SECOND EDITION

NUCLEAR SYSTEMS

VOLUME 1

THERMAL HYDRAULIC FUNDAMENTALS

SECOND EDITION

NUCLEAR SYSTEMS

VOLUME 1

THERMAL HYDRAULIC FUNDAMENTALS

NEIL E. TODREAS • MUJID S. KAZIMI

CRC Press
Taylor & Francis Group
Boca Raton London New York

CRC Press is an imprint of the
Taylor & Francis Group, an **informa** business

CRC Press
Taylor & Francis Group
6000 Broken Sound Parkway NW, Suite 300
Boca Raton, FL 33487-2742

Printed in the United States of America on acid-free paper
Version Date: 20110816

International Standard Book Number: 978-1-4398-0887-0 (Hardback)

Visit the Taylor & Francis Web site at
http://www.taylorandfrancis.com

and the CRC Press Web site at
http://www.crcpress.com

To our families for their support in this endeavor
Carol, Tim, and Ian
Nazik, Yasmeen, Marwan, and Omar

Contents

Preface

In the 22 years since the first edition appeared, nuclear power systems have regained favor worldwide as a scalable, dependable, and environmentally desirable energy supply technology. While water-cooled reactor types promise to dominate the deployment of nuclear power systems well into this century, several alternate cooled reactors among the six Generation IV proposed designs are now under intensive development for deployment in the next decades.

Consequently, while this new edition continues to emphasize pressurized and boiling light water reactor technologies, the principal characteristics and analysis approaches for the Generation IV designs of most international interest, the liquid-metal- and gas-cooled options, are introduced as well.

To accomplish this, the content of most chapters has been amplified by introducing both new analysis approaches and correlation methods which have gained the favor of experts over the past several decades. Nevertheless, these enhancements have been introduced into the sequence of chapters of the first edition, a sequence which has proven to be pedagogically effective. This organization of text chapters is shown in Table 0.1.

The content of these chapters emphasizing new material in particular is summarized below:

- Chapter 1 surveys the characteristics of all current principal reactor types: operating and advanced light water-cooled designs as well as the six Generation IV designs.
- Chapter 2 emphasizes the relations that integrate the performance of the core nuclear steam supply system and balance of plant. In doing so, the performance measures of the multiple disciplines—neutronics, thermal hydraulics, fuel behavior, structural mechanics, and reactor operations—are introduced and interrelated. In particular, power density and specific power measures are thoroughly developed for standard fuel pin arrays as well as inverted (fuel and coolant are spatially interchanged) assembly configurations.
- Chapter 3 presents a thorough description of the magnitude and spatial distribution of the generation and deposition of energy within the reactor vessel assembly. Emphasis is given to the evolution of the American Nuclear Society Decay Heat Generation Standard up to 2005. Multiple other sources of stored energy are present in reactor systems such as those associated with elevated temperature and pressure conditions of the primary and secondary reactor coolants and the chemical reactions of materials of construction under extreme temperatures characteristic of accident conditions. The stored energy and the energy liberated by the chemical reactions characteristic of the light water reactor materials are now presented in detail.

TABLE 0.1
Text Contents by Chapter and Subject

1—Reactor-Type Overview

2—Core and Plant Performance Measures

3—Fission Energy Generation and Deposition

	Conservation Equations	Thermodynamics	Fluid Flow	Heat Transfer
Single-Phase Coolant	4—Differential and Integral Formulations		8—Fuel Pins and Assemblies	
			9—Single Channels	10—Single Channels
Two-Phase Coolant	5—Differential and Integral Formulations; 2φ Parameter Definitions		11—Single Channels	12—Pool Boiling 13—Flow Boiling in Single Channel
Single-Phase and Two-Phase Coolants		6—Power Cycles 7—Components and Containment	14—Single-Channel Examples	

- Chapters 4 and 5, which present the mass, momentum, energy, and entropy conservation equations for single- and two-phase coolants in differential and control volume formulations, are unchanged from the first edition.
- Chapter 6 presents the analysis of power generation cycles, both Rankine and Brayton types. The analysis of the supercritical carbon dioxide (SCO_2) cycle, which has recently been introduced as a performance and cost-efficient option for Generation IV systems, is presented. Of note, this presentation is designed to clearly illustrate how this recompression cycle must be designed and analyzed, a subtlety which otherwise easily confounds the inexperienced analyst.
- Chapter 7 applies thermodynamic principles for analysis of the fuel and coolant mixture in a severe accident, for analysis of containment design and accident performance, and for analysis of PWR pressurizer performance. The chapter remains as in the first edition with the fundamentals of pressurizer performance becoming the focus now by deletion of the more complex multiregion pressurizer model introduced in the first edition.
- Chapter 8 significantly amplifies the discussion of fuel pin materials and thermal analysis of various fuel geometries in the first edition. Extensive fuel and cladding material properties for thermal and fast neutron spectrum reactors are now included. Additionally, the case of annular fuel geometry, which is cooled inside and outside, is fully analyzed. Finally, the correlations for cladding-coolant surface oxidation are now included.
- Chapters 9 and 10 present the fluid flow and thermal analysis techniques for single-phase coolant in laminar and turbulent flow. Noteworthy new

material presents the analysis of geometries configured for enhanced heat transfer albeit with pressure loss penalty.

- Chapters 11 and 12 along with Chapter 13 present fluid flow and thermal analysis techniques for two-phase flow. Chapter 11 now has expanded coverage of critical flow and new coverage of flow instabilities and condensation. Boiling heat transfer is now separated into pool boiling and flow boiling in Chapters 12 and 13, respectively. The entire two-phase flow treatment in these three chapters has been expanded to include recent correlations of current usage, particularly for void fraction, pressure loss, and the critical condition.
- Chapter 14 has been thoroughly restructured to present a systematic analysis of heated flow channel performance for single- and two-phase flow utilizing the methods and correlations of Chapters 9 through 13. Specific focus is directed to conditions of equilibrium and nonequilibrium for both equal/unequal phasic temperatures (thermal equilibrium) and equal phasic flow velocities (mechanical equilibrium).

Finally, throughout the text we refer as appropriate to further elaboration of relevant technical information which appears in the companion *Volume II* of this text.*

* Todreas, N. E. and Kazimi, M. S., *Nuclear Systems II: Elements of Thermal Hydraulic Design.* New York: Taylor & Francis, 2001.

Preface to the First Edition

This book can serve as a textbook for two to three courses at the advanced undergraduate and graduate student level. It is also suitable as a basis for the continuing education of engineers in the nuclear power industry who wish to expand their knowledge of the principles of thermal analysis of nuclear systems. The book, in fact, was an outgrowth of the course notes used for teaching several classes at MIT over a period of nearly 15 years.

The book is meant to cover more than thermal hydraulic design and analysis of the core of a nuclear reactor. Thus, in several parts and examples, other components of the nuclear power plant, such as the pressurizer, the containment, and the entire primary coolant system, are addressed. In this respect the book reflects the importance of such considerations in thermal engineering of a modern nuclear power plant. The traditional concentration on the fuel element in earlier textbooks was appropriate when the fuel performance had a higher share of the cost of electricity than in modern plants. The cost and performance of nuclear power plants has proved to be more influenced by the steam supply system and the containment building than previously anticipated.

The desirability of providing in one book basic concepts as well as complex formulations for advanced applications has resulted in a more comprehensive textbook than those previously authored in the field. The basic ideas of both fluid flow and heat transfer as applicable to nuclear reactors are discussed in *Volume I*. No assumption is made about the degree to which the reader is already familiar with the subject. Therefore, various reactor types, energy source distribution, and fundamental laws of conservation of mass, momentum, and energy are presented in early chapters. Engineering methods for analysis of flow hydraulics and heat transfer in single-phase as well as two-phase coolants are presented in later chapters. In *Volume II*, applications of these fundamental ideas to multi-channel flow conditions in the reactor are described as well as specific design considerations such as natural convection and core thermal reliability. They are presented in a way that renders it possible to use the analytical development in simple exercises and as the bases for numerical computations similar to those commonly practiced in the industry.

A consistent nomenclature is used throughout the text and a table of the nomenclature is included in the Appendix. Each chapter includes problems identified as to their topic and the section from which they are drawn. While the SI unit system is principally used, British Engineering Units are given in brackets for those results commonly still reported in the United States in this system.

Acknowledgments

The authors are indebted to several fellow faculty, professional associates, staff, and students who provided significant assistance in preparing and reviewing this second edition of *Nuclear Systems Volume I.* Besides those named here, there were many who sent us notes and comments over the years, which we took into consideration in this second edition.

Our faculty colleague, Jacopo Buongiorno, provided essential insights on a number of technical topics introduced or expanded upon in this new edition. Also, he provided a collection of solved problems drawn from MIT courses taught using this text, which have been integrated into the exercises at the end of each chapter. Our faculty colleagues, Arthur Bergles, Michael Driscoll, and Eugene Shwageraus, as well as our professional associates, Mahmoud Massoud, Tom Newton, Aydin Karadin, and Charles Kling, carefully proofed major portions of the text and offered materials and comments for our consideration.

The following students greatly assisted in the preparation of the text by researching the technical literature as well as preparing figures and example solutions: Muhammed Ayanoglu, Tom Conboy, Jacob DeWitte, Paolo Ferroni, Giancarlo Lenci, and Wenfeng Liu. Bryan Herman prepared the manual of solutions to all homework problems in the text. In the proofreading of the final manuscript we were greatly assisted by the following students: Nathan Andrews, Tyrell Arment, Ramsey Arnold, Jacob DeWitte, Mihai Diaconeasa, You Ho Lee, Giancarlo Lenci, Alexander Mieloszyk, Stefano Passerini, Joshua Richard, Koroush Shirvan, John Stempien, and Francesco Vitillo.

The CD-ROM accompanying the text contains the Problemsolver software developed by Dr. Massoud and several Excel spreadsheets developed by Giancarlo Lenci for critical heat flux application. We convey our additional thanks to both for providing these useful problem-solving tools to enhance our text.

The preparation of the manuscript was expertly and conscientiously done by Richard St. Clair, who tirelessly worked on the iterations of insertions and deletions of material in the text. We also extend our gratitude to Paula Cornelio who prepared the majority of the new figures for this edition as she equally did for the original text.

Finally, we acknowledge the review of our proposed plan for the technical content of this second edition by Professors Samim Anghaie, Fred Best, Larry Hockreiter, and Michel Giot. Their observations and suggestions were influential in our final selection of topics to be rewritten, added, and deleted, although the final selection was made by us.

Authors

Neil E. Todreas is Professor Emeritus in the Departments of Nuclear Science and Engineering and Mechanical Engineering at the Massachusetts Institute of Technology. He held the Korea Electric Power Corporation (KEPCO) chair in Nuclear Engineering from 1992 until his retirement to part-time activities in 2006. He served an 8-year period from 1981 to 1989 as the Nuclear Engineering Department Head. Since 1975 he has been a codirector of the MIT Nuclear Power Reactor Safety summer course, which presents current issues of reactor safety significant to an international group of nuclear engineering professionals. His area of technical expertise includes thermal and hydraulic aspects of nuclear reactor engineering and safety analysis. He started his career at Naval Reactors working on submarine and surface nuclear vessels after earning the B.Eng. and the M.S. in mechanical engineering from Cornell University. Following his Sc.D. in Nuclear Engineering at MIT, he worked for the Atomic Energy Commission (AEC) on organic cooled/heavy water-moderated and sodium-cooled reactors until he returned as a faculty member to MIT in 1970. He has an extensive record of service for government (Department of Energy (DOE), U.S. Nuclear Regulatory Commission (USNRC), and national laboratories) and utility industry review committees including INPO, and international scientific review groups. He has authored more than 200 publications and a reference book on safety features of light water reactors. He is a member of the U.S. National Academy of Engineering and a fellow of the American Nuclear Society (ANS) and the American Society of Mechanical Engineers (ASME). He has received the American Nuclear Society Thermal Hydraulics Technical Achievement Award, its Arthur Holly Compton Award in Education, and its jointly conferred (with the Nuclear Energy Institute) Henry DeWolf Smyth Nuclear Statesman Award.

Mujid S. Kazimi is Professor in the Departments of Nuclear Science and Engineering and Mechanical Engineering at the Massachusetts Institute of Technology. He is the director of the Center for Advanced Nuclear Energy Systems (CANES) and holds the Tokyo Electric Power Company (TEPCO) chair in Nuclear Engineering at MIT. Prior to joining the MIT faculty in 1976, Dr. Kazimi worked for a brief period at the Advanced Reactors Division of Westinghouse Electric Corporation and at Brookhaven National Laboratory. At MIT he was head of the Department of Nuclear Engineering from 1989 to 1997 and Chair of the Safety Committee of the MIT Research Reactor from 1998 to 2009. He has served since 1990 as the codirector of the MIT Nuclear Power Reactor Safety summer course. He is active in the development of innovative designs of fuel and other components of nuclear power plants and in analysis of the nuclear fuel cycle options for sustainable nuclear energy. He cochaired the 3-year MIT interdisciplinary study on the *Future of the Nuclear Fuel Cycle*, published in 2011. He has served on scientific advisory committees at the U.S. National Academy of Engineering and several other national agencies and laboratories in the United States, Japan, Spain, Switzerland, Kuwait, the United

Arab Emirates, and the International Atomic Energy Agency. He has authored more than 200 articles and papers that have been published in journals and presented at international conferences. Dr. Kazimi holds a B.Eng. degree from Alexandria University in Egypt, and M.S. and Ph.D. degrees from MIT, all in nuclear engineering. He is a fellow of the American Nuclear Society and the American Association for the Advancement of Science. Among his honors is the Technical Achievement Award in Thermal Hydraulics by the American Nuclear Society.

1 Principal Characteristics of Power Reactors

1.1 INTRODUCTION

This chapter presents the basic characteristics of power reactors. These characteristics, along with more detailed thermal hydraulic parameters presented in further chapters, enable the student to apply the specialized techniques presented in the remainder of the text to a range of reactor types. Water-, gas-, and sodium-cooled reactor types, identified in Table 1.1, encompass the principal nuclear power reactor designs that have been employed in the world. The thermal hydraulic characteristics of these reactors are presented in Sections 1.2 through 1.5 as part of the description of the power cycle, primary coolant system, core, and fuel assembly design of these reactor types. Three classes of advanced reactors are also presented in subsequent sections, the Generation III, III+, and IV designs. The Generation III designs are advanced water reactors that have already been brought into operation (ABWR) or are under construction (EPR). The Generation III+ designs are advanced water- and gas-cooled reactors, several of which are being licensed and brought into service in the 2010 decade [12]. These Generation III and III+ designs are discussed in Section 1.6. The Generation IV reactors described in Section 1.7 were selected by an international roadmapping process and are being pursued through an internationally coordinated research and development activity for deployment in the period 2020–2040 [13]. Figure 1.1 presents the evolution and categorization by the generation of the world's reactor types. Tables in Chapters 1 and 2 and Appendix K provide detailed information on reactor characteristics useful for application to specific illustrative examples and homework problems in the text.

1.2 POWER CYCLES

In these plants, a primary coolant is circulated through the reactor core to extract energy for ultimate conversion to electricity in a turbine connected to an electric generator. Depending on the reactor design, the turbine may be driven directly by the primary coolant or by a secondary coolant that has received energy from the primary coolant. The number of coolant systems in a plant equals the sum of the one primary and one or more secondary systems. For the boiling water reactor (BWR) and the high-temperature gas reactor (HTGR) systems, which produce steam and hot helium by passage of a primary coolant through the core, direct use of these primary coolants in the turbine is possible, leading to a single-coolant system. The BWR

TABLE 1.1
Basic Features of Major Power Reactor Types

Reactor Type	Neutron Spectrum	Moderator	Coolant	Fuel: Chemical Form	Fuel: Approximate Fissile Content (All ^{235}U Except the Sodium-Cooled Reactors)
Water-cooled	Thermal				
PWR		H_2O	H_2O	UO_2	3–5% enrichment
BWR		H_2O	H_2O	UO_2	3–5% enrichment
PHWR (CANDU)		D_2O	D_2O	UO_2	Natural
SGHWR[a]		D_2O	H_2O	UO_2	~3% enrichment
Gas-cooled	Thermal	Graphite			
Magnox			CO_2	U metal	Natural
AGR			CO_2	UO_2	~3% enrichment
HTGR			Helium	UO_2	~7–20% enrichment[b]
Sodium-cooled	Fast	None	Sodium		
SFBR[c]				UO_2/PuO_2	~15–20% of HM is Pu[e]
SFR[d]				NU–TRU–Zr[f] metal or oxide	~15% of HM is TRU

[a] Steam-generating heavy water reactor.
[b] Older operating plants have enrichments of more than 90% and used a variety of thorium and carbide fuel forms.
[c] Sodium-cooled fast-breeder reactor.
[d] Sodium fast reactor operating on a closed cycle.
[e] Heavy metal (HM).
[f] Natural uranium (NU), transuranic elements (TRU), and zirconium (Zr).

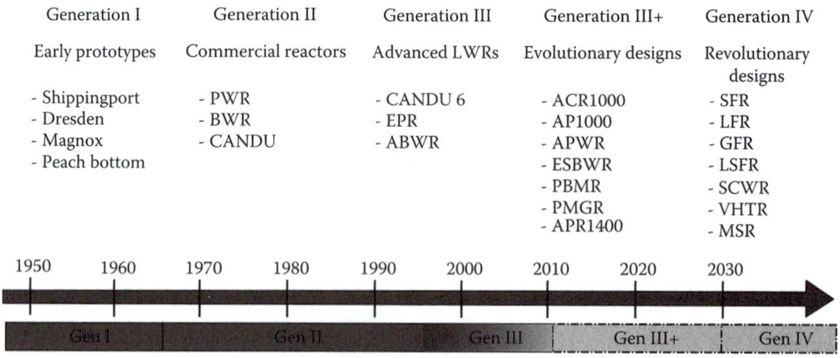

FIGURE 1.1 The evolution of nuclear power. (Adopted from U.S. Department of Energy, http://www.gen-4.org/Technology/evolution.htm.)

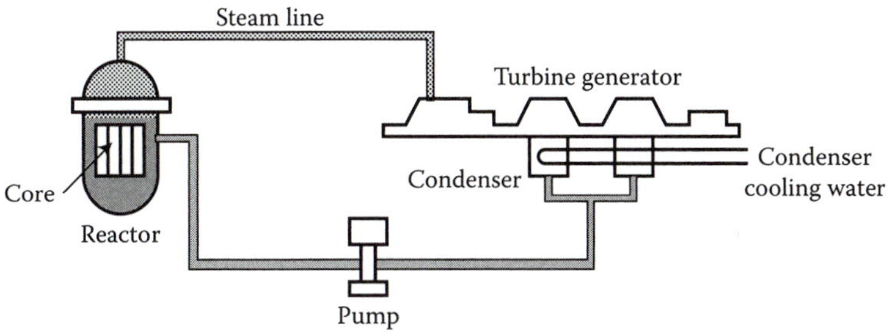

FIGURE 1.2 Direct, single-coolant Rankine cycle. (Adopted from U.S. Department of Energy.)

single-coolant system, based on the Rankine cycle (Figure 1.2), is in common use. The Fort St. Vrain HTGR plant used a secondary water system in a Rankine cycle because the technology did not exist to produce a large, high-temperature, helium-driven turbine. Although the HTGR direct turbine system has not yet been built for a commercial reactor, it would use the Brayton cycle, as illustrated in Figure 1.3. Thermodynamic analyses for typical Rankine and Brayton cycles are presented in Chapter 6.

The pressurized water reactor (PWR) and the pressurized heavy water reactor (PHWR) are two-coolant systems. This design is necessary to maintain the primary coolant conditions at a nominal subcooled liquid state while the turbine is driven by steam in the secondary system. Figure 1.4 illustrates the PWR two-coolant steam cycle.

The sodium-cooled fast reactors (both SFRs and SFBR) employ three-coolant systems: a primary sodium coolant system, an intermediate sodium coolant system, and a steam–water, turbine–condenser coolant system (Figure 1.5). The sodium-to-sodium heat exchange is accomplished in an intermediate heat exchanger (IHX), and the sodium-to-water/steam heat exchange in a steam generator. Three-coolant systems were specified to isolate the radioactive primary sodium coolant from the

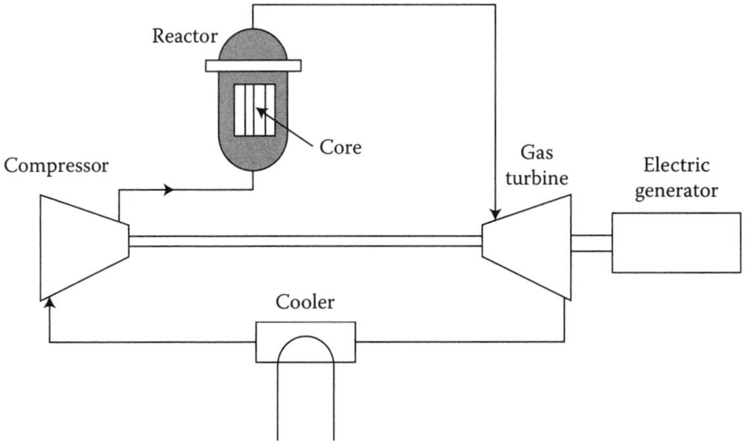

FIGURE 1.3 Direct, single-coolant Brayton cycle. (Adopted from U.S. Department of Energy.)

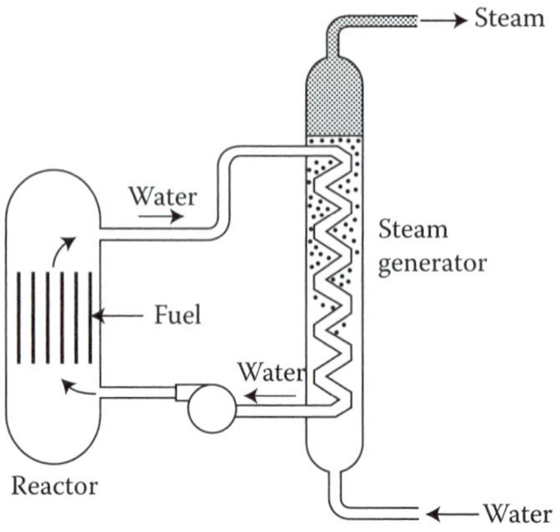

FIGURE 1.4 Two-coolant system steam cycle. (From Shultis, J. K. and Faw, R. E., *Fundamentals of Nuclear Science and Engineering*, 2nd Ed. CRC Press, Boca Raton, FL, 2008.)

steam–water circulating through the turbine, condenser, and associated conventional plant components. The SFR concept being developed in the United States draws on this worldwide SFR technology and the operational experience base, but it is not designed as a breeder. Sodium-cooled reactor characteristics and examples presented in this chapter are for both the SFBRs, which were built in the late 1900s, and the SFR, which is currently under development and design.

The significant characteristics of the thermodynamic cycles and coolant systems used in these reference reactor types are summarized in Table 1.2.

1.3 PRIMARY COOLANT SYSTEMS

The Generation II BWR single-loop primary coolant system is illustrated in Figure 1.6, while Figure 1.7 highlights the flow paths within the reactor vessel. The

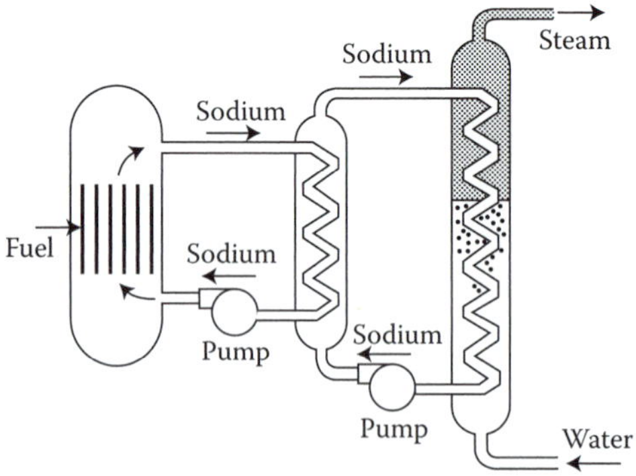

FIGURE 1.5 Three-coolant system steam cycle. (From Shultis, J. K. and Faw, R. E., *Fundamentals of Nuclear Science and Engineering*, 2nd Ed. CRC Press, Boca Raton, FL, 2008.)

TABLE 1.2
Typical Characteristics of the Thermodynamic Cycle for Six Reference Power Reactor Types

Characteristics	BWR	PWR	PHWR	HTGR	AGR	SFBR
			Reference Design			
Manufacturer	General Electric	Westinghouse	Atomic Energy of Canada, Ltd.	General Atomic	National Nuclear Corp.	Novatome
System (reactor station)	BWR/5 (NMP2)	(Seabrook)	CANDU-600	(Fulton)[a]	(Heysham 2)	(Superphenix-1)
Steam cycle						
No. coolant systems	1	2	2	2	2	3
Primary coolant	H_2O	H_2O	D_2O	He	CO_2	Liq. Na
Secondary coolant	[b]	H_2O	H_2O	H_2O	H_2O	Liq. Na/H_2O
			Energy Conversion			
Thermal power, MW_t	3323	3411	2180	3000	1551	3000
Electric power, MW_e	1062	1148	638	1160	660	1240
Efficiency (%)	32.0	33.7	29.3	38.6	42.5	41.3
			Heat Transport System			
No. of primary loops/pumps	2/2	4/4	2/2	6/6	8/8	Pool with 4 pumps
No. of IHXs	0	0	0	0	0	8
No. of steam generators	[b]	4	4	6	8	4
Steam generator type	[b]	U tube	U tube	Helical coil	Helical coil	Helical coil

continued

TABLE 1.2 (continued)
Typical Characteristics of the Thermodynamic Cycle for Six Reference Power Reactor Types

Characteristics	BWR	PWR	PHWR	HTGR	AGR	SFBR
Thermal Hydraulics						
Primary coolant						
Pressure (MPa)	7.14	15.51	10.0	5.0	4.27	~0.1
Inlet temp. (°C)	278.3	293.1	267	318	292	395
Ave. outlet temp. (°C)	286.1	326.8	310	741	638	545
Core flow rate (Mg/s)	13.671	17.476	7.6	1.41	3.92	15.7
Volume or mass	—	336 m^3	120 m^3	7850 kg	5300 m^3	3.2×10^6 kg Na/H_2O
Secondary coolant						
Pressure (MPa)	[b]	6.89	4.7	17.3	17.0	~0.1/17.7
Inlet temp. (°C)	[b]	227	187	188	157.0	345/237
Outlet temp. (°C)	[b]	285	260	513	543.0	525/490

Source: BWR and PWR: Adopted from *Seabrook Power Station Updated Safety Analysis Report*, Revision 8, Seabrook Station, Seabrook, NH, 2002; Appendix K. PHWR: Adopted from Knief, R. A. *Nuclear Engineering: Theory and Technology of Commercial Nuclear Power*, pp. 707–717. American Nuclear Society, La Grange Park, IL, 2008. HTGR: Adopted from Breher, W., Neyland, A., and Shenoy, A. *Modular High-Temperature Gas-Cooled Reactor (MHTGR) Status*. GA Technologies, GA-A18878, May 1987. AGR-Heysham 2: Adopted from AEAT/R/PSEG/0405 Issue 3. *Main Characteristics of Nuclear Power Plants in the European Union and Candidate Countries*. Report for the European Commission, September 2001; *Nuclear Engineering International*. Supplement. August 1982; Alderson, M. A. H. G. (UKAEA, pers. comm., October 6, 1983 and December 6, 1983). SuperPhenix-1: Adopted from IAEA-TECDOC-1531. *Fast Reactor Database 2006 Update*. International Atomic Energy Agency. December 2006. The Russian VVER is similar to the US PWR while their RBMK, which is no longer being built, is a low-enriched uranium oxide-fueled, light water-cooled, graphite-moderated pressure tube design.

[a] Designed but not built.

[b] Not applicable.

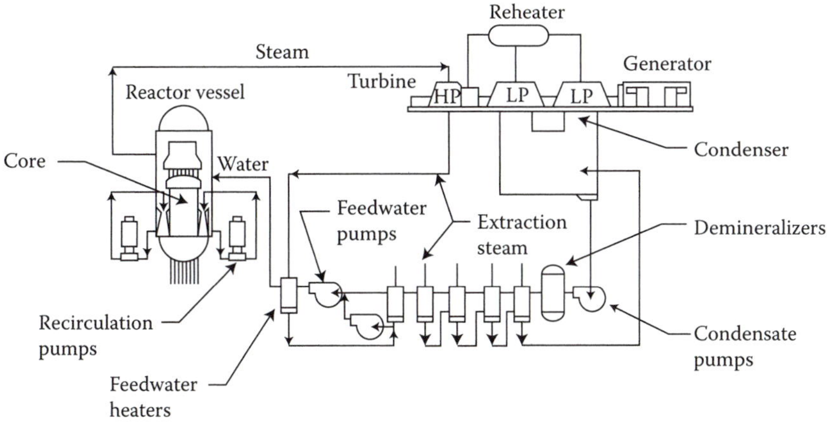

FIGURE 1.6 BWR single-loop primary coolant system. (From Shultis, J. K. and Faw, R. E., *Fundamentals of Nuclear Science and Engineering*, 2nd Ed. CRC Press, Boca Raton, FL, 2008.)

steam–water mixture first enters the steam separators after exiting the core. After subsequent passage through a steam separator and dryer assembly located in the upper portion of the reactor vessel, dry saturated steam flows directly to the turbine. Saturated water, which is separated from the steam, flows downward in the periphery of the reactor vessel and mixes with the incoming main feed flow from the condenser. This combined flow stream is pumped into the lower plenum through jet pumps mounted around the inside periphery of the reactor vessel. The jet pumps are driven

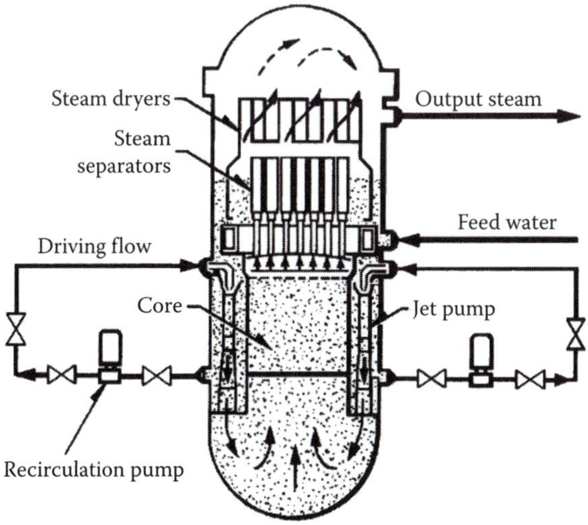

FIGURE 1.7 Steam and recirculation water flow paths in the Generation II BWR. (From Shultis, J. K. and Faw, R. E., *Fundamentals of Nuclear Science and Engineering*, 2nd Ed. CRC Press, Boca Raton, FL, 2008.)

by flow from recirculation pumps located in relatively small-diameter (~50 cm) external recirculation loops, which draw flow from the plenum just above the jet pump discharge location. In the ABWR, all external recirculation loops are eliminated and replaced with recirculation pumps placed internal to the reactor vessel. In the economic simplified boiling water reactor (ESBWR), all jet pumps as well as external recirculation pumps were eliminated by the natural circulation flow design.

In all BWRs the core flowrate is much greater than the feed water flowrate, reflecting the fact that the average core exit quality $\bar{x}_e$ is about 15%. Hence, the recirculation ratio (RR) is obtained as

$$RR \equiv \frac{\text{Mass flowrate of recirculated liquid}}{\text{Mass flowrate of vapor produced}} = \frac{1-\bar{x}_{exit}}{\bar{x}_{exit}} = \frac{0.85}{0.15} = 5.7 \tag{1.1}$$

The primary coolant system of a PWR consists of a multiloop arrangement arrayed around the reactor vessel. Higher power reactor ratings are achieved by adding loops of identical design. Designs of two, three, and four loops have been built with three- and four-loop reactors being the most common. In a typical four-loop configuration (Figure 1.8), each loop has a vertically oriented steam generator* and

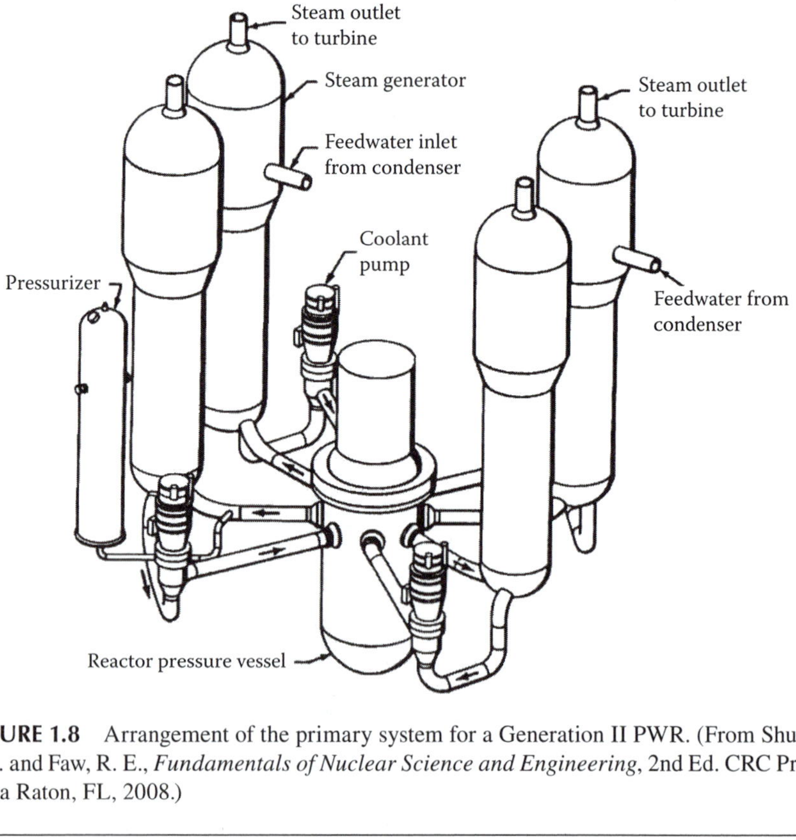

FIGURE 1.8 Arrangement of the primary system for a Generation II PWR. (From Shultis, J. K. and Faw, R. E., *Fundamentals of Nuclear Science and Engineering*, 2nd Ed. CRC Press, Boca Raton, FL, 2008.)

* Russian VVERs employ horizontal steam generators.

coolant pump. The coolant flows through the steam generator within an array of U tubes that connect the inlet and outlet plena located at the bottom of the steam generator. The system's single pressurizer is connected to the hot leg of one of the loops. The hot (reactor vessel to steam generator inlet) and cold (steam generator outlet to reactor vessel) leg pipes are typically 31–42 and 29–30 in. (78.7–106.7 and 73.7–76.2 cm) in diameter, respectively.

The flow path through the PWR reactor vessel is illustrated in Figure 1.9. The inlet nozzles communicate with an annulus formed between the inside of the reactor vessel and the outside of the core support barrel. The coolant entering this annulus flows downward into the inlet plenum formed by the lower head of the reactor vessel. Here it turns upward and flows through the core into the upper plenum that communicates with the reactor vessel's outlet nozzles.

The HTGR primary system is composed of several loops, each housed within a large cylinder of prestressed concrete. A compact HTGR arrangement as embodied in the modular high-temperature gas-cooled reactor (MHTGR) is illustrated in

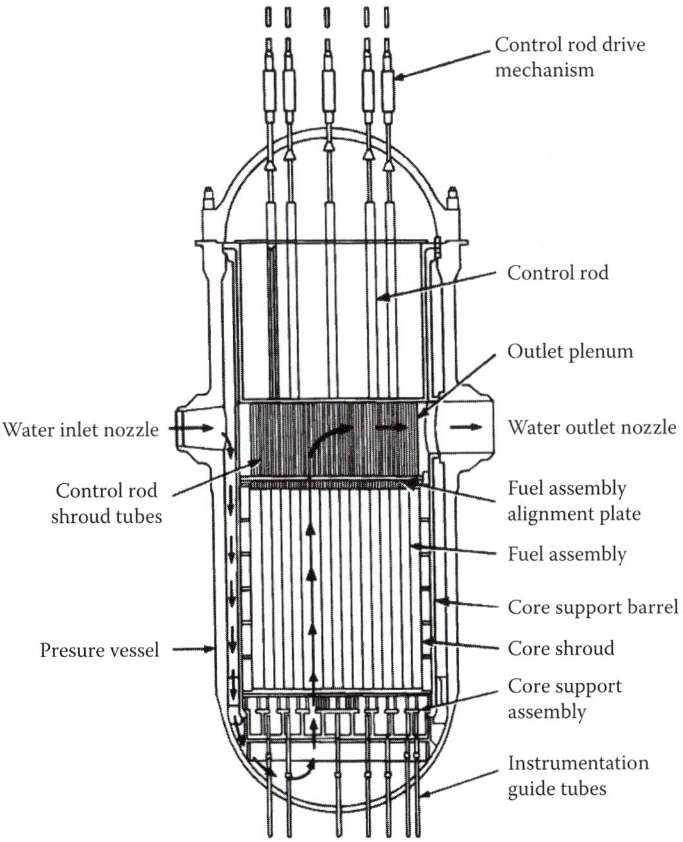

FIGURE 1.9 Flow path through a PWR reactor vessel. (From Shultis, J. K. and Faw, R. E., *Fundamentals of Nuclear Science and Engineering*, 2nd Ed. CRC Press, Boca Raton, FL, 2008.)

Figure 1.10. In this 588 MWe MHTGR arrangement [2], the flow is directed downward through the core by a circulator mounted above the steam generator in the cold leg. The reactor vessel and steam generator are connected by a short, horizontal cross duct, which channels two oppositely directed coolant streams. The coolant from the core exit plenum is directed laterally through the 47 in. (119.4 cm) interior diameter region of the cross duct into the inlet of the steam generator. The coolant from the steam generator and circulator is directed laterally through the outer annulus (equivalent pipe diameter of approximately 46 in. [116.8 cm]) of the cross duct into the core inlet plenum and then upward through the reactor vessel's outer annulus into the inlet core plenum at the top of the reactor vessel.

SFBR and SFR primary systems have been of the loop and pool types. The pool-type configuration of the Superphenix reactor [14] is shown in Figure 1.11. Its characteristics are detailed in Table 1.2. The coolant flow path is upward through the reactor core into the upper sodium pool of the main vessel. The coolant from this pool flows downward by gravity through the IHX and discharges into a low-pressure

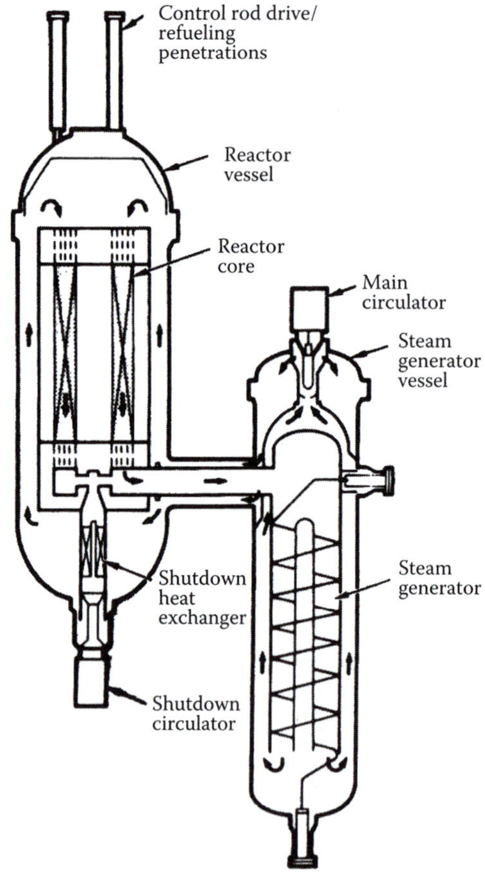

FIGURE 1.10 Modular HTGR primary coolant flow path. (Courtesy of U.S. Department of Energy.)

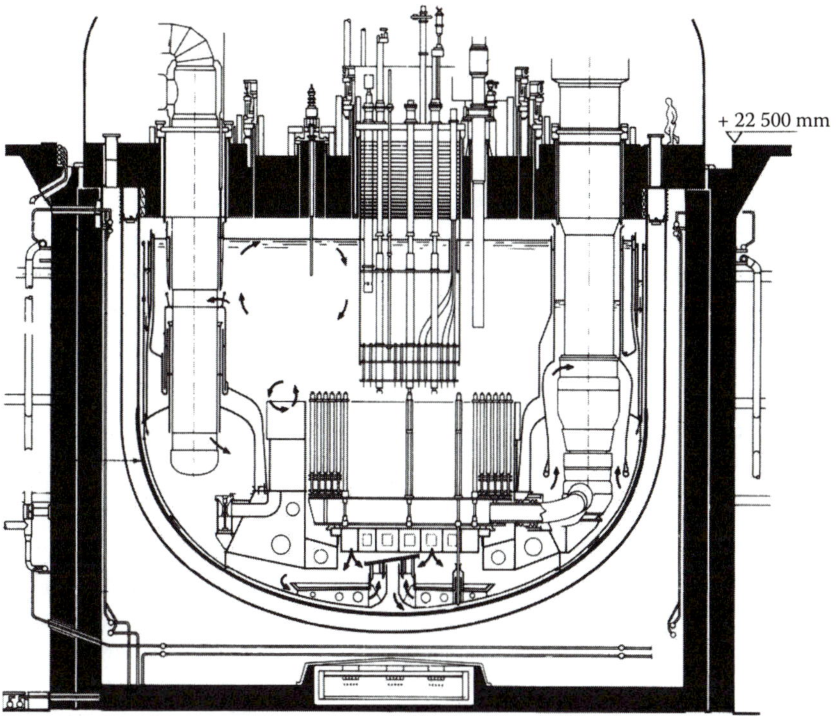

FIGURE 1.11 Primary system sodium flow path in the Superphenix reactor. (Courtesy of Électricité de France.)

toroidal plenum located in the periphery of the lower portion of the main vessel. Vertically oriented primary pumps draw the coolant from this low-pressure plenum and discharge it into the core inlet plenum.

1.4 REACTOR CORES

The reactor cores of all these reactors, except for the HTGR, are composed of assemblies of cylindrical fuel rods surrounded by the coolant that flows along the rod length. The prismatic HTGR core consists of graphite moderator hexagonal blocks that function as fuel assemblies. The blocks or assemblies are described in detail in Section 1.5.

There are two design features that establish the principal thermal hydraulic characteristics of reactor cores: the orientation and the degree of hydraulic isolation of an assembly from its neighbors. It is simple to adopt a reference case and describe the exceptions. Let us take as the reference case a vertical array of assemblies that communicate only at inlet and exit plena. This reference case describes the BWR, SFBR, and the advanced gas reactor (AGR) systems. The HTGR is nominally configured in this manner also, although leakage between the graphite blocks that are stacked to create the proper core length creates a substantial degree of communication between coolant passages within the core. The PHWR core consists of horizontal pressure tubes penetrating a low-pressure calandria tank filled with a heavy water moderator. The fuel assemblies housed within the pressure tubes are cooled by high-pressure heavy water, which is directed to and from the tubes by an array of inlet and outlet headers. The more advanced Canadian reactors use light water for cooling within the pressure tubes but retain heavy water in the calandria tank. Both the PHWR and the AGR are designed for online refueling.

The PWR and BWR assemblies are vertical, but unlike the BWR design, the PWR assemblies are not isolated hydraulically by enclosing the fuel rod array within ducts (called fuel channels in the BWR) over the core length. Hence, PWR fuel rods are grouped into assemblies only for handling and other structural purposes.

1.5 FUEL ASSEMBLIES

The principal characteristics of power reactor fuel bundles are the array (geometric layout and rod spacing) and the method of fuel pin separation and support along their span. The light water reactors (BWR and PWR), PHWR, AGR, and SFBR/SFR all use fuel rods. The HTGR has graphite moderator blocks in which adjacent penetrating holes for fuel and flowing helium coolant exist.

Light water reactors (LWRs), where the coolant also serves as the moderator, have small fuel-to-water volume ratios (commonly called the *metal-to-water ratio*) and consequently rather large fuel rod centerline-to-centerline spacing (commonly called the *rod pitch, P*). This moderate packing fraction permits the use of a simple square array and requires a rod support scheme of moderately small frontal area to yield low-pressure drops. The one LWR exception is the VVER, which uses a hexagonal array. A variety of grid support schemes have evolved for these applications.

Heavy water reactors and advanced gas reactors are designed for online refueling and consequently consist of fuel assemblies stacked within circular pressure tubes. This circular boundary leads to an assembly design with an irregular geometric array of rods. The online refueling approach has led to short fuel bundles in which the rods are supported at the assembly ends and by a center brace rather than by LWR-type grid spacers.

SFRs require no moderator and achieve high-power densities by compact hexagonal fuel rod array packing. With this tight rod-to-rod spacing, a lower pressure drop is obtained using spiral wire wrapping around each rod than could be obtained with a grid-type spacer. This wire wrap serves a dual function: as a spacer and as a promoter of coolant mixing within the fuel bundle. However, some SFR assemblies do use grid spacers.

The principal characteristics of the fuel for the six reference power reactor types are summarized in Table 1.3. The HTGR does not consist of an array of fuel rods within a coolant continuum. Rather, the HTGR blocks that contain fuel compacts, a coolant, and a moderator are designated as inverted fuel assemblies. In these blocks, the fuel–moderator combination is the continuum that is penetrated by isolated, cylindrically shaped coolant channels.

The LWRs (PWR and BWR), PHWR, AGR, and SFBR utilize an array of fuel rods surrounded by the coolant. For each of these arrays, the useful geometric characteristics are given in Table 1.3 and typical subchannels identified in Figure 1.12.

TABLE 1.3
Typical Characteristics of the Fuel for Six Reference Power Reactor Types

Characteristics	BWR	PWR	PHWR	HTGR	AGR	SFBR
			Reference Design			
Manufacturer	General Electric	Westinghouse	Atomic Energy of Canada, Ltd.	General Atomic	National Nuclear Corp.	Novatome
System (reactor station)	BWR/5 (NMP2)	(Seabrook)	CANDU-600	(Fulton)	(Heysham 2)	(Superphenix 1)
Moderator	H_2O	H_2O	D_2O	Graphite	Graphite	None
Neutron energy	Thermal	Thermal	Thermal	Thermal	Thermal	Fast
Fuel production	Converter	Converter	Converter	Converter	Converter	Breeder
			Fuel[b]			
Geometry	Cylindrical pellet	Cylindrical pellet	Cylindrical pellet	Microspheres[c]	Cylindrical pellet	Cylindrical pellet
Dimensions (mm)	$9.60D \times 10.0L$	$8.192D \times 9.8L$	$12.2D \times 16.4L$	400–800 μm D	$14.51D \times 14.51L$	7.14 D
Chemical form	UO_2	UO_2	UO_2	UC/ ThO_2	UO_2	PuO_2 /UO_2
Fissile (first core avg. wt% unless designated as equilibrium core)	^{235}U (3.5 eq. core)	^{235}U (3.57 avg. eq. core)	^{235}U (0.711)	^{235}U (93)	^{235}U (2 zones at 2.1 and 2.7)	^{239}Pu (2 zones at 16 and 19.7)
Fertile	^{238}U	^{238}U	^{238}U	Th	^{238}U	Depleted U
			Fuel Rods			
Geometry	Pellet stack in clad tube	Pellet stack in clad tube	Pellet stack in clad tube	Cylindrical fuel compacts	Pellet stack in clad tube	Pellet stack in clad tube

continued

TABLE 1.3 (continued)

Typical Characteristics of the Fuel for Six Reference Power Reactor Types

Characteristics	BWR	PWR	PHWR	HTGR	AGR	SFBR
Dimensions (mm)	$11.20D \times 3.588$ m H ($\times 4.09$ m L)	$9.5D \times 3.658$ m H ($\times 3.876$ m L)	$13.1D \times 493L$	$15.7D \times {\sim}742H$ ($\times 793L$)	$15.3D \times 987H$ ($\times 1.04$ m L)	$8.5D \times 2.7$ m H(C)[d] $15.8D \times 1.94$ m H(RB)
Clad material	Zircaloy-2	Zirlo™	Zircaloy-4	No clad	Stainless steel	Stainless steel
Clad thickness (mm)	0.71	0.572	0.42	[g]	0.37	0.56
Fuel Assembly						
Geometry	9 × 9 square rod array[a]	17 × 17 square rod array	Concentric circles of rods	Cylindrical fuel compacts within a hex. graphite block[e]	Concentric circles of rods surrounded by graphite sleeve 25.7	Hexagonal rod array
Rod pitch (mm)	14.37	12.60	14.6	23	37	9.8 (C)/17.0 (RB)
No. rod locations	81	289	37	132 (SA)/76 (CA)[f]	36	271 (C)/91 (RB)
No. fuel rods	74 (8 part length)	264	37	132 (SA)/76 (CA)[f]	190.4 (sleeve inner D)	271 (C)/91 (RB)

Outer dimensions (mm)	139 × 139	214 × 214	102*D* × 495*L*	359*F* × 793*L*	Yes	173*F* (inner) × 5.4 m *L*
Channel	Yes	No	No	No	395	Yes
Total weight (kg)	273	~640	~25	—		—

Source: BWR and PWR. Data from Appendix K. The BWR/5 is the Nine Mile Point 2 unit (NMP2) and the PWR is the four-loop Seabrook unit. PHWR: Adopted from Knief, R. A. *Nuclear Engineering: Theory and Technology of Commercial Nuclear Power*, pp. 707–717. American Nuclear Society, La Grange Park, IL, 2008. HTGR: Breher, W., Neyland, A., and Shenoy, A. *Modular High-Temperature Gas-Cooled Reactor (MHTGR) Status*. GA Technologies, GA-A18878, May 1987. AGR-Heysham-2: Adopted from AEAT/R/PSEG/0405 Issue 3. *Main Characteristics of Nuclear Power Plants in the European Union and Candidate Countries*. Report for the European Commission, September 2001; *Nuclear Engineering International*. Supplement. August 1982. Alderson, M. A. H. G. (UKAEA, pers. comm., October 6, 1983 and December 6, 1983); Getting the most out of the AGRs, *Nucl. Eng. Int.*, 28(358):8–11, August 1984. Superphenix-1: Adopted from IAEA-TECDOC-1531. *Fast Reactor Database 2006 Update*. International Atomic Energy Agency. December 2006.

[a] The most recent General Electric BWR fuel bundle designs, which have 9 × 9 and 10 × 10 rod arrays, have been introduced but data for many parameters are held proprietary. The table presents available data for a typical 9 × 9 GE11 design.

[b] Fuel and fuel rod dimensions: diameter (*D*), heated length (*H*), total length (*L*), (across the) flats (*F*).

[c] Blends of fuel microspheres are molded to form fuel cylinders each having a diameter of 15.7 mm and a length of 5–6 cm

[d] SFBR-core (C), SFBR-radial blanket (RB).

[e] Early HTGRs (the German AVR and THTR) employed pebble fuel rather than prismatic graphite blocks.

[f] HTGR-standard assembly (SA), HTGR-control assembly (CA).

[g] Not applicable.

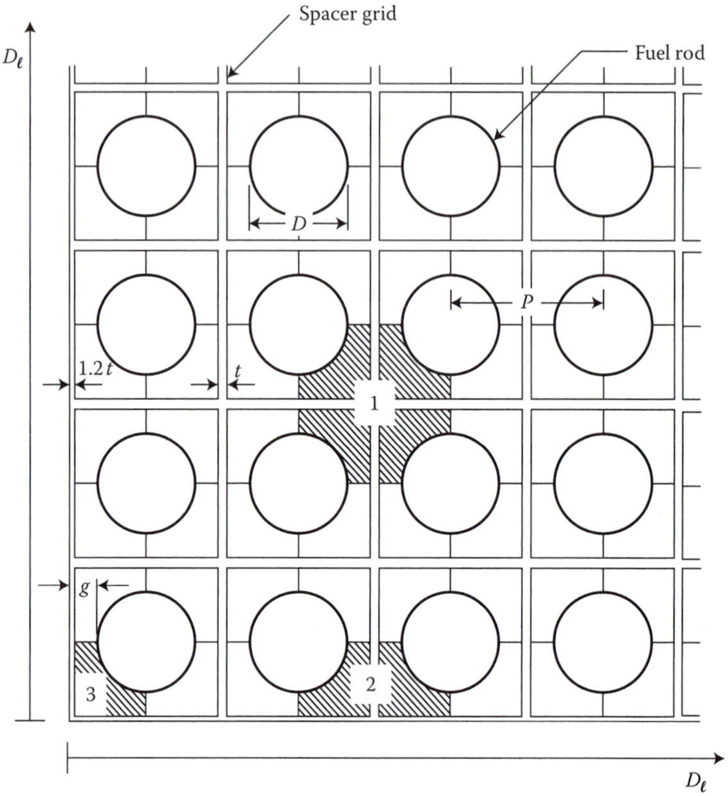

FIGURE 1.12 Typical fuel array for a light water reactor. Subchannel designation: 1, interior; 2, edge; 3, corner.

These three typical types of subchannels are defined as coolant regions between fuel rods and hence are "coolant-centered" subchannels. Alternately, a "rod-centered" subchannel has been defined as that coolant region surrounding a fuel rod. This alternate definition is infrequently used. All arrays of LWR fuel assemblies are held in place by above-assembly "springs." Tubes that may normally contain water or control rods provide structural flexural rigidity by virtue of their enhanced diameter and cladding thickness.

1.5.1 LWR Fuel Bundles: Square Arrays

A typical PWR fuel assembly, including its grid-type spacer, is shown in Figure 1.13. The spring clips of the spacer contact and support the fuel rods. A variety of spacer designs are now in use. Table 1.4 summarizes the number of subchannels of various-sized square arrays. Modern BWRs utilize assemblies of 64–100 rods, whereas PWR assemblies are typically composed of 225–289 rods. The formulae for subchannel and bundle dimensions, based on a PWR-type ductless assembly, are presented in Appendix J.

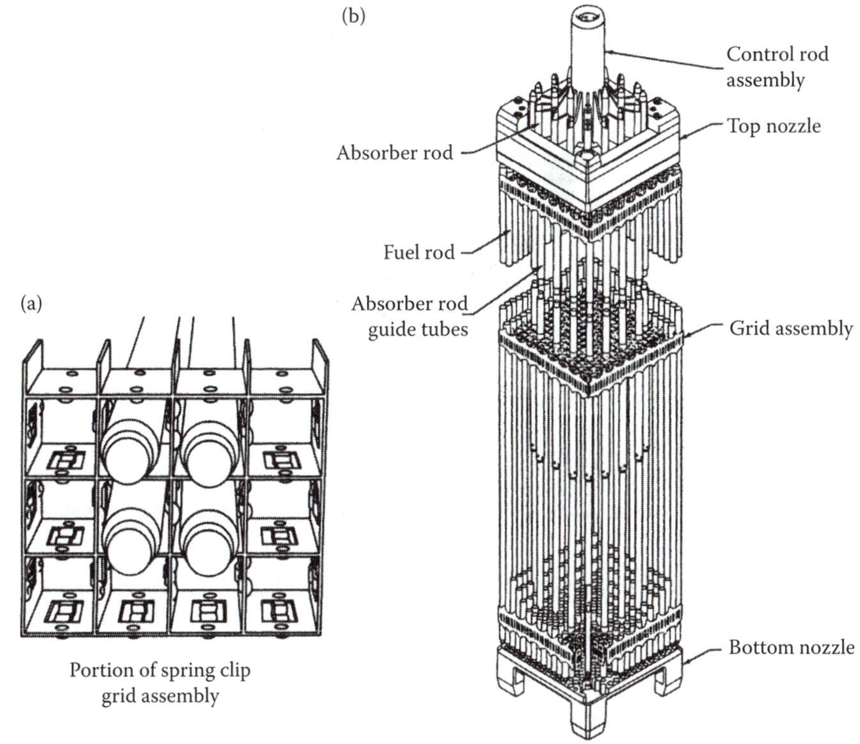

FIGURE 1.13 (a) Typical spacer grid for a light water reactor fuel assembly. (Courtesy of Westinghouse Electric Corporation.) (b) Typical light water reactor fuel assembly. (From Shultis, J. K. and Faw, R. E., *Fundamentals of Nuclear Science and Engineering*, 2nd Ed. CRC Press, Boca Raton, FL, 2008.)

TABLE 1.4
Subchannels for Square Arrays

Rows of Rods	N_p Total No. of Rods	N_1 No. of Interior Subchannels	N_2 No. of Edge Subchannels	N_3 No. of Corner Subchannels
1	1	0	0	4
2	4	1	4	4
3	9	4	8	4
4	16	9	12	4
5	25	16	16	4
6	36	25	20	4
7	49	36	24	4
8	64	49	28	4
N_{rows}	N_{rows}^2	$(N_{rows} - 1)^2$	$4(N_{rows} - 1)$	4

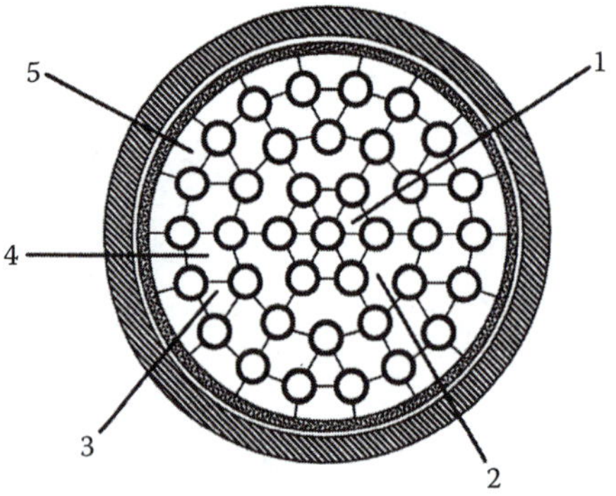

FIGURE 1.14 Fuel array of AGRs and PHWRs. Subchannel types: 1, interior first row (triangular); 2, interior second row (irregular); 3, interior third row (triangular); 4, interior third row (rhombus); 5, edge outer row. *Note:* Center pin is fueled in the PHWR and unfueled in the AGR. Some PHWRs now use the new Canadian Flexible Fuel Bundle (CANFLEX®) in which the central pin contains natural uranium with about 4% dysprosium burnable poison.

1.5.2 PHWR and AGR Fuel Bundles: Mixed Arrays

The geometry and subchannel types for the PHWR and AGR fuel bundles are shown in Figure 1.14. Because these arrays are arranged in a circular sleeve, the geometric characteristics are specific to the number of rods in the bundle. Therefore, the exact number of rods in the PHWR and the AGR bundle is shown.

1.5.3 SFBR/SFR Fuel Bundles: Hexagonal Arrays

A typical hexagonal array for a sodium-cooled reactor assembly with the rods wire wrapped is shown in Figure 1.15. As with the light water reactor, a different number of rods are used to form bundles for various applications. A typical fuel assembly has about 271 rods. However, arrays of 7–331 rods have been designed for irradiation and out-of-pile simulation experiments of fuel, blanket, and absorber materials. The axial distance over which the wire wrap completes a helix of 360° is called the lead length or axial pitch. Therefore, axially averaged dimensions are based on averaging the wires over one lead length. Table 1.5 summarizes the number of subchannels of various-sized hexagonal arrays and Appendix J presents the dimensions for unit subchannels and the overall array.

Additional useful relations between N_p, N_{ps}, N_1, and N_2 are as follows:

$$N_{ps} = \frac{\left[1+\sqrt{1+\frac{4}{3}(N_p-1)}\right]}{2}$$

$$N_p = 3N_{ps}(N_{ps}-1)+1 \quad \text{or alternately } 3N_{rings}^2+3N_{rings}+1$$

$$N_1 = 6(N_{ps}-1)^2$$

$$N_2 = 6(N_{ps}-1)$$

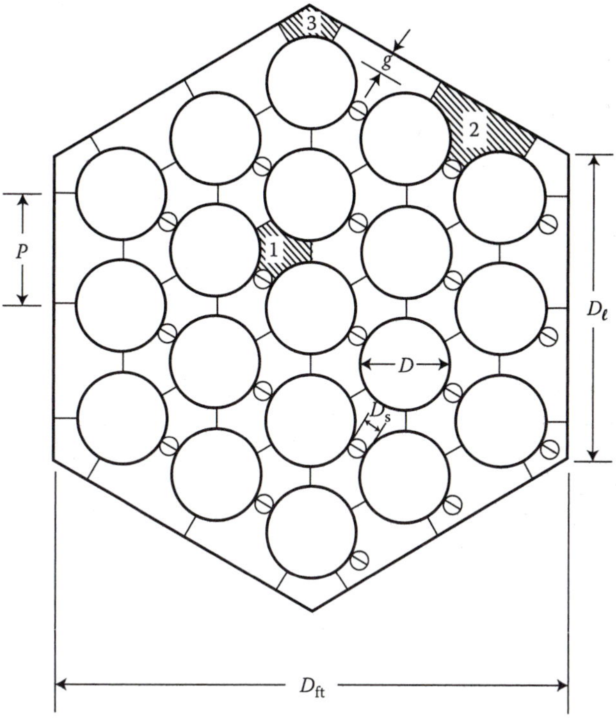

FIGURE 1.15 Typical fuel array for a liquid-metal-cooled fast breeder reactor. Subchannel designation: 1, interior; 2, edge; 3, corner. $N = 19$ in this example. *Note:* The sectional view of the wire should strictly be elliptical.

TABLE 1.5
Subchannels for Hexagonal Arrays

Rings of Rods	N_p Total No. of Rods	N_{ps} No. of Rods Along a Side	N_1 No. of Interior Subchannels	N_2 No. of Edge Subchannels	N_3 No. of Corner Subchannels
1	7	2	6	6	6
2	19	3	24	12	6
3	37	4	54	18	6
4	61	5	96	24	6
5	91	6	150	30	6
6	127	7	216	36	6
7	169	8	294	42	6
8	217	9	384	48	6
9	271	10	486	54	6
N_{rings}	$\sum_{n=1}^{N_{rings}} 6n+1$	$N_{rings}+1$	$6N_{rings}^2$	$6N_{rings}$	6

1.6 ADVANCED WATER- AND GAS-COOLED REACTORS (GENERATIONS III AND III+)

These reactor designs retain the basic technologies of the operating reactors but are improved in reliability of their operational and safety systems. They also have features to enhance the maintenance and repair efficiencies to reduce the operating costs.

The reactors are of two classes: (1) designs using actively powered safety systems that are evolutionary improvements over existing operating reactors, and (2) designs using passive safety system features that rely on natural forces, for example, gravity-driven flows, thermal conduction, and single explosive discharge-activated valves.

In the first class, the principal PWR plants are the European-pressurized reactor or internationalized as the evolutionary power reactor (EPR) and the advanced-pressurized water reactor (APWR). The corresponding principal BWR plant is the ABWR*.

In the second class, the principal PWR plants are the light water-cooled and moderated Advanced Passive 1000 (AP 1000) and the light water-cooled and heavy water-moderated Advanced Canadian Deuterium Uranium Reactor (ACR 1000). The corresponding BWR plant is the ESBWR.

The advanced gas-cooled reactors use helium as the coolant, graphite as the moderator, and the refractory-coated particle fuel described in Section 2.4.2 of Chapter 2. Two designs exist—the pebble-bed modular reactor (PBMR) and the prismatic modular gas reactor (PMGR)—also called the gas turbine-modular helium reactor (GT-MHR). As the terms pebble bed and prismatic connote, the designs differ in the form of the graphite within which the particle fuel is placed. In the PBMR, the fuel particles are contained in tennis-ball-sized spheres that are refueled online. In the PMGR, the fuel particles are contained in cylindrical graphite compacts that are in turn embedded in prismatic-shaped graphite blocks. These blocks are removed from the core and replaced at periodic-refueling shutdown intervals.

Tables 1.6 and 1.7 summarize the principal characteristics of the advanced water- and gas-cooled reactors. Table 1.6, analogous to Table 1.2, covers the thermodynamic cycle and coolant system characteristics. Table 1.7, analogous to Table 1.3, covers the fuel characteristics.

1.7 ADVANCED THERMAL AND FAST NEUTRON SPECTRUM REACTORS (GENERATION IV)

These reactors are cooled by a wide variety of coolants—liquid metal sodium, lead or lead–bismuth, helium, and supercritical carbon dioxide gas, liquid salt, and supercritical water. They were identified through an international evaluation [12] based on a set of sustainability, safety, economic, proliferation resistance, and physical protection criteria. Their individual development is led by the interested countries that coordinate their efforts through the Generation IV International Forum.

* The ABWR is unique among advanced water reactors in having been deployed prior to 2000. Various units have been built by General Electric in combination with Hitachi and others with Toshiba.

TABLE 1.6
Typical Characteristics of the Thermodynamic Cycle for Generation III+ Reactors

	PWR			BWR		HTGR	
Characteristics	**EPR**[c]	**US-APWR**[d]	**AP1000**[e]	**ABWR**[f]	**ESBWR**[g]	**PBMR**[h]	**PMGR**[i]**(GT-MHR)**
				Reference Design			
Manufacturer	Areva	Mitsubishi	Westinghouse	General Electric	General Electric	PBMR Pty. Ltd. of S. Africa	General Atomics
Power conversion cycle	Rankine	Rankine	Rankine	Rankine	Rankine	Rankine	Brayton
No. coolant systems	2	2	2	1	1	1	1
Primary coolant	H_2O	H_2O	H_2O	H_2O	H_2O	He	He
Secondary coolant	H_2O	H_2O	H_2O	[k]	[k]	H_2O	None
				Energy Conversion[b]			
Gross thermal power, MW_{th}	4590	4451	3400	3926	4500	200	600
Net electrical power, MW_e	1580	1700	1117	1371	1550	80	288
Efficiency (%)	34.4	38.0	32–34%[j]	34.4	34.4	40	48
				Heat Transport System			
No. primary loops/pumps	4/4	4/4	2/4	Vessel only/10 internal pumps	Vessel only/0	1/(2 circulators)	1/(1 circulator)
No. steam generators	4	4	2	[k]	[k]	1	[k]
Steam generator type	U-tube	U-tube	U-tube	[k]	[k]	Once-through	[k]
				Thermal Hydraulics			
Primary coolant							
Pressure (MPa)	15.5	15.5	15.5	7.17	7.17	6.1	7.07

continued

TABLE 1.6 (continued)

Typical Characteristics of the Thermodynamic Cycle for Generation III+ Reactors

	PWR			BWR		HTGR	
Characteristics	**EPR**[c]	**US-APWR**[d]	**AP1000**[e]	**ABWR**[f]	**ESBWR**[g]	**PBMR**[h]	**PMGR**[i]**(GT-MHR)**
Inlet temp. (°C)	295.3	289	280.7	278[a]	270–272[a]	250	490
Ave. outlet temp. (°C)	329.2	323.7	321.1	14.5% quality	17.0% quality	750	850
Core flow rate (Mg/s)	21.0	~22	13.46	14.5	10	0.077	0.321
Volume or mass	460 m³	—	271.8 m³	—	—	—	6000–7000 kg
Secondary Coolant							
Pressure (MPa)	7.65	6.9	5.67	[k]	[k]	19	[k]
Inlet temp. (°C)	230	235.9	226.7	[k]	[k]	170	[k]
Outlet temp. (°C)	291.9	282.8	272.9	[k]	[k]	530	[k]

[a] Feed water temperature: 215.6°C.

[b] Electrical output and efficiency are strongly dependent on site-specific characteristics, which especially impact the condenser backpressure and the amount of house loads.

[c] M. Parece, 2008, pers. comm.

[d] Design Certification Application: Design Review Document, December 2007.

[e] L. Oriani, 2010, pers. comm.

[f] ABWR Plant General Description, October 2006.

[g] ESBWR Plant General Description, October 2006, and P. Saha, 2008, pers. comm.

[h] M. Anness, 2010, pers. comm.

[i] A. Shenoy, 2008, pers. comm.

[j] The plant efficiency range provided is representative of different site conditions and site-specific design solutions.

[k] Not applicable.

TABLE 1.7
Typical Characteristics of the Fuel for Generation III+ Reactors[a]

	PWR			BWR		HTGR	
Characteristics	**EPR**	**US-APWR**	**AP1000**	**ABWR**	**ESBWR**	**PBMR**	**PMGR (GT-MHR)**
Reference Design							
Manufacturer	Areva	Mitsubishi	Westinghouse	General Electric	General Electric	Pty. Ltd. of South Africa	General Atomics
Moderator	H_2O	H_2O	H_2O	H_2O	H_2O	Graphite	Graphite
Neutron energy	Thermal	Thermal	Thermal	Thermal	Thermal	Thermal	Thermal
Fuel production	Converter	Converter	Converter	Converter	Converter	Converter	Converter
Fuel[b]							
Geometry	Cylindrical pellet	Cylindrical pellet	Cylindrical pellet	Cylindrical pellet	Cylindrical pellet	Coated TRISO microspheres	Coated TRISO microspheres
Dimensions (mm)	8.19D	8.19D	8.19$D \times$ 9.83L	8.76D	8.76D	[c]	Fissile: 770 μm D; Fertile: 850 μm D
Chemical form	UO_2	UO_2	UO_2	UO_2	UO_2	UO_2	UCO
Fissile (first core avg. wt%)	^{235}U (<5%)	^{235}U (<5%)	^{235}U (4.8 at eq.core)	^{235}U (2)	^{235}U (<5%)	^{235}U (9.6)	^{235}U
Fertile	^{238}U	^{238}U	^{238}U	^{238}U	^{238}U	^{238}U	^{238}U
Fuel Rods							
Geometry	Pellet stack in clad tube	Pellet stack in clad tube	Pellet stack in clad tube	Pellet stack in clad tube	Pellet stack in clad tube	Pebbles	Cylindrical fuel compact
Dimensions (mm)	9.50$D \times$ 4.20 m H ($\times$ 4.55 m L)	9.50$D \times$ 4.27m H	9.50$D \times$ 4.267 m H ($\times$4.795 m L)	10.26$D \times$ 3.7m H ($\times$4.47 m L)	10.26$D \times$ 3.05m H ($\times$3.79 mL)	60D	12.446$D \times$ 49.276H

continued

TABLE 1.7 (continued)

Typical Characteristics of the Fuel for Generation III+ Reactors[a]

	PWR			BWR		HTGR	
Characteristics	**EPR**	**US-APWR**	**AP1000**	**ABWR**	**ESBWR**	**PBMR**	**PMGR (GT-MHR)**
Clad material	M5™	Zirlo™	Zirlo™	Zr-2	Zr-2	No clad	No clad
Clad thickness (mm)	0.57	0.56	0.572	0.66	~0.66	[d]	[d]
				Fuel Assembly			
Geometry	17 × 17 square rod array	17 × 17 square rod array	17 × 17 square rod array	10 × 10 square rod array	10 × 10 square rod array	Pebble bed	Cylindrical fuel compacts within a hex. graphite block
Rod pitch (mm)	12.60	12.60	12.60	12.95	12.95	[d]	18.8
							210 fuel holes/block
No. rod locations	289	289	264	289	100	100	4.52×10^5
No. fuel rods	265	214 × 214	264	92	92	[d]	15 fuel compacts/fuel hole/block
Outer dimensions (mm)	214 × 214	No	214 × 214	140 × 140	140 × 140	[d]	360F × 800L
Channel	No	~760	No	Yes	Yes	[d]	[d]
Total weight (kg)	~760		786	183 (UO_2 only)	144 (UO_2 only)	~0.21 per pebble	160

[a] Data sources as given in Table 1.6.

[b] Fuel and fuel rod dimensions: diameter (*D*), heated length (*H*), total length (*L*), (across the) flats (*F*).

[c] 0.92-mm particle diameter; 0.5-mm kernel diameter.

[d] Not applicable.

The missions of these reactors include production of electricity, process heat including hydrogen as well as waste management by transmutation of actinides (neptunium, plutonium, americium, and curium). Principal reactor characteristics are therefore the average neutron energy and primary system outlet temperature. The Generation IV reactor concepts are displayed in terms of these variables in Figure 1.16. These fast and thermal reactors are, respectively, the

- Sodium-cooled fast reactor (SFR)
- Lead-cooled fast reactor (LFR)
- Gas-cooled fast reactor (GFR)
- Supercritical water-cooled reactor (SCWR)
- Very-high-temperature (gas-cooled) reactor (VHTR)
- Molten salt reactor (MSR)

The salt and supercritical water reactors that were emphasized as thermal systems were also cited as potential fast reactor systems. Of these, significant work has been done on the salt fast reactor which is called the liquid salt-cooled fast reactor (LSFR) [11]. The LSFR uses solid fuel and salt coolant, whereas the MSR uses salt as the coolant and fuel carrier. The SCWR can be based on either light or heavy water coolant. The SCWR designed to operate as a fast spectrum reactor adopts a tighter fuel lattice pitch than the thermal spectrum design. Also, it does not adopt the special moderator water rods or other special moderator materials of the thermal spectrum design.

Table 1.8 summarizes the principal characteristics of the Generation IV reactors. Achievable fuel burnup is not specified since it depends on pin geometry, for example, gas plenum length, specific cladding material, and fuel pellet density and operating temperature, all of which are under optimization. However, because of longer fission gas plenum length, steel versus zirconium alloy cladding, and higher fuel pellet-operating temperature, oxide fuel in fast reactors can achieve a burnup

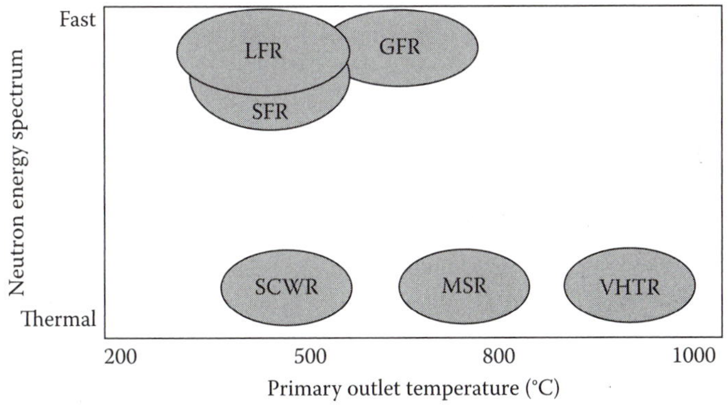

FIGURE 1.16 Conceptual mapping of Generation IV nuclear reactor concepts. (Adopted from Corradini, M. L. *Rohsenow Symposium on Future Trends in Heat Transfer*, Massachusetts Institute of Technology, MA, May 16, 2003.)

TABLE 1.8
Typical Characteristics of Generation IV Reactors

Parameters	LFR	SFR	GFR	LSFR	VHTR[b]	SCWR[a]	MSR
Mission (products)	Electricity, hydrogen	Electricity	Electricity, hydrogen	Electricity, hydrogen	Hydrogen	Electricity	Electricity, hydrogen
Power (range), (MW_{th})	125–3000	400–4000	1500–3000	2400	600	4000	2500
Net plant efficiency (%)	43	43	48	44	>50	42	44–50
Coolant inlet/outlet temperature (°C) and pressure (MPa)	479/550/0.1	371/530–550/0.1	490/850/9.1	496/581/0.1	640/1000/5–7	280/510/25	565/700/4.8×10^{-6}
Fuel[c]	NU–TRU–Zr[e] (metal or oxide) or UN Steel-clad rods	NU–TRU–Zr (metal or oxide) Steel-clad rods	UC–SiC Cercer plates, blocks, or rods[d]	U–TRU–Zr (metal or oxide) Steel-clad rods	Triso particles	UO_2, steel-clad rods	UF_4 dissolved in coolant[b]
Moderator	None	None	None	None	Graphite pebble or hex block	Water rods	Graphite
Spectrum	Fast	Fast	Fast	Fast	Thermal	Thermal	Thermal
Coolant	Pb or Pb · Bi	Na	He or CO_2	Molten chloridesalt	He	Supercritical water	Molten fluoride salts
Power density (kW/L)	100	300	100	100	4–8	70	20 (2/3 of fuel is ex-core)
Fuel enrichment (%)	15 (Pu)	16 (Pu)	16 (Pu)	10–16.5 (Pu)	8	6.3	3.3
Specific power (kW/kg_{IHM})	30	80	38	22	100	30	30

Source: Adopted from Driscoll, M. J. and Hejzlar, P. *Nucl. Eng. Technol.*, 37(1): February 2005; Todreas, N. E. et al., *Nucl. Eng. Des.* 239(12):2582–2595, December 2009.

[a] There is also a CANDU version of the SCWR; a fast reactor version of the SCWR is also an option.
[b] There is also a (fuel-free) molten salt-cooled version of the VHTR.
[c] Fuel shown as uranium-based actually employs plutonium with or without minor actinides as the fissile loading.
[d] UN and $U^{15}N$ are also candidates.
[e] Naturan uranium (NU), transuranic elements (TRU), and zirconium (Zr).

about double that of thermal reactor utilization, that is, ~100–150 GWd/Mt_{IHM}* versus 50–60 GWd/Mt_{IHM}. Achievable burnup of metal fuels in fast reactor application also strongly depends on the selected cladding material and fission gas plenum length but is in the range around 120 GWd/Mt_{IHM} as well. Further details will be forthcoming as conceptual designs for these concepts are developed by interested countries.

TABLE 1.9
Worldwide Utilization of Power Reactor Technology, Thermal Reactor Types

			Coolant						
			Light water		Heavy water		Organic HB-40 Santowax-OM	Gas hydrogen, nitrogen, CO_2, helium	Liquid metal NaK Na
			Pressurized	Boiling	Pressurized	Boiling			
Moderators	Light water								
	Heavy water	Vessel							
		Tube							
	Graphite								
	Beryllium								
	Organic								

References for thermal reactor types:
1. List of operational nuclear power plants. *Nucl. News*, August 1992.
2. Dietrich, J. R. and Zinn, W. H. *Solid Fuel Reactors*, Reading, MA: Addison-Wesley Pub. Co., 1958.
3. *Directory of Nuclear Reactors*, Vienna: International Atomic Energy Commission, published annually.
4. Kuljian, H. A. *Nuclear Power Plant Design*, Cranbury, NJ: A.S. Barnes & Co., 1968.
5. Meserve, R., Chairman, *Safety Issues at the Defense Production Reactors: A Report to the U.S. Department of Energy*, National Academy Press, Washington, DC, 1987.
6. Zinn, W. H., Pittman, F. K., and Hogerton, J. F. *Nuclear Power, USA*, McGraw-Hill, New York, NY, 1964.

* The symbol t signifies tonne (metric ton). Hence the prefix M in Mt which signifies metric ton or tonne is redundant. Use of the prefix M before t is, however, retained in this chapter to clearly identify that t is not English tons but metric tons or tonnes.

PROBLEM

1.1. Worldwide utilization of power reactor technology
   1. For each position of Table 1.9, identify either:
      a. The principal technical reason for which this moderator–coolant combination cannot be exploited, or
      b. The name of one (or more) terrestrial power reactor plants that have been built using this combination.
   2. Which of these combinations would be best for submarine propulsion? Explain your choice.

REFERENCES

1. AEAT/R/PSEG/0405 Issue 3, *Main Characteristics of Nuclear Power Plants in the European Union and Candidate Countries.* Report for the European Commission, September 2001.
2. Breher, W., Neyland, A., and Shenoy, A., *Modular High-Temperature Gas-Cooled Reactor (MHTGR) Status.* GA Technologies, GA-A18878, May 1987.
3. Corradini, M. L., Advanced nuclear energy systems: Heat transfer issues and trends. *Rohsenow Symposium on Future Trends in Heat Transfer*, Massachusetts Institute of Technology, MA, May 16, 2003.
4. Driscoll, M. J. and Hejzlar, P., Reactor physics challenges in Gen-IV reactor design. *Nucl. Eng. Technol.*, 37(1):1–10, February 2005.
5. IAEA-TECDOC-1531, *Fast Reactor Database 2006 Update.* International Atomic Energy Agency, December 2006.
6. Knief, R. A., *Nuclear Engineering: Theory and Technology of Commercial Nuclear Power*, pp. 707–717. La Grange Park, IL: American Nuclear Society, 2008.
7. *Nuclear Engineering International.* Supplement, August 1982.
8. Getting the most out of the AGRs, *Nucl. Eng. Int.*, 28(358):8–11, August 1984.
9. *Seabrook Power Station Updated Safety Analysis Report*, Revision 8, Seabrook Station, Seabrook, NH, 2002.
10. Shultis, J. K. and Faw, R. E., *Fundamentals of Nuclear Science and Engineering*, 2nd Ed. Boca Raton, FL: CRC Press, 2008.
11. Todreas, N. E., Hejzlar, P., Nikiforova, A., Petroski, R., Shwageraus, E., Fong, C. J., Driscoll, M. J., Elliott, M. A., and Apostolakis, G., Flexible conversion ratio fast reactors: Overview. *Nucl. Eng. Des.*, 239(12):2582–2595, December 2009.
12. U.S. Department of Energy, *A Roadmap to Deploy New Nuclear Power Plants in the United States by 2010*, Vol. I: *Summary Report.* Office of Nuclear Energy, Science and Technology and Nuclear Energy Research Advisory Committee, Subcommittee on Generation IV Technology Planning, October 31, 2001.
13. U.S. Department of Energy, *A Technology Roadmap for Generation IV Nuclear Energy Systems.* Nuclear Energy Research Advisory Committee and the Generation IV International Forum, December 2002.
14. Vendryes, G. A., Superphenix: A full-scale breeder reactor. *Sci. Am.*, 236:26–35, 1977.

2 Thermal Design Principles and Application

2.1 INTRODUCTION

The general principles of reactor thermal design are introduced in this chapter, with the focus on the operating parameters, design limits, and figures of merit by which the thermal design process is characterized. The energy of a power reactor originates from the fission process within the fuel elements. Energy deposited in the fuel is transferred to the coolant by conduction, convection, and radiation. A small fraction of the fission energy is also directly deposited in the coolant and structures. We do not attempt to present a procedure for thermal design because nuclear, thermal, and structural aspects are intertwined in a complicated, interactive process. Specific design analysis techniques are detailed in Volume II. Table 2.1 presents typical core thermal performance characteristics for the reference power reactor types introduced in Chapter 1.

2.2 OVERALL PLANT CHARACTERISTICS INFLUENCED BY THERMAL HYDRAULIC CONSIDERATIONS

Thermal hydraulic considerations are important when selecting overall plant characteristics. Core coolant temperature and pressure are key characteristics related to both the coolant selection and plant thermal performance. In particular, thermal efficiency is dictated by the bounds of the maximum allowable primary coolant outlet temperature and the minimum achievable temperature of the heat sink. Because the ambient temperature is relatively fixed, improved thermal efficiency requires increased reactor coolant outlet temperature. Figure 2.1 illustrates the relation among reactor plant temperatures for a typical PWR.

The analogous relations for both liquid- and gas-cooled fast reactors are illustrated in Figure 2.2. For liquid-cooled fast reactors, IHXs couple the primary coolant to the power conversion cycle coolant, whereas for gas-cooled fast reactors the direct power conversion cycle is used.

Bounds on the achievable primary outlet temperature depend on the coolant type. For liquid metals and molten salts, in contrast to water, the saturated vapor pressure is low, that is, less than atmospheric pressure at outlet temperatures of interest of 500–650°C. Thus, the outlet temperature for SFRs is not limited by the boiling point of the sodium coolant but, rather, by the creep lifetime characteristics of the

TABLE 2.1
Typical Core Thermal Performance Characteristics for Six Reference Power Reactor Types

Characteristics	BWR	PWR(W)	PHWR	HTGR	AGR	SFBR[a]
Core						
Axis	Vertical	Vertical	Horizontal	Vertical	Vertical	Vertical
No. of assemblies						
Axial	1	1	12	8	8	1
Radial	764	193	380	493	332	364 (C) 233 (BR)
Assembly pitch (mm)	152	214	286	361	460	179
Active fuel height (m)	3.588	3.658	5.94	6.30	8.296	1.0 (C) 1.6 (C + BA)
Equivalent diameter (m)	4.75	3.37	6.29	8.41	9.458	3.66
Total fuel weight (MT)	160 UO_2	101 UO_2	89.3 UO_2	1.56 U 34.0 Th	103.0 UO_2	29 UO_2
Reactor vessel						
Inside dimensions (m)	$6.05D \times 21.6H$	$4.83D \times 13.4H$	$7.6D \times 4L$	$11.3D \times 14.4H$	$20.25D \times 21.87H$	$21D \times 19.5H$
Wall thickness (mm)	152	224	28.6	4720	5800	25
Material[b]	SS-clad carbon steel	SS-clad carbon steel	Stainless steel	Prestressed concrete	Concrete helical prestressed	Stainless steel
Other features			Pressure tubes	Steel liners	Steel lined	Pool type
Power density core average (kW/L)	52.3	104.5	12	8.4	2.66	280
Linear heat rate						
Core average (kW/m)	17.6	17.86	25.7	7.87	17.0	29
Core maximum (kW/m)	47.24	44.62	44.1	23.0	29.8	45

Performance						
Equilibrium burnup (MWd/MT)	50,000	50,000	8300	105,000	20,000	110,000
Average assembly residence (full-power days)	2192	1644	470	1170	1320	640 (C) 320 (BR row 1) 640 (BR row 2)
Refueling						
Sequence	$\frac{1}{4}$ per year	$\frac{1}{3}$ per year	Continuous on-line	$\frac{1}{2}$ per year	Continuous on-line	Variable
Outage time (days)	25	25	None for refueling	Unavailable	None for refueling	32

Source: Principally Appendix K (Knief, R. A. *Nuclear Engineering: Theory and Technology of Commercial Nuclear Power*, 2nd Ed. La Grange Park, IL: American Nuclear Society, 2008.) except for AGR data from M. A. H. G. Alderson (pers. comm., October 6, 1983) and A. A. Debenham (pers. comm., August 5, 1988).

[a] SFBR: core (C), radial blanket (BR), axial blanket (BA).

[b] SS = stainless steel.

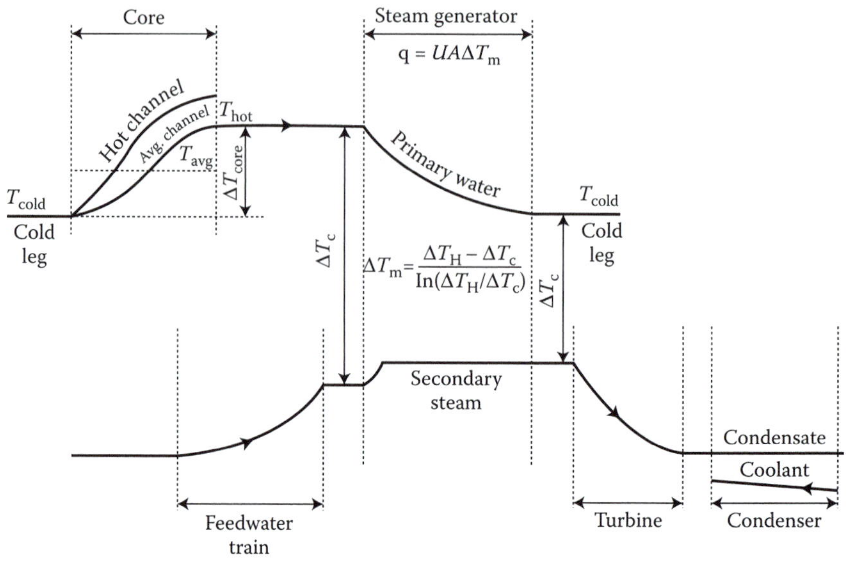

FIGURE 2.1 Relations among reactor plant temperatures for a typical PWR. (Adopted from Tong, L. S., *Nucl. Eng. Des.*, 6:301–324, 1967.)

stainless-steel primary system material and the allowable cladding temperature. Because the cladding operates for a limited in-core residency period, allowable cladding temperature limits are larger than the primary system material. Additionally, for liquid metal and molten salt coolants their freezing temperature imposes a constraint on the core inlet temperature as Figure 2.2 illustrates.

For water-cooled reactors, on the other hand, high core outlet coolant temperatures require correspondingly high system pressures (7–15.5 MPa), which increases

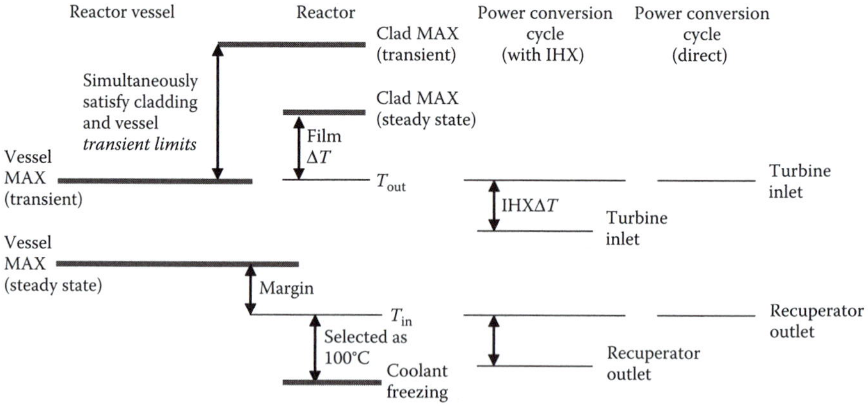

FIGURE 2.2 Thermal design logic (operating temperatures constrained by material limits for steady-state and transient conditions) for liquid- and gas-cooled fast reactors.

TABLE 2.2
Light-Water Reactor Thermal Conditions

Condition	PWR	BWR/5
Primary coolant outlet temperature	326.8°C	286.1°C
Primary coolant system pressure	15.51 MPa	7.14 MPa
Turbine steam saturation conditions		
Pressure	5.7 MPa	7.14 MPa
Temperature	272.3°C	287.2°C
Plant thermal efficiency	33.7%	32.0%

Source: Appendix K.

the stored energy in the coolant and requires increased structural piping and component wall thicknesses. Single-phase gas coolants offer the potential for high outlet temperatures without such inherently coupled high pressures. For these reactors, the system pressure is dictated by the desired core heat transfer capabilities, since gas properties that enter the heat transfer correlations are strongly dependent on pressure. The resulting pressures are moderate, that is, 4–8 MPa, whereas achievable outlet temperatures are high, that is, 635–850°C for the AGR and HTGR, respectively. The supercritical CO_2-cooled GFR, however, is designed to operate on a direct cycle at 20 MPa. This high pressure is dictated by the desire to operate the compressor at supercritical CO_2 conditions* and the cycle at a modest pressure ratio but sufficient to achieve a high thermodynamic cycle efficiency [4].

The plant thermal efficiency depends on the maximum temperature in the power conversion system. For indirect cycles, this temperature is lower than the reactor coolant outlet temperature, owing to the temperature difference needed to transfer heat between the primary and secondary systems in the steam generator or IHX. When a direct cycle is employed, the reactor outlet temperature is identical (neglecting losses) to the inlet temperature of the turbine. This outlet temperature is also limited to the saturation condition, since no reactors operate under primary coolant superheat conditions.† In a typical BWR, however, the average outlet enthalpy achieved corresponds to an average quality of 15%.‡ The PWRs and BWRs achieve approximately equal thermal efficiencies, because the turbine steam conditions are comparable even though the primary system pressure and temperature conditions significantly differ (Table 2.2). Note that, because of detailed differences in thermodynamic cycles, the example PWR plant achieves a slightly higher plant thermal efficiency than does the BWR plant even though its steam temperature is lower.

* The critical point for CO_2 is 304.1 K, 7.3773 MPa.

† Superheated reactor concepts were tested at GE Valecitos, California site and in the BONUS reactor in Puerto Rico, but not found practical due to cladding material deterioration. However, Babcock and Wilcox designed PWRs with once-through steam generators which did produce slightly superheated secondary coolant to the turbine.

‡ The ESBWR has a higher exit quality of 24.3% (see Table 1.6).

Other plant characteristics are strongly affected by thermal hydraulic conditions. Some notable examples are as follows:

1. Primary coolant temperature
   a. Corrosion behavior, though strongly dependent on coolant chemistry control, is also temperature dependent.
   b. The reactor vessel resistance to brittle fracture degrades with fluence (accumulated neutron exposure). Vessel behavior under low-temperature, high-pressure transients from operating conditions is carefully evaluated to ensure that the vessel retains the required material toughness over its lifetime.
2. Primary system inventory
   a. The time response during accidents and less severe transients strongly depends on coolant inventories. The reactor vessel inventory above the core is important to behavior in primary system rupture accidents. For a PWR, in particular, the pressurizer and steam generator inventories dictate the transient response for a large class of events.
   b. During steady-state operation, the inlet plenum serves as a mixing chamber to homogenize the coolant flow into the reactor. The upper plenum serves a similar function in plants with multiple cooling systems with regard to the IHX/steam generator while at the same time protecting the reactor vessel nozzles from thermal shock in transients.
3. System arrangement
   a. The arrangement of reactor core and IHX/steam generator thermal centers is crucial to the plant's capability to remove decay heat by natural circulation.
   b. Orientation of pump shafts and heat exchanger tubes coupled with support designs and impingement velocities is important relative to the prevention of troublesome vibration problems.

The system arrangement issues are sometimes little appreciated if the design of a reactor type has been established for some years. For example, note that the dominant characteristics of the primary coolant system configuration for a typical U.S.-operating PWR (see Figure 1.7) are cold leg pump placement, vertical orientation of pumps, vertical orientation of the steam generators, and elevation of the steam generator thermal center of gravity above that of the core. The pumps are located in the cold leg to take advantage of the increased subcooling, which increases the available net positive suction head and thus the margin to the onset of pump cavitation. Vertical orientation of the pumps provides for convenient accommodation of loop thermal expansion in a more compact layout and facilitates their serviceability, and the vertical steam generator orientation eases the structural support design as well as eliminates a possible flow stratification problem.* The relative vertical arrangement of the thermal center of

* However, horizontal steam generators have been successfully employed in the Russian VVER design. Unlike vertical steam generators, in horizontal steam generators, sludge does not accumulate on the tubesheet.

steam generators above the core is established to provide the capability for natural circulation heat removal in the primary loops. These characteristics lead to the creation of a loop seal (a section of U-shaped piping) of varying size in each arrangement between the steam generator exit and the pump inlet. This loop seal is of significance when considering reactor system response in natural circulation and loss-of-coolant circumstances. In Generation III+ PWRs, the loop seal has been eliminated by redesign of the primary system arrangement. Also, in later versions of Generation III BWRs, the external recirculation pumps have been eliminated.

2.3 ENERGY PRODUCTION AND TRANSFER PARAMETERS

Energy production in a nuclear reactor core stems from neutron–nucleus interaction. Unlike neutrons whose characteristics are a function of many variables (space, direction, energy, and time), energy production is a function of only space and time. The various terms used in this text to describe the analysis of core energy production follow.

Parameter	Units
Volumetric energy (or heat) generation rate	$q'''(\vec{r})$ W/cm^3
Surface heat flux	$\vec{q}''(S)$ kW/m^2
Linear heat generation rate or linear power	$q'(z)$ kW/m
Energy generation rate per pin	$\dot{q}$ kW
Core power	$\dot{Q}$ MW

Two additional parameters below are also commonly used figures of merit of core thermal performance. They are discussed later (see Section 2.6) after design limits and design margin considerations are presented.

$$\text{Core power density: } Q''' \equiv \frac{\dot{Q}_{\text{th}}}{V_{\text{core}}} \quad (\text{kW/L}) \tag{2.1}$$

$$\text{Core-specific power: } Q^{\text{M}} \equiv \frac{\dot{Q}_{\text{th}}}{M_{\text{IHM}}} \quad (\text{kW/kg}_{\text{IHM}}) \tag{2.2}$$

where the superscript M reminds us that specific power is per unit fuel mass and the subscript IHM reflects that the mass of interest is the initial heavy metal portion of the fuel.

The relations among these terms must be well understood to ensure communication between the reactor physicist, the thermal designer, and the metallurgist or ceramicist. The reactor physicist deals with fission reaction rates, which lead to the volumetric energy generation rate: $q'''(\vec{r})$. The triple-prime notation represents the fact that it is an energy generation rate per unit volume of the fuel material. Normally in this text, a dot above the symbol is added for a rate, but for simplicity it is deleted here and in other energy quantities with primes. Hence, the energy generation rate is expressed per unit length cubed, as $q'''(\vec{r})$.

The thermal designer must calculate the fuel element surface heat flux, $\vec{q}''(S)$, which (at steady state) is related to $q'''(\vec{r})$ as

$$\iint_S \vec{q}''(S)\cdot\vec{n}\,\mathrm{d}S = \iiint_V q'''(\vec{r})\,\mathrm{d}V \tag{2.3}$$

where S is the surface area that bounds the volume V within which the heat generation occurs.

Both the thermal and the metallurgical designers express some fuel performance characteristics in terms of a linear power rating (alternately called the linear energy generation rate) $q'(z)$ as

$$\int_L q'(z)\,\mathrm{d}z = \iiint_V q'''(\vec{r})\,\mathrm{d}V \tag{2.4}$$

where L is the length of the volume V bounded by the surface S within which the heat generation occurs.

If the volume V is taken as the entire heat generating volume of a fuel pin, the quantity $\dot{q}$, unprimed, is the energy generation rate in a pin, that is,

$$\dot{q} = \iiint_V q'''(\vec{r})\,\mathrm{d}V \tag{2.5}$$

Finally, the core thermal power $\dot{Q}_{\mathrm{th}}$ is obtained by summing the energy deposited in each fuel pin summed over all the pins in the core (N_{p}) plus the energy generation that occurs in the moderator/coolant and noncladding structure yielding:

$$\dot{Q}_{\mathrm{th}} = \sum_{n=1}^{N_{\mathrm{p}}} \dot{q}_n + \dot{Q}_{\text{coolant and noncladding structure}} \tag{2.6}$$

Actually, as discussed in Chapter 3, approximately 2.5% of the reactor power is generated directly in the moderator/coolant and noncladding structural materials for a typical PWR while the corresponding value for a BWR is about 3.5%.

These general relations take many specific forms, depending on the size of the region over which an average is desired and the specific shape (plate, cylindrical, spherical) of the fuel element. For example, considering a core with N_{p} cylindrical fuel pins each having an active fuel length L, core average values of the thermal parameters that can be obtained from Equations 2.3 through 2.6 are related as

$$\begin{aligned}\dot{Q}_{\mathrm{th}} - \dot{Q}_{\text{coolant and noncladding structure}} &= N_{\mathrm{p}}\langle \dot{q}\rangle = N_{\mathrm{p}}L\langle q'\rangle = N_{\mathrm{p}}L\pi D_{\mathrm{co}}\langle q''_{\mathrm{co}}\rangle \\ &= N_{\mathrm{p}}L\pi R_{\mathrm{fo}}^2\langle q'''\rangle \end{aligned} \tag{2.7}$$

where D_{co} = outside cladding diameter; R_{fo} = fuel pellet radius; and $\langle\ \rangle$ = a core volume-averaged value. A more detailed examination of the above relations and the various factors affecting the heat generation distribution in the reactor core are discussed in Chapter 3.

2.4 THERMAL DESIGN LIMITS

The principal design limits for the power reactors discussed in Chapter 1 are detailed here. All reactors except for the thermal and fast gas-cooled reactors employ a metallic cladding to hermetically seal the cylindrical fuel pellets. The helium gas-cooled reactors use graphite and silicon carbide barriers around the fuel particles to contain the fission products within the fuel particles.

2.4.1 Fuel Pins With Metallic Cladding

For hermetically sealed fuel pins, thermal design limits are imposed to maintain the integrity of the cladding. In theory, these limits should all be expressed in terms of structural design parameters, that is, strain and fatigue limits for both steady-state and transient operation. However, the complete specification of limits in these terms is currently impractical because of the complex behavior of materials in radiation and thermal environments characteristic of power reactors. For this reason, design limits in power reactors have been imposed directly on certain temperatures and heat fluxes, although the long-term trend should be to transform these limits into more specific structural design terms.

The design limits for LWRs and the SFR which employ cylindrical, metallic cladding, oxide fuel pellets are summarized in Table 2.3. The distinction between conditions that would cause damage (loss of cladding integrity) and those that would exceed the design limits is highlighted. Also, both PWRs and BWRs have hydrodynamic stability limits. The inherent characteristics of light-water reactors limit cladding temperatures to a narrow band above the coolant saturation temperature and thus preclude the necessity for a steady-state limit on cladding midwall temperature. However, a significant limit on cladding average temperature does exist in transient situations, specifically in the loss of coolant accident (LOCA). For this accident a number of design criteria are being imposed, key among which is maintenance of the Zircaloy cladding below 2200°F (1204°C) to prevent extensive metal–water reaction from occurring.

The particular design limit that governs the reactor design varies with the reactor type and the continually evolving state of design methods. For example, for LWRs, the fuel centerline temperature is typically maintained well within its design limit due to restrictions imposed by the critical heat flux (CHF) limit. Furthermore, with the application of improved LOCA analysis methods during the mid-1980s, the LOCA-imposed limit on cladding temperature has not been the dominant limit. Finally, for LWRs, the occurrence of excessive mechanical interaction between the pellet and cladding (pellet–cladding interaction) has led to operational restrictions on allowable rates of change of reactor power and the extensive development of fuel and cladding materials to alleviate these restrictions on reactor load-following ability.

The CHF phenomenon results from a relatively sudden reduction of the heat transfer capability of the two-phase coolant. The resulting thermal design limit is expressed in terms of the departure from nucleate boiling (DNB) condition for PWRs and the critical power condition for BWRs. For fuel rods, where the volumetric energy-generation rate $q'''(r,z,t)$ is the independent parameter, reduction in surface

TABLE 2.3
Typical Fuel Pin Thermal Design Limits

Characteristics	PWR	BWR	SFR
Damage limit	1% cladding strain or MDNBR[a] ≤ 1.0	1% cladding strain or MCPR[a] ≤ 1.0	0.7% cladding strain
Design limits			
Fuel centerline temperature			
Steady state	—	—	—
Transient	No incipient melt	No incipient melt	No incipient melt
Clad average temperature			
Steady state	—	—	649–704°C (1200–1300°F)
Transient	<1204°C (2200°F) (LOCA)[a]	<1204°C (2200°F) (LOCA)	788°C (1450°F) for anticipated transients; 871°C (1600°F) for unlikely events
Surface heat flux			
Steady state	—	MCPR ≥ 1.2	—
Transient	MDNBR ≥ 1.3[b] at 112% power	—	—

[a] LOCA = loss of coolant accident; MDNBR = minimum departure from nucleate boiling ratio; MCPR = minimum critical power ratio.

[b] Corresponding value of minimum departure from nucleate boiling ratio is dependent on the particular correlation used, and can be as high as approximately 1.9.

heat transfer capability for nominally fixed bulk coolant temperature (T_b) and heat flux causes the cladding temperature to rise, that is,

$$T_{co} - T_b = \frac{q''}{h} = \frac{q'''R_{fo}^2}{hD_{co}} \tag{2.8}$$

where h = heat transfer coefficient. Physically, this reduction occurs because of a change in the liquid–vapor flow patterns at the heated surface. At low-void fractions typical of PWR-operating conditions, the heated surface, which is normally cooled by nucleate boiling, becomes vapor blanketed, resulting in a cladding surface temperature excursion by DNB. At high-void fractions typical of BWR-operating conditions, the heated surface, which is normally cooled by a liquid film, overheats owing to film dryout (Dryout). The Dryout phenomenon depends significantly on channel thermal hydraulic conditions upstream of the Dryout location rather than on the local conditions at the Dryout location. Because DNB is a local condition and Dryout depends on channel history, the correlations and graphical representations for DNB are in terms of heat flux ratios, whereas for Dryout they are in terms of channel power ratios.

These two mechanisms for the generic CHF phenomenon are shown in Figure 2.3. Correlations have been established for both these conditions in terms of different operating parameters (see Chapter 13). Because these parameters change over the fuel length, different margins exist between the actual operating heat flux and the

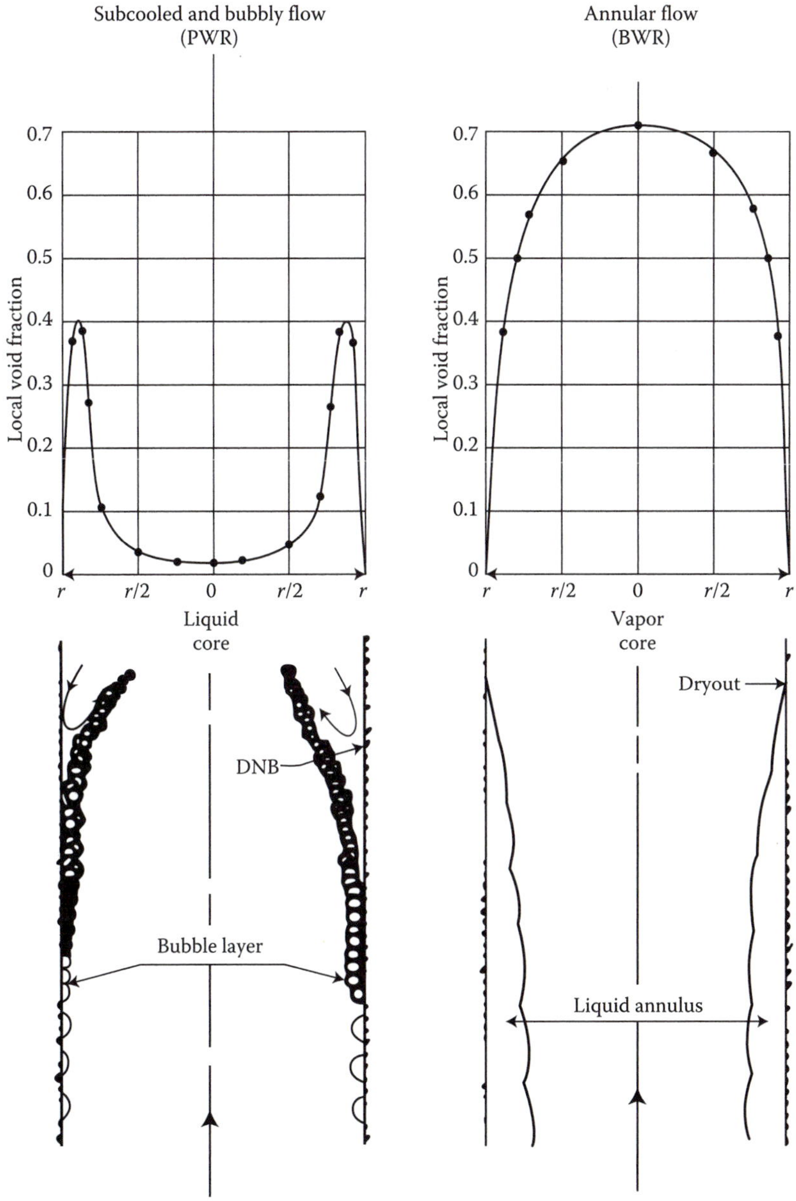

FIGURE 2.3 Critical heat flux mechanisms for PWR- and BWR-operating conditions. (From Tong, L. S., *Nucl. Eng. Des.*, 6:301–324, 1967.)

limiting heat flux for the occurrence of DNB or Dryout. These differences are illustrated in Figure 2.4 for a typical DNB case. The ratio between the predicted correlation heat flux and the actual operating heat flux is called the departure from nucleate boiling ratio (DNBR). This ratio changes over the fuel length and reaches a minimum value, as shown in Figure 2.4, somewhere downstream of the peak-operating heat flux location. An alternative representation in terms of bundle average conditions, which depend on total power input, exists for BWR Dryout conditions. This representation is expressed as the critical power ratio (CPR) and is presented in Chapter 13 along with the DNBR.

Critical condition limits are established for this minimum value of the appropriate ratio, that is, MDNBR or MCPR (Table 2.3). Furthermore, prior to Brown's Ferry Unit 1, the BWR MCPR limit was applied to operational transient conditions, as is the present PWR practice. Subsequently, it was applied to the 100% power condition. These limits for a BWR at 100% power and a PWR at 112% power allow for consistent overpower margin, as can be demonstrated for any specific case of prescribed axial heat flux distribution and coolant channel conditions.

For the SFR, the present practice is to require a level of subcooling such that the sodium temperature does not exceed its boiling temperature for transient conditions. Additionally, considerable effort is being applied to ensure that coolant voiding in accident situations can be satisfactorily accommodated. Hence, the SFR limits are placed on fuel and cladding temperatures. At present, no incipient fuel melting is allowed. However, work is being directed at developing cladding strain criteria. These criteria will reflect the fact that the stainless-steel cladding is operating in the creep regimen and that some degree of center fuel melting can be accommodated without cladding failure.

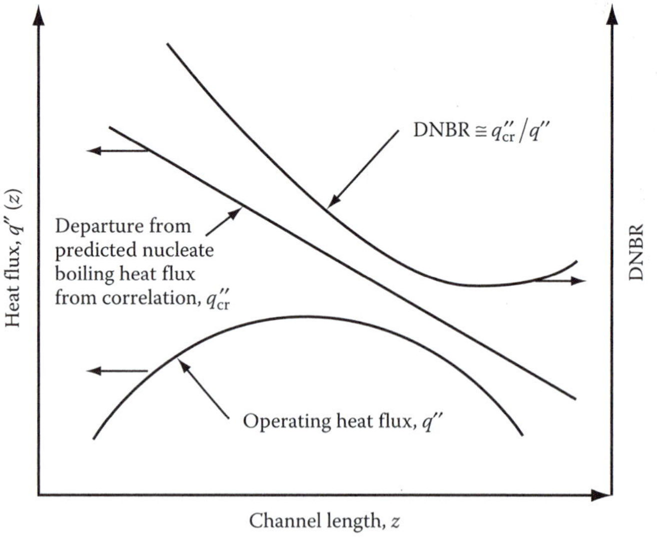

FIGURE 2.4 Axial variation of the DNBR.

2.4.2 Graphite-Coated Fuel Particles

The HTGR fuel in a block-type core is in the form of coated particles distributed within cylindrical graphite compacts. These compacts and coolant channels are symmetrically placed within the graphite matrix blocks. In an HTGR of the pebble-bed type, the coated particles are dispersed within a tennis-ball-sized graphite matrix. Two types of coated particles are used (Figure 2.5). The BIstructural ISOtropic-type (BISO) has a fuel kernel surrounded by a low-density pyrolitic carbon buffer region, which is itself surrounded by a high-density, high-strength pyrolitic carbon layer. The TRIstructural ISOtropic-type (TRISO) sandwiches a layer of silicon carbide between the two high-density pyrolitic carbon layers. In both fuel types, the inner layer and the fuel kernel are designed to accommodate expansion of the particle and to trap gaseous fission products. The buffer layer acts to attenuate the fission fragment recoils. The laminations described also help to prevent crack propagation. Silicon carbide is used to supply dimensional stability and low-diffusion rates. It has a greater thermal expansion rate than the surrounding pyrolitic carbon coating and thus is normally in compression. In both kinds of particles, the total coating thickness is about 150 μm. In early HTGRs, the TRISO coating was used for uranium fuel particles that are enriched to 93% ^{235}U. The BISO coating was used for thorium oxide particles, which comprise the fertile material. The TRISO particle is adopted for use in recent designs.

Because fission gas release occurs in this system by diffusion through the coatings and directly from rupture of the coatings, limits are imposed on steady-state fission product concentrations within the primary circuit. These activity levels are established to ensure that radiation doses resulting from accidental release of the

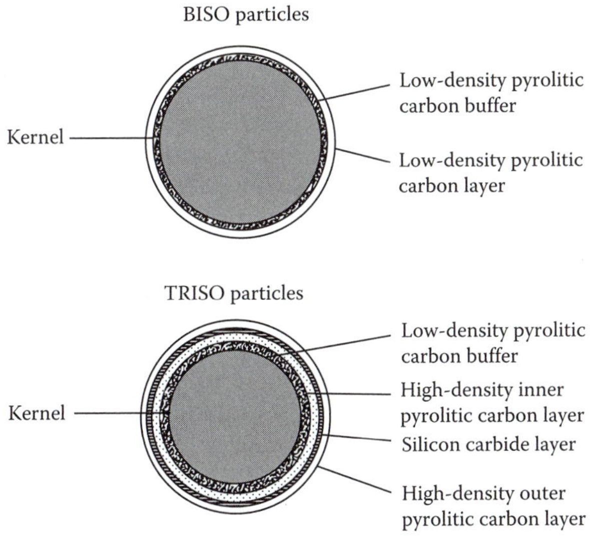

FIGURE 2.5 High-temperature gas reactor fuel particles. Typical dimensions: kernel diameters, 100–300 μm; total coating thickness, 50–190 μm.

primary circuit inventory to the atmosphere are within regulations. Given the dependence of fission gas release on temperature, these limits are also imposed on fuel particle center temperature.

Steady-state 100% power ≃ 1300°C (2372°F)
Peak transient ≃ 1600°C (2822°F) for short term

Since the particles are very small, the average temperature is very close to the center temperature. The full power limit minimizes steady-state diffusion, whereas the transient limit, based on in-pile tests, is imposed to minimize cracking of the protective coatings.

2.5 THERMAL DESIGN MARGIN

A striking characteristic of thermal conditions existing in any core design is the large difference among conditions in different spatial regions of the core. The impact of the variation in thermal conditions for a typical core is shown in Figure 2.6. Starting with the nominal core average condition characterized by the linear power rating $\langle q' \rangle$, defined by Equation 2.4, nuclear power-peaking factors, engineering, and process parameter uncertainty factors are sequentially applied, leading to a more limiting $\langle q' \rangle$ value. Each condition, appearing in Figure 2.6, is clearly defined by a combination of the terms nominal, average, peak, and maximum. No consistent set of definitions exists for these terms. However, a consensus of usage, which we adopt, is the following:

Nominal: Normally, the value of a parameter calculated using variables at their design-specified values.

Average: Usually, a core or assembly or pin averaged value. The distinction must be made or inferred from the application.

Peak: Sometimes referred to as hot spot; refers exclusively to a physical location at which the extreme value occurs; that is, peak pin is that pin located

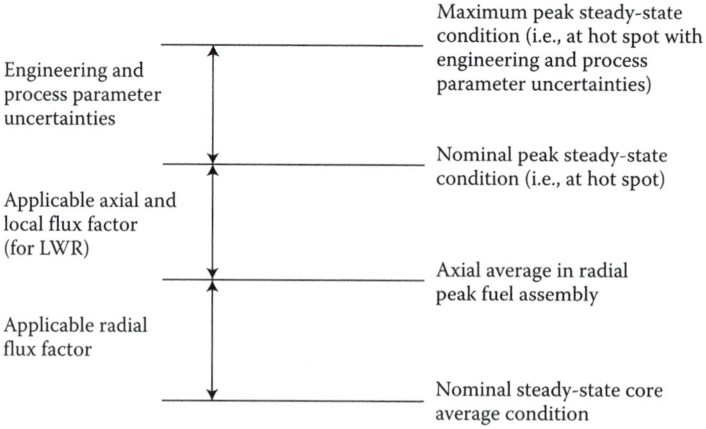

FIGURE 2.6 Thermal design nomenclature.

where the radial power profile is greatest; peak linear power rating occurs at an axial location along the peak pin.

Maximum: Value of a parameter calculated with allowance provided for deviation of the variables from their design-specified values.

These differences stem from the neutronic-driven spatial power distribution and engineering and process parameter uncertainties. The progression from the nominal steady-state core average condition to the transient-operating limit involves additional parameters beyond those shown in Figure 2.6. When expressed in terms of the limiting CHF condition, margins include those for correlation uncertainties, engineering and process parameter uncertainties, core anomalies, and transients. The fuel pin failure limit is unity in MDNBR terms and operation at this condition would correspond to the highest power rating in Figure 2.7. As margin is progressively introduced to accommodate uncertainties, core anomalies, and transients, the allowable core power decreases to ultimately define the allowable-rated power. The numerical values for MDNBR shown are all specific to individual plant design and the DNB correlation utilized. The radial and axial power-peaking factors are due to the neutron distribution in a classic core that follows a cosine (axial) shape and a Bessel function (radial) shape. The values shown in Figure 2.7 are representative of the PWR reactor values for typical fuel cells from the South Texas Plant's *Final Safety Analysis Report* [9].

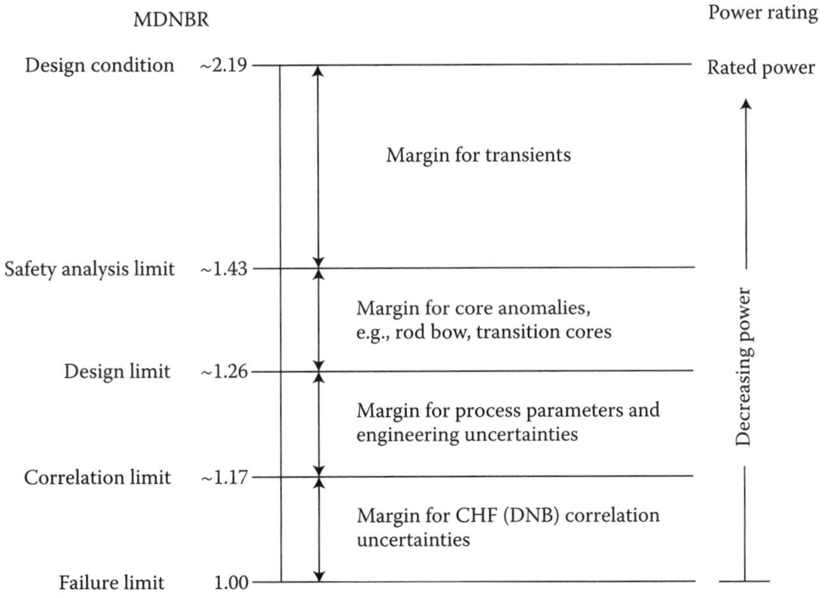

FIGURE 2.7 *Thermal analysis in practice. (MDNBR numerical values are from the South Texas PWR Plant Final Safety Analysis Report.)

* A transition core is composed of fuel assemblies of different designs. This typically occurs as fuel assemblies of a new design are gradually added in replacement of an existing design.

Core power can be enhanced by flattening the shape of the power generation rate, optimizing the power-to-flow ratio in every radial zone of the core, or both. Power flattening is achieved by a combination of control and fissile material distribution steps involving reflector regions, enrichment zoning, shuffling of fuel assemblies with burnup, and burnable poison and shim control placement. In water reactors, local power-peaking effects due to water regions comprise an additional factor that has been addressed by considerable attention to the detailed layout of the fuel/coolant lattice. However, power flattening may not be desirable if neutron leakage from the core is to be minimized.

Optimizing the power-to-flow ratio in the core is important for achieving the desired reactor vessel outlet temperature at maximum net reactor power, that is, reactor power generation minus pumping power, where pumping power for an ideal pump is expressed as

$$\begin{aligned}\text{Pumping power} &= \text{Work to raise coolant pressure per unit time expressed as} \\ &\quad \text{force through distance per unit time } (FL/T) \\ &= \text{Force times Velocity} = (\Delta p A_\text{f})\upsilon = \Delta p Q = \dot{m}\Delta p/\rho \end{aligned} \tag{2.9}$$

where Δp = pressure drop through the circuit (F/L^2); A_f = cross-sectional area of the flow path (L^2); υ = average coolant velocity (L/T); and Q = volumetric flow rate (L^3/T).

Flow control involves establishing coolant flow rates across the core at the levels necessary to achieve equal coolant exit conditions in all assemblies and minimizing the amount of inlet coolant that bypasses the heated core regions. Some bypass flow is required, however, to maintain certain regions (e.g., the inner reactor vessel wall) at low temperatures. Variations in energy generation within assemblies with burnup do make it difficult to maintain the power-to-flow ratio near unity. The assembly flow rates are normally adjusted by orificing. Orificing is accomplished by restricting the coolant flow in the assembly, usually at the inlet. Because the orifice devices are not typically designed to be changed during operation, deviations in the power-to-flow ratio from optimum over a fuel cycle are inevitable. Alternatively, large spatial variations in flow rate can also be accomplished using multiple inlet plena. This design approach, however, is complicated.

Example 2.1: Work Requirement of a Pump

PROBLEM

A typical primary system pump in a three-loop PWR operates at a head of 85 m and a flow rate of 7000 kg/s. How much power is required for the pump in steady state if the pump efficiency η is 85%? Does it constitute a significant fraction of the work produced by a power plant?

SOLUTION

$$\text{Pump power} = (\Delta p A_\text{f})(\upsilon)/\eta = (\Delta p)(\dot{m}/\rho\eta) \tag{2.9}$$

Here $\Delta p = 85(\text{m}) \times 9.81(\text{m/s}^2) \times 1000(\text{kg/m}^3) = 8.34 \times 10^5\ \text{kg/ms}^2$

For an 85% efficient pump, we can estimate the pumping power from

$$\text{Pump power} = \frac{(\Delta p)(\dot{m})}{\eta\rho} = \frac{(8.34\times10^5\,\text{kg/ms}^2)(7000\,\text{kg/s})}{(0.85)(1000\,\text{kg/m}^2)}$$
$$= 6.9\times10^6\,\text{W}$$

Hence pump work is 6.9 MWe/pump or about 21 MWe/plant.

The typical electrical output of a large PWR power plant is of the order of 1000 MWe, and so the work required to run the pumps in a three-loop PWR is a small fraction of the station power output, for example, of order 2% contrasted to HTGR circulators that consume about 30% of plant output.

2.6 FIGURES OF MERIT FOR CORE THERMAL PERFORMANCE

The design performance of a power reactor can be characterized by two figures of merit: the power density (Q''') and the specific power (Q^M). Table 2.4 tabulates the power density, specific power, and other parameters that are important factors in establishing the cost of electricity from a nuclear power plant.*

2.6.1 Power Density

Core power density as defined by Equation 2.1 is a measure of the energy-generated relative to the core volume. Because the size of the reactor vessel and hence the capital cost are nominally related to the core size, the power density is an indicator of the

TABLE 2.4
Core Parameters Important to the Cost of Electricity for Typical LWRs

Parameter	Symbol	Defining Equation	PWR	BWR-5
Core power (MW_{th})	Q_{th}		3411	3323
Net electric output (MW_e)	Q_e		1148	1062
Coolant			Water	Water
Fuel			UO_2	UO_2
Core power density (kW_{th}/L)	Q'''	2.10	104.5	52.3
Enrichment equil core average (wt%)	r	2.15	3.57	3.5
Initial heavy metal (IHM) loading in core (Mt)	M_{IHM}		89	141
Specific power (kW_{th}/kg_{IHM})	Q^M	2.11	38.3	23.6
Number of batches	n	2.25	3	3
Operating cycle length or refueling interval (years)	T_c	226	1.5	2
Capacity factor (fraction)	L_c	2.28	0.9	0.9

Source: Appendix K.

* For simplified relations for calculating the components of electricity cost which involve these parameters, see Tester et al. [11].

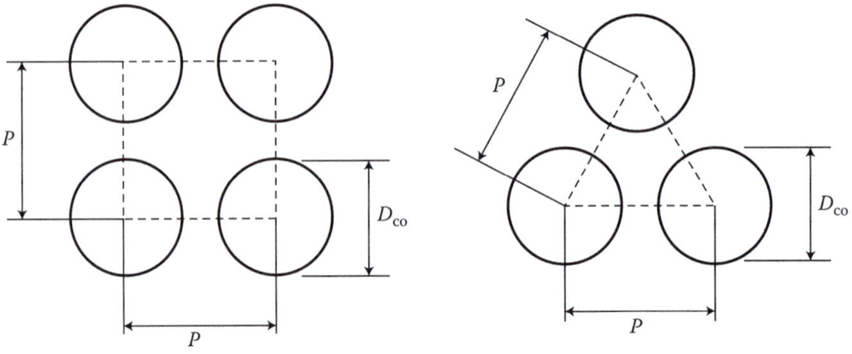

FIGURE 2.8 Square and triangular rod arrays.

capital cost of a concept. For propulsion reactors', where weight and hence, size are at a premium, power density is a directly relevant figure of merit.

The power density can be varied by changing the fuel pin arrangement in the core. For an infinite square array, shown in Figure 2.8, the power density (Q''') is related to the array pitch (P) as

$$(Q''')_{\text{square array}} = \frac{4(1/4\pi R_{\text{fo}}^2)q'''\text{d}z}{P^2\text{d}z} = \frac{q'}{P^2} \tag{2.10a}$$

whereas for an infinite triangular array, the comparable result is

$$(Q''')_{\text{triangular array}} = \frac{3(1/6\pi R_{\text{fo}}^2)q'''\text{d}z}{P/2\left[(\sqrt{3}/2)P\right]\text{d}z} = \frac{q'}{\left(\sqrt{3}/2\right)P^2} \tag{2.10b}$$

Comparing Equations 2.10a and 2.10b, we observe that the power density of a triangular array is 15.5% greater than that of a square array for a given pitch and linear power. For this reason, reactor concepts such as the SFR adopt triangular arrays, which, however, are more complicated mechanically than square arrays. For water-cooled reactors, on the other hand, the simpler square array is more desirable,* since the necessary neutron moderation can be provided by the looser-packed square array.

2.6.2 Specific Power

Core specific power as defined by Equation 2.2 is a measure of the energy deposited per unit mass of fuel material. It is usually expressed as kilowatts per kilogram of initial heavy atoms. This parameter has direct implications on core inventory

* Nevertheless, the Russian VVER-pressurized water reactor and the Canadian CANDU reactor use the triangular array with grid spacers.

requirements and hence the fuel cycle cost. For the fuel pellet shown in Figure 2.9, the specific power is

$$Q^{\mathrm{M}} = \frac{\dot{Q}_{\mathrm{th}}}{\text{mass of initial heavy atoms}} = \frac{\langle q' \rangle / \gamma}{\pi R_{\mathrm{fo}}^2 \rho_{\mathrm{pellet}} f_{\mathrm{IHM}}} = \frac{\langle q' \rangle / \gamma}{\pi \left(R_{\mathrm{fo}} + \delta_{\mathrm{g}}\right)^2 \rho_{\mathrm{smeared}} f_{\mathrm{IHM}}} \quad (2.11)$$

where

$\langle q' \rangle$ = core volume average linear power rating

γ = percentage of energy deposited in the fuel rods

$$\rho_{\mathrm{smeared}} = \frac{\pi R_{\mathrm{fo}}^2 \rho_{\mathrm{pellet}}}{\pi \left(R_{\mathrm{fo}} + \delta_{\mathrm{g}}\right)^2} \quad (2.12)$$

$$\delta_{\mathrm{g}} = \text{Fuel to clad gap spacing}$$

and

$$f_{\mathrm{IHM}} = \text{Mass fraction of initial heavy atoms in the fuel}$$
$$= \frac{\text{Grams of initial heavy atoms}}{\text{Grams of fuel}} \quad (2.13)$$

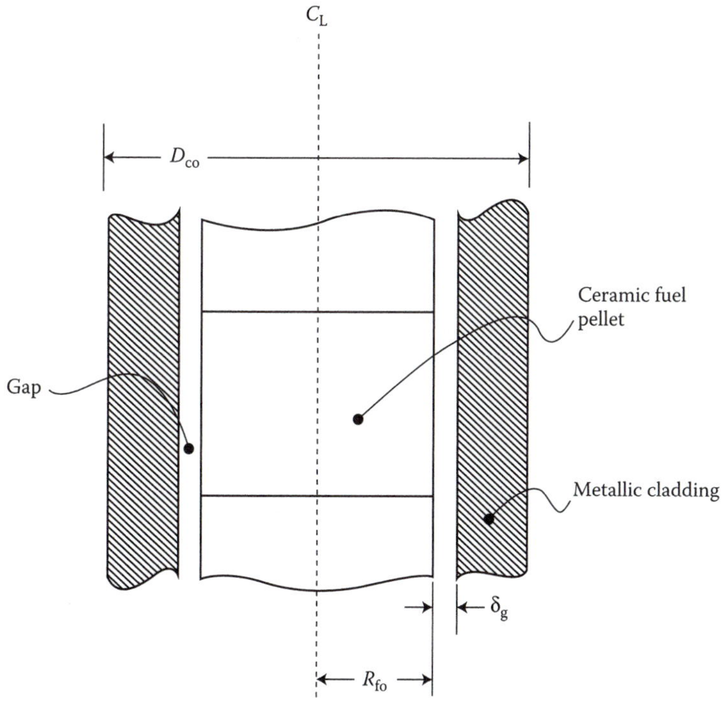

FIGURE 2.9 Typical components of light-water power reactor fuel. Actual fuel pellets are dished and ends are champered to enhance fuel irradiation performance. These features are not shown.

The specific power in Equation 2.11 is expressed in terms of both the pellet and the smeared densities. The *pellet density* includes the void present within the pellet. The *smeared density* adds the void that is present as the gap between the fuel pellet and the cladding inside diameter. The smeared density can be a cold or hot density, depending on whether the gap spacing δ_g, is taken at the cold or hot condition. Smeared density is an important parameter associated with accommodation of fuel swelling with burnup.

The fuel-cladding gap spacing is extensively discussed in Section 8.7 of Chapter 8.

The heavy metal mass fraction (f_{HM}) is evaluated based on the following definitions of the relevant terms:

1. Heavy atoms include all the U, Pu, or Th isotopes and are therefore composed of fissile atoms (M_{ff}) and nonfissile atoms (M_{nf}), where M is the molar mass.
2. Fuel is the entire fuel-bearing material, that is, UO_2 but not the cladding.

Thus for oxide fuel:

$$f_{HM} = \frac{N_{ff}M_{ff} + N_{nf}M_{nf}}{N_{ff}M_{ff} + N_{nf}M_{nf} + N_{O_2}M_{O_2}} \tag{2.14}$$

where N = atomic density.

The enrichment (r) is the mass ratio of fissile atoms to total heavy atoms, that is,

$$r = \frac{N_{ff}M_{ff}}{N_{ff}M_{ff} + N_{nf}M_{nf}} \tag{2.15}$$

and for later convenience

$$1 - r = \frac{N_{nf}M_{nf}}{N_{ff}M_{ff} + N_{nf}M_{nf}} \tag{2.16}$$

It is useful to express f_{HM} in terms of the molar mass and enrichment. It follows from the observation that for UO_2

$$N_{O_2} \equiv N_U = N_{ff} + N_{nf} \tag{2.17}$$

and manipulation of Equations 2.15 and 2.16 to yield

$$N_{ff} = \left[r\frac{M_{nf}}{M_{ff}}\right]\frac{N_{nf}}{1-r}; \quad N_{nf} = \left[(1-r)\frac{M_{ff}}{M_{nf}}\right]\frac{N_f}{r} \tag{2.18a,b}$$

$$N_{ff} + N_{nf} = \left[r\frac{M_{nf}}{M_{ff}} + (1-r)\right]\frac{N_{nf}}{1-r} \quad \text{and equally using Equation 2.18b}$$

$$= \left[(1-r)\frac{M_{ff}}{M_{nf}} + r\right]\frac{N_{ff}}{r} \tag{2.19a,b}$$

For UO_2 fuel, division of each term of Equation 2.14 by N_{O_2} and substitution of Equations 2.18a,b and 2.19a,b yields the desired result, that is,

$$f_{HM} = \frac{r/r+(1-r)(M_{ff}/M_{nf})M_{ff}+(1-r)/r(M_{nf}/M_{ff})+(1-r)M_{nf}}{r/r+(1-r)(M_{ff}/M_{nf})M_{ff}+(1-r)/r(M_{nf}/M_{ff})+(1-r)M_{nf}+M_{O_2}} \quad (2.20)$$

which, for the case where $M_{ff} \simeq M_{nf}$, simplifies to

$$f_{HM} \simeq \frac{rM_{ff}+(1-r)M_{nf}}{rM_{ff}+(1-r)M_{nf}+M_{O_2}} \quad (2.21)$$

2.6.3 Power Density and Specific Power Relationship

The relationship between core power density and core specific power is

$$Q''' = 10^{-3}Q^{M}\frac{M_{IHM}/V_f}{V_{core}/V_f} = 10^{-3}Q^{M}\rho_{IHM}\nu_f \quad (2.22)$$

where

Q'''	core power density	kW/L or MW/m^3
Q^M	core specific power	kW/kg$_{IHM}$
V_f	volume of fuel in the core	m^3_{fuel}
ρ_{IHM}	heavy metal density in the fuel	kg_{IHM}/m^3_{fuel}
ν_f	volume fraction of fuel in the core	$m^3{}_{fuel}/m^3{}_{core}$
10^{-3}	conversion factor m^3/L	

Values of these last two parameters for a typical PWR and BWR are given in Table 2.5.

Example 2.2: Power Density and Specific Power for a PWR

PROBLEM

Confirm the power density listed in Table 2.1 for the PWR case. Calculate the specific power of the PWR of Tables 2.1 and 1.3.

TABLE 2.5
PWR and BWR Fuel Parameters

Parameter		PWR	BWR-5
Heavy metal density in UO_2 fuel ($kg_{IHM}/m^3{}_{fuel}$)	ρ_{HM}	9.67×10^3	9.67×10^3
Volume fraction of fuel in the core ($m^3{}_{fuel}/m^3{}_{core}$)	ν_f	0.285	0.22

SOLUTION

a. Power density: The power density based on a square unit cell of side P is given as

$$Q'''_{\text{square array}} = \frac{\langle q' \rangle}{P^2}$$

From Table 2.1

$$\langle q' \rangle_{\text{core average}} = 17.86 \text{ kW/m}$$

Take $q' = \langle q' \rangle_{\text{core average}}$.

From Table 1.3, $P = 12.6$ mm.

$$Q'''_{\text{PWR}} = \frac{17.86}{(12.6 \times 10^{-3})^2} = 1.13 \times 10^5 \text{ kW/m}^3 = 113 \text{kW/L or MW/m}^3 \text{ or W/cm}^3$$

Observe that Table 2.1 lists the average power density as

$$Q'''_{\text{PWR}} = 104.5 \text{kW/L}$$

The difference arises for several reasons. First, the $\langle q' \rangle$ value used of 17.86 kW/m includes only energy deposited in the fuel rods. Actually, energy is deposited also directly in core structures and the coolant, as will be discussed in Chapter 3. This adjustment would make the calculated value higher than 113 kW/L. Second, volumes used are inconsistent. Our calculation is based on the core as an infinite square array, whereas in practice each fuel assembly has side and corner subchannels, control finger guide tubes, and there is a finite space between assemblies. On the other hand, the power density in Table 2.1 is based on the actual core volume. This effect shows that considering only a unit cell gives an overestimate of the power density. Hence, the two effects compensate with the second being dominant.

b. Specific power:

1. Solution using Equation 2.11. The specific power is given by Equation 2.11 as

$$Q^{\text{M}} = \frac{\langle q' \rangle / \gamma}{\pi R_{\text{fo}}^2 \, \rho_{\text{pellet}} \, f_{\text{IHM}}}$$

where γ = percent of energy deposited in the fuel rods = 97.4% as given in Appendix K and f_{HM} is expressed as f_{IHM} since specific power is defined based on initial heavy metal mass.

Evaluating f_{HM} using Equation 2.21 for the enrichment of 3.57% listed in Table 1.3 yields

$$f_{\text{HM}} = \frac{0.0357(235.0439) + 0.9643(238.0508)}{0.0357(235.0439) + 0.9643(238.0508) + 2(15.9944)} = 0.8815$$

For a pellet density of 95% of the UO_2 theoretical density of 10.97 g/cm^3 and a fuel pellet diameter of 8.192 mm, the specific power is

$$Q^{M} = \frac{[(1.786\times 10^{4})/0.974]}{\pi[(8.192\times 10^{-3}/2)]^{2}[0.95(10.97)/10^{-6}]0.8815}\frac{\text{W/m}}{\text{m}^{2}\text{g/m}^{3}} = 37.9\frac{\text{kW}}{\text{kg}_{\text{IHM}}(\text{uranium})}$$

In this method, the pellet volume should actually be reduced by about 1% to account for pellet dishing and chamfering done to safeguard fuel mechanical behavior on irradiation. This factor would increase the estimation of Q^M to about 38.3 W/g uranium.

2. Solution using total fuel-loading value directly. Table 2.1 lists the total core loading of fuel material as 101 × 10^3 kg of UO_2, that is, M_{fc}. In this case

$$Q^{M}_{\text{initial}} = \frac{\text{Core power}}{\text{Initial heavy metal fuel loading}} = \frac{\dot{Q}_{\text{th}}}{f_{\text{IHM}}M_{\text{fc}}} \tag{2.23}$$

If $\dot{Q}$ is evaluated from the PWR dimensions as in Tables 2.1 and 1.3:

$$\dot{Q}_{\text{th}} = \frac{\langle q'\rangle}{\gamma}LN = \frac{0.01786}{0.974}\frac{\text{MW}}{\text{m}}\left[3.658\text{m}\right]\left[193(264)\right] = 3418\,\text{MW}_{\text{th}}$$

which is the thermal power value given in Table 1.2.

Then,

$$Q^{M} = \frac{3418\text{ MW}_{\text{th}}}{0.8815(101\times 10^{3}\text{ kg})} = 38.3\frac{\text{kW}}{\text{kg}_{\text{IHM}}(\text{uranium})}$$

2.6.4 Specific Power in Terms of Fuel Cycle Operational Parameters

The specific power is also related to fuel cycle operational parameters as follows:

$$Q^{M} = \frac{Bu_{\text{d}}}{0.365L_{\text{c}}T_{\text{res}}} \tag{2.24}$$

where

Q^M	specific power	kW_{th}/kg_{IHM}
Bu_d	fuel discharge burnup	MWd/kg_{IHM}
T_{res}	fuel in-core residence time	years
0.365	conversion factor	days/year × 10^{-3}
L_c	plant capacity factor	fraction

The relationship between the parameters of Equation 2.24 is illustrated in Figure 2.10 for a plant capacity factor of 90%. For a current operating PWR on an 18-month cycle with a three-batch fuel management scheme, the fuel in-core residence time is 4.5 years. Hence, at a fuel discharge burnup of 55 MWd/kg, the specific power is 36.7 kW/kg_{IHM}.

In Chapter 3, it will be noted that 5% of the fission power is lost as kinetic energy of the small but neutrally charged "neutrino" particles. However, an equivalent

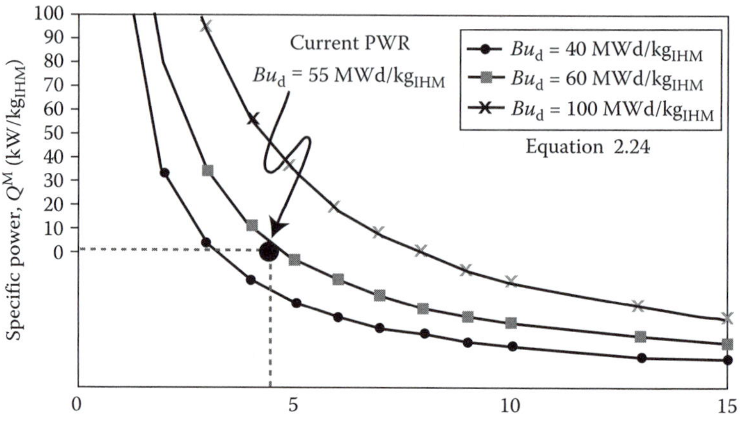

FIGURE 2.10 The specific power fuel residence time trade-off for plant capacity factor of 90%. (From Saccheri, J., Todreas, N. E., and Driscoll, M. J., *Nucl. Tech.*, 158(3):315–347, 2007.)

amount of energy is being deposited in the fuel as decay heat from fission products that were produced earlier. Thus, the relationship between the core-specific power and the burnup (Equation 2.24) is fairly accurate.

In order to complete the discussion on the determination of core-specific power from fuel cycle operational parameters, by Equation 2.24, we next demonstrate the manner in which the following needed parameters Bu_d, T_{res}, and L_c are determined

1. Discharge burnup, Bu_d, and in-core residence time, T_{res}: Achievement of discharge burnup in a single versus a multibatch fuel management scheme requires increased enrichment. It also results in uneconomical discharge of peripheral fuel assemblies at considerably lower burnup because neutron leakage lowers the neutron flux at the periphery and determines the core radial (and axial) power profiles. Hence, multibatch fuel management schemes are used that shuffle fuel within the core as well as add fresh fuel to $1/n$th of the core at refueling outages. Assuming equal-sized batches, the discharge burnup Bu_d, operating cycle burnup Bu_c, and single batch ($n = 1$) loaded core burnup Bu_1 are related* as follows:

$$Bu_d = nBu_c = \frac{2n}{n+1} Bu_1 \tag{2.25}$$

where n is the number of fuel batches. Note that in general, n need not be an integer.

* Based on linear reactivity theory from Driscoll, M. J., Downar, T. J., and Pilat, E. E., *The Linear Reactivity Model for Nuclear Fuel Management*. La Grange Park, IL: American Nuclear Society, 1991.

The corresponding relevant time periods are related as

$$T_{res} = nT_c \tag{2.26}$$

where T_c is the operating cycle length or refueling interval.

The fuel in-core residence time, and hence the number of batches, is also limited by allowable coolant-side oxidation of the fuel cladding and fast fluence exposure to the cladding and assembly structures.

It has been found by Xu et al. [13] that while single-batch burnup, Bu_1, depends on many variables, it can be related to core-average reload fuel enrichment, r_p from 3 to 20 w/o (20 w/o is normally taken as the maximum enrichment to achieve acceptable proliferation-resistant fuel).

$$Bu_1 = 64.6\sqrt{r_p + 13.4} - 240.4\,\text{MWd/kg}_{\text{IHM}} \tag{2.27}$$

(multiply by 1000 to obtain MWd/MTU, the unit plotted in Figure 2.11).

Hence, solving Equations 2.24 through 2.27 simultaneously, we can obtain the useful map of Figure 2.11 relating discharge burnup to operating cycle length for a typical PWR. Note that the specific power and capacity factor in Equation 2.24 have been fixed at prescribed values cited in Figure 2.11.

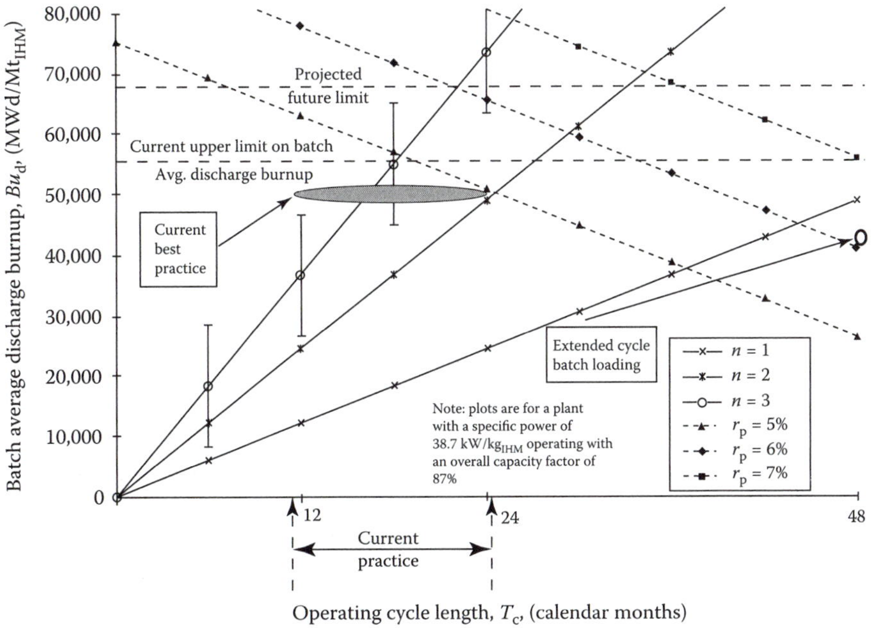

FIGURE 2.11 Burnup-cycle length map for a representative PWR. *Note:* Plots are for a plant with a specific power of 38.7 kW/kg$_{IHM}$ operating with an overall capacity factor of 87%. (From Handwerk, C. S. et al., *Economic Analysis of Implementing a Four-Year Extended Operating Cycle in Existing PWRs.* Cambridge, MA: Massachusetts Institute of Technology, CANES Report MIT-ANP-TR-049, January 1997.)

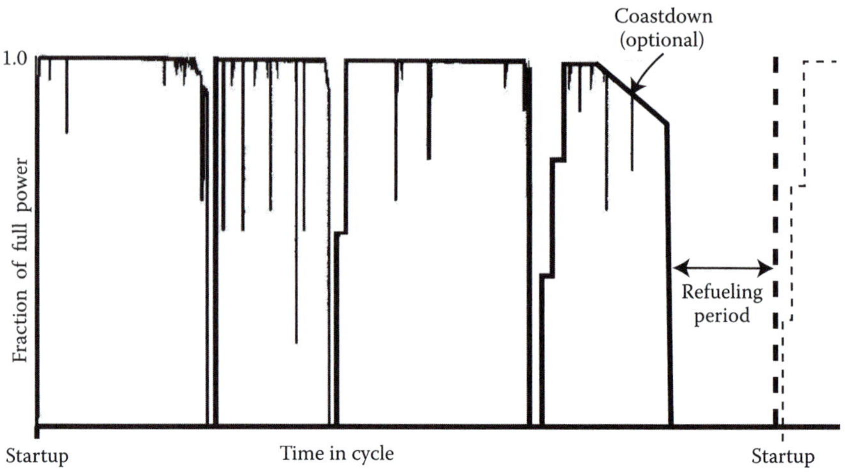

FIGURE 2.12 Typical nuclear plant-operating power history (the skyline chart).

2. Plant Capacity Factor, L_C: Considerable management attention is paid to the enhancement of plant capacity factors throughout the operating fleet. Central has been control of the duration of planned plant shutdowns for refueling and maintenance and reduction of unplanned shutdowns from outage extensions and forced outages.

   Figure 2.12 is an illustration of the power history of a hypothetical plant during an operating cycle that encompasses the period from startup to shutdown to perform core refueling. Figure 2.12 is typically called a plant skyline reflective of the imagery of the illustration. As shown, the power losses can be partial load reductions or full shutdowns. Some are planned, some unplanned, and some not under control of the plant operator. Also shown is a power coastdown period prior to the refueling period at the end of the cycle, which is an option sometimes selected to allow the refueling period to coincide with reduced seasonal power demand.

Various parameters have been defined to characterize this operational history, all of which report some time-averaged electricity production as a fraction of full power operation. They differ principally in the components of operating power losses they account for. Here we will first focus on the plant capacity factor L_C, and plant unit capability factor L_{CB}. Plant availability, L_A, will be introduced later.

The plant capacity factor, L_C, accounts for downtime from all causes, that is, due to causes under plant management control (such as planned outages) as well as energy losses not considered to be under the control of plant management (such as lack of demand, grid instability or failure, environmental limitations, and seasonal variations). Plant unit capability factor, L_{CB}, in contrast, accounts only for downtime from causes under plant management control.

Hence, plant capacity factor is

$$L_C = \frac{\text{Reference energy generation minus all energy losses}}{\text{Reference energy generation}}$$

$$= \frac{\text{Actual generation}}{\text{Reference energy generation}} \tag{2.28}$$

while the plant unit capability factor, L_{CB}, is a measure of the energy that could have been produced but was lost due to factors under plant operator control:

$$L_{CB} = \frac{\text{Reference energy generation minus energy losses within operator control}}{\text{Reference energy generation}} \tag{2.29}$$

We desire to transform these factors to effective times (for operation or loss) at full power so that corresponding times (e.g., downtime due to a series of forced load reductions) are equally weighted and therefore can be summed. We express effective time in terms of actual time and power rating as follows:

$$T_{\text{eff}}\, P_{\text{rated}} = \sum_i T_{\text{ACT}}\, (P_{\text{rated}} - P_{\text{actual operating}}) \tag{2.30}$$

where $(P_{\text{rated}} - P_{\text{actual operating}})$ is the power lost for the actual time T_{ACT} for occurrence i.

So, if we are interested in forced outage loss, that is, the effective time for loss of full power, we simply interpret T_{eff} as forced outage time T_{FO} and rewrite Equation 2.30 as

$$T_{\text{FO}} = \frac{\sum_i T_{\text{ACT}} (P_{\text{rated}} - P_{\text{actual operating}})_{\text{forced outages}}}{P_{\text{rated}}} \tag{2.31}$$

Table 2.6 illustrates the effective time periods for full power operation (EFPP), unplanned outages, T_{EO} and $T_{\text{FO,}}$ planned outages, T_{RO} and T_{MO}, and idle time, T_{I}.

Now we can express the plant capacity factor and the unit capability factor in terms of EFPP, T_{CB} (capability factor equivalent cycle length), and T_{C} (cycle length). Figure 2.13 illustrates these relationships where the terms L_{A} (plant availability) and T_{A} (availability time) have also been introduced. Note in particular that the individual times T_{PO}, T_{UO}, and T_{I} are illustrated only as their sum that equals T_{C}—EFPP.

The shaded areas in Figure 2.13 are energy quantities since the ordinate is percent of full power and the abscissa is time. Since the actual energy generated over a cycle, T_{C}, is fixed from the energy balance of Figure 2.13, we can write

$$100\%\ \text{EFPP} = L_{\text{A}} T_{\text{A}} = L_{\text{CB}} T_{\text{CB}} = L_{\text{C}} T_{\text{C}} \tag{2.32a}$$

Hence,

$$\frac{L_{\text{C}}}{L_{\text{CB}}} = \frac{T_{\text{CB}}}{T_{\text{C}}} \tag{2.32b}$$

TABLE 2.6
Time Periods in an Operating Cycle

	Outages Due to Plant Condition (Within Operator Control)				Outages Due to Grid (Demand) Conditions or Regulatory Limits (Outside Operator Control)
Operation	**Unplanned Outages (UO)**		**Planned Outages (PO)**		**Idle Outages (I)**
At power	T_{UO}		T_{PO}		T_I
	Outage Extension (EO)	**Forced Outage (FO)**	**Refueling Outage (RO)**	**Maintenance Outage (MO)**	
Effective full power period (EFPP)	T_{EO}	T_{FO}	T_{RO}	T_{MO}	T_I

It is this ratio between capacity and capability factors that we seek to derive. Expressing Equation 2.28 in terms of effective times at reference power, we write the plant capacity factor, L_C, as

$$L_C = \frac{T_C - T_I - T_{PO} - T_{UO}}{T_C} \tag{2.33}$$

Similarly, from Equation 2.29, we can express the plant capability factor, L_{CB}, as

$$L_{CB} = \frac{T_C - T_{PO} - T_{UO}}{T_C} \tag{2.34}$$

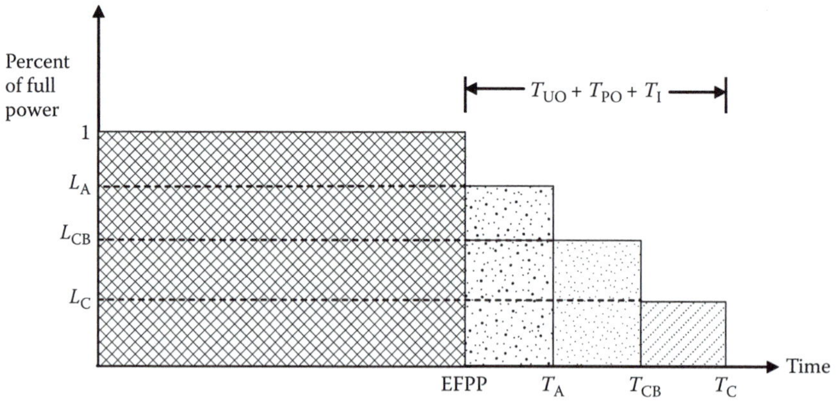

FIGURE 2.13 Plant-operating characteristics.

For simplicity, let us next take T_{EO} and T_{MO} equal to zero. From the definitions of Table 2.6

$$T_{PO} = T_{RO} + T_{MO} \Rightarrow T_{RO} \quad \text{and} \quad T_{UO} = T_{EO} + T_{FO} \Rightarrow T_{FO} \qquad (2.35a,b)$$

Now T_{FO} is typically expressed in terms of the forced loss rate (FLR), defined as*

$$\text{FLR} = \frac{\text{Unplanned forced energy losses}}{\text{Reference energy generation} - \text{Planned plus unplanned outage extension energy losses}} \qquad (2.36)$$

Using the effective times introduced earlier, we obtain

$$\text{FLR} = \frac{T_{FO}}{T_C - T_{PO} - T_{EO}} \qquad (2.37a)$$

For our assumption that T_{EO} and T_{MO} are zero, T_{PO} reduces to T_{RO} and Equation 2.37a becomes

$$\text{FLR} = \frac{T_{FO}}{T_C - T_{RO}} \qquad (2.37b)$$

Hence, T_{FO} is available from Equation 2.37b in terms of the assumed known parameters FLR, T_C, and T_{RO} as

$$T_{FO} = \text{FLR}\,(T_C - T_{RO}) \qquad (2.38)$$

Now the plant capacity factor can be expressed from Equations 2.33, 2.35a, 2.35b, and 2.38 as

$$L_C = \frac{T_C - T_I - T_{RO} - T_{FO}}{T_C} = \frac{(T_C - T_{RO})(1 - \text{FLR}) - T_I}{T_C} \qquad (2.39)$$

Similarly, the plant capability factor, L_{CB}, can be expressed from Equations 2.34, 2.35a, 2.35b, and 2.38 as

$$L_{CB} = \frac{T_C - T_{RO} - T_{FO}}{T_C} = \left[\frac{T_C - T_{RO}}{T_C}\right](1 - \text{FLR}) \qquad (2.40)$$

If, for example, a 30-day idle time T_I is assumed along with a 30-day refueling outage length T_{RO}, the plant capacity factor L_C, and the plant capability factor L_{CB}, can be calculated from Equation 2.40 for given values of FLR. The results are displayed in

* This indicator has been constructed by the industry to monitor the progress in minimizing the outage time and power reductions that result from unplanned equipment failures, human errors, or other conditions during the operating period. The operating period which is the denominator of the definition has therefore been defined to exclude planned outages and their possible unplanned extensions.

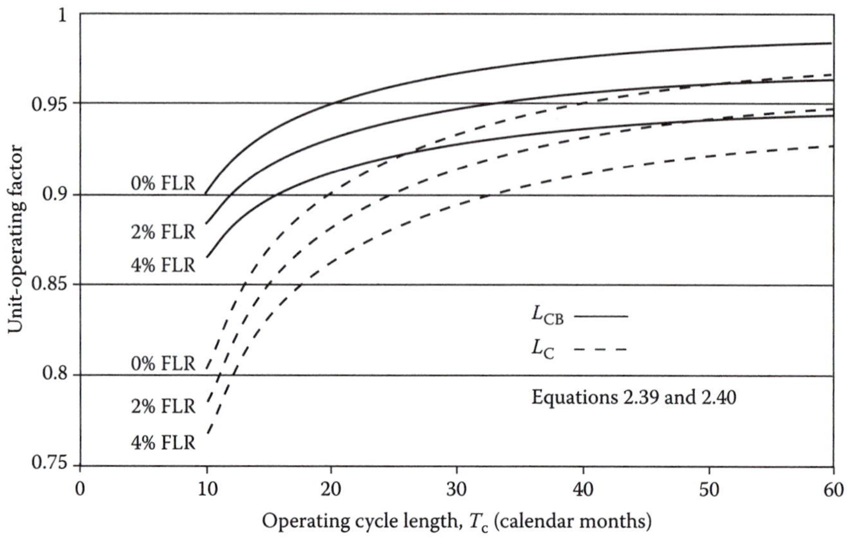

FIGURE 2.14 Effect of cycle length on plant-operating factors (assuming a 30-day refueling outage length, T_{RO}) and 30-day idle time period, T_I, outside the plant operator's control.

Figure 2.14. Note that the ratio of the capability and capacity factors as illustrated in Figure 2.14 for the same values of T_{RO} and FLR is given by

$$\frac{L_C}{L_{CB}} = 1 - \frac{T_I}{(T_C - T_{RO})(1 - \text{FLR})} \tag{2.41}$$

which is obtained by dividing Equation 2.39 by Equation 2.40. From Figure 2.14, at assumed values of 4% FLR, $T_I = 30$ days, $T_{RO} = 30$ days, and $T_C = 24$ months, the capability factor is 0.925, while the capacity factor is lower at 0.885. The ratio between these two factors is 0.955 as can be confirmed using Equation 2.41.

Plant management can minimize the effect of forced outages by seeking to operate at longer cycle lengths as well as by eliminating the root causes for the forced outages.

2.7 THE INVERTED FUEL ARRAY

In addition to the traditional fuel-pin array geometry on which attention has been focused, core designs in which a fueled matrix is penetrated by a regular array of coolant channels can also be utilized. Several reactor concepts utilize this fuel arrangement.

The General Atomic thermal HTGR reactor employs a version of this inverted geometry in which both cylindrical fuel compacts and coolant channels are in intermixed hexagonal arrays within a graphite-only matrix. The fuel compacts are graphite slugs loaded with uniformly dispersed fuel particles of the type illustrated in Figure 2.5. The HTGR matrix fuel assembly is illustrated in Figure 2.15. The fuel

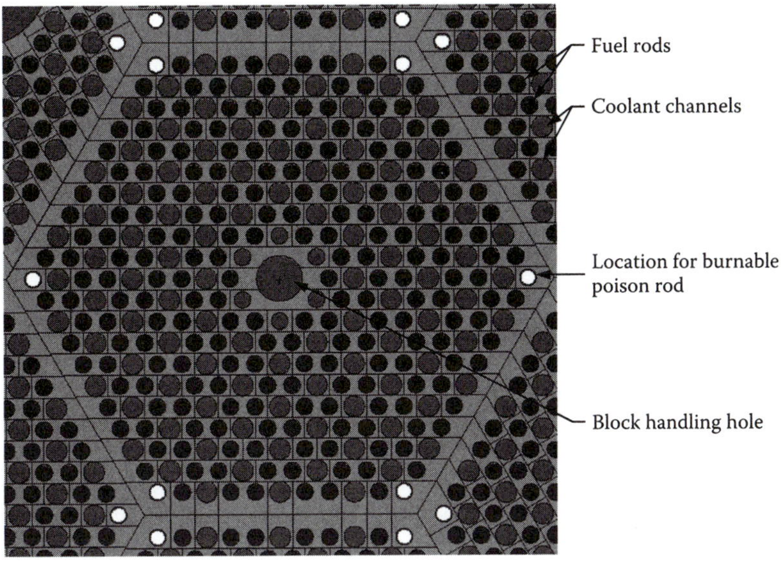

FIGURE 2.15 HTGR matrix fuel assembly. (From Sterbentz, J. W. et al., *Reactor Physics Parametric and Depletion Studies in Support of TRISO Particle Fuel Specification for the Next Generation Nuclear Plant.* INEEL/EXT-04-02331, Idaho National Engineering and Environmental Laboratory, 2004.)

compact/coolant channel array design in Figure 2.15 has been optimized such that the diameters of the compacts and coolant channels differ and the array pattern, while regular, has fewer coolant channels than fuel compacts. The supercritical CO_2-cooled GFR concept [1] proposed a matrix of hexagonal-shaped pellets penetrated by circular coolant holes formed into a hexagonal subassembly. The fuel is a homogeneous blend of (U, TRU) UO_2 with a BeO diluent. This GFR-inverted fuel assembly is illustrated in Figure 2.16.

Further, note that for the HTGR design the fuel-bearing region is distinct from the fuel region. The fuel-bearing region is the graphite of the compact and the graphite and silicon carbide layers surrounding the fuel kernels. The fuel region is the uranium or thorium-containing kernel of the BISO- or TRISO-type particle. The volume fractions of these regions are defined as

$$\text{Volume fraction of fuel-bearing region: } \nu_{\text{fb}} = \frac{V_{\text{fb}}}{V \text{ unit cell}} \tag{2.42a}$$

$$\text{Volume fraction of fuel region (within the fuel bearing): } \nu_{\text{f}} = \frac{V_{\text{f}}}{V_{\text{fb}}} \tag{2.42b}$$

These two different regions will be separately accounted for in the analysis that follows.

The inverted geometry is most clearly presented by contrasting it to the fuel pin array geometry. Figure 2.17 illustrates both fuel pin and inverted geometries for the triangular array.

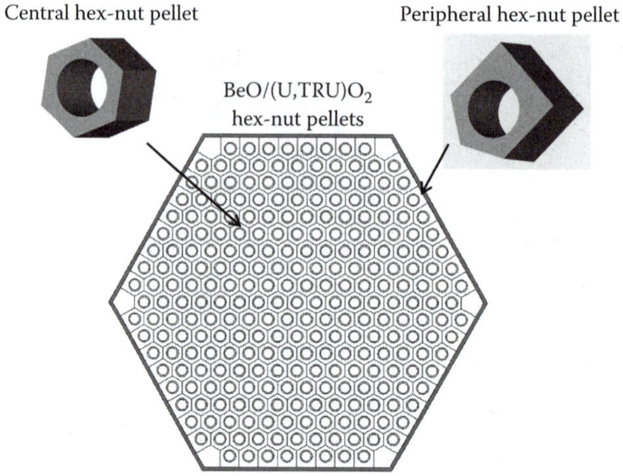

FIGURE 2.16 Horizontal cross section of fuel assembly with hex-nut pellets. (From Pope, M. A., *Nucl. Eng. Des.*, 239:840–854, 2009.)

Observing Figure 2.17, the fuel pin geometry results when the continuous region is the coolant and discontinuous region—the circles—is the fuel. Conversely, the fuel-inverted geometry results when the continuous region is the fuel and the circles are coolant channels. The terminology "inverted" follows directly from the inversion of fuel and coolant placement within these arrays.

Example 2.3: Comparison of Fuel-to-Coolant Ratio in Triangular-Inverted Array Versus the Square Pin Array

PROBLEM

Determine the comparative behavior of the fuel-to-coolant ratio as a function of pitch to diameter for the following two fuel arrays:

Square pin array
Triangular-inverted array

Assume that the fuel-cladding gap regions are occupied by a liquid–metal bond.

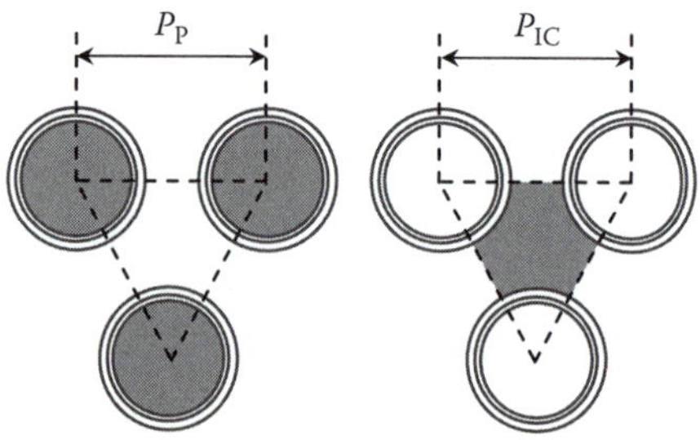

FIGURE 2.17 Pin geometry (*left*) versus inverted geometry (*right*). Shaded regions are fuel.

SOLUTION

The unit cell geometries for the two arrays are illustrated in Figure 2.18.

For the square pin array, the coolant channel area $A_{C,SC}$, total unit cell area $A_{T,SC}$, and fuel area $A_{F,SC}$ are defined as

$$A_{C,SC} = P_{SC}^2 - \frac{\pi D_{SC}^2}{4}; \quad A_{T,SC} = P_{SC}^2; \quad A_{F,SC} = \beta\frac{\pi D_{SC}^2}{4} \qquad (2.43a,b,c)$$

where the subscript SC signifies the standard core. The equivalent diameter of the standard design is defined as follows:

$$D_{e,SC} = \frac{4A_{C,SC}}{P_{W,SC}} = \frac{4\left(P_{SC}^2 - (\pi D_{SC}^2/4)\right)}{\pi D_{SC}} \qquad (2.44)$$

FCR is defined as the ratio of fuel cross-sectional area to coolant cross-sectional area, and is expressed in terms of P_{SC} and D_{SC} as follows from Equations 2.43:

$$\text{FCR} = \frac{A_{F,SC}}{A_{C,SC}} = \beta\left[\frac{4}{\pi}\left(\frac{P_{SC}}{D_{SC}}\right)^2 - 1\right]^{-1} \qquad (2.45)$$

where β is the area ratio of fuel to fuel plus gap and cladding, that is,

$$\beta = \frac{A_F}{A_{FGC}} \qquad (2.46)$$

For the triangular-inverted array, the coolant channel area $A_{C,IC}$, total unit cell area $A_{T,IC}$, and fuel area $A_{F,IC}$ are defined as

$$A_{C,IC} = \frac{\pi D_{IC}^2}{4}; \quad A_{T,IC} = \frac{\sqrt{3}}{2}P_{IC}^2; \quad A_{F,IC} = \beta\left(A_{T,IC} - A_{C,IC}\right) \qquad (2.47a,b,c)$$

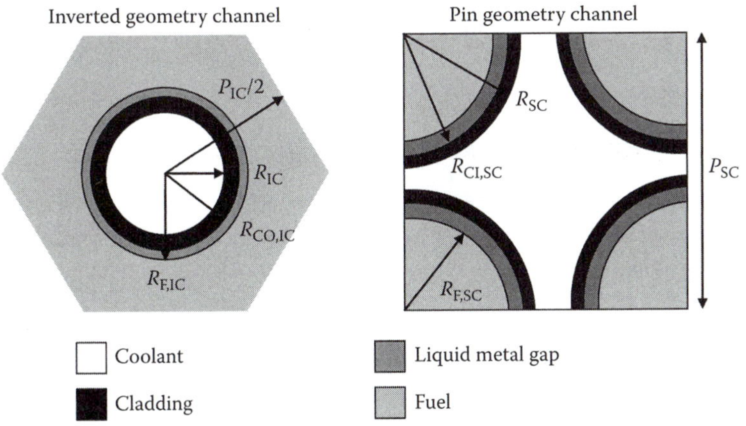

FIGURE 2.18 Comparison of standard pin bundle and inverted designs.

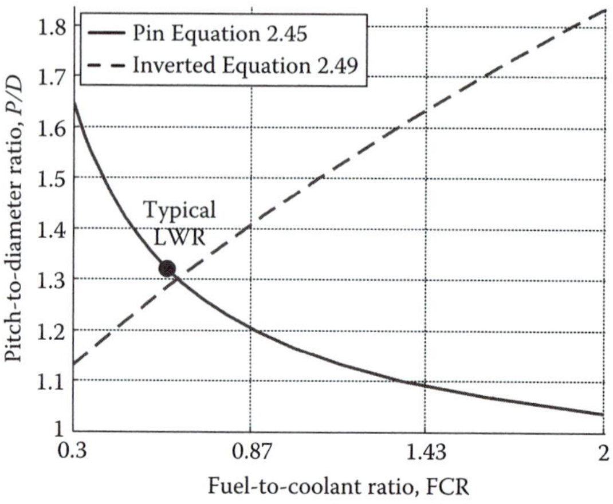

FIGURE 2.19 Relationship of governing geometric parameters defining fuel and coolant channel arrays for $\beta = 0.736$. (From Malen, J. et al., *Nucl. Eng. Des.*, 239:1471–1480, 2009.)

$D_{e,IC}$ is equal to the inverted channel diameter (D_{IC}) in the case of the IC design.

$$D_{e,IC} = \frac{4A_{C,IC}}{P_{W,IC}} = D_{IC} \tag{2.48}$$

Hence, FCR can be expressed in terms of β, P_{IC}, and D_{IC} as follows:

$$\text{FCR} = \frac{A_{F,IC}}{A_{C,IC}} = \beta\left(\frac{2\sqrt{3}}{\pi}\left(\frac{P_{IC}}{D_{IC}}\right)^2 - 1\right) \tag{2.49}$$

Since the gap and cladding thicknesses are normally scaled to the pellet diameter, β is a constant value for which we take a typical value of 0.736. Utilizing Equations 2.45 and 2.49, the desired comparison is illustrated in Figure 2.19.

2.8 THE EQUIVALENT ANNULUS APPROXIMATION

The unit cells to be approximated are either of square or triangular geometry. These cells are further simplified by formulating an equivalent annulus in which the area of the material bounding the cell's central cylinder is replaced by an annulus.

For the square array under the condition of equal areas of the equivalent annulus and the region around the cylinder but bounded by the hexagon (the coolant region for the fuel pin array while the fuel region for the inverted fuel array), we get

$$\pi(R_{EA}^2 - r_o^2) = P^2 - \pi r_o^2 \tag{2.50}$$

Hence,

$$R_{EA}\Big|_{\text{sq array}} = \frac{P}{\sqrt{\pi}} \tag{2.51a}$$

or

$$d_{cell}\Big|_{\text{sq array}} \equiv 2R_{EA} = \frac{2}{\sqrt{\pi}}P \tag{2.51b}$$

Analogously for the triangular array

$$R_{EA}\Big|_{\text{tri array}} = \sqrt{\frac{\sqrt{3}}{2\pi}}P \tag{2.52a}$$

$$d_{cell}\Big|_{\text{tri array}} = \sqrt{\frac{2\sqrt{3}}{\pi}}P \tag{2.52b}$$

Figure 2.20 illustrates the pin and inverted fuel geometries where each is now represented as a unit cell—hence, they will be called the pin cell (for fuel-pin array) and the inverted cell (for the inverted element array). The equivalent annulus approximation is an accurate representation of the temperature fields for the coolant in the pin cell and the fuel in the inverted cell only for sufficiently large *P/D* ratios, typically greater than *P/D* = 1.2, as demonstrated in Chapter 8. The relevant geometric relationships for these unit cells are given in Table 2.7.

The relevant thermal performance parameters for these two cells are based on the following definitions.

The fuel-bearing region contains the fuel material and the matrix material (if any). The fuel material is composed of heavy metal and diluent, that is, UO_2 or UC. The matrix material is typically material in which the fuel is embedded which resides within cladding or, if unclad, within the region whose temperature is controlled by the

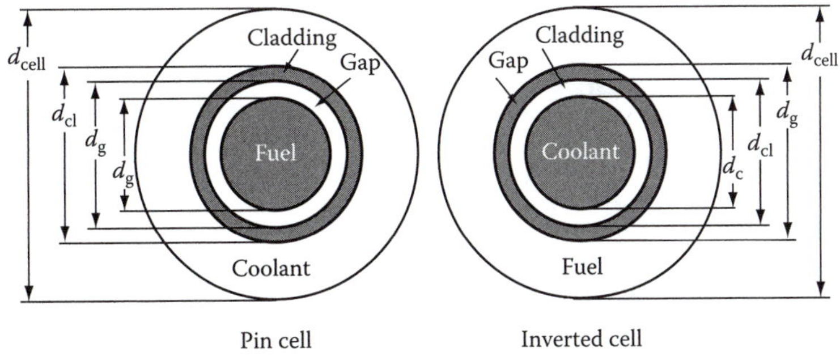

FIGURE 2.20 Equivalent annulus representations of the pin cell geometry and the inverted cell geometry.

TABLE 2.7
Geometric Relationships for Equivalent Annulus Representations of Fuel Arrays

	Symbol	Pin Cell	Inverted Cell
Equivalent gap thickness	t_g	$\frac{d_g - d_f}{2}$	$\frac{d_g - d_{c\ell}}{2}$
Cladding thickness	$t_{c\ell}$	$\frac{d_{c\ell} - d_g}{2}$	$\frac{d_{c\ell} - d_c}{2}$
Fuel-bearing region volume fraction	v_{fb}	$\frac{d_f^2}{d_{cell}^2}$	$\frac{d_{cell}^2 - d_g^2}{d_{cell}^2}$
Gap volume fraction	v_g	$\frac{d_g^2 - d_f^2}{d_{cell}^2}$	$\frac{d_g^2 - d_{c\ell}^2}{d_{cell}^2}$
Cladding volume fraction	$v_{c\ell}$	$\frac{d_{c\ell}^2 - d_g^2}{d_{cell}^2}$	$\frac{d_{c\ell}^2 - d_c^2}{d_{cell}^2}$
Coolant volume fraction	v_c	$\frac{d_{cell}^2 - d_{c\ell}^2}{d_{cell}^2}$	$\frac{d_c^2}{d_{cell}^2}$
Volume fraction summation		$\nu_{fb} + \nu_g + \nu_{c\ell} + \nu_c = 1$	

coolant. The heavy metal in turn is composed of fissile and nonfissile (including possibly fertile) isotopes. Hence, for example, for the HTGR TRISO fuel of Figure 2.5.

Fuel-bearing material = Fuel particles plus matrix (UO_2, pyrolitic carbon, silicon carbide plus graphite)
Fuel = Heavy metal and diluent (U and O)
Heavy metal = Fissile and fertile (U^{235} and U^{238})

The relevant densities, mass fractions, and volume fractions for these materials are

$$\rho_{fb} = \frac{M_{fb}}{V_{fb}}; \quad \rho_f = \frac{M_f}{V_f}; \quad \rho_{HM} = \frac{M_{HM}}{V_{HM}} \tag{2.53a,b,c}$$

The actual fuel density, ρ_f, also equals

$$\rho_f = \rho_{fTD} \times \% \text{ of TD} \tag{2.53d}$$

$$f_{fb} = \frac{M_{fb}}{M_{cell}}; \quad f_f = \frac{M_f}{M_{fb}}; \quad f_{HM} = \frac{M_{HM}}{M_f} \quad \text{(2.54a,b; analogous to 2.13)}$$

$$\nu_{fb} = \frac{V_{fb}}{V_{cell}}; \quad \nu_f = \frac{V_f}{V_{fb}}; \quad \nu_{HM} = \frac{V_{HM}}{V_f} \tag{2.42a,b; 2.55}$$

(Note that unlike the volume fractions defined in Table 2.7, ν_f and ν_{HM} are not defined as fractions of the fuel cell.)

Express the specific power in terms of unit cell parameters and initial heavy metal loading, and hence

$$Q^M = \frac{\dot{Q}}{M_{IHM}} = \frac{q'''_{cell} V_{cell}}{V_{IHM}\rho_{IHM}} \tag{2.56}$$

From Equations 2.53b, 2.53c, 2.13, and 2.42b

$$\rho_{HM} = \rho_f \frac{M_{HM}}{V_{HM}} \frac{V_f}{M_f} = \rho_f f_{HM} \nu_f \frac{V_{fb}}{V_{HM}} \tag{2.57}$$

Now expressing this result in terms of IHM and substituting it into Equation 2.56 and applying Equation 2.42a yields

$$Q^M = \frac{q'''_{cell}}{\nu_{fb}\nu_f f_{IHM}\rho_f} \tag{2.58}$$

Since by energy balance

$$q'''_{cell} V_{cell} = q'''_{fb} V_{fb} \tag{2.59}$$

Equation 2.58 can be expressed in terms of the volumetric energy generation rate of the fuel-bearing region by applying Equations 2.59 and 2.42a, that is

$$Q^M = \frac{q'''_{fb}}{\nu_f f_{IHM}\rho_f} \tag{2.60}$$

Table 2.8 summarizes the thermal performance parameters for equivalent annular representations of the inverted and pin fuel arrays.

Example 2.4: Comparative Pin and Inverted Fuel Cell Thermal Performance

PROBLEM

Assume a pin cell and an inverted cell of identical diameters, $d_{cell} = 2.214$ cm operate at the same power density, $q'''_{cell} = 50\,\text{W/cm}^3$ using uranium carbide (UC) fuel. Take the inverted cell coolant diameter, d_c, as 1.4 cm. Compare the pin cell and inverted cell performance (q', q'', Q^M) for the following set of conditions.

Geometry	Fuel Properties
$d_{cell} = 2.214$ cm	$\rho_{fTD} = 13$ g/cm³ (theoretical density)
$t_g = 0.005$ cm	ρ_f, density of fuel = 94% of theoretical density
$t_{c\ell} = 0.038$ cm	$f_{IHM} = \frac{235}{250} = 0.952 g_{IHM}/g_{fuel}$
$d_c = 1.4$ cm	—
$d_{c\ell} = 1.72$ cm	—

TABLE 2.8
Thermal Performance Parameters for Equivalent Annulus Representations of Fuel Arrays

	Pin Cell	Inverted Cell
q'	$q'''_{cell}\frac{\pi}{4}d^2_{cell}$	
q''	$\frac{q'}{\pi d_{c\ell}}=\frac{q'''_{cell}d^2_{cell}}{4d_{c\ell}}$	$\frac{q'}{\pi d_c}=\frac{q'''_{cell}d^2_{cell}}{4d_c}$
Q^M	$\frac{q'''_{cell}}{v_{fb}v_f f_{IHM}\rho_f}$	
	$\frac{q'''_{fb}}{v_f f_{IHM}\rho_f}$	

SOLUTION

These performance characteristics can be expressed as

$$q' = q'''_{cell}\frac{\pi}{4}d^2_{cell} \tag{2.61}$$

$$q'' = \frac{q'''_{cell}d^2_{cell}}{4d_c} \quad \text{(inverted cell)} \tag{2.62a}$$

$$q'' = \frac{q'''_{cell}d^2_{cell}}{4d_{c\ell}} = \frac{q'''_{cell}d^2_{cell}}{4[d_f + 2(t_g + t_{c\ell})]} \quad \text{(pin cell)} \tag{2.62b}$$

where d_f must be determined

$$q'''_{fb} = \frac{q'''_{cell}}{v_{fb}} = \frac{q'''_{cell}}{\left((d^2_{cell} - d^2_g)/d^2_{cell}\right)} \quad \text{(inverted cell)} \tag{2.63a}$$

$$= \frac{q'''_{cell}}{\left(d^2_f/d^2_{cell}\right)} \quad \text{(pin cell)} \tag{2.63b}$$

So again d_f must be determined

$$Q^M = \frac{q'''_{cell}}{v_{fb}v_f f_{IHM}\rho_f} = \frac{q'''_{cell}}{(v_{fb})^2 f_{IHM}\rho_f} \tag{2.58}$$

since in this case the fuel-bearing volume and the fuel volumes are identical.

Hence, the problem solution requires determination of d_f for the pin cell as discussed next.

The volume fraction terms in the volume fraction balance of Table 2.7, that is

$$v_{fb} + v_g + v_{c\ell} + v_c = 1 \tag{2.64}$$

TABLE 2.9
Thermal Performance Cell Characteristics

Equation		Pin Cell (Both Square and Triangular)	Inverted Cell
(2.61)	$q' = q'''_{\text{cell}} \frac{\pi}{4} d^2_{\text{cell}}$	19.25 kW/m	19.25 kW/m
(2.62a)	$q'' = \frac{q'''_{\text{cell}} d^2_{\text{cell}}}{4d_c}$	—	43.8 W/cm^2
(2.62b)	$q'' = \frac{q'''_{\text{cell}} d^2_{\text{cell}}}{4d_{\text{cl}}}$	35.7 W/cm^2	—
(2.63a,b)	$q'''_{\text{fb}} = \frac{q'''_{\text{cell}}}{v_{\text{fb}}}$	92.4 W/cm^3	90.9 W/cm^3
(2.58)	$Q^M = \frac{q'''_{\text{cell}}}{(v_{\text{fb}})^2 f_{\text{IHM}} \rho_f} = \frac{q'''_{\text{fb}}}{v_{\text{fb}} f_{\text{IHM}} \rho_f}$	14.7 kW/kg$_{\text{IHM}}$	14.2 kW/kg$_{\text{IHM}}$

can be expressed in terms of the given data and the unknown fuel pin diameter, d_f. Performing this operation on Equation 2.64, the fuel pin diameter, d_f, can be obtained as 1.629 cm. Then, the volume fractions of the pin cell can be calculated as tabulated below. Those of the inverted cell are directly available from the given data and also are listed, since v_{fb} (inverted cell) is needed for calculating q'''_{fb} and Q^M for that cell.

	Pin Cell	Inverted Cell
v_g	0.007	0.006
v_{cl}	0.052	0.044
v_c	0.400	0.400
v_{fb}	0.541	0.550

The performance characteristics can now be determined as presented in Table 2.9. Note that these cells have been sized for the given q'''_{cell}-operating condition so that q' is identical for each geometry. The parameters q'''_{fb} and Q^M differ for each geometry since the fuel-bearing region fractions, v_{fb}, differ as shown above.

PROBLEMS

2.1. *Relations among fuel element thermal parameters in various power reactors* (Section 2.3)
Compute the core average values of the volumetric energy-generation rate in the fuel (q''') and outside surface heat flux (q'') for the reactor types of Table 2.1. Use the core average linear power levels in Table 2.1 and the geometric parameters in Table 1.3.

Answer (for BWR): $q''' = 243.2$ MW/m^3
$q'' = 500.2$ kW/m^2

2.2. *Relationships between assemblies of different pin arrays* (Section 2.3)
A utility wishes to replace the fuel in its existing PWR from 15 × 15 fuel pin array assemblies to 17 × 17 fuel pin array assemblies. What is the ratio of the core average linear power, q', in the new core to the old core, assuming that reactor power, length, and number of fuel assemblies are maintained constant?

Answer:

a. $\dfrac{\overline{q'_{17\times17}}}{\overline{q'_{15\times15}}} = 0.779$

b. $\dfrac{q''_{17\times17}}{q''_{15\times15}} = 0.882$

2.3. *Minimum CHF ratio in a PWR for a flow coastdown transient* (Section 2.4)
Describe how you would determine the minimum CHF ratio versus time for a flow coastdown transient by drawing the relevant channel-operating curves and the CHF limit curves for several time values. Draw your sketches in relative proportion and be sure to state all assumptions.

2.4. *Minimum CPR in a BWR* (Section 2.4)
Calculate the minimum CPR for a typical 1062 MWe BWR operating at 100% power using the data in Appendix K. Assume that:

1. The axial linear power shape can be expressed as

$$q'(z) = q'_{\text{ref}} \exp(-\alpha z/L) \sin \pi z/L$$

where $\alpha = 1.96$. Determine q'_{ref} such that $q'_{\text{max}} = 47.24$ kW/m

2. The critical bundle power is 9319 kW.

Answers:
$q'_{\text{ref}} = 104.75\,\text{kW/m}$
MCPR = 1.28

2.5. *Primary cooling system pumping power for a PWR reactor* (Section 2.5)
Calculate the pumping power under steady-state operating conditions for a typical PWR reactor–coolant system using only the following operating conditions:

Core power = 3411 MWth
$\Delta T_{\text{core}} = 33.7°C$
$T_{\text{IN}} = 293°C$
$p = 15.5$ MPa
Reactor–coolant system pressure drop = 778 kPa
Pump efficiency = 85%

Answer:
Pumping power = 21.8 MW_e

2.6. *Relations among thermal design conditions in a PWR* (Section 2.5)
Compute the margin as defined in Figure 2.5 for a typical PWR having a core average linear power rate of 17.86 kW/m. Assume that the failure

limit is established by centerline melting of the fuel at 70 kW/m. Use the following multiplication factors:

Radial flux = 1.55
Axial and local flux factor = 1.70
Engineering uncertainty factor = 1.05
Overpower factor = 1.15

Answer:
Margin = 1.23

REFERENCES

1. Dostal, V., Hejzlar, P., and Driscoll, M. J., High performance supercritical carbon dioxide cycle for next generation nuclear reactors. *Nucl. Tech.*, 154:265–282, 2006.
2. Driscoll, M. J., Downar, T. J., and Pilat, E. E., *The Linear Reactivity Model for Nuclear Fuel Management.* La Grange Park, IL: American Nuclear Society, 1991.
3. Handwerk, C. S., Driscoll, M. J., Todreas, N. E., and McMahon, M. V., *Economic Analysis of Implementing a Four-Year Extended Operating Cycle in Existing PWRs.* Cambridge, MA: Massachusetts Institute of Technology, CANES Report MIT-ANP-TR-049, January 1997.
4. Hejzlar, P., Pope, M. J., Williams, W. C., and Driscoll, M. J., Gas cooled fast reactor for Generation IV service. *Prog. Nucl. Energy*, 47(1–4):271–282, 2005.
5. Knief, R. A., *Nuclear Engineering: Theory and Technology of Commercial Nuclear Power*, 2nd Ed. La Grange Park, IL: American Nuclear Society, 2008.
6. Malen, J., Todreas, N., Hejzlar, P., Ferroni, P., and Bergles, A., Thermal hydraulic design of a hydride-fueled inverted PWR core. *Nucl. Eng. Des.*, 239:1471–1480, 2009.
7. Pope, M. A., Lee, J. I., Hejzlar, P., and Driscoll, M. J., Thermal hydraulic challenges of gas cooled fast reactors with passive safety features. *Nucl. Eng. Des.*, 239:840–854, 2009.
8. Saccheri, J., Todreas, N. E., and Driscoll, M. J., Design and economic evaluation of an advanced tight-lattice core for the IRIS integral primary system reactor. *Nucl. Tech.*, 158(3):315–347, 2007.
9. South Texas Plant. *Final Safety Analysis Report*, Revision 12.
10. Sterbentz, J. W., Phillips, B., Sant, R. L., Chang, G. S., and Bayless, P. D., *Reactor Physics Parametric and Depletion Studies in Support of TRISO Particle Fuel Specification for the Next Generation Nuclear Plant.* INEEL/EXT-04-02331, Idaho National Engineering and Environmental Laboratory, 2004.
11. Tester, J. W., Drake, E. M., Golay, M. W., Driscoll, M. J., and Peters, W. A., *Sustainable Energy: Choosing among Options*. Cambridge, MA: MIT Press, 2005.
12. Tong, L. S., Heat transfer in water-cooled nuclear reactors. *Nucl. Eng. Des.,* 6:301–324, 1967.
13. Xu, Z., Driscoll, M. J., and Kazimi, M. S., *Design Strategies for Optimizing High Burnup Fuel in Pressurized Water Reactors.* Cambridge, MA: Massachusetts Institute of Technology, CANES Report MIT-NFC-TR-053, July 2003.

3 Reactor Energy Distribution

3.1 INTRODUCTION

The energy generation distribution throughout the nuclear reactor is achieved via a neutronic analysis of the reactor. Accurate knowledge of the energy source is a prerequisite for analysis of the temperature distribution, which in turn is required for definition of the nuclear and physical properties of the fuel, coolant, and structural materials. Therefore, coupling the neutronic and thermal analyses of a nuclear core is required for accurate prediction of its steady state as well as its transient conditions. For simplicity, the neutronic and thermal analyses performed during thermal design evaluations may not be coupled, in which case the magnitude and distribution of the energy generation rate are assumed to be fixed, and thermal analysis is carried out to predict the temperature distribution in the reactor core.

As discussed in Chapter 2, the operational power of the core is limited by thermal, not nuclear, considerations. That is, in practice, the allowable core power is limited by the rate at which heat can be transported from fuel to coolant without reaching, either at steady-state or during specified transient conditions, excessively high temperatures that would cause degradation of the fuel, structures, or both. The design limits are discussed in Chapter 2.

Sections 3.2 through 3.8 of this chapter describe the fission energy generation and deposition during reactor operation.

Section 3.9 describes the shutdown energy generation, that is, the decay power* and energy values after reactor shutdown.

Section 3.10 describes the sources of energy generation from chemical reactions among reactor materials at off-normal conditions that could be encountered in accident scenarios.

3.2 ENERGY GENERATION AND DEPOSITION

3.2.1 Forms of Energy Generation

The energy generated in a reactor is produced by exothermic nuclear reactions in which part of the nuclear mass is transformed to energy. Most of the energy is released when nuclei of heavy atoms split as they absorb neutrons. The splitting of

* The ANS Standards use the more complete terminology of "decay heat power."

these nuclei is called *fission*. A small fraction of the reactor energy comes from nonfission neutron capture in the fuel, moderator, coolant, and structural materials.

Splitting a fissile atom produces two smaller atoms called fission products and on average two or more neutrons. A fissile atom is one prone to fission on absorption of a neutron of any energy. Uranium 235 is the principal core fissile material. Its abundance is roughly 0.7% of naturally occurring uranium. Other fissile materials are generated by neutron capture in the so-called fertile atoms. Thus, fissile plutonium 239 and plutonium 241 are produced by neutron absorption in atoms of uranium 238 and plutonium 240, respectively. Another practical fissile material is uranium 233, which is produced by neutron capture in thorium 232.

The probability of a fission event from neutron absorption in an atom strongly depends on the specific isotope and the neutron energy. Generally, the ratio of fission to nonfission probability increases with increasing neutron energy. Most actinides* (usually the elements 90 [thorium] through 103 [lawrencium]) are prone to fission when absorbing a fast neutron. Even ^{238}U and ^{232}Th may undergo fission when absorbing fast neutrons with energies higher than 1 MeV. Hence, in thermal and fast reactor cores undergoing multiple recycles, the higher-atomic-weight plutonium isotope, ^{241}Pu as well as minor actinide isotopes of neptunium, americium, and curium are also fissioned.

The energy release on fission, E_f, depends slightly on the different fissile materials. Unik and Gindler [23] suggest the following relation for energy release due to fission by a thermal neutron:

$$E_f(\text{MeV/fission}) = 1.29927 \times 10^{-3}(Z^2 A^{0.5}) + 3.12 \quad (3.1)$$

where Z and A are the atomic number and atomic mass of the fissioning nuclide. For nuclides between ^{232}Th and ^{242}Pu, the authors state that this equation yields values within 1% of experimental data. This relation is used in the code ORIGEN2 by Croff [6]. The more recent nuclear data libraries JEFF 3.1 [11] and ENDF-VII [5] report such data for all isotopes. The fission energy is roughly 200 MeV (or 3.2×10^{11} J) per fission.

In a typical LWR only about one-half of the neutrons are absorbed in fissile isotopes; the other one-half are captured by the fertile isotopes, control, and structural materials. The LWR fresh fuel is composed of uranium-based fuel, that is, UO_2, slightly enriched to between 3.5 and 5.0 wt/% in uranium 235. The plutonium produced in the LWR core during operation also participates in the energy-release process and may contribute up to 50% of the fission energy release near the end of a fuel cycle.

The fission energy appears as kinetic and decay energy of the fission products, kinetic energy of the newborn neutrons, and energy of emitted γ-rays and neutrinos. Many of the fission fragments are radioactive, undergoing β-decay accompanied by neutron release. The β-emission makes certain isotopes unstable, with resultant delayed neutron and γ-ray emission.

* According to the International Union of Pure and Applied Chemistry, actinide is now referred to as actinoid.

Immediately on capture, the neutron-binding energy, which ranges from 2.2 MeV in hydrogen to 6–8 MeV in heavy materials, is released in the form of γ-rays. Many capture products are unstable and undergo decay by emitting β-particles and γ-rays. An approximate accounting of the forms of the released energy is given in Figure 3.1, and the energy distribution among these forms is outlined in Table 3.1.

The neutrons produced in the fission process have a relatively high kinetic-energy and are therefore called *fast neutrons.* Most of these prompt neutrons have energies between 1 and 2 MeV, although a small fraction may emerge at energies as high as 10 MeV. The potential for a neutron to be absorbed into an atom is improved if its energy is reduced to levels comparable to the surroundings by a slowing process called *neutron moderation.* The slow neutrons, referred to as *thermal neutrons*, have energy less than 1 eV. Since the most energy loss per collision occurs with light nuclei atoms, the best moderating materials are those with low atomic masses. Hence,

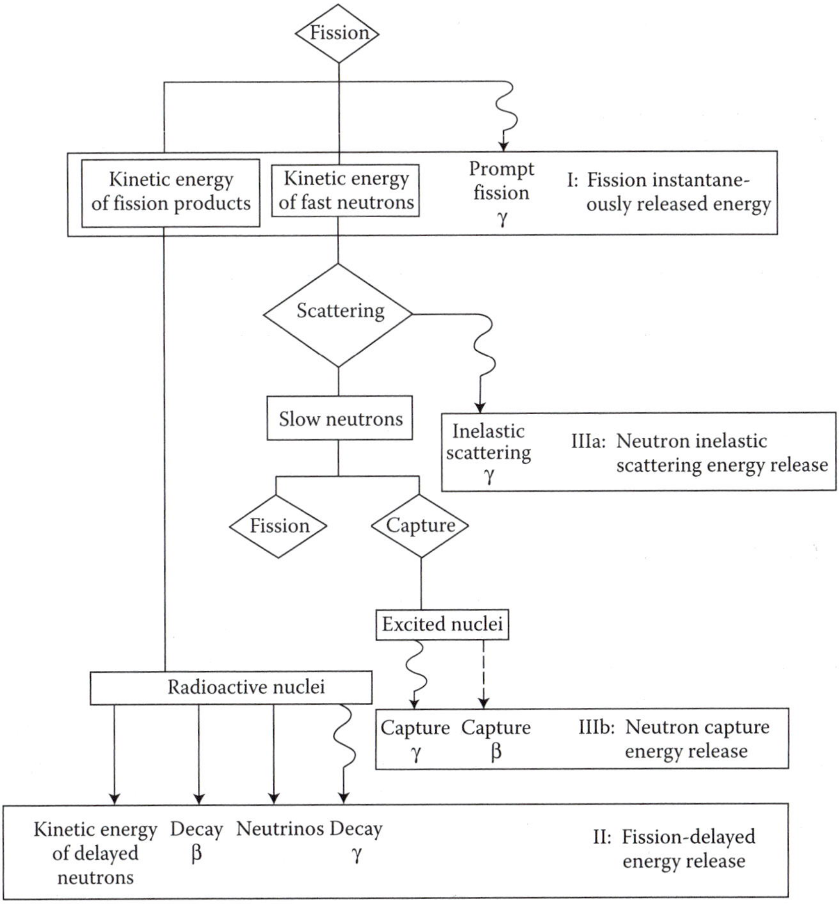

FIGURE 3.1 Forms of energy release in a reactor from fission.

TABLE 3.1
Approximate Distribution of Energy Release and Deposition in Thermal Reactors

Type	Process	Percent of Total Released Energy	Principal Position of Energy Deposition
	Fission		
I. Instantaneously released energy	Kinetic energy of fission products	80.5	Fuel
	Kinetic energy of newly born fast neutrons	2.5	Moderator
	γ-Energy released at time of fission	2.5 (1.3/1.2)	Fuel/structures
II. Delayed energy	β-Decay energy of fission products	3.0	Fuel
	Neutrinos associated with β-decay	5.0	Nonrecoverable
	γ-Decay energy of fission products	3.0 (1.6/1.4)	Fuel/structures
	Kinetic energy of delayed neutrons	0.02	Moderator
Subtotal fission		96.5	
	Inelastic Scattering and Neutron Capture		
III. Instantaneously released and delayed energy	Nonfission reactions due to excess neutrons plus β- and γ-decay energy of (*n*, γ) products	3.5 (2.0/1.5)	Fuel and structures
Total		100	

Source: Adopted from El-Wakil, M. M. *Nuclear Heat Transport*. International Textbook Co., Scranton, PA, 1971.

moderators such as carbon, hydrogen, and deuterium have been used in power reactors that rely mostly on fission from slow (thermal) neutrons.

3.2.2 Energy Deposition

The fission products moving through the core material lose their energy through interaction with the surrounding matter. The energy loss rate depends mainly on the penetrating ability of the fission products. The fission fragments that carry most of the fission energy have a short range (<0.25 mm). The γ-rays' energy is released in structural as well as fuel materials. A considerable amount of the kinetic energy of neutrons is released in the moderator and the structure. The neutrino energy is unrecoverable, since neutrinos do not interact with the surrounding materials.

Let us consider the energy deposition rate per unit volume at any position due to fission, $q_f'''(\vec{r})$, hence the subscript f. This quantity is written as $q_f'''(\vec{r})$ and conventionally called a volumetric energy generation rate although as just explained its source is not the fission energy released at that same position $(\vec{r})$. It should be recognized that $q_f'''(\vec{r})$ is due to products of reaction events at all neighboring positions that pass through position $\vec{r}$. Let the volumetric energy generation rate at position $\vec{r}$ due to reaction products of type i and energy E be $q_i'''(\vec{r},E)$. In order to obtain the total volumetric energy generation rate at position $\vec{r}$, we must sum over all particle (and photon) types and energy spectrum:

$$q_f'''(\vec{r}) = \sum_i \int_0^\infty q_{fi}'''(\vec{r},E)\,dE \tag{3.1}$$

Thus, to exactly calculate the energy generation at a particular point of the reactor is difficult. However, the energy generation can be well approximated in various parts of the reactor by established reactor physics analysis methods.

The typical distribution of energy deposition in fuel and nonfuel materials in an LWR is also given in Table 3.1. From Table 3.1 for the LWR, about 88.4% of the total energy released per fission is recovered in the fuel, 2.5% in the moderator, and 4.1% in the structure; 5% is unrecoverable (neutrino energy). These fractions depend on the exact range of the particles identified in Table 3.1 and, hence, on the materials, their arrangement within the reactor, and thus on the reactor type.

3.3 FISSION POWER AND CALORIMETRIC (CORE THERMAL) POWER

Take the total energy generated in the core due to fission as $\dot{Q}_f$. This includes all the various forms of energy release from fission, inelastic scattering, and neutron capture, which are illustrated in Figure 3.1 and Table 3.1. Since 5% is carried away by neutrinos, the amount of energy deposited in the core is $0.95\dot{Q}_f$. Further, since core thermal power rating is established by calorimetric means, for example, the product of measured primary coolant mass flowrate, $\dot{m}$, and primary coolant temperature rise, ΔT_p, the core thermal power rating $\dot{Q}_{th}$ equals

$$\dot{Q}_{th} = 0.95\dot{Q}_f \tag{3.2}$$

Now considering the spatial distribution of energy deposition in the core, we have

$$\dot{Q}_{th} = \sum_{n=1}^{N} \dot{q}_n + \dot{Q}_{\text{moderator and noncladding structure}} \tag{2.6}$$

where $\dot{q}_n$ is the average energy deposition in a fuel rod and N_p is the number of fuel rods in the core.

From the previous section, the amount of energy deposited in the fuel is $0.884\dot{Q}_f$. However, the fraction of the energy deposited in the cladding is unspecified. From Tong and Weisman [22], for a PWR no more than about 1% of the total energy released is deposited within the thermal shield. Hence, of the 4.1% deposited in the structures, 3% resides in the cladding, so that 91.4% can be assumed deposited in the fuel rods of a PWR. Applying Equation 3.2, the amount of core thermal power deposited in the fuel is

$$0.914\dot{Q}_{\mathrm{f}} = 0.914\left(\frac{\dot{Q}_{\mathrm{th}}}{0.95}\right) = 0.962\,\dot{Q}_{\mathrm{th}} \tag{3.3}$$

or

$$\dot{Q}_{\mathrm{th}} = \frac{1}{0.962}\sum_{n=1}^{N_{\mathrm{p}}}\dot{q}_n \equiv \frac{1}{\gamma}\sum_{n=1}^{N_{\mathrm{p}}}\dot{q}_n \tag{3.4}$$

where $\gamma \equiv$ fraction of core thermal power deposited in the fuel rods. Since $\gamma\dot{Q}_{\mathrm{th}}$ equals $\sum_{n=1}^{N_{\mathrm{p}}}\dot{q}_n$ from Equation 3.4, Equation 2.6 yields

$$\frac{\dot{Q}_{\text{moderator and non-cladding structure}}}{\dot{Q}_{\mathrm{th}}} = 1-\gamma \tag{3.5}$$

Now commonly these thermal energy deposition rates, $\gamma\dot{Q}_{\mathrm{th}}$ and $(1-\gamma)\dot{Q}_{\mathrm{th}}$, are designated as the heat generation rates in fuel rods and in the moderator and noncladding structure. This is convenient since we can more easily express heat generation rate quantities as fractions of core thermal power versus core fission energy rates. Table 3.2 summarizes these power definitions and the associated numerical values for typical PWR conditions.

3.4 ENERGY GENERATION PARAMETERS

3.4.1 Energy Generation and Neutron Flux in Thermal Reactors

Analogous to Equation 3.1, the energy generation rates will be summed over all particle (and photon) types, energy spectrum, and space but now labeled simply as

$$q'''(\vec{r})$$

and called the volumetric energy generation rate. The volumetric energy generation rate in the fuel, $q'''(\vec{r})$, is typically computed by assuming that the energy released by a fission reaction is recovered at the position of the fission event, except for the fraction carried away by neutrinos and the fraction deposited in nonfuel materials.* In other words, the spatial distribution of energy deposition from the fission fragments, γ's and β's, is assumed to follow the spatial distribution of the fission reaction rate $RR_{\mathrm{f}}(\vec{r})$.

We define the recoverable energy per fission reaction of isotope j as χ_{f}^{j}. For a typical thermal reactor, χ_{f}^{25} is about 190 MeV per fission.†

* In addition to the loss of the neutrino energy and the energy deposited outside the fuel, the magnitude of the energy released per reaction, which appears in the fuel, should theoretically be reduced by conversion of a small part to potential energy of various metallurgical defects. It is, however, a negligible effect.

† Historically, the superscript j for ^{235}U has been written as 25 where the 2 represents the element $_{92}U$ and the 5, the isotope ^{235}U.

TABLE 3.2
Power Definitions and Typical PWR Conditions[a]

Fission Power		**Calorimetric Core Power**	
Fission energy, $\dot{Q}_f$: $\frac{3411}{0.95} = 3591\,\text{MW}_{th}$		Core thermal power, $\dot{Q}_{th}$: 3411 MW_{th}	
% Deposition in		% Deposition in	
Fuel rods	91.4%	Fuel rods	96.2%[b] (91.4%/0.95)
Moderator	2.5%	Moderator	2.6%
Noncladding structure	1.1%	Noncladding structure	1.2%
Neutrinos	5%	Neutrinos	—
Core average volumetric energy generation rate		Core average volumetric energy generation rate in the fuel	
$\langle q_f''' \rangle = \frac{\dot{Q}_f}{V_{core}} = \frac{3591}{32.63} = 110\frac{\text{MW}}{\text{m}^3}$		$\langle q''' \rangle \equiv \frac{\gamma \dot{Q}_{th}/N_p}{V_f} = \frac{(0.962)3411/50952}{1.93\times10^{-4}} = 334\,\text{MW/m}^3$	
		Core average volumetric energy generation rate (the power density)	
		$\langle q''' \rangle_{core} \equiv \frac{\dot{Q}_{th}}{V_{core}} = \frac{3411}{32.63} = 104.5\frac{\text{MW}}{\text{m}^3}$	

[a] All PWR data from Appendix K. Core volume, $V_{core} = 32.63\ \text{m}^3$; Pins per core, $N_p = 50952$;

$$\text{Pin fuel volume} \equiv V_f = \frac{\pi D_{fo}^2}{4}L_f = \frac{\pi\left(8.192\times10^{-3}\right)^2}{4}\times 3.658 = 1.93\times10^{-4}\ \text{m}^3$$

[b] The best estimate thermal fuel rod performance characteristics of LWRs depend on the value adopted for the percent of energy released which is deposited in the fuel rods. *This value is 97.4% for the Seabrook plant, a typical Westinghouse Electric Co. PWR, and 97.5% for the Calvert Cliffs plant, a typical Combustion Engineering Co. designed PWR. These values are slightly above the 96.2% in Table 3.2. For the BWR with more noncladding core structure as channels and its unique core internal structures, the value analogous to 97.4% is about 96.5%.* These plant values are reflected in Appendix K together with the associated values of core average linear heat generation rate (LHGR), $\langle q' \rangle$ in kW/m.

Evaluation of the heat source distribution requires the knowledge of fission reaction rates, RR_f^j, which, when summed over all atom types, yield the total fission reaction rate, $RR_f(\vec{r})$:

$$RR_f(\vec{r}) = \sum_j RR_f^j(\vec{r}) \tag{3.6}$$

Let $\sigma_a^j(E)$ be the microscopic absorption cross section of isotope j, which is the equivalent projected area of an atom for an absorption reaction. The cross section $\sigma_a^j(E)$ is proportional to the probability that one atom of type j absorbs an incident neutron of energy E. Because the absorption of a neutron by an atom can lead to either a fission process or a nonfission (or capture) process, the absorption cross section is the sum of the fission cross section σ_f and the capture cross section σ_c. The units of the microscopic cross section are normally given as square centimeters or barns (1 barn = 10^{-24} cm^2). Typical values of the absorption and fission cross sections of thermal neutrons are given in Table 3.3.

The macroscopic fission cross section is the sum of all microscopic cross sections for a fission reaction due to all atoms of type j within a unit volume interacting with an incident neutron per unit time. The macroscopic cross section is defined by

$$\Sigma_f^j(\vec{r},E) \equiv N^j(\vec{r})\,\sigma_f^j(E) \tag{3.7a}$$

where N^j is the atomic density of isotope j that can be obtained from the mass density of isotope j, ρ_j, using the relation

$$N^j = \frac{A_v \rho_j}{M_j} \tag{3.7b}$$

where A_v = Avogadro's number of atoms (and hence also nuclei) in 1 mol (0.60225×10^{24} atoms/mol); M_j = atomic mass of the isotope j (g/mol): ρ_j = grams of isotope j/cm^3.

The fission rate of isotope j at position $\vec{r}$ due to the neutron flux $\phi(\vec{r},E)$ within the interval of neutron energy of E to $E + dE$ is obtained as

$$RR_f^j(\vec{r},E)\,dE = \Sigma_f^j(\vec{r},E)\phi(\vec{r},E)\,dE \tag{3.8a}$$

TABLE 3.3
Thermal (0.0253 eV) Neutron Cross Sections

	Cross Section (Barns)	
Material	**Fission σ_f**	**Absorption σ_a**
Uranium 233	531	579
Uranium 235	582	681
Uranium 238	—	2.70[a]
Uranium, natural	4.2	7.6
Plutonium 239	743	1012
Boron	—	759
Cadmium	—	2450
Carbon	—	0.0034
Deuterium	—	0.0005
Helium	—	<0.007
Hydrogen	—	0.33
Iron	—	2.55
Oxygen	—	0.00027
Sodium	—	0.53
Zirconium	—	0.19

[a] The effective absorption cross section of ^{238}U in a typical LWR is substantially higher owing to the larger cross section at epithermal energies.

The fission reaction rate for isotope j due to neutrons of all energies is

$$RR_{\mathrm{f}}^{j}(\vec{r}) = \int_{0}^{\infty} RR_{\mathrm{f}}^{j}(\vec{r}, E)\,\mathrm{d}E \tag{3.8b}$$

The energy generation rate per unit volume at $\vec{r}$ due to isotope j is

$$q_{\mathrm{j}}'''(\vec{r}) = \int_{0}^{\infty} \chi_{\mathrm{f}}^{j} RR_{\mathrm{f}}^{j}(\vec{r}, E)\,\mathrm{d}E \tag{3.9}$$

Summing over all isotopes yields the total volumetric energy generation rate:

$$q'''(\vec{r}) = \sum_{j} \int_{0}^{\infty} \chi_{\mathrm{f}}^{j} RR_{\mathrm{f}}^{j}(\vec{r}, E)\,\mathrm{d}E \tag{3.10}$$

In practice, the energy range is subdivided into a few intervals or groups. A multienergy group model is then used to calculate the neutron fluxes; thus,

$$q'''(\vec{r}) = \sum_{j} \sum_{k=1}^{K} \chi_{\mathrm{f}}^{j} \Sigma_{\mathrm{f}k}^{j}(\vec{r}) \phi_{k}(\vec{r}) \Delta E_{k} \tag{3.11}$$

where K = number of energy groups; $\Sigma_{\mathrm{f}k}$ and ϕ_{k} = macroscopic fission cross section and neutron flux, respectively, for the energy group k.

If we use a one energy group approximation, which gives good results for homogeneous thermal reactors at locations far from the reactor core boundaries, we have

$$q'''(\vec{r}) = \sum_{j} \chi_{\mathrm{f}}^{j} \Sigma_{\mathrm{f1}}^{j}(\vec{r}) \phi_{1}(\vec{r}) \tag{3.12}$$

If we also assume uniform fuel material composition, then

$$q'''(\vec{r}) = \sum_{j} \chi_{\mathrm{f}}^{j} \Sigma_{\mathrm{f1}}^{j} \phi_{1}(\vec{r}) \tag{3.13}$$

where Σ_{f1}^{j} is independent of position $\vec{r}$.

Assuming that $\chi_{\mathrm{f}}^{j} = \chi_{\mathrm{f}}$ for all fissile material, the volumetric heat generation rate in the fuel may be given by

$$q'''(\vec{r}) = \chi_{\mathrm{f}} \Sigma_{\mathrm{f1}} \phi_{1}(\vec{r}) \tag{3.14a}$$

where

$$\Sigma_{\mathrm{f1}} = \sum_{j} \Sigma_{\mathrm{f1}}^{j} \tag{3.14b}$$

It should be noted that the thermal neutron flux in a typical LWR is only 10–15% of the total neutron flux. However, the fission cross section of ^{235}U and ^{239}Pu is so large at thermal energies that thermal fissions constitute 70–85% of all fissions. The lower percentage value corresponds to beginning of cycle conditions for current core enrichments of about 5%, due to the harder spectrum.

Example 3.1: Determination of the Neutron Flux at a Given Power in a Thermal Reactor

PROBLEM

The Seabrook PWR (Appendix K) designed to produce power at a rating of 3411 MW_{th} has 193 fuel assemblies each loaded with 558.5 kg of UO_2. If the average isotopic content of the fuel is 2.78 weight percent ^{235}U, what is the average thermal neutron flux in the reactor?

Assume uniform fuel composition, and take the ^{235}U value of χ_f = 190 MeV/fission (3.04×10^{-11} J/fission). Also assume that 97.4% of the reactor energy, that is, of the recoverable fission energy, is generated in the fuel. The spectrum averaged, single group thermal fission cross section of ^{235}U $\left(\sigma_f^{25}\right)$ for this reactor is 35 barns. Note that this effective value of σ_f^{25} is smaller than that appearing in Table 3.3, because it is the average value of $\sigma_f^{25}(E)$ over the energy spectrum of the neutron flux, including the epithermal neutrons.

SOLUTION

Use Equation 3.14a to find $\phi_I(\vec{r})$.

$$\phi_I(\vec{r}) = \frac{q'''(\vec{r})}{\chi_f \Sigma_f} \quad (3.15a)$$

but because $\Sigma_f = \Sigma_f^{25}$ in this reactor, Σ_f^{25} is obtained from Equation 3.7a, yielding

$$\phi_I(\vec{r}) = \frac{q'''(\vec{r})}{\chi_f \sigma_f^{25} N^{25}} \quad (3.15b)$$

Consequently, the core-average value of the neutron flux in a reactor with uniform density of ^{235}U is obtained from the average energy generation rate in the fuel by

$$\langle \phi_I \rangle = \frac{\langle q''' \rangle}{\chi_f \sigma_f^{25} N^{25}} \quad (3.15c)$$

Multiplying both the numerator and the denominator by the UO_2 volume (V_{UO_2}), we get

$$\langle \phi_I \rangle = \frac{\langle q''' \rangle V_{UO_2}}{\chi_f \sigma_f^{25} N^{25} V_{UO_2}} = \frac{\dot{Q}_{th}}{\chi_f \sigma_f^{25} N^{25} V_{UO_2}} \quad (3.15d)$$

In order to use the above equation, only N^{25} needs to be calculated. All other values are given. The value of N^{25}, the atomic density of isotope ^{235}U, can be obtained from the uranium atomic density if the ^{235}U atomic fraction (a) is known

$$N^{25} = aN^{U} \tag{3.16a}$$

The uranium atomic density is equal to the molecular density of UO_2, because each molecule of UO_2 contains one uranium atom:

$$N^{U} = N^{UO_2} \tag{3.16b}$$

Expressing Equation 3.7b in terms of molecules of UO_2 and substituting Equations 3.16a and 3.16b, we obtain

$$N^{25} = a\frac{A_v \rho_{UO_2}}{M_{UO_2}} \tag{3.17a}$$

Multiplying each side of Equation 3.17a by V_{UO_2}, we get

$$N^{25}V_{UO_2} = aN^{UO_2}V_{UO_2} = \frac{aA_v m_{UO_2}}{M_{UO_2}} \tag{3.17b}$$

Now,

$$m_{UO_2} = 193\ (\text{assemblies}) \times 558.5\left[\frac{\text{kg}\,UO_2}{\text{assembly}}\right] \times \frac{1000\,\text{g}}{\text{kg}} = 1.0779 \times 10^8\ \text{g}$$

The molar mass of UO_2 is calculated from the atomic mass of uranium isotopes 235, M_{25}, and 238, M_{28}, and the molar mass of oxygen, M_O, as*

$M_{25} = 235.0439$ g/mol for ^{235}U
$M_{28} = 238.0508$ g/mol for ^{238}U
$M_O = 15.9994$ g/mol for oxygen

$$M_{UO_2} = M_U + 2M_O \tag{3.18a}$$

$$M_U = aM_{25} + (1 - a)M_{28} \tag{3.18b}$$

$$\therefore\ M_{UO_2} = aM_{25} + (1 - a)M_{28} + 2M_O \tag{3.18c}$$

In order to obtain the value of a, we use the known ^{235}U weight fraction, or enrichment (r):

$$r = \frac{aM_{25}}{M_U} = \frac{aM_{25}}{aM_{25} + (1 - a)M_{28}} \tag{3.19}$$

Equation 3.19 can be solved for a yielding

$$a = \frac{r}{r + M_{25}/M_{28}(1 - r)} = \frac{0.0278}{0.0278 + (235.0439/238.0508)(0.9722)}$$

$$= 0.028146\frac{\text{atoms }^{235}U}{\text{atoms U}} \tag{3.20}$$

* The 0.0055% abundance of ^{234}U in natural uranium is neglected.

From Equation 3.18c

$$M_{UO_2} = 0.028146(235.0439) + (0.971854)(238.0508) + 2(15.9994)$$
$$= 237.9657 + 2(15.9994) = 269.9645 \text{ g/mol}$$

Then from Equation 3.17b

$$N^{25}V_{UO_2} = \frac{0.028146(\text{atoms } ^{235}\text{U/atoms U})\left[0.60225\times10^{24}(\text{atoms U/g}\cdot\text{mol})\right](1.0779\times10^{8}\text{ g})}{269.9645\text{ g/mol}}$$

$$= 6.768\times10^{27}\text{ atoms } ^{235}\text{U}$$

Now using Equation 3.15d, the value for the average flux is calculated

$$\langle\phi_l\rangle = \frac{0.974(3411\text{ MW}_{th})\left(10^{6}(\text{W/MW})\right)}{\left(3.04\times10^{-11}\text{ J/fission}\right)\left(35\times10^{-24}\text{ cm}^2/\text{atom U}^{25}\right)\left(6.768\times10^{27}\text{ atom } ^{235}\text{U}\right)}$$

Answer: $\langle\phi_l\rangle = 4.62\times10^{14}$ neutrons/(cm^2 s)

3.4.2 Relation between Heat Flux, Volumetric Energy Generation, and Core Power

3.4.2.1 Single Pin Parameters

Three thermal parameters, introduced in Chapter 2, are related to the volumetric heat generation in the fuel: (1) the fuel pin power or rate of heat generation $(\dot{q})$; (2) the heat flux $(\vec{q}'')$, normal to any heat transfer surface of interest that encloses the fuel (e.g., the heat flux may be defined at the inner and outer surfaces of the cladding or the surface of the fuel itself); and (3) the power rating per unit length (linear energy generation rate or linear power) of the pin (q').

At steady state, the three quantities are related by Equations 2.3 through 2.5, which can be combined and applied to the nth fuel pin to yield

$$\dot{q}_n = \iiint_{V_{fn}} q'''(\vec{r})\,dV = \iint_{S_n} \vec{q}''\cdot\vec{n}\,dS = \int_L q'\,dz \tag{3.21}$$

where V_{fn} = volume of the energy generating region of a fuel element; $\vec{n}$ = outward unity vector normal to the surface S_n surrounding V_{fn}; L = length of the active fuel element.

It is also useful to define the mean heat flux through the surface of our interest:

$$\{q''\}_n = \frac{1}{S_n}\iint_{S_n} \vec{q}''(\vec{r})\cdot\vec{n}\,dS = \frac{\dot{q}_n}{S_n} \tag{3.22}$$

where { } = a surface averaged quantity.

The mean linear power rating of the fuel element is obtained from

$$q'_n = \frac{1}{L}\int_L q'\,dz = \frac{\dot{q}_n}{L} \tag{3.23}$$

Let us apply the relations of Equation 3.21 to a practical case. For a cylindrical fuel rod of pellet radius R_{fo}, outer cladding radius R_{co}, and length L, the total rod power is related to the volumetric energy generation rate by

$$\dot{q}_n = \int_{-L/2}^{L/2}\int_{o}^{R_{fo}}\int_{o}^{2\pi} q'''(r,\theta,z)r\,\mathrm{d}\theta\,\mathrm{d}r\,\mathrm{d}z \quad (3.24a)$$

The pin power can be related to the heat flux at the cladding outer surface (q''_{co}) by

$$\dot{q}_n = \int_{-L/2}^{L/2}\int_{o}^{2\pi} q''_{co}(\theta,z)R_{co}\,\mathrm{d}\theta\,\mathrm{d}z \quad (3.24b)$$

Here we have neglected axial heat transfer through the ends of the rod and heat generation in the cladding and gap. Finally, the rod power can be related to the linear power by

$$\dot{q}_n = \int_{-L/2}^{L/2} q'(z)\,\mathrm{d}z = q'_n L \quad (3.24c)$$

The mean heat flux through the outer surface of the clad is according to Equation 3.22:

$$\{q''_{co}\}_n = \frac{1}{2\pi R_{co}L}\int_{-L/2}^{L/2}\int_{o}^{2\pi} q''_{co}(\theta,z)\,R_{co}\,\mathrm{d}\theta\,\mathrm{d}z = \frac{\dot{q}_n}{2\pi R_{co}L} \quad (3.25)$$

It should be recognized that the linear power at any axial position is equal to the heat flux integrated over the perimeter:

$$q'(z) = \int_{o}^{2\pi} q''_{co}(\theta,z)R_{co}\,\mathrm{d}\theta \quad (3.26a)$$

The linear power can also be related to the volumetric energy generation rate by

$$q'(z) = \int_{o}^{R_{fo}}\int_{o}^{2\pi} q'''(r,\theta,z)r\,\mathrm{d}\theta\,\mathrm{d}r \quad (3.26b)$$

Then, for any cylindrical fuel rod

$$\dot{q}_n = Lq'_n = L2\pi R_{co}\{q''_{co}\}_n = L\pi R_{fo}^2\langle q'''\rangle_n \quad (3.27)$$

where the average volumetric energy generation rate in the pin is

$$\langle q'''\rangle_n = \frac{\dot{q}_n}{\pi R_{fo}^2 L} = \frac{\dot{q}_n}{V_{fn}} \quad (3.28)$$

3.4.2.2 Core Power and Fuel Pin Parameters

Consider a core consisting of N_p fuel pins. The overall thermal power in the core has been written as

$$\dot{Q}_{th} = \sum_{n=1}^{N_p} \dot{q}_n + \dot{Q}_{\text{moderator and noncladding structure}} \tag{2.6}$$

With γ defined as the fraction of energy deposited in the fuel and 1 − γ the fraction deposited in nonfuel regions, we obtain

$$\dot{Q}_{th} = \frac{1}{\gamma}\sum_{n=1}^{N_p} \dot{q}_n; \quad \frac{\dot{Q}_{\text{moderator and noncladding structure}}}{\dot{Q}_{th}} = 1-\gamma \tag{3.4; 3.5}$$

From Equations 3.27 and 3.28, for N_p fuel pins of identical dimensions

$$\dot{Q}_{th} = \frac{1}{\gamma}\sum_{n=1}^{N_p} \dot{q}_n = \frac{1}{\gamma}\sum_{n=1}^{N_p} Lq'_n = \frac{1}{\gamma}\sum_{n=1}^{N_p} L2\pi R_{co}\{q''_{co}\}_n = \frac{1}{\gamma}\sum_{n=1}^{N_p} L\pi R_{fo}^2 \langle q''' \rangle_n \tag{3.29}$$

We can define core-wide thermal parameters for an average pin as

$$\frac{\dot{Q}_{th}}{N_p} = \frac{1}{\gamma}\langle \dot{q} \rangle = \frac{L}{\gamma}\langle q' \rangle = \frac{L}{\gamma}2\pi R_{co}\langle q''_{co} \rangle = \frac{L}{\gamma}\pi R_{fo}^2 \langle q''' \rangle \tag{3.30}$$

The core-average volumetric energy generation rate in the fuel is given by

$$\langle q''' \rangle = \frac{\gamma \dot{Q}_{th}}{V_{fuel}} = \frac{\gamma \dot{Q}_{th}}{N_p V_{fn}} \tag{3.31}$$

The core-wide average fuel volumetric energy generation rate $\langle q''' \rangle$ should not be confused with the core power density Q''', defined in Chapter 2 as

$$Q''' = \frac{\dot{Q}_{th}}{V_{core}} \tag{3.32}$$

which takes into account V_{core}, the volume of all the core constituents: fuel, moderator, and structures.

Example 3.2: Heat Transfer Parameters in Various Power Reactors

PROBLEM

For the set of reactor parameters of Table 3.4, calculate for each reactor type:

1. Equivalent core diameter and core length.
2. Average core power density Q'''(MW/m^3).

TABLE 3.4
Reactor Parameters for Example 3.2

Quantity	PWR[a]	BWR[a]	PHWR[b] (CANDU)	SFBR[c]	HTGR[d]
Core power level (MW_{th})	3411	3323	2140	780	3000
% of power deposited in fuel rods	97.4	96.5	91.6	94.5	100
Fuel assemblies/core	193	764	12 × 380 = 4560	198	8 × 493 = 3944
Assembly lateral spacing (mm)	215 (square pitch)	152 (square pitch)	280 (square pitch)	144 (across hexagonal flats)	361 (across hexagonal flats)
Fuel rods/assembly	264	74	37	217	72
Fuel rods length (mm)	3658	3588	480	914	787
Fuel rod diameter (mm)	9.5	11.2	13.1	5.8	21.8

[a] Data taken from Appendix K.
[b] Twelve horizontal fuel assemblies are stacked end to end at 380 locations.
[c] Blanket assemblies are excluded from the SFBR calculation.
[d] Eight fuel assemblies are stacked end to end at 493 locations. Fuel rod dimensions given actually refer to coolant holes.

3. Core-wide average linear energy generation rate of a fuel rod, $\langle q' \rangle$(kW/m).
4. Core-wide average heat flux at the interface between the rod and the coolant, $\langle q''_{co} \rangle$(MW/m^2)

SOLUTION

Only the PWR case is considered in detail here; the results for the other reactors are summarized in Table 3.5.

1. Equivalent core diameter and length calculation

Fuel assembly area = (0.215 m)^2 = 0.046 m^2
Core area = (0.046 m^2) (193 fuel assemblies) = 8.92 m^2
Equivalent circular diameter: $\pi D^2/4 = 8.92$ m^2. Hence, $D = 3.37$ m (corresponds to Appendix K).
Core length (L) = 3.658 m
Total core volume = 3.658 $\pi(3.37)^2/4$ = 32.63 m^3

2. Average power density in the core from Equation 3.32:

$$Q''' = \frac{\dot{Q}_{th}}{\pi R^2 L} = \frac{\dot{Q}_{th}}{V_{core}} = \frac{3411\,\text{MW}}{32.63\,\text{m}^3} = 104.5\,\text{MW/m}^3$$

(corresponds to Appendix K)

3. Average linear energy generation rate in a fuel rod can be obtained from Equation 3.30 as

TABLE 3.5
Solutions to Example 3.2

Quantity	BWR	CANDU	SFBR	HTGR
Equivalent core diameter (m)	4.75[a]	6.16	2.13	8.42
Core length (m)	3.588	5.76	0.914	6.3
Core power density (MW/m^3)	52.3[a]	12.1	230.9	8.55
Average linear energy generation rate of a fuel rod (kW/m)	17.6	24.2	19.2	13.4
Average heat flux at the interface of fuel rod and coolant (MW/m^2)	0.50	0.59	1.05	0.20

[a] corresponds with Appendix K.

$$\langle q' \rangle = \frac{\gamma \dot{Q}_{th}}{N_p L} = \frac{0.974(3411\,\text{MW})}{(264\ \text{rods/assembly})(193\ \text{assemblies})(3.658\,\text{m/rod})}$$
$$= 17.86\,\text{kW/m}$$

4. Average heat flux at the interface between a rod and the coolant: From Equation 3.30, we can also obtain $\langle q''_{co} \rangle$ as

$$\langle q''_{co} \rangle = \frac{\gamma \dot{Q}_{th}}{N_p L 2\pi R_{co}} = \frac{\gamma \dot{Q}_{th}}{N_p L \pi D_{co}} = \frac{\langle q' \rangle}{\pi D_{co}}$$
$$= \frac{(17.86\,\text{kW/m})(10^{-3}\,\text{MW/kW})}{\pi(0.0095\,\text{m})} = 0.60\,\text{MW/m}^2$$

Calculations for the other reactor types are left as an exercise for the reader. The solutions are given in Table 3.5.

Note: Area for hexagon-shaped assembly (Figure 3.2) is given by

$$\text{Area hexagon} = \frac{12}{2}(D_{ft}/2)(D_{ft}/2)(\tan 30°) = \frac{\sqrt{3}}{2} D_{ft}^2 \tag{3.33}$$

where D_{ft} = distance across hexagonal flats.

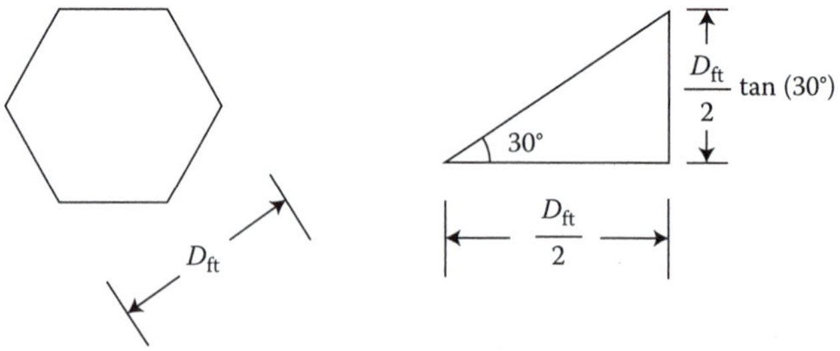

FIGURE 3.2 Hexagonal assembly.

TABLE 3.6
Distribution of Energy Generation Rate in a Homogeneous Unreflected Core

Geometry	Coordinate	$q'''(\vec{r})/q'''_{max}$ or $F(\vec{r})$	$q'''_{max}/\langle q'''\rangle$ (Ignoring Extrapolation Lengths)
Infinite slab	x	$\cos \pi x/L_e$	$\pi/2$
Rectangular parallelepiped	x, y, z	$\cos(\pi x/L_{xe})\cos(\pi y/L_{ye})\cos(\pi z/L_{ze})$	$\pi^3/8$
Sphere	r	$\dfrac{\sin(\pi r/R_e)}{\pi r/R_e}$	$\pi^2/3$
Finite cylinder	r, z	$J_o(2.405r/R_e)\cos(\pi z/L_e)$	$2.32(\pi/2)$

Note: $L_e = L + 2\delta L$; $R_e = R + \delta R$; L_e, R_e = extrapolated dimensions; L, R = fuel physical dimensions. δL and δR are typically taken as $0.71\lambda_{tr}$ where λ_{tr} is the transport mean free path of neutrons in the reactor core.

Source: Adopted from Rust, J. H., *Nuclear Power Plant Engineering*. Buchanan, GA: Haralson Publishing, 1979.

3.5 POWER PROFILES IN REACTOR CORES

We shall consider simple cases of reactor cores to form an appreciation of the overall power distribution in various geometries. The simplest core is one in which the fuel is homogeneously mixed with the moderator and uniformly distributed within the core volume. Consideration of such a core is provided here as a means to establish the general tendency of neutron flux behavior. In a neutronically heterogeneous power reactor, the fuel material is dispersed in lumps within the moderator (see Chapter 1).

In practice, different strategies may be sought for the fissile material distribution in the core. For example, to burn the fuel uniformly in the core, uniform energy generation is desired. Therefore, various enrichment zones may be introduced, with the highest enrichment located at the low neutron flux region near the core periphery.

3.5.1 Homogeneous Unreflected Core

In the case of a homogeneous unreflected core, the whole core can be considered as one fuel element. Using the one energy group scheme, it is clear from Equation 3.14a that $q'''(\vec{r})$ is proportional to $\phi_1(\vec{r})$.

Solving the appropriate one-group neutron diffusion equation, simple analytical expressions for the neutron flux, and hence the volumetric energy generation rate, have been obtained for simple geometries. The general distribution is given by*

$$q'''(\vec{r}) = q'''_{max}F(\vec{r}) \tag{3.34}$$

where q'''_{max} is the energy generation at the center of the homogeneous core. Expressions for $F(\vec{r})$ are given in Table 3.6. Thus, for a cylindrical core

* Note that in a homogeneous reactor $\langle q'''\rangle$ and Q''' are identical.

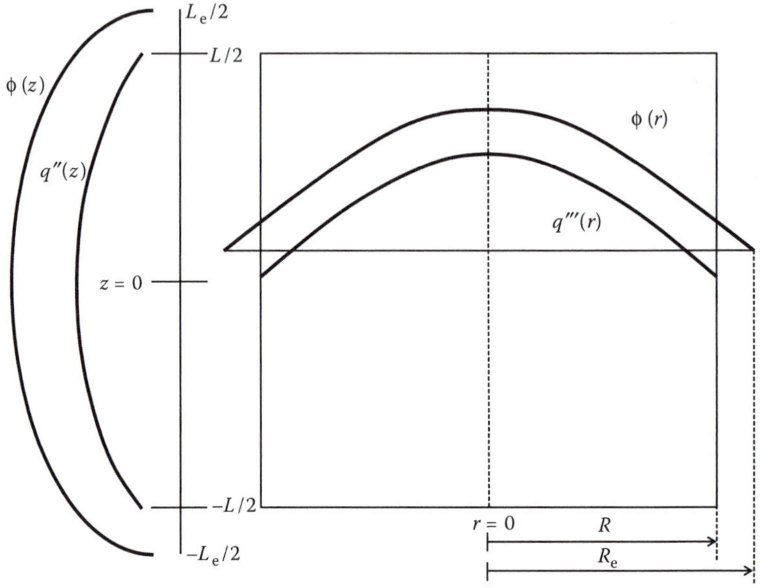

FIGURE 3.3 Neutron flux and energy generation rate profiles in a homogeneous cylindrical reactor.

$$q'''(r,z) = q'''_{max} J_o\left(2.4048\frac{r}{R_e}\right)\cos\left(\frac{\pi z}{L_e}\right) \quad \text{for } -L/2 \le z \le L/2;\ r \le R_o \tag{3.35}$$

where r and z are measured from the center of the core.

The shape of q''' as a function of r is shown in Figure 3.3. It is seen that the neutron flux becomes zero at a small distance δR from the actual core boundary. The distances δR and δL are called the *extrapolation lengths* and are usually small relative to R and L, respectively.

The overall core energy generation rate is given by

$$\dot{Q} = q'''_{max}\iiint_{V_{core}} F(\vec{r})\,dV \tag{3.36}$$

In real reactors, the higher burnup of fuel at locations of high neutron fluxes leads to flattening of radial and axial power profiles.

3.5.2 Homogeneous Core with Reflector

For a homogeneous core with a reflector, it is also possible to use a one-group scheme inside the reactor. For the region near the boundary between the core and the reflector, a two-group approximation is usually required. An analytical expression for q''' in this case is more difficult. The radial shape of the thermal neutron flux is shown in Figure 3.4.

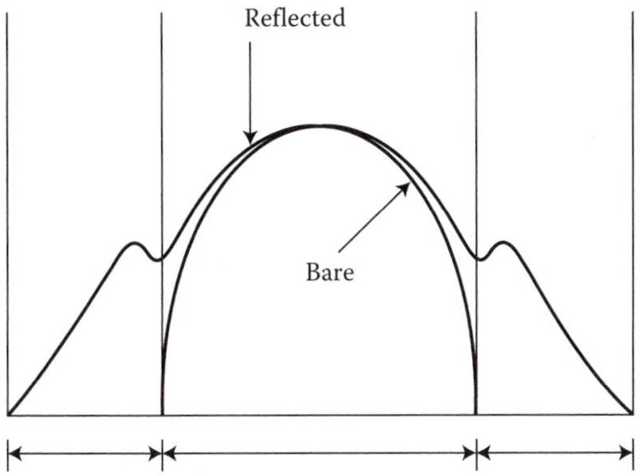

FIGURE 3.4 Effect of neutron reflector on the thermal neutron flux radial distribution.

3.5.3 Heterogeneous Core

In the case of a heterogeneous thermal reactor, energy is produced mainly in the fuel elements, and the thermal neutron flux is generated in the moderator.

In most power reactors, there are large numbers of rods. Thus, with little error we can approximate the profile of the energy generation rate in a fresh core with uniform enrichment by the previous expressions for a homogeneous core, provided that now $q'''(r)$ is understood to represent the energy generation rate in a fuel rod that is a distance r from the center of the reactor core. However, in practice, a fresh core may have zones of variable enrichment, which negates the possibility of using the homogeneous core expressions. With fuel burnup, various amounts of plutonium and fission products are introduced. A typical power distribution in a PWR mid-burnup core is shown in Figure 3.5. It is clear that no analytic expression can easily describe the spatial distribution.

3.5.4 Effect of Control Rods

The control rods depress the neutron flux radially and axially. Thus, the radial power profile is also depressed near the rods (Figure 3.6). In addition, soluble poison is the primary means for burnup cycle control in PWRs. In many reactors, some control material is uniformly mixed with the fuel in several fuel pins so that it is burned up as the fuel burns, thereby readily compensating for the loss of fuel fissile content.

Example 3.3: Local Pin Power for a Given Core Power

PROBLEM

For a heavy water-moderated reactor with uniform distribution of enriched UO_2 fuel in a cylindrical reactor core, calculate the power generated in a fuel rod located

Center-line	A	B	C	D	E	F	G	H
A	7FG 27,841 35,997 0.72	9AH 10,506 20,929 0.94	8GB 22,820 31,824 0.81	9FG 8021 18,967 1.03	9FF 12,007 22,721 1.03	8AH 20,927 29,456 0.82	8BG 23,452 32,601 0.93	10AH 0 10,505 1.12
B	9HA 10,506 20,928 0.94	9BG 12,827 23,454 0.97	9CH 9884 20,535 0.98	8DH 19,247 28,621 0.86	9DH 8086 19,249 1.08	8EG 21,118 30,125 0.88	10BG 0 12,826 1.36	10BH 0 10,847 1.16
C	8GB 22,820 31,820 0.81	9HC 9884 20,530 0.98	9GB 12,827 22,827 0.91	8FG 18,968 27,840 0.81	9 EG 10,514 21,117 1.02	8CH 20,532 29,348 0.86	8DG 23,409 32,506 0.92	10CH 0 9883 1.03
D	9FG 8021 18,956 1.02	8HD 19,236 28,594 0.86	8GF 19,382 28,183 0.80	8FF 22,722 31,183 0.79	9BH 10,847 21,660 1.07	9DG 12,829 23,410 1.07	10DG 0 12,830 1.37*	10DH 0 8086 0.83
E	9FF 12,007 22,714 1.03	9HD 8086 19,238 1.08	9GE 10,514 21,098 1.02	9HB 10,847 21,656 1.07	9GF 8021 19,382 1.15	8BH 21,656 31,071 0.95	10EG 0 10,516 1.10	
F	8HA 20,926 29,454 0.82	8GE 21,099 30,105 0.88	8HC 20,527 29,340 0.86	9GD 12,829 23,408 1.07	8HB 21,651 31,067 0.95	10FF 0 12,010 1.25	10FG 0 8022 0.82	
G	8BG 23,452 32,601 0.93	10GB 0 12,825 1.36	8GD 23,407 32,504 0.92	10GD 0 12,829 1.37*	10GE 0 10,516 1.10	10GF 0 8022 0.82		
H	10HA 0 10,505 1.12	10HB 0 10,847 1.16	10HC 0 9883 1.03	10HD 0 8086 0.83				

* = Maximum relative power

Fuel lots 7, 8, 9, 10 initially 3.20 w/o U-235

Cycle average burnup = 10,081 MWd/MT

Cycle thermal energy = 896.8 GWd

Key

1AA	Assembly number
0	BOC burnup, MWd/MT
17302	EOC burnup, MWd/MT
1.04	BOC relative power (Assembly/average)

FIGURE 3.5 Typical PWR assembly power and burnup distribution, assuming fresh fuel is introduced at the outer core location. (From Benedict, M., Pigford, T. H., and Levi, H. W., *Nuclear Chemical Engineering*, 2nd Ed. New York, NY: McGraw-Hill, 1981.)

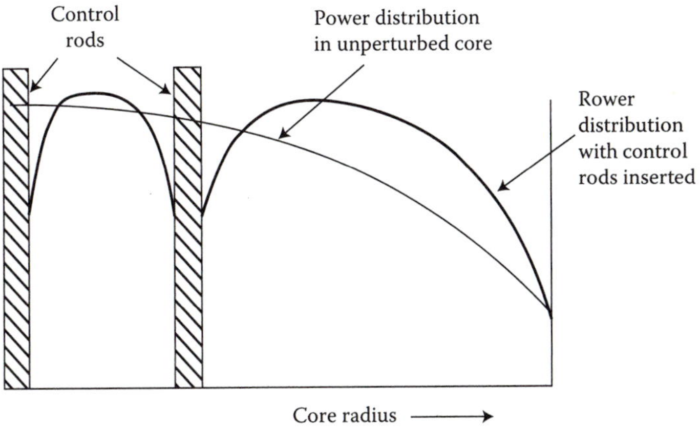

FIGURE 3.6 Radial power profile in a cylindrical reactor with inserted control rods.

half-way between the core centerline and its outer boundary. The important parameters for the core are as follows.

1. Core radius (R) = 2.44 m
2. Core height (L) = 6.10 m
3. Fuel pellet outside diameter = 1.524 cm
4. Maximum thermal neutron flux = 10^{13} neutrons/(cm^2 s)

Assume that the extrapolated dimensions can be approximated by the physical dimensions and the enrichment and average moderator temperature are such that, at every position

$$q'''\left(\text{W/m}^3\right) = 7.27 \times 10^{-6}\phi$$

with ϕ in neutrons/(cm^2 s). (This neglects the actual CANDU segmentation of fuel rods in short axial fuel bundles.)

SOLUTION

This reactor can be approximated as a homogeneous unreflected core in the form of a finite cylinder. From Table 3.6:

$$q'''(r,z) = q'''_{\max} J_o\left(2.4048\frac{r}{R_e}\right)\cos\left(\frac{\pi z}{L_e}\right) \tag{3.35}$$

The value of $q'''_{\max}$ is readily established as

$$q'''_{\max} = 7.27 \times 10^{-6}\phi_{\max} = 7.27 \times 10^{-6}(10^{13})\,\text{W/m}^3 = 72.7\,\text{MW/m}^3$$

For $r/R_e = 0.5$:

$$q'''(r,z) = q'''_{\max} J_o(1.202)\cos\left(\frac{\pi z}{L_e}\right)$$

For the power generated from a single fuel rod, Equation 3.24a yields

$$\dot{q}_n = \frac{L}{2}\int_{-L/2}^{L/2} q'''_n(r,z)A\,dz \tag{3.37}$$

where $A = \frac{\pi D_{fo}^2}{4}$.

Taking L_e as L

$$\begin{aligned}\dot{q}_n &= \int_{-L/2}^{L/2} q'''_{max} J_o(1.202)\cos\left(\frac{\pi z}{L}\right)A\,dz \\ &= q'''_{max} J_o(1.202)\frac{L}{\pi}\sin\frac{\pi z}{L}\Big|_{-L/2}^{L/2} A \\ &= q'''_{max} J_o(1.202)\frac{2L}{\pi}A\end{aligned}$$

where $J_o(1.202) = 0.6719$ [19]. Hence:

$$\dot{q}_n = 72.7(0.6719)\frac{2}{\pi}(6.1)\frac{\pi(1.524\times 10^{-2})^2}{4}$$

Answer: $\dot{q}_n = 34.6\,\text{kW}$

3.6 ENERGY GENERATION RATE WITHIN A FUEL PIN

3.6.1 Fuel Pins of Thermal Reactors

Consider a cylindrical fuel pin inside the reactor. The energy generation rate at a particular point of the rod, $q'''(r,\theta,z)$, depends on the position of the rod in the reactor and the concentration of the various fissionable materials at this point.

In thermal reactors with uniform fuel enrichment, the profile of the energy generation rate follows approximately the thermal neutron flux. For single-phase cooled reactors, in many cases the axial profile of q''' can be approximated by a cosine function, that is

$$q'''(z) \sim \cos\left(\frac{\pi z}{L_e}\right) \tag{3.38}$$

As mentioned earlier, for fresh fuel this formula gives adequate results for illustrative purposes.

Radially within the fuel element, q''' is expected to be reduced at the center because of thermal neutron flux depression. With burnup, plutonium buildup at the pellet rim causes further increase in the radial peak-to-average thermal flux. This depression is often neglected for small-diameter low-absorbing fuel rods but should not be ignored for thick, highly absorbing rods and for rods with high burnup (>40 MWd/kg_{HM}).

3.6.2 Fuel Pins of Fast Reactors

The dependence of q''' on z is similar to that of thermal reactors. However, the shape of q''' as a function of r within a rod is different because in fast reactors energetic neutrons, even fission neutrons, contribute to the fission reaction directly without being slowed. The radial profile of q''' tends to be flatter than in thermal and epithermal systems. Fortunately, in the real cases the diameter of the fuel rod is relatively small and the mean free path of the neutrons relatively large. Therefore, the assumption that q''' is independent of r and θ within a fuel pin is a good approximation.

3.7 ENERGY DEPOSITION RATE WITHIN THE MODERATOR

The energy deposition rate in the moderator mainly comes from (1) neutron slowing down by scattering due to collisions with the nuclei of the moderator material, and (2) γ-ray absorption. The dominant mechanism of energy production is neutron slowing down due to elastic scattering. Neutrons also lose energy through inelastic collisions as a result of excitation of the target nuclei. However, for light nuclei, moderation by inelastic scattering is less important than by elastic scattering. With heavy nuclei, such as those involved in structures, the inelastic scattering is the principal mechanism for neutron moderation.

With elastic scattering, the energy lost from the neutrons appears as kinetic energy of the struck nucleus. With inelastic scattering, the energy lost by the neutrons appears as γ-rays.

Let $\Sigma_{s,e\ell}(\vec{r},E)$ be the macroscopic elastic scattering cross section of neutrons of energy E at position $\vec{r}$, and let $\phi(\vec{r},E)$ be the neutron flux at this position and energy. Hence, the elastic scattering reaction rate $RR_{s,e\ell}(\vec{r},E)\mathrm{d}E$ within the energy interval E to $E + \mathrm{d}E$ is given by

$$RR_{s,e\ell}(\vec{r},E)\mathrm{d}E = \Sigma_{s,e\ell}(\vec{r},E)\phi(\vec{r},E)\mathrm{d}E \tag{3.39}$$

If $\overline{\Delta E(E)}$ is the mean energy loss per collision at neutron energy E, the energy deposition rate due to $\phi(r,\vec{E})$ is

$$q'''_{e\ell}(\vec{r},E)\mathrm{d}E = \overline{\Delta E(E)}\Sigma_{s,e\ell}(\vec{r},E)\phi(\vec{r},E)\mathrm{d}E \tag{3.40}$$

Now, as shown by Glasstone and Sesonske [8],

$$\overline{\Delta E(E)} = E\frac{2A}{(A+1)^2}\left[1-\mu_o(E)\right] \tag{3.41}$$

where μ_o (E) is the average cosine of the scattering angle for neutrons of energy E.

Note that

$$2A/(A+1)^2 = 1-\alpha/2 = \text{The arithmetic average value of } E_1 - E_2/E_1 \text{ per collision} \tag{3.42}$$

where E_1 = the initial neutron energy
E_2 = the neutron energy after the collision

$$\alpha = \left(\frac{A-1}{A+1}\right)^2 \tag{3.43}$$

The ratio of E_2/E_1 is equal to

$$\frac{E_2}{E_1} = \frac{1}{2}\left[(1+\alpha)+(1-\alpha)\cos\theta\right] \tag{3.44}$$

For the minimum loss of energy in a collision, $\theta = 0$, $E_2/E_1 = 1$, while for the maximum energy loss in a collision, $\theta = \pi$, $E_2/E_1 = \alpha$.

This ratio, E_2/E_1, is not to be confused with the familiar average logarithmic energy decrement per collision:

$$\xi \equiv \ell n\frac{E_1}{E_2} = 1 + \frac{\alpha}{1-\alpha}\ell n\,\alpha \tag{3.45}$$

Applying Equation 3.41 to Equation 3.40 yields

$$q'''_{e\ell}(\vec{r},E)\,\mathrm{d}E = E\frac{2A}{(A+1)^2}\left[1-\mu_o(E)\right]\Sigma_{s,e\ell}(\vec{r},E)\phi(\vec{r},E)\,\mathrm{d}E \tag{3.46}$$

Usually, the moderator can be considered a homogeneous material, so that Σ_s is independent of $\vec{r}$ within the moderator. The thermal energy generation rate from neutron elastic scatterings at all energies is then

$$q'''_{e\ell}(\vec{r}) = \frac{2A}{(A+1)^2}\int_{E_c}^{\infty} E\left[1-\mu_o(E)\right]\Sigma_{s,e\ell}(\vec{r})\,\phi(\vec{r},E)\,\mathrm{d}E \tag{3.47}$$

where E_c = an energy level below which the energy loss by neutrons is negligible (e.g., $E_c = 0.1$ eV).

In order to evaluate $q'''_{e\ell}(\vec{r})$ of Equation 3.47, the required parameters are

- $\Sigma_{s,e\ell}(\vec{r})$—the macroscopic elastic neutron scattering cross-section of the moderator material.
- μ_o (E)—for various materials and neutron energies given as f_1(E) in Table 7.2.1 of Jaeger [10].

Note that the factor $2A/(A + 1)^2$ decreases dramatically with increasing mass number A of the scattering material. For $A = 1$ (hydrogen), the factor is 0.5 while for $A = 40$ (calcium) that is a dominant constituent of concrete, the factor is 0.05. Hence, for large A materials, heating by neutron elastic scattering is small.

3.8 ENERGY DEPOSITION IN THE STRUCTURE

The main mechanisms of energy deposition in the structure are (1) γ-ray absorption, (2) elastic scattering of neutrons, and (3) inelastic scattering of neutrons.

3.8.1 γ-Ray Absorption

The photon "population" at the particular point $\vec{r}$ of the structure is mainly due to (1) γ-rays born somewhere in the fuel, which arrive without scattering at position $\vec{r}$

within the structure; (2) γ-rays born within the structural materials, which arrive unscattered at $\vec{r}$; and (3) scattered photons (Compton effect).

Consider the quantity $N_\gamma(\vec{r},E)\mathrm{d}E$, which is the photon density at a particular position r within the structure having energy between E and $E+\mathrm{d}E$. The energy flux is defined as

$$I_\gamma(\vec{r},E)\mathrm{d}E \equiv EN_\gamma(\vec{r},E)\mathrm{d}E \tag{3.48}$$

The absorption rate is described by application of the linear energy absorption coefficient $\mu_a\,(E)$ as follows:

$$q_\gamma'''(\vec{r},E)\mathrm{d}E = \mu_a(E)I_\gamma(\vec{r},E)\mathrm{d}E \tag{3.49}$$

where $q_\gamma''(\vec{r},E)$ = the absorbed energy density per unit time from the γ-ray energy flux within the interval E to E + dE; $\mu_a\,(E)$ = a function of the material, as can be found in Table 3.7.

TABLE 3.7
Linear γ-Ray Attenuation and Absorption Coefficients

	Coefficient (m^{-1})			
γ-Ray Energy (MeV)	**Water**	**Iron**	**Lead**	**Concrete**
0.5				
μ	9.66	65.1	164	20.4
μ_a	3.30	23.1	92.4	7.0
1.0				
μ	7.06	46.8	77.6	14.9
μ_a	3.11	20.5	37.5	6.5
1.5				
μ	5.74	38.1	58.1	12.1
μ_a	2.85	19.0	28.5	6.0
2.0				
μ	4.93	33.3	51.8	10.5
μ_a	2.64	18.2	27.3	5.6
3.0				
μ	3.96	28.4	47.7	8.53
μ_a	2.33	17.6	28.4	5.08
5.0				
μ	3.01	24.6	48.3	6.74
μ_a	1.98	17.8	32.8	4.56
10.0				
μ	2.19	23.1	55.4	5.38
μ_a	1.65	19.7	41.9	4.16

Source: From Templin, L. T. (ed.), *Reactor Physics Constants*, 2nd Ed. ANL-5800, Argonne National Laboratory, Argonne, IL, 1963.

The total energy deposition rate then is

$$q'''_\gamma(\vec{r}) = \int_0^{E_\infty} q'''_\gamma(\vec{r},E)\mathrm{d}E = \int_0^{E_\infty} \mu_a(E)I_\gamma(\vec{r},E)\mathrm{d}E \tag{3.50}$$

where E_∞ is selected at a sufficiently high value.

$I_\gamma(\vec{r},E)$ can be found by solving the appropriate transport equation. However, practical calculation of γ-ray attenuation is often greatly simplified by the use of the so-called buildup factors and the uncollided γ-ray flux. With this procedure, the simplified transport equation is first solved neglecting the scattering process to yield the uncollided flux $I^*_\gamma(\vec{r},E)$. For example, for a plane geometry, the energy of the uncollided γ-ray flux is obtained from

$$I^*_\gamma = I^\mathrm{o}_\gamma\, \mathrm{e}^{-\mu(E)x} \tag{3.51}$$

where μ (E) = linear attenuation coefficient for photons at energy E due to absorption and scattering (Table 3.7); I^o_γ = unattenuated γ-ray flux. If $I_\gamma(\vec{r},E')$ is the real energy flux at a point $\vec{r}$ resulting from I^o_γ, the buildup factor (B) is defined as

$$B(\vec{r},\mu,E) = \frac{\int_0^{E_\infty} \mu_a(E')I_\gamma(\vec{r},E')\mathrm{d}E'}{\mu_a(E)I^*_\gamma(\vec{r},E)} \tag{3.52}$$

Note that by definition $B(\vec{r},E)$ is greater than unity and depends primarily on the boundary conditions and the energy level of the uncollided photons (through the linear attenuation coefficient and the scattered photon source distribution). The values for the cases of interest are tabulated elsewhere [9,12]. Utilizing this definition of B, Equation 3.49 becomes

$$q'''_\gamma(\vec{r},E)\mathrm{d}E = B\mu_a(E)I^*_\gamma(\vec{r},E)\mathrm{d}E \tag{3.53}$$

The following are mathematical expressions for simplified cases for the evaluation of Equation 3.53 when $I^*_\gamma(\vec{r},E)$ can be analytically evaluated.

1. Point isotropic source emitting S photons of energy E_0 per second. In this case

$$q'''_\gamma(r) = S\,B\mu_a(E_0)E_0\,\frac{\mathrm{e}^{-\mu r}}{4\pi r^2} \tag{3.54}$$

   where r is the distance from the point source.

2. Infinite plane source emitting S photons with energy E_0 per unit time per unit surface in the positive direction of the x-axis:

$$q'''_\gamma(x) = S\,B\mu_a(E_0)E_0\mathrm{e}^{-\mu x} \tag{3.55}$$

3. Plane isotropic source emitting S photons of energy E_0 per unit source area, per second in all directions:

$$q_\gamma'''(x) = S\,B\mu_a(E_0)\frac{E_0}{2}\int_{\mu x}^{\infty}\frac{e^{-t}}{t}\,dt \tag{3.56}$$

3.8.2 Neutron Slowing Down

In the structure, the neutrons slow down by (1) elastic scattering and (2) inelastic scattering. For elastic scattering, we can use the approximation of

$$q_{e\ell}'''(\vec{r}) = \frac{2A}{(A+1)^2}\int_{E_c}^{\infty} E\left[1-\mu_o(E)\right]\Sigma_{s,e\ell}(\vec{r})\phi(\vec{r},E)\,dE \tag{3.47}$$

Taking into consideration that (1) inelastic scattering heating is not large compared with γ-heating, and (2) the γ-rays due to inelastic scattering are of moderate energies and therefore are absorbed in relatively short distances [8], the energy can be assumed to be released at the point of the inelastic scattering event. Then $q_{il}'''(\vec{r})$ is given by an expression similar to that of Equation 3.47:

$$q_{i\ell}'''(\vec{r}) = \int_{E_c}^{\infty} Ef(E)\Sigma_{s,i\ell}(E)\phi(\vec{r},E)\,dE \tag{3.57}$$

where $f(E)$ = the fraction of the neutron energy E lost in the collision. The parameter f is a function of E and the material composition.

Finally, the heat deposition within the structure is

$$q'''(\vec{r}) = q_\gamma'''(\vec{r}) + q_{e\ell}'''(\vec{r}) + q_{i\ell}'''(\vec{r}) \tag{3.58}$$

Example 3.4: Energy Deposition Rate in a Thermal Shield

PROBLEM

In a PWR, the core is surrounded by a thermal shield (Figure 3.7) to protect the pressure vessel from γ-ray heating and neutron-induced radiation damage. For an iron thermal shield with the radiation values given below, calculate the volumetric energy deposition rate in the shield at its outermost position.

Assume that the core of the reactor is equivalent to an infinite plane source and that the shield can be treated as a slab owing to the small thickness-to-radius ratio.

For steel, the inelastic scattering rate may be assumed to be equal to the elastic scattering rate:

$$\Sigma_{s,e\ell} \approx \Sigma_{s,i\ell} \tag{3.59}$$

The total neutron cross section can be approximated by the sum of both scatterings:

$$\Sigma_T \approx \Sigma_{s,e\ell} + \Sigma_{s,i\ell} \tag{3.60}$$

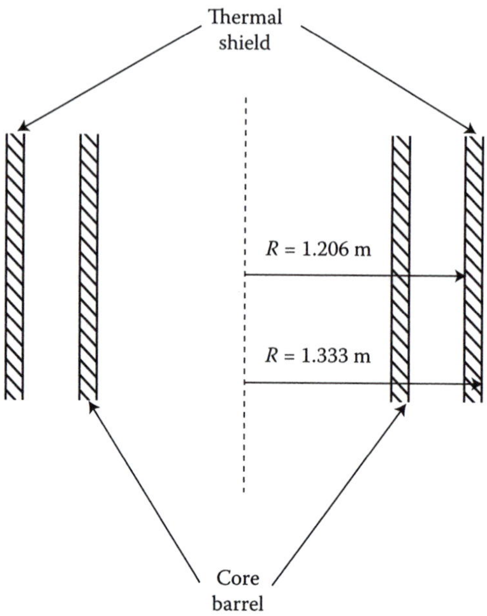

FIGURE 3.7 Thermal shield and core barrel of a PWR.

Use the values for the radiation flux parameters of Table 3.8.

SOLUTION

Equation 3.55 may be used to find $q_\gamma'''(x)$:

$$q_\gamma'''(x) = SB\mu_a(E_0)E_0 e^{-\mu x}$$

where x is measured from the inner wall:

$$x = 1.333\,\text{m} - 1.206\,\text{m} = 0.127\,\text{m} = 12.7\,\text{cm}$$

From Table 3.7, μ_a for iron at $E_0 = 2.0$ MeV is 0.182 cm^{-1} and μ for iron at $E_0 = 2.0$ MeV is 0.333 cm^{-1}. Hence,

$$q_\gamma''' = (10^{14})(4.212)(0.182)(2)e^{-0.333(12.7)} = 2.23 \times 10^{12}\,\text{MeV/cm}^3\,\text{s}$$

TABLE 3.8
Radiation Flux Parameters for Example 3.4

γ-Radiation	**Neutron Radiation**
$E_0 = 2$ MeV	$\phi_{fast} = 10^{14}$ neutrons/cm^2 s
$S = 10^{14}$ γ/cm^2 s	Effective neutron energy = 0.6 MeV
$B_{Fe} = 4.212$	Neutron total scattering cross section = σ_T $(0.1 < E < 15) = 3$ barns
	$f(2\text{ MeV}) = 0.1$

In order to solve for the elastic and inelastic scattering of the neutrons, Equations 3.47 and 3.57 must be solved

$$q'''_{e\ell}(\vec{r}) = \frac{2A}{(A+1)^2}\int_{E_c}^{\infty} E[1-\mu_o(E)]\Sigma_{s,e\ell}(\vec{r})\phi(\vec{r},E)\,dE \tag{3.47}$$

$$q'''_{i\ell}(\vec{r}) = \int_{E_c}^{\infty} Ef(E)\Sigma_{s,i\ell}(E)\phi(\vec{r},E)\,dE \tag{3.57}$$

Using a one-group approximation for the integral and defining ϕ_{fast} as the fast neutron flux, the above equations reduce to

$$q'''_{e\ell}(\vec{r}) = \frac{2A}{(A+1)^2}[1-\mu_o(E)]\bar{E}\Sigma_{s,e\ell}\phi_{fast} \tag{3.61}$$

$$q'''_{i\ell}(\vec{r}) = f(\bar{E})\bar{E}\,\Sigma_{s,i\ell}\phi_{fast} \tag{3.62}$$

where $\bar{E}$ is the effective energy of the flux.

Now for iron $A = 55.85$ so that

- $\frac{2A}{(A+1)^2} = 0.0346$
- $\mu_o(E) = 0.205$ from Table 7.2.1 of Jaeger [10] for the specified effective energy of 0.6 MeV.

The total cross section can be obtained from

$$\begin{aligned}\Sigma_T &= \frac{\rho_{Fe}A_v}{M_{Fe}}\sigma_T \\ &= \frac{7.87\,\text{g/cm}^3(0.6022\times 10^{24}\,\text{atom/mole})}{55.85\,\text{g/mole}}(3\times 10^{-24}\,\text{cm}^2) \\ &= 0.254\,\text{cm}^{-1}\end{aligned} \tag{3.63}$$

Therefore, with the assumption that $\Sigma_{s,e\ell} = \Sigma_{s,i\ell}$, each would equal one-half of Σ_T:

$$\Sigma_{s,i\ell} = \Sigma_{s,e\ell} \sim 0.5\Sigma_T = 0.127\,\text{cm}^{-1} \tag{3.64}$$

Now the energy deposition rate due to neutron scattering from this mono-energetic neutron flux can be calculated

$$\begin{aligned}q'''_{e\ell} &= \frac{2A}{(A+1)^2}[1-\mu_o(E)]\bar{E}\,\Sigma_{s,e\ell}\phi_{fast} \\ &= (0.0346)(1-0.205)(0.6\,\text{MeV})(0.127\,\text{cm}^{-1})(10^{14}\,\text{neutron/cm}^2\,\text{s}) \\ &= 0.21\times 10^{12}\,\text{MeV/cm}^3\,\text{s}\end{aligned} \tag{3.61}$$

$$\begin{aligned}q'''_{i\ell} &= f(\bar{E})\bar{E}\,\Sigma_{s,i\ell}\phi_{fast} \\ &= (0.1)(0.6\,\text{MeV})(0.127\,\text{cm}^{-1})(10^{14}\,\text{neutron/cm}^2\,\text{s}) \\ &= 0.76\times 10^{12}\,\text{MeV/cm}^3\,\text{s}\end{aligned} \tag{3.62}$$

$$\therefore q''' = q'''_{\gamma} + q'''_{e\ell} + q'''_{i\ell}$$
$$= 2.23 \times 10^{12} + 0.21 \times 10^{12} + 0.76 \times 10^{12}\ \text{MeV/cm}^3\,\text{s}$$
$$= 3.2 \times 10^{12}\ \text{MeV/cm}^3\,\text{s} = 0.51\,\text{W/cm}^3 \quad (3.58)$$

Note that γ-rays are the principal source of energy deposition in the structure. In fact, because of the high buildup factor (*B*) for iron, the calculated energy deposition rate due to γ-rays may exceed the incident flux of photons. A more refined transport calculation in which the photon energy and scattering properties are accounted for in detail is needed if exact prediction of the energy deposition rate is desired.

3.9 SHUTDOWN ENERGY GENERATION RATE

It is important to evaluate the rate of energy generated in a reactor after shutdown for determining cooling requirements under normal conditions and accident consequences following abnormal events. In order to do so, it is instructive to first examine the energy sources for a reactor at power and then illustrate how these individual energy sources disappear or change on shutdown.

Let us consider a typical LWR reactor startup to 100% power. The energy source components, neglecting the neutrino energy, are

- Prompt neutron fissions
- Delayed neutron fissions
- Fission product decay
- Actinide decay

The energy generation rate, or power, from these components is illustrated in Figure 3.8 for startup to 100% power of an LWR fueled with 3% enriched UO_2 by a 0.065%

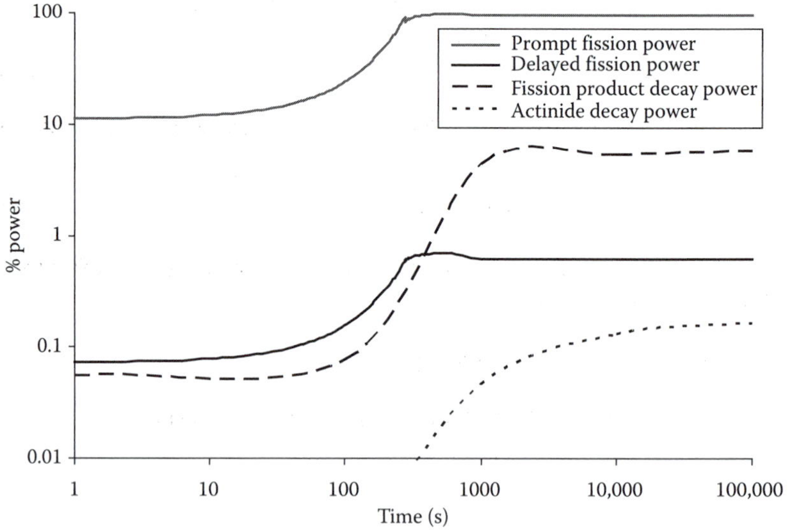

FIGURE 3.8 Reactor startup and return to criticality given by Equation 3.66 with component power contributions displayed.

dk/k positive step reactivity insertion at time zero, where k is the effective neutron multiplication factor. (Note that a 0.065% dk/k negative step reactivity insertion made at 77 s was sufficient to level power at 100%—no feedback effects were modeled.) The relative magnitudes of the power contribution and the time to achieve saturation of each of these components are well illustrated in Figure 3.8. Prompt fissions dominate both the magnitude and prompt response of the total power characteristic. Delayed neutrons, which constitute about 0.67% of all neutrons, follow the same time profile, but are seconds behind.

On shutdown, the contribution of prompt fissions decays very rapidly. Energy generation on reactor shutdown becomes the sum of the remaining source components listed above, that is, (1) fissions from delayed neutron or photoneutron emissions, and (2) decay of fission products and actinides, fertile materials, and other activation products from neutron capture. These sources initially contribute equal amounts to the shutdown power. However, within minutes after shutdown, fissions from delayed neutron emission are reduced to a negligible amount.

The first source, fission energy, is described next in Section 3.9.1. The second source will be characterized in detail in two subsections. Initially in Section 3.9.2, it is presented by historical relations that are useful in illustrating (1) the physical dependence of decay power on the extent of reactor operation preceding reactor shutdown and (2) the enhancement of decay power due to the presence of actinides beyond ^{235}U and neutron capture in fission products. In Section 3.9.3, the characterization is from standards developed by the American Nuclear Society (ANS) under procedures accredited by the American National Standards Institute (ANSI). The most recent 2005 standard version [3] is focused on light water reactors.

3.9.1 Fission Power after Reactivity Insertion

The energy generated from fissions by delayed neutrons is obtained by solving the neutron kinetic equations after a step insertion of reactivity. Assuming a single "effective" group of delayed neutrons, the time-dependent neutron flux can be given by [20]:

$$\phi(t) = \phi_0 \left[\frac{\beta}{\beta - \rho} \exp\left(\frac{\lambda\rho}{\beta - \rho} t_s \right) - \frac{\rho}{\beta - \rho} \exp\left(-\frac{\beta - \rho}{\Lambda} t_s \right) \right] \tag{3.65}$$

where

ϕ_0	= steady-state neutron flux prior to shutdown;
$\beta = \sum_{i=1}^{6} \beta_i$	= effective delayed neutron yield fraction obtained from the more general 6-group treatment of delayed neutrons;
β_i	= yield fraction of the ith delayed neutron group;
ρ	= step reactivity change;
$\lambda = \left(\sum_{i=1}^{6} \beta_i \lambda_i \right) / \beta$	= effective decay constant of delayed neutron precursors;
λ_i	= decay constant of the ith delayed neutron precursor group;
t_s	= time after initiation of the insertion of negative reactivity;

Λ	$= lP/k$ = mean generation time between the birth of a fission neutron and the subsequent absorption leading to another fission;
lP	= prompt neutron lifetime;
k	= neutron multiplication factor.

Substituting typical values for a ^{235}U-fueled, water-moderated reactor of β = 0.0067, $\lambda = 0.08\ \text{s}^{-1}$ and $\Lambda = 6 \times 10^{-5}\ \text{s}^{-1}$ into Equation 3.65 for a positive step reactivity insertion of $\rho = 0.1\beta$ (often referred to as \$0.10), the fractional power, which is proportional to the fractional flux, is given by

$$\frac{\dot{Q}}{\dot{Q}_0} = 1.11e^{+0.0089t_s} - 0.11e^{-100.5t_s} \tag{3.66}$$

where t_s is in seconds. Contributions of prompt fissions, delayed fissions, fission product decay, and actinide decay to reactor power for startup of a reactor of these characteristics due to a step positive \$0.10 reactivity insertion at time zero followed by negative reactivity insertion at 77 s to return to criticality are shown in Figure 3.8. It is seen that the steady state is reached at about 250 s, at a power level ten times the initial value. In power reactors, the rise in fuel temperature introduces negative reactivity that may be sufficient to stabilize the power.

3.9.2 Power from Fission Product Decay

The major source of shutdown power is fission product decay. Simple, empirical formulas for the rate of energy release due to β and γ emissions from decaying fission products are given by [7]

$$\begin{aligned} \beta\text{-energy release rate} &= 1.40t'^{-1.2}\ \text{MeV/fission s} \\ \gamma\text{-energy release rate} &= 1.26t'^{-1.2}\ \text{MeV/fission s} \end{aligned} \tag{3.67}$$

where t' = time after the occurrence of fission in seconds.

The equations above are accurate within a factor of 2 for $10\,\text{s} < t' < 100$ days. Integrating the above equations over the reactor operation time yields, the rate of decay energy released from fission products after a reactor has shutdown.

Assuming 200 MeV are released for each fission, 3.1×10^{10} fissions per second would be needed to produce 1 W of operating power. Thus, a fission rate of $3.1 \times 10^{10} q_o'''$ fissions/cm$^3 \cdot$s is needed to produce q_o''' W/cm^3. The decay heat at a time τ seconds after reactor startup due to fissions occurring during the time interval between τ' and $\tau' + d\tau'$ is given by the following (see Figure 3.9 for time relations):

$$dP_\beta = 1.40(\tau - \tau')^{-1.2}(3.1 \times 10^{10})q_o'''\,d\tau'\ \text{MeV/cm}^3\,\text{s} \tag{3.68a}$$

$$dP_\gamma = 1.26(\tau - \tau')^{-1.2}(3.1 \times 10^{10})q_o'''\,d\tau'\ \text{MeV/cm}^3\,\text{s} \tag{3.68b}$$

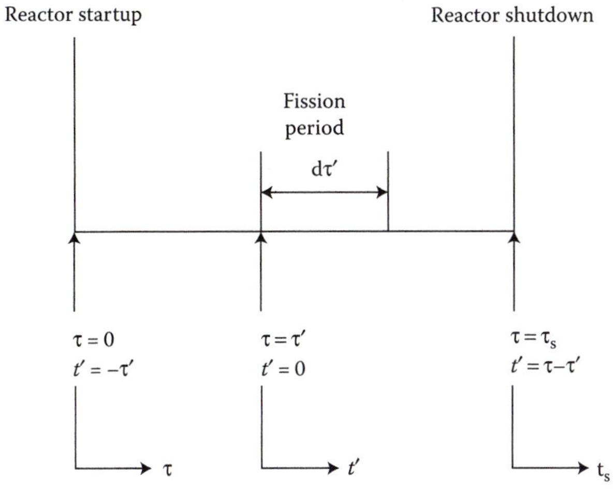

FIGURE 3.9 Time intervals. τ = time after reactor startup; $t' = \tau - \tau'$ = time after fission; $t_s = \tau - \tau_s$ = time after shutdown.

For a reactor operating at a constant power level over the period τ_s, we integrate Equations 3.68a and 3.68b to get the decay power from all fissions:

$$P_\beta = 2.18 \times 10^{11} q_o'''[(\tau - \tau_s)^{-0.2} - \tau^{-0.2}]\,\text{MeV/cm}^3\,\text{s} \tag{3.69a}$$

$$P_\gamma = 1.95 \times 10^{11} q_o'''[(\tau - \tau_s)^{-0.2} - \tau^{-0.2}]\,\text{MeV/cm}^3\,\text{s} \tag{3.69b}$$

The decay power level may be expressed as a fraction of the constant operating power level (P_0)—that is associated with the steady-state volumetric heat generation rate (q_o''')—by multiplying P_γ and P_β in Equations 3.69a and 3.69b by 1.602×10^{-13} to convert the units to W/cm^3—and rearranging to obtain

$$\frac{P_\beta}{P_0} = 0.035[(\tau - \tau_s)^{-0.2} - \tau^{-0.2}] \tag{3.70a}$$

$$\frac{P_\gamma}{P_0} = 0.031[(\tau - \tau_s)^{-0.2} - \tau^{-0.2}] \tag{3.70b}$$

The total fission product decay power (P) fraction is then given by

$$\frac{P}{P_0} = 0.066[(\tau - \tau_s)^{-0.2} - \tau^{-0.2}] \tag{3.70c}$$

Although almost all the energy from the β-particles is deposited in the fuel material, depending on the reactor configuration, only a fraction of the γ energy is deposited in the fuel material. The rest is deposited within the structural materials of the core and the surrounding supporting structures.

The decay power fraction predicted by Equation 3.70c is plotted in Figure 3.10 as a function of time after reactor shutdown for various times of reactor operation. Note

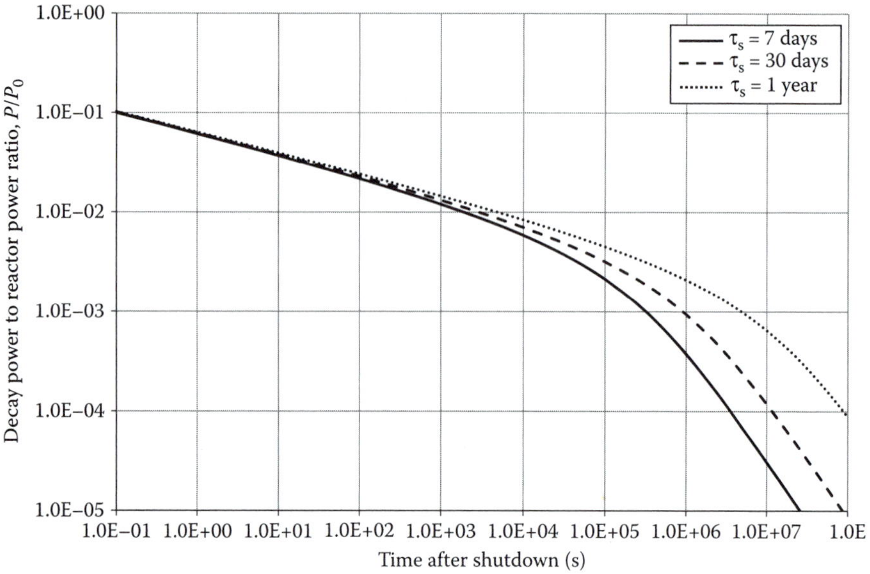

FIGURE 3.10 Fractional decay power from empirical relations as a function of shutdown time.

that for operating times (τ_s) of a few days or longer, the decay power shortly after shutdown is independent of reactor operating time. However, the reactor operating time is relatively important for determining the long-term decay power.

This decay power behavior of Figure 3.10 can be demonstrated by rewriting Equation 3.70c as

$$\frac{P}{P_0} = 0.066[t_s^{-0.2} - (t_s + \tau_s)^{-0.2}] \quad (3.70d)$$

and taking appropriate limits, for example, for short times after shutdown compared to the operation time, that is, $t_s << \tau_s$:

$$\frac{P}{P_0} \cong 0.066\left[t_s^{-0.2} - \tau_s^{-0.2}\right] \cong 0.066 t_s^{-0.2}$$

Hence, $P(t_s)$ does not depend on operating time, τ_s, whereas for a long time after shutdown compared to the operation time, $t_s > \tau_s$, Equation 3.70c must be evaluated in full and $P(t_s)$ increases with longer operating time, τ_s.

For comparison, another equation was later experimentally obtained from a 2.54 cm-diameter uranium rod. The resulting equation is given by [7]:

$$\frac{P}{P_0} = 0.1[(\tau - \tau_s + 10)^{-0.2} - (\tau + 10)^{-0.2} + 0.87(\tau + 2\times10^7)^{-0.2} - 0.87(\tau - \tau_s + 2\times10^7)^{-0.2} \quad (3.71)$$

This equation predicts higher decay powers due to, among other factors, the inclusion of the decay heat of the actinides ^{239}U and ^{289}Np along with decay of ^{235}U fission products. The effect of neutron capture in fission products is to increase the decay power on the order of a few percent, depending on the level of burnup and the operating time.

3.9.3 ANS Standard Decay Power

3.9.3.1 UO_2 in Light Water Reactors

The prediction of decay power and its codification for design use has undergone a progressive series of refinements through the publication of ANSI/ANS standards.

In 1961, the data from several experiments were combined to provide a more accurate method for predicting fission product decay power [17]. The results were adopted in 1971 by the American Nuclear Society (ANS) as the basis for a draft standard [1] for reactor shutdown cooling requirements. The revised standards issued subsequently as summarized below have progressively reduced the conservatism in the predicted level of decay power.

Investigation of the inaccuracies in the 1971 ANS proposed standard due to the assumptions that (1) fission product decay power from different fissile isotopes are equal and (2) neutron capture effects are negligible led to a new standard that was developed in 1979 and reaffirmed in 1985. The revised ANS standard, which consisted of equations based on summation calculations, explicitly accounts for decay power from ^{235}U, ^{238}U, and ^{239}Pu fission products. Neutron capture in fission products is included through a correction factor multiplier. The new standard is also capable of accounting for changes in fissile nuclides with fuel life. Accuracy within the first 10^4 s after shutdown was emphasized in the new standard's development for accident consequence evaluation.

Subsequent Standard revisions were issued in 1994 [2] and 2005 [3]. These revisions have principally improved the accounting for neutron capture in fission products, improved the prediction of decay power from fission products of major fissionable nuclides present in LWRs (^{235}U, ^{239}Pu, $^{241}Pu_{thermal}$, $^{238}U_{fast}$), included decay power from additional actinide isotopes (^{239}U, ^{239}Np), and extended results to longer times, 10^{10} s, after reactor shutdown (referred to as cooling time).

Figures 3.11a and b present the decay power derived [18] from the tabular results of the 2005 ANS Standard [3], which is for a UO_2-fueled PWR with 4.2% enrichment after 1350 EFPD* and a corresponding discharge burnup of 51 MWd/kg_{IHM}. Each curve includes fission products, ^{239}U and ^{239}Np.

For times greater than 10^8 s, the ANS Standard increasingly underestimates the decay power due to neglect of the other actinides, principally ^{238}Pu(~40% of total actinide decay power between 10^8 and 10^9 s), ^{244}Cm (~40% of total actinide decay power at 10^8 s, ~17% at 10^9 s) and ^{241}Am (~6% of total actinide decay power at 10^8 s, ~30% at 10^9 s). At 10^{10} s the ANS FP + ^{239}U + ^{239}Np curve underestimates the total decay power by slightly more than two orders of magnitude.

Figures 3.11a and 3.11b, which span a time range from 1 s to 10^{10} s, refer to reactors initially fueled with uranium and operated at a constant power (P_0) for an infinite

* 1350 EFPD is a sufficient period for all fission products to have reached saturation levels. This period can be considered as an infinite period of operation.

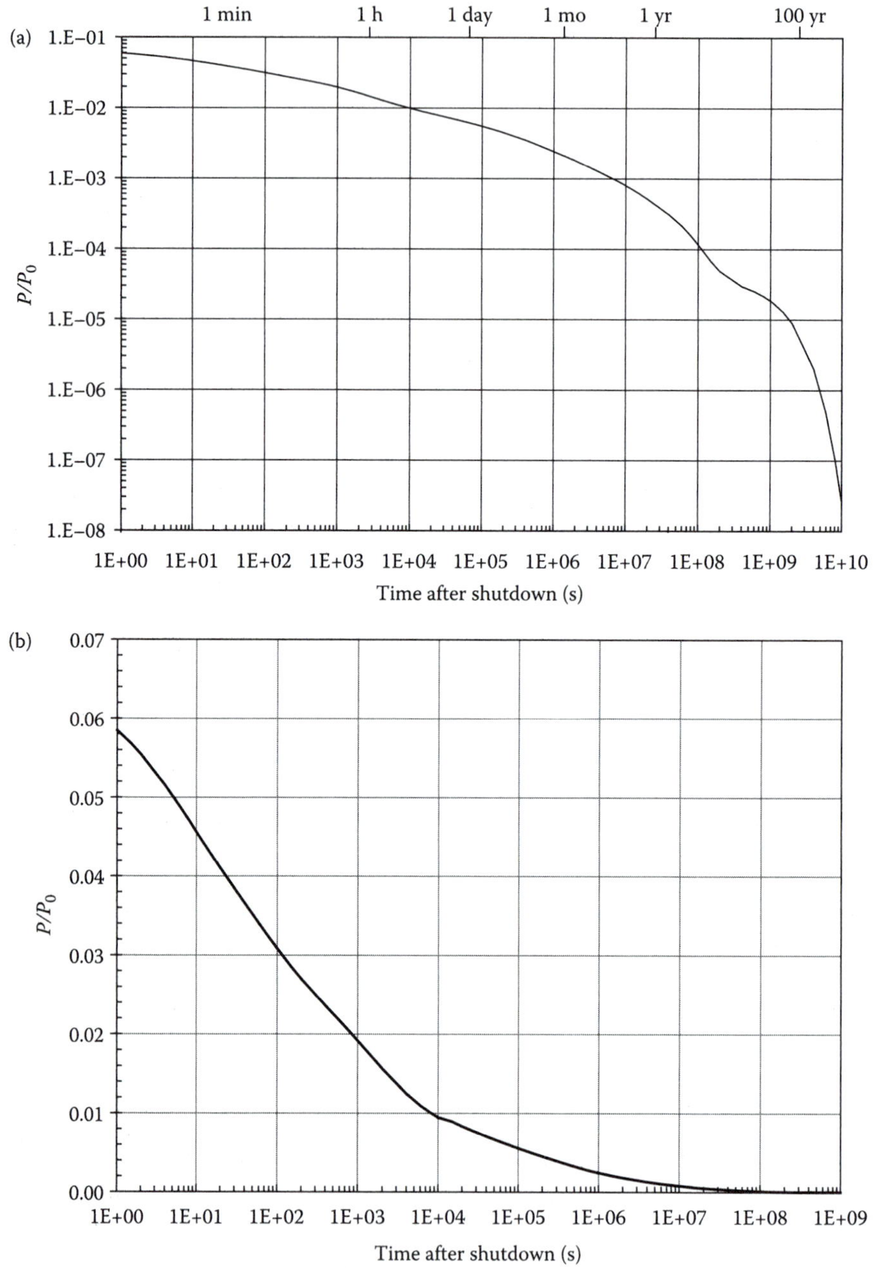

FIGURE 3.11 Fractional decay power from the ANS 2005 Standard as a function of time after shutdown for a burnup of 51 MWd/kg_{IHM}: (a) log-log coordinates and (b) linear-log coordinates. (From Shwageraus, E. and Fridman, E., *Proceedings of International Youth Nuclear Congress, IYNC08*, Interlaken, Switzerland, September 20–26, 2008.)

period before being instantaneously shut down. For the time range $10^2 \leq t_s \leq 10^6$ s, a reasonable approximation of the ANS FP + ^{239}U + ^{239}Np decay power curve is

$$\frac{P(t_s)}{P_0} = 1.250 \times 10^{-1} \times t_s^{-0.2752} \quad (3.72a)$$

with t_s expressed in seconds. This relation approximates the decay power curve with a relative error ranging between −5% and +1% for the period $400 < t_s < 4 \times 10^5$ s; it is instead conservative by as much as 14% for the periods $100 \leq t_s \leq 400$ s and $4 \times 10^5 \leq t_s \leq 10^6$ s. A more accurate fit to the ANS FP + ^{239}U + ^{239}Np curve of Figures 3.11a and b over the full time range is given by the following relations:

$$\frac{P(t_s)}{P_0} = -6.14575 \times 10^{-3}\, \ell n t_s + 0.060157 \quad \text{for } 1.5 \leq t_s \leq 400 \text{ s} \quad (3.72b)$$

$$\frac{P(t_s)}{P_0} = 1.40680 \times 10^{-1} \times t_s^{-0.286} \quad \text{for } 400 < t_s \leq 4 \times 10^5 \text{ s} \quad (3.72c)$$

$$\frac{P(t_s)}{P_0} = 8.70300 \times 10^{-1} \times t_s^{-0.4255} \quad \text{for } 4 \times 10^5 < t_s \leq 4 \times 10^6 \text{ s} \quad (3.72d)$$

$$\frac{P(t_s)}{P_0} = 1.28420 \times 10^{1} \times t_s^{-0.6014} \quad \text{for } 4 \times 10^6 < t_s \leq 4 \times 10^7 \text{ s} \quad (3.72e)$$

$$\frac{P(t_s)}{P_0} = 4.03830 \times 10^{4} \times t_s^{-1.0675} \quad \text{for } 4 \times 10^7 < t_s \leq 4 \times 10^8 \text{ s} \quad (3.72f)$$

$$\frac{P(t_s)}{P_0} = 3.91130 \times 10^{-5} \exp(-7.3541 \times 10^{-10} t_s) \quad \text{for } 4 \times 10^8 < t_s \leq 10^{10} \text{ s} \quad (3.72g)$$

Equations 3.72b through 3.72g approximate the decay power curve of Figure 3.11a with a relative error of about

$$1.5 \leq t_s \leq 2 \times 10^7 \text{ s} \quad +/-\ 6\%$$

$$2 \times 10^7 < t_s \leq 4 \times 10^8 \text{ s} \quad +/-\ 15\%$$

$$4 \times 10^8 < t_s \leq 10^{10} \text{ s from 0 to } -\ 6\%$$

TABLE 3.9
Decay Power After Shutdown

	Fraction of Thermal Operating Power (P/P_0) at Various Times After Shutdown (Cooling Period)				
	Cooling Period				
Operating Time	**1 h**	**12 h**	**1 day**	**10 days**	**100 days**
20 days	0.0116	4.7×10^{-3}	3.6×10^{-3}	1.0×10^{-3}	9×10^{-5}
200 days	0.0129	6.1×10^{-3}	4.9×10^{-3}	2.0×10^{-3}	4×10^{-4}
Infinite	0.0135	6.6×10^{-3}	5.5×10^{-3}	2.6×10^{-3}	9×10^{-4}

The value of P/P_0 for reactors operated for a finite period (τ_s) may be obtained from Figure 3.11 (a or b) by subtracting the value of P/P_0 at the time $\tau_s + t_s$ from the value of P/P_0 at the time t_s, where t_s is the cooling time after shutdown.

Some important observations may be derived from Figures 3.11a and b. Consider the values of the ANS FP + ^{239}U + ^{239}Np curve of P/P_0 for reactor operating times of 20 days, 200 days, and infinity and for various cooling times. These values are given in Table 3.9. They show that the longer the cooling period, the stronger is the dependence of the fractional decay power on the period of operation. This situation is due to the rapid decay of fission products of short half-lives, which reach their saturation values after short periods of operation. The decay heat at longer cooling times is due to fission products with longer half-lives. Because the amount of these products present at shutdown is dependent on the reactor operation time (τ_s), the decay heat after cooling times longer than 1 day is also dependent on the operating time.

The uncertainty associated with the 2005 proposed ANS Standard was given only for fission product decay as 2–3% to 10^9 s and up to 5% between 10^9 and 10^{10} s. The uncertainty for actinide power decay and the effect of neutron capture in the fission products are not given.

The ANS FP + ^{239}U + ^{239}Np decay power curve of Figure 3.11a can be integrated from shutdown, $t_s = 0$, to obtain the total decay energy generated to time t_s. Figure 3.12 presents this energy curve in terms of energy/steady state power (Q/P_0). From this curve, the important information on the total decay energy that must be managed

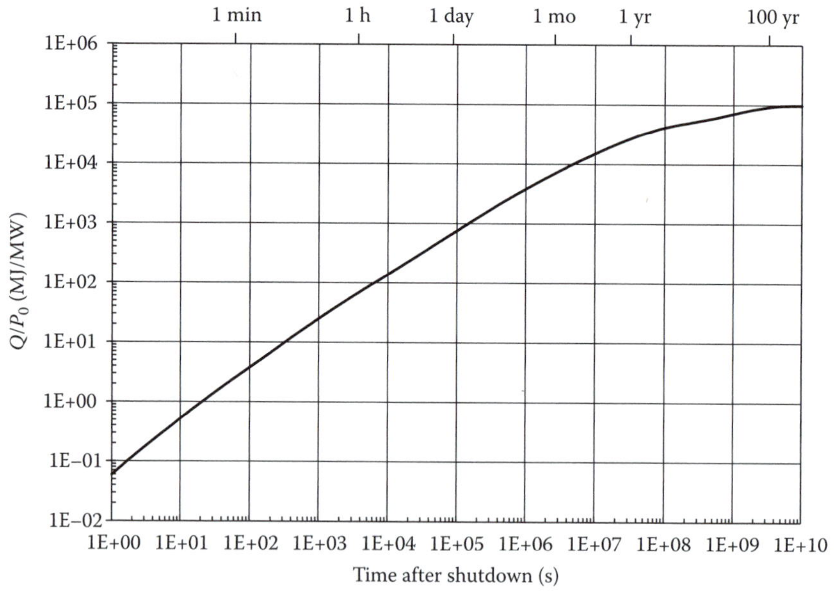

FIGURE 3.12 Normalized decay energy from the ANS 2005 Standard (from ANS FP + ^{239}U + ^{239}Np decay power curve) as a function of time after shutdown for a burnup of 51 MWd/kg_{IHM}. (From Shwageraus, E. and Fridman, E., *Proceedings of International Youth Nuclear Congress, IYNC08*, Interlaken, Switzerland, September 20–26, 2008.)

following reactor shutdown can be calculated. The following relations approximate the decay energy Q (relative errors are indicated in parenthesis):

$$Q = P_0\left[6.28300\times10^{-2}\times t_s^{0.91}\right] \quad \text{for} \quad 1.5\le t_s\le 400\,\text{s} \quad (+/-\ 3\%) \quad \text{s} \tag{3.73a}$$

$$Q = P_0\left[9.87400\times10^{-2}\times t_s^{0.79378}\right] \quad \text{for} \quad 400< t_s\le 10^4\,\text{s} \quad (+/-\ 7\%) \tag{3.73b}$$

$$Q = P_0\left[2.76780\times10^{-1}\times t_s^{0.68455}\right] \quad \text{for} \quad 10^4< t_s\le 6\times10^6\,\text{s} \,(+/-\ 10\%) \tag{3.73c}$$

$$Q = P_0\left[8.77910\times t_s^{0.46183}\right] \quad \text{for} \quad 6\times10^6< t_s\le 10^8\,\text{s} \quad (+/-\ 5\%) \tag{3.73d}$$

$$Q = P_0\left[1.36824\times10^4\ln(t_s)-2.1202\times10^5\right] \quad \text{for} \quad 10^8< t_s\le 10^{10}\,\text{s} \quad (-8\%\text{ to }+1\%) \tag{3.73e}$$

with Q in MJ, P_0 in MW and t_s in seconds. As already noted, while for $t_s \le 10^8$ s the curve accurately reproduces the decay energy, for $t_s > 10^8$ s it increasingly underestimates such energy, up to about −40% at 10^{10} s, due to the neglect of actinides such as ^{238}Pu, ^{241}Am, and ^{244}Cm.

Example 3.5: Decay Power after a Specified Operating Power-Shutdown History

QUESTION

A reactor operates at 40% power for t_1 days and then at full power for t_2 days after t_1 before it is shut down. What is the decay power at a time t_s after shutdown? Assume $t_1 = 10$ days, $t_2 = 50$ days, and $t_s = 100$ days.

SOLUTION

The complete power-shutdown history is illustrated in Figure 3.13. In order to find the decay power during reactor operation or after shutdown, the power histories corresponding to infinite operation of Figure 3.14 may be superposed.

Figure 3.11a, corresponding to the 2005 ANS Standard for decay power, yields the ratio of decay power to total reactor power, $F(t_s, \tau)$ where t_s is the time after shutdown and τ is the operating period at constant power. This ratio represents the sum of all fission product contributions to the total decay power.

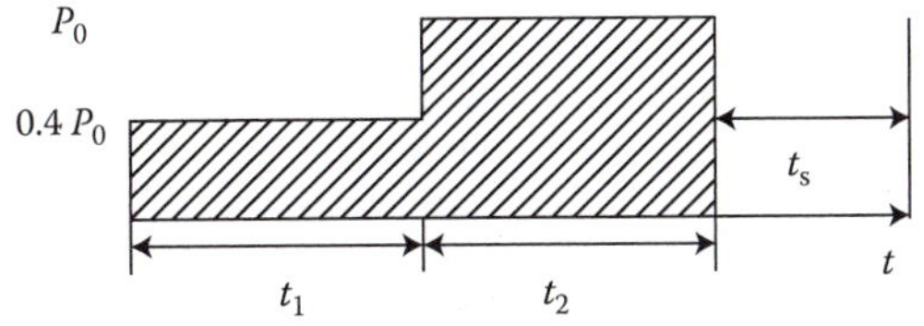

FIGURE 3.13 Operating power and shutdown time history of Example 3.5.

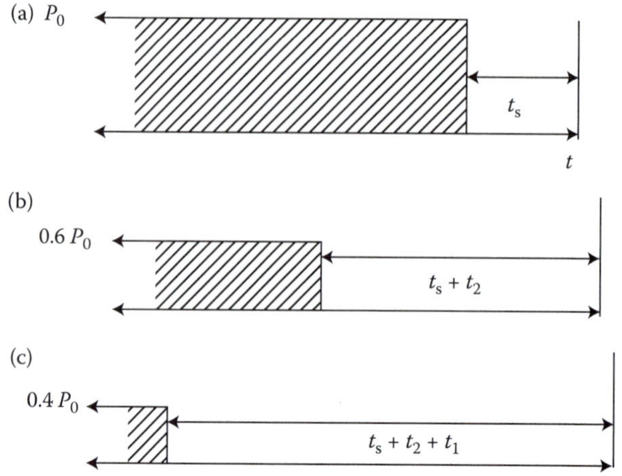

FIGURE 3.14 Infinite operating histories to be superimposed for Example 3.5. (a) Full power, P_0, operation prior to time t_s; (b) 60% power operation prior to time $t_s + t_2$; and (c) 40% power operation prior to time $t_s + t_2 + t_1$.

For (a): $F(t_s, \tau) = F(t_s, \infty)$
For (b): $F(t_s, \tau) = 0.6 \times F(t_s + t_2, \infty)$
For (c): $F(t_s, \tau) = 0.4 \times F(t_s + t_2 + t_1, \infty)$
Hence:

$$\frac{P}{P_0} = 1.0F(t_s,\infty) - 0.6F(t_s + t_2,\infty) - 0.4(t_s + t_2 + t_1,\infty) \tag{3.74a}$$

Evaluating the terms from Figure 3.11a yields

$t_s = 100$ days	or	8.6400×10^6 s	$F = 8.7 \times 10^{-4}$
$t_s + t_2 = 150$ days	or	1.2960×10^7 s	$F = 6.8 \times 10^{-4}$
$t_s + t_2 + t_1 = 160$ days	or	1.3824×10^7 s	$F = 6.6 \times 10^{-4}$

Hence,

$$\frac{P}{P_0} = 1.0\left(8.7\times10^{-4}\right) - 0.6\left(6.8\times10^{-4}\right) - 0.4\left(6.6\times10^{-4}\right) \cong 2\times10^{-4} \tag{3.74b}$$

The preceding decay power and energy results are for typical PWR fuel at a burnup of 51 MWd/kg_{IHM}. Spent fuel at burnup of 33 MWd/kg_{IHM} dominates the existing PWR-spent fuel inventory due to historical accumulation, while 55 MWd/kg_{IHM} fuel is more representative of current PWR fuel discharge burnup. Even higher burnup fuel can be contemplated based on ongoing research. The decay power dependence on burnup for times after discharge is shown in Figure 3.15. The burnup dependence is normalized to 33 MWd/kg_{IHM}. Note that to achieve the higher burnups increased enrichments are required, although for a given assembly design the same fuel load, that is, heavy metal mass is required. The decay power per unit mass of initial heavy metal is related to the local limits for spent fuel assemblies, which determines the spent fuel geological repository capacity given a fixed volume. The corresponding decay energy dependence on burnup is shown in Figure 3.16.

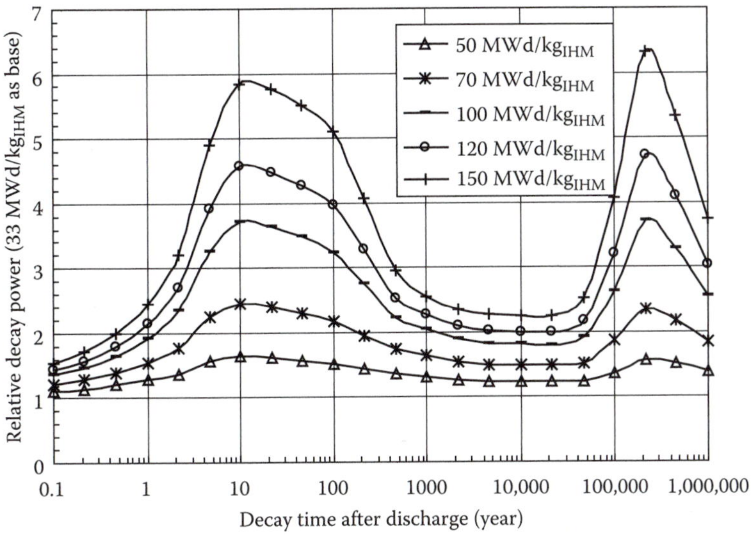

FIGURE 3.15 Relative magnitude of decay power per kg$_{IHM}$ after discharge. (From Xu, Z., Kazimi, M.S., and Driscoll, M.J., *Nucl. Sci. Eng.*, 151:261–273, 2005.)

As expected, higher burnup leads to a higher decay heat generation rate per kg$_{IHM}$ (or per assembly if the initial heavy metal loading is the same). Interestingly, however, if the decay power is taken relative to the energy produced, that is, per GW-yr(e), as Figure 3.17 shows, the decay power of high-burnup spent fuel is only larger than the reference fuel within two time periods: (1) from 4 years

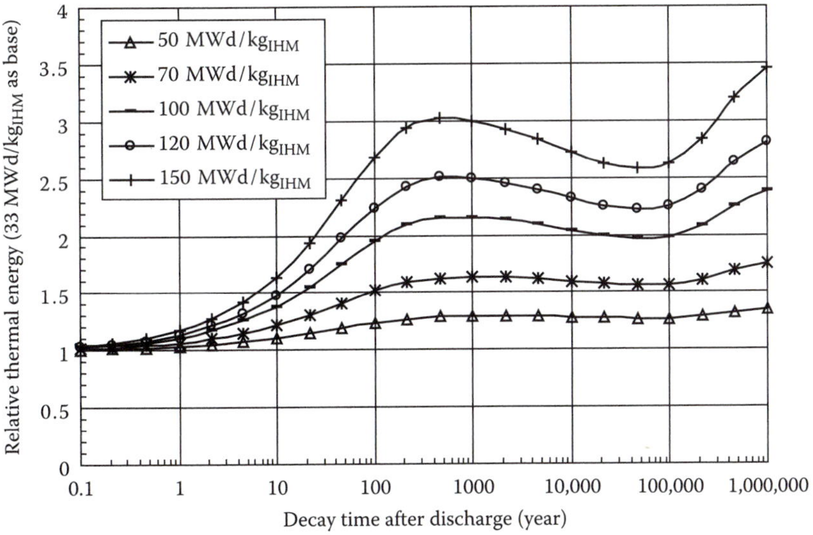

FIGURE 3.16 Relative magnitude of accumulated decay heat per kg$_{IHM}$ after discharge. (From Xu, Z., Kazimi, M.S., and Driscoll, M.J., *Nucl. Sci. Eng.*, 151:261–273, 2005.)

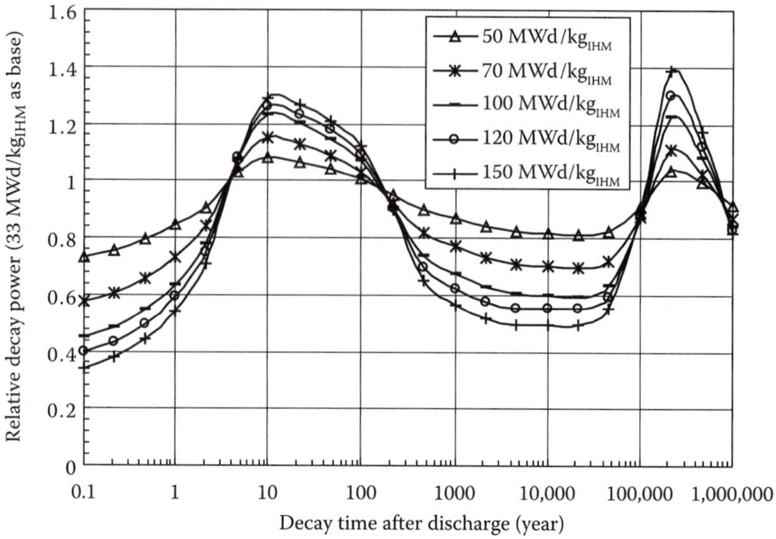

FIGURE 3.17 Relative magnitude of decay power per GW-yr(e) after discharge. (From Xu, Z., Kazimi, M.S., and Driscoll, M.J., *Nucl. Sci. Eng.*, 151:261–273, 2005.)

to 150 years, and (2) from 150,000 years to 700,000 years. The short time frame difference suggests that high-burnup spent fuel needs a longer time in storage before final disposal. In storage and in disposal, there will be less volume to handle per unit electricity generated. Because a repository is likely to be limited by the thermal load after 100 years from discharge, this value based on per unit electricity generation is a global measure quantifying the overall spent fuel characteristics, which indicates the overall repository requirement and hence is directly proportional to the total repository cost.

If X_1 is per kg$_{IHM}$ and X_2 is per GW-yr(e), the transformation between the two bases can be expressed as

$$X_2 = \frac{3.65 \times 10^5}{\eta B_d} X_1 \tag{3.75}$$

where η is plant thermal to electrical conversion efficiency, that is

$$1150\ MW_e/3411 MW_{th} = 33.7\%$$

and B_d is the discharge burnup, MWd/kg$_{IHM}$ (same as GWd/Mt$_{IHM}$). The corresponding decay energy dependence on burnup for this unit energy basis is shown in Figure 3.18.

3.9.3.2 Alternative Fuels in Light Water and Fast Reactors

It is important to keep in mind that the ANS Standards have only been developed for LWRs, hence the typical low-enrichment uranium core loadings. Other core fuel compositions will have different decay power characteristics particularly at long time periods when different actinides of long half-lives make significant contributions to the decay power.

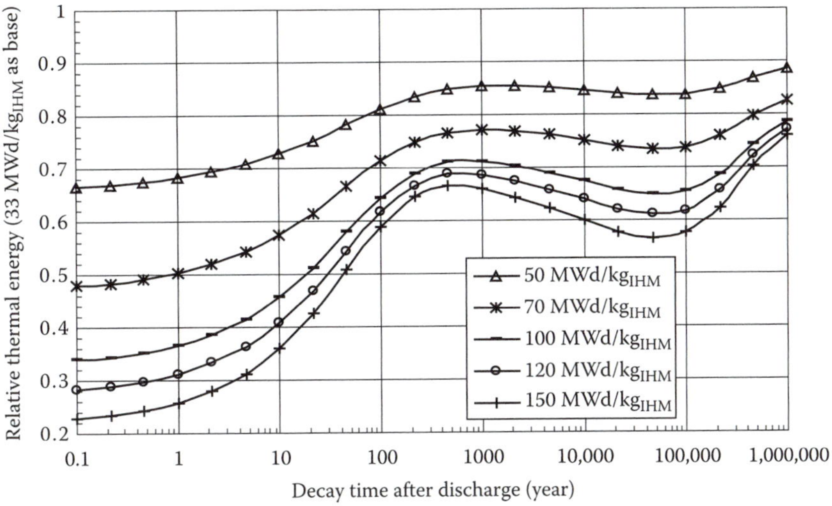

FIGURE 3.18 Relative magnitude of accumulated decay heat per GW-yr(e) after discharge. (From Xu, Z., Kazimi, M.S., and Driscoll, M.J., *Nucl. Sci. Eng.*, 151:261–273, 2005.)

Figure 3.19 illustrates the decay power for mixed oxide (UO_2 and PuO_2) and plutonium/thorium oxide (MOX and TOX) cores. The fuel compositions for the two cores illustrated are given in Table 3.10 [15]. Fast reactors of near unity conversion ratios which operate in a sustainable closed cycle have considerable actinide loadings. Figure 3.20 illustrates the decay power for lead-cooled (LFR) and sodium-cooled (SFR) Generation IV reactors described in Chapter 1. Two core cases are illustrated for the lead coolant—a sustainable closed fuel cycle of conversion ratio unity (CR = 1) and a fuel cycle burning actinides from LWR spent fuel (CR = 0). For the sodium coolant, the CR = 1 case is shown. The fuel composition for each of these reactors is also presented in Table 3.10. As Figure 3.20 illustrates, the actinides in these cores, which significantly differ in quantity from those in a typical LWR core, cause the decay power to be significantly higher at longer times than for the LWR core. Also contributing to this decay power deviation are differences in (1) thermal and fast fission product yields, (2) fission product yields for actinides other than thermal fissions of ^{235}U, ^{239}Pu, ^{241}Pu and fast fission of $^{238}U_{fast}$, and (3) differences in the levels of burnup of the fuel. The plots of Figures 3.19 and 3.20 show the relative difference in the decay heat as a fraction of power:

$$\frac{\left[(P/P_0)-(P/P_0)_{UO_2}\right]}{(P/P_0)_{UO_2}}$$

However, the amount of fuel in each reactor will not necessarily be the same.

3.10 STORED ENERGY SOURCES

In accident situations, reactor materials can undergo chemical reactions that release stored energy in addition to the decay heat generation discussed in Section 3.9.

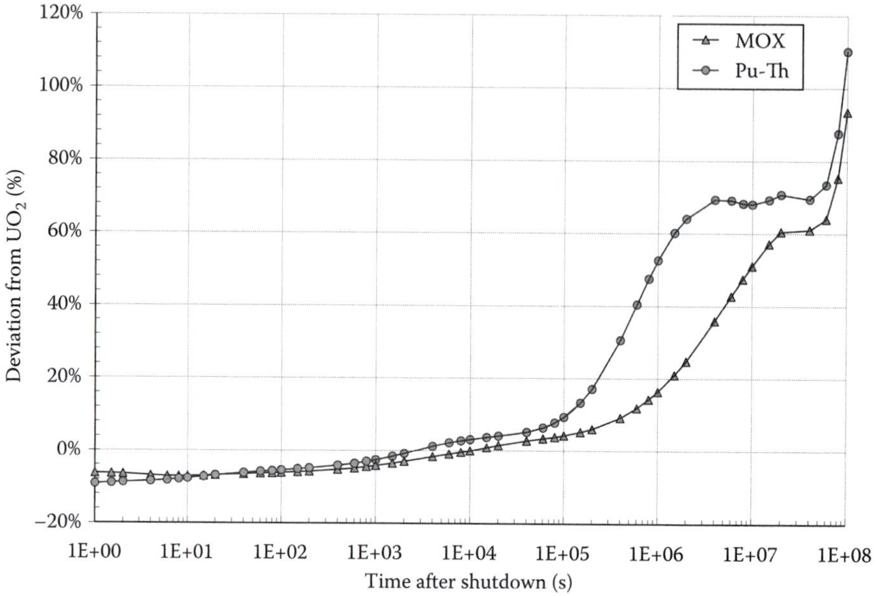

FIGURE 3.19 Decay power of advanced fuels in a thermal spectrum compared to UO_2 in a LWR from 2005 ANS Standard for 1000-MWe plants. (From Shwageraus, E. and Fridman, E., *Proceedings of International Youth Nuclear Congress, IYNC08*, Interlaken, Switzerland, September 20–26, 2008.)

Principal among these exothermic reactions are

- For water-cooled reactors—oxidation of the zircaloy and steel core cladding and structures by the primary water coolant.
- For sodium-cooled reactors—oxidation of the primary sodium coolant (a) by secondary water coolant leakage from steam generator tubing (for the Rankine cycle power conversion system), or (b) by secondary carbon dioxide coolant leakage from intermediate heat exchanger tubing (for the Brayton cycle power conversion system).
- For carbon dioxide-cooled reactors—oxidation of the graphite moderator by the carbon dioxide coolant.
- For graphite-moderated reactors—release of Wigner-stored energy.
- For all reactors—oxidation of metals that may exist in molten core material (called corium) by water and carbon dioxide released from thermal decomposition of the concrete containment basemat on contact with corium. Corium contact with the basemat could only occur if the reactor vessel failed.

In many of these reactions, hydrogen and carbon monoxide are produced which themselves can be oxidized in subsequent exothermal reactions.* Mixtures of dry air and H_2 are flammable in a composition range between 4% H_2, the lower

* Additional sources of hydrogen production are radiolysis in the core during normal operation and water interaction with metals in the containment during accident conditions.

TABLE 3.10
Description of Fuel Types in Decay Power Calculations

Fuel Type	Fuel Composition	Burnup Before Shutdown
PWR UO_2	4.2 wt% ^{235}U enrichment	50 MWd/kg_{IHM}
PWR UO_2-PuO_2 MOX	7.5 vol.% RG[c]PuO_2 + Nat. UO_2	51 MWd/kg_{IHM}
PWR ThO_2-PuO_2 TOX	9.2 vol.% RG PuO_2 + ThO_2	55 MWd/kg_{IHM}
Lead-cooled fast reactor, CR = 1 (LFR1)	Metallic Zr-NU-TRU (16.7wt% in HM)[a]	77 MWd/kg_{IHM}
Lead-cooled fast reactor, CR = 0 (LFR0)	Metallic Zr-TRU (34.0wt% in HM)	120/240/323 MWd/kg_{IHM}[b]
Sodium-cooled fast reactor, CR = 1 (SFR1)	Metallic Zr-NU-TRU (15.2wt% in HM)	78 MWd/kg_{IHM}

Source: Adopted from Shwageraus, E. and Fridman, E. *Proceedings of International Youth Nuclear Congress, IYNC08*, Interlaken, Switzerland, September 20–26, 2008.

[a] Fixed TRU enrichment relative to total HM but variable Zr content in three radial core zones.

[b] Individual batch burnups in three batch core.

[c] Pu isotope wt%: 238, 2.5; 239, 54.3; 240, 23.8; 241, 12.6; 242, 6.8.

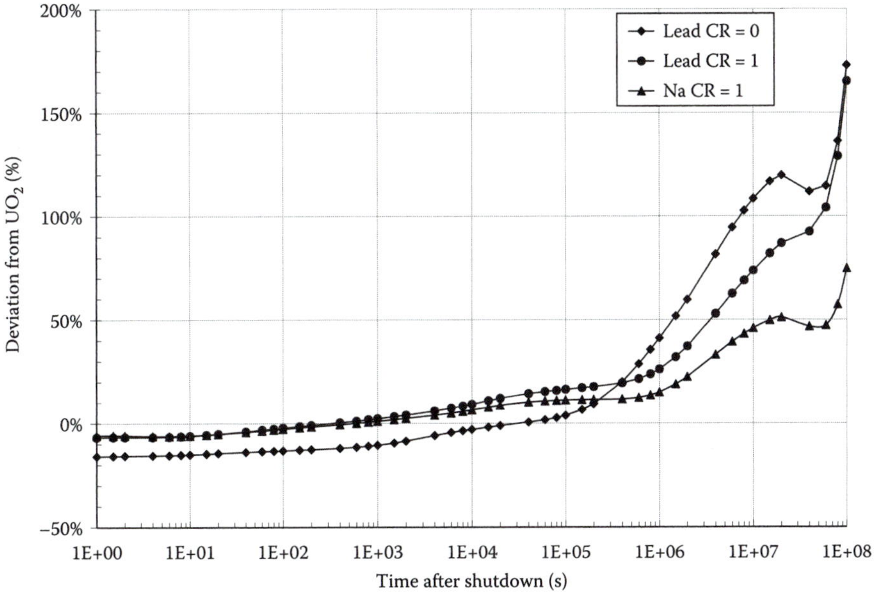

FIGURE 3.20 Decay power of advanced fuels in a fast spectrum compared to UO_2 in an LWR from 2005 ANS Standard for 1000 MWe plants. (From Shwageraus, E. and Fridman, E., *Proceedings of International Youth Nuclear Congress, IYNC08*, Interlaken, Switzerland, September 20–26, 2008.)

flammabililty limit (LFL), and 75% H_2, the upper flammability limit (UFL). Importantly in reactor applications, the presence of steam and carbon dioxide can suppress this reaction, so that the LFL increases and the UFL decreases as these inert gases are progressively added to a fuel–air mixture. Also in reactor applications, oxygen in air may be partially depleted by these or other reactions, and the excess nitrogen has a similar inertant effect on the LFL and UFL. For a sufficiently large mole concentration of inertant, the LFL and UFL merge, and the mixture is not flammable for any fuel concentration. This occurs for about 55% steam or carbon dioxide, and for about 70% excess nitrogen. Flammability data are represented in a flammability limit diagram as shown in Figure 3.21 for three different fuels, H_2, CO, and an equimolar mixture of H_2 and CO. In this diagram, the inertant gas is an equimolar mixture of N_2 and ($CO_2 + H_2O$), and about 62% of the inertant mixture is sufficient to suppress flammability for any fuel concentration. The flammability envelope expands with increasing temperature, which may also be important in reactor applications.

Example 3.6: Rendering an Atmosphere Containing Hydrogen Nonflammable

PROBLEM

To the atmosphere defined below, 2 moles of steam are added. Is this steam addition sufficient to render the mixture nonflammable?

Hydrogen	2.0 moles
Nitrogen	5.2 moles
Oxygen	0.8 moles

SOLUTION

First examine the flammability of the original atmosphere. In order to use Figure 3.19, the excess nitrogen content must be determined. In air, neglecting the small argon content, the oxygen-to-nitrogen ratio is about 20%:80% or 1:4. Hence, the 0.8 moles of oxygen in normal air would be accompanied by 3.2 moles of nitrogen. Since the atmosphere is postulated to contain 5.2 moles of nitrogen, it is oxygen deficient (compared to normal air) and contains 2.0 moles of excess nitrogen. (Note that this reflects a common situation where hydrogen recombiners act to combine a portion of oxygen with hydrogen.)

Moles of mixture $= 2H_2 + 0.8O_2 + 5.2N_2 = 8$ moles
Moles of fuel $= 2H_2$
Moles of inertant $= 2N_2$ (excess)
Hence the fuel fraction $= \frac{2}{8} = \frac{1}{4}$
and the inertant fraction $= \frac{2}{8} = \frac{1}{4}$

Entering Figure 3.19, observe that the mixture is flammable.

Now compute the coordinates of Figure 3.19 for the addition of 2 moles of steam.

Moles of mixture $= 2H_2 + 0.8O_2 + 5.2N_2 + 2H_2O = 10$ moles
Moles of fuel $= 2H_2$
Moles of inertant $= 2N_2$ (excess) $+ 2H_2O$

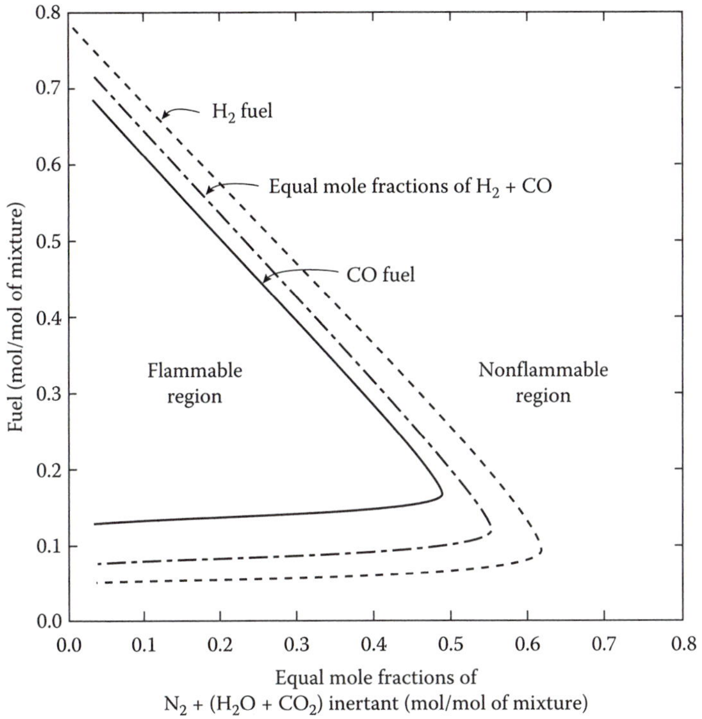

FIGURE 3.21 Generalized flammability limit diagram. The inert gas mixture contains equal mole fractions of N_2 and (CO_2 and H_2O), and separate curves are shown for three fuels: H_2, CO, and H_2 + CO combined at equal mole fractions. (Adopted from Plys, M. G., *Nucl. Tech.*, 101:400–410, 1993.)

$$\text{Hence the fuel fraction} = \frac{2}{10}$$

$$\text{and the inertant fraction} = \frac{4}{10}$$

Consequently, entering Figure 3.19, observe that the mixture is still flammable.

The rate and nature of the gas phase combustion reaction also depends on gas composition and temperature [13]. For low concentrations of hydrogen in air, flame propagation is subsonic and the term deflagration is applied to the reaction propagation. The reaction is generally incomplete for hydrogen concentrations less than 8%. For hydrogen concentrations above about 10–12%, flame acceleration is observed, and a transition is possible from subsonic deflagration to a supersonic detonation, a phenomenon known as deflagration to detonation transition, or DDT. Factors that influence flame speed and potential for DDT are the fuel and inertant concentrations, temperature, and geometry, particularly the presence or absence of obstacles that promote turbulence and thereby promote flame acceleration. The nature of the reaction is important because the final pressure and temperature for a deflagration are bounded by thermodynamic ideal values. In contrast, dynamic pressure loads during a detonation can have the equivalent impact of approximately double the ideal value.

We next return to the principal exothermic reactions which have been listed for the various reactor types. For each reaction, the relevant equations are presented below.* We focus on the reaction energy characteristics rather than the extent or kinetics of the reactions.

The reaction energy is expressed as the change in enthalpy, that is, total energy of the products minus reactants at states and a temperature of interest:

$$\Delta H^{\circ}{}_{\text{f Temperature}}$$

This is called the Heat of Formation and varies with the state (gas, liquid, solid) of the reactants and products and, to a lesser extent, with their temperatures and pressures. For crystalline solids, care is necessary to specify the specific crystal form of the compound involved. In the reactions listed in Sections 3.10.1 through 3.10.4 below, typical temperatures at which they might occur under reactor conditions are assumed. The heat of formation is given directly on the right-hand side (RHS) of each equation as an energy quantity. For an exothermic reaction, the energy term on the RHS is preceded by a positive sign conforming to a negative value of $\Delta H^{\circ}{}_{\text{f Temperature}}$ since Δ reflects enthalpy of products minus reactants.

3.10.1 The Zircaloy–Water Reaction

$$\mathrm{Zr(s)} + 2\mathrm{H_2O}(l) \rightarrow \mathrm{ZrO_2}(s) + 2\mathrm{H_2}(g) + 537.8\,\mathrm{kJ/(mol\ Zr)}\ (500\,\mathrm{K}) \tag{3.76}$$

This represents the normally occurring zircaloy cladding corrosion process which deposits a protective zirconium oxide film on the cladding under normal LWR- and HWR-operating conditions. In the loss of coolant scenario, when the cladding temperature undergoes an excursion due to undercooling, the maximum cladding temperature is limited to 1204°C (2200°F) to prevent an autocatalytic excursion due to the exothermic nature of the reaction. In this case, the reaction is with steam yielding

$$\mathrm{Zr}(s) + 2\mathrm{H_2O}(g) \rightarrow \mathrm{ZrO_2}(s) + 2\mathrm{H_2}(g) + 583.6\,\mathrm{kJ/(mol\ Zr)}\ (1477\,\mathrm{K}) \tag{3.77}$$

The hydrogen produced can be oxidized as

$$\mathrm{H_2}(g) + \frac{1}{2}\mathrm{O_2}(g) \rightarrow \mathrm{H_2O}(g) + 241.8\,\mathrm{kJ/(mol\ H2)}\ (298\,\mathrm{K}) \tag{3.78}$$

The potential amount of hydrogen generation for LWRs depends on zircaloy inventory and an estimation of the percentage of reaction of that inventory. Table 3.11 gives these amounts for typical PWRs and BWRs, assuming 100% of the inventory is reacted, a conservative assumption in safety assessments of degraded cores.

* Letters in parentheses in the reaction equations, for example, (*s*), (*l*), (*g*), signify the state of the substance.

TABLE 3.11
Potential for Hydrogen Generation in Typical Light-Water Reactors

	TMI-2	Browns Ferry 2
Reactor type	PWR	BWR
Containment type	Large dry	Mark I
Thermal power, MW_{th}	2270	3300
Zircaloy inventory, kg		
Cladding	24,000	37,000
Channel box	—	25,000
Total	24,000	62,000
Potential H_2 generation, kg (for 100% zircaloy inventory reaction)	1061	2740
Power specific H_2, kg H_2/MW_{th}	0.47	0.83

3.10.2 The Sodium–Water Reaction

$$\mathrm{Na}(l) + \mathrm{H_2O}(g) \rightarrow \mathrm{NaOH}(l) + \frac{1}{2}\mathrm{H_2}(g) + 160.1\,\mathrm{kJ/(mol\ Na)}\ (798\ \mathrm{K}) \tag{3.79}$$

$$\mathrm{Na}(l) + \mathrm{NaOH}(l) \rightarrow \mathrm{Na_2O}(s) + \frac{1}{2}\mathrm{H_2}(g) + 13.3\,\mathrm{kJ/(mol\ Na)}\ (798\ \mathrm{K}) \tag{3.80}$$

$$\mathrm{Na}(l) + \frac{1}{2}\mathrm{H_2}(g) \rightarrow \mathrm{NaH}(s) + \ 57.3\,\mathrm{kJ/(mol\ Na)}\ (798\ \mathrm{K}) + 57.3\,\mathrm{kJ/(mol\ Na)}\ (798\,\mathrm{K}) \tag{3.81}$$

3.10.3 The Sodium–Carbon Dioxide Reaction

Four reactions can occur, three of which are exothermic:

$$\mathrm{Na}(l) + \frac{3}{4}\mathrm{CO_2}(g) \rightarrow \frac{1}{2}\mathrm{Na_2CO_3}(s) + \frac{1}{4}\mathrm{C}(s) + 266.4\,\mathrm{kJ/(mol\ Na)}\ (798\,\mathrm{K}) \tag{3.82}$$

$$\mathrm{Na}(l) + \frac{1}{4}\mathrm{CO_2}(g) \rightarrow \frac{1}{2}\mathrm{Na_2O}(s) + \frac{1}{4}\mathrm{C}(s) - 84.8\,\mathrm{kJ/(mol\ Na)}\ (798\,\mathrm{K}) \tag{3.83}$$

$$\mathrm{Na}(l) + \frac{1}{2}\mathrm{CO_2}(g) \rightarrow \frac{1}{2}\mathrm{Na_2O}(s) + \frac{1}{2}\mathrm{CO}(g) + 68.3\,\mathrm{kJ/(mol\ Na)}\ (798\,\mathrm{K}) \tag{3.84}$$

$$\mathrm{Na}(l) + \frac{3}{2}\mathrm{CO}(g) \rightarrow \frac{1}{2}\mathrm{Na_2CO_3}(s) + \mathrm{C}(s) + 397.3\,\mathrm{kJ/(mol\ Na)}\ (798\,\mathrm{K}) \tag{3.85}$$

Subsequent reaction of the carbon, albeit endothermic, produces carbon monoxide, which can undergo a subsequent exothermic reaction with air (an oxygen source):

$$\mathrm{C}(s) + \mathrm{CO_2}(g) \rightarrow 2\mathrm{CO}(g) - 174.6\,\mathrm{kJ/(mol\ C)}\ (798\,\mathrm{K}) \tag{3.86}$$

$$\mathrm{CO}(g) + \frac{1}{2}\mathrm{O_2}(g) \rightarrow \mathrm{CO_2}(g) + 283.3\,\mathrm{kJ/(mol\ CO)}\ (798\,\mathrm{K}) \tag{3.87}$$

3.10.4 The Corium–Concrete Interaction

Corium is the mixture of molten materials that would be formed within the reactor vessel from melting of core and structure materials in the hypothetical case of a nuclear reactor severe accident. The primary features of the interaction of corium with concrete are illustrated in Figure 3.22. The interaction proceeds as follows: all reactions are taken at 2500 K, a typical temperature of the corium as it reaches the concrete:

1. Thermal decomposition of concrete is an endothermic reaction that yields

$$H_2O \quad \text{and} \quad CO_2$$

   The amount of these products which can be released depends on the type of concrete. These products, together with unoxidized metals (zircaloy, iron, chromium) that may exist in the corium, react to produce the combustible gases H_2 and CO, which are themselves subject to the oxidation reactions already discussed.
2. Oxidation of metal (remaining metal in the core melt):

$$Zr(l) + 2H_2O(g) \rightarrow ZrO_2(s) + 2H_2(g) + 598.2\,\text{kJ/(mol Zr)}\ (2500\,\text{K}) \quad (3.88)$$

$$Zr(l) + CO_2(g) \rightarrow ZrO_2(s) + C(s) + 729.5\,\text{kJ/(mol Zr)}\ (2500\,\text{K}) \quad (3.89)$$

   Similar reactions for iron and chromium occur as well as reduction of SiO_2 and Fe_2O_3 by zirconium.

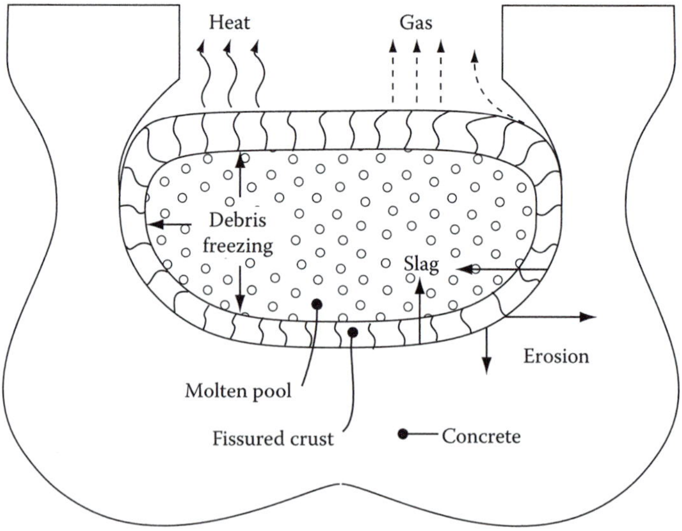

FIGURE 3.22 Core–concrete interaction representation.

3. After the zircaloy is depleted, free carbon reacts with gases:

$$C(s) + H_2O(g) \rightarrow CO(g) + H_2(g) - 154.2\,\text{kJ/(mol C)}\ (2500\,\text{K}) \tag{3.90}$$

$$C(s) + CO_2(g) \rightarrow 2CO(g) - 177.0\,\text{kJ/(mol C)}\ (2500\ \text{K}) \tag{3.91}$$

   Chromium and iron also oxidize with H_2O to produce H_2 and CO_2 and subsequently CO.
4. While the metals are reacting, the following gas phase reaction also takes place for chemical equilibrium among the oxidizing gases and their reduced forms:

$$CO_2(g) + H_2(g) \rightarrow CO(g) + H_2O(g) - 22.9\,\text{kJ/(mol } CO_2)\ (2500\,\text{K}) \tag{3.92}$$

These reactions illustrate the significance of the CO_2 and H_2O content of the concrete. Concrete exists in a range of forms from limestone to siliceous, which have a wide range of CO_2 content (33–4%) and about 5–6 w/o H_2O. The large CaO content in limestone is responsible for its high liquidus temperature. Table 3.12 presents properties for these concretes plus an intermediate limestone–sand form.

Example 3.7: Energy Sources in a Typical PWR

QUESTION

Compute the energy sources in a typical PWR (Seabrook—Table 1.2) which the containment might have to accommodate.

1. Stored energy in primary coolant. Should enthalpy or internal energy be used? Explain.
2. Stored energy in the secondary side of the steam generator (assume 89 m^3 of saturated liquid at secondary system conditions).

TABLE 3.12
Approximate Compositions of Typical Concrete

	Siliceous (wt%)	Limestone–Common Sand (wt%)	Limestone (wt%)
SiO_2	70	28	6
CaO	13	26	45
MgO	1	10	4
Al_2O_3	3	3	2
H_2O	6	6	5
CO_2	4	20	33
Others	3	7	5
Solidus/liquidus (K)	1400/1500	1400/1600	1400/2600

Source: From Seiler, J. M. and Froment, K., *Sci. Tech.*, 12(2):117–257, 2000.

3. Decay heat-integrated release over a 1 day shutdown period after infinite operation.
4. Chemical reactions.
   a. Metal–water reaction of 75% of the zircaloy cladding.
   b. Reaction of 25% of the zircaloy clad with CO_2 to produce C which then reacts with CO_2 to produce CO. Assume reactions are constrained only by amount of zircaloy available.
5. Combustion of hydrogen and carbon monoxide product in (4) above.
6. Neglect carbon–water reaction.

Relevant reactions:

$$\mathrm{Zr}(l) + 2\mathrm{H_2O}(g) \rightarrow \mathrm{ZrO_2}(s) + 2\mathrm{H_2}(g) + 598.2\,\mathrm{kJ/(mol\ Zr)}$$

$$\mathrm{Zr}(l) + \mathrm{CO_2}(g) \rightarrow \mathrm{ZrO_2}(s) + \mathrm{C}(s) + 598.2\,\mathrm{kJ/(mol\ Zr)}$$

$$\mathrm{C}(s) + \mathrm{CO_2}(g) \rightarrow 2\mathrm{CO}(g) + 598.2\,\mathrm{kJ/(mol\ Zr)} - 177.0\,\mathrm{kJ/(mol\ C)}$$

$$\mathrm{H_2}(g) + \frac{1}{2}\mathrm{O_2}(g) \rightarrow \mathrm{H_2O}(g) + 241.8\,\mathrm{kJ/(mol\ H_2)}$$

$$\mathrm{CO}(g) + \frac{1}{2}\mathrm{O_2}(g) \rightarrow \mathrm{CO_2}(g) + 283.3\,\mathrm{kJ/(mol\ CO)}$$

SOLUTION

To begin this solution, the mass of zirconium in the core must be determined. Unless otherwise noted, the data used in the calculation of the amount of zirconium in the core of the Westinghouse Seabrook PWR comes from Table 1.3.

1. Stored energy in primary coolant:

$$p = 15.5\ \mathrm{MPa} \qquad T_{av} = 310°\mathrm{C} \qquad V = 336\ \mathrm{m^3}$$

   Internal energy and density at $P = 15.5$ MPa and $T_{av} = 310°$C

$$u = 1371.79\ \mathrm{kJ/kg} \quad \rho = 704.76\ \mathrm{kg/m^3}$$
$$U_p = M_p u = \rho V u$$
$$= (704.76\ \mathrm{kg/m^3}) \times (336\ \mathrm{m^3}) \times (1371.79\ \mathrm{kJ/kg})$$
$$= 3.248 \times 10^8\ \mathrm{kJ}$$

2. Stored energy in the secondary side of the steam generator (secondary coolant):

$$p = 6.89\ \mathrm{MPa} \qquad V = 89\ \mathrm{m^3}$$

   Internal energy and density of saturated liquid at $p = 6.89$ MPa

$$u = 1252.7\ \mathrm{kJ/kg}$$
$$\rho = 741.69\ \mathrm{kg/m^3}$$
$$U_s = M_s u = \rho V_u$$
$$= (741.69\ \mathrm{kg/m^3}) \times (89\ \mathrm{m^3}) \times (1252.7\ \mathrm{kJ/kg})$$
$$= 8.27 \times 10^7\ \mathrm{kJ}$$

3. Decay heat:
   Equations 3.72b and c can be integrated to obtain the decay heat generated in the day following shutdown:

$$Q_{1\text{day}} = \int_0^{86400} P(t_s)dt_s \cong \int_1^{86400} P(t_s)dt_s$$

$$= P_0 \left\{ \int_1^{400} \left(-6.14575 \times 10^{-3} \ln t_s + 0.060157\right) dt_s \right.$$

$$\left. + \int_{400}^{86400} 0.14068 \times t_s^{-0.286} \, dt_s \right\}$$

$$= P_0 \left\{ \left[-6.14575 \times 10^{-3} \left(t_s \ln t_s - t_s\right) + 0.060157 \times t_s\right]_1^{400} \right.$$

$$\left. + \left[\frac{0.14068}{0.714} t_s^{0.714}\right]_{400}^{86400} \right\}$$

$$= P_0 \times 657 = 2.2410 \times 10^9 \text{ kJ}$$

4. Chemical reaction:
   Total amount of Zr = 24,000 kg (from Table 3.11)
   Atomic weight of Zr = 91.22 g/mol

   a. Zr and H_2O reaction:

$$Zr(l) + 2H_2O(g) \rightarrow ZrO_2(s) + 2H_2(g) + 598.2 \text{ kJ/(mol Zr)}$$

$$Q = 0.75 \times (24{,}000 \text{ kg Zr}) \times (1000 \text{ g/kg Zr}) \times (1 \text{ mol Zr}/91.22 \text{ g}) \times (598.2 \text{ kJ/mol Zr}) = 1.180{\times}10^8 \text{ kJ}$$

   b. Zr and CO_2 reaction:

$$Zr(l) + CO_2(g) \rightarrow ZrO_2(s) + C(s) + 729.5 \text{kJ/(mol Zr)} \quad (3.93)$$

$$C(s) + CO_2(g) \rightarrow 2CO(g) - 177.0 \text{kJ/(mol C)} \quad (3.94)$$

Adding Equations 3.93 and 3.94 gives

$$Zr(l) + 2CO_2(g) \rightarrow ZrO_2(s) + 2CO(g) + 552.5 \text{kJ/(mol Zr)}$$

$$Q = 0.25 \times (24{,}000 \text{ kg Zr}) \times (1000 \text{ g/kg Zr}) \times (1 \text{ mol Zr}/91.22 \text{ g}) \times (552.5 \text{ kJ/mol Zr}) = 3.634{\times}10^7 \text{ kJ}$$

5. Combustion:
   a. Hydrogen combustion:
      Amount of H_2 produced = 0.75 × (24,000 kg Zr) × (1000 g/kg Zr) × (1 mol Zr/91.22 g)
      × (2 mol H_2/1 mol Zr)
      = 3.9465×10^5 mol H_2

TABLE 3.13
Summary of Energy Sources of Example 3.7

Sources	Energy (GJ)	Fraction (%)
1. Primary coolant	324.8	11.1
2. Secondary coolant	82.7	2.8
3. Decay heat (1 day)	2241.0	76.3
4. Chemical reaction	118.0	4.0
– Zr and H_2O	36.3	1.2
– Zr and CO_2	95.4	3.3
5. Combustion	37.3	1.3
– H_2 combustion		
– CO combustion		
Total	2935.5	100.0

H_2 combustion reaction:

$$H_2(g) + \frac{1}{2}O_2(g) \rightarrow H_2O(g) + 241.8\,\text{kJ/(mol H}_2)$$

Total energy, Q:

$Q = (3.9465 \times 10^5 \text{ mol H}_2) \times (241.8 \text{ kJ/mol H}_2) = 9.54\times10^7 \text{ kJ}$

b. CO combustion amount of CO produced:

$$= 0.25 \times (24{,}000 \text{ kg Zr}) \times (1000 \text{ g/kg Zr}) \times (1 \text{ mol Zr}/91.22 \text{ g}) \times (2 \text{ mol CO/mol Zr}) = 1.3155 \times 10^5 \text{ mol CO}$$

CO combustion reaction:

$$CO(g) + \frac{1}{2}O_2(g) \rightarrow CO_2(g) + 241.8\,\text{kJ/(mol H}_2) + 283.3\,\text{kJ/(mol CO)}$$

Total energy, Q:

$$Q = (1.3155 \times 10^5 \text{ mol CO}) \times (283.3 \text{ kJ/mol CO}) = 3.73 \times 10^7 \text{ kJ}$$

Table 3.13 summarizes all the energy sources of this Example.

PROBLEMS

3.1. *Thermal design parameters for a cylindrical fuel pin* (Section 3.4)
Consider the PWR reactor of Example 3.1.

1. Evaluate the average neutron flux if the enrichment of the fuel is 3.25 wt%.
2. Evaluate the average volumetric heat generation rate in the fuel in MW/m^3. Assume that the fuel density is 90% of theoretical density.
3. Calculate the average linear power of the fuel, assuming there are 264 fuel rods per assembly. Assume the fuel rod length to be 3658 mm.

4. Calculate the average heat flux at the cladding outer radius, when the cladding diameter is 9.5 mm.

Answers:

$\langle\phi\rangle = 3.56\times10^{13}$ neutrons/cm^2 s

$\langle q'''\rangle = 289.3$ MW/m^3

$\langle q'\rangle = 16.95$ kW/m

$\langle q''\rangle = 567.8$ kW/m^2

3.2. *Power profile in a homogeneous reactor* (Section 3.5)

Consider an ideal core with the following characteristics: The U-235 enrichment is uniform throughout the core, and the flux distribution is characteristic of an unreflected, uniformly fueled cylindrical reactor, with extrapolation distances δz and δR of 10 cm. How closely do these assumptions allow prediction of the following characteristics of a PWR?

1. Ratio of peak to average power density and heat flux?
2. Maximum heat flux?
3. Maximum linear heat generation rate of the fuel rod?
4. Peak-to-average enthalpy rise ratio, assuming equal coolant mass flow rates in every fuel assembly?
5. Temperature of water leaving the central fuel assembly?

Calculate the heat flux on the basis of the area formed by the cladding outside diameter and the active fuel length. Use as input only the following values from Tables 1.2, 1.3, 2.1, and Appendix K.

Total power = 3411 MW_{th}
Equivalent core diameter = 3.37 m
Active length = 3.658 m
Fraction of energy deposited in fuel = 0.962
Total number of rods = 50,952
Rod outside diameter = 9.5 mm
Total flow rate = 17.476×10^3 kg/s
Inlet temperature = 293.1°C
Core average pressure = 15.5 MPa

Answers:

1. $\phi_o/\bar{\phi} = 3.11$
2. $q''_{max} = 1.84$ MW/m^2
3. $q'_{max} = 54.8$ kW/m
4. $(\Delta h)max/\overline{\Delta h} = 2.08$
5. $(\Delta Tout)max = 344.8$°C

3.3. *Power generation in thermal shield* (Section 3.8)

Consider the heat generation rate in the PWR core thermal shield discussed in Example 3.4.

1. Calculate the total power generation in the thermal shield if it is 4.0-m high.
2. How would this total power change if the thickness of the shield is increased from 12.7 to 15 cm?

Assume uniform axial power profile.

Answers:

1. $\dot{Q} = 23.21\ MW_{th}$
2. $\dot{Q} = 23.53\ MW_{th}$

3.4. *Decay heat energy* (Section 3.9)
Using Equation 3.70c or 3.70d, evaluate the energy generated in a 3411 MW_{th} PWR after the reactor shuts down. The reactor operated for 1 year at the equivalent of 75% of total power.

1. Consider the following time periods after shutdown:
a. 1 h
b. 1 day
c. 1 month
2. How would your answers be different if you had used Equation 3.71 (i.e., would higher or lower values be calculated)?

Answers:
1a. 0.128 TJ (1TJ = 1012 J)
1b. 1.42 TJ
1c. 14.8 TJ
2. Higher

3.5. *Decay heat from a PWR fuel rod* (Section 3.9)
A decay heat cooling system is capable of removing 1 kW from the surface of a typical PWR (Seabrook) fuel rod (Appendix K). Assume the rod has operated for an essentially infinite period before shutdown.

1. At what time will the decay energy generation rate be matched by the cooling capability?
2. What is the maximum amount of decay heat energy that will be stored in the rod following shutdown?

Answers:
1. $t = 1490.6$ s
2. $Q_{max} = 373 \times 10^5$ J

3.6. *Decay power calculations of a 3-batch PWR core* (Section 3.9)
A PWR core has been operated on a three-batch fuel management scheme on an 18-month refueling cycle, for example, at every 18 months, one-third the core loading is replaced with fresh fuel. A new batch is first loaded into the core in a distributed fashion such that it generates 43% of the core power. After 18 months of operation, it is shuffled to other core locations where it generates 33% of the core power. After another 18 months, it is moved to other core locations where it generates 24% of the core power.

Question

The plant rating is 3411 MW_{th}. Assume it is shutdown after an 18-month operating cycle. What is the decay power of the plant 1 h after shutdown if it has operated continuously at 100% power during each of the preceding three 18-month operating cycles and the shutdown periods for refueling were each of 35-day duration?

Solve this problem in two ways:

1. Consider the explicit operating history of each of the three batches to the core decay power.
2. Assume the whole core had been operating for an infinite period before shutdown.

Answers:

1. $\dot{Q} = 38.0\ MW_{th}$
2. $\dot{Q} = 43.8\ MW_{th}$

3.7. *Effect of continuous refueling on decay heat* (Section 3.9)
Using Equation 3.70c or 3.70d, estimate the decay heat rate in a 3000 MW_{th} reactor in which 3.2% ^{235}U-enriched UO_2 assemblies are being fed into the core. The burned-up fuel stays in the core for 3 years before being replaced. Consider two cases:

1. The core is replaced in two batches every 18 months.
2. The fuel replacement is so frequent that refueling can be considered a continuous process. (*Note:* The PHWR reactors and some of the water-cooled graphite-moderated reactors in former Soviet Union are effectively continuously refueled.)

Compare the two situations at 1 min, 1 h, 1 month, and 1 year.

Answers:

	Case 1	Case 2
1 min	$P = 81.9\ MW_{th}$	$P = 81.0\ MW_{th}$
1 h	$P = 33.1\ MW_{th}$	$P = 32.2\ MW_{th}$
1 day	$P = 15.0\ MW_{th}$	$P = 14.1\ MW_{th}$
1 month	$P = 4.95\ MW_{th}$	$P = 4.24\ MW_{th}$
1 year	$P = 1.29\ MW_{th}$	$P = 0.965\ MW_{th}$

REFERENCES

1. ANSI/ANS-5.1–1971, *Decay Heat Power in Light Water Reactors*. American Nuclear Society, La Grange Park, IL, 1971.
2. ANSI/ANS-5.1–1994, *Decay Heat Power in Light Water Reactors*. American Nuclear Society, La Grange Park, IL, 1994.
3. ANSI/ANS-5.1–2005, *Decay Heat Power in Light Water Reactors*. American Nuclear Society, La Grange Park, IL, 2005.
4. Benedict, M., Pigford, T. H., and Levi, H. W., *Nuclear Chemical Engineering*, 2nd Ed. New York, NY: McGraw-Hill, 1981.
5. Chadwork, M. B., Oblozinsky, P. L., Herman, M. et al., ENDF/B-VII.0: Next generation evaluated nuclear data library for nuclear science and technology. *Nucl. Data Sheets*, 107(12):2931–3060, 2006.
6. Croff, A. G., ORIGEN-2: A versatile computer code for calculating the nuclide compositions and characteristics of nuclear materials. *Nucl. Tech.*, 62:335, 1983.

7. El-Wakil, M. M., *Nuclear Heat Transport*. International Textbook Co., Scranton, PA, 1971.
8. Glasstone, S. and Sesonske, A., *Nuclear Reactor Engineering*. Van Nostrand Reinhold, New York, NY, 1967 (2nd Ed.), 1981 (3rd Ed.).
9. Goldstein, H., *Fundamental Aspects of Reactor Shielding*. Addison Wesley, Reading, MA, 1959.
10. Jaeger, R. G. Ed.-in-Chief, *Engineering Compendium on Radiation Shielding, Vol. I, Shielding Fundamentals and Methods*, p. 232. Springer-Verlag, New York, NY, 1968.
11. Koning, A., Forrest, R., Kellett, M., Mills, R., Henriksson, H., and Rugama, Y. (Eds.), The JEFF-3.1 Nuclear Data Library, JEFF Report 21, An OECD Publication, Paris, ISBN 92-64-02314-3, 2006.
12. Leipunskii, O., Noroshicov, B. V., and Sakharov, V. N., *The Propagation of Gamma Quanta in Matter*. Pergamon Press, Oxford, UK, 1965.
13. Marshall, B. W., Hydrogen:Air:Steam Flammability Limits and Combustion Characteristics in the FITS Vessel. NUREG/CR-3468, SAND84–0383, Sandia National Laboratories, December 1986.
14. Plys, M. G., Hydrogen production and combustion in severe reactor accidents: An integral assessment perspective. *Nucl. Tech.*, 101:400–410, 1993.
15. Rust, J. H., *Nuclear Power Plant Engineering*. Haralson Publishing, Buchanan, GA, 1979.
16. Seiler, J. M. and Froment, K., Material effects on multiphase phenomena in late phases of severe accidents of nuclear reactors. *Multiphase Sci. Tech.*, 12(2):117–257, 2000.
17. Shure, I., *Fission Product Decay Energy,* pp. 1–17. USAEC report WAPD-BT-24, 1961, Pittsburgh, PA, 1961.
18. Shwageraus, E. and Fridman, E., Decay Power Calculation for Safety Analysis of Innovative Reactor Systems. *Proceedings of International Youth Nuclear Congress, IYNC08*, Interlaken, Switzerland, September 20–26, 2008.
19. Spiegel, U. S., *Mathematical Handbook*, p. 111. Outline Series. New York, NY: McGraw-Hill, 1968.
20. Stacey, W. M., *Nuclear Reactor Physics*. John Wiley & Sons, Inc., New York, NY, 2001.
21. Templin, L. T. (ed.), *Reactor Physics Constants*, 2nd Ed. ANL-5800, Argonne National Laboratory, Argonne, IL,1963.
22. Tong, L. S. and Weisman, J., *Thermal Analysis of Pressurized Water Reactors*, 3rd Ed. American Nuclear Society, La Grange Park, IL, 1996.
23. Unik, J. P. and Gindler, J. E., *A Critical Review of the Energy Released in Nuclear Fission*. ANL-7748, Argonne National Laboratory, Argonne, IL, 1971.
24. Xu, Z., Kazimi, M. S., and Driscoll, M. J., High-burnup impact on PWR spent fuel. *Nucl. Sci. Eng.*, 151:261–273, 2005.

4 Transport Equations for Single-Phase Flow

4.1 INTRODUCTION

Thermal analyses of power-conversion systems involve the solution of transport equations of mass, momentum, and energy in forms that are appropriate for the system conditions. Engineering analysis often starts with suitably tailored transport equations. These equations are achieved by simplifying the general equations, depending on the necessary level of resolution of spatial distributions, the nature of the fluids involved (e.g., compressibility), and the numerical accuracy required for the analysis. The general forms of the transport equations for single-phase flow and many of the simplifying assumptions are discussed in this chapter.

The basic assumption made here is that the medium can be considered a continuum. That is, the smallest volume of concern contains enough molecules to allow each point of the medium to be described on the basis of average properties of the molecules. Hence, unique values for temperature, velocity, density, and pressure (collectively referred to as field variables) can be assumed to exist at each point in the medium of consideration. Differential equations of conservation of mass, momentum, and energy in a continuum can then be developed to describe the average molecular values of the field variables. The equations of a continuum do not apply when the mean free path of molecules is of a magnitude comparable to the dimension of the volume of interest. Under these conditions few molecules would exist in the system, and averaging loses its meaning when only a few molecules are involved. In that case statistical distribution of the molecular motion should be used for the description of the macroscopic behavior of the medium. Because a cube of gas whose side is 1 μm contains 2.5×10^7 molecules at normal temperature and pressure, the continuum condition is readily met in most practical systems. An example of a situation where the continuum equations should not be applied is that of rarefield gases, as in fusion vacuum equipment or for space vehicles flying at the edge of the atmosphere. (The mean free path of air at atmospheric standard conditions is 6×10^{-8} m, whereas at 100 miles' elevation it is 50 m.)

4.1.1 Equation Forms

Two approaches are used to develop the transport equations: an integral approach and a differential approach. The integral approach can be further subdivided into two categories: the lumped parameter approach and the distributed parameter integral approach.

The *integral approach* addresses the behavior of a system of a specific region or mass. With the *lumped parameter integral approach*, the medium is assumed to occupy one or more compartments, and within each compartment the spatial distributions of the field variables and transport parameters of the material are ignored when setting up the integral equations. Conversely, with the *distributed parameter integral approach* the spatial dependence of the variables within the medium is taken into consideration when obtaining the equations.

The *differential approach* is naturally a distributed parameter approach but with balance equations for each point and not for an entire region. Integrating the differential equations over a volume yields integral-distributed parameter equations for the volume. However, integral-distributed parameter equations for a volume may also be formulated on the basis of conveniently assumed spatial behavior of the parameters in the volume. In that case the information about the point-by-point values of the parameter would not be as accurate as predicted by a differential approach.

The integral equations can be developed for two types of system: a control mass or a control volume (Table 4.1). With the *control mass approach*, the boundary of the system is the boundary of the mass, and no mass is allowed to cross this boundary (a closed-system formulation). With the *control volume approach*, mass is allowed to cross the boundaries of the system (an open-system formulation), as illustrated in Figure 4.1. In general, the boundary surface surrounding a control mass or a control volume may be deformable, although in practice rigid boundaries are encountered for most engineering applications.

The system of differential equations used to describe the control mass is called Lagrangian. With the *Lagrangian system* the coordinates move at the flow velocity (as if attached to a particular mass), and thus the spatial coordinates are not independent of time. Eulerian equations are used to describe transport equations as applied to a control volume. The *Eulerian system* of equations can be derived for a coordinate frame moving at any velocity. When the coordinate frame's origin is stationary in space at a particular position, the resulting equations are the most often used form of the Eulerian equations.

Applications of the integral forms of these equations are found in Chapters 6 and 7, and the differential forms are in Chapters 8, 9, 10, and 14.

TABLE 4.1
Classification of Transport Equations

Integral	Differential
Lumped parameter	
Control mass	
Control volume	
Distributed parameter	
Control mass	Lagrangian equations
Control volume	Eulerian equations

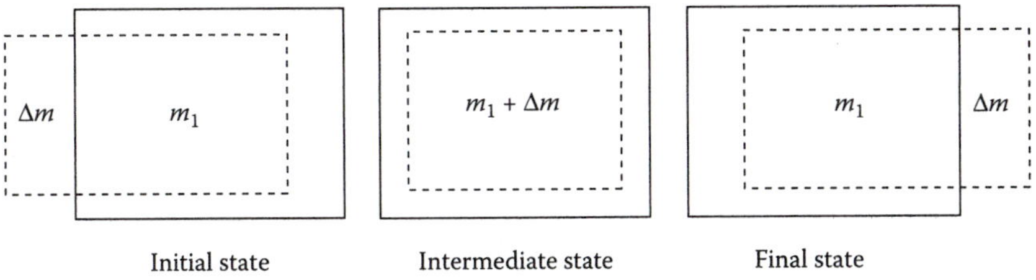

FIGURE 4.1 Control mass and control volume boundaries. *Solid line* indicates the control volume. *Dotted line* indicates the control mass. Note that at the intermediate stage the control mass boundary is identical to the control volume boundary.

4.1.2 Intensive and Extensive Properties

System properties whose values are obtained by the summation of their values in the components of the system are called *extensive properties*. These properties depend on the extent of a system. Properties that are independent of the size of the system are called *intensive properties*. Table 4.2 lists some of the properties appearing in the transport equations.

For a control mass (m), let C be an extensive property of the medium (e.g., volume, momentum, or internal energy). Also, denote the specific value per unit mass of the extensive property as c. The lumped parameter approach is based on the assumption that the medium is of uniform properties, and thus $C = mc$. For a distributed parameter approach, the mass density (ρ) and c are not necessarily uniform, and $C = \iiint_V \rho c \, dV$.

4.2 MATHEMATICAL RELATIONS

It is useful to remind the reader of some of the basic mathematics encountered in this chapter.

4.2.1 Time and Spatial Derivative

Consider a property c as a function of time and space. The time rate of change of the property c as seen by an observer at a fixed position in space is denoted by the partial time derivative $\partial c/\partial t$.

The time rate of change of a property c as seen by an observer moving at a velocity $\vec{\upsilon}_o$ within a medium at rest is given by the total time derivative dc/dt. In Cartesian coordinates, the total derivative is related to the partial derivative by

$$\frac{dc}{dt} = \frac{\partial c}{\partial t} + \frac{\partial x}{\partial t}\frac{\partial c}{\partial x} + \frac{\partial y}{\partial t}\frac{\partial c}{\partial y} + \frac{\partial z}{\partial t}\frac{\partial c}{\partial z} = \frac{\partial c}{\partial t} + \upsilon_{ox}\frac{\partial c}{\partial x} + \upsilon_{oy}\frac{\partial c}{\partial y} + \upsilon_{oz}\frac{\partial c}{\partial z} \tag{4.1}$$

TABLE 4.2
Extensive and Intensive Properties[a]

Property	Total Value	Specific Value Per Unit Mass	Specific Value Per Unit Volume
Mass	m	1	ρ
Volume	V	$v \equiv \frac{V}{m} \equiv \frac{1}{\rho}$	1
Momentum	$m\vec{\upsilon}$	$\vec{\upsilon}$	$\rho\vec{\upsilon}$
Kinetic energy	$\frac{1}{2}m\ \upsilon^2$	$\frac{1}{2}\upsilon^2$	$\frac{1}{2}\rho\upsilon^2$
Potential energy	mgz	gz	ρgz
Internal energy	$U = mu$	u	ρu
Stagnation internal energy	$U^\circ = mu^\circ$	$u^\circ \equiv u + \frac{1}{2}\upsilon^2$	ρu°
Enthalpy	$H = mh$	$h = u + p\upsilon$	ρh
Stagnation enthalpy	$H^\circ = mh^\circ$	$h^\circ = h + \frac{1}{2}\upsilon^2$	ρh°
Total energy	$E = me$	$e \equiv u^\circ + gz$	ρe
Entropy	$S = ms$	s	ρs
Temperature	T	T	T
Pressure	p	p	p
Velocity	$\vec{\upsilon}$	$\vec{\upsilon}$	$\vec{\upsilon}$

[a] Properties that are independent of the size of the system are called intensive properties. Intensive properties include $\vec{\upsilon}$, p, and T, as well as the specific values of the thermodynamic properties.

where υ_{ox}, υ_{oy}, and υ_{oz} are the components of $\vec{\upsilon}_o$ in the x, y, and z directions. The general relation between the total and the partial derivatives for a property c is given by

$$\frac{\mathrm{d}c}{\mathrm{d}t} = \frac{dc}{dt} + \vec{\upsilon}_o \cdot \nabla c \tag{4.2}$$

If the frame of reference for the coordinates is moving at the flow velocity $\vec{\upsilon}$, the rate of change of c as observed at the origin (the observer) is given by the substantial derivative Dc/Dt. A better name for this derivative is *material derivative*, as suggested by Whitaker [7]. It is given by

$$\frac{\mathrm{D}c}{\mathrm{D}t} = \frac{\partial c}{\partial t} + \vec{\upsilon} \cdot \nabla c \tag{4.3}$$

The right-hand side of Equation 4.3 describes the rate of change of the variable in Eulerian coordinates, and the left-hand side describes the time rate of change in Lagrangian coordinates.

In a steady-state flow system, the rate of change of c at a fixed position $\partial c/\partial t$ is zero. Hence at the fixed position an observer does not see any change in c with time. However, the fluid experiences a change in its property c as it moves because of the nonzero values of $\vec{\upsilon}\cdot\nabla c$. Thus in the Lagrangian system $\mathrm{D}c/\mathrm{D}t$ is nonzero, and the observer moving with the flow does see a change in c with time.

Note that

$$\frac{\mathrm{D}c}{\mathrm{D}t} = \frac{\mathrm{d}c}{\mathrm{d}t} + (\vec{\upsilon} - \vec{\upsilon}_{\mathrm{o}})\cdot\nabla c \tag{4.4}$$

Thus, the total derivative becomes identical to the substantial derivative when $\vec{\upsilon}_{\mathrm{o}} = \vec{\upsilon}$.

For a vector, $\vec{c}$, the substantial derivative is obtained by applying the operator to each component so that when they are summed we get

$$\frac{\mathrm{D}\vec{c}}{\mathrm{D}t} = \frac{\mathrm{D}c_i}{\mathrm{D}t}\vec{i} + \frac{\mathrm{D}c_j}{\mathrm{D}t}\vec{j} + \frac{\mathrm{D}c_k}{\mathrm{D}t}\vec{k} \tag{4.5}$$

where $\vec{i}, \vec{j}$, and $\vec{k}$ = the unit vectors in the Cartesian directions x, y, and z.

Example 4.1: Various Time Derivatives

PROBLEM

Air bubbles are being injected at a steady rate into a water channel at various positions such that the bubble population along the channel is given by

$$N_{\mathrm{b}} = N_{\mathrm{bo}}\left[1 + \left(\frac{z}{L}\right)^2\right]$$

where L = channel length along the z axis; N_{bo} = bubble density at the channel inlet ($z = 0$).

What is the observed rate of change of the bubble density by

1. A stationary observer at $z = 0$?
2. An observer moving in the channel with a constant speed υ_0?

SOLUTION

The time rate of change of the bubble density as observed by a stationary observer at $z = 0$ is

$$\frac{\partial N_{\mathrm{b}}}{\partial t} = 0$$

However, for the observer moving along the channel with a velocity υ_o m/s, the rate of change of the bubble density is

$$\frac{dN_b}{dt} = \frac{\partial N_b}{\partial t} + \upsilon_o \frac{\partial N_b}{\partial z} = \upsilon_o N_{bo}\left(\frac{2z}{L^2}\right)$$

4.2.2 Gauss's Divergence Theorem

If V is a closed region in space, totally bounded by the surface S, the volume integral of the divergence of a vector $\vec{c}$ is equal to the total flux of the vector at the surface S:

$$\iiint_V (\nabla \cdot \vec{c})\mathrm{d}V = \oiint_S \vec{c} \cdot \vec{n}\,\mathrm{d}S \tag{4.6}$$

where $\vec{n}$ = unit vector directed normally outward from S.

This theorem has two close corollaries for scalars and tensors:

$$\iiint_V \nabla c\,\mathrm{d}V = \oiint_S c\vec{n}\,\mathrm{d}S \tag{4.7}$$

and

$$\iiint_V \nabla \cdot \bar{\bar{c}}\,\mathrm{d}V = \oiint_s (\bar{\bar{c}} \cdot \vec{n})\,\mathrm{d}S \tag{4.8}$$

The last equation is also applicable to a dyadic product of two vectors $\vec{c}_1\vec{c}_2$.

4.2.3 Leibnitz's Rules

The general Leibnitz rule for differentiation of an integral of a function (f) is given by [3]

$$\frac{\mathrm{d}}{\mathrm{d}\lambda}\int_{a(\lambda)}^{b(\lambda)} f(x,\lambda)\,\mathrm{d}x = \int_{a(\lambda)}^{b(\lambda)} \frac{\partial f(x,\lambda)}{\partial \lambda}\,\mathrm{d}x + f(b,\lambda)\frac{\mathrm{d}b}{\mathrm{d}\lambda} - f(a,\lambda)\frac{\mathrm{d}a}{\mathrm{d}\lambda} \tag{4.9}$$

It should be noted that for this rule to be applicable the functions f, a, and b must be continuously differentiable with respect to λ. Also the function f and $\partial f/\partial \lambda$ should be continuous with respect to x between $a(\lambda)$ and $b(\lambda)$.

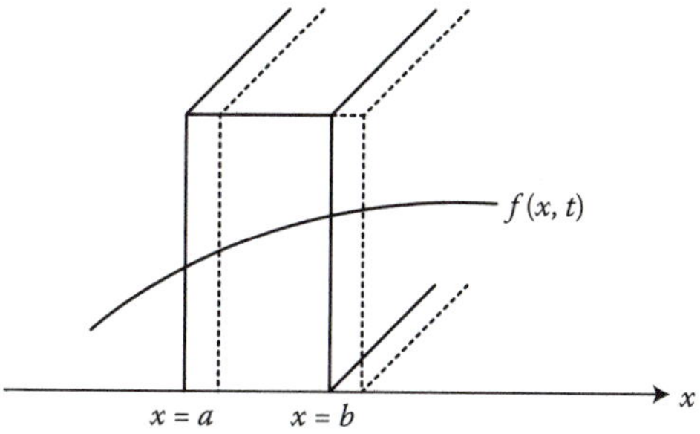

FIGURE 4.2 Plate moving in the x direction.

This rule can be used to yield a general relation between the time differential of a volume integral and the surface velocity of the volume as follows. Let x be the Cartesian coordinate within a plate of a surface area A and extending from $x = a(t)$ to $x = b(t)$, as seen in Figure 4.2. According to Equation 4.9, the time derivative of an integral of the function $f(x, t)$ over x is given by

$$\frac{\mathrm{d}}{\mathrm{d}t}\int_{a(t)}^{b(t)} f(x,t)A\,\mathrm{d}x = \int_{a(t)}^{b(t)} \frac{\partial f(x,t)}{\partial t}A\,\mathrm{d}x + f(b,t)\,A\frac{\mathrm{d}b}{\mathrm{d}t} - f(a,t)\,A\frac{\mathrm{d}a}{\mathrm{d}t} \tag{4.10}$$

The extension of Equation 4.10 to three-dimensional integration of a function $f(\vec{r},t)$ over a volume V surrounded by the surface S yields [5]:

$$\frac{\mathrm{d}}{\mathrm{d}t}\iiint_V f(\vec{r},t)\mathrm{d}V = \iiint_V \frac{\partial f(\vec{r},t)}{\partial t}\mathrm{d}V + \oiint_S f(\vec{r},t)\vec{\upsilon}_s\cdot\vec{n}\,\mathrm{d}S \tag{4.11}$$

where $\vec{\upsilon}_s$ = local and instantaneous velocity of the surface S.

The velocity $\vec{\upsilon}_s$ may be a function of the spatial coordinate (if the surface is deforming) and time (if the volume is accelerating or decelerating). For a deformable volume, whether the center of the volume is stationary or moving, $\vec{\upsilon}_s$ is not equal to zero.

Equation 4.11 is the *general transport theorem* [6]. A special case of interest is when the volume under consideration is a material volume V_m (i.e., a volume encompassing a certain mass) within the surface S_m. In that case the surface velocity represents the fluid velocity, that is, $\vec{\upsilon}_s = \vec{\upsilon}$, and the left-hand side represents the substantial derivative of a volume integral (refer to Equation 4.4). In this case, Equation 4.11 becomes

$$\frac{\mathrm{D}}{\mathrm{D}t}\iiint_{V_m} f(\vec{r},t)\mathrm{d}V = \iiint_{V_m} \frac{\partial f(\vec{r},t)}{\partial t}\mathrm{d}V + \oiint_{S_m} f(\vec{r},t)\vec{\upsilon}\cdot\vec{n}\,\mathrm{d}S \tag{4.12}$$

This equation is the *Reynolds transport theorem.* Note that the integration in Equation 4.12 is performed over a material volume V_m bounded by the surface S_m. The Reynolds transport theory is useful for transforming the derivatives of integrals from material-based coordinates (Lagrangian left-hand side of Equation 4.12) into spatial coordinates (Eulerian right-hand side of Equation 4.12).

Another special case of Equation 4.11 is that of a fixed volume (i.e., a stationary and nondeformable volume) where $\vec{\upsilon}_s = 0$, in which case Equation 4.11 reduces to

$$\frac{d}{dt}\iiint_V f(\vec{r},t)\,dV = \iiint_V \frac{\partial f(\vec{r},t)}{\partial t}\,dV = \frac{\partial}{\partial t}\iiint_V f(\vec{r},t)\,dV \quad (4.13)$$

The last equality is obtained because the volume in this case is time-independent.

The total rate of change of the integral of the function $f(\vec{r},t)$ can be related to the material derivative at a particular instant when the volume boundaries of V and V_m are the same, so that by subtracting Equation 4.11 from Equation 4.12:

$$\frac{D}{Dt}\iiint_V f(\vec{r},t)dV = \frac{d}{dt}\iiint_V f(\vec{r},t)dV + \oiint_S f(\vec{r},t)(\vec{\upsilon}-\vec{\upsilon}_s)\cdot\vec{n}\,dS \quad (4.14)$$

where $\vec{\upsilon}-\vec{\upsilon}_s$ = relative velocity of the material with respect to the surface of the control volume:

$$\vec{\upsilon}_r = \vec{\upsilon} - \vec{\upsilon}_s \quad (4.15)$$

The first term on the right-hand side of Equation 4.14 is the "time rate of change" term, and the second term is the "flux" term, both referring to the function $f(\vec{r},t)$ within the volume V. The time rate of change term may be nonzero for two distinct conditions: if the integrand is time-dependent or the volume is not constant. A de-formable control surface is not a sufficient condition to produce a change in the total volume, as expansion of one region may be balanced by contraction in another. Note also that only the component of $\vec{\upsilon}_r$ normal to the surface S contributes to the flux term.

The general transport theorem can be applied to a vector as well as to a scalar. The extension to vectors is easily obtained if Equation 4.11 is applied *to* three scalar functions (f_x, f_y, f_z), then multiplied by $\vec{i}, \vec{j}$, and $\vec{k}$, respectively, and summed to obtain

$$\frac{d}{dt}\iiint_V \vec{f}(\vec{r},t)dV = \iiint_V \frac{\partial \vec{f}(\vec{r},t)}{\partial t}dV + \oiint_S \vec{f}(\vec{r},t)(\vec{\upsilon}_s\cdot\vec{n})\,dS \quad (4.16)$$

The reader should consult the basic fluid mechanics textbooks for further discussion and detailed derivation of the transport theorems given here [e.g., 2,4,6].

Example 4.2: Difference in Total and Substantial Derivatives

PROBLEM

Consider loss of coolant from a pressure vessel. The process can be described by observing the vessel as a control volume or the coolant as a control mass. Assume the coolant leaves the vessel at an opening of area A_1 with velocity υ_1. Evaluate the rate of change of the mass remaining in the vessel.

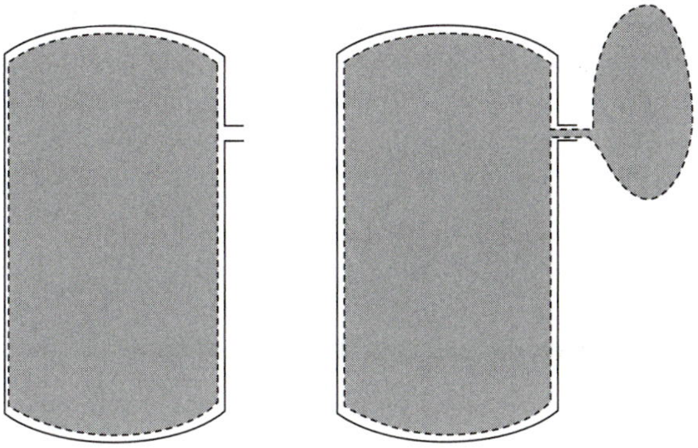

FIGURE 4.3 Boundaries of the vessel control volume (*solid line*) and the coolant control mass (*dashed line*) after a loss of coolant accident.

SOLUTION

The solid line in Figure 4.3 defines the control volume around the vessel, and the dashed line surrounds the coolant mass (i.e., material volume).

The observed rate of change of mass in the material volume is $Dm/Dt = 0$. We can obtain dm/dt from Dm/Dt by considering Equation 4.14 and the fact that the total mass is the integral of the coolant density (ρ) over the volume of interest to get

$$\frac{dm}{dt} = \frac{Dm}{Dt} - \iint \rho \vec{\upsilon}_r \cdot \vec{n}\, dS$$

However $\vec{\upsilon}_r = \vec{\upsilon}_1 - 0$ over the opening A_1 and zero elsewhere around the surface of the vessel. Therefore if the coolant density at the opening is p_1, we get

$$\frac{dm}{dt} = 0 - \rho_1 \vec{\upsilon}_1 \cdot \vec{A}_1$$

4.3 LUMPED PARAMETER INTEGRAL APPROACH

4.3.1 Control Mass Formulation

4.3.1.1 Mass

By definition, the control mass system is formed such that no mass crosses the boundaries. Hence, the conservation of mass implies that the mass (m) in the system volume, (V_m), is constant, or

$$\frac{Dm}{Dt} = \frac{D}{Dt} - \iiint_{V_m} \rho\, dV = 0 \tag{4.17a}$$

Because we are treating mass and energy as distinct quantities, this statement ignores any relativistic effects such as those involved in nuclear reactions, where mass is transformed into energy. From Equation 4.14, taking $f(\vec{r},t)$ equal to $\rho(\vec{r},t)$, we can recast Equation 4.17a into

$$\frac{d}{dt}\iiint_V \rho \, dV + \iint_S \rho\vec{\upsilon}_r \cdot \vec{n} \, dS = 0 \tag{4.17b}$$

For a control mass, the material expands at the rate of change of the volume it occupies, so that $\vec{\upsilon}_r = 0$ and

$$\frac{Dm}{Dt} = \frac{d}{dt}\iiint_{V_m} \rho \, dV = \left(\frac{dm}{dt}\right)_{c.m.} = 0 \tag{4.17c}$$

where c.m. denotes the control mass.

4.3.1.2 Momentum

The momentum balance (Newton's second law of motion) applied to the control mass equates the rate of change of momentum to the net externally applied force. Thus when all parts of the mass (m) have the same velocity ($\vec{\upsilon}$):

$$\frac{Dm\vec{\upsilon}}{Dt} = \left(\frac{dm\vec{\upsilon}}{dt}\right)_{c.m.} = \sum_k \vec{F}_k = \sum_k m\vec{f}_k \tag{4.18}$$

The forces $\vec{F}_k$ may be body forces arising from gravitational, electrical, or magnetic effects, or surface forces such as those exerted by pressure. The forces are positive if acting in the positive direction of the coordinates. In each case, $\vec{f}_k$ is the force per unit mass.

4.3.1.3 Energy

The *energy equation*, the first law of thermodynamics, states that the rate of change of the stored energy in a control mass should equal the difference between the rate of energy addition (as heat or as work) to the control mass and the rate of energy extraction (as heat or as work) from the control mass. For systems that do not involve nuclear or chemical reactions, stored energy is considered to consist of the thermodynamic internal energy and the kinetic energy. Whereas stored energy is a state property of the control mass, heat and work are the forms of energy transfer to or from the mass. By convention, heat *added to* a system is considered positive, but work *provided by* the system to the surroundings (i.e., extracted from the system) is considered positive. Thus,

$$\frac{DU^\circ}{Dt} = \left(\frac{dU^\circ}{dt}\right)_{c.m.} = \left(\frac{dQ}{dt}\right)_{c.m.} - \left(\frac{dW}{dt}\right)_{c.m.} \tag{4.19a}$$

The change in the stored energy associated with internally distributed chemical and electromagnetic interactions are not treated in this book. Stored energy changes due to nuclear interactions are treated as part of the internal energy in Chapter 6 only. However, as the nuclear internal energy changes lead to volumetric heating of the fuel, it is more convenient to account for this energy source by an explicit heat source term on the right-hand side of Equation 4.19a:

$$\left(\frac{dU^\circ}{dt}\right)_{c.m.} = \left(\frac{dQ}{dt}\right)_{c.m.} + \left(\frac{dQ}{dt}\right)_{gen} - \left(\frac{dW}{dt}\right)_{c.m.} \tag{4.19b}$$

where, for the lumped parameter approach, the internal energy is assumed uniform in the control mass and is given by:

$$U^\circ = mu^\circ = m(u + \upsilon^2/2)$$

We may subdivide the work term into shaft work, expansion work by surface forces (which can be normal forces such as those exerted by pressure or tangential forces, e.g., shear forces), and work associated with the body forces (e.g., gravity, electrical, or magnetic forces). Thus if the body forces consist of a single force ($\vec{f}$) per unit mass, the work term can be decomposed into

$$\left(\frac{dW}{dt}\right)_{c.m.} = \left(\frac{dW}{dt}\right)_{shaft} + \left(\frac{dW}{dt}\right)_{normal} + \left(\frac{dW}{dt}\right)_{shear} + \vec{\upsilon}\cdot m\vec{f} \tag{4.20}$$

The energy effects arising from a body force may be included in Equations 4.19a and 4.19b as part of the stored energy term instead of the work term only when the force is associated with a time-independent field, for example, gravity. To illustrate, consider $\vec{f}$ the force per unit mass due to spatial change in a force field ψ per unit mass such that

$$\vec{f} = -\nabla\psi \tag{4.21}$$

Using Equation 4.3, we get

$$\vec{\upsilon}\cdot m\vec{f} = -m\vec{\upsilon}\cdot\nabla\psi = -m\frac{D\psi}{Dt} + m\frac{\partial\psi}{\partial t} \tag{4.22a}$$

When the field is time-independent, Equation 4.22a reduces to

$$\vec{\upsilon}\cdot m\vec{f} = -m\frac{D\psi}{Dt} = -\frac{D[m\psi]}{Dt}\left(\text{because } \frac{Dm}{Dt} = 0\right) \tag{4.22b}$$

Thus for a control mass:

$$\vec{\upsilon}\cdot m\vec{f} = -m\left(\frac{d\psi}{dt}\right)_{c.m.} = -\left(\frac{d[m\psi]}{dt}\right)_{c.m.} \tag{4.22c}$$

For gravity:

$$\psi = gz \quad \text{and} \quad \vec{f} = \vec{g} \tag{4.23}$$

If the only body force is gravity, Equations 4.22c and 4.23 can be used to write the energy equation 4.19b in the form:

$$\left(\frac{d}{dt}[mu^\circ]\right)_{c.m.} = \left(\frac{dQ}{dt}\right)_{c.m.} + \left(\frac{dQ}{dt}\right)_{gen} - \left(\frac{dW}{dt}\right)_{shaft} - \left(\frac{dW}{dt}\right)_{normal} - \left(\frac{dW}{dt}\right)_{shear} - \left(\frac{d(mgz)}{dt}\right)_{c.m.}$$

which can be rearranged to give

$$\left(\frac{d}{dt}[m(u^\circ + gz)]\right)_{c.m.} = \left(\frac{dQ}{dt}\right)_{c.m.} + \left(\frac{dQ}{dt}\right)_{gen} - \left(\frac{dW}{dt}\right)_{shaft} - \left(\frac{dW}{dt}\right)_{normal} - \left(\frac{dW}{dt}\right)_{shear} \tag{4.24a}$$

or equivalently in terms of the total energy (E):

$$\left(\frac{d}{dt}[E]\right)_{c.m.} = \left(\frac{dQ}{dt}\right)_{c.m.} + \left(\frac{dQ}{dt}\right)_{gen} - \left(\frac{dW}{dt}\right)_{shaft} - \left(\frac{dW}{dt}\right)_{normal} - \left(\frac{dW}{dt}\right)_{shear} \tag{4.24b}$$

These forms of the energy equation hold for reversible and irreversible processes. For a reversible process, when the only normal force on the surface is due to pressure (p), $(dW/dt)_{normal}$ can be replaced by $p(dV/dt)_{c.m.}$, the work involved in changing the volume of this mass. In this case, Equation 4.24b takes the form:

$$\left(\frac{d}{dt}[E]\right)_{c.m.} = \left(\frac{dQ}{dt}\right)_{c.m.} + \left(\frac{dQ}{dt}\right)_{gen} - \left(\frac{dW}{dt}\right)_{shaft} - p\left(\frac{dV}{dt}\right)_{c.m.} - \left(\frac{dW}{dt}\right)_{shear} \tag{4.24c}$$

4.3.1.4 Entropy

The *entropy equation*, an expression of the second law of thermodynamics, states that the net change of the entropy of a control mass interacting with its surroundings and the entropy change of the surroundings of the system should be equal to or greater than zero. Because there is no mass flow across the control mass boundary, the entropy exchanges with the surroundings are associated with heat interaction with the surroundings. The entropy equation then is

$$\frac{DS}{Dt} = \left(\frac{dS}{dt}\right)_{c.m.} \geq \frac{(dQ/dt)_{c.m.}}{T_s} \tag{4.25a}$$

where T_s = temperature at the location where the energy (Q) is supplied as heat.

The equality holds if the control mass undergoes a reversible process, whereas the inequality holds for an irreversible process. For a reversible process, the control mass temperature (T) would be equal to heat supply temperature (T_s). For the inequality case, the excess entropy is that produced within the control mass by irreversible action. Thus, we can write an entropy balance equation as

$$\left(\frac{dS}{dt}\right)_{c.m.} = \dot{S}_{gen} + \frac{(dQ/dt)_{c.m.}}{T_s} \tag{4.25b}$$

where $\dot{S}_{gen}$ = rate of entropy generation due to irreversibilities.

For a reversible adiabatic process, Equation 4.25b indicates that the entropy of the control mass is unchanged, that is, that the process is isentropic:

$$\left(\frac{dS}{dt}\right)_{c.m.} = 0 \quad \text{reversible adiabatic process} \tag{4.26}$$

It is also useful to introduce the definition of the availability function (A). The availability function is a characteristic of the state of the system and the environment which acts as a reservoir at constant pressure (p_o) and constant temperature (T_o). The availability function of the control mass is defined as

$$A \equiv E + p_o V - T_o S \tag{4.27}$$

The usefulness of this function derives from the fact that the change in the availability function from state to state yields the maximum useful work ($W_{u,max}$), obtainable from the specified change in state, that is

$$\left(\frac{dW}{dt}\right)_{u,max} \equiv -\frac{dA}{dt} = -\frac{d}{dt}(E + p_o V - T_o S) \tag{4.28}$$

For a change from state 1 to 2

$$W_{u,max\ 1\to 2} \equiv A_1 - A_2 = E_1 - E_2 + p_o(V_1 - V_2) - T_o(S_1 - S_2) \tag{4.29}$$

4.3.2 Control Volume Formulation

4.3.2.1 Mass

The conservation of mass in the control volume approach requires that the net mass flow rate into the volume equals the rate of change of the mass in the volume. As already illustrated in Example 4.2, Equation 4.14 can be specified for the mass balance, that is, $f(\vec{r},t) = \rho$, in a lumped parameter system to get

$$\frac{\mathrm{D}m}{\mathrm{D}t} = \left(\frac{\mathrm{d}m}{\mathrm{d}t}\right)_{\text{c.v.}} - \sum_{i=1}^{I} \dot{m}_i \tag{4.30a}$$

Because $\mathrm{D}m/\mathrm{D}t = 0$:

$$\left(\frac{\mathrm{d}m}{\mathrm{d}t}\right)_{\text{c.v.}} = \sum_{i=1}^{I} \dot{m}_{\mathrm{i}} \tag{4.30b}$$

where c.v. = control volume; I = number of the gates i through which flow is possible.

Note that $\dot{m}_{\mathrm{i}}$ is defined positive for mass flow into the volume and is defined negative for mass flow out of the volume as follows:

$$\dot{m}_{\mathrm{i}} = -\iint_{S_{\mathrm{i}}} \rho \vec{\upsilon}_{\mathrm{r}} \cdot \vec{n}\, \mathrm{d}S \tag{4.31a}$$

For uniform properties at S_{i}:

$$\dot{m}_{\mathrm{i}} = -\rho(\vec{\upsilon} - \vec{\upsilon}_{\mathrm{s}})_{\mathrm{i}} \cdot \vec{S}_{\mathrm{i}} \tag{4.31b}$$

4.3.2.2 Momentum

The momentum law, applied to the control volume, accounts for the rate of momentum change because of both the accumulation of momentum due to net influx and the external forces acting on the control volume. By specifying Equation 4.14 to the linear momentum $[f(r,t) = \rho\vec{\upsilon}]$ of a lumped parameter system and using Equation 4.31, we get

$$\left(\frac{\mathrm{d}m\vec{\upsilon}}{\mathrm{d}t}\right)_{\text{c.m.}} = \frac{\mathrm{D}m\vec{\upsilon}}{\mathrm{D}t} = \left(\frac{\mathrm{d}m\vec{\upsilon}}{\mathrm{d}t}\right)_{\text{c.v.}} - \sum_{i=1}^{I} \dot{m}_i \vec{\upsilon}_i \tag{4.32}$$

Equations 4.18 and 4.32 can be combined to give

$$\left(\frac{\mathrm{d}m\vec{\upsilon}}{\mathrm{d}t}\right)_{\text{c.v.}} = \sum_{i=1}^{I} \dot{m}_i \vec{\upsilon}_i + \sum_{k} m \vec{f}_k \tag{4.33}$$

4.3.2.3 Energy

The energy equation (first law of thermodynamics) applied to a control volume takes into consideration the rate of change of energy in the control volume due to net influx and any sources or sinks within the volume.

By specifying Equation 4.14 to the stagnation energy, that is, $f(\vec{r},t) = \rho u^\circ$, and applying Equation 4.31a, we get the total change in the stagnation energy in the volume due to the flow in and out of the volume:

$$\frac{DU^\circ}{Dt} = \left(\frac{dU^\circ}{dt}\right)_{c.v.} - \sum_{i=1}^{I} \dot{m}_i u_i^\circ \tag{4.34}$$

Hence, the first law of thermodynamics applied to this system (the system here is the control volume) is

$$\left(\frac{dU^\circ}{dt}\right)_{c.v.} = \sum_{i=1}^{I} \dot{m}_i u_i^\circ + \left(\frac{DQ}{Dt}\right) - \left(\frac{DW}{Dt}\right) \tag{4.35}$$

The heat addition term does not have a component that depends on the flow into the volume. This fact is emphasized here by dropping the subscript c.v. for this term:

$$\frac{DQ}{Dt} = \left(\frac{dQ}{dt}\right)_{c.v.} = \frac{dQ}{dt} \tag{4.36}$$

The work term in Equation 4.35 should account for the work done by the mass in the control volume as well as the work associated with mass flow into and out of the volume. Thus

$$\frac{DW}{Dt} = \left(\frac{dW}{dt}\right)_{c.v.} - \sum_{i=1}^{I} \dot{m}_i (p\mathrm{v})_i \tag{4.37}$$

The shaft, normal, and shear components of the work must be evaluated independently (Figure 4.4). For a time-independent force field, as in the case of gravity, the work associated with the body force can be added to the stored internal energy. Now, similar to Equation 4.24b, we can write an equation for the total energy change in the volume as

$$\left(\frac{dE}{dt}\right)_{c.v.} = \sum_{i=1}^{I} \dot{m}_i \left(u_i^\circ + \frac{p_i}{\rho_i} + gz_i\right) + \frac{dQ}{dt} + \left(\frac{dQ}{dt}\right)_{gen} - \left(\frac{dW}{dt}\right)_{shaft} - \left(\frac{dW}{dt}\right)_{normal}^{c.v.} - \left(\frac{dW}{dt}\right)_{shear} \tag{4.38}$$

When the only surface normal force is that exerted by pressure and for a reversible process, $(dW/dt)_{normal}^{c.v.}$ can be replaced by $p(dV/dt)_{c.v.}$.

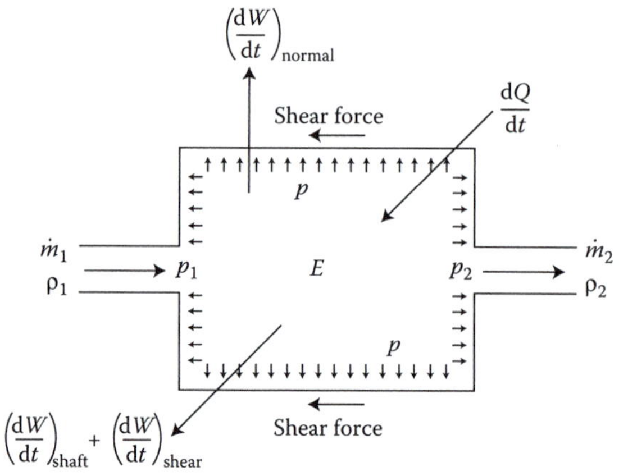

FIGURE 4.4 Lumped parameter control volume with a moving boundary.

For a stationary control volume, and given that $u_i^o + (p_i/\rho_i) = h_i^o$, we get

$$\left(\frac{\partial E}{\partial t}\right)_{\text{c.v.}} = \sum_{i=1}^{I} \dot{m}_i (h_i^o + gz_i) + \frac{dQ}{dt} + \left(\frac{dQ}{dt}\right)_{\text{gen}} - \left(\frac{dW}{dt}\right)_{\text{shaft}} - \left(\frac{\partial W}{\partial t}\right)_{\text{normal}}^{\text{c.v.}} - \left(\frac{dW}{dt}\right)_{\text{shear}} \tag{4.39}$$

Example 4.3: Determining the Pumping Power

PROBLEM

Obtain an expression for the power requirements of a PWR pump in terms of the flow rate.

SOLUTION

The PWR pump is modeled here as an adiabatic, steady-state pump with no internal heat sources. The control volume is taken as the pump internal volume and is thus fixed, that is, stationary and nondeformable. Pump inlet and outlet are at the same elevation, and the difference in kinetic energy between inlet and outlet streams is negligible.

With these assumptions, Equation 4.39 becomes

$$0 = \dot{m}(h_{in} - h_{out}) - \dot{W}_{shaft} - \dot{W}_{shear}$$

In order to evaluate the pumping work (i.e., W_{shaft} and W_{shear}), we need to compute the $h_{in} - h_{out}$ term. It is done by determining the thermodynamic behavior of the fluid undergoing the process. Pumping is assumed to be isentropic and the fluid to be incompressible. Because $h(T,p) = u(T,p) + p/\rho(T,p)$, $h(T,p) = u(T) + p/\rho(T)$ when $\partial\rho/\partial p = 0$. Thus, $h_{in} - h_{out} = u_{in}(T) - u_{out}(T) + (p/\rho)_{in} - (p/\rho)_{out} = (p_{in} - p_{out})/\rho$.

Hence,

$$\text{Pump power} = -\dot{W}_{\text{shaft}} - \dot{W}_{\text{shear}} = \dot{m}(h_{\text{out}} - h_{\text{in}}) = \frac{\dot{m}}{\rho}(p_{\text{out}} - p_{\text{in}})$$

4.3.2.4 Entropy

The entropy equation (second law of thermodynamics) applied to a control volume is obtained from examination of the condition already introduced for a control mass (Equation 4.25b):

$$\left(\frac{\mathrm{d}S}{\mathrm{d}t}\right)_{\text{c.m.}} = \dot{S}_{\text{gen}} + \frac{(\mathrm{d}Q/\mathrm{d}t)_{\text{c.m.}}}{T_{\text{s}}}$$

Now by Equation 4.14, when $f(\vec{r},t)$ is taken as ρs, we get

$$\left(\frac{\mathrm{d}S}{\mathrm{d}t}\right)_{\text{c.m.}} = \left(\frac{\mathrm{d}S}{\mathrm{d}t}\right)_{\text{c.v.}} - \sum_{i=1}^{I}\dot{m}_i s_i \tag{4.40a}$$

For a stationary control volume:

$$\left(\frac{\mathrm{d}S}{\mathrm{d}t}\right)_{\text{c.m.}} = \left(\frac{\partial S}{\partial t}\right)_{\text{c.v.}} - \sum_{i=1}^{I}\dot{m}_i s_i \tag{4.40b}$$

Hence, applying Equations 4.36 and 4.40a, Equation 4.25b can be recast in the form

$$\left(\frac{\partial S}{\partial t}\right)_{\text{c.v.}} = \sum_{i=1}^{I}\dot{m}_i s_i + \dot{S}_{\text{gen}} + \frac{(\mathrm{d}Q/\mathrm{d}t)}{T_{\text{s}}} \tag{4.41}$$

It is also desirable to define the maximum useful work and the lost work (or irreversibility) for a control volume with respect to the environment acting as a reservoir at pressure p_{o} and temperature T_{o}. (Useful work may be defined with respect to any reservoir. Most often the reservoir is chosen to be the environment.) The actual work is given by

$$\left(\frac{\mathrm{d}W}{\mathrm{d}t}\right)_{\text{actual}} \equiv \left(\frac{\mathrm{d}W}{\mathrm{d}t}\right)_{\text{shaft}} + \left(\frac{\mathrm{d}W}{\mathrm{d}t}\right)_{\text{normal}} \tag{4.42}$$

Note that a certain amount of the work is required to displace the environment. Therefore, the useful part of the actual work is given by

$$\left(\frac{\mathrm{d}W}{\mathrm{d}t}\right)_{\text{u,actual}} = \left(\frac{\mathrm{d}W}{\mathrm{d}t}\right)_{\text{actual}} - p_{\text{o}}\frac{\mathrm{d}V}{\mathrm{d}t} \tag{4.43}$$

Realizing that for a stationary volume $(dV/dt) = (\partial V/\partial t)$, the useful actual work can be obtained by substituting from Equations 4.39 and 4.42 into Equation 4.43 as

$$\left(\frac{dW}{dt}\right)_{u,actual} = \sum_{i=1}^{I} \dot{m}_i (h_i^o + gz_i) + \frac{dQ}{dt} + \left(\frac{dQ}{dt}\right)_{gen} - \left(\frac{\partial(E + p_o V)}{\partial t}\right)_{c.v.} - \left(\frac{dW}{dt}\right)_{shear} \tag{4.44}$$

Now multiplying Equation 4.41 by T_o and subtracting from Equation 4.44, we obtain after rearrangement:

$$\left(\frac{dW}{dt}\right)_{u,actual} = -\left[\frac{\partial(E + p_o V - T_o S)}{\partial t}\right]_{c.v.} + \sum_{i=1}^{I} \dot{m}_i (h^o - T_o s + gz)_i + \left(1 - \frac{T_o}{T_s}\right)\frac{dQ}{dt} + \left(\frac{dQ}{dt}\right)_{gen} - \left(\frac{dW}{dt}\right)_{shear} - T_o \dot{S}_{gen} \tag{4.45}$$

The maximum useful work $(dW/dt)_{u,max}$ is given by Equation 4.45 when $\dot{S}_{gen}$ is zero:

$$\left(\frac{dW}{dt}\right)_{u,max} = -\left[\frac{\partial(E + p_o V - T_o S)}{\partial t}\right]_{c.v.} + \sum_{i=1}^{I} \dot{m}_i (h^o - T_o s + gz)_i + \left(1 - \frac{T_o}{T_s}\right)\frac{dQ}{dt} + \left(\frac{dQ}{dt}\right)_{gen} - \left(\frac{dW}{dt}\right)_{shear} \tag{4.46}$$

The irreversibility (lost work) is then:

$$\dot{I} \equiv \left(\frac{dW}{dt}\right)_{u,max} - \left(\frac{dW}{dt}\right)_{u,actual} \equiv T_o \dot{S}_{gen} \tag{4.47}$$

Evaluating $(dW/dt)_{u,max}$ from Equation 4.46 and $(dW/dt)_{u,actual}$ from Equation 4.44, the irreversibility becomes

$$\dot{I} = T_o \left(\frac{\partial S}{\partial t}\right)_{c.v.} - T_o \sum_{i=1}^{I} \dot{m}_i s_i - \frac{T_o}{T_s}\frac{dQ}{dt} \tag{4.48}$$

where $\partial S/\partial t$ involves only the temporal variations in the entropy (zero at steady state).

Example 4.4: Setting Up the Energy Equation for a PWR Pressurizer

PROBLEM

Consider the pressurizer of a PWR plant that consists of a large steel container partially filled with water (liquid); the rest of the volume is filled with steam (vapor) (Figure 4.5). The pressurizer is connected at the bottom to the hot leg of the plant

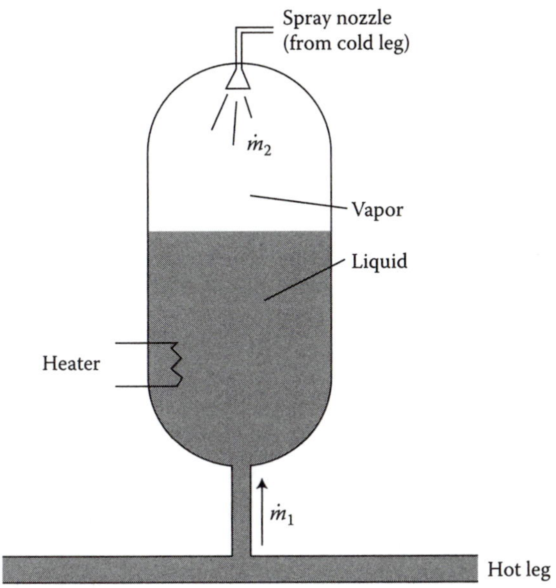

FIGURE 4.5 PWR pressurizer.

primary system and draws spray water from the cold leg for injection at the top. A heater is installed in the lower part to vaporize water, if needed, for pressure regulation.

Consider the energy equations for a control volume and answer the following questions:

1. If the pressurizer is the control volume, can we apply Equation 4.39?
   Answer: Yes, because the pressurizer is a stationary volume.
2. How many "gates" do the pressurizer have for flow into it (or out of it).
   Answer: Two: the nozzle for the spray and the connection to the primary.
3. What are the contributing factors to the heat input term dQ/dt?
   Answer: Any heat provided by the electric heater; also the heat extracted from or conducted into the vessel walls.
4. Does the value of $\partial E^{\circ}/\partial t$ equal the value of $\partial U^{\circ}/\partial t$ in this case?
   Answer: Not necessarily; because $E^{\circ} = U^{\circ} + mgz$, both m and z (the effective height of the center of mass) may be changing.
5. Should the work terms $\left(\frac{dW}{dt}\right)_{shaft}$, $p\left(\frac{\partial V}{\partial t}\right)_{c.v.}$ and $\left(\frac{dW}{dt}\right)_{shear}$ be included in this pressurizer analysis?
   Answer:
   $(dW/dt)_{shaft} = 0$, as no motor exists.
   $p(\partial V/\partial t)_{c.v.} = 0$; the volume considered is nondeformable and stationary.
   $(dW/dt)_{shear} < 0$, because some work has to be consumed to cause motion. This term is small for slow-moving media.

4.4 DISTRIBUTED PARAMETER INTEGRAL APPROACH

With the distributed parameter integral approach, the properties of the fluid are not assumed uniform within the control volume. As a general law, the rate of change of an extensive property (C) is manifested in changes in the local distribution of the mass, the property, or both and is governed by the net effects of the external factors contributing to changes in C.

Recall from Equation 4.12 that the rate of change of the extensive property C can be written as the sum of the volume integral of the local rate of change of the property and the net efflux from the surface:

$$\frac{\mathrm{D}C}{\mathrm{D}t} = \iiint_{V_\mathrm{m}} \frac{\partial}{\partial t}(\rho c)\,\mathrm{d}V + \oiint_{S_\mathrm{m}} (\rho c)\vec{\upsilon}\cdot\vec{n}\,\mathrm{d}S \tag{4.49}$$

where c is the specific value of the property per unit mass; ρ is the mass density; $\vec{\upsilon}$ is the mass velocity at the boundary S_m, which surrounds the material volume V_m.

The rate of change of the quantity C due to external volumetric effects as well as surface effects can be expressed as

$$\frac{\mathrm{D}C}{\mathrm{D}t} = \iiint_{V_\mathrm{m}} \rho\Phi\,\mathrm{d}V + \oiint_{S_\mathrm{m}} \vec{J}\cdot\vec{n}\,\mathrm{d}S \tag{4.50}$$

where Φ is the rate of introduction of c per unit mass within the volume V_m (a "body" factor); $\vec{J}\cdot\vec{n}$ = rate of loss of C per unit area of S_m due to surface effects (a "surface" factor).

Combining Equations 4.49 and 4.50, we get the general form for the integral transport equation as

$$\iiint_{V_\mathrm{m}} \frac{\partial}{\partial t}(\rho c)\mathrm{d}V + \oiint_{S_\mathrm{m}} (\rho c)\vec{\upsilon}\cdot\vec{n}\,\mathrm{d}S = \iiint_{V_\mathrm{m}} \rho\Phi\,\mathrm{d}V + \oiint_{S_\mathrm{m}} \vec{J}\cdot\vec{n}\,\mathrm{d}S \tag{4.51}$$

It is now assumed that this general integral balance is applicable to any arbitrary volume. Therefore, Equation 4.51 can be written as

$$\iiint_{V} \frac{\partial(\rho c)}{\partial t}\mathrm{d}V + \oiint_{S} (\rho c)\vec{\upsilon}\cdot\vec{n}\,\mathrm{d}S = \iiint_{V} \rho\Phi\,\mathrm{d}V + \oiint_{S} \vec{J}\cdot\vec{n}\,\mathrm{d}S \tag{4.52}$$

Substituting for the first term on the left-hand side of Equation 4.52 from Equation 4.11, we obtain

$$\frac{\mathrm{d}}{\mathrm{d}t}\iiint_{V} (\rho c)\mathrm{d}V + \oiint_{S} (\rho c)(\vec{\upsilon}-\vec{\upsilon}_\mathrm{s})\cdot\vec{n}\,\mathrm{d}S = \iiint_{V} \rho\Phi\,\mathrm{d}V + \oiint_{S} \vec{J}\cdot\vec{n}\,\mathrm{d}S \tag{4.53}$$

When the body force is due only to gravity, the general Equations 4.52 and 4.53 can be specified for mass, momentum, and energy by specifying $c, \vec{J}$, and Φ as follows:

For the mass equation:

$$c = 1, \quad \vec{J} = 0, \quad \Phi = 0 \tag{4.54}$$

For the momentum equation:

$$c = \vec{\upsilon}, \quad \vec{J} = \bar{\bar{\tau}} - p\bar{\bar{I}}, \quad \Phi = \vec{g} \tag{4.55}$$

For the energy equation:

$$c = u^{\circ} = u + \frac{\upsilon^2}{2}, \quad \vec{J} = -\vec{q}'' + (\bar{\bar{\tau}} - p\bar{\bar{I}}) \cdot \vec{\upsilon}, \quad \Phi = \frac{q'''}{\rho} + \vec{g} \cdot \vec{\upsilon} \tag{4.56}$$

where $\bar{\bar{\tau}}$ = a stress tensor; $\bar{\bar{I}}$ = a unity tensor; $\vec{g}$ = gravitational acceleration; $\vec{q}''$ = surface heat flux; q''' = volumetric heat-generation rate.

Substituting from Equation 4.54 into Equation 4.53, we get the integral mass balance equation:

$$\frac{\mathrm{d}}{\mathrm{d}t} \iiint_V \rho \,\mathrm{d}V + \oint\!\!\oint_S \rho(\vec{\upsilon} - \vec{\upsilon}_s) \cdot \vec{n} \,\mathrm{d}S = 0 \tag{4.57}$$

Note that

$$\iiint_V \rho \,\mathrm{d}V = (m)_{\text{c.v.}} \tag{4.58}$$

and from Equation 4.31a:

$$\iint_{S_i} \rho(\vec{\upsilon} - \vec{\upsilon}_s) \cdot \vec{n} \,\mathrm{d}S_i = -\dot{m}_i \tag{4.59}$$

where i identifies a localized opening at the boundary such that $(\vec{\upsilon} - \vec{\upsilon}_s)_i \neq 0$. Thus, Equation 4.57 can be cast in a rearranged form of Equation 4.30b:

$$\left(\frac{\mathrm{d}}{\mathrm{d}t} m\right)_{\text{c.v.}} - \sum_i \dot{m}_i = 0$$

For momentum balance, we substitute from Equation 4.55 into Equation 4.53 to get

$$\frac{\mathrm{d}}{\mathrm{d}t} \iiint_V \rho\vec{\upsilon} \,\mathrm{d}V + \oint\!\!\oint_S \rho\vec{\upsilon}(\vec{\upsilon} - \vec{\upsilon}_s) \cdot \vec{n} \,\mathrm{d}S = \oint\!\!\oint_S (\bar{\bar{\tau}} - p\bar{\bar{I}}) \cdot \vec{n} \,\mathrm{d}S + \iiint_V \rho\vec{g} \,\mathrm{d}V \tag{4.60}$$

If $\vec{\upsilon}$ is uniform within V, we can recast Equation 4.60 in a form similar to Equation 4.33 as

$$\left(\frac{\mathrm{d}}{\mathrm{d}t}m\vec{\upsilon}\right)_{\text{c.v.}} - \sum_i \dot{m}_i\vec{\upsilon}_i = \sum_j m\vec{f}_j + m\vec{g} = \sum_k m\vec{f}_k = \sum_k \vec{F}_k \tag{4.61}$$

where $\vec{f}_j$ is the force at the surface portion j and is obtained from

$$\vec{f}_j = \frac{1}{m}\iint_{S_j}(\bar{\bar{\tau}} - p\bar{\bar{I}})\cdot\vec{n}\,\mathrm{d}S \tag{4.62}$$

For energy balance, the parameters c, Φ, and $\vec{J}$ from Equation 4.56 are substituted into Equation 4.53 to get

$$\frac{\mathrm{d}}{\mathrm{d}t}\iiint_V \rho u^{\circ}\mathrm{d}V + \oiint_S \rho u^{\circ}(\vec{\upsilon} - \vec{\upsilon}_{\mathrm{s}})\cdot\vec{n}\,\mathrm{d}S$$
$$= \oiint_S [-\vec{q}'' + (\bar{\bar{\tau}} - p\bar{\bar{I}})\cdot\vec{\upsilon}]\cdot\vec{n}\,\mathrm{d}S + \iiint_V (q''' + \rho\vec{g}\cdot\vec{\upsilon})\mathrm{d}V \tag{4.63}$$

If u° is uniform within V, then Equation 4.63 can be recast as

$$\left(\frac{\mathrm{d}}{\mathrm{d}t}mu^{\circ}\right)_{\text{c.v.}} - \sum_i \dot{m}_i u_i^{\circ} = \frac{\mathrm{d}Q}{\mathrm{d}t} - \frac{\mathrm{d}W}{\mathrm{d}t} \tag{4.64}$$

where

$$\frac{\mathrm{d}Q}{\mathrm{d}t} = \iiint_V q'''\,\mathrm{d}V - \oiint_S \vec{q}''\cdot\vec{n}\,\mathrm{d}S \tag{4.65}$$

and

$$\frac{\mathrm{d}W}{\mathrm{d}t} = \oiint_S [(p\bar{\bar{I}} - \bar{\bar{\tau}})\cdot\vec{\upsilon}]\cdot\vec{n}\,\mathrm{d}S - \iiint_V \rho\vec{g}\cdot\vec{\upsilon}\,\mathrm{d}V \tag{4.66}$$

If the pressure-related energy term is moved to the left-hand side of Equation 4.63, we get

$$\frac{\mathrm{d}}{\mathrm{d}t}\iiint_V \rho u^{\circ}\,\mathrm{d}V + \oiint_S \rho\left(u^{\circ} + \frac{p}{\rho}\right)(\vec{\upsilon} - \vec{\upsilon}_{\mathrm{s}})\cdot\vec{n}\,\mathrm{d}S + \oiint_S p\vec{\upsilon}_{\mathrm{s}}\cdot\vec{n}\,\mathrm{d}S$$
$$= \frac{\mathrm{d}Q}{\mathrm{d}t} + \oiint_S \bar{\bar{\tau}}\cdot\vec{\upsilon}\cdot\vec{n}\,\mathrm{d}S + \iiint_V \rho\vec{g}\cdot\vec{\upsilon}\,\mathrm{d}V \tag{4.67}$$

TABLE 4.3
Various Forms of Integral Transport Equations of Mass

Deformable control volume
$\frac{d}{dt}\iiint_V \rho dV + \oiint_S \rho\vec{\upsilon}_r \cdot \vec{n}\, dS = 0$
$\iiint_V \frac{\partial\rho}{\partial t}\, dV + \oiint_S \rho\vec{\upsilon}_s \cdot \vec{n}\, dS + \oiint_S \rho\vec{\upsilon}_r \cdot \vec{n}\, dS = 0$
Nondeformable control volume ($\vec{\upsilon}_s = 0,\ \vec{\upsilon}_r = \vec{\upsilon}$)
$\frac{d}{dt}\iiint_V \rho dV + \oiint_S \rho\vec{\upsilon} \cdot \vec{n}\, dS = 0$
or
$\iiint_V \frac{\partial\rho}{\partial t}\, dV + \oiint_S \rho\vec{\upsilon} \cdot \vec{n}\, dS = 0$
Steady flow
$\oiint_S \rho\vec{\upsilon} \cdot \vec{n}\, dS = 0$
Steady uniform flow (single inlet and outlet systems)
$\rho_1\vec{\upsilon}_1 \cdot \vec{A}_1 = \rho_2\vec{\upsilon}_2 \cdot \vec{A}_2 = \dot{m}$

$\vec{\upsilon}_r$ = velocity of fluid relative to control volume surface = $\vec{\upsilon} - \vec{\upsilon}_s$
$\vec{\upsilon}_s$ = velocity of control surface boundary.
$\vec{\upsilon}$ = velocity of fluid in a fixed coordinate system.

The various forms of the integral transport equations are summarized in Tables 4.3 through 4.5.

4.5 DIFFERENTIAL CONSERVATION EQUATIONS

It is possible to derive the differential transport equations using the integral transport equation (Equation 4.51) as follows. Apply Gauss's theorem to the material volume (V_m) so that the last term on the right-hand side of Equation 4.51 is transformed to a volume integral:

$$\oiint_{S_m} \vec{J} \cdot \vec{n}\, dS = \iiint_{V_m} \nabla \cdot \vec{J}\, dV \tag{4.68}$$

as well as the last term on the left-hand side of Equation 4.51:

$$\oiint_{S_m} \rho c\vec{\upsilon} \cdot \vec{n}\, dS = \iiint_{V_m} \nabla \cdot (\rho c\vec{\upsilon})\, dV \tag{4.69}$$

TABLE 4.4
Various Forms of Integral Transport Equations of Momentum

Deformable control volume

$$\frac{d}{dt}\iiint_V \rho\vec{\upsilon}\,dV + \oiint_S \rho\vec{\upsilon}(\vec{\upsilon}_r \cdot n)\,dS = \Sigma \vec{F}$$

or

$$\iiint_V \frac{\partial}{\partial t}(\rho\vec{\upsilon})\,dV + \oiint_S \rho\vec{\upsilon}(\vec{\upsilon}_s \cdot \vec{n})\,dS + \oiint_S \rho\vec{\upsilon}(\vec{\upsilon}_r \cdot \vec{n})\,dS = \Sigma \vec{F}$$

Nondeformable control volume ($\vec{\upsilon}_s = 0,\ \vec{\upsilon}_r = \vec{\upsilon}$)

$$\frac{d}{dt}\iiint_V \rho\vec{\upsilon}\,dV + \oiint_S \rho\vec{\upsilon}(\vec{\upsilon} \cdot \vec{n})\,dS = \Sigma \vec{F}$$

or

$$\iiint_V \frac{\partial}{\partial t}(\rho\vec{\upsilon})\,dV + \oiint_S \rho\vec{\upsilon}(\vec{\upsilon} \cdot \vec{n})\,dS = \Sigma \vec{F}$$

Steady flow

$$\oiint_S \rho\vec{\upsilon}(\vec{\upsilon} \cdot \vec{n})\,dS = \Sigma \vec{F}$$

Steady uniform flow (single inlet and outlet systems)

$$\rho_2(\vec{\upsilon}_2 \cdot \vec{A}_2)\vec{\upsilon}_2 - \rho_1(\vec{\upsilon}_1 \cdot \vec{A}_1)\vec{\upsilon}_1 = \dot{m}(\vec{\upsilon}_2 - \vec{\upsilon}_1) = \Sigma \vec{F}$$

$\vec{\upsilon}_r$, $\vec{\upsilon}_s$, and $\vec{\upsilon}$ are defined in the footnotes to Table 4.3.

Substituting from Equations 4.68 and 4.69 into Equation 4.51, we get

$$\iiint_{V_m} \left[\frac{\partial(\rho c)}{\partial t} + \nabla \cdot (\rho c\vec{\upsilon})\right] dV = \iiint_{V_m} [\nabla \cdot \vec{J} + \rho\Phi]\,dV \tag{4.70}$$

For Equation 4.70 to be true for any arbitrary volume (V_m) and if ρ, c, $\vec{\upsilon}$, and Φ are continuous functions of time and space, the integrands of both sides of Equation 4.70 must be equal. Hence for a continuum, the local instantaneous transport equation can be written as

$$\frac{\partial}{\partial t}[\rho c] + \nabla \cdot [\rho c\vec{\upsilon}] = \nabla \cdot \vec{J} + \rho\Phi \tag{4.71}$$

where c, $\vec{J}$, and Φ can be specified from Equations 4.54 through 4.56.

The differential formulations can also be derived by considering an infinitesimal elementary control volume. The shape of this volume should be selected according to the system of coordinates. In the derivation here, only Cartesian coordinates are considered. In Cartesian coordinates, the control volume is an elementary cube (Figure 4.6). Also, the control volume is small enough that the variables can be considered uniform over the control surfaces that bound it.

TABLE 4.5
Various Forms of Integral Transport Equations of Energy

Deformable control volume

$$\frac{dQ}{dt}-\frac{dW^+}{dt}=\frac{d}{dt}\iiint_V \rho u^o \, dV+\oiint_S \rho\left[u^o+\frac{p}{\rho}\right]\vec{\upsilon}_r\cdot\vec{n}\,dS+\oiint_S p\,\vec{\upsilon}_s\cdot\vec{n}\,dS$$

$$\frac{dQ}{dt}-\frac{dW^+}{dt}=\iiint_V \frac{\partial}{\partial t}\rho u^o \, dV+\oiint_S \rho\left[u^o+\frac{p}{\rho}\right]\vec{\upsilon}_r\cdot\vec{n}\,dS+\oiint_S \rho\left[u^o+\frac{p}{\rho}\right]\vec{\upsilon}_s\cdot\vec{n}\,dS$$

Nondeformable control volume $\vec{\upsilon}_s=0$, $\vec{\upsilon}_r=\vec{\upsilon}$

$$\frac{dQ}{dt}-\frac{dW^+}{dt}=\frac{d}{dt}\iiint_V \rho u^o \, dV+\oiint_S \rho\left[u^o+\frac{p}{\rho}\right]\vec{\upsilon}\cdot\vec{n}\,dS$$

$$\frac{dQ}{dt}-\frac{dW^+}{dt}=\iiint_V \frac{\partial}{\partial t}\rho u^o \, dV+\oiint_S \rho\left[u^o+\frac{p}{\rho}\right]\vec{\upsilon}\cdot\vec{n}\,dS$$

Steady flow

$$\frac{dQ}{dt}-\frac{dW^+}{dt}=\oiint_S \rho\left[u^o+\frac{p}{\rho}\right]\vec{\upsilon}\cdot\vec{n}\,dS$$

Steady uniform flow (single inlet and outlet systems)[a]

$$\frac{dQ}{dt}-\left(\frac{dW}{dt}\right)_{\text{shaft}}-\left(\frac{dW}{dt}\right)_{\text{shear}}=\left[\left(\frac{\upsilon_2^2}{2}+\frac{p_2}{\rho_2}+gz_2+u_2\right)-\left(\frac{\upsilon_1^2}{2}+\frac{p_1}{\rho_1}+gz_1+u_1\right)\right]\dot{m}$$

$\vec{\upsilon}_r$, $\vec{\upsilon}_s$, $\vec{\upsilon}$ are defined in the footnotes to Table 4.3

$$\frac{dW^+}{dt}=\left(\frac{dW}{dt}\right)_{\text{shaft}}+\left(\frac{dW}{dt}\right)_{\text{shear}}+\left(\frac{dW}{dt}\right)_{\text{body force}}$$

[a] Gravity is the only body force.

4.5.1 Conservation of Mass

Denote υ_x, υ_y, and υ_z the three components of the velocity vector $\vec{\upsilon}$. The mass equation can be written as

[Rate of change of mass in control volume] = [mass flow rate into control volume] – [mass flow rate out of control volume]

$$\frac{\partial\rho}{\partial t}(dx\,dy\,dz)=\rho\upsilon_x(dy\,dz)-\left[\rho\upsilon_x+\frac{\partial}{\partial x}(\rho\upsilon_x)dx\right](dy\,dz)$$
$$+\rho\upsilon_y(dx\,dz)-\left[\rho\upsilon_y+\frac{\partial}{\partial y}(\rho\upsilon_y)dy\right](dx\,dz)$$
$$+\rho\upsilon_z(dx\,dy)-\left[\rho\upsilon_z+\frac{\partial}{\partial z}(\rho\upsilon_z)dz\right](dx\,dy)$$

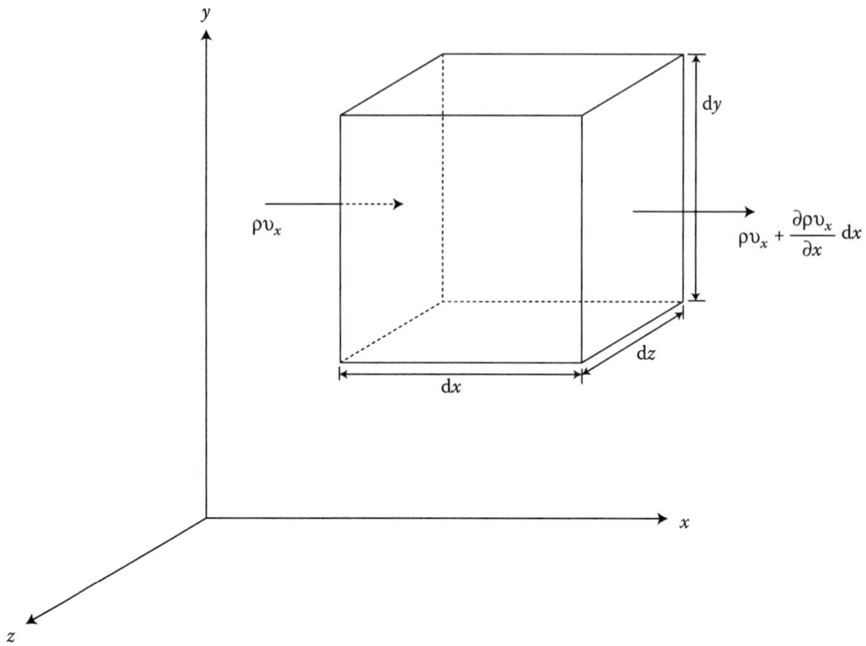

FIGURE 4.6 Components of mass flow in the x direction.

or, after simplification:

$$\frac{\partial \rho}{\partial t} = -\frac{\partial}{\partial x}(\rho \upsilon_x) - \frac{\partial}{\partial y}(\rho \upsilon_y) - \frac{\partial}{\partial z}(\rho \upsilon_z) \tag{4.72}$$

which when rearranged and written in vector-algebra notation becomes

$$\frac{\partial \rho}{\partial t} + \nabla \cdot (\rho \vec{\upsilon}) = 0 \tag{4.73}$$

This expression is the Eulerian form of the mass conservation equation.

Expanding the second term of Equation 4.73 as

$$\nabla \cdot (\rho \vec{\upsilon}) = \rho(\nabla \cdot \vec{\upsilon}) + \vec{\upsilon} \cdot \nabla \rho$$

and substituting for $\nabla \cdot (\rho \vec{\upsilon})$ in Equation 4.73, we get

$$\frac{D\rho}{Dt} + \rho(\nabla \cdot \vec{\upsilon}) = 0 \tag{4.74}$$

where

$$\frac{D\rho}{Dt} = \frac{\partial \rho}{\partial t} + \vec{\upsilon} \cdot \nabla \rho$$

Equation 4.74 is the Lagrangian form of the mass conservation equation.

When the density does not change appreciably in the domain of interest, and when small density changes do not affect appreciably the behavior of the system, we can assume that the density is constant.

For such an incompressible fluid, the continuity Equation 4.74 simplifies to

$$\nabla \cdot \vec{\upsilon} = 0; \quad \rho = \text{constant} \tag{4.75}$$

4.5.2 Conservation of Momentum

The momentum equation expresses mathematically the fact that the rate of change of momentum in the control volume equals the momentum flow rate into the control volume minus the momentum flow rate out of the control volume plus the net external force on the control volume. Both body forces and surface forces are to be included (Figures 4.7 and 4.8).

The forces that must be accounted for include, in addition to gravitational, electrical, or magnetic forces, three surface forces on each face: one normal and two tangential.

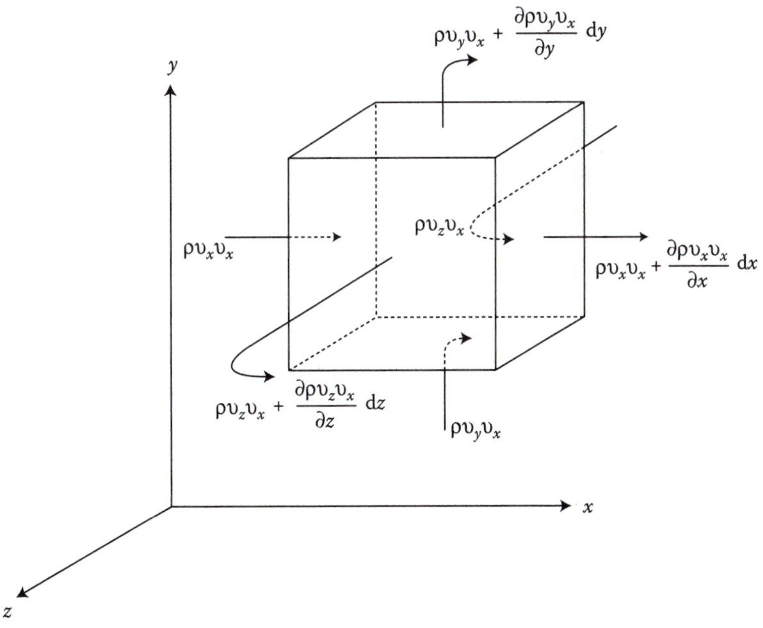

FIGURE 4.7 Components of the momentum efflux in the x direction.

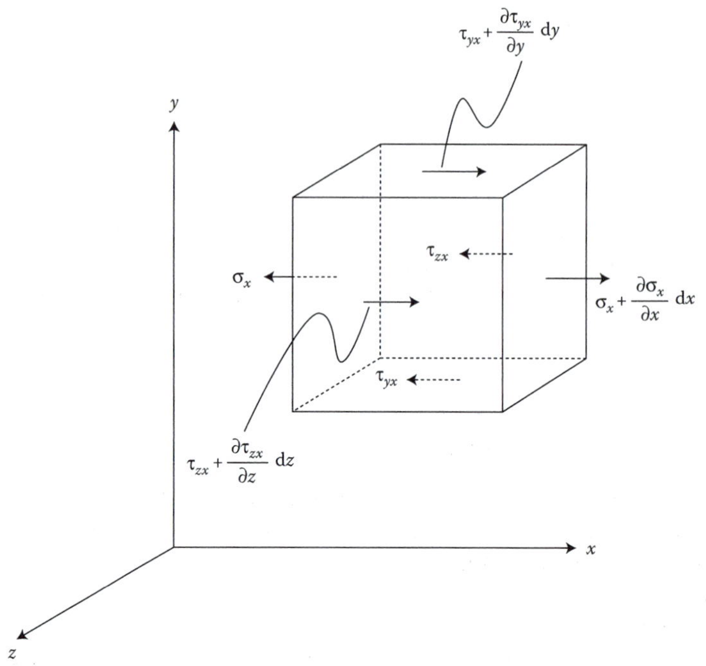

FIGURE 4.8 Components of stress tensor in the x direction. (The shear stress is positive when pointing in a positive direction on a positive surface or in a negative direction on a negative surface.)

The normal forces are caused by the pressure and internal frictional effects that act to elongate the fluid element. The shear (tangential) forces are due to internal friction, which attempts to rotate the fluid element. Thus for the momentum in the x direction:

$$\begin{aligned}
\frac{\partial \rho \upsilon_x}{\partial t}(dx\,dy\,dz) = {} & \rho \upsilon_x \upsilon_x dy\,dz - \left(\rho \upsilon_x \upsilon_x + \frac{\partial \rho \upsilon_x \upsilon_x}{\partial x} dx \right) dy\,dz \\
& + \rho \upsilon_y \upsilon_x dx\,dz - \left(\rho \upsilon_y \upsilon_x + \frac{\partial \rho \upsilon_y \upsilon_x}{\partial y} dy \right) dx\,dz \\
& + \rho \upsilon_z \upsilon_x dy\,dx - \left(\rho \upsilon_z \upsilon_x + \frac{\partial \rho \upsilon_z \upsilon_x}{\partial z} dz \right) dy\,dx \\
& + \left(\sigma_x + \frac{\partial \sigma_x}{\partial x} dx \right) dy\,dz - \sigma_x\, dy\,dz \\
& + \left(\tau_{yx} + \frac{\partial \tau_{yx}}{\partial y} dy \right) dx\,dz - \tau_{yx}\, dx\,dz \\
& + \left(\tau_{zx} + \frac{\partial \tau_{zx}}{\partial z} dz \right) dy\,dx - \tau_{zx}\, dy\,dx \\
& + \rho f_x\, dx\,dy\,dz
\end{aligned}$$

which upon simplification and rearrangement leads to

$$\frac{\partial}{\partial t}(\rho \upsilon_x) + \frac{\partial}{\partial x}(\rho \upsilon_x \upsilon_x) + \frac{\partial}{\partial y}(\rho \upsilon_x \upsilon_y) + \frac{\partial}{\partial z}(\rho \upsilon_x \upsilon_z) = \frac{\partial \sigma_x}{\partial x} + \frac{\partial \tau_{yx}}{\partial y} + \frac{\partial \tau_{zx}}{\partial z} + \rho f_x \quad (4.76)$$

The left-hand side of Equation 4.76 represents the rate of momentum change for any fluid packet as it flows with the stream in the x direction. The three-dimensional change in the momentum can be written as

$$\text{Rate of momentum change} = \frac{\partial}{\partial t}\rho \vec{\upsilon} + \nabla \cdot \rho \vec{\upsilon}\vec{\upsilon} \quad (4.77)$$

where $\vec{\upsilon}\vec{\upsilon}$ is a dyadic product of all the velocity components and is given in Cartesian coordinates by

$$\vec{\upsilon}\vec{\upsilon} = \begin{pmatrix} \upsilon_x \upsilon_x & \upsilon_x \upsilon_y & \upsilon_x \upsilon_z \\ \upsilon_y \upsilon_x & \upsilon_y \upsilon_y & \upsilon_y \upsilon_z \\ \upsilon_z \upsilon_x & \upsilon_z \upsilon_y & \upsilon_z \upsilon_z \end{pmatrix} \quad (4.78)$$

The normal stress σ_x can be expanded into a pressure component and an internal friction component so that

$$\begin{aligned} \sigma_x &= -p + \tau_{xx} \\ \sigma_y &= -p + \tau_{yy} \\ \sigma_z &= -p + \tau_{zz} \end{aligned} \quad (4.79)$$

Substituting Equation 4.79 into Equation 4.76 and utilizing Equation 4.77, it is possible to write the three-dimensional equation of the momentum in vector form as:

$$\frac{\partial}{\partial t}\rho \vec{\upsilon} + \nabla \cdot \rho \vec{\upsilon}\vec{\upsilon} = -\nabla p + \nabla \cdot \bar{\bar{\tau}} + \rho \vec{f} = -\nabla \cdot (p\bar{\bar{I}} - \bar{\bar{\tau}}) + \rho \vec{f} \quad (4.80)$$

where the tensor $\bar{\bar{\tau}}$ is given by

$$\bar{\bar{\tau}} = \begin{pmatrix} \tau_{xx} & \tau_{xy} & \tau_{xz} \\ \tau_{yx} & \tau_{yy} & \tau_{yz} \\ \tau_{zx} & \tau_{zy} & \tau_{zz} \end{pmatrix} \quad (4.81)$$

By taking moments about the axis at the position (x, y, z), it is found that for equilibrium conditions:

$$\tau_{xy} = \tau_{yx}; \quad \tau_{xz} = \tau_{zx}; \quad \tau_{yz} = \tau_{zy}$$

That is, the tensor $\bar{\bar{\tau}}$ is symmetric.

By expanding the left-hand side of Equation 4.80 and utilizing the continuity Equation 4.73, we get

$$\frac{\partial}{\partial t}\rho\vec{\upsilon} + \nabla\cdot\rho\vec{\upsilon}\vec{\upsilon} = \rho\frac{\partial}{\partial t}\vec{\upsilon}+\vec{\upsilon}\frac{\partial\rho}{\partial t} + \rho\vec{\upsilon}\cdot\nabla\vec{\upsilon}+\vec{\upsilon}\nabla\cdot\rho\vec{\upsilon}$$
$$=\rho\frac{\partial\vec{\upsilon}}{\partial t}+\rho\vec{\upsilon}\cdot\nabla\vec{\upsilon}=\rho\frac{\mathrm{D}\vec{\upsilon}}{\mathrm{D}t} \quad (4.82)$$

Equation 4.80 can therefore be written as

$$\rho\frac{\mathrm{D}\vec{\upsilon}}{\mathrm{D}t}=-\nabla p+\nabla\cdot\bar{\bar{\tau}}+\rho\vec{f} \quad (4.83)$$

This expression is the Lagrangian form of the momentum equation. It is easily seen that it is a restatement of Newton's law of motion, whereby the left-hand side is the mass times the acceleration, and the right-hand side is the sum of the forces acting on that mass. To solve the momentum equation and obtain the velocity field $\vec{\upsilon}(\vec{r},t)$, the density ρ and forces ∇p and $\vec{f}$ must be specified, and the internal stress force $\nabla\cdot\bar{\bar{\tau}}$ must be stated in terms of the velocity field, that is, the velocity or velocity gradients and fluid properties. In other words, a constitutive relation for $\bar{\bar{\tau}}$ is needed for mathematic closure. In practice, the mass Equation 4.74 and a state relation of pressure and density are solved together with the momentum equation, thus providing the means to specify $\rho(\vec{r},t)$ and $\nabla p(\vec{r},t)$. This subject is discussed in Chapter 9.

For the reason that all fluids are isotropic with respect to stress–strain behavior, the stresses in the three directions must be related to material properties that are independent of coordinates. Thus if the fluids are assumed to follow the Newtonian laws of viscosity, the friction terms are given by [1]:

$$\tau_{ii}=2\mu\left(\frac{\partial\upsilon_i}{\partial x_i}\right)-\left(\frac{2}{3}\mu-\mu'\right)(\nabla\cdot\vec{\upsilon}) \quad (4.84)$$

$$\tau_{ij}=\tau_{ji}=\mu\left(\frac{\partial\upsilon_i}{\partial x_j}+\frac{\partial\upsilon_j}{\partial x_i}\right) \quad (4.85)$$

where μ = ordinary "dynamic viscosity"; μ' = "bulk viscosity."

For dense gases and fluids, μ' is negligible. It is identically zero for low-density monoatomic gases. It is only important for acoustic or shock-wave problems in gases where it produces viscous decay of the high-frequency waves. Thus, in general, we can take $\mu' = 0$ so that Equation 4.84 becomes

$$\tau_{ii}=2\mu\frac{\partial\upsilon_i}{\partial x_i}-\frac{2}{3}\mu\nabla\cdot\vec{\upsilon} \quad (4.86)$$

When Equations 4.79, 4.85, and 4.86 are substituted into Equation 4.76, we get

$$\frac{\partial}{\partial t}(\rho\upsilon_x)+\frac{\partial}{\partial x}(\rho\upsilon_x\upsilon_x)+\frac{\partial}{\partial y}(\rho\upsilon_x\upsilon_y)+\frac{\partial}{\partial z}(\rho\upsilon_x\upsilon_z)$$
$$=-\frac{\partial p}{\partial x}+\frac{\partial}{\partial x}\left[2\mu\frac{\partial \upsilon_x}{\partial x}-\frac{2}{3}\mu(\nabla\cdot\vec{\upsilon})\right]$$
$$+\frac{\partial}{\partial y}\left[\mu\left(\frac{\partial \upsilon_x}{\partial y}+\frac{\partial \upsilon_y}{\partial x}\right)\right]+\frac{\partial}{\partial z}\left[\mu\left(\frac{\partial \upsilon_x}{\partial z}+\frac{\partial \upsilon_z}{\partial x}\right)\right]+\rho f_x \tag{4.87}$$

This expression is the *Navier–Stokes equation* for the momentum balance in the x direction. The momentum equation in the y and z directions can be obtained by permutation of x, y, and z in Equation 4.87. The resultant equations can be written in vector form as

$$\frac{\partial}{\partial t}(\rho\vec{\upsilon})+\nabla\cdot\rho\vec{\upsilon}\vec{\upsilon}=-\nabla p-\nabla\cdot[\mu\nabla\cdot\vec{\upsilon}]+\nabla\left[\frac{4}{3}\mu\nabla\cdot\vec{\upsilon}\right]+p\vec{f} \tag{4.88}$$

The momentum equations, the continuity equation, the equations specifying the pressure–density and viscosity–density relation, and the initial and boundary conditions are sufficient to define the velocity, density, and pressure distributions in the medium.

For an incompressible fluid ($\nabla\cdot\vec{\upsilon}=0$) with a constant viscosity, the Navier–Stokes equation in the x direction simplifies to

$$\rho\frac{\partial \upsilon_x}{\partial t}+\rho\nabla\cdot\upsilon_x\vec{\upsilon}=-\frac{\partial p}{\partial x}+\mu\nabla^2\upsilon_x+\rho f_x;\quad \rho=\text{constant},\ \mu=\text{constant} \tag{4.89}$$

where

$$\nabla^2\upsilon_x=\frac{\partial^2\upsilon_x}{\partial x^2}+\frac{\partial^2\upsilon_x}{\partial y^2}+\frac{\partial^2\upsilon_x}{\partial z^2}$$

Note that

$$\nabla\cdot\rho\vec{\upsilon}\vec{\upsilon}=\rho\vec{\upsilon}\cdot\nabla\vec{\upsilon}+\vec{\upsilon}(\nabla\cdot\rho\vec{\upsilon})$$

so that for a constant density fluid:

$$\nabla\cdot\rho\vec{\upsilon}\vec{\upsilon}=\rho\vec{\upsilon}\cdot\nabla\vec{\upsilon}$$

However,

$$\rho\nabla\cdot\upsilon_x\vec{\upsilon}=\rho\vec{\upsilon}\cdot\nabla\upsilon_x+\rho\upsilon_x(\nabla\cdot\vec{\upsilon})$$

so that for a constant-density fluid:

$$\rho\nabla\cdot\upsilon_x\vec{\upsilon} = \rho\vec{\upsilon}\cdot\nabla\upsilon_x$$

In vector form, the momentum balance equation for a fluid with constant viscosity and density is then given by simplification of Equation 4.88:

$$\rho\frac{\partial\vec{\upsilon}}{\partial t} + \rho\vec{\upsilon}\cdot\nabla\vec{\upsilon} = -\nabla p + \mu\nabla^2\vec{\upsilon} + \rho\vec{f}; \quad \rho = \text{constant} \quad \mu = \text{constant} \tag{4.90a}$$

or

$$\rho\frac{D\vec{\upsilon}}{Dt} = -\nabla p + \mu\nabla^2\vec{\upsilon} + \rho\vec{f}; \quad \rho = \text{constant} \quad \mu = \text{constant} \tag{4.90b}$$

For flow that can be considered inviscid (i.e., of negligible viscosity effects), the second term on the right-hand side in Equations 4.90a and 4.90b disappears. Hence, Equation 4.90b becomes

$$\rho\frac{D\vec{\upsilon}}{Dt} = -\nabla p + \rho\vec{f}; \quad \text{inviscid flow} \tag{4.91}$$

Equation 4.91 is the equation derived by Euler in 1755 for inviscid flow.

Note that if the only body force present is the gravity,

$$\vec{f} = \vec{g} \tag{4.92}$$

Example 4.5: Shear Stress Distribution in One-Dimensional, Newtonian Fluid Flow

PROBLEM

Consider uniaxial flow of an incompressible Newtonian fluid in a duct of rectangular cross section (Figure 4.9). The flow is in the axial direction and is given by

$$\vec{\upsilon} = \upsilon_x\vec{i} = \upsilon_{max}\left[1-\left(\frac{y}{L_y}\right)^2\right]\left[1-\left(\frac{z}{L_z}\right)^2\right]\vec{i}$$

Evaluate the maximum shear stress in the fluid.

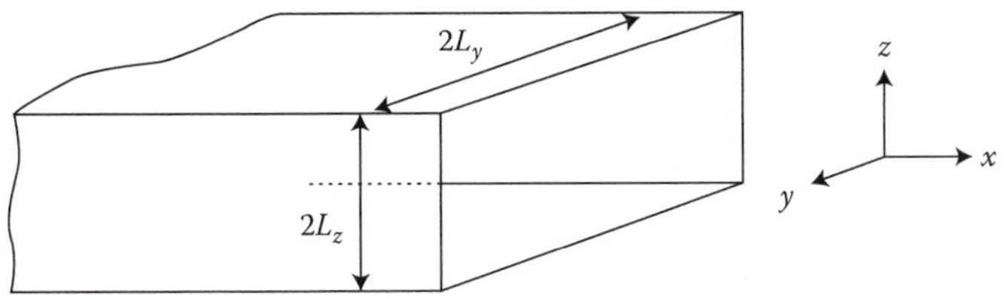

FIGURE 4.9 Flow channel for Example 4.5.

SOLUTION

From Equation 4.85, the shear stress components are given by

$$\tau_{xy} = \mu\left(\frac{\partial \upsilon_x}{\partial y} + 0\right) = -\mu\, \upsilon_{max}\left[1 - \left(\frac{z}{L_z}\right)^2\right]\left[\frac{2y}{L_y^2}\right]$$

$$\tau_{xz} = \mu\left(\frac{\partial \upsilon_x}{\partial z} + 0\right) = -\mu\, \upsilon_{max}\left[1 - \left(\frac{y}{L_y}\right)^2\right]\left[\frac{2z}{L_z^2}\right]$$

$$\tau_{yz} = 0$$

From Equation 4.84:

$$\tau_{xx} = 0 \quad \text{because} \quad \frac{\partial \upsilon_x}{\partial x} = 0 \ \text{and} \ \nabla \cdot \vec{\upsilon} = 0$$

Therefore, the maximum values of the shear components are

$$(\tau_{xy})_{max} = \pm\frac{2\mu\, \upsilon_{max}}{L_y} \quad \text{at } y = \mp L_y \text{ and } z = 0$$

$$(\tau_{xz})_{max} = \pm\frac{2\mu\, \upsilon_{max}}{L_z} \quad \text{at } y = 0 \text{ and } z = \mp L_z$$

4.5.3 Conservation of Energy

4.5.3.1 Stagnation Internal Energy Equation

The energy equation expresses the fact that the rate of change of total internal energy in an infinitesimal volume must be equal to the rate at which internal energy is brought into the volume by the mass inflow, minus that removed by mass outflow, plus the heat transported diffusively or generated, minus the work performed by the medium in the volume and the work needed to put the flow through the volume. Again the stagnation energy (u^o) refers to the internal energy and the kinetic energy. (The potential energy is included in the work term.)

Now for the volume dx dy dz:

$$
\begin{aligned}
\left(\frac{\partial}{\partial t} pu^o\right)\mathrm{d}x\,\mathrm{d}y\,\mathrm{d}z = &-\left(\frac{\partial \rho \upsilon_x u^o}{\partial x}\,\mathrm{d}x\right)\mathrm{d}y\,\mathrm{d}z - \left(\frac{\partial \rho \upsilon_y u^o}{\partial y}\,\mathrm{d}y\right)\mathrm{d}x\,\mathrm{d}z \\
&-\left(\frac{\partial \rho \upsilon_z u^o}{\partial z}\,\mathrm{d}z\right)\mathrm{d}x\,\mathrm{d}y - \left(\frac{\partial q_x''}{\partial x}\,\mathrm{d}x\right)\mathrm{d}y\,\mathrm{d}z \\
&-\left(\frac{\partial q_y''}{\partial y}\,\mathrm{d}y\right)\mathrm{d}x\,\mathrm{d}z - \left(\frac{\partial q_z''}{\partial z}\,\mathrm{d}z\right)\mathrm{d}x\,\mathrm{d}y \\
&+ (q''')\,\mathrm{d}x\,\mathrm{d}y\,\mathrm{d}z + \left(\frac{\partial}{\partial x}(\sigma_x \upsilon_x + \tau_{xy}\upsilon_y + \tau_{xz}\upsilon_z)\,\mathrm{d}x\right)\mathrm{d}y\,\mathrm{d}z \\
&+ \left(\frac{\partial}{\partial y}(\sigma_y \upsilon_y + \tau_{yx}\upsilon_x + \tau_{yz}\upsilon_z)\,\mathrm{d}y\right)\mathrm{d}x\,\mathrm{d}z \\
&+ \left(\frac{\partial}{\partial z}(\sigma_z \upsilon_z + \tau_{zx}\upsilon_x + \tau_{zy}\upsilon_y)\,\mathrm{d}z\right)\mathrm{d}x\,\mathrm{d}y \\
&+ (\upsilon_x \rho f_x + \upsilon_y \rho f_y + \upsilon_z \rho f_z)\,\mathrm{d}x\,\mathrm{d}y\,\mathrm{d}z
\end{aligned}
\tag{4.93}
$$

Dividing both sides of Equation 4.93 by dx dy dz and using the vector notation, we get:

$$\frac{\partial}{\partial t}\rho u^o = -\nabla\cdot\rho u^o\vec{\upsilon} - \nabla\cdot\vec{q}'' + q''' - \nabla\cdot p\vec{\upsilon} + \nabla\cdot(\bar{\bar{\tau}}\cdot\vec{\upsilon}) + \vec{\upsilon}\cdot\rho\vec{f} \tag{4.94}$$

The first term on the left-hand side is the local rate of change of the stagnation internal energy. The first term on the right represents the net change in the internal energy per unit time due to convection; the second term is the net heat transport rate by conduction and radiation (if present); the third term is the internal heat-generation rate. The fourth, fifth, and sixth terms are the work done on the fluid by the pressure, viscous forces, and body forces, respectively, per unit time.

Grouping the first term on the right side and that on the left side of Equation 4.94 and expanding both terms, we get

$$
\begin{aligned}
\frac{\partial}{\partial t}\rho u^o + \nabla\cdot\rho u^o\vec{\upsilon} &= \rho\frac{\partial u^o}{\partial t} + u^o\frac{\partial \rho}{\partial t} + \rho\vec{\upsilon}\cdot\nabla u^o + u^o\nabla\cdot\rho\vec{\upsilon} \\
&= \rho\frac{\partial u^o}{\partial t} + \rho\vec{\upsilon}\cdot\nabla u^o = \rho\frac{\mathrm{D}u^o}{\mathrm{D}t}
\end{aligned}
\tag{4.95}
$$

Now, Equation 4.94 may be rewritten as

$$\rho\frac{\mathrm{D}u^o}{\mathrm{D}t} = -\nabla\cdot\vec{q}'' + q''' - \nabla\cdot p\vec{\upsilon} + \nabla\cdot(\bar{\bar{\tau}}\cdot\vec{\upsilon}) + \vec{\upsilon}\cdot\rho\vec{f} \tag{4.96a}$$

Example 4.6: Energy Equation for One-Dimensional Newtonian Fluid Flow

PROBLEM

For the velocity profile of Example 4.5, determine the rate of the volumetric work production in the coolant and compare it to the power density in the PWR core at steady state.

SOLUTION

Because

$$\vec{\upsilon} = \upsilon_x \vec{i} = \upsilon_{\max}\left[1-\left(\frac{y}{L_y}\right)^2\right]\left[1-\left(\frac{z}{L_z}\right)^2\right]\vec{i}$$

the values of the three work terms in Equation 4.96a can be determined as follows:

$$\nabla \cdot p\vec{\upsilon} = \frac{\partial}{\partial x} p\upsilon_x = \upsilon_x \frac{\partial p}{\partial x}$$

$$\nabla \cdot (\bar{\bar{\tau}} \cdot \vec{\upsilon}) = \nabla \cdot \begin{pmatrix} 0 & \tau_{xy} & \tau_{xz} \\ \tau_{xy} & 0 & 0 \\ \tau_{xz} & 0 & 0 \end{pmatrix} \cdot \begin{pmatrix} \upsilon_x \\ 0 \\ 0 \end{pmatrix} = \frac{\partial}{\partial x}(0) + \frac{\partial}{\partial y}(\tau_{xy}\upsilon_x) + \frac{\partial}{\partial z}(\tau_{xz}\upsilon_x)$$

$$\vec{\upsilon} \cdot \rho \vec{f} = \upsilon_x \rho g_x$$

At steady state, Equation 4.96a specified to our system leads to

$$\rho\upsilon_x \frac{\partial u^\circ}{\partial x} = -\nabla \cdot \vec{q}'' + q''' - \upsilon_x \frac{\partial p}{\partial x} + \frac{\partial}{\partial y}(\tau_{xy}\upsilon_x) + \frac{\partial}{\partial z}(\tau_{xz}\upsilon_x) + \upsilon_x \rho g_x$$

The shear stresses τ_{xy} and τ_{xz} have been evaluated in Example 4.5. Applying them here and differentiating, we get

$$\begin{aligned}
\rho\upsilon_x \frac{\partial u^\circ}{\partial x} &= -\nabla \cdot \vec{q}'' + q''' - \upsilon_x \frac{\partial p}{\partial x} - \mu\,\upsilon_{\max}^2 \left[1-\left(\frac{z}{L_z}\right)^2\right]^2 \frac{\partial}{\partial y}\left[\frac{2y}{L_y^2}\left\{1-\left(\frac{y}{L_y^2}\right)^2\right\}\right] \\
&\quad - \mu\,\upsilon_{\max}^2 \left[1-\left(\frac{y}{L_y}\right)^2\right]^2 \frac{\partial}{\partial z}\left[\frac{2z}{L_z^2}\left\{1-\left(\frac{z}{L_z^2}\right)^2\right\}\right] + \upsilon_x \rho g_x \\
&= -\nabla \cdot \vec{q}'' + q''' - \upsilon_x \frac{\partial p}{\partial x} \\
&\quad - \mu\upsilon_{\max}^2 \left\{\left[1-\left(\frac{z}{L_z}\right)^2\right]^2\left[\frac{2}{L_y^2} - \frac{6y^2}{L_y^4}\right] + \left[1-\left(\frac{y}{L_y}\right)^2\right]^2\left[\frac{2}{L_z^2} - \frac{6z^2}{L_z^4}\right]\right\} + \upsilon_x \rho g_x
\end{aligned} \tag{4.96b}$$

Now consider two points within the channel:

1. $y = L_y$ and $z > 0$
2. $y = z = 0$

For $y = L_y$ and $z > 0$, $\upsilon_x = 0$ and Equation 4.96b becomes

$$0 = -\nabla \cdot \vec{q}'' + q''' - 0 - \mu\upsilon_{max}^2 \left\{ \left[1 - \left(\frac{z}{L_z} \right)^2 \right] \left[\frac{2}{L_y^2} - \frac{6}{L_y^2} \right] - 0 \right\} + 0$$

For $y = z = 0$, $\upsilon_x = \upsilon_{max}$, and Equation 4.96b becomes

$$\rho\upsilon_{max} \frac{\partial u^\circ}{\partial x} = -\nabla \cdot q'' + q''' - \upsilon_{max} \frac{\partial p}{\partial x} - \mu\upsilon_{max}^2 \left\{ \frac{2}{L_y^2} + \frac{2}{L_z^2} \right\} + \upsilon_{max}\rho g_x$$

Hence, there is internal heat generation and consequent molecular heat diffusion due to the shear work term as well as to the other work terms. However, in energy conversion systems this work term is small. For example if $\upsilon_{max} = 10$ m/s, $\mu = 100\ \mu\text{Pa} \cdot \text{s}$ (water at 300°C and 15 MPa), and $L_y = L_z = 0.01$ m (representative of subchannels in PWRs), the largest value of the work term is

$$\frac{4\mu\upsilon_{max}^2}{L_y^2} = \frac{4 \times 100 \times 10^{-6} \times (10)^2}{(0.01)^2} \frac{(\text{Pa} \cdot \text{s})(\text{m}^2/\text{s}^2)}{(\text{m}^2)} = 400\,\text{Pa/s} = 400\,\text{W/m}^3$$

This amount is small relative to the power density in a PWR of about 100 MW/m³ (see Example 3.2). It should be mentioned that the numbers given here are for illustration only. In a real PWR at the given velocity, the shape of the velocity profile in a subchannel does not correspond to the parabolic shape.

4.5.3.2 Stagnation Enthalpy Equation

The explicit term $\nabla \cdot p\vec{\upsilon}$ can be eliminated if the energy equation is cast in terms of the rate of change of enthalpy instead of internal energy. If we expand this term as:

$$\nabla \cdot p\vec{\upsilon} = \nabla \cdot \left(\frac{p}{\rho} \right) \rho\vec{\upsilon} = \left(\frac{p}{\rho} \right) \nabla \cdot \rho\vec{\upsilon} + \rho\vec{\upsilon} \cdot \nabla \left(\frac{p}{\rho} \right)$$

and apply the continuity equation, Equation 4.73:

$$\frac{\partial \rho}{\partial t} + \nabla \cdot \rho\vec{\upsilon} = 0$$

we get

$$\nabla \cdot p\vec{\upsilon} = -\left(\frac{p}{\rho} \right) \frac{\partial \rho}{\partial t} + \rho\vec{\upsilon} \cdot \nabla \left(\frac{p}{\rho} \right) \tag{4.97}$$

Considering that

$$\frac{\partial((p/\rho)\rho)}{\partial t} = \left(\frac{p}{\rho}\right)\frac{\partial \rho}{\partial t} + \rho\frac{\partial(p/\rho)}{\partial t} \tag{4.98}$$

we get

$$\left(\frac{p}{\rho}\right)\frac{\partial \rho}{\partial t} = \frac{\partial p}{\partial t} - \rho\frac{\partial}{\partial t}\left(\frac{p}{\rho}\right) \tag{4.99}$$

Substituting from Equation 4.99 into Equation 4.97, we get

$$\nabla \cdot p\vec{\upsilon} = -\left(\frac{\partial p}{\partial t}\right) + \rho\frac{\partial}{\partial t}\left(\frac{p}{\rho}\right) + \rho\vec{\upsilon}\cdot\nabla\left(\frac{p}{\rho}\right) = -\frac{\partial p}{\partial t} + \rho\frac{\mathrm{D}}{\mathrm{D}t}\left(\frac{p}{\rho}\right) \tag{4.100}$$

Substituting for $\nabla \cdot p\vec{\upsilon}$ in Equation 4.96a and rearranging, we get

$$\rho\frac{\mathrm{D}h^{\mathrm{o}}}{\mathrm{D}t} = -\nabla\cdot\vec{q}'' + q''' + \frac{\partial p}{\partial t} + \nabla\cdot(\bar{\bar{\tau}}\cdot\vec{\upsilon}) + \vec{\upsilon}\cdot\rho\vec{f} \tag{4.101}$$

where $h^{\mathrm{o}} \equiv u^{\mathrm{o}} + \dfrac{p}{\rho}$.

4.5.3.3 Kinetic Energy Equation

It is possible to relate the change in kinetic energy to the work done on the fluid by multiplying both sides of Equation 4.83 by $\vec{\upsilon}$ to obtain the mechanical energy equation:

$$\rho\vec{\upsilon}\cdot\frac{\mathrm{D}\vec{\upsilon}}{\mathrm{D}t} = -\vec{\upsilon}\cdot\nabla p + \vec{\upsilon}\cdot(\nabla\cdot\bar{\bar{\tau}}) + \vec{\upsilon}\cdot\rho\vec{f}$$

or

$$\rho\frac{\mathrm{D}}{\mathrm{D}t}\left(\frac{1}{2}\upsilon^2\right) = -\vec{\upsilon}\cdot\nabla p + \vec{\upsilon}\cdot(\nabla\cdot\bar{\bar{\tau}}) + \vec{\upsilon}\cdot\rho\vec{f} \tag{4.102}$$

However, for a symmetrical stress tensor:

$$(\bar{\bar{\tau}}:\nabla\vec{\upsilon}) \equiv (\bar{\bar{\tau}}\cdot\nabla)\cdot\vec{\upsilon} = \nabla\cdot(\bar{\bar{\tau}}\cdot\vec{\upsilon}) - \vec{\upsilon}\cdot(\nabla\cdot\bar{\bar{\tau}}) \tag{4.103}$$

Therefore, Equation 4.102 may be written as

$$\rho\frac{\mathrm{D}}{\mathrm{D}t}\left(\frac{1}{2}\upsilon^2\right) = -\vec{\upsilon}\cdot\nabla p + \nabla\cdot(\bar{\bar{\tau}}\cdot\vec{\upsilon}) - (\bar{\bar{\tau}}:\nabla\vec{\upsilon}) + \vec{\upsilon}\cdot\rho\vec{f}$$

which can be rearranged as

$$\rho \frac{D}{Dt}\left(\frac{1}{2}\upsilon^2\right) = \nabla \cdot [(\bar{\bar{\tau}} \cdot p\bar{\bar{I}}) \cdot \vec{\upsilon} + p\nabla \cdot \vec{\upsilon} + \vec{\upsilon} \cdot \rho \vec{f} - (\bar{\bar{\tau}} : \nabla \vec{\upsilon}) \tag{4.104}$$

where $\bar{\bar{I}}$ is the unity tensor.

4.5.3.4 Thermodynamic Energy Equations

By subtracting Equation 4.104 from Equation 4.101, we get the energy equation in terms of the enthalpy (h)—not the stagnation enthalpy (h^o):

$$\rho \frac{Dh}{Dt} = -\nabla \cdot \vec{q}'' + q''' + \frac{Dp}{Dt} + (\bar{\bar{\tau}} : \nabla \vec{\upsilon}) \tag{4.105}$$

By a manipulation similar to that of Equation 4.95, it is seen that Equation 4.105 may also be written in the form:

$$\rho \frac{Dh}{Dt} = \frac{\partial}{\partial t}(\rho h) + \nabla \cdot (\rho h \vec{\upsilon}) = -\nabla \cdot \vec{q}'' + q''' + \frac{Dp}{Dt} + (\bar{\bar{\tau}} : \nabla \vec{\upsilon}) \tag{4.106}$$

If we define,

$$\Phi \equiv (\bar{\bar{\tau}} : \nabla \vec{\upsilon}) = \text{dissipation function} \tag{4.107}$$

then Equation 4.106 may be written in a form often encountered in thermal analysis textbooks:

$$\rho \frac{Dh}{Dt} = \frac{\partial}{\partial t}(\rho h) + \nabla \cdot (\rho h \vec{\upsilon}) = -\nabla \cdot \vec{q}'' + q''' + \frac{Dp}{Dt} + \Phi \tag{4.108}$$

By applying the definition of the enthalpy,

$$h \equiv u + \frac{p}{\rho}$$

and rearranging Equation 4.108, we get a similar energy equation for the internal energy u:

$$\frac{\partial}{\partial t}(\rho u) + \nabla \cdot (\rho u \vec{\upsilon}) = -\nabla \cdot \vec{q}'' + q''' - p\nabla \cdot \vec{\upsilon} + \Phi \tag{4.109}$$

or

$$\rho \frac{Du}{Dt} = -\nabla \cdot \vec{q}'' + q''' - p\nabla \cdot \vec{\upsilon} + \Phi \tag{4.110}$$

4.5.3.5 Special Forms

To solve Equations 4.96a, 4.101, 4.106, or 4.109, the viscosity term must be explicitly written in terms of the velocity field and fluid properties. In most cases, the rate of heat addition due to viscous effects can be neglected compared to the other terms, and the energy Equation 4.105 may be given by

$$\rho \frac{\mathrm{D}h}{\mathrm{D}t} = -\nabla \cdot \vec{q}'' + q''' + \frac{\mathrm{D}p}{\mathrm{D}t}; \quad \text{inviscid flow} \tag{4.111}$$

Furthermore, for incompressible as well as inviscid flow, the energy equation can be simplified as Φ and $\nabla \cdot \vec{\upsilon}$ equal 0. Thus, Equation 4.110 becomes

$$\rho \frac{\mathrm{D}h}{\mathrm{D}t} = q''' - \nabla \cdot \vec{q}''; \quad \text{incompressible and inviscid flow} \tag{4.112}$$

Note also that the heat flux $\vec{q}''$ may be due to both conduction and radiation, so that in general:

$$\vec{q}'' = \vec{q}_c'' + \vec{q}_r'' \tag{4.113}$$

where $\vec{q}_c''$ and $\vec{q}_r''$ denote the conduction and radiation heat fluxes, respectively. Within dense materials in most engineering analysis $\vec{q}_r''$ can be neglected, although the transmissivity of radiation within a given medium is dependent on the wavelength of the radiation.

Fourier's law of conduction gives the molecular heat flux as linearly proportional to the gradient of temperature.

$$\vec{q}_c'' = -k\nabla T \tag{4.114}$$

In heat transfer problems, it is often convenient to cast Equation 4.108 in terms of temperature rather than enthalpy. For a pure substance, the enthalpy is a function of only two thermodynamic properties, for example, temperature and pressure:

$$\mathrm{d}h = \left.\frac{\partial h}{\partial T}\right|_p \mathrm{d}T + \left.\frac{\partial h}{\partial p}\right|_T \mathrm{d}p = c_p\, \mathrm{d}T + \left.\frac{\partial h}{\partial p}\right|_T \mathrm{d}p \tag{4.115}$$

Also, for a pure substance, the first law of thermodynamics as applied to a unit mass can be written as

$$\mathrm{d}h = T\,\mathrm{d}s + \mathrm{d}p/\rho$$

so that

$$\left.\frac{\partial h}{\partial p}\right|_T = T \left.\frac{\partial s}{\partial p}\right|_T + \frac{1}{\rho} \tag{4.116}$$

Defining β as the volumetric thermal expansion coefficient, we have

$$\beta \equiv -\frac{1}{\rho}\frac{\partial \rho}{\partial T}\bigg|_p \tag{4.117}$$

From Maxwell's thermodynamic relations, we know that

$$\frac{\partial s}{\partial p}\bigg|_T = -\frac{\partial(1/\rho)}{\partial T}\bigg|_p = \frac{1}{\rho^2}\frac{\partial \rho}{\partial T}\bigg|_p = -\frac{\beta}{\rho} \tag{4.118}$$

Substituting from Equation 4.118 into Equation 4.116, we get

$$\frac{\partial h}{\partial p}\bigg|_T = -\frac{\beta T}{\rho} + \frac{1}{\rho} \tag{4.119}$$

Now we substitute Equation 4.119 into Equation 4.115 to get

$$dh = c_p dT + (1 - \beta T)\frac{dp}{\rho} \tag{4.120a}$$

Note that for an ideal gas, Equation 4.120a reduces to

$$dh = c_p dT \tag{4.120b}$$

Applying Equation 4.120a to a unit mass in the Lagrangian formulation, we get

$$\rho\frac{Dh}{Dt} = \rho c_p \frac{DT}{Dt} + (1 - \beta T)\frac{Dp}{Dt} \tag{4.121}$$

If we introduce Equations 4.113, 4.114, and 4.121 into Equation 4.108, the energy equation takes a form often used for solving heat transfer problems:

$$\rho c_p \frac{DT}{Dt} = -\nabla \cdot \vec{q}'' + q''' + \beta T \frac{Dp}{Dt} + \Phi \tag{4.122a}$$

or

$$\rho c_p \frac{DT}{Dt} = \nabla \cdot k \nabla T - \nabla \cdot \vec{q}_r'' + q''' + \beta T \frac{Dp}{Dt} + \Phi \tag{4.122b}$$

The energy term due to thermal expansion of fluids is often small compared to the other terms. An exception is near the front of a shock wave, where Dp/Dt can be of a high magnitude.

For stagnant fluids or solids, neglecting compressibility and thermal expansion (ρ is constant) and the dissipation function, the energy equation is given by

$$\rho c_p \frac{\partial T}{\partial t} = \nabla \cdot k \nabla T + q''' \tag{4.123}$$

Note also that, for incompressible materials, $c_p = c_v$. This is the reason behind neglecting the difference between c_p and c_v when solving heat transfer problems for liquids and solids but not for gases.

4.5.4 Summary of Equations

Useful differential forms of the relevant conservation equations are compiled in Tables 4.6 through 4.12.

4.6 TURBULENT FLOW

In turbulent flow, it can be assumed that the moving medium properties exhibit fluctuations in time, which are the result of random movement of eddies (or pockets) of fluid. The local fluid motion can still be described instantaneously by the equations

TABLE 4.6
Differential Fluid Transport Equations in Vector Form[a]

Generalized form

Unsteady term + convectional term = diffusion term + source term (Equation 4.71)

$$\frac{\partial}{\partial t}[\rho c] + \nabla \cdot \rho c \vec{\upsilon} = \nabla \cdot \vec{J} + \rho \Phi$$

Equation	c	$\vec{J}$	Φ[a]	Equation
Continuity	1	0	0	4.73
Linear momentum	$\vec{\upsilon}$	$\bar{\bar{\tau}} - p\bar{\bar{I}}$	$\vec{g}$	4.80
Stagnation internal energy	u^o	$(\bar{\bar{\tau}} - p\bar{\bar{I}}) \cdot \vec{\upsilon} - \vec{q}''$	$\dfrac{q'''}{\rho} + \vec{\upsilon} \cdot \vec{g}$	4.96a
Energy				
Internal energy	u	$-\vec{q}''$	$\dfrac{1}{\rho}(q''' - p\nabla \cdot \vec{\upsilon} + \Phi)$	4.109
Enthalpy	h	$-\vec{q}''$	$\dfrac{1}{\rho}\left(q''' + \dfrac{Dp}{Dt} + \Phi\right)$	4.108
Kinetic energy	$\dfrac{1}{2}\upsilon^2$	$(\bar{\bar{\tau}} - p\bar{\bar{I}}) \cdot \vec{\upsilon}$	$\dfrac{1}{\rho}(\vec{\upsilon} \cdot \rho \vec{g} - \Phi + p\nabla \cdot \vec{\upsilon})$	4.104

Note: $\frac{\partial}{\partial t}[\rho c] + \nabla \cdot \rho c \vec{\upsilon} = \rho \frac{Dc}{Dt}$

[a] Assuming gravity is the only body force.

TABLE 4.7
Differential Equation of Continuity

Vector form

$$\frac{\partial \rho}{\partial t} + \nabla \cdot (\rho \vec{\upsilon}) = 0 \quad (4.73)$$

or

$$\frac{D\rho}{Dt} + \rho(\nabla \cdot \vec{\upsilon}) = 0 \quad (4.74)$$

Cartesian

$$\frac{\partial \rho}{\partial t} + \frac{\partial}{\partial x}(\rho \upsilon_x) + \frac{\partial}{\partial y}(\rho \upsilon_y) + \frac{\partial}{\partial z}(\rho \upsilon_z) = 0$$

Cylindrical

$$\frac{\partial \rho}{\partial t} + \frac{1}{r}\frac{\partial}{\partial r}(\rho r \upsilon_r) + \frac{1}{r}\frac{\partial}{\partial \theta}(\rho \upsilon_\theta) + \frac{\partial}{\partial z}(\rho \upsilon_z) = 0$$

Spherical

$$\frac{\partial \rho}{\partial t} + \frac{1}{r^2}\frac{\partial}{\partial r}(\rho r^2 \upsilon_r) + \frac{1}{r \sin\theta}\frac{\partial}{\partial \theta}(\rho \upsilon_\theta \sin\theta) + \frac{1}{r \sin\theta}\frac{\partial}{\partial \phi}(\rho \upsilon_\varphi) = 0$$

of Section 4.5; but in fact, the practical characteristics of the moving fluid require time-averaging manipulation of these equations. It is also possible to space-average the local equations to obtain volume-average or area-average properties. Such space averaging is particularly useful for the two-phase flow condition, described in Chapter 5. Furthermore, applications to reactor core analysis by the porous media approach (see Volume II) draw on such space averaging.

TABLE 4.8
Differential Equation of Continuity for Incompressible Materials

Vector form

$$\nabla \cdot \vec{\upsilon} = 0 \quad (4.75)$$

Cartesian

$$\frac{\partial}{\partial x}(\upsilon_x) + \frac{\partial}{\partial y}(\upsilon_y) + \frac{\partial}{\partial z}(\upsilon_z) = 0$$

Cylindrical

$$\frac{1}{r}\frac{\partial}{\partial r}(r\upsilon_r) + \frac{1}{r}\frac{\partial(\upsilon_\theta)}{\partial \theta} + \frac{\partial(\upsilon_z)}{\partial z} = 0$$

Spherical

$$\frac{1}{r^2}\frac{\partial}{\partial r}(r^2 \upsilon_r) + \frac{1}{r \sin\theta}\frac{\partial}{\partial \theta}(\upsilon_\theta \sin\theta) + \frac{1}{r \sin\theta}\frac{\partial}{\partial \phi}(\upsilon_\varphi) = 0$$

TABLE 4.9
Differential Equation of Momentum[a]

Vector form

$$\rho\frac{D\vec{\upsilon}}{Dr}=-\nabla p+\nabla\cdot\bar{\bar{\tau}}+\rho\vec{f} \tag{4.83}$$

$$\rho\frac{D\vec{\upsilon}}{Dt}=-\nabla p-\nabla\cdot[\mu\nabla\cdot\vec{\upsilon}]+\nabla\left[\frac{4}{3}\mu\nabla\cdot\vec{\upsilon}\right]+\rho\vec{f} \tag{4.88}$$

Note:

$$\rho\frac{D\vec{\upsilon}}{Dt}=\frac{\partial}{\partial t}\rho\vec{\upsilon}+\nabla\cdot\rho\vec{\upsilon}\vec{\upsilon} \tag{4.82}$$

Cartesian

$$\frac{D}{Dt}=\frac{\partial}{\partial t}+\upsilon_x\frac{\partial}{\partial x}+\upsilon_y\frac{\partial}{\partial y}+\upsilon_z\frac{\partial}{\partial z}$$

$$\rho\frac{D\upsilon_x}{Dt}=-\frac{\partial p}{\partial x}+\frac{\partial}{\partial x}\left[2\mu\frac{\partial\upsilon_x}{\partial x}-\frac{2}{3}\mu\nabla\cdot\vec{\upsilon}\right]+\frac{\partial}{\partial y}\left[\mu\left(\frac{\partial\upsilon_x}{\partial y}+\frac{\partial\upsilon_y}{\partial x}\right)\right]+\frac{\partial}{\partial z}\left[\mu\left(\frac{\partial\upsilon_x}{\partial z}+\frac{\partial\upsilon_z}{\partial x}\right)\right]+\rho f_x$$

$$\rho\frac{D\upsilon_y}{Dt}=-\frac{\partial p}{\partial y}+\frac{\partial}{\partial y}\left[2\mu\frac{\partial\upsilon_y}{\partial y}-\frac{2}{3}\mu\nabla\cdot\vec{\upsilon}\right]+\frac{\partial}{\partial z}\left[\mu\left(\frac{\partial\upsilon_y}{\partial z}+\frac{\partial\upsilon_z}{\partial y}\right)\right]+\frac{\partial}{\partial x}\left[\mu\left(\frac{\partial\upsilon_x}{\partial y}+\frac{\partial\upsilon_y}{\partial x}\right)\right]+\rho f_y$$

$$\rho\frac{D\upsilon_z}{Dt}=-\frac{\partial p}{\partial y}+\frac{\partial}{\partial z}\left[2\mu\frac{\partial\upsilon_z}{\partial z}-\frac{2}{3}\mu\nabla\cdot\vec{\upsilon}\right]+\frac{\partial}{\partial x}\left[\mu\left(\frac{\partial\upsilon_z}{\partial x}+\frac{\partial\upsilon_x}{\partial z}\right)\right]+\frac{\partial}{\partial y}\left[\mu\left(\frac{\partial\upsilon_y}{\partial z}+\frac{\partial\upsilon_z}{\partial y}\right)\right]+\rho f_z$$

Cylindrical

$$\frac{D}{Dt}=\frac{\partial}{\partial t}+\upsilon_r\frac{\partial}{\partial r}+\frac{\upsilon_\theta}{r}\frac{\partial}{\partial\theta}+\upsilon_z\frac{\partial}{\partial z}$$

$$\rho\left[\frac{D\upsilon_r}{Dt}-\frac{\upsilon_\theta^2}{r}\right]=-\frac{\partial p}{\partial r}+\frac{\partial}{\partial r}\left[2\mu\frac{\partial\upsilon_r}{\partial r}-\frac{2}{3}\mu\nabla\cdot\vec{\upsilon}\right]+\frac{1}{r}\frac{\partial}{\partial\theta}\left[\mu\left(\frac{1}{r}\frac{\partial\upsilon_r}{\partial\theta}+\frac{\partial\upsilon_\theta}{\partial r}-\frac{\upsilon_\theta}{r}\right)\right]$$
$$+\frac{\partial}{\partial z}\left[\mu\left(\frac{\partial\upsilon_r}{\partial z}+\frac{\partial\upsilon_z}{\partial r}\right)\right]+\frac{2\mu}{r}\left(\frac{\partial\upsilon_r}{\partial r}-\frac{1}{r}\frac{\partial\upsilon_\theta}{\partial\theta}-\frac{\upsilon_r}{r}\right)+\rho f_r$$

$$\rho\left[\frac{D\upsilon_\theta}{Dt}-\frac{\upsilon_r\upsilon_\theta}{r}\right]=-\frac{1}{r}\frac{\partial p}{\partial\theta}+\frac{1}{r}\frac{\partial}{\partial\theta}\left[\frac{2\mu}{r}\frac{\partial\upsilon_\theta}{\partial\theta}-\frac{2}{3}\mu\nabla\cdot\vec{\upsilon}\right]+\frac{\partial}{\partial z}\left[\mu\left(\frac{1}{r}\frac{\partial\upsilon_z}{\partial\theta}+\frac{\partial\upsilon_\theta}{\partial z}\right)\right]$$
$$+\frac{\partial}{\partial r}\left[\mu\left(\frac{1}{r}\frac{\partial\upsilon_r}{\partial\theta}+\frac{\partial\upsilon_\theta}{\partial r}-\frac{\upsilon_\theta}{r}\right)\right]+\frac{2\mu}{r}\left(\frac{1}{r}\frac{\partial\upsilon_r}{\partial\theta}+\frac{\partial\upsilon_\theta}{\partial r}-\frac{\upsilon_\theta}{r}\right)+\rho f_\theta$$

$$\rho\frac{D\upsilon_z}{Dt}=-\frac{\partial p}{\partial z}+\frac{\partial}{\partial z}\left[2\mu\frac{\partial\upsilon_z}{\partial z}-\frac{2}{3}\mu\nabla\cdot\vec{\upsilon}\right]+\frac{1}{r}\frac{\partial}{\partial r}\left[\mu r\left(\frac{\partial\upsilon_r}{\partial z}+\frac{\partial\upsilon_z}{\partial r}\right)\right]$$
$$+\frac{1}{r}\frac{\partial}{\partial\theta}\left[\mu\left(\frac{1}{r}\frac{\partial\upsilon_z}{\partial\theta}+\frac{\partial\upsilon_\theta}{\partial z}\right)\right]+\rho f_z$$

Spherical

$$\frac{D}{Dt}=\frac{\partial}{\partial t}+\upsilon_r\frac{\partial}{\partial r}+\frac{\upsilon_\theta}{r}\frac{\partial}{\partial\theta}+\frac{\upsilon_\phi}{r\sin\theta}\frac{\partial}{\partial\phi}$$

continued

TABLE 4.9 (continued)
Differential Equation of Momentum[a]

$$\rho\left[\frac{D\upsilon_r}{Dt}-\frac{\upsilon_\theta^2+\upsilon_\phi^2}{r}\right]=-\frac{\partial p}{\partial r}+\frac{\partial}{\partial r}\left[2\mu\frac{\partial\upsilon_r}{\partial r}-\frac{2}{3}\mu\nabla\cdot\vec{\upsilon}\right]+\frac{1}{r}\frac{\partial}{\partial\theta}\left[\mu\left\{r\frac{\partial}{\partial r}\left(\frac{\upsilon_\theta}{r}\right)+\frac{1}{r}\frac{\partial\upsilon_r}{\partial\theta}\right\}\right]$$
$$+\frac{1}{r\sin\theta}\frac{\partial}{\partial\phi}\left[\left\{\frac{1}{r\sin\theta}\frac{\partial\upsilon_r}{\partial\phi}+r\frac{\partial}{\partial r}\left(\frac{\upsilon_\phi}{r}\right)\right\}\right]$$
$$+\frac{\mu}{r}\left[4\frac{\partial\upsilon_r}{\partial r}-\frac{2}{r}\frac{\partial\upsilon_\theta}{\partial\theta}-\frac{4\upsilon_r}{\mathrm{r}}-\frac{2}{r\sin\theta}\frac{\partial\upsilon_\phi}{\partial\phi}-\frac{2\upsilon_\theta\cot\theta}{r}\right.$$
$$\left.+r\cot\theta\frac{\partial}{\partial r}\left(\frac{\upsilon_\theta}{r}\right)+\frac{\cot\theta}{r}\frac{\partial\upsilon_r}{\partial\theta}\right]+\rho f_r$$

$$\rho\left[\frac{D\upsilon_\theta}{Dt}-\frac{\upsilon_r\upsilon_\theta}{r}-\frac{\upsilon_\phi^2\cot\theta}{r}\right]=-\frac{1}{r}\frac{\partial p}{\partial\theta}+\frac{1}{r}\frac{\partial}{\partial\theta}\left[\frac{2\mu}{r}\left(\frac{\partial\upsilon_\theta}{\partial\theta}+\upsilon_r\right)-\frac{2}{3}\mu\nabla\cdot\vec{\upsilon}\right]$$
$$+\frac{1}{r\sin\theta}\frac{\partial}{\partial\phi}\left[\mu\left\{\frac{\sin\theta}{r}\frac{\partial}{\partial\theta}\left(\frac{\upsilon_\phi}{\sin\theta}\right)+\frac{1}{r\sin\theta}\frac{\partial\upsilon_\theta}{\partial\phi}\right\}\right]$$
$$+\frac{\partial}{\partial r}\left[\mu\left\{r\frac{\partial}{\partial r}\left(\frac{\upsilon_\theta}{r}\right)+\frac{1}{r}\frac{\partial\upsilon_r}{\partial\theta}\right\}\right]$$
$$+\frac{\mu}{r}\left[2\left(\frac{1}{r}\frac{\partial\upsilon_\theta}{\partial\theta}-\frac{1}{r\sin\theta}\frac{\partial\upsilon_\phi}{\partial\phi}-\frac{\upsilon_\theta\cot\theta}{r}\right)\cdot\cot\theta\right.$$
$$\left.+3\left\{r\frac{\partial}{\partial r}\left(\frac{\upsilon_\theta}{r}\right)+\frac{1}{r}\frac{\partial\upsilon_r}{\partial\theta}\right\}\right]+\rho f_\theta$$

$$\rho\left[\frac{D\upsilon_\phi}{Dt}+\frac{\upsilon_\phi\upsilon_r}{r}+\frac{\upsilon_\theta\upsilon_\phi\cot\theta}{r}\right]=-\frac{1}{r\sin\theta}\frac{\partial p}{\partial\phi}$$
$$+\frac{1}{r\sin\theta}\frac{\partial}{\partial\phi}\left[\frac{2\mu}{r}\left(\frac{1}{\sin\theta}\frac{\partial\upsilon_\phi}{\partial\phi}+\upsilon_r+\upsilon_\theta\cot\theta\right)-\frac{2}{3}\mu\nabla\cdot\vec{\upsilon}\right]$$
$$+\frac{\partial}{\partial r}\left[\mu\left\{\frac{1}{r\sin\theta}\frac{\partial\upsilon_r}{\partial\phi}+r\frac{\partial}{\partial r}\left(\frac{\upsilon_\phi}{r}\right)\right\}\right]$$
$$+\frac{1}{r}\frac{\partial}{\partial\theta}\left[\mu\left\{\frac{\sin\theta}{r}\frac{\partial}{\partial\theta}\left(\frac{\upsilon_\phi}{\sin\theta}\right)+\frac{1}{r\sin\theta}\frac{\partial\upsilon_\theta}{\partial\phi}\right\}\right]$$
$$+\frac{\mu}{r}\left[3\left\{\frac{1}{r\sin\theta}\frac{\partial\upsilon_r}{\partial\phi}+r\frac{\partial}{\partial r}\left(\frac{\upsilon_\phi}{r}\right)\right\}\right.$$
$$\left.+2\cot\theta\left\{\frac{\sin\theta}{r}\frac{\partial}{\partial\theta}\left(\frac{\upsilon_\phi}{\sin\theta}\right)+\frac{1}{r\sin\theta}\frac{\partial\upsilon_\theta}{\partial\phi}\right\}\right]+\rho f_\phi$$

[a] $\mu'=0$ in Equation 4.84.

TABLE 4.10
Differential Equation of Momentum for Incompressible Fluids[a]

Vector form

$$\rho\frac{D\vec{\upsilon}}{Dt} = -\nabla p + \mu\nabla^2\vec{\upsilon} + \rho\vec{f} \tag{4.90b}$$

Cartesian

$$\rho\left(\frac{\partial \upsilon_x}{\partial t} + \upsilon_x\frac{\partial \upsilon_x}{\partial x} + \upsilon_y\frac{\partial \upsilon_x}{\partial y} + \upsilon_z\frac{\partial \upsilon_x}{\partial z}\right) = -\frac{\partial p_x}{\partial x} + \mu\left(\frac{\partial^2 \upsilon_x}{\partial x^2} + \frac{\partial^2 \upsilon_x}{\partial y^2} + \frac{\partial^2 \upsilon_x}{\partial z^2}\right) + \rho f_x$$

$$\rho\left(\frac{\partial \upsilon_y}{\partial t} + \upsilon_x\frac{\partial \upsilon_y}{\partial x} + \upsilon_y\frac{\partial \upsilon_y}{\partial y} + \upsilon_z\frac{\partial \upsilon_y}{\partial z}\right) = -\frac{\partial p}{\partial y} + \mu\left(\frac{\partial^2 \upsilon_y}{\partial x^2} + \frac{\partial^2 \upsilon_y}{\partial y^2} + \frac{\partial^2 \upsilon_y}{\partial z^2}\right) + \rho f_y$$

$$\rho\left(\frac{\partial \upsilon_z}{\partial t} + \upsilon_x\frac{\partial \upsilon_z}{\partial x} + \upsilon_y\frac{\partial \upsilon_z}{\partial y} + \upsilon_z\frac{\partial \upsilon_z}{\partial z}\right) = -\frac{\partial p}{\partial z} + \mu\left(\frac{\partial^2 \upsilon_z}{\partial x^2} + \frac{\partial^2 \upsilon_z}{\partial y^2} + \frac{\partial^2 \upsilon_z}{\partial z^2}\right) + \rho f_z$$

Cylindrical

$$\rho\left[\frac{\partial \upsilon_r}{\partial t} + \upsilon_r\frac{\partial \upsilon_r}{\partial r} + \frac{\upsilon_\theta}{r}\frac{\partial \upsilon_r}{\partial \theta} + \upsilon_z\frac{\partial \upsilon_r}{\partial z} - \frac{\upsilon_\theta^2}{r}\right]$$

$$= -\frac{\partial p}{\partial r} + \mu\left[\frac{\partial^2 \upsilon_r}{\partial r^2} + \frac{1}{r}\frac{\partial \upsilon_r}{\partial r} + \frac{1}{r^2}\frac{\partial^2 \upsilon_r}{\partial \theta^2} + \frac{\partial^2 \upsilon_r}{\partial z^2} - \frac{\upsilon_r}{r_2} + \frac{2}{r^2}\frac{\partial \upsilon_\theta}{\partial \theta}\right] + \rho f_r$$

$$\rho\left[\frac{\partial \upsilon_\theta}{\partial t} + \upsilon_r\frac{\partial \upsilon_\theta}{\partial r} + \frac{\upsilon_\theta}{r}\frac{\partial \upsilon_\theta}{\partial \theta} + \upsilon_z\frac{\partial \upsilon_r}{\partial z} - \frac{\upsilon_r\upsilon_\theta}{r}\right]$$

$$= -\frac{1}{r}\frac{\partial p}{\partial \theta} + \mu\left[\frac{\partial^2 \upsilon_\theta}{\partial r^2} + \frac{1}{r}\frac{\partial \upsilon_\theta}{\partial r} + \frac{1}{r^2}\frac{\partial^2 \upsilon_\theta}{\partial \theta^2} + \frac{\partial^2 \upsilon_\theta}{\partial z^2} + \frac{2}{r^2}\frac{\partial \upsilon_r}{\partial \theta} - \frac{\upsilon_\theta}{r^2}\right] + \rho f_\theta$$

$$\rho\left[\frac{\partial \upsilon_z}{\partial t} + \upsilon_r\frac{\partial \upsilon_z}{\partial r} + \frac{\upsilon_\theta}{r}\frac{\partial \upsilon_z}{\partial \theta} + \upsilon_z\frac{\partial \upsilon_z}{\partial z}\right] = -\frac{\partial p}{\partial z} + \mu\left[\frac{\partial^2 \upsilon_z}{\partial r^2} + \frac{1}{r}\frac{\partial \upsilon_z}{\partial r} + \frac{1}{r^2}\frac{\partial^2 \upsilon_z}{\partial \theta^2} + \frac{\partial^2 \upsilon_z}{\partial z^2}\right] + \rho f_z$$

Spherical

$$\rho\left[\frac{\partial \upsilon_r}{\partial t} + \upsilon_r\frac{\partial \upsilon_r}{\partial r} + \frac{\upsilon_\theta}{r}\frac{\partial \upsilon_r}{\partial \theta} + \frac{\upsilon_\phi}{r\sin\theta}\frac{\partial \upsilon_r}{\partial \phi} - \frac{\upsilon_\theta^2 + \upsilon_\phi^2}{r}\right] = -\frac{\partial p}{\partial r} + \mu\left[\frac{1}{r^2}\frac{\partial}{\partial r}\left(r^2\frac{\partial \upsilon_r}{\partial r}\right)\right.$$

$$\left. + \frac{1}{r^2\sin\theta}\left(\sin\theta\frac{\partial \upsilon_r}{\partial \theta}\right) + \frac{1}{r^2\sin^2\theta}\frac{\partial^2 \upsilon_r}{\partial \phi^2} - \frac{2\upsilon_r}{r^2} - \frac{2}{r^2}\frac{\partial \upsilon_\theta}{\partial \theta} - \frac{2\upsilon_\theta\cot\theta}{r^2} - \frac{2}{r^2\sin\theta}\frac{\partial \upsilon_\phi}{\partial \phi}\right] + \rho f_r$$

$$\rho\left[\frac{\partial \upsilon_\theta}{\partial t} + \upsilon_r\frac{\partial \upsilon_\theta}{\partial r} + \frac{\upsilon_\theta}{r}\frac{\partial \upsilon_\theta}{\partial \theta} + \frac{\upsilon_\phi}{r\sin\theta}\frac{\partial \upsilon_\theta}{\partial \phi} + \frac{\upsilon_r\upsilon_\theta}{r} - \frac{\upsilon_\phi^2\cot\theta}{r}\right] = -\frac{1}{r}\frac{\partial p}{\partial \theta} + \mu\left[\frac{1}{r^2}\frac{\partial}{\partial r}\left(r^2\frac{\partial \upsilon_\theta}{\partial r}\right)\right.$$

$$\left. + \frac{1}{r^2\sin\theta}\frac{\partial}{\partial \theta}\left(\sin\theta\frac{\partial \upsilon_\theta}{\partial \theta}\right) + \frac{1}{r^2\sin^2\theta}\frac{\partial^2 \upsilon_\theta}{\partial \phi^2} + \frac{2}{r^2}\frac{\partial \upsilon_r}{\partial \theta} - \frac{\upsilon_\theta}{r^2\sin^2\theta} - \frac{2\cos\theta}{r^2\sin^2\theta}\frac{\partial \upsilon_\phi}{\partial \phi}\right] + \rho f_\theta$$

$$\rho\left[\frac{\partial \upsilon_\phi}{\partial t} + \upsilon_r\frac{\partial \upsilon_\phi}{\partial r} + \frac{\upsilon_\theta}{r}\frac{\partial \upsilon_\phi}{\partial \theta} + \frac{\upsilon_\phi}{r\sin\theta}\frac{\partial \upsilon_\phi}{\partial \phi} + \frac{\upsilon_\phi\upsilon_r}{r} + \frac{\upsilon_\theta\upsilon_\phi\cot\theta}{r}\right] = -\frac{1}{r\sin\theta}\frac{\partial p}{\partial \phi}$$

$$+ \mu\left[\frac{1}{r^2}\frac{\partial}{\partial r}\left(r^2\frac{\partial \upsilon_\phi}{\partial r}\right) + \frac{1}{r^2\sin\theta}\frac{\partial}{\partial \theta}\left(\sin\theta\frac{\partial \upsilon_\phi}{\partial \theta}\right) + \frac{1}{r^2\sin^2\theta}\frac{\partial^2 \upsilon_\phi}{\partial \phi^2} - \frac{\upsilon_\phi}{r^2\sin^2\theta}\right.$$

$$\left. + \frac{2}{r^2\sin^2\theta}\frac{\partial \upsilon_r}{\partial \theta} + \frac{2\cos\theta}{r^2\sin^2\theta}\frac{\partial \upsilon_\theta}{\partial \phi}\right] + \rho f_\phi$$

[a] $\mu' = 0$; μ = constant; ρ = constant.

TABLE 4.11

Differential Equations of Energy[a]

Vector form

$$\rho \frac{Dh}{Dt} = -\nabla \cdot \vec{q}'' + q''' + \frac{Dp}{Dt} + \Phi \tag{4.108}$$

$$\rho \frac{Du}{Dt} = -\nabla \cdot \vec{q}'' + q''' - p\nabla \cdot \vec{\upsilon} + \Phi \tag{4.110}$$

$$\rho c_p \frac{DT}{Dt} = \nabla \cdot k\nabla T - \nabla \cdot \vec{q}_r'' + q''' + \beta T \frac{Dp}{Dt} + \Phi \tag{4.122b}$$

$$\Phi = \bar{\bar{\tau}} : \nabla \vec{\upsilon} = \nabla \cdot (\bar{\bar{\tau}} \cdot \vec{\upsilon}) - \vec{\upsilon} \cdot (\nabla \cdot \bar{\bar{\tau}}) \tag{4.103}$$

Cartesian

$$\rho \frac{Dh}{Dt} = \frac{\partial}{\partial x}\left(k \frac{\partial T}{\partial x}\right) + \frac{\partial}{\partial y}\left(k \frac{\partial T}{\partial y}\right) + \frac{\partial}{\partial z}\left(k \frac{\partial T}{\partial z}\right) - \nabla \cdot \vec{q}_r'' + q''' + \frac{Dp}{Dt} + \Phi$$

Cylindrical

$$\rho \frac{Dh}{Dt} = \frac{1}{r}\frac{\partial}{\partial r}\left(rk \frac{\partial T}{\partial r}\right) + \frac{1}{r^2}\frac{\partial}{\partial \theta}\left(k \frac{\partial T}{\partial \theta}\right) + \frac{\partial}{\partial z}\left(k \frac{\partial T}{\partial z}\right) - \nabla \cdot \vec{q}_r'' + q''' + \frac{Dp}{Dt} + \Phi$$

Spherical

$$\rho \frac{Dh}{Dt} = \frac{1}{r^2}\frac{\partial}{\partial r}\left(r^2 k \frac{\partial T}{\partial r}\right) + \frac{1}{r^2 \sin\theta}\frac{\partial}{\partial \theta}\left(k \sin\theta \frac{\partial T}{\partial \theta}\right) + \frac{1}{r^2 \sin\theta}\frac{\partial}{\partial \phi}\left(k \frac{\partial T}{\partial \phi}\right) - \nabla \cdot \vec{q}_r'' + q''' + \frac{Dp}{Dt} + \Phi$$

[a] $\mu' = 0$; ρ = constant.

TABLE 4.12

Differential Equations of Energy for Incompressible Materials[a]

Vector form

$$\rho \frac{Dh}{Dt} = -\nabla \cdot \vec{q}'' + q''' + \Phi$$

$$\rho \frac{Du}{Dt} = -\nabla \cdot \vec{q}'' + q''' + \Phi$$

$$\rho c_p \frac{DT}{Dt} = -\nabla \cdot q'' + q''' + \Phi$$

$$\Phi = \bar{\bar{\tau}} : \nabla \vec{\upsilon} = \nabla \cdot (\bar{\bar{\tau}} \cdot \vec{\upsilon}) - \vec{\upsilon} \cdot (\nabla \cdot \bar{\bar{\tau}})$$

Special case for k = constant

$$\rho \frac{Dh}{Dt} = k\nabla^2 T - \nabla \cdot \vec{q}_r'' + q''' + \Phi$$

Cartesian

$$\rho \frac{Dh}{Dt} = k\left(\frac{\partial^2 T}{\partial x^2} + \frac{\partial^2 T}{\partial y^2} + \frac{\partial^2 T}{\partial z^2}\right) - \nabla \cdot \vec{q}_r'' + q''' + \Phi$$

TABLE 4.12 (continued)

Differential Equations of Energy for Incompressible Materials[a]

$$\Phi = 2\mu\left[\left(\frac{\partial \upsilon_x}{\partial x}\right)^2 + \left(\frac{\partial \upsilon_y}{\partial y}\right)^2 + \left(\frac{\partial \upsilon_z}{\partial z}\right)^2 + \frac{1}{2}\left(\frac{\partial \upsilon_x}{\partial y} + \frac{\partial \upsilon_y}{\partial x}\right)^2 + \frac{1}{2}\left(\frac{\partial \upsilon_x}{\partial z} + \frac{\partial \upsilon_z}{\partial x}\right)^2 + \frac{1}{2}\left(\frac{\partial \upsilon_y}{\partial z} + \frac{\partial \upsilon_z}{\partial y}\right)^2\right]$$

Cylindrical

$$\rho\frac{Dh}{Dt} = k\left(\frac{\partial^2 T}{\partial r^2} + \frac{1}{r}\frac{\partial T}{\partial r} + \frac{1}{r^2}\frac{\partial^2 T}{\partial \theta^2} + \frac{\partial^2 T}{\partial z^2}\right) - \nabla\cdot\vec{q}_r'' + q''' + \Phi$$

$$\Phi = \mu\left[2\left\{\left(\frac{\partial \upsilon_r}{\partial r}\right)^2 + \left(\frac{1}{r}\frac{\partial \upsilon_\theta}{\partial \theta} + \frac{\upsilon_r}{r}\right)^2 + \left(\frac{\partial \upsilon_z}{\partial z}\right)^2\right\} + \left(\frac{\partial \upsilon_z}{\partial \theta} + \frac{\partial \upsilon_\theta}{\partial z}\right)^2 + \left(\frac{\partial \upsilon_r}{\partial z} + \frac{\partial \upsilon_z}{\partial r}\right)^2 + \left(\frac{1}{r}\frac{\partial \upsilon_r}{\partial \theta} + \frac{\partial \upsilon_\theta}{\partial r} - \frac{\upsilon_\theta}{r}\right)^2\right]$$

Spherical

$$\rho\frac{Dh}{Dt} = k\left[\frac{1}{r^2}\frac{\partial}{\partial r}\left(r^2\frac{\partial T}{\partial r}\right) + \frac{1}{r^2\sin\theta}\frac{\partial}{\partial \theta}\left(\sin\theta\frac{\partial T}{\partial \theta}\right) + \frac{1}{r^2\sin^2\theta}\frac{\partial^2 T}{\partial \phi^2}\right] - \nabla\cdot\vec{q}_r'' + q''' + \Phi$$

$$\Phi = \mu\left[2\left\{\left(\frac{\partial \upsilon_r}{\partial r}\right)^2 + \left(\frac{1}{r}\frac{\partial \upsilon_\theta}{\partial \theta} + \frac{\upsilon_r}{r}\right)^2 + \left(\frac{1}{r\sin\theta}\frac{\partial \upsilon_\phi}{\partial \phi} + \frac{\upsilon_r}{r} + \frac{\upsilon_\theta\cot\theta}{r}\right)^2\right\} + \left\{\frac{1}{r\sin\phi}\frac{\partial \upsilon_\theta}{\partial \phi} + \frac{\sin\theta}{r}\frac{\partial}{\partial \theta}\left(\frac{\upsilon_\phi}{\sin\theta}\right)\right\}^2 + \left\{\frac{1}{r\sin\theta}\frac{\partial \upsilon_r}{\partial \phi} + r\frac{\partial}{\partial r}\left(\frac{\upsilon_\phi}{r}\right)\right\}^2 + \left\{r\frac{\partial}{\partial r}\left(\frac{\upsilon_\theta}{r}\right) + \frac{1}{r}\frac{\partial \upsilon_r}{\partial \theta}\right\}^2\right]$$

[a] $\mu' = 0$; ρ = constant.

For any property c, a time-averaged value can be obtained by

$$\bar{c} \equiv \frac{1}{\Delta t}\int_{t-\Delta t/2}^{t+\Delta t/2} c\,dt \tag{4.124}$$

Thus, the instantaneous value of the property may be written as

$$c \equiv \bar{c} + c' \tag{4.125}$$

The period Δt is chosen to be sufficiently large to smooth out c but sufficiently small with respect to the transient time constants of the flow system under consideration.

Let us average 4.71 over the period Δt to get

$$\overline{\frac{\partial[\rho c]}{\partial t}} + \overline{\nabla \cdot [\rho c \vec{\upsilon}]} = \overline{\nabla \cdot \vec{J}} + \overline{\rho \Phi} \tag{4.126}$$

Because Δt is sufficiently small:

$$\overline{\frac{\partial[\rho c]}{\partial t}} = \frac{\partial\overline{[\rho c]}}{\partial t} \tag{4.127a}$$

For a stationary system of coordinates, time and space are independent variables, so that from Equation 4.11, because $\upsilon_s = 0$, we get

$$\overline{\nabla \cdot [\rho c \vec{\upsilon}]} = \nabla \cdot \overline{[\rho c \vec{\upsilon}]} \tag{4.127b}$$

and

$$\overline{\nabla \cdot \vec{J}} = \nabla \cdot \overline{\vec{J}} \tag{4.127c}$$

Applying Equations 4.127a through 4.127c to Equation 4.126, we get

$$\frac{\partial\overline{[\rho c]}}{\partial t} + \nabla \cdot \overline{[\rho c \vec{\upsilon}]} = \nabla \cdot \overline{\vec{J}} + \overline{\rho \Phi} \tag{4.128}$$

Assume that the density, velocity, and stagnation internal energy can be represented as a sum of a time-averaged value and a perturbation such that

$$\rho \equiv \bar{\rho} + \rho' \tag{4.129a}$$

$$\vec{\upsilon} \equiv \overline{\vec{\upsilon}} + \vec{\upsilon}' \tag{4.129b}$$

$$u^{\circ} \equiv \overline{u^{\circ}} + u^{\circ\prime} \tag{4.129c}$$

It is clear that

$$\overline{c'} = 0 \quad \text{and} \quad \overline{\bar{c}\, c'} = 0 \tag{4.130}$$

Therefore, Equation 4.128 can be expanded to

$$\frac{\partial[\bar{\rho}\,\bar{c}]}{\partial t} + \frac{\partial\overline{[\rho' c']}}{\partial t} + \nabla \cdot [\bar{\rho}\,\bar{c}\,\overline{\vec{\upsilon}}]$$
$$+ \nabla \cdot [\overline{\rho' c'}\,\overline{\vec{\upsilon}} + \overline{\rho' \vec{\upsilon}'}\,\bar{c} + \bar{\rho}\overline{c' \vec{\upsilon}'} + \overline{\rho' c' \vec{\upsilon}'}] = \nabla \cdot \overline{\vec{J}} + \bar{\rho}\,\bar{\Phi} + \overline{\rho' \Phi'} \tag{4.131}$$

Equation 4.131 can be rearranged to collect the fluctuation terms on the right-hand side:

$$\frac{\partial[\bar{\rho}\,\bar{c}]}{\partial t} + \nabla\cdot[\bar{\rho}\,\bar{c}\,\bar{\vec{\upsilon}}] = \nabla\cdot\bar{\vec{J}} + \bar{\rho}\,\bar{\Phi} + \left\{\overline{\rho'\Phi'} - \frac{\partial[\overline{\rho' c'}]}{\partial t} - \nabla\cdot\vec{J}^t\right\} \tag{4.132}$$

where

$$\vec{J}^t = \overline{\rho' c'}\;\bar{\vec{\upsilon}} + \overline{\rho'\vec{\upsilon}'}\;\bar{c} + \bar{\rho}\;\overline{c'\vec{\upsilon}'} + \overline{\rho' c'\vec{\upsilon}'} \tag{4.133}$$

If it is assumed that the density fluctuations are small or that the fluid is incompressible, $\rho' = 0$ and the turbulent transport Equation 4.132 reduce to

$$\frac{\partial[\rho\bar{c}]}{\partial t} + \nabla\cdot[\rho\bar{c}\;\bar{\vec{\upsilon}}] = \nabla\cdot\bar{\vec{J}} + \rho\bar{\Phi} - \nabla\cdot[\rho\overline{c'\vec{\upsilon}'}];\quad \rho' = 0 \tag{4.134}$$

The mass equation can be obtained by specifying c, $\vec{J}$, and Φ from Table 4.6, in Equation 4.134 to get the turbulent flow equations of an incompressible fluid.
The *mass equation* becomes

$$\frac{\partial[\rho]}{\partial t} + \nabla\cdot\left[\rho\bar{\vec{\upsilon}}\right] = 0 \tag{4.135}$$

The *momentum equation* becomes

$$\frac{\partial[\rho\bar{\vec{\upsilon}}]}{\partial t} + \nabla\cdot[\rho\bar{\vec{\upsilon}}\,\bar{\vec{\upsilon}}] = \nabla\cdot[\,\bar{\bar{\tau}} - \bar{p}\,\bar{\bar{I}}\,] + \rho\vec{g} - \nabla\cdot[\rho\overline{\vec{\upsilon}'\vec{\upsilon}'}] \tag{4.136}$$

The *stagnation energy equation* becomes

$$\frac{\partial[\rho\,\bar{u}^{\circ}]}{\partial t} + \nabla\cdot[\rho\bar{u}^{\circ}\bar{\vec{\upsilon}}] = \nabla\cdot[-\bar{\vec{q}}'' + \overline{(\bar{\bar{\tau}} - p\,\bar{\bar{I}})\cdot\vec{\upsilon}}]$$
$$+\bar{q}''' + \rho\vec{g}\cdot\bar{\vec{\upsilon}} - \nabla\cdot[\rho\overline{u^{\circ\prime}\vec{\upsilon}'}] \tag{4.137}$$

The solution of Equations 4.136 and 4.137 requires identification of constitutive relations between the terms, including the fluctuating variables and the behavior of the average variables. That is, it is required that $[\rho\overline{\vec{\upsilon}'\vec{\upsilon}'}]$, $\overline{(\bar{\bar{\tau}} - p\,\bar{\bar{I}})\cdot\vec{\upsilon}}$ and $[\rho\overline{u^{\circ\prime}\vec{\upsilon}'}]$ be specified in terms of ρ, $\bar{\vec{\upsilon}}$, and $\bar{u}^{\circ}$ for the above set of equations to be solvable. The constitutive equations are discussed in Chapters 9 and 10.

PROBLEMS

4.1. *Various time derivatives* (Section 4.2)
Air bubbles are being injected at a steady rate into the bottom of a vertical water channel of height L. The bubble rise velocity with respect to the water is υ_b. The water itself is flowing vertically at a velocity υ_ℓ

What is the velocity of the bubbles at the channel exit as observed by the following:

1. A stationary observer
2. An observer who moves upward with velocity υ_e
3. An observer who moves upward with velocity $2\upsilon_f$

Answers:

1. $\upsilon_\ell + \upsilon_b$
2. υ_b
3. $\upsilon_b - \upsilon_\ell$.

4.2. *Conservation of energy in a control volume* (Section 4.3)
A mass of 9 kg of gas with an internal energy of 1908 kJ is at rest in a rigid cylinder. A mass of 1.0 kg of the same gas with an internal energy of 95.4 kJ and a velocity of 30 m/s flows into the cylinder. In the absence of heat transfer to the surroundings and with negligible change in the center of gravity, find the internal energy of the 10 kg of gas finally at rest in the cylinder. The absolute pressure of the flowing gas crossing the control surface is 0.7 MPa, and the specific volume of the gas is 0.00125 m³/kg:

Answer:
$U_f = 2004.7$ kJ

4.3. *Process-dependent heat addition to a control mass* (Section 4.3)
Two tanks, A and B (Figure 4.10), each with a capacity of 0.566 m³, are perfectly insulated from the surroundings. A diatomic perfect gas is initially confined in tank A at a pressure of 1.013 MPa abs and a temperature of 21.1°C. Valve C is initially closed, and tank B is completely evacuated.

1. If valve *C* is opened and the gas is allowed to reach the same temperature in both tanks, what are the final pressure and temperature?
2. If valve *C* is opened just until the pressure in the two tanks is equalized and is then closed, what are the final temperature and pressure in tanks A and B? Assume no transfer of heat between tank A and tank B. Note that for a diatomic gas, $c_v = (5/2)\,R$.

Answers:

1. $p = 0.507$ MPa; $T = 21.1$°C
2. $p = 0.507$ MPa; $T_A = -31.8$°C; $T_B = 103.6$°C

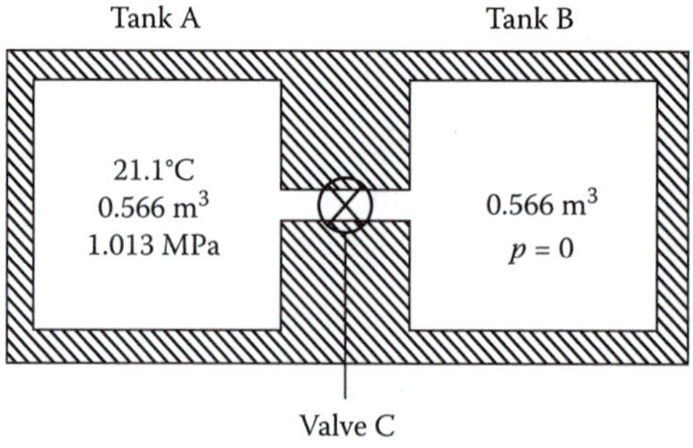

FIGURE 4.10 Initial state of perfectly insulated tanks for Problem 4.3.

4.4. *Control mass energy balance* (Section 4.3)
A perfectly insulated vessel contains 9.1 kg of water at an initial temperature of 277.8 K. An electric immersion heater with a mass of 0.454 kg is also at an initial temperature of 277.8 K. The water is slowly heated by passing an electric current through the heater until both water and heater attain a final temperature of 333.3 K. The specific heat of the water is 4.2 kJ/kg K and that of the heater is 0.504 kJ/kg K. Disregard the volume changes and assume that the temperatures of water and heater are the same at all stages of the process.

Calculate $\int \frac{dQ}{T}$ and ΔS for

1. The water as a system
2. The heater as a system
3. The water and heater as the system

Answers:

1. $\Delta S = \int \frac{dQ}{T} = 6.92\,\text{kJ/K}$

2. $\Delta S = 41.6$ J/K

$$\int \frac{dQ}{T} = -6.92\,\text{kJ/K}$$

3. $\Delta S = 6.97$ kJ/K

$$\int \frac{dQ}{T} = 0$$

4.5. *Qualifying a claim against the first and second laws of thermodynamics* (Section 4.3)
An engineer claims to have invented a new compressor that can be used with a small gas-cooled reactor. The CO_2 used for cooling the reactor enters the compressor at 1.378 MPa and 48.9°C and leaves the compressor at 2.067 MPa and −6.7°C. The compressor requires no input power but operates simply by transferring heat from the gas to a low-temperature reservoir surrounding the compressor. The inventor claims that the compressor can handle 0.908 kg of CO_2 per second if the temperature of the reservoir is −95.6°C and the rate of heat transfer is 63.6 kJ/s. Assuming that the CO_2 enters and leaves the device at very low velocities and that no significant elevation changes are involved

1. Determine if the compressor violates the first or the second law of thermodynamics.
2. Draw a schematic of the process on a *T-s* diagram.
3. Determine the change in the availability function (A) of the fluid flowing through the compressor.
4. Using the data from items 1 through 3, determine if the compressor is theoretically possible.

Answers:

1. The compressor does not violate either the first or the second law.
2. $\Delta A = 16.9$ kJ/kg

4.6. *Determining rocket acceleration from an energy balance* (Section 4.3) A rocket ship is traveling in a straight line through outer space, beyond the range of all gravitational forces. At a certain instant, its velocity is V, its mass is M, the rate of consumption of propellant is P, the rate of energy liberation by the chemical reaction is $\dot{Q}_R$. From the first law of thermodynamics alone (i.e., without using a force balance on the rocket), derive a general expression for the rate of change of velocity with time as a function of the above parameters and the discharged gas velocity V_d and enthalpy h_d.

Answer:

$$\frac{dV}{dt} = \frac{1}{MV}\left[\dot{Q}_R - M\frac{du}{dt} + \left(u + \frac{V^2}{2} - h_d - \frac{V_d^2}{2}\right)P\right]$$

4.7. *Momentum balance for a control volume* (Section 4.3) A jet of water is directed at a vane (Figure 4.11) that could be a blade in a turbine. The water leaves the nozzle with a speed of 15 m/s and mass flow of 250 kg/s; it enters the vane tangent to its surface (in the x direction). At the point the water leaves the vane, the angle to the x direction of 120°. Compute the resultant force on the vane if:

1. The vane is held constant.
2. The vane moves with a velocity of 5 m/s in the x direction.

Answers:

1. $F_x = 5625$ N, $F_y = -3248$ N
2. $F_x = 2500$ N, $F_y = -1443$ N

4.8. *Internal conservation equations for an extensive property* (Section 4.4) Consider a coolant in laminar flow in a circular tube. The one-dimensional velocity is given by

$$\vec{\upsilon} = \upsilon_{max}\left[1 - \left(\frac{r}{R}\right)^2\right]\vec{i}_z$$

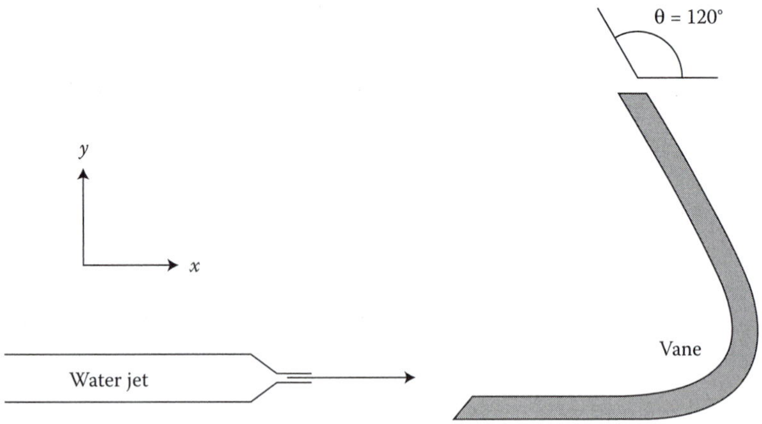

FIGURE 4.11 Jet of water directed at a vane.

where $\upsilon_{max} = 2.0$ m/s; R = radius of the tube = 0.05 m; $\vec{i}_z$ = a unit vector in the axial direction. Assume the fluid density is uniform within the tube ($\rho_0 = 300$ kg/m^3).

1. What is the coolant flow rate in the rube?
2. What is the coolant average velocity (V) in the tube?
3. What is the true kinetic head of this flow? Does the kinetic head equal $\frac{1}{2}\rho_0 V^2$?

Answers:

1. $Q = 7.854 \times 10^{-3}$ m^3/s
2. $V = 1$ m/s
3. Kinetic head $= 1.33\dfrac{\rho_o V^2}{2}$

4.9. *Differential transport equations* (Section 4.5)
Can the following sets of velocities belong to possible incompressible flow cases?

1. $\upsilon_x = x + y + z^2$
   $\upsilon_y = x - y + z$
   $\upsilon_z = 2xy + y^2$
2. $\upsilon_x = xyzt$
   $\upsilon_y = -xyzt^2$
   $\upsilon_z = \dfrac{z^2}{2}(xt^2 - yt)$

Answers:

1. Yes
2. Yes

REFERENCES

1. Bird, R. B., Stewart, W. E., and Lightfoot, E. N., *Transport Phenomena*. New York, NY: Wiley, 1960.
2. Carrie, I. G., *Fundamental Mechanics of Fluids*. New York, NY: McGraw-Hill, 1974.
3. Hildebrand, F. B., *Advanced Calculus for Applications*. Englewood Cliffs, NJ: Prentice Hall, 1962.
4. Potter, M. C. and Foss, J. R., *Fluid Mechanics*. New York, NY: Wiley, 1975.
5. Slattery, J. C. and Gaggioli, R. A., The macroscopic angular momentum balance. *Chem. Eng. Sci.*, 17:873–895, 1962.
6. Whitaker, S., *Introduction to Fluid Mechanics*. Huntington, NY: Krieger, 1981.
7. Whitaker, S., Laws of continuum physics for single phase, single component systems. In G. Hetsroni (ed.). *Handbook of Multiphase Systems*. Washington, DC: Hemisphere, 1982.

5 Transport Equations for Two-Phase Flow

5.1 INTRODUCTION

In principle, transport of mass, momentum, and energy can be formulated by taking balances over a control volume or by integrating the point equations over the desired region. It has been found that integrating (averaging) the point continuum equations over the desired volume is a more accurate and insightful, although more lengthy, approach to the formulation. Because of its simplicity, the balance approach has been useful in well-known one-dimensional applications, for example, the separate flow model of Martinelli and Nelson [4]. However, the averaging approach, which was developed later, can better account for the interfacial conditions at the gas–liquid boundaries.

The formulation of appropriate models for two-phase flow in nuclear, chemical, and mechanical systems has been an area of substantial interest since the mid-1960s. However, the proliferation of two-phase modeling efforts has made it difficult for newcomers to identify the basic assumptions and the limitations of the abundant models. This chapter provides a framework for defining the relations between the more established approach of taking balances to derive control volume equations that the reader may find in Wallis [7] or Collier [1] and the more recent approaches of averaging the local and instantaneous balance equations in time and/or space as described by Ishii [3] or Delhaye et al. [2].

5.1.1 Macroscopic versus Microscopic Information

The fundamental difficulty of two-phase flow description arises from the multiplicity of internal configurations that must be taken into consideration. A sensor at a localized position in the flow channel may feel the presence of one phase continuously, as with annular flow, or the two phases intermittently, as with bubbly flow. On the other hand, the viewers of an x-ray snapshot of a cross section of a two-phase flow channel can observe only the space-averaged behavior of the mixture of two phases and cannot identify the local behavior of each point in the field. It is this macroscopic or averaged two-phase flow behavior that is of interest in practice. However, the interaction between the two phases or between the fluids and the structures depends to a great extent on the microscopic behavior. Therefore, the basic question is how to predict the practically needed (and measurable) macroscopic behavior on the basis of whatever microscopic behavior may exist—and to the degree of accuracy required for the application.

With a single-phase flow, bridging between the microscopic phenomena and macroscopic description of the flow is easier. The observer of a point moving within the single-phase volume of consideration sees one continuous material. With two-phase flow, the local observer encounters interphase surfaces that lead to jump conditions between the two phases. The observer of a large volume may be able to consider the two phases as a single fluid but only after the properties of the two phases are properly averaged and the jump conditions are accounted for. These jump conditions describe mass, momentum, or energy exchange between the two phases.

5.1.2 Multicomponent versus Multiphase Systems

It is also beneficial to note that a two-phase system can be classified as a one-component or a multicomponent system, where "component" refers to the chemical species of the substance. Thus, steam–water is a one-component, two-phase system, whereas nitrogen–water is a two-component, two-phase system. In the latter case mass transfer between the two phases can often be ignored, but momentum and energy transfer must be accounted for. It is possible that a mixture of various chemical components constitutes one of the phases, for example, air and steam mixture in a reactor containment following a loss of coolant accident (LOCA). A multicomponent phase is often treated as a homogeneous mixture. Obviously, a multicomponent phase may be treated as a multifluid problem at the expense of greater complexity.

As chemical reactions between the phases are the exception in nuclear engineering applications, they are not considered here. Thus, the two-phase flow equations formulated here are still a special case of a more general formulation that may consider chemically reactive two phases.

5.1.3 Mixture versus Multifluid Models

A wide variety of possibilities exist for choosing the two-phase flow model. They range from describing the two-phase flow as a pseudo single-phase fluid (mixture) to a multifluid flow (e.g., liquid film, vapor, and droplets). Generally, as the two-phase flow model becomes more complex, more constitutive equations are required to represent the interactions between the fluids.

The homogeneous equilibrium model (HEM) is the simplest of the mixture models. It assumes that there is no relative velocity between the two phases (i.e., homogeneous flow) and that the vapor and liquid are in thermodynamic equilibrium. In this case the mass, momentum, and energy balance equations of the mixture are sufficient to describe the flow. The HEM assumptions are clearly limiting but may be adequate for certain flow conditions. Extensions of this model to include relative velocity, or slip, between the two phases and thermal nonequilibrium (e.g., subcooled boiling) effects are possible using externally supplied, usually empirical, constitutive relations. The HEM model has the advantage of being useful in analytical studies, as demonstrated in Chapters 11 and 14 in this book.

Mixture models other than HEM add some complexity to the two-phase flow description. By allowing the vapor and liquid phases to have different velocities but constraining them to thermal equilibrium, these methods allow for more accurate velocity predictions. Alternatively, by allowing one phase to depart from thermal equilibrium, enthalpy prediction can be improved. In this case, four or even five transport equations may be needed as well as a number of externally supplied relations to specify the interaction between the two phases. A well-known example for such mixture models is the thermal equilibrium drift flux model, which allows the vapor velocity to be different from that of the liquid by providing an algebraic relation for the velocity differential between the two phases (three-equation drift flux model). By considering two mass equations but using only mixture momentum and energy equations, one of the phases may depart from thermal equilibrium (i.e., a four-equation drift flux model).

In the two-fluid model, three conservation equations are written for both the vapor and the liquid phases. Hence, the model is often called the six-equation model. This model allows a more general description of the two-phase flow. However, it also requires a larger number of constitutive equations. The most important relations are those that represent the transfer of mass (Γ), transfer of energy (Q_s), and transfer of momentum (F_s) across liquid–vapor interfaces. The advantage of using this model is that the two phases are not restricted to prescribed temperature or velocity conditions. Extensions of the two-fluid model to multifluid models—in which vapor bubbles, a continuous liquid, a continuous vapor, and liquid droplets are described by separate sets of conservation equations—are also possible but have not been as widely applied as the simpler two-fluid model.

The combination of a mixture transport equation and one separate phase equation can also be used to replace two separate phase equations. Table 5.1 summarizes a variety of possible models to describe the two-phase flow problem. Note that there are six conservation equations and imposed restrictions for all models except for the three-fluid model. Note also that models requiring a larger number of constitutive relations, though not reducing the number of imposed restrictions, are generally not worth exploiting. Thus, the four-equation model B is less useful than the four-equation model A. For the five-equation models, model C is less useful than model B.

The choice among the alternative models discussed above depends on the nature of the problem to be solved. In LWR applications, the three-equation HEM model may be adequate for predicting the pressure drop in a flow channel under high-pressure steady-state conditions. Calculating the void distribution requires a specified relative velocity or a four-equation mixture model because of the tendency of the vapor to move faster than the liquid. Fast transients with sudden pressure changes are best handled by the six-equation, two-fluid model because of the expected large departure from equilibrium conditions.

5.2 AVERAGING OPERATORS FOR TWO-PHASE FLOW

Given the nature of two-phase flow, the field parameters of interest are usually average quantities in either space or time and often in both. Thus, it is important to introduce certain averaging mathematical operators here as well as define the more basic parameters that appear in the two-phase flow transport equations.

TABLE 5.1
Two-Phase Flow Models: $p_\ell = p_v = p_s$

	Conservation Equations				Imposed Restrictions			Constitutive Laws					
								Wall		Interphase			
Two-Phase Flow Model	Mass	Energy	Mom.	Total No.	Phase Enthalpy or Temperatures	Phase Velocities	Total No.	Mom. F_w	Energy Q_w	Mass Γ	Energy Q_s	Mom. F_s	Total No.
General 3-equation models	1	1	1	3	T_v and T_ℓ specified	Specified	3	1	1	0	0	0	2
Homogeneous equilibrium	1	1	1	3	T_v and T_ℓ equilibrium	Equal	3	1	1	0	0	0	2
Equilibrium drift flux	1	1	1	3	T_v and T_ℓ equilibrium	Specified drift flux	3	1	1	0	0	0	2
4-Equation models													
A	2	1	1	4	T_v or T_ℓ	Slip relation	2	1	1	1	0	0	3
B	1	2	1	4	T_v or T_ℓ equilibrium	Slip relation	2	1	2	1[a]	1	0	5
C	1	1	2	4	T_v and T_ℓ equilibrium	None	2	2	1	1[a]	0	1	5
5-Equation models													
A	2	2	1	5	None	Slip relation	1	1	2	1	1	0	5
B	2	1	2	5	T_v or T_ℓ equilibrium	None	1	2	1	1	0	1	5
C	1	2	2	5	T_v or T_ℓ equilibrium	None	1	2	2	1[a]	1	1	7
Two-fluid	2	2	2	6	None	None	0	2	2	1	1	1	7
Three-fluid: continuous liquid, vapor, and liquid drops	3	3	3	9	None	None	0	3	3	2	2	2	12

[a] Γ is needed whenever Q_s or F_s is needed.

5.2.1 Phase Density Function

If a phase (k) is present at a point ($\vec{r}$), a phase density function (α_k) can be defined as follows:

$$\left.\begin{aligned}\alpha_k(\vec{r},t) &\equiv 1 \text{ if point } \vec{r} \text{ is occupied by phase } k\\ \alpha_k(\vec{r},t) &\equiv 0 \text{ if point } \vec{r} \text{ is not occupied by phase } k\end{aligned}\right\} \quad (5.1)$$

5.2.2 Volume-Averaging Operators

Any volume (V) can be viewed as divided into two domains (V_k) (where $k = \upsilon$ or ℓ), each containing one of the two phases. We can then define two instantaneous volume-averaging operators acting on any parameter (c); one over the entire volume:

$$\langle c\rangle \equiv \frac{1}{V}\iiint_V c\,\mathrm{d}V \quad (5.2)$$

and the other over the volume occupied by the phase k:

$$\langle c\rangle_k \equiv \frac{1}{V_k}\iiint_{V_k} c\,\mathrm{d}V_k = \frac{1}{V_k}\iiint_V c\alpha_k\,\mathrm{d}V \quad (5.3)$$

5.2.3 Area-Averaging Operators

The description of the two-phase flow conditions at the boundary of a control volume requires the space- and time-averaged values of a property on the surface area surrounding the control volume. Thus, we introduce here the definition of such an area average for a parameter c:

$$\{c\} \equiv \frac{1}{A}\iint_A c\,\mathrm{d}A \quad (5.4)$$

It should also be clear that the average of parameter c over the area occupied only by the phase k may be obtained from

$$\{c\}_k \equiv \frac{1}{A_k}\iint_{A_k} c\,\mathrm{d}A_k \equiv \frac{1}{A_k}\iint_A c\alpha_k\,\mathrm{d}A \quad (5.5)$$

5.2.4 Local Time-Averaging Operators

As the two phases may intermittently pass through a point $\vec{r}$, a time-averaged process can be defined as follows:

$$\tilde{c} \equiv \frac{1}{\Delta t^*}\int_{t-\Delta t^*/2}^{t+\Delta t^*/2} c\,\mathrm{d}t \quad (5.6)$$

where Δt^* is chosen large enough to provide a meaningful statistical count but short enough such that the flow conditions are not substantially altered during the observation of a transient flow condition. Defining a suitable Δt^* can be a practical problem in fast transients.

It is also possible to define a phase time-averaging operator by averaging only over that portion of time within Δt^* when a single phase occupies the position $\vec{r}$ so that

$$\tilde{c}_k \equiv \int_{t-\Delta t^*/2}^{t+\Delta t^*/2} c\alpha_k \mathrm{d}t \div \int_{t-\Delta t^*/2}^{t+\Delta t^*/2} \alpha_k \, \mathrm{d}t \tag{5.7}$$

Note that Δt^* is long compared to a short time period Δt that may be chosen for averaging the turbulent effects (i.e., high-frequency fluctuations) within each single phase as was done in Section 4.6 in Chapter 4. To filter out the high-frequency single-phase fluctuation, an average value of c has been defined by

$$\overline{c} \equiv \frac{1}{\Delta t} \int_{t-\Delta t/2}^{t+\Delta t/2} c \, \mathrm{d}t \tag{4.124}$$

5.2.5 Commutativity of Space- and Time-Averaging Operations

Vernier and Delhaye [6] argued that the volumetric average of $\tilde{c}$ must be identical to the time average of $\langle c \rangle$, so that

$$\langle \tilde{c} \rangle = \widetilde{\langle c \rangle} \tag{5.8}$$

Sha et al. [5] suggested that Equation 5.8 is strictly valid only for one-dimensional flow with uniform velocity. Their restriction has not yet been widely evaluated. Here, we adopt the view that commutativity of the time and space operators is valid for the Eulerian coordinate system where the time and space act as independent variables.

Similarly, for averaging the turbulent equations of a single phase

$$\langle \overline{c} \rangle = \overline{\langle c \rangle} \tag{5.9}$$

5.3 VOLUME-AVERAGED PROPERTIES

5.3.1 Void Fraction

5.3.1.1 Instantaneous Space-Averaged Void Fraction

The fraction of the control volume (V) that is occupied by the phase k at a given time can be obtained from

$$\langle \alpha_k \rangle = \frac{1}{V} \iiint_V \alpha_k \, \mathrm{d}V = \frac{V_k}{V} = \frac{V_k}{\sum_1^N V_k} \tag{5.10}$$

The volume fraction of the gaseous phase is normally called the *void fraction* and is simply referred to as $\langle\alpha\rangle$, that is

$$\langle\alpha\rangle \equiv \langle\alpha_v\rangle \tag{5.11a}$$

Similarly, $\langle 1-\alpha\rangle$ or $\langle\alpha_\ell\rangle$ is used to refer to the liquid volume fraction.

$$\langle 1-\alpha\rangle \equiv \langle\alpha_\ell\rangle \tag{5.11b}$$

A spatial void fraction may be defined for an area or a line, as well as for a volume.

5.3.1.2 Local Time-Averaged Void Fraction

Since any point is instantaneously occupied only by one phase, the phase fraction at point $\vec{r}$ in a flow channel is present only in a time-averaged sense and is defined by

$$\tilde{\alpha}_k = \frac{1}{\Delta t^*}\int_{t-\Delta t^*/2}^{t+\Delta t^*/2} \alpha_k \mathrm{d}t \tag{5.12}$$

where $\tilde{\alpha}_k$ = a local time-averaged phase fraction. For the gaseous phase $\tilde{\alpha}_k \equiv \tilde{\alpha}_v$.

5.3.1.3 Space- and Time-Averaged Void Fraction

The two-phase flow is sufficiently turbulent to make the value of $\langle\alpha\rangle$ a fluctuating value, even at steady-state flow conditions. It is useful to introduce the time-averaged value as the basis for describing the measurable quantities such as pressure gradient and mass fluxes. Thus, the time- and volume-averaged void fraction is defined as

$$\widetilde{\langle\alpha\rangle} = \frac{1}{\Delta t^*}\int_{t-\Delta t^*/2}^{t+\Delta t^*/2} \langle\alpha\rangle \, \mathrm{d}t \tag{5.13}$$

From Equation 5.8, the time average of $\langle\alpha_k\rangle$ is equal to the volume average of $\tilde{\alpha}_k$ where $\langle\alpha_k\rangle$ and $\tilde{\alpha}_k$ have been defined by Equations 5.10 and 5.12, respectively. Therefore,

$$\widetilde{\langle\alpha\rangle} = \widetilde{\langle\alpha_v\rangle} = \langle\tilde{\alpha}_v\rangle \tag{5.14a}$$

and

$$\widetilde{\langle 1-\alpha\rangle} = \widetilde{\langle\alpha_\ell\rangle} = \langle\tilde{\alpha}_\ell\rangle \tag{5.14b}$$

5.3.2 Volumetric Phase Averaging

5.3.2.1 Instantaneous Volumetric Phase Averaging

It is common to assume that the spatial variation of the phase properties can be ignored within a control volume, so that average properties within the phase can be used to derive the balance equations, as done later in Sections 5.5 and 5.6. Therefore, it is desirable to derive an expression for the average value of a property, that is, $\langle c_k \rangle_k$ where the phasic property c_k has been averaged over only that volume occupied by phase k (V_k). Note that the parameter c of Equation 5.2 is now considered as a property of phase k. We express $\langle c_k \rangle_k$ as

$$\left\langle c_k \right\rangle_k \equiv \frac{1}{V_k} \iiint_{V_k} c_k \, dV_k \equiv \frac{1}{V_k} \iiint_{V} \alpha_k c_k \, dV \tag{5.15}$$

However, the average of a phase property c_k over the entire volume is given by

$$\left\langle c_k \right\rangle \equiv \frac{1}{V} \iiint_{V} \alpha_k c_k \, dV_k \equiv \frac{1}{V} \iiint_{V_k} c_k \, dV_k \tag{5.16}$$

From Equations 5.15, 5.16, and 5.10, it is seen that

$$\left\langle c_k \right\rangle = \left\langle c_k \right\rangle_k \frac{V_k}{V} = \left\langle c_k \right\rangle_k \left\langle \alpha_k \right\rangle \tag{5.17}$$

Thus, for any phasic property c_k, the instantaneous phasic-averaged value $\langle c_k \rangle_k$ is related to the instantaneous volume-averaged value $\langle c_k \rangle$ by the volumetric phase fraction $\langle \alpha_k \rangle$. For example, if the property c_k is the liquid phase or the vapor phase density (ρ_ℓ or ρ_v), Equation 5.17 yields

$$\left\langle \rho_\ell \right\rangle = \left\langle \rho_\ell \right\rangle_\ell \left\langle 1 - \alpha \right\rangle \tag{5.18a}$$

$$\left\langle \rho_v \right\rangle = \left\langle \rho_v \right\rangle_v \left\langle \alpha \right\rangle \tag{5.18b}$$

Note that $\langle \rho_\ell \rangle_\ell$ and $\langle \rho_v \rangle_v$ are merely the thermodynamic state values found in handbooks, whereas $\langle \rho_\ell \rangle$ and $\langle \rho_v \rangle$ depend also on the values of $\langle \alpha \rangle$ and $\langle 1 - \alpha \rangle$ in the volume of interest.

5.3.2.2 Time Averaging of Volume-Averaged Quantities

The time average of the product of two variables does not equal the product of the time average of the two variables. Assuming that the instantaneous values of $\langle c_k \rangle_k$

and $\langle\alpha_k\rangle$ can be given as the sum of a time-averaged component and a fluctuation component, we get, using Equation 5.17:

$$\widetilde{\langle c_k \rangle} = \overline{\widetilde{\langle c_k \rangle_k \langle \alpha_k \rangle}} = \overline{\left[\widetilde{\langle c_k \rangle_k} + \langle c_k \rangle_k'\right]\left[\widetilde{\langle \alpha_k \rangle} + \langle \alpha_k \rangle'\right]}$$

Because $\left\langle \widetilde{\widetilde{c_k}} \right\rangle_k = \left\langle \widetilde{c_k} \right\rangle_k$ and $\left\langle \widetilde{\widetilde{\alpha_k}} \right\rangle = \left\langle \widetilde{\alpha_k} \right\rangle$

$$\widetilde{\langle c_k \rangle} = \widetilde{\langle c_k \rangle_k} \widetilde{\langle \alpha_k \rangle} + \psi' \tag{5.19}$$

where

$$\psi' = \overline{\widetilde{\langle c_k \rangle_k} \langle \alpha_k \rangle'} + \overline{\langle c_k \rangle_k' \widetilde{\langle \alpha_k \rangle}} + \overline{\langle c_k \rangle_k' \langle \alpha_k \rangle'} \tag{5.20}$$

But $\left\langle \widetilde{c_k} \right\rangle_k' = 0$ and $\left\langle \widetilde{\alpha_k} \right\rangle' = 0$

$$\psi' = \overline{\langle c_k \rangle_k' \langle \alpha_k \rangle'} \tag{5.21}$$

Clearly, the term ψ' is the result of the time fluctuations in $\langle c_k\rangle$ and $\langle\alpha_k\rangle$ and may become important in high flows or rapidly varying flow conditions. Therefore, experimentally obtained values of $\widetilde{\langle c_k \rangle}$ should not always be readily assumed to equal $\widetilde{\langle c_k \rangle_k}\ \widetilde{\langle \alpha_k \rangle}$ unless fluctuations in $\langle c_k\rangle_k$ or $\langle\alpha_k\rangle$ can be ignored.

5.3.3 Static Quality

An important variable in many two-phase situations is the mass fraction of vapor in a fixed volume. This fraction, called the *static quality*, is typically written without the brackets for volume averages as x_{st}. It is given at any instant by

$$x_{st} = \frac{m_v}{m_v + m_\ell} \tag{5.22}$$

$$x_{st} = \frac{\langle \rho_v \rangle V}{(\langle \rho_v \rangle + \langle \rho_\ell \rangle) V} \tag{5.23}$$

Note that x_{st} is a volume-averaged property by definition; therefore, $\langle x_{st}\rangle \equiv x_{st}$.

Using Equation 5.17, we can recast Equation 5.23 as

$$x_{st} = \frac{\langle \rho_v \rangle_v \langle \alpha \rangle}{\langle \rho_v \rangle_v \langle \alpha \rangle + \langle \rho_\ell \rangle_\ell \langle 1 - \alpha \rangle} \tag{5.24}$$

5.3.4 Mixture Density

The two-phase mixture density in a volume can be given by

$$\langle \rho \rangle = \frac{m_{\mathrm{v}} + m_{\ell}}{V}$$

and by using Equations 5.16 and 5.17 we get

$$\langle \rho \rangle = \langle \rho_{\mathrm{v}} \rangle_{\mathrm{v}} \langle \alpha \rangle + \langle \rho_{\ell} \rangle_{\ell} \langle 1 - \alpha \rangle \tag{5.25}$$

Consequently, the average phasic density may be given by (using Equations 5.24 and 5.25):

$$\langle \rho_{\mathrm{v}} \rangle_{\mathrm{v}} = \frac{x_{\mathrm{st}} \langle \rho \rangle}{\langle \alpha \rangle} \tag{5.26a}$$

or

$$\langle \rho_{\ell} \rangle_{\ell} = \frac{(1 - x_{\mathrm{st}}) \langle \rho \rangle}{\langle 1 - \alpha \rangle} \tag{5.26b}$$

5.4 AREA-AVERAGED PROPERTIES

5.4.1 Area-Averaged Phase Fraction

The instantaneous fraction of an area occupied by phase k is given by

$$\{\alpha_k\} \equiv \frac{A_k}{A} = \frac{1}{A} \iint_A \alpha_k \, \mathrm{d}A \tag{5.27}$$

As the area not occupied by phase k must be occupied by the other phase k', the fraction of area occupied by phase k can also be given as

$$\{\alpha_k\} \equiv \frac{A_k}{A} = \frac{A_k}{A_k + A_{k'}} \tag{5.28}$$

If space- and time-averaging commutativity are applied, the time- and area-averaged phase fraction can be given by

$$\widetilde{\{\alpha_k\}} \equiv \{\tilde{\alpha}_k\} \tag{5.29}$$

Similar to Equation 5.17, it is possible to relate the phase property average to the total area property average by

$$\{c_k\} = \{c_k\}_k\{\alpha_k\} \tag{5.30}$$

Similar to Equation 5.19, the time- and area-averaged property $\widetilde{\{c_k\}}$ may be obtained from the following manipulations:

$$\{c_k\} = \widetilde{\{c_k\}_k\{\alpha_k\}} = \widetilde{\widetilde{\{c_k\}_k}\widetilde{\{\alpha_k\}}} + \widetilde{\widetilde{\{c_k\}_k}\{\alpha_k\}'} + \widetilde{\{c_k\}'_k\widetilde{\{\alpha_k\}}} + \widetilde{\{c_k\}'_k\{\alpha_k\}'} \tag{5.31a}$$

However,

$$\widetilde{\{\alpha_k\}'} = \widetilde{\{c\}'_k} = 0 \text{ and } \widetilde{\widetilde{\{c_k\}}} = \widetilde{\{c_k\}}$$

Therefore, the last equation becomes

$$\widetilde{\{c_k\}} = \widetilde{\{c_k\}_k}\widetilde{\{\alpha_k\}} + \widetilde{\{c_k\}'_k\{\alpha_k\}'} \tag{5.31b}$$

So that when turbulent fluctuations in either of $\{\alpha_k\}$ or $\{c_k\}$ are ignored

$$\widetilde{\{c_k\}_k} = \frac{\widetilde{\{c_k\}}}{\widetilde{\{\alpha_k\}}} \tag{5.32}$$

Example 5.1: Time Average of Area-Averaged Void Fraction

PROBLEM

Consider a series of cylindrical bubbles, each of length ℓ_2 and diameter d_b, moving at a velocity V_b in a cylindrical pipe. The pipe diameter is D, and its length is L.

If the bubbles are separated by a distance $\ell_1 = 0.5\ell_2$, and the bubble diameter $d_b = 0.4D$, evaluate the area-averaged void fraction at cross sections 1 and 2 at the instant illustrated in Figure 5.1. Also, evaluate the time- and area-averaged void fraction at positions 1 and 2.

SOLUTION

The instantaneous area-averaged void fractions at the positions indicated in Figure 5.1 are obtained by applying Equation 5.28:

$$\{\alpha\}_1 = 0$$

$$\{\alpha\}_2 = \frac{\pi d_b^2/4}{\pi D^2/4} = (0.4)^2 = 0.16$$

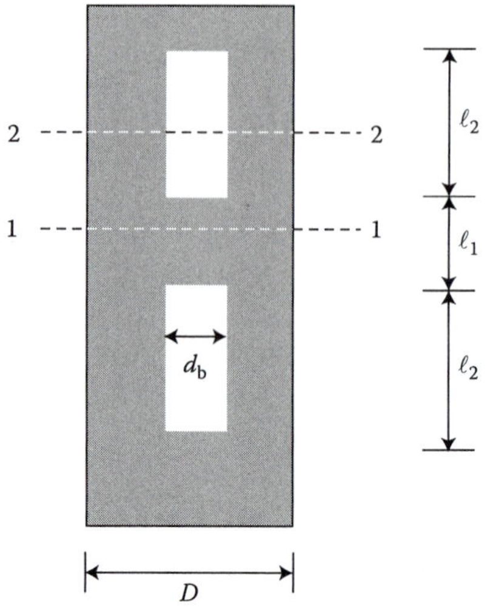

FIGURE 5.1 Bubble and pipe characteristics for Example 5.1.

The time- and area-averaged values are equal: $\widetilde{\{\alpha\}}_1 = \widetilde{\{\alpha\}}_2 = \widetilde{\{\alpha\}}$ and are obtained by applying Equation 5.6:

$$\widetilde{\{\alpha\}} = \frac{1}{\Delta t^*} \int_{t-\Delta t^*/2}^{t+\Delta t^*/2} \{\alpha\}\, dt = \frac{1}{\Delta t^*} \int_0^{\Delta t^*} \{\alpha\}\, dt$$

$$\widetilde{\{\alpha\}} = \frac{(0)(\ell_1/V_b) + (0.16)(\ell_2/V_b)}{(\ell_1 + \ell_2)/V_b}$$

$$\widetilde{\{\alpha\}} = 0.16\left(\frac{\ell_2}{\ell_1 + \ell_2}\right) = 0.16\left(\frac{1}{0.5 + 1}\right) = 0.1067$$

Note that Δt^* is chosen such that

$$\Delta t^* > \frac{\ell_1 + \ell_2}{V_b}$$

to properly average the two-phase properties. Under transient conditions, it is important to select an averaging time less than the characteristic bubble residence time in the tube so that

$$\Delta t^* < \frac{L}{V_b}$$

For steady-state conditions this condition is immaterial, as the behavior of the two-phase system is periodic.

Example 5.2: Equivalence of Time–Area and Area–Time Sequence of Operators

PROBLEM

Consider vertical flow of a gas–liquid mixture between two parallel plates with two types of gas bubble with plate geometry as shown in Figure 5.2a. The size and spacing of the bubbles are indicated on the figure. The bubble velocities are related by $V_A = 2V_B$. Six probes measure the phase–time relations: r_1 to r_6. The measured signals appear in Figure 5.2b.

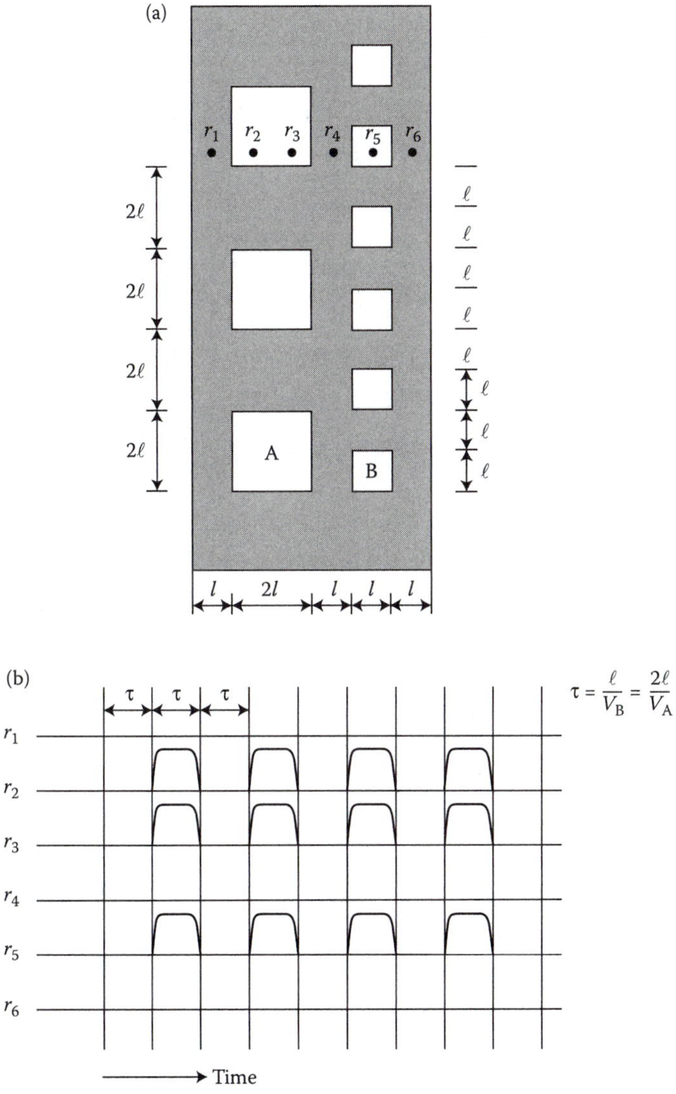

FIGURE 5.2 Bubble configuration (a) and vapor sensor output (b) for Example 5.2.

Determine the area average of the time-averaged local void fractions, at r_1 to r_6, that is, $\{\tilde{\alpha}_v\}$. Also, determine the time average of the instantaneous area-averaged void fraction at the plane of the probes, that is, $\widetilde{\{\alpha_v\}}$. Are they equal for the stated conditions?

SOLUTION

The local time-averaged void fraction can be written from Equation 5.12 as

$$\tilde{\alpha}_k = \frac{1}{\Delta t^*}\int_0^{\Delta t^*} \alpha_k \, dt; \quad \alpha_k = 1 \text{ if phase } k \text{ is present}$$

where $\Delta t^* \gg \tau$, the time it takes one bubble to pass by the probe. Therefore,

$$\tilde{\alpha}_{v1} = 0$$
$$\tilde{\alpha}_{v2} = 0.5$$
$$\tilde{\alpha}_{v3} = 0.5$$
$$\tilde{\alpha}_{v4} = 0$$
$$\tilde{\alpha}_{v5} = 0.5$$
$$\tilde{\alpha}_{v6} = 0$$

At the plane of the sensors, the area average of the local time-averaged void fraction is

$$\{\tilde{\alpha}_v\} = \frac{1}{6\ell}\int_0^{6\ell} \tilde{\alpha}_v \, d\ell = \frac{1}{6\ell}\sum_j \tilde{\alpha}_{vj}\ell_j = [0.5(2\ell + \ell)]\frac{1}{6\ell} = 0.25$$

Next consider the average of the instantaneous void fraction at the plane of the sensors for the time shown ($t = 0$):

$$\{\alpha_v\} = \frac{\sum_j \alpha_v \ell_j}{6\ell} = 0.5$$

If we observe the time behavior of $\{\alpha_v\}$, we get

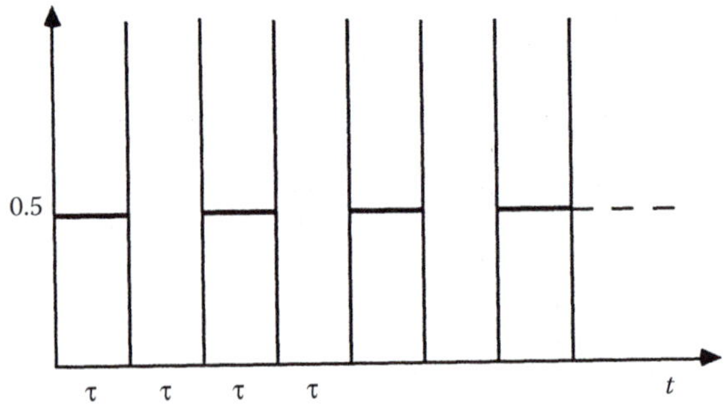

which implies that

$$\widetilde{\{\alpha_v\}} = \frac{1}{\Delta t^*}\int_0^{\Delta t^*} \{\alpha_v\} dt = \frac{0.5\tau + 0\tau}{2\tau} = 0.25$$

Therefore,

$$\{\tilde{\alpha}_v\} = \widetilde{\{\alpha_v\}}$$

which is what Vernier and Delhaye [6] proved mathematically.

5.4.2 Flow Quality

The instantaneous mass flow rate of phase k through an area $\vec{A}_j$ can be obtained from

$$\dot{m}_{kj} = \iint_{A_j} \alpha_k \rho_k \vec{\upsilon}_k \cdot \vec{n}\, dA_j = \{\alpha_k \rho_k \vec{\upsilon}_k\}_j \cdot \vec{A}_j \tag{5.33a}$$

Using Equation 5.30, the instantaneous mass flow rate can be written as

$$\dot{m}_{kj} = \{\rho_k \vec{\upsilon}_k\}_{kj} \{\alpha_k\}_j \cdot \vec{A}_j \tag{5.33b}$$

The time-averaged flow rate of phase k is then obtained by integrating Equation 5.33b over the time Δt^*. Performing this integration we get, according to Equation 5.31b:

$$\widetilde{\dot{m}_{kj}} = \widetilde{\{\rho_k \vec{\upsilon}_k\}_{kj}}\, \widetilde{\{\alpha_k\}_j} \cdot \vec{A}_j \text{ (turbulent fluctuation in } \{\alpha_k\} \text{ or } \{\rho_k \vec{\upsilon}_k\}_k \text{ is ignored)} \tag{5.34}$$

The vapor mass flow fraction of the total flow is called the *flow quality* (x). In order to define the flow quality in a one-dimensional flow, for example, in the z direction, we write

$$x_z \equiv \frac{\dot{m}_{vz}}{(\dot{m}_{vz} + \dot{m}_{\ell z})} \tag{5.35}$$

Let

$$\dot{m} = \dot{m}_{vz} + \dot{m}_{\ell z} \tag{5.36}$$

If the total flow rate is assumed nonfluctuating, that is

$$\widetilde{\dot{m}_z} = \dot{m}_z$$

for the time-averaged value of the quality, we get

$$\tilde{x}_z = \frac{\tilde{\dot{m}}_{vz}}{\tilde{\dot{m}}_{vz} + \tilde{\dot{m}}_{\ell z}} = \frac{\tilde{\dot{m}}_{vz}}{\dot{m}_z} \tag{5.37}$$

The flow quality is often used in the analysis of predominantly one-dimensional two-phase flow. Many correlations that are extensively used in two-phase flow and heat transfer are given in terms of the flow quality. The flow quality becomes particularly useful when thermodynamic equilibrium between the two phases is assumed. In that case the flow quality can be obtained from the energy balances of the flow, although additional information or correlations are needed to calculate the void fraction. With two-dimensional flow, the quality at a given plane has two components and is not readily defined by a simple energy balance. Therefore, for two- or three-dimensional flow, the flow quality becomes less useful because of the scarcity of literature on the effect of the vectorial nature of the quality on the flow conditions.

The concept of flow quality breaks down when the net flow rate is zero or when there is countercurrent flow of two phases (extensions to these cases are possible but not very useful).

5.4.3 Mass Fluxes

The superficial mass fluxes or mass velocities of the phases are defined as the phase flow rates per unit cross-sectional area, for example

$$G_{\ell z} \equiv \frac{\dot{m}_{\ell z}}{A_z} = \frac{\dot{m}_z(1-x_z)}{A_z} = G_{mz}(1-x_z) \tag{5.38a}$$

and

$$G_{vz} = \frac{\dot{m}_{vz}}{A_z} = \frac{\dot{m}_z x_z}{A_z} = G_{mz} x_z \tag{5.38b}$$

The superficial mass velocity can also be expressed in terms of the actual phasic velocity by specifying $\dot{m}_{vz}$ from Equation 5.33b:

$$G_{kz} \equiv \frac{\{\rho_k \vec{v}_k\}_{kz}\{\alpha_k\}_z \cdot \vec{A}_z}{A_z} = \{\rho_k v_{kz}\}_{kz}\{\alpha_k\}_z \tag{5.39}$$

The mass flux (G_m) is defined as the average mass flow rate per unit flow area. Thus, in the z direction

$$G_{mz} \equiv \frac{\dot{m}_z}{A_z} = \frac{\dot{m}_{vz} + \dot{m}_{\ell z}}{A_z} = G_{vz} + G_{\ell z} \tag{5.40a}$$

Utilizing Equation 5.39, the mass flux of the mixture can also be given by

$$G_{mz} = \{\rho_v \upsilon_{vz}\}_{vz}\{\alpha_v\}_z + \{\rho_\ell \upsilon_{\ell z}\}_{\ell z}\{\alpha_\ell\}_z \tag{5.40b}$$

which, using Equation 5.30, can be written as

$$G_{mz} = \{\alpha_v \rho_v \upsilon_{vz} + \alpha_\ell \rho_\ell \upsilon_{\ell z}\}_z \tag{5.40c}$$

Time-averaging G_{mz}, using Equation 5.40b and ignoring time fluctuations in $[\alpha_k]$, we obtain

$$\tilde{G}_{mz} = \frac{\tilde{\dot{m}}_z}{A_z} = \overline{\{\rho_v \upsilon_{vz}\}_{vz}\{\alpha\}_z} + \overline{\{\rho_\ell \upsilon_{\ell z}\}_{\ell z}\{1-\alpha\}_z} \tag{5.41}$$

5.4.4 Volumetric Fluxes and Flow Rates

The instantaneous volumetric flow rate of phase k may be related to the volumetric flux (or superficial velocity) over the area A_j by

$$Q_{kj} = \{\vec{j}_k\}_j \cdot \vec{A}_j \tag{5.42}$$

Note that for one-dimensional flow in the z-direction, when the spatial variation in ρ_k can be neglected over the area A_z, we get

$$\{j_v\}_z = \frac{Q_{vz}}{A_z} = \frac{\dot{m}_{vz}}{\rho_v A_z} = \frac{G_{vz}}{\rho_v} = \frac{G_{mz} x_z}{\rho_v} \tag{5.43a}$$

Similarly

$$\{j_\ell\}_z = \frac{Q_{\ell z}}{A_z} = \frac{\dot{m}_{\ell z}}{\rho_\ell A_z} = \frac{G_{\ell z}}{\rho_\ell} = \frac{G_{mz}(1-x_z)}{\rho_\ell} \tag{5.43b}$$

and

$$\{j\}_z = \frac{Q_{\ell z} + Q_{vz}}{A_z} = \frac{Q_z}{A_z} = G_m\left[\frac{1-x_z}{\rho_\ell} + \frac{x_z}{\rho_v}\right] \tag{5.44}$$

Because the volumetric flow rates Q_ℓ and Q_v are simply related to the mass flow rates and the phasic densities from Equations 5.43a and b, the volumetric fluxes $[j_k]$ can be considered known if the mass flow rates of the two phases—or, alternatively, the total mass flow rate and the quality—are known.

The local volumetric flux (or superficial velocity) of phase k is obtained from

$$\vec{j}_k = \alpha_k \vec{\upsilon}_k \tag{5.45}$$

Space- and time-averaged values of the volumetric fluxes can be obtained starting with either operation, as space and time commutation can be assumed

$$\{\widetilde{\vec{j}_k}\} = \widetilde{\{\vec{j}_k\}} = \widetilde{\{\alpha_k \ \vec{\upsilon}_k\}} \tag{5.46}$$

Similar to Equation 5.31b, by ignoring time fluctuations in $\{a_k\}$ or $\{\vec{\upsilon}_k\}$ Equation 5.46 can be shown to yield

$$\widetilde{\{\vec{j}_k\}} = \widetilde{\{\alpha_k\}}\widetilde{\{\vec{\upsilon}_k\}_k} \tag{5.47}$$

5.4.5 Velocity (Slip) Ratio

The time- and space-averaged, or macroscopic, velocity ratio, often referred to as *slip ratio* (S), is defined as

$$S_i \equiv \frac{\{\tilde{\upsilon}_{vi}\}_v}{\{\tilde{\upsilon}_{\ell i}\}_\ell} \tag{5.48}$$

S_i is a directional value, as it depends on the velocity in the direction i.

For spatially uniform liquid and vapor densities and when time fluctuations in $\{\alpha_k\}$ or $\{\upsilon_{ki}\}$ are ignored, Equations 5.43a, 5.43b, and 5.47 can be used to obtain

$$S_i = \frac{\{\tilde{j}_v\}_i}{\{\tilde{j}_\ell\}_i} \frac{\{1-\alpha\}_i}{\{\alpha\}_i} = \frac{x_i}{1-x_i} \frac{\rho_\ell}{\rho_v} \frac{\{1-\alpha\}_i}{\{\alpha\}_i} \tag{5.49}$$

This equation is a commonly used relation between x_i, $\{\alpha_v\}_i$, and S_i. If any two are known, the third can be determined. Experimental void fraction data for one-dimensional flows have been used in the past to calculate the velocity ratio and then correlate S_i in terms of the relevant parameters. However, $\vec{S}$ cannot be easily related to meaningful parameters in three-dimensional flows. For one-dimensional flows, several models can be used to estimate S_i, as discussed in Chapter 11.

5.4.6 Mixture Density Over an Area

Analogous to Equation 5.25, we define the mixture average density of an area by

$$\rho_m \equiv \{\rho\} = \{\alpha\}\{\rho_v\}_v + \{1-\alpha\}\{\rho_\ell\}_\ell \tag{5.50a}$$

Note that this density can be called the *static density* of the mixture because it is not affected by the flow velocity. Using Equation 5.30 and the area averaging analogy to Equation 5.16, we can write ρ_m as

$$\rho_m = \{\alpha_v \rho_v\} + \{\alpha_\ell \rho_\ell\} \tag{5.50b}$$

5.4.7 Volumetric Flow Ratio

The volumetric flow ratio of vapor to the total mixture is given by

$$\{\beta_z\} = \frac{Q_{vz}}{Q_{vz} + Q_{\ell z}} = \frac{\{j_v\}_z}{\{j\}_z} \tag{5.51}$$

5.4.8 Flow Thermodynamic Quality

The flow (mixing-cup) enthalpy is defined for flow in the direction "i" as

$$(h_m^+)_i \equiv x_i h_v + (1 - x_i) h_\ell \tag{5.52}$$

The flow thermodynamic (or equilibrium) quality is given by relating the flow enthalpy to the saturation liquid and vapor enthalpies as

$$(x_e)_i = \frac{(h_m^+)_i - h_f}{h_g - h_f} \tag{5.53}$$

where h_f and h_g = saturation-specific enthalpies of the liquid and vapor, respectively.

Thus, the flow quality can be used to define the equilibrium quality:

$$(x_e)_i = \frac{x_i h_v + (1 - x_i) h_\ell - h_f}{h_g - h_f} \tag{5.54}$$

Under thermal equilibrium conditions

$$h_\ell = h_f \text{ and } h_v = h_g; \text{ therefore } (x_e)_i = (x)_i$$

5.4.9 Summary of Useful Relations for One-Dimensional Flow

Table 5.2 gives a summary of useful relations between parameters in one-dimensional flow. It may be helpful to summarize the relations between the void fraction $\{\alpha\}$ and the static quality (x_{st}) at an area (*static parameters*), with the volumetric flow vapor fraction $\{\beta\}$ and the flow quality (x) (i.e., the dynamic parameters). Assuming uniform phase densities across an area, Equation 5.49 leads to

$$\{\alpha\} = \frac{1}{1 + [(1 - x)/x](\rho_v/\rho_\ell)S} \tag{5.55}$$

TABLE 5.2
One-Dimensional Relation between Two Phase Parameters

	$\dot{m}$	G	Q	$\{\alpha\}$	$\{\beta\}$	$\{\upsilon\}$	$\{j\}$	x_{st}	x
$\dot{m}$	$\dot{m} = \dot{m}_k + \dot{m}_{k'}$								
G	$G_k = \frac{\dot{m}_k}{A}$	$G_m = G_k + G_{k'}$							
Q	$Q_k = \frac{\dot{m}_k}{\rho_k}$	$Q_k = \frac{G_k A}{\rho_k}$	$Q = Q_k + Q_{k'}$						
$\{\alpha\}$	$\{\alpha_k\} = \left(1 + \frac{\dot{m}_{k'}\{\rho\upsilon\}_k}{\dot{m}_k\{\rho\upsilon\}_{k'}}\right)^{-1}$	$\{\alpha_k\} = \left(1 + \frac{G_{k'}\{\rho\upsilon\}_k}{G_k\{\rho\upsilon\}_{k'}}\right)^{-1}$	$\{\alpha_k\} = \left(1 + \frac{Q_{k'}}{Q_k}\frac{\{\upsilon\}_k}{\{\upsilon\}_{k'}}\right)^{-1}$	—					
$\{\beta\}$	$\{\beta_k\} = \frac{(\dot{m}/\rho)_k}{(\dot{m}/\rho)_k + (\dot{m}/\rho)_{k'}}$	$\{\beta_k\} = \frac{(G/\rho)_k}{(G/\rho)_k + (G/\rho)_{k'}}$	$\{\beta_k\} = \frac{Q_k}{Q}$	$\{\beta_k\} = \left(1 + \frac{\{\alpha\upsilon\}_{k'}}{\{\alpha\upsilon\}_k}\right)^{-1}$	—				
$\{\upsilon\}$	$\{\upsilon_k\} = \frac{\dot{m}_k}{\{\rho A\}_k}$	$\{\upsilon_k\} = \frac{G_m x_k}{\{\rho\alpha\}_k}$	$\{\upsilon_k\} = \frac{Q_k}{A_k}$	$\frac{\{\upsilon_k\}}{\{\upsilon_{k'}\}} = \frac{\{\alpha\rho\}_{k'}}{\{\alpha\rho\}_k}\frac{x_k}{x_{k'}}$	$\frac{\{\upsilon_k\}}{\{\upsilon_{k'}\}} = \frac{\{\beta_k\}}{\{\beta_{k'}\}}\frac{\{\alpha_{k'}\}}{\{\alpha_k\}}$	—			
$\{j\}$	$\{j_k\} = \frac{\dot{m}_k}{\rho_k A}$	$\{j_k\} = \frac{G_k}{\rho_k}$	$\{j_k\} = \frac{Q_k}{A}$	$\{j_k\} = \{(\alpha\upsilon)_k\}$	$\{j_k\} = \{\beta_k\}\{j\}$	$\{j_k\} = \{(\upsilon\alpha)_k\}$	$\{j\} = \{j_k\} + \{j_k'\}$		
x_{st}	$\{x_{st_k}\} = \left(1 + \frac{\dot{m}_{k'}}{\dot{m}_k}\frac{\{\upsilon_k\}}{\{\upsilon_{k'}\}}\right)^{-1}$	$\{x_{st_k}\} = \left(1 + \frac{G_{k'}}{G_k}\frac{\{\upsilon_k\}}{\{\upsilon_{k'}\}}\right)^{-1}$	$\{x_{st_k}\} = \frac{Q_k}{\{\upsilon_k\}A}\frac{\rho_k}{\rho_m}$	$\{x_{st_k}\} = \frac{\{(\alpha\rho)_k\}}{\rho_m}$	$\{x_{st_k}\} = \frac{\{\beta_k\}\{j\}}{\{\upsilon_k\}}\frac{\rho_k}{\rho_m}$	See $\{x_{st_k}\}$ vs. Q, β, or j	$\{x_{st_k}\} = \frac{\{j_k\}}{\{\upsilon_k\}}\frac{\rho_k}{\rho_m}$	—	
x	$x_k = \frac{\dot{m}_k}{\dot{m}}$	$x_k = \frac{G_k}{G_m}$	$x_k = \left[1 + \frac{(Q\rho)_{k'}}{(Q\rho)_k}\right]^{-1}$	$x_k = \left(1 + \frac{\{(\alpha\rho\upsilon)_{k'}\}}{\{(\alpha\rho\upsilon)_k\}}\right)^{-1}$	$x_k = \left(1 + \frac{\{(\beta\rho)_{k'}\}}{\{(\beta\rho)_k\}}\right)^{-1}$	See x_k vs. α or $\{x_{st_k}\}$	$x_k = \left(1 + \frac{\{(j\rho)_{k'}\}}{\{(j\rho)_k\}}\right)^{-1}$	$x_k = \left[1 + \frac{\{(x_{st}\upsilon)_{k'}\}}{\{(x_{st}\upsilon)_k\}}\right]^{-1}$	—

Note: All entries above the diagonal can be obtained by inverting corresponding existing entries in this table.

and

$$\frac{x}{1-x} = \frac{\{\alpha\}\,\rho_v}{(1-\{\alpha\})\rho_\ell}S \tag{5.56}$$

However, the area-averaged static quality (x_{st}) can be defined (see Equation 5.26a) for space-independent density ρ_v and ρ_ℓ as

$$x_{st} \equiv \frac{\{\alpha\}\rho_v}{\rho_m} \tag{5.57}$$

which yields

$$\frac{x_{st}}{1-x_{st}} = \frac{\{\alpha\}\rho_v}{(1-\{\alpha\})\rho_\ell} \tag{5.58}$$

Comparing Equations 5.56 and 5.58, it is obvious that if $S = 1$, $x = x_{st}$.
Consider also the volumetric flow fraction given by

$$\{\beta\} = \frac{\{j_v\}}{\{j\}} = \frac{1}{1+\{j_\ell\}/\{j_v\}} \tag{5.59}$$

Since for space-independent ρ_v and ρ_ℓ

$$\{j_\ell\} = \frac{(1-x)G_m}{\rho_\ell} \text{ and } \{j_v\} = \frac{xG_m}{\rho_v}$$

we get

$$\{\beta\} = \frac{1}{1+\left[(1-x)/x\right](\rho_v/\rho_\ell)} \tag{5.60}$$

By comparing Equations 5.55 and 5.60, we get

$$\{\alpha\} = \{\beta\} \quad \text{if } S = 1$$

The assumption of $S = 1$ is referred to as the *homogeneous flow assumption*. Generally, $S \neq 1$, as is discussed in Chapter 11.

A summary of the relations of x, x_{st}, $\{\alpha\}$, and $\{\beta\}$ appears in Figure 5.3.

5.5 MIXTURE EQUATIONS FOR ONE-DIMENSIONAL FLOW

The one-dimensional case is presented first as it is often encountered when describing energy equipment, and because it is easier to establish than the multidimensional cases. A formal derivation of the one-dimensional equations as a special case of the three-dimensional equations is given in Section 5.7.

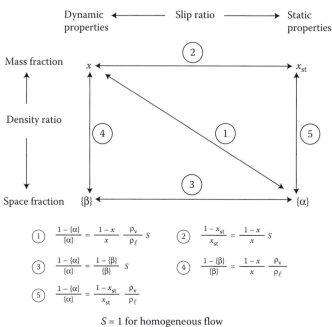

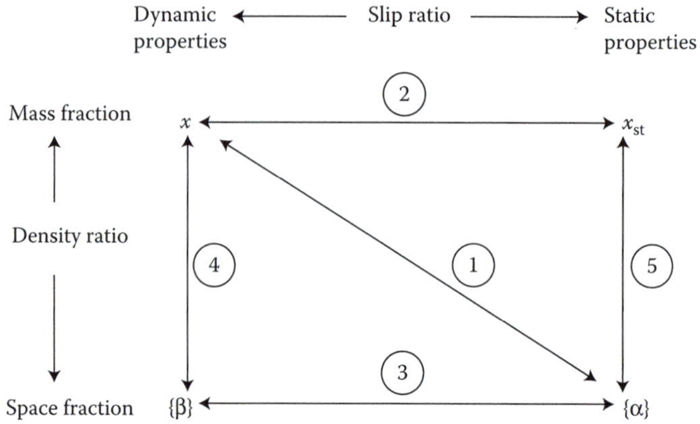

$$\text{(1)}\quad \frac{1-\{\alpha\}}{\{\alpha\}} = \frac{1-x}{x}\,\frac{\rho_v}{\rho_\ell}\,S \qquad \text{(2)}\quad \frac{1-x_{st}}{x_{st}} = \frac{1-x}{x}\,S$$

$$\text{(3)}\quad \frac{1-\{\alpha\}}{\{\alpha\}} = \frac{1-\{\beta\}}{\{\beta\}}\,S \qquad \text{(4)}\quad \frac{1-\{\beta\}}{\{\beta\}} = \frac{1-x}{x}\,\frac{\rho_v}{\rho_\ell}$$

$$\text{(5)}\quad \frac{1-\{\alpha\}}{\{\alpha\}} = \frac{1-x_{st}}{x_{st}}\,\frac{\rho_v}{\rho_\ell}$$

$S = 1$ for homogeneous flow

FIGURE 5.3 Useful relations for one-dimensional flow and spatially uniform phase densities.

5.5.1 Mass Continuity Equation

Consider the one-dimensional flow through a plane of an area A_z at a position z along a channel as illustrated in Figure 5.4. The mass balance equation can be written as

$$\frac{\partial}{\partial t}\iint_{A_z} \rho\, dA_z + \frac{\partial}{\partial z}\iint_{A_z} \rho \upsilon_z\, dA_z = 0 \tag{5.61}$$

or

$$\frac{\partial}{\partial t}\{\rho_v \alpha + \rho_\ell (1-\alpha)\}A_z + \frac{\partial}{\partial z}\{\rho_v \alpha \upsilon_{vz} + \rho_\ell (1-\alpha) v_{\ell z}\}A_z = 0 \tag{5.62}$$

where υ_{vz} and $\upsilon_{\ell z}$ = vapor and liquid velocities in the axial direction. Equation 5.62 can also be written as

$$\frac{\partial}{\partial t}(\rho_m A_z) + \frac{\partial}{\partial z}(G_m A_z) = 0 \tag{5.63}$$

where ρ_m and G_m = average mixture density and mass flux over the area A_z, defined in Equations 5.50b and 5.40c, respectively.

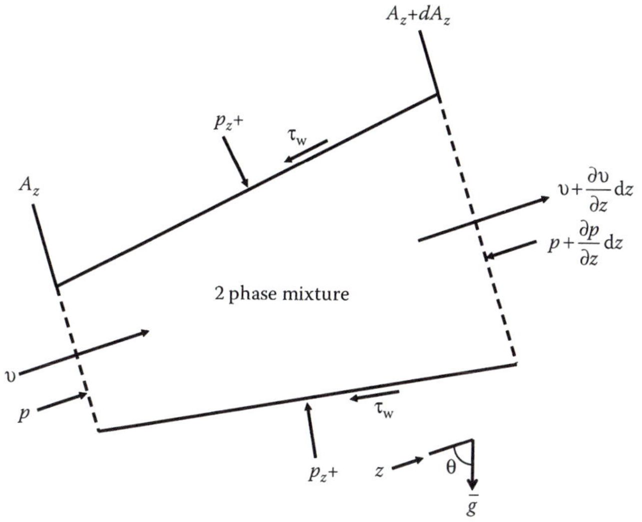

FIGURE 5.4 One-dimensional flow in a channel.

5.5.2 Momentum Equation

Consider next the momentum equation, which can be obtained from equating the rate of momentum change to the net forces (shown in Figure 5.4 in the axial direction):

$$\frac{\partial}{\partial t}\iint_{A_z}\rho\upsilon_z\,dA_z+\frac{\partial}{\partial z}\iint_{A_z}\rho\upsilon_z^2\,dA_z=-\iint_{A_z}\frac{\partial p}{\partial z}dA_z+\frac{\partial A_z}{\partial z}\frac{1}{P_z}\int_{P_z}p\,dP_z$$
$$-\int_{P_z}\tau_w\,dP_z-\iint_{A_z}\rho g\cos\theta\,dA_z \tag{5.64}$$

where P_z = perimeter of the channel at the location z, and θ = angle between the axis of the channel and the upward vertical direction. The correct form of the τ_w term in Equation 5.64 should be τ_w cos α, where α is the angle formed by the slope of the channel walls with respect to the channel axis. If this slope is small, the use of τ_w only, as written above in Equation 5.64, is sufficiently accurate. Thus

$$\frac{\partial}{\partial t}\{\rho_v\alpha\upsilon_{vz}+\rho_\ell(1-\alpha)\upsilon_{\ell z}\}A_z+\frac{\partial}{\partial z}\{\rho_v\alpha\upsilon_{vz}^2+\rho_\ell(1-\alpha)\upsilon_{\ell z}^2\}A_z=$$
$$-\left(\frac{\partial\{p\}A_z}{\partial z}\right)+\frac{\partial A_z}{\partial z}\frac{1}{P_z}\int_{P_z}p\,dP_z-\int_{P_z}\tau_w\,dP_z-\{\rho_v\alpha+\rho_\ell(1-\alpha)\}g\cos\theta A_z \tag{5.65}$$

If it is assumed that the pressure p is uniform within the area A_z and the pressure averaged over the perimeter can simply be taken as p, then we can write Equation 5.65 as

$$\frac{\partial}{\partial t}(G_m A_z)+\frac{\partial}{\partial z}\left(\frac{G_m^2}{\rho_m^+}A_z\right)=-\frac{\partial(pA_z)}{\partial z}+p\frac{\partial A_z}{\partial z}-\int_{P_z}\tau_w\,dP_z-\rho_m g\cos\theta A_z \tag{5.66}$$

where ρ_m is defined in Equations 5.50a and 5.50b, and ρ_m^+ = dynamic (or mixing cup) density given by

$$\frac{1}{\rho_m^+} \equiv \frac{1}{G_m^2}\left[\{\rho_v \upsilon_{vz}^2 \alpha\} + \{\rho_\ell \upsilon_{\ell z}^2 (1-\alpha)\}\right] \quad (5.67)$$

Note that because the velocity in a general case has three components, the dynamic density is, in effect, a directional quantity.

5.5.3 Energy Equation

Finally, for the energy balance, we get

$$\frac{\partial}{\partial t}\iint_{A_z} \rho u^o \, dA_z + \frac{\partial}{\partial z}\iint_{A_z} \rho h^o \upsilon_z \, dA_z = q' + \iint_{A_z} q''' \, dA_z \quad (5.68)$$

where q' = linear heat addition rate from the walls, q''' = volumetric heat generation rate in the coolant, and the axial heat conduction and the work terms are ignored. Note that in this formulation there is no expansion work term, since the control volume is fixed and the surface forces due to pressure and wall shear, p and τ_w, are acting on a stationary surface. In reality, the internal shear leads to conversion of some of the mechanical energy into an internal heat source, which is often negligibly small. The wall shear stress τ_w indirectly affects u^o by influencing the pressure in the system. Finally, the work due to the gravity force is considered negligible. A much more elaborate derivation is given in Section 5.7.

Equation 5.68 may be written as

$$\frac{\partial}{\partial t}\iint_{A_z} \rho h^o \, dA_z + \frac{\partial}{\partial z}\iint_{A_z} \rho h^o \, \upsilon_z \, dA_z = \frac{\partial}{\partial t}\iint_{A_z} \rho (pv) \, dA_z + q' + \iint_{A_z} q''' \, dA_z \quad (5.69)$$

or

$$\frac{\partial}{\partial t}(\{\rho_v \alpha h_v^o + \rho_\ell (1-\alpha) h_\ell^o\} \, A_z) + \frac{\partial}{\partial z}(\{\rho_v \alpha h_v^o \upsilon_{vz} + \rho_\ell (1-\alpha) h_\ell^o v_{\ell z}\} A_z)$$
$$= \left(\frac{\partial p}{\partial t}\right) A_z + q' + \iint_{A_z} q''' \, dA_z \quad (5.70)$$

where p was assumed uniform over A_z. Ignoring the kinetic energy of both phases, Equation 5.70 may be written as

$$\frac{\partial}{\partial t}(\rho_m h_m A_z) + \frac{\partial}{\partial z}(G_m h_m^+ A_z) = \left(\frac{\partial p}{\partial t}\right) A_z + q' + \iint_{A_z} q''' \, dA_z \quad (5.71)$$

where

$$h_m = \frac{1}{\rho_m}\{\rho_v \alpha h_v + \rho_\ell (1-\alpha) h_\ell\} \tag{5.72}$$

is the static mixture enthalpy averaged over the area A, and

$$h_m^+ = \frac{1}{G_m}(\{\rho_v \alpha h_v \upsilon_{vz} + \rho_\ell (1-\alpha) h_\ell \upsilon_{\ell z}\}) \tag{5.73}$$

is the dynamic (or mixing cup) average enthalpy of the flowing mixture (similar to Equation 5.52), which is a directional quantity.

The balance equations presented above do not impose restrictions on the velocities or the temperatures of the two phases. However, various simplifications can be imposed if deemed accurate for the case of interest. For example, if υ_{vz} and $\upsilon_{\ell z}$ were taken to be equal

$$\rho_m^+ = \rho_m \text{ for equal velocities} \tag{5.74}$$

and

$$h_m^+ = h_m \text{ for equal velocities} \tag{5.75}$$

Furthermore, when saturated conditions are assumed to apply, Equations 5.63, 5.66, and 5.71 in ρ_m, G_m, h_m, and p can be solved when: (1) q', q''', and τ_w are specified in relation to the other variables, and (2) the state equation for the fluid $\rho_m(p,h_m)$ is used. Note that these equations can be extended to apply to the single-phase liquid, saturated mixture, and single-phase vapor flow sections of the channel.

Various approaches to the specification of τ_w are discussed in Chapter 11.

5.6 CONTROL-VOLUME INTEGRAL TRANSPORT EQUATIONS

A two-fluid (separate flow) approach is used to derive the instantaneous transport equations for liquid and vapor. Considering each of the phases to be a continuum material, we can write the mass, momentum, and energy equations by applying the relations derived for deformable volumes in Section 4.4 of Chapter 4.

We shall consider two volumes V_v and V_ℓ (Figure 5.5), which are joined to obtain the two-fluid equations in the total volume, that is

$$V = V_\ell + V_v \tag{5.76}$$

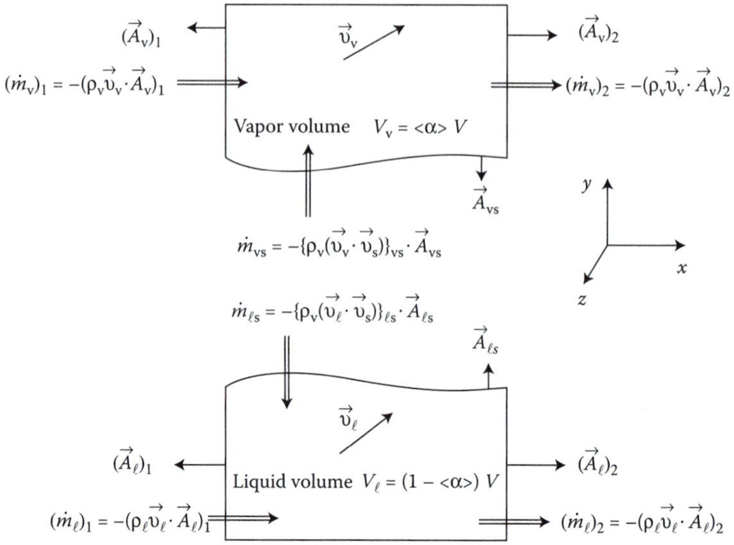

FIGURE 5.5 Mass balance components in a two-phase volume. The geometric control volume is composed of the phasic subvolumes separated by the wavy lines. *Single arrows* represent vectors, which are positive in the direction of the coordinate axes. *Double arrows* are scalars, which are positive when adding to the control volume content.

5.6.1 Mass Balance

5.6.1.1 Mass Balance for Volume V_k

The rate of change of the mass in the control volume V_k depicted in Figure 5.5 should be equal to the net convection of phase k mass into the volume and due to the change of phase. Applying Equation 4.57 to the control volume of the phase k, we obtain

$$\frac{d}{dt}\iiint_{V_k} \rho_k \, dV + \iint_{S_k} \rho_k (\vec{\upsilon}_k - \vec{\upsilon}_s) \cdot \vec{n} \, dS = 0$$

Let us divide the surface S_k into a fixed surface A_{kj} at the control volume boundary (across which convection may occur) and the deformable surface A_{ks} at the liquid–vapor interface (across which change of phase occurs). Thus, we get

$$\frac{d}{dt}\iiint_{V_k} \rho_k \, dV + \iint_{A_{kj}} \rho_k (\vec{\upsilon}_k - \vec{\upsilon}_s) \cdot \vec{n} \, dS + \iint_{A_{ks}} \rho_k (\vec{\upsilon}_k - \vec{\upsilon}_s) \cdot \vec{n} \, dS = 0$$

Rate of mass increase in the control volume	+	Rate of mass loss by convection	+	Rate of mass loss by change of phase	=	0	(5.77)

Note that $\vec{\upsilon}_s = 0$ at the A_{kj} surface, as the total control volume is taken as stationary and nondeformable. Thus, using Equations. 5.15 and 5.5, Equation 5.77 may be written in the form

$$\frac{d}{dt}\left[\langle\rho_k\rangle_k\ V_k\right]+\sum_j\{\rho_k\vec{\upsilon}_k\}_{kj}\cdot\vec{A}_{kj}+\{\rho_k(\vec{\upsilon}_k-\vec{\upsilon}_s)\}_{ks}\cdot\vec{A}_{ks}=0 \tag{5.78}$$

where $\vec{\upsilon}_k$ in the last term = phase k velocity at the vapor–liquid interface, and $\vec{\upsilon}_s$ = velocity of the interface itself. The vector designating an area $\vec{A}_k$ is normal to the area pointing outward of the subvolume containing the phase k.

Denoting the rate of mass addition due to phase change by $\dot{m}_{ks}$, it is seen that

$$\dot{m}_{ks}=-\{\rho_k(\vec{\upsilon}_k-\vec{\upsilon}_s)\}_{ks}\cdot\vec{A}_{ks} \tag{5.79}$$

If $(\vec{\upsilon}_k-\vec{\upsilon}_s)_{ks}$ is positive, its dot product with $\vec{A}_{ks}$ takes the sign of $\vec{A}_{ks}$. The area $\vec{A}_{ks}$ is considered positive if it points in a positive coordinate direction. On a positive surface, the right-hand side of Equation 5.79 renders $\dot{m}_{ks}$ negative, which implies that mass is being lost through the surface.

5.6.1.2 Mass Balance in the Entire Volume V

Let us now write the mass equation of phase k in terms of the total volume and surface area, recalling that

$$V_k=\langle\alpha_k\rangle V \text{ and } \vec{A}_{kj}=\{\alpha_k\}_j\vec{A}_j$$

We can rewrite Equation 5.78 as

$$\frac{d}{dt}\left[\langle\rho_k\rangle_k\langle\alpha_k\rangle V\right]+\sum_j\{\rho_k\vec{\upsilon}_k\}_{kj}\cdot\{\alpha_k\}_j\vec{A}_j=\dot{m}_{ks} \tag{5.80}$$

Let m_k be the mass of phase k in the control volume V:

$$m_k=\langle\rho_k\rangle_k\langle\alpha_k\rangle V \tag{5.81}$$

and $\dot{m}_{kj}$ be the mass flow rate of phase k through the area A_j:

$$\dot{m}_{kj}=-\{\rho_k\vec{\upsilon}_k\}_{kj}\cdot\vec{A}_{kj}=-\{\rho_k\vec{\upsilon}_k\}_{kj}\cdot\{\alpha_k\}_j\vec{A}_j \tag{5.82}$$

Note that in Figure 5.5 the vector $\vec{A}_{k1}$ is pointing in a negative direction, whereas the vector $\vec{A}_{k2}$ is in a positive direction. Hence, $\vec{\upsilon}_k\cdot\vec{A}_{k1}$ is negative and $\vec{\upsilon}_k\cdot\vec{A}_{k2}$ is positive. The use of the negative sign on the right-hand side of Equation 5.82 makes

$\dot{m}_{kj}$ positive for inflow and negative for outflow. We can recast Equation 5.80 into the form

$$\frac{d}{dt}m_k - \sum_j \dot{m}_{kj} = \dot{m}_{ks} \tag{5.83}$$

For a mixture, we get

$$\frac{d}{dt}(m_v + m_\ell) - \sum_j (\dot{m}_{\ell j} + \dot{m}_{vj}) = \dot{m}_{vs} + \dot{m}_{\ell s} \tag{5.84}$$

It should also be remembered that because the frame of reference of the control volume is stationary, by using Equation 4.13, Equation 5.83 reduces to

$$\frac{\partial}{\partial t}m_k - \sum_j \dot{m}_{kj} = \dot{m}_{ks} \tag{5.85a}$$

or

$$\frac{\partial}{\partial t}m_k = \sum_j \dot{m}_{kj} + \dot{m}_{ks} \tag{5.85b}$$

Recognizing that x is the flow quality, Equation 5.85b may be written for a stationary control volume as follows:

For vapor

$$\frac{\partial}{\partial t}m_v = \sum_j (x\dot{m})_j + \dot{m}_{vs} \tag{5.86a}$$

For liquid

$$\frac{\partial}{\partial t}m_\ell = \sum_j \left[(1-x)\dot{m}\right]_j + \dot{m}_{\ell s} \tag{5.86b}$$

5.6.1.3 Interfacial Jump Condition

As there is no mass source or sink at the interface, the entire vapor added by change of phase should appear as a loss to the liquid and vice versa. This describes the interphase "jump condition." If $\dot{m}_{\ell v}$ is the rate of change of phase of liquid into vapor

$$\dot{m}_{vs} = -\dot{m}_{\ell s} \equiv \dot{m}_{\ell v} \tag{5.87}$$

It is also useful to define the rate of phase addition by a change in state per unit volume (i.e., vaporization or condensation) so that

$$\Gamma_k = \frac{\dot{m}_{ks}}{V} \tag{5.88}$$

From Equation 5.87, it is clear that the jump condition can also be written as

$$\Gamma_v = -\Gamma_\ell = \Gamma \tag{5.89}$$

5.6.1.4 Simplified Form of the Mixture Equation

The mass balance for the two-phase mixture is obtained by combining Equations 5.86a and 5.86b, and 5.87 to obtain an equation for the rate of change of the mixture density $\langle\rho\rangle$:

$$\frac{\partial}{\partial t}[m_v + m_\ell] = \sum_j \dot{m}_j \tag{5.90}$$

or

$$\frac{\partial}{\partial t}[\langle\rho\rangle V] = \sum_j \dot{m}_j \tag{5.91}$$

where

$$\langle\rho\rangle = \langle\rho_v\rangle_v \langle\alpha\rangle + \langle\rho_\ell\rangle_\ell \langle 1-\alpha\rangle \tag{5.25}$$

When the phase densities are assumed to be uniform in the volume, we get

$$\langle\rho\rangle = \alpha\rho_v + (1-\alpha)\rho_\ell \tag{5.92}$$

5.6.2 Momentum Balance

5.6.2.1 Momentum Balance for Volume V_k

The momentum balance for each phase is an application of Newton's law of motion to the deformable volumes of the vapor and liquid phases. The forces acting on each volume are illustrated in Figure 5.6.

Applying Equation 4.60 to the control volume V_k, we obtain

$$\frac{d}{dt}\iiint_{V_k} \rho_k \vec{\upsilon}\, dV + \oiint_{S_k} \rho_k \vec{\upsilon}_k (\vec{\upsilon}_k - \vec{\upsilon}_s)\cdot \vec{n}\, dS = \oiint_{S_k} (\overline{\overline{\tau}}_k - p_k \overline{\overline{I}})\cdot \vec{n}\, dS + \iiint_{V_k} \rho_k \vec{g}\, dV \tag{5.93}$$

Again, consider the surface S_k to be divided into a number of fixed bounding areas (A_{kj}) and a deformable area (A_{ks}). Applying the volume- and area-averaging operations, and rearranging the above equation, we get

$$\begin{aligned}\frac{d}{dt}\left[\langle\rho_k \vec{\upsilon}_k\rangle_k V_k\right] = &-\sum_j \{\rho_k \vec{\upsilon}_k \vec{\upsilon}_k\}_{kj}\cdot \vec{A}_{kj} - \{\rho_k \vec{\upsilon}_k (\vec{\upsilon}_k - \vec{\upsilon}_s)\}_{ks}\cdot \vec{A}_{ks} \\ &+ \sum_j \vec{F}_{jk} - \sum_j \{p_k\}_{kj} \vec{A}_{kj} + \vec{F}_{sk} + \langle\rho_k\rangle_k \vec{g} V_k\end{aligned} \tag{5.94}$$

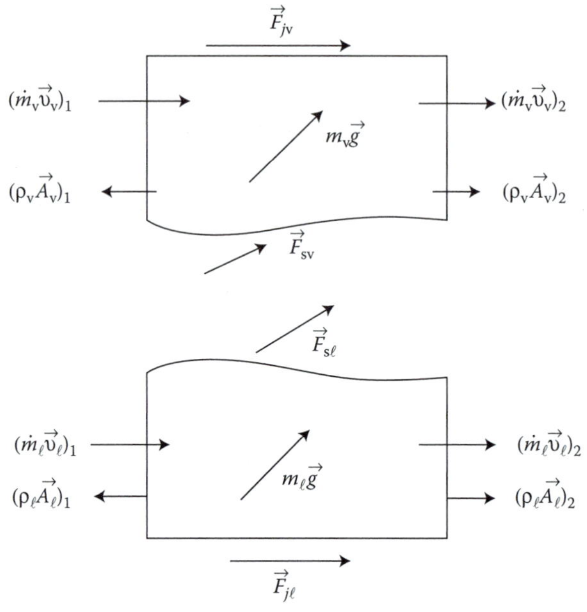

FIGURE 5.6 Control volume momentum balance components. The arrows represent vectors which are positive in the direction of the coordinate axes.

where

$$\vec{F}_{jk} = \iint_{A_{kj}} \overline{\overline{\tau}}_k \cdot \vec{n}\, \mathrm{d}S \tag{5.95}$$

represents the shear forces at all surfaces surrounding the phase except for the vapor–liquid interface, and

$$\vec{F}_{sk} = \iint_{A_{ks}} (\overline{\overline{\tau}}_k - p_k \overline{\overline{I}}) \cdot \vec{n}\, \mathrm{d}S \tag{5.96}$$

represents the shear and normal forces at the vapor–liquid interface. Equation 5.94 represents the following balance:

Rate of change of momentum within control volume	=	Rate of momentum addition by convection	+	Rate of momentum addition by change of phase	+	Net shear forces at the fixed boundaries A_{kj}	+	Net pressure forces at the fixed boundaries A_{kj}	+	Net shear forces and pressure forces at the vapor–liquid interface	+	Gravity force

For the stationary control volume V_k, when the velocities $\vec{\upsilon}_k$ are assumed position-independent within the volume and at each area A_{kj} and A_{sk} of the surrounding surface, we can reduce Equation 5.94 to the lumped parameter form:

$$\frac{\partial}{\partial t}[(m_k\vec{\upsilon}_k)] = \sum_j (\dot{m}_k\vec{\upsilon}_k)_j + \dot{m}_{ks}\vec{\upsilon}_{ks} + \sum_j \vec{F}_{jk} - \sum_j (p_k\vec{A}_k)_{kj} + \vec{F}_{sk} + m_k\vec{g} \quad (5.97)$$

where $\vec{\upsilon}_{ks}$ = velocity of the vaporized mass at the phase k side of the interface S.

5.6.2.2 Momentum Balance in the Entire Volume V

When the two volumes V_v and V_ℓ are added together to form the volume V, we can recast Equation 5.94 for the phase k in the form

$$\frac{d}{dt}\left[\langle\rho_k\vec{\upsilon}_k\rangle_k\langle\alpha_k\rangle V\right] = -\sum_j \{\rho_k\vec{\upsilon}_k\vec{\upsilon}_k\}_{kj}\cdot\{\alpha_k\}_j\vec{A}_j - \{\rho_k\vec{\upsilon}_k(\vec{\upsilon}_k - \vec{\upsilon}_s)\}_{ks}\cdot\vec{A}_{ks}$$
$$+\sum_j \vec{F}_{jk} - \sum_j \{p_k\}_{kj}\{\alpha_k\}_j\vec{A}_j + \vec{F}_{sk} + \langle\rho_k\rangle_k\langle\alpha_k\rangle\vec{g}\,V \quad (5.98)$$

where

$$\sum_{k=1}^{2} V_k = V$$

$$\sum_{k=1}^{2} \vec{A}_{kj} = \vec{A}_j$$

$$\sum_{k=1}^{2} \vec{A}_{ks} = 0$$

(Because the interface areas are equal but have directionally opposite normal vectors.)

Using Equations 5.79 and 5.88,

$$-\{\rho_k\vec{\upsilon}_k(\vec{\upsilon}_k - \vec{\upsilon}_s)\}_{ks}\cdot\vec{A}_{ks} = \dot{m}_{ks}\vec{\upsilon}_{ks} = \Gamma_k\vec{\upsilon}_{ks}V \quad (5.99)$$

then by Equation 5.82, we get

$$-\{\rho_k\vec{\upsilon}_k\vec{\upsilon}_k\}_{kj}\cdot\{\alpha_k\}_j\vec{A}_j = (\dot{m}_k\vec{\upsilon}_k)_j \quad (5.100)$$

Finally, we can simplify the situation by assuming uniform ρ_k and $\vec{\upsilon}_k$ within each volume, so that the averaging operations are not needed. Equation 5.98 can then be written for a fixed control volume as

$$\frac{\partial}{\partial t}(\alpha_k \rho_k \vec{\upsilon}_k)V = \sum_j (\dot{m}_k \vec{\upsilon}_k)_j + \Gamma_k \vec{\upsilon}_{ks} V + \sum_j \vec{F}_{jk} - \sum_j (\alpha_k p_k \vec{A})_j + \vec{F}_{sk} + \alpha_k \rho_k \vec{g} V \tag{5.101}$$

The mixture equation is obtained by adding the vapor and liquid momentum equations (Equation 5.101):

$$\begin{aligned}\frac{\partial}{\partial t}[\alpha \rho_v \vec{\upsilon}_v + (1-\alpha)\rho_\ell \vec{\upsilon}_\ell]V = &\sum_j [\dot{m}_v \vec{\upsilon}_v + \dot{m}_\ell \vec{\upsilon}_\ell]_j \\ &+ \Gamma(\vec{\upsilon}_{vs} - \vec{\upsilon}_{\ell s})V + \sum_j (\vec{F}_{jv} + \vec{F}_{j\ell}) \\ &- \sum_j [\alpha p_v + (1-\alpha)p_\ell]_j \vec{A}_j + \vec{F}_{sv} \\ &+ \vec{F}_{s\ell} + [\alpha \rho_v + (1-\alpha)\rho_\ell]\vec{g}V\end{aligned} \tag{5.102}$$

5.6.2.3 Interfacial Jump Condition

If the surface tension force (and hence the surface deformation energy) is negligible, no net momentum is accumulated at the interface. Thus, the net momentum change plus the interface forces can be expected to vanish:

$$\Gamma(\vec{\upsilon}_{vs} - \vec{\upsilon}_{\ell s})V + \vec{F}_{sv} + \vec{F}_{s\ell} = 0 \tag{5.103}$$

Several assumptions have been used in the literature with regard to $\vec{\upsilon}_{vs}$ and $\vec{\upsilon}_{\ell s}$. The most common approaches are as follows. Assume either

$$\vec{\upsilon}_{vs} = \vec{\upsilon}_{\ell s}; \quad \text{hence } \vec{F}_{sv} = -\vec{F}_{s\ell} \tag{5.104a}$$

or

$$\vec{\upsilon}_{vs} = \vec{\upsilon}_v \text{ and } \vec{\upsilon}_{\ell s} = \vec{\upsilon}_\ell; \quad \text{hence } \vec{F}_{sv} = -\vec{F}_{s\ell} - \Gamma(\vec{\upsilon}_v - \vec{\upsilon}_\ell)V \tag{5.104b}$$

In Equation 5.104a, it is often assumed that

$$\vec{\upsilon}_{vs} = \vec{\upsilon}_{\ell s} = \eta \vec{\upsilon}_v + (1-\eta)\vec{\upsilon}_\ell \tag{5.104c}$$

where $\eta = 0$ if $\Gamma < 0$; or $\eta = 1$ if $\Gamma > 0$.

The last formulation avoids penetration of the interface in a direction not compatible with Γ.

5.6.2.4 Common Assumptions

It is common to assume that if the flow area A_j is horizontal, the pressure would be uniform across the area and hence

$$(p_{\text{v}})_j = (p_\ell)_j = p_j \tag{5.105}$$

Although this assumption is often extended to all areas irrespective of their inclination, it may be a poor approximation for a vertical flow area with stratified flow where significant pressure variations due to gravity may exist.

The shear forces at the boundary areas A_j are usually assumed to constitute friction terms with the wall* that can be considered additive:

$$\sum_j (\vec{F}_{j\text{v}} + \vec{F}_{j\ell}) = \vec{F}_{\text{w}} \tag{5.106}$$

5.6.2.5 Simplified Forms of the Mixture Equation

Imposing the conditions of Equations 5.103, 5.105, and 5.106 on Equation 5.102, we get a simplified mixture momentum balance equation:

$$\frac{\partial}{\partial t}[\alpha\rho_{\text{v}}\vec{\upsilon}_{\text{v}} + (1-\alpha)\rho_\ell\vec{\upsilon}_\ell]V = \sum_j [x\vec{\upsilon}_{\text{v}} + (1-x)\vec{\upsilon}_\ell]_j \dot{m}_j + \vec{F}_{\text{w}} - \sum_j p_j\vec{A}_j + \rho_{\text{m}}\vec{g}V \tag{5.107}$$

where x_j = flow quality in the direction perpendicular to $\vec{A}_j$.

Because the mass flux $\vec{G}_{\text{m}}$ can be defined for spatially uniform properties as

$$\vec{G}_{\text{m}} = \alpha\rho_{\text{v}}\vec{\upsilon}_{\text{v}} + (1-\alpha)\rho_\ell\vec{\upsilon}_\ell \tag{5.108}$$

and

$$\dot{m}_j = (\vec{G}_{\text{m}} \cdot \vec{A})_j \tag{5.109}$$

It is possible to write Equation 5.107 in terms of the mass flux $\vec{G}_{\text{m}}$ as

$$\frac{\partial}{\partial t}(\vec{G}_{\text{m}}V) = \sum_j [x\vec{\upsilon}_{\text{v}} + (1-x)\vec{\upsilon}_\ell]_j (\vec{G}_{\text{m}} \cdot \vec{A})_j + \vec{F}_{\text{w}} - \sum_j p_j\vec{A}_j + \rho_{\text{m}}\vec{g}V \tag{5.110}$$

---

* It is assumed that the shear forces due to velocity gradients in the fluid at open portions of the surface area are much smaller than those at fluid–solid surfaces.

Example 5.3: Modeling of a BWR Suppression Pool Transient

PROBLEM

Consider the simplified diagram of a BWR suppression pool shown in Figure 5.7. The function of the suppression pool is to act as a means of condensing the steam emerging from the primary system during an LOCA. In a postulated accident, a sudden rupture of a BWR recirculation pipe causes a jet of somewhat high-pressure steam–water to enter into the suppression pool from a connecting pipe. This jet is preceded by the air initially in the pipe, which is noncondensible. Because of the rapid nature of this transient, there is some concern that the water slug impacted by the jet may exert a large pressure spike on the bottom of the suppression pool.

Describe how you would use Equation 5.94 to obtain an approximate answer for the value of the pressure spike. Consider a control volume drawn around the water slug between the exit of the connecting pipe and the bottom of the suppression pool, as shown in Figure 5.8, and make the following assumptions:

1. The nature of this transient is sufficiently fast such that a water slug mass does not have time to deform from a roughly cylindrical configuration assumed to have the same diameter as the connecting pipe.
2. Within the time interval of interest, the momentum of the water within the control volume remains fairly constant.
3. All the water in the slug is carried to the bottom of the pool, where it is deflected to move parallel to the bottom of the pool.
4. The distance between the connecting pipe outlet and the bottom of the suppression pool is on the order of meters.

SOLUTION

Let us consider the terms in Equation 5.94 for the water slug momentum in the vertical direction.

1. By assumption 2, $\frac{d}{dt}[\langle \rho_k \vec{v}_k \rangle V_k] = 0$.
2. $-\sum_j \{\rho_k \vec{v}_k \vec{v}_k\}_{kj} \cdot \vec{A}_{kj} = \rho_w V_w^2 A_p = \dot{m}\vec{V}_w$, where the subscript w = water, and $\dot{m}_w$ = downward flow rate of the water. Note here that

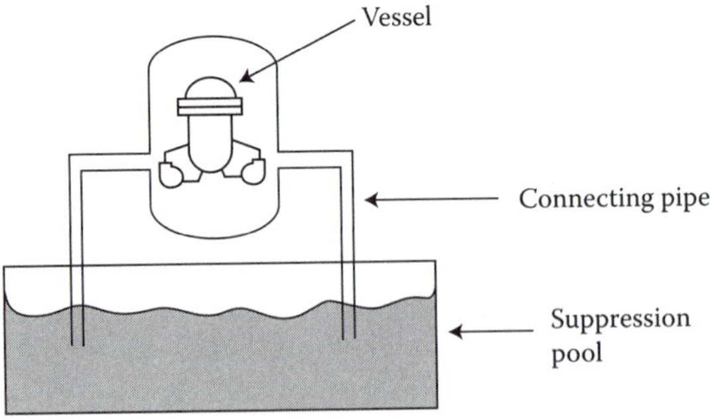

FIGURE 5.7 Simplified diagram of the suppression pool in a BWR.

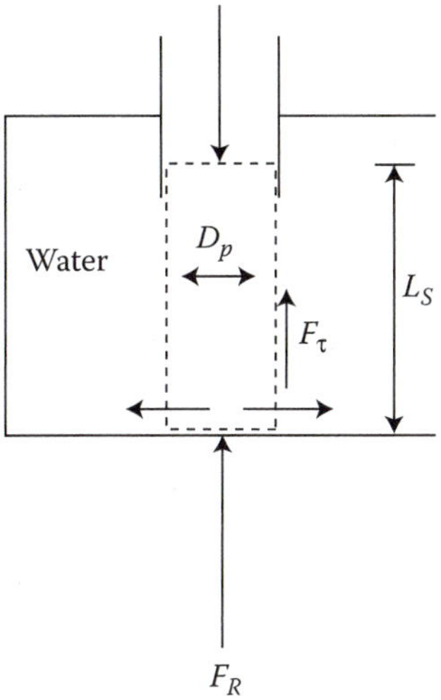

FIGURE 5.8 Control volume for analysis of BWR suppression pool pressure spike.

a. We consider the flow at the pipe outlet because the flow at this part of the control volume surface is in the vertical direction.
b. The sign convention makes the flow positive because it is at an inlet to the control volume.

3. $-\{\rho_k \vec{\upsilon}_k(\vec{\upsilon}_k - \vec{\upsilon}_s)\}_{ks} \cdot \vec{A}_{ks} = 0$ because this term represents momentum transfer by change of phase, and there is no phase change between water and air. If the air were replaced by steam, this term may become important because of the expected condensation.
4. $\sum_j \vec{F}_{jk} \simeq 0$ = the force exerted on the control volume by shear at the ends of the control volume.
5. $\sum_j \{p_k\}_{kj} \vec{A}_{kj} = \vec{F}_R - p_{air}\pi \frac{D_p^2}{4} \vec{n}_z$, where $\vec{F}_R$ = force sought at the bottom and $\vec{n}_z$ = vertical unity vector.
6. $\vec{F}_{sk} = -\vec{F}_z = -\tau_s(\pi D_p L_s)\vec{n}_z$ is the shear force at the interface of the water slug and the pool.
7. $\langle \rho_k \rangle_k \vec{g}\, V_k \simeq 0$, as the column is relatively short by assumption 4.

Equation 5.94 then becomes

$$0 = \dot{m}\, \vec{V}_w - 0 + 0 - \vec{F}_R + p_{air} \frac{\pi D_p^2}{4} \vec{n}z - \vec{F}_{\ddot{A}} + 0$$

$$\vec{F}_R = \dot{m}\, \vec{V}_w + p_{air}\pi \frac{D_p^2}{4} \vec{n}_z - \tau_s(\pi D_p L_s)\vec{n}_z$$

5.6.3 Energy Balance

5.6.3.1 Energy Balance for the Volume V_k

The energy balance for each phase is an application of the first law of thermodynamics to the phase subvolume. Equation 4.63 can be applied to the vapor volume shown in Figure 5.9 to obtain

$$\frac{d}{dt}\iiint_{V_k} \rho_k u_k^o \, dV + \iint_{S_k} \rho_k u_k^o (\vec{\upsilon}_k - \vec{\upsilon}_s)\cdot \vec{n}\, dS =$$

$$\oiint_{S_k} \left[-\vec{q}_k'' + (\bar{\bar{\tau}}_k - p_k \bar{\bar{I}})\cdot \vec{\upsilon}_k\right]\cdot \vec{n}\, dS + \iiint_{V_k} \rho_k \vec{g}\cdot \vec{\upsilon}_k \, dV + \iiint_{V_k} q_k''' \, dV \quad (5.111)$$

A heat addition term due to both internal (to the fluid k) and surface heat addition can be defined as

$$\left(\frac{dQ}{dt}\right)_k = \iiint_{V_k} q_k''' \, dV - \oiint_{S_k} \vec{q}_k''\cdot \vec{n}\, dS \quad (5.112)$$

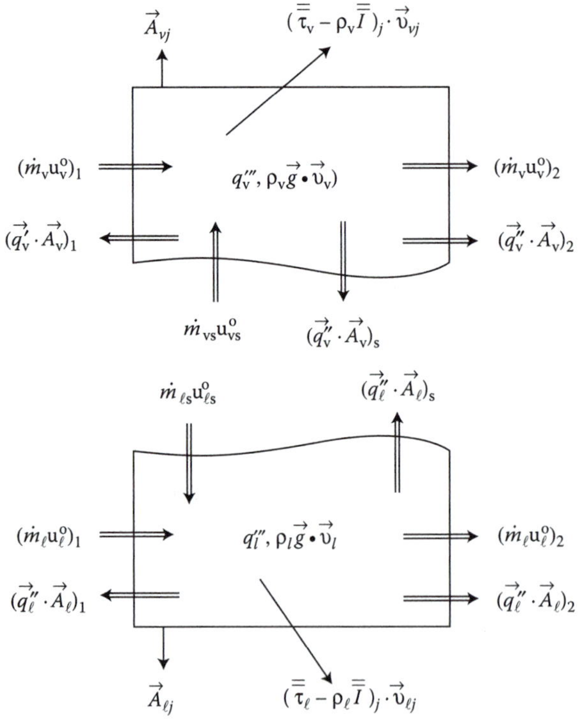

FIGURE 5.9 Forces and heat fluxes at the vapor–liquid interface. *Single arrows* represent vectors, which are positive in the positive direction of the coordinate axis. *Double arrows* represent scalars, which are positive when adding to the control volume content.

where $\vec{q}_k'' \cdot \vec{n}$ = outward heat flux from the surface.

A work term due to shear forces can be defined as

$$\left(\frac{\mathrm{d}W}{\mathrm{d}t}\right)_{k,\text{shear}} = -\oiint_{S_k} (\bar{\bar{\tau}}_k \cdot \vec{\upsilon}_k) \cdot \vec{n}\, \mathrm{d}S \tag{5.113}$$

A work term due to the gravity (body) force can be defined as

$$\left(\frac{\mathrm{d}W}{\mathrm{d}t}\right)_{k,\text{gravity}} = -\iiint_{V_k} (\rho_k \vec{g} \cdot \vec{\upsilon}_k \mathrm{d}V) \tag{5.114}$$

Recalling that the surface S can be divided into a number of fixed bounding surfaces A_{kj} and the deformable interface A_{ks}, Equation 5.111 can be recast using Equations 5.112 through 5.114 in the form

$$\begin{aligned}\frac{d}{\mathrm{d}t}\left[\langle \rho_k u_k^{\mathrm{o}} \rangle_k V_k\right] = &-\sum_j \{\rho_k u_k^{\mathrm{o}} (\vec{\upsilon}_k - \vec{\upsilon}_{\mathrm{s}}) \cdot \vec{n}\}_{kj} A_{kj} \\ &- \{\rho_k u_k^{\mathrm{o}} (\vec{\upsilon}_k - \vec{\upsilon}_{\mathrm{s}}) \cdot \vec{n}\}_{ks} A_{ks} \\ &- \sum_j \{p_k \vec{\upsilon}_k \cdot \vec{n}\}_{kj} A_{kj} - \{p_k \vec{\upsilon}_k \cdot \vec{n}\}_{ks} A_{ks} \\ &+ \left(\frac{\mathrm{d}Q}{\mathrm{d}t}\right)_k - \left(\frac{\mathrm{d}W}{\mathrm{d}t}\right)_{k,\text{shear}} - \left(\frac{\mathrm{d}W}{\mathrm{d}t}\right)_{k,\text{gravity}}\end{aligned} \tag{5.115}$$

Note that

$$\{p_k \vec{\upsilon}_k \cdot \vec{n}\}_{ks \text{ or } kj} = \{p_k \vec{\upsilon}_{\mathrm{s}} \cdot \vec{n}_{\mathrm{s}}\}_{ks \text{ or } kj} + \{p_k (\vec{\upsilon}_k - \vec{\upsilon}_{\mathrm{s}}) \cdot \vec{n}_j\}_{ks \text{ or } kj} \tag{5.116}$$

The energy balance for V_k (Equation 5.114) can then be written in terms of the total enthalpy:

$$\begin{aligned}\frac{d}{\mathrm{d}t}\left[\langle \rho_k u_k^{\mathrm{o}} \rangle_k V_k\right] = &-\sum_j \{\rho_k h_k^{\mathrm{o}} (\vec{\upsilon}_k - \vec{\upsilon}_{\mathrm{s}}) \cdot \vec{n}\}_{kj} A_{kj} - \sum_j \{p_k \vec{\upsilon}_{\mathrm{s}} \cdot \vec{n}\}_{kj} A_{kj} \\ &- \{\rho_k h_k^{\mathrm{o}} (\vec{\upsilon}_k - \vec{\upsilon}_{\mathrm{s}}) \cdot \vec{n}\}_{ks} A_{ks} - \{p_k \vec{\upsilon}_{\mathrm{s}} \cdot \vec{n}\}_{ks} A_{ks} \\ &+ \left(\frac{\mathrm{d}Q}{\mathrm{d}t}\right)_k - \left(\frac{\mathrm{d}W}{\mathrm{d}t}\right)_{k,\text{shear}} - \left(\frac{\mathrm{d}W}{\mathrm{d}t}\right)_{k,\text{gravity}}\end{aligned} \tag{5.117}$$

For a fixed control volume $\{\vec{\upsilon}_s\}_{kj} = 0$, then we have

$$\frac{\partial}{\partial t}\left[\left\langle \rho_k u_k^o \right\rangle_k V_k\right] = -\sum_j \{\rho_k h_k^o (\vec{\upsilon}_k)\cdot\vec{n}\}_{kj} A_{kj} - \{\rho_k h_k^o(\vec{\upsilon}_k - \vec{\upsilon}_s)\cdot\vec{n}\}_{ks} A_{ks} - \{p_k \vec{\upsilon}_s \cdot \vec{n}\}_{ks} A_{ks} + \left(\frac{dQ}{dt}\right)_k - \left(\frac{dW}{dt}\right)_{k,\text{shear}} - \left(\frac{dW}{dt}\right)_{k,\text{gravity}} \quad (5.118)$$

For uniform properties within the volume and at each gate area, we can write the lumped parameter form of the energy equation (Equation 5.118) as

$$\frac{\partial}{\partial t}(m_k u_k^o) = \sum_j (\dot{m}_k h_k^o)_j + (\dot{m}_k h_k^o)_s + \left(\frac{\partial Q}{\partial t}\right)_k - \left(\frac{\partial W}{\partial t}\right)_{k,\text{c,v.}} - \left(\frac{dW}{dt}\right)_{k,\text{gravity}} \quad (5.119)$$

where

$$\left(\frac{\partial W}{\partial t}\right)_{k,\text{e,v.}} = \left(\frac{\partial W}{\partial t}\right)_{k,\text{shear}} + \{p_k \vec{\upsilon}_s \cdot \vec{n}\}_{ks} A_{ks} \quad (5.120)$$

Note that $(\partial W/\partial t)_{k,\text{c,v.}}$ can be written in terms of the work required to change the vapor volume given the various forces surrounding the vapor volume. The control volume surface and the phase interface area are moving at velocity $\vec{\upsilon}_j$ and $\vec{\upsilon}_s$, respectively. Thus

$$\left(\frac{\partial W}{\partial t}\right)_{k,\text{c,v.}} = \sum_j \left(\frac{\partial W}{\partial t}\right)_{k,\text{c,v.},j} + \left(\frac{\partial W}{\partial t}\right)_{k,\text{c,v.,s}} \quad (5.121)$$

where for uniform properties over each area

$$\left(\frac{\partial W}{\partial t}\right)_{k,\text{c,v.},j} = -\sum_j (\bar{\bar{\tau}}_k \cdot \vec{\upsilon}_k)_j \cdot \vec{n}_{kj} A_{kj} \quad (5.122)$$

and

$$\left(\frac{\partial W}{\partial t}\right)_{k,\text{c,v.,s}} = -(\bar{\bar{\tau}}_k \cdot \vec{\upsilon}_k)_s \cdot \vec{n}_{ks} A_s + (p_k \vec{\upsilon}_s)\cdot \vec{n}_{ks} A_s \quad (5.123)$$

Also, the heat addition term can be divided into three parts:

$$\left(\frac{\partial Q}{\partial t}\right)_k = \dot{Q}_k - \sum_j (\vec{q}_k'' \cdot \vec{n}_k)_j A_{kj} - (\vec{q}_k'' \cdot \vec{n}_k)_s A_s \tag{5.124}$$

where $\vec{q}_{kj}''$ and $\vec{q}_{ks}''$ = heat transfer rates at the boundary surfaces j and s surrounding the phasic volume, respectively, in the positive coordinate direction, and $\dot{Q}_k$ = rate of heat generation in the fluid k.

Let us now substitute from Equations 5.122 and 5.124 into 5.119 to obtain, after rearranging the terms

$$\begin{aligned}\frac{\partial}{\partial t}(m_k u_k^{\circ}) = {} & \sum_j (\dot{m}_k h_k^{\circ})_j + \dot{Q}_k - \sum_j (\vec{q}_k'' \cdot \vec{n}_k)_j A_{kj} \\ & - \sum_j (\bar{\bar{\tau}}_k \cdot \vec{\upsilon}_k)_j \cdot \vec{n}_{kj} A_{kj} - \left(\frac{\mathrm{d}W}{\mathrm{d}t}\right)_{k,\text{gravity}} + (\dot{m}_k h_k^{\circ})_s \\ & - (\vec{q}_k'' + \bar{\bar{\tau}}_k \cdot \vec{\upsilon}_k - p_k \vec{\upsilon}_s) \cdot \vec{n}_{ks} A_{ks}\end{aligned} \tag{5.125}$$

Note that the last two terms represent the energy exchange rate between the interior of a phase and the vapor–liquid interface due to change of phase, heat transport, and work of surface forces.

5.6.3.2 Energy Equations for the Total Volume *V*

The two-phase mixture energy equation can be obtained by adding the vapor and liquid equations of 5.125:

$$\begin{aligned}\frac{\partial}{\partial t}(m_v u_v^{\circ} + m_\ell u_\ell^{\circ}) = {} & \sum_j [\dot{m}_\ell h_\ell^{\circ} + \dot{m}_v h_v^{\circ}]_j \\ & + \dot{Q}_\ell + \dot{Q}_v - \sum_j (\vec{q}_\ell''\alpha_\ell + \vec{q}_v''\alpha_v)_j \cdot \vec{n}_j A_j \\ & - \sum_j [(\bar{\bar{\tau}}_\ell \cdot \vec{\upsilon}_\ell)\alpha_\ell + (\bar{\bar{\tau}}_v \cdot \vec{\upsilon}_v)\alpha_v]_j \cdot \vec{n}_j A_j \\ & - \left(\frac{\mathrm{d}W}{\mathrm{d}t}\right)_{v,\text{gravity}} - \left(\frac{\mathrm{d}W}{\mathrm{d}t}\right)_{\ell,\text{gravity}} + (\dot{m}_\ell h_\ell^{\circ})_s \\ & + (\dot{m}_v h_v^{\circ})_s - \{[\vec{q}_v'' + \bar{\bar{\tau}}_v'' \cdot \vec{\upsilon}_v - p_v \vec{\upsilon}_s]\} \cdot \vec{n}_{vs} A_{vs} \\ & + \{[\vec{q}_\ell'' + \bar{\bar{\tau}}_\ell \cdot \vec{\upsilon}_\ell - p_\ell \vec{\upsilon}_s] \cdot \vec{n}_{\ell s} A_{\ell s}\end{aligned} \tag{5.126}$$

where it has been noted that

$$(\vec{q}_k'' \cdot \vec{n}_k)_j A_{kj} = \vec{q}_k''\alpha_k \cdot \vec{n}_j A_j$$

and

$$\bar{\bar{\tau}}_k \cdot \vec{\upsilon}_k \cdot \vec{n}_{kj} A_{kj} = \bar{\bar{\tau}}_k \cdot \vec{\upsilon}_k \alpha_k \cdot \vec{n}_j A_j$$

5.6.3.3 Jump Condition

It should be clear that no energy sources or sinks should exist at the interface (when the surface tension is ignored), that is, at the interface all energy exchange should add up to zero. Applying the conditions

$$\dot{m}_{\mathrm{vs}} = -\dot{m}_{\ell s} = \dot{m}_{\ell \mathrm{v}}, \vec{n}_{\mathrm{vs}} A_{\mathrm{vs}} = -\vec{n}_{\ell s} A_{\ell s}, \text{ and } A_{\mathrm{vs}} = A_{\ell s} \equiv A_s$$

we get

$$\dot{m}_{\ell \mathrm{v}} (h_{\mathrm{v}}^{\mathrm{o}} - h_{\ell}^{\mathrm{o}})_{\mathrm{s}} - [(\vec{q}_{\mathrm{v}}'' - \vec{q}_{\ell}'')_s + (\bar{\bar{\tau}}_{\mathrm{v}} \cdot \vec{\upsilon}_{\mathrm{v}} - \bar{\bar{\tau}}_{\ell} \cdot \vec{\upsilon}_{\ell})_s - (p_{\mathrm{v}} - p_{\ell})_s \vec{\upsilon}_s] \cdot \vec{n}_{\mathrm{vs}} A_s = 0 \quad (5.127)$$

Several simplifications can be made. The most common assumption used is that the term related to the interfacial forces can be ignored, which reduces Equation 5.127 to

$$\dot{m}_{\ell \mathrm{v}} = (\vec{q}_{\mathrm{v}}'' - \vec{q}_{\ell}'')_{\mathrm{s}} \cdot \frac{\vec{n}_{\mathrm{vs}} A_s}{(h_{\mathrm{v}}^{\mathrm{o}} - h_{\ell}^{\mathrm{o}})_s} \quad (5.128)$$

The terms $(\vec{q}_{\mathrm{v}}'')_{\mathrm{s}}$ and $(\vec{q}_{\ell}'')_{\mathrm{s}}$ have to be externally supplied as constitutive equations.

Substituting from Equation 5.127 into Equation 5.126, the mixture energy equation can be written as

$$\frac{\partial}{\partial t}(m u_{\mathrm{m}}^{\mathrm{o}}) = \sum_j [\{(1-x) h_{\ell}^{\mathrm{o}} + x h_{\mathrm{v}}^{\mathrm{o}}\} \dot{m}]_j + \dot{Q} - \sum_j \vec{q}_j'' \cdot \vec{n}_j A_j$$
$$-\sum_j \bar{\bar{\tau}}_{\mathrm{eff}} \cdot \vec{j} \cdot \vec{A}_j - \left(\frac{\mathrm{d}W}{\mathrm{d}t}\right)_{\mathrm{gravity}} \quad (5.129)$$

where

$$\dot{Q} = \dot{Q}_{\ell} + \dot{Q}_{\mathrm{v}}$$
$$\vec{q}_j'' = (\vec{q}_{\ell}'' \alpha_{\ell} + \vec{q}_{\ell}'' \alpha_{\mathrm{v}})_j$$
$$\bar{\bar{\tau}}_{\mathrm{eff}} \cdot \vec{j} \cdot \vec{A}_j = [\bar{\bar{\tau}}_{\ell} \cdot \vec{\upsilon}_{\ell} \alpha_{\ell} + \bar{\bar{\tau}}_{\mathrm{v}} \cdot \vec{\upsilon}_{\mathrm{v}} \alpha_{\mathrm{v}}]_j \cdot \vec{n}_j A_j$$

5.7 ONE-DIMENSIONAL SPACE-AVERAGED TRANSPORT EQUATIONS

In this section, the differential form of the space-averaged equations is derived from the integral transport equations. No time averaging is invoked.

5.7.1 Mass Equations

From Equation 5.80, when V is considered equal to a small volume $A_z\Delta z$, we get

$$\frac{d}{dt}\left[\langle\rho_k\rangle_k\langle\alpha_k\rangle A_z\Delta z\right]+\{\rho_k\vec{\upsilon}_k\}_{kz^+}\cdot\{\alpha\}_{z^+}\vec{A}_{z^+}-\{\rho_k\vec{\upsilon}_k\}_{kz^-}\cdot\{\alpha_k\}_{z^-}\vec{A}_{z^-}=\dot{m}_{ks}=\Gamma_k A_z\Delta z \tag{5.130}$$

where z^+ and z^- refer to flows entering and leaving the volume $A_z\Delta z$.

In the limit of infinitesimal Δz and for a fixed frame of coordinates, we get

$$\frac{\partial}{\partial t}(\{\rho_k\alpha_k\}A_z)+\frac{\partial}{\partial z}(\{\rho_k\upsilon_{kz}\}_k\{\alpha_k\}A_z)=\Gamma_k A_z \tag{5.131}$$

Because Δz has been taken infinitesimally small, the volume-averaged quantities are equivalent to the area-averaged ones over the area A_z. Note that Γ_k is then also the average over A_z. Now, with the application of Equation 5.30, Equation 5.131 can be written as

$$\frac{\partial}{\partial t}(\{\rho_k\alpha_k\}A_z)+\frac{\partial}{\partial z}(\{\rho_k\upsilon_{kz}\alpha_k\}A_z)=\Gamma_k A_z \tag{5.132}$$

By writing Equation 5.132 for each phase, adding them, and applying the condition $\Gamma_v+\Gamma_\ell=0$, we get the mixture equation

$$\frac{\partial}{\partial t}(\rho_m A_z)+\frac{\partial}{\partial z}(G_m A_z)=0 \tag{5.63}$$

where

$$\rho_m=\{\rho_v\alpha\}+\{(1-\alpha)\rho_\ell\} \tag{5.50b}$$

and

$$G_{mz}=\{\rho_v\upsilon_{vz}\}_{vz}\{\alpha_v\}_z+\{\rho_\ell\upsilon_{\ell z}\}_{\ell z}\{\alpha_\ell\}_z \tag{5.40b}$$

For a constant area channel, Equation 5.132 leads to the following equations for vapor and liquid:

$$\text{Vapor: }\frac{\partial}{\partial t}\{\rho_v\alpha\}+\frac{\partial}{\partial z}\{\rho_v\upsilon_{vz}\alpha\}=\Gamma_v \tag{5.133a}$$

$$\text{Liquid: } \frac{\partial}{\partial t}\{\rho_\ell(1-\alpha)\} + \frac{\partial}{\partial z}\{\rho_\ell v_{\ell z}(1-\alpha)\} = \Gamma_\ell \tag{5.133b}$$

5.7.2 Momentum Equations

From Equation 5.98, when V is considered equal to $A_z\Delta z$ and the limit of infinitesimal volume is applied, we get for a fixed frame of reference:

$$\begin{aligned}\frac{\partial}{\partial t}(\{\rho_k \upsilon_{kz}\}_k\{\alpha_k\}A_z) + \frac{\partial}{\partial z}(\{\rho_k \upsilon_{kz}^2\}_k\{\alpha_k\}A_z) &= \iint_{A_z} \Gamma_k(\vec{\upsilon}_{ks})_{\text{av}} \cdot d\vec{A}_z \\ + \iint_{A_z} \vec{F}_{\text{wk}}''' \cdot d\vec{A}_z - \frac{\partial}{\partial z}(\{\text{p}_k\}_z\{\alpha_k\})A_z &+ \iint_{A_z} \vec{F}_{\text{sk}}''' \cdot d\vec{A}_z + \{\rho_k\alpha_k\}\vec{g}A_z\end{aligned} \tag{5.134}$$

where

$$\Gamma_k(\vec{\upsilon}_{ks})_{\text{av}} = -\frac{1}{V}\{\rho_k\vec{\upsilon}_k(\vec{\upsilon}_k - \vec{\upsilon}_s)\}_{ks} \cdot \vec{A}_{ks} \tag{5.135}$$

$$\vec{F}_{\text{wk}}''' = \frac{1}{V}\sum_j \vec{F}_{kj} \tag{5.136}$$

$$\vec{F}_{\text{sk}}''' = \frac{1}{V}\vec{F}_{\text{sk}} \tag{5.137}$$

If we again apply Equation 5.30, we get

$$\begin{aligned}\frac{\partial}{\partial t}(\{\rho_k \upsilon_{kz}\alpha_k\}A_z) + \frac{\partial}{\partial z}(\{\rho_k \upsilon_{kz}^2\alpha_k\}A_z) &= \{\Gamma_k\vec{\upsilon}_{ks}\cdot\vec{n}_z\}A_z \\ + \{\vec{F}_{\text{wk}}'''\cdot\vec{n}_z\}A_z - \frac{\partial}{\partial z}(\{p_k\alpha_k\}_z A_z) &+ \{\vec{F}_{\text{sk}}'''\cdot\vec{n}_z\}\cdot A_z + \{\rho_k\alpha_k\}\vec{g}\cdot\vec{n}_z A_z\end{aligned} \tag{5.138}$$

By adding Equation 5.138 as applied to each phase and applying the jump condition at the interface,

$$\Gamma_{\text{v}}\vec{\upsilon}_{\text{vs}}\cdot\vec{n}_z + \vec{F}_{\text{sv}}'''\cdot\vec{n}_z + \Gamma_\ell\vec{\upsilon}_{\ell\text{s}}\cdot\vec{n}_z + \vec{F}_{\text{s}\ell}'''\cdot\vec{n}_z = 0 \tag{5.139}$$

we get

$$\frac{\partial}{\partial t}(G_{\text{m}}A_z) + \frac{\partial}{\partial z}\left(\frac{G_{\text{m}}^2 A_z}{\rho_{\text{m}}^+}\right) = -F_{\text{wz}}'''A_z - \frac{\partial}{\partial z}(\{p\}A_z) - \rho_{\text{m}}gA_z\cos\theta \tag{5.140}$$

where

$$\frac{1}{\rho_{\mathrm{m}}^{+}} \equiv \frac{1}{G_{\mathrm{m}}^{2}}\left[\left\{\rho_{\mathrm{v}}\upsilon_{\mathrm{vz}}^{2}\alpha\right\}+\left\{\rho_{\ell}\upsilon_{\ell z}^{2}(1-\alpha)\right\}\right] \tag{5.67}$$

$$F_{\mathrm{wz}}''' \equiv -(\vec{F}_{\mathrm{wv}}''' \cdot \vec{n}_z + \vec{F}_{\mathrm{w}\ell}''' \cdot \vec{n}_z) \tag{5.141}$$

$$\{p\} \equiv \{p_{\mathrm{v}}\alpha\} + \{p_{\ell}(1-\alpha)\} \tag{5.142}$$

and θ is the angle between the flow direction and the vertical direction.

The assumption of $p_{\mathrm{v}} = p_{\ell} = p$ can usually be applied in well-mixed one-dimensional two-phase flow. Special relations are developed to relate F_{wz}''' to the average flow parameters such as G_{m} and ρ_{m}. It should be noted that F_{wz}''' is the net force per unit volume due to shear forces at the walls. Thus, it can be obtained from

$$F_{\mathrm{wz}}''' = \frac{1}{A_z}\int \tau_{\mathrm{w}}\, \mathrm{d}P_z \tag{5.143}$$

where the shear stress at the walls (τ_{w}) is a function of the flow conditions, as is discussed in Chapter 11.

5.7.3 Energy Equations

Again considering $V = A_z\Delta z$ and the limit of infinitesimal Δz, we can get from Equation 5.118 a one-dimensional energy equation. Let us recall that

$$\left(\frac{\mathrm{d}Q}{\mathrm{d}t}\right)_k = \iiint_{V_k} q_k'''\, \mathrm{d}V - \oiint_{S_k} \vec{q}_k'' \cdot \vec{n}\, \mathrm{d}S \tag{5.112}$$

and

$$\left(\frac{\mathrm{d}W}{\mathrm{d}t}\right)_{k,\mathrm{shear}} = -\oiint_{S_k} (\bar{\bar{\tau}}_k \cdot \vec{\upsilon}_k) \cdot \vec{n}\, \mathrm{d}S \tag{5.113}$$

Note that

$$\frac{1}{V}(p_k \vec{\upsilon}_{\mathrm{s}} \cdot \vec{n}_{\mathrm{s}})_{k\mathrm{s}} A_{k\mathrm{s}} = p_k \frac{\partial}{\partial t}(\alpha_k) \tag{5.144}$$

Then for a fixed frame of reference, we get

$$\frac{\partial}{\partial t}(\{\rho_k u_k^{\mathrm{o}}\alpha_k\}A_z)+\frac{\partial}{\partial z}(\{\rho_k h_k^{\mathrm{o}}\upsilon_{kz}\alpha_k\}A_z)=\{\Gamma_k h_{ks}^{\mathrm{o}}\}A_z-\left\{p_k\frac{\partial\alpha_k}{\partial t}\right\}A_z$$
$$+\{q_k'''\alpha_k\}A_z-(q_{wk}''\alpha_{wk}P_w)-(q_{sk}''P_s)-\frac{\partial}{\partial z}(q_{kz}''\alpha_k A_z)$$
$$+\frac{\partial}{\partial z}(\{(\tau_{xz}\upsilon_x)_k+(\tau_{yz}\upsilon_y)_k+(\tau_{zz}\upsilon_z)_k\}_k\{\alpha_k\}A_z)-\{\rho_k g\upsilon_{kz}\alpha_k\}A_z \quad (5.145)$$

where q_{wk}'' = heat flux from phase k to the wall in the A_z plane, P_w = wall perimeter in the A_z plane, and P_s = interphase perimeter in the A_z plane.

For one-dimensional flow in a uniform area channel, A_z is constant. The axial heat conduction and the shear effect are small so that both can be neglected. In addition, p_k may be assumed constant in the channel, that is, $p_v = p_\ell = p$. Thus, Equation 5.145 can be reduced to

$$\frac{\partial}{\partial t}\{\rho_k u_k^{\mathrm{o}}\alpha_k\}+\frac{\partial}{\partial z}\{\rho_k h_k^{\mathrm{o}}\upsilon_{kz}\alpha_k\}=\Gamma_k h_{ks}^{\mathrm{o}}-p\frac{\partial\alpha_k}{\partial t}+\{q_k'''\alpha_k\}-q_{wk}''\alpha_{wk}\frac{P_w}{A_z}$$
$$-\{\rho_k g\upsilon_{kz}\alpha_k\}+\{Q_{sk}^*\} \quad (5.146)$$

where Q_{sk}^* is given by

$$Q_{sk}^* = -\{\vec{q}_k''\}_s\cdot\frac{\vec{n}_{ks}A_{sk}}{V}=\{\vec{q}_k''\}_s\cdot\frac{\vec{n}P_S}{A_z} \quad (5.147)$$

and where the jump condition is given by

$$\sum_{k=1}^{2}(\Gamma_k h_{ks}^{\mathrm{o}}+Q_{sk}^*)=0 \quad (5.148)$$

For the one-dimensional mixture equation, we add the phasic equations to obtain

$$\frac{\partial}{\partial t}\left\{\rho_m\left[h_m+\frac{1}{2}(\upsilon^2)_m\right]-p\right\}+\frac{\partial}{\partial z}\left\{G_m\left[h_m^+ +\frac{1}{2}(\upsilon^2)_m^+\right]\right\}=$$
$$q_m'''-q_w''\frac{P_w}{A_z}-gG_m\cos\theta \quad (5.149)$$

where

$$\rho_m=\{\alpha_v\rho_v\}+\{\alpha_\ell\rho_\ell\} \quad (5.50b)$$

$$G_{mz} = \{\alpha_v\rho_v\upsilon_{vz} + \alpha_\ell\rho_\ell\upsilon_{\ell z}\}_z \tag{5.40c}$$

$$h_m = \frac{\{\rho_v h_v \alpha_v + \rho_\ell h_\ell \alpha_\ell\}}{\rho_m} \tag{5.150}$$

$$h_m^+ = \frac{\{\rho_v h_v \upsilon_{vz}\alpha_v + \rho_\ell h_\ell \upsilon_{\ell z}\alpha_\ell\}}{G_m} \tag{5.151}$$

$$(\upsilon^2)_m = \frac{\{\alpha_v\rho_v\upsilon_v^2 + \alpha_\ell\rho_\ell\upsilon_\ell^2\}}{\rho_m} \tag{5.152}$$

$$(\upsilon^2)_m^+ = \frac{\{\alpha_v\rho_v\upsilon_v^3 + \alpha_\ell\rho_\ell\upsilon_\ell^3\}}{G_m} \tag{5.153}$$

$$q_m''' = q_v'''\alpha_v + q_\ell'''\alpha_\ell \tag{5.154}$$

When υ_{kz} is uniform within A_z, it is possible to obtain an equation for the kinetic energy of each phase by multiplying the one-dimensional momentum equation for the k phase (Equation 5.138) by υ_{kz} to get

$$\upsilon_{kz}\left(\upsilon_{kz}\frac{\partial\rho_k\alpha_k}{\partial t}A_z + \rho_k\alpha_k A_z\frac{\partial\upsilon_{kz}}{\partial t} + \upsilon_{kz}\frac{\partial\rho_k\alpha_k\upsilon_{kz}A_z}{\partial z} + \rho_k\alpha_k\upsilon_{kz}A_z\frac{\partial\upsilon_{kz}}{\partial z}\right) =$$
$$-\upsilon_{kz}F_{wkz}'''A_z - \upsilon_{kz}\frac{\partial p\alpha_k}{\partial z}A_z - \rho_k\alpha_k\upsilon_{kz}gA_z + \vec{\upsilon}_{kz}\cdot\left(\Gamma_k\vec{\upsilon}_{ks} + \vec{F}_{sk}'''\right)A_z \tag{5.155}$$

The left-hand side (LHS) can be simplified using the mixture mass balance equation to

$$\rho_k\alpha_k A_z\frac{\partial(\upsilon_{kz}^2/2)}{\partial t} + \rho_k\alpha_k\upsilon_{kz}A_z\frac{\partial(\upsilon_{kz}^2/2)}{\partial z} + \upsilon_{kz}^2\Gamma_k A_z \tag{5.156}$$

Hence, by adding to the left-hand side the term

$$\frac{\upsilon_{kz}^2}{2}\left(\frac{\partial}{\partial t}\rho_k\alpha_k + \frac{\partial}{\partial z}\rho_k\alpha_k\upsilon_k - \Gamma_{vk}\right)A_z$$

which equals zero (from the mass balance), we get

$$\frac{\partial}{\partial t}\left[\left(\frac{\rho_k\alpha_k\upsilon_{kz}^2}{2}\right)A_z\right] + \frac{\partial}{\partial z}\left[\rho_k\alpha_k\upsilon_{kz}\left(\frac{\upsilon_{kz}^2}{2}\right)A_z\right] + \frac{\upsilon_{kz}^2}{2}\Gamma_k A_z =$$
$$-\upsilon_{kz}F_{wkz}'''A_z - \upsilon_{kz}\frac{\partial p\alpha_k}{\partial z}A_z - \rho_k\alpha_k\upsilon_{kz}gA_z + \vec{\upsilon}_{kz}\cdot(\Gamma_k\vec{\upsilon}_{ks} + \vec{F}_{sk}''')A_z \tag{5.157}$$

If we add the equations of kinetic energy for vapor and liquid, we get

$$\frac{\partial}{\partial t}\rho_m\left[\frac{(\upsilon_m^2)}{2}\right]A_z+\frac{\partial}{\partial z}G_m\left[\frac{(\upsilon^2)_m^+}{2}\right]A_z=-\upsilon_{vz}F'''_{wvz}A_z+\upsilon_{\ell z}F'''_{w\ell z}A_z$$
$$-\left(\upsilon_{vz}\frac{\partial p\alpha_v}{\partial z}+\upsilon_{\ell z}\frac{\partial p\alpha_\ell}{\partial z}\right)A_z-(\rho_v\alpha_v\upsilon_{vz}g+\rho_\ell\alpha_\ell\upsilon_{\ell z}g)A_z$$
$$+\vec{\upsilon}_{vz}\cdot[\Gamma_v(\vec{\upsilon}_{vs}-\vec{\upsilon}_{vz})+\vec{F}'''_{sv}]A_z+\vec{\upsilon}_{\ell z}\cdot[\Gamma_\ell(\vec{\upsilon}_{\ell s}-\vec{\upsilon}_{\ell z})+\vec{F}'''_{s\ell}]A_z \quad (5.158)$$

Let us simplify by assuming that the last two terms can be dropped owing to cancelation of each other. Then, by substracting Equation 5.158 from Equation 5.149, we obtain an energy equation for the mixture enthalpy:

$$\frac{\partial}{\partial t}(\rho_m h_m-p)A_z+\frac{\partial}{\partial z}(G_m h_m^+ A_z)=q'''_m A_z-q''_w P_w+\upsilon_{vz}\left(F'''_{wvz}+\frac{\partial p\alpha_v}{\partial z}\right)A_z$$
$$+\upsilon_{\ell z}\left(F'''_{w\ell z}+\frac{\partial p\alpha_\ell}{\partial z}\right)A_z \quad (5.159)$$

Equation 5.159 is an approximate one, as a term pertaining to the kinetic energy exchange at the liquid–vapor interface was neglected. However, it is interesting to note that the gravity term has dropped from the equation. The last two terms are the heat addition terms, included because of wall friction and static pressure changes. When these terms are small, and for models where the vapor and liquid velocities are not easily identifiable, it is possible to approximate the energy equation as

$$\frac{\partial}{\partial t}[(\rho_m h_m-p)A_z]+\frac{\partial}{\partial z}(G_m h_m^+ A_z)=q'''_m A_z-q''_w P_w+\frac{G_m}{\rho_m}\left(F'''_{wz}+\frac{\partial p}{\partial z}\right)A_z \quad (5.160)$$

Table 5.3 summarizes the one-dimensional relations for a single-phase situation as well as for the two-phase mixture.

PROBLEMS

5.1. *Area-averaged parameters* (Section 5.4)
In a BWR assembly, it is estimated that the exit quality is 0.15 and the mass flow rate is 17.5 kg/s. If the pressure is 7.2 MPa, and the slip ratio can be given as $S = 1.5$, determine $\{\alpha\}$, $\{\beta\}$, $\{j_v\}$, G_v, and G_ℓ. The flow area of the assembly is 1.2×10^{-2} m^2.

Answers:
$\{\alpha\} = 0.6968$
$\{\beta\} = 0.7751$
$\{j_v\} = 5.80$ m/s
$G_v = 218.75$ kg/m^2 s
$G_\ell = 1239.6$ kg/m^2 s

TABLE 5.3
One-Dimensional Transport Equations for Uniform Density Within Each Phasic Region

	Mass Equations
Phase	$\frac{\partial}{\partial t}\{\rho_k\alpha_k\}A_z + \frac{\partial}{\partial z}\{\rho_k\upsilon_{kz}\alpha_k\}A_z = \Gamma_k A_z$
Jump condition	$\sum_{k=1}^{2}\Gamma_k = 0$
Mixture	$\frac{\partial}{\partial t}(\rho_m A_z) + \frac{\partial}{\partial z}(G_m A_z) = 0$
where	$\rho_m = \{\rho_v\,\alpha\}+\{\rho_\ell(1-\alpha)\}$ $G = \{\rho_v\upsilon_{vz}\,\alpha\} + \{\rho_\ell\upsilon_{\ell z}(1-\alpha)\}$
	Momentum Equations
Phase	$\frac{\partial}{\partial t}\{\rho_k\upsilon_{kz}\alpha_k\}A_z + \frac{\partial}{\partial z}\{\rho_k\upsilon_{kz}^2\alpha_k\}A_z = \{\Gamma_k\vec{\upsilon}_{ks}\cdot\vec{n}_z\}A_z + \{\vec{F}'''_{wk}\cdot\vec{n}_z\}A_z - \frac{\partial}{\partial z}\{\rho_k\alpha_k\}A_z + \{\vec{F}'''_{sk}\cdot\vec{n}_z\}A_z + \{\rho_k\alpha_k\}\vec{g}\cdot\vec{n}_z A_z$
Jump condition	$\sum_{k=1}^{2}(\Gamma_k\vec{\upsilon}_{ks}\cdot\vec{n}_z + \vec{F}'''_{sk}\cdot\vec{n}_z) = 0$
Mixture	$\frac{\partial}{\partial t}(G_m A_z) + \frac{\partial}{\partial z}\left(\frac{G_m^2 A_z}{\rho_m^+}\right) = -F'''_{wz}A_z - \frac{\partial}{\partial z}\{p\}A_z - \rho_m g A_z \sin\theta$
where	$\frac{G_m^2}{\rho_m^+} \equiv \{\rho_v\upsilon_{vz}^2\alpha\} + \{\rho_\ell\upsilon_{\ell z}^2(1-\alpha)\}$ $\{\rho\} \equiv \rho_v\,\alpha + \rho_\ell\,(1-\alpha)$ $F'''_{wz} = -(\vec{F}_{wv} + \vec{F}_{w\ell})\cdot\vec{n}_z$
	Energy Equations
Phase	$\frac{\partial}{\partial t}\{\rho_k u_k^o\alpha_k\}A_z + \frac{\partial}{\partial z}\{\rho_k h_k^o\upsilon_{kz}\alpha_k\}A_z = \Gamma_k h_{ks}^o A_z - p\frac{\partial\alpha_k}{\partial z}A_z + \{q_k'''\alpha_k\}A_z - q''_{wk}\alpha_{wk}P_w - \{\rho_k g\upsilon_{kz}\alpha_k\}A_z + \{Q_{sk}^a\}A_z$
Jump	$\sum_{k=1}^{2}(\Gamma_k h_{ks}^o + Q_{ks}^*) = 0$
Mixture	$\frac{\partial}{\partial t}(\rho_m h_m - p)A_z + \frac{\partial}{\partial z}(G_m h_m^+)A_z = q_w'''A_z - q_w'''P_w + \frac{G_m}{\rho_m}\left(F'''_{wz} + \frac{\partial p}{\partial z}\right)$
where	$h_m = \{\rho_v\,h_v\,\alpha_v + \rho_\ell\,h_\ell\,\alpha_\ell\}/\rho_m$ $h_m{}^+ = \{\rho_v\,h_v\,\upsilon_{vz}\,\alpha_v + \rho_\ell\,h_\ell\upsilon_{\ell z}\,\alpha_\ell\}/G_m$

5.2. *Momentum balance for a two-phase jet load* (Section 5.5)
Calculate the force on a wall subjected to a two-phase jet (Figure 5.10) that has the following parameters:
Mass flux at exit (G_m) = 10.75×10^3 kg/m² s
Exit diameter (D) = 0.3 m
Upstream pressure (p) = 7.2 MPa
Pressure at throat (p_o) = 3.96 MPa
Exit quality (x_o) = 0.68
Slip ratio (S_o) = 1.5

Answer:
Force = 0.416 MN

5.3. *Estimating phase velocity differential* (Section 5.4)
A vertical tube is operating at high-pressure conditions as follows (Figure 5.11):

Operating Conditions
Pressure (p) = 7.4 MPa
Mass flux (G) = 2000 kg/m² s
Exit quality (x_e) = 0.0693

Geometry
D = 10.0 mm
L = 3.66 m

Saturated water properties at 7.4 MPa
h_f = 1331.33 kJ/kg
h_g = 2759.60 kJ/kg
h_{fg} = 1448.27 kJ/kg
v_f = 0.001381 m³/kg
v_g = 0.02390 m³/kg
v_{fg} = 0.02252 m³/kg

Assuming that thermal equilibrium between steam and water has been attained at the tube exit, find:

1. The tube exit cross-sectional averaged true and superficial vapor velocities; that is, find $\{\upsilon_v\}_v$ and $\{j_v\}$.

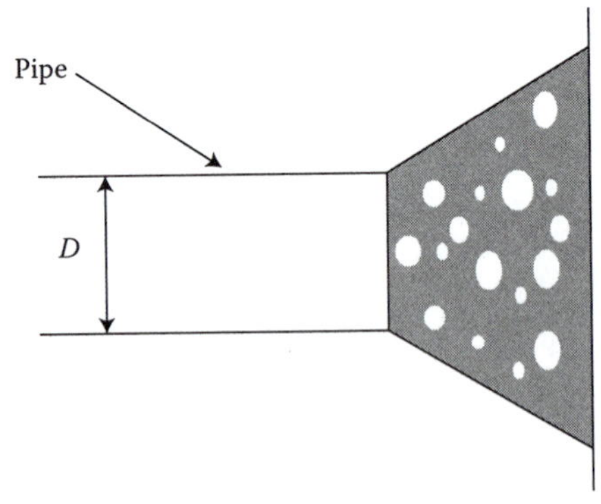

FIGURE 5.10 Two-phase jet impacting a wall.

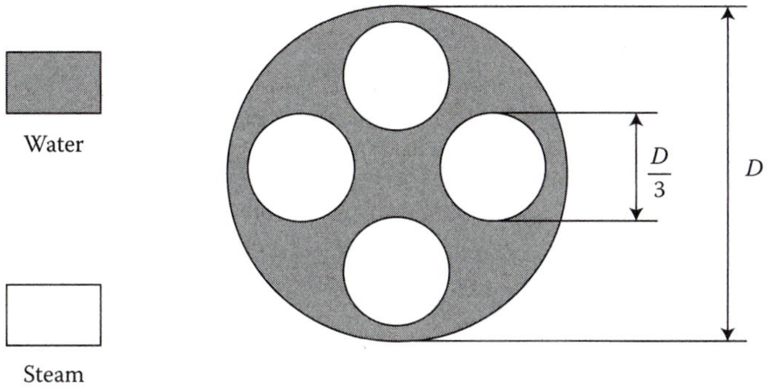

FIGURE 5.11 Cross-section of a high-pressure tube.

2. The difference between the tube exit cross-sectional averaged vapor and liquid velocities; that is, find $\{\upsilon_v\}_v - \{\upsilon_\ell\}_\ell$ at the exit.
3. The difference between the tube exit cross-sectional averaged vapor and liquid superficial velocities; that is, find $\{j_v\} \sim \{j_\ell\}$.

Answers:
1. $\{\upsilon_v\}_v = 7.45$ m/s, $\{j_v\} = 3.31$ m/s
2. 2.83 m/s
3. 0.74 m/s

5.4. *Torque on vessel due to jet from hot leg break* (Section 5.5)
Calculate the torque on a pressure vessel when a break develops in the hot leg as shown in Figure 5.12. The conditions of the two-phase emerging jet are:
$G_{cr} = 10.75 \times 10^3$ kg/m^2 s
$x_{cr} = 0.68$
$L = 5$ m
Angle $\theta = 70°$
$S = 1.5$
$p_o = 3.96$ MPa
$p = 7.2$ Mpa
Flow area at the rupture ≈ 0.08 m^2

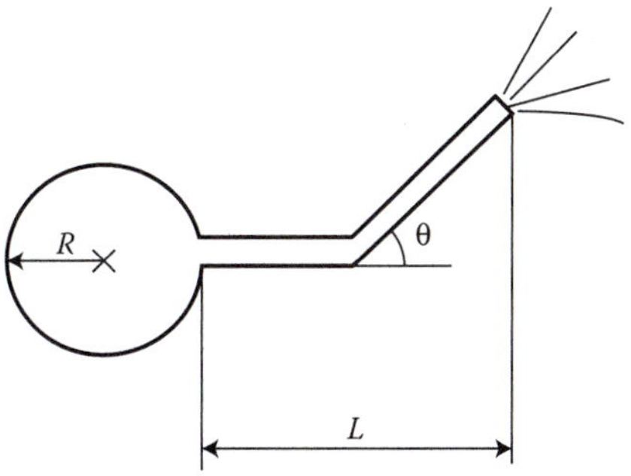

FIGURE 5.12 Two-phase jet at a pipe break.

Answer:
Torque ≈ 2.8 MN m

5.5. *Interfacial term in the momentum equation* (Section 5.6)
In a two-fluid model, the momentum equation of the vapor in one-dimensional flow may be written as

$$\frac{\partial}{\partial t}(\alpha\rho_v V_v A) = -\frac{\partial}{\partial z}(\alpha G_v V_v A) + \Gamma A V_{vs} - \frac{\partial p}{\partial z}A - F_{wv} - F_{sv} - \alpha\rho_v gA\cos\theta$$

where Γ = rate of mass exchange between vapor and liquid; F_{wv} = rate of momentum loss at the wall due to friction; A = flow area; V_{vs} = vapor velocity at the liquid–vapor interface; and F_{sv} = rate of momentum exchange between vapor and liquid.
Given the following:
Steady-state flow condition in a tube
Tube diameter is D
Uniform axial heat flux q''
Annular flow conditions prevail

1. Write appropriate mass balance and energy balance equations for vapor to complete the model. State and assumption you made.
2. Provide an expression for Γ. Justify your answer.

Answer: 2. $\Gamma = \dfrac{4q''}{Dh_{fg}}$

REFERENCES

1. Collier, J. G., *Convective Boiling and Condensation*, 2nd Ed. New York, NY: McGraw-Hill, 1980.
2. Delhaye, J. M., Giot, M., and Reithmuller, M. L., *Thermodynamics of Two-Phase Systems for Industrial Design and Nuclear Engineering*. New York, NY: McGraw-Hill, 1981.
3. Ishii, M., *Thermo-fluid Dynamic Theory of Two-Phase Flow*. Eyrolles, Paris: Scientific and Medical Publications of France, 1975.
4. Martinelli, R. C., and Nelson, D. B., Prediction of pressure drop during forced circulation boiling of water. *Trans. ASME*, 49:695–702, 1948.
5. Sha, W. T., Chao, B. T., and Soo, S. L., *Time Averaging of Local Volume-Averaged Conservation Equations of Multiphase Flow*. ANL-83–49, July 1983.
6. Vernier, P., and Delhaye, J. M., General two-phase flow equations applied to the thermo-hydraulics of boiling water nuclear reactors. *Energie Primaire* 4:5–46, 1968.
7. Wallis, G. B., *One Dimensional Two-Phase Flow*. New York, NY: McGraw-Hill, 1969.

6 Thermodynamics of Nuclear Energy Conversion Systems

Nonflow and Steady Flow: First- and Second-Law Applications

6.1 INTRODUCTION

The working forms of the first and second laws for the control mass and control volume approaches are summarized in Table 6.1. There are many applications of these laws to the analysis of nuclear systems. It is of prime importance that the reader not only develops the proficiency to apply these laws to new situations, but also recognizes which approach (control mass or control volume) is more convenient for the formulation of the solution of specific problems.

The elementary application of these laws avoids the time-dependent prediction of parameters. Usually, processes are either modeled as nontransient by specifying the initial and end states or as steady-state processes. This choice is not dictated by inherent limitations in the first and second laws but, rather, by the complexity involved when describing the heat and work rate terms, $\dot{Q}$ and $\dot{W}$, which appear in these laws. For example, the analytic description of $\dot{Q}$ requires definition of the heat transfer rates which, in many processes, are complex and available in empirical form only.

Nonflow and steady-flow processes are discussed here. Variable-flow processes are discussed in Chapter 7. The examples analyzed in this chapter as nonflow processes are essentially transient processes. However, as for any engineering analysis, it is important to consider the question posed and then to model the process in the simplest form, consistent with the required objective and the information available. Therefore when illustrating the control mass and control volume approaches, the examples are modeled in different ways to obviate or include the time-dependent description of the relevant rate processes. In doing so, the differences in the results achieved when complex processes are modeled in these fundamentally different ways are illustrated. Table 6.2 summarizes the approaches and examples covered here and in Chapter 7.

TABLE 6.1
Summary of the Working Forms of the First and Second Laws of Thermodynamics

<table>
<tr><th>Parameter</th><th>First Law</th><th>Second Law</th></tr>
<tr><td>Control mass</td><td>$\dot{U}^{\circ}_{\text{c.m.}} = \dot{Q}_{\text{c.m.}} - \dot{W}_{\text{c.m}}$ (Equation 4.19a)
For a process involving finite changes between states 1 and 2, Equation 4.19a becomes
$U_2 - U_1 = Q_{1+2} - W_{1+2}$ (Equation 6.1) if kinetic energy differences are negligible
Convention: W_{out} and Q_{in} are positive
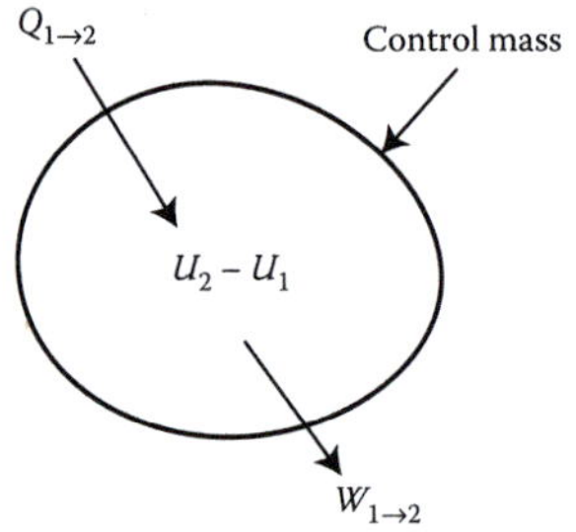
</td><td>$\dot{S}_{\text{c.m.}} = \dot{S}_{\text{gen}} + \frac{\dot{Q}_{\text{c.m.}}}{T_s}$
(Equation 4.25b)
where T_s is the temperature at which heat is supplied</td></tr>
<tr><td>Control volume (stationary)</td><td>$\dot{E}_{\text{c.v.}} = \sum_{i=1}^{I} \dot{m}_i (h_i^{\circ} + gz_i) + \dot{Q} + \dot{Q}_{\text{gen}} - \dot{W}_{\text{shaft}} - \dot{W}_{\text{normal}} - \dot{W}_{\text{shear}}$ (Equation 4.39)
Neglecting shear work and differences in kinetic and potential energy, and treating $\dot{Q}_{\text{gen}}$ as part of $\dot{U}$, Equation 4.39 becomes
$\dot{U}_{\text{c.v.}} = \sum_{i=1}^{I} \dot{m}_i h_i + \dot{Q} - \dot{W}_{\text{shaft}} - \dot{W}_{\text{normal}}$ (Equation 6.2)
Convention: W_{out} and Q_{in} are positive
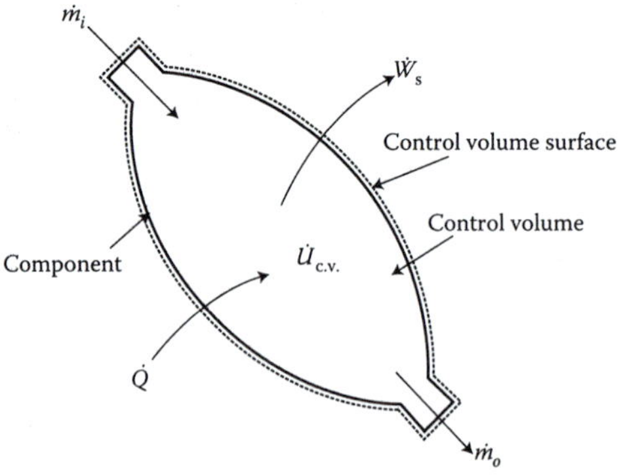
</td><td>$\dot{S}_{\text{c.v.}} = \sum_{i=1}^{I} \dot{m}_i s_i + \dot{s}_{\text{gen}} + \frac{\dot{Q}}{T_s}$
(Equation 4.41)</td></tr>
</table>

TABLE 6.2
Examples Covered in Chapters 6 and 7

Process	Control Mass	Control Volume
Nonflow	Chapter 6, Section 6.2: expansion work from a fuel–coolant interaction process	—
Steady flow	—	Chapter 6, Section 6.4: reactor system thermodynamic efficiency and irreversibility analysis
Nonsteady flow	Chapter 7, Section 7.2: reactor containment pressurization from loss of coolant accident	Chapter 7, Section 7.2: reactor containment pressurization from loss of coolant accident
	—	Chapter 7, Sections 7.3 and 7.4: pressurizer response to load change

6.2 NONFLOW PROCESS

The essential step in the nonflow approach is to carefully define the control mass and the sign convention for the interactions. A sketch of this control mass, complete with the relevant energy flows, is useful. Also desirable is a state diagram using suitable thermodynamic properties of the initial and final equilibrium states of the process the control mass undergoes. Finally, if the control mass undergoes a known reversible process, it is useful to represent the process path on the state diagram.

To determine the energy flows, it is necessary to specify the time base over which the analysis is applicable. Usually, a fixed time period is not specified; rather, it is stated or implicitly assumed that the analysis applies over a period sufficient for a desired transition between fixed thermodynamic states for the control mass. In that way, we do not explicitly involve the rate processes that would require detailed heat transfer and fluid mechanics information. The price for this simplification is that we do not learn anything of the transient aspects of the process, only the relation between the end states and, if the process path is defined, the work and heat interactions of the control mass.

The first law (Table 6.1) for a control mass undergoing a process involving finite changes between state "1" and state "2" is

$$U_2 - U_1 = Q_{1\to 2} - W_{1\to 2} \tag{6.1}$$

where positive W = work done by the control mass, and positive Q = heat transferred into the control mass. Note that for a control volume Equation 6.2 of Table 6.1 is the applicable equation.

As an illustration of application of this law to a control mass, let us examine the thermal interaction between a hot liquid and a more volatile cold liquid. This phenomenon, a molten fuel–coolant interaction, is of interest when evaluating the integrity of the containment under hypothetical accident conditions. It can be postulated to occur in both light-water- and sodium-cooled reactors only in unlikely situations. In both reactors, the hot liquid would consist of the molten fuel, cladding, and structural materials from the partially melted core; and the volatile, cold liquid would be the coolant, water, or sodium. We are interested in evaluating the effect of this interaction on the containment by calculating the expansion work resulting from the mixing of these hot and cold liquids. The relevant properties for these materials are presented in Table 6.3.

To estimate this expansion work, assume that the process occurs in two steps. In the first step, an equilibrium temperature is found for the fuel and coolant under the condition of no expansion by the more volatile coolant or the fuel. This constant-volume condition is reasonable if the thermal equilibration time is small (≤1 ms), which, however, is not the case if film boiling occurs when the fuel mixes with the coolant. In the second step, work occurs when the coolant and fuel are assumed to expand reversibly to a prescribed state. This expansion may take place: (1) without heat transfer between the fuel and the coolant, or (2) with heat transfer between the fuel and the coolant so that thermal equilibrium is maintained during the expansion.

With these somewhat artificial prescriptions for the interaction, we have implicitly prescribed time periods for the two steps by defining the intermediate state (i.e., that corresponding to the equilibrium temperature at the initial volume) and the final state. In this manner, the truly transient fuel–coolant interaction process has been idealized as a nonflow process. In fact, these time periods are not real because the processes occurring during the two steps are not physically distinct. Furthermore, there are heat transfer and expansion work rates that characterize this process. However, for convenience, our idealization of this process between end states allows us to establish conservative bounds on the expansion work. The analysis of this process that follows generally adopts the approach of Hicks and Menzies [3].

6.2.1 A Fuel–Coolant Thermal Interaction

6.2.1.1 Step I: Coolant and Fuel Equilibration at Constant Volume

We may define either one control mass, consisting of the combined mass of the fuel and the coolant, or two control masses, one consisting of the mass of the fuel and the other consisting of the mass of the coolant. We choose the second representation because it is easier to define the *T–s* diagrams for a one-component (two-phase) fluid (i.e., the coolant or fuel alone) then it is to define the *T–s* diagram for the two-component coolant–fuel mixture.

The control masses do not provide work to the environment during the constant-volume thermal equilibration process. To analyze the interaction conservatively, we assume that all the heat transferred from the fuel during the process is transferred to the coolant. Finally, because the thermal equilibration process is so fast, the decay heat generation rate is low enough to be neglected during the process. With these assumptions, the control mass and the process representation for the coolant alone may be sketched as in Figure 6.1.

TABLE 6.3
Properties for Fuel–Coolant Interaction Examples

Parameter	Symbol	Units	Sodium	Water	Fuel: UO_2 or Mixed Oxide
Mass (typically primary system inventory)	m_c or m_f	kg	3500	4000	40,000
Initial temperature	T_1	K	600	400	3100
Initial density	ρ_1	kg/m^3	835	945	~ 8000
Saturation pressure	p_{sat}	MPa	$\ln p_{sat}(\mathrm{MPa}) = 8.11 - \frac{12{,}016}{T(K)}$	$\ln p_{sat}(\mathrm{MPa}) = 10.55 - \frac{4798}{T(K)}$	—
Gas constant	R	J/kg K	361	462	31
Vapor specific heat ratio	Γ	—	1.15	1.3	~ 1.06–1.07
Liquid specific heat at constant volume[a]	c_{v_c} or c_{v_l}	J/kg K	1300	4184	560
Vapor specific heat at constant pressure	$c_{p_{vc}}$ or $c_{p_{vl}}$	J/kg K	$\frac{\gamma R}{\gamma - 1} = 2767$	2003	~ 500
Vapor specific heat at constant volume	$c_{v_{vc}}$ or $c_{v_{vl}}$	J/kg K	$\frac{R}{\gamma - 1} = 2410$	1540	~ 475
Latent heat of vaporization	h_{fg}	J/kg	2.9×10^6	1.9×10^6	~ 1.9×10^6
Coolant critical point properties					
Pressure	p_{crit}	MPa	40.0	22.1	—
Temperature	T_{crit}	K	2733	647.3	—
Density	ρ_{crit}	kg/m^3	818	317	—
Internal energy	u_{crit}	J/kg	4.29×10^6	2.03×10^6	—

[a] For nearly incompressible liquids, the constant pressure and constant volume specific heats can be taken as equal.

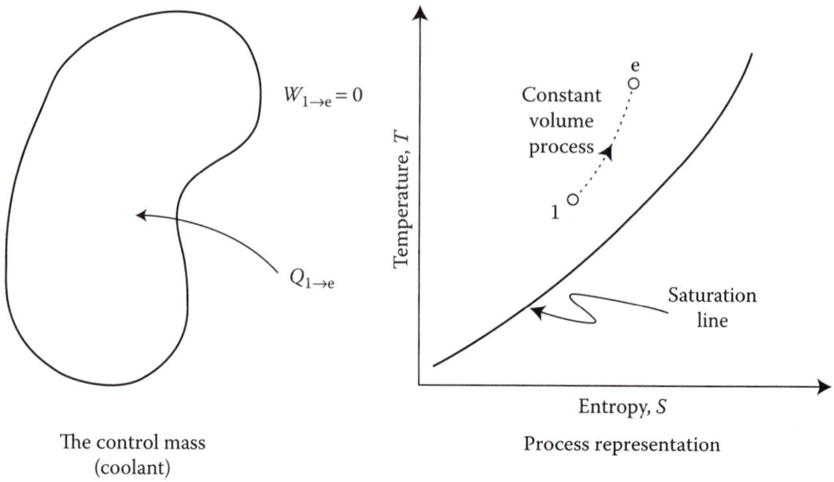

FIGURE 6.1 Coolant control mass behavior in the constant-volume thermal interaction with fuel: Step I.

To evaluate Q, we identify the fuel as another control mass undergoing a constant-volume cooling process to the equilibrium temperature. Analogous sketches for this fuel control mass are shown in Figure 6.2.

Expressing the first law for each control mass, undergoing a change of state from 1 to e, neglecting potential and kinetic energy changes, we obtain

$$\text{Coolant} \quad \Delta E_c = \Delta U_c = Q_{1\to e} \tag{6.3}$$

$$\text{Fuel} \quad \Delta E_f = \Delta U_f = -Q_{1\to e} \tag{6.4}$$

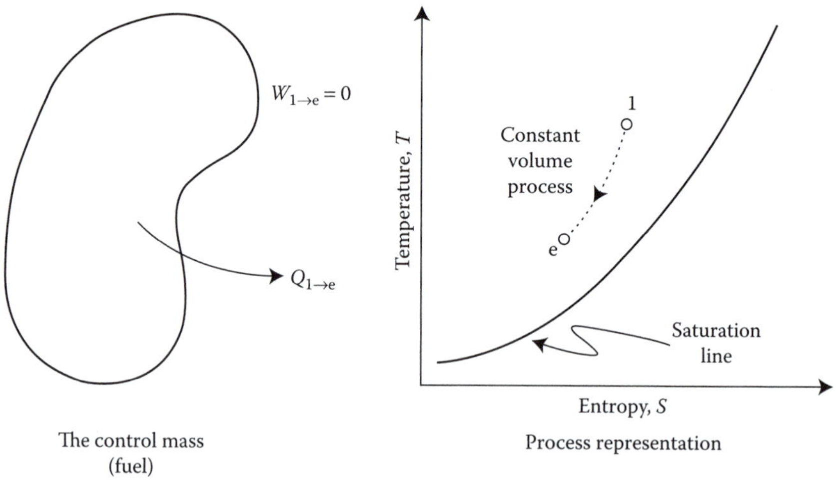

FIGURE 6.2 Fuel control mass behavior in the constant-volume thermal interaction with coolant: Step I.

Using the state equation that expresses internal energy as a function of temperature, we obtain

$$\Delta U_{\mathrm{c}} = m_{\mathrm{c}} c_{\mathrm{v_c}} (T_{\mathrm{e}} - T_{1_{\mathrm{c}}})$$
$$\Delta U_{\mathrm{f}} = m_{\mathrm{f}} c_{\mathrm{v_f}} (T_{\mathrm{e}} - T_{1_{\mathrm{f}}})$$

The interaction process of the combined coolant–fuel system is adiabatic. Hence

$$m_{\mathrm{c}} c_{\mathrm{v_c}} = (T_{\mathrm{e}} - T_{1_{\mathrm{c}}}) = m_{\mathrm{f}} c_{\mathrm{v_f}} (T_{1_{\mathrm{f}}} - T_{\mathrm{e}})$$

or

$$T_{\mathrm{e}} = \frac{(m_{\mathrm{f}} c_{\mathrm{v_f}} / m_{\mathrm{c}} c_{\mathrm{v_c}}) T_{1_{\mathrm{f}}} + T_{1_{\mathrm{c}}}}{1 + (m_{\mathrm{f}} c_{\mathrm{v_f}} / m_{\mathrm{c}} c_{\mathrm{v_c}})} \tag{6.5}$$

Because this equilibrium state is to the left of the saturated liquid line on a T–s plot for each component, the static quality of each fluid is by definition either 1.0 or 0 depending on whether T_{e} is above or below the critical temperature.

Example 6.1: Determination of the Equilibrium Temperature

PROBLEM

Compute the equilibrium temperature achieved by constant volume mixing of (1) sodium and mixed oxide (UO_2-15 weight percent PuO_2), and (2) water and UO_2 for the parameters of Table 6.3.

SOLUTION

For the sodium and mixed oxide, Equation 6.4 yields

$$T_{\mathrm{e}} = \frac{[40{,}000\ (560)/3500\ (1300)]3100 + 600}{1 + (40{,}000\ (560)/3500\ (1300))} = 2678\mathrm{K}$$

Because the critical temperature of sodium is 2733 K, $(x_{\mathrm{st}})_{\mathrm{e}} = 0$. The quality is unsubscripted with regard to coolant versus fuel because the fuel remains a subcooled liquid in the examples of the fuel–coolant interaction in this chapter. For simplicity, the subscript st is deleted in the remainder of this section. For the water and UO_2 combination:

$$T_{\mathrm{e}} = \frac{[40{,}000\ (560)/4000\ (4184)]3100 + 400}{1 + (40{,}000\ (560)/4000\ (4184))} = 1945\mathrm{K}$$

6.2.1.2 Step II: Coolant and Fuel Expanded as Two Independent Systems, Isentropically and Adiabatically

Consider the fuel and the coolant each undergoing an adiabatic isentropic, and hence reversible expansion. Let us first estimate the work done by the coolant when

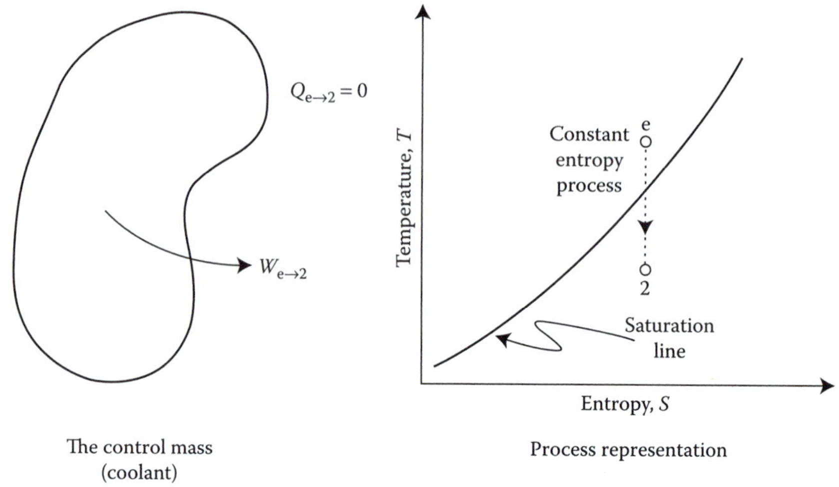

FIGURE 6.3 Coolant control mass behavior in the adiabatic, isentropic expansion: Step II.

expanding to 1 atmosphere. The control mass and the process representation for the coolant are shown in Figure 6.3.

Note that when expanding to the final state (state 2) some coolant remains in liquid form. Assume that this liquid occupies negligible volume and is incompressible.

From the first law for our control mass neglecting potential and kinetic energy changes, we obtain

$$W_{e\to 2} = -\Delta U_{e2} \equiv U_e - U_2 \tag{6.6}$$

In order to obtain an analytic solution for the $W_{e\to 2}$, the internal energy is next expressed by an approximate equation of state, which allows for expansion into the two-phase region. Thus, the change in internal energy is expressed as

$$\Delta u \equiv c_v \Delta T + \Delta(x u_{fg}) \equiv c_v \Delta T + \Delta[x(h_{fg} - p v_{fg})] \tag{6.7}$$

where $u_{fg} = u_g - u_f$, $h_{fg} = h_g - h_f$, and $v_{fg} = v_g - v_f$; c_v and c_p without further subscript refer to liquid specific heats, whereas vapor specific heats are designated c_{v_v} and c_{p_v}. Considering the change in specific internal energy (Δu) as defined in Equation 6.6, the work $W_{e\to 2}$ can be written as

$$\begin{aligned} W_{e\to 2} &= -m_c \Delta u = -m_c(u_2 - u_e) \\ W_{e\to 2} &= m_c[c_v(T_e - T_2) + x_e(h_{fg} - p v_{fg})_e - x_2(h_{fg} - p v_{fg})_2] \end{aligned} \tag{6.8}$$

If we neglect the liquid volume at state 2, this equation can be written as

$$W_{e\to 2} = m_c\{c_v(T_e - T_2) + [(x h_{fg})_e - (x h_{fg})_2] - [(x p v_{fg})_e - (x p v_g)_2]\} \tag{6.9}$$

To evaluate $W_{e \to 2}$, we must first determine x_2, which is done as follows. For a pure substance, the following relation exists between thermodynamic properties for two infinitesimally close equilibrium states:

$$T\mathrm{d}s = \mathrm{d}h - v\mathrm{d}p \tag{6.10a}$$

which for an isentropic process can be rewritten as

$$\left(\frac{\mathrm{d}h}{T}\right)_{\mathrm{s}} = v\left(\frac{\mathrm{d}p}{T}\right)_{\mathrm{s}} \tag{6.10b}$$

Analogous to the treatment of the internal energy by an approximate equation of state (Equation 6.6), take h_{fg} and c_{p} independent of temperature over the range of interest and express the enthalpy change for this isentropic expansion as

$$\mathrm{d}h = c_{\mathrm{p}}\mathrm{d}T + h_{\mathrm{fg}}\mathrm{d}x \tag{6.11}$$

Neglecting the liquid specific volume and applying the perfect gas result:

$$v \equiv v_{\mathrm{f}} + xv_{\mathrm{fg}} \approx xv_{\mathrm{g}} \approx x\frac{RT}{p} \tag{6.12}$$

In an isentropic process, the differential of $p(T,s)$ is

$$\mathrm{d}p = \left(\frac{\partial p}{\partial T}\right)_{\mathrm{s}} \mathrm{d}T$$

If we assume state e is almost at the saturation line, so that the expansion is almost totally under the saturation dome, we can express $(\partial p/\partial T)_{\mathrm{s}}$ utilizing the Clausius–Clapeyron relation, which links parameters along the saturation line to the enthalpy and volume of vaporization. Hence

$$\mathrm{d}p = \left(\frac{\partial p}{\partial T}\right)_{\mathrm{s}} \mathrm{d}T = \left(\frac{\partial p}{\partial T}\right)_{\mathrm{sat}} \mathrm{d}T = \frac{h_{\mathrm{fg}}}{Tv_{\mathrm{fg}}}\mathrm{d}T = \frac{ph_{\mathrm{fg}}}{RT^2}\mathrm{d}T \tag{6.13}$$

where the last step utilizes the assumptions of Equation 6.12. Substituting Equations 6.11 through 6.13 into Equation 6.10 and rearranging, we obtain

$$h_{\mathrm{fg}}\frac{\mathrm{d}x}{T} + c_{\mathrm{p}}\frac{\mathrm{d}T}{T} - h_{\mathrm{fg}}x\frac{\mathrm{d}T}{T^2} = 0$$

which can be written as

$$d\left(\frac{x}{T}\right) = -\frac{c_{\mathrm{p}}}{h_{\mathrm{fg}}}\frac{\mathrm{d}T}{T}$$

where h_{fg} and c_{p} have been taken constant.

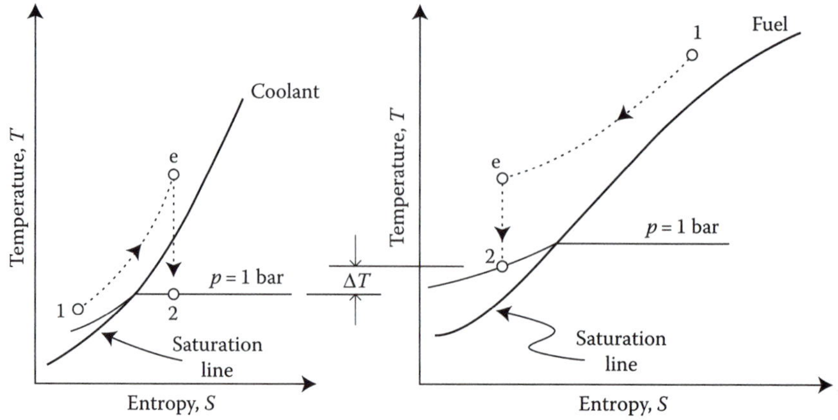

FIGURE 6.4 Fuel control mass and coolant control mass behavior in the adiabatic isentropic expansion: Step II.

Integrating this result between the initial equilibrium temperature of the fuel and coolant and the final coolant temperature, we obtain the expression defining x_2 in terms of the fuel/coolant equilibrium temperature and the final coolant temperature.

$$x_2 = T_2\left(\frac{x_e}{T_e} + \frac{c_p}{h_{fg}}\ell n\frac{T_e}{T_2}\right) \tag{6.14}$$

The work has been calculated for expansion of only the coolant. Next consider the work from the fuel as it reversibly changes volume. The fuel cooldown to a partial pressure which is negligible consistent with a total pressure of 1 atmosphere would terminate at a temperature different from that achieved in the coolant-only expansion, as Figure 6.4 illustrates. However, the work associated with fuel cooldown is small because the fuel remains a subcooled liquid. Hence, it is not evaluated here.

Example 6.2: Determination of Final Quality and Work Done by Sodium Coolant

PROBLEM

Evaluate the final quality (x_2) and the work done by the coolant for the sodium case taking the final state (state 2) as the saturation temperature (1154 K) at 1 atmosphere. Use the properties of Table 6.3.

SOLUTION

Utilizing Equation 6.14 and $x_e = 0$ from Example 6.1 yields

$$x_2 = 1154\left[0 + \frac{1300}{2.9(10^6)}\ln\frac{2678}{1154}\right] = 0.435$$

Returning to Equation 6.9, the expansion work done by the coolant is

$$W = 3500[1300(2678 - 1154) - 0.435(2.9)(10^6) + 0.435(1.013)(10^5)4.11] = 3153\ \text{MJ}$$

where

$$v_{g2} \approx \left(\frac{RT}{p}\right)_2 = \frac{361(1154)}{1.013 \times 10^5} = 4.11\text{m}^3/\text{kg}$$

and

$$x_e = 0.$$

Examination of the water–fuel case for the properties given in Table 6.3 using Equation 6.14 shows that the final state is superheated. Because the equilibrium condition has been already shown to be supercritical, use of Equation 6.7 is inappropriate and Equation 6.9 for the expansion work must be rederived.

6.2.1.3 Step III: Coolant and Fuel Expanded as One System in Thermal Equilibrium, Adiabatically and Isentropically

Let us now calculate the work considering that the sodium and the fuel remains in thermal equilibrium as the mixture expands adiabatically and isentropically to the final pressure. This process furnishes the upper bound of the expansion work. In this expansion process, the coolant–fuel system expands adiabatically and isentropically in thermal equilibrium. Consequently, heat is being removed from the fuel and added to the coolant so that the coolant entropy increases and the fuel entropy decreases an equal amount. Hence the coolant passes into the two-phase region and may achieve a superheated vapor state. On the other hand, the fuel remains a subcooled liquid throughout the expansion process because the equilibrium temperature is below the fuel boiling point even at 1 atm. The control mass and process representations for the coolant and for the fuel as separate control masses are shown in Figure 6.5, which illustrates the entropy increase of the coolant and the equal entropy decrease of the fuel. For illustration, the final coolant state is shown as superheated vapor. The work ($W_{e\to 2}$) is the expansion work performed by the coolant–fuel system. Although the entropy of each component of the mixture changes, the entropy of the mixture is constant, as Figure 6.5 illustrates.

In the analysis, we assume that the liquid phase of each component occupies negligible volume and is incompressible. Equation 6.6 is the relevant first-law formulation but now is applied to the mixture. Considering the mixture-specific internal energy (u_m), Equation 6.6 is written as

$$W_{e\to 2} = -m_m \Delta u_m = -(m_c \Delta u_c + m_f \Delta u_f) \tag{6.15}$$

To express the coolant internal energy change (Δu_c), the final state of the coolant (i.e., whether it is a two-phase mixture or a superheated vapor) must be known. Therefore, we must evaluate the final coolant quality (x_2) to see if it is less than or greater than unity. (Note that there is no need to subscript the quality as x_{c_2}, as the fuel is always a subcooled liquid.) To do so, we repeat the procedure of Section 6.2, but now consider the mixture expansion as adiabatic and isentropic.

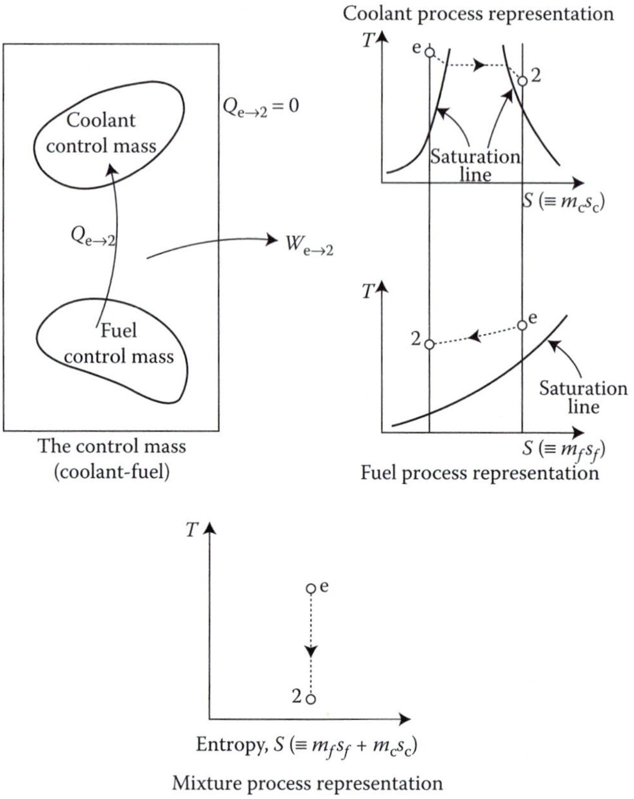

FIGURE 6.5 Coolant–fuel control mass and coolant–fuel interaction during the adiabatic, isentropic, isentropic control mass expansion: Step II.

Expressing Equation 6.10a for the mixture in this isentropic process:

$$T_m ds_m = dh_m - v_m dp_m = 0 \tag{6.16}$$

From the definition of mixture enthalpy

$$\Delta h_m = \frac{(m_c c_{p_c} + m_f c_{p_f})\Delta T + m_c \Delta(x h_{fg_c})}{m_c + m_f} \tag{6.17}$$

Differentiating, assuming h_{fg} constant:

$$dh_m = \frac{(m_c c_{p_c} + m_f c_{p_f})dT + m_c h_{fg_c}\, dx}{m_c + m_f} \tag{6.18}$$

Neglecting the fuel volume and the liquid coolant volume:

$$v_m = \frac{x m_c v_{g_c}}{m_c + m_f} = \frac{x R_c T}{p}\left(\frac{m_c}{m_c + m_f}\right) \tag{6.19}$$

Utilizing the Clausius–Clapeyron relation in the same manner as previously, reexpress Equation 6.13 for the mixture as follows noting that the fuel remains as a subcooled liquid:

$$\mathrm{d}p_{\mathrm{m}} = \left(\frac{\partial p}{\partial T}\right)_{\mathrm{sat}} \mathrm{d}T = \frac{ph_{\mathrm{fg_c}}}{R_{\mathrm{c}}} \frac{\mathrm{d}T}{T^2} \tag{6.20}$$

Substituting the above three relations into the state principle (Equation 6.16) and rearranging, we obtain

$$(m_{\mathrm{c}}c_{\mathrm{p_c}} + m_{\mathrm{f}}c_{\mathrm{p_f}})\frac{\mathrm{d}T}{T} + m_{\mathrm{c}}h_{\mathrm{fg_c}}\,\mathrm{d}\left(\frac{x}{T}\right) = 0 \tag{6.21}$$

Integrating this result between the initial equilibrium temperature of the fuel and coolant and final mixture temperature, we obtain

$$x_2 = T_2\left[\frac{x_{\mathrm{e}}}{T_{\mathrm{e}}} + \left(\frac{m_{\mathrm{c}}c_{\mathrm{p_c}} + m_{\mathrm{f}}c_{\mathrm{p_f}}}{m_{\mathrm{c}}h_{\mathrm{fg_c}}}\right)\ln\frac{T_{\mathrm{e}}}{T_2}\right] \tag{6.22}$$

If evaluation of x_2 indicates that the final state is a superheated vapor, that is, $x_2 > 1$, Δu_{c} is written as

$$\Delta u_{\mathrm{c}} = \Delta u_{\mathrm{c}}\ (x_{\mathrm{e}} = 0 \text{ to } x = 1) + \Delta u_{\mathrm{c}}\ (x = 1 \text{ to the actual state 2})$$

which can be written as follows utilizing Equation 6.7 and the numerical values of x_{e} and x:

$$\Delta u_{\mathrm{c}} = [c_{\mathrm{v}}(T - T_{\mathrm{e}}) + h_{\mathrm{fg}} - p\mathrm{v}_{\mathrm{fg}}]_{\mathrm{c}} + c_{\mathrm{v_{vc}}}(T_2 - T) \tag{6.23}$$

where T = temperature corresponding to x equal to unity, and the subscript on c_{v} in the second term explicitly indicates that the coolant is in vapor form.

The change in fuel internal energy is

$$\Delta u_{\mathrm{f}} = c_{\mathrm{v_f}}(T - T_{\mathrm{e}}) + c_{\mathrm{v_f}}(T_2 - T) \tag{6.24}$$

where $c_{\mathrm{v_f}}$ is the specific heat at constant volume for liquid fuel.

To evaluate Δu_{c} and Δu_{f}, the final temperature (T_2) must be determined in the superheat region. It is done again using the state principle (Equation 6.16) but now considering only that portion of the expansion process in which the coolant state is a superheated vapor. Therefore, Equation 6.18 is written as

$$\mathrm{d}h_{\mathrm{m}} = \frac{(m_{\mathrm{c}}c_{\mathrm{p_{vc}}} + m_{\mathrm{f}}c_{\mathrm{p_f}})\mathrm{d}T}{m_{\mathrm{c}} + m_{\mathrm{f}}} \tag{6.25}$$

where again $c_{p_{vc}}$ explicitly indicates that the coolant is in vapor form, and Equation 6.19 is written for $x = 1$ as

$$v_m = \frac{R_c T}{p}\left(\frac{m_c}{m_c + m_f}\right) \tag{6.26}$$

Substituting Equations 6.25 and 6.26 into Equation 6.16, we obtain the following relation between mixture temperature and pressure:

$$\left(\frac{m_c c_{p_{vc}} + m_f c_{p_f}}{m_c R_c}\right)\frac{dT}{T} = \frac{dp}{p} \tag{6.27}$$

Integration of Equation 6.27 to the unknown final state p_2, T_2 yields

$$T_2 = T\left(\frac{p_2}{p}\right)^{1/n} \tag{6.28}$$

where n = the term in parentheses in Equation 6.27, which can be written in terms of coolant vapor properties as

$$n \equiv \frac{m_c(\gamma R_c/\gamma - 1) + m_f c_{p_f}}{m_c R_c} = \frac{m_c c_{p_{vc}} + m_f c_{p_f}}{m_c R_c} \tag{6.29}$$

Although T ($x = 1$) can be determined by expressing Equation 6.22 for $x = 1$ and this corresponding T, the corresponding p is the coolant partial pressure, not the mixture pressure. However, because the fuel is liquid, its partial pressure is negligible. Therefore we can evaluate T_2 from Equation 6.28 with negligible error. Finally, returning to evaluation of the expansion work, it can be expressed by applying Equations 6.23 and 6.24 to 6.15, yielding:

$$\begin{aligned} W = W(\text{to } x = 1) + W(x = 1 \text{ to } T_2) &= m_c[c_v(T_e - T) - h_{fg} + R_c T]_c \\ &+ m_f c_{v_f}(T_e - T) + m_c c_{v_{vc}}(T - T_2) + m_f c_{v_f}(T - T_2) \end{aligned} \tag{6.30}$$

where the coolant vapor is taken as perfect gas allowing (pv) to be replaced by $R_c T$.

Example 6.3: Determination of Final Quality, Temperature, and Expansion Work for the Sodium-Mixed Oxide Combination

PROBLEM

For the sodium–mixed oxide combination, evaluate the final quality and temperature (x_2 and T_2) and the expansion work using the parameters of Table 6.3.

SOLUTION

First evaluate x_2 from Equation 6.22, assuming initially that state 2 is a two-phase state at 1 atmosphere. For this assumption T_2 is the sodium saturation temperature corresponding to 1 atmosphere or 1154 K. The quality (x_2) is obtained from Equation 6.22 as

$$x_2 = 1154\left[0 + \frac{40{,}000(560) + 3500(1300)}{(3500)2.9(10^6)}\ln\frac{2678}{1154}\right] = 2.58$$

Because $x_2 > 1$, this final state is superheated, and our original assumption is false. Therefore it becomes necessary to find the state at which $x = 1$. Returning to Equation 6.22 but defined for the state $x = 1$ and the corresponding temperature T, yielding:

$$1 = T\left[0 + \frac{40{,}000(560) + 3500(1300)}{(3500)2.9(10^6)}\ln\frac{2678}{T}\right]$$

Solving for T yields 2250 K, which has a corresponding sodium vapor pressure of 158 bars.

T_2 can now be evaluated from Equation 6.28 utilizing n obtained from Equation 6.29. These steps yield

$$T_2 = 2250\left(\frac{1}{158}\right)^{(1/25.4)} = 1843\text{K}$$

because

$$n = \frac{3.5(2767) + 40(560)}{3.5(361)} = 25.4$$

Finally, evaluating the expansion work by Equation 6.30 yields

$$\begin{aligned} W &= 3500[1300(2678 - 2250) - 2.9 \times 10^6 + (361)(2250)] \\ &\quad + 40{,}000(560)(2678 - 2250) + 3500(2410)(2250 - 1843) \\ &\quad + 40{,}000(560)(2250 - 1843) \\ &= -5360 + 9587 + 3433 + 9117 \\ &= 16{,}777\,\text{MJ} \end{aligned}$$

Note that the work evaluated by Example 6.3 is greater than that from Example 6.2 because work is dependent on the process path and the initial and end states. Although both processes have the same initial fuel and coolant states and are reversible, the process paths as well as the final states of the fuel and coolant are different. Consequently, there is no reason to expect that expansion work integrals performed to different final states should be the same. For the adiabatic and thermal equilibrium cases, the work is evaluated by Equation 6.9 and 6.30, respectively. A comparison of the numerical values of the terms of each equation is given in Table 6.4 to illustrate the origin of the differences in net work. In the thermal equilibrium case, the terms involving coolant properties and those involving fuel properties should not be interpreted as the work contributions due to the coolant and the fuel separately.

TABLE 6.4
Comparison between Results of Examples 6.2 and 6.3

Parameter	Fuel and Coolant Expansion: Two Independent Systems (Equation 6.9)	Fuel-Coolant Expansion: One-System (Equation 6.30)
Terms involving coolant properties	+ 3153 MJ	− 5360 + 3433 = −1927 MJ
Terms involving fuel properties	0	+ 9587 + 9117 = +18,704 MJ
Net work	+ 3153 MJ	+ 16,777 MJ

6.3 THERMODYNAMIC ANALYSIS OF NUCLEAR POWER PLANTS

The analysis of nuclear power plants represents a prime application of the methods for thermodynamic analysis of steady-flow processes. The results of such analyses determine the relation between the mixed mean outlet coolant temperatures (or enthalpy for BWR) of the core through the primary and secondary systems to the generation of electricity at the turbine. The variety of possible reactor systems, with their associated coolants, leads to a corresponding multiplicity of primary and secondary system configurations. In addition, because gaseous reactor coolants can be used directly to drive electric turbines, the Brayton cycle can be considered for gas-cooled reactor systems, whereas systems that use steam-driven electric turbines employ a Rankine cycle. The various cycles employed with the principal reactor types have been described in Chapter 1. In this section, the cycles used for the various reactor types and the methods of thermodynamic analysis of these cycles are described.

The Rankine and Brayton cycles are constant-pressure heat addition and rejection cycles for steady-flow operation. They differ regarding the phase changes the working fluid undergoes. In the Rankine cycle the working fluid is vaporized and condensed, whereas in the Brayton cycle the working fluid remains a single gaseous phase as it is heated and cooled. In central station nuclear Rankine cycles, water is employed as the working fluid, whereas helium is used in proposed Brayton cycles.

The PWR and BWR employ the water Rankine cycle. Because the PWR limits the reactor coolant to a nominal saturated mixed mean core outlet condition, the vapor that drives the turbine must be generated in a steam generator in a secondary system. A simplified pressurized water reactor two-coolant system is illustrated in Figure 6.6. A steam generator links these primary and secondary systems. Figure 6.7 illustrates the temperature distribution within such a recirculating PWR steam generator. The mixed mean core outlet condition (state 5) is dictated by the allowable core performance, particularly material corrosion limits.

As the temperature is raised, the primary loop pressure must also be raised to maintain state 5 at the nominally saturated condition. Establishment of the optimum primary pressure in current PWRs at 2250 psia (15.5 MPa) has involved many considerations, among them the piping and reactor vessel wall dimensions and the pressure dependence of the critical heat flux limit. The secondary side temperature and

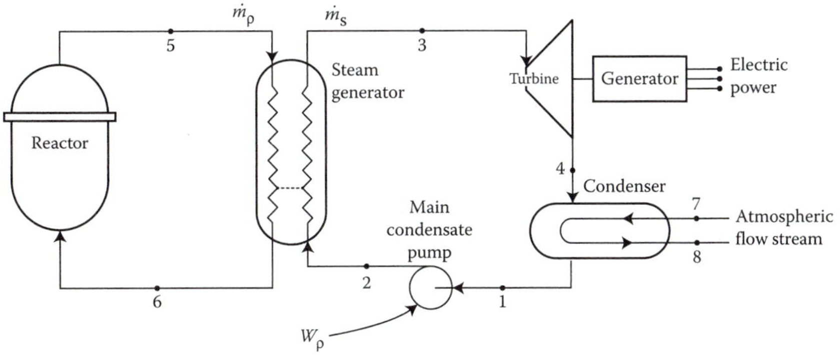

FIGURE 6.6 Simplified PWR plant.

hence pressure are related to these primary side conditions, as suggested in Figure 6.7. This relation is demonstrated in Section 6.4.

The BWR employs a direct Rankine cycle. The reactor is itself the steam generator, so that the mixed mean core outlet is also nominally saturated. The outlet temperature and pressure conditions are thus established by both the allowable core and secondary loop design conditions. As Table 2.2 illustrates, the Rankine working fluid conditions (at turbine entrance) for both PWRs and BWRs are close, that is, 5.7 and 7.14 MPa, respectively.

The simple Brayton cycle is illustrated in Figure 6.8. There are four components, as in the secondary system of the PWR plant operating under the Rankine cycle. The pictured Brayton cycle compressor and heat exchanger perform functions analogous to those of the Rankine cycle condensate pump and condensor. In practice, the

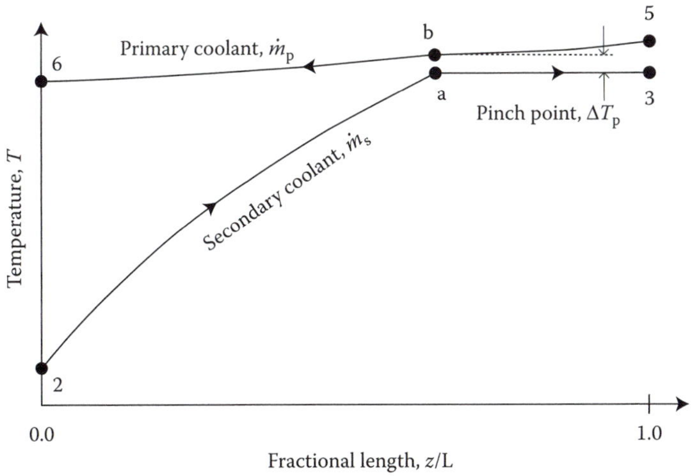

FIGURE 6.7 Temperature distribution within the steam generator of the simplified PWR plant of Example 6.4.

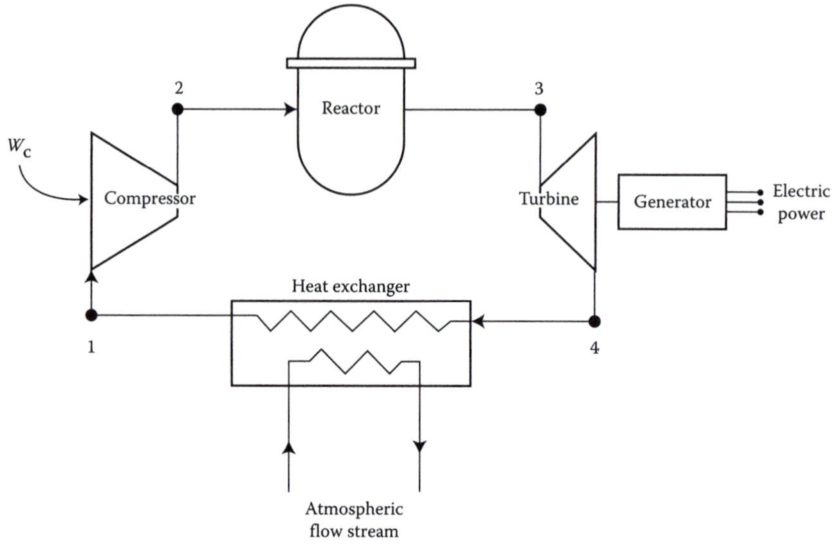

FIGURE 6.8 Simple Brayton cycle.

maximum temperature of the Brayton cycle (T_3) is set by turbine blade and gas-cooled reactor core material limits far higher than those for the Rankine cycle, which is set by liquid-cooled reactor core materials limits. The Brayton cycle in its many possible variations is discussed in Section 6.6.

It is useful to assess the thermodynamic performance of components and cycles using nondimensional ratios called *efficiencies*. Three definitions for efficiency are considered

- Thermodynamic efficiency (or effectiveness) (ζ)
- Isentropic efficiency (η_s)
- Thermal efficiency (η_{th})

The *thermodynamic efficiency* or effectiveness is defined as

$$\zeta \equiv \frac{\dot{W}_{u,actual}}{\dot{W}_{u,max}} \tag{6.31}$$

where $\dot{W}_{u,max}$ is defined by Equation 4.46, which for the useful case of a fixed, nondeformable control volume with zero shear work and negligible kinetic and potential energy differences between the inlet and outlet flow streams takes the form:

$$\dot{W}_{u,max} = -\left[\frac{\partial(U - T_o S)}{\partial t}\right] + \sum_{i=1}^{I} \dot{m}_i (h - T_o s)_i + \left(1 - \frac{T_o}{T_s}\right)\dot{Q} \tag{6.32}$$

where T_s = temperature at which heat is supplied and $(dQ/dt)_{gen}$ is treated as part of dU/dt.

The *isentropic efficiency* is defined as

$$\eta_s = \left(\frac{\dot{W}_{u,actual}}{\dot{W}_{u,max}}\right)_{\dot{Q}=0} \tag{6.33}$$

Obviously, for an adiabatic control volume, $\zeta = \eta_s$. The quantity $\dot{W}_{u,max|\dot{Q}=0}$ is the useful, maximum work associated with a reversible adiabatic (and hence isentropic) process for the control volume. It can be expressed from Equation 6.32 as

$$\dot{W}_{u,max|\dot{Q}=0} = -\left[\frac{\partial U}{\partial t}\right]_{c.v.} + \sum_{i=1}^{I} \dot{m}_i h_{is} \tag{6.34}$$

which for steady-state conditions becomes

$$\dot{W}_{u,max|\dot{Q}=0} = \sum_{i=1}^{I} \dot{m}_i h_{is} \tag{6.35}$$

The useful, actual work of an adiabatic nondeformable control volume with zero shear and negligible kinetic and potential energy differences between the inlet and outlet flow streams for steady-state conditions from Equation 4.44 (treating $\dot{Q}_{gen}$ as part of $\dot{E}$) is

$$\dot{W}_{u,actual} = \sum_{i=1}^{I} \dot{m}_i h_i \tag{6.36}$$

and the isentropic efficiency becomes

$$\eta_s = \frac{\sum_{i=1}^{I} \dot{m}_i h_i}{\sum_{i=1}^{I} \dot{m}_i h_{is}} \tag{6.37}$$

Finally, the *thermal efficiency* is defined as

$$\eta_{th} = \frac{\dot{W}_{u,actual}}{\dot{Q}_{in}} \tag{6.38}$$

where $\dot{Q}_{in}$ = rate of heat addition to the control volume. For adiabatic systems, the thermal efficiency is not a useful measure of system performance.

These efficiencies are now evaluated for nuclear plants. Typical plants employing the Rankine and Brayton cycles have been presented in Figures 6.6 and 6.8, respectively. Such plants have two interactions with their surroundings: a net work output as electricity and a flow cooling stream that is in mutual equilibrium with the atmosphere. Hence, the availability of the inlet and the outlet streams are zero, that is,

$$\sum_{i=1}^{I} \dot{m}_i (h - T_o s)_i = 0$$

Note, however, that the energy and entropy of the stream change as the stream passes through the plant condenser or heat exchanger, but the stream remains at pressure p_0 and temperature T_0.

In Section 6.4, each component of typical plants is analyzed using a stationary, nondeformable control volume with zero shear work and negligible kinetic and potential energy differences between the inlet and the outlet flow streams. These results are then utilized for evaluating the entire nuclear plant modeled by one adiabatic control volume, as illustrated in Figure 6.9a. Two other control volume representations of the complete nuclear plant are illustrated in Figures 6.9b and c.

Let us now contrast the computed effectiveness and thermal efficiencies for the three nuclear plant representations of Figure 6.9. In Section 6.4, the maximum useful work for control volume 1 of Figure 6.9a is shown equal to the fission rate, which equals the coolant enthalpy rise across the reactor. Hence,

$$\dot{W}_{\text{u,max}} = \dot{m}_{\text{p}} (h_{\text{out}} - h_{\text{in}})_{\text{reactor}}$$

(see Equation 6.53). The effectiveness of the nuclear plant is then:

$$\zeta = \frac{\dot{W}_{\text{u,actual}}}{\dot{m}_{\text{p}} (h_{\text{out}} - h_{\text{in}})_{\text{reactor}}} \tag{6.39}$$

In Figure 6.9b, the same nuclear plant is modeled by two control volumes. Control volume 1 encompasses the reactor plant except for the fuel loading, which is segregated into the second control volume. The energy liberated by fission is transferred from the second control volume to the first control volume as heat ($\dot{Q}$) at the temperature of the fuel. Because $\dot{Q}$ equals the fission rate:

$$\dot{Q} = \dot{m}_{\text{p}} (h_{\text{out}} - h_{\text{in}})_{\text{reactor}} \tag{6.40}$$

and

$$\underset{\text{(Fig. 6.9b)}}{\eta_{\text{th}}\big|_{\text{c.v.1}}} = \frac{\dot{W}_{\text{u,actual}}}{\dot{Q}} = \underset{\text{(Fig. 6.9a)}}{\zeta_{\text{c.v.1}}} \tag{6.41}$$

Finally, the same nuclear plant is alternatively modeled as shown in Figure 6.9c, with the reactor (including the fuel) and the primary side of the steam generator located within control volume 2. In this case, the same heat transfer rate ($\dot{Q}$) to control volume 1 occurs across the steam generator but now at the coolant temperature, so that

$$\underset{\text{(Fig. 6.9c)}}{\eta_{\text{th}}\big|_{\text{c.v.1}}} = \underset{\text{(Fig. 6.9b)}}{\eta_{\text{th}}\big|_{\text{c.v.1}}} = \underset{\text{(Fig. 6.9a)}}{\zeta_{\text{c.v.1}}} \tag{6.42}$$

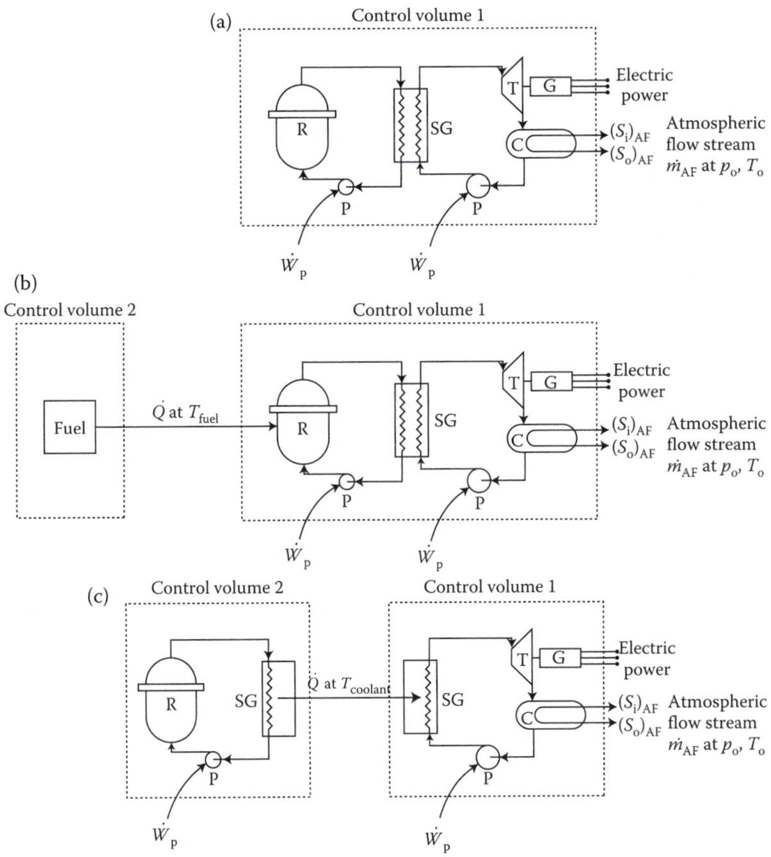

FIGURE 6.9 Alternative control volume representations of a batch-fueled reactor plant. (a) Entire nuclear plant in one control volume, (b) fuel contained in a separate control volume, and (c) primary and secondary coolant systems in separate control volumes.

Figure 6.9a is the only control volume configuration in which the entire nuclear plant is contained within one control volume. The effectiveness of the portions of the reactor plant contained within the three control volumes designated c.v.1 is not identical because the maximum, useful work associated with each of these control volumes is different. In Section 6.4.2, we show how transfer of the fission energy to the reactor coolant may be modeled as a series of three heat interactions that bring the fission energy from $T_{fission}$ to $T_{coolant}$. Consequently:

$$\underset{\text{(Fig. 6.9a)}}{\dot{W}_{u,max|c.v.1}} > \underset{\text{(Fig. 6.9b)}}{\dot{W}_{u,max|c.v.1}} > \underset{\text{(Fig. 6.9c)}}{\dot{W}_{u,max|c.v.1}} \tag{6.43}$$

Hence, the effectiveness of the portion of the plant within control volume 1 in each of the three configurations is related as

$$\underset{\text{(Fig. 6.9c)}}{\zeta_{|c.v.1}} > \underset{\text{(Fig. 6.9b)}}{\zeta_{|c.v.1}} > \underset{\text{(Fig. 6.9a)}}{\zeta_{|c.v.1}} \tag{6.44}$$

In this text, we often refer to the entire nuclear plant, and therefore the representation in Figure 6.9a appears and the term thermodynamic efficiency or effectiveness is used. However, by virtue of the equalities in Equation 6.42, it is also possible and useful to alternatively refer to the thermal efficiency of the cycle, which is also done. The previous examples illustrate the conditions under which the computation of various efficiencies lead to the same numerical values.

Finally, it is of interest to recall a case in which the effectiveness and the thermal efficiency do not coincide: For the case of reversible cycles operating between a heat source at T_{hjgh} and a reservoir at T_{low}, the thermal efficiency equals the Carnot efficiency. On the other hand, the effectiveness equals unity, as $\dot{W}_{\text{u,max}} = \dot{W}_{\text{u,actual}}$ for these cycles.

6.4 THERMODYNAMIC ANALYSIS OF A SIMPLIFIED PWR SYSTEM

The steady-state thermodynamic analysis of power plant systems is accomplished by considering in turn the components of these systems and applying to each the control volume form of the first and second laws. Let us examine the simplified PWR two-coolant system shown in Figure 6.6. We wish to develop the procedures for evaluating the thermodynamic states and system flow rates so that the overall plant thermodynamic efficiency and sources of lost work or irreversibilities can be computed from a combined first and second law analysis. At the outset, assume for simplicity that pressure changes occur only in the turbine and condensate pumps.

6.4.1 First Law Analysis of a Simplified PWR System

Consider any one component with multiple inlet and outlet flow streams operating at steady state and surround it with a nondeformable, stationary control volume (Figure 6.10). Applying the first law (Equation 6.2) to this control volume, we obtain

$$\sum_{k=1}^{I} (\dot{m}h)_{\text{in,k}} - \sum_{k=1}^{I} (\dot{m}h)_{\text{out,k}} = \dot{W}_{\text{shaft}} - \dot{Q} \tag{6.45}$$

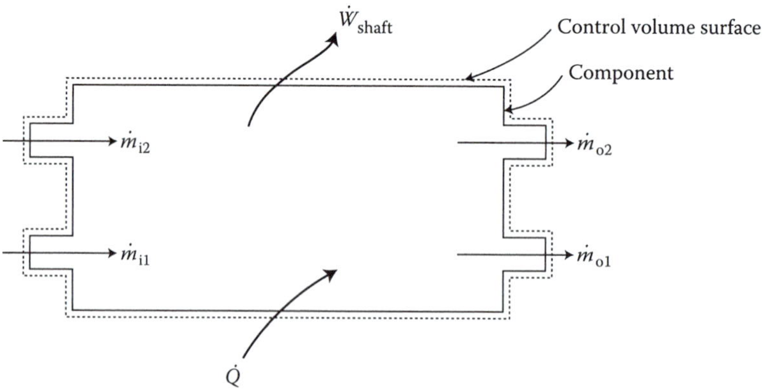

FIGURE 6.10 Generalized fixed-volume component with surrounding control volume.

where the notation of the summations has been generalized from that of Equation 4.30b to specifically identify the multiple inlet and outlet flow streams, where for steady state:

$$\dot{m} = \sum_{k=1}^{I} \dot{m}_{\text{in},k} - \sum_{k=1}^{I} \dot{m}_{\text{out},k}$$

Equation 6.45 can be further specialized for the components of interest. In particular for the components of Figure 6.6, heat addition from an external source can be neglected. For all components except the turbine and the pump, no shaft work exists. Table 6.5 summarizes the results of specializing Equation 6.45 to the components of Figure 6.6. The adiabatic turbine actual work is related to the ideal work by the component isentropic efficiency defined by Equation 6.33. From Equation 6.37, which is a specialized form of Equation 6.33, the turbine efficiency is

$$\eta_T = \frac{h_{\text{in}} - h_{\text{out}}}{h_{\text{in}} - h_{\text{out,s}}} \tag{6.46}$$

where

$$h_{\text{out,s}} = f[p(h_{\text{out}}), s_{\text{in}}]. \tag{6.47}$$

TABLE 6.5
Forms of First Law (Equation 6.2) for PWR Plant Components

Component	Desired Quantity	Assumptions	Resultant Equation	Consequence of Sign Convention on the Desired Parameter
Turbine	$\dot{W}_{\text{shaft}}$	$\dot{Q}_2 = 0$ $\Delta\left(\frac{1}{2}\upsilon^2\right) = 0$	$\dot{W}_T^{a} = [\dot{m}(h_{\text{in}} - h_{\text{out}})]_T$ $= [\dot{m}\eta_T(h_{\text{in}} - h_{\text{out,s}})]_T$	$\dot{W}_T$ positive (work out)
Pump	$\dot{W}_{\text{shaft}}$	$\dot{Q}_2 = 0$ $\Delta\left(\frac{1}{2}\upsilon^2\right) = 0$	$\dot{W}_P = [\dot{m}(h_{\text{in}} - h_{\text{out}})]_P$ $= \left[\frac{\dot{m}}{\eta_P}(h_{\text{in}} - h_{\text{out,s}})\right]_P$	$\dot{W}_P$ negative (work in)
Condensor steam generator, any feed water heater	h_o of one stream, for example, h_{o1}	$\dot{Q}_2 = 0$ $\dot{W}_{\text{shaft}} = 0$ $\Delta\left(\frac{1}{2}\upsilon^2\right) = 0$	$h_{\text{out},1} = \dfrac{\sum_{k=1}^{I} (\dot{m}h)_{\text{in},k} - \sum_{k=2}^{I} (\dot{m}h)_{\text{out},k}}{\dot{m}_{\text{out},1}}$	$h_{\text{out},1} — h_{\text{in},1}$ positive (stream 1 is heated)

[a] Because the control volumes are nondeformable, the shaft is dropped on $\dot{W}$ and replaced by a subscript describing the component.

Because work is supplied to the pump, the isentropic efficiency of the pump is analogously defined as

$$\eta_P = \frac{\text{Ideal work required}}{\text{Actual work required}} = \frac{h_{in} - h_{out,s}}{h_{in} - h_{out}} \tag{6.48}$$

The primary and secondary flow rates are related by application of the first law to the component they both flow through, that is, the steam generator. Considering only one primary and one secondary flow stream, the first law can be used to express the ratio of these two flow rates. The first law (Equation 6.45) for the steam generator yields

$$\frac{\dot{m}_P}{\dot{m}_s} = \frac{(h_{out} - h_{in})_s}{(h_{in} - h_{out})_P} \tag{6.49}$$

where subscripts p and s = primary and secondary flow streams, respectively.

It is often useful to apply Equation 6.49 to several portions of the steam generator. In Figure 6.7, temperatures are plotted versus fractional length assuming constant pressure for a recirculating-type PWR steam generator producing saturated steam. The minimum primary to secondary temperature difference occurs at the axial location of the onset of bulk boiling in the secondary side. All temperatures are related to this minimum, or pinch point, temperature difference. The steam generator heat transfer area is inversely related to this minimum temperature difference for a given heat exchange capacity. Also the irreversibility of the steam generator is directly related to this temperature difference. Therefore, the specification of the pinch-point temperature difference is an important design choice based on tradeoff between cost and irreversibility.

The ratio of flow rates in terms of the pinch-point temperature difference (ΔT_p) is from the first law:

$$\frac{\dot{m}_P}{\dot{m}_s} = \frac{h_3 - h_a}{\bar{c}_p[T_5 - (T_a + \Delta T_p)]} = \frac{h_a - h_2}{\bar{c}_p[(T_a + \Delta T_p) - T_6]} \tag{6.50}$$

where $\bar{c}_p$ = average coolant-specific heat over the temperature range of interest.

For the reactor plant, $\dot{W}_{u,actual}$ is the net rate of work generated, that is,

$$\dot{W}_{u,actual} = \dot{W}_T + \dot{W}_P = [\dot{m}_s(h_{in} - h_{out})]_T + [\dot{m}_s(h_{in} - h_{out})]_P \tag{6.51}$$

$$= [\eta_T \dot{m}_s(h_{in} - h_{out,s})]_T + \left[\frac{\dot{m}_s}{\eta_P}(h_{in} - h_{out,s})\right]_P \tag{6.52}$$

For a pump $h_{in} < h_{out,s}$, whereas for a turbine $h_{in} > h_{out,s}$. Hence, $\dot{W}_p$ is negative, indicating, per our sign convention presented in Tables 6.1 and 6.5, that work is supplied to the pump. Analogously, $\dot{W}_T$ is positive, and the turbine delivers work. The familiar expression of the maximum useful work of the nuclear plant that is derived in Section 6.4.2 is

$$\dot{W}_{u,max} = \dot{m}_p(h_{out} - h_{in})_R = \dot{m}_p(h_5 - h_6) \tag{6.53}$$

for Figure 6.6. It is desirable to reexpress $\dot{W}_{u,max}$ in terms of secondary side conditions. To do it, apply the first law (Equation 6.45) to the PWR steam generator of Figure 6.6, yielding:

$$\dot{m}_p(h_{in} - h_{out})_{SGp} = \dot{m}_s(h_{out} - h_{in})_{SGs} \tag{6.54}$$

From Figure 6.6 observe that the primary pump work is neglected so that $(h_{out})_R$ is identically $(h_{in})_{SGp}$ and $(h_{in})_R$ is identically $(h_{out})_{SGp}$. Hence, from Equations 6.53 and 6.54:

$$\dot{W}_{u,max} = \dot{m}_p(h_{out} - h_{in})_R = [\dot{m}_s(h_{out} - h_{in})]_{SGs} \tag{6.55}$$

or

$$\dot{W}_{u,max} = \dot{m}_p(h_5 - h_6) = \dot{m}_s(h_3 - h_2) \tag{6.56}$$

for Figure 6.6. Hence utilizing Equations 6.31, 6.51, and 6.55, the overall nuclear plant thermodynamic efficiency or effectiveness (ζ) is

$$\zeta = \frac{[\dot{m}_s(h_{in} - h_{out})]_T + [\dot{m}_s(h_{in} - h_{out})]_P}{[\dot{m}_s(h_{out} - h_{in})]_{SGs}} \tag{6.57}$$

For the plant of Figure 6.6, with the entire secondary flow passing through a single turbine, Equation 6.57 becomes

$$\zeta = \frac{h_3 - h_4 + h_1 - h_2}{h_3 - h_2} \tag{6.58}$$

Example 6.4: Thermodynamic Analysis of a Simplified PWR Plant

PROBLEM

The PWR plant of Figure 6.6 operates under the conditions given in Table 6.6. Assume that the turbine and pump have isentropic efficiencies of 85%.

1. Draw the temperature–entropy (T–s) diagram for this cycle.
2. Compute the ratio of the primary to secondary flow rates.

TABLE 6.6
PWR Operating Conditions for Example 6.4

State	Temperature °R (K)	Pressure psia (kPa)	Condition
1	—	1 (6.89)	Saturated liquid
2	—	1124 (7750)	Subcooled liquid
3	—	1124 (7750)	Saturated vapor
4	—	1 (6.89)	Two-phase mixture
5	1078.2 (599)	2250 (15,500)	Subcooled liquid
6	1016.9 (565)	2250 (15,500)	Subcooled liquid
7			Subcooled liquid
8			Subcooled liquid
A	—	1124 (7750)	Saturated liquid
B	$T_a + 26$ ($T_a + 14.4$)	2250 (15,500)	Subcooled liquid

3. Compute the nuclear plant thermodynamic efficiency.
4. Compute the cycle thermal efficiency.

SOLUTION

1. The T–s diagram is sketched in Figure 6.11. Note that states 4 and 2 reflect the fact that the turbine and the pump are not 100% efficient. Also, states on the primary side of the steam generator are shown at higher temperatures than the corresponding (same x/L position) states on the second side.

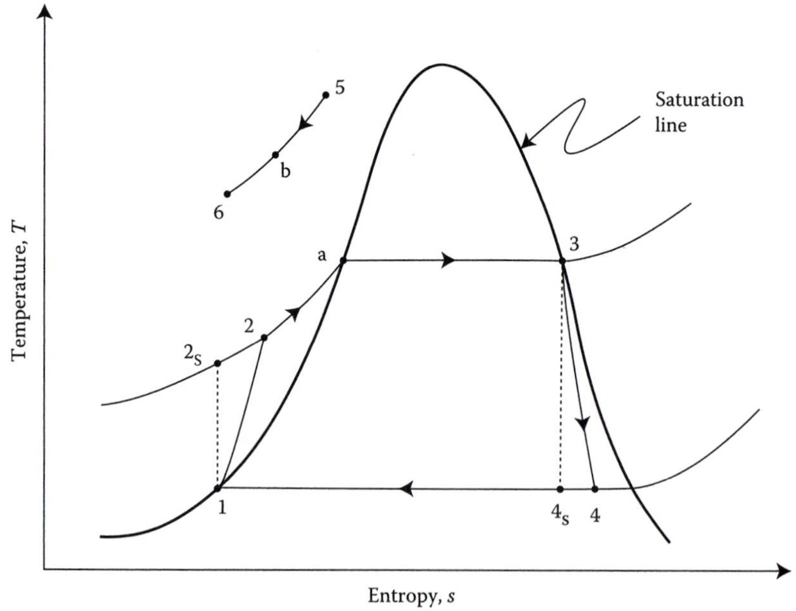

FIGURE 6.11 T–s diagram for PWR cycle: Example 6.4.

The T versus x/L diagram is the same as Figure 6.7.

If a primary system main coolant pump were to be included, the reactor system and its representation on a T–s diagram would be as sketched in Figure 6.12, where the primary system pressure loss is taken as 0.48 MPa (70 psia).

2. From Equation 6.50:

$$\frac{\dot{m}_p}{\dot{m}_s} = \frac{h_3 - h_a}{\bar{c}_p[T_5 - (T_a + \Delta T_p)]}$$

From steam tables:

$h_3 = h_g$(sat. at 1124 psia) = 1187.29 Btu/lb (2.771 MJ/kg)

$h_a = h_f$(sat. at 1124 psia) = 560.86 Btu/lb (1.309 MJ/kg)

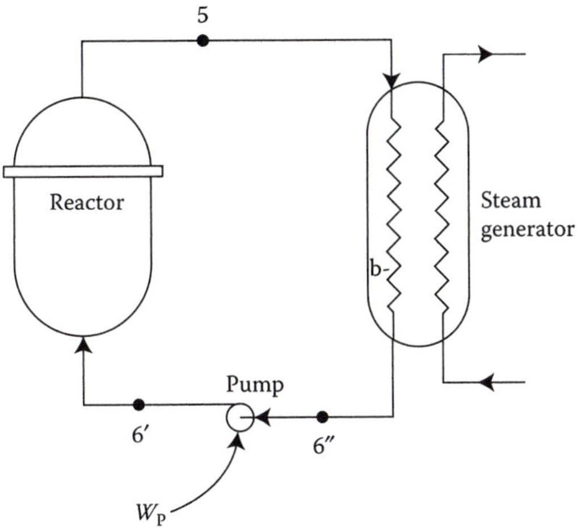

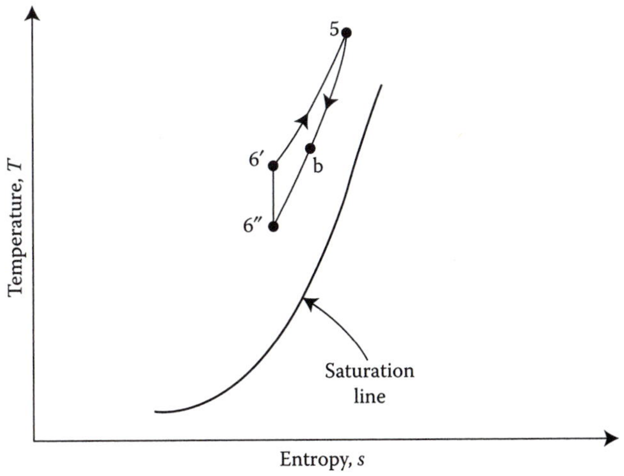

FIGURE 6.12 Primary system condition of Example 6.4.

T_a = sat. liquid at 1124 psia = 1018.8 R° (566.0 K)
Hence

$$\frac{\dot{m}_p}{\dot{m}_s} = \frac{1187.29 - 560.86}{1.424\,[1078.2 - (1018.8 + 26)]} = 13.18$$

in SI

$$\frac{\dot{m}_p}{\dot{m}_s} = \frac{2.77 \times 10^6 - 1.309 \times 10^6}{5941\,[599 - (566 + 14.4)]} = 13.18$$

3. The nuclear plant thermodynamic efficiency (ζ) for the plant within control volume 1 of Figure 6.9a is given by Equation 6.39. Draw a control volume around the PWR plant of interest in Figure 6.6. The result is an arrangement identical to that of Figure 6.9a. For this PWR plant, Equation 6.39 reduces to

$$\zeta = \frac{h_3 - h_4 + h_1 - h_2}{h_3 - h_2} \tag{6.58}$$

Proceed to obtain the required enthalpies.
From the steam tables:

$$h_1 = h_f\,(\text{sat. at 1 psia}) = 69.74 \text{ Btu/lb } (0.163 \text{ MJ/kg})$$

From the definition of component isentropic efficiencies:

$$h_2 = h_1 + \frac{h_{2s} - h_1}{\eta_P}$$

$$h_4 = h_3 - \eta_T\,(h_3 - h_{4s})$$

State 2:

$$s_{2s} = s_1 = 0.13266 \text{ Btu/lb°R } (557 \text{ J/kg K})$$

$$p_{2s} = p_2 = 1124 \text{ psia } (7.75 \text{ MP}_a)$$

From the subcooled liquid tables by interpolation:

$$h_{2s} = 73.09 \text{ Btu/lb } (0.170 \text{ MJ/kg})$$

$$h_2 = 69.74 + \frac{73.09 - 69.74}{0.85} = 73.69 \text{Btu/lb}$$

$$= 0.163 + \frac{0.170 - 0.163}{0.85} = 0.171 \text{MJ/kg}$$

Alternately, considering the liquid-specific volume constant and the compression process isothermal, h_2 can be evaluated as

$$h_2 = h_1 + v_1 \frac{(p_2 - p_1)}{\eta_P}$$

$$= 69.74 + \frac{0.016136(1123)}{0.85}\frac{144}{778} = 73.69 \text{Btu/lb}$$

$$= 0.163 + \frac{7.75 - 0.0069}{0.85(993)} = 0.171 \text{MJ/kg}$$

State 4:

$$s_{4s} = s_3 = 1.3759 \text{ Btu/lb } (5756 \text{ J/kg°K})$$

$$x_{4s} = \frac{s_{4s} - s_f}{s_{fg}} = \frac{1.3759 - 0.13266}{1.8453} = 0.674$$

$$h_{4s} = hf + x_{4s}h_{fg} = 69.74 + 0.674(1036.0) = 768.00 \text{ Btu/lb } (1.79 \text{ MJ/kg})$$

$$h_4 = h_3 - \eta_T(h_3 - h_{4s}) = 1187.29 - 0.85(1187.29 - 768.00)$$
$$= 830.9 \text{ Btu/lb } (1.94 \text{ MJ/kg})$$

Substituting into Equation 6.58, the thermodynamic efficiency can be evaluated as

$$\zeta = \frac{1187.29 - 830.9 + 69.74 - 73.69}{1187.29 - 73.69} \text{(English units)} = 0.317$$
$$= \frac{2.771 - 1.94 + 0.163 - 0.171}{2.771 - 0.171} \text{(SI units)} = 0.317$$

4. The plant thermal efficiency (η_{th}) is defined by Equation 6.38 and may be determined for the portion of the plant within control volumes 1 of Figure 6.9b and c. In either case, from Equation 6.40, $\dot{Q}$ in these figures equals the enthalpy rise of the coolant across the reactor. From Equation 6.42, the thermal efficiencies of the portions of the plants within control volumes 1 of Figure 6.9b and c are equal to the thermodynamic efficiency of the plant within control volume 1 of Figure 6.9a:

$$\underset{\text{(Fig. 6.9c)}}{\eta_{th|c.v.1}} = \underset{\text{(Fig. 6.9b)}}{\eta_{th|c.v.1}} = \underset{\text{(Fig. 6.9a)}}{\zeta_{c.v.1}} \tag{6.42}$$

The thermal efficiencies of these portions of the plant are commonly referred to as the *cycle thermal efficiency*. (Because, they are numerically equal.) This nomenclature is used in the remainder of the chapter for the various reactor plants considered. Because the PWR plant of Figure 6.6 is just a specific case of the general plant illustrated in Figure 6.9, we can directly state that the thermal efficiency of the PWR cycle equals the thermodynamic efficiency of the plant; that is, $\eta_{th} = \zeta$ for the PWR plant.

Improvements in thermal efficiency for this simple saturated cycle are achievable by (1) increasing the pressure (and temperature) at which energy is supplied to the working fluid in the steam generator (i.e., path 2-3, the steam generator inlet and turbine inlet conditions), and (2) decreasing the pressure (and temperature) at which energy is rejected from the working fluid in the condenser (i.e., path 4-1, the turbine outlet and pump inlet conditions). For the cycle of Figure 6.6, assuming an ideal turbine and condenser, the insets in Figures 6.13 and 6.14 illustrate the cycles in which each of these strategies is demonstrated.

The effect on cycle thermal efficiency of increasing the pressure at which energy is supplied as heat and of decreasing the pressure at which energy is rejected as heat is illustrated in Figures 6.13 and 6.14, respectively. The cross-hatched areas in the insets represent changes in energy supplied to and rejected from the working fluid.

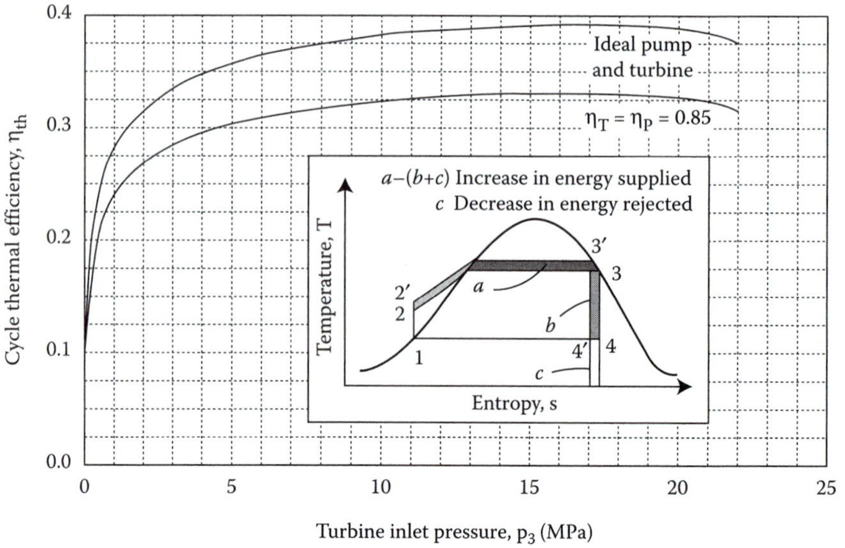

FIGURE 6.13 Thermal efficiency of Rankine cycle using saturated steam for varying turbine inlet pressure. Turbine inlet: saturated vapor. Exhaust pressure: 7 kPa.

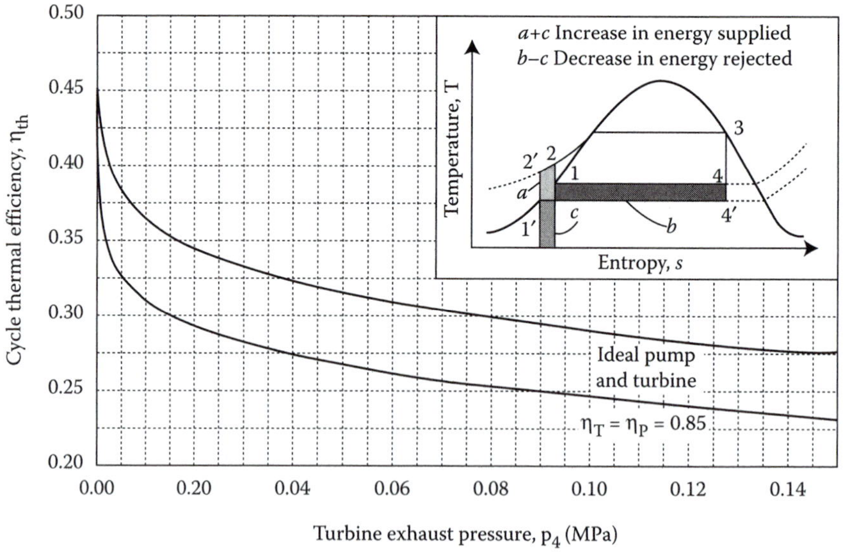

FIGURE 6.14 Thermal efficiency of Rankine cycle for a saturated turbine inlet state for varying turbine outlet pressure. Turbine inlet: 7.8 MPa saturated vapor.

6.4.2 Combined First and Second Law or Availability Analysis of a Simplified PWR System

Consider again the stationary, nondeformable control volume of Figure 6.10 operating at steady state with inlet and outlet flow streams at prescribed states. Neglect shear work and differences between kinetic and potential energies of the inlet and outlet streams. The maximum useful work obtainable from this control volume, which is in the form of shaft work because the control volume is nondeformable, can be expressed from Equation 4.46 as follows:

$$\dot{W}_{\text{u,max}} = \sum_{i=1}^{I} \dot{m}_i (h - T_o s)_i + \left(1 - \frac{T_o}{T_s}\right) \dot{Q} \tag{6.59}$$

where T_s = temperature at which $\dot{Q}$ is supplied and $(\mathrm{d}Q/\mathrm{d}t)_{\text{gen}}$ is treated as part of $\mathrm{d}E/\mathrm{d}t$.

For the same control volume under the same assumptions, the *lost work* or the *irreversibility* is given by Equations. 4.47 and 4.48 as

$$\dot{W}_{\text{lost}} \equiv \dot{I} = -T_o \sum_{i=1}^{I} \dot{m}_i s_i - \frac{T_o}{T_s} \dot{Q} = T_o \dot{S}_{\text{gen}} \tag{6.60}$$

Let us now restrict the above results to adiabatic conditions, that is, $\dot{Q} = 0$, and examine two cases: a control volume with shaft work and a control volume with no shaft work. Also recall that because the control volume is nondeformable, $(\mathrm{d}W/\mathrm{d}t)_{\text{normal}} = 0$.

Case I

$\dot{W}_{\text{shaft}} \neq 0$

From the first law, Equation 4.39

$$\dot{W}_{\text{shaft}} = \sum_{i=1}^{I} \dot{m}_i h_i \tag{6.61a}$$

From Equation 4.42

$$\dot{W}_{\text{actual}} = \dot{W}_{\text{shaft}} \tag{6.62a}$$

Now the irreversibility is given by Equation 6.60 as

$$\dot{I} = -T_o \sum_{i=1}^{I} \dot{m}_i s_i = T_o \dot{S}_{\text{gen}} \tag{6.63a}$$

However, the irreversibility is also equal from Equation 4.47 to

$$\dot{I} = \dot{W}_{\text{u,max}} - \dot{W}_{\text{actual}} = \sum_{i=1}^{I} \dot{m}_i (h - T_o s)_i - \dot{W}_{\text{shaft}} \tag{6.63b}$$

where the maximum useful work has been given by Equation 6.59 with $\dot{Q} = 0$ as

$$\dot{W}_{\text{u,max}} = \sum_{i=1}^{I} \dot{m}_i (h - T_o s)_i \tag{6.63c}$$

Case II
$\dot{W}_{\text{shaft}} \equiv 0$
From the first law, Equation 4.39

$$\sum_{i=1}^{I} \dot{m}_i h_i = 0 \tag{6.61b}$$

From Equation 4.42

$$\dot{W}_{\text{actual}} = 0 \tag{6.62b}$$

Hence the maximum useful work and the irreversibility, expressed by Equations 6.59 and 6.60, respectively, are equal and given by

$$\dot{W}_{\text{u,max}} = \dot{I} = -T_o \sum_{i=1}^{I} \dot{m}_i s_i = T_o \dot{S}_{\text{gen}} \tag{6.64}$$

Let us now apply these results to PWR system components.

6.4.2.1 Turbine and Pump

The turbine and pump have finite shaft work (case I) and single inlet and exit streams. Hence, the irreversibilities and the maximum useful work can be expressed from Equations 6.63a and 6.63c, respectively, as

$$\dot{I} = \dot{m}\, T_o (s_{\text{out}} - s_{\text{in}}) \tag{6.65}$$

and

$$\dot{W}_{\text{u,max}} = \dot{m}[(h_{\text{in}} - h_{\text{out}}) - T_o (s_{\text{in}} - s_{\text{out}})] \tag{6.66}$$

6.4.2.2 Steam Generator and Condenser

The steam generator and condenser have zero shaft work (case II) and two inlet and exit streams. Hence, the irreversibilities and the maximum useful work can be expressed from Equation 6.64 as

$$\dot{I} = \dot{W}_{\text{u,max}} = T_o \left(\sum_{k=1}^{2} \dot{m}_{\text{out},k} s_{\text{out},k} - \sum_{k=1}^{2} \dot{m}_{\text{in},k} s_{\text{in},k} \right) \tag{6.67}$$

It is useful to expand on this result for the condenser. Figure 6.15 illustrates a condenser cooled by an atmospheric flow stream ($\dot{m}_{AF}$) that is drawn from and returns to the environment, which acts as a reservoir at p_o, T_o. For the condenser conditions specified in Figure 6.15, the irreversibility of the condenser can be expressed from Equation 6.67 as

$$\dot{I} = \dot{m}_{AF} T_o \Delta s_{AF} + \dot{m}_s T_o \Delta s_s \tag{6.68}$$

For this constant pressure atmospheric flow stream interaction, Equation 6.10a can be written as

$$T\,ds = dh \tag{6.10b}$$

which can be expressed for the atmospheric coolant stream as

$$\dot{m}_{AF} T_o \Delta s_{AF} = \dot{m}_{AF} \Delta h_{AF} \tag{6.69}$$

From an energy balance on the condenser:

$$\dot{m}_{AF} \Delta h_{AF} = -\dot{m}_s \Delta h_s \tag{6.70}$$

Substituting Equation 6.70 into Equation 6.69, we obtain the desired result:

$$\dot{m}_{AF} T_o \Delta s_{AF} = -\dot{m}_s \Delta h_s \tag{6.71}$$

The irreversibility can be rewritten using Equation 6.71 as

$$\dot{I} = T_o \dot{m}_s \left[-\frac{\Delta h_s}{T_o} + \Delta s_s \right] \tag{6.72a}$$

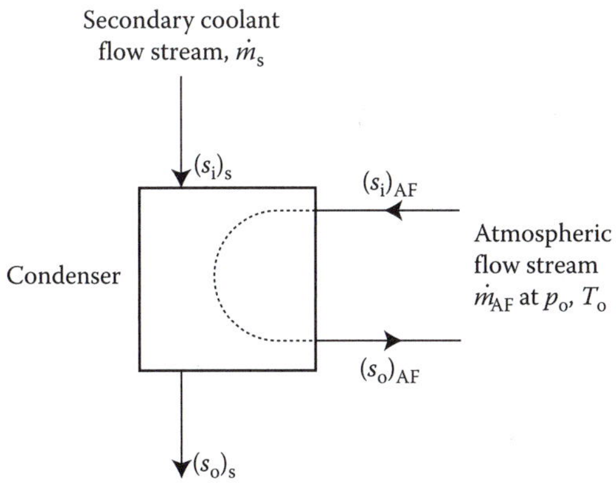

FIGURE 6.15 Flow streams for a condenser.

which, referring to Figure 6.6, becomes

$$\dot{I} = T_o \dot{m}_s \left[-\left(\frac{h_1 - h_4}{T_o} \right) + s_1 - s_4 \right] \quad (6.72b)$$

The condenser irreversibility also can be obtained directly from an availability balance. The irreversibility of the condenser is simply the difference between the availability of the inlet and exit secondary coolant flow streams because there is no change of availability in the atmospheric flow stream. Note that there is, however, a change of energy and entropy in the atmospheric flow stream. This availability difference is evaluated from the definition of the availability function A given by Equation 4.27 as

$$\dot{I} = \dot{m}_{in}(h_{in} - T_o s_{in}) - \dot{m}_{out}(h_{out} - T_o s_{out})$$

For the state of the secondary coolant flow stream, this result is identical to Equation 6.72b.

6.4.2.3 Reactor Irreversibility

To proceed, we first compute the maximum work that could be done by the fissioning fuel. Upon fissioning and with respect to fission products, each fission fragment at steady state may be regarded as having a kinetic energy of about 100 MeV. If it were a perfect gas in a thermodynamic equilibrium state, the fragment would have a temperature T_a such that $\frac{3}{2} kT_a = 100\,\text{MeV}$. Maximum work is done by the fragment in a reversible process that brings the fragment to a state of mutual equilibrium with the environment at $T_o = 298$ K or $kT_o = 0.025$ eV. At this state, the energy of the fragment is $u_o = 0.0375$ eV. Thus, the maximum work per unit mass of fission fragments is

$$W_{u,max} = h_a - h_o - T_o(s_a - s_o)$$

which can be approximated as

$$W_{u,max} \simeq u_a - u_o - cT_o \ln \frac{T_a}{T_o} = (u_a - u_o)\left[1 - \frac{kT_o \ln(T_a/T_o)}{k(T_a - T_o)}\right] \quad (6.73)$$

For $T_a \gg T_o$, Equation 6.73 becomes

$$W_{u,max} \simeq u_a$$

The maximum work is thus approximately equal to the fission energy (u_a). In other words, all the fission energy may be considered to be available as work.

We compute the rate of lost work in several steps. The maximum useful power available as a result of fission equals the fission power ($\dot{u}_a$) and is delivered to the fuel rods considered at a constant fuel temperature (T_{fo}). Because this fuel temperature is far below T_a, for the purpose of this computation the fuel temperature and the clad surface temperature are taken to be T_{fo}. The transfer of power from the fission fragments to the fuel rods is highly irreversible, as $T_{fo} \ll T_a$. The maximum power available from the fuel rod at fixed surface temperature T_{fo} is

$$(\dot{W}_{u,max})_{fuel} = (\dot{W}_{u,max})_{fission}\left(1 - \frac{T_o}{T_{fo}}\right) \tag{6.74}$$

and therefore the irreversibility is

$$\dot{I}_{fuel} = (\dot{W}_{u,max})_{fission} - (\dot{W}_{u,max})_{fuel} = (\dot{W}_{u,max})_{fission}\frac{T_o}{T_{fo}} \tag{6.75}$$

From the fuel rods, all the fission power is transferred to the coolant. The maximum work available from the coolant is

$$(\dot{W}_{u,max})_{coolant} = \dot{m}_p[(h_{out} - h_{in})_R - T_o(s_{out} - s_{in})_R] \tag{6.76}$$

so here the irreversibility is

$$\dot{I}_{coolant} = (\dot{W}_{u,max})_{fuel} - (\dot{W}_{u,max})_{coolant} = (\dot{W}_{u,max})_{fission}\left[\frac{T_o(s_{out} - s_{in})_R}{(h_{out} - h_{in})_R} - \frac{T_o}{T_{fo}}\right] \tag{6.77}$$

because $(\dot{W}_{u,\,max})_{fission}$ equals the fission power $\dot{U}_a$, which in turn equals $\dot{m}_p\,(h_{out} - h_{in})_R$, by application of the first law (Equation 4.39) to the reactor control volume, which is adiabatic, is nondeformable, and has no shaft work—and where kinetic energy, potential energy, and shear work terms are neglected. Hence, the irreversibility of the reactor is

$$\dot{I}_R = \dot{I}_{fission} + \dot{I}_{fuel} + \dot{I}_{coolant} \approx 0 + (\dot{W}_{u,max})_{fission}\frac{T_o}{T_{fo}}$$
$$+ (\dot{W}_{u,max})_{fission}\left[\frac{T_o(s_{out} - s_{in})_R}{(h_{out} - h_{in})_R} - \frac{T_o}{T_{fo}}\right]$$

Again, because $(\dot{W}_{u,\,max})_{fission}$ equals $\dot{m}_p\,(h_{out} - h_{in})_R$, the above result becomes

$$\dot{I}_R = \dot{m}_p T_o(s_{out} - s_{in})_R \tag{6.78}$$

Equation 6.78 is a particular form of the general control volume entropy generation equation as expressed by Equation 6.60. Equation 6.78 states that for a steady-state, adiabatic reactor control volume the change of entropy of the coolant stream equals the entropy generation within the reactor.

6.4.2.4 Plant Irreversibility

Finally, consider the entire nuclear plant within one control volume as illustrated in Figure 6.9a. For this adiabatic, nondeformable, stationary control volume, the maximum useful work, given by Equation 6.34, is

$$\dot{W}_{\text{u,max}_{\substack{\text{reactor}\\\text{plant}}}} = -\left[\frac{\partial(U - T_o S)}{\partial t}\right]_{\substack{\text{reactor}\\\text{plant}}} + \dot{m}_{\text{AF}}[(h_{\text{in}} - h_{\text{out}}) - T_o(s_{\text{in}} - s_{\text{out}})]_{\text{AF}} \tag{6.79}$$

because $\dot{Q} = 0$ and differences in kinetic and potential energy as well as shear work are neglected. The first term of the right-hand side of Equation 6.79 involves only the fission process within the reactor, because for all other components the time rate of change of $U - TS$ is zero. Hence

$$-\left[\frac{\partial(U - T_o S)}{\partial t}\right]_{\substack{\text{reactor}\\\text{plant}}} = -\left[\frac{\partial(U - T_o S)}{\partial t}\right]_{\text{reactor}} = -\left(\frac{\partial U}{\partial t}\right)_{\text{reactor}} \tag{6.80a}$$

because for the reactor $\dot{I}_{\text{R}} = T_o \dot{S}_{\text{gen}}$ and $\dot{Q} = 0$; hence Equation 6.79 becomes

$$\left(\frac{\partial S}{\partial t}\right)_{\text{reactor}} = \dot{m}_{\text{p}}(s_{\text{in}} - s_{\text{out}})_{\text{R}} + \frac{\dot{I}_{\text{R}}}{T_o} = 0$$

utilizing Equation 6.78 for $\dot{I}_{\text{R}}$.

Again, the first law (Equation 4.39) for the reactor control volume, which is adiabatic, is nondeformable, and has no shaft work, becomes (treating $\dot{Q}_{\text{gen}}$ as part of $\dot{E}$):

$$\left(\frac{\partial U}{\partial t}\right)_{\text{reactor}} = \dot{m}_{\text{in}} h_{\text{in}} - \dot{m}_{\text{out}} h_{\text{out}} \tag{6.80b}$$

Hence, the change in reactor internal energy can be expressed in terms of steam generator stream enthalpy differences as

$$-\left(\frac{\partial U}{\partial t}\right)_{\text{reactor}} = \dot{m}_{\text{p}}(h_{\text{in}} - h_{\text{out}})_{\text{SGp}} = \dot{m}_{\text{s}}(h_{\text{out}} - h_{\text{in}})_{\text{SGs}} \tag{6.81}$$

Furthermore, from Equation 6.69, the second term on the right-hand side of Equation 6.79 is zero. Hence, applying Equation 6.81 to Equation 6.79 yields

$$[\dot{W}_{\text{u,max}}]_{\substack{\text{reactor}\\\text{plant}}} = \dot{m}_{\text{p}}(h_{\text{in}} - h_{\text{out}})_{\text{SGp}} = \dot{m}_{\text{s}}(h_{\text{out}} - h_{\text{in}})_{\text{SGs}} \tag{6.55}$$

which is the result earlier presented in Equations 6.53 through 6.55. The irreversibility from Equation 4.48 is

$$\dot{I}_{\substack{\text{reactor}\\\text{plant}}} = -T_o \sum_{i=1}^{I} \dot{m}_i s_i = T_o \dot{m}_{AF} (s_{out} - s_{in})_{AF} \quad (6.82)$$

which by Equations 6.69 and 6.70 yields

$$\dot{I}_{\substack{\text{reactor}\\\text{plant}}} = -\dot{m}_s \Delta h_s = \dot{m}_s (h_{in} - h_{out})_s \quad (6.83)$$

Referring to Figure 6.6, Equation 6.83 becomes

$$\dot{I}_{\substack{\text{reactor}\\\text{plant}}} = \dot{m}_s (h_4 - h_1) \quad (6.84)$$

Example 6.5: Second Law Thermodynamic Analysis of a Simplified PWR Cycle

PROBLEM

Consider the PWR plant of Figure 6.6, which was analyzed in Example 6.4. The plant-operating conditions from Example 6.4 are summarized in Table 6.7 together with additional state conditions that can be obtained from the results of Example 6.4. We wish to determine the magnitude of the irreversibility for each component. This analysis is a useful step for assessing the incentive for redesigning components to reduce the irreversibility.

SOLUTION

First evaluate the irreversibility of each component. These irreversibilities sum to the irreversibility of the reactor plant, which may be independently determined by

TABLE 6.7
PWR Operating Conditions for Example 6.5

Point	Flow Rate[a]	Pressure psia (kPa)	Temperature °F (°C)	Temperature °R (K)	Enthalpy Btu/lb (kJ/kg)	Entropy Btu/lb °R (kJ/kg K)
1	1.0	1 (6.9)	101.7 (38.72)	561.3 (311.8)	69.74 (162.2)	0.13266 (0.55542)
2	1.0	1124 (7750)	103.0 (39.44)	562.6 (312.6)	73.69 (171.4)	0.13410 (0.56144)
3	1.0	1124 (7750)	559.1 (292.8)	1018.7 (565.9)	1187.29 (2761.6)	1.37590 (5.76062)
4	1.0	1 (6.9)	101.7 (38.72)	561.3 (311.8)	830.70 (1932.2)	1.48842 (6.23172)
5	13.18	2250 (15,513)	618.5 (325.8)	1078.1 (598.9)	640.31 (1489.4)	0.83220 (3.48425)
6	13.18	2250 (15,513)	557.2 (291.8)	1016.8 (564.9)	555.80 (1292.8)	0.75160 (3.14680)

[a] Flow rate relative to 1 lb of steam through turbine.

considering the entire plant within one control volume. T_o is the reservoir temperature taken at 539.6°R (298.4 K).

For the steam generator, utilizing Equation 6.67 and dividing by $\dot{m}_s$:

$$
\begin{aligned}
I_{SG} &= \frac{\dot{I}_{SG}}{\dot{m}_s} = T_o \frac{\dot{m}_p}{\dot{m}_s}(s_6 - s_5) + T_o(s_3 - s_2) \\
&= 539.6(13.18)(0.7516 - 0.8322) + (539.6)(1.3759 - 0.1341) \\
&= -573.22 + 670.08 \\
&= 96.85 \text{ Btu/lb steam } (0.226\,\text{MJ/kg})
\end{aligned} \tag{6.85}
$$

For the turbine, using Equation 6.65 and dividing by $\dot{m}_s$:

$$
\begin{aligned}
I_T &= \frac{\dot{I}_T}{\dot{m}_s} = T_o(s_4 - s_3) \\
&= 539.6(1.4884 - 1.3759) \\
&= 60.71 \text{ Btu/lb steam } (0.142\,\text{MJ/kg})
\end{aligned} \tag{6.86}
$$

For the condenser, using Equation 6.72b and dividing by $\dot{m}_s$:

$$
\begin{aligned}
I_{CD} &= \frac{\dot{I}_{CD}}{\dot{m}_s} = T_o\left(\frac{h_4 - h_1}{T_o} + s_1 - s_4\right) \\
&= 539.6\left(\frac{830.7 - 69.74}{539.6} + 0.13266 - 1.48842\right) \\
&= 29.39 \text{ Btu/lb steam } (6.86 \times 10^{-2}\,\text{MJ/kg})
\end{aligned} \tag{6.87}
$$

For the pump, using Equation 6.65 and dividing by $\dot{m}_s$:

$$
\begin{aligned}
I_P &= \frac{\dot{I}_P}{\dot{m}_s} = T_o(s_2 - s_1) \\
&= 539.6\ (0.1341 - 0.13266) \\
&= 0.78 \text{ Btu/lb steam } (1.81 \times 10^{-2}\,\text{MJ/kg})
\end{aligned} \tag{6.88}
$$

For the reactor, using Equation 6.78 and dividing by $\dot{m}_s$:

$$
\begin{aligned}
I_R &= \frac{\dot{I}_R}{\dot{m}_s} = \frac{\dot{m}_p}{\dot{m}_s} T_o(s_5 - s_6) \\
&= 13.18(539.6)\ (0.8322 - 0.7516) \\
&= 573.22 \text{ Btu/lb steam } (1.33\,\text{MJ/kg})
\end{aligned} \tag{6.89}
$$

This reactor irreversibility can be broken into its two nonzero components, $\dot{I}_{fuel}$ and $\dot{I}_{coolant}$, that is, irreversibility associated with fission power transfer from the fission products to the fuel rod and from the fuel rod to the coolant. These irreversibility components are expressed using Equations. 6.75 and 6.77 assuming a fuel surface temperature of 1760°R (977.6 K) as

$$
\begin{aligned}
I_{fuel} &= \frac{\dot{I}_{fuel}}{\dot{m}_s} = \frac{(\dot{W}_{u,max})_{fission}}{\dot{m}_s}\frac{T_o}{T_{fo}} \\
&= \left(\frac{\dot{m}_p}{\dot{m}_s}\right)(h_{out} - h_{in})_R \frac{T_o}{T_{fo}} \\
&= (13.18)\,(640.31 - 555.8)\frac{539.6}{1760} \\
&= 34.14 \text{ Btu/lb steam } (7.97 \times 10^{-2} \text{ MJ/kg})
\end{aligned}
$$

$$
\begin{aligned}
I_{coolant} &= \frac{\dot{I}_{coolant}}{\dot{m}_s} \\
&= \frac{(\dot{W}_{u,max})_{fission}}{\dot{m}_s}\left[\left[\frac{T_o(s_{out} - s_{in})_R}{(h_o - h_{in})_R}\right] - \frac{T_o}{T_{fo}}\right] \\
&= \left(\frac{\dot{m}_p}{\dot{m}_s}\right) T_o \left[(s_{out} - s_{in})_R - \frac{(h_{out} - h_{in})_R}{T_{fo}}\right] \\
&= (13.18)\;539.6\left[0.8322 - 0.7516 - \left(\frac{640.31 - 555.80}{1760}\right)\right] \\
&= 539.08 \text{ Btu/lb steam } (1.25 \text{ MJ/kg})
\end{aligned}
$$

The sum of the numerical values of the irreversibilities of all components is
$I_R = 573.22$
$I_{SG} = 96.85$
$I_T = 60.71$
$I_{CD} = 29.39$
$I_P = 0.78$
Total = 760.95 Btu/lb steam
which equals (considering round-off error) the irreversibility of the reactor plant evaluated in Equation 6.84 as

$$I_{RP} = \frac{\dot{I}_{RP}}{\dot{m}_s} = h_4 - h_1 = 830.70 - 69.74 = 760.96 \text{ Btu/lb steam}$$

The generality of this result may be illustrated by summing the expressions for the irreversibility of each component and showing that the sum is identical to the irreversibility of the reactor plant. Proceeding in this way, we obtain

$$
\begin{aligned}
\dot{I}_{RP} &= \dot{W}_{u,max} - \dot{W}_{net} \\
&= \dot{m}_s[h_3 - h_2 - [h_3 - h_4 - (h_2 - h_1)]] \qquad (6.84)\\
&= \dot{m}_s(h_4 - h_1)
\end{aligned}
$$

$$\dot{I}_R = T_o\dot{m}_p(s_5 - s_6) \qquad (6.89)$$

$$\dot{I}_{SG} = T_o\dot{m}_p(s_6 - s_5) + T_o\dot{m}_s(s_3 - s_2) \tag{6.85}$$

$$\dot{I}_T = T_o\dot{m}_s(s_4 - s_3) \tag{6.86}$$

$$\dot{I}_P = T_o\dot{m}_s(s_2 - s_1) \tag{6.88}$$

$$\dot{I}_{CD} = T_o\dot{m}_s(s_1 - s_4) + \dot{m}_s(h_4 - h_1) \tag{6.87}$$

Hence

$$\dot{I} = \dot{m}_s(h_4 - h_1) = \dot{I}_{RP} \tag{6.84}$$

6.5 MORE COMPLEX RANKINE CYCLES: SUPERHEAT, REHEAT, REGENERATION, AND MOISTURE SEPARATION

Improvements in overall cycle performance can be accomplished by superheat, reheat, and regeneration, as each approach leads to a higher average temperature at which heat is received by the working fluid. Furthermore, to minimize erosion of the turbine blades, the turbine expansion process can be broken into multiple stages with intermediate moisture separation. The superheat and reheat options also allow maintenance of higher exit steam quality, thereby decreasing the liquid fraction in the turbine.

The superheat, reheat, regeneration, and moisture separation options are illustrated in subsequent figures for ideal turbines. In actual expansion processes, the coolant entropy increases, and the exit steam is dryer, although the shaft work generated per unit mass is decreased. Although each process is discussed independently, in practice, they are utilized in varying combinations. Superheat is accomplished by heating the working fluid into the superheated vapor region. A limited degree of superheat is being achieved in PWR systems employing once-through versus recirculating steam generators. Figure 6.16 illustrates the heat exchanger and turbine processes in a superheated power cycle.

Reheat is a process whereby the working fluid is returned to a heat exchanger after partial expansion in the turbine. As shown in Figure 6.17, the fluid can be reheated to the maximum temperature and expanded. Actual reheat design conditions may differ with application.

Regeneration is a process by which the colder portion of the working fluid is heated by the hotter portion of the working fluid. Because this heat exchange is internal to the cycle, for single-phase fluids (e.g., Brayton cycle) the average temperature of external heat rejection is reduced and the average temperature of external heat addition is increased. For a two-phase fluid (e.g., Rankine cycle) only the average temperature of external heat addition is changed, that is, increased. In either case, the cycle thermal efficiency increases. For the limiting case of this heat exchange

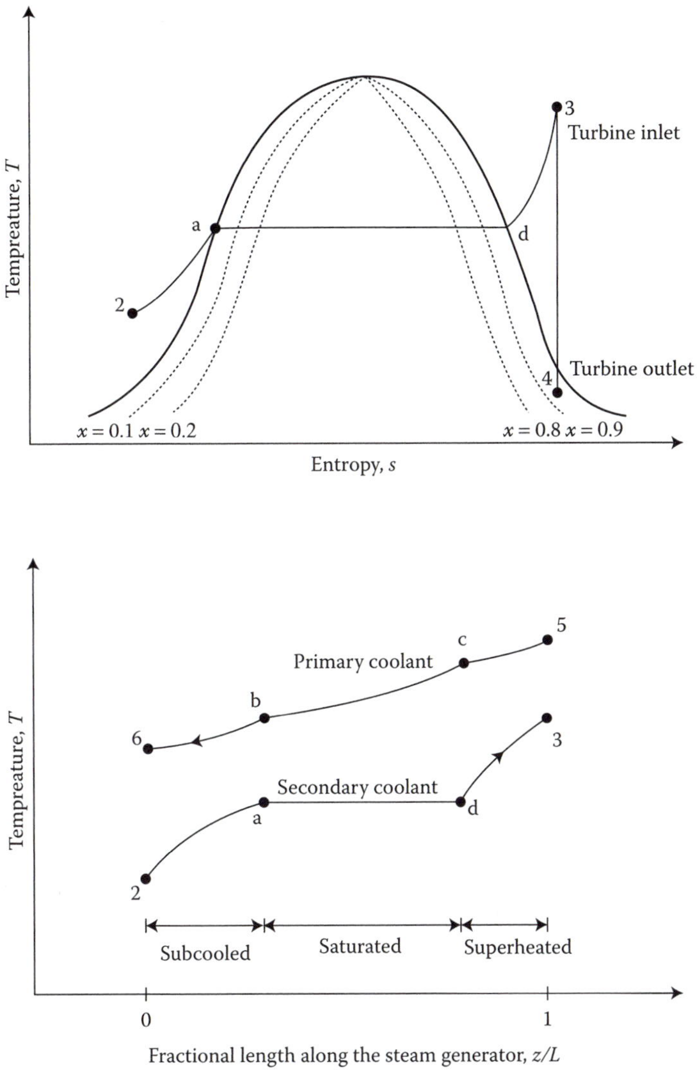

FIGURE 6.16 Heat exchanger and turbine processes in a superheated power cycle.

occurring at infinitesimal temperature differences—an ideal regeneration treatment—it can be shown that the cycle thermal efficiency equals the Carnot efficiency of a cycle operating between the same two temperature limits. In practice for a Rankine cycle, this regenerative process is accomplished by extracting steam from various turbine stages and directing it to a series of heaters where the condensate or feedwater is preheated. These feedwater heaters may be open (OFWH), in which the streams are directly mixed, or closed (CFWH) in which the heat transfer occurs through tube walls. Figure 6.18 illustrates a system and associated T–s diagram for

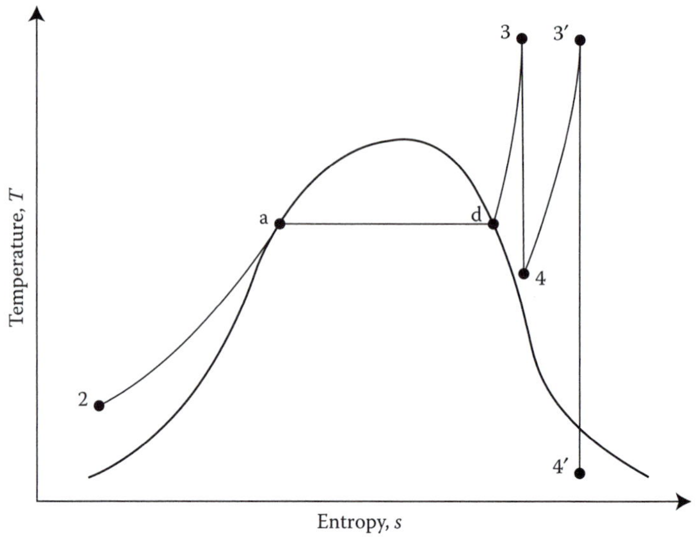

FIGURE 6.17 Heat exchanger and turbine processes in a power cycle with superheat and reheat.

the case of two open feedwater heaters. In this case, a progressively decreasing flow rate passes through later stages of the turbine. Hence, the total turbine work is expressed as

$$\dot{W}_{\mathrm{T}} = \dot{m}_{\mathrm{s}}(h_3 - h_{3'}) + (\dot{m}_{\mathrm{s}} - \dot{m}_{3'})(h_{3'} - h_{3''}) + \dot{m}_4(h_{3''} - h_4) \tag{6.90}$$

The exit enthalpy from each feedwater heater can be expressed by specializing the relations of Table 6.5, yielding:

For OFWH No. 1:

$$h_z = \frac{\dot{m}_{3''}h_{3''} + \dot{m}_4 h_{1'}}{\dot{m}_{3''} + \dot{m}_4} = \frac{\dot{m}_{3''}h_{3''} + \dot{m}_4 h_{1'}}{\dot{m}_{\mathrm{s}} - \dot{m}_{3'}} \tag{6.91}$$

For OFWH No. 2:

$$h_y = \frac{\dot{m}_{3'}h_{3'} + (\dot{m}_{\mathrm{s}} - \dot{m}_{3'})h_{z'}}{\dot{m}_{\mathrm{s}}} \tag{6.92}$$

Figures 6.19 and 6.20 illustrate a system and the associated T–s diagram for the case of two closed feedwater heaters. Steam traps at the bottom of each heater allow the passage of saturated liquid only. The condensate from the first heater is returned to the main condensate line by a drip pump. For the last heater (No. 2), the condensate drip is flashed back to the adjacent upstream heater (No. 1). Observe on the T–s

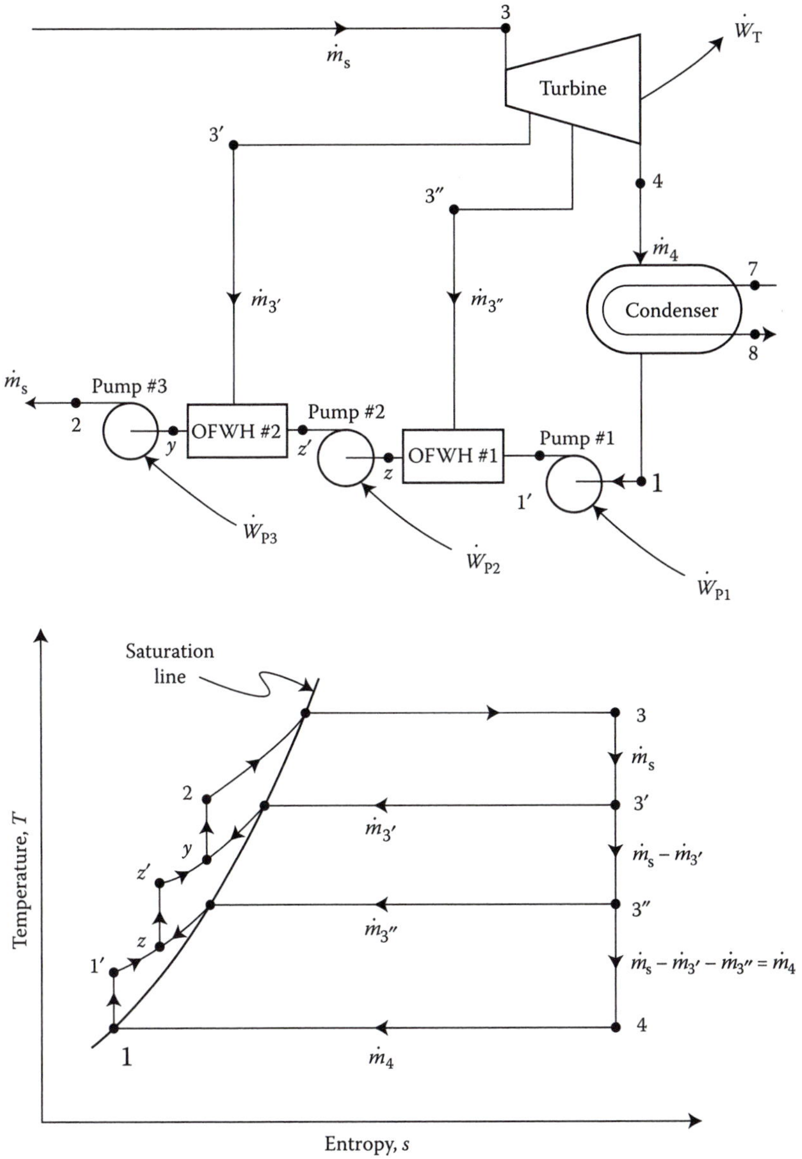

FIGURE 6.18 Power cycle with open feedwater heaters.

diagram that certain temperature differentials must be maintained. In this case, exit enthalpies of each CFWH and the mixing tee can be written

For CFWH No. 1:

$$h_y = \frac{\dot{m}_{3''}h_{3''} + \dot{m}_4 h_{1'} + \dot{m}_{3'}h_{x'} - (\dot{m}_{3'} + \dot{m}_{3''})h_z}{\dot{m}_4} \tag{6.93}$$

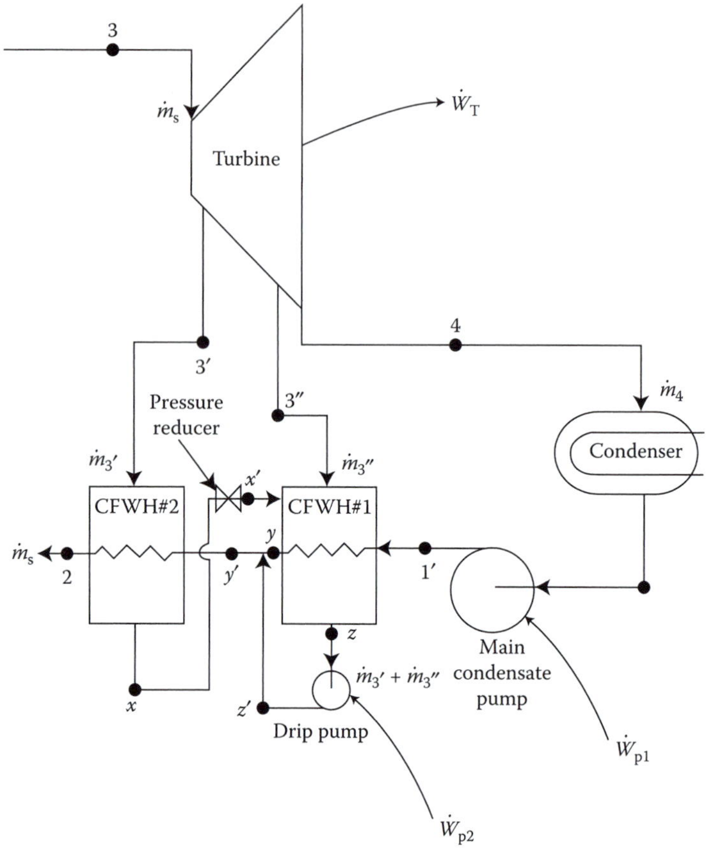

FIGURE 6.19 Portion of Rankine power cycle with closed feedwater heaters.

For the mixing tee, at position y':

$$h_{y'} = \frac{\dot{m}_4 h_y + (\dot{m}_{3'} + \dot{m}_{3''})h_{z'}}{\dot{m}_s} \tag{6.94}$$

For CFWH No. 2:

$$h_2 = \frac{\dot{m}_{3'}(h_{3'} - h_x) + \dot{m}_s h_{y'}}{\dot{m}_s} \tag{6.95}$$

From these feedwater heater illustrations, it can be observed that the OFWH approach requires a condensate pump for each heater, whereas the CFWH approach requires only one condensate pump plus a smaller drip pump. However, higher heat transfer rates are achievable with the OFWH. Additionally, the OFWH permits deaeration of the condensate. For most applications closed heaters are favored, but for purposes of feedwater deaeration at least one open heater is provided.

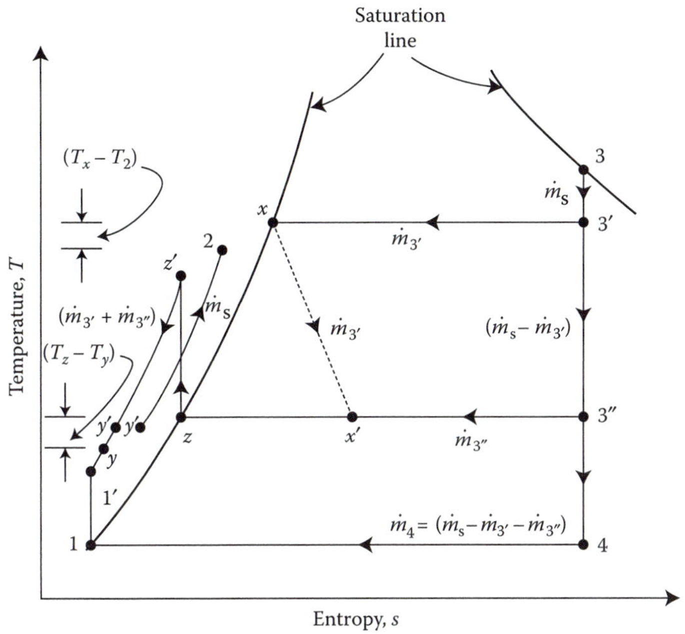

FIGURE 6.20 Portion of Rankine power cycle with closed feedwater heaters. *Note*: Process y' to 2 offset from constant pressure line $1'yy'z'2$ for clarity.

The final process considered is moisture separation. The steam from the high-pressure turbine is passed through a moisture separator, and the separated liquid is diverted to a feedwater heater while the vapor passes to a low-pressure turbine. The portion of the power cycle with a moisture separator and the associated T–s diagram are shown in Figure 6.21. The moisture separator is considered ideal, producing two streams of fluid, a saturated liquid flow rate of magnitude $(1-x_{3'})\dot{m}_s$ and enthalpy $h_{3''} = h_f$ (at $p_{3'}$), and a saturated vapor flow rate of magnitude $x_{3'}\dot{m}_s$ and enthalpy $h_{3'''} = h_g$(at $p_{3'}$). In this case, the feedwater heater exit enthalpy is given as

$$h_{1'} = \frac{h_{3''}(1-x_{3'})\dot{m}_s + h_1 x_{3'}\dot{m}_s}{\dot{m}_s} \tag{6.96}$$

Example 6.6: Thermodynamic Analysis of a PWR Cycle with Moisture Separation and One Stage of Feedwater Heating

PROBLEM

This example demonstrates the advantage in cycle performance gained by adding moisture separation and open feedwater heating to the cycle of Example 6.4. The cycle, as illustrated in Figure 6.22, operates under the conditions given in Table 6.8. All states are identical to those in Example 6.4 except state 2, as states

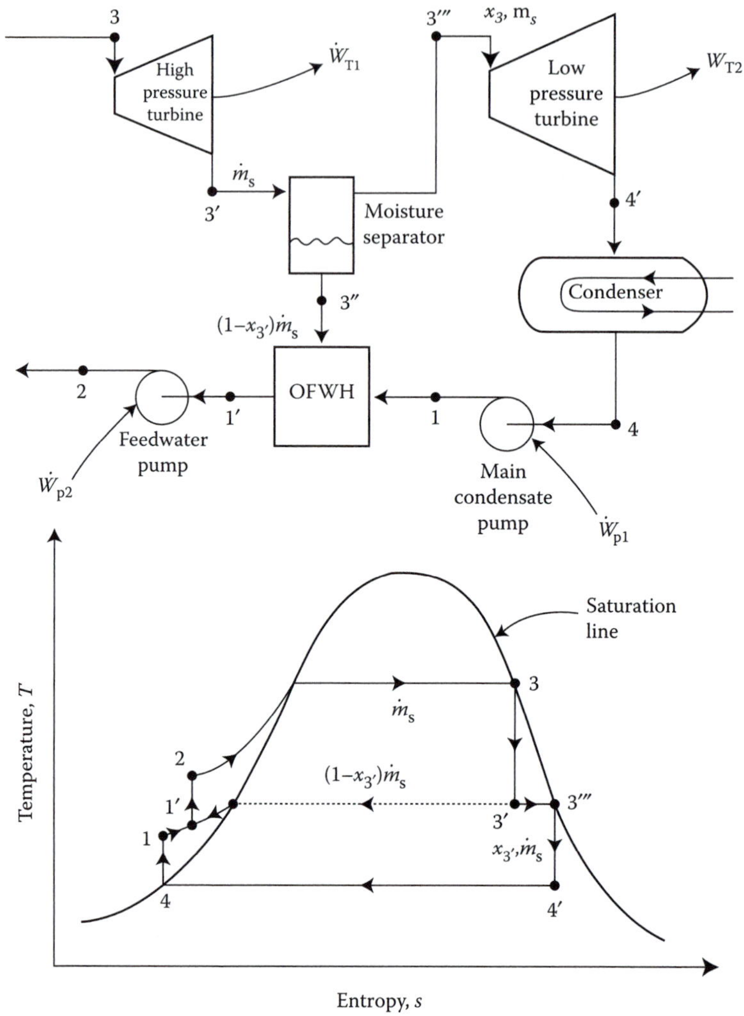

FIGURE 6.21 Portion of power cycle with ideal moisture separation.

9 through 13 have been added. The flow rate at state 11, which is diverted to the open feedwater heater, is sufficient to preheat the condensate stream at state 13 to the conditions of state 10. Assume that all turbines and pumps have isentropic efficiencies of 85%.

1. Draw the temperature–entropy (T–s) diagram for this cycle, as well as the temperature–fractional length (T versus z/L) diagram for the steam generator.
2. Compute the ratio of primary to secondary flow rates.
3. Compute the cycle thermal efficiency.

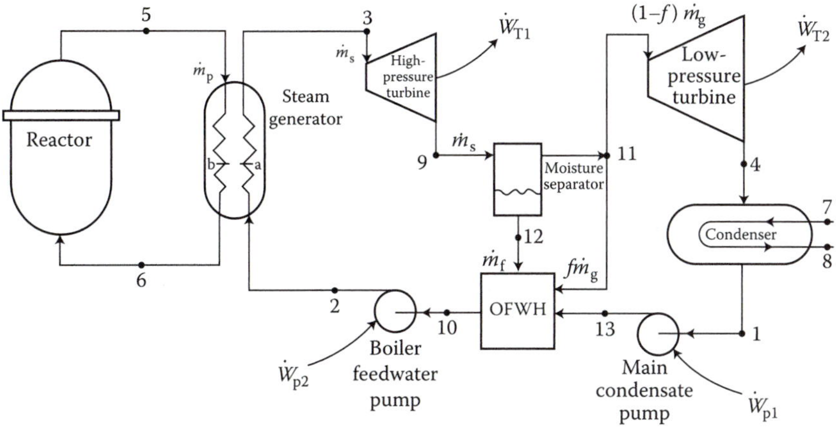

FIGURE 6.22 PWR cycle analyzed in Example 6.6.

SOLUTION

1. The T–s diagram is sketched in Figure 6.23. The T versus z/L diagram is the same as that in Figure 6.7.
2. $\dfrac{\dot{m}_p}{\dot{m}_s} = \dfrac{h_3 - h_a}{\bar{c}_p[T_5 - (T_a - \Delta T_p)]} = 13.18$, as states 3, 5, a, and b were specified the same as in Example 6.4.
3. $\eta_{th} = \dfrac{\dot{m}_s(h_3 - h_9) + (1-f)\dot{m}_g(h_{11} - h_4)}{\dot{m}_s(h_3 - h_2)}$, neglecting pump work.

TABLE 6.8
PWR Operating Conditions for Example 6.6

State	Temperature °F (K)	Pressure psia (MPa)	Condition
1		1 (6.89×10^{-3})	Saturated liquid
2		1124 (7.75)	Subcooled liquid
3		1124 (7.75)	Saturated vapor
4		1 (6.89×10^{-3})	Two-phase mixture
5	618.5 (599)	2250 (15.5)	Subcooled liquid
6	557.2 (565)	2250 (15.5)	Subcooled liquid
7			Subcooled liquid
8			Subcooled liquid
a		1124 (7.75)	Saturated liquid
b	$T_a + 26$ (14.4)	2250 (15.5)	Subcooled liquid
9		50 (0.345)	Two-phase mixture
10		50 (0.345)	Saturated liquid
11		50 (0.345)	Saturated vapor
12		50 (0.345)	Saturated liquid
13		50 (0.345)	Subcooled liquid

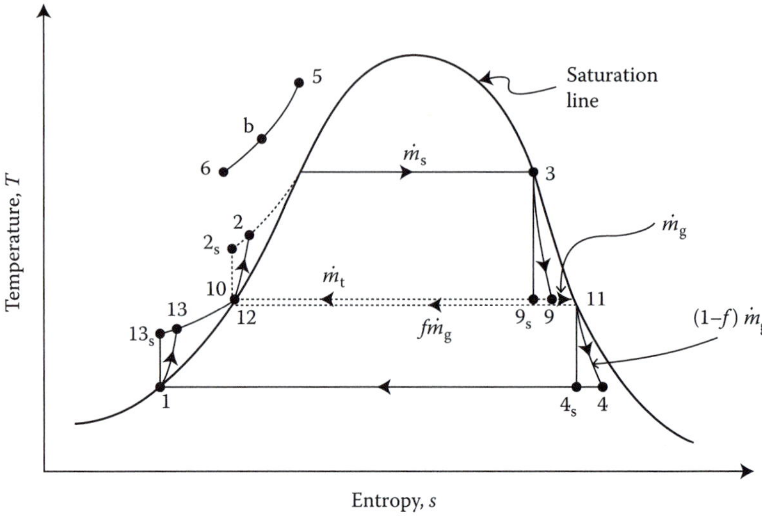

FIGURE 6.23 PWR cycle with moisture separation and one stage of feedwater heating.

From the problem statement, the following enthalpies can be directly determined:

$h_3 = 1187.29$ Btu/lb or 2.77 MJ/kg (Example 6.4)
$h_{11} = 1174.40$ Btu/lb (2.74 MJ/kg)
$h_{12} = h_{10} = 250.24$ Btu/lb (0.584 MJ/kg)

The following parameters must be calculated: h_9, $\dot{m}_g$, h_4, f, and h_2.

$$h_9 = h_3 - \eta_T(h_3 - h_{9s})$$

where

$$h_{9s} = h_f + x_{9s}h_{fg} = h_f + \left(\frac{s_{9s} - s_f}{s_{fg}}\right)h_{fg}$$

Hence:

$$h_{9s} = 250.24 + \left(\frac{1.3759 - 0.41129}{1.2476}\right)924.2$$
$$= 964.81\,\text{Btu/lb}\ (2.25\,\text{MJ/kg})$$

$$h_9 = 1187.29 - 0.85(1187.29 - 964.81) = 998.18\ \text{Btu/lb}\ (2.330\ \text{MJ/kg})$$

$$\dot{m}_g = x_9\dot{m}_s = \frac{h_9 - h_f}{h_{fg}}\dot{m}_s = \frac{998.18 - 250.24}{924.2}\dot{m}_s = 0.81\dot{m}_s\ \text{(English units)}$$
$$= \frac{2.331 - 0.584}{2.156}\dot{m}_s = 0.81\dot{m}_s\ \text{(SI units)}$$

$$\dot{m}_f = \left(1 - \frac{\dot{m}_g}{\dot{m}_s}\right)\dot{m}_s = 0.19\,\dot{m}_s$$

$$\begin{aligned} h_4 &= h_{11} - \eta_T(h_{11} - h_{4s}) = 1174.4 - 0.85(1174.4 - 926.61) \\ &= 963.77\ \text{Btu/lb}\ (2.25\ \text{MJ/kg}) \end{aligned}$$

where

$$\begin{aligned} h_{4s} &= h_f + x_{4s}h_{fg} = h_f + \left(\frac{s_{4s} - s_f}{s_{fg}}\right)h_{fg} \\ &= 69.74 + \left(\frac{1.6589 - 0.13266}{1.8453}\right)1036.0 \\ &= 926.61\,\text{Btu/lb}\ (2.16\,\text{MJ/kg}) \end{aligned}$$

$$\begin{aligned} h_{13} &= h_1 + \frac{v_1(p_{13} - p_1)}{\eta_p} \\ &= 69.91\,\text{Btu/lb}\ (0.163\,\text{MJ/kg})\ \text{(same as state 2, Example 6.4)} \end{aligned}$$

Applying the first law to the feedwater heater:

$$\dot{m}_f h_{12} + f\dot{m}_g h_{11} + (1 - f)\dot{m}_g h_{13} = \dot{m}_s h_{10}$$

$$f = \frac{\dot{m}_s h_{10} - \dot{m}_f h_{12} - \dot{m}_g h_{13}}{\dot{m}_g(h_{11} - h_{13})} = \frac{h_{10} - h_{13}}{h_{11} - h_{13}} = \frac{250.24 - 69.91}{1174.4 - 69.91} = 0.16$$

$$\begin{aligned} h_2 &= h_{10} + \frac{v_{10}(p_2 - p_{10})}{\eta_p} \\ &= 250.24 + \frac{0.017269(1124 - 50)}{0.85}\left(\frac{144\,\text{in}^2/\text{ft}^2}{778\,\text{lb}_f - \text{ft/Btu}}\right) \\ &= 254.28\,\text{Btu/lb}\ (0.594\,\text{MJ/kg}) \end{aligned}$$

Hence

$$\begin{aligned} \eta_{th} &= \left[\frac{(1187.29 - 998.18) + (1 - 0.16)0.81(1174.4 - 963.77)}{1187.29 - 254.28}\right]100\ \text{English units} \\ &= \left[\frac{2.77 - 2.330 + (1 - 0.16)\,0.81\,(2.74 - 2.25)}{2.79 - 0.594}\right]100\ \text{SI units} \\ &= 35.6\% \end{aligned}$$

6.6 SIMPLE BRAYTON CYCLE

Reactor systems that employ gas coolants offer the potential for operating as direct Brayton cycles by passing the heated gas directly into a turbine. This Brayton cycle is ideal for single-phase, steady-flow cycles with heat exchange and therefore is the basic cycle for modern gas turbine plants as well as proposed nuclear gas-cooled reactor plants. As with the Rankine cycle, the remaining processes and components are, sequentially, a heat exchanger for rejecting heat, a compressor, and a heat source, which in our case is the reactor. The ideal cycle is composed of two reversible constant-pressure heat-exchange processes and two reversible, adiabatic work processes. However, because the working fluid is single phase, no portions of the heat-exchange processes are carried out at constant temperature as in the Rankine cycle. Additionally, because the volumetric flow rate is higher, the compressor work, or "backwork," is a larger fraction of the turbine work than is the pump work in a Rankine cycle. This large backwork has important ramifications for the Brayton cycle.

The *T–s* plot of the simple Brayton cycle of Figure 6.8 is shown in Figure 6.24. The pressure or compression ratio of the cycle is defined as

$$r_{\mathrm{p}} \equiv \frac{p_2}{p_1} = \frac{p_3}{p_4} \tag{6.97}$$

For isentropic processes with a perfect gas, the thermodynamic states are related in the following manner:

$$Tv^{\gamma-1} = \text{constant} \tag{6.98}$$

$$Tp^{1-\gamma/\gamma} = \text{constant} \tag{6.99}$$

where $\gamma \equiv c_{\mathrm{p}}/c_{\mathrm{v}}$. For a perfect gas, because enthalpy is a function of temperature only and the specific heats are constant:

$$\Delta h = c_{\mathrm{p}}\Delta T \tag{6.100}$$

Applying this result for Δh (Equation 6.100) and the isentropic relations for a perfect gas (Equations 6.98 and 6.99) to the specialized forms of the first law of

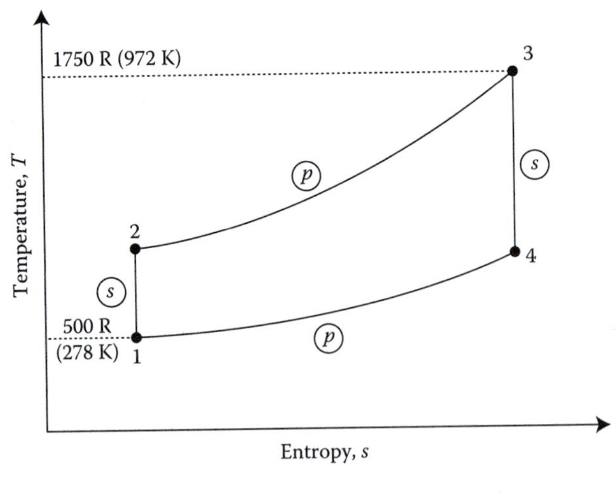

FIGURE 6.24 Temperature–entropy (*T–s*) plot of the simple Brayton cycle. (*p*) Constant pressure process. (*s*) Constant entropy process.

Table 6.5, the Brayton cycle can be analyzed. The following equations are in terms of the pressure ratio (r_p) and the lowest and highest cycle temperatures (T_1 and T_3) both of which are generally known quantities.

The turbine and compressor works are given as follows. For a perfect gas

$$\dot{W}_T = \dot{m}c_p(T_3 - T_4) = \dot{m}c_pT_3\left(1 - \frac{T_4}{T_3}\right) \tag{6.101a}$$

Furthermore, for an isentropic process, application of Equation 6.99 to Equation 6.101a yields

$$\dot{W}_T = \dot{m}c_pT_3\left[1 - \frac{1}{(r_p)^{\gamma-1/\gamma}}\right] \tag{6.101b}$$

Analogously for the compressor:

$$\dot{W}_{CP} = \dot{m}c_p(T_2 - T_1) = \dot{m}c_pT_1\left[\frac{T_2}{T_1} - 1\right] = \dot{m}c_pT_1[(r_p)^{\gamma-1/\gamma} - 1] \tag{6.102}$$

The heat input from the reactor and the heat rejected by the heat exchanger are

$$\dot{Q}_R = \dot{m}c_p(T_3 - T_2) = \dot{m}c_pT_1\left[\frac{T_3}{T_1} - (r_p)^{\gamma-1/\gamma}\right] \tag{6.103}$$

$$\dot{Q}_{HX} = \dot{m}c_p(T_4 - T_1) = \dot{m}c_pT_3\left[\frac{1}{(r_p)^{\gamma-1/\gamma}} - \frac{T_1}{T_3}\right] \tag{6.104}$$

As in Equation 6.56, the maximum useful work of the reactor plant is equal to the product of the system flow rate and the coolant enthalpy rise across the core. Hence

$$\dot{W}_{u,\,max} \equiv \dot{Q}_R = \dot{m}c_pT_1\left[\frac{T_3}{T_1} - (r_p)^{\gamma-1/\gamma}\right] \tag{6.105}$$

The Brayton nuclear plant thermodynamic efficiency is then:

$$\begin{aligned}\zeta &= \frac{\dot{W}_T - \dot{W}_{CP}}{\dot{W}_{u,max}} = \frac{T_3\left[1-(1/(r_p)^{\gamma-1/\gamma})\right] - T_1(r_p)^{\gamma-1/\gamma}\left[1-(1/(r_p)^{\gamma-1/\gamma})\right]}{T_1\left[(T_3/T_1) - (r_p)^{\gamma-1/\gamma}\right]} \\ &= 1 - \frac{1}{(r_p)^{\gamma-1/\gamma}}\end{aligned} \tag{6.106}$$

It can be further shown that the optimum pressure ratio for maximum net work is

$$(r_p)_{\text{optimum}} = \left(\frac{T_3}{T_1}\right)^{\gamma/2(\gamma-1)} \tag{6.107}$$

The existence of an optimum compression ratio can also be seen on a T–s diagram by comparing the enclosed areas for cycles operating between fixed temperature limits of T_1 and T_3.

Example 6.7: First Law Thermodynamic Analysis of a Simple Brayton Cycle

PROBLEM

Compute the cycle efficiency for the simple Brayton cycle of Figures 6.8 and 6.24 for the following conditions:

1. Helium as the working fluid taken as a perfect gas with
   $c_p = 1.25$ Btu/lb °R (5230 J/kg K)
   $\gamma = 1.658$
   $\dot{m}$ in lb/s (English units) or kg/s (SI units)
2. Pressure ratio of 4.0
3. Maximum and minimum temperatures of 1750°R (972 K) and 500°R (278 K), respectively

SOLUTION

$$\frac{\dot{W}_T}{\dot{m}} = c_p T_3\left[1 - \frac{1}{(r_p)^{\gamma-1/\gamma}}\right] = 1.25\,(1750)\left[1 - \frac{1}{(4.0)^{0.397}}\right]$$
$$= 925.9\,\text{Btu/lb}\ (2.150\,\text{MJ/kg})$$

$$\frac{\dot{W}_{CP}}{\dot{m}} = c_p T_1[(r_p)^{\gamma-1/\gamma} - 1] = 1.25\,(500)\ [(4.0)^{0.397} - 1]$$
$$= 458.67\,\text{Btu/lb}\ (1.066\,\text{MJ/kg})$$

$$\frac{\dot{W}_{u,\max}}{\dot{m}} = c_p T_1\left[\frac{T_3}{T_1} - (r_p)^{\gamma-1/\gamma}\right] = 1.25\,(500)\left[\frac{1750}{500} - (4.0)^{0.397}\right]$$
$$= 1103.8\,\text{Btu/lb}\ (2.560\,\text{MJ/kg})$$

$$\zeta = \frac{(\dot{W}_T - \dot{W}_{CP})/\dot{m}}{\dot{W}_{u,\max}/\dot{m}} = \left(\frac{925.9 - 458.7}{1103.8}\right) 100\ \text{(English units)}$$
$$= \left(\frac{2.15 - 1.066}{2.56}\right) 100\ \text{(SI units)}$$
$$= 42.3\%$$

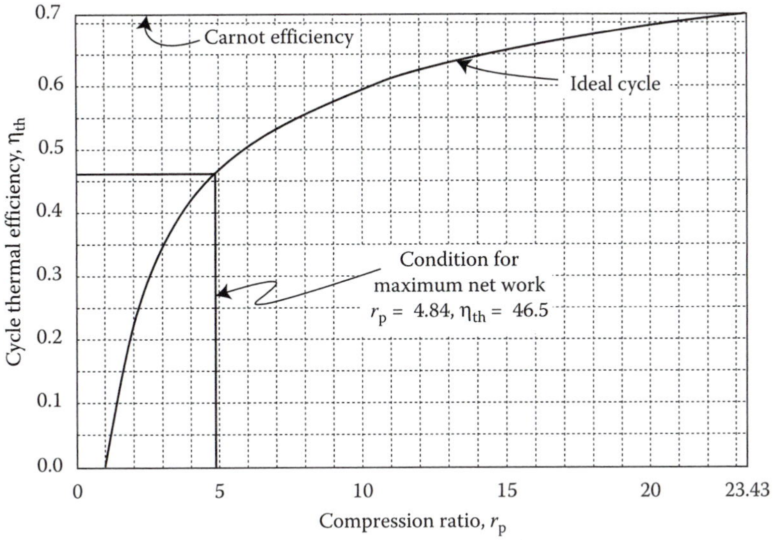

FIGURE 6.25 Thermal efficiency of an ideal Brayton cycle as a function of the compression ratio. $\gamma = 1.658$.

Analogous to the PWR cycle demonstrated in Example 6.5, the Brayton cycle thermal efficiency equals the Brayton nuclear plant thermodynamic efficiency, that is,

$$\eta_{th} = \zeta$$

For ideal Brayton cycles, as Equation 6.106 demonstrates, the thermodynamic efficiency increases with the compression ratio. The thermal efficiency of a Carnot cycle operating between the maximum and minimum temperatures specified is

$$\eta_{Carnot} = \frac{T_{max} - T_{min}}{T_{max}} = \frac{972\text{K} - 278\text{K}}{972\text{K}} = 0.714$$

Figure 6.25 illustrates the dependence of cycle efficiency on the compression ratio for the cycle of Example 6.7. The pressure ratio corresponding to the maximum net work is also identified. This optimum pressure ratio is calculated from Equation 6.107 for a perfect gas as

$$(r_p)_{optimum} = \left(\frac{T_3}{T_1}\right)^{\gamma/2(\gamma-1)} = \left(\frac{972}{278}\right)^{1.658/2(0.658)} = 4.84$$

6.7 MORE COMPLEX BRAYTON CYCLES

In this section, the various realistic considerations are included in the analysis and are illustrated through a number of examples.

Example 6.8: Brayton Cycle with Real Components

PROBLEM

Compute the thermal efficiency for the cycle depicted in Figure 6.26 if the isentropic efficiencies of the compressor and the turbine are each 90%. All other conditions of Example 6.7 apply.

SOLUTION

For $\dot{W}_T$:

$$\eta_T = \frac{\text{Actual work out of turbine}}{\text{Ideal turbine work}} = \frac{\dot{W}_T}{\dot{W}_{Ti}} = \frac{\dot{m}c_p(T_3 - T_4)}{\dot{m}c_p(T_3 - T_{4s})}$$

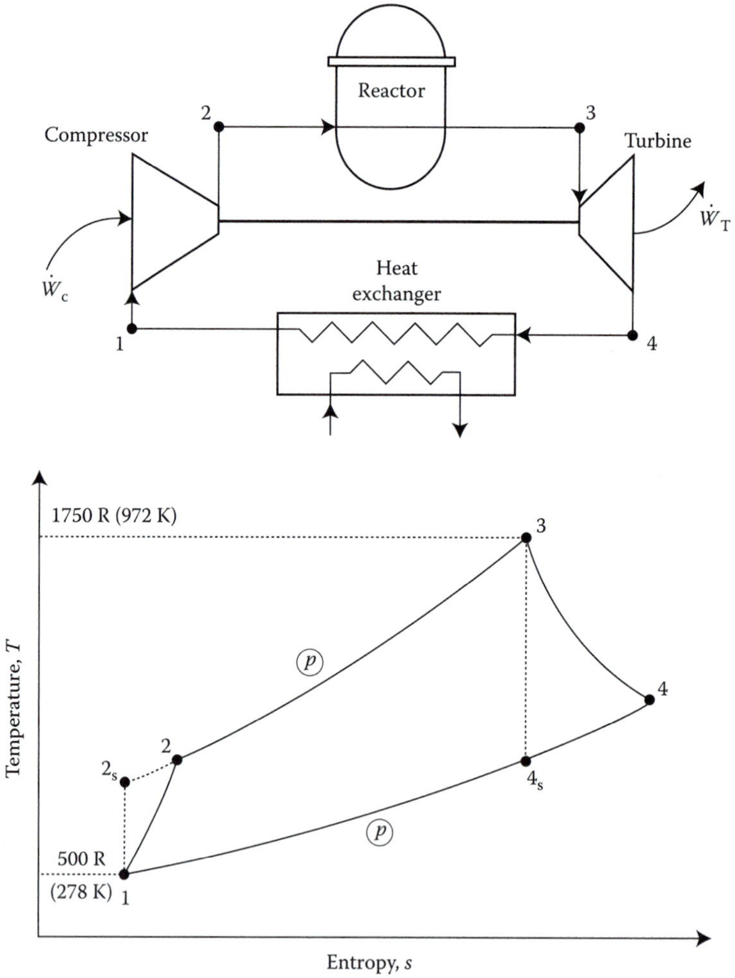

FIGURE 6.26 Brayton cycle with real components: Example 6.8.

$$\therefore \dot{W}_T = \eta_T \dot{W}_{Ti} = \eta_T \dot{m} c_p (T_3 - T_{4s}) = \eta_T \dot{m} c_p T_3 \left(1 - \frac{T_{4s}}{T_3}\right)$$

$$= \eta_T \dot{m} c_p T_3 \left[1 - \frac{1}{(r_p)^{\gamma-1/\gamma}}\right] = \eta_T \dot{m}\ 925.9 = (0.9)(925.9)\ \dot{m}$$

$$= 833.3\ \dot{m}\ \text{Btu/s}\ (1.935\ \dot{m}\ \text{MJ/s or MW})$$

For $\dot{W}_{CP}$:

$$\eta_{CP} = \frac{\text{Ideal compressor work}}{\text{Actual compressor work}} = \frac{\dot{W}_{CPi}}{\dot{W}_{CP}} = \frac{\dot{m} c_p (T_{2s} - T_1)}{\dot{m} c_p (T_2 - T_1)}$$

$$\dot{W}_{CP} = \frac{\dot{m}}{\eta_{CP}} c_p (T_{2s} - T_1) = \frac{\dot{m}}{\eta_{CP}} c_p T_1 \left(\frac{T_{2s}}{T_1} - 1\right) = \frac{\dot{m}\ 458.7}{0.9}$$

$$= 509.7\ \dot{m}\ \text{Btu/s}\ (1.184\ \dot{m}\ \text{MW})$$

$$\dot{W}_{NET} = \dot{W}_T - \dot{W}_{CP} = \dot{m}\ (833.3 - 509.7) = 323.6\ \dot{m}\ \text{Btu/s}\ (0.752\ \dot{m}\ \text{MW})$$

$$\dot{Q}_R = \dot{m} c_p (T_3 - T_2)$$

To evaluate $\dot{Q}_R$, it is necessary to find T_2.

In Example 6.7, we found $\dot{W}_{CPi} = \dot{m} c_p (T_{2s} - T_1) = 458.7\ \dot{m}$ Btu/s $= 1.066\ \dot{m}$ MW, so that from the expressions above for $\dot{W}_{CP}$ and η_{CP}:

$$T_2 - T_1 = \frac{\dot{W}_{CP}}{\dot{m} c_p} = \frac{\dot{W}_{CPi}}{\dot{m} c_p \eta_{CP}} = \frac{458.7}{1.25(0.9)} = 407.7°\text{R}\ (226.5\text{K})$$

$$T_2 = 407.7 + T_1 = 407.7 + 500 = 907.7°\text{R}\ (504.3\ \text{K})$$

$$\dot{Q}_R = (1.25)(T_3 - T_2)\ \dot{m} = 1.25(1750 - 907.7)\dot{m}$$

$$= 1052.9\ \dot{m}\ \text{Btu/s}\ (2.45\ \dot{m}\ \text{MW})$$

$$\eta_{th} = \frac{\dot{W}_{NET}}{\dot{Q}_R} = \left(\frac{323.6}{1052.9}\right) 100\ \text{(English units)}$$

$$= \left(\frac{0.752}{2.45}\right) 100\ \text{(SI units)}$$

$$= 30.7\%$$

Example 6.9: Brayton Cycle Considering Duct Pressure Losses

PROBLEM

Compute the cycle thermal efficiency considering pressure losses in the reactor and heat exchanger processes as well as 90% isentropic turbine and compressor efficiencies. The cycle is illustrated in Figure 6.27. The pressure losses are characterized by the parameter β where

$$\beta \equiv \left(\frac{p_4}{p_1} \frac{p_2}{p_3} \right)^{\gamma - 1/\gamma} = 1.05$$

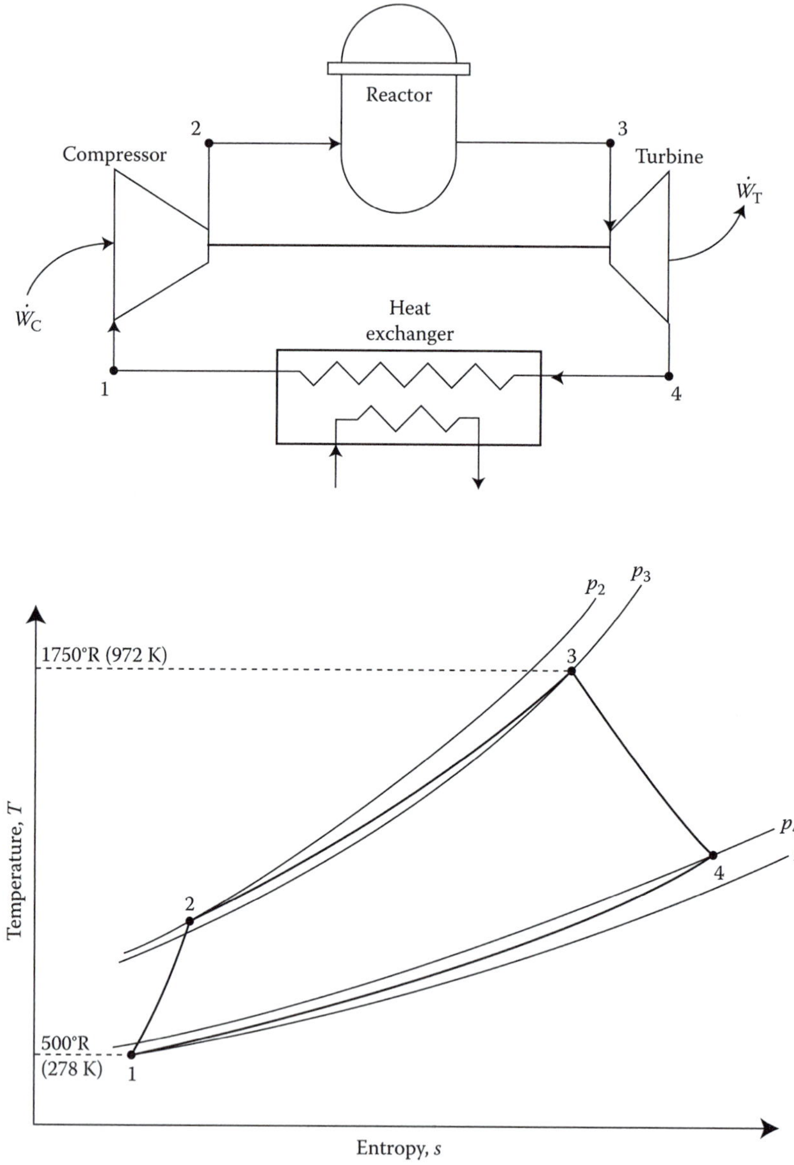

FIGURE 6.27 Brayton cycle considering duct pressure losses: Example 6.9.

All other conditions of Example 6.7 apply.

$$\dot{W}_T = \eta_T \dot{m} c_p T_3 \left(1 - \frac{T_{4s}}{T_3}\right) = \eta_T \dot{m} c_p T_3 \left[1 - \frac{1}{(p_3/p_4)^{\gamma-1/\gamma}}\right]$$

SOLUTION

Because β is defined as

$$\left(\frac{p_4}{p_1} \cdot \frac{p_2}{p_3}\right)^{\gamma-1/\gamma}$$

$$\therefore \left(\frac{p_4}{p_3}\right)^{\gamma-1/\gamma} = \frac{(p_4/p_1 \cdot p_2/p_3)^{\gamma-1/\gamma}}{(p_2/p_1)^{\gamma-1/\gamma}} = \frac{\beta}{(r_p)^{\gamma-1/\gamma}}$$

$$\begin{aligned}\therefore \dot{W}_T &= \eta_T \dot{m} c_p T_3 \left[1 - \frac{\beta}{(r_p)^{\gamma-1/\gamma}}\right] \\ &= 0.9\dot{m}(1.25)(1750)\left[1 - \frac{1.05}{(4)^{0.397}}\right] \\ &= 776.5\dot{m} \text{ Btu/s } (1.803\dot{m} \text{ MW})\end{aligned}$$

Again as in Example 6.8:

$$\begin{aligned}\dot{W}_{CP} &= \frac{\dot{m} c_p}{\eta_{CP}}(T_{2s} - T_1) = \dot{m}\frac{c_p T_1}{\eta_{CP}}\left(\frac{T_{2s}}{T_1} - 1\right) = \frac{\dot{m} c_p T_1}{\eta_{CP}}\left[\left(\frac{p_2}{p_1}\right)^{\gamma-1/\gamma} - 1\right] \\ &= \frac{\dot{m}(1.25)(500)}{0.9}(1.7338 - 1.0) \\ &= 509.7\dot{m} \text{ Btu/s } (1.184\dot{m} \text{ MW})\end{aligned}$$

Now $\dot{Q}_R = \dot{m} c_p (T_3 - T_2)$
We know that

$$\eta_{CP} = \frac{\dot{m} c_p (T_{2i} - T_1)}{\dot{m} c_p (T_2 - T_1)} = \frac{\dot{W}_{CPi}}{\dot{W}_{CP}}$$

where $\dot{W}_{CPi}$ was calculated in Example 6.7.

$$\therefore T_2 - T_1 = \frac{\dot{W}_{CPi}}{c_p \eta_{CP}} = \frac{458.7}{(1.25)(0.9)} = 407.7°\text{R } (226.5\text{K})$$

$$\therefore T2 = 500 + 407.7 = 907.7°\text{R } (504.3 \text{ K})$$

$$\therefore \dot{Q}_R = \dot{m}c_p(1750 - 907.7) = 1052.9\dot{m}\ \text{Btu/s}\ (2.45\dot{m}\ \text{MW})$$

$$\dot{W}_{NET} = \dot{W}_T - \dot{W}_{CP} = \dot{m}(776.5 - 509.6) = 266.9\dot{m}\ \text{Btu/s}\ (0.620\dot{m}\ \text{MW})$$

$$\eta_{th} = \frac{\dot{W}_{NET}}{\dot{Q}_R} = \left(\frac{266.9}{1052.9}\right) 100\ \text{(English units)}$$
$$= \left(\frac{0.620}{2.45}\right) 100\ \text{(SI units)}$$
$$= 25.3\%$$

Examples 6.8 and 6.9 demonstrate that consideration of real component efficiencies and pressure losses cause the cycle thermal efficiency to decrease dramatically. Employment of regeneration, if the pressure ratio allows, reverses this trend, as demonstrated next in Example 6.10, which considers both ideal and real turbines and compressors.

Example 6.10A: Brayton Cycle with Regeneration for Ideal Turbines and Compressors

PROBLEM

Compute the cycle thermal efficiency first for ideal turbines and compressors but with the addition of a regenerator of effectiveness 0.95. The cycle is illustrated in Figure 6.28. Regenerator effectiveness is defined as the actual preheat temperature change over the maximum possible temperature change, that is,

$$\xi = \frac{T_5 - T_2}{T_4 - T_2}$$

All other conditions of Example 6.7 apply.
Solution

$$\dot{W}_{C_p} = \dot{m}c_p(T_2 - T_1) = \dot{m}c_pT_1\left[\left(\frac{p_2}{p_1}\right)^{\gamma-1/\gamma} - 1\right]$$
$$= 458.6\dot{m}\ \text{Btu/ s}\ (1.066\dot{m}\ \text{MW})$$

(as in Example 6.7). Likewise

$$\dot{W}_T = \dot{m}c_p(T_3 - T_4) = \dot{m}c_pT_3\left[1 - \frac{1}{(r_p)^{\gamma-1/\gamma}}\right]$$
$$= 925.9\dot{m}\ \text{Btu/s}\ (2.150\dot{m}\ \text{MW})$$
$$\dot{Q}_R = \dot{m}c_p(T_3 - T_5)$$
$$\xi\ \text{(effectiveness of regenerator)} = \frac{T_5 - T_2}{T_4 - T_2} = 0.95$$

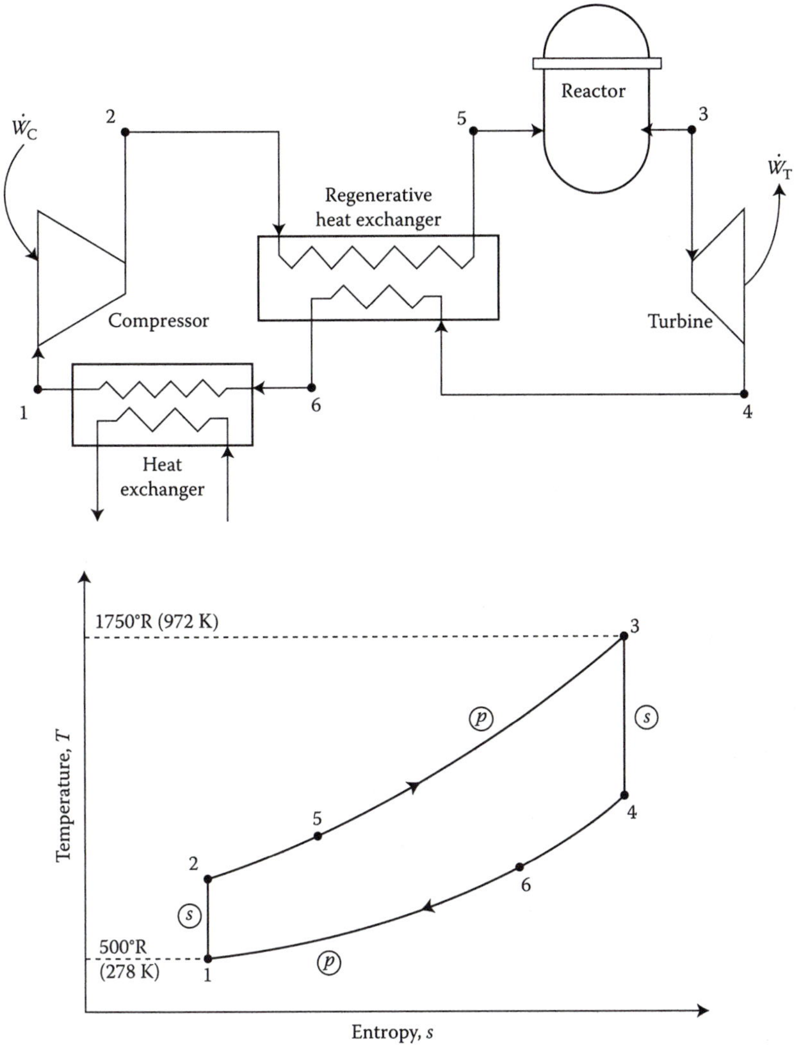

FIGURE 6.28 Brayton cycle with ideal components and regeneration: Example 6.10A.

Hence

$$T_5 = (T_4 - T_2)(0.95) + T_2 = 0.95T_4 + 0.05T_2$$

Writing T_4 in terms of T_3 and r_p, T_2 in terms of T_1 and r_p, and noting that processes 1–2 and 3–4 are isentropic, we obtain

$$\begin{aligned} T_5 &= (0.95)\left[\frac{T_3}{(r_p)^{\gamma-1/\gamma}}\right] + 0.05T_1(r_p)^{\gamma-1/\gamma} \\ &= (0.95)(0.5767)(1750) + (0.05)(500)(1.7338) \\ &= 1002.1\,°\text{R}(556.7\,\text{K}) \end{aligned}$$

Hence

$$\dot{Q}_R = \dot{m}c_p(1750 - 1002.1) = 934.9\dot{m}\ \text{Btu/s}\ (2.172\dot{m}\ \text{MW})$$

$$\dot{W}_{NET} = \dot{W}_T - \dot{W}_{CP} = \dot{m}(925.9 - 458.7) = 467.2\ \dot{m}\ \text{Btu/s}\ (1.084\dot{m}\ \text{MW})$$

$$\eta_{th} = \frac{\dot{W}_{NET}}{\dot{Q}_R}$$

$$= \left(\frac{467.2}{934.9}\right)100\,(\text{English units})$$

$$= \left(\frac{1.084}{2.172}\right)100\,(\text{SI units})$$

$$= 50.0\%$$

Example 6.10B: Brayton Cycle with Regeneration for Real Turbines and Compressors

Repeating these calculations for real compressors and turbines with component isentropic efficiencies of 90%, the thermal efficiency is reduced to 38.3%. Table 6.9 lists the intermediate calculations under Example 6.10B.

It is interesting to compare these results with those of Examples 6.7 and 6.8, which are identical except for regeneration. One sees that the use of regeneration has increased the efficiency for the cases of ideal components (50.0% versus 42.3%) and real components (38.3% versus 30.7%). In each situation, regeneration has caused the thermal efficiency of the cycle to increase by decreasing $\dot{Q}_R$ ($\dot{W}_{NET}$ being the same), a result that is possible only if T_4, the turbine exit, is hotter than T_2, the compressor exit.

Example 6.11: Brayton Cycle with Regeneration for Ideal Turbines and Compressors at Elevated Pressure Ratio

The condition above is not the case for a sufficiently high pressure ratio (e.g., $p_2 = 8\ p_1$). In this case (Example 6.11 in Table 6.9), the cycle thermal efficiency is reduced to 35.5%.

A loss of thermal efficiency has resulted from the introduction of a regenerator compared to the same cycle operated at a lower compression ratio, that is, Example 6.10A (35.5% versus 50.1%). An additional comparison is between this case and the case of an ideal cycle without regenerator but operated at the same high compression ratio of 8. Again the efficiency of this case is less than this ideal cycle without regeneration (35.5% versus 56.2%—where the value of 56.2% can be obtained from Equation 6.106 or, equivalently, Figure 6.25). Physically both comparisons reflect the fact that for this case (Example 6.11) r_p is so high that T_2 is higher than T_4; that is, the exhaust gases are cooler than those after compression. Heat would thus be transferred to the exhaust gases, thereby requiring $\dot{Q}_R$ to increase and η_{th} to decrease. Obviously, for this high compression ratio, inclusion of a regenerator in the cycle is not desirable.

Example 6.12: Brayton Cycle with Intercooling (See Table 6.9)

TABLE 6.9
Results of Brayton Cycle Cases of Examples 6.7 through 6.14

Parameter	Ex. 6.7	Ex. 6.8	Ex. 6.9	Ex. 6.10A	Ex. 6.10B	Ex. 6.11	Ex. 6.12	Ex. 6.13	Ex. 6.14
$\beta = \left(\frac{p_2 p_4}{p_3 p_1}\right)^{\gamma-1/\gamma}$	1.0	1.0	1.05	1.0	1.0	1.0	1.0	1.0	1.0
Component isentropic efficiency (η_s)	1.0	0.9	0.9	1.0	0.9	1.0	1.0	1.0	1.0
Regenerator effectiveness (ξ)	—	—	—	0.95	0.95	0.95	—	—	—
Pressure ratio (r_p)	4	4	4	4	4	8	4	4	4
Intercooling	—	—	—	—	—	—	$\frac{p_1'}{p_1} = \frac{1}{2}\frac{p_2}{p_1}$ $T_1'' = T_1$	—	$\frac{p_1'}{p_1} = \frac{1}{2}\frac{p_2}{p_1}$ $T_1'' = T$
Reheat	—	—	—	—	—	—	—	$\frac{p_3'}{p_4} = \frac{1}{2}\frac{p_3}{p_4}$ $T_3'' = T_3$	$\frac{p_3'}{p_4} = \frac{1}{2}\frac{p_3}{p_4}$ $T_3'' = T_3$
Turbine work ($\dot{W}_T/\dot{m}$) Btu/lb MJ/kg	925.9	833.3	776.5	925.9	833.3	1229.4	925.9	1052.5	1052.5
	2.150	1.935	1.803	2.150	1.935	2.855	2.150	2.444	2.444
Compressor work ($\dot{W}_c/\dot{m}$) Btu/lb MJ/kg	458.7	509.7	509.7	458.7	509.7	801.9	395.96	458.7	395.96
	1.066	1.184	1.184	1.066	1.184	1.862	0.920	1.066	0.920
Net work ($\dot{W}_{NET}/\dot{m}$ Btu/lb MJ/kg	467.2	323.6	266.9	467.2	323.6	427.5	529.9	593.8	656.5
	1.084	0.752	0.620	1.084	0.752	0.993	1.23	1.378	1.524
Heat in ($\dot{Q}_R/\dot{m}$) Btu/lb MJ/kg	1103.0	1052.9	1052.9	934.9	844.4	1205.7	1364.5	1630.1	1890.8
	2.560	2.45	2.45	2.172	1.961	2.800	3.169	3.786	4.391
Cycle thermal efficiency (η_{th})(%)	42.3	30.7	25.3	50.0	38.3	35.46	38.8	36.4	34.7

Example 6.13: Brayton Cycle with Reheat (See Table 6.9)

Major process variations possible for cycles with high compression ratios are the intermediate extraction and cooling of gases undergoing compression (to reduce compressor work) and the intermediate extraction and heating of gases undergoing expansion (to increase turbine work). These processes are called *intercooling* and *reheat*, respectively. Cases of intercooling only and reheat only are included in Table 6.9 as Examples 6.12 and 6.13, respectively.

Example 6.14: Brayton Cycle with Reheat and Intercooling

PROBLEM

Calculate the thermal efficiency for the cycle employing both intercooling and reheat as characterized below. The cycle is illustrated in Figure 6.29. All other conditions of Example 6.7 apply.

$$\text{Intercooling:} \quad \frac{p_1'}{p_1} = \frac{p_2}{p_1'} = r_p' \quad T_1'' = T_1$$

$$\text{Reheat:} \quad \frac{p_3'}{p_4} = \frac{p_3}{p_3'} = r_p' \quad T_3'' = T_3$$

SOLUTION

$$\dot{W}_{CP} = \dot{m}c_p(T_1' - T_1) + \dot{m}c_p(T_2' - T_1'')$$

$$\dot{W}_{CP} = \dot{m}c_pT_1\left(\frac{T_1'}{T_1} - 1\right) + \dot{m}c_pT_1''\left(\frac{T_2'}{T_1''} - 1\right)$$

$$= \dot{m}c_pT_1[(r_p')^{\gamma-1/\gamma} - 1] + \dot{m}c_pT_1''\,[(r_p')^{\gamma-1/\gamma} - 1]$$

$$= 2\dot{m}c_pT_1[(r_p')^{\gamma-1/\gamma} - 1] = 2\dot{m}c_pT_1[(2)^{0.397} - 1]$$

$$= 395.96\,\dot{m}\ \text{Btu/s}\ (0.920\,\dot{m}\ \text{MW})$$

$$\dot{W}_T = \dot{m}c_p(T_3 - T_3') + \dot{m}c_p(T_3'' - T_4') = \dot{m}c_pT_3\left(1 - \frac{T_3'}{T_3}\right) + \dot{m}c_pT_3''\left(1 - \frac{T_4'}{T_3''}\right)$$

Again for the isentropic case:

$$\dot{W}_T = 2\dot{m}c_pT_3\left[1 - \frac{1}{(r_p')^{\gamma-1/\gamma}}\right]$$

$$= 2\dot{m}(1.25)1750\left[1 - \frac{1}{(2)^{0.397}}\right]$$

$$= 1052.5\,\dot{m}\ \text{Btu/s}\ (2.444\,\dot{m}\ \text{MW})$$

$$\frac{T_3'}{T_3} = \frac{1}{(r_p')^{\gamma-1/\gamma}}$$

$$\therefore T_3' = \frac{T_3}{(r_p')^{\gamma-1/\gamma}} = \frac{1750}{(2)^{0.397}} = \frac{1750}{1.317}$$

$$= 1329°\text{R}(738.3\text{K})$$

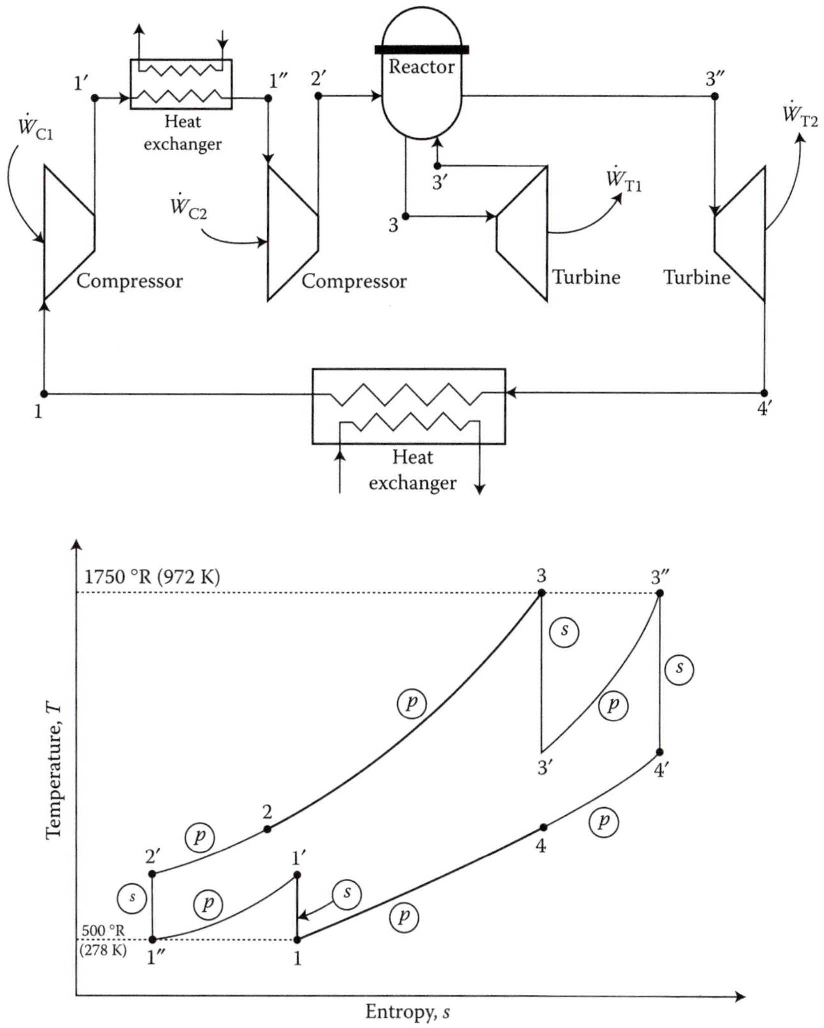

FIGURE 6.29 Brayton cycle with reheat and intercooling: Example 6.14.

$$\text{where } T_2' = T_1''(r_p')^{\gamma-1/\gamma} \quad \text{and} \quad T_1'' = T_1 = 500°\text{R (278K)}$$

$$\therefore\ T_2' = (500°\text{R})2^{0.397} = 658.4°\text{R}(365.8\text{K})$$

$$\dot{Q}_R = \dot{m}c_p(T_3 - T_2) + \dot{m}c_p(T_3'' - T_3')$$

$$\dot{Q}_R = \dot{m}c_p[(1750 - 658.4) + (1750 - 1329.0)]$$

$$= 1890.8\,\dot{m}\ \text{Btu/s}\ (4.391\dot{m}\ \text{MW})$$

$$\dot{W}_{NET} = \dot{W}_T - \dot{W}_{CP} = \dot{m}\,(1052.5 - 395.96) = 656.5\,\dot{m}\ \text{Btu/s}\ (1.524\,\dot{m}\ \text{MW})$$

$$\eta_{th} = \frac{\dot{W}_{NET}}{\dot{Q}_R}$$

$$= \left(\frac{656.5}{1890.8}\right)100\,(\text{English units})$$
$$= \left(\frac{1.524}{4.391}\right)100\,(\text{SI units})$$
$$= 34.7\%$$

Compared with Example 6.7, both the net work and the maximum useful work have increased, but the thermal efficiency has decreased. An increase in the thermal efficiency would result if regeneration were employed. These changes are the combined result of the intercooling and reheating processes. They can best be understood by examining each separately with reference to the results of Examples 6.12 and 6.13, respectively.

In comparison with Example 6.7 (which is the same except for intercooling), we see that intercooling has decreased the compressor work and hence raised the $\dot{W}_{NET}$. However, the cycle efficiency has decreased to 38.8% from 42.3%, as the cycle with intercooling has a smaller pressure ratio than the basic cycle. However, by lowering inlet temperature to the reactor, that is, T_2 to T_2', a portion of this added energy can be supplied regeneratively, thereby offering the possibility to recoup some of the decrease in efficiency.

As we see in comparison with Example 6.7, reheat also decreases cycle thermal efficiency (36.4% versus 42.3%) because the cycle with reheat has a smaller pressure ratio than the basic cycle itself. However, the turbine exit temperature has been raised from T_4 to T_4', offering the possibility of supplying some of the energy addition from states 2 to 3 regeneratively with its associated improvement in cycle thermal efficiency.

6.8 SUPERCRITICAL CARBON DIOXIDE BRAYTON CYCLES

In the previous two sections, Brayton cycles using perfect gas with constant specific heats were presented. Here we consider supercritical carbon dioxide (S-CO_2), which must be treated as a real gas and as well has specific heats which vary considerably with the state conditions typical of the Brayton cycles considered.

We will show that S-CO_2 Brayton cycles operating at lower temperatures can outperform helium Brayton cycles operating at the higher temperatures already considered. However, to achieve this performance the S-CO_2 cycle design must carefully accommodate the significant numerical changes which are encountered in values of specific heat values at constant pressure.*

* For the student performing calculations for supercritical CO_2 systems, extreme care must be taken when near the critical point. Linear interpolation with tabulated values will not accurately reflect the large changes in property values within 2–3 degrees of the critical point. Thus, rough calculations can be performed with a table, but a reliable correlation must be used for more accurate results. The National Institute of Standards and Technology publishes REFPROP, a software program capable of producing values of all properties for a number of fluids, including CO_2. REFPROP uses a large data set, with corresponding correlations to produce very accurate values. The software is available via the NIST Website at http://www.nist.gov/srd/nist23.htm.

TABLE 6.10
State Conditions for the Simple S-CO_2 Brayton Cycle

State Number	Pressure (MPa)	Temperature (K)	Entropy (kJ/kg K)	Enthalpy (kJ/kg)	Specific Heat at Constant Pressure (kJ/kg K)
1	7.7	305	1.3394	304.11	13.3
2	20	332.2	1.3394	322.20	2.4901
3	20	823	2.7408	1034.9	1.2408
4	7.7	697.6	2.7408	896.18	1.1653

6.8.1 Simple S-CO_2 Brayton Cycle

Figure 6.24 for the helium Brayton cycle also represents the temperature entropy plot of the simple S-CO_2 Brayton cycle. Let us, however, assume the less optimistic limits for the lowest and highest cycle temperatures (T_1 and T_3) of 305 and 823 K. As well select the pressure ratio of 2.6. However, take the minimum pressure as 7.7 MPa so that the cycle operates in the supercritical region (for CO_2 the critical point is 304.1 K, 7.3773 MPa).

Table 6.10 summarizes the state conditions for the simple S-CO_2 Brayton cycle derived from the above given conditions, that is,

State 1 $p_1 = 7.7$ MPa; $T_1 = 305$ K
State 3 $p_3 = 2.6\ \mathrm{p}_1$; $T_3 = 823$ K

The turbine and compressor works are given as follows. Note that enthalpies must be used versus temperatures because the specific heat values at constant pressure change significantly for the cycle state conditions of each process.

$$\frac{\dot{W}_T}{\dot{m}} = (h_3 - h_4) = (1034.9 - 896.18) = 138.72\ \text{kJ/kg} \tag{6.108}$$

$$\frac{\dot{W}_{CP}}{\dot{m}} = (h_2 - h_1) = (322.2 - 304.11) = 18.09\ \text{kJ/kg} \tag{6.109}$$

$$\frac{\dot{Q}_R}{\dot{m}} = (h_3 - h_2) = (1034.9 - 322.2) = 712.7\ \text{kJ/kg} \tag{6.110}$$

The Brayton nuclear plant thermodynamic efficiency is then:

$$\eta_{th} = \frac{\dot{W}_T - \dot{W}_{CP}}{\dot{Q}_R} \tag{6.111}$$

$$= \left(\frac{138.72 - 18.09}{712.7}\right)100 = 16.9\%$$

This is indeed a far lower thermodynamic efficiency than that of the comparable helium Brayton cycle (Ex. 6.7, 42.3%) so that the S-CO_2 cycle design must be enhanced. The obvious step is to add regeneration.

6.8.2 S-CO_2 Brayton Cycle with Ideal Components and Regeneration

Figure 6.28 also represents this cycle. Again, however, the temperature limits are changed to 305 and 823 K. The achievable thermodynamic efficiency for this cycle hinges on the assumed regenerator effectiveness value, ξ. As for the helium cycle, let us take this as 0.95. This assumption has significant implications which we will revisit after completing the analysis.

For this regeneration cycle, the thermodynamic efficiency increases because the reactor heat input, $\dot{Q}_R$, decreases dramatically. For this cycle, the reactor heat input is now:

$$\dot{Q}_R = \dot{m}(h_3 - h_5) \tag{6.112}$$

so that h_5 must be determined. From the assumed effectiveness value, T_5 is obtained as

$$\xi \equiv \frac{T_5 - T_2}{T_4 - T_2} \tag{6.113}$$

Hence

$$\begin{aligned} T_5 &= \xi T_4 + (1-\xi)T_2 \\ &= 0.95(697.6) + 0.05(332.2) \\ &= 679.3\,\text{K} \end{aligned} \tag{6.114}$$

With the pressure (20 MPa) and temperature (679.3 K) established, all needed properties of state 5 can be determined and are listed in Table 6.11.

State 6 conditions are determined from a heat balance in the regenerator yielding:

$$\dot{m}(h_4 - h_6) = \dot{m}(h_5 - h_2) \tag{6.115a}$$

TABLE 6.11
Additional State Conditions for the S-CO_2 Brayton Cycle with Regeneration ($\xi = 0.95$)

State Number	Pressure (MPa)	Temperature (K)	Entropy (kJ/kg K)	Enthalpy (kJ/kg)	Specific Heat at Constant Pressure (kJ/kg K)
5	20	679.3	2.5047	858.10	1.2246
6	7.7	306.5	1.5256	360.28	29.692

Importantly for the design, the mass flow rate is the same for both streams. Solving for the unknown state 6 enthalpy:

$$h_6 = h_4 - (h_5 - h_2) = 896.18 - (858.10 - 322.2) = 360.28 \text{ kJ/kg} \quad (6.115\text{b})$$

With this enthalpy and pressure of 7.7 MPa, state 6 is defined and its properties are added to Table 6.11.

We immediately observe, however, that $T_6 < T_2$, an impossibility for the operation of the regenerative heat exchanger. Let us examine why this occurs. Observe from Figure 6.28 that in the counter flow regenerator hot, low pressure S-CO_2 entering at state 4 and exiting at state 6 heats the cold, high pressure S-CO_2 stream entering at state 2 and exiting at state 5. Each stream experiences a significant variation in specific heat value as the state conditions in Tables 6.10 and 6.11 illustrate. Specifically for the hot stream, the constant pressure specific heat changes from 1.1653 (state 4) to 29.692 kJ/kg K (state 6), while for the cold stream the change is from 2.4901 (state 2) to 1.2246 kJ/kg K (state 5). These large variations in constant pressure specific heat dictate that an enthalpy-based rather than a temperature-based regeneration effectiveness definition be employed.

The effectiveness relations now need to account for the fact that the specific heat at constant pressure is a strong function of temperature. Since effectiveness is a ratio of heat transfer rates, we generalize its form from the simple temperature ratio of Example 6.10 as follows:

$$\xi = \frac{\dot{Q}_{\text{actually transferred}}}{\dot{Q}_{\text{max possible}}} \quad (6.116)$$

The numerator expresses the heating of the cold stream:

$$\dot{Q}_{\text{actually transferred}} = \dot{m}_{\text{c}}(h_{\text{c out}} - h_{\text{c in}}) \quad (6.117)$$

The denominator expresses the maximum heat transfer from the hot stream, that is, the exit of this stream at the pressure of the hot stream and the cold inlet temperature.

$$\dot{Q}_{\text{max possible}} = \dot{m}_{\text{h}}(h_{\text{h in}} - h_{\text{p hot},T_{\text{c in}}}) \quad (6.118)$$

Hence for the regenerator of Figure 6.28, we express the effectiveness as follows:

$$\xi = \frac{h_5 - h_2}{h_4 - h(p_6, T_2)} \quad (6.119\text{a})$$

Solving for h_5, we obtain

$$h_5 = \xi(h_4 - h(p_6, T_2)) + h_2 \quad (6.119\text{b})$$

From property tables find:

$$h(p_6,T_2) = 460.64 \text{ kJ/kg}$$

and hence

$$h_5 = 0.95(896.18 - 460.64) + 322.20 = 735.96 \text{ kJ/kg}$$

From the heat balance of Equation 6.115b:

$$h_6 = h_4 - (h_5 - h_2)$$

or

$$h_6 = h_4 - \xi(h_4 - h(p_6,T_2)) = 896.18 - 0.95(896.18 - 460.64) = 482.17 \text{ kJ/kg}$$

With this value for enthalpy and $p_6 = 7.7$ MPa, we find $T_6 = 345.09$ K. Now $T_6 > T_2$, that is, 345.09 K > 332.23 K as it should be.

However, the resulting cycle thermodynamic efficiency, while higher than the 16.9% for the simple S-CO_2 cycle (Equation 6.111), is only 40.5% even though $\dot{Q}_R/\dot{m}$ is decreased, for example

$$\frac{\dot{Q}_R}{\dot{m}} = h_3 - h_5 = 1034.9 - 735.96 = 298.94 \text{ kJ/kg}$$

yielding from Equation 6.111 the cited efficiency as

$$\eta_{th} = \left(\frac{138.72 - 18.09}{298.94}\right)100 = 40.5\%$$

The reason for this low thermodynamic efficiency is the less than desirable performance of the single regenerator of this cycle. Figures 6.30 and 6.31 illustrate the temperature and specific heat profiles through this regenerator.

The adoption of a recompression cycle which splits the S-CO_2 flow by introducing two regenerators offers an approach to increase this efficiency as presented next.

6.8.3 S-CO_2 Recompression Brayton Cycle with Ideal Components

The Brayton cycle with regenerator must be reconfigured in light of the specific heat values presented in the previous section. From Figure 6.31, observe that the specific heat difference in the low-temperature region of the regenerator ending at states 2 and 6 is generally larger than that in the high-temperature region, starting with states 4 and 5. These regions correspond to the last and first 50% of the regenerator length.

Hence to achieve a heat balance, the flow streams in the regenerator must be made different particularly where the specific heat differences are greatest, that is, the lower temperature region. This redesigned Brayton cycle is reflected in Figure 6.32.

The regenerator is partitioned into two segments, a low- and a high-temperature component and importantly the low-temperature stream in the low-temperature

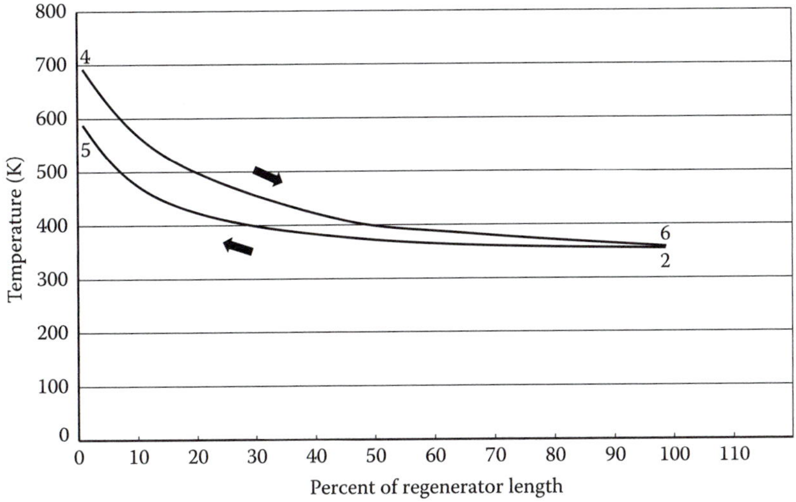

FIGURE 6.30 Temperature profiles in regenerator.

component, state 2 to 2′, has a fraction f of the flow rate of the high-temperature stream in the low-temperature regenerator, state 4′ to 6. The streams in the high-temperature regenerator remain with equal flow rates. This cycle is called by its designers a recompression cycle [1,2].

In this cycle, the assumed state conditions are the same as the simple S-CO_2 cycle. These conditions are restated in Table 6.12: State 6 and State 2′ have the same entropy.

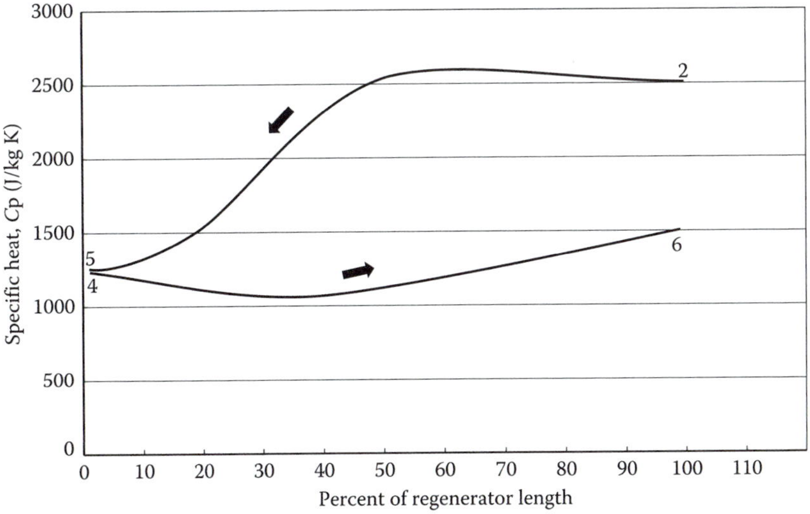

FIGURE 6.31 Specific heat capacity in regenerator.

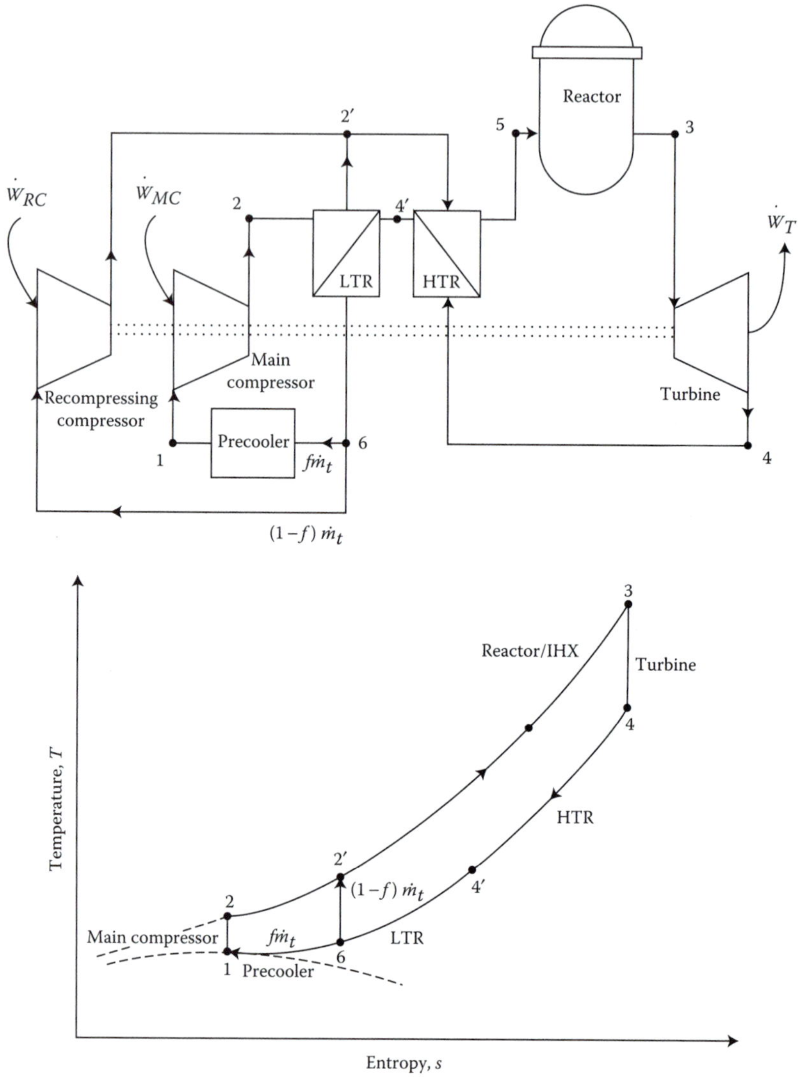

FIGURE 6.32 Recompression Brayton cycle with ideal components.

TABLE 6.12
State Conditions for the Recompression Brayton Cycle of Figure 6.32

Assumed	State 1	$p_1 = 7.7$ MPa	$T_1 = 305$ K
Assumed	State 3	$p_3 = 2.6\ P_1$	$T_3 = 823$ K
Determined	State 2	$p_2 = 20.0$ MPa	$s_2 = s_1$
Determined	State 4	$p_4 = 7.7$ MPa	$s_4 = s_3$

The calculation of the cycle thermodynamic efficiency requires determination of the cycle states and the flow fractions throughout the cycle. It is accomplished by expressing the applicable effectiveness and heat balance relations.

For the recompression cycle regenerators of Figure 6.32, we express the effectiveness in the enthalpy-based form of Equations 6.116 through 6.118 and the heat balances as follows. We take the low-temperature regenerator effectiveness as 0.92 and the high-temperature regenerator effectiveness as 0.98.

For the low-temperature regenerator:

$$\xi_{\text{LTR}} = 0.92 = \frac{h_{4'} - h_6}{h_{4'} - h(p_6, T_2)} \tag{6.120}$$

$$\dot{Q}_{\text{LTR}} = (h_{2'} - h_2) f \dot{m}_{\text{T}} = \dot{m}_{\text{T}} (h_{4'} - h_6) \tag{6.121}$$

For the high-temperature regenerator:

$$\xi_{\text{HTR}} = 0.98 = \frac{h_4 - h_{4'}}{h_4 - h(p_{4'}, T_{2'})} \tag{6.122}$$

$$\dot{Q}_{\text{HTR}} = (h_5 - h_{2'}) = (h_4 - h_{4'}) \tag{6.123}$$

Further, given state 6, state 2′ can be determined since $p_{2'}$ is known, and $s_{2'} = s_6$.

Hence there are four equations (Equations 6.120 through 6.123) but five unknowns ($h_{4'}$, f, $T_{2'}$, $h_{2'}$, h_5, and $h_{2'}$).

$T_{2'}$ and $h_{2'}$ are related since $p_{2'}$ is known, so these reduce to only one unknown. Hence the unknowns are now four—*f*, state 2′ (for which we know one property, $p_{2'}$), $h_{4'}$ and h_5. The solution is achieved by assuming a value for $T_{2'}$, say 169°C, and iterating to convergence. The final state conditions obtained for this recompression cycle are given in Table 6.13. The flow split factor is 0.59. The temperature and specific

TABLE 6.13
State Points for Recompression S-CO_2 Brayton Cycle

State	*T* (K)	*p* (MPa)	*h* (kJ/kg)	*s* (kJ/kgK)	c_p (kJ/kgK)
1	305.00	7.7	304.11	1.3394	13.3000
2	332.23	20	322.20	1.3394	2.4901
3	823.00	20	1034.90	2.7408	1.2408
4	697.60	7.7	896.18	2.7408	1.1653
2′	421.27	20	520.09	1.8723	1.6554
6	337.96	7.7	470.80	1.8723	1.7002
4′	426.73	7.7	587.61	2.1813	1.1652
5	655.29	20	828.66	2.4606	1.2251

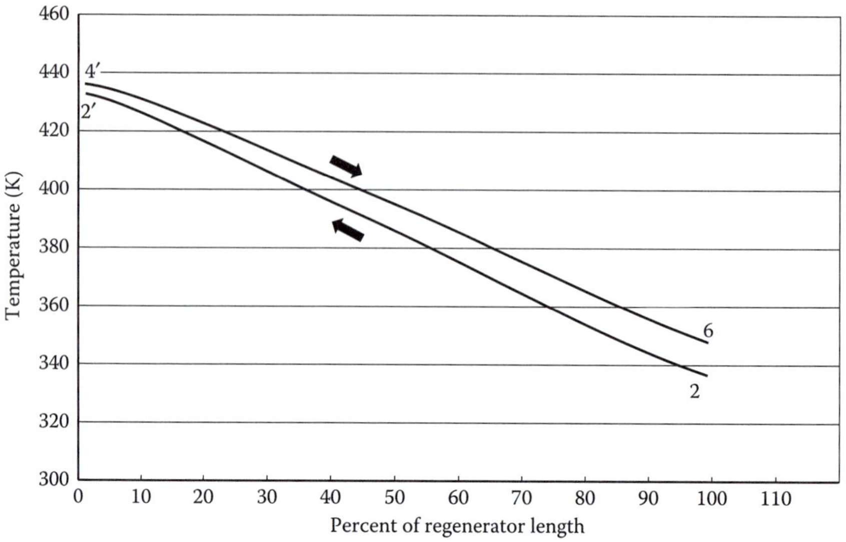

FIGURE 6.33 Temperature in low-temperature regenerator.

heat profiles throughout these regenerators are shown in Figures 6.33 through 6.36. The thermodynamic efficiency of this cycle is 52.2%. This high efficiency is principally due to the high temperature of heat addition (between states 5 and 3), the low temperature of heat rejection (between states 6 and 1) and the small main compressor work (between states 1 and 2).

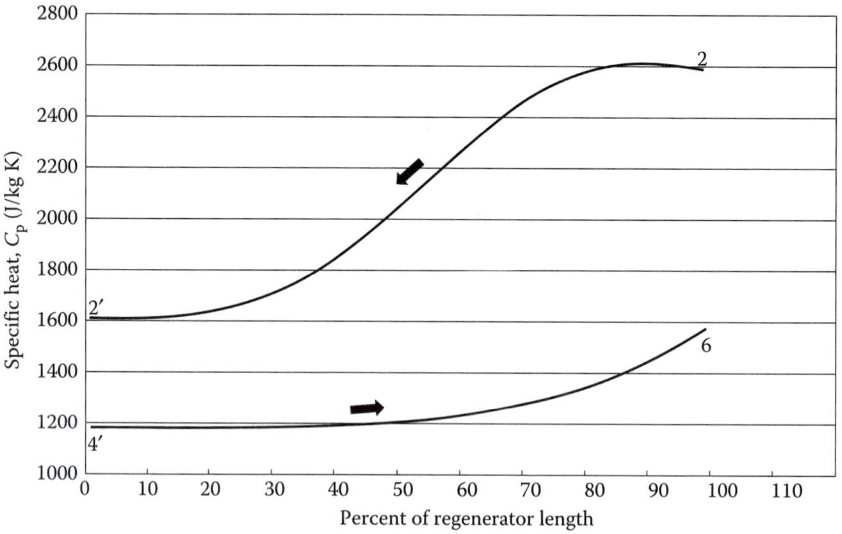

FIGURE 6.34 Specific heat in low-temperature regenerator.

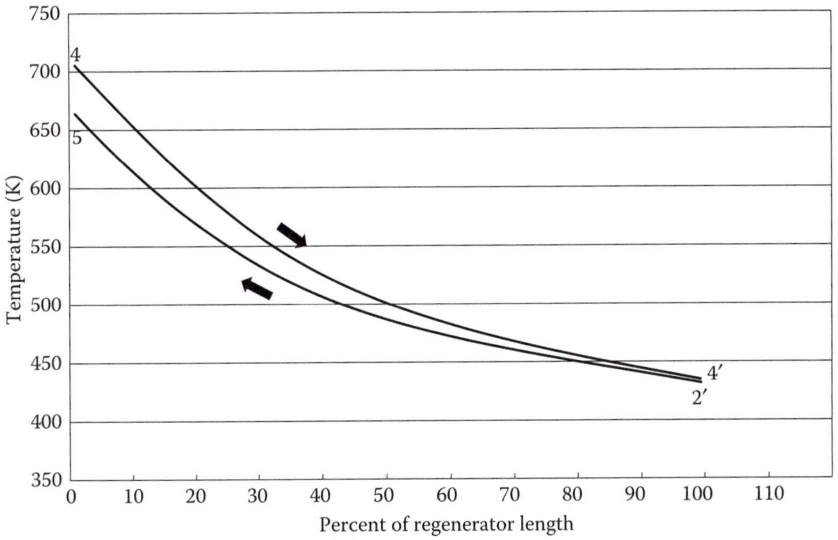

FIGURE 6.35 Temperature in high-temperature regenerator.

6.8.4 S-CO_2 Recompression Brayton Cycle with Real Components and Pressure Losses

It is of interest to investigate the effect of adding real component and system losses on this recompression cycle. We do this next for four new cases with sequential addition of real losses (case 1 is the ideal case just presented).

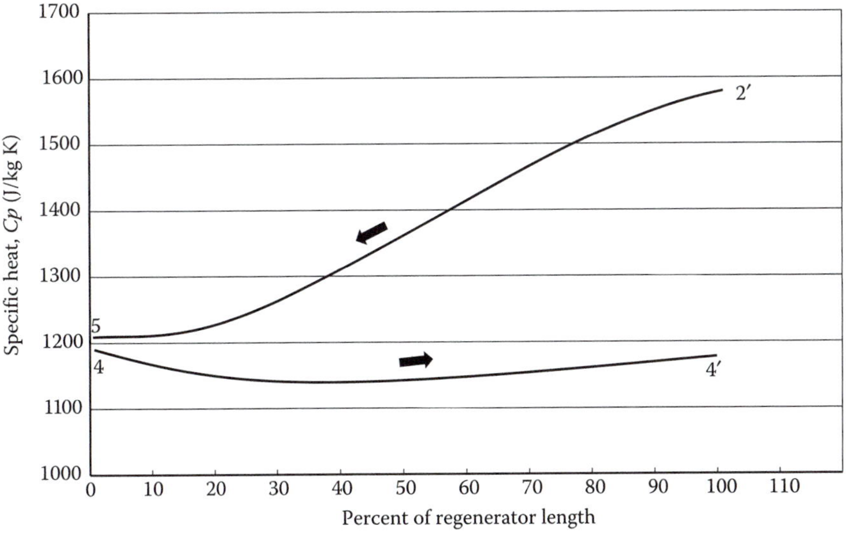

FIGURE 6.36 Specific heat in high-temperature regenerator.

TABLE 6.14
Results for S-CO_2 Recompression Brayton Cycle Cases[a]

Case Number	Machines	Fractional Δp in IHX	Fractional Δp in Cycle HXs	Fractional Δp in Pipes	Cycle Efficiency (%)
1	Ideal	None	None	None	52.2
2	T: 92%, C: 89%	None	None	None	46.5
3	T: 92%, C: 89%	2.6%	None	None	45.6
4	T: 92%, C: 89%	2.6%	2.9%	None	44.8
5	T: 92%, C: 89%	2.6%	2.9%	6.1%	43.0

[a] Turbine inlet temperature of 550°C.

Case 2. Real turbine and compressor efficiencies of 92% and 89%, respectively.

Case 3. Typical pressure losses in the IHX added.

Case 4. Typical pressure losses in the regenerators and the precooler added.

Case 5. Typical pressure losses in the piping of the power conversion system added.

Table 6.14 presents these results. It shows the cycle efficiency is reduced from 52.2% to 43.0% by considering these unavoidable real conditions. The state conditions for Cases 2, 3, 4, and 5 are presented in Tables 6.15 through 6.18. Note the smallest temperature for state 1 is in Case 5. This is because of the large sensitivity of CO_2 properties near the critical point to small temperature changes, which requires consideration of kinetic energy in the energy conservation equation. As the flow coming from the precooler outlet is accelerated in the pipe at the main compressor inlet, the increase of kinetic energy is accompanied by reduction of the static enthalpy and a corresponding reduction of static temperature. As a result, compressor inlet temperature is reduced from 32°C at the precooler outlet to 31.87°C at the main compressor inlet even though there is adiabatic flow through the piping (assumption of perfect insulation). In Table 6.14, the pressure drops are defined as a percent of the

TABLE 6.15
State Conditions for Case 2

State	T (°C)	T (K)	p (MPa)	h (kJ/kg)	s (kJ/kg K)	c_p (kJ/kg K)
1	31.99	305.14	7.69	306.66	1.3477	15.658
2	61.09	334.24	20.00	327.23	1.3545	2.5154
3	550.00	823.15	20.00	1035.13	2.7411	1.2409
4	431.57	704.72	7.69	904.54	2.7530	1.1674
2′	154.74	427.89	20.00	530.87	1.8977	1.6030
6	67.35	340.50	7.69	475.14	1.8853	1.6447
4′	163.98	437.13	7.69	599.70	2.2095	1.1534
5	388.59	661.74	20.00	836.57	2.4726	1.2248

TABLE 6.16
State Conditions for Case 3

State	*T* (°C)	*T* (K)	*p* (MPa)	*h* (kJ/Kg)	*s* (kJ/kg K)	c_p (kJ/kg K)
1	31.99	305.14	7.69	306.66	1.3477	15.658
2	61.09	334.24	20.00	327.23	1.3545	2.5154
3	550.00	823.15	20.00	1035.13	2.7411	1.2409
4	434.61	707.76	7.69	908.09	2.7580	1.1684
2′	154.64	427.79	20.00	530.71	1.8973	1.6037
6	67.31	340.46	7.69	475.06	1.8851	1.6455
4′	164.11	437.26	7.69	599.85	2.2098	1.1533
5	391.28	664.43	20.00	839.86	2.4776	1.2247

TABLE 6.17
State Conditions for Case 4

State	*T* (°C)	*T* (K)	*p* (MPa)	*h* (kJ/kg)	*s* (kJ/kg K)	c_p (kJ/kg K)
1	32.00	305.15	7.69	306.65	1.3483	15.813
2	61.09	334.24	20.00	327.23	1.3545	2.5154
3	550.00	823.15	19.36	1035.58	2.7480	1.2390
4	437.89	710.37	7.86	911.70	2.7590	1.1703
2′	153.54	427.50	19.94	529.19	1.8942	1.6101
6	66.83	340.09	7.72	473.92	1.8812	1.6617
4′	163.22	437.15	7.81	598.30	2.2036	1.1576
5	394.15	666.88	19.91	843.49	2.4839	1.2242

TABLE 6.18
State Conditions for Case 5

State	*T* (°C)	*T* (K)	*p* (MPa)	*h* (kJ/kg)	*s* (kJ/kg K)	c_p (kJ/kg K)
1	31.87	305.02	7.66	306.65	1.3475	16.140
2	61.12	334.27	20.00	327.30	1.3547	2.5158
3	549.96	823.11	18.95	1035.84	2.7524	1.2379
4	444.20	717.35	8.11	918.76	2.7629	1.1734
2′	152.75	425.90	19.69	529.00	1.8956	1.6085
6	66.52	339.67	7.71	473.59	1.8802	1.6658
4′	163.24	436.39	7.93	597.77	2.1998	1.1611
5	399.96	673.11	19.66	850.92	2.4974	1.2228

absolute pressure in the subject component and are therefore called fractional pressure drops. For the regenerators, which have a high pressure side and a low pressure side, the fractional pressure drops from each side are summed. The recuperator pressure drops include losses in the inlet and outlet distribution plena, which are essentially a part of the heat exchanger.

An important point to note is the relatively large pressure loss in piping. This is due to the relatively low specific heat of CO_2, which requires larger flow rates to transport a given heat rate. Increasing the size of the pipes is not always a feasible option. Considering the fact that high pressure pipes operate at 20 MPa and at temperatures up to 550°C, very large pipes would require extremely thick walls and would become impractical. If 1 m pipe diameter is taken as a limit, the rating of the S-CO_2 power cycle is limited to 250–300 MWe per unit. Therefore, large plants will typically require several trains of S-CO_2 power conversion systems. In our example, the cycle rating was 500 MWth, producing 215 MWe. Because the system power rating is close to the limit and the maximum pipe diameter of 1 m is utilized, no further reduction of pressure losses through pipe diameter increase can be achieved. As a result, the fractional pressure drop in pipes is relatively high. Power conversion systems with smaller ratings would have smaller fractional pressure losses in pipes.

PROBLEMS

6.1. *Work output of a fuel–water interaction* (Section 6.2)
Compute the work done by a fuel–water interaction assuming that the 40,000 kg of mixed oxide fuel and 4000 kg of water expand independently and isentropically to 1 atmosphere. Assume that the initial fuel and water conditions are such that equilibrium mixture temperature (T_e) achieved is 1945 K. Other water conditions are as follows: $T_{\text{initial}} = 400$ K; $\rho_{\text{initial}} = 945$ kg/m^3; $c_v = 4184$ J/kg K. Caution: Equation 6.9 is inappropriate for these conditions since the coolant at state *e* is supercritical.

Answer: 1.67×10^{10} J

6.2. *Evaluation of alternate ideal Rankine cycles* (Section 6.3)
Three alternative steam cycles illustrated in Figure 6.37 are proposed for a nuclear power station capable of producing either saturated steam or superheated steam at a temperature of 293°C. The condensing steam temperature is 33°C.

1. Assuming ideal machinery, calculate the cycle thermal efficiency and steam rate (kg steam/kWe-h) for each cycle using the steam tables.
2. For each cycle, compare the amount of heat added per unit mass of working fluid in the legs $3' \rightarrow 4$ and $4 \rightarrow 1$.
3. Briefly compare advantages and disadvantages of each of these cycles. Which would you use?

Answers:

1. $\eta_{th} = 38.2\%$, 45.9%, 36.8%; steam rate = 3.60, 5.38, 3.54 kg steam/(kWe-h)

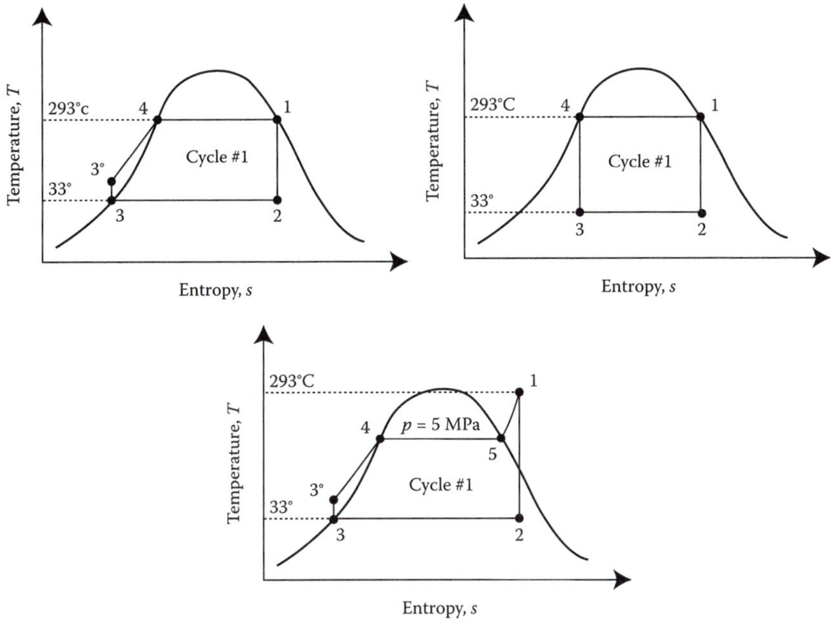

FIGURE 6.37 Alternate ideal Rankine cycles.

2. $\dot{Q}_{3'\rightarrow 4} = 1.163 \times 10^3$; 0; 1.011×10^3 kJ/kg

   $\dot{Q}_{4\rightarrow 1} = 1.456 \times 10^3$; 1.456×10^3; 1.748×10^3 kJ/kg

6.3. *Availability analysis of a simplified BWR* (Section 6.4)
A BWR system with one stage of moisture separation is shown in Figure 6.38. The conditions in Table 6.19 may be used.
Turbine isentropic efficiency = 90%
Pump isentropic efficiency = 85%
Environmental temperature = 30°C

1. Calculate the cycle thermal efficiency.
2. Recalculate the thermal efficiency of the cycle assuming that the pumps and turbines have isentropic efficiency of 100%.
3. Calculate the lost work due to the irreversibility of each component in the cycle and show numerically that the available work equals the sum of the lost work and the net work.

Answers:

1. $\eta_{th} = 34.2\%$
2. $\eta_{thmax} = 37.7\%$
3. $\dot{W}_{u,max} = 2510.9$ kJ/kg

   $\dot{W}_{NET} + \dot{I}_{TOT} = 2510.9$ kJ/kg

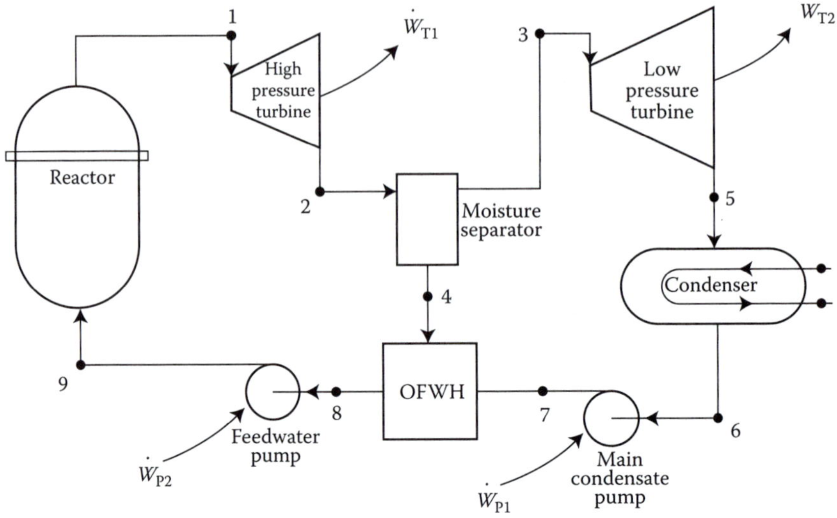

FIGURE 6.38 BWR plant.

6.4. *Thermodynamic analysis of a gas turbine* (Section 6.4)
Is transformation $1 \rightarrow 2$, shown in Figure 6.39, thermodynamically possible for a gas turbine? If so, under what conditions? If not, why? (Assume steady-state)

6.5. *Analysis of a steam turbine* (Section 6.4)
In the test of a steam turbine, the following data were observed:
$h_1 = 3000$ kJ/kg; $p_1 = 10$ MPa; $\upsilon_1 = 150$ m/s
$h_2 = 2600$ kJ/kg; υ_2 is negligible, $p_2 = 0.5$ MPa
$z_2 = z_1$ and $\dot{W}_{1,2} = 384.45\,\text{kJ/kg}$

TABLE 6.19
State Conditions for Problem 6.3

Points	p (kPa)	Condition
1	6890	Saturated vapor
2	1380	
3	1380	Saturated vapor
4	1380	Saturated liquid
5	6.89	
6	6.89	Saturated liquid
7	1380	
8	1380	
9	6890	

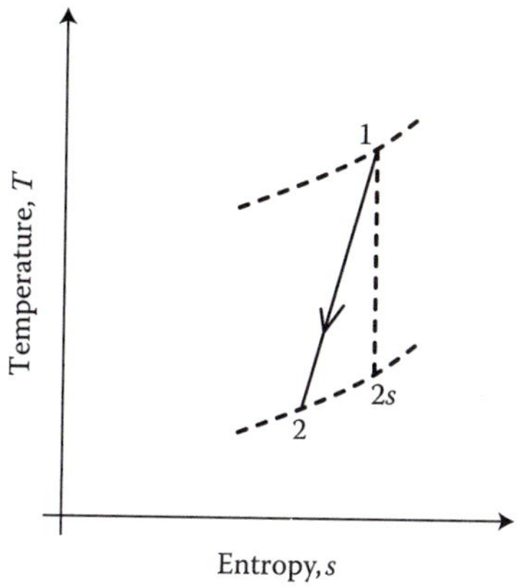

FIGURE 6.39 Temperature–entropy diagram for transformation in the turbine.

1. Assume steady flow, and determine the heat transferred to the surroundings per kilogram of steam.
2. What is the quality of the exit steam?

Answers:
$\dot{Q} = -26.8\,\text{kJ/kg}$
$x = 92.9\%$

6.6. *Irreversibility problems involving the Rankine cycle* (Section 6.4) Consider the Rankine cycles given in the T–s diagram of Figure 6.40 and defined by operating conditions of Table 6.20. The cycles differ in the temperature and pressure of the condensation process. What

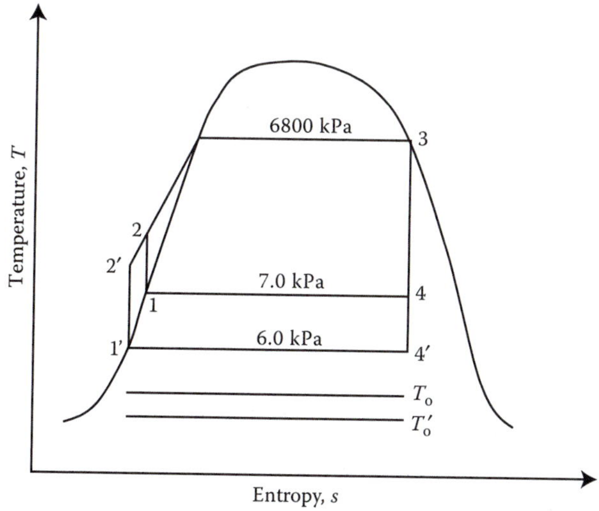

FIGURE 6.40 Temperature–entropy characteristics of two cycles.

TABLE 6.20
Operating Conditions of Cycles of Problem 6.6

Points	Pressure (kPa)	Condition
1	7.0	Saturated liquid
2	6800.0	
3	6800.0	Saturated vapor
4	7.0	
1′	6.0	Saturated liquid
2	6800.0	
3	6800.0	Saturated vapor
4′	6.0	

are the differences in cycle irreversibilities between the two cases for irreversibilities defined as

1. Irreversibility per unit mass flow rate of working fluid, $\dot{I}/\dot{m}_s$, and
2. Irreversibility per unit mass flow rate of working fluid and energy output, that is, $\dot{I}/\dot{Q}_{in}\dot{m}_s$.

Answers:

1. $I \equiv \dot{I}/\dot{m}_s$ for cycle $1234 = 1645.1\,\text{kJ/kg}$
   $I \equiv \dot{I}/\dot{m}_s$ for cycle $1'234' = 1642.0\,\text{kJ/kg}$
2. $I/\dot{Q}_{IN} \equiv \dot{I}/\dot{Q}_{IN}\dot{m}_s$ for cycle $1234 = 0.632$
   $I/\dot{Q}_{IN} \equiv \dot{I}/\dot{Q}_{IN}\dot{m}_s$ for cycle $1'234' = 0.627$

6.7. *Replacement of a steam generator in a PWR with a flash tank* (Sections 6.4 and 6.5)
Consider a "direct" cycle plant with a pressurized water-cooled reactor. This proposed design consists of using most of the typical PWR plant components except the steam generator. In place of the steam generator, a large "flash tank" is incorporated with the capability to take the primary coolant and reduce the pressure to the typical secondary side pressure. The resulting steam is separated, dried, and taken to the balance of plant. The feedwater from the condenser returns to this flash tank. The primary water from the flash tank is repressurized and circulated back to the core.
Make a schematic drawing of this direct cycle plant, and discuss the benefits and/or problems with this design. Also, compare a typical PWR plant design with this direct cycle design with respect to

- Plant thermal efficiencies (perform a numerical comparison and explain your results), and
- Nuclear plant safety (perform a qualitative comparison).

6.8. *Advantages of moisture separation and feedwater heating* (Section 6.5)
A simplified BWR system with moisture separation and an open feedwater heater is described in Problem 6.3. Compute the improvement

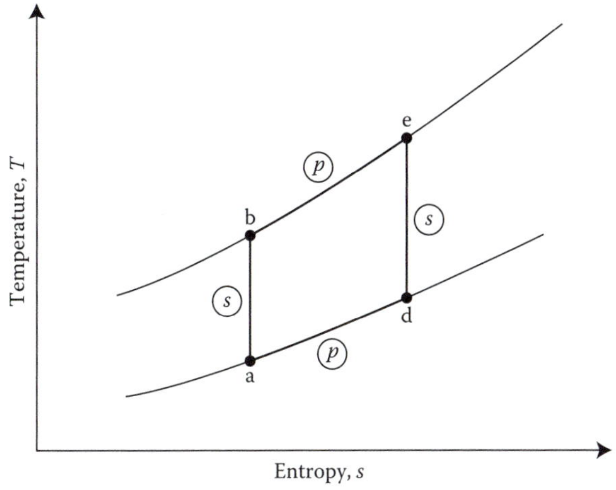

FIGURE 6.41 Ideal Brayton cycle (State a: $p = 137.9$ kPa, $T = 32.2$°C, State c: $p = 689.5$ kPa, $T = 537.8$°C).

in thermal efficiency that results from the inclusion of these two components in the power cycle. Do you think the thermal efficiency improvement from these components is sufficient to justify the capital investment required? Are there other reasons for having moisture separation?

Answer: η_{th} changes from 36.9% to 37.7%

6.9. *Ideal Brayton cycle* (Section 6.6)
The Brayton cycle shown in Figure 6.41 operates using CO_2 as a working fluid with compressor and turbine isentropic efficiencies of 1.0. Calculate the thermal efficiency of this cycle when the working fluid is modeled as

1. A perfect gas of $\gamma = 1.30$.
2. A real fluid (see below for extracted values from Keenan and Kaye's gas tables).
3. A real fluid and the compressor and turbine both have isentropic efficiencies of 0.95.

The parameters needed for a real fluid (from Keenan and Kaye's gas tables) are shown in Table 6.21, where T = temperature, °C; p_r = relative pressure; $\bar{h}$ = enthalpy per mole, kJ/kg-mole. The ratio of the pressures p_a and p_b corresponding to the temperatures T_a and T_b, respectively, for an isentropic process is equal to the ratio of the relative pressures p_{ra} and p_{rb} as tabulated for T_a and T_b, respectively. Thus

$$\left(\frac{p_a}{p_b}\right)_{s=\text{constant}} = \frac{p_{ra}}{p_{rb}}$$

Answers: $\eta_{th} = 31.0\%$, 25.5%, and 21.9%, respectively.

6.10. *Complex real Brayton cycle* (Section 6.7)

TABLE 6.21
Parameters for Problem 6.9

Parameter	a	b	c	d
T	32.2		537.8	
p	137.9	689.5	689.5	137.9
p_r	0.16108	0.8054	31.5	6.3
$\bar{h}$	9643.6	14513.8	33038.5	23439.1

A gas-cooled reactor is designed to heat helium gas to a maximum temperature of 540°C. The helium flows through a gas turbine, generating work to run the compressors and an electric generator, and then through a regenerative heat exchanger and two stages of compression with precooling to 40°C before entering each compressor. Each compressor and the turbine have an isentropic efficiency of 85%, and the exchanger drop factor β is equal to 1.05. Each compression stage has a pressure ratio (r_p) of 1.27. The heat exchanger effectiveness (ξ) is 0.90.

Determine the cycle thermal efficiency. The Brayton cycle system is illustrated in Figure 6.42. The pressure drop factor β is defined as

$$\beta \equiv \left(\frac{p_4}{p_6} \cdot \frac{p_7}{p_1} \cdot \frac{p_2}{p_3} \right)^{\gamma-1/\gamma} = 1.05$$

Answer: $\eta_{th} = 13.8\%$

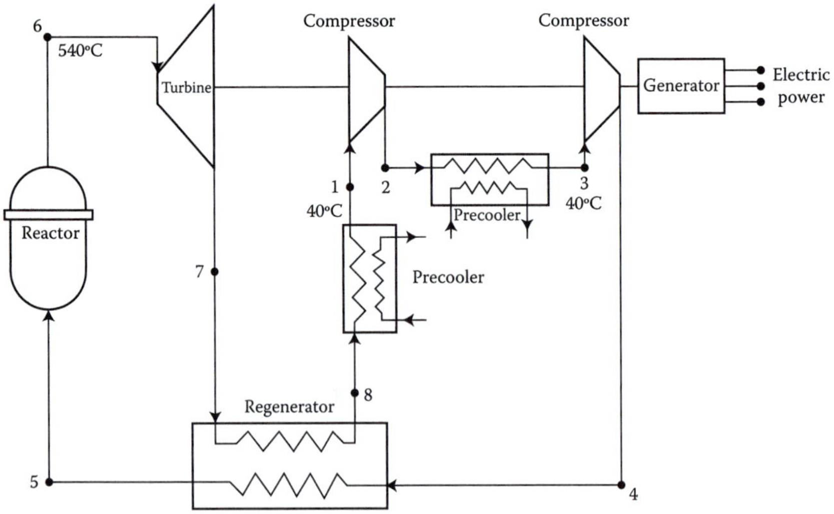

FIGURE 6.42 Complex Brayton cycle.

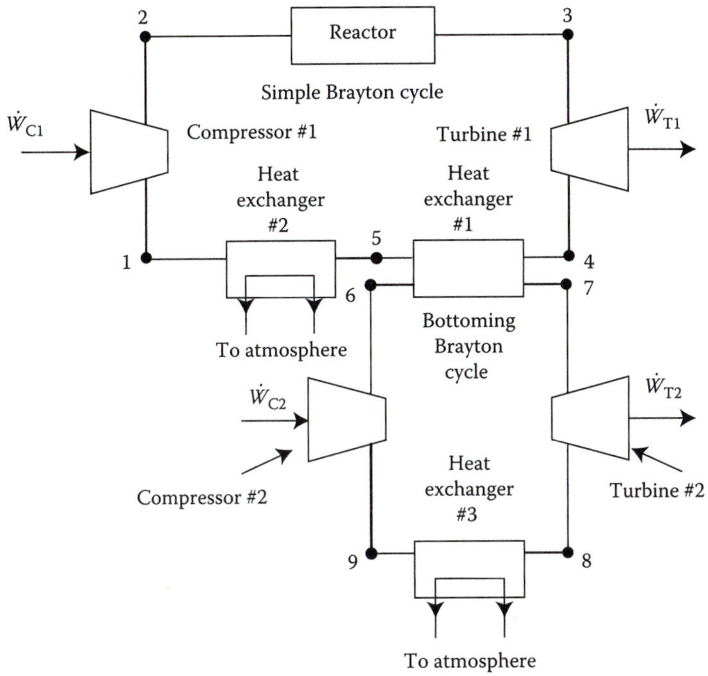

FIGURE 6.43 A simple Brayton cycle with a Brayton bottoming cycle.

6.11. *Cycle thermal efficiency problem involving a bottoming cycle* (Section 6.7)
In Example 6.10A, it is shown that the cycle thermal efficiency of the simple Brayton cycle shown in Figure 6.24 can be increased by utilizing regeneration. Specifically, it was found that, with the addition of a regenerator of effectiveness 0.95, the cycle thermal efficiency was increased from 42.3% to 50%. Another way of improving the efficiency of the simple Brayton cycle is to use a bottoming cycle. To this end, consider the system shown in Figure 6.43. It shows the simple Brayton

TABLE 6.22
Parameters and Assumptions for Problem 6.11

$T_1 = 278$ K	r_p for the simple Brayton cycle = 4.0
$T_3 = 972$ K	c_p for both cycles = 5230 J/kg K
$T_9 = T_1$	γ for both cycles = 1.658
$(\Delta T_p)_1$ = pinch point of heat exchanger #1= 15°C	Mass flow rate for the simple Brayton cycle = twice the mass flow rate for the Brayton bottoming cycle
All turbine and compressors in both cycles are ideal	No duct pressure losses in either cycle

cycle with a Brayton bottoming cycle. Parameters and assumptions for this Problem are given in Table 6.22.

Questions

1. Draw the T–s diagram for the entire system.
2. What must be the pressure ratio of Turbine #2 and Compressor #2 such that the cycle thermal efficiency of the entire system is maximized?
3. What is the maximum cycle thermal efficiency?

Answers:
 2. $r_p = 2.34$
 3. $\eta_{th} = 46.9\%$

6.12 *Nuclear cogeneration plant* (Section 6.8)
A High-Temperature Gas Reactor (HTGR) is being considered for cogeneration of electricity and heat for residential heating. This HTGR uses the direct Brayton cycle shown in Figure 6.44, which comprises a turbine, a regenerator, a cogeneration heat exchanger ($3 \rightarrow 4$), and a compressor. The cogeneration heat exchanger is used to generate steam, which is then sent to the residential area served by the plant. The helium temperature and pressure at the turbine inlet are 1000 K and 9 MPa, respectively. The minimum temperature in the cycle is 373 K. The cycle operates with a compression ratio equal to 2. The isentropic efficiency for the turbo-machines (turbine and compressor) is 0.9. Assume negligible pressure losses throughout the cycle.

1. Sketch the T–s diagram for the cycle.

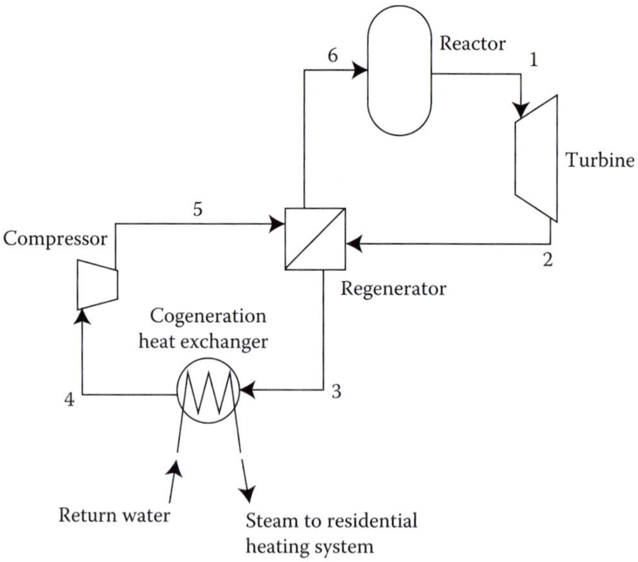

FIGURE 6.44 Schematic of the nuclear cogeneration plant.

2. An important parameter to select is the cogeneration temperature T_3. If T_3 is too high, regeneration is minimal and the cycle thermal efficiency becomes too low. If T_3 is too low, the amount of heat delivered to the residential heating system may be too low. Find the value of T_3 that will give a cycle thermal efficiency equal to 30%.
3. What is the energy utilization factor (EUF) of this cycle? The EUF is defined as the ratio of the energy utilized (net work + cogeneration heat) to the heat input (reactor heat).
4. What is the reactor thermal power if the plant is to produce 100 kg/s of saturated steam at 0.5 MPa from the return water at 80°C?
5. Nuclear cogeneration for residential heating has been rarely done although the Agesta reactor performed this function in Stockholm for a decade starting in 1963. What are in your opinion the drawbacks of this approach?

Useful properties

Helium: Treat as an ideal gas with $c_p = 5193$ J/kg K, $R = 2077$ J/kg K, $\gamma = 1.667$.
Water at 0.5 MPa ($T_{sat} = 152$°C): specific heat = 4.24 kJ/kg K, enthalpy of vaporization = 2109 kJ/kg.

Answers:

2. $T_3 = 572.5$ K
3. EUF = 1
4. $\dot{Q}_R = 344.9\,\text{MW}_{th}$

REFERENCES

1. Dostal, V., Hejzlar, P., and Driscoll, M. J., High performance supercritical carbon dioxide cycle for next generation nuclear reactors. *Nucl. Technol.*, 154:265–282, 2006.
2. Dostal, V., Hejzlar, P., and Driscoll, M. J., The supercritical carbon dioxide power cycle: Comparison to other advanced power cycles. *Nucl. Technol.*, 154:283–301, 2006.
3. Hicks, E. P. and Menzies, D. C., Theoretical studies on the fast reactor maximum accident. In: *Proceedings of the Conference on Safety, Fuels, and Core Design in Large Fast Power Reactors.* ANL-7120, Argonne, IL, 1965, pp. 654–670.

7 Thermodynamics of Nuclear Energy Conversion Systems *Nonsteady Flow First Law Analysis*

7.1 INTRODUCTION

Clear examples of time-varying flow processes relevant to nuclear technology are (1) pressurization of the containment due to postulated rupture of the primary or secondary coolant systems, (2) response of a PWR pressurizer to turbine load changes, and (3) BWR suppression pool heatup by addition of primary coolant. Unlike the steady-flow analysis, variable-flow analyses can be performed with equal ease by either the control mass or the control volume approach. These approaches are demonstrated for the containment example. The pressurizer example is solved using the control volume approach. The suppression pool case is given as Problem 7.5.

7.2 CONTAINMENT PRESSURIZATION PROCESS

The analysis of rapid mixing of a noncondensible gas and a flashing liquid has application in reactor safety; for example, for the light-water reactor, one postulated accident is the release of primary or secondary coolant within the containment. The magnitude of the peak pressure and the time to peak pressure are of interest for structural considerations of the containment.

The fluid released in the containment can be due to the rupture of either the primary or secondary coolant loops. In both cases, the assumed pipe rupture begins the blowdown. The final state of the water–air mixture depends on several other factors: (1) the initial thermodynamic state and mass of water in the reactor and the air in the containment; (2) the rate of release of fluid into the containment and the possible heat sources or sinks involved; (3) the likelihood of exothermic chemical reactions; and (4) the core decay heat. Table 7.1 lists the various external factors in the blowdown process that could affect the peak pressure. Figure 7.1 is a general pictorial representation of the factors involved in the containment response to the loss of primary coolant. We include only heat loss to structure and heat gain from the core in the following discussion. Furthermore, potential and kinetic energy effects are neglected.

TABLE 7.1
Factors to Consider during Analysis of Coolant System Ruptures

Possible Heat Sinks	Possible Heat Sources	Possible Fluid Added from External Sources
	Primary System Rupture	
Containment walls and other cool surfaces	Stored heat	Emergency core cooling water
	Decay heat	
Active containment heat removal systems—air coolers, sprays, and heat exchangers	Other energy sources in core (e.g., Zr–H_2O reactions, H_2 explosion)	Feedwater (BWR)
Steam generator secondary side	Steam generator secondary side	
	Secondary System Rupture	
Containment walls and other cool surfaces	Primary coolant through steam generator	Condensate makeup (PWR)
Active containment heat removal systems—air coolers, sprays, heat exchangers		

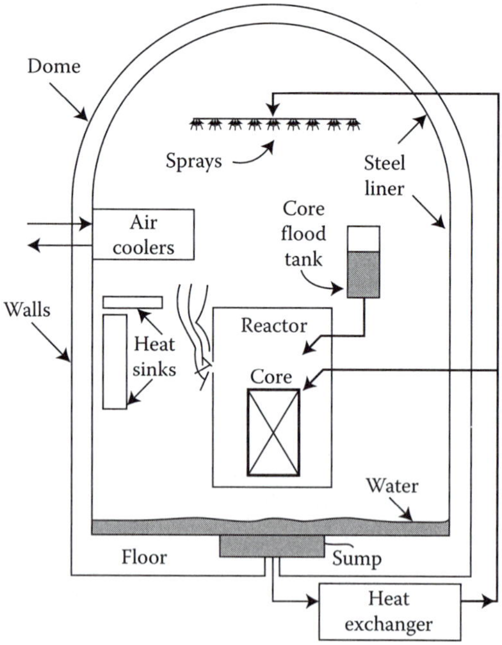

FIGURE 7.1 Containment features in a loss of primary coolant accident.

7.2.1 Analysis of Transient Conditions

We perform the analysis of transient conditions using the control mass and control volume approaches to illustrate that the two techniques are equally applicable to variable-flow processes.

7.2.1.1 Control Mass Approach

Let us define the thermodynamic system of interest as composed of three subsystems: mass of air in containment (m_a), water vapor initially in the air of the containment (m_{wc_1}), and water initially in the primary (or secondary) system depending on rupture assumption (m_{wp}). At any given time, of the mass m_{wp}, the portion m_{wpd} has discharged into the containment and the portion m_{wpr} remains in the primary system. Hence, the total primary system inventory $m_{wp} = m_{wpd} + m_{wpr}$.

For calculation purposes, it can be assumed that each mass in the containment exists at the total containment pressure (p_T) and therefore at a partial volume as seen in Figure 7.2, where the free containment volume (V_c) is given by

$$V_c = V_a + V_{wc_1} + V_{wpd} \tag{7.1}$$

Application of the first law (Equation 6.1) to each of these subsystems (m_a, m_{wc_1}, m_{wp}), neglecting potential and kinetic energy effects, leads to the following:

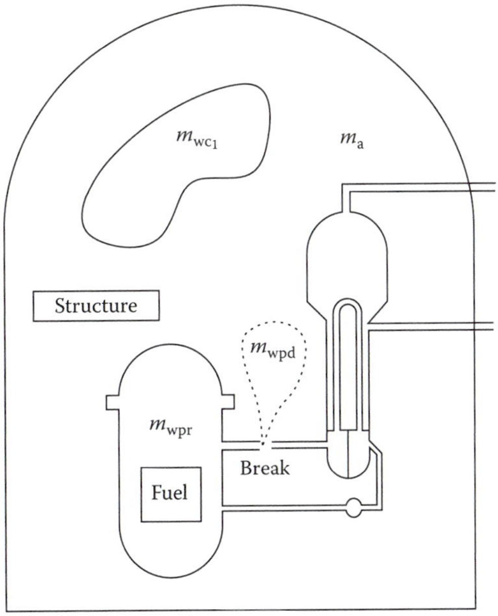

FIGURE 7.2 Control mass for transient analysis of containment conditions.

For m_a:

$$\frac{d(m_a u_a)}{dt} = \dot{Q}_{wc_1-a} + \dot{Q}_{wpd-a} - \dot{Q}_{a-st} - p_T \frac{dV_a}{dt} \quad (7.2a)$$

For m_{wc_1}:

$$\frac{d(m_{wc_1} u_{wc_1})}{dt} = \dot{Q}_{wpd-wc_1} - \dot{Q}_{wc_1-a} - \dot{Q}_{wc_1-st} - p_T \frac{dV_{wc_1}}{dt} \quad (7.2b)$$

For m_{wp}:

$$\frac{d(m_{wpd} u_{wpd} + m_{wpr} u_{wpr})}{dt} = \dot{Q}_{n-wpr} - \dot{Q}_{wpd-wc_1} - \dot{Q}_{wpd-a} - \dot{Q}_{wpd-st} - p_T \frac{dV_{wpd}}{dt} \quad (7.2c)$$

where $\dot{Q}_{i-j}$ = rate of heat transfer from subsystem i to subsystem j. p_T = total containment pressure; V_{wpd} = instantaneous partial volume of the portion of m_{wp} (i.e., m_{wpd}) discharged into the containment; V_{wpr} = volume of the portion of m_{wp} (i.e., m_{wpr}) remaining in the primary system. Subscripts a, wc_1, wpd, wpr, n, and st refer to air, initial containment water, coolant discharged from primary system, coolant remaining in primary system, core fuel, and structures, respectively.

These equations do not explicitly reflect the sign of the time derivative of the volume terms, that is, dV_i/dt. This determination is obtained by applying the volume constraint for the total free volume available to the subsystem, that is, the free containment and the primary system volumes:

$$\frac{d(V_a + V_{wc_1} + V_{wpd} + V_{wpr})}{dt} = 0 \quad (7.3)$$

The time derivative of the volume of the coolant remaining in the primary system is zero, although the relative volumes of liquid and vapor of this subsystem do change with time.

$$\frac{d(V_{wpr})}{dt} = 0 \quad (7.4)$$

On the other hand, the sign of the heat transfer rate terms do explicitly indicate whether heat flow is into or out of the control mass or system. When Equations 7.2a through 7.2c are added together, and the volume constraint of Equation 7.3 is applied, we obtain

$$\frac{d}{dt}(m_a u_a + m_{wc_1} u_{wc_1} + m_{wpd} u_{wpd} + m_{wpr} u_{wpr}) = \dot{Q}_{n-wpr} - \sum_i \dot{Q}_{i-st} \quad (7.5)$$

where $\sum_i$ represents the three subsystems that comprise the containment atmosphere, that is, a, wc_1, and wpd.

Upon integration of Equation 7.5 from the time of break occurrence (1) to a later time (2) during the discharge process, we get

$$U_2 - U_1 = Q_{n-wpr} - \sum_i Q_{i-st} \tag{7.6}$$

where $U_2 = m_a u_{a_2} + (m_{wc_1} + m_{wpd_2})u_{wc_2} + m_{wpr_2} u_{wp_2}$

$$U_1 = m_a u_{a_1} + m_{wc_1} u_{wc_1} + m_{wpr_1} u_{wp_1}$$

because at the time of break occurrence m_{wpr_1}, which is identical to m_{wp}, exists at internal energy u_{wp_1}. If time 1 is taken as an arbitrary later time after the break, U_1 should be expressed as

$$U_1 = m_a u_{a_1} + (m_{wc_1} + m_{wpd_1})u_{wc_1} + m_{wpr_1} u_{wp_1} \tag{7.7}$$

Equation 7.6 is the desired result for transient analysis of containment conditions. Specification of $m_{wpd}(t)$ and $u_{wp}(t)$ are required. The discharged primary mass, $m_{wpd}(t)$, is obtained by integrating the break flow rate, $\dot{m}(t)$, over the interval 1 to 2, that is

$$m_{wpd_2} = m_{wpd_1} + \int_1^2 \dot{m}(t)\,dt \tag{7.8}$$

where the break rate is obtained from a critical flow analysis as presented in Chapter 11.

The primary system internal energy $u_{wp}(t)$ is obtained from the integrated discharge rate, the primary system volume, and an assumption on how to calculate the state of the coolant remaining in the primary system. For example, if the remaining coolant is assumed homogenized and undergoing a reversible, adiabatic expansion, the internal, energy $u_{wp}(t)$ can be obtained from the initial known entropy s_{wp_1} and the calculable time-dependent specific volume $v_{wp}(t)$. Finally, Q_{n-wpr} and $\sum_i Q_{i-st}$ are obtained by transient thermal analyses.

7.2.1.2 Control Volume Approach

Consider now the control volume of Figure 7.3. Upon rupture, the system coolant flows into the control volume at the rate $\dot{m}(t)$. The control volume shape remains constant with time, and there is no shaft work. Therefore, the first law for a control volume (Equation 6.2 of Table 6.1) is written as

$$\dot{U}_{c.v.} = \dot{m}(t)h_p(t) + \dot{Q}_{wpr-c} - \dot{Q}_{c-st} \tag{7.9}$$

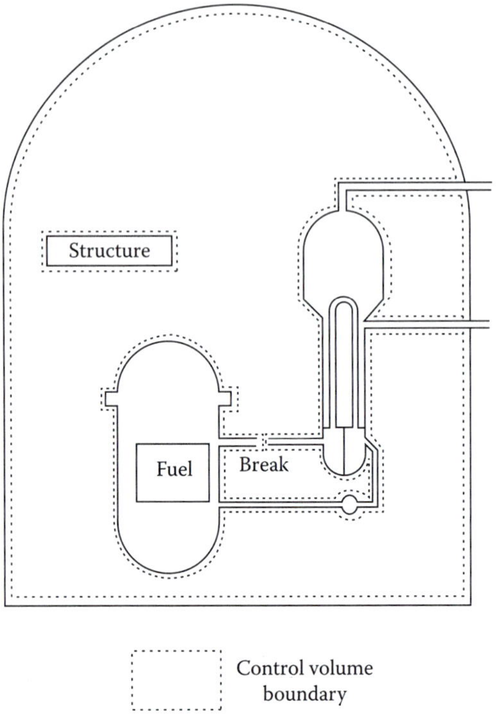

FIGURE 7.3 Control volume for transient analysis of containment conditions.

where the second term on the RHS side represents heat flow into the control volume at the system discharge plane, and the third term on the RHS is heat flow from the control volume into structures.

Integrating between times 1 and 2, Equation 7.9 becomes

$$U_2 - U_1 = +\int_1^2 h_p(t)\dot{m}(t)\,dt + Q_{wpr-c} - Q_{c-st} \tag{7.10}$$

where now for the control volume:

$$U_2 = m_a u_{a_2} + (m_{wc_1} + m_{wpd_2})u_{wc_2}$$
$$U_1 = m_a u_{a_1} + m_{wc_1} u_{wc_1}$$

Equation 7.10 becomes

$$m_a u_{a_2} + (m_{wc_1} + m_{wpd_2})u_{wc_2} = m_a u_{a_1} + m_{wc_1} u_{wc_1} + \int_1^2 h_p(t)\dot{m}(t)\,dt + Q_{wpr-c} - Q_{c-st} \tag{7.11}$$

Equation 7.6 should be identical to Equation 7.11, as both the control mass and the control volume approaches should give identical results. To show this equivalence, we must show that the following terms are equal, that is

$$m_{\mathrm{wpr}_2} u_{\mathrm{wp}_2} - m_{\mathrm{wpr}_1} u_{\mathrm{wp}_1} = -\int_1^2 h_{\mathrm{p}}(t)\dot{m}(t)\,\mathrm{d}t + Q_{\mathrm{n-wpr}} - Q_{\mathrm{wpr-c}} \tag{7.12}$$

as $\sum_i Q_{\mathrm{i-st}}$ is equal to $Q_{\mathrm{c-st}}$ by definition.

This result follows from applying the first law to a new control volume, the primary coolant system volume. This volume remains constant with time.

For this control volume, the first law (Equation 6.2) becomes

$$\dot{U}_{\mathrm{c.v.}} = -h_{\mathrm{p}}(t)\dot{m}(t) + \dot{Q}_{\mathrm{n-wpr}} - \dot{Q}_{\mathrm{wpr-c}} \tag{7.13}$$

where the signs on the RHS of this equation follow from the observations that the discharge flow is out of the control volume, and heat flows into this control volume from the core fuel and out of this control volume into the containment. Integrating over the time interval 1 to 2 yields

$$m_{\mathrm{wpr}_2} u_{\mathrm{wp}_2} - m_{\mathrm{wpr}_1} u_{\mathrm{wp}_1} = -\int_1^2 h_{\mathrm{p}}(t)\dot{m}(t)\,\mathrm{d}t + Q_{\mathrm{n-wpr}} - Q_{\mathrm{wpr-c}}$$

This result is identical to Equation 7.12.

Containment condition histories can be evaluated by Equation 7.11 in the following manner. First obtain the break flow rate, $\dot{m}(t)$, from a critical flow analysis and the heat transferred from the coolant remaining in the vessel to the containment and the heat transferred to containment structures $Q_{\mathrm{wpr-c}}$ and $Q_{\mathrm{c-st}}$, respectively, by transient thermal analyses. Primary system enthalpy is obtained in a manner analogous to that described for primary system internal energy after Equation 7.8.

7.2.2 Analysis of Final Equilibrium Pressure Conditions

Determination of the transient pressure must be based on a variable-flow analysis such as that just described. If we wish to simplify the analysis, we can ask a simpler question and accept a more approximate answer. For example, consider only final conditions upon completion of the blowdown process and establishment of pressure equilibrium between the contents of the containment vessel and the primary system. We retain provision for heat transfer to the containment but only as the total heat transferred rather than the rate of heat transfer.

7.2.2.1 Control Mass Approach

As in Section 7.2.1.1, we define our thermodynamic system of interest as the mass of air and water vapor in the containment and all the water coolant in the primary system, as Figure 7.4 illustrates.

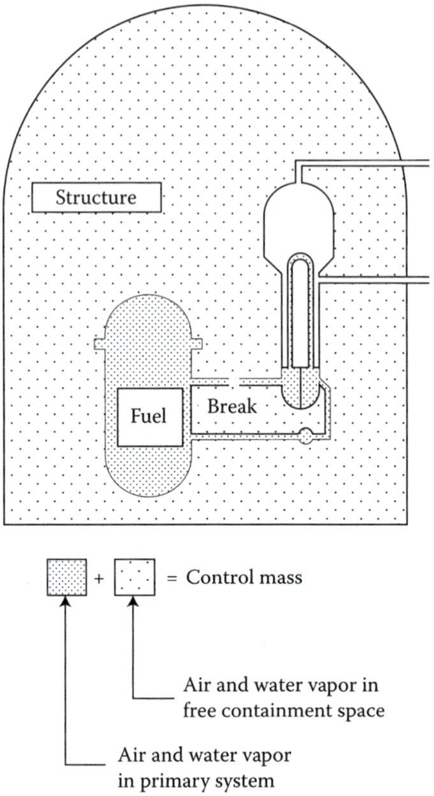

FIGURE 7.4 Control mass for final containment conditions.

Equation 7.6 is still applicable, but now state 2 is the state after completion of blowdown and achievement of pressure equilibrium. Hence u_{wc_2} and u_{wp_2} are identical, and at state 1 all the primary coolant is in the primary system so that Equation 7.6 becomes

$$U_2 - U_1 = Q_{\text{n-wpr}} - \sum_i Q_{\text{i-st}} \tag{7.14a}$$

where $U_2 = m_a u_{a_2} + (m_{wc_1} + m_{wp}) u_{wc_2}$

$$U_1 = m_a u_{a_1} + m_{wc_1} u_{wc_1} + m_{wp} u_{wp_1}$$

7.2.2.2 Control Volume Approach

For the control volume approach, we simply draw the control volume around the containment and primary system, as illustrated in Figure 7.5. For this control volume, there are no entering or exiting flow streams, no shaft work, and no expansion

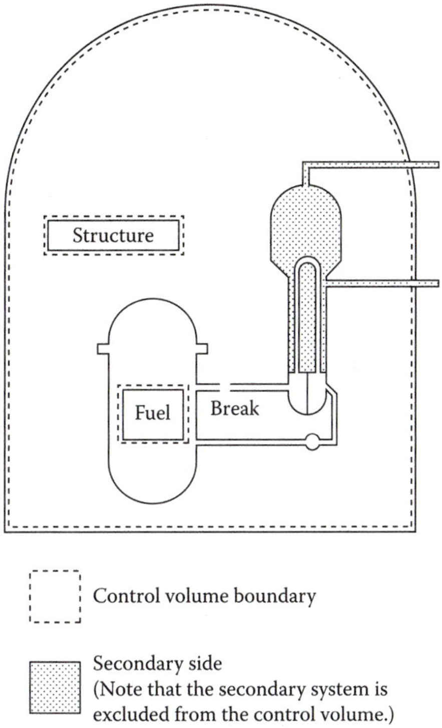

FIGURE 7.5 Control volume for final containment conditions.

work. The first law for control volume (Equation 6.2 of Table 6.1) reduces directly to Equation 7.14a, where the heat loss to the structures is reexpressed as $Q_{\text{c-st}}$, that is

$$U_2 - U_1 = Q_{\text{n-wpr}} - Q_{\text{c-st}} \tag{7.14b}$$

7.2.2.3 Governing Equations for Determination of Final Conditions

Although Equation 7.14b has been identified as the governing equation, by specifying the example more fully we can transform this result and introduce subsidiary relations in working forms suitable for numerical manipulation. Let us assume that the following initial and final conditions are known

1. The initial state (designated "1") for the primary or secondary system fluid: For example, for a PWR, the primary system water at operating conditions is subcooled at a known pressure (p_{w_1}) and temperature (T_{w_1}), with a known mass (m_{w}).
2. The initial mass (m_{a}) and thermodynamic state of the air. The amount of water vapor (relative humidity) initially in the air is known and can be included in the analysis, although its effect is usually small.

3. The final assumed containment condition, which depends on the reason for the analysis. If we are analyzing a preexisting plant, the containment volume is known and the peak pressure is sought. If, on the other hand, we are designing a new plant, we can specify the peak pressure as a design limit and seek the containment volume needed to limit the pressure to this peak value.

We can now elaborate on the governing equations. For simplicity, redesignate the water initially in the containment air as m_{wa}. Also note that the ruptured system can be either the primary or secondary system, so that for these general equations replace the subscript p with sys. Finally, treat the internal energy of the air in terms of its specific heat-temperature product.

Reexpressing the energy balance of Equation 7.14b in this nomenclature yields

$$m_{\text{w}}(u_{\text{w}_2} - u_{\text{w}_1}) + m_{\text{a}}c_{\text{va}}(T_2 - T_{\text{a}_1}) = Q_{\text{n-wsysr}} - Q_{\text{c-st}} \tag{7.15}$$

and

$$m_{\text{w}}u_{\text{w}_1} \equiv m_{\text{wa}}u_{\text{wa}_1} + m_{\text{wsys}}u_{\text{wsys}_1}$$

$$m_{\text{w}}u_{\text{w}_2} \equiv (m_{\text{wa}} + m_{\text{wsys}})u_{\text{w}_2}$$

where m_{w} = mass of water, which is composed of water vapor initially in the air and water or water and steam initially in the failed system, that is, $m_{\text{wa}} + m_{\text{wsys}}$; m_{a} = mass of air in containment; $u = u(T, v)$ = internal energy per unit mass defined with respect to a reference internal energy $u_0(T_0, v_0)$ per unit mass; u_{w_1} = internal energy of the water initially in the containment air and the water initially in the failed system, that is, u_{wa_1} and u_{wsys_1}; c_{va} = specific heat of air at constant volume; T_{a_1} = initial air temperature; and T_2 = final temperature for the air–water mixture in the containment. The initial air temperature and initial water internal energy are known, whereas the final equilibrium state (T_2, u_2) is unknown.

Additional relations are needed that relate the properties of the fluids being mixed—water and air—to the total pressure and volume of the mixture. If the small volume occupied by liquid water is neglected, we can assume that the air occupies the same total volume (V_{T}) the liquid water plus water vapor occupy, that is, V_{T} equals the free containment volume (V_{c}) plus the system, which is either the secondary system (V_{s}) or the primary system (V_{p}) depending on the problem definition. Further assume that the water vapor and liquid exist at the partial pressure of the saturated water vapor. Actually, whereas the water vapor and air are intermingled gases, each exerting its partial pressure, the liquid is agglomerated and at a pressure equal to the total pressure. Then from Dalton's law of partial pressures:

$$p_2 = p_{\text{w}_2}(T_2) + p_{\text{a}_2}(T_2) \tag{7.16}$$

where p_2 = final equilibrium pressure of the mixture; p_{w_2} = partial pressure of the saturated water vapor corresponding to T_2; p_{a_2} = partial pressure of air corresponding to T_2; and from the associated fact that each mixture component occupies the total volume:

$$V_{\mathrm{T}} = m_{\mathrm{w}_2} v_{\mathrm{w}_2} (T_{2,\mathrm{sat}}) \simeq m_{\mathrm{a}} v_{\mathrm{a}} (T_2, p_{\mathrm{a}_2}) \tag{7.17}$$

Introducing the definition of the steam static quality (x_{st}) in the containment and treating air as a perfect gas, Equation 7.17 becomes

$$V_{\mathrm{T}} = m_{\mathrm{w}_2} [v_{\mathrm{f}_2} + x_{\mathrm{st}} v_{\mathrm{fg}_2} (T_{2,\mathrm{sat}})] \simeq \frac{m_{\mathrm{a}} R_{\mathrm{a}} T_2}{p_{\mathrm{a}_2}} \tag{7.18}$$

Equations 7.15, 7.16, and 7.18 are used to find the final equilibrium state.

Establishment of the initial air pressure (p_{a_1}) in the containment should consider the water vapor present. This correction on p_{a_1} is minor but illustrates the use of Dalton's law of partial pressures. The initial conditions are characteristically stated in terms of a relative humidity (ϕ), the dry bulb temperature (T_{a_1}), and the total pressure (p_1). From the definition of relative humidity (ϕ), the saturated water vapor pressure for the given initial condition (p_{wa_1}) is given by

$$p_{\mathrm{wa}_1} = \phi p_{\mathrm{sat}} (T_{\mathrm{a}_1}) \tag{7.19}$$

Figure 7.6 illustrates this relation. Therefore, by Dalton's law of partial pressures

$$p_{\mathrm{a}_1} = p_1 - p_{\mathrm{wa}_1} \tag{7.20}$$

and by the perfect gas law

$$m_{\mathrm{a}} = \frac{p_{\mathrm{a}_1} V_{\mathrm{c}}}{R_{\mathrm{a}} T_{\mathrm{a}_1}} \tag{7.21}$$

Finally, we relate the final partial pressure of air to this initial air pressure by neglecting the difference in volume available to the air at the final versus the initial condition. Hence, again from the gas law

$$p_{\mathrm{a}_2} \simeq p_{\mathrm{a}_1} \frac{T_2}{T_1} \tag{7.22}$$

7.2.2.4 Individual Cases

We can now treat two general cases for the final condition in the containment vessel.

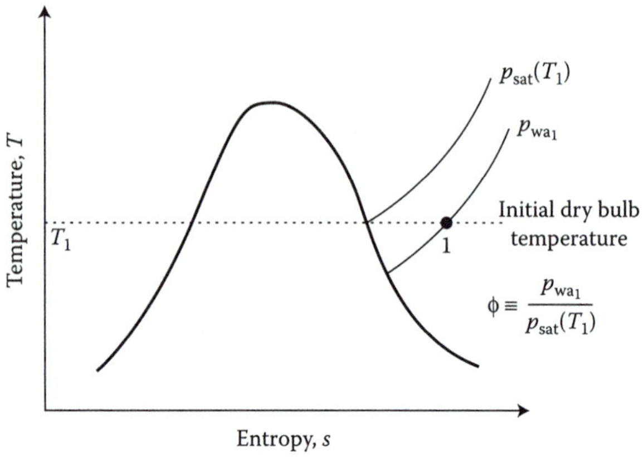

FIGURE 7.6 Initial water vapor pressure from relative humidity.

1. Saturated water mixture in equilibrium with the air. It is the expected result in the containment after a postulated large primary system pipe rupture. Because the heat addition from the core is relatively small, we neglect it here for simplicity. The process representation for this case is shown in Figure 7.7.
2. Superheated steam in equilibrium with the air. This case requires that heat be added to the thermodynamic system. Such a situation could occur upon rupture of a PWR main steam line, as the intact primary system circulates through the steam generator and adds significant heat to the secondary coolant, which is blowing down into the containment. The process representation for this case is shown in Figure 7.8, where the water vapor path is illustrated as one of entropy increase.

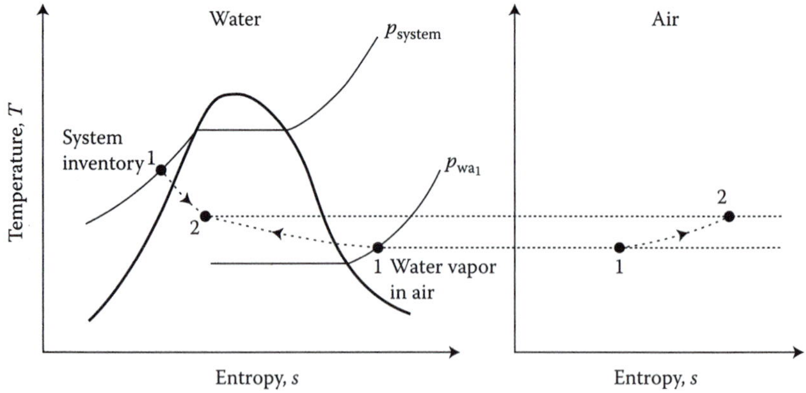

FIGURE 7.7 Process representation: saturated water mixture in equilibrium with air.

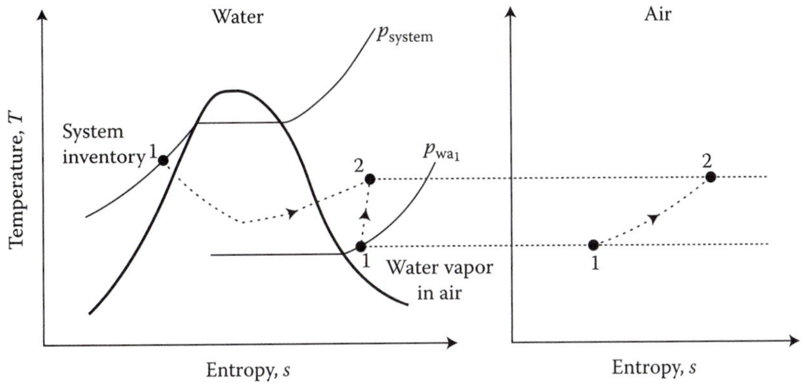

FIGURE 7.8 Process representation: superheated steam in equilibrium with air.

In both cases, heat losses to active heat removal systems and to the structure within containment are neglected. Therefore, the results obtained from this simplified analysis should be conservative, that is, overprediction of peak pressure or containment volume. If heat transfer to structures is included, they need to be input in practice in a manner that takes into account their transient character.

Example 7.1 Containment Pressurization: Saturated Water Mixture in Equilibrium with Air Resulting from a PWR Primary System Rupture

PROBLEM A

Find the peak pressure, given the containment volume.

SOLUTION

Equations 7.15 and 7.18 are the governing equations, and numerical values for containment conditions are drawn from Table 7.2. There are several ways in which the final pressure can be determined. One method is a trial-and-error solution using the steam tables.

The approach is to assume a final temperature (T_2) and from Equation 7.18 calculate the static quality (x_{st}). This quality–temperature pair is checked in Equation 7.15, and the search is continued until these equations are simultaneously satisfied. A quality greater than 1 indicates that the equilibrium water condition is superheated, and this search technique fails. This result is unlikely for realistic reactor containment conditions.

Let us now elaborate this procedure. To start, assume that the final temperature (T_2) is 415 K. The first value needed is the mass of air in the containment (m_a), which can be found by

1. Using Equation 7.19 to find the water vapor partial pressure (p_{w_1}):

$$p_{w_1} = \Phi p_{sat}(T_1) = 0.8(3498 \text{ Pa}) = 2798 \text{ Pa}$$

TABLE 7.2
Conditions for Containment Examples

Fluid	Heat Addition during Blowdown (Joules)	Volume (m³)	Pressure (MPa)	Temperature (K)	Quality (x_{st}) or Relative Humidity (Φ)
Example 7.1: Saturated Water Mixture in Equilibrium with Air as Final State					
Primary coolant water (initial)		$V_p = 354$	15.5	617.9	Assumed saturated liquid
Containment vessel air (initial)		$V_c = 50{,}970$	0.101	300.0	$\Phi = 80\%$
Mixture (final)	$Q = 0$	$V_T = 51{,}324$	0.523	415.6	$x_{st} = 50.5\%$
Example 7.2: Superheated Steam in Equilibrium with Air as Final State					
Secondary coolant water (initial)		$V_s = 89$	6.89	558	Assumed saturated liquid
Containment vessel air (initial)		$V_c = 50{,}970$	0.101	300	$\Phi = 80\%$
Mixture (final)	$Q = 10^{11}$	$V_T = 51{,}059$	0.446 (64.7 psia)	478	$\Phi = 17\%$

2. Using Equation 7.20 to find the air partial pressure (p_{a_1}):

$$p_{a_1} = p_1 - p_{w_1} = 101{,}378 - 2798 = 98{,}580 \text{ Pa}$$

3. Using Equation 7.21 to find the air mass (m_a):

$$m_a = \frac{p_{a_1} V_c}{R_a T_{a_1}} = \frac{(98{,}580)\text{Pa}\,(50{,}970)\text{m}^3}{(286)\text{ J/kg °K}\,(300)\text{°K}} = 5.9(10^4)\text{kg}$$

Using Equation 7.17, we find the mass of water initially in the containment:

$$m_{wa} = \frac{V_c}{v_{wa_1}} = \frac{50{,}970\text{m}^3}{50.02\text{m}^3/\text{kg}} = 1019\text{kg}$$

because v_{wa_1} is the specific volume of superheated water vapor at p_{w_1} and T_{a_1}, which numerically equals 50.02 m³/kg.

Now we can find the quality (x_{st}) at state 2 from Equation 7.18 as

$$x_{st} = \frac{(V_T/m_w) - v_{f_2}}{v_{fg2}} = \frac{(51{,}324)\text{m}^3/2.11(10^5)\text{kg} - 0.00108\text{m}^3/\text{kg}}{0.485\text{m}^3/\text{kg}} = 0.499$$

where by the steam tables:

v_{f_2} = saturated liquid specific volume at 415 K = 0.00108 m³/kg
v_{fg2} = volume of vaporization specific volume at 415 K = 0.485 m³/kg
v_{wp} = saturated liquid specific volume at primary system pressure (15.5 MPa) = 0.00168 m³/kg

$$m_{wp} = \frac{V_p}{v_{wp}} = \frac{354\text{ m}^3}{1.68(10^{-3})\text{ m}^3/\text{kg}} = 2.1(10^5)\text{ kg}$$

$$m_w = m_{wp} + m_{wa} = 2.1(10^5) + 0.01(10^5) = 2.11(10^5)\text{ kg}$$

and where the failed system is the primary so that subscript sys is written as p.

The quality (x_{st}) is checked by using Equation 7.15 rewritten to express the water conditions separately as primary water and water in air.

$$m_{wp}(u_{f_2} + x_{st}u_{fg_2} - u_{wp_1}) + m_{wa}(u_{f_2} + x_{st}u_{fg_2} - u_{wa_1}) + m_a c_{va}(T_2 - T_{a_1}) = 0$$

$$2.1(10^5)[595{,}380 + x_{st}1.95(10^6) - 1.6(10^6)] + 1.02(10^3)$$
$$\times [595{,}380 + x_{st}1.95(10^6) - 2.41(10^6)] + 5.9(10^4)(719)[415 - 300] = 0$$

Solving for x_{st}, we get

$$x_{st} = 0.505$$

where

u_{wp_1} = saturated liquid u at 15.5 MPa = 1.6(10⁶) J/kg
u_{f_2} = saturated liquid u at 415 K = 595,380 J/kg
u_{fg2} = heat of vaporization u at 415 K = 1.95(10⁶) J/kg
u_{wa_1} = superheated vapor u at 300 K, 2798 Pa = 2.41(10⁶) J/kg

Continuing the iteration on x_{st}, the final result is

$$T_2 = 415.6\text{K}, \quad p_{w_2} = 0.386\text{MPa}$$

Now the final air partial pressure (p_{a_2}) can be calculated using Equation 7.18 or 7.22, yielding, respectively

$$p_{a_2} = \frac{m_a R_a T_a}{V_T} = \frac{5.9(10)^4\text{ kg}(286\text{ J/kgK})(415\text{K})}{51{,}324\text{m}^3} = 1.37(10^5)\text{Pa}$$

$$p_{a_2} = p_{a_1}\left(\frac{T_{a2}}{T_{a_1}}\right) = 0.099\frac{(415.6)}{300} = 0.137\text{MPa}$$

This gives us the total pressure from Equation 7.16 as

$$p_2 = p_{w_1} + p_{a_2} = 0.386 + 0.137 = 0.523\,\text{MPa}$$

PROBLEM B

Find the containment volume, given a design limit for the peak pressure (p_2).

SOLUTION

For this problem we would know the initial air pressure (p_{a_1}), but the mass of air is dependent on the containment volume (Equation 7.21). Therefore, we can use p_{a_1} from Equation 7.22 directly in Equation 7.16 as

$$p_2 = p_{w_2}(T_2) + p_{a_1}\frac{T_2}{T_1} \tag{7.23}$$

Now we can simply iterate on Equation 7.23 by assuming T_2 and finding $p_{w2} = p_{sat}(T_2)$ and then comparing the RHS to p_2. Once T_2 is found, water–steam properties are available, that is, $v_2, v_{fg2}, u_{f_2}, u_{fg2}$. These values can be substituted into Equations 7.15 and 7.18, which can be solved simultaneously for the two unknowns x_{st} and V_c (when V_T is expressed as V_c plus V_p).

Proceeding numerically for the design condition of $p_2 = 0.523$ MPa, we find that Equation 7.23 is satisfied when T_2 equals 415.6 K, that is

$$p_2 = p_{w_2}(T_2) + p_{a_1}\left(\frac{T_2}{T_1}\right) = 0.386 + 0.099\left(\frac{415.6}{300}\right) = 0.523\,\text{MPa}$$

Then Equations 7.15 and 7.18 in the unknowns x_{st} and V_c become, respectively

$$\left(m_{wp} + \frac{V_c}{v_{wa_1}}\right)(u_{f_2} + x_{st}u_{fg2}) - m_{wp}u_{wp_1} - \frac{V_c}{v_{wa_1}}u_{wa_1} + p_{a_1}\frac{V_c}{R_a T_{a_1}}c_{va}(T_2 - T_{a_1}) = 0 \tag{7.24}$$

and

$$V_c + V_p = \left(m_{w_p} + \frac{V_c}{v_{wa_1}}\right)(v_{f_2} + x_{st}v_{fg2}) \tag{7.25}$$

where v_{wa_1} has been obtained from superheated steam tables.

Substituting numerical values into these two relations yields

$$\left[2.1(10^5) + \frac{V_c}{50}\right][595{,}380 + x_{st}1.95(10^6)] - 2.1(10^5)[1.6(10^6)] - \frac{V_c}{50}[2.41(10^6)] + \frac{0.099(10^6)V_c}{286(300)}(719)(415 - 300) = 0 \tag{7.26}$$

$$V_c + 354 = \left[2.1(10^5) + \frac{V_c}{50}\right][0.00108 + x_{st}0.485] \quad (7.27)$$

Upon simultaneous solution, x_{st} and V_c are found as

$$x_{st} = 0.505, \quad V_c = 51{,}593\,\text{m}^3$$

Example 7.2 Containment Pressurization: Superheated Steam in Equilibrium with Air Resulting from a PWR Secondary System Rupture

A rupture of a main steam line adds water to the containment while the intact primary system circulates water through the steam generator, transferring energy to the secondary water that is discharging into the containment. The amount of water added is primarily dependent on the size of the steam generator. For this example, typical PWR four-loop plant values have been used for the amount of secondary water added and energy transferred via the steam generator. These assumed values are listed in Table 7.2.

PROBLEM A

Find the peak pressure, given the containment volume.

If the steam is superheated we have a situation where the relative humidity has increased, as has the equilibrium temperature (T_2). Equations 7.16 and 7.17 are not linked by the steam table-specific volumes but are equivalently given by

$$p_2 = \frac{m_w R_w T_2}{V_T} + \frac{m_a R_a T_2}{V_T} \quad (7.28)$$

where $m_w = m_{wa} + m_{ws}$ (i.e., water in containment air and the failed secondary system, respectively). Equation 7.28 treats the superheated steam as a perfect gas that deviates from reality by only a few percent.

There are now three unknowns (T_2, p_2, u_{w_2}) and three equations (Equations 7.15, 7.28, and superheat steam tables). Assume T_2, use Equation 7.28 to calculate p_2, and find u_{w_2} by using the steam tables. Now use Equation 7.15 to find u_{w_2} and compare with the previous value and iterate until convergence. As before, the initial water internal energy (u_{w_1}) is composed of two parts: (1) the system's liquid water, and (2) water vapor initially in the air. A condition may exist such that the final equilibrium state is above the critical temperature of the water. However, it does not affect the analysis, and the same procedure is used.

First we assume a final temperature, $T_2 = 450$ K, and substitute into Equation 7.28.

$$\begin{aligned} p_2 &= \frac{m_w R_w T_2}{V_T} + \frac{m_a R_a T_2}{V_T} \\ &= \frac{(67{,}223\,\text{kg})(462\,\text{J/kg K})(450\,\text{K})}{(51{,}059\,\text{m}^3)} + \frac{[5.9(10^4)\,\text{kg}](286\,\text{J/kg K})(450\,\text{K})}{(51{,}059\,\text{m}^3)} \\ &= 0.274 + 0.149 = 0.42\,\text{MPa} \end{aligned}$$

where m_a and m_{wa} are found in a similar fashion as before using an equation of the form of Equation 7.21 for each component.

From the steam tables for $p_{w_2} = 0.274$ MPa and $T_2 = 450$ K, we find $u_{w_2} = 2.61(10^6)$ J/kg. Using Equation 7.15 with the initial conditions given in Table 7.2, we can solve for u_{w_2} and compare with the steam table value. Expressing Equation 7.15 for this case and substituting numerical values, we have

$$m_{ws}(u_{w_2} - u_{ws_1}) + m_{wa}(u_{w_2} - u_{wa_1}) + m_a c_{va}(T_2 - T_{a_1}) = Q \tag{7.29}$$

$$(66,204 \text{ kg})[u_{w_2} - 1.25(10^6) \text{ J/kg}] + 1019.0 \text{ kg } [u_{w_2} - 2.41(10^6) \text{ J/kg}] \\ + 5.9(10^4) \text{ kg } [719(450 - 300)] = 10^{11} \text{ J}$$

or

$$u_{w_2} = 2.66(10^6) \text{ J/kg}$$

where

u_{ws_1} = sat. liquid u at 6.89 MPa = 1.25(10⁶) J/kg
u_{ws_1} = superheated vapor u at 2798 Pa
= 2.41(10⁶) J/kg ($T_1 = 300$ K)

We can iterate on T_2 using the calculated u_{w_2} to adjust the guess on T_2; the final result is

$$T_2 = 478 \text{K}, \quad p_2 = 0.4463 \text{MPa}$$

The final relative humidity is 17%, that is

$$\Phi_2 = \frac{p_{w2}}{p_{sat}(T_2)} = \frac{0.291 \text{ MPa}}{1.725 \text{ MPa}} = 0.17$$

where

$$p_{w2} = \frac{(67,233 \text{ kg})(462 \text{ J/kg K})(478 \text{ K})}{51,059 \text{ m}^3} = 0.291 \text{MPa}$$

PROBLEM B

Find the containment volume, given the peak pressure.

The same procedure can be used here, except that Equation 7.28 can be solved in terms of V_c.

$$V_T = \frac{m_a R_a T_2}{p_2} + \frac{m_w R_w T_2}{p_2} \tag{7.30}$$

where $V_T = V_c + V_s$ and $m_w = m_{wa} + m_{ws}$.

The initial air mass (m_a) is normally not known if the initial volume is unknown (V_c). Therefore, we can substitute for m_a and m_{wa} by Equation 7.21 written for each component, yielding

$$V_c + V_s = \left[\left(\frac{p_{a1}}{p_2}\right)\left(\frac{T_2}{T_{a_1}}\right) + \left(\frac{p_{wa}}{p_2}\right)\left(\frac{T_2}{T_{a_1}}\right)\right]V_c + \frac{m_{ws}R_w T_2}{p_2} \tag{7.31}$$

Now we can assume T_2 and solve for V_c and thus m_a and m_{wa}. These values can be substituted into Equation 7.15 to solve for u_{w_2} and compare it to $u_{w2}(T_2, p_{w2})$ from the steam tables. Now, we simply iterate until u_{w_2} from Equation 7.15 matches the steam table value.

We can proceed in this example for the assumed design condition $p_2 = 0.4463$. Equation 7.31 can be used to find V_c, assuming $T_2 = 478$ K.

$$V_c + 89\text{m}^3 = \left(\frac{0.0986}{0.4463}\frac{478}{300} + \frac{0.0027}{0.4463}\frac{478}{300}\right)V_c + \frac{(66,204\,\text{kg})(462\,\text{J/kgK})478\text{K}}{0.4463 \times 10^6\,\text{Pa}}$$

giving $V_c = 51,093$ m^3.

Using V_c, we can find u_{w_2} for $p_{w_2} = 0.291$ MPa and $T_2 = 478$ K from the steam tables:

$$u_{w_2} = 2.66(10^6)\,\text{J/kg}$$

Using Equation 7.15 and solving for u_{w_2}, we get

$$m_{ws}(u_{w_2} - u_{ws_1}) + m_{wa}(u_{w_2} - u_{wa_1}) + m_a c_{va}(T_2 - T_{a_1}) = Q$$

where

$$m_{wa} = \frac{V_c}{v_{w_1}} = \frac{51,093\text{m}^3}{50\text{m}^3/\text{kg}} = 1022\text{kg}$$

$$m_a = \frac{p_{a_1}V_c}{R_a T_{a_1}} = \frac{[0.099(10^6)\text{Pa}][51,093\text{m}^3]}{(286\,\text{J/kgK})(300\text{K})} = 5.9(10^4)\text{kg}$$

and the other properties are known from the steam tables as before.

$$(66,204\text{kg})[u_{w_2} - 1.25(10^6)] + 1022[u_{w_2} - 2.41(10^6)] + 5.9(10^4)(719)(478 - 300) = 10^{11}\,\text{J}.$$

$$u_{w_2} = 2.643(10^6)\,\text{J/kg}$$

Thus

$$T_2 = 478\text{K}, \quad V_c = 51,093\text{m}^3$$

7.3 RESPONSE OF A PWR PRESSURIZER TO LOAD CHANGES

The pressurizer vessel employed to control system pressure in a PWR provides another example of the analysis of a transient process by the control volume approach. The pressurizer vessel has an upper steam region and a lower water region. Water surges in out of the lower region as a result of temperature changes in the system connected to the pressurizer. The lower pressurizer region is typically connected to a hot leg.

Upon an insurge, a water spray, which is typically taken from a cold leg, is actuated in the upper steam region to condense steam. Heaters are actuated to restore the initial saturated condition. In an outsurge, electric heaters in the lower water region are actuated to maintain the system pressure, which is otherwise decreased owing to the departure of liquid and subsequent expansion of the vapor volume. Figure 7.9 summarizes the response of the pressurizer to insurge and outsurge.

7.3.1 Equilibrium Single-Region Formulation

We start the analysis of pressurizer behavior with a simple formulation in which the entire pressurizer is represented as a single region at equilibrium conditions. The energy quantities being added to or departing the pressurizer owing to inlet spray at the top from the cold leg; insurge and outsurge through the bottom to the hot leg and

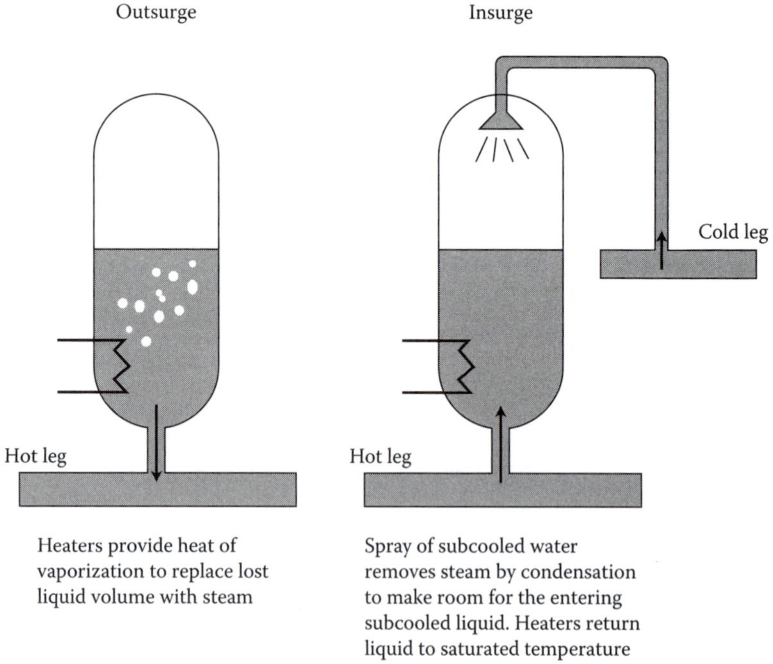

FIGURE 7.9 Pressurizer operation (without relief valves).

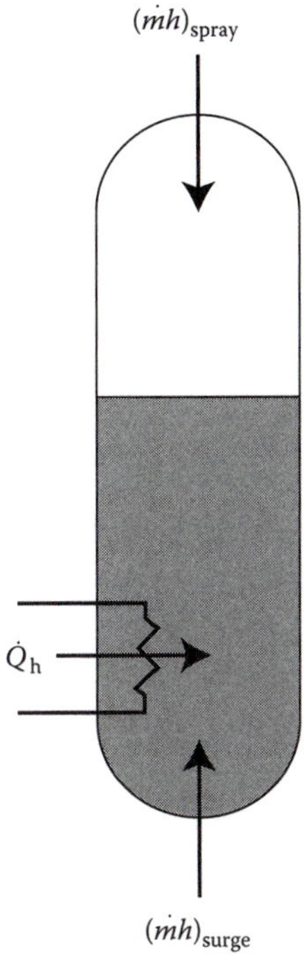

FIGURE 7.10 Externally supplied mass flow rate, enthalpy, and heat to the one-region pressurizer.

heater inputs are illustrated in Figure 7.10. The pressurizer wall is assumed to be perfectly insulated with negligible heat capacity.

The general transient mass and energy equations are as follows:

$$\frac{d}{dt}m = \dot{m}_{surge} + \dot{m}_{spray} \tag{7.32}$$

$$\frac{d}{dt}(mu) = \dot{m}_{surge}h_{surge} + \dot{m}_{spray}h_{spary} + \dot{Q}_h - p\frac{d}{dt}(mv) \tag{7.33}$$

where

$$m = m_v + m_\ell \tag{7.34}$$

$$mu = m_v u_v + m_\ell u_\ell \tag{7.35}$$

$$mv = m_v v_v + m_\ell v_\ell \tag{7.36}$$

Additionally, a constraint exists on the total volume, which is fixed, that is

$$\frac{\mathrm{d}}{\mathrm{d}t}(mv) = \frac{\mathrm{d}}{\mathrm{d}t}(m_\mathrm{v}v_\mathrm{v} + m_\ell v_\ell) = 0 \tag{7.37}$$

There are five prescribed input parameters: $\dot{m}_\mathrm{spray}$, h_spray, $\dot{m}_\mathrm{surge}$, h_surge, and $\dot{Q}_\mathrm{h}$.

There are seven unknowns, that is, p, m_v, u_v, v_v, m_ℓ, u_ℓ, and v_ℓ. These unknowns are so far related by only three equations (Equations 7.32, 7.33, and 7.37). Closure of the problem is obtained by use of four equations of state, reflecting the assumption that all vapor and liquid conditions are maintained at saturation:

$$u_\mathrm{v} \equiv u_\mathrm{g} = f(p) \tag{7.38}$$

$$u_\ell \equiv u_\mathrm{f} = f(p) \tag{7.39}$$

$$v_\mathrm{v} \equiv v_\mathrm{g} = f(p) \tag{7.40}$$

$$v_\ell \equiv v_\mathrm{f} = f(p) \tag{7.41}$$

This set of equations and prescribed inputs is sufficient to perform a transient analysis to determine the unknowns for this simple one-region equilibrium formulation.

7.3.2 Analysis of Final Equilibrium Pressure Conditions

If we wish to ask simpler questions than the transient nature of pressurizer behavior, the analysis can be simplified to consider only the end states of the transient process. For example, let us determine the size of the pressurizer necessary to accommodate specified insurge and outsurge events based on accommodating the end states only. In this idealized illustration, the initial and final pressurizer states are prescribed as saturated states at a fixed pressure.

In this case, the unknowns become the liquid and vapor masses and the heater power, and the equations are those for continuity, energy conservation, and volume constraint. The inputs are spray mass and enthalpy, surge mass and enthalpy, and pressure with the associated saturation internal energy and specific volume properties.

The heaters must always be entirely liquid covered, which for a fixed pressurizer geometry prescribes the minimum required liquid volume. The insurge is accommodated by providing sufficient initial vapor volume such that the final state is a liquid-filled pressurizer. In this case the spray completely condenses the initial vapor, leading to a liquid-filled pressurizer, and sufficient heater input is provided to restore the initial pressure. In practice, the limiting condition would be the relief valve pressure set point.

The outsurge is to be accommodated by providing sufficient initial liquid volume that in the final state the heaters remain submerged after having operated to provide

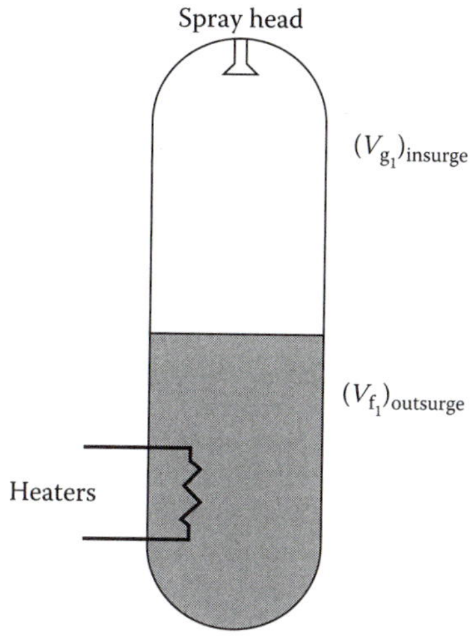

FIGURE 7.11 Initial state of pressurizer awaiting insurge or outsurge.

enough energy to restore the initial pressure. In practice, the limiting condition for the outsurge could be the minimum-allowed primary system pressure.

Figure 7.11 illustrates the initial state of the pressurizer, and Figure 7.12 shows the end states for the insurge and outsurge events. The required total pressurizer volume is the sum of the initial vapor volume to accommodate the insurge, $(V_{g_1})_{insurge}$, and the initial liquid volume to keep the heaters submerged in an outsurge, $(V_{f_1})_{outsurge}$. The sum is required for the pressurizer to operate with sufficient liquid level and vapor space to accommodate either the insurge or the outsurge event.

$$V_T = (V_{g_1})_{insurge} + (V_{f_1})_{outsurge} \tag{7.42}$$

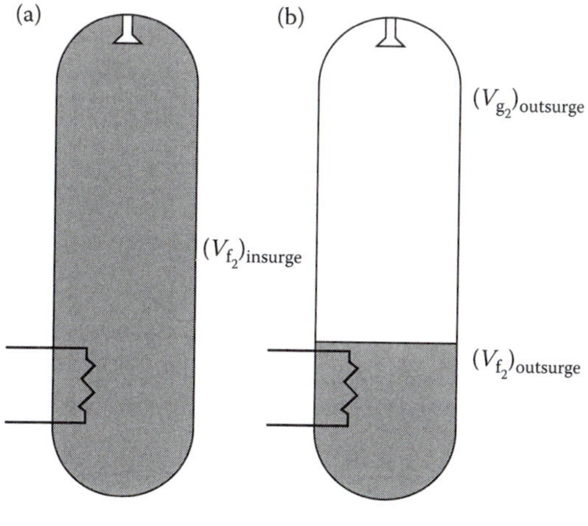

FIGURE 7.12 Final states of pressurizer. (a) Insurge and (b) outsurge.

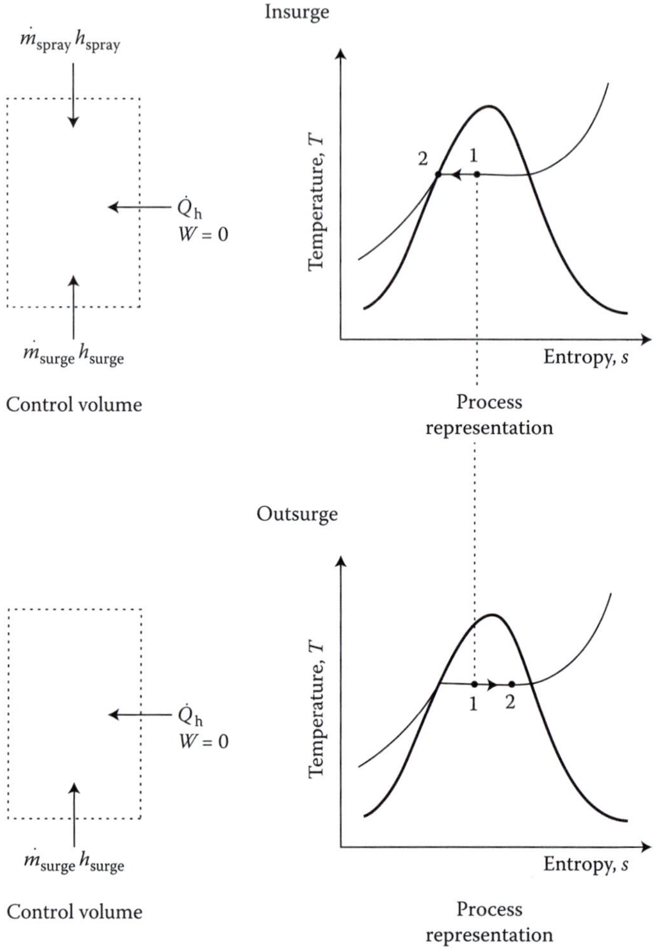

FIGURE 7.13 Thermodynamic representation of an insurge and an outsurge.

This example is handled most easily by taking the interior of the entire pressurizer as the control volume. The control volume and process representations for the insurge and outsurge are shown in Figure 7.13.

The continuity, energy, and volume constraint equations for the control volume are Equations 7.32, 7.33, and 7.37. Integrating these equations between the initial (1) and the final (2) state, assuming the flow rates and surge and spray enthalpies are constant with time, we obtain

$$m_2 - m_1 = m_{surge} + m_{spray} \tag{7.43}$$

$$m_2u_2 - m_1u_1 = m_{surge}h_{surge} + m_{spray}h_{spray} + Q_h \tag{7.44}$$

$$m_2v_2 = m_1v_1 \tag{7.45}$$

At this point, the insurge and outsurge cases must be treated separately to specialize the equations for each case. For the insurge, the final state is taken to be a liquid-filled pressurizer, which, after the water has been completely mixed, is saturated at the initial pressure. For this case, we can express the governing equations in the following form:

$$m_{f_2} = m_{surge} + m_{spray} + m_{f_1} + m_{g_1} \tag{7.46}$$

or

$$m_{f_2} = m_{surge}(1+f) + m_{f_1} + m_{g_1} \tag{7.47}$$

where the initial mass m_1 is split up into its liquid (m_{f_1}) and steam (m_{g_1}) components, and f is defined as the ratio of spray to surge flow:

$$fm_{surge} = m_{spray} \tag{7.48}$$

Note that, in practice, it is not necessary that $\dot{m}_{surge}$ and $\dot{m}_{spray}$ are related by a time-invariant proportionality constant; that is, f in Equation 7.48 can be a function of time. The energy balance likewise becomes

$$m_{f_2}u_{f_2} = m_{surge}(h_{surge} + fh_{spary}) + m_{g_1}u_{g_1} + m_{f_1}u_{f_1} + Q_h \tag{7.49}$$

with the volume constraint being

$$m_{f_2}v_{f_2} = m_{g_1}v_{g_1} + m_{f_1}v_{f_1} \tag{7.50}$$

Because states 1 and 2 are saturated at the same pressure, the properties appear in subsequent equations without state subscripts. The volume constraint (Equation 7.50) can be substituted into the mass and energy balance (Equations 7.47 and 7.49) to eliminate the final mass (m_{f_2}); the resulting equations are

$$m_{g_1} = \frac{m_{surge}(1+f)v_f}{v_g - v_f} \tag{7.51}$$

and

$$m_{g_1} = \frac{v_f[m_{surge}(h_{surge} + fh_{spray}) + Q_h]}{v_g u_f - v_f u_g} \tag{7.52}$$

Note that the unknown (m_{f_1}), the initial liquid mass in the pressurizer, does not appear in Equations 7.51 or 7.52. This finding is reasonable, as this liquid does not participate in accommodating the insurge. Combining these equations yields the desired solution for Q_h as

$$(Q_h)_{\text{insurge}} = \frac{m_{\text{surge}}(1+f)[v_g u_f - v_f u_g]}{v_g - v_f} - m_{\text{surge}}(h_{\text{surge}} + f h_{\text{spray}}) \tag{7.53}$$

The steam volume, $(V_{g_1})_{\text{insurge}}$, is obtained as

$$(V_{g_1})_{\text{insurge}} = m_{g_1} v_g \tag{7.54}$$

To compute the desired total volume, we must now consider the outsurge case and compute $(V_{f_1})_{\text{outsurge}}$. This task is accomplished by specialization of Equations 7.43 through 7.45 to the outsurge process. In an outsurge, the final state has a liquid level sufficient to cover the pressurizer heaters. The mass balance (Equation 7.43) becomes

$$m_{f_2} + m_{g_2} - m_{f_1} - m_{g_1} = -m_{\text{surge}} \tag{7.55}$$

The energy balance equation becomes

$$m_{f_2} u_{f_2} + m_{g_2} u_{g_2} - m_{f_1} u_{f_1} - m_{g_1} u_{g_1} = -m_{\text{surge}} h_{\text{surge}} + Q_h \tag{7.56}$$

The volume constraint is

$$m_{f_2} v_{f_2} + m_{g_2} v_{g_2} = m_{f_1} v_{f_1} - m_{g_1} v_{g_1} \tag{7.57}$$

The volume constraint (Equation 7.57) can be substituted into the mass and energy balances (Equations 7.55 and 7.56) to eliminate the final vapor mass (m_{g_1}); the resulting equations are

$$m_{f_1} = m_{f_2} + m_{\text{surge}} \frac{v_g}{v_g - v_f} \tag{7.58}$$

and

$$m_{f_1} = m_{f_2} + \frac{Q_h - m_{\text{surge}} h_{\text{surge}}}{v_f / v_g (u_g - u_f)} \tag{7.59}$$

Note that the unknown (m_{g_1}), the initial vapor mass in the pressurizer, does not appear in Equation 7.58 or 7.59. This omission is reasonable, as this vapor does not participate in accommodating the outsurge.

Combining these equations yields the desired solution for Q_h as

$$(Q_h)_{outsurge} = m_{surge}h_{surge} - \left(u_f - \frac{v_f}{v_g}u_g\right)\left(m_{surge}\frac{v_g}{v_g - v_f}\right) \quad (7.60)$$

The initial liquid mass (m_{f_1}) can now be obtained from either Equation 7.58 or Equation 7.59.

Hence

$$(V_{f_1})_{outsurge} = m_{f_1}v_f \quad (7.61)$$

The total volume (V_T) can now be obtained by utilizing the results of Equations 7.54 and 7.61 in Equation 7.42, that is

$$V_T = (V_{g_1})_{insurge} + (V_{f_1})_{outsurge} \quad (7.42)$$

Example 7.3: Pressurizer Sizing Example

PROBLEM

Determine the size of the pressurizer that can accommodate a maximum outsurge of 14,000 kg and a hot leg insurge of 9500 kg for the conditions of Table 7.3.

SOLUTION

The value of $(Q_h)_{insurge}$ is obtained from Equation 7.53 as

$$\begin{aligned}(Q_h)_{insurge} &= \frac{9500(1+0.03)[(9.81\times10^{-3})(1.60\times10^{6}) - (1.68\times10^{-3})(2.44\times10^{6})]}{(9.81-1.68)\times10^{-3}} \\ &\quad - 9500[1.43\times10^{6} + 0.03(1.27\times10^{6})] \\ &= 1.06\times10^{7}\ \mathrm{J}\end{aligned}$$

The value of m_{g_1} is obtained from Equation 7.51 as

$$m_{g_1} = \frac{9500(1+0.03)1.68\times10^{-3}}{(9.81-1.68)\times10^{-3}} = 2022\ \mathrm{kg}$$

The steam volume needed for the insurge from Equation 7.54 is

$$(V_{g_1})_{insurge} = (2.022\times10^{3})(9.81\times10^{-3}) = 19.84\ \mathrm{m^3}$$

TABLE 7.3
Conditions for Pressurizer Design Problem

Saturation pressure	15.5 MPa	2250 psia
Saturation temperature	618.3 K	652.9°F
Saturation properties		
u_f	1.60×10^6 J/kg	689.9 B/lb
u_g	2.44×10^6 J/kg	1050.6 B/lb
v_f	1.68×10^{-3} m³/kg	0.02698 ft³/lb
v_g	9.81×10^{-3} m³/kg	0.15692 ft³/lb
Mass of maximum outsurge	14,000 kg	
Mass of maximum insurge	9500 kg	
Hot leg insurge enthalpy	1.43×10^6 J/kg	612.8 B/lb
Cold leg spray enthalpy	1.27×10^6 J/kg	546.8 B/lb
Cold leg spray expressed as a fraction of hot leg insurge (f)	0.03	
Outsurge enthalpy	1.63×10^6 J/kg	701.1 B/lb
Mass of liquid water necessary to cover the heaters (requires an assumption about the pressurizer configuration)	1827 kg	

Proceeding similarly for the outsurge, obtain $(Q_h)_{outsurge}$ from Equation 7.60 as

$$\begin{aligned}(Q_h)_{outsurge} &= 14{,}000(1.63 \times 10^6) \\ &\quad - \left(1.60 \times 10^6 - \frac{1.68 \times 10^{-3}}{9.81 \times 10^{-3}}(2.44 \times 10^6)\right) \\ &\quad \times \left(14{,}000\left[\frac{9.81 \times 10^{-3}}{(9.81 - 1.68) \times 10^{-3}}\right]\right) \\ &= 2.2820 \times 10^{10} - (1.1821 \times 10^6)(1.6893 \times 10^4) \\ &= 2.851 \times 10^9 \text{ J}\end{aligned}$$

The value of m_{f_1} from Equation 7.58 is

$$m_{f_1} = 1827 + 14{,}000\left[\frac{9.81 \times 10^{-3}}{(9.81 - 1.68) \times 10^{-3}}\right] = 1.8720 \times 10^4 \text{ kg}$$

Hence, from Equation 7.61

$$(V_{f_1})_{outsurge} = (1.8720 \times 10^4)(1.68 \times 10^{-3}) = 31.45 \text{m}^3$$

Utilizing Equation 7.42, the total volume is

$$V_T = (V_{g_1})_{insurge} + (V_{f_1})_{outsurge} = 19.84 + 31.45 = 51.29 \text{m}^3 \text{ or } 51.3 \text{m}^3$$

PROBLEMS

7.1. *Containment pressure analysis* (Section 7.2)

For the plant analyzed in Example 7.1, Problem A, the peak containment pressure resulting from primary system blowdown is 0.523 MPa. Assume that the primary system failure analyzed in that accident sends an acoustic wave through the primary system that causes a massive failure of steam generator tubes. Although the main and auxiliary feedwater to all steam generators are shut off promptly, the entire secondary system inventory of 89 m^3 at 6.89 MPa is also now released to containment by blowdown through the primary system. Assume for this case that the secondary system inventory is all at saturated liquid conditions.

1. What is the new containment pressure?
2. Can it be less than that resulting from only primary system failure?

Answers:

1. 0.6 MPa.
2. Yes, if the secondary fluid properties are such that this fluid acts as an effective heat sink.

7.2. *Ice condenser containment analysis* (Section 7.2)

Calculate the minimum mass of ice needed to keep the final containment pressure below 0.4 MPa, assuming that the total volume consists of a 5.05×10^4 m^3 containment volume and a 500 m^3 primary volume. Neglect the initial volume of the ice. Additionally, assume the following initial conditions:

Containment pressure = 1.013×10^5 Pa (1 atm)
Containment temperature = 300 K
Ice temperature = 263 K (−10°C)
Ice pressure = 1.013×10^5 Pa (1 atm)
Primary pressure = 15.5 MPa (saturated liquid conditions)
Relevant properties are:
c_v for air = 719 J/kg K
R for air = 286 J/kg K
c for ice = 4.23×10^3 J/kg K
Heat of fusion for water = 3.33×10^5 J/kg (at 1 atm)

Answer: 1.78×10^5 kg of ice

7.3. *Containment pressure increase due to residual core heat* (Section 7.2)

Consider the containment system as shown in Figure 7.14 after a loss of coolant accident (LOCA). Assume that the containment is filled with saturated liquid, saturated vapor, and air and nitrogen gas released from rupture of one of the accumulators, all at thermal equilibrium. The containment spray system has begun its recirculating mode, during which the sump water is pumped by the residual heat removal (RHR) pump through the RHR heat exchanger and sprayed into the containment.

Assume that after 1 h of operation the RHR pump fails and the containment is heated by core decay heat. The decay heat from the core is assumed constant at 1% of the rated power of 2441 MW_{th}.

Find the time when the containment pressure reaches its design limit, 0.827 MPa (121.6 psia). Additional necessary information is given below.

Conditions after 1 h:

Water mixture mass (m_w) = 1.56×10^6 kg

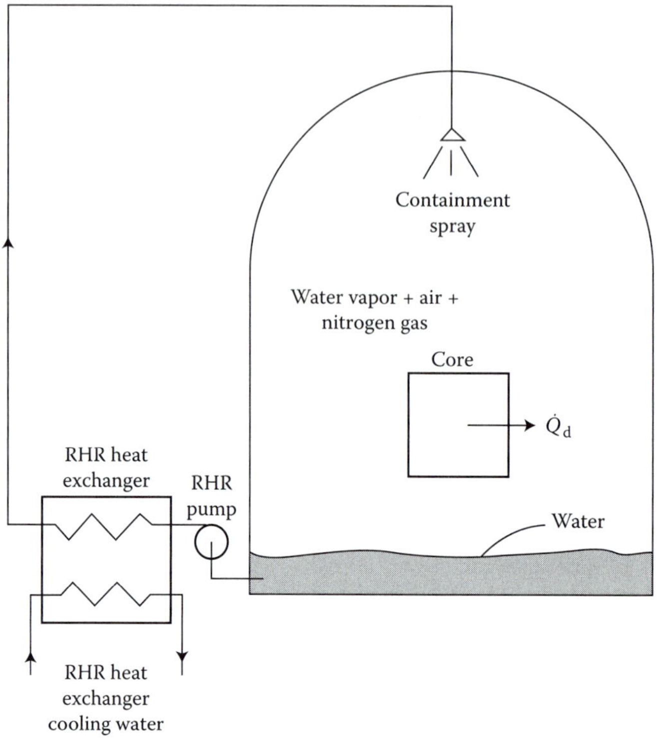

FIGURE 7.14 Containment.

Water mixture quality (x_{st}) = 0.0249
Air mass (m_a) = 5.9×10^4 kg
Nitrogen mass (m_{N_2}) = 1.0×10^3 kg
Initial temperature (T) = 381.6 K

Thermodynamic properties of gases:

$R_{air} = 0.286$ kJ/kg K
$c_{va} = 0.719$ kJ/kg K
$R_{N_2} = 0.296$ kJ/kg K
$c_{vN} = 0.742$ kJ/kg K

Answer: 7.08 h

7.4. *Loss of heat sink in a sodium-cooled reactor* (Section 7.2)
A 1000 MW_{th} SFR has three identical coolant loops. The reactor vessel is filled with sodium to a prescribed level, with the remained of the vessel being occupied by an inert cover gas, as shown in Figure 7.15. Under the steady operating condition, the ratio of the volume of cover gas to that of sodium in the primary system is 0.1. At time $t = 0$, the primary system of this reactor suffers a complete loss of heat sink accident, and the reactor power instantly drops to 2% of full power. Using a lumped parameter approach, calculate the pressure of the primary system at $t = 60$ s. At this time, check if the sodium is boiling.

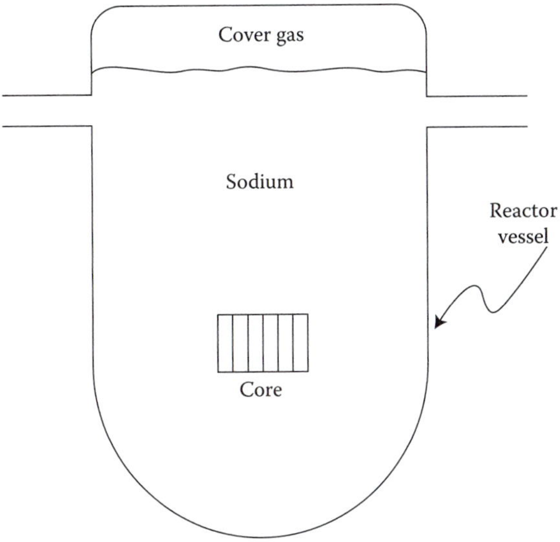

FIGURE 7.15 Reactor vessel in a sodium-cooled reactor.

Useful data:

Initial average primary system temperature = 527°C

Initial primary system pressure = 345 kPa

Total primary system coolant mass = 8165 kg

β (volumetric coefficient of thermal expansion of sodium) = 2.88×10^{-4}/°C

c_p for sodium = 1256 J/kg K (at 527°C)

ρ for sodium = 823.3 kg/m^3 (527°C)

Sodium-saturated vapor pressure:

$$p = \exp\left[18.832 - (13{,}113/T) - 1.0948 \ln T + 1.9777(10^{-4})\, \mathrm{T}\right]$$

where p is in atmospheres and T is in K.

Answers:

$p = 607$ kPa

$T = 644$°C, thus no boiling occurs.

7.5. *Response of a BWR suppression pool to safety/relief valve discharge* (Section 7.2)

Compute the suppression pool temperature after 5 min for a case in which the reactor is scrammed and steam is discharged from the reactor pressure vessel (RPV) into the suppression pool such that the temperature of the RPV coolant is reduced at a specific cooldown rate. During this process, makeup water is supplied to the RPV. Heat input to the RPV is only from long-term decay energy generation.

Numerical parameters applicable to this problem are given in Table 7.4.

Answer: 34.6°C

TABLE 7.4
Conditions for Suppression Pool Heat-Up Analysis

Parameter	Value
Specified cooldown rate	38°C/h (68.4°F/h)
Reactor power level prior to scram	3434 MW_{th}
RPV initial pressure	7 MPa (1015.3 psia)
Saturation properties at 7 MPa	$T = 285.88$°C
	$v_f = 1.3513 \times 10^{-3}$ m³/kg
	$v_{fg} = 26.0187 \times 10^{-3}$ m³/kg
	$u_f = 1257.55$ kJ/kg
	$u_{fg} = 1323.0$ kJ/kg
	$s_f = 3.1211$ kJ/kg K
	$s_{fg} = 2.6922$ kJ/kg K
Discharge period	5 min
Makeup water flow rate	32 kg/s (70.64 lb_m/s)
Makeup water enthalpy	800 kJ/kg (350 BTU/lb_m)
RPV free volume	656.3 m³ (2.3184×10^4 ft³)
RPV initial liquid mass	0.303×10^6 kg (0.668×10^6 lb_m)
RPV initial steam mass	9.0264×10^3 kg (19.9×10^3 lb_m)
Suppression pool initial temperature	32°C (90°F)
Suppression pool initial pressure	0.1 MPa (14.5 psia)
Suppression pool water mass	3.44×10^6 kg (7.6×10^6 lb_m)

RHR heat exchanger is actuated at high suppression pool temperature: 43.4°C (110°F)

7.6. *Containment sizing for a gas-cooled reactor with passive emergency cooling* (Section 7.2)
An advanced helium-cooled graphite-moderated reactor generates a nominal thermal power of 300 MW. To prevent air ingress in the core during an LOCA, the reactor containment is filled with helium at atmospheric pressure and room temperature (Figure 7.16a). The reactor also features an emergency cooling system to remove the decay heat from the containment during an LOCA. To function properly, this system, which is passive and based on natural circulation of helium inside the containment, requires a minimum containment pressure of 1.3 MPa.

1. Find the containment volume, so that the pressure in the containment is 1.3 MPa immediately after a large-break LOCA occurs (Figure 7.16b). (Assume that thermodynamic equilibrium within the containment is achieved instantaneously after the break.)
2. Assuming that the emergency cooling system removes 2% of the nominal reactor thermal power, calculate at what time the pressure in the containment reaches its peak value after the LOCA as well as the peak temperature and pressure. (Calculate the decay heat rate assuming infinite operation time.)
3. To reduce the peak pressure in the containment, a nuclear engineer suggests venting the containment gas to the atmosphere through

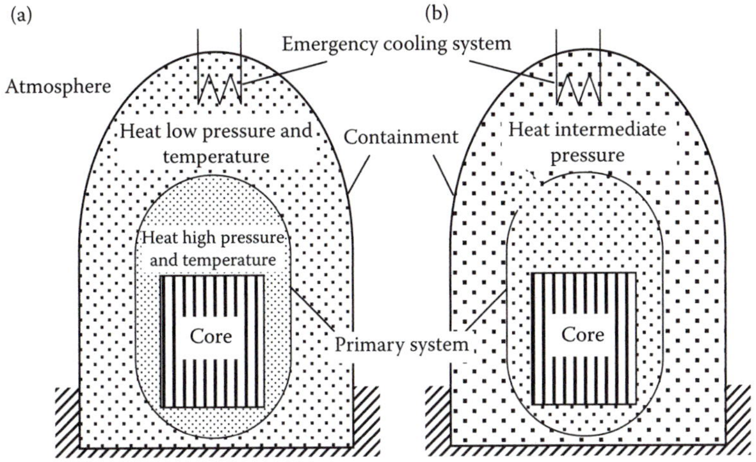

FIGURE 7.16 Helium-cooled reactor with helium-filled containment: (a) normal operating conditions and (b) post-LOCA situation.

a filter. What would be the advantages and disadvantages of this approach?

Assumptions:

- Treat helium as an ideal gas.
- Neglect the heat contribution from fission and chemical reactions.
- Neglect the thermal capacity of the structures.

Data:

Gas volume in the primary system: 200 m^3
Initial primary system temperature and pressure: 673 K, 7.0 MPa
Initial containment temperature and pressure: 300 K, 0.1 MPa
Helium-specific heat at constant volume: $c_v = 12.5$ J/mol K
Helium atomic weight = 0.004 kg/mol
Gas constant: $R = 8.31$ J/mol K

Answers:

1. 950 m^3
2. 787 K and 1.6 MPa at 391 s

7.7. *Containment problem involving an LOCA* (Section 7.2)
Upon a loss of primary coolant accident (LOCA), the primary system flashes as it discharges into the containment. At the resulting final equilibrium condition, the containment and primary system are filled with a mixture of steam and liquid. A containment is being designed as shown in Figure 7.17 which directs the liquid portion of this mixture to flood into a reactor cavity in which primary system is located. The condensate which passes back into the core through the break can satisfactorily cool the core if it can submerge it, that is, if the condensate level is high enough.

Find the containment volume which will yield a final equilibrium pressure following primary system rupture sufficient to create the 125 m^3 of liquid required to fill the cavity and submerge the core.

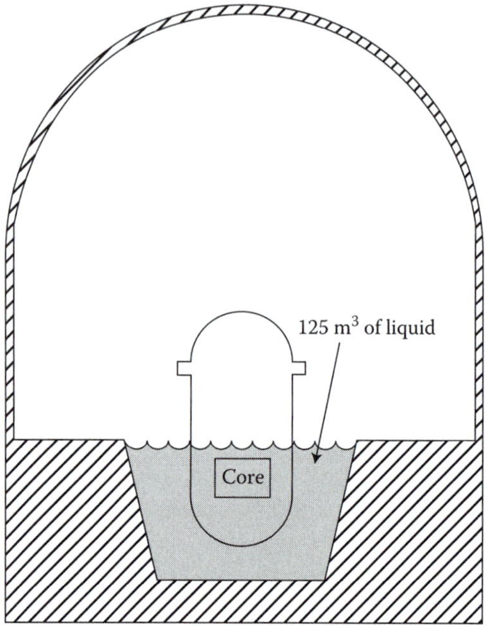

FIGURE 7.17 Containment with liquid fraction of discharge from LOCA directed to core cooling.

The pressure and volume of the primary system are 15.5 MPa and 354 m^3, respectively. The containment initial conditions are 300 K and 0.101 MPa.

- Neglect the initial relative humidity.
- Neglect $\dot{Q}_{c-st}$ and $\dot{Q}_{c-atm}$

Answers:

$$V_T = 2.32 \times 10^4 \, m^3$$

$$V_C = 2.27 \times 10^4 \, m^3$$

$$p_T = 0.97 \, MPa$$

7.8. *Analysis of a transient overpower in the PWR steam generator* (Section 7.2)

The steam generator of a large PWR delivers dry saturated steam at 5.7 MPa to the turbine. Consider the steam generator secondary side, which has a volume of 100 m^3 and receives a thermal power $\dot{Q}$ from the primary coolant flowing in the U-tubes (Figure 7.18).

At steady state, the operating conditions for the secondary coolant are as follows:

- Inlet mass flow rate $\dot{m}_i = 456$ kg/s
- Inlet temperature $T_i = 267°C$ ($h_i = 1170$ kJ/kg)
- Mass of steam 880 kg
- Mass of liquid 54,000 kg

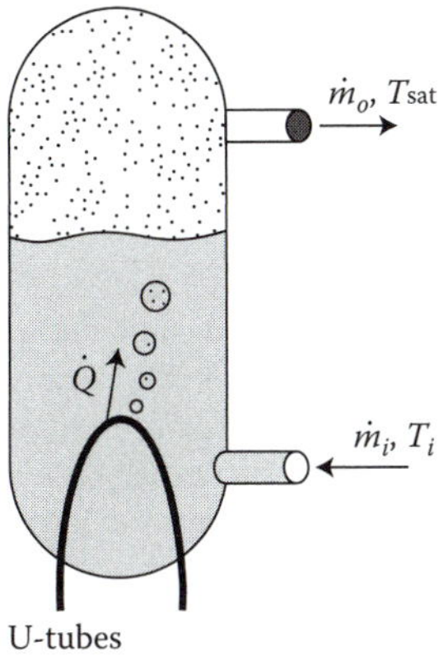

FIGURE 7.18 Schematic of the steam generator.

Properties of saturated water at 5.7 MPa are given in Table 7.5.

1. Calculate $\dot{Q}$.

   At one point in time, the operator maneuvers the reactor so that the thermal power supplied to the secondary coolant increases to $1.2\dot{Q}$. Assume that the secondary coolant pressure, inlet mass flow rate, and inlet temperature do not change during the transient.
2. Write a complete set of equations that would allow you to find how the secondary coolant mass ($M_{SC}(t)$) in the steam generator changes during the transient. Clearly identify all known and unknown parameters in the equations. You may neglect kinetic and gravitational terms. State all your assumptions.
3. Does the secondary coolant outlet mass flow rate increase, decrease, or stay the same during the transient?
4. Now imagine that after 2 min both the secondary coolant inlet and outlet are suddenly and simultaneously closed shut, while the thermal power remains at $1.2\dot{Q}$. Does the secondary coolant pressure increase or decrease during this transient? Write a complete set of

TABLE 7.5
Properties of Saturated Water at 5.7 MPa

Parameter	Value
T_{sat}	272°C
v_f	1.3×10^{-3} m³/kg
v_g	0.034 m³/kg
h_f	1196 kJ/kg
h_g	2788 kJ/kg
$c_{p,f}$	5.2 kJ/kg °C
$c_{p,g}$	4.7 kJ/kg °C
u_f	1189 kJ/kg
u_g	2592 kJ/kg

equations that would allow you to find the pressure change in the secondary coolant during this transient.

Answers:
1. 737.8 MW_{th}
3. Increases
4. Increases

7.9. *Drain tank pressurization problem* (Section 7.2)
A drain tank is used to temporarily store water discharged from the pressurizer through the pressure-operated relief valve (PORV) (Figure 7.19). The drain tank has a burst disk on it which ruptures if the pressure inside the drain tank becomes too large. For this problem, assume that the PORV at the top of the pressurizer is stuck open, and saturated water at 15.4 MPa leaves the presssurizer at a constant flow rate of 3 kg/s and enters a perfectly insulated drain tank of total volume 12 m^3. In addition, assume that the initial conditions (before the water due to the stuck-open PORV has entered the drain tank) in the drain tank are
No air present Initial vapor volume = 10 m^3
Initial pressure = 3 MPa Initial liquid volume = 2 m^3
Also assume that the liquid and the water vapor are in thermal equilibrium at all times in the drain tank, and that the burst disk on the drain tank ruptures at 10 MPa.

Questions:
a. Define the control mass or control volume you will use and the equation set you will develop.
b. Solve for the elapsed time to burst disk rupture.
c. Now assume 11.93 kg of air is present in the drain tank along with the liquid water and water vapor (p_{1w}) initial = 3 MPa and that the

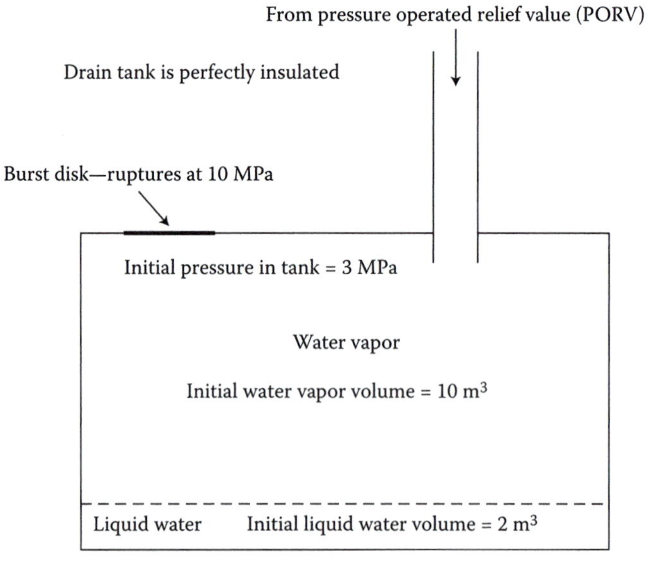

FIGURE 7.19 Drain tank to store discharge from the PORV.

change in volume of the liquid water from the initial state to the final state is large. What is the new time to rupture?

Answers:
b. 1128 s
c. 1044 s

7.10. *Containment pressurization following zircaloy–hydrogen reaction* (Section 7.2)
Consider an LOCA in a typical PWR in which the emergency cooling system is insufficient to prevent metal–water reaction of 75% of the Zircaloy clad and the hydrogen produced subsequently combusts. Using the results of Problem 3.6, this sequence of events yields the following material changes and energy releases relevant to the containment pressurization:
Primary coolant released = 2.1×10^5 kg
Zr reacted = $0.75 \times 24{,}000$ kg
Energy released from Zr–H_2O reaction = 1.18×10^{11} J
H_2 produced and reacted = 394.7 mols
Energy released from H_2 combustion = 9.54×10^{10} J
O_2 consumed = *you must determine*
Net H_2O change = *you must determine*
Take the initial primary coolant and containment vessel geometry and conditions the same as Table 7.2. Also, assume that nitrogen has the same properties as air.

Question:
For the sequence described (e.g., LOCA, 75% Zircaloy clad reaction and subsequent complete combustion of the hydrogen produced):
a. Demonstrate that the final equilibrium temperature is 450 K, neglecting containment heat sinks using the initial conditions of Table 7.2.
b. Find the final equilibrium pressure.
Hint: Is the final state likely saturated water or superheated steam in equilibrium with the air?
Consider the energy releases compared to those of Example 7.2.

Answer:
b. p_2 (450 K) = 1.06×10^6 Pa

7.11. *Effect of noncondensable gas on pressurizer response to an insurge* (Section 7.3)
Compute the pressurizer and heater input resulting from an insurge of liquid from the primary system to a pressurizer containing a mass of air (m_a). Use the following initial, final, and operating conditions.
Initial conditions
Mass of liquid = m_{f1}
Mass of steam = m_{g1}
Mass of air (in steam space) = m_a
Total pressure = p_1
Equilibrium temperature = T_1
Operating conditions
Mass of surge = m_{surge}
Mass of spray = fm_{surge}

Enthalpy of surge = h_{surge}
Enthalpy of spray = h_{spray}
Heater input = Q_h

Final condition

Equilibrium temperature = $T_2 = T_1$

You may make the following assumptions for the solution:

1. Perfect phase separation
2. Thermal equilibrium throughout the pressurizer
3. Liquid water properties that are independent of pressure

Answers:

$$p_2 = p_w(T_{1\text{sat}}) + \frac{(m_a R T_1)(v_{g_1} - v_{f_1})}{(v_{g_1})(m_{g_1} v_{g_1} - (1+f) m_{\text{surge}} v_{f_1} - m_{g_1} v_{f_1})}$$

$$Q_h = \frac{(1+f)}{V_{fg}} m_{\text{surge}} (u_f V_g - u_g V_f) - m_{\text{surge}} (h_{\text{surge}} + f h_{\text{spray}}) \quad (7.53)$$

That is, the same as Equation 7.53.

7.12. *Pressurizer-sizing analysis* (Section 7.3)

The size of a pressurizer is determined by the criteria that the vapor volume must be capable of accommodating the largest insurge and the liquid volume must handle the outsurge. The important limitations of the design are that the pressurizer should not be totally liquid filled or the immersion heaters should not be uncovered after possible transients. To size the vapor volume, a maximum insurge is assumed to completely fill the pressurizer with liquid with some of the insurge being diverted to the spray to condense the vapor. Treating the entire pressurizer volume, V_t, as the control volume, find the vapor volume, V_{g1}, which will accommodate the insurge given below.

Data:

Initial pressurizer conditions

Saturation at 15.51 MPa and 345°C
Initial liquid mass = 1827
Maximum insurge (includes spray)
Mass = 2740 kg
Enthalpy = 1.2×10^6 J/kg

Final pressurizer condition

Assume completely filled with liquid at 15.51 MPa

Answers:

$$V_{g\ell} = 5.57\,\text{m}^3$$

$$V_t = 8.64\,\text{m}^3$$

7.13. *Pressurizer insurge problem* (Section 7.3)

For insurge case, why is latent heat of vaporization of vapor which is condensed insufficient to heat insurge mass to saturation?

7.14. *Behavior of a fully contained pressurized pool reactor under decay power conditions* (Section 7.3)

A 1600 MW_{th} pressurized pool reactor has been proposed in which the entire primary coolant system is submerged in a large pressurized pool of cold water with a high boric acid content. The amount of water in the pool is sufficient to provide for core decay heat removal for at least 1 week following any incident, assuming no cooling systems are operating. In this mode, the pool water boils and is vented to the atmosphere. The vessel geometry is illustrated in Figure 7.20. The core volume can be neglected.

Assume that at time $t=0$ with an initial vessel pressure of 0.10135 MPa (1 atm), the venting to the atmosphere fails. Does water cover the core for all t? Plot the water level measured from the top of the vessel versus time. You may assume that the vessel volume can be subdivided into an upper saturated vapor and a lower saturated liquid volume. Also, assume that the decay heat rate is constant for the time interval of interest at 25 MW_{th}.

Answer: Water always covers the core.

7.15. *Depressurization of a primary system* (Section 7.3)

The pressure of the primary system of a PWR is controlled by the pressurizer via the heaters and spray. A simplified drawing of the primary system of a PWR is shown in Figure 7.21. If the spray valve were to fail in the open position, depressurization of the primary system would result.

Calculate the time to depressurize from 15.5 to 12.65 MPa if the spray rate is 30.6 kg/s with an enthalpy of 1252 kJ/kg (constant with time). Also calculate the final liquid and vapor volumes. You may assume that

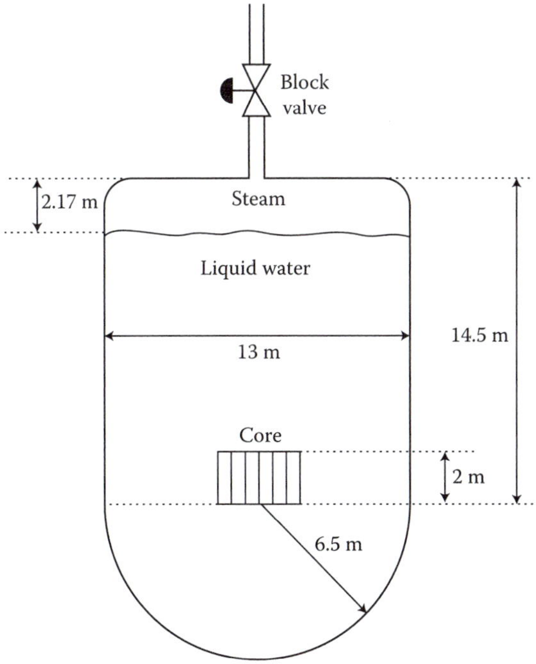

FIGURE 7.20 Pressurized pool reactor.

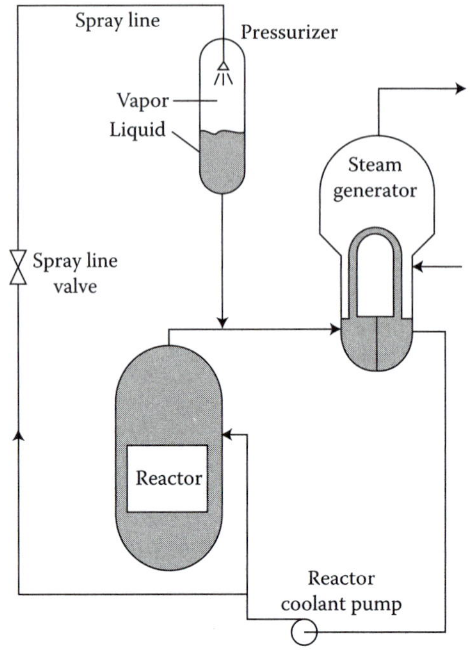

FIGURE 7.21 Simplified drawing of the primary system of a PWR.

1. Heaters do not operate.
2. Pressurizer wall is adiabatic.
3. Pressurizer vapor and liquid are in thermal equilibrium and occupy initial volumes of 20.39 and 30.58 m^3, respectively.
4. Complete phase separation occurs in the pressurizer.
5. The subcooled liquid in the primary system external to the pressurizer is incompressible so that the spray mass flow rate is exactly balanced by the pressurizer outsurge.

Answers:
225 s
22.13 m^3 vapor
28.84 m^3 liquid

8 Thermal Analysis of Fuel Elements

8.1 INTRODUCTION

The fuel matrix itself and the hermetically sealed cladding of the fuel pin are the first two barriers to the release of radioactivity in a nuclear reactor (the third and fourth being the reactor coolant system boundary and the containment structure, respectively). Consequently, an accurate description of the temperature distributions in fuel elements and reactor structures is essential to the prediction of the lifetime behavior of these components. The temperature gradients, which control the thermal stress levels in the materials, together with the mechanical loads contribute to the determination of the potential for plastic deformation at high temperatures or cracking at low temperatures. The temperature level at coolant-solid surfaces controls chemical reactions and diffusion processes, thus profoundly affecting the corrosion process. Furthermore, the impact of the fuel and coolant temperatures on neutron reaction rates provides an incentive for accurate modeling of the temperature behavior under transient as well as steady-state operating conditions. In this chapter, the focus is on the steady-state temperature field in the fuel elements. Many of the principles presented are also useful for describing the temperature field in the structural components.

The temperature in the fuel material depends on the energy deposition rate, the fuel material properties, and the coolant and cladding temperature conditions. The rate of heat generation and deposition in a fuel pin depends on the neutron reaction rates within the fuel, β-ray absorption, and γ-ray attenuation in the fuel pin, as described in Chapter 3. In return, the neutron reaction rate depends on the fuel material (both the initial composition and burnup level) and the moderator material (if present) and their temperatures. Hence, an exact prediction of the fuel material temperature requires simultaneous determination of the neutronic and temperature fields, although for certain conditions it is possible to decouple the two fields. Thus, it is assumed that the energy deposition rate is known as we proceed here to obtain the fuel temperature field.

Uranium dioxide (UO_2) has been used as fuel material in light water power reactors ever since it was used in the Shippingport PWR in 1955. Uranium metal and its alloys have been used in research reactors. Early liquid-metal-cooled reactors relied on plutonium as a fuel and more recently on a mixture of UO_2 and PuO_2. This mixture called mixed oxide (MOX) for MOX fuel, albeit in different proportions, is also now being used to fuel about a third of the core loading in a significant number of LWRs in Europe and some in Russia and Japan. The mid-1980s saw resurgence in the interest in the metal as fuel in U.S.-designed fast reactors. The properties of these materials are highlighted in this chapter. UO_2 and MOX use in LWRs has been characterized by satisfactory chemical and irradiation tolerance. This tolerance has overshadowed the disadvantages of low-thermal conductivity and uranium atom density relative to other materials, for example, the nitrides and carbides or even the metal itself. The carbides and nitrides, if proved not to swell excessively under irradiation, may be used in future reactors.

The general fuel assembly geometry and operating characteristics are given in Chapters 1 and 2. Tables 8.1 and 8.2 compare the thermal properties of the various fuel and cladding materials, respectively. Table 8.1a presents thermal properties of ceramic fuels, while Table 8.1b gives metallic fuel thermal properties. Tables 8.2a and 8.2b present thermal properties of stainless-steel- and zirconium-based cladding and non-stainless-steel- and non-zirconium-based cladding materials, respectively. Cladding corrosion behavior is discussed in Section 8.7.2.

8.2 HEAT CONDUCTION IN FUEL ELEMENTS

8.2.1 General Equation of Heat Conduction

The energy transport equation (Equation 4.123) describes the temperature distribution in a solid (which is assumed to be an incompressible material with negligible thermal expansion as far as the effects on temperature distribution are concerned). If Equation 4.123 is written with explicit dependence on the variables $\vec{r}$ and t, it becomes

$$\rho c_p(\vec{r},T)\frac{\partial T(\vec{r},t)}{\partial t} = \nabla\cdot\left[k(\vec{r},T)\nabla T(\vec{r},t)\right] + q'''(\vec{r},t) \tag{8.1}$$

Note that for incompressible materials $c_p = c_v$.

At steady state, Equation 8.1 reduces to

$$\nabla\cdot k(\vec{r},T)\nabla T(\vec{r}) + q'''(\vec{r}) = 0 \tag{8.2}$$

Because by definition the conduction heat flux is given by $\vec{q}'' \equiv -k\nabla T$, Equation 8.2 can be written as

$$-\nabla\cdot\vec{q}''(\vec{r},T) + q'''(\vec{r}) = 0 \tag{8.3}$$

8.2.2 Thermal Conductivity Approximations

In a medium that is isotropic with regard to heat conduction, k is a scalar quantity that depends on the material, temperature, and pressure of the medium. In a nonisotropic medium, thermal behavior is different in different directions. Highly oriented crystalline-like materials can be significantly anisotropic. For example, thermally deposited pyrolytic graphite can have a thermal conductivity ratio as high as 200:1

TABLE 8.1A
Thermal Properties of Ceramic Fuel Materials (Unirradiated)

Property	UO_2	PuO_2	MOX (94% UO_2–6% PuO_2)[a]	ThO_2	UC	UN	U_3Si_2
Theoretical density at room temperature (kg/m^3)	10.97×10^3	11.45×10^3	11.00×10^3	9.60×10^3	13.63×10^3	14.32×10^3	12.20×10^3
Heavy metal density (kg/m^3)	9.67×10^3	10.09×10^3	9.69×10^3	8.45×10^3	12.97×10^3	13.60×10^3	11.29×10^3
Melting point (°C)	2800	2374	2774	3378	2390	2800	1665
Stability range	Up to M.P.	Up to M.P.	Up to M.P.	Up to M.P.	Up to M.P.	Up to M.P.	Up to M.P.
Thermal conductivity (W/m°C) [average between 200°C and 1000°C]	3.6	4.3	3.7[b]	5.76	23	21	15
Specific heat (J/kg °C) [100°C]	247	286	261	301	146	206[c]	209
Linear thermal expansion coefficient (/°C) [1000°C]	10.1×10^{-6}	13.1×10^{-6}	12.5×10^{-6}	9.7×10^{-6}	11.1×10^{-6}	9.4×10^{-6}	15.5×10^{-6}
Crystal structure	Face-centered cubic	Face-centered cubic	Face-centered cubic	Face-centered cubic	Face-centered cubic	Face-centered cubic	Varies
Tensile strength (MPa)	110	—	—	—	62	Not well defined	255

[a] Thermal properties of MOX fuel are evaluated as weighted fraction of the contribution from PuO_2 and UO_2, that is, $X_{MOX} = X_{UO_2}\,(1–6\%) + X_{PuO_2}\,*6\%$, where X is the thermal property except for thermal conductivity which is not modeled as a function of weight fraction of PuO_2. The linear thermal coefficient of expansion is calculated using material properties in [34]. Other thermal properties are evaluated referring to [19].

[b] The thermal conductivity of PuO_2 is evaluated using [14].

[c] The specific heat of UN is calculated using the model of [39].

TABLE 8.1B
Thermal Properties of Metallic Fuel Materials (Unirradiated)

Property	U	U-Zr Metal (U-10% Zr)	U-Pu-Zr Metal (U-20% Pu-10% Zr)	U-TRU-Zr Metal (U-20% TRU-10% Zr)	$UZrH_{1.6}$
Theoretical density at room temperature (kg/m^3)	19.04×10^3	15.97×10^3	16.07×10^3	15.98×10^3	8.26×10^3
Heavy metal density (kg/m^3)	19.04×10^3[a]	14.37×10^3	14.46×10^3	14.38×10^3	3.72×10^3
Melting point (°C)	1133	1380	1310	<1310	1855[d]
Stability range	Up to 665°C[b]	Up to M.P.	Up to M.P.	Up to M.P.	600[e]
Thermal conductivity (W/m°C) (average between 200°C and 1000°C)	32	28	25	24	17.6
Specific heat (J/kg°C) (100°C)	116	131	129	129	295
Linear thermal expansion coefficient (/°C)	1.4×10^{-5} (25°C)	1.3×10^{-5} (25°C)	1.7×10^{-5} (25°C)	1.6×10^{-5} (25°C)	1.5×10^{-5} (500°C)
Crystal structure	Below 655°C: α, orthorhombic Above 770°C: γ, body-centered cubic	Orthorhombic	Orthorhombic	Orthorhombic	Face-centered cubic
Tensile strength (MPa)	344–1380[c]	403	398	397	120–165

[a] Uranium metal density in the compound at its theoretical density.
[b] Addition of a small amount of Mo, Nb, Ti, or Zr extends stability up to the melting point.
[c] The higher values apply to cold-worked metal.
[d] Not well defined. Hydrogen release starts at about 600°C (H_2 equilibrium pressure at 600°C = 0.01 atm). Once all hydrogen has been released, the remaining U-Zr fuel melts between 1135°C (U melting point) and 1855°C (Zr melting point).
[e] Stable as long as H is not released. Hydrogen release, other than causing the change of the fuel composition, induces crystal structure modification of ZrH_x phase (from δ-$ZrH_{1.6}$ to α-ZrH_x or β-ZrH_x depending on the temperature).

TABLE 8.2A
Thermal Properties of Stainless-Steel- and Zirconium-Based Cladding

Property	Zirconium	Zircaloy 2[a]	Zr-1% Nb [37]	HT9	Type 316 SS	ODS
Density at room temperature (kg/m^3)	6.50×10^3	6.5×10^3	6.55×10^3	8.1×10^3	7.8×10^3	7.25×10^3
Melting point (°C)	1852	1850	1860	1470	1400	1480
Thermal conductivity (W/m °C) [400°C]	20.4	13	21	18	23	16.9
Specific heat (J/kg °C) [400°C]	328	330	380	614	580	575
Linear thermal expansion coefficient (/°C) [400°C]	5.7×10^{-6}[c]	4.4×10^{-6}[b] 6.72×10^{-6}[c]	5.4×10^{-6}[b] 1.02×10^{-5}[c]	1.6×10^{-5}	2.0×10^{-5}	1.2×10^{-5}

[a] Linear thermal expansion coefficient of Zry-2 is evaluated using the material properties in [20].
[b] Axial direction.
[c] Diametral direction.

TABLE 8.2B
Thermal Properties of Non-Stainless-Steel- and Non-Zirconium-Based Cladding

Property	Beryllium	Magnesium	Aluminum	Inconel 718	SiC[a] [36,38]
Density at room temperature (kg/m^3)	1.842	1.746	2.707	8.19	3.21
Melting point (°C)	1287	650	660	1260	2700
Thermal conductivity (W/m °C) [400°C]	200	156	237	17	70
Specific heat (J/kg °C) [400°C]	1820	1067	963	502	1070
Linear thermal expansion coefficient (/°C) [400°C]	1.1×10^{-5}	2.5×10^{-5}	2.3×10^{-5}	1.3×10^{-5}	4.6×10^{-6}

[a] All SiC properties are for β-phase.

in directions parallel and normal to basal planes. For anisotropic and nonhomogeneous materials, k is a tensor, which in Cartesian coordinates can be written as

$$\bar{\bar{k}} = \begin{pmatrix} k_{xx} & k_{xy} & k_{xz} \\ k_{yx} & k_{yy} & k_{yz} \\ k_{zx} & k_{zy} & k_{zz} \end{pmatrix} \quad (8.4)$$

For anisotropic homogeneous solids, the tensor is symmetric, that is, $k_{ij} = k_{ji}$. In most practical cases, k can be taken as a scalar quantity. We restrict ourselves to this particular case of a scalar k for the remainder of this text.

As mentioned before, thermal conductivity is different for different media and generally depends on the temperature and pressure. The numerical value of k varies from practically zero for gases under extremely low pressures to about 400 W/m K or 7000 Btu/ft h F for a natural copper crystal at very low temperatures.

The change of k with pressure depends on the physical state of the medium. Whereas in gases there is a strong pressure effect on k, in solids this effect is negligible. Therefore, the conductivity of solids is mainly a function of temperature, $k = k(T)$, and can be determined experimentally. For most metals, the empirical formula [2]

$$k = k_0[1 + \beta_0(T - T_0)] \quad (8.5)$$

gives a good fit to the data in a relatively large temperature range. The values of k_0 and β_0 are constants for the particular metal. It is evident that k_0 corresponds to the reference temperature (T_0). The value of β_0 can be positive or negative. In general, β_0 is negative for pure homogeneous metals, whereas for metallic alloys β_0 becomes positive.

In the case of nuclear fuels, the situation is more complicated because k becomes a function of the irradiation as a result of changes in the chemical and physical composition (porosity changes due to temperature and fission products).

Even when k is assumed to be a scalar, it may be difficult to solve Equation 8.2 analytically because of its nonlinearity. The simplest way to overcome the difficulties is to transform Equation 8.2 to a linear one, which can be done by four techniques:

1. In the case of small changes of k within a given temperature range, assume k is constant. In this case, Equation 8.2 becomes

$$k\nabla^2 T(\vec{r}) + q'''(\vec{r}) = 0 \quad (8.6)$$

2. If the change in k over the temperature range is large, define a mean $\bar{k}$ as follows:

$$\bar{k} = \frac{1}{T_2 - T_1}\int_{T_1}^{T_2} k\mathrm{d}T \quad (8.7)$$

and use Equation 8.6 with $\bar{k}$ replacing k.

3. If an empirical formula for k exists, it may be used to obtain a single variable differential equation, which in many cases can be transformed to a relatively simple linear differential equation. For example, Equation 8.5 can be used to write

$$T - T_0 = \frac{k - k_0}{\beta_0 k_0} \tag{8.8}$$

so that

$$k\nabla T = \frac{k\nabla k}{\beta_0 k_0} = \frac{\nabla k^2}{2\beta_0 k_0}$$

and

$$\nabla \cdot (k\nabla T) = \frac{\nabla^2 k^2}{2\beta_0 k_0} \tag{8.9}$$

Substituting from Equation 8.9 into Equation 8.2, we get

$$\nabla^2 k^2 + 2\beta_0 k_0 q''' = 0 \tag{8.10}$$

which is a linear differential equation in k^2.

4. Finally, the heat conduction equation can be linearized by Kirchhoff's transformation, as briefly described here: in many cases, it is useful to know the integral

$$\int_{T_1}^{T_2} k(T)\,\mathrm{d}T$$

where $T_2 - T_1$ is the temperature range of interest. Kirchhoff's method consists of finding such integrals by solving a modified heat conduction equation. Define

$$\theta \equiv \frac{1}{k_0} \int_{T_0}^{T} k(T)\,\mathrm{d}T \tag{8.11}$$

The new variable θ can be used to give

$$\nabla\theta = \frac{1}{k_0} \nabla \int_{T_0}^{T} k(T)\mathrm{d}T = \frac{1}{k_0}\left[\nabla T \frac{d}{\mathrm{d}T} \int_{T_0}^{T} k(T)\mathrm{d}T\right] = \frac{k(T)}{k_0}\nabla T \tag{8.12}$$

From this equation, we find

$$k\nabla T = k_0 \nabla\theta \tag{8.13}$$

and

$$\nabla \cdot [k\nabla T] = k_0 \nabla^2\theta \quad (8.14)$$

Then at steady state, Equation 8.2 becomes

$$k_0\nabla^2\theta + q''' = 0 \quad (8.15)$$

which is a linear differential equation that can generally be solved more easily than Equation 8.2.

In practice, nuclear computer programs have allowed common use of temperature-dependent conductivity in the numerical solutions. Equation 8.2 can be readily solved in one-dimensional geometries, as illustrated later in this chapter. Therefore, the above approaches are useful only if one is interested in analytic solutions of multidimensional problems.

8.3 THERMAL PROPERTIES OF UO_2 AND MOX

LWR fuel is principally UO_2; hence, the focus of this section is on the properties of this most widely used fuel material. In particular, the thermal conductivity, melting point, specific heat, and fractional gas release are discussed. MOX fuel {$(U,Pu)O_2$} properties are also presented but to a more limited degree.

8.3.1 Thermal Conductivity

Many factors affect UO_2 thermal conductivity. The major factors are temperature, porosity, and oxygen-to-metal atom ratio, PuO_2 content, pellet cracking, and burnup. A brief description of the change in k with each of these factors is discussed.

8.3.1.1 Temperature Effects

It has been experimentally observed that $k(T)$ decreases with increasing temperature until $T = 1750°C$ and then starts to increase. Figure 8.1 illustrates this behavior for fresh UO_2 as well as irradiated UO_2 with and without the typical-added burnable poison, 5% by weight of gadolinium oxide (Gd_2O_3) called gadolinia. The often-used integral of $k(T)$ is given in Figure 8.2. The fuel density for all cases in Figures 8.1 and 8.2 is 95% of theoretical density. Figure 8.3 presents both UO_2 and MOX {$(U,Pu)O_2$} fresh fuel performance for a range of theoretical fuel densities. The equation for $k(T)$ used in the COPERNIC code [4] is given in Table 8.3. The typical value of the conductivity integral for unirradiated UO_2 from 0°C to melting is 90 W/cm.

$$\int_{0°C}^{\text{melting}} k\, dT$$

For an ionic solid, thermal conductivity can be derived by assuming that the solid is an ideal gas whose particles are the quantized elastic wave vibrations in a crystal

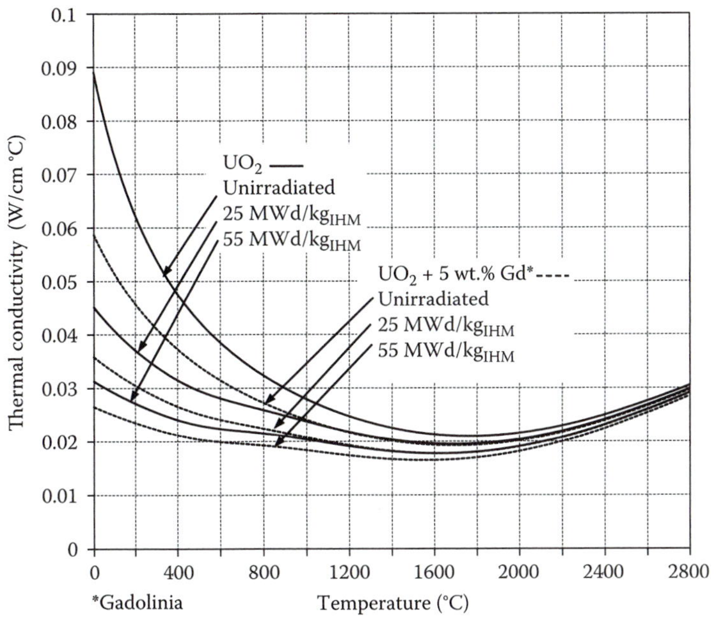

FIGURE 8.1 Thermal conductivity of UO_2 and UO_2 with 5% gadolinia—all cases at 95% of theoretical density. (Based on results of Lanning, D. D., Beyer, C. E., and Geelhood, K. J., *FRAPCON-3 updates, including mixed-oxide fuel properties*. Tech. Rep. NUREG/CR-6534 Vol. 4, PNNL-11513, 2005.)

(referred to as *phonons*). It can be shown that the behavior of the UO_2 conductivity with temperature can be predicted by such a model [30].

8.3.1.2 Porosity (Density) Effects

The oxide fuel is generally fabricated by sintering-pressed powdered UO_2 or MOX $\{(U,Pu)O_2\}$ at high temperature. By controlling the sintering conditions, material of any desired density, usually around 90% of the maximum possible or theoretical density of the solid, can be produced.

Generally, the conductivity of a solid decreases with increasing presence of voids (pores) within its structure. Hence, low porosity is desirable to maximize the conductivity. However, fission gases produced during operation within the fuel result in internal pressures that may swell, and hence deform, the fuel. Thus, a certain degree of porosity is desirable to accommodate the fission gases and limit the swelling potential. This is particularly true for fast reactors where the specific power level is higher, and hence the rate at which gases are produced per unit fuel volume is higher than in thermal reactors.

Let us define the porosity (P) as

$$P = \frac{\text{Volumes of pores } (V_p)}{\text{Total volumes of pores } (V_p) \text{ and solids } (V_s)} = \frac{V_p}{V} = \frac{V - V_s}{V}$$

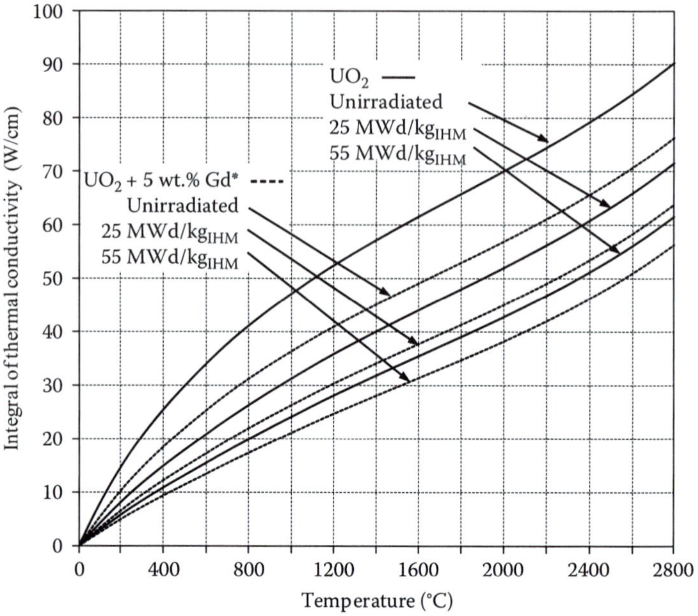

FIGURE 8.2 Integral of thermal conductivity of UO_2 and UO_2 with 5% gadolinia—all cases at 95% of theoretical density. (Based on results of Lanning, D. D., Beyer, C. E., and Geelhood, K. J., *FRAPCON-3 updates, including mixed-oxide fuel properties*. Tech. Rep. NUREG/CR-6534 Vol. 4, PNNL-11513, 2005.) *Note:* 32.8 W/cm = 1kW/ft.

or

$$P = 1 - \frac{\rho}{\rho_{TD}} \tag{8.17}$$

where ρ_{TD} is the theoretical density of the poreless solid. The effect of porosity on $\int kdT$ for UO_2 and MOX expressed in terms of theoretical density is shown in Figure 8.3. The analytic expression for these $\int kdT$ curves is given in Equation 8.22d.

By considering the linear porosity to be $P^{1/3}$ and the cross-sectional porosity to be $P^{2/3}$, Kampf and Karsten [17] derived an equation for negligible pore conductance

$$k = (1 - P^{2/3})k_{TD} \tag{8.18}$$

Earlier, the analysis of Loeb [23] was used by Francl and Kingery [11] to derive the equation referred to as the Loeb equation for this condition of negligible pore conductance

$$k = (1 - P)k_{TD} \tag{8.19}$$

Equation 8.19 was found to underestimate the porosity effect.

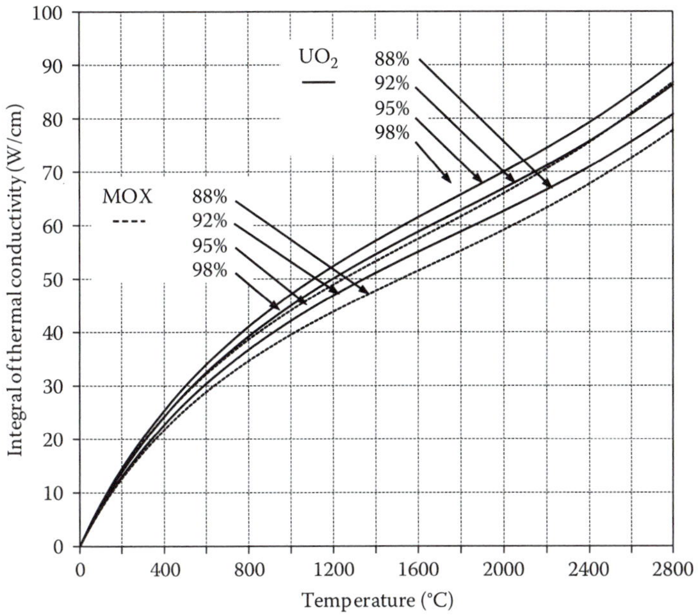

FIGURE 8.3 Integral of thermal conductivity of UO_2 and $(U,Pu)O_2$ fresh fuel (up to 7 wt.% PuO_2 content) at different densities. (Based on results of Lanning, D. D., Beyer, C. E., and Geelhood, K. J., *FRAPCON-3 updates, including mixed-oxide fuel properties*. Tech. Rep. NUREG/CR-6534 Vol. 4, PNNL-11513, 2005.)

A modified Loeb equation often used to fit the UO_2 conductivity measurements is expressed as

$$k = (1 - \alpha_1 P)k_{TD} \tag{8.20}$$

where α_1 is between 2 and 5 [7].

Biancharia [5] derived the following formula for the porosity effect, which accounts for the shape of the pores

$$k = \frac{(1-P)}{1+(\alpha_2 - 1)P} k_{TD} \text{ where } \quad \alpha_2 = 1.5 \quad \text{for spherical pores} \tag{8.21}$$

For axisymmetric shapes (e.g., ellipsoids), α_2 is larger. This formula has been often used in liquid metal fast reactor applications.

The following fit for the temperature and porosity effects on conductivity is used in the FRAPCON fuel analysis package for irradiated fuel with gadolinia, both UO_2 and $(U,Pu)O_2$ [19]

$$k_{0.95} = \frac{1}{A + BT + a \cdot \text{gad} + f(Bu) + [1 - 0.9\exp(-0.04Bu)]g(Bu)h(T)} + \frac{E}{T^2}\exp(-F/T) \tag{8.22a}$$

TABLE 8.3
Thermal Conductivity Model Used by AREVA

AREVA (formerly Framatome ANP) thermal conductivity model of the fuel performance code COPERNIC

$$k_{TD} = \frac{1}{A_1 + A_2 T + A_3 Bu + A_4 f(T)} + g(T) \tag{8.16}$$

where:

$A_1 = 0.0375$ m K/W
$A_2 = 2.165\text{e}{-4}$ m/W
$A_3 = 1.70\text{e}{-3}$ m K/W/(GWd/Mt_{IHM})
$A_4 = 0.058$ m K/W

Bu is burnup in GWd/$Mt_{IHM} \equiv$ MWd/kg_{IHM}

$$f(T) = \left[1 + \exp\left(\frac{T - 900}{80}\right)\right]^{-1} \text{ dimensionless}$$

$$g(T) = 4.715 \times 10^9 T^{-2} \exp\left(-\frac{16{,}361}{T}\right) \text{W/m K}$$

k_{TD} thermal conductivity at the theoretical density in W/m K, that is, TD density
T is in K

Source: From Bernard, L. C., Jacoud, J. L., and Vesco, P., *J. Nucl. Mater.*, 302:2–3, 2002.

which is applicable to the following range of conditions:

Temperature: 300–3000 K
Rod-average burnup: 0–62 GWd/Mt_{IHM}
Gadolinia content: 0–10 wt.%

As-fabricated density: 92–97% TD (accounted for by multiplying $k_{0.95}$ by the porosity factor η given in Equation 8.22b below) where

a = 115.99 in cm K/W
Bu = burnup in MWd/kg_{IHM}
$f(Bu)$ = effect of fission products in crystal matrix (solution)
= 0.187 Bu in cm K/W
gad = mass fraction of gadolinia
$g(Bu)$ = effect of irradiation defects = 3.8 $Bu^{0.28}$ in cm K/W
$h(T)$ = temperature dependence of annealing of irradiation defects
= $\frac{1}{1}$ + 396 exp($-Q/T$) dimensionless
$k_{0.95}$ = thermal conductivity at 95% theoretical density in W/cm K
Q = temperature of annealing parameter = 6380 K
T = temperature in kelvin

Constants A, B, E, and F for UO_2 and $(U,Pu)O_2$ are given in Table 8.4.

TABLE 8.4
Values of Constants in FRAPCON Correlations for Thermal Conductivity[a]

	A (cm K/W)	B (cm/W)	E (W K/cm)	F (K)
UO_2	4.52	2.46×10^{-2}	3.5×10^7	16,361
$(U, Pu)O_2$[b]	$285x + 3.5$	$(2.86 - 7.15x) \times 10^{-2}$	1.5×10^7	13,520

[a] Equations 8.22a, 8.22c, and 8.22d.
[b] $x = 2.00 - O/M$ (oxygen-to-metal atomic ratio).

For thermal conductivity of as-fabricated fuel density between 92% to 97% TD, the model of Equation 8.22a that is normalized to $\rho_{.95}$ is multiplied by a porosity correction factor η given by Equation 8.22b.

$$\eta = \frac{1.0789\rho/\rho_{TD}}{1.0 + 0.5(1 - \rho/\rho_{TD})} \tag{8.22b}$$

where ρ/ρ_{TD} is the as-fabricated density given as a fraction of theoretical density.

The thermal conductivity of fresh UO_2 and $(U,Pu)O_2$ without addition of gadolinia is given by Equation 8.22c, which is a reduced form of Equation 8.22a.

For $300 \leq T \leq 3000$ *K*:

$$k_{0.95} = \frac{1}{A + BT} + \frac{E}{T^2}\exp\left(-\frac{F}{T}\right) \tag{8.22c}$$

where again constants *A*, *B*, *E*, and *F* for UO_2 and $(U,Pu)O_2$ are given in Table 8.4.

The integral of Equation 8.22c is

$$\int_0^T k_{0.95}dT = \frac{1}{B}\ell n\left(\frac{A + BT}{A}\right) + \frac{E}{F}\exp\left(-\frac{F}{T}\right) \tag{8.22d}$$

Curves of this integral for fresh UO_2 and $(U,Pu)O_2$ fuel with the appropriate porosity correction factor η for different densities are given in Figure 8.3.

8.3.1.3 Oxygen-to-Metal Atomic Ratio

The oxygen-to-metal ratio of the uranium and plutonium oxides can vary from the theoretical (or stoichiometric) value of 2. This variation affects almost all the physical properties of the fuel. The departure from the initial stoichiometric condition occurs during burnup of fuel. In general, the effect of both the hyper- and hypostoichiometry is to reduce the thermal conductivity, as shown in Figure 8.4.

8.3.1.4 Plutonium Content

Thermal conductivity of the MOX fuel decreases as the plutonium oxide content increases, as can be seen in Figure 8.5.

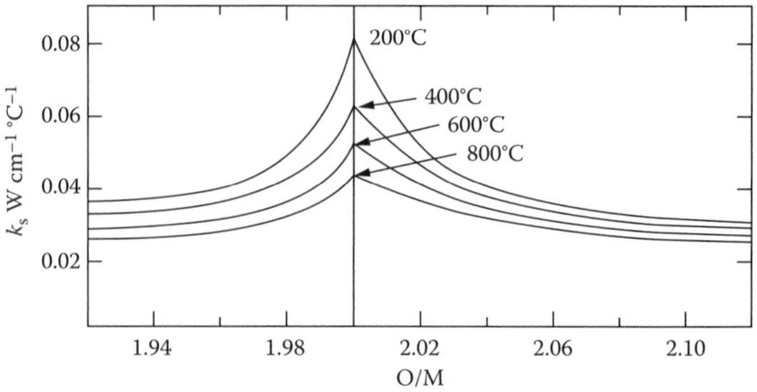

FIGURE 8.4 Thermal conductivity of $UO_{0.8}Pu_{0.2}O_{2\pm x}$ as a function of the O/(U + Pu) ratio. (Adopted from Schmidt, H. E. and Richter, J., The influence of stoichiometry on the thermal conductivity of $(U_{0.8}Pu_{0.2})O_2x$. Presented at the *Symposium on Oxide Fuel Thermal Conductivity*, Stockholm, Sweden, 1967.)

8.3.1.5 Effects of Pellet Cracking

Fuel pellet cracking and fragment relocation into the pellet-cladding gap during operation alter the fuel thermal conductivity and the gap conductance. A series of tests at the Idaho National Laboratory has led to an empirical formula for the decrease in the UO_2 thermal conductivity due to cracking. For a fresh, helium-filled LWR fuel rod with cracked and broken fuel pellets, this relation for effective conductivity is [24]

$$k_{\text{eff}} = k_{UO_2} - (0.0002189 - 0.050867X + 5.6578X^2) \tag{8.23a}$$

where

$$X = (\delta_{\text{hot}} - 0.014 - 0.14\,\delta_{\text{cold}})\left(\frac{0.0545}{\delta_{\text{cold}}}\right)\left(\frac{\rho}{\rho_{TD}}\right)^8 \tag{8.23b}$$

where k is in kW/m K; δ_{hot} = calculated hot radial gap width (mm) for the uncracked fuel; δ_{cold} = cold radial gap width (mm); ρ_{TD} = theoretical density of UO_2.

The effect of cracking on fuel conductivity is illustrated in Figure 8.6.

A semiempirical approach by MacDonald and Weisman [25] yielded the following relation between the effective and theoretical conductivity

$$k_{\text{eff}}\left(\text{W/m K}\right) = \frac{k_{UO_2}\left(\text{W/m K}\right)}{\left[\dfrac{2\delta_{\text{hot}} - A}{D_{\text{fo}}\left(B\,k_{\text{gas}}/k_{UO_2} + C\right)}\right] + 1} \tag{8.24}$$

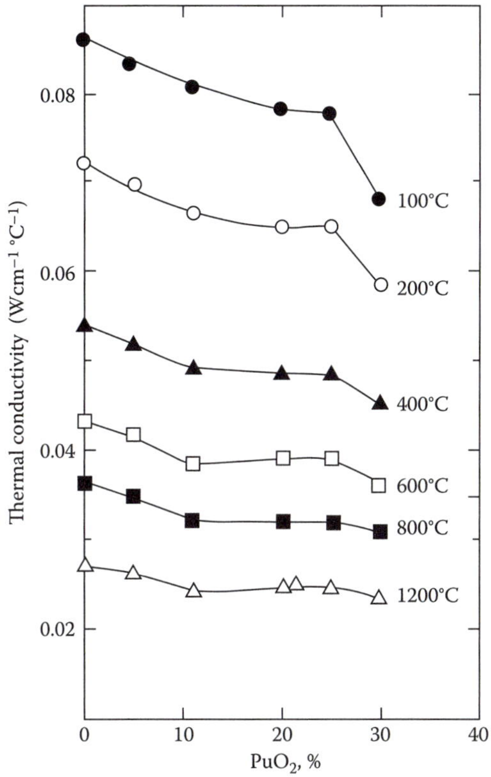

FIGURE 8.5 Thermal conductivity of (U,Pu)O_2 solid solutions as a function of PuO_2 content. (From Gibby, R. L., *J. Nucl. Mater.*, 38:163, 1971.)

where $A = 6.35 \times 10^{-5}$ m; $B = 0.077$; $C = 0.015$; D_{fo} = hot pellet radial diameter in meters; k_{gas} = thermal conductivity of the gas in the gap; and δ_{hot} = hot gap in meters.

8.3.1.6 Burnup

The irradiation of fuel induces several changes in the porosity, composition, and stoichiometry of the fuel. These changes, however, are generally small in LWRs, where the burnup is only of the order of 5% of the initial uranium atoms. In fast reactors, this effect would be larger, as the expected burnup is of the order of 10% of the initial uranium and plutonium atoms.

Introduction of fission products into the fuel with burnup leads to a slight decrease in the conductivity.

Finally, the oxide material operating at temperatures higher than a certain temperature, about 1400°C, undergoes a sintering process that leads to an increase in the fuel density. This increase in density, which occurs in the central region of the fuel, affects the conductivity and the fuel temperature distribution, which is discussed in detail in Section 8.6.

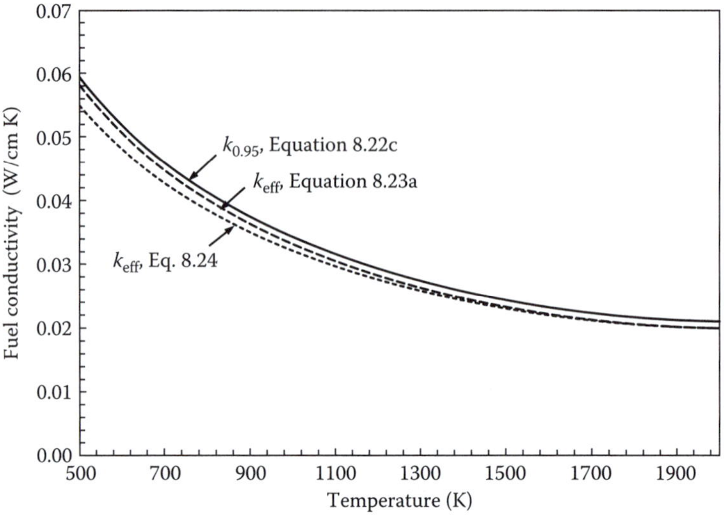

FIGURE 8.6 Representative comparison between FRAPCON fuel thermal conductivity and calculated effective fuel thermal conductivity for typical BWR fuel operating conditions (pellet OD 10 mm, cold gap 0.22 mm, hot gap 0.08 mm, and linear power 30 kW/m).

8.3.2 Fission Gas Release

It is important for the design of the fuel pin to calculate the gas released to the fuel pin plenum. Some of the fission gases are released from the UO_2 pellet at low temperature. Accompanying the change in the structure of the fuel at high temperatures is a significant additional release of fission gases to the fuel boundaries. For high-burnup fuel, the fission gas accumulated at the grain boundaries of UO_2 can contribute a large fraction of release during fast transient conditions [1]. The accumulated gases lead to pressurization of the cladding. For engineering analysis, a simple scheme is often used in which a certain fraction (f) of the gas is assumed to be released depending on the fuel temperature. An empirically based formula is given as [28]

$$\begin{aligned}
f &= 0.05 & T &< 1400°\text{C} \\
f &= 0.10 & 1500 > T &< 1400°\text{C} \\
f &= 0.20 & 1600 > T &< 1500°\text{C} \\
f &= 0.40 & 1700 > T &< 1600°\text{C} \\
f &= 0.60 & 1800 > T &< 1700°\text{C} \\
f &= 0.80 & 2000 > T &< 1800°\text{C} \\
f &= 0.98 & T &< 2000°\text{C}
\end{aligned} \tag{8.25}$$

More complicated approaches to fission gas release have been proposed based on various physical mechanisms of gas migration. Most designers, however, still prefer the simple empirically based models.

8.3.3 Melting Point

The melting point of UO_2 is in the vicinity of 2840°C (5144°F). The melting process for the oxide starts at a solidus temperature but is completed at a higher temperature called the *liquidus point*. The melting range is affected by the oxygen-to-metal ratio (Figure 8.7) and by the Pu content. Thus, in LWR designs the conservatively low value of 2600°C (4700 F) is often used. Olsen and Miller [31] fitted the melting point for the MATPRO code as

$$T(\text{solidus}) = 113 - 5.414\xi + 7.468 \times 10^{-3}\,\xi^2 K \tag{8.26}$$

where ξ = mole percent of PuO_2 in the oxide. The effect of increased PuO_2 content is shown in Figure 8.8, which plots Equation 8.26.

8.3.4 Specific Heat

The specific heat of the fuel plays a significant role in determining the sequence of events in many transients. It varies greatly over the temperature range of the fuel as Figure 8.9 based on Equation 8.27 illustrates.

Equation 8.27 proposed by Fink [10] is applicable to UO_2 and PuO_2 using the appropriate constants from Table 8.5:

$$c_p = c_2 + 2c_3t + 3c_4t^2 + 4c_5t^3 + 5c_6t^4 - c_7t^{-2} \tag{8.27}$$

where c_p is expressed in units of J/mol K.

$$t = T/1000 \quad \text{for } T \text{ in } K$$

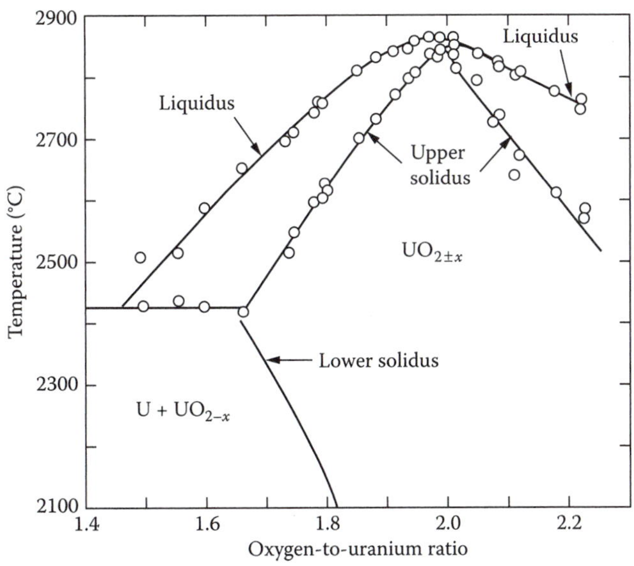

FIGURE 8.7 Partial phase diagram for uranium from $UO_{1.5}$ to $UO_{2.23}$. (From Latta, R. E. and Frysell, R. E., *J. Nucl. Mater.*, 35:195, 1970.)

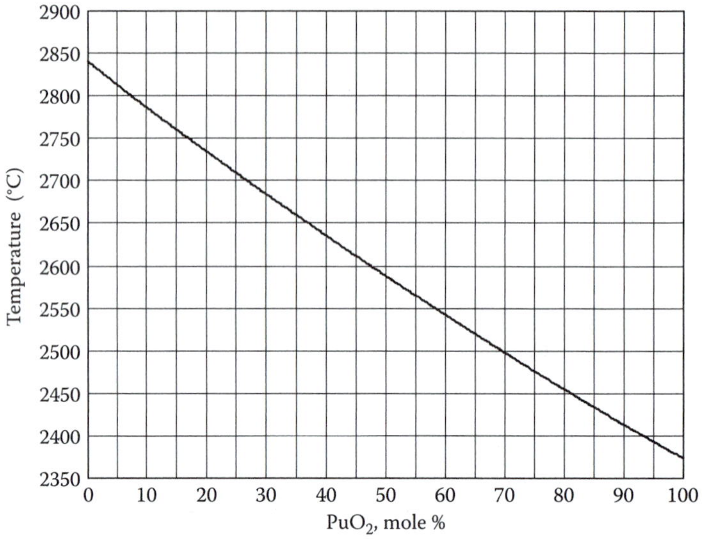

FIGURE 8.8 Melting points of mixed uranium–plutonium oxides. (From Olsen, C. S. and Miller, R. L., *MATPRO, Vol. II: A Handbook of Materials Properties for Use in the Analysis of Light Water Reactor Fuel Behavior.* NUREG/CR-0497, USNRC, 1979.)

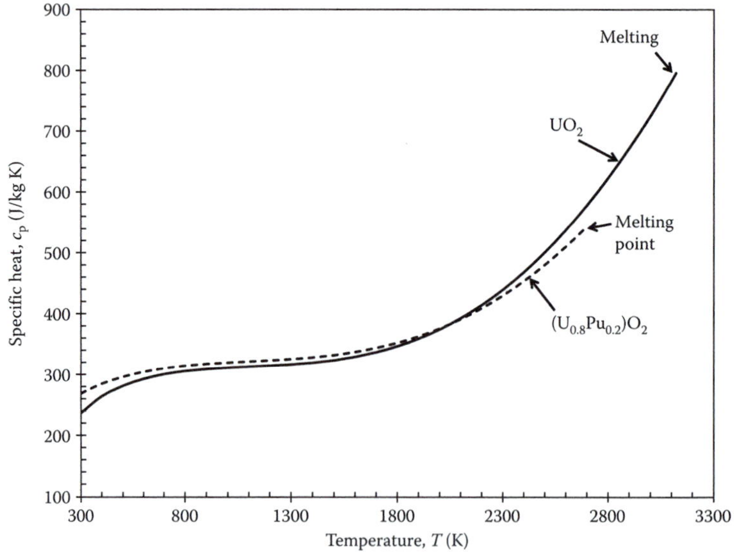

FIGURE 8.9 Temperature dependence of the specific heat capacity of UO_2 and $(U,Pu)O_2$. (From Olsen, C. S. and Miller, R. L., *MATPRO, Vol. II: A Handbook of Materials Properties for Use in the Analysis of Light Water Reactor Fuel Behavior.* NUREG/CR-0497, USNRC, 1979.)

TABLE 8.5
Constants Used in Heat Capacity Correlations (Equation 8.27)

Constant	UO_2	PuO_2	Units
c_2[a]	52.1743	84.495	J/mol K
	193.238	311.7866	J/kg K
c_3	43.9735	5.3195	J/mol K^2
	162.8647	19.629	J/kg K^2
c_4	– 28.0804	– 0.20379	J/mol K^3
	– 104.0014	– 0.752	J/kg K^3
c_5	7.88552	0	J/mol K^4
	29.2056	0	J/kg K^4
c_6	– 0.52668	0	J/mol K^5
	– 1.9507	0	J/kg K^5
c_7	0.71391	1.90056	J K/mol
	2.6441	7.0131	J K/kg

[a] For the burnup effect replace c_2 with c_2^* given by $c_2^* = c_2(1 + 0.011B)$ for burnup B in at %.

valid from 298.15 to 3120 K for UO_2; and 298.15 to 2701 K for PuO_2 where the upper limits are the melting points.

This equation employs the variable $t = T/1000$ instead of T to minimize the size of the coefficients.

8.3.5 The Rim Effect

At high burnup in LWRs, the buildup of plutonium near the outer surface of a fuel pellet due to neutron absorption in ^{238}U leads to significantly higher local volumetric heat generation rates near the surface. Thus, the assumption of uniform heat generation rate in the fuel becomes less accurate. However, this rim effect is confined to a narrow ring near the surface and therefore has limited influence on the bulk temperature of the fuel. However, at very high burnup, the rim structure becomes highly porous and retains a lot of fission gases which add to potential fission gas release during transients. This is important to pressure buildup during a transient and fuel-cladding interaction in response to transients.

Example 8.1: Effect of Cracking on Fuel Conductivity

PROBLEM

Evaluate the effective conductivity of the fuel after cracking at 1000°C for the geometry and operating conditions given below:

1. *Geometry and materials:* BWR fuel rod with UO_2 solid fuel pellet and zircaloy cladding. Cold fuel rod dimensions (at 27°C) are
   a. Cladding outside diameter = 11.2 mm.

   b. Cladding thickness = 0.71 mm.
   c. Gap thickness = 90 μm.
2. *Assumptions*:
   a. Initial fuel density = 0.88 ρ_{TD}.
   b. UO_2 conductivity is predicted by the AREVA correlation in Table 8.3.
   c. Porosity correction factor for the conductivity is given by Equation 8.21 assuming spherical pores.
   d. Fuel conductivity of the cracked fuel is given by Equation 8.23a.
3. *Operating conditions*:
   a. Fuel temperature = 1000°C.
   b. Cladding temperature = 295°C.c.Burnup = 26 MWd/kg.

SOLUTION

Consider first the conductivity of the uncracked fuel pellet. From Equation 8.16 (Table 8.3), the UO_2 conductivity at 1000°C and theoretical density is

$$\begin{aligned} f(1273) &= \left[1+\exp\left(\frac{1273-900}{80}\right)\right]^{-1} \\ &= [1+105.9]^{-1} \\ &= 0.0093545 \end{aligned}$$

$$\begin{aligned} g(1273) &= 4.715\times 10^{9}\times 1273^{-2}\times \exp(-16{,}361/1273) \\ &= 4.715\times 10^{9}\times 6.17\times 10^{-7}\times 2.62\times 10^{-6} \\ &= 0.007622 \text{ W/mK} \end{aligned}$$

$$\begin{aligned} A_1 + A_2T + A_3Bu + A_4f(T) &= 0.0375 + 2.165\times 10^{-4}\times 1273 \\ &\quad + 1.7\times 10^{-3}\times 26 + 0.058\times 0.0093545 \\ &= 0.0375 + 0.2756 + 0.0442 + 5.4256\times 10^{-4} \\ &= 0.35784 \text{ mK/W} \end{aligned}$$

$$\begin{aligned} k_{TD} &= \frac{1}{A_1 + A_2T + A_3Bu + A_4f(T)} + g(T) \\ &= \frac{1}{0.35784} + 0.007622 \\ &= 2.79454 + 0.007622 \\ &= 2.802 \text{ W/mK} \end{aligned} \tag{8.16}$$

Applying the porosity correction factor of Equation 8.21:

$$\frac{k}{k_{TD}} = \frac{1-P}{1+0.5P} = \frac{\rho/\rho_{TD}}{1+0.5(1-\rho/\rho_{TD})}$$

Because $\rho/\rho_{TD} = 0.88$ for the 88% theoretical density fuels, we get

$$\frac{k_{88}}{k_{TD}} = \frac{0.88}{1+0.5(0.12)} = 0.83$$

Therefore, the conductivity of the 88% theoretical density uncracked fuel pellet is

$$k_{88} = 0.83(0.002802) = 2.326 \times 10^{-3} \text{ kW/m °C}$$

Consider the cracked fuel effective conductivity. From Equations 8.23a and 8.23b

$$k = k_{UO_2} - (0.0002189 - 0.050867X + 5.6578X^2)\text{kW/m°C} \quad (8.23a)$$

$$X = (\delta_{hot} - 0.014 - 0.14\delta'_{cold})\left(\frac{0.0545}{\delta_{cold}}\right)\left(\frac{\rho}{\rho_{TD}}\right)^8 \quad (8.23b)$$

In order to evaluate δ_{hot}, we must evaluate the change in the radius of the fuel and the cladding. We neglect the fission products swelling and relocation of UO_2 pellet. If the fuel pellet radius is R_{fo} and the cladding inner radius is R_{ci}

$$\begin{aligned}\delta_{hot} &= (R_{ci})_{hot} - (R_{fo})_{hot} \\ &= (R_{ci})_{cold}\left[1 + \alpha_c(T_c - 27)\right] - (R_{fo})_{cold}\left[1 + \alpha_f(T_f - 27)\right]\end{aligned}$$

where α = linear thermal expansion coefficient.

However

$$(R_{ci})_{cold} = \frac{11.2}{2} - 0.71 = 4.89\text{mm}$$

and

$$(R_{fo})_{cold} = \frac{11.2}{2} - 0.71 - 0.09 = 4.80\text{mm}$$

From Table 8.1, $\alpha_f = 10.1 \times 10^{-6}$/°C
From Table 8.2, $\alpha_c = 6.72 \times 10^{-6}$/°C
Therefore

$$\begin{aligned}\delta_{hot} &= 4.89\left[1 + 6.72\times 10^{-6}(295 - 27)\right] - 4.80\left[1 + 10.1 \times 10^{-6}(1000 - 27)\right] \\ &= 4.89881 - 4.84717 \\ &= 0.0516\text{mm}\end{aligned}$$

Now we can determine the parameter X from Equation 8.23b

$$\begin{aligned}X &= [0.0516 - 0.014 - 0.14(0.09)]\left[\frac{0.0545}{0.09}\right][0.88]^8 \\ &= 0.0250(0.6056)(0.359635) \\ &= 0.00545\end{aligned}$$

Hence, the cracked fuel conductivity for the 88% theoretical density fuel is given by Equation 8.23a as

$$\begin{aligned} k_{\text{eff}} &= 0.002326 - [0.0002189 - 0.050867(0.00545) + 5.6587(0.00545)^2] \\ &= 0.002326 - (0.0002189 - 0.0002772 + 0.0001681) \\ &= 0.002326 - 0.0001098 \\ &= 0.002216 \text{ kW/m °C} \end{aligned}$$

Thus, the effect of cracking is to reduce the fuel thermal conductivity in this fuel from 2.326 to the effective value of 2.216 W/m °C.

8.4 TEMPERATURE DISTRIBUTION IN PLATE FUEL ELEMENTS

Assume a fuel plate is operating with a uniform heat generation rate, q'''. The fuel is clad in thin metallic sheets, with perfect contact between the fuel and the cladding as shown in Figure 8.10.

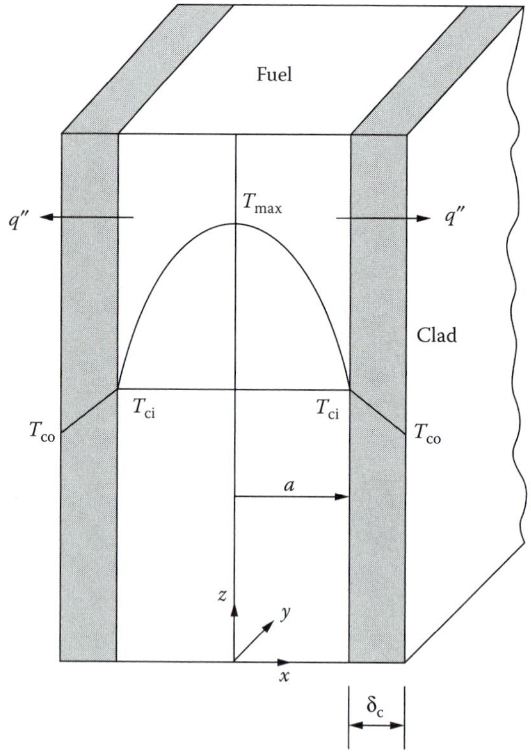

FIGURE 8.10 Plate fuel element.

If the fuel plate is thin and extends in the y and z directions considerably more than it does in the x direction, the heat conduction equation (Equation 8.2)

$$\frac{\partial}{\partial x}k\frac{\partial T}{\partial x}+\frac{\partial}{\partial y}k\frac{\partial T}{\partial y}+\frac{\partial}{\partial z}k\frac{\partial T}{\partial z}+q'''=0 \tag{8.28a}$$

can be simplified by assuming the heat conduction in the y and z directions to be negligible, that is

$$k\frac{\partial T}{\partial y}\simeq k\frac{\partial T}{\partial z}\simeq 0 \tag{8.29}$$

Hence, we need only to solve the one-dimensional equation

$$\frac{\mathrm{d}}{\mathrm{d}x}k\frac{\mathrm{d}T}{\mathrm{d}x}+q'''=0 \tag{8.28b}$$

By integrating once, we get

$$k\frac{\mathrm{d}T}{\mathrm{d}x}+q'''x=C_1 \tag{8.30}$$

8.4.1 Heat Conduction in Fuel

Because q''' is uniform, and if the temperatures at both interfaces between cladding and fuel are equal, the fuel temperature should be symmetric around the center plane. In the absence of any heat source or sink at $x = 0$, no heat flux should cross the plane at $x = 0$. Hence

$$k_{\mathrm{f}}\left.\frac{\mathrm{d}T}{\mathrm{d}x}\right|_{x=0}=0 \tag{8.31}$$

Applying the condition of Equation 8.31 to Equation 8.30 leads to

$$C_1=0 \tag{8.32}$$

and

$$k_{\mathrm{f}}\frac{\mathrm{d}T}{\mathrm{d}x}+q'''x=0 \tag{8.33}$$

Integrating Equation 8.33 between $x = 0$ and any position x, and applying the condition that $T = T_{max}$ at $x = 0$, we get

$$\int_{T_{max}}^{T} k_f \, dT + q''' \frac{x^2}{2}\Big|_0^x = 0$$

or

$$\int_{T}^{T_{max}} k_f \, dT = q''' \frac{x^2}{2} \tag{8.34}$$

Three conditions may exist, depending on whether q''', T_{ci}, or T_{max} are known.

1. q''' Specified: A relation between T_{max} and T_{ci} can be determined when q''' is specified by substituting $x = a$ in Equation 8.34 to get

$$\int_{T_{ci}}^{T_{max}} k_f \, dT = q''' \frac{a^2}{2} \tag{8.35}$$

If k_f is constant

$$(T_{max} - T_{ci}) = q''' \frac{a^2}{2k_f} \tag{8.36}$$

2. T_{ci} specified: If T_{ci} is known, Equation 8.34 can be used to specify the relation between q''' and T_{max} in the form

$$q''' = \frac{2}{a^2} \int_{T_{ci}}^{T_{max}} k_f \, dT \tag{8.37}$$

3. T_{max} specified: Equation 8.37 can be used to specify the relation between q''' and T_{ci} given the value of T_{max}.

8.4.2 Heat Conduction in Cladding

The heat conduction in the cladding can also be assumed to be a one-dimensional problem, so that Equation 8.29 also applies in the cladding. Furthermore, the heat generation in the cladding is negligible (mainly owing only to absorption of γ-rays and both elastic and inelastic scattering of neutrons). Hence, in the cladding the heat conduction equation is given by

$$\frac{\mathrm{d}}{\mathrm{d}x} k_{\mathrm{c}} \frac{\mathrm{d}T}{\mathrm{d}x} = 0 \tag{8.38}$$

Integrated once, it leads to the equation

$$k_{\mathrm{c}} \frac{\mathrm{d}T}{\mathrm{d}x} = \text{constant} \tag{8.39}$$

which implies that the heat flux is the same at any position in the cladding. Let q'' be the heat flux in the cladding in the positive or negative outward x direction. Therefore

$$-k_{\mathrm{c}} \frac{\mathrm{d}T}{\mathrm{d}x} = q'' \tag{8.40}$$

Integrating Equation 8.40 between $x = a$ and any positive position x leads to

$$\int_{T_{\mathrm{ci}}}^{T} k_{\mathrm{c}}\, \mathrm{d}T = -q''(x-a) \tag{8.41}$$

or for k_{c} constant

$$k_{\mathrm{c}}\,(T - T_{\mathrm{ci}}) = -q''(x - a) \tag{8.42}$$

Thus, the right-hand side (RHS) external-cladding surface temperature is given by

$$k_{\mathrm{c}}(T_{\mathrm{co}} - T_{\mathrm{ci}}) = -q''_{\mathrm{RHS}}(a + \delta_{\mathrm{c}} - a) \tag{8.43}$$

or

$$T_{\mathrm{co}} = T_{\mathrm{ci}} - \frac{q''_{\mathrm{RHS}}\delta_{\mathrm{c}}}{k_{\mathrm{c}}} \tag{8.44a}$$

where δ_{c} = cladding thickness. Integration of Equation 8.40 in the negative outward x direction yields

$$T_{\mathrm{co}} = T_{\mathrm{ci}} + \frac{q''_{\mathrm{LHS}}\delta_{\mathrm{c}}}{k_{\mathrm{c}}} \tag{8.44b}$$

The heat flux q''_{LHS} is negative since the temperature gradient within the left-hand side (LHS) cladding region is positive. By symmetry, the LHS and RHS external-cladding surface temperatures are identical.

8.4.3 Thermal Resistances

Note that q'' is equal to the heat generated in one-half of the fuel plate. Thus, for the LHS fuel region:

$$q'' = q'''a \tag{8.45}$$

Therefore, the fuel temperature drop may also be obtained by substituting for $q'''a$ from Equation 8.45 into Equation 8.36 and rearranging the result

$$T_{ci} = T_{max} - q''\frac{a}{2k_f} \tag{8.46}$$

Substituting T_{ci} from Equation 8.46 into Equation 8.44a, we get

$$T_{co} = T_{max} - q''\left(\frac{a}{2k_f} + \frac{\delta_c}{k_c}\right) \tag{8.47}$$

By simple manipulation of Equation 8.47, the heat flux q'' can be given as

$$q'' = \frac{T_{max} - T_{co}}{(a/2k_f) + (\delta_c/k_c)} \tag{8.48}$$

Thus, the temperature difference acts analogously to an electrical potential difference that gives rise to a current (q'') whose value is dependent on two thermal resistances in series (Figure 8.11). This concept of resistances proves useful for simple transient fuel temperature calculations.

8.4.4 Conditions for Symmetric Temperature Distributions

The temperature field symmetry in the preceding discussion enabled us to solve for the temperature by considering only one-half the plate. It is useful to reflect on the required conditions to produce such symmetry.

Consider the general case for a plate fuel element with internal heat generation that is cooled on both sides (Figure 8.12). In this case, some of the heat is removed from the RHS, and the rest is removed from the LHS. Therefore, a plane exists within the fuel plate through which no heat flux passes. Let the position of this plane be x_0. Thus

$$q''\Big|_{x_0} = -k\frac{\partial T}{\partial x}\Big|_{x_0} = 0 \tag{8.49}$$

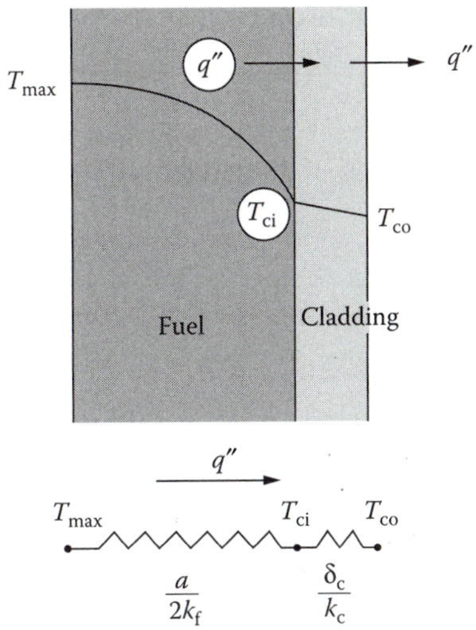

FIGURE 8.11 Electrical current equivalence with the heat flux.

The value of x_0 is zero if symmetry of the temperature distribution exists. This symmetry can be *a priori* known under specific conditions, all of which should be present simultaneously. These conditions are the following:

1. Symmetric distribution of heat generation in the fuel plate.
2. Equal resistances to heat transfer on both sides of the plate, which translates in the simplest case into similar material and geometric configurations on both sides (i.e., uniform fuel, fuel-cladding gap, and cladding material thicknesses).

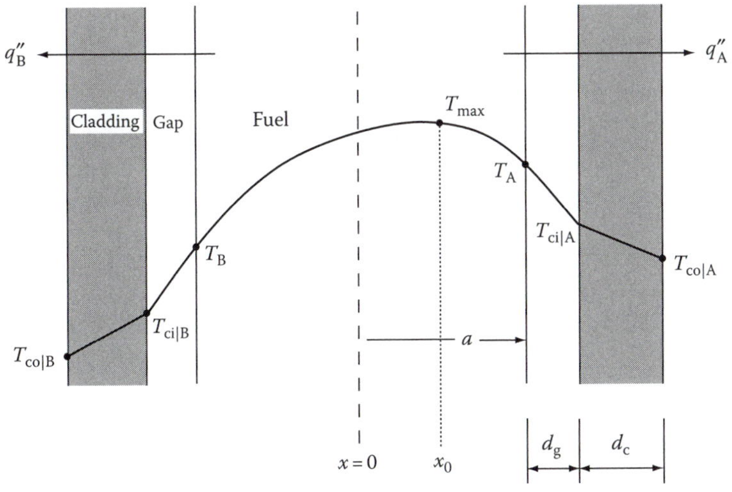

FIGURE 8.12 Plate with asymmetric temperature distribution.

3. Equal temperatures of the boundary of the plate on both sides, that is $T_{\mathrm{colA}} = T_{\mathrm{colB}}$.

If the three conditions exist in a fuel element, symmetry of the temperature field can be assumed and the condition of zero heat flux at the midplane can be applied, that is, $x_0 = 0$ in Equation 8.49:

$$k\left.\frac{\partial T}{\partial x}\right|_{x_0} = 0 \tag{8.50}$$

The following examples violate the conditions of symmetry mentioned above:

1. Nonuniform heat generation due to absorption of an incident flux of γ-rays, as occurs in a core barrel or thermal shield (see Section 3.8).
2. Nonidentical geometry of the regions surrounding the fuel as happens in an off-center fuel element.
3. Unequal temperatures of the bounding surfaces of the plate as shown in Figure 8.12.

In a general coordinate system with a volumetric heat source, the condition for the temperature gradient at the origin of the coordinate system is as follows:

$$\begin{aligned} \nabla T = 0 \quad & \text{at } x = 0 \text{ (plate)} \\ & \text{at } r = 0 \text{ (cylinder)} \\ & \text{at } r = 0 \text{ (sphere)} \end{aligned} \tag{8.51}$$

For condition 8.51 to be valid, it is implied that the fuel elements are solid. If there is a central void, condition 8.51 is not useful for solution of the temperature profile within the fuel region. When an inner void exists, and if a symmetric condition of the temperature field applies, no heat flux would exist at the void boundary. This situation is examined for cylindrical pins in Section 8.5 which follows.

Example 8.2: Maximum Temperature in a Fuel Plate under Various Coolant Temperature Conditions

PROBLEM

For the metallic fuel plate of Figure 8.12, eliminate the gap and take the cladding bonded to the fuel, the fuel half-thickness a is 0.38 mm, the cladding thickness δ_c is 0.38 mm, the fuel and cladding have constant conductivities k_f and k_c equal to 186 and 41 W/mK, respectively, and the volumetric energy generation rate is constant at 1150 MW/m^3.

Find the maximum fuel temperature and its location if

a. The boundary-cladding temperatures T_{co} are both equal to 350 K.
b. The left boundary, cladding temperature $T_{\mathrm{co|B}}$ is 350 K while the right boundary-cladding temperature $T_{\mathrm{co|A}}$ is 355 K.

Note that if the right boundary-cladding temperature becomes sufficiently elevated, the heat flux out of that surface is zero and all the energy generated in the plate flows out the left boundary.

SOLUTION

Case a: For this case, the fuel temperature is symmetric since the conditions of Section D are satisfied. Hence, the location of the maximum fuel temperature is at $x = 0$ and its magnitude is given by Equation 8.47 with $q'' = q'''a$ since the temperature distribution is symmetric.

$$
\begin{aligned}
T_{max} &= T_{co} + q'''a\left(\frac{a}{2k_f} + \frac{\delta_c}{k_c}\right) \\
&= 350 + 1150 \times 10^6(0.38 \times 10^{-3})\left(\frac{0.38 \times 10^{-3}}{2(186)} + \frac{0.38 \times 10^{-3}}{41}\right) \\
&= 350 + 0.437[1.02 + 9.26] \\
&= 354.5\text{K}
\end{aligned}
\tag{8.47}
$$

Case b: For this case, the temperature distribution through the fuel plate is asymmetric since the plate boundary temperatures differ, that is, $T_{co|A} > T_{co|B}$. The location of the maximum fuel temperature is displaced toward the higher-temperature boundary (see Figure 8.12 but with $\delta_g = 0$) and consequently $q''_A < |q''_B|$.

First let us express the inside-cladding temperature on both sides A and B from Equation 8.44 as

$$T_{ci|A} = T_{co|A} + \frac{q''_A \delta_c}{k_c} \tag{8.44a}$$

$$T_{ci|B} = T_{co|B} - \frac{q''_B \delta_c}{k_c} \tag{8.44b}$$

Now the temperature distribution through the fuel is obtained by integrating Equation 8.28b twice yielding

$$T_x = -\frac{q'''x^2}{2k_f} + C_1 x + C_2 \tag{8.52a}$$

and applying the boundary conditions that

$$\text{at } x = a, \quad T_x = T_{ci|A} \tag{8.52b}$$

$$\text{at } x = -a, \quad T_x = T_{ci|B} \tag{8.52c}$$

Obtain

$$C_1 = \frac{T_{ci|A} - T_{ci|B}}{2a} \tag{8.53a}$$

and

$$C_2 = \frac{T_{ci|A} + T_{ci|B}}{2} + \frac{q'''a^2}{2k_f} \tag{8.53b}$$

Inserting these constants into Equation 8.52a, we obtain

$$T_x = \frac{q'''}{2k_f}(a^2 - x^2) + \left(T_{ci|A} - T_{ci|B}\right)\frac{x}{2a} + \frac{T_{ci|A} + T_{ci|B}}{2} \tag{8.54}$$

The location x_0 of the maximum fuel temperature is obtained by setting the differential of Equation 8.54 with respect to x equal to zero and solving for x_0, that is

$$\left.\frac{dT_x}{dx}\right|_{x0} = 0 = \frac{-q'''x_0}{k_f} + \frac{\left(T_{ci|A} - T_{ci|B}\right)}{2a}$$

or

$$x_0 = \frac{k_f}{q'''}\left[\frac{T_{ci|A} - T_{ci|B}}{2a}\right] \tag{8.55}$$

Hence, from Equation 8.54

$$T_{max} = \frac{q'''}{2k_f}\left(a^2 - x_0^2\right) + \left(T_{ci|A} - T_{ci|B}\right)\frac{x_0}{2a} + \frac{\left(T_{ci|A} + T_{ci|B}\right)}{2} \tag{8.56}$$

We desire a solution for T_{max} and the position x_0 in terms of the known boundary condition temperatures $T_{co|A}$ and $T_{co|B}$. Hence we need further manipulation of Equations 8.55 and 8.56, since they are expressed in terms of inner-cladding temperatures $T_{ci|A}$ and $T_{ci|B}$.

To complete the solution, express q''_A and q''_B from an energy balance as

$$q''_A = q'''\left(a - x_0\right) \tag{8.57a}$$

$$q''_B = -q'''\left(a + x_0\right) \tag{8.57b}$$

Equations 8.44 can now be reexpressed as

$$T_{ci|A} = T_{co|A} + \frac{q'''(a - x_0)\delta_c}{k_c} \tag{8.58a}$$

$$T_{ci|B} = T_{co|B} + \frac{q'''(a + x_0)\delta_c}{k_c} \tag{8.58b}$$

Subtracting and adding Equations 8.58a and b, we obtain

$$T_{ci|A} - T_{ci|B} = T_{co|A} - T_{co|B} - \frac{q'''\delta_c}{k_c}(2x_0) \quad (8.59a)$$

$$T_{ci|A} + T_{ci|B} = T_{co|A} + T_{co|B} + \frac{q'''\delta_c}{k_c}(2a) \quad (8.59b)$$

In order to find x_0 in terms of the known temperatures $T_{co|A}$ and $T_{co|B}$, insert Equation 8.59a into Equation 8.55 to obtain

$$x_0 = \frac{k_f/q'''2a\left(T_{co|A} - T_{co|B}\right)}{1+(k_f/k_c)(\delta_c/a)}$$

and applying the numerical values we get the value of position x_0 as

$$x_0 = \frac{(186/1150\times10^6(2)0.38\times10^{-3})(355-350)}{1+(186/41)(0.38\times10^{-3}/0.38\times10^{-3})} = \frac{1.06\times10^{-3}}{5.54} = 0.191\,\text{mm}$$

Thus, the position x_0 is very nearly equal to half the distance between the fuel surface and the fuel centerline.

The maximum fuel temperature is given by Equation 8.56. Hence, first obtain $T_{ci|A} - T_{ci|B}$ and $T_{ci|A} + T_{ci|B}$ from Equations 8.59

$$\begin{aligned} T_{ci|A} - T_{ci|B} &= 355 - 350 - \frac{1150\times10^6\left(0.38\times10^{-3}\right)}{41}(2)\left(0.191\times10^{-3}\right) \\ &= 5.0 - 4.1 = 0.9\text{K} \end{aligned}$$

$$\begin{aligned} T_{ci|A} + T_{ci|B} &= 355 + 350 + \frac{1150\times10^6\left(0.38\times10^{-3}\right)}{41}(2)\left(0.38\times10^{-3}\right) \\ &= 705 + 8.1 = 713.1\text{K} \end{aligned}$$

Finally, from Equation 8.56 obtain the maximum fuel temperature as

$$\begin{aligned} T_{max} &= \frac{1150\times10^6}{2(186)}\left[\left(0.38\times10^{-3}\right)^2 - \left(0.191\times10^{-3}\right)^2\right] \\ &\quad + 0.9\left[\frac{0.191\times10^{-3}}{2\left(0.38\times10^{-3}\right)}\right] + \frac{713.1}{2} \\ &= 0.34 + 0.23 + 356.55 \\ &= 357.1\,\text{K} \end{aligned}$$

8.5 TEMPERATURE DISTRIBUTION IN CYLINDRICAL FUEL PINS

We first derive the basic relations for cylindrical (solid and annular) fuel pellets. Two annular cases are considered—the inner surface either insulated or cooled by an axial flow stream. In both cases, the outer surface is cooled by an independent axial flow stream. Then conclusions are drawn with regard to (1) the maximum fuel temperature, T_{max}, for a given linear heat rate, q', and (2) the maximum possible heat rate, q'_{max}, for a given T_{max}.

Cylindrical fuel pellets are nearly universally used as the fuel form in power reactors. Dimensions of the fuel pellet and cladding are given in Table 1.3.

8.5.1 General Conduction Equation for Cylindrical Geometry

If the neutron flux is assumed to be uniform within the fuel pellet, the heat generation rate can be assumed uniform. The coolant turbulent flow conditions in a typical LWR fuel assembly of a pin pitch-to-diameter ratio of more than 1.2 are such that the azimuthal flow conditions can be taken to be essentially the same around the fuel rod. (More information on the azimuthal heat flux distribution is available in Chapter 7, Volume II.) The above two conditions lead to the conclusion that no significant azimuthal temperature gradients exist in the fuel pellet. Also, for a fuel pin of a length-to-diameter ratio of more than 10, it is safe to neglect the axial heat transfer within the fuel relative to the radial heat transfer for most of the pin length. However, near the top and bottom ends, axial heat conduction plays a role in determining the temperature field.

Thus, at steady state, the heat conduction equation sufficiently far from the rod ends reduces to a one-dimensional equation in the radial direction

$$\frac{1}{r}\frac{\mathrm{d}}{\mathrm{d}r}\left(kr\frac{\mathrm{d}T}{\mathrm{d}r}\right)+q'''=0 \tag{8.60}$$

Integrating Equation 8.60 once, we get

$$kr\frac{\mathrm{d}T}{\mathrm{d}r}+q'''\frac{r^2}{2}+C_1=0 \tag{8.61}$$

which can be divided by r to obtain

$$k\frac{\mathrm{d}T}{\mathrm{d}r}+q'''\frac{r}{2}+\frac{C_1}{r}=0 \tag{8.62}$$

Equation 8.62 can be integrated between r and R_v to yield, after rearrangement

$$-\int_{T_{max}}^{T} k\,\mathrm{d}T=\frac{q'''}{4}\left[r^2-R_v^2\right]+C_1\ell n\left(\frac{r}{R_v}\right) \tag{8.63}$$

where R_v is the radius of the central void region.

8.5.2 Solid Fuel Pellet

The general heat flux condition that can be applied in cylindrical geometry at $r = R_v$ is

$$q''\big|_{r=R_v} = -k\frac{dT}{dr}\bigg|_{r=R_v} = 0 \tag{8.64}$$

Applying this condition to Equation 8.62 leads to

$$C_1 = -\frac{q'''R_v^2}{2} \tag{8.65}$$

But since $R_v = 0$ for the solid fuel pellet, then from Equation 8.65

$$C_1 = 0 \tag{8.66}$$

Hence, Equation 8.63 becomes

$$\int_{T}^{T_{max}} k\,dT = \frac{q'''r^2}{4} \tag{8.67}$$

A relation among T_{max}, T_{fo}, and R_{fo} can be obtained from Equation 8.67 by taking $r = R_{fo}$ to get

$$\int_{T_{fo}}^{T_{max}} k\,dT = \frac{q'''R_{fo}^2}{4} \tag{8.68}$$

The linear heat rate is given by an energy balance as

$$q' = \pi R_{fo}^2 q''' \tag{8.69}$$

Therefore

$$\int_{T_{fo}}^{T_{max}} k\,dT = \frac{q'}{4\pi} \tag{8.70}$$

It is interesting to note that the temperature difference across a solid fuel pellet is fixed by q' and is independent of the pellet radius (R_{fo}). Thus, a limit on q' is directly implied by a design requirement on the maximum fuel temperature.

It should also be mentioned, and the student can verify on his or her own, that for a constant conductivity the average temperature* in a fuel pin is given by

$$T_{\text{avg}} - T_{\text{fo}} = \frac{1}{2}(T_{\text{max}} - T_{\text{fo}}) = \frac{q'}{8\pi k} \tag{8.71a}$$

compared to maximum temperature in the fuel pin of

$$T_{\text{max}} - T_{\text{fo}} = \frac{q'}{4\pi k} \tag{8.71b}$$

8.5.3 Annular Fuel Pellet (Cooled Only on the Outside Surface R_{fo})

For the annular pellet of Figure 8.13, in this case no heat flux exists at R_{v}. Hence, C_1 from Equation 8.65 can be substituted into Equation 8.63 to get

$$-\int_{T_{\text{max}}}^{T} k\text{d}T = \frac{q'''}{4}\left[r^2 - R_{\text{v}}^2\right] - \frac{q''' R_{\text{v}}^2}{2} \ell n\left(\frac{r}{R_{\text{v}}}\right) \tag{8.72a}$$

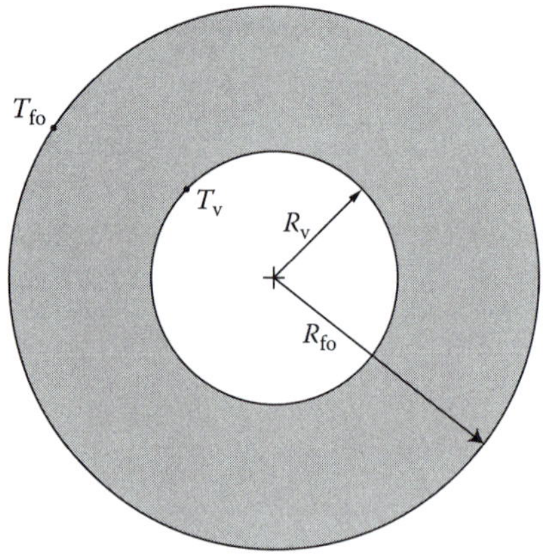

FIGURE 8.13 Cross section of an annular pellet.

---

* Average fuel temperature $\equiv \dfrac{\int_0^R T(r) 2\pi r \, dr}{\int_0^R 2\pi r \, dr}$

Equation 8.72a can be rearranged as

$$\int_{T}^{T_{max}} k\, dT = \frac{q''' r^2}{4}\left\{\left[1-\left(\frac{R_v}{r}\right)^2\right]-\left(\frac{R_v}{r}\right)^2 \ell n\left(\frac{r}{R_v}\right)^2\right\} \tag{8.72b}$$

Equation 8.72b can be used to provide a relation among T_{max}, T_{fo}, R_v, and R_{fo} when the condition $T = T_{fo}$ at $r = R_{fo}$ is applied. Thus, we get

$$\int_{T_{fo}}^{T_{max}} k\, dT = \frac{q''' R_{fo}^2}{4}\left\{\left[1-\left(\frac{R_v}{R_{fo}}\right)^2\right]-\left(\frac{R_v}{R_{fo}}\right)^2 \ell n\left(\frac{R_{fo}}{R_v}\right)^2\right\} \tag{8.73}$$

Note that for uniform heat generation the linear heat rate is given by

$$q' = \pi\left(R_{fo}^2 - R_v^2\right)q''' \tag{8.74a}$$

so that

$$q''' R_{fo}^2 = \frac{q'}{\pi\left[1-\left(R_v/R_{fo}\right)^2\right]} \tag{8.74b}$$

Substituting for $q''' R_{fo}^2$ from Equation 8.74b into Equation 8.73, we get

$$\int_{T_{fo}}^{T_{max}} k\, dT = \frac{q'}{4\pi}\left[1-\frac{\ell n\left(R_{fo}/R_v\right)^2}{\left(R_{fo}/R_v\right)^2 - 1}\right] \tag{8.75}$$

If a void factor for the outside surface-cooled case is defined as

$$F_{V_{OC}} \equiv 1-\frac{\ell n(\alpha^2)}{\beta(\alpha^2-1)} \equiv F_{V_{OC}}(\alpha,\beta) \tag{8.76}$$

Equation 8.75 can be written as

$$\int_{T_{fo}}^{T_{max}} k\, dT = \frac{q'}{4\pi}\left[F_{V_{OC}}\left(\frac{R_{fo}}{R_v},1\right)\right] \tag{8.77}$$

Figure 8.14 provides a plot of $F_{V_{OC}}$ in terms of α. Note that $F_{V_{OC}}$ is always less than 1. Here the parameter $\beta = 1$. This parameter can be used to reflect the effect of nonuniform radial heat generation rate in large pellets.

8.5.4 Annular Fuel Pellet (Cooled on Both Surfaces)

Consider that the annular fuel pellet of Figure 8.12 is cooled both on surfaces R_v and R_{fo} such that the surface temperatures are T_v and T_{fo}, respectively. Let us examine the performance of this pellet for the general case where these temperatures differ.

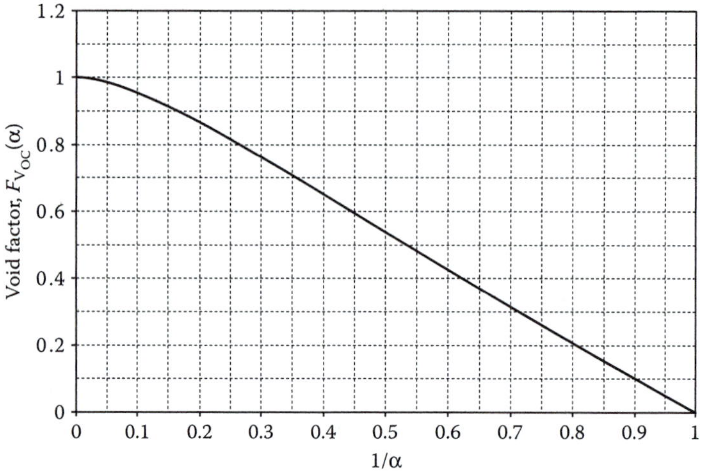

FIGURE 8.14 Void factor function, Equation 8.76. For an annular region, α is the ratio of the outer to the inner radius. Here β = 1.

Integrating Equation 8.62 with reference temperature T_v to obtain

$$\int_{T_v}^{T} k\,dT + \frac{q'''r^2}{4} + C_1 ln\, r + C_2 = 0 \tag{8.78}$$

Applying the surface temperature boundary conditions stated above

$$\frac{q'''R_v^2}{4} + C_1 lnR_v + C_2 = 0 \tag{8.79a}$$

and

$$\int_{T_v}^{T_{fo}} k\,dT + \frac{q'''R_{fo}^2}{4} + C_1 lnR_{fo} + C_2 = 0 \tag{8.79b}$$

Subtracting Equation 8.79b from 8.79a to find C_1 and then substituting the result into Equation 8.79a to find C_2, we obtain

$$C_1 = -\left[\int_{T_v}^{T_{fo}} k\,dT + \frac{q'''}{4}\left(R_{fo}^2 - R_v^2\right)\right]\left(\frac{1}{ln(R_{fo}/R_v)}\right) \tag{8.80a}$$

$$C_2 = -\frac{q'''R_v^2}{4} + \left[\int_{T_v}^{T_{fo}} k\,dT + \frac{q'''}{4}\left(R_{fo}^2 - R_v^2\right)\right]\frac{ln\,R_v}{ln(R_{fo}/R_v)} \tag{8.80b}$$

Hence, the temperature distribution becomes

$$\int_{T_v}^{T} k\,dT = \frac{q'''\left(R_v^2 - r^2\right)}{4} + \left[\int_{T_v}^{T_{fo}} k\,dT + \frac{q'''}{4}\left(R_{fo}^2 - R_v^2\right)\right]\frac{ln\,(r/R_v)}{ln\,(R_{fo}/R_v)} \tag{8.81}$$

To find the location, R_o, of the maximum temperature differentiate Equation 8.81 with respect to r and set the temperature gradient equal to zero yielding

$$k(T)\frac{\delta T}{\delta r} = 0 = -\frac{q'''R_o}{2} + \left[\int_{T_v}^{T_{fo}} k\,dT + \frac{q'''}{4}\left(R_{fo}^2 - R_v^2\right)\right]\frac{1/R_o}{ln(R_{fo}/R_v)} \tag{8.82}$$

$$R_o = \sqrt{\frac{\left[\int_{T_v}^{T_{fo}} k dT + \frac{q'''}{4}\left(R_{fo}^2 - R_v^2\right)\right]2}{q''' ln(R_{fo}/R_v)}} \tag{8.83a}$$

For the special case of $T_{fo} = T_v$, then

$$R_o = \sqrt{\frac{\left(R_{fo}^2 - R_v^2\right)}{ln\left(R_{fo}/R_v\right)^2}} = R_v\sqrt{\frac{\left(R_{fo}/R_v\right)^2 - 1}{ln\left(R_{fo}/R_v\right)^2}} \tag{8.83b}$$

The maximum temperature T_{R_o} is given by

$$\int_{T_v}^{T_{R_o}} k\,dT = \frac{q'''\left(R_v^2 - R_o^2\right)}{4} + \left[\int_{T_v}^{T_{fo}} k\,dT + \frac{q'''}{4}\left(R_{fo}^2 - R_v^2\right)\right]\frac{ln(R_o/R_v)}{ln(R_{fo}/R_v)} \tag{8.84a}$$

Now for an annular pellet geometry, the linear power at a uniform q''' is given as

$$q' = q'''\pi\left(R_{fo}^2 - R_v^2\right) \tag{8.85}$$

Hence

$$\int_{T_v}^{T_{R_o}} k\,dT = \frac{q'}{4\pi}\left(\frac{R_v^2 - R_o^2}{R_{fo}^2 - R_v^2}\right) + \left[\left(\int_{T_v}^{T_{fo}} k\,dT\right) + \frac{q'}{4\pi}\right]\frac{ln\,(R_o/R_v)}{ln\,(R_{fo}/R_v)} \tag{8.84b}$$

For the case of equal surface temperatures, that is, $T_{fo} = T_v$, then

$$\int_{T_v}^{T_{R_o}} k\, dT = \frac{q'}{4\pi} F_{v_{DC}} \tag{8.86}$$

where

$$F_{v_{DC}} = f\left(R_{fo}/R_v \equiv \alpha\right) = \left[\frac{\left[1-\left(R_o/R_v\right)^2\right]}{\left(R_{fo}/R_v\right)^2 - 1} + \frac{ln\left(R_o/R_v\right)}{ln\left(R_{fo}/R_v\right)}\right] \tag{8.87}$$

and the ratio R_o/R_v is given by Equation 8.83b as a function of α.

The parameters R_o/R_v and $F_{v_{DC}}$ are plotted in Figures 8.15 and 8.16 using Equations 8.83b and 8.87, respectively. Again note that $F_{v_{DC}}$ is always both less than unity and less than $F_{v_{OC}}$ for $\beta = 1$ for the annular pellet with only the outer surface cooled.

The heat fluxes on the outside and inside pellet surfaces are

$$q''_{fo} = \frac{q'''}{2}\left(\frac{R_{fo}^2 - R_o^2}{R_{fo}}\right) \tag{8.88a}$$

$$q''_v = \frac{q'''}{2}\left(\frac{R_o^2 - R_v^2}{R_v}\right) \tag{8.88b}$$

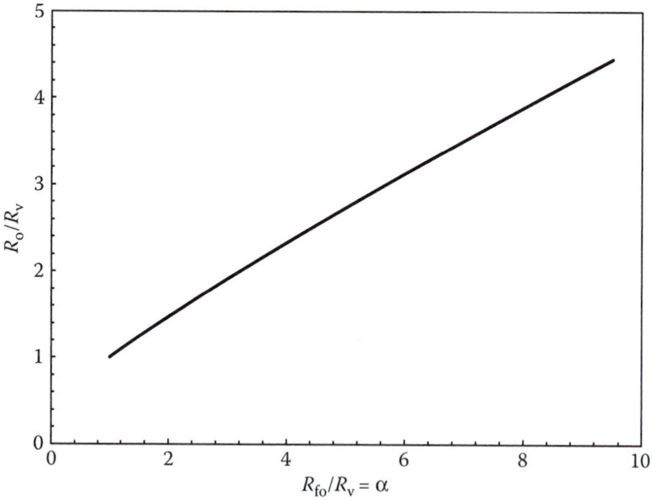

FIGURE 8.15 Location of the maximum temperature T_{R_o} for an annular pellet cooled on both the inside and outside surfaces with equal surface temperatures per Equation 8.83b. Note that for $R_{fo}/R_v \geq 6$, $R_o/R_v \approx 0.5\ R_{fo}/R_v$.

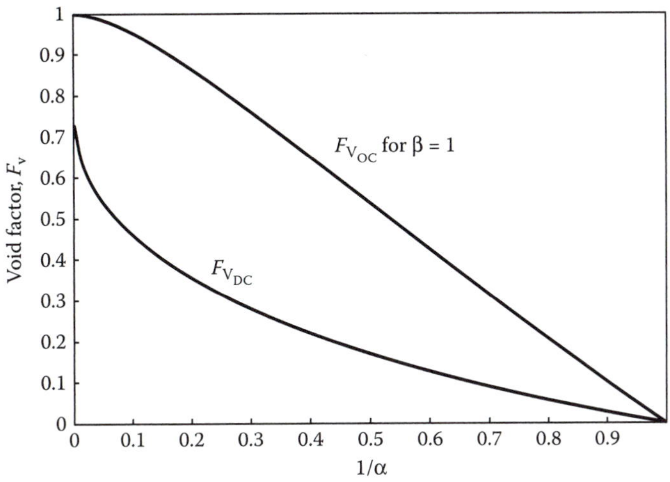

FIGURE 8.16 Void factor functions for annular pellets. $F_{V_{OC}}$ is the void factor for outside-cooled pellets, Equation 8.76. $F_{V_{DC}}$ is the void factor for dual-cooled pellets with equal surface temperatures, Equation 8.87.

8.5.5 Solid versus Annular Pellet Performance

Hence, when comparing the solid and annular pellets, the following conclusions can be made by observing Equations. 8.70, 8.77, and 8.86:

1. For the same temperature limit T_{max}

$$q'_{\text{Ann}_{DC}} F_{v_{DC}} = q'_{\text{Ann}_{OC}} F_{v_{OC}} = q'_{\text{solid}} \tag{8.89}$$

provided that T_{fo} and k are the same in the annular elements as in the solid. Since $F_{V_{DC}}$ is less than $F_{V_{OC}}$, then

$$q'_{\text{Ann}_{DC}} > q'_{\text{Ann}_{OC}} > q'_{\text{solid}} \tag{8.90}$$

That is, annular pellets can operate at a higher linear heat rate than a solid pellet if T_{max}, T_{fo}, and k are the same while the dual-cooled surface pellet can operate at higher linear power than the outside surface-cooled pellet again for T_{max}, T_{fo}, and k the same.

2. For the same heat rate (q') the temperature integrals of the conductivity are related by

$$\frac{\int_{T_{fo}}^{T_{max}} k_{\text{Ann}_{DC}} \mathrm{d}T}{F_{v_{DC}}} = \frac{\int_{T_{fo}}^{T_{max}} k_{\text{Ann}_{OC}} \mathrm{d}T}{F_{v_{OC}}} = \int_{T_{fo}}^{T_{max}} k_{\text{solid}} \mathrm{d}T \tag{8.91}$$

so that if the fuel material thermal conductivities are the same, that is, the same $k(T)$, and T_{fo} is the same, we get

$$T_{\max \text{lannular}_{DC}} < T_{\max \text{lannular}_{OC}} < T_{\max_{\text{solid}}} \tag{8.92}$$

In this case, the maximum operating temperatures of the annular fuel cases are less than that of the solid fuel pellet and the dual-surface-cooled pellet operates at a lower maximum temperature than the outside-only-cooled surface pellet.

Note that the conductivity integral is a function of the fuel pellet density, which depends on the initial fuel manufacturing conditions and irradiation conditions in the reactor. This dependence was discussed in Section 8.3. The conditions of Equations 8.89 and 8.92 do not necessarily apply if $k(T)$ of the annular fuel does not equal $k(T)$ of the solid fuel.

Example 8.3: Linear Power of Solid and Annular Cylindrical Fuel Pellets

PROBLEM

For the geometry of the PWR pellets described in Table 8.6, evaluate the linear power of the solid and annular (cooled on both surfaces) pellet-fueled pins when the fuel outer temperature is 495°C and the maximum fuel temperature is 1400°C. Assume that the fuel is unirradiated UO_2 of density 95% ρ_{TD}, and ignore the effect of cracking on fuel conductivity.

SOLUTION

Solid Fuel Pellet

The linear power is related to the temperature difference by Equation 8.70

$$\int_{T_{fo}}^{T_{max}} k\,dT = \frac{q'}{4\pi} \tag{8.70}$$

TABLE 8.6
Dimensions (cm) of Cylindrical Fuel Elements of Various Shapes[a]

	Array	D_{cii}	D_{cio}	D_{fi}	D_{fo}	D_{ci}	D_{co}	Pitch
Annular[b]	13 × 13	0.863	0.978	0.990	1.410	1.422	1.537	1.651
Solid	17 × 17	—	—	—	0.819	0.836	0.95	1.26

[a] Subscripts, ci, f, and co designate inner cladding, fuel, and outer cladding, respectively; an additional subscript designates outer (o) or inner (i) surface.
[b] Dual surface cooled.

which can be used to write

$$q' = 4\pi\left(\int_{0°C}^{T_{max}} k\,dT - \int_{0°C}^{T_{fo}} k\,dT\right)$$

Applying the conditions specified in the problem, using Figure 8.2 we get

$$\begin{aligned} q' &= 4\pi\left(\int_{0°C}^{1400°C} k_{0.95}\,dT - \int_{0°C}^{495°C} k_{0.95}\,dT\right) \\ &= 4\pi(58 - 30)\ \text{W/cm} \\ &= 351.9\ \text{W/cm} \\ &= 35.2\ \text{kW/m} \end{aligned}$$

Annular Pellet Cooled on Both Surfaces

For this pellet, the linear power is related to the temperature difference by Equation 8.86

$$\int_{T_v}^{T_{RO}} k\,dT = \frac{q'}{4\pi}F_{VDC} \tag{8.86}$$

where F_{VDC} is given by Equation 8.87 and Figure 8.16.

From Table 8.6

$$R_{fo}/R_v = 1.410/0.99 = 1.424.$$

Hence from Figure 8.16, $F_{VDC} = 0.09$

Therefore, from Equation 8.86, q' is obtained as

$$\begin{aligned} q' &= \frac{4\pi}{F_{VDC}}\int_{T_v}^{T_{RO}} k\,dT \\ &= \frac{4\pi}{0.09}(58 - 30) \\ &= 3909\ \text{W/cm} \\ &= 390.9\ \text{kW/m} \end{aligned}$$

Hence, the achievable linear power for the dual surface cooled pellet is an order of magnitude larger than that of the solid pellet.

8.5.6 Annular Fuel Pellet (Cooled Only on the Inside Surface R_v)

This configuration yields the approximate solution for the maximum fuel temperature of the inverted fuel array. The unit cells of square- and hexagonal-inverted fuel arrays have been given in Figure 2.17. The maximum fuel temperature occurs at the vertices since the bounding edges of the hexagon and square cells are adiabatic. However, because the bounding surfaces of the fuel region do not coincide with any

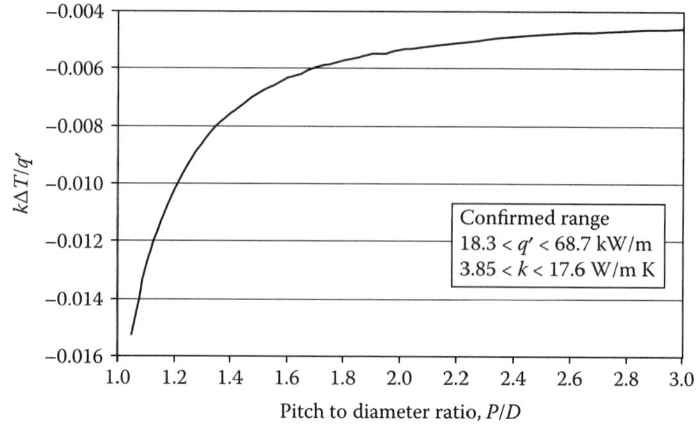

FIGURE 8.17 Accuracy of the equivalent annulus prediction of maximum temperature.

orthogonal coordinate system, an exact solution of the fuel temperature field is not possible.* Therefore, it is customary to perform numerical calculations to obtain this maximum fuel temperature.

Alternatively for inverted arrays of sufficiently large *P/D*, an equivalent annulus, as illustrated in Figure 2.20, is constructed and analyzed. As Figure 8.17 demonstrates, the equivalent annulus solution approximates the actual maximum temperature increase well as the *P/D* ratio increases.

$$(\Delta T \equiv T_{\text{equivalent annulus solution}} - T_{\text{exact solution}})$$

The outer fuel radius of the equivalent annulus, R_{EA}, is chosen to create an equivalent fuel cross-sectional area as for the inverted array unit cell. This condition of equal flow areas yields

$$R_{EA} = P\sqrt{\frac{1}{\pi}} \text{ for the square array; } P\sqrt{\frac{\sqrt{3}}{2\pi}} \text{ for the hexagonal array} \quad (2.51a, 2.52a)$$

In the solution which follows, the outer fuel radius of the equivalent annulus, R_{EA}, is written as R_{fo} to be consistent with the nomenclature of the fuel pellet cases of Sections 8.5.2 through 8.5.4.

For this geometry, the boundary condition is

$$q''\Big|_{r=R_{fo}} = -k\frac{dT}{dr}\Big|_{r=R_{fo}} = 0 \quad (8.93)$$

* An approximate solution by the point-matching technique is demonstrated in Chapter 7 of Volume II.

Applying this condition to Equation 8.62 leads to C_1, as

$$C_1 = -\frac{q'''R_{fo}^2}{2} \tag{8.94}$$

Substituting this result into Equation 8.62 and integrating between r and R_{fo}, the location of T_{max} yields

$$-\int_{T_{max}}^{T} k\,dT = \frac{q'''}{4}\left[r^2 - R_{fo}^2\right] - \frac{q'''R_{fo}^2}{2}\,\ell n\left(\frac{r}{R_{fo}}\right) \tag{8.95}$$

The above equation can be rearranged as

$$\int_{T}^{T_{max}} k\,dT = \frac{q'''r^2}{4}\left\{\left[1-\left(\frac{R_{fo}}{r}\right)^2\right]-\left(\frac{R_{fo}}{r}\right)^2 \ell n\left(\frac{r}{R_{fo}}\right)^2\right\} \tag{8.96}$$

Applying the condition $T = T_v$ at $r = R_v$, we get

$$\int_{T_v}^{T_{max}} k\,dT = \frac{q'''R_v^2}{4}\left\{\left[1-\left(\frac{R_{fo}}{R_v}\right)^2\right]-\left(\frac{R_{fo}}{R_v}\right)^2 \ell n\left(\frac{R_v}{R_{fo}}\right)^2\right\} \tag{8.97}$$

The linear heat rate is given by

$$q' = \pi\left(R_{fo}^2 - R_v^2\right)q''' \tag{8.98}$$

so that

$$q'''R_v^2 = \frac{q'}{\pi\left[\left(R_{fo}/R_v\right)^2 - 1\right]} \tag{8.99}$$

Substituting for $q'''R_v^2$ from Equation 8.99 into Equation 8.97, we get

$$\int_{T_v}^{T_{max}} k\,dT = \frac{q'}{4\pi}\left[\frac{\left(1-\left(R_{fo}/R_v\right)^2\right)-\left(R_{fo}/R_v\right)^2 \ell n\left(R_v/R_{fo}\right)^2}{\left(R_{fo}/R_v\right)^2 - 1}\right] \tag{8.100}$$

We express Equation 8.100 in the form below in terms of a void factor for this inside-cooled case, $F_{V_{IC}}$, analogous to Equation 8.87 for the outside-cooled, $F_{V_{OC}}$, and dual-cooled, $F_{V_{DC}}$, pellet cases, respectively

$$\int_{T_v}^{T_{max}} k\,dT = \frac{q'}{4\pi} F_{V_{IC}} \tag{8.101}$$

where

$$F_{V_{IC}} = f\left(\frac{R_{fo}}{R_v}\right) \equiv f(\alpha) = \frac{\left(1-(R_{fo}/R_v)^2\right)-(R_{fo}/R_v)^2\,\ell n(R_v/R_{fo})^2}{(R_{fo}/R_v)^2-1} \tag{8.102}$$

Figure 8.18 plots the behavior of the inside-cooled void factor, $F_{V_{IC}}$. Note the contrast of this behavior to that of the outside- and dual-cooled cases shown in Figure 8.16. This behavior of $F_{V_{IC}}$ is explained by the following considerations. For $1/\alpha = 0$, the geometry of the inside-cooled pellet tends to that of an insulated solid pellet with uniform heat generation. Hence $F_{V_{IC}} \rightarrow \infty$ yielding $T_{MAX} \rightarrow \infty$. Conversely, for $1/\alpha = 1$, the inside-cooled pellet geometry tends to that of an infinitely thin-insulated annulus. Hence, $F_{V_{IC}} \rightarrow 0$ yielding $T_{MAX} \rightarrow T_v$.

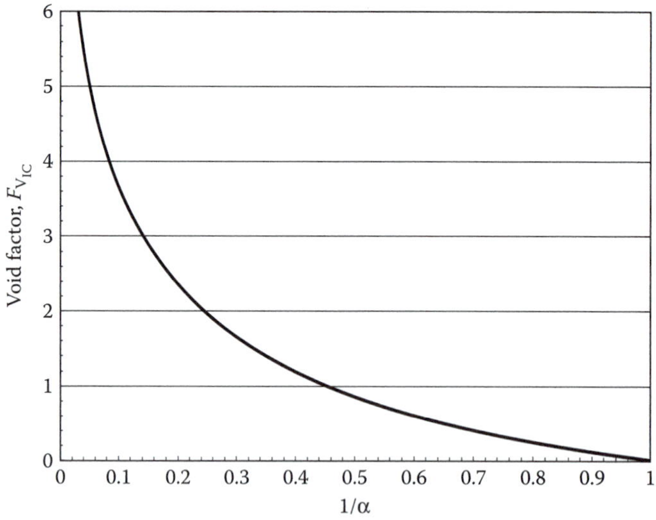

FIGURE 8.18 Void factor function for inside-cooled annular geometry from Equation 8.102. Note the void factor approaches ∞ as α approaches ∞.

8.6 TEMPERATURE DISTRIBUTION IN RESTRUCTURED FUEL ELEMENTS

Operation of an oxide fuel material at a high temperature leads to alterations of its morphology. The fuel region in which the temperature exceeds a certain sintering temperature experiences a loss of porosity. In a solid cylindrical fuel pellet, the inner region is restructured to form a void at the center, surrounded by a dense fuel region. In fast reactors, where the fuel may have a higher temperature near the center, restructuring has been found to lead to three distinct regions (Figure 8.19). In the outermost ring, where no sintering (i.e., no densification) occurs, the fuel density remains equal to the original (as fabricated) density, whereas the intermediate and inner regions have densities of 95–97% and 98–99%, respectively. It should be noted that most of the restructuring occurs within the first few days of operation, with slow changes afterward. In LWRs, where the fuel temperature is not as high as in fast reactors, two-region pellet restructuring may occur in the core regions operating at high powers. The sintering temperatures as well as the density in each fuel structure are not universally agreed on, as seen in Table 8.7.

In this section, the heat conduction problem in cylindrical fuel elements that have undergone some irradiation, and hence developed sintered (densified) regions, is solved. Temperature distributions are obtained on the assumption that the fuel element may be represented by three zones (Figure 8.20). The two-zone fuel temperature distribution is obtained by reduction of the equations of the three-zone treatment.

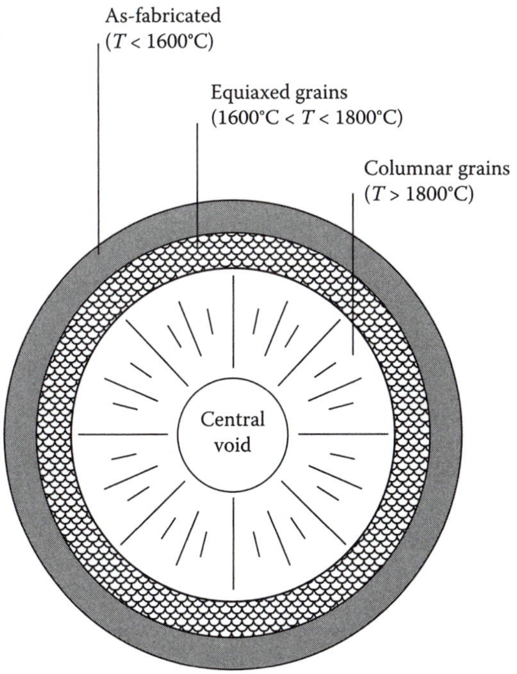

FIGURE 8.19 Restructuring of an oxide fuel pellet during high-temperature irradiation.

TABLE 8.7
Fuel Sintering Temperature and Densities

	Columnar Grains		Equiaxed Grains	
Recommendation Source	T_1(°C)	ρ_1/ρ_{TD}	T_2(°C)	ρ_2/ρ_{TD}
Atomics international	1800	0.98	1600	0.95
General electric	2150	0.99	1650	0.97
Westinghouse	2000	0.99	1600	0.97

Source: Adopted from Marr, W. M. and Thompson, D. H., *Trans. Am. Nucl. Soc.*, 14:150, 1971.

8.6.1 Mass Balance

From conservation of mass across a section in the fuel rod before and after restructuring, we conclude that the original mass is equal to the sum of the fuel mass in the three rings. Hence, when the pellet length is assumed unchanged

$$\pi R_{fo}^2 \rho_0 = \pi\left(R_1^2 - R_v^2\right)\rho_1 + \pi\left(R_2^2 - R_1^2\right)\rho_2 + \pi\left(R_{fo}^2 - R_2^2\right)\rho_3 \qquad (8.103)$$

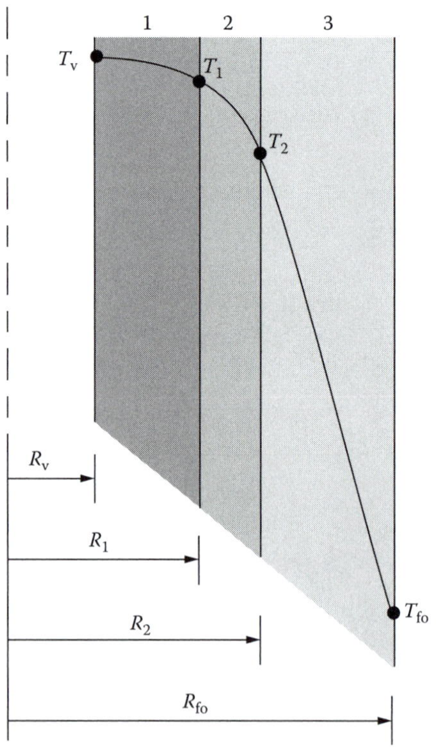

FIGURE 8.20 Temperature profile in a three-zone restructured fuel pellet.

However, the initial density ρ_0 is equal to ρ_3, so that an explicit expression for R_v can be obtained from Equation 8.103 as

$$R_v^2 = \left(\frac{\rho_1 - \rho_2}{\rho_1}\right) R_1^2 + \left(\frac{\rho_2 - \rho_3}{\rho_1}\right) R_2^2 \tag{8.104}$$

8.6.2 Power Density Relations

For a uniform neutron flux, the volumetric heat generation rate is proportional to the mass density:

$$q_2''' = \frac{\rho_2}{\rho_3} q_3''' \tag{8.105}$$

and

$$q_1''' = \frac{\rho_1}{\rho_3} q_3''' \tag{8.106}$$

However, the linear heat rate is given by the summation of the heat generation rate in the three rings. Therefore, the linear heat generation rate in the restructured fuel (q'_{res}) is given by

$$q'_{res} = q_3'''\pi\left(R_{fo}^2 - R_2^2\right) + q_2'''\pi\left(R_2^2 - R_1^2\right) + q_1'''\pi\left(R_1^2 - R_v^2\right) \tag{8.107}$$

or

$$q'_{res} = q_3'''\left[\pi\left(R_{fo}^2 - R_2^2\right) + \frac{\rho_2}{\rho_3}\pi\left(R_2^2 - R_1^2\right) + \frac{\rho_1}{\rho_3}\pi\left(R_1^2 - R_v^2\right)\right] \tag{8.108}$$

Combining Equations 8.103 and 8.108, we obtain

$$q'_{res} = \pi R_{fo}^2 q_3''' \tag{8.109a}$$

that is

$$q_3''' = \frac{q'_{res}}{\pi R_{fo}^2} \tag{8.109b}$$

That is, the power density in the as-fabricated region can be obtained from the assumption that the mass is uniformly distributed in the fuel pellet. It is equivalent to expecting the power density in the outer region not to be affected by the redistribution of the fuel within the other two zones.

8.6.3 Heat Conduction in Zone 3

The differential equation is

$$\frac{1}{r}\frac{\mathrm{d}}{\mathrm{d}r}\left(rk_3\frac{\mathrm{d}T}{\mathrm{d}r}\right) = -q_3''' \tag{8.110}$$

By integrating once, we get

$$k_3\frac{\mathrm{d}T}{\mathrm{d}r} = -q_3'''\frac{r}{2} + \frac{C_3}{r} \tag{8.111}$$

and integrating again between the zone boundaries

$$\int_{T_2}^{T_{\mathrm{fo}}} k_3\,\mathrm{d}T = -q_3'''\frac{R_{\mathrm{fo}}^2 - R_2^2}{4} + C_3\ell n\left(\frac{R_{\mathrm{fo}}}{R_2}\right) \tag{8.112}$$

Using Equation 8.109b, we can evaluate the temperature gradient at R_{fo} from the heat flux as follows:

$$q''_{R_{\mathrm{fo}}} = -k_3\left.\frac{\mathrm{d}T}{\mathrm{d}r}\right|_{R_{\mathrm{fo}}} = \frac{q'_{\mathrm{res}}}{2\pi R_{\mathrm{fo}}} = q_3'''\frac{\pi R_{\mathrm{fo}}^2}{2\pi R_{\mathrm{fo}}} = \frac{q_3'''R_{\mathrm{fo}}}{2} \tag{8.113}$$

Equation 8.113 provides a boundary condition to be satisfied by Equation 8.111. This condition leads to

$$C_3 = 0 \tag{8.114}$$

Hence, Equation 8.112 reduces to

$$\int_{T_{\mathrm{fo}}}^{T_2} k_3\,\mathrm{d}T = \frac{q_3'''}{4}R_{\mathrm{fo}}^2\left[1-\left(\frac{R_2}{R_{\mathrm{fo}}}\right)^2\right] \tag{8.115}$$

which, using Equation 8.109b, can also be written as

$$\int_{T_{\mathrm{fo}}}^{T_2} k_3\,\mathrm{d}T = \frac{q'_{\mathrm{res}}}{4\pi}\left[1-\left(\frac{R_2}{R_{\mathrm{fo}}}\right)^2\right] \tag{8.116}$$

8.6.4 Heat Conduction in Zone 2

The heat conduction equation integration for zone 2 leads to an equation similar to Equation 8.111 but applicable to zone 2:

$$k_2 \frac{dT}{dr} = -q_2''' \frac{r}{2} + \frac{C_2}{r} \tag{8.117}$$

At R_2, continuity of heat flux leads to

$$-k_2 \left.\frac{dT}{dr}\right|_{R_2} \text{ from zone } 2 = -k_3 \left.\frac{dT}{dr}\right|_{R_2} \text{ from zone } 3 \tag{8.118}$$

Substituting from Equations 8.111 (with $C_3 = 0$) and 8.117 into Equation 8.118 leads to

$$+q_2''' \frac{R_2}{2} - \frac{C_2}{R_2} = q_3''' \frac{R_2}{2}$$

or

$$C_2 = \frac{R_2^2}{2}\left[q_2''' - q_3'''\right] \tag{8.119}$$

By integrating Equation 8.117 between $r = R_1$ and $r = R_2$, we obtain

$$\int_{T_1}^{T_2} k_2 \, dT = -\frac{q_2'''}{4}\left(R_2^2 - R_1^2\right) + \frac{R_2^2}{2}\left(q_2''' - q_3'''\right)\ell n\left(\frac{R_2}{R_1}\right) \tag{8.120}$$

which, using Equations 8.105 and 8.109b, can be written as

$$\int_{T_2}^{T_1} k_2 \, dT = \frac{q'_{\text{res}}}{4\pi}\left(\frac{\rho_2}{\rho_3}\right)\left[1 - \left(\frac{R_1}{R_2}\right)^2\right]\left(\frac{R_2}{R_{\text{fo}}}\right)^2 - \frac{q'_{\text{res}}}{4\pi}\left(\frac{\rho_2}{\rho_3}\right)\left(\frac{R_2}{R_{\text{fo}}}\right)^2\left(1 - \frac{\rho_3}{\rho_2}\right)\ell n\left(\frac{R_2}{R_1}\right)^2 \tag{8.121}$$

Hence

$$\int_{T_2}^{T_1} k_2 \, dT = \frac{q'_{\text{res}}}{4\pi}\left(\frac{\rho_2}{\rho_3}\right)\left(\frac{R_2}{R_{\text{fo}}}\right)^2\left[1 - \left(\frac{R_1}{R_2}\right)^2 - \left(\frac{\rho_2 - \rho_3}{\rho_2}\right)\ell n\left(\frac{R_2}{R_1}\right)^2\right] \tag{8.122a}$$

Using the notation of Equations 8.76 and 8.122a can be recast in the form

$$\int_{T_2}^{T_1} k_2 \, dT = \frac{q'_{res}}{4\pi}\left(\frac{\rho_2}{\rho_3}\right)\left(\frac{R_2^2 - R_1^2}{R_{fo}^2}\right)F_{V_{oc}}\left(\frac{R_1}{R_2}, \frac{\rho_2}{\rho_2 - \rho_3}\right) \tag{8.122b}$$

where

$$F_{V_{oc}}(\alpha,\beta) \equiv 1 - \frac{\ell n \alpha^2}{\beta(\alpha^2 - 1)}$$

8.6.5 Heat Conduction in Zone 1

Again the integration of the heat conduction equation once leads to

$$k_1 \frac{dT}{dr} = -q_1''' \frac{r}{2} + \frac{C_1}{r} \tag{8.123}$$

To obtain the value of C_1, we can either use the zero heat flux condition at R_v or the continuity of heat flux at R_1. Applying the first boundary condition

$$k_1 \left.\frac{dT}{dr}\right|_{R_v} = 0 \tag{8.124}$$

leads to

$$C_1 = q_1''' \frac{R_v^2}{2} \tag{8.125}$$

By integrating Equation 8.123 again, we get

$$\int_{T_v}^{T_1} k_1 \, dT = \frac{-q_1'''}{4}(R_1^2 - R_v^2) + C_1 \ell n\left(\frac{R_1}{R_v}\right) \tag{8.126}$$

Substituting from Equations. 8.106 and 8.109b for q_1''' and q_3''' and Equation 8.125 for C_1, we get

$$\int_{T_v}^{T_1} k_1 \, dT = \frac{q'_{res}}{4\pi}\left(\frac{\rho_1}{\rho_3}\right)\left(\frac{R_1}{R_{fo}}\right)^2\left[1 - \left(\frac{R_v}{R_1}\right)^2 - \left(\frac{R_v}{R_1}\right)^2 \ell n\left(\frac{R_1}{R_v}\right)^2\right] \tag{8.127a}$$

Using the notation of Equation 8.76, the last equation can be written as

$$\int_{T_v}^{T_1} k_1\, dT = \frac{q'_{res}}{4\pi}\left(\frac{\rho_1}{\rho_3}\right)\frac{R_1^2 - R_v^2}{R_{fo}^2}\left[1 - \frac{\ell n(R_1/R_v)^2}{(R_1/R_v)^2 - 1}\right]$$

$$= \frac{q'_{res}}{4\pi}\left(\frac{\rho_1}{\rho_3}\right)\left(\frac{R_1^2 - R_v^2}{R_{fo}^2}\right)F_{V_{oc}}\left(\frac{R_1}{R_v}, 1\right) \quad (8.127b)$$

8.6.6 Solution of the Pellet Problem

Equations 8.104, 8.116, 8.122a (or 8.122b), and 8.127a (or 8.127b) provide a set of four equations in terms of six parameters, namely R_1, R_2, R_v, T_{fo}, T_v, and q'_{res}. Thus, if any two are specified, the rest can be evaluated. Note that the values of T_1 and T_2 are assumed known from Table 8.7.

Generally, two conditions are of interest.

1. A linear heat rate (q'_{res}) and the outer surface temperature (T_{fo}) are specified; so that T_v, along with R_v, R_1, and R_2 can be evaluated.
2. The maximum temperatures (T_v and T_{fo}) are specified; q'_{res} along with R_v, R_1, and R_2 are to be evaluated.

8.6.7 Two-Zone Sintering

For two-zone representation of the fuel, $R_1 = R_2 = R_s$, $\rho_1 = \rho_2 = \rho_s$, and $T_1 = T_2 = T_s$, Equation 8.104 reduces to

$$R_v^2 = \frac{\rho_s - \rho_3}{\rho_s} R_s^2 \quad (8.128)$$

Equation 8.126 takes the form

$$\int_{T_{fo}}^{T_1} k_3\, dT = \frac{q'_{res}}{4\pi}\left[1 - \left(\frac{R_s}{R_{fo}}\right)^2\right] \quad (8.129)$$

Equation 8.122a is not needed. Equation 8.127a takes the form

$$\int_{T_s}^{T_v} k_s\, dT = \frac{q'_{res}}{4\pi}\left(\frac{\rho_s}{\rho_3}\right)\left(\frac{R_s}{R_{fo}}\right)^2\left\{1 - \left(\frac{R_v}{R_s}\right)^2\left[1 + \ell n\left(\frac{R_s}{R_v}\right)^2\right]\right\} \quad (8.130)$$

Example 8.4: Linear Power of a Two-Zone Pellet Under Temperature Constraint

PROBLEM

Consider an initially solid SFR fresh UO_2 fuel pellet under fixed maximum temperature constraint. Evaluate the linear power (q'_{res}) for the sintered pellet for the same temperature constraint. The sintered pellet may be represented by two zones. Given

Solid pellet	$q' = 14.4$ kW/ft $= 472.4$ W/cm
	$\rho_0 = 88\%\ \rho_{TD}$
	$T_{fo} = 960°C$
Two-zone-sintered pellet	$\rho_s = 98\%\ \rho_{TD}$
	$T_s = 1800°C$
	$T_{fo} = 960°C$

SOLUTION

1. Evaluate the maximum temperature of the solid pellet using Equation 8.70 for $k = k_{0.88}$:

$$\int_{T_{fo}}^{T_{max}} k_{0.88}\, dT = \frac{q'}{4\pi} = \frac{14.4}{4\pi} = 1.15\,\text{kW/ft} = 37.7\ \text{W/cm}$$

   Expressing the conductivity integral in the form represented in Figure 8.3:

$$\int_{0°C}^{T_{max}} k_{0.88} dT - \int_{0°C}^{T_{fo}=960°C} k_{0.88}\, dT = 37.7\ \text{W/cm}$$

   which yields T_{max} as follows:

$$\int_{0°C}^{T_{max}} k_{0.88}\, dT = 37.7 + 41 = 78.70\ \text{W/cm} \rightarrow T_{max} = 2700°C$$

2. Obtain the relation between q'_{res} and R_s for the restructured pellet. Using Equation 8.129 for $k_3 = k_{0.88}$, we get

$$\int_{T_{fo}=960°C}^{T_s=1800°C} k_{0.88}\, dT = \frac{q'_{res}}{4\pi}\left[1 - \left(\frac{R_s}{R_{fo}}\right)^2\right]$$

or

$$59 - 41 = 18\ \text{W/cm} = \frac{q'_{res}}{4\pi}\left[1 - \left(\frac{R_s}{R_{fo}}\right)^2\right] \tag{8.131}$$

3. Obtain the relation between q'_{res}, R_{fo}, R_s, and R_v from Equation 8.120 for $k_s = k_{0.98}$:

$$\int_{T_s=1800°C}^{T_{max}=2700°C} k_{0.98}\, dT = \frac{q'_{res}}{4\pi}\left(\frac{\rho_s}{\rho_0}\right)\left(\frac{R_s}{R_{fo}}\right)^2 \left\{1 - \left(\frac{R_v}{R_s}\right)^2\left[1 + \ell n\left(\frac{R_s}{R_v}\right)^2\right]\right\}$$

Hence

$$(78.7 - 59) = 19.7 = \frac{q'_{res}}{4\pi}\frac{0.98}{0.88}\left(\frac{R_s}{R_{fo}}\right)^2 \left\{1 - \left(\frac{R_v}{R_s}\right)^2\left[1 + \ell n\left(\frac{R_s}{R_v}\right)^2\right]\right\} \tag{8.132}$$

4. Eliminate q'_{res} between Equations 8.131 and 8.132 and rearrange to obtain

$$1.0175\left\{1 - \left(\frac{R_v}{R_s}\right)^2\left[1 + \ell n\left(\frac{R_s}{R_v}\right)^2\right]\right\} = \left(\frac{R_{fo}}{R_s}\right)^2 - 1 \tag{8.133}$$

5. From the mass balance Equation 8.128 in the solid and sintered pellets

$$R_v^2 = \frac{\rho_s - \rho_0}{\rho_s}R_s^2 = \frac{0.98 - 0.88}{0.98}R_s^2 = 0.102 R_s^2 \tag{8.128}$$

Thus

$$\left(\frac{R_s}{R_v}\right)^2 = 9.8 \tag{8.134}$$

6. Substituting the value of $(R_v/R_s)^2$ from Equation 8.134 into Equation 8.133:

$$1.0175\{1 - 0.102[1 + \ell n(9.8)]\} = \left(\frac{R_{fo}}{R_s}\right)^2 - 1$$

or

$$\frac{R_s}{R_{fo}} = 0.772 \tag{8.135}$$

7. Substituting the value of R_s/R_{fo} from Equation 8.135 into Equation 8.131, we get

$$q'_{res} = \frac{4\pi(18)}{1-(0.772)^2} = 560\ \text{W/cm} = 18.37\ \text{kW/ft}$$

that is, $q'_{res} > q'$. Hence, under the same temperature limits, the linear power of a sintered pellet is higher than that of a solid pellet. Conversely, at the same linear power, the maximum temperature of a sintered pellet is lower than that of a solid pellet.

Example 8.5: Comparison between Three-Zone and Two-Zone Fuel Pellet Maximum Temperature

PROBLEM

Consider a fresh UO_2 SFR fuel pin operating at a fixed linear power q' of 14.4 kW/ft. Obtain the maximum fuel temperature if the fuel pin is in one of the following restructured conditions.

Three-zone condition:

Columnar: $\rho_1 = 98\%\ \rho_{TD}$, $T_1 = 1800°C$
Equiaxed : $\rho_2 = 95\%\ \rho_{TD}$, $T_2 = 1600°C$
Unrestructured: $\rho_3 = 88\%\ \rho_{TD}$, $T_{fo} = 1000°C$

Two-zone condition:

Sintered: $\rho_s = 98\%\ \rho_{TD}$, $T_s = 1700°C$
Unrestructured: $\rho_{fo} = 88\%\ \rho_{TD}$, $T_{fo} = 1000°C$

SOLUTION

We note that $q'_{res} = q' = 14.4\ \text{kW/ft} = 472.4\ \text{W/cm}$.

Case 1

1. We first obtain the value of (R_2/R_{fo}) from Equation 8.116

$$\int_{T_{fo}}^{T_2} k_3\, dT = \frac{q'_{res}}{4\pi}\left[1-\left(\frac{R_2}{R_{fo}}\right)^2\right] \tag{8.116}$$

where $k_3 = k_{0.88\ TD}$.
From Equation 8.22d (Figure 8.3 has the equivalent information but the accuracy limitation from use of the curves of Figure 8.3 obscures the point being made in this example of the maximum fuel temperature difference between the two and three zone models.), we obtain

$$\int_{1000°C}^{1600°C} k_{0.88}\, dT = \int_{0K}^{1873K} k_{0.88}\, dT - \int_{0K}^{1273K} k_{0.88}\, dT$$

$$\int_{0\text{K}}^{1873\text{K}} k_{0.88}\,dT = \eta_{0.88}\left[\frac{1}{B}\ell n\frac{A+BT}{A} + \frac{E}{F}\exp\left(-\frac{F}{T}\right)\right]$$

$$= \eta_{0.88}\left[\frac{1}{2.46\times10^{-4}}\ell n\frac{4.52\times10^{-2}+2.46\times10^{-4}(1873)}{4.52\times10^{-2}} + \frac{3.5\times10^{9}}{1.636\times10^{4}}\exp\left(-\frac{1.6361\times10^{4}}{1873}\right)\right]$$

$$= 0.8957\left[0.9817\times10^{4} + 0.00428\times10^{4}\right]$$

$$= 88.32\ \text{W/cm}$$

since from Equation 8.22b:

$$\eta_{0.88} = \frac{1.0789(0.88)}{1.0+0.5(1-0.88)} = 0.8957$$

Similarly evaluating the integral $\int_{0\text{K}}^{1273\text{K}} k_{0.88}\,dT$, we obtain 75.37 W/cm. Hence

$$\int_{1000°\text{C}}^{1600°\text{C}} k_{0.88}\,dT = 88.32 - 75.37 = 12.95\ \text{W/cm}$$

Consequently

$$\left(\frac{R_2}{R_{\text{fo}}}\right)^2 = 1 - \frac{4\pi}{q'}12.95 = 1 - \frac{51.8\pi}{472.4} = 1 - 0.344 = 0.656$$

Therefore, $R_2 = 0.8096\ R_{\text{fo}}$.

2. Obtain a value for R_1/R_2 from Equation 8.122a:

$$\int_{T_2}^{T_1} k_2\,dT = \frac{q'}{4\pi}\left(\frac{\rho_2}{\rho_3}\right)\left(\frac{R_2}{R_{\text{fo}}}\right)^2\left[1-\left(\frac{R_1}{R_2}\right)^2 - \left(1-\frac{\rho_3}{\rho_2}\right)\ell n\left(\frac{R_2}{R_1}\right)^2\right] \quad (8.122\text{a})$$

where $k_2 = k_{0.95\,\text{TD}}$.
From Equation 8.22d:

$$\int_{1600°\text{C}}^{1800°\text{C}} k_{0.95}\,dT = 102.81 - 98.60 = 4.21\ \text{W/cm}$$

Substituting in Equation 8.122a:

$$4.21 = \frac{472.4}{4\pi}\left(\frac{0.95}{0.88}\right)(0.656)\left[1-\left(\frac{R_1}{R_2}\right)^2 - \left(1-\frac{0.88}{0.95}\right)\ell n\left(\frac{R_2}{R_1}\right)^2\right]$$

$$4.21 = 26.62\left[1-\left(\frac{R_1}{R_2}\right)^2 - 0.074\ell n\left(\frac{R_2}{R_1}\right)^2\right]$$

$$0.8418 = \left(\frac{R_1}{R_2}\right)^2 + 0.074\ell n\left(\frac{R_2}{R_1}\right)^2$$

Solving iteratively, we get

$$\left(\frac{R_1}{R_2}\right)^2 = 0.8278 \text{ and } \left(\frac{R_1}{R_{fo}}\right)^2 = \left(\frac{R_1}{R_2}\right)^2\left(\frac{R_2}{R_{fo}}\right)^2 = (0.8278)(0.65) = 0.538$$

or $R_1 = 0.910$, R_2 and $R_1 = 0.734\ R_{fo}$.

3. Determine R_v from the mass balance equation:

$$R_v^2 = \left(\frac{\rho_1 - \rho_2}{\rho_1}\right)R_1^2 + \left(\frac{\rho_2 - \rho_3}{\rho_1}\right)R_2^2 \quad (8.104)$$

or

$$\left(\frac{R_v}{R_2}\right)^2 = \frac{0.98 - 0.95}{0.98}(0.8278) + \frac{0.95 - 0.88}{0.98}$$
$$= 0.0253 + 0.0714 = 0.0967$$

or

$$R_v = 0.311,\ R_2 = 0.342,\ R_1 = 0.251 R_{fo}$$

4. Determine T_v from the integral of the heat conduction equation over zone 1 (Equation 8.127a):

$$\int_{T_1}^{T_v} k_1\, dT = \frac{q'}{4\pi}\left(\frac{\rho_1}{\rho_3}\right)\left(\frac{R_1}{R_{fo}}\right)^2\left[1-\left(\frac{R_v}{R_1}\right)^2 - \left(\frac{R_v}{R_1}\right)^2 \ell n\left(\frac{R_1}{R_v}\right)^2\right]$$
$$= \frac{472.4}{4\pi}\left(\frac{0.98}{0.88}\right)(0.538)\left[1-(0.344)^2 - (0.344)^2 \ell n\left(\frac{1}{0.344}\right)^2\right]$$
$$= 22.52[1 - 0.118 - 0.2518]$$
$$= 14.19 \text{ W/cm}$$

Because $k_1 = k_{0.98}$, we now have

$$\int_{1800°C}^{T_v} k_{0.98}\,dT = 14.19 \text{ W/cm}$$

or, using Equation 8.22d

$$\int_{0K}^{T_v} k_{0.98}\,dT = 14.19 + \int_{0K}^{T_1=2073K} k_{0.98}\,dT = 14.19 + 107.62$$
$$= 121.8 \text{ W/cm} \rightarrow T_v \simeq 2405°C$$

Case 2
Consider the two-zone-sintered fuel, under the condition $q'_{res} = q'$ and $k_3 = k_{0.88}$.

1. Obtain R_s from the integral of the heat conduction equation over the unrestructured zone (Equation 8.129):

$$\int_{T_{fo}=1000°C}^{T_s=1700°C} k_{0.88}\,dT = \frac{q'}{4\pi}\left[1-\left(\frac{R_s}{R_{fo}}\right)^2\right] = 90.14 - 75.37 = 14.77 \text{ W/cm}$$

Hence

$$\left(\frac{R_s}{R_{fo}}\right)^2 = 1 - \frac{4\pi}{472.4}(14.77) = 1 - 0.393 = 0.607$$
$$\therefore R_s = 0.779 R_{fo}$$

2. Obtain R_v from the mass balance Equation 8.128:

$$R_v^2 = \left(\frac{\rho_s - \rho_o}{\rho_s}\right) R_s^2 = \frac{0.98 - 0.88}{0.98} R_s^2$$

that is

$$R_v^2 = 0.102 R_s^2$$
$$R_v = 0.319 R_s = 0.249 R_{fo}$$

3. Obtain T_v from the heat conduction integral (Equation 8.130) when $k_s = k_{0.98}$:

$$\int_{1700°C}^{T_v} k_s\,dT = \frac{q'}{4\pi}\left(\frac{\rho_s}{\rho_o}\right)\left(\frac{R_s}{R_{fo}}\right)^2\left[1-\left(\frac{R_v}{R_s}\right)^2 - \left(\frac{R_v}{R_s}\right)^2 \ell n\left(\frac{R_s}{R_v}\right)^2\right] \quad (8.130)$$

or

$$\int_{0°C}^{T_v} k_{0.98}\,dT = \int_{0°C}^{1700°C} k_{0.98}\,dT + \frac{472.4}{4\pi}\left(\frac{0.98}{0.88}\right)(0.607)$$

$$\left[1 - 0.102 - 0.102\ell n\left(\frac{1}{0.102}\right)\right]$$

$$= 105.35 + 25.4(1 - 0.102 - 0.232)$$

$$= 105.35 + 25.4(0.666) = 105.35 + 16.92$$

$$= 122.27\,\text{W/cm}$$

$$\text{Hence} \quad T_v = 2423°C$$

Although the three-region model is a more exact one, there is only a small advantage in using the three-zone approach, in that T_v from the three-zone model is only about 18°C lower than that from the two-zone approach. While a simpler solution to this example using Figure 8.3 is possible, the uncertainty in determining numerical values from Figure 8.3 is significant.

8.6.8 Design Implications of Restructured Fuel

The result of fuel sintering is to reduce the effective thermal resistance between the highest fuel temperature (T_{max}) and the pellet outer temperature (T_{fo}). Thus, two operational options exist after restructuring.

1. If the fuel maximum temperature is kept constant, the linear power can be increased. This condition is applicable to liquid-metal-cooled fast reactors where the heat flux to the coolant is not a limiting factor. This case is depicted graphically as Case A in Figure 8.21. The assumption of a constant T_{fo} in Figure 8.21 implies that the coolant conditions have been adjusted.
2. If the linear power is kept constant, the fuel maximum temperature is reduced. This case may be applicable to LWRs, as other considerations limit the operating heat flux of a fuel pin. This case is depicted graphically as Case B in Figure 8.21. If the fuel maximum temperature is calculated ignoring the sintering process, a conservative value is obtained.

8.7 THERMAL RESISTANCE BETWEEN THE FUEL AND COOLANT

The overall thermal resistance between the fuel and coolant consists of (1) the resistance of the fuel itself, (2) the resistance of the gap between the fuel and the cladding, (3) the resistance of the cladding, (4) the resistance of the zirconium oxide corrosion film on the cladding surface, and (5) the resistance of the coolant.

For typical LWR and fast reactor oxide fuel rods, the resistance of the UO_2 fuel is by far the largest, as can be inferred from the temperature profile of Figure 8.22. The next largest resistance is that of the gap. Hence, models for the gap conductance with increasing sophistication have been developed over the years. In this section, the gap resistance models are presented first, and then the overall resistance is discussed.

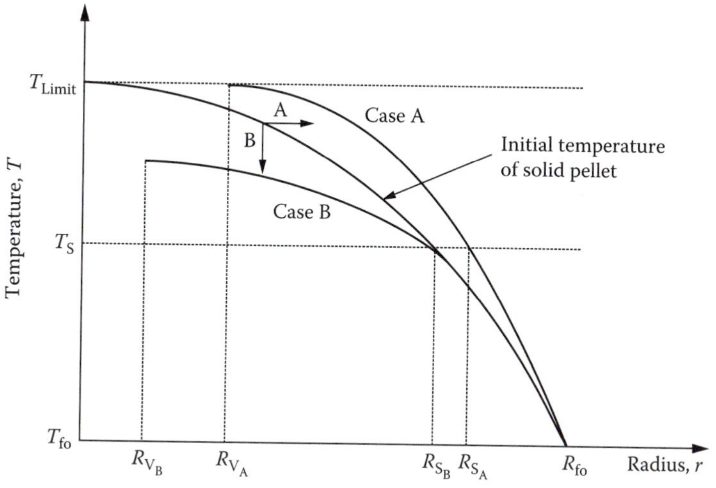

Case A	$T_{max,A} = T_{max,initial} = T_{limit}$
Case B	$q'_{initial} = q'_B$

FIGURE 8.21 Design and irradiation strategies for initially solid fuel pellets. Note the change in slope for both Case A and Case B at the radius between the original and restructured fuel where sintering is initiated.

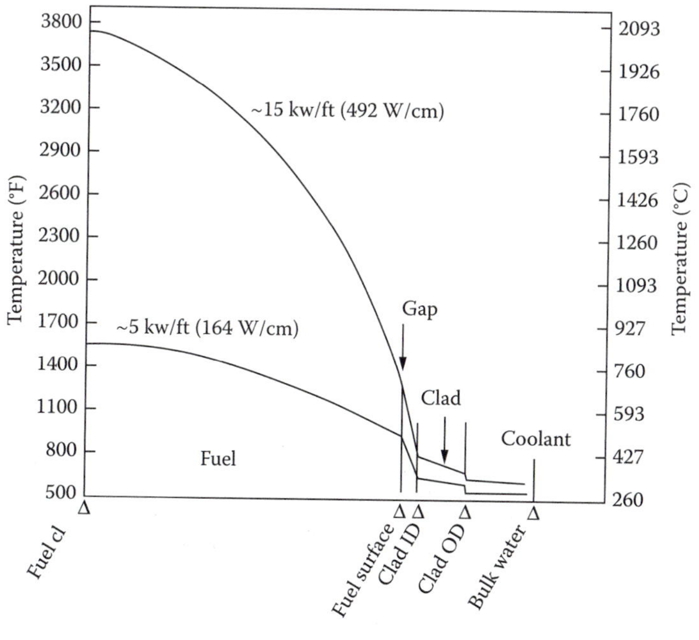

FIGURE 8.22 Typical PWR fuel rod temperature profile for two LHGRs. (From Jordan, R., MIT Reactor Safety Course, 1979.)

8.7.1 Gap Conductance Models

It is usually assumed that the gap consists of an annular space occupied by gases. The gas composition is initially the fill gas, which is an inert gas such as helium, but is gradually altered with burnup by the addition of gaseous fission products such as xenon and krypton. (Sodium-cooled reactors also use sodium bonding with metal fuel.) However, this simple picture does not reflect the real conditions of the fuel pin after some irradiation. The fuel pellets usually crack upon irradiation, as shown in Figure 8.23, and this situation leads to circumferential variation in the gap. In addition, thermal expansions of the fuel and cladding are often different and the fuel swells while the cladding creeps down with lifetime, the result being substantial pellet-cladding contact at the interface. This contact reduces the thermal resistance and hence effectively increases the "gap" conductance with the burnup. It occurs despite the lower conductivity of the fission gas products, compared to the initial conductivity of helium. A typical change of gap conductance with burnup is shown in Figure 8.24.

8.7.1.1 As-Fabricated Gap

The gap conductance at the as-fabricated conditions of the fuel can be modeled as due to conduction through an annular space as well as to radiation from the fuel. Thus, the heat flux at an intermediate gap position can be given by

$$q''_{g} = h_{g}(T_{fo} - T_{ci}) \tag{8.136}$$

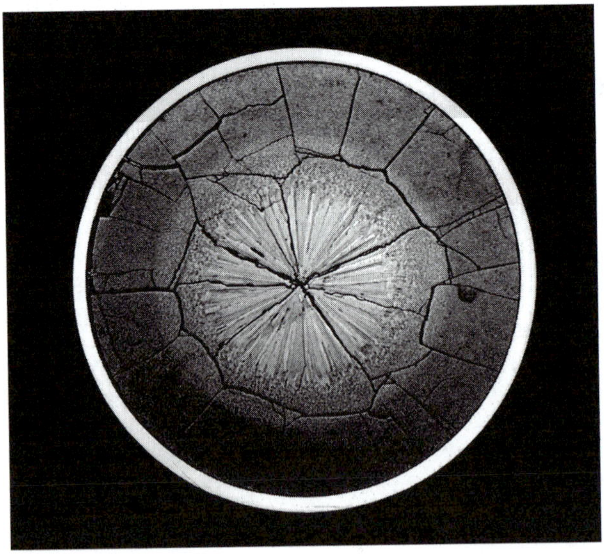

FIGURE 8.23 Example of a cracked fuel cross section at a burnup of 9.8 MWd/kg$_{IHM}$. (From Nuclear Energy Agency, NEA-1596 IFPE/AECL-BUNDLE, *NEA Data Bank Computer Programs: Category Y. Integral Experiments Data, Databases, Benchmarks*, 2000. Provided courtesy of AECL.)

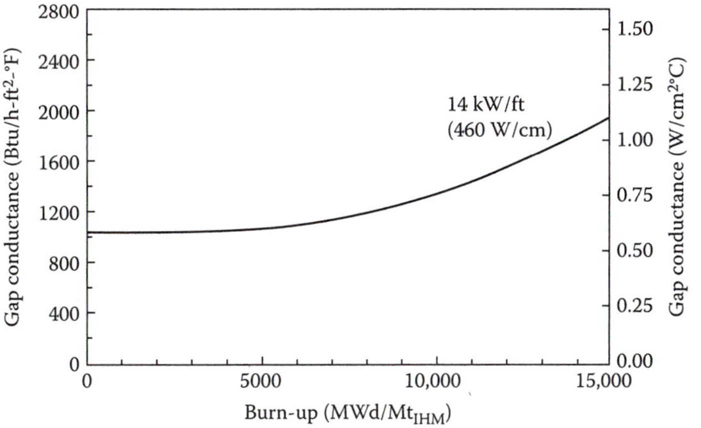

FIGURE 8.24 Variations of gap conductance with burnup for a PWR fuel rod (pressurized with helium and operating at 14 kW/ft (460 W/cm)). (From Fenech, H. In H. Fenech (ed.), *Heat Transfer and Fluid Flow in Nuclear Systems*. Pergamon Press, Oxford, UK, 1981.)

where for an open gap

$$h_{g,open} = \frac{k_{gas}}{\delta_{eff}} + \frac{\sigma}{(1/\varepsilon_f)+(1/\varepsilon_c)-1}\frac{T_{fo}^4 - T_{ci}^4}{T_{fo} - T_{ci}} \quad (8.137a)$$

where $h_{g,open}$ = gap conductance for an open gap; T_{fo} = fuel surface temperature; T_{ci} = cladding inner surface temperature; k_{gas} = thermal conductivity of the gas; δ_{eff} = effective gap width; σ = Stefan–Boltzman constant; and ε_f, ε_c = surface emissivities of the fuel and cladding, respectively.

Equation 8.137a can often be approximated as

$$h_{g,open} = \frac{k_{gas}}{\delta_{eff}} + \frac{\sigma T_{fo}^3}{1/\varepsilon_f + 1/\varepsilon_c - 1} \quad (8.137b)$$

It should be noted that the effective gap width is larger than the real gap width because of the temperature discontinuities at the gas–solid surface. The temperature discontinuities arise near the surface owing to the small number of gas molecules present near the surface. Thus, it is possible to relate δ_{eff} to the real gap width (δ_g), as illustrated in Figure 8.25, by

$$\delta_{eff} = \delta_g + \delta_{jump1} + \delta_{jump2} \quad (8.138)$$

At atmospheric pressure, ($\delta_{jump1} + \delta_{jump2}$), was found to equal 10 μm in helium and 1 μm in xenon [33].

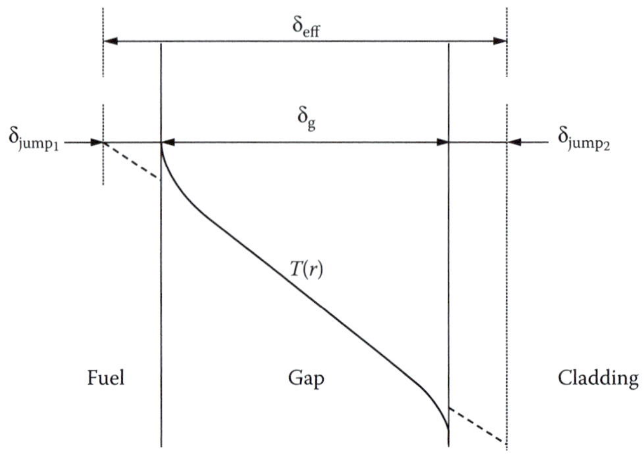

FIGURE 8.25 Temperature profile across the fuel-cladding gap.

The gas conductivity of a mixture of two gases is given by [17]

$$k_{gas} = (k_1)^{x_1} (k_2)^{x_2} \tag{8.139}$$

where x_1 and x_2 are the mole fractions of gases 1 and 2, respectively. For rare gases, the conductivity dependence on temperature is given by [40]

$$k(\text{pure gas}) = A \times 10^{-6} T^{0.79} \text{ W/cm K} \tag{8.140}$$

where T is in K and $A = 15.8$ for helium, 1.97 for argon, 1.15 for krypton, and 0.72 for xenon.

8.7.1.2 Gap Closure Effects

Calza-Bini et al. [6] observed that the model of Equations 8.136 through 8.138 provides a reasonable estimate for the gap conductance on the first rise to power. Subsequently, the measured gap conductance increased and was attributed to a portion of the fuel contacting the cladding and, after fuel cracking, remaining in contact. Additionally, cracking has a negative effect on thermal conductivity of the fuel (see Section 8.3.1.5).

When gap closure occurs because of fuel swelling and thermal expansion, the contact area with the cladding is proportional to the surface contact pressure between the fuel and cladding (Figure 8.26). Thus, the contact-related heat transfer coefficient can be given by

$$h_{contact} = C \frac{2k_f k_c}{k_f + k_c} \frac{p_i}{H\sqrt{\delta_g}} (\text{Btu/ft}^2\ \text{h}\ ^\circ\text{F}) \tag{8.141}$$

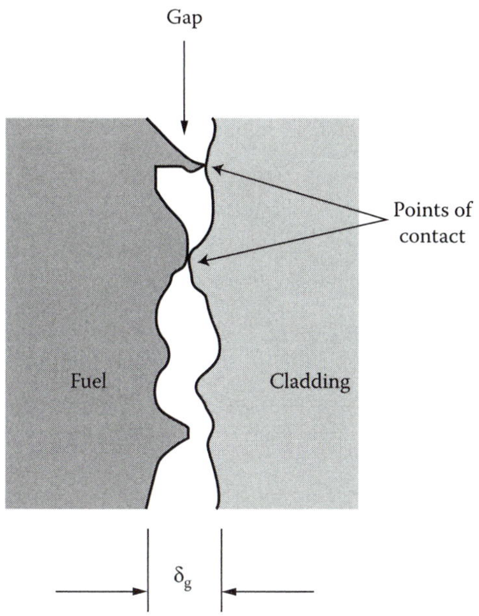

FIGURE 8.26 Close-up view of fuel-cladding contact.

where $C = 10\,\text{ft}^{-1/2}$ (a constant); p_i = surface contact pressure (psi) (usually calculated based on the relative thermal expansion of the fuel and the cladding, but ignoring the elastic deformation of the cladding); H = Meyer's hardness number of the softer material (typical values are, for steel, 13×10^4 psi and, for zircaloy, 14×10^4 psi); δ_g = mean thickness of the gas space (feet) (calculated based on the roughness of the materials in contact); k_f and k_c = thermal conductivities of fuel and cladding (Btu/h ft F).

The total gap conductance on contact may be given by

$$h_g = h_{g,open} + h_{contact} \tag{8.142}$$

Ross and Stoute [33] have shown that the general features of Equations 8.141 and 8.142 can be observed in experiments. From their data they concluded that δ_{jump} is 1–10 μm at atmospheric pressure, which is 10–30 times the gas molecule mean free path. Jacobs and Todreas [15] used the work of Cooper et al. [8] to formulate a more realistic model by recognizing the nonuniform distribution of the surface protrusions. Yet, even their improved model could not explain all the experimental observations. Thus, inadequacies in modeling remain to be resolved.

The effect of increased swelling at higher linear powers is to increase the gap conductance unless the initial gap thickness is so small as to lead to saturation of the contact effect, as can be observed in Figure 8.27. Table 8.8 provides the design and operating parameters used in the construction of Figure 8.27. The reader should consult the empirically based relations available in the database of FRAPCON for additional information [20].

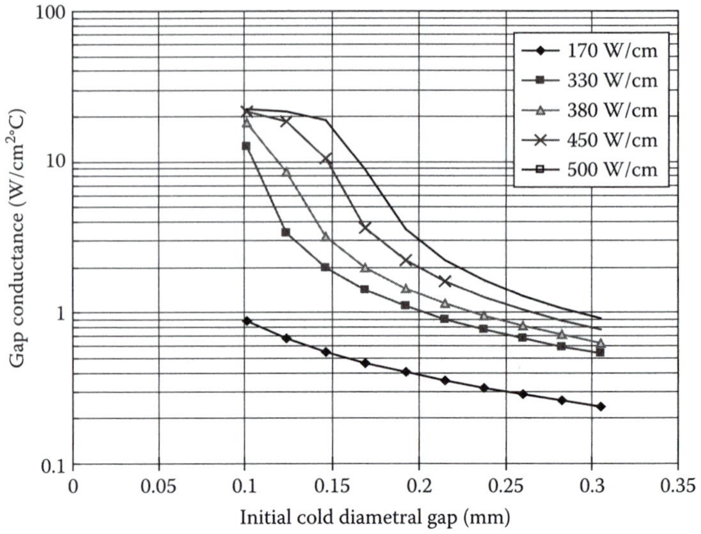

FIGURE 8.27 Gap conductance as a function of cold diametral gap in a typical PWR fuel rod calculated by FRAPCON (beginning of life, 1 h of operation).

8.7.2 Cladding Corrosion: Oxide Film Buildup and Hydrogen Consequences

During operation of the reactor, the surface of the cladding reacts with water to form a layer of oxide film at the surface. The reaction producing the zirconium dioxide film is

$$Zr + H_2O \rightarrow ZrO_2 + 2H_2 + 6.5 \times 10^6 \text{ J/kg}_{Zr} \tag{8.143}$$

This reaction also produces hydrogen, which has significant consequences for steady-state and accident conditions. The hydrogen, due to its small atom size, can diffuse into the matrix from the metal–oxide interface. Above its solubility limit in zircaloy, hydride precipitation occurs. Hydrides are brittle in nature and reduce the ductility of the cladding. The exothermic character of the reaction at high temperatures leads to more rapid reaction rate. Consequently, a total local oxidation of less than 17% of the cladding thickness and a maximum cladding temperature of 1204°C (2200°F) are current regulatory-imposed limits. The limit on film thickness has recently been considered an indirect limit on the presence of hydrogen in the clad. During loss of coolant accident, the hydrogen presence is critical to effects of thermal shock on quenching with cold emergency water. Hydrogen is a potential explosive energy source when present in large quantities mixed with oxygen or air. (Steam tends to suppress such potential.) Due to enhanced kinetics under loss of coolant accident conditions, hydrogen formation and accumulation in reactor containment have also been of concern.

TABLE 8.8
Design and Operating Parameters of a Reference Input Case for FRAPCON 3.3 Calculations of Figure 8.27

	Parameter	Unit	Value
Design	Cladding type		Zry-4
	Cladding OD	Mm	9.44–9.64
	Cladding thickness	Mm	0.57
	Fuel OD	Mm	8.2
	Diametral gap thickness	Mm	0.1–0.3
	Fuel stack length	M	3.66
	Pellet height	Mm	13.5
	Fill gas		He
	Fill gas pressure	MPa	2
	Fuel density	% TD	94
	Fuel enrichment	%	2.6
Operating	Reactor type		PWR
	Coolant pressure	MPa	15.5
	Coolant inlet temperature	K	559
	Mass flow rate	kg/m²-s	3460
	Peak linear power	W/cm	170–500
	Irradiation time	Hr	1

Corrosion control of zircaloy during normal operation is provided by alloying element addition to the zirconium matrix, such as Fe, Cr, Ni, and Nb and adjusting the microstructure via specific heat treatments and cold working. These added elements are typically insoluble and form precipitates such as $Zr(Fe,Cr)_2$ and $Zr_2(Ni, Fe)$. The size distribution of the precipitates strongly affects the corrosion behavior by means of anodic protection. For uniform corrosion observed in a PWR environment, the large precipitates are favored, while fine precipitates are better for a BWR environment to gain resistance against nodular corrosion [22]. The kinetics of the reaction distinguish between two regimes, or phases, for the corrosion rate. In the first phase, oxidation is relatively slow and the oxide formed is relatively dense until a transition thickness (typically 2.0 microns) is reached. In contrast, the oxide forms faster in the second phase (or post transition) and has a higher porosity. In general, the corrosion rate is sensitive to the temperature:

$$\frac{dS^3}{dt} = K_{pre} \exp\left(\frac{-Q_{pre}}{RT_{co}}\right) \quad \text{for pretransition} \tag{8.144a}$$

and

$$\frac{dS}{dt} = K_{post} \exp\left(\frac{-Q_{post}}{RT_{co}}\right) \quad \text{for posttransition} \tag{8.144b}$$

where S is the oxide layer thickness (μm), t is time (d), K is the frequency factor (μm^3/d or μm/d), T_{co} is the metal–oxide layer interface temperature (K), R is the universal gas constant (1.986 cal/K/mol), and Q is the activation energy (cal/mol).

It is well known that the in-reactor corrosion rate of zirconium is much higher than that in tests outside the reactor, most likely as a result of the presence of free oxygen due to radiolysis. However, other concepts have been offered to explain the increase.

For the PWR environment, several empirical models exist for prediction of the corrosion rate using Equations 8.144a and 8.144b. These models and their associated parameters for use in these equations are given in Table 8.9. Note that the criterion for transition from Equation 8.144a to 8.144b also depends on the model adopted. For the Electric Power Research Institute (EPRI) model the transition occurs at a critical value of oxide film thickness, S_c, whereas for the Asea Brown Boveri (ABB) and Commissariat à l'Énergie Atomique (CEA) models the criterion is at a transition time, t_t. Numerical values for all these criteria are also given in Table 8.9.

For the BWR for zircaloy-2 corrosion rate can be expressed by the model of Equation 8.144c [20].

$$\frac{dS}{dt} = K\exp\left(\frac{-Q}{RT}\right)\left[1 + Cq''\exp\left(\frac{-Q}{RT}\right)\right] \tag{8.144c}$$

TABLE 8.9
Constants for Corrosion Rate Calculation (Equations 8.144a and 8.144b)

	EPRI [32]	ABB [3]	CEA [3]
K_{pre} (μm^3/d)	18.9×10^9	4.0×10^{10}	11.4×10^{10}
Q_{pre} (cal/mol)	32,289	32,289	34,119
K_{post} (μm/d)	$K_o + U(M\varphi_n)^{0.24} = 8.04 \times 10^7 + 2.59 \times 10^8(7.46 \times 10^{-15}\varphi_n)^{0.24}$	$K_oT[1 + (1 + u\varphi_n)(S - S_c)] = 1.04 \times 10^8[1 + (1 + 5 \times 10^{-14}\varphi_n)(S - 6)]$	$2.15\ K_o = 2.15(1.86 \times 10^{11})$
Q_{post} (cal/mol)	27,354	27,354	36,542
S_c (μm)	$2.1410^7 \left\{\exp\left[\frac{-10{,}763}{RT_{co}} - 1.17 \times 10^{-2}T_{co}\right]\right\}$	—	—
t_t (h)	—	$6.5 \times 10^{11}\exp(-0.035\,T_{co})$	$8.857 \times 10^{10}\exp\left[\frac{1830}{RT_{co}} - 0.035T_{co}\right]$

K_{pre} and K_{post} are the frequency factors, Q_{pre} and Q_{post} are the energies of activation, K_o is the out-of-reactor constant, φ_n is the fast neutron flux in n/cm^2 s, S_c is a critical oxide thickness beyond which corrosion accelerates in-reactor, t_t is a transition time beyond which corrosion accelerates in-reactor, T_{co} is also here the metal interface temperature with the oxide film, and U, u, and M are numerical constants.

where S is the oxide layer thickness (μm), t is the time (days or seconds consistent with the units of K), K is $8.04 \times 10^{+7}$ (μm/day) or 930.6 (μm/s), Q is 27,350 (cal/mol), R is 1.987 (cal/K/mol), T is the temperature of the metal–oxide interface (K), C is 2.5×10^{-16} (m^2/W), and q'' is the surface heat flux (W/m^2).

Accelerated corrosion has been reported in both PWR and BWR environments at high burnup, and possible reasons were suggested. For the PWR, dense hydride precipitation at the metal/oxide interface at high burnup deteriorates the lattice coherency leading to destabilization of the oxide film [18]. In addition, excessive buildup of LiOH concentration at locations of subcooled boiling may also lead to destabilization of the oxide film and accelerated corrosion [26]. Hence, correlations given in Table 8.9 do not reflect these effects since wide agreement on such effects does not yet exist. For the BWR, accelerated corrosion was suggested to result on loss of secondary particle precipitates, such as Zr_2(Ni, Fe), due to irradiation-induced dissolution. The relation given in 8.144c was suggested to be limited to a fluence less than 1.0×10^{26} n/m^2 above which accelerated corrosion results [13].

The alloying of Zr with a small amount of Nb, from 1% to 2%, together with careful adjustment of impurity concentration, microstructure, and texture formation slows down the corrosion reaction considerably. Reducing the calculated corrosion rate of Zircaloy-4 by a factor of 2–2.3 can achieve a good fit with experimental data [25]. These reduction factors reflect the corrosion resistant behavior of advanced cladding materials such as ZIRLO™ and M5™, respectively, in PWR environment. Achievement of low corrosion resistance in a BWR environment at high burnup remains as a challenge today [12].

8.7.3 Overall Thermal Resistance

The linear power of a cylindrical fuel pin can be related to the temperature drop between the maximum fuel temperature, T_{max}, and the mean coolant temperature of the coolant at the axial position of interest, T_m, by considering the series of thermal resistances posed by the fuel, the gap, the cladding, the oxide corrosion layer, and the coolant. Consider, for simplicity, a solid fuel pellet with a constant thermal conductivity, k_f. From Equation 8.70, we get

$$T_{max} - T_{fo} = \frac{q'}{4\pi k_f} \tag{8.145}$$

Across the gap, the temperature drop is predicted from Equation 8.136 by

$$T_{fo} - T_{ci} = \frac{q''_g}{h_g} = \frac{2\pi R_g q''_g}{2\pi R_g h_g} = \frac{q'}{2\pi R_g h_g} \tag{8.136}$$

where R_g = mean radius in the gap; h_g = effective gap conductance.

The temperature drop across the cladding can be determined as

$$T_{ci} - T_{co} = \frac{q'}{2\pi k_c} \ell n\left(\frac{R_{co}}{R_{ci}}\right) \approx \frac{q'}{2\pi R_c k_c/\delta_c} \text{(for thin cladding)} \tag{8.146}$$

where R_c = mean radius in the cladding and δc is the cladding thickness.

The temperature drop across the thin oxide corrosion layer is given by

$$T_{co} - T_{oo} = \frac{q''}{k_o/\delta_o} = \frac{q'}{2\pi R_{oo}(k_o/\delta_o)} \tag{8.147}$$

where R_{oo} = outside oxide film radius, T_{oo} = outside oxide film temperature contacting the coolant, k_o = oxide film thermal conductivity, and δ_o = oxide film thickness.

The thermal conductivity of the reference oxide film, ZrO_2, formed in reactor is [28]

$$k_o\,(\text{W/mK}) = 0.835 + 1.81 \times 10^{-4}\, T(\text{K}) \tag{8.148}$$

The oxide corrosion layer thermal conductivity is about 0.020 W/cm °C. The various modelers reported in Table 8.8 use the following values of conductivity: 0.016 W/cm °C, EPRI, and CEA; 0.022 W/cm °C, ABB.

The heat flux emerging from the cladding oxide film is given by

$$q''_{oo} = h(T_{oo} - T_m) \tag{8.149}$$

so that

$$T_{oo} - T_m = \frac{2\pi R_{oo} q''_{oo}}{2\pi R_{oo} h} = \frac{q'}{2\pi R_{oo} h} \tag{8.150}$$

The heat transfer coefficient, h, is dependent on the coolant flow conditions. The linear power is then given by

$$q' = 2\pi R_{oo} q''_{oo} = 2\pi R_{oo} h(T_{oo} - T_m) \tag{8.151}$$

Hence, the linear power of the pin can be related to an overall thermal resistance and the temperature drop $T_{max} - T_m$ (from Equations 8.145, 8.136, 8.146, 8.147, and 8.150):

$$T_{max} - T_m = \frac{q'}{2\pi}\left[\frac{1}{2k_f} + \frac{1}{R_g h_g} + \frac{1}{k_c}\ell n\left(\frac{R_{co}}{R_{ci}}\right) + \frac{1}{R_{oo} k_o/\delta_o} + \frac{1}{R_{oo} h}\right] \tag{8.152}$$

The linear power is then analogous to an electrical current driven by the potential $T_{max} - T_m$ across a series of resistances, as depicted in Figure 8.28. This concept can be used to analyze the rate of temperature change in the fuel during a transient, provided the transient is sufficiently slow to allow the quasi-steady-state treatment of the thermal resistance.

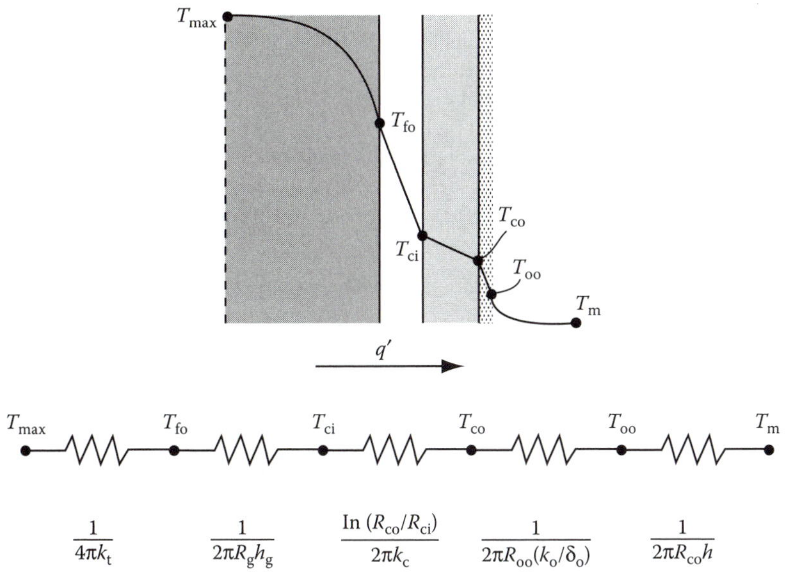

FIGURE 8.28 Thermal resistance circuit using the electrical analogy for a cylindrical fuel pin.

PROBLEMS

8.1. *Application of Kirchhoff's law to pellet temperature distribution* (Section 8.2)

For a PWR cylindrical solid fuel pellet operating at a heat flux equal to 1.7 MW/m^2 and a surface temperature of 400°C, calculate the maximum temperature in the pellet for two assumed values of conductivities.

1. $k = 3$ W/m °C independent of temperature
2. $k = 1 + 3e^{-0.0005T}$ where T is in °C

UO_2 pellet diameter = 8.192 mm
UO_2 density 95% theoretical density

Answers:
1. $T_{max} = 1560.5$°C
2. $T_{max} = 1627.9$°C

8.2. *Conductivity integral* (Section 8.2)

Describe an experiment by which you would obtain the results of Figure 8.2, that is, the value of the conductivity integral. Be sure to explicitly state what measurements and observations are to be made and how the conductivity integral is to be determined from them.

8.3. *Effect of cracking on UO_2 conductivity* (Section 8.3)

For the conditions given in Example 8.1, evaluate the effective conductivity of the UO_2 pellet after cracking using the empirical relation of Equation 8.24. Assume the gas is helium at a temperature $T_{gas} = 0.7\ T_{c\ell} + 0.3\ T_f$. Compare the results to those obtained in Example 8.1.

Answer: $k_{eff} = 1.83$ W/m K

8.4. *Temperature fields in fresh and irradiated fuel* (Section 8.3)
Consider two conditions for heat transfer in the pellet and the pellet-cladding gap of a BWR fuel pin.
Initial uncracked pellet with no relocation.
Cracked and relocated fuel.

1. For each combination, find the temperatures at the cladding inner surface, the pellet outer surface, and the pellet centerline.
2. Find for each case the volume-weighted average temperature of the pellet.

Geometry and material information:
Cladding outside diameter = 11.20 mm
Cladding thickness = 0.71 mm
Fuel-cladding gap thickness = 90 μm
Initial solid pellet with density = 88%

Basis for heat transfer calculations:
Cladding conductivity is constant at 17 W/m K
Gap conductance
Without fuel relocation, 4300 W/m² K;
With fuel relocation, 31,000 W/m² K;
Fuel conductivity (average) at 95% density
Uncracked, 2.7 W/m K,
Cracked, 2.4 W/m K;
Volumetric heat deposition rate: uniform in the fuel and zero in the cladding
Do not adjust the pellet conductivity for restructuring.
Use Biancharia's porosity correction factor (Equation 8.21).
Operating conditions:
Cladding outside temperature = 295°C
Linear heat generation rate = 44 kW/m

Answers:

	Uncracked	Cracked
1. T_{max}(°C)	2135	2026
T_{fo}(°C)	687	398
T_{ci}(°C)	351	351
2. T_{ave}(°C)	1411	1212

8.5. *Comparison of UO_2 and UC fuel temperature fields* (Section 8.4)
A fuel plate is of half-width $a = 10$ mm and is contained in a sheet of zircaloy cladding with thickness $\delta_c = 2$ mm. The heat is generated uniformly in the fuel.
Compare the temperature drop across the fuel plate when the fuel is UO_2 with that of a UC fuel for the same heat generation rate (i.e., calculate the ratio of $T_{max} - T_{co}$ for the UC plate to that of the UO_2 plate).

Answer: $\Delta T_{UO_2}/\Delta T_{UC} = 4.16$

8.6. *Thermal conduction problem involving design of a BWR core* (Section 8.5)
A core design is proposed in which BWR type UO_2 pins are located in holes within graphite hexagonal blocks (Figure 8.29). These blocks

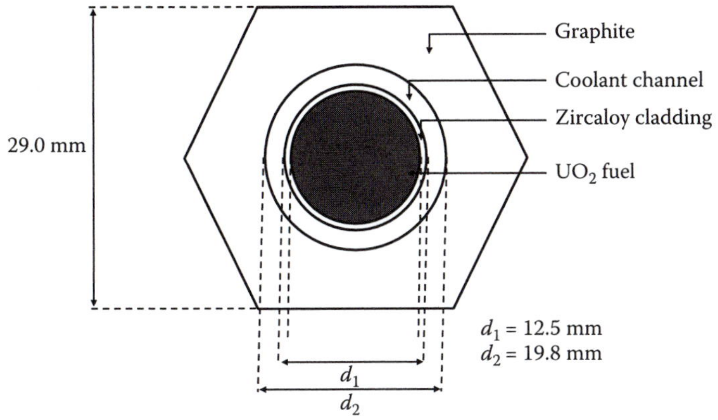

FIGURE 8.29 Unit cell dimensions.

then form a core of radius R_o. Constants and constraints are defined in Figure 8.30 and Table 8.10. Basically these constraints exist under decay power conditions where the outside of the core radiates its energy to a passive air chimney. However, the outside of the core which is in touch with a vessel at the same temperature is limited to 500°C. The cladding outside temperature, T_{co}, which radiates to the graphite, T_{gi}, is also constrained, here to a temperature 649°C.

Calculate the achievable linear heat power of the core (MW/m) as a function of core radius, R_o(m) for constraints of Table 8.10. Present the result as a plot.

Notes for Figure 8.29:

- Unit cell using BWR fuel pin in MHTGR prismatic block holding the ratio of fuel to graphite constant.
- Coolant channel size established by taking the area of water normally associated with a pin in conventional BWR.

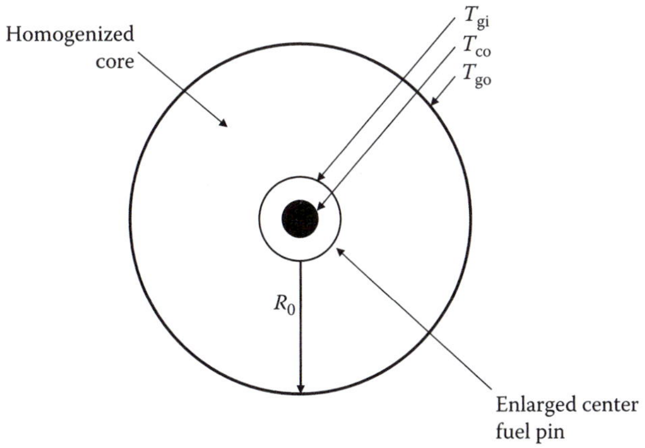

FIGURE 8.30 Configuration of a solid core and variables.

TABLE 8.10
Constants and Constraints for Homogenized Core Power Analysis

Constraints	Constants
$T_{co} < 649°C$	$A_{cell} = 7.30 \times 10^{-4}$ m^2
$T_{go} < 500°C$	$d_1 = 12.5$ mm
	$d_2 = 19.8$ mm
	$\in_1 = 0.6$ (cladding emissivity)
	$\in_2 = 0.7$ (graphite emissivity)
	$k_g = 60$ W/m
	$\sigma = 5.669 \times 10^{-8}$ W/m$^2 \cdot$ K^4

Notes for Figure 8.30:

T_{gi} is the temperature at the inner surface of the matrix graphite surrounding the fuel pin.
T_{go} is the temperature of the matrix graphite at the core outer surface.
T_{co} is the temperature at the cladding outer surface.
d_1 and d_2 are shown in Figure 8.29.

8.7. *Comparison of thermal energy that can be extracted from a spherical hollow fuel pellet versus a cylindrical annular fuel pellet* (Section 8.5) Consider a cylindrical annular fuel pellet of length L, inside radius R_V, and outside radius R_{foc}. It is operating at q_c''', such that for a given outside surface temperature, T_{fo}, the inside surface temperature, T_V, is just at the fuel melting limit T_{melt}.

A fellow engineer claims that if the same volume of fuel is arranged as a sphere with an inside-voided region of radius R_V and operated between the same two surface temperature limits, that is, T_V and T_{fo}, more power can be extracted from the spherical fuel volume then from a cylindrical fuel pellet. In both cases volumetric generation rate is radially constant. Is the claim correct? Prove or disprove it. Use the nomenclature of Figure 8.31. Assume no sintering occurs.

Useful relations:
The one-dimensional heat conduction equation in the radial direction in spherical coordinates is:

$$\frac{1}{r^2}\frac{\mathrm{d}}{\mathrm{d}r}\left(kr^2\frac{\mathrm{d}T}{\mathrm{d}r}\right) + q''' = 0$$

For a sphere: $V_S = 4/3\pi R^3$ and $A_S = 4\pi R^2$

Cylindrical Annular Fuel Pellet	Spherical Hollow Fuel Pellet
q_c'''	q_s'''
$R_V = 0.25$ mm	$R_V = 0.25$ mm
$R_{foc} = 1$ cm	$R_{fos} = ?$
$L = 1$ cm	(to be determined from the problem statement)

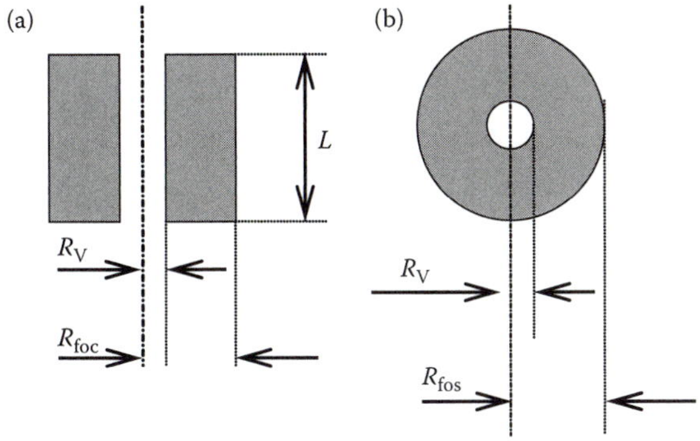

FIGURE 8.31 Fuel Pellet geometry. (a) cylindrical annular pellet. (b) spherical hollow pellet.

Answer: The sphere does generate more power.

8.8. *Fuel pin problem* (Section 8.5)

A fuel pin is operating with solid pellets of 88% theoretical density and outside radius 5 mm such that at the axial location of maximum fuel temperature, the fuel centerline temperature, T_{CL}, is 2500°C and the fuel surface temperature, T_{fo}, is 700°C. It is desired to raise the pin linear power by 10% by employing one of the following alternative strategies (In each case, all the other conditions except for the one cited are held constant.):

a. Raise the maximum allowable fuel temperature.
b. Use an annular pellet with the center void of dimension R_v, or
c. Increase the pellet density.

For each strategy find the new value of the cited parameter necessary to achieve the desired 10% increase of linear pin power. Sintering effects may be neglected.

Answers:

a. $T_{max} = 2610°C$
b. $R_V = 0.75$ mm
c. $\rho = 0.96\ \rho_{TD}$

8.9. *Radially averaged fuel temperature and stored energy in solid and annular pellet* (Section 8.5)

Consider a solid pellet of radius b and an annular pellet of inside radius a, and outside radius b, each operating at the same linear power rate, q'. Define

$$\Delta T(r) \equiv T(r) - T_b$$

and

$$\overline{\Delta T(r)} \equiv \overline{T(r)} - T_b$$

1. Find across each pellet, the value of $\overline{\Delta T}/\Delta T$. Use the subscript "s" for solid and the subscript "a" for annular.

2. What is the ratio of the stored energy in the solid to the annular pellet?

Answers:

1. Solid Pellet -

$$\frac{\overline{\Delta T_s}}{\Delta T_s} = \frac{1}{2\left(1 - \frac{r^2}{b^2}\right)}$$

Annular Pellet -

$$\frac{\overline{\Delta T_a}}{\Delta T_a} = \frac{(2/F)(b^4/4 - a^2b^2 + 3/4\, a^4 + a^4 \ell n\, b/a)}{(b^2 - a^2)}$$

2. $$\frac{c_{p_s}}{c_{p_a}} \frac{M_s \overline{\Delta T_s}}{M_a \overline{\Delta T_a}} = \frac{c_{p_s}(\overline{\Delta T_s})}{c_{p_a}(\Delta T_a)} \frac{\rho_s}{\rho_a} \frac{b^4}{4}\left(\frac{b^4}{4} - a^2b^2 + \frac{3}{4}a^4 + a^4 \ell n \frac{b}{a}\right)^{-1}$$

8.10. *Maximum linear power from a duplex fuel pellet* (Section 8.5)
To uprate the power of its LWR fleet, an electric utility is considering the use of duplex fuel pellets. A duplex fuel pellet consists of two radial zones, one loaded with UO_2 and one with PuO_2 (Figure 8.32). The pellet outer temperature is fixed at 400°C.

1. Assuming that centerline fuel melting is the design limit for this pellet, which oxide would you put in Zone 1 (i.e., the inner zone)?
2. Using only the properties in Table 8.11 and assuming that the volumetric heat generation rate in PuO_2 is 50% higher than in UO_2, calculate the maximum linear power at which the pellet can be operated without melting the fuel. (Neglect the thermal conductivity dependence on temperature.)
3. How does the maximum linear power compare with the maximum linear power for an all-UO_2 pellet?

Answers:

1. UO_2
2. 95.6 kW/m
3. 90.5 kW/m

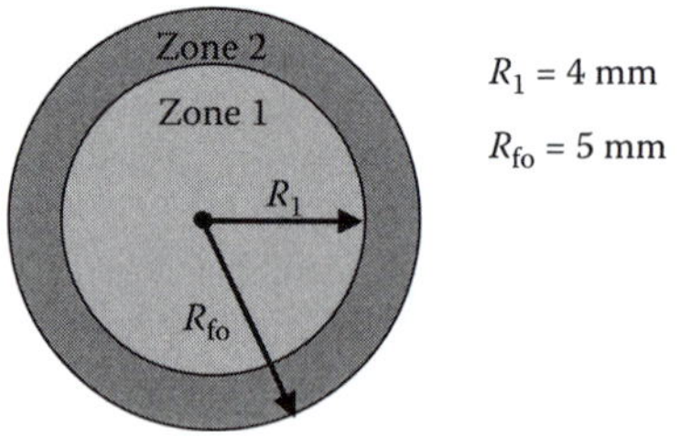

FIGURE 8.32 Cross-sectional view of a duplex fuel pellet.

TABLE 8.11
Properties of Oxide Fuels

Parameter	UO_2	PuO_2
Density (g/cm^3)	10.5	10.9
Thermal conductivity (W/m °C)	3.0	2.5
Melting point (°C)	2800	2300
Specific heat (J/kg °C)	410	380

8.11. *Effect of internal cooling on fuel temperature* (Section 8.5)
Consider the following three UO_2 pellet configurations:
- Solid pellet
- Annular pellet with only external cooling
- Annular pellet with simultaneous internal and external cooling

The dimensions for all three pellets are in Table 8.12 below. Assume that the fuel thermal conductivity is k_f=3 W/m K (independent of temperature), the pellet surface temperature is 700°C, and the linear power is q' = 40 kW/m in all three cases.
1. Calculate the maximum temperature for the solid pellet.
2. Calculate the maximum temperature for the annular pellet with only external cooling.
3. Calculate the maximum temperature for the annular pellet with simultaneous external and internal cooling.
4. For the annular pellet with simultaneous internal and external cooling calculate also the heat flux at the inner and outer surfaces.
5. What are the advantages and drawbacks of the annular fuel pellet with simultaneous internal and external cooling?

Answers:
1. T_{max} = 1761°C
2. T_{max} = 1579°C
3. T_{max} = 793°C
4. Inner: q'' = −567.9 kW/m^2 Outer: q'' = 504.3 kW/m^2

8.12. *Temperature field in a restructured fuel pin* (Section 8.6)
Using the conditions of Problem 8.4 for the uncracked fuel, calculate the maximum fuel temperature for the given operating conditions. Assume two-zone sintering, with $T_{sintering}$ = 1700 °C and $\rho_{sintered}$ = 98% TD.

Answer:
T_v = 1961°C

TABLE 8.12
Geometry of the Three UO_2 Pellets

	ID (mm)	OD (mm)
Solid pellet	N/A	8.2
Annular pellet with only external cooling	2.0	8.44
Annular pellet with internal and external cooling	9.9	14.1

8.13. *Eccentricity effects in a plate-type fuel* (Section 8.7)
A nuclear fuel element is of plate geometry (Figure 8.33). It is desired to investigate the effects of fuel offset within the cladding. For simplicity, assume uniform heat generation in the fuel, temperature-independent fuel conductivity, and heat conduction only in the gap. Calculate:
1. The temperature difference between the offset fuel and the concentric fuel maximum temperatures:

$$T_{MO} - T_{MC}$$

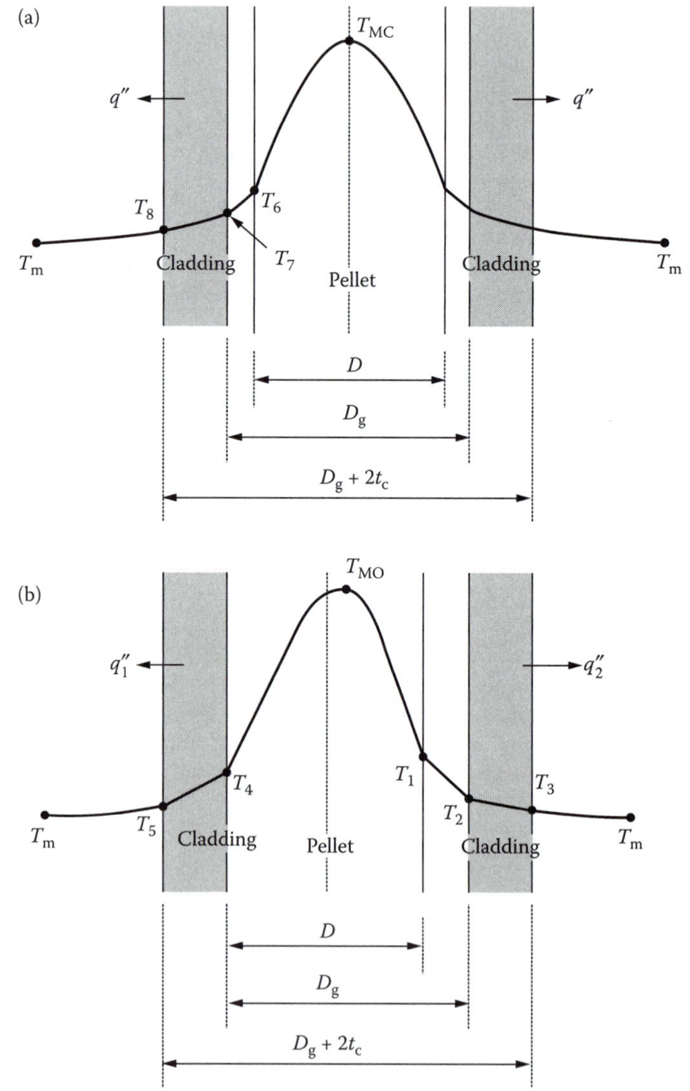

FIGURE 8.33 Effect of fuel offset. (a) Centered fuel pellet. (b) Effect of fuel pellet offset.

2. The temperature difference between the cladding maximum temperatures:

$$T_7 - T_4$$

3. The ratio of heat fluxes to the coolant.

$$\frac{q''}{q_1''} \text{ and } \frac{q''}{q_2''}$$

$T_{coolant}$ and heat transfer coefficient to coolant may be assumed constant on both sides and for both cases. Neglect interface contact resistance for fuel and cladding.

k_f = 3.011 W/m°C ($PuO_2 - UO_2$)
k_g = 0.289 W/m°C (He)
k_c = 21.63 W/m°C (SS)
D = 6.352 mm
D_g = 6.428 mm
t_c = 0.4054 mm
q''' = 9.313×10^5 kW/m³
$h_{coolant}$ = 113.6 kW/m²°C (Na)

Answers:
1. $T_{MO} - T_{MC} = -23.8$°C
2. $T_7 - T_4 = -8.83$°C
3. $\frac{q''}{q_1''} = 0.902$ and $\frac{q''}{q_2''} = 1.121$

8.14. *Determining the linear power given a constraint on the fuel average temperature* (Section 8.7)
For a PWR fuel pin with pellet radius of 4.096 mm, cladding inner radius of 4.178 mm, and outer radius of 4.75 mm, calculate the maximum linear power that can be obtained from the pellet such that the mass average temperature in the fuel does not exceed 1204°C (2200°F). Take the bulk fluid temperature to be 307.5°C and the coolant heat transfer coefficient to be 28.4 kW/m²°C. Consider only conduction based on the effective gap width using the gap conductance model.
Fuel conductivity $k_f = 3.011$ W/m°C
Cladding conductivity $k_c = 18.69$ W/m°C
Helium gas conductivity $k_g = 0.277$ W/m°C

Answer:
$q' = 31.7$ kW/m

REFERENCES

1. Amaya, M., Sugiyama, T., and Fuketa, T., Fission gas release in irradiated UO_2 fuel at burnup of 45 GWd/t during simulated reactivity initiated accident (RIA) condition. *J. Nucl. Sci. Technol.*, 41(10):966–972, 2004.
2. Arpaci, V. S., *Conduction Heat Transfer*. Addison-Wesley, Menlo Park, CA, 1966.
3. Bailly, H., Ménessier, D., and Prunier, C., *The Nuclear Fuel of Pressurized Water Reactors and Fast Neutron Reactors*. Lavoisier Publishing, Paris, France, 1999.

4. Bernard, L. C., Jacoud, J. L., and Vesco, P., An efficient model for the analysis of fission gas release. *J. Nucl. Mater.*, 302:2–3, 2002.
5. Biancharia, A., The effect of porosity on thermal conductivity of ceramic bodies. *Trans. ANS*, 9:15, 1966.
6. Calza-Bini, A., Cosoli, G., Filacchioni, G., Lanchi, M., Nobili, A., Pesce, E., Rocca, U., Rotoloni, P. L. In-pile measurement of fuel cladding conductance for pelleted and vipac zircaloy-2 sheathed fuel pin. *Nucl. Technol.*, 25:103, 1975.
7. Clark, P. A. E., Clough, D. J., Smith, R. C. *Post Irradiation Examination of Two High Burnup Fuel Rods Irradiated in the Halden BWR.* AEREG3207, AERE Harwell Report. May 1985.
8. Cooper, M. G., Mikic, B. B., and Yovanovich, M. M., Thermal contact conductance. *Int. J. Heat Mass Transf.*, 12:279, 1969.
9. Fenech, H., General considerations on thermal design and performance requirements of nuclear reactor cores. In H. Fenech (ed.). *Heat Transfer and Fluid Flow in Nuclear Systems.* Pergamon Press, Oxford, UK, 1981.
10. Fink, J. K., Thermophysical properties of uranium dioxide. *J. Nucl. Mater.*, 279:1–18, 2000.
11. Francl, J. and Kingery, W. D., Thermal conductivity. IX. Experimental investigation of porosity on thermal conductivity. *J. Am. Ceram. Soc.*, 37:99, 1954.
12. Garzarolli, F., Adamson, R., and Ruddling, P., Optimization of BWR Fuel Rod Cladding Condition at High Burnups, *2010 LWR Fuel Performance Meeting*, Orlando, FL, September 26–29, 2010.
13. Garzarolli, F., Ruhmann, H., and Van Swam, L., Alternative Zr alloys with irradiation resistant precipitates for high burnup BWR application. *Zirconium in the Nuclear Industry: Thirteenth International Symposium*, ASTM STP 1423:119–132, 2002.
14. Gibby, R. L., The effect of plutonium content on the thermal conductivity of (U, Pu)O_2 solid solutions. *J. Nucl. Mater.*, 38:163, 1971.
15. Jacobs, G. and Todreas, N., Thermal contact conductance in reactor fuel elements. *Nucl. Sci. Eng.*, 50:282, 1973.
16. Jordan, R., *Massachusetts Institute of Technology Reactor Safety Course*, Cambridge, MA, July 1979.
17. Kampf, H. and Karsten, G., Effects of different types of void volumes on the radial temperature distribution of fuel pins. *Nucl. Appl. Technol.*, 9:288, 1970.
18. Kim, Y. S., Woo, H. K., Im, K. S., and Kwun, S. I., The cause for enhanced corrosion of zirconium alloys by hydrides. *Zirconium in the Nuclear Industry: Thirteenth International Symposium*, ASTM STP 1423:274–296, 2002.
19. Lanning, D. D., Beyer, C. E., and Geelhood, K. J., *FRAPCON-3 updates, including mixed-oxide fuel properties.* Tech. Rep. NUREG/CR-6534 Vol. 4, PNNL-11513, 2005.
20. Lanning, D. D., Beyer, C. E., and Painter, C. L., *FRAPCON-3: Modifications to fuel rod material properties and performance models for high-burnup application.* NUREG/CR-6534 Vol. 1, PNNL-11513, 1997.
21. Latta, R. E. and Frysell, R. E., Determination of solidus-liquidus temperatures in the UO_{2+n} system ($-0.50 < n < 0.20$). *J. Nucl. Mater.*, 35:195, 1970.
22. Lemaignan, C., Physical phenomena concerning corrosion under irradiation of Zr alloys. *Zirconium in the Nuclear Industry: Thirteenth International Symposium*, ASTM STP 1423:20–29, 2002.
23. Loeb, A. L., Thermal conductivity. VIII. A theory of thermal conductivity of porous material. *J. Am. Ceram. Soc.*, 37:96, 1954.
24. MacDonald, P. E. and Smith, R. H., An empirical model of the effects of pellet cracking on the thermal conductivity of UO_2 light water reactor fuel. *Nucl. Eng. Des.*, 61:163, 1980.
25. MacDonald, P. E. and Weisman, J., Effect of pellet cracking on light water reactor fuel temperatures. *Nucl. Technol.*, 31:357, 1976.

26. McDonald, S. G., Sabol, G. P. and Sheppard, S. P., Effect of lithium hyroxide on the corrosion behavior of zircaloy-4. *Zirconium in the Nuclear Industry: Thirteenth International Symposium*, ASTM STP 824:519–530, 2002.
27. Marr, W. M. and Thompson, D. H., Prediction of fuel melting in an LMFBR fuel element. *Trans. Am. Nucl. Soc.*, 14:150, 1971.
28. Notley, M. J. F., A computer model to predict the performance of UO_2 fuel element irradiated at high power output to a burnup of 10,000 MWD/MTU. *Nucl. Appl. Technol.*, 9:195, 1970.
29. Nuclear Energy Agency, NEA-1596 IFPE/AECL-BUNDLE, *NEA Data Bank Computer Programs: Category Y. Integral Experiments Data, Databases, Benchmarks*, 2000.
30. Olander, D. R., *Fundamental Aspects of Nuclear Reactor Fuel Elements.* T1D-26711-P1, 1976.
31. Olsen, C. S. and Miller, R. L., *MATPRO, Vol. II: A Handbook of Materials Properties for Use in the Analysis of Light Water Reactor Fuel Behavior.* NUREG/CR-0497, USNRC, 1979.
32. Pyecha, T. D., Bain, G. M., McInteer, W. A., and Pham, C. H., Waterside corrosion of PWR fuel rods through burnups of 50,000 MWD/MTU. *ANS Topical Meeting on Light Water Reactor Fuel Performance*. Orlando, FL, 1985.
33. Ross, A. M. and Stoute, R. L., *Heat Transfer Coefficient between UO_2 and Zircaloy-2.* AECL-1552, 1962.
34. SCDAP/RELAP5-3D© Code Development Team. SCDAP/RELAP5-3D© CODE Manual Volume 4: *MATPRO – A Library of Materials Properties for Light-Water-Reactor Accident Analysis.* INEEL/EXT-02-00589-V4-R2.2, 2003.
35. Schmidt, H. E. and Richter, J., The influence of stoichiometry on the thermal conductivity of $(U_{0.8}Pu_{0.2})O_2x$. Presented at *Oxide Fuel Thermal Conductivity Symposium*, Stockholm, Sweden, June 1, 1967.
36. Shackelford, J. F. (ed.), *CRC Materials Science and Engineering Handbook,* 2nd Ed. CRC Press, Boca Raton, FL, 1994.
37. Shestopalov, A., Lioutov, K., and Yegorova, L., Adaptation of USNRC's FRAPTRAN and IRSN's SCANAIR Transient Codes and Updating of MATPRO Package for Modeling of LOCA and RIA Validation Cases with Zr-1% Nb (VVER type) Cladding, NUREG/IA-0209 IRSN 2002-33, NSI RRC KI 3067, 2003.
38. Snead, L., Nozawa, T., Katoh, Y., Byun, T., Kondo, S., and Petti, D., Handbook of SiC properties for fuel performance modeling. *J. Nucl. Mater.*, 371:329–337, 2007.
39. Thetford, R. and Mignanelli, M., The chemistry and physics of modeling nitride fuels for transmutation. *J. Nucl. Mater.*, 320(1–2):44–53, 2003.
40. Von Ubisch, H., Hall, S., and Srivastov, R., Thermal conductivities of mixtures of fission product gases with helium and argon. Presented at the *2nd U.N. International Conference on Peaceful Uses of Atomic Energy*. Sweden, 1958.

9 Single-Phase Fluid Mechanics

9.1 APPROACH TO SIMPLIFIED FLOW ANALYSIS

The objective of the fluid mechanics analysis is to provide the velocity and pressure distributions in a given geometry for specified boundary and initial conditions. The transport equations of mass, momentum, and energy were derived for both a control volume and at a local point in Chapter 4. Theoretically, we need to solve these detailed equations simultaneously to obtain the velocity, pressure, and temperature distributions in the flow system of interest. Practically, however, we first simplify the equations to be solved by eliminating the insignificant terms for the situation of interest.

Furthermore, our objective can often be achieved by applying the accumulated engineering experience in a manner that empirically relates macroscopic quantities, for example, the pressure drop and the flow rate through a tube, without obtaining the detailed distribution of the fluid velocity or density in the tube. This engineering approach can be used whenever the flow characteristics fall within the range of previously established empiric relations.

The analytic approach and the empiric engineering relations for single-phase flow analysis are discussed in this chapter. The heat transport analysis is dealt with in Chapter 10, and the fluid mechanics of two-phase flow are considered in Chapter 11.

9.1.1 Solution of the Flow Field Problem

Determination of the velocity field in a moving fluid requires simultaneous solution of the mass, momentum, and energy equations:
Mass:

$$\frac{\partial \rho}{\partial t} + \nabla \cdot (\rho \vec{v}) = 0 \tag{4.73}$$

Linear momentum:

$$\frac{\partial}{\partial t} \rho \vec{v} + \nabla \cdot \rho \vec{v}\vec{v} = -\nabla \cdot (p\bar{\bar{I}} - \bar{\bar{\tau}}) + \rho \vec{f} \tag{4.80}$$

Energy, in one of its various forms, for example:

$$\rho \frac{Du^\circ}{Dt} = -\nabla \cdot \vec{q}'' + q''' - \nabla \cdot p\vec{v} + \nabla \cdot (\bar{\bar{\tau}} \cdot \vec{v}) + \vec{v} \cdot \rho \vec{f} \tag{4.96}$$

or equivalently:

$$\rho c_p \frac{DT}{Dt} = -\nabla \cdot \vec{q}'' + q''' + \beta T \frac{Dp}{Dt} + \Phi \tag{4.122a}$$

There are six unknowns in the above equations: ρ, $\vec{\upsilon}$, u°(or T), p, $\overline{\overline{\tau}}$, and $\vec{q}''$. Two additional quantities are *a priori* given, q''' and $\vec{g}$. Therefore, the three transport equations need to be supplemented with three additional equations: the equation of state for the fluid:

$$\rho = \rho(p, T) \tag{9.1}$$

and two constitutive equations that relate shear stress and heat flux to the unknown or given quantities (in magnitude as well as spatial gradients):

$$\overline{\overline{\tau}} = \overline{\overline{\tau}}(\rho, \vec{\upsilon}, T) \tag{9.2}$$

$$\vec{q}'' = \vec{q}''(\rho, \vec{\upsilon}, T) \tag{9.3}$$

Note that T in Equations 9.1 through 9.3 can be replaced by u° if the energy equations in the form 4.96 are used.

Finally, initial and boundary conditions necessary for completely specifying the solution of the differential transport equations should be supplied.

Thus given q''' and $\vec{g}$, the above six equations can be solved to obtain the six variables, ρ, $\vec{\upsilon}$, T(or u°), p, $\overline{\overline{\tau}}$, and $\vec{q}''$.

Often, however, we tend to use simplified forms that represent acceptable approximations to the physical conditions. Thus, the complexity of the problem can be reduced, and even analytic forms for all the variables may be obtained.

9.1.2 Possible Simplifications

1. The most significant of these approximations is the assumption of temperature-independent physical properties, which leads to decoupling of the velocity field solution from the energy equation, as both ρ and $\overline{\overline{\tau}}$ no longer depend on T. Thus Equations 4.73 and 4.80 can then be solved along with the simplified equations:

$$\rho = \rho(p) \tag{9.1a}$$

$$\overline{\overline{\tau}} = \overline{\overline{\tau}}(\rho, \vec{\upsilon}) \tag{9.2a}$$

to obtain the unknowns, ρ, $\vec{\upsilon}$, p, $\overline{\overline{\tau}}$. This assumption removes the energy equation from the system of equations to be solved, as illustrated in Figure 9.1.

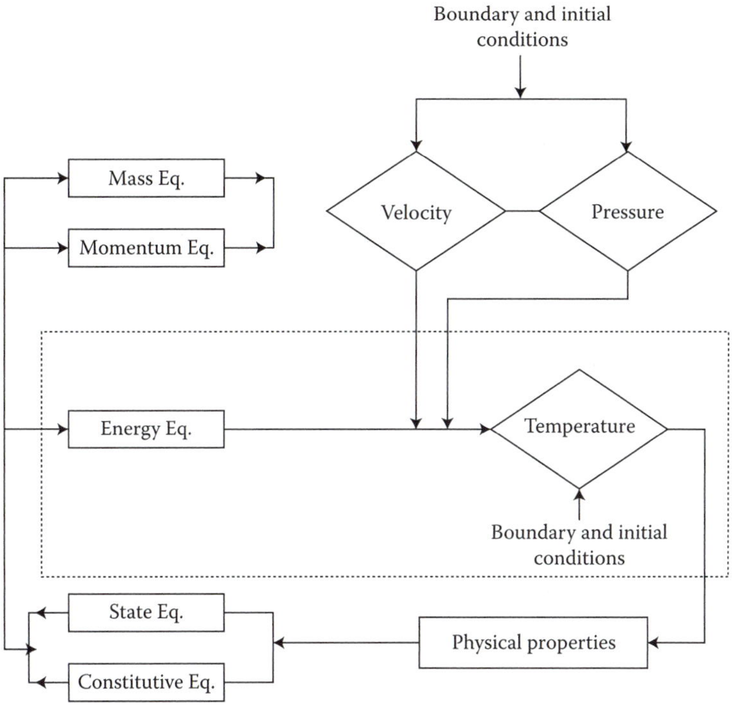

FIGURE 9.1 Fluid mechanics analysis schemes. The region within the *dotted line* is considered separately when the physical properties are assumed temperature independent. Note that in a transient situation the initial conditions for each time period (one time step) are the values of the variable at the end of the preceding time period.

This assumption is good in flow fields that do not span a wide range of temperatures, provided the selection for $\rho(p)$ and $\bar{\bar{\tau}}(\rho,\vec{\upsilon})$ is made to represent their values at an appropriate temperature within the range of interest.

Other possible simplifying assumptions are

2. The density (ρ) is constant, which is a valid assumption when the effect of pressure as well as temperature is small. This is an incompressible flow problem with constant-temperature properties. It is a reasonable assumption for nearly all practical problems involving liquids. At high pressure, it may also be applied to gases if the pressure variation within the system is small.
3. The effects of viscosity are negligible, so that $\nabla\cdot\bar{\bar{\tau}}$ is a negligible term in the momentum equation. This is called an inviscid flow problem and is appropriate for flows where the momentum effects dominate, such as the case at high flow velocities in large compartments or even with open channel flow. In some nuclear reactor components, for example, large pipes, or within a large reactor vessel plenum, the flow may be considered inviscid.

4. The problem can be solved in the fewest number of dimensions. For example, a problem can be solved as one-dimensional if the flow in the other dimensions is either nonexistent or very small. In that case, the point equation is to be integrated over the directions that are to be eliminated, which results in the one-dimensional problem.
5. Finally, if the information desired is for the endpoint conditions (e.g., the pressure drop across a channel), the momentum integral equation can be solved, rather than the local momentum equation. For steady-state one-dimensional flows, this approach is often used, and the change in momentum flux between the inlet and the exit is related to the static pressure change as well as the shear forces and gravity forces. The laws representing shear forces depend on the flow velocity and material properties in a manner described in the next few sections.

The relevant categories of single-flow situations are given in Figure 9.2. Some of the categories that appear are further explained in this chapter.

9.2 INVISCID FLOW

9.2.1 Dynamics of Inviscid Flow

In the absence of viscosity (or shear forces), the momentum equation (Equation 4.80) and the mass balance equation (Equation 4.73) can be combined to yield

$$\rho \frac{D\vec{\upsilon}}{Dt} \equiv \rho \frac{\partial \vec{\upsilon}}{\partial t} + \rho \vec{\upsilon} \cdot \nabla \vec{\upsilon} = -\nabla p + \rho \vec{f} \tag{4.91}$$

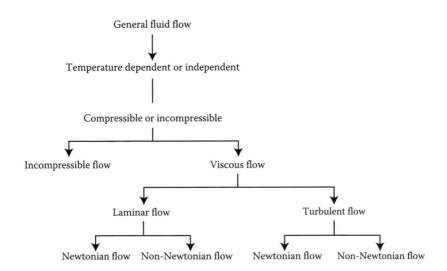

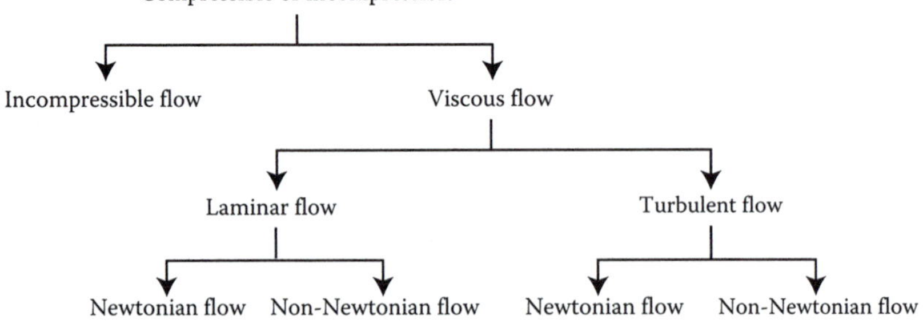

FIGURE 9.2 Categorization of fluid flow situations.

The above equation can be rewritten in the form of gradient and curl operations with the help of the definition of $\vec{\upsilon}\cdot\nabla\vec{\upsilon}$ as

$$\rho\vec{\upsilon}\cdot\nabla\vec{\upsilon} = \frac{\rho}{2}\nabla\upsilon^2 - [\rho\vec{\upsilon}\times(\nabla\times\vec{\upsilon})] \tag{9.4a}$$

so that the inviscid flow momentum equation takes the form:

$$\frac{\partial\vec{\upsilon}}{\partial t} + \nabla\left(\frac{\upsilon^2}{2}\right) - [\vec{\upsilon}\times(\nabla\times\vec{\upsilon})] = -\frac{1}{\rho}\nabla p + \vec{f}a \tag{9.4b}$$

The term $\nabla\times\vec{\upsilon}$ represents an angular rotation of the fluid element and is referred to as the *vorticity* of the fluid ($\vec{\omega}$). It can be shown [3] that the vorticity is twice the angular velocity of any two perpendicular lines intersecting at a point (Figure 9.3). Thus, the inviscid flow momentum equation takes the form:

$$\frac{\partial\vec{\upsilon}}{\partial t} + \nabla\left(\frac{\upsilon^2}{2}\right) - \vec{\upsilon}\times\vec{\omega} = -\frac{1}{\rho}\nabla p + \vec{f} \tag{9.4c}$$

9.2.2 Bernoulli's Integral

9.2.2.1 Time-Dependent Flow

The body forces can often be expressed as the gradient of a scalar function called the potential function (ψ):

$$\vec{f} = -\nabla\psi \tag{4.21}$$

For example, in a gravity field $\psi = gz$, where g = the gravitational constant and z = the coordinate in the vertical direction:

$$\vec{f} = -\left(\frac{\partial}{\partial z}gz\right)\vec{k} = -g\vec{k} = \vec{g} \tag{9.5}$$

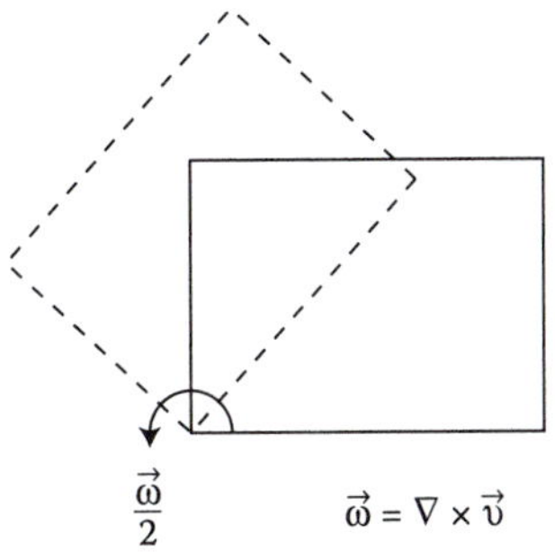

FIGURE 9.3 Rotation of a fluid element about a point.

When the body forces are derived from a field potential, they are said to be conservative. For conservative body forces, the inviscid fluid momentum (Equation 9.4) can be written as

$$\frac{\partial \vec{\upsilon}}{\partial t} + \frac{1}{\rho}\nabla p + \nabla\left(\psi + \frac{\upsilon^2}{2}\right) = \vec{\upsilon} \times \vec{\omega} \tag{9.6a}$$

Let $d\vec{r}$ be an elementary displacement (dx, dy, dz). Note that for an arbitrary scalar quantity, X, $\nabla X \cdot d\vec{r} = dX$, as

$$\begin{aligned}\nabla X \cdot d\vec{r} &= \left(\vec{i}\frac{\partial X}{\partial x} + \vec{j}\frac{\partial X}{\partial y} + \vec{k}\frac{\partial X}{\partial z}\right) \cdot (\vec{i}dx + \vec{j}dy + \vec{k}dz) \\ &= \frac{\partial X}{\partial x}dx + \frac{\partial X}{\partial y}dy + \frac{\partial X}{\partial z}dz = dX\end{aligned} \tag{9.7}$$

Thus when the terms in Equation 9.6a are multiplied by $d\vec{r}$, we get

$$\frac{\partial \vec{\upsilon}}{\partial t} \cdot d\vec{r} + \frac{dp}{\rho} + d\left(\psi + \frac{\upsilon^2}{2}\right) = (\vec{\upsilon} \times \vec{\omega}) \cdot d\vec{r} \tag{9.6b}$$

The right-hand side term is zero if $\vec{\omega} = 0$, that is, if the flow is irrotational, or if $d\vec{r} = d\vec{s}$, when $d\vec{r}$ is directed along a stream line $d\vec{s}$ and hence in the same direction as $\vec{\upsilon}$.

The second case is true because of the identity:

$$(\vec{\upsilon} \times \vec{\omega}) \cdot d\vec{r} = (\vec{\omega} \times d\vec{r}) \cdot \vec{\upsilon} = (d\vec{r} \times \vec{\upsilon}) \cdot \vec{\omega} = 0 \tag{9.8}$$

as $d\vec{r}$ is parallel to $\vec{\upsilon}$ and $d\vec{r} \times \vec{\upsilon} = 0$. Thus, either when the flow is irrotational or when we are considering changes along a stream line:

$$\frac{\partial \vec{\upsilon}}{\partial t} \cdot d\vec{r} + \frac{dp}{\rho} + d\left(\psi + \frac{\upsilon^2}{2}\right) = 0 \tag{9.9}$$

For a barotropic fluid, where ρ is only a function of p, we can integrate Equation 9.9 along a stream line, or between any two points if the flow is irrotational, to obtain the general form of the Bernoulli integral equation:

$$\int \frac{\partial \vec{\upsilon}}{\partial t} \cdot d\vec{r} + \int \frac{dp}{\rho} + \left(\psi + \frac{\upsilon^2}{2}\right) = f(t) \tag{9.10a}$$

Note that $f(t)$, the constant of integration, is the same for any two points at a given time but might change with time. Equation 9.10a is generally written as the integral between two positions along the stream line:

$$\int_1^2 \frac{\partial \vec{\upsilon}}{\partial t} \cdot d\vec{r} + \int_1^2 \frac{dp}{\rho} + \Delta\left(\psi + \frac{\upsilon^2}{2}\right) = 0 \tag{9.10b}$$

where

$$\Delta\left(\psi + \frac{\upsilon^2}{2}\right) = \left(\psi + \frac{\upsilon^2}{2}\right)_2 - \left(\psi + \frac{\upsilon^2}{2}\right)_1$$

9.2.2.1.1 Time-Dependent Flow through a Variable-Geometry Channel

Consider the time-dependent flow of an incompressible, inviscid fluid created by the pressure difference between two ends of a channel. To make this case as general as possible, we consider a channel consisting of several sections of variable geometry. Typical transient situations arise during normal reactor operation, such as the startup of a hydraulic loop or incidents such as inadvertent closing of a valve. The inviscid-flow assumption in this case must be interpreted as meaning that the frictional pressure drop is negligible compared to pressure drops created by temporal and spatial acceleration of the fluid. Figure 9.4 illustrates the system under consideration.

We seek to determine the variation of the flow rate, $\dot{m}(t)$, through the system of N sections for an arbitrary pressure drop between the inlet and the exit, $p_{in} - p_{out}$.

Assuming that the flow can be considered incompressible and one-dimensional (irrotational assumption implied) everywhere in the flow duct, Equation 9.10b yields

$$\int_1^N \frac{\partial \vec{\upsilon}}{\partial t} \cdot d\vec{r} + \frac{p_{out} - p_{in}}{\rho} + g(z_N - z_1) + \frac{1}{2}(\upsilon_N^2 - \upsilon_1^2) = 0 \tag{9.11}$$

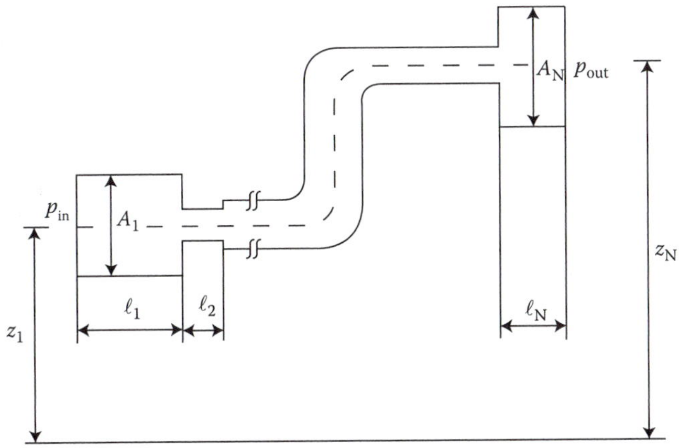

FIGURE 9.4 Hydraulic system configuration.

The integral can be performed by recalling that for incompressible flow the flow rate ($\dot{m}$) is not space-dependent

$$\int_1^N \frac{\partial \vec{\upsilon}}{\partial t}\cdot d\vec{r} = \frac{\partial}{\partial t}\int_1^N \vec{\upsilon}\cdot d\vec{r} = \frac{\partial}{\partial t}\int_1^N \frac{\dot{m}}{\rho A}\, d\ell$$
$$= \frac{1}{\rho}\frac{d\dot{m}}{dt}\int_1^N \frac{d\ell}{A} = \frac{1}{\rho}\frac{d\dot{m}}{dt}\sum_{n=1}^{N}\frac{\ell_n}{A_n} \tag{9.12}$$

The term $\sum_{n=1}^{N}(\ell_n/A_n)$ represents an equivalent inertia length $(\ell/A)_{\text{T}}$ for the system and can be computed knowing the system dimensions. Thus, after expressing the velocity term in Equation 9.11 in terms of $\dot{m}$, A, and ρ, and multiplying Equation 9.11 by ρ, the Bernoulli integral becomes

$$\left(\frac{\ell}{A}\right)_{\text{T}}\frac{d\dot{m}}{dt} + (p_{\text{out}} - p_{\text{in}}) + \rho g(z_{\text{N}} - z_1) + \frac{\dot{m}^2}{2\rho}\left(\frac{1}{A_{\text{N}}^2} - \frac{1}{A_1^2}\right) = 0 \tag{9.13}$$

Note that for incompressible flow only the inertia term (first term alone) involves a channel integral quantity, whereas all other terms represent conditions at the endpoints. If $p_{\text{in}} - p_{\text{out}}$ is prescribed, this equation can be solved to determine the variation of the flow rate in time.

Example 9.1: Pump Start-Up in an Inviscid Flow Loop

PROBLEM

Consider a reactor flow loop, shown in Figure 9.5. We are interested in determining the time it takes the coolant, initially at rest, to reach a steady-state flow level once the pump is turned on. From a hydraulic standpoint, the loop can be modeled as a one-dimensional flow path where a pressure head $\Delta p = p_{\text{in}} - p_{\text{out}}$ is provided by the pump. Assume the pump inlet and outlet are at the same elevation. Equation 9.13 can be used, but observe that it does not account for the inertia of the pump, only for the coolant in the flow.

Derive an expression for $\dot{m}(t)$. Note in particular what happens when $A_5 = A_1$. Does it seem reasonable? What has been neglected in deriving Equation 9.13 that, if included, would give a more reasonable answer for the case $A_1 = A_5$?

Given that the pump provides a pressure head equivalent to 85.3 m of water, and that the density of water is approximately $\rho_w \cong 1000$ kg/m³, evaluate the various parameters appearing in the expression for $\dot{m}(t)$. What is the expected steady-state value of the flow rate? How long does it take the system to reach 90% of this value?

(Note that in Example 9.2 we estimate the flow rate through a venturi meter located at the pump outlet piping of a PWR system with hydraulic characteristics similar to those of this reactor loop.)

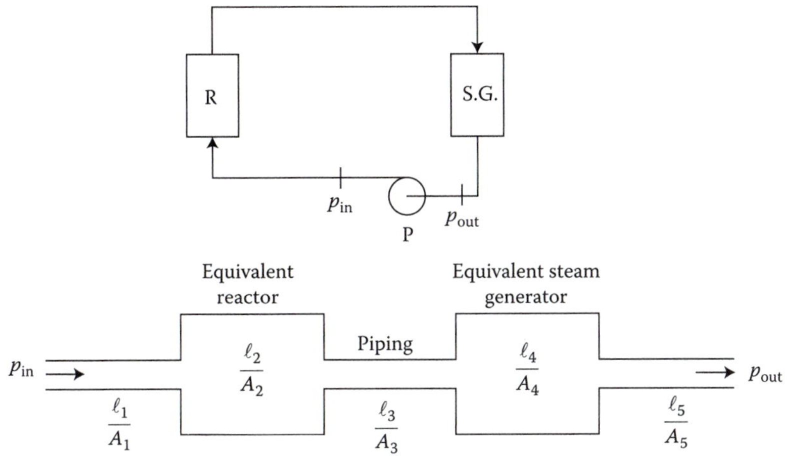

Table of characteristic dimensions

Component no.	Component name	ℓ(m)	A(m^2)
1	Pump outlet piping	8.0	0.4
2	Reactor	14.5	20.9
3	Reactor outlet piping	17.0	0.4
4	Steam generator	16.5	1.5
5	Pump inlet piping	10.0	0.35

FIGURE 9.5 Simplified reactor loop with a table of characteristic dimensions.

SOLUTION

First, because the pump inlet and outlet are at the same elevation, there is no net change in the gravity head. Thus, the gravity term can be dropped from Equation 9.13. Then, because $\Delta p = p_{in} - p_{out}$, we have

$$\left(\frac{\ell}{A}\right)_T \frac{d\dot{m}}{dt} - \Delta p + \frac{\dot{m}^2}{2\rho}\left(\frac{1}{A_5^2} - \frac{1}{A_1^2}\right) = 0$$

$$\left(\frac{\ell}{A}\right)_T \frac{d\dot{m}}{dt} = \Delta p - \dot{m}^2\left[\frac{1}{2\rho}\left(\frac{1}{A_5^2} - \frac{1}{A_1^2}\right)\right]$$

Then

$$\frac{(\ell/A)_T(1/\Delta p)d\dot{m}}{1 - \dot{m}^2 C^2} = dt$$

where

$$C^2 = \frac{1}{2\rho\Delta p}\left(\frac{1}{A_5^2} - \frac{1}{A_1^2}\right)$$

Hence

$$\frac{d\dot{m}}{(1-\dot{m}C)(1+\dot{m}C)} = \Delta p\left(\frac{A}{\ell}\right)_T dt$$

Expanding the left-hand side in partial fractions:

$$\frac{1/2\,d\dot{m}}{1-\dot{m}C} + \frac{1/2\,d\dot{m}}{1+\dot{m}C} = \Delta p\left(\frac{A}{\ell}\right)_T dt$$

Integrating the last equation, we get

$$-\frac{1}{2C}\ell n(1-\dot{m}C) + \frac{1}{2C}\ell n(1+\dot{m}C) = \Delta p\left(\frac{A}{\ell}\right)_T t + C_o$$

Because at $t = 0, \dot{m} = 0$, we find that $C_o = 0$, and

$$\frac{1+\dot{m}C}{1-\dot{m}C} = \exp\left[2C\Delta p\left(\frac{A}{\ell}\right)_T t\right]$$

or

$$\dot{m} = \frac{1}{C}\left\{\frac{\exp\left[2C\,\Delta p\left(\frac{A}{\ell}\right)_T t\right]-1}{\exp\left[2C\,\Delta p\left(\frac{A}{\ell}\right)_T t\right]+1}\right\}$$

Thus, we see that as $t \to \infty$, $\dot{m} \to 1/C$. Hence, at steady state there is a balance between the pressure driving force and the acceleration pressure drop.

We can evaluate C using the information given above:

$$\Delta p = \rho g \Delta z$$

$$\Delta p = (1000\text{ kg/m}^3)(9.81\text{ m/s}^2)(85.3\text{ m})$$

$$= 8.37 \times 10^5\text{ kg/ms}^2 \text{ or Pa}$$

$$C^2 = \frac{1}{2(1000\text{ kg/m}^3)(8.37\times 10^5\text{ kg/ms}^2)} \times \left(\frac{1}{(0.35\text{m}^2)^2} - \frac{1}{(0.4\text{m}^2)^2}\right) = 1.143\times 10^{-9}\text{ s}^2/\text{kg}^2$$

$$\dot{m} \text{ at steady state} = \frac{1}{C} = 2.96\times 10^4\text{ kg/s}$$

To obtain the transient time constant, we evaluate the term appearing in the exponent:

$$\left(\frac{\ell}{A}\right)_T = \sum_{n=1}^{5}\left(\frac{\ell}{A}\right)_n$$
$$= \left(\frac{8}{0.4}\right)+\left(\frac{14.5}{20.9}\right)+\left(\frac{17.0}{0.4}\right)+\left(\frac{16.5}{1.5}\right)+\left(\frac{10.0}{0.35}\right)$$
$$= 102.8\ \mathrm{m^{-1}}$$

$$2C\,\Delta p\left(\frac{A}{\ell}\right)_T = 2(1.143\times10^{-9})^{1/2}\ \mathrm{s/kg}\,(8.37\times10^5)\,\mathrm{kg/ms^2}\left(\frac{1}{102.8}\ \mathrm{m}\right)$$
$$= 0.551\ \mathrm{s^{-1}}$$

To find the time, it takes for the system to reach 90% of its full flow value, let

$$0.9 = \frac{e^{0.551t}-1}{e^{0.551t}+1}$$
$$0.9(e^{0.551t}+1) = e^{0.551t}-1$$
$$1.9 = 0.1e^{0.551t}$$
$$t = \left(\frac{1}{0.551}\ell n 19\right) = 5.34\,\mathrm{s}$$

We note some interesting observations here. First, as will be shown in Example 9.2, the flow rate through the PWR loop (measured by a venturi meter located at the pump outlet piping) is approximately 4250 kg/s. However, the result of this calculation shows that the inviscid flow rate would be 29,000 kg/s and that the flow rate is sensitive to the values of A_5 and A_1.

When $A_5 = A_1$, we can no longer use the above expressions for $\dot{m}(t)$ because $C \to 0$. The equation governing the flow rate is still Equation 9.13, but with both the gravity and acceleration terms set to zero (Remember the friction term does not show up because the flow is inviscid.):

$$\left(\frac{\ell}{A}\right)_T \frac{d\dot{m}}{dt} - \Delta p = 0$$

or

$$\dot{m} = \Delta p\left(\frac{A}{\ell}\right)_T t$$

Hence there is no limit to the flow rate, as there is no resistance to the applied force in such a system. In reality, form and friction forces are encountered in a flow loop with a viscous fluid. This subject is discussed in Example 9.8. In addition, the pressure head of the pump decreases with the flow rate. Thus, if $\Delta p = A - B\dot{m}$, the maximum achievable flow rate is limited to $\dot{m}_{max} = A/B$ even in the absence of friction forces. Pump head representation is described in Chapter 3, Volume II.

9.2.2.2 Steady-State Flow

At steady state, Equation 9.10 reduces to

$$\int_1^2 \frac{dp}{\rho} + \Delta\left(\psi + \frac{\upsilon^2}{2}\right) = 0 \tag{9.14}$$

which for an incompressible flow is given by

$$\Delta\left(\frac{p}{\rho} + \psi + \frac{\upsilon^2}{2}\right) = 0 \tag{9.15a}$$

In most cases, the body forces consist only of gravitational forces, $\psi = gz$; and a total head (z°) is usually defined as the actual height (z) plus the static head and the velocity head:

$$z^\circ = \frac{p}{\rho g} + \frac{\upsilon^2}{2g} + z \tag{9.16}$$

Thus, the total head of an incompressible and inviscid flow in a channel is constant along a stream line or at any location if the flow is also irrotational. Hence, between channel locations we can write

$$\Delta\left(\frac{p}{\rho g} + \frac{\upsilon^2}{2g} + z\right) = 0 \tag{9.15b}$$

Equations 9.15a and 9.15b are called the steady-flow Bernoulli equation.

Summarizing, the underlying assumptions of the Bernoulli equation 9.15 are that the flow is

1. Along a single stream line
2. Steady
3. Incompressible
4. Inviscid
5. Involves no shaft work transfer—there is a flow work transfer across areas A_1 and A_2
6. Involves no heat transfer—applies if the fluid is a thermodynamically coupled system such as an ideal gas. When the fluid is a liquid in which the thermal and mechanical energy storage modes are uncoupled, this constraint can be relaxed.

9.2.2.2.1 Incompressible Flow in a Constant-Area Duct

Steady-state flow of a constant density fluid in a constant area has a constant velocity. For the fully developed flow condition, there are no form or acceleration pressure drops. Additionally, the frictionless flow case has no friction pressure drop, whereas the viscous flow case has the familiar friction pressure drop formulations developed later in Section 9.3.

9.2.2.2.2 Flow through a Nozzle

The flow of an incompressible, inviscid fluid through a nozzle can be predicted by application of Equation 9.15. Consider a nozzle of diameter D_2 attached to a pipe of diameter D_1 (Figure 9.6). From Equation 9.15b, when gravity is the only body force, we get

$$\left(\frac{p}{\rho g}+\frac{\upsilon^2}{2g}+z\right)_1=\left(\frac{p}{\rho g}+\frac{\upsilon^2}{2g}+z\right)_2 \tag{9.17}$$

For a horizontal nozzle, however

$$z_1=z_2 \tag{9.18}$$

From the mass balance consideration, we get

$$\rho\upsilon_1\frac{\pi D_1^2}{4}=\rho\upsilon_2\frac{\pi D_2^2}{4}$$

which leads to

$$\upsilon_1=\upsilon_2\left(\frac{D_2}{D_1}\right)^2 \tag{9.19}$$

Thus, substituting from Equations 9.18 and 9.19 into Equation 9.17, we get

$$\frac{\upsilon_2^2}{2}\left[\left(\frac{D_2}{D_1}\right)^4-1\right]=\frac{p_2-p_1}{\rho}$$

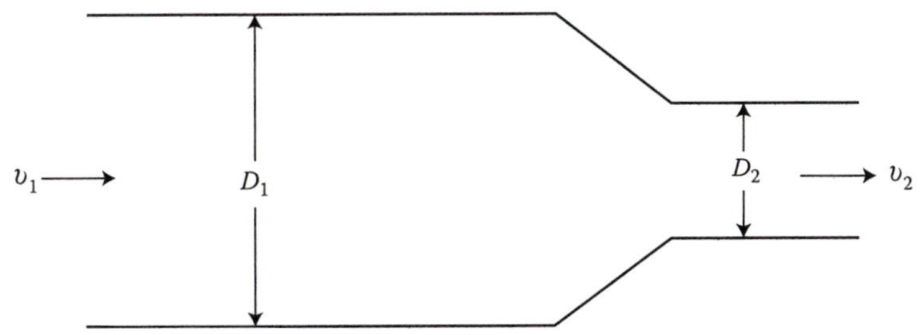

FIGURE 9.6 Nozzle.

or

$$\upsilon_2 = \sqrt{\frac{2(p_1 - p_2)}{\rho[1-(D_2/D_1)^4]}} \tag{9.20a}$$

It should be noted that under real conditions the velocity in the nozzle is less than that predicted by Equation 9.20a. The departure from ideal conditions due to fluid viscous forces is accounted for by introducing a nozzle coefficient (C_D) such that the velocity v_2 is given by

$$\upsilon_2 = C_D\sqrt{\frac{2(p_1 - p_2)}{\rho[1-(D_2/D_1)^4]}} \tag{9.20b}$$

The values C_D are typically 0.7 to 0.95 depending on the nozzle geometry [18]. The flow rate is given by

$$\dot{m} = \rho\upsilon_2\left(\frac{\pi D_2^2}{4}\right) = C_D\left(\frac{\pi D_2^2}{4}\right)\sqrt{\frac{2\rho(p_1 - p_2)}{[1-(D_2/D_1)^4]}} \tag{9.21}$$

Example 9.2: Venturi Meter for Flow Measurement

PROBLEM

A venturi meter is inserted into one flow loop of a PWR as shown in Figure 9.7. The dimensions are $D_1 = 0.711$ m (28 in.) and $D_2 = 0.686$ m (27 in.). The venturi meter is mostly filled with stagnant water that is separated from the primary flow by an air bubble. This setup enables visual determination of the difference in water elevation levels (h) and hence the pressure difference between the contraction and the loop. Given that the height (h) is 0.924 m (3 ft.) and the water density is approximately $\rho_w \cong 1000$ kg/m^3, what are the velocity and the mass flow rate in the loop?

SOLUTION

From Figure 9.7, the loop velocity (υ_1) is related to throat velocity (υ_2) from a mass balance such that

$$\upsilon_1 = \upsilon_2\left(\frac{D_2}{D_1}\right)^2$$

Using this equation and Equation 9.20a, υ_1 can be found from

$$\upsilon_1 = \sqrt{\frac{2(p_1 - p_2)}{\rho_w[(D_1/D_2)^4 - 1]}}$$

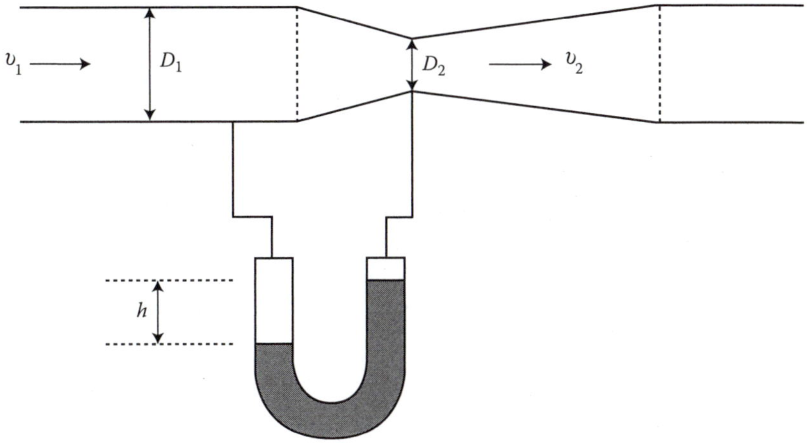

FIGURE 9.7 Venturi meter.

Because $p_1 - p_2 = \rho_w gh$,

$$\upsilon_1 = \sqrt{\frac{2gh}{[(D_1/D_2)^4 - 1]}} = \sqrt{\frac{2(9.81\ \text{m/s}^2)(0.914\ \text{m})}{\left[\left(\frac{0.711}{0.686}\right)^4 - 1\right]}} = 10.71\ \text{m/s}$$

The mass flow rate through the loop is

$$\dot{m} = \rho_w \upsilon_1 A_1 = (1000\ \text{kg/m}^3)(10.71\ \text{m/s})\left[\frac{\pi}{4}(0.711\ \text{m})^2\right] = 4252\ \text{kg/s}$$

9.2.2.2.3 Flow through a Sudden Area Change

Again apply Equation 9.15 and assume a horizontal flow geometry yielding:

$$p_1 - p_2 = \rho\left(\frac{\upsilon_2^2}{2} - \frac{\upsilon_1^2}{2}\right) \tag{9.22a}$$

which for our case of steady flow is also equal to

$$p_1 - p_2 = \frac{\dot{m}^2}{2\rho}\left(\frac{1}{A_2^2} - \frac{1}{A_1^2}\right) \tag{9.22b}$$

Hence if the flow area expands ($A_2 > A_1$), the static pressure increases ($p_2 > p_1$), whereas the inverse holds for an area contraction.

This is the same result as in the variable geometry channel analysis of Equation 9.13, where the flow cross-sectional area A_n corresponds to A_2. The result above for incompressible, inviscid flow through a sudden area change, Equation 9.22, will be contrasted in later sections with results for a variety of combinations of flow area geometries and flow conditions as summarized in Table 9.1.

9.2.3 Compressible Inviscid Flow

In real fluids, the effects of compressibility and viscosity may have to be considered. Let us qualitatively discuss the effects of compressibility here. If the pressure at one end of the system changes abruptly, this effect is not experienced immediately at the other end, because pressure waves are transmitted with finite velocities in compressible fluids. Thus, the flow rate at the other end of the system cannot change until the pressure wave arrives there.

Because the sonic wave propagation velocity is high in liquids (more than 1000 m/s), it is generally acceptable to assume that pressure changes propagate instantaneously and that the flow rate responds instantaneously to pressure transients. This assumption is of course equivalent to assuming incompressible flow. The assumption of incompressible flow, however, breaks down as the time scale of interest becomes comparable to the sonic transit time in the system. (Sonic effects on flow transient calculations are discussed in Chapter 2, Volume II in some detail.)

At the other end of the scale, as the transients become slower, the transient (or inertia) term in Equation 9.13, $(\ell/A)_T\, d\dot{m}/dt$, diminishes in magnitude, and for slow enough transients it can be completely neglected. The problem can then be solved in a quasi-steady-state fashion, that is, by calculating the instantaneous flow rate using the steady-state relation between momentum and the driving pressure difference.

9.2.3.1 Flow in a Constant-Area Duct

Compressible flow in a constant-area duct is of interest in a nuclear reactor with a gaseous coolant that is being heated as it flows through the core. For the steady-state frictionless case, the governing equation can be derived from both momentum and energy balance conservation principles including the more general assumption of variable flow area. Proceeding first from energy considerations, write the energy balance for heat addition in this steady-state case from Equation 4.39 as

$$0 = d\dot{Q} + \dot{m}\left(h + \frac{\upsilon^2}{2}\right)\Bigg|_z - \dot{m}\left(h + \frac{\upsilon^2}{2}\right)\Bigg|_{z+dz} \tag{9.23a}$$

The second law for inviscid flow can be expressed for steady state from reduction of Equation 4.41 as

$$0 = \frac{d\dot{Q}}{T_s} - \dot{m}ds \tag{4.41b}$$

Hence, Equation 9.23a reduces to

$$0 = \dot{m}T_s ds - \dot{m}dh - \dot{m}d\left(\frac{\upsilon^2}{2}\right) \tag{9.23b}$$

Now from Equation 6.10a:

$$\mathrm{d}h = T\mathrm{d}s + \frac{\mathrm{d}p}{\rho} \tag{6.10a}$$

Substituting Equation 6.10a into Equation 9.23b and assuming an internally reversible process (T equal to T_s) yields the desired result:

$$\mathrm{d}p = -\rho \upsilon \mathrm{d}\upsilon \tag{9.24a}$$

which is exactly the steady-state Bernoulli Equation 9.9 for negligible gravity effects.

The same result follows from the momentum balance on a variable area flow channel, $A(z) = \pi\, R^2\,(z)$, between planes z and $z + \mathrm{d}z$ as

$$0 = \rho\upsilon^2 A\big|_z - \rho\upsilon^2 A\big|_{z+\mathrm{d}z} + pA\big|_z - pA\big|_{z+\mathrm{d}z} + p\mathrm{d}A \tag{9.25a}$$

where the term

$$p\mathrm{d}A = p2\pi R\mathrm{d}Rs \tag{9.26}$$

is the axial component of the pressure force on the channel wall.

Equation 9.25a can be written as

$$0 = -\mathrm{d}(\rho\upsilon^2 A) - \mathrm{d}(pA) + p\mathrm{d}A \tag{9.25b}$$

Now

$$-d(pA) = -p\mathrm{d}A - A\mathrm{d}p \tag{9.25c}$$

and

$$\mathrm{d}(\rho\upsilon^2 A) = \rho\upsilon A\mathrm{d}\upsilon \tag{9.25d}$$

since the second term from the differentiation $\upsilon\mathrm{d}(\rho\upsilon A) = 0$ by continuity.

Hence when the results of Equations 9.25c and 9.25d are substituted into Equation 9.25b, which has been obtained from momentum principles, Equation 9.25b reduces to the form of Equation 9.24 derived earlier from energy considerations:

$$\mathrm{d}p = -\rho \upsilon \mathrm{d}\upsilon \tag{9.24a}$$

Now let us evaluate the pressure change between stations 1 and 2 for the case of interest here, inviscid compressible flow in a constant-area duct. Since the fluid has a temperature, pressure and density, all a function of length ℓ in the flow direction, for clarity we rewrite Equation 9.24 as

$$dp(\ell) = -\rho(\ell)\upsilon(\ell)d\upsilon(\ell) \tag{9.24b}$$

For a constant-area duct, however, $\rho(\ell)\upsilon(\ell) = \text{constant}$, so when we integrate between station 1 and 2 we obtain

$$\begin{aligned} p_2 - p_1 &= -\frac{\dot{m}}{A}(\upsilon_2 - \upsilon_1) \\ p_1 - p_2 &= \rho_2\upsilon_2^2 - \rho_1\upsilon_1^2 \end{aligned} \tag{9.27}$$

Note that for the case of incompressible, frictionless flow with an abrupt area change treated in Section 9.2.2.2.3, Equation 9.24b can be integrated as follows:

$$p_1 - p_2 = \rho\int_1^2 \upsilon\, d\upsilon \tag{9.28}$$

to obtain the result of Section 9.2.2.2.3:

$$p_1 - p_2 = \rho\left(\frac{\upsilon_2^2}{2} - \frac{\upsilon_1^2}{2}\right) \tag{9.22a}$$

For flow with friction, the usual approximation is to simply introduce an integrated friction pressure drop. This treatment implicitly assumes that the friction pressure drop is small enough not to perturb the frictionless solutions for $\rho(\ell)$ and $v(\ell)$.

9.2.3.2 Flow through a Sudden Expansion or Contraction

The problem of compressible flow through a sudden expansion or contraction is complicated. Two simplifying assumptions can be made: (1) the flow has a small Mach number (so the pressure field does not greatly affect the density), and (2) the expansion or contraction region is short enough that there is little heating or cooling of the gas for the case of adiabatic change in pressure. (Hence, the temperature field does not greatly affect the density.) In this case, the results of Section 9.2.2.2.3 are applicable without modification.

9.3 VISCOUS FLOW

9.3.1 Viscosity Fundamentals

The shear stress $\bar{\bar{\tau}}$ in the momentum balance Equation 4.80 arises from the resistance of a fluid to move with a uniform velocity when only a portion of it is subjected to an externally imposed velocity. Consider as an example a fluid layer initially at rest between two plates. If the upper plate begins to move to the right at a constant

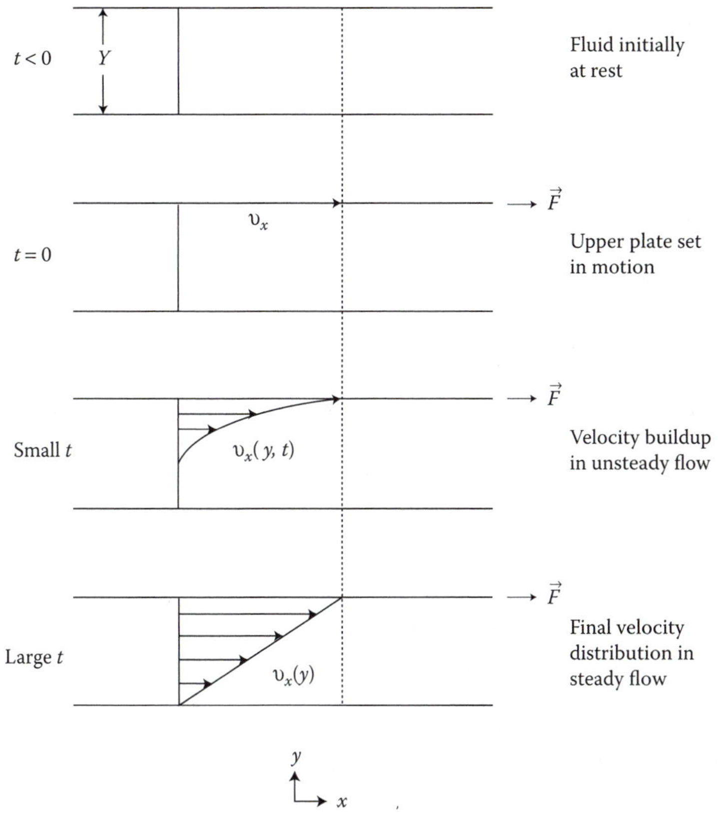

FIGURE 9.8 Laminar velocity profile in a fluid between two plates when the upper plate moves at a constant velocity (υ_x).

velocity (υ_x), the rest of the fluid, with time, acquires a finite velocity. Thus, the x momentum is said to be transportable in the y direction, as indicated in Figure 9.8.

When eventually a steady-state velocity profile is established, a force (F_x) is required to maintain the motion of the upper plate. The value of F_x is generally a function of the velocity (υ_x), the distance Y between the two plates, and the fluid properties, so that

$$F_x = F_x(\upsilon_x, Y, \text{fluid properties}) \tag{9.29}$$

For most ordinary fluids, the velocity profile is linear in y, and the force can be expressed as

$$F_x = \tau_{yx} A = \mu\left(\frac{\upsilon_x}{Y}\right) A \tag{9.30}$$

where A = plate area, τ = shear force per unit area (i.e., shear stress), and μ = dynamic viscosity of the fluid. Such fluid behavior is called Newtonian, and the stress tensor components can be written following Equation 4.84 for an incompressible flow and Equation 4.85 as*

$$\tau_{ij} = \mu\left(\frac{\partial \upsilon_i}{\partial x_j} + \frac{\partial \upsilon_j}{\partial x_i}\right) \tag{4.85}$$

where j here can differ from i or be identically i.

Fluids that do not follow the Newtonian mechanics include slurries and aerosol-carrying gases.

9.3.2 Viscosity Changes with Temperature and Pressure

Molecular movement can be used to predict the dynamic viscosity of gases and liquids as well as the dependence of viscosity on the pressure and temperature [3]. The data on the dynamic viscosity of ordinary fluids, that is water and air, can be generalized as a function of the critical (thermodynamic) pressure (p_c) and temperature (T_c). In Figure 9.9, the reduced viscosity (μ/μ_c; where μ_c = viscosity at the critical point) is plotted against the reduced temperature (T/T_c) and the reduced pressure (p/p_c). It is seen from Figure 9.9 that at a given temperature the viscosity of a gas approaches a limit as the pressure approaches zero. For ideal gases, the dynamic viscosity is independent of pressure. (Note that gases can be considered ideal at pressures much lower than their critical pressures.) Also seen are an increase in gas viscosity and a decrease in liquid viscosity with increasing temperature. This picture is also clear in the plot of viscosity of various materials given in Figure 9.10. In Figure 9.11, the trend of dependence of the kinematic viscosity (ν) on temperature is found to be similar to that of the dynamic viscosity (μ). The kinematic viscosity (ν) is defined as

$$\nu \equiv \frac{\mu}{\rho} \tag{9.31a}$$

In the absence of empirical data, μ_c for ordinary fluids (gases and liquids) can be estimated from

$$\mu_c = 7.7\,\mathrm{M}^{1/2} p_c^{2/3} T_c^{-1/6} \tag{9.31b}$$

where μ_c is in micropoise (1 micropoise = 10^{-7} Pa s), p_c is in atmospheres, and T_c is in K; M = molecular weight of the material in grams/mole.

For a mixture of gases, pseudocritical properties can be used

$$p_c = \sum_n x_n p_{cn}; \quad T_c = \sum_n x_n T_{cn}; \quad \mu_c = \sum_n x_n \mu_{cn} \tag{9.32}$$

where x_n = mole fraction of the component n. These pseudocritical properties can be used along with Figure 9.9 to determine the mixture viscosity, following the same procedure as that for pure fluids.

---

* See Chapter 4 for the convention on direction and sign of the shear stresses.

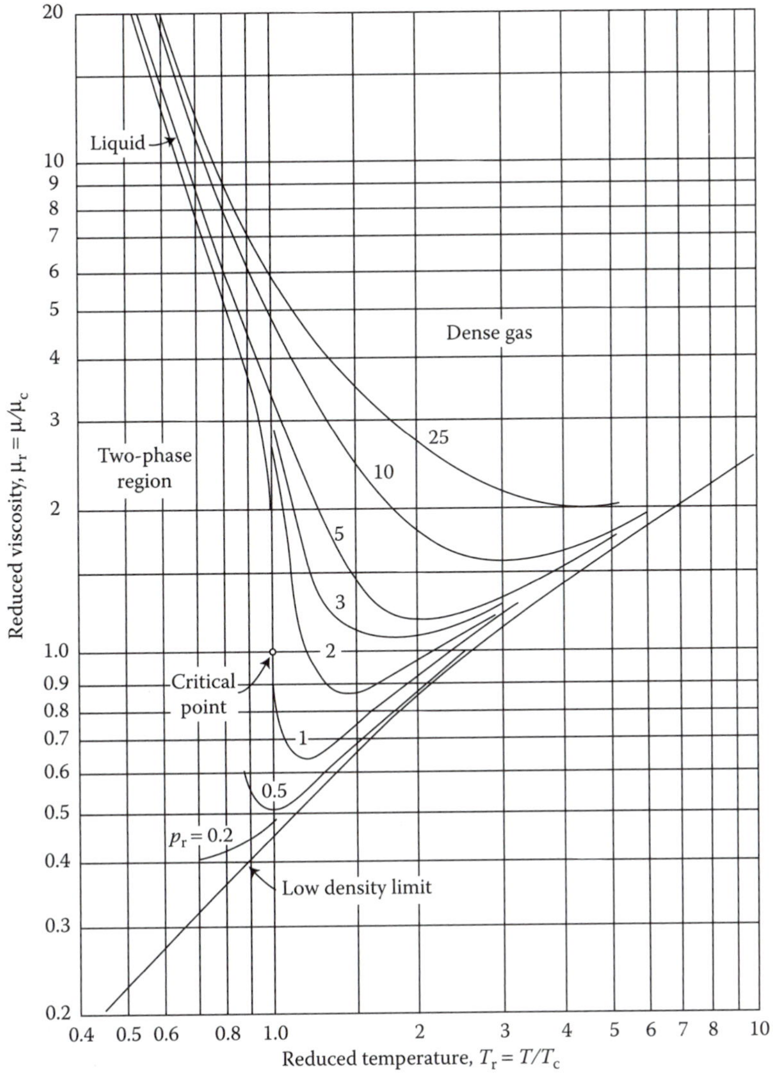

FIGURE 9.9 Reduced viscosity as a function of temperature for various values of reduced pressure. (From Bird, R. B., Stewart, W. E., and Lightfoot, E. N., *Transport Phenomena*. New York, NY: Wiley, 1960.)

9.3.3 Boundary Layer

For the most practical flow conditions, the effects of viscosity on the flow over a surface can be assumed as confined to a "thin" region close to the surface. This region is called the *boundary layer.* The velocity of the fluid at the surface is taken to be zero, which is referred to as the no-slip boundary condition. For external flows, the flow away from the layer can be treated as inviscid (Figure 9.12). Thus, by

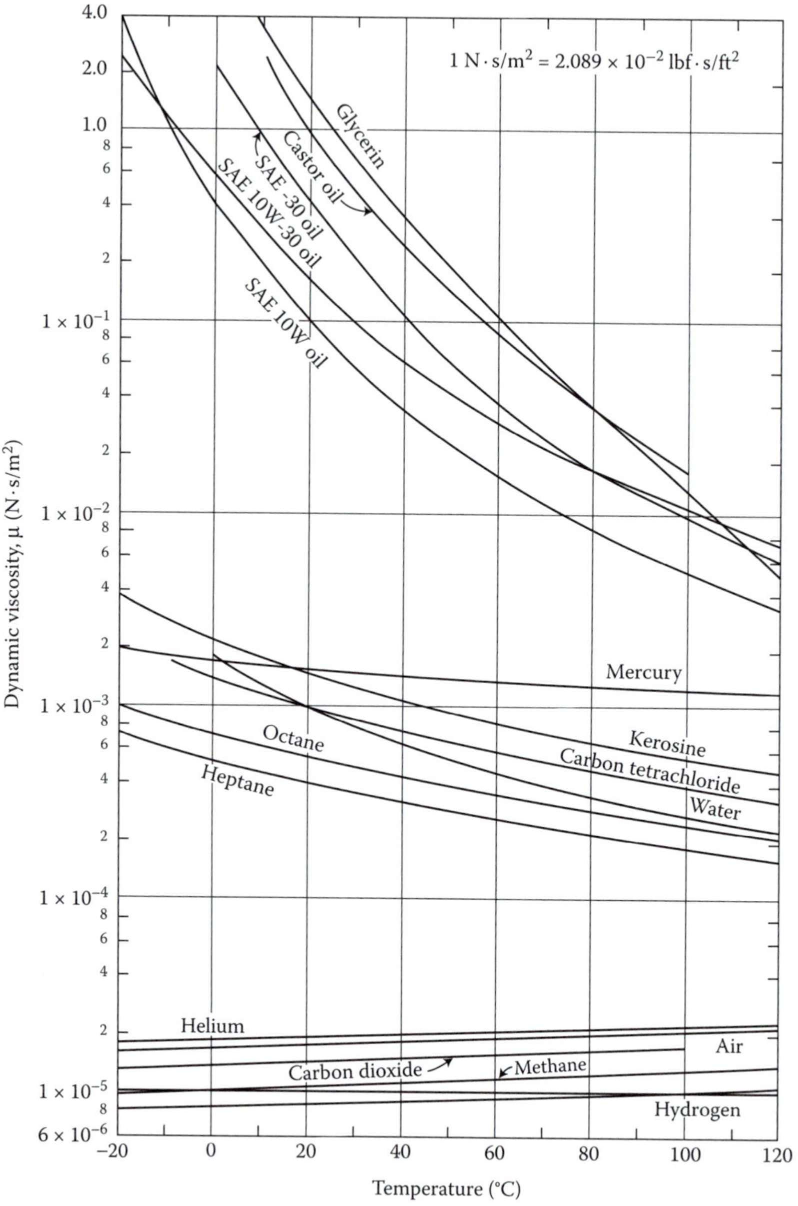

FIGURE 9.10 Dynamic viscosity of fluids. (From Fox, R. W. and McDonald, A. T., *Introduction to Fluid Mechanics*, 2nd Ed. New York, NY: Wiley, 1978.)

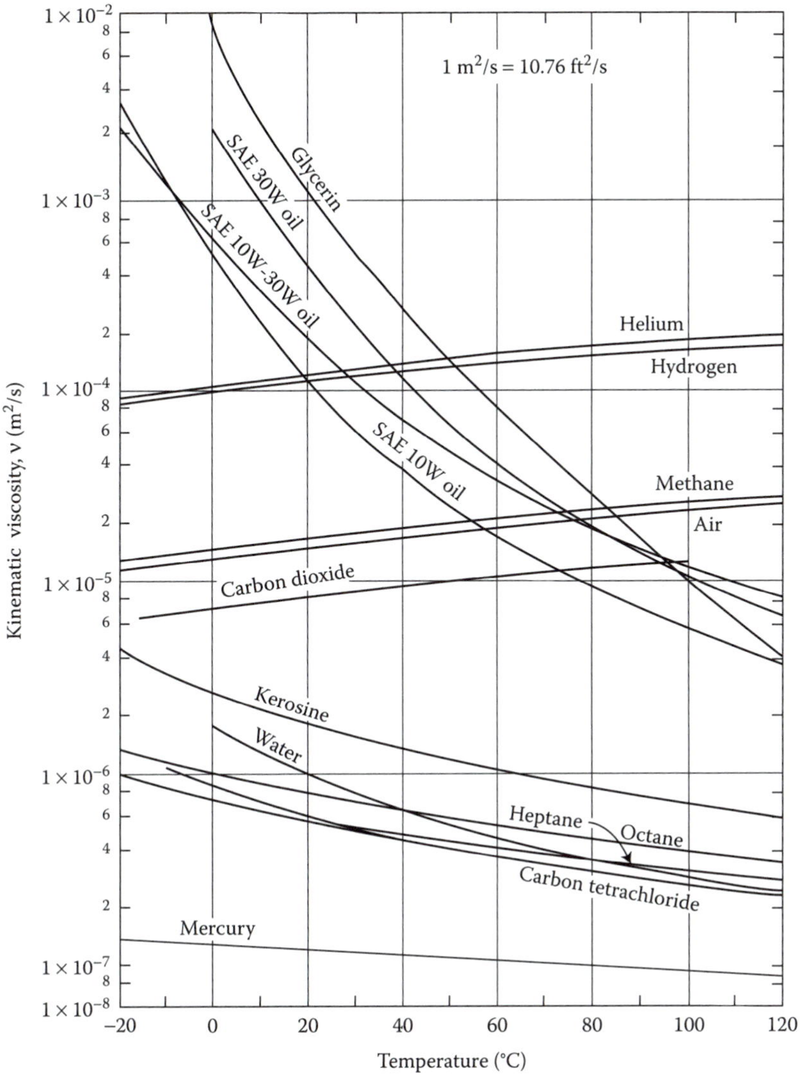

FIGURE 9.11 Kinematic viscosity of fluids. (From Fox, R. W. and McDonald, A. T., *Introduction to Fluid Mechanics*, 2nd Ed. New York, NY: Wiley, 1978)

definition the boundary layer thickness is taken as the region in which the velocity changes from a free stream (inviscid) velocity to zero at the surface. In reality, the velocity at the fluid side of the boundary layer is taken to be about 99% of the free stream velocity to account for the presence of a weak effect of the viscosity even in the bulk of the flow.

In the case of internal flows, for example, flow inside a tube, the boundary layer is assumed to start developing at the entrance of the channel and to grow from the surface until it reaches the tube centerline (Figure 9.13).

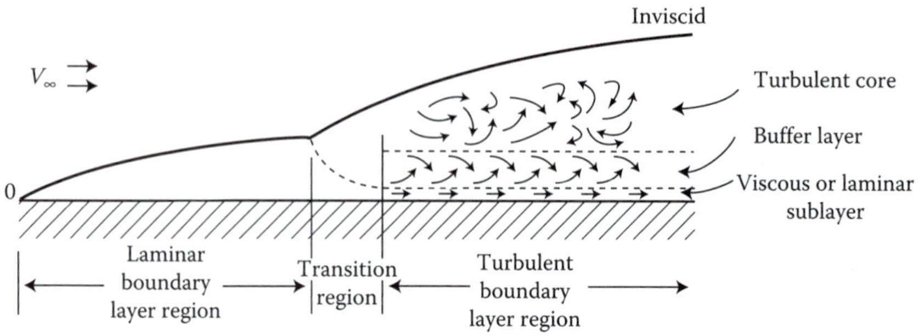

FIGURE 9.12 Boundary layer velocity distribution for flow on an external surface.

The length from the channel entrance required for the boundary layer to grow to occupy the entire flow area is called the *hydrodynamic entrance* or *developing length*. The flow is deemed fully developed once it is past the developing length.

When heat is transferred between the flowing fluid and the surface, the temperature gradient can be similarly treated as occurring mostly across a thermal boundary layer. The hydrodynamic and thermal layers are not necessarily of the same dimensions.

The boundary layer concept is of great value because it allows for simplification of the flow equations, as is shown later in this section. The greatest simplification arises from the ability to assume that the flow in the boundary layer is predominantly parallel to the surface. Thus, the pressure gradient in the perpendicular directions to the flow can be taken as zero. For a flow in a tube, within the boundary layer the pressure gradient conditions are

$$\frac{\partial p}{\partial r} \simeq 0 \quad \text{and} \quad \frac{\partial p}{r\partial \theta} = 0 \tag{9.33}$$

The velocity conditions are

$$\upsilon_z \gg \upsilon_r$$

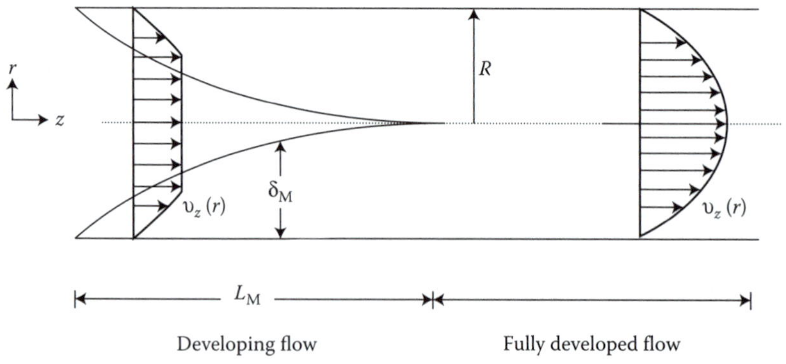

FIGURE 9.13 Boundary layer development in a tube.

and

$$\frac{\partial v_z}{\partial r} \gg \frac{\partial v_z}{\partial z} \tag{9.34}$$

The second condition arises because the velocity changes from zero at the wall to the free stream value across the boundary.

9.3.4 Turbulence

At low velocities, the flow within a boundary layer proceeds along stream lines and hence is of laminar type (Figure 9.12). However, with sufficient length, disturbances within the flow appear, leading to time-variable velocity (in direction as well as magnitude) at any position; thus, the flow becomes turbulent. The higher the flow velocity, the shorter is the purely laminar flow length. With turbulent flow, eddies are formed that have a random velocity, which destroys the laminar flow lines. However, even for turbulent flow, the flow near the wall (where the velocity is small) appears to have a laminar region or sublayer, as illustrated in Figure 9.12.

The eddies substantially increase the rate at which momentum and energy can be transferred across the main (or mean) flow direction. The enhancement, by the eddies, of the transport processes above the level possible by molecular effects leads to the practical preference for the use of turbulent flow in most heat transfer equipment. Treatment of the turbulent flow equations is more complicated than the laminar flow equations, and several models for turbulence exist. Experimentally based correlations for the macroscopic turbulent flow characteristics have been developed for most applications

9.3.5 Dimensionless Analysis

It is useful to relate the hydraulic characteristics of flow to the ratio of the various forces encountered in the flow. A list of dimensionless groups that have been applied to fluid mechanics analysis appears in Appendix G.

The groupings originate from the various ways by which the flow equations can be nondimensionalized. As an example, the fluid equations for Newtonian fluids of constant density and viscosity can be nondimensionalized as follows:

Let

$$\vec{v}^* \equiv \frac{\vec{v}}{V}$$

where V = some characteristic velocity

$$x^*, y^*, z^* \equiv \frac{x}{D_e}, \frac{y}{D_e}, \frac{z}{D_e}$$

where D_e = a reference length

$$t^* \equiv \frac{tV}{D_e} \quad \text{and} \quad p^* \equiv \frac{p}{\rho V^2}$$

The Navier–Stokes equation for the body force as gravity only

$$\rho \frac{D\vec{\upsilon}}{Dt} = -\nabla p + \mu \nabla^2 \vec{\upsilon} + \rho \vec{g} \tag{4.90b}$$

can then be written as:

$$\rho \frac{V}{D_e} \frac{D}{Dt^*}(V\vec{\upsilon}^*) = -\frac{\nabla^*}{D_e}(\rho V^2 p^*) + \mu \frac{\nabla^{*2}}{D_e^2}(V\vec{\upsilon}^*) + \rho \vec{g}$$

which can be rearranged as:

$$\frac{D\vec{\upsilon}^*}{Dt^*} = -\nabla^* p^* + \left(\frac{\mu}{\rho V D_e}\right) \nabla^{*2} \vec{\upsilon}^* + \left(\frac{D_e g}{V^2}\right) \frac{\vec{g}}{g} \tag{9.35}$$

where g is the gravitational acceleration.

Equation 9.35 can be written, using the definitions of the *Reynolds number* (Re) and *Froude number* (Fr), as:

$$\frac{D\vec{\upsilon}^*}{Dt^*} = -\nabla^* p^* + \frac{1}{\text{Re}} \nabla^{*2} \vec{\upsilon}^* + \frac{1}{\text{Fr}} \frac{\vec{g}}{g} \tag{9.36}$$

Thus when two flow systems with the same Reynolds and Froude numbers and initial and boundary conditions are found, they can be described by the same dimensionless velocity and pressure. Such systems are said to be dynamically similar. This similarity is used for identifying the flow patterns in large systems by constructing less expensive, smaller prototypes.

9.3.6 Pressure Drop in Channels

This section presents the approach to calculation of pressure loss in single-phase flow in channels. Incompressible and compressible flows are considered.

The effect of viscosity introduces shear within the fluid (internal friction) as well as friction with the confining wall. A difficulty arises because we do not know how the shear stress varies under transient conditions, for example, due to a change in the velocity profile. For engineering applications this difficulty is often ignored, and the "quasi-steady-state" values are assumed to hold for the transient conditions. This approximation is strictly valid for relatively slow transients for which it can be argued that dynamic effects in the governing equations are small.

The shear effects lead to a loss of the driving pressure, so that a pressure loss term must be added to the Bernoulli equation. Equation 9.13 takes then the form:

$$\left(\frac{\ell}{A}\right)_{\mathrm{T}} \frac{\mathrm{d}\dot{m}}{\mathrm{d}t} + p_{\mathrm{out}} - p_{\mathrm{in}} + \rho g(z_{\mathrm{N}} - z_1) + \frac{\dot{m}^2}{2\rho}\left(\frac{1}{A_{\mathrm{N}}^2} - \frac{1}{A_1^2}\right) + \Delta p_{\mathrm{loss}} = 0 \tag{9.37}$$

where Δp_{loss} is to be evaluated either from a viscous flow analysis or empirically. It is usually decomposed into the pressure losses due to wall friction ($\Delta p_{\mathrm{friction}}$) and flow form losses ($\Delta p_{\mathrm{form}}$) as described in the following Equation 9.39.

Note that Equation 9.37 can be rearranged into the form:

$$p_{\mathrm{in}} - p_{\mathrm{out}} = \Delta p_{\mathrm{inertia}} + \Delta p_{\mathrm{acceleration}} + \Delta p_{\mathrm{gravity}} + \Delta p_{\mathrm{form}} + \Delta p_{\mathrm{friction}} \tag{9.38}$$

where

$$\Delta p_{\mathrm{inertia}} = \left(\frac{\ell}{A}\right)_{\mathrm{T}} \frac{\mathrm{d}\dot{m}}{\mathrm{d}t} \tag{9.39a}$$

$$\Delta p_{\mathrm{acceleration}} = \frac{\dot{m}^2}{2\rho}\left(\frac{1}{A_{\mathrm{N}}^2} - \frac{1}{A_1^2}\right) \tag{9.39b}$$

$$\Delta p_{\mathrm{gravity}} = \rho g(z_{\mathrm{N}} - z_1) \tag{9.39c}$$

The pressure loss due to an abrupt change in flow direction and/or geometry is usually called a *form loss*. The pressure head loss due to form losses is, in practice, related to the kinetic pressure, that is the pressure arising if the flow is brought to zero velocity. So the pressure loss is given by

$$\Delta p_{\mathrm{form}} \equiv K\left(\frac{\rho \upsilon_{\mathrm{ref}}^2}{2}\right) \tag{9.39d}$$

The reference velocity in Equation 9.39d is usually the higher of the two velocities on both sides of the abrupt flow area change. The evaluation of loss coefficients, K, for a variety of flow restriction geometries is discussed in Section 9.4.4 for laminar and Section 9.5.7 for turbulent flow. Loss coefficient values for abrupt area changes in multichannel systems are given in Section 9.6.4.

The frictional pressure losses within a pipe of constant flow area can be written in the form of Equation 9.39d. The frictional pressure drop coefficient is, however, conveniently taken to be proportional to the length of the flow channel and inversely proportional to the channel equivalent diameter, so the pressure drop due to wall friction is, in practice, given by

$$\Delta p_{\text{friction}} = \bar{f}\frac{L}{D_e}\left(\frac{\rho \upsilon_{\text{ref}}^2}{2}\right) \tag{9.39e}$$

Since Equation 9.39e is written in terms of a finite channel length, the friction factor is an average value for that channel length. The average friction factor $\bar{f}$ depends on the actual channel geometry for laminar flow while for turbulent flow the channel geometry can be characterized by the equivalent diameter. The reference velocity appearing directly in Equation 9.39e, which is also used to establish the Reynolds number for determination of the friction factor, is the bulk velocity in the channel.

The modified Bernoulli's equation (Equation 9.37) can be used to obtain the effective pressure loss between points 1 and 2 at steady state:

$$\left(\frac{p}{\rho} + gz + \frac{\upsilon^2}{2}\right)_1 - \left(\frac{p}{\rho} + gz + \frac{\upsilon^2}{2}\right)_2 = K_T\frac{\upsilon_{\text{ref}}^2}{2} + \bar{f}\frac{L}{D_e}\frac{\upsilon_{\text{ref}}^2}{2} \tag{9.40}$$

where

$$K_T\frac{\upsilon_{\text{ref}}^2}{2} = \sum_i K_i\frac{(\upsilon_{\text{ref}}^2)_i}{2} \tag{9.41}$$

and L and D_e = length and equivalent diameter, respectively, of the pipe connecting the two points 1 and 2.

For incompressible fluids and steady-state flow, the velocity varies inversely to the flow area; thus

$$\frac{K_T}{A_{\text{ref}}^2} = \sum_i \frac{K_i}{A_i^2} \tag{9.42}$$

and

$$K_T = \sum_i K_i\left(\frac{A_{\text{ref}}}{A_i}\right)^2 \tag{9.43}$$

The relation in Equation 9.43 allows all the partial pressure losses to be referred to the velocity at some reference cross section (A_{ref}).

9.3.7 Summary of Pressure Changes in Inviscid/Viscid and in Compressible/Incompressible Flows

The preceding sections have presented the components of pressure change in situations with and without viscosity and compressibility (for subsonic flows). These results are summarized in Table 9.1. The presentation of loss coefficients for incompressible flow through abrupt area changes in multi channel systems is given later in Sections 9.5.7 and 9.6.4 as noted in Table 9.1.

9.4 LAMINAR FLOW INSIDE A CHANNEL

In this section, we examine the velocity and pressure relations for laminar flow inside a channel. The approach to the solution of laminar flow problems is illustrated via a

TABLE 9.1
Summary of Pressure Changes for Multiple Flow Conditions

Flow Geometry and Condition	Incompressible	Applicable to Both Incompressible and Compressible	Compressible
	Constant Area Channels		
Frictionless	$\Delta p_{acc} = 0$ (Section 9.2.2.2.1)	$\Delta p_{form} = 0$ $\Delta p_{fric} = 0$	$\Delta p_{acc} = \rho_2 \upsilon_2^2 - \rho_1 \upsilon_1^2$ (9.27) (Section 9.2.3.1)
Viscous	$\Delta p_{acc} = 0$ $\Delta p_{fric} = \bar{f}\dfrac{L}{D_e}\rho\dfrac{\upsilon_{ref}^2}{2}$ (9.39e) (Sections 9.2.2.2.1 and 9.3.5)	$\Delta p_{form} = 0$	$\Delta p_{acc} = \rho_2 \upsilon_2^2 - \rho_1 \upsilon_1^2$ (9.27) $\Delta p_{fric} \approx \int_0^L \dfrac{f(\ell)}{2D_e}\rho(\ell)\upsilon^2(\ell)\, d\ell$ (Section 9.2.3.1)
	Abrupt Area Change[a]		
Frictionless	$\Delta p_{acc} = \rho\left(\dfrac{\upsilon_2^2}{2} - \dfrac{\upsilon_1^2}{2}\right)$ (9.22a) (Section 9.2.2.2.3)	$\Delta p_{form} = 0$ $\Delta p_{fric} = 0$	$\Delta p_{acc} \approx \rho\left(\dfrac{\upsilon_2^2}{2} - \dfrac{\upsilon_1^2}{2}\right)$ (9.22a) (Section 9.2.3.2)
Viscous	$\Delta p_{acc} = \rho\left(\dfrac{\upsilon_2^2}{2} - \dfrac{\upsilon_1^2}{2}\right)$ (9.22a) $\Delta p_{form,c} = K_c\rho\dfrac{\upsilon_2^2}{2}$ (9.22c) $\Delta p_{form,e} = K_e\rho\dfrac{\upsilon_1^2}{2}$ (9.22d) $\Delta p_{fric} = 0$ (Sections 9.5.7 and 9.6.4)		Not covered

[a] $\Delta p_{fric} = 0$ in this part for viscous flow because the effect of friction is included in Δp_{form}.

detailed analysis for a channel of circular geometry. However, results for other geometries are also presented. As mentioned before, the analysis can be split into two regions:

1. The developing flow region, where subdivision between an inviscid core flow and a laminar boundary layer is needed to determine the velocity variation across the entire channel.
2. The fully developed flow region, where the viscous laminar flow analysis can be applied over the entire flow region.

9.4.1 Fully Developed Laminar Flow in a Circular Tube

Consider steady-state viscous flow in a horizontal circular tube of radius R. The flow enters the tube, at $z = 0$, with a uniform velocity. The boundary layer thickness is zero at the inlet but eventually grows to the centerline, and the layer occupies the entire flow channel. Because of symmetry in the θ direction, the velocity ($\vec{\upsilon}$) and the pressure (p) are independent of θ. The applicable momentum equation in the z direction is

$$\rho\upsilon_z \frac{\partial \upsilon_z}{\partial z} + \rho\upsilon_r \frac{\partial \upsilon_z}{\partial r} = -\frac{\partial p}{\partial z} + \mu \frac{\partial^2 \upsilon_z}{\partial z^2} + \frac{\mu}{r}\frac{\partial}{\partial r}\left(r \frac{\partial \upsilon_z}{\partial r}\right) \quad \text{(from Table 4.10)}$$

Fully developed laminar flow implies that there is no change in the velocity along the longitudinal flow direction; therefore

$$\frac{\partial \vec{\upsilon}}{\partial z} = 0 \quad \text{which yields} \quad \frac{\partial \upsilon_z}{\partial z} = 0 \tag{9.44a,b}$$

For constant ρ and μ, the continuity equation can be written as

$$\nabla \cdot \vec{\upsilon} = 0 \tag{9.45}$$

Since $\upsilon_\theta = 0$ from symmetry and $(\partial \upsilon_z/\partial z)$, then the continuity equation yields

$$\upsilon_r = \text{constant}$$

which reduces to

$$\upsilon_r = 0 \tag{9.44c}$$

when applying the boundary condition $\upsilon_r = 0$ at the wall.

Applying the conditions of Equations 9.44b and 9.44c yields

$$\frac{\mu}{r}\frac{\partial}{\partial r}\left(r \frac{\partial \upsilon_z}{\partial r}\right) = \frac{\partial p}{\partial z} \tag{9.46}$$

Because from the conditions of Equation 9.33 $\partial p/\partial r = 0$ and $\partial p/\partial \theta = 0$; however,

$$\frac{\partial p}{\partial z} = \frac{dp}{dz} \tag{9.47}$$

Hence the equation to be solved for υ_z as a function of r is:

$$\frac{\mu}{r}\frac{d}{dr}\left(r\frac{d\upsilon_z}{dr}\right) = \frac{dp}{dz} \tag{9.48}$$

Note that the total derivatives are used because υ_z and p are only functions of r and z, respectively.

Equation 9.48 can be directly integrated twice over r, as the pressure is not a function of r and dp/dz is not a function of r. Applying the boundary conditions at

$$r = R \quad \upsilon_z = 0 \tag{9.49a}$$

and

$$r = 0 \quad \upsilon_z \text{ remains finite} \tag{9.49b}$$

Condition 9.49b yields

$$\left.\frac{\partial \upsilon_z}{\partial r}\right|_r = 0 \text{ (i.e., } \upsilon_z \text{ is maximum at } r = 0) \tag{9.49c}$$

Hence Equation 9.48 yields

$$\upsilon_z = \frac{R^2}{4\mu}\left(-\frac{dp}{dz}\right)\left(1 - \frac{r^2}{R^2}\right) \tag{9.50}$$

Hence, when dp/dz is negative, the velocity is positive in the axial direction.

It is useful to obtain the mean (mass weighted) velocity (V_m):

$$V_m = \frac{\int_0^R \rho\upsilon_z(2\pi r)\, dr}{\int_0^R \rho(2\pi r)\, dr} \tag{9.51}$$

For a constant density:

$$V_m = \frac{\int_0^R \upsilon_z(2\pi r)\, dr}{\pi R^2}$$

Substituting υ_z from Equation 9.50 and performing the integration, we obtain

$$V_m = \frac{R^2}{8\mu}\left(-\frac{dp}{dz}\right) \tag{9.52}$$

From Equations 9.50 and 9.52, the local velocity can be written as

$$\frac{\upsilon_z}{V_m} = 2\left(1 - \frac{r^2}{R^2}\right) \tag{9.53}$$

The parabolic nature of the velocity profile leads to a linear profile of the shear stress τ_{rz}*, as

$$\tau_{rz} = +\mu\frac{d\upsilon_z}{dr} = -4\mu\frac{V_m}{R}\left(\frac{r}{R}\right) \tag{9.54}$$

Note that τ_{rz} is opposite in sign to V_m, which means that it acts opposite to the flow direction, as expected. The maximum shear stress is given by specifying the value of $r = R$ in Equation 9.54 to get

$$\tau_{rz}\big|_{max} = -4\mu\frac{V_m}{R} = \frac{R}{2}\left(+\frac{dp}{dz}\right) \tag{9.55}$$

In Figure 9.14, we illustrate the shear stress distribution and the wall shear stress. Since the wall shear stress, τ_w, is always opposite to the direction of flow, it is convenient to define it as positive in the opposite direction of the flow. Thus

$$\tau_w = \left|\tau_{rz}\right|_{max}$$

Note that the wall shear stress can also be obtained from an integral momentum balance over an infinitesimal δz (Figure 9.14) in the pipe such that the net pressure

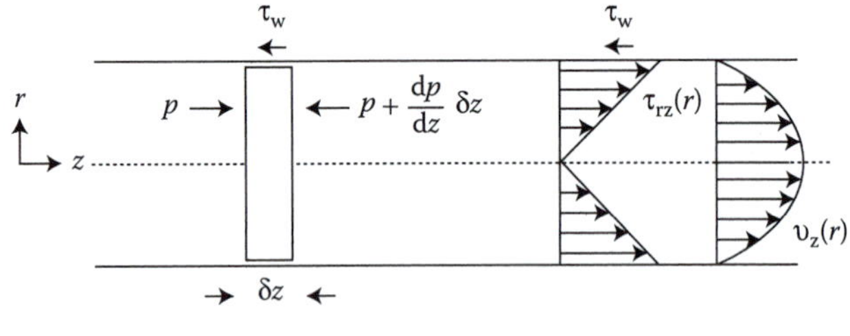

FIGURE 9.14 Shear stress distribution and velocity profile in fully developed pipe flow.

* τ_{ij} = stress in j direction on a face normal to the i axis.

force is balanced by the wall shear stress force since the net momentum flow here is zero, hence yielding

$$\tau_w(2\pi R)(\delta z)+\frac{dp}{dz}(\pi R^2)(\delta z)=0$$

or

$$\tau_w=-\frac{R}{2}\left(\frac{dp}{dz}\right) \quad (9.56)$$

Note dp/dz has negative values for flow in the positive z direction.

Of great practical importance is the definition of a friction factor (f) such that the pressure gradient is related to the kinetic pressure based on the average velocity and the diameter of the pipe. From Equation 9.52, we can see that

$$-\frac{dp}{dz}=\frac{8\mu}{R^2}V_m=\frac{64\mu}{\rho D^2 V_m}\left(\frac{\rho V_m^2}{2}\right) \quad (9.57)$$

which can be recast in the form originally proposed by Darcy and defined in Equation 9.39e as

$$-\frac{dp}{dz}=\frac{f}{D}\frac{\rho V_m^2}{2} \quad (9.58)$$

where f, the friction factor, in this case is given by

$$f=\frac{64}{\rho D V_m/\mu}=\frac{64}{\text{Re}} \quad (9.59)$$

The result of this analysis leads to a condition usually observed for laminar flow, that is, that the product fRe is a constant dependent only on the geometry of the flow. Experience shows that laminar flow in a tube exists up to a Reynolds number of about 2100. By combining Equations 9.56 and 9.58, we get a relation for the wall shear stress and the friction factor f:

$$\tau_w=\frac{R}{2}\frac{f}{2R}\frac{\rho V_m^2}{2}=\frac{f}{4}\frac{\rho V_m^2}{2} \quad (9.60)$$

Unfortunately, there is another friction factor that appears in the literature, the Fanning friction factor, which is defined in terms of the shear stress (τ_w) relation to the kinetic pressure, such that

$$\tau_w=f'\frac{\rho V_m^2}{2} \quad (9.61)$$

Thus, the Darcy factor f is related to the Fanning factor f' as

$$f = 4f' \tag{9.62}$$

The friction pressure drop across a pipe of length L, when the developing flow region can be ignored, is given by

$$\Delta p_{\text{friction}} = \int_{z_{\text{in}}}^{z_{\text{out}}} \left(-\frac{dp}{dz}\right) dz = p_{\text{in}} - p_{\text{out}} = \frac{\bar{f}L}{D_e}\frac{\rho V_m^2}{2} \tag{9.39e}$$

9.4.2 Fully Developed Laminar Flow in Noncircular Geometries

By following a procedure similar to the one outlined above, solutions for the velocity distribution and wall friction factor coefficient have been obtained for a variety of flow geometries. The values of f' Re for fully developed flow in rectangular and annular channels are given in Figures 9.15 and 9.16. In those cases, the *Reynolds number* is defined by

$$\text{Re} = \frac{\rho V_m D_e}{\mu} \tag{9.63}$$

where D_e = the equivalent or hydraulic diameter defined by

$$D_e = \frac{4A_f}{P_w} \tag{9.64}$$

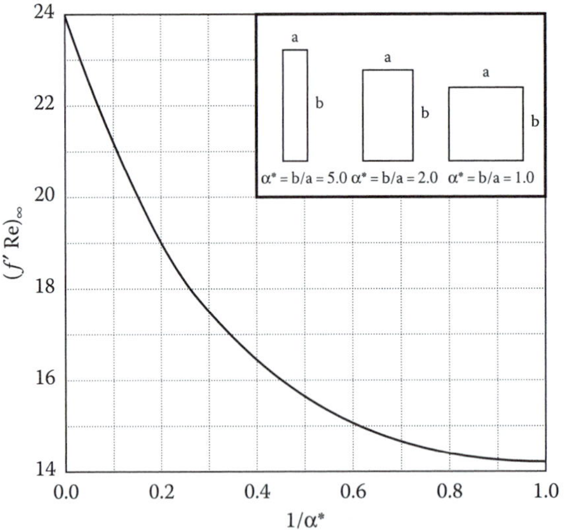

FIGURE 9.15 Product of laminar friction factor (Fanning) and Reynolds number for fully developed flow with rectangular geometry. (From Kays, W. M. and Crawford, M. E., *Convective Heat and Mass Transfer*, 3rd Ed. New York, NY: McGraw-Hill, 1993.)

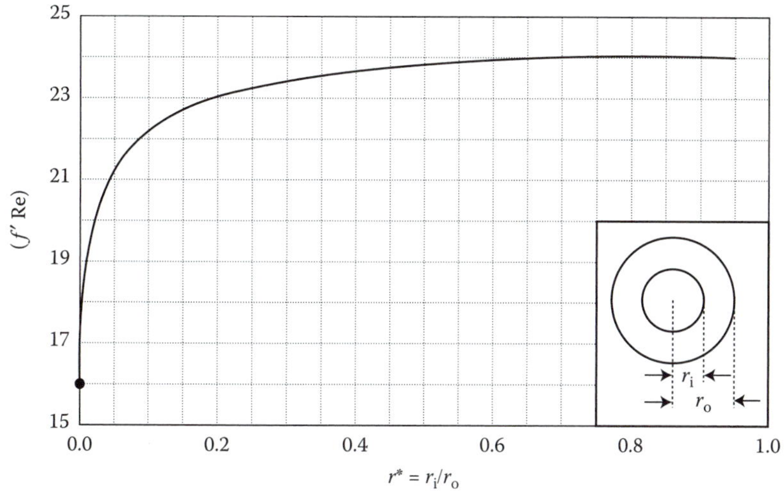

FIGURE 9.16 Product of laminar friction factor (Fanning) and Reynolds number for fully developed flow in an annular channel. (Adopted from Kays, W. M. and Crawford, M. E., *Convective Heat and Mass Transfer*, 3rd Ed. New York, NY: McGraw-Hill, 1993.)

where A_f = flow area, and P_w = wetted perimeter (or surface per unit length) of the channel.

Although the values of f' Re for these cases are expressed in terms of the equivalent hydraulic diameter, these results are not equivalent to simply transforming the circular tube results utilizing the equivalent diameter concept. It can be seen directly by noting from Equation 9.58 that such a transformation would yield

$$f'\,\mathrm{Re} = 16 \quad (\text{or } f\,\mathrm{Re} = 64) \tag{9.65}$$

where Re is defined in Equation 9.63, and f' is used in place of f. Note that the plots in Figures 9.15 and 9.16 yield the value of f' Re = 24 for infinite parallel plates (i.e., for $1/\alpha^* = 0$ and $r_i/r_o = 1$) and illustrate that values of f'Re vary with flow geometry. This result is to be expected with laminar flow, where the molecular shear effects are significant throughout the flow cross section so that the governing equations have to be solved for each specific geometry. As discussed in Chapter 10, the need to solve for each specific geometry is also the case for heat transfer in laminar flow.

9.4.3 Laminar Developing Flow Length

In the developing (or entrance) region of flow, the steady-state velocity distribution and friction factor coefficients can be obtained from the equation:

$$\frac{\mu}{r}\frac{\partial}{\partial r}\left(r\frac{\partial \upsilon_z}{\partial r}\right) = \rho\upsilon_z\frac{\partial \upsilon_z}{\partial z} + \frac{\mathrm{d}p}{\mathrm{d}z} \tag{9.66}$$

because $\partial \upsilon_z/\partial z$ is not zero in that region. However, υ_r remains approximately zero, and $\partial p/\partial r$ is also approximately zero. Momentum integral solutions to this problem lead to determination of the dependence of $f\text{Re}$ on $\text{Re}/(z/D)$ as the nondimensional parameter of significance. Langhaar and No [24] calculated the value of both the local value of the friction factor f_z' as well as the effective value over any length for a circular tube. Their results are shown in Figure 9.17. It is seen that the local value for $f'\text{Re}$ approaches 16 as $\text{Re}/(z/D)$ approaches 20. Hence, it can be said that the flow becomes fully developed at that distance, such that

$$\frac{z}{D} \simeq \frac{\text{Re}}{20} \tag{9.67}$$

Three friction coefficients are indicated in Figure 9.17: f', f_z', and f_{app}'. The local friction coefficient is described as f_z' and is based on the actual local wall shear stress at z. The mean friction coefficient from $z = 0$ to z is described by f'. Part of the pressure drop in the entrance region of a tube is attributable to an increase in the total fluid momentum flux, which is associated with the development of the velocity profile. The combined effects of surface shear and momentum flux have led to identifying an apparent friction factor (f_{app}'), which is the mean friction factor from $z = 0$ to z. If the momentum flux variation is accounted for explicitly in the momentum equations (via the acceleration Δp terms), f'(not f_{app}') should be used to calculate the pressure drop due to friction alone.

It should be noted that the friction factor, or pressure gradient, is higher in the developing region than in the fully developed region. Note that for large values of z/D, f_z' and f' have essentially the same value. Thus, in channels with multiple flow

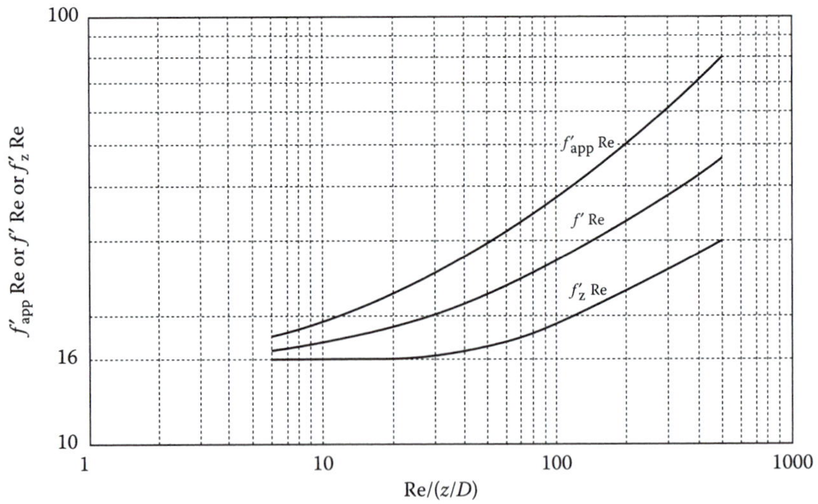

FIGURE 9.17 Developing laminar flow friction factor. (From Langhaar, H. L. and No, P. H., *J. Appl. Mech.*, Sect. 3, 9:1942.)

obstructions, the pressure drop is higher than that of unobstructed open channels for two reasons: (1) the form losses imposed by the obstructions, and (2) the destruction of the fully developed flow pattern that occurs both upstream and downstream of the obstructions. The solution of the entrance region flow for concentric annuli, including flow between parallel plates, can be found in Heaton et al. [17] and Fleming and Sparrow [13].

Example 9.3: Laminar Flow Characteristics in a Steam Generator Tube

PROBLEM

During a shutdown condition in a PWR, the flow is driven through the loop by natural circulation at a rate corresponding to about 1% of the full flow rate provided by the pumps. Assuming that the total flow rate is $\dot{m}_T = 4686$ kg/s in the full flow condition and that there are approximately 3800 tubes of 0.0222 m (7/8 in.) inside diameter in the steam generator of average length 16.5 m, determine

1. Whether the flow is turbulent or laminar.
2. The value of the friction factor.
3. The friction pressure loss between the inlet and outlet of one tube (neglecting the pressure loss due to the tube bend of 180°).

Use $\rho = 1000\ \text{kg/m}^3$ and $\mu = 0.001\ \text{kg/m}\cdot\text{s}$ or Pa s

SOLUTION

1. The value of the Reynolds number at 1% of full flow is

$$\text{Re} = \frac{\rho V_m D}{\mu} = \frac{\dot{m} D}{\mu A} = \frac{4\dot{m}}{\pi \mu D} = \frac{4(0.01)(4686/3800)\ \text{kg/s}}{\pi(0.001\ \text{kg/m}\cdot\text{s})(0.0222\ \text{m})} = 706.5$$

   Flow in pipes is laminar below a Re of about 2100, so the flow is laminar.

2. For these tubes where $L/D \gg 1$, the fully developed laminar flow friction factor can be used for the whole tube length. Hence, from Equation 9.59:

$$f = \frac{64}{\text{Re}} = \frac{64}{706.5} = 0.0906$$

3. The pressure drop due to friction in one tube is then

$$\Delta p = f\left(\frac{L}{D}\right)\frac{\rho V_m^2}{2}$$

$$V_m = \frac{\dot{m}}{\rho A} = \frac{(0.01)4686/3800\,\text{kg/s}}{(1000\,\text{kg/m}^3)\frac{\pi}{4}(0.0222\,\text{m})^2} = 0.0318\,\text{m/s}$$

$$\Delta p = 0.0906\left(\frac{16.5\,\text{m}}{0.0222\,\text{m}}\right)\frac{(1000\,\text{kg/m}^3)(0.0318\,\text{m/s})^2}{2} = 34.0\,\text{Pa}$$

9.4.4 Form Losses in Laminar Flow

Form losses in laminar flow are important contributors to pressure loss in reactors under decay power and accident conditions when the flow can approach the laminar regime. This is particularly the case for the helium-cooled gas fast reactor (GFR), which operates in laminar flow under post loss of coolant accident (LOCA) situation. For this helium system, the post-LOCA system pressure must be maintained at 1.5–2.5 MPa during the first minutes after LOCA to achieve the needed natural circulation conditions. The significant increase in form loss with decrease in Reynolds number can impair decay heat removal and result in substantially higher needed system pressure.

Laminar form losses vary considerably with the geometry of the flow restriction and often are expressed in cumbersome equations or graphic representations. Hence, the reader is advised to consult handbooks (e.g., Idelchik [18]) for specific needs. In general, form losses in the laminar regime can become an order of magnitude larger than turbulent values as the Reynolds number value decreases below 100 and a factor of 2 or so larger in the laminar range above Reynolds number equals 100.

9.5 TURBULENT FLOW INSIDE A CHANNEL

9.5.1 Turbulent Diffusivity

As mentioned in Section 9.3.4, at sufficiently long distances or high velocities (i.e., high Re values), the smooth flow of the fluid is disturbed by the irregular appearance of eddies. For flow in pipes, the laminar flow becomes unstable at Re values above 2100. However, the value of Re that is needed to stabilize a fully turbulent flow is about 10,000. For Re values between 2100 and 10,000, the flow is said to be in transition from laminar to turbulent flow.

The turbulent enhancement of the momentum and energy lateral transport above the rates possible by molecular effects alone flattens the velocity and temperature profiles (Figure 9.18).

In Chapter 4, it was shown that the instantaneous velocity can be divided into a time-averaged component and a fluctuating component:

$$\vec{\upsilon} = \bar{\vec{\upsilon}} + \vec{\upsilon}' \tag{4.129b}$$

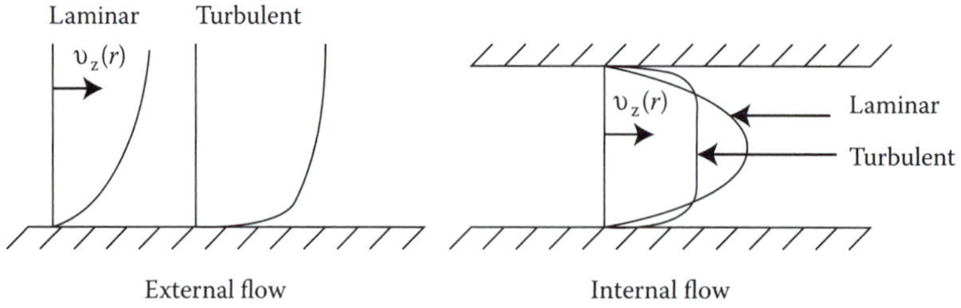

FIGURE 9.18 Velocity profiles of fully developed laminar and turbulent flows.

The time-averaged mass and momentum equations (Equations 4.135 and 4.136) can be reduced for negligible density fluctuations to the form:

$$\nabla \cdot \left[\rho \bar{\vec{\upsilon}}\right] = 0 \tag{9.68}$$

$$\frac{\partial \rho \bar{\vec{\upsilon}}}{\partial t} + \nabla \cdot \left[\rho \bar{\vec{\upsilon}} \bar{\vec{\upsilon}}\right] = -\nabla p + \nabla \cdot \left[\bar{\bar{\tau}} - \rho \overline{\vec{\upsilon}' \vec{\upsilon}'}\right] + \rho \vec{g} \tag{9.69}$$

From Equation 9.69, it is seen that the lateral and much smaller axial momentum transport rates can be considered to be composed of two parts: one due to the molecular effects and the other due to the fluctuations of the eddies, that is

$$\bar{\bar{\tau}}_{\text{eff}} = \bar{\bar{\tau}} - \rho \overline{\vec{\upsilon}' \vec{\upsilon}'} \tag{9.70}$$

Thus for flow inside a pipe, the effective shear in the flow direction can be given by

$$(\tau_{rz})_{\text{eff}} = \tau_{rz} - \rho \overline{\upsilon_z' \upsilon_r'} \tag{9.71}$$

The most common approach to analyzing the turbulent flow problems is to relate the additional term to the time-averaged behavior by defining a new momentum diffusivity (ε_M) in the plane of interest. For example

$$(\varepsilon_M)_{zr} \equiv \frac{-\overline{\upsilon_z' \upsilon_r'}}{\left(\partial \bar{\upsilon}_z / \partial r + \partial \bar{\upsilon}_r / \partial z\right)} \tag{9.72}$$

so that from Equations 9.71 and 9.72 the effective shear can be given by

$$(\tau_{rz})_{\text{eff}} = (\mu + \rho \varepsilon_M)\left(\frac{\partial \bar{\upsilon}_z}{\partial r} + \frac{\partial \bar{\upsilon}_r}{\partial z}\right) \tag{9.73}$$

since

$$(\varepsilon_M)_{zr} = (\varepsilon_M)_{rz}$$

(Under steady-state flow conditions in a fully developed pipe flow, $\bar{\upsilon}_r = 0$.) In general, the turbulent diffusivity is assumed to be independent of the orientation of flow, so that

$$(\varepsilon_M)_{zr} = (\varepsilon_M)_{r\theta} = (\varepsilon_M)_{\theta z} \tag{9.74}$$

The quantity $\rho\varepsilon_M$ is clearly analogous to the dynamic viscosity (μ); but whereas μ is a property of the fluid, $\rho\varepsilon_M$ depends on the velocity and geometry as well.

Other approaches have been introduced to determine the effect of turbulence on the momentum transport and energy transport. A brief discussion of two common approaches is found in Section 10.4.

9.5.2 Turbulent Velocity Distribution

It has been found useful to express the turbulent velocity profile in a dimensionless form that depends on the wall shear:

$$\upsilon_z^+ = \frac{\upsilon_z}{\sqrt{\tau_w/\rho}} \tag{9.75}$$

υ_z^+ is often called the universal turbulent velocity. The term $\sqrt{\tau_w/\rho}$ is referred to as the "shear velocity." Also, the distance from the wall (y) can be nondimensionalized as

$$y^+ = y\frac{\sqrt{\tau_w/\rho}}{\nu} \tag{9.76}$$

If a linear shear stress distribution is assumed near the wall where molecular effects dominate, we have

$$\tau_w \approx \rho\nu\frac{d\upsilon_z}{dy} \tag{9.77}$$

where y = distance from the wall (in a pipe, $y = R - r$). After integrating Equation 9.77:

$$\upsilon_z = \frac{\tau_w}{\rho\nu}y + C \tag{9.78}$$

Because $\upsilon_z = 0$ at $y = 0$, we get

$$\upsilon_z = \frac{\tau_w}{\rho\nu}y \tag{9.79a}$$

or

$$\upsilon_z^+ = y^+ \tag{9.79b}$$

Various investigators have developed expressions for the universal velocity. Often the boundary layer is subdivided into a laminar sublayer near the wall (where $\varepsilon_M = 0$), an intermediate or buffer sublayer, and a fully turbulent sublayer (where $\mu << \rho\varepsilon_M$). Martinelli [31] described the resulting distribution by

$$y^+ < 5 \quad \upsilon_z^+ = y^+ \tag{9.80a}$$

$$5 < y^+ < 30 \quad \upsilon_z^+ = -3.05 + 5.00\,\ell n(y^+) \tag{9.80b}$$

$$y^+ > 30 \quad \upsilon_z^+ = 5.5 + 2.5\,\ell n(y^+) \tag{9.80c}$$

An important equation for ε_M (for flow in a pipe) in the turbulent sublayer is that reported by Reichardt [44]. He proposed that

$$\frac{\varepsilon_M}{\nu} = \frac{kR^+}{6}\left[1-\left(\frac{r}{R}\right)^2\right]\left[1+2\left(\frac{r}{R}\right)^2\right] \tag{9.81}$$

where

$$R^+ = R\frac{\sqrt{\tau_w/\rho}}{\nu} \tag{9.82}$$

and $k \approx 0.4$. This equation leads to the following expression for the velocity in the turbulent sublayer:

$$\upsilon_z^+ = 5.5 + 2.5\,\ell n\left[y^+\frac{1.5(1+r/R)}{1+2(r/R)^2}\right] \tag{9.83}$$

This equation appears to satisfy the experiments at all values of r, including at the center line.

A fully developed turbulent velocity profile for a pipe has been sketched in Figure 9.18. Note that the velocity profile is flatter than in laminar flow. The velocity profile can be approximated by

$$\frac{\upsilon_z}{\upsilon_{c\ell}} = \left(\frac{y}{R}\right)^{1/7} = \left(\frac{R-r}{R}\right)^{1/7} \tag{9.84}$$

where $\upsilon_{c\ell}$ is the velocity at the pipe centerline.

Consequently the average velocity is given by

$$V_m = 0.817\,\upsilon_{c\ell} \tag{9.85}$$

9.5.3 Turbulent Friction Factors in Adiabatic and Diabatic Flows

9.5.3.1 Turbulent Friction Factor: Adiabatic Flow

A commonly encountered expression for the friction factor is the Karman–Nikuradse equation:

$$\frac{1}{\sqrt{f}} = -0.8 + 0.87\,\ell n\left(\mathrm{Re}\sqrt{f}\right) \tag{9.86}$$

This transcendental equation is difficult to use in practice, and simplified relations are often applied. An approximate equation for a smooth tube where $3 \times 10^4 < \mathrm{Re} < 10^6$ is the McAdams relation:

$$f = 0.184\,\mathrm{Re}^{-0.20} \tag{9.87}$$

For $4 \times 10^3 < \mathrm{Re} < 10^5$, the turbulent friction factor for a smooth tube may be given by the Blasius relation:

$$f = 0.316\,\mathrm{Re}^{-0.25} \tag{9.88}$$

The tube roughness, characterized by the ratio of depth of surface protrusions to the tube diameter (λ/D), increases the effective friction factor. The Moody chart [33] given in Figure 9.19 is commonly used to obtain Darcy friction factors. In Moody's chart, the effect of the roughness depends on the pipe diameter (D), as is reasonable

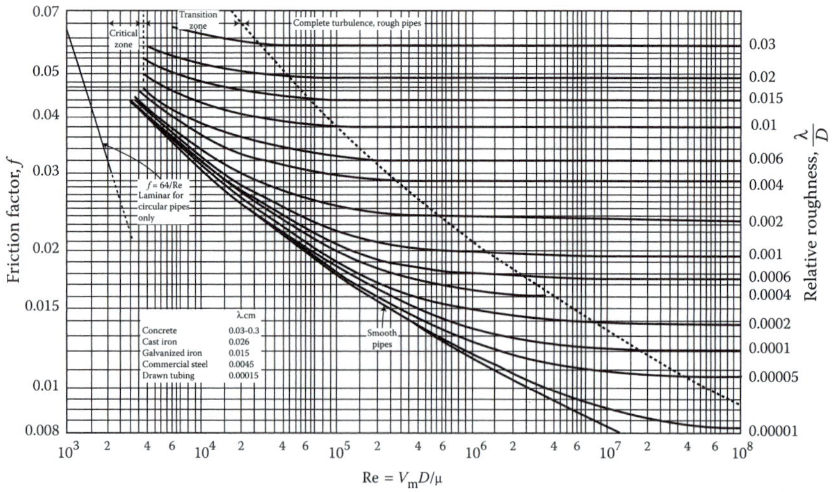

FIGURE 9.19 Moody's chart for Darcy friction factors for flow inside circular pipes. (From Moody, L. F., *Trans. ASME*, 66:671, 1944.)

to expect. The Moody diagram is a graphic representation of the Colebrook equation [6], which covers both rough and smooth walls and is given by

$$\frac{1}{\sqrt{f}} = -2\log_{10}\left[\frac{\lambda/D}{3.70} + \frac{2.51}{\mathrm{Re}\sqrt{f}}\right] \tag{9.89}$$

Note that roughness has a negligible effect on the laminar friction factor as shown in Figure 9.19. The iteration to calculate f required by use of the transcendental Colebrook equation is eliminated in the following explicit relations by Haaland [16] and Churchill [5], which yield friction factors within 2% and 1%, respectively, of that from the Colebrook Equation 9.89:

$$f_{\text{Churchill}} = \frac{8}{6.0516}\left\{\ell n\left[\frac{\lambda/D}{3.7} + \left(\frac{7}{\mathrm{Re}}\right)^{0.9}\right]\right\}^{-2} \tag{9.90}$$

and

$$\frac{1}{\sqrt{f_{\text{Haaland}}}} \approx -1.8\log_{10}\left[\frac{6.9}{\mathrm{Re}} + \left(\frac{\lambda/D}{3.7}\right)^{1.11}\right] \tag{9.91}$$

9.5.3.2 Turbulent Friction Factor: Diabatic Flow

For heated or cooled fluids, the temperature dependence of the fluid properties affects the friction factor and heat transfer coefficient through their influence on variation of velocity and temperature over the cross-sectional area of flow. Liquid and gas properties exhibit different behaviors with temperature. For most liquids, density, specific heat, and thermal conductivity are nearly independent of temperature while the viscosity changes inversely with temperature, that is, decreases as temperature increases. For gases, only the specific heat varies slightly with temperature. The other three properties vary at about the same rate as the absolute temperature. Hence for gases, the viscosity decreases as temperature decreases. This behavior is correlated relative to the friction factor based on constant properties evaluated at the bulk temperature, T_b, as

For liquids:

$$\frac{f}{f_{T_b}} = g\left(\frac{\mu_b}{\mu_w}\right) \tag{9.92a}$$

where g indicates a generic function.

For gases:

$$\frac{f}{f_{T_b}} = \left(\frac{T_b}{T_w}\right)^n \tag{9.92b}$$

where T_w and T_b are absolute wall and bulk temperatures, respectively.

Correlations for liquids are relatively consistent, whereas for gases recommendations vary. The recommended values of the parameters in Equations 9.92a and 9.92b for liquids and gases, respectively, are those by Petukhov [36] presented in Table 9.2 from Lienhard and Lienhard [26], who considered but did not explicitly adopt the formulations for gases of McEligot [32] and Taylor [49]. The needed friction factor f_{T_b} is evaluated using the following equation from Petukhov [36]:

$$f_{T_b} = \frac{1}{(1.82\log_{10}\mathrm{Re_D} - 1.64)^2} \tag{9.93}$$

Example 9.4: Turbulent Flow in a Steam Generator Tube

PROBLEM

For the condition of full flow through a U-tube steam generator and assuming bulk properties for the primary coolant ($p = 15.5$ MPa; $T_b = 309°C$), find the same quantities asked for in Example 9.3:

1. Determine whether the flow is turbulent or laminar.
2. Find the value of the friction factor.
3. Find the average pressure loss between the inlet and outlet of a tube (neglecting the pressure loss due to the tube bend of 180°).

Then, repeat the calculation accounting for the difference in water properties between the bulk ($T_b = 309°C$) and the tube inner wall ($T_w = 295°C$) using the Petukhov correlations shown in Table 9.2.

SOLUTION

1. Water density and viscosity at 15.5 MPa and 309°C are 707 kg/m³ and 8.497×10^{-5} Pa s, respectively. At 15.5 MPa and 295°C, the viscosity is

TABLE 9.2
Coefficients for Evaluation of Friction Factors in Diabatic Flow for Use in Equations 9.92a and 9.92b

	Liquids $0.5 \le \mu_b/\mu_w \le 3$	Gases $0.14 \le T_b/T_w \le 3.3$
Heating ($T_w > T_b$)	$g\left(\frac{\mu_b}{\mu_w}\right) = \frac{1}{6}\left(7 - \frac{\mu_b}{\mu_w}\right)$	$n = 0.23$
Cooling ($T_w < T_b$)	$g\left(\frac{\mu_b}{\mu_w}\right) = \left(\frac{\mu_b}{\mu_w}\right)^{-0.24}$	$n = 0.23$

Source: From Lienhard IV, J. H. and Lienhard V, J. H., *A Heat Transfer Textbook*. Cambridge, MA: Phlogiston Press, 2008.

9.047×10^{-5} Pa s. These data, together with the total primary flow rate to the steam generator and the geometric parameters from Example 9.3, can be used to calculate the Reynolds number:

$$\text{Re} = \frac{\rho V_m D}{\mu} = \frac{4\dot{m}}{\pi \mu D} = \frac{4(4686/3800)}{\pi(8.497 \times 10^{-5})(0.0222)} = 832{,}358$$

which is clearly in the turbulent range (Re > 2100).

2. If bulk-averaged water properties are used, the appropriate expression for the friction factor is now Equation 9.87:

$$f = 0.184\text{Re}^{-0.2} = 0.184(832{,}358)^{-0.2} = 0.0120$$

If the radial variation of the water viscosity across the tube is instead accounted for, the friction factor can be calculated through Equation9.92a as

$$f = f_{T_b}\left(\frac{\mu_b}{\mu_w}\right)^n = 0.0120\left(\frac{8.497 \times 10^{-5}}{9.047 \times 10^{-5}}\right)^{-0.24} = 0.0122$$

where the value $n = -0.24$ has been used for the exponent according to Table 9.2. Accounting for the higher viscosity at the wall causes only a minor correction in the friction factor, which is well within the overall accuracy of the basic correlation.

3. Using bulk average properties, the friction pressure drop in one of the tubes is now:

$$V_m = \frac{\dot{m}}{\rho A} = \frac{4686/3800 \text{ kg/s}}{(707 \text{ kg/m}^3)\frac{\pi}{4}(0.0222 \text{ m})^2} = 4.51 \text{ m/s}$$

$$\Delta p = f\frac{L}{D}\frac{\rho V_m^2}{2}$$

$$= 0.0120\frac{16.5}{0.0222}\frac{707 \times (4.51)^2}{2}$$

$$= 64.0 \text{ kPa}$$

With the friction factor obtained through the Petukhov correlation, the pressure drop is only slightly different, that is, 65.1 kPa.

Relative to Example 9.3, the friction factor is lower by a factor of about 7, but the velocity has increased by a factor of about 140, and the pressure drop has therefore increased by a factor of about 1880.

9.5.4 Fully Developed Turbulent Flow with Noncircular Geometries

In the turbulent flow case, the velocity gradient is principally near the wall. Hence, the flow channel geometry does not have an important influence on the friction factor.

Therefore, the equivalent or hydraulic diameter concept can be used for predicting the friction factor. Specifically, the hydraulic diameter of noncircular channels characterized as

$$D_e = \frac{4A_f}{P_w} \tag{9.64}$$

can be used in place of the circular channel diameter D in circular channel friction factor correlations. As demonstrated in Section 9.4, this is not the case for laminar flow in noncircular channels.

9.5.5 Turbulent Developing Flow Length

The entrance length of laminar flow was found to extend to a maximum axial distance z such as

$$\frac{z}{D} \simeq \frac{\text{Re}}{20} \quad \text{for Re} < 2000 \tag{9.67}$$

For turbulent flow, the boundary layer can develop faster than in the high Re laminar region, so that

$$\frac{z}{D} = 25 \text{ to } 40 \tag{9.94}$$

Example 9.5: Comparative Pin and Inverted Fuel Array Geometries under Viscous Flow

PROBLEM

Consider pin and inverted fuel geometries (see Figure 9.20) for a hexagonal array of the same axial length L.

It is desired for neutronic reasons to establish the fuel volume fractions for these arrays as $v_f = 75\%$ and hence the coolant volume fractions as $v_c = 25\%$.

What are the dimensions of the geometries of each fuel type—pin and inverted—if they are to have the same pressure drop under turbulent flow conditions at equal mass flow rates?

SOLUTION

Friction pressure drop is given as

$$\Delta p = f\frac{L}{D_e}\frac{\rho V_m^2}{2} = f\frac{L}{D_e}\frac{1}{2\rho}\left(\frac{\dot{m}}{A_F}\right)^2$$

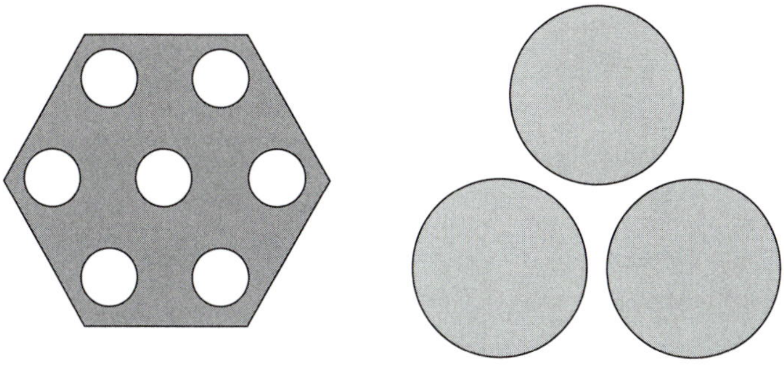

FIGURE 9.20 Inverted and pin fuel lattices having equivalent coolant volume fraction and hydraulic diameter. (From Pope, M. A., Hejzlar, P., and Driscoll, M. J., *Thermal Hydraulics of a 2400 MWth Supercritical CO_2-Direct Cycle GFR*. Cambridge: Massachusetts Institute of Technology, CANES Report MIT-ANP-TR-112, September 2006.)

Equal pressure drop requires that

$$\left[f\frac{L}{D_e}\frac{\dot{m}^2}{2\rho A_F^2}\right]_{pin} = \left[f\frac{L}{D_e}\frac{\dot{m}^2}{2\rho A_f^2}\right]_{inverted} \tag{9.95a}$$

Now since

$$f = \frac{C}{Re^n} = \frac{C}{(\rho V_m D_e/\mu)^n} = \frac{C}{(\dot{m}D_e/\mu A_F)^n}$$

the equal pressure drop condition becomes

$$\left[\frac{C\mu^n}{2\rho}\left(\frac{L}{D_e^{1+n}}\right)\left(\frac{\dot{m}}{A_f}\right)^{2-n}\right]_{pin} = \left[\frac{C\mu^n}{2\rho}\left(\frac{L}{D_e^{1+n}}\right)\left(\frac{\dot{m}}{A_f}\right)^{2-n}\right]_{inverted} \tag{9.95b}$$

Since $\nu_c \equiv (A_F/A_T)$, where A_T is the total cross-sectional area; a constant ν_c implies that

$$(A_F)_{pin} = (A_F)_{inverted}.$$

For conditions of equal mass flow rate, properties, and coolant void fraction, Equation 9.95b reduces to

$$D_{e_{pin}}^{1+n} = D_{e_{inverted}}^{1+n}$$

where $n = 0.2$ for turbulent flow.

Hence for turbulent flow:

$$D_{e_{pin}} = D_{e_{inverted}} \tag{9.96a}$$

Now for the inverted geometry:

$$D_{e_{inverted}} \equiv D_{channel}$$

whereas for the pin geometry:

$$D_{e_{pin}} = \frac{4A_F}{P_w} = \frac{4\left[(\sqrt{3}/4)P_{pin}^2 - (\pi/8)D_{pin}^2\right]}{(1/2)\pi D_{pin}} = D_{pin}\left[\frac{2\sqrt{3}}{\pi}\left(\frac{P}{D}\right)_{pin}^2 - 1\right]$$

Hence the condition of Equation 9.96a requires that

$$D_{channel} = D_{pin}\left[\frac{2\sqrt{3}}{\pi}\left(\frac{P}{D}\right)_{pin}^2 - 1\right] \tag{9.96b}$$

Now P_{pin}/D_{pin} is obtained as

$$\nu_{c_{pin}} \equiv \frac{A_F}{A_T} = \frac{(\sqrt{3}/4)P_{pin}^2 - (\pi/8)D_{pin}^2}{(\sqrt{3}/4)P_{pin}^2} = 1 - \frac{\pi\sqrt{3}}{6(P/D)_{pin}^2} \tag{9.97}$$

so that for

$$\nu_{c_{pin}} = 0.25, \quad \left(\frac{P}{D}\right)_{pin} = 1.10$$

Substituting this result into Equation 9.96b yields the desired ratio of channel to pin diameters as

$$\frac{D_{pin}}{D_{channel}} = 3$$

Further, the $(P/D)_{inverted}$ also follows from the known value of $\nu_c = 0.25$ as

$$\nu_c = \frac{A_F}{A_T} = \frac{(\pi D_{chan}^2/8)}{(\sqrt{3}/4)P_{inv}^2} = \frac{\pi\sqrt{3}}{6(P/D)_{inv}^2} = 0.25 \tag{9.98}$$

or

$$\left(\frac{P}{D}\right)_{inv} = 1.905$$

Hence, the resulting geometries of the inverted and pin fuel arrays are illustrated to scale in Figure 9.20.

9.5.6 Turbulent Friction Factors—Geometries for Enhanced Heat Transfer

The augmentation of thermal performance by the use of extended surfaces for heat transfer and twisted tapes within the flow channel has been an area of extensive investigation. The work of Bergles [2] and that of Webb and Kim [51] have been the most comprehensive in this area. This increased heat transfer performance, which is described by correlations in Chapter 10, is accompanied, however, by increased pressure loss. The friction factors are presented in this section for both extended surfaces and twisted tape inserts.

9.5.6.1 Extended Surfaces

The first nuclear application was the adoption of fins on the Magnox cladding (abbreviation for magnesium nonoxidizing; magnesium with small amounts of aluminum and other metals) of the graphite-moderated, natural uranium-fueled Magnox reactors to enhance the heat transfer to the carbon dioxide coolant maintained at relatively low temperatures. The prime recent nuclear application has been in the design of tubing for heat exchangers in Generation IV reactors in which heat transfer occurs between high Nusselt number fluids (e.g., lead, sodium) and low Nusselt number fluids (e.g., carbon dioxide) [34]. In this application, the low Nusselt number flow channel has an augmented surface area by virtue of internal spiral or longitudinal fins which increase heat transfer through disturbance of the surface sublayer. The primary surface variables are the number of starters, ratio of rib height to inner tube diameter, and helical rib angle (rib axial pitch).

The correlation of Ravigururajan and Bergles [38] with Re range as extended by Ravigururajan [37] is presented here because of its applicability to a wide range of hydraulic and geometric parameters for internally augmented tubes. The correlation is based on statistical analysis of a previously available extensive database and validated by water testing of a variety of commercially available tubes. The friction factor expressed by this correlation is defined as a Fanning factor for the augmented surface with respect to that of the smooth tube, meaning that, once this factor is calculated, the computation of the friction pressure drop has to be performed according to the formula:

$$\Delta p = 4f'\left(\frac{L}{D}\right)\rho\left(\frac{V_0^2}{2}\right) \tag{9.99}$$

in which L is the nominal length of the tube, D is the inside diameter of the smooth tube (equal to the maximum inside diameter of the ribbed tube), and V_0 is the mean (bulk) velocity that the fluid would have if the tube was smooth. The correlation* is in terms of the ratio of augmented f'_{a} to smooth tube f'_{sm} Fanning friction factors as

$$\frac{f'_{\mathrm{a}}}{f'_{\mathrm{sm}}} = \left\{1+\left[29.1\mathrm{Re}^{Y1}\times\left(\frac{e}{D}\right)^{Y2}\left(\frac{p}{D}\right)^{Y3}\times\left(\frac{\alpha}{90}\right)^{Y4}\times\left(1+\frac{2.94}{n}\right)\sin\beta\right]^{15/16}\right\}^{16/15} \tag{9.100a}$$

* This form of the correlation contains corrections from A. Bergles (pers. comm., June 2011).

$$Y1 = 0.67 - 0.06\frac{p}{D} - 0.49\frac{\alpha}{90}$$

$$Y2 = 1.37 - 0.157\frac{p}{D}$$

$$Y3 = -1.66 \cdot 10^{-6}\,\text{Re} - 0.33\frac{\alpha}{90}$$

$$Y4 = 4.59 + 4.11 \times 10^{-6}\,\text{Re} - 0.15\frac{p}{D}$$

where f'_{sm} is given by Filonenko [12] as a Fanning friction factor:

$$f'_{\text{sm}} = \left(1.58\,\ell n\,\text{Re} - 3.28\right)^{-2} \quad (9.100\text{b})$$

and the Reynolds number shown in Equations 9.100a and 9.100b is that of the fluid assumed to flow in the smooth tube.

This correlation developed on air, water, hydrogen, n-butyl alcohol data is applicable over the following fluid, operational, and geometric conditions:

$$0.66 \le \Pr \le 37.6$$

$$3 \times 10^3 \le \text{Re} \le 5 \times 10^5$$

and

$$0.01 \le e/D \le 0.2$$
$$0.1 \le p/D \le 7.0$$
$$0.3 \le \alpha/90 \le 1.0$$

where

e is the ib height
D is the maximum inside tube diameter
p is the rib separation, that is, separation between adjacent ribs which for one start is the traditional axial pitch
α is the helix angle (deg)
β is the contact angle profile (deg)

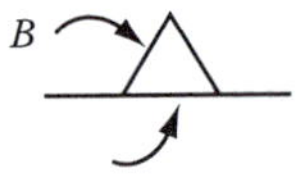

n is the number of sharp corners facing the flow that characterize the rib profile (e.g., two for both triangular and rectangular profiles)

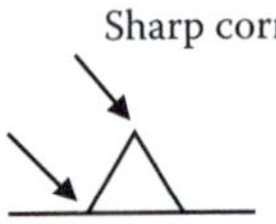

ns is the number of starts (which affects p value)

This friction factor correlation predicts 96% of the database to within 50% and 77% within 20%.

Example 9.6: Friction Factor for an Extended Surface Intermediate Heat Exchanger Tube

PROBLEM

Consider a tube-and-shell heat exchanger with lead-coolant on the shell side and supercritical CO_2 (S-CO_2) on the tube side. The heat exchanger is desired to be as compact as possible. Heat transfer augmentation is considered for the S-CO_2 side because otherwise it exhibits a heat transfer coefficient three to five times lower than that of the lead side. One of the possible methods to increase the heat transfer coefficient is through roughening of the inside tube surface. Repeated helical ribs are introduced on the inner tube side, which increase the S-CO_2 heat transfer through disturbance of the surface sublayer. For a 300 MW_{th} heat exchanger, compare

a. The friction factor
b. The friction pressure drop if the tube length is 5 m

for the case of smooth tubes and the case of square-shaped ribbed tubes.

For predicting the friction factor, use the correlation developed by Ravigururajan and Bergles of Equations 9.100a and 9.100b. Input parameters for the IHX are given in Table 9.3.

SOLUTION

a. *Calculate the friction factor for smooth tubes:*

$$V_{o_{CO_2}} = \frac{\dot{m}}{\rho A_{\text{total}}} = \frac{\dot{m}}{\rho\left(N(\pi D_i^2/4)\right)} = \frac{1594}{138.5 \times 10{,}000 \times 3.14\left(1.60 \times 10^{-2}\right)^2/4}$$

$$= 5.73 \text{ m/s}$$

$$\text{Re}_{CO_2} = \left(\frac{\rho V_o D_e}{\mu}\right)_{CO_2} = \frac{138.5 \times 5.73 \times 1.60 \times 10^{-2}}{3.52 \times 10^{-5}} = 3.607 \times 10^5$$

The flow is clearly turbulent. Using the Fanning friction factor correlation of Equation 9.100b developed by Filonenko [12] for smooth tubes, we obtain

$$f'_{sm} = (1.58 \ln \text{Re} - 3.28)^{-2} = (1.58 \ln (3.607 \times 10^5) - 3.28)^{-2} = 3.49 \times 10^{-3}$$

Convert it to Darcy friction factor by multiplying by 4:

$$f_{sm} = 4 \times 3.49 \times 10^{-3} = 1.39 \times 10^{-2}$$

TABLE 9.3
Intermediate Heat Exchanger Characteristics

Target power transmitted in the IHX, $\dot{Q}$ (MW_{th})	300
Lead mass flow rate (kg/s)	21,700
S-CO_2 mass flow rate (kg/s)	1594
Lead inlet temperature (°C)	600
Lead outlet temperature (°C)	503
S-CO_2 inlet temperature (°C)	397
S-CO_2 target outlet temperature (°C)	550
S-CO_2 pressure (MPa)	19.7
Tube Geometry	
Lattice	Triangular
Number of tubes (per IHX)	10,000
Outer tube diameter (mm)	20.0
Inner tube diameter, D (mm)	16.0
Pitch to diameter ratio	1.23
Rib Geometry	
Rib height, e (mm)	0.35
Number of starts, ns	5
Helix angle, α (°)	27
Contact angle profile, β (°)	90
Fin base thickness, b (mm)	0.35
Rib separation, p (mm)	20.0
Useful Fluid Properties (Average)	
CO_2 density (kg/m^3)	138.5
CO_2 viscosity (Pa s)	3.52E − 05
CO_2 thermal conductivity (W/m K)	5.77E − 02
CO_2 specific heat (J/kg K)	1.23E + 03

b. *Calculate the friction factor for enhanced or augmented surface tubes:* Even though the effect of the rib addition to the tube inner surface is to reduce the equivalent diameter on the S-CO_2 side and to increase the fluid velocity, there is no need to compute these parameters for the pressure drop calculation. This is because the Fanning friction factor in the Ravigururajan–Bergles correlation (Equation 9.100a) is defined with respect to the empty tube and the smooth tube parameters need therefore to be used. Let us check the validity of the Ravigururajan–Bergles correlation:

$$\mathrm{Re}_{CO_2} = 3.607 \times 10^5$$

$$\mathrm{Pr}_{CO_2} = \left(\frac{c_p\mu}{k}\right)_{CO_2} = \frac{1230 \times 3.52 \times 10^{-5}}{5.77 \times 10^{-2}} = 7.50 \times 10^{-1}$$

Check geometric conditions:

$$\frac{e}{D} = \frac{0.35}{16.0} = 0.022 \quad \frac{p}{D} = \frac{20.0}{16.0} = 1.25 \quad \frac{\alpha}{90} = \frac{27}{90} = 0.3$$

Hence, Re_{CO_2}, Pr_{CO_2}, and rib geometry are within correlation limits.

$$Y1 = 0.67 - 0.06 \times 1.25 - 0.49 \times 0.3 = 4.48 \times 10^{-1}$$
$$Y2 = 1.37 - 0.157 \times 1.25 = 1.174$$
$$Y3 = -1.66 \times 10^{-6} \times (3.607 \times 10^5) - 0.33 \times 0.3 = -6.98 \times 10^{-1}$$
$$Y4 = 4.59 + 4.11 \times 10^{-6} \times (3.607 \times 10^5) - 0.15 \times 1.25 = 5.886$$

$$f_a' = 3.49 \times 10^{-3} \left\{ 1 + \left[\begin{array}{l} 29.1 \times \left(3.607 \times 10^5\right)^{0.448} \times (0.022)^{1.174} \\ \times (1.25)^{-0.698} \times (0.3)^{5.886} \times \left(1 + \dfrac{2.94}{2}\right) \sin 90^\circ \end{array} \right]^{15/16} \right\}^{16/15}$$
$$= 4.23 \times 10^{-3}$$

The corresponding Darcy friction factor is

$$f_a = 4 \times 4.23 \times 10^{-3} = 1.69 \times 10^{-2}$$

c. The friction pressure drop in one of the tubes
Smooth:

$$\Delta p = f_{sm} \left(\frac{L}{D}\right) \frac{\rho V_o^2}{2} = 1.39 \times 10^{-2} \frac{5}{1.60 \times 10^{-2}} \frac{138.5 \times 5.73^2}{2} = 9.90 \text{ kPa}$$

Enhanced:

$$\Delta p = f_a \left(\frac{L}{D}\right) \frac{\rho V_o^2}{2} = 1.69 \times 10^{-2} \frac{5}{1.60 \times 10^{-2}} \frac{138.5 \times 5.73^2}{2} = 12.02 \text{ kPa}$$

Note that the pressure drop increases when ribs are introduced in the tubes. However, the tube length will decrease because of improved heat transfer as demonstrated in Example 10.3, which will result in a smaller pressure drop than calculated in this example.

9.5.6.2 Twisted Tape Inserts

Twisted tape inserts are the most widely used class of swirl-flow producing devices. They combine heat transfer performance enhancement with ease of manufacture and installation, as well as the possibility of easily retrofitting existing heat exchangers. Twisted tapes increase heat transfer by generating an enhanced fluid mixing centrifugal field in which colder, heavier fluid is pushed toward the tube walls and by increased axial flow velocity due to the partial area blockage by the tape. Minor contributions come from the fin-effect and by the longer (helical) flow length relative to a flow channel without such tube inserts. Twisted tapes can be used to enhance performance by increasing the heat transfer coefficient, as described in Section 10.5.2.2, and by also increasing the critical heat flux performance. This latter strategy has been applied to an inverted fuel assembly design as described by Malen et al. [27].

The friction factor correlations for twisted tape inserts, for laminar and turbulent flow, are reviewed in [30]. In the development of these correlations, the friction factor was defined as a Fanning factor with respect to the empty tube: this means that, once this factor is calculated, the computation of the friction pressure drop has to be performed according to Equation 9.99.

For laminar flow, Manglik and Bergles [28] recommended the following correlation:

$$f' = (f'\mathrm{Re})_s \frac{1}{\mathrm{Re}}\left(\frac{\pi}{\pi-(4\delta/D)}\right)\left[1+\left(\frac{\pi}{2y}\right)^2\right] \tag{9.101a}$$

for the following range of fluid conditions:

$43 < \mathrm{Re} < 2720$
$1.5 \le y \le \infty$
$0.019 \le \delta/D \le 0.119$

where

- f' is the Fanning friction factor;

$$(f'\mathrm{Re})_s = \left\{15.767\left[\frac{\pi+2-2(\delta/D)}{\pi-4(\delta/D)}\right]^2\right\}[1+10^{-6}\,\mathrm{Sw}^{2.55}]^{1/6} \tag{9.101b}$$

- Sw: swirl parameter, which accounts for the tape thickness δ, the tube inside diameter D, the twist ratio y and the absolute (resulting) fluid velocity:

$$V_s = V_0\left[1+\left(\frac{\pi}{2y}\right)^2\right]^{1/2} \tag{9.101c}$$

$$\mathrm{Sw} = \left(\frac{\mathrm{Re}}{\sqrt{y}}\right)\left\{\frac{\pi}{[\pi-4(\delta/D)]}\right\}\left[1+\left(\frac{\pi}{2y}\right)^2\right]^{1/2} \tag{9.101d}$$

- Re = Reynolds number of the tube assumed to be empty (= $\rho V_0 D/\mu$)
  H: tape half-pitch (axial length required for a 180° twist)
  y: twist ratio (= H/D)

In turbulent flow, Manglik and Bergles [28] recommended the following correlation:

$$f' = \frac{0.0791}{\mathrm{Re}^{0.25}}\left[1+\frac{2.752}{y^{1.29}}\right]\left[\frac{\pi}{\pi-(4\delta/D)}\right]^{1.75}\left[\frac{\pi+2-(2\delta/D)}{\pi-(4\delta/D)}\right]^{1.25} \tag{9.102}$$

where f' is again the Fanning friction factor for the following range of conditions:

$\mathrm{Re} \geq 10^4$
$2 \leq y \leq \infty$
$0.03 \leq \delta/D \leq 0.83$

For transition flows, Manglik and Bergles [29] recommended the following expression to provide a smooth transition between the results of Equations 9.101a and 9.102:

$$f' = \left(f_{\mathrm{L}}'^{10} + f_{\mathrm{T}}'^{10}\right)^{0.1} \tag{9.103}$$

They further provided [30] the following recommendation for bulk to wall viscosity (for liquids) and temperature (for gases) ratio factors to be applied to the adiabatic flow results above to accommodate diabatic flow conditions*:

For liquids:

$$f'_{\text{diabatic}} = f'_{\text{adiabatic}}\left(\frac{\mu_{\mathrm{w}}}{\mu_{\mathrm{b}}}\right)^{m} \tag{9.104a}$$

For constant heat flux conditions: $m = 0.61$ for heating, $m = 0.54$ for cooling. For constant wall temperature conditions: $m = 0.65$ for heating, $m = 0.58$ for cooling.

For gases:

$$f'_{\text{diabatic}} = f'_{\text{adiabatic}}\left(\frac{T_{\mathrm{w}}}{T_{\mathrm{b}}}\right)^{-0.1} \tag{9.104b}$$

where T_{w} and T_{b} are absolute wall and bulk temperatures, respectively.

---

* Typographical error in the exponent in Equation 9.104a of paper [30] corrected from $-m$ to m.

Example 9.7: Friction Pressure Loss for a Twisted Tape Insert

PROBLEM

In this example and in Chapter 10, we will explore the performance of a twisted tape in place of the ribbed surface of Example 9.6. The ribbed surface yielded a Darcy friction factor value of 1.69×10^{-2} and a pressure drop for the 5 m length tube of 12.02 kPa. What twist tape ratio must be maintained in the same smooth tube of inside diameter, $D = 16$ mm to match the ribbed tube pressure drop? Take the tape thickness as $\delta = 1$ mm.

SOLUTION

Solve Equation 9.102 for the twist tape ratio, y. Needed parameters from Example 9.6 and this example question are

$D = 16$ mm
$\delta/D = 1/16 = 0.0625$
$Re_{CO_2} = 3.607 \times 10^5$ (Reynolds number for an empty tube)
$f_a = 1.69 \times 10^{-2}$ (Darcy friction factor of ribbed-surface tubes)

Since the friction factors given by Equations 9.101a and 9.102, as well as that by Equation 9.100a are defined with respect to the empty tube (see discussion earlier in this section), the comparison between the friction pressure drop of ribbed-surface tubes and twisted tape tubes can be reduced to the comparison between the corresponding friction factors. Therefore, we need to set the Darcy friction factor of the twisted tape tubes equal to 1.69×10^{-2} calculated for the ribbed or augmented surface in Example 9.6:

$$1.69 \times 10^{-2} = \frac{4 \times 0.0791}{Re^{0.25}}\left(1 + \frac{2.752}{y^{1.29}}\right)\left(\frac{\pi}{\pi - (4\delta/D)}\right)^{1.75}\left(\frac{\pi + 2 - (2\delta/D)}{\pi - (4\delta/D)}\right)^{1.25}$$

$$= \frac{4 \times 0.0791}{\left(3.607 \times 10^5\right)^{0.25}}\left(1 + \frac{2.752}{y^{1.29}}\right)\left(\frac{\pi}{\pi - 4 \times 0.0625}\right)^{1.75} \times \left(\frac{\pi + 2 - 2 \times 0.0625}{\pi - 4 \times 0.0625}\right)^{1.25}$$

$$= \frac{4 \times 0.0791}{\left(3.607 \times 10^5\right)^{0.25}}\left(1 + \frac{2.752}{y^{1.29}}\right)\left(\frac{\pi}{\pi - 4 \times 0.0625}\right)^{1.75} \times \left(\frac{\pi + 2 - 2 \times 0.0625}{\pi - 4 \times 0.0625}\right)^{1.25}$$

$$= 2.97 \times 10^{-2}\left(1 + \frac{2.752}{y^{1.29}}\right)$$

From the above equation it can be seen that, regardless of the value of the twist ratio y, a twisted tape-provided tube cannot have a Darcy friction factor matching that of the ribbed-surface tube examined in Example 9.6. Particularly, the smallest possible Darcy friction factor for the present case, which is that corresponding to a straight tape ($y = \infty$), would be 2.97×10^{-2}.

9.5.7 Turbulent Form Losses

Loss coefficients for turbulent flows in typical flow geometry changes are given in Table 9.4. As noted, the coefficients given are approximate values. More precise values for most cases involve extended empirical relations which are dependent on flow configuration geometry, the ratio of the channel upstream to downstream areas, and Reynolds number. The reader is directed to handbook compilations (e.g., Idelchick [18]), should accurate coefficient values be desired.

The major pressure drop components for a channel with an inlet contraction and an exit expansion are shown in Figure 9.21. An incompressible flow condition is shown. For the abrupt area change regions, the subscript 1 denotes an upstream value

TABLE 9.4
Form Loss Coefficients for Turbulent Flow through Various Flow Geometries* for Evaluating Irrecoverable Pressure Loss

Parameter	K	Reference Velocity
Pipe entrance from a plenum		
Well-rounded entrance to pipe	0.04	In pipe
Slightly rounded entrance to pipe	0.23	In pipe
Sharp-edged entrance	0.50	In pipe
Projecting pipe entrance	0.5–1	In pipe
Pipe exit to a plenum	1.0	In pipe
Any pipe exit	$0.50\,(1-\beta)^{0.75}$	Downstream
Sudden changes in cross-sectional area	$(1-\beta)^2$	Upstream
Sudden contraction		
Sudden expansion		

where $\beta \equiv \dfrac{\text{small pipe cross-sectional area}}{\text{large pipe cross-sectional area}}$

Parameter	$(L^b/D)_{equiv}$	K	Reference Velocity
Fittings[a]			
90° standard elbow	30	0.35–0.9	In pipe
90° long-radius elbow	20	0.2–0.6	In pipe
45° standard elbow	16	0.17–0.45	In pipe
Standard tee (flow through run)	20	0.2–0.6	Downstream
Standard tee (flow through branch)	60	0.65–1.70	Downstream
Valves (various types)			
Fully open		0.15–15	In pipe
Half-closed		13–450	In pipe

* Approximate values; consult Idelchik [18] for extensive tabulation. Also, Section 9.6 gives an accounting for the theoretical basis for obtaining K for sudden contraction and expansion.

[a] Values of K depend on the pipe diameter and surface condition.

[b] Upstream of fitting.

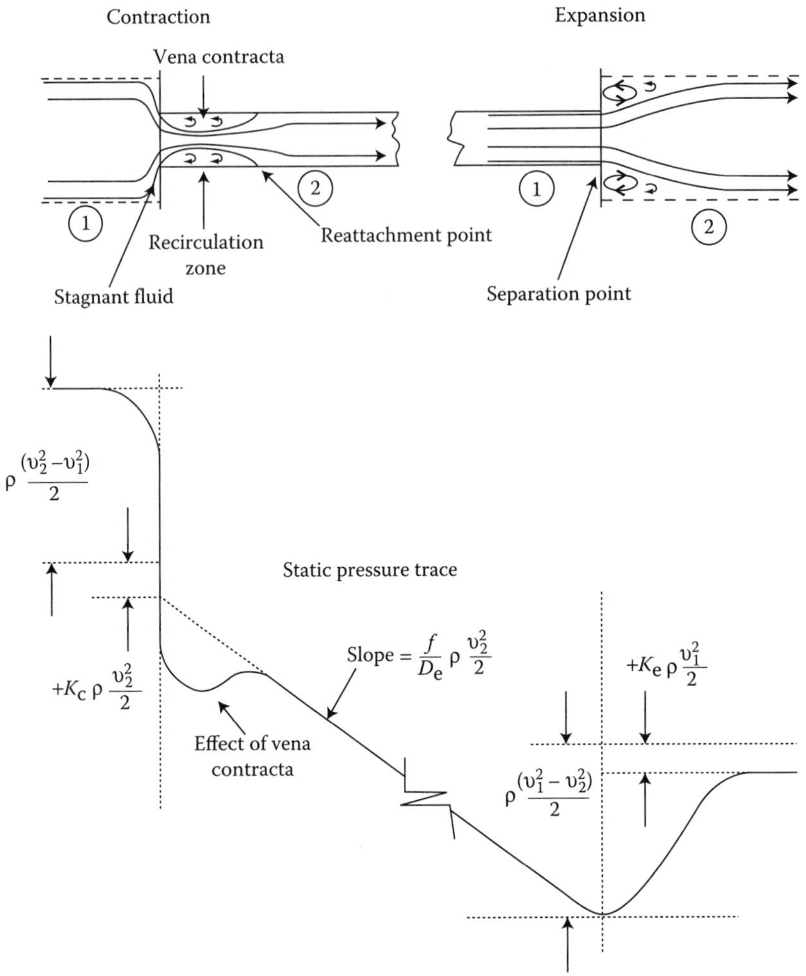

FIGURE 9.21 Flow of a viscous, incompressible fluid (which could be laminar or turbulent) through a sudden expansion or contraction.

and the subscript 2, a downstream value. Unsubscripted variables are constants in the flow.

As Figure 9.21 illustrates, the total static pressure change at sudden channel area changes is composed of flow acceleration and deceleration components and irrecoverable losses. The unrecoverable losses can be added to the acceleration/deceleration static pressure loss and gain terms re-expressed from Equation 9.22 in terms of the flow area changes as:

For contraction:

$$\rho\left(\frac{\upsilon_2^2-\upsilon_1^2}{2}\right)+K_c\frac{\rho\upsilon_2^2}{2}=\frac{\rho\upsilon_2^2}{2}\left[\left(1-\beta^2\right)+K_c\right] \tag{9.22c}$$

For expansion:

$$\rho\left(\frac{\upsilon_1^2 - \upsilon_2^2}{2}\right) - K_e\frac{\rho\upsilon_1^2}{2} = \frac{\rho\upsilon_1^2}{2}\left[\left(1-\beta^2\right) - K_e\right] \tag{9.22d}$$

where

$$\beta \equiv \frac{\text{small pipe cross-sectional area}}{\text{large pipe cross-sectional area}}$$

The irrecoverable losses are characterized by the coefficients K_c and K_e, which are listed in Table 9.4 for single channels. For multichannel systems connected to headers, the irrecoverable loss coefficients are presented in Section 9.6.4.

Example 9.8: Pump Start-Up for a Viscous Flow Loop

PROBLEM

Let us return to the problem of Example 9.1 and solve for the time-dependent flow rate, taking friction and form losses into account. Equation 9.37 is now specified for our horizontal system, so it becomes

$$\left(\frac{\ell}{A}\right)_T \frac{d\dot{m}}{dt} - \Delta p + \frac{\dot{m}^2}{2\rho}\left(\frac{1}{A_5^2} - \frac{1}{A_1^2}\right) + K_R\frac{\rho\upsilon_R^2}{2} + K_{SG}\frac{\rho\upsilon_{SG}^2}{2} + \sum_i f\frac{L_i}{D_i}\frac{\rho\upsilon_i^2}{2} = 0$$

where K_R and K_{SG} = form pressure loss coefficients for the reactor and steam generator, respectively, and the friction pressure loss term is a summation over the different pipes in the system assuming a constant friction factor (f). Writing it in terms of the total mass flow rate, we find

$$\left(\frac{\ell}{A}\right)_T \frac{d\dot{m}}{dt} - \Delta p + \frac{\dot{m}^2}{2\rho}\left(\frac{1}{A_5^2} - \frac{1}{A_1^2} + \frac{K_R}{A_R^2} + \frac{K_{SG}}{A_{SG}^2} + \sum_i f\frac{L_i}{D_i}\frac{1}{A_i^2}\right) = 0$$

Taking $K_R = 18$, $K_{SG} = 52$, $f = 0.015$, $\rho = 1000\ \text{kg/m}^3$, and the pump head as 85.3 m, and using the information on lengths and diameters from the previous problem, what is the asymptotic value of the flow rate?

SOLUTION

Note that this differential equation is of the same form as the one solved in the previous problem. The solution to the equation is still:

$$\dot{m}(t) = \frac{1}{C}\left[\frac{\exp\left[2C\Delta p\left(\frac{A}{\ell}\right)_\tau t\right] - 1}{\exp\left[2C\Delta p\left(\frac{A}{\ell}\right)_\tau t\right] + 1}\right]$$

where now

$$\left(\frac{A}{\ell}\right)_T = \left(\frac{1}{102.8}\text{m}\right) \quad \text{from Example 9.1.}$$

$$\begin{aligned}\Delta p &= 85.3\,\text{m} \times 9.81\,\text{m/s}^2 \times 1000\,\text{kg/m}^3 \\ &= 8.37 \times 10^5\,\text{kg/ms}^2\end{aligned}$$

again as in Example 9.1.

Evaluating C, we find

$$C^2 = \frac{1}{2\rho\Delta p}\left(\frac{1}{A_5^2} - \frac{1}{A_1^2} + \frac{K_R}{A_R^2} + \frac{K_{SG}}{A_{SG}^2} + \sum_i f\frac{L_i}{D_i}\frac{1}{A_i^2}\right)$$

$$\begin{aligned}C^2 &= \frac{1}{2(1000)(8.37\times 10^5)} \\ &\quad \times\left\{\frac{1}{(0.35)^2} - \frac{1}{(0.4)^2} + \frac{18}{(20.9)^2} + \frac{52}{(1.5)^2} + 0.015\right. \\ &\quad \left.\times\left[\frac{8}{0.714}\frac{1}{(0.4)^2} + \frac{17}{0.714}\left(\frac{1}{0.4}\right)^2 + \frac{10}{0.668}\left(\frac{1}{0.35}\right)^2\right]\right\} \\ &= 1.803 \times 10^{-8}\,\text{s}^2/\text{kg}^2\end{aligned}$$

The time-dependent flow rate, $\dot{m}(t)$, is now available, but again note the mass of the pump rotor has been assumed to be zero. At steady state, the flow rate is

$$\dot{m}(\text{steady state}) = \frac{1}{C} = 7447\,\text{kg/s}$$

The steady-state flow rate is now in more reasonable agreement with that of a typical PWR primary pump.

9.6 PRESSURE DROP IN ROD BUNDLES

The principal pressure drop components along a reactor cooling channel under nominal operating conditions include (1) the friction pressure drop along the fuel rods, (2) the form losses due to the presence of spacers, and (3) entrance and exit pressure losses between the vessel plena and the core internals. The inlet loss can include orifice pressure loss for those assemblies orificed to achieve a balance of power to flow radially across the core. These losses are treated next in that sequence.

9.6.1 Friction Loss along Bare Rod Bundles

9.6.1.1 Laminar Flow

In Section 9.4.2, it was demonstrated that friction factors for noncircular geometries in laminar flow could not be obtained by transforming circular tube results using the equivalent diameter concept. However, an alternate approximate method follows from the observation that the coolant region in rod arrays can be represented as an array of equivalent annuli around the rods. The equivalent annulus geometry is defined in Figure 9.22. As Figure 9.23 suggests, the equivalent annulus approximation improves as the rod spacing increases.

The solutions of the momentum equations for the exact bare rod array geometry and for the equivalent annulus approximation have been obtained for laminar and turbulent flow. The solution procedure and a number of significant special cases are presented in Chapter 7, Volume II. Here, the most commonly encountered cases are summarized.

In this section, we use the superscripted symbol Re′ to refer to the Reynolds number in a bare (spacerless) rod bundle. In Section 9.6.2, we use Re to refer to the Reynolds number in a bundle with spacers.

Figure 9.23 presents the product $f\,\mathrm{Re}'_{\mathrm{De}}$ for fully developed laminar flow in a triangular array derived by Sparrow and Loeffler [46] and for the equivalent annulus. The equivalent annulus approximation is seen as satisfactory for pitch-to-diameter (*P*/*D*) ratios greater than about 1.3. A complete set of laminar results are available from Rehme [40]. His results have been fit by Cheng and Todreas [4] with polynomials for each subchannel type. The polynomials, which have the form:

$$C'_{\mathrm{fiL}} = a + b_1(P/D - 1) + b_2(P/D - 1)^2 \tag{9.105a}$$

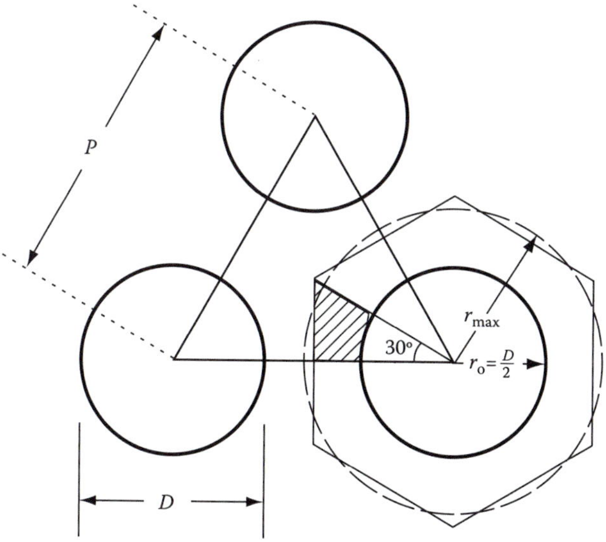

FIGURE 9.22 Definition of an equivalent annulus in a triangular array. *Cross-hatched area* represents an elemental coolant flow section. Circle of radius r_{max} represents the equivalent annulus with equal flow area.

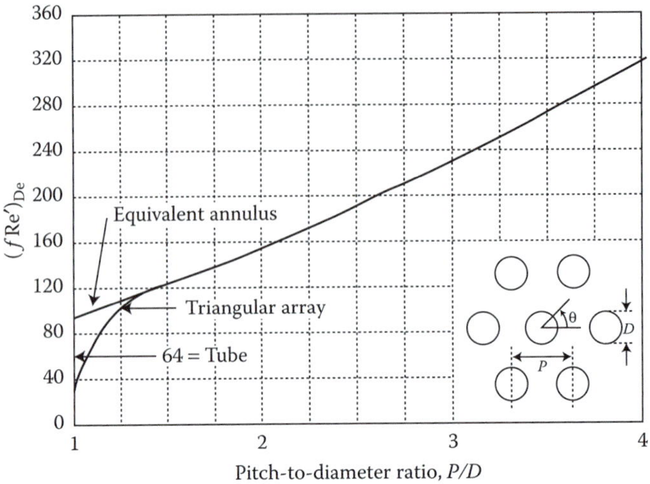

FIGURE 9.23 Product of laminar friction factors and Reynolds number for parallel flow in an infinite triangular array of rods. (From Sparrow, E. M. and Loeffler, A. L. Jr., *A.I.Ch.E. J.*, 5:325, 1959.)

can then be used to obtain the friction factor corresponding to the subchannel type "i" to which they are referred, as

$$f_{\mathrm{iL}} \equiv \frac{C'_{\mathrm{fiL}}}{\mathrm{Re}'^{n}_{\mathrm{iL}}} \tag{9.105b}$$

where $n = 1$ for laminar flow. When Equation 9.105a is used for edge and corner subchannels, P/D is replaced by W/D, where W = rod diameter plus gap between rod and bundle wall. The effect of P/D (or W/D) was separated into two regions; $1.0 \le P/D \le 1.1$ and $1.1 \le P/D \le 1.5$. Tables 9.5 and 9.6 present the coefficients a, b_1, and b_2 for the subchannels of hexagonal and square arrays, respectively.

Bundle average friction factors are obtained from the subchannel friction factors by assuming that the pressure difference across all subchannels is equal and applying the mass balance condition for total bundle flow in terms of the subchannel flow. This procedure, which is demonstrated in Chapter 4, Volume II, yields

$$C'_{\mathrm{bL}} = D'_{\mathrm{eb}}\left[\sum_{i=1}^{3} S_{\mathrm{i}}\left(\frac{D'_{\mathrm{ei}}}{D'_{\mathrm{eb}}}\right)^{\frac{n}{2-n}}\left(\frac{C'_{\mathrm{fi}}}{D'_{\mathrm{ei}}}\right)^{\frac{1}{n-2}}\right]^{n-2} \tag{9.106}$$

where S_{i} is the ratio of the total flow area of subchannels of type i to the bundle flow area, while D_{eb} and D_{ei} are the equivalent diameters of the bundle and of the subchannel type "i," respectively. Figure 9.24 compares Equation 9.106 to the available data for laminar flow ($n = 1$) in a 37-pin triangular array.

TABLE 9.5
Coefficients in Equations 9.105a and 9.109a for Bare Rod Subchannel Friction Factor Constants C'_n in Hexagonal Array

	$1.0 \le P/D \le 1.1$			$1.1 \le P/D \le 1.5$		
Subchannel	a	b_1	b_2	a	b_1	b_2
			Laminar Flow			
Interior	26.00	888.2	−3334	62.97	216.9	−190.2
Edge	26.18	554.5	−1480	44.40	256.7	−267.6
Corner	26.98	1636	−10050	87.26	38.59	−55.12
			Turbulent Flow			
Interior	0.09378	1.398	−8.664	0.1458	0.03632	−0.03333
Edge	0.09377	0.8732	−3.341	0.1430	0.04199	−0.04428
Corner	0.1004	1.625	−11.85	0.1499	0.006706	−0.009567

9.6.1.2 Turbulent Flow

Early work in the area of turbulent flow includes that of Deissler and Taylor [7], who derived friction factors that depend on a universal velocity profile obtained from early measurements. Their approach was compared, along with the equivalent diameter concept, to measurements in square and triangular rod bundles with $P/D = 1.12$, 1.20, and 1.27. The results show that the circular tube prediction of Equation 9.88 for $\text{Re} > 100{,}000$ provides an answer that lies within the scatter of the data. However, the data show a dependence on P/D that cannot be reproduced by the equivalent diameter concept and the circular tube correlation. LeTourneau et al. [25] tested rod

TABLE 9.6
Coefficients in Equations 9.105a and 9.109a for Bare Rod Subchannel Friction Factor Constants C'_{fi} in Square Array

	$1.0 \le P/D \le 1.1$			$1.1 \le P/D \le 1.5$		
Subchannel	a	b_1	b_2	a	b_1	b_2
			Laminar Flow			
Interior	26.37	374.2	−493.9	35.55	263.7	−190.2
Edge	26.18	554.5	−1480	44.40	256.7	−267.6
Corner	28.62	715.9	−2807	58.83	160.7	−203.5
			Turbulent Flow			
Interior	0.09423	0.5806	−1.239	0.1339	0.09059	−0.09926
Edge	0.09377	0.8732	−3.341	0.1430	0.04199	−0.04428
Corner	0.09755	1.127	−6.304	0.1452	0.02681	−0.03411

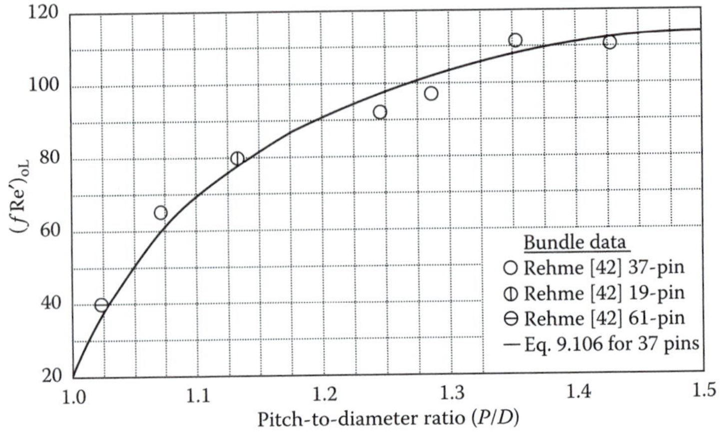

FIGURE 9.24 Laminar flow results in triangular array bare rod bundles. (From Cheng, S. K., and Todreas, N. E., *Nucl. Eng. Des.*, 92(2):227–251, 1986.)

bundles of square lattice with *P/D* ratios of 1.12 and 1.20 and of triangular lattice with a *P/D* ratio of 1.12. These data fall within a band between the smooth tube prediction and a curve 10% below that for Re = 3×10^3 to 3×10^5.

Later, Trupp and Azad [50] obtained velocity distributions, eddy diffusivities, and friction factors with airflow in triangular array bundles. These data indicated friction factors somewhat higher than Deissler and Taylor's predictions. For Reynolds numbers between 10^4 and 10^5, their data at *P/D* = 1.2 were about 17% higher than the circular tube data. The data at *P/D* = 1.5 were about 27% higher than the circular tube data.

For the turbulent flow situation, solution of both the exact and the equivalent annulus geometry requires, assumptions about the turbulent velocity distribution. For a triangular array, Rehme [42] obtained the following equivalent annulus solutions:

For $\mathrm{Re}'_{\mathrm{De}} = 10^4$:

$$\frac{f}{f_{\mathrm{c.t.}}} = 1.045 + 0.071(P/D - 1) \tag{9.107}$$

For $\mathrm{Re}'_{\mathrm{De}} = 10^5$:

$$\frac{f}{f_{\mathrm{c.t.}}} = 1.036 + 0.054(P/D - 1) \tag{9.108}$$

where $f_{\mathrm{c.t.}}$ = circular tube friction factor.

Rehme [43] also proposed a method for solving the turbulent flow case in the actual geometry. Cheng and Todreas [4] fitted results of this method with the polynomial of the form of Equation 9.105a, where now:

$$C'_{\mathrm{fiT}} = a + b_1(P/D - 1) + b_2(P/D - 1)^2 \tag{9.109a}$$

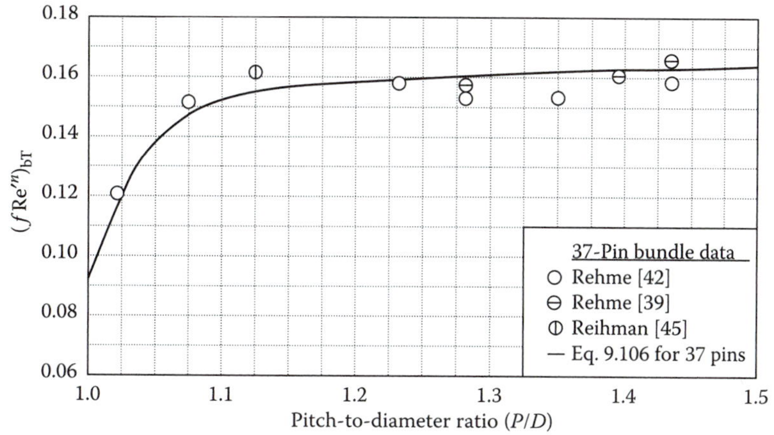

FIGURE 9.25 Turbulent flow results in triangular array bare rod bundles. (From Cheng, S. K. and Todreas, N. E., *Nucl. Eng. Des.*, 92(2):227–251, 1986.)

where

$$f_{\mathrm{iT}} \equiv \frac{C'_{\mathrm{fiT}}}{(\mathrm{Re}'_{\mathrm{iT}})^{n}} \tag{9.109b}$$

and $n = 0.18$. Tables 9.5 and 9.6 list the coefficients a, b_1, and b_2 for subchannels of hexagonal and square arrays. The turbulent bundle friction factor constant C'_{bT} can be obtained from an equation of the form of Equation 9.106 but where C'_{fi} and n are turbulent values, C'_{fi} from Equation 9.109a and $n = 0.20$ or 0.25 from Equation 9.87 or 9.88. Figure 9.25 compares this bundle friction factor for a 37-pin triangular array with available data.

9.6.2 Pressure Loss at Fuel Pin Spacer and Support Structures

Pressure losses across spacer grids or wires (Figure 9.26) are form drag-type, pressure losses that can be calculated using pressure-loss coefficients. The pressure drop across the spacers can be comparable in magnitude to the friction along the bare rod bundle.

9.6.2.1 Grid Spacers

De Stordeur [8] measured the pressure drop characteristics of a variety of spacers and grids. He correlated his results in terms of a modified drag coefficient (C_{s}). The pressure drop (Δp_{s}) across the grid or spacer is given by

$$\Delta p_{\mathrm{s}} = C_{\mathrm{s}}(\rho \upsilon_{\mathrm{s}}^{2}/2)(A_{\mathrm{s}}/A_{\mathrm{v}}) \tag{9.110}$$

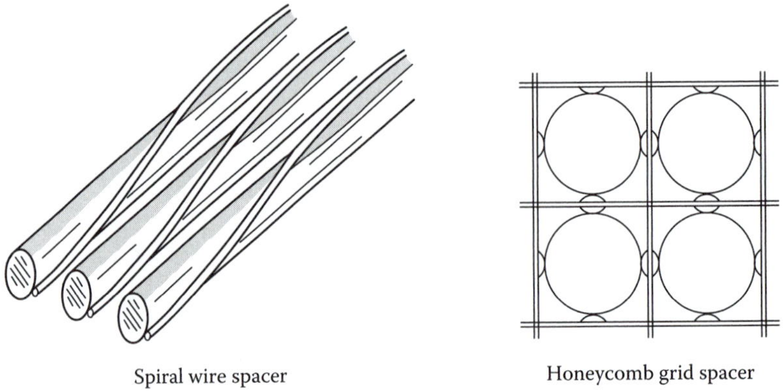

FIGURE 9.26 Rod bundle fuel element spacers.

where υ_s = velocity in the spacer region; A_v = unrestricted flow area away from the grid or spacer; and A_s = projected frontal area of the spacer.

The grid drag coefficient is a function of the Reynolds number for a given spacer or grid type. At high Reynolds number ($Re \cong 10^5$) as shown in Figure 9.27, honeycomb grids showed drag coefficients of ≈1.65. The test results indicated that the loss coefficient proved independent of the grid strap height. The grid strap thickness which directly affects the form loss characteristic appears to have been used for the parameter D_{es} in defining the applicable Reynolds number in Figure 9.27. The pressure drop across the grid made of circular wires was ≈10% lower.

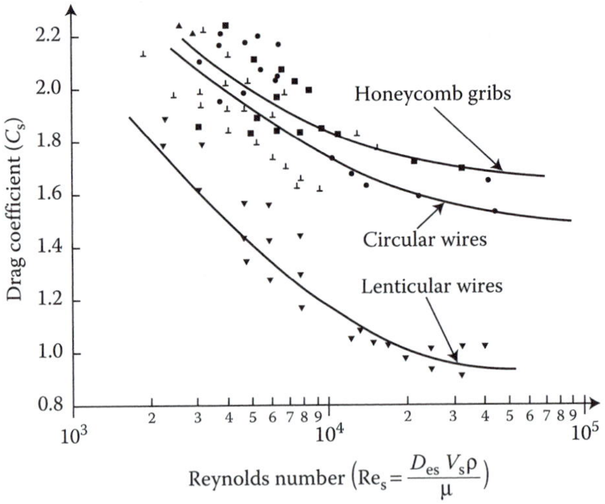

FIGURE 9.27 Drag coefficients for rod bundle transverse grid spacers. D_{es}: wire diameter for circular and lenticular wire grids; and grid strap thickness for honeycomb grids. (From De Stordeur, A. M., *Nucleonics*, 19(6):74, 1961.)

On the basis of tests of several grids, Rehme [41] found that the effect of the ratio A_s/A_v is more pronounced than was indicated by de Stordeur. Rehme concluded that grid pressure drop data are better correlated by

$$\Delta p_s = C_v(\rho \upsilon_v^2/2)(A_s/A_v)^2 \tag{9.111}$$

where C_v = modified drag coefficient, and υ_v = average bundle fluid velocity away from the grid.

The drag coefficient, C_v, is a function of the average bundle, unrestricted area Reynolds number. Rehme's data indicated that for square arrays $C_v = 9.5$ at $Re = 10^4$ and $C_v = 6.5$ at $Re = 10^5$ (Figure 9.28). The data range of Spengos [47] shown in Figure 9.28 was used by De Stordeur in establishing his correlation. Hence, it can be

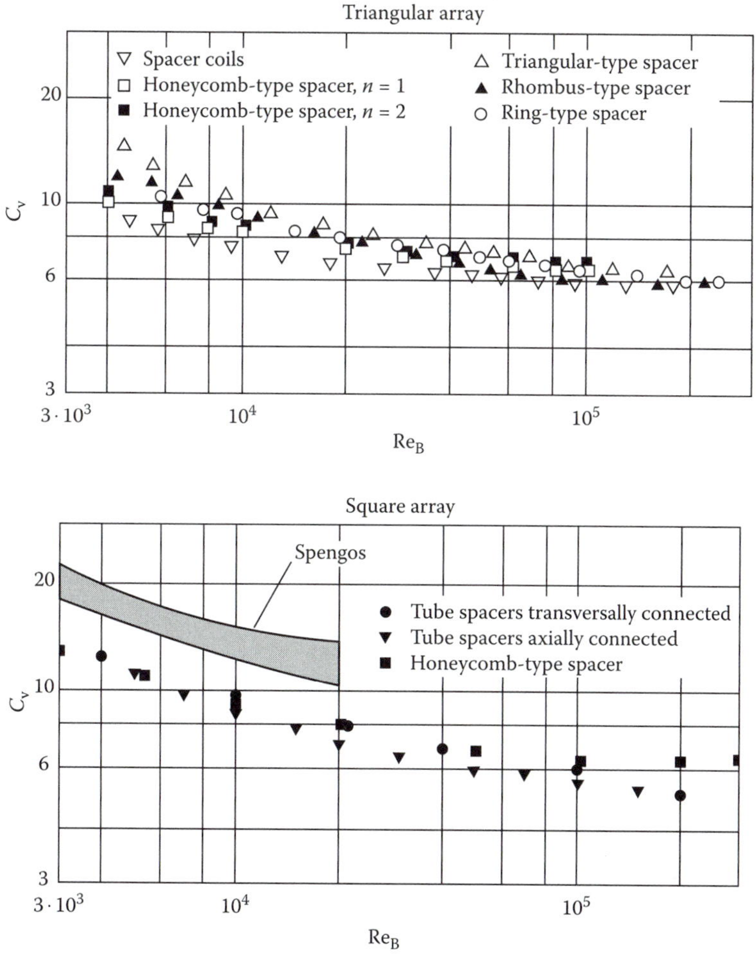

FIGURE 9.28 Modified drag coefficients. (From Rehme, K., *Nucl. Technol.*, 17:15, 1973.)

concluded that predicted pressure drops from De Stordeur are somewhat larger than those of Rehme under the same flow and area ratio, A_s/A_v, conditions.

More recently, pressure loss correlations for PWR grid spacers have been published by In et al. [19]. In his work, In defines the grid spacer loss coefficient with respect to the fluid velocity away from the grid:

$$\Delta p_{\text{grid}}^{\text{In}} = K_{\text{grid}}^{\text{In}} \frac{\rho \upsilon_{\text{v}}^2}{2} \tag{9.112}$$

where the grid spacer loss coefficient $K_{\text{grid}}^{\text{In}}$ is computed as the sum of three terms: grid form loss (term A), grid friction loss (term B) and rod friction loss within the spacer region (term C). As will be shown in Example 9.9 for a honeycomb grid of typical geometry under PWR conditions, terms A and C are about equal and an order of magnitude greater than term B.

The formulation for the loss coefficient $K_{\text{grid}}^{\text{In}}$ is

$$K_{\text{grid}}^{\text{In}} = \left[C_{\text{grid}}^{\text{form}} \frac{\varepsilon}{(1-\varepsilon)^2} \right]_{\text{A}} + \left[C_{\text{grid}}^{\text{fric}} \frac{A_{\text{grid, wetted}}}{A_{\text{flow, bundle}}} \frac{1}{(1-\varepsilon)^2} \right]_{\text{B}} + \left[C_{\text{rod}}^{\text{fric}} \frac{A_{\text{rods, wetted@grid}}}{A_{\text{flow, bundle}}} \frac{1}{(1-\varepsilon)^2} \right]_{\text{C}} \tag{9.113}$$

where:

- Term A:

$$C_{\text{grid}}^{\text{form}} = 2.75 - 0.27 \log_{10}(\text{Re})_{\text{away from grid}} \tag{9.114a}$$

  - ε: ratio between the total projected grid cross section (frontal grid area) and the bundle flow area away from the grid:

$$\varepsilon = \left(\frac{A_{\text{grid, frontal}}}{A_{\text{flow}}} \right)_{\text{bundle}} \tag{9.114b}$$

- Term B:

$$C_{\text{grid}}^{\text{fric}} = C_{\text{grid,lam}}^{\text{fric}} \frac{3 \times 10^4 \mu_{\text{avg}}}{G_{\text{@grid}} H} + C_{\text{grid, turb}}^{\text{fric}} \frac{\left(H - (3 \times 10^4 \mu_{\text{avg}} / G_{\text{@grid}}) \right)}{H} \quad \text{for } H \geq \frac{3 \times 10^4 \mu_{\text{avg}}}{G_{\text{@grid}}} \tag{9.115a}$$

$$= C_{\text{grid, lam}}^{\text{fric}} \frac{3 \times 10^4 \mu_{\text{avg}}}{G_{\text{@grid}} H} \quad \text{for } H < \frac{3 \times 10^4 \mu_{\text{avg}}}{G_{\text{@grid}}} \tag{9.115b}$$

where H is the grid strap height in meters

$$C^{\text{fric}}_{\text{grid,lam}} = 1.328\left\{G_{@\text{grid}}\frac{\left(H-(3\times10^{4}\mu_{\text{avg}}/G_{@\text{grid}})\right)}{\mu_{\text{avg}}}\right\}^{-1/2} \tag{9.115c}$$

$$C^{\text{fric}}_{\text{grid,turb}} = 0.523\left\{\ln\left[0.06\times G_{@\text{grid}}\frac{\left(H-(3\times10^{4}\mu_{\text{avg}}/G_{@\text{grid}})\right)}{\mu_{\text{avg}}}\right]\right\}^{-2} \tag{9.115d}$$

- $A_{\text{grid,wetted}}$: total wetted area of the grid straps
- $A_{\text{flow,bundle}}$: total flow area of the bundle, away from the grid

- Term C:

$$C^{\text{fric}}_{\text{rod}} = 0.184\,\text{Re}^{-0.2}_{@\text{grid}} \tag{9.116}$$

- $A_{\text{rod,wetted @ grid}}$: total wetted area of rods at the grid ($= N_{\text{rods}} \times \pi D \times$ grid height)

The effect of the mixing vanes on the pressure drop in the mechanistic correlation was formulated by means of an additional term that, if mixing vanes are present, needs to be added to the right-hand side of Equation 9.113. This term is formulated by multiplication of an empirical form drag coefficient for abrupt flow blockage and a relative plugging of the flow cross-section by the mixing vanes. It is assumed by In that the pressure drop by the mixing vane largely depends on the plugging of the flow cross-section. Hence, similar to the form loss coefficient of the spacer grid, the pressure loss coefficient of the mixing vanes is given by

$$C_{\text{mv}} = C_{\text{d,mv}}\frac{\varepsilon_{\text{mv}}}{(1-\varepsilon_{\text{mv}})^2} \tag{9.117}$$

where

$C_{\text{d,}}\text{m}_v$: empirical form drag coefficient which, according to In, does not show noticeable dependence on the Reynolds number but rather a small variation in the range 0.6–0.8. A value of 0.72 is recommended by In.

$\varepsilon_{\text{mv}} = A_{\text{mv}}/A_{\text{o}}$ (ratio of the total plugging area of the mixing vanes to the bundle flow area away from the grid).

The pressure loss coefficients estimated by the correlations are compared with the measured data for spacer grids with and without mixing vanes. Since the measurements provided the spacer loss coefficients for a Reynolds number < 100,000, the measured data were also extrapolated to a Reynolds number of 500,000 to show comparisons at a nominal PWR-operating condition.

In's empirical correlations without mixing vanes (Equation 9.113) and with mixing vanes (Equation 9.113 plus Equation 9.117) result in pressure loss coefficients that agree within 10% of the experimental data. The form loss due to the mixing vane accounts for about 15% of the grid loss.

Example 9.9: Pressure Loss at Spacers

PROBLEM

For the PWR fuel assembly conditions of Tables 1.2, 1.3, and 2.1, calculate the pressure drop across the spacers. Use In's, de Stordeur's, and Rehme's correlations for spacer pressure drop and compare the results.

Assume the spacer type is honeycomb, and there are eight identical spacers along the fuel length. Take the grid strap thickness (t) equal to 0.45×10^{-3} m and the grid strap height (H) equal to 40×10^{-3} m. Figure 9.29 illustrates the unit cell to be considered. Relevant data:

$D = 9.5$ mm
$P = 12.60$ mm
$T_{\text{coolant,avg}} = 307.5°\text{C}$
Average pressure = 15.5 MPa
$\rho_{\text{avg}} = 710.4 \text{ kg/m}^3$
$\mu_{\text{avg}} = 8.557 \times 10^{-5}$ Pa s

From Tables 2.1, 1.3, and 1.2, respectively, find the PWR parameters of 193 assemblies, 17 × 17 rod array, total core flow 17476 kg/s. Assume

- Guide thimbles as solid rods having same diameter as fuel rods.
- Grid spacers without mixing vanes.*
- Same coolant properties at all the axial elevations corresponding to the grids, equal to the average properties given above.

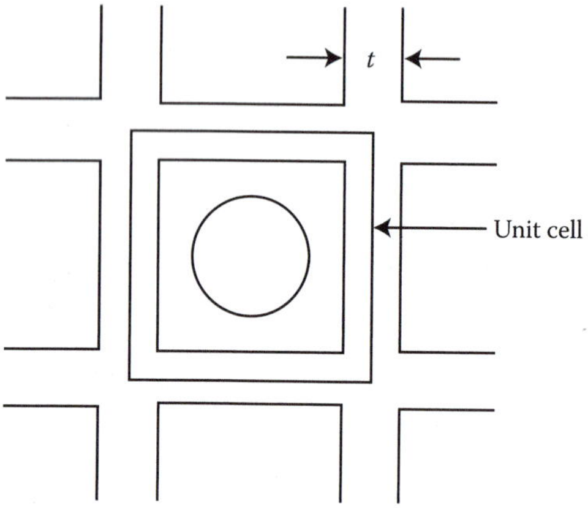

FIGURE 9.29 Unit cell analysis in Example 9.9.

* While In's model allows accounting for mixing vanes, the de Stordeur and Rehme models were developed for grid spacers not provided with mixing vanes.

SOLUTION

Some general parameters need first to be computed. The subscript "@grid" is used to distinguish parameters calculated at the grid from those calculated away from the grid.

Grid projected frontal area per unit cell:

$$A_{\text{grid,frontal}} = P^2 - (P-t)^2 = 1.11375 \times 10^{-5}\ \text{m}^2$$

Subchannel area:

$$A_{\text{subch}} = P^2 - 0.25\pi D^2 = 8.78778 \times 10^{-5}\ \text{m}^2$$

$$A_{\text{subch@grid}} = A_{\text{subch}} - A_{\text{grid,frontal}} = 7.67403 \times 10^{-5}\ \text{m}^2$$

Subchannel equivalent diameter:

$$D_{\text{eq}} = \frac{4A_{\text{subch}}}{P_{\text{w}}} = \frac{4A_{\text{subch}}}{\pi D} = 11.778 \times 10^{-3}\ \text{m}$$

$$D_{\text{eq@grid}} = \frac{4A_{\text{subch@grid}}}{P_{\text{w@grid}}} = 4\frac{A_{\text{subch@grid}}}{\pi D + 4(P-t)} = 3.913 \times 10^{-3}\ \text{m}$$

Average mass flux:

$$G = \frac{\dot{m}_{\text{subch}}}{A_{\text{subch}}} = \frac{(\dot{m}_{\text{core}}/N_{\text{subch}})}{A_{\text{subch}}} = \frac{(17{,}476/193 \times 17^2)}{8.78778 \times 10^{-5}} = 3565$$

$$G_{\text{@grid}} = \frac{\dot{m}_{\text{subch}}}{A_{\text{subch@grid}}} = \frac{(\dot{m}_{\text{core}}/N_{\text{subch}})}{A_{\text{subch@grid}}} = \frac{(17{,}476/193 \times 17^2)}{7.67403 \times 10^{-5}} = 4083\ \text{kg/sm}^2$$

Reynolds number:

$$\text{Re} = \frac{GD_{\text{eq}}}{\mu_{\text{avg}}} = \frac{3565 \times 11.778 \times 10^{-3}}{8.557 \times 10^{-5}} = 4.907 \times 10^5$$

$$\text{Re}_{\text{@grid}} = \frac{G_{\text{@grid}} D_{\text{eq@grid}}}{\mu_{\text{avg}}} = \frac{4083 \times 3.913 \times 10^{-3}}{8.557 \times 10^{-5}} = 1.867 \times 10^5$$

Criteria of Equation 9.115:

$$\frac{3 \times 10^4 \mu_{\text{avg}}}{G_{\text{@grid}}} = \frac{3 \times 10^4 (8.557 \times 10^{-5})}{4083} = 0.63 \times 10^{-3} < H$$

In's model

In [19] defines the grid spacer loss coefficient based on fluid velocity, undisturbed by the grid so that

$$\Delta p_{\text{grid}}^{\text{In}} = K_{\text{grid}}^{\text{In}} \frac{G^2}{2\rho_{\text{avg}}} \tag{9.112}$$

Particularly, for spacer grids not provided with mixing vanes In's correlation computes the grid spacer loss coefficient $K_{\text{grid}}^{\text{In}}$ as sum of three terms: grid form loss, grid friction loss, and rod friction loss within the grid region, following Equation 9.113:

$$K_{\text{grid}}^{\text{In}} = \left[C_{\text{grid}}^{\text{form}} \frac{\varepsilon}{(1-\varepsilon)^2} \right] + \left[C_{\text{grid}}^{\text{fric}} \frac{A_{\text{grid,wetted}}}{A_{\text{subch}}} \frac{1}{(1-\varepsilon)^2} \right] + \left[C_{\text{rod}}^{\text{fric}} \frac{A_{\text{rod,wetted@grid}}}{A_{\text{subch}}} \frac{1}{(1-\varepsilon)^2} \right] \tag{9.113}$$

The following parameters are inputs to Equation 9.113.

$$C_{\text{grid}}^{\text{form}} = 2.75 - 0.27\log_{10}\text{Re} = 2.75 - 0.27\log_{10}\left(4.907 \times 10^5\right) = 1.21348 \tag{9.114a}$$

ε: ratio between the total projected grid frontal area and the bundle flow area. Since the problem asks, for simplicity, for a unit cell calculation:

$$\varepsilon = \frac{A_{\text{grid,frontal}}}{A_{\text{subch}}} = 0.12674 \tag{9.114b}$$

$$(1-\varepsilon)^2 = 0.7626$$

$$H > \frac{3 \times 10^4 \mu_{\text{avg}}}{G_{@\text{grid}}}$$

since $40 \times 10^{-3} > 0.63 \times 10^{-3}$.

Hence $C_{\text{grid}}^{\text{fric}}$ is given by Equation 9.115a requiring calculation of $C_{\text{grid,lam}}^{\text{fric}}$ and $C_{\text{grid,turb}}^{\text{fric}}$

$$C_{\text{grid,lam}}^{\text{fric}} = 1.328\left[4083\frac{\left(40 \times 10^{-3} - 0.63 \times 10^{-3}\right)}{8.557 \times 10^{-5}}\right]^{-1/2} \tag{9.115c}$$

$$= 9.6892 \times 10^{-4}$$

$$C_{\text{grid,turb}}^{\text{fric}} = 0.523\left\{\ell n\left[0.06(4083)\frac{\left(40 \times 10^{-3} - 0.63 \times 10^{-3}\right)}{8.557 \times 10^{-5}}\right]\right\}^{-2} \tag{9.115d}$$

$$= 3.8650 \times 10^{-3}$$

yielding

$$C_{grid}^{fric} = 9.6892 \times 10^{-4}\left(\frac{0.63 \times 10^{-3}}{40 \times 10^{-3}}\right) + 3.8650 \times 10^{-3}\left(\frac{40 \times 10^{-3} - 0.63 \times 10^{-3}}{40 \times 10^{-3}}\right) = 3.8194 \times 10^{-3} \quad (9.115a)$$

$$C_{rod}^{fric} = 0.184 \mathrm{Re}_{@grid}^{-0.2} \quad (9.116)$$

$$A_{grid,wetted} = 4(P - t)H = 1.9440 \times 10^{-3}\,\mathrm{m}^2$$

$$A_{rod,wetted@grid} = \pi DH = 1.1938 \times 10^{-3}\,\mathrm{m}^2$$

By introducing the numerical values just calculated into the formula for K_{grid}^{In}, we get

$$K_{grid}^{In} = \left[1.21348\frac{0.1267}{0.7626} + 3.8194 \times 10^{-3}\frac{1.9440 \times 10^{-5}}{8.7878 \times 10^{-5}}\frac{1}{0.7626} + 1.624 \times 10^{-2}\frac{1.1938 \times 10^{-3}}{8.7878 \times 10^{-5}}\frac{1}{0.7626}\right] = 0.2016 + 1.108 \times 10^{-3} + 0.2893 = 0.4920 \quad (9.113)$$

The total pressure drop across the eight grids can be calculated as

$$\Delta p_{grids}^{In} = 8K_{grid}^{In}\frac{G^2}{2\rho_{avg}} = 8(0.4920)\left[\frac{(3565)^2}{2(710.4)}\right] = 35.2\,\mathrm{kPa}$$

De Stordeur's Model

Unlike In, de Stordeur [8] defines the grid spacer loss coefficient using the coolant velocity at the grid (Equation 9.110):

$$\Delta p_{grid}^{deStordeur} = \left(C_s\frac{A_{grid,frontal}}{A_{subch}}\right)\frac{\rho_{avg}\upsilon_s^2}{2} = \left(C_s\frac{A_{grid,frontal}}{A_{subch}}\right)\frac{G_{@grid}^2}{2\rho_{avg}}$$

To perform a consistent comparison among models, not only on the basis of the pressure drop but also of the grid spacer loss coefficient, it is important to have the form loss coefficient always defined with respect to the same velocity. Consistent with the calculations made for the In's model, the velocity away from the grid should be chosen. Therefore, the above equation needs to be rewritten by applying a simple mass balance:

$$\Delta p_{\text{grid}}^{\text{deStordeur}} = \left(C_s \frac{A_{\text{grid,frontal}}}{A_{\text{subch}}}\right)\frac{G_{@\text{grid}}^2}{2\rho_{\text{avg}}} = \left(C_s \frac{A_{\text{grid,frontal}}}{A_{\text{subch}}}\right)\frac{G^2}{2\rho_{\text{avg}}}\left(\frac{A_{\text{subch}}}{A_{\text{subch@grid}}}\right)^2$$

$$= K_{\text{grid}}^{\text{deStordeur}} \frac{G^2}{2\rho_{\text{avg}}}$$

where

$$K_{\text{grid}}^{\text{deStordeur}} = \left(C_s \frac{A_{\text{grid,frontal}}}{A_{\text{subch}}}\right)\left(\frac{A_{\text{subch}}}{A_{\text{subch@grid}}}\right)^2$$

The value for C_s can be obtained from the bottom plot of Figure 9.26 using, as a Reynolds number, that calculated using the grid spacer strap thickness as characteristic dimension (consistent with de Stordeur procedure):

$$\text{Re}_{\text{deStordeur}} = \frac{tV_s\rho_{\text{avg}}}{\mu_{\text{avg}}} = \frac{tG_{@\text{grid}}}{\mu_{\text{avg}}} = \frac{0.45 \times 10^{-3} \times 4083}{8.557 \times 10^{-5}} = 21{,}472$$

The value for C_s corresponding to this Reynolds number is about 1.7. Therefore

$$K_{\text{grid}}^{\text{deStordeur}} = \left(1.7\frac{1.11375}{8.78788}\right)\left(\frac{8.78788}{7.67403}\right)^2 = 0.2825$$

and

$$\Delta p_{\text{grids}}^{\text{deStordeur}} = 8K_{\text{grid}}^{\text{deStordeur}} \frac{G^2}{2\rho_{\text{avg}}} = 20.2 \text{ kPa}$$

Rehme's Model

According to Rehme's model (Equation 9.111):

$$\Delta p_{\text{grid}}^{\text{Rehme}} = C_V\left(\frac{A_{\text{grid,frontal}}}{A_{\text{subch}}}\right)^2 \frac{\rho_{\text{avg}}V_V^2}{2} = C_V\left(\frac{A_{\text{grid,frontal}}}{A_{\text{subch}}}\right)^2 \frac{G^2}{2\rho_{\text{avg}}} = K_{\text{grid}}^{\text{Rehme}} \frac{G^2}{2\rho_{\text{avg}}}$$

where

$$K_{\text{grid}}^{\text{Rehme}} = C_V\left(\frac{A_{\text{grid,frontal}}}{A_{\text{subch}}}\right)^2$$

The value to be used for C_V can be obtained from Figure 9.26 using, as Reynolds number, that corresponding to the conditions away from the grid spacers, that is, $Re = 4.907 \times 10^5$. A value of about 6.5 can be obtained. Therefore

$$K_{grid}^{Rehme} = C_V \left(\frac{A_{grid,frontal}}{A_{subch}} \right)^2 = 6.5 \left(\frac{1.11375 \times 10^{-5}}{8.78788 \times 10^{-5}} \right)^2 = 0.1044$$

The pressure drop due to the eight grid spacers is therefore

$$\Delta p_{grids}^{Rehme} = 8K_{grid}^{Rehme} \frac{G^2}{2\rho_{avg}} = 7.5 \text{ kPa}$$

It can be concluded that the three correlations yield different spacer-pressure drop in the sequence In (35.2 kPa), de Stordeur (20.2 kPa), and Rehme (7.5 kPa). For this case of spacer grid without mixing vane, the Rehme correlation results in much lower loss coefficient because the correlation coefficient was determined from the loss coefficient of a short and simply supported spacer. Nevertheless, these dramatically different results point to the need for experimental determination of the pressure drop for any proposed spacer design.

9.6.2.2 Wire Wrap Spacers

Rehme [41] also correlated the total pressure losses in wire-wrapped bundles. More recently Cheng and Todreas [4] correlated wire-wrapped pressure losses utilizing the much larger database existing in the literature up to 1984. Their correlations covered the laminar, transition, and turbulent flow regimes and reduced smoothly to the bare rod correlations presented in Section 9.4.1. Friction factors and flow split factors are presented for each subchannel for the hexagonal array. The friction factor for the rod bundle is also presented, as given below. It predicts most of the data within 10% except at the extreme ends of the P/D and H/D range (where H = the lead length or axial pitch of the wire wrap).

Turbulent region ($Re \geq Re_T$):

$$f = \frac{C_{fT}}{Re^{0.18}} \tag{9.118a}$$

Transition region ($Re_L < Re < Re_T$):

$$f = \frac{C_{fT}}{Re^{0.18}} \varphi^{1/3} + \left(\frac{C_{fL}}{Re} \right) (1 - \varphi)^{1/3} \tag{9.118b}$$

Laminar region ($Re \leq Re_L$):

$$f = \frac{C_{fL}}{Re} \tag{9.118c}$$

where

$$\begin{aligned}\varphi &= \log_{10}(\mathrm{Re}/\mathrm{Re_L})/\log_{10}(\mathrm{Re_T}/\mathrm{Re_L}) \\ &= [\log_{10}(\mathrm{Re}) - (1.7P/D + 0.78)](2.52 - P/D)\end{aligned} \tag{9.119}$$

$$\begin{aligned}C_{\mathrm{fT}} = \{&0.8063 - 0.9022[\log_{10}(H/D)] \\ &+ 0.3526[\log_{10}(H/D)]^2\}(P/D)^{9.7}(H/D)^{1.78-2.0(P/D)}\end{aligned} \tag{9.120a}$$

$$C_{\mathrm{fL}} = [-974.6 + 1612.0(P/D) - 598.5(P/D)^2](H/D)^{0.06-0.085(P/D)} \tag{9.120b}$$

and all parameters are bundle average values.

The range of applicability of the correlation is

$1.025 \le P/D \le 1.42$
$8.0 \le H/D \le 50.0$

The flow region boundary definitions are

$$\text{Turbulent}: \quad \mathrm{Re} \ge \mathrm{Re_T} = 10^{(0.7P/D+3.3)} \tag{9.121a}$$

$$\text{Laminar}: \quad \mathrm{Re} < \mathrm{Re_L} = 10^{(1.7P/D+0.78)} \tag{9.121b}$$

9.6.2.3 Grid versus Wire Wrap Pressure Loss

We are now in a position to illustrate the relative magnitudes of grid versus wire wrap pressure losses. Fundamentally, it is expected that wire wraps will cause less pressure drop in tightly packed fuel arrays since the wire diameter decreases as the fuel pin spacing, $P - D$, decreases whereas the fixed grid strap thickness, t, will occupy an increasing fraction of the spacing as the spacing decreases. Conversely, as the bundle array spacing loosens, the grid spacers outperform the increasing large diameter wire spacer.

Figure 9.29 compares the pressure drop performance of these two spacer types for an interior subchannel over a wide range of hexagonal fuel array geometries for typical PWR reactor operating conditions. The wire wrap axial pitch is taken equal to the typical PWR grid axial separation length (0.4064 m for 8 equally spaced grids over a 12-f long core), a condition which should yield comparable lateral support to the fuel pins for the two spacer types.

The pressure drop calculation methodology of the preceding section was utilized in Figure 9.30 for a hexagonal array with 8 grids spaced at 0.406 m versus the same array with wire wraps with an axial pitch set to the grid spacing. The correlations used were the In [19] loss coefficients for a honeycomb grid of height 40 mm and strap

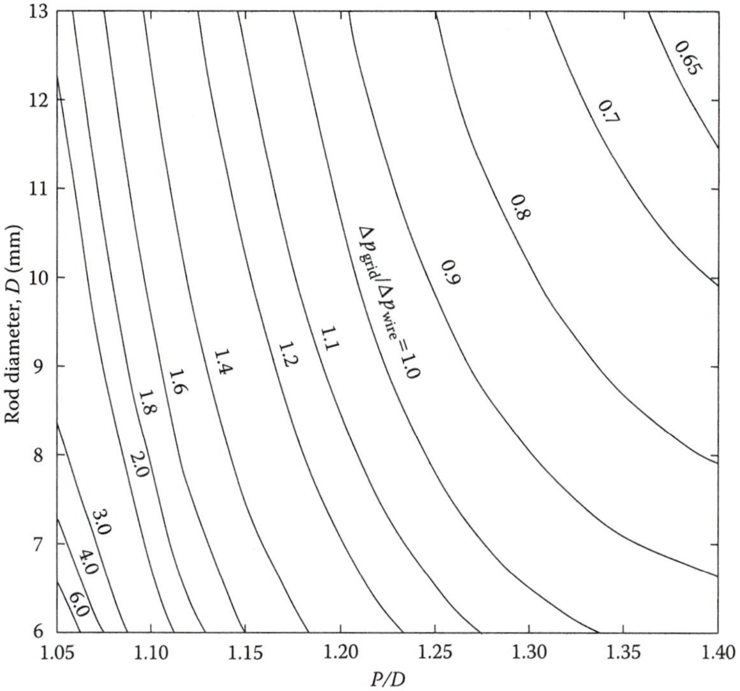

FIGURE 9.30 Comparative grid to wire rod support pressure drop for interior subchannels over a large rod hexagonal array geometry range, that is, $\Delta p_{grid}/\Delta p_{wire}$. (Adopted from Diller, P., Todreas, N., and Hejzlar, P. *Nucl. Eng. Des.*, 239:1461–1470, 2009.)

thickness 0.45 mm with mixing vanes, the Cheng–Todreas correlation for bare rod friction factor and the Cheng–Todreas correlation for wire wrap friction coefficients.

Figure 9.30 illustrates that the pressure drop is the wire support is less than that for the grid over most of the geometry range displayed. The slope of the curves of constant grid/wire pressure drop is similar to those of constant rod-to-rod gap spacing, suggesting that this ratio is related to the magnitude of this spacing.

9.6.3 Pressure Loss for Cross Flow

9.6.3.1 Across Bare Rod Arrays

For heat exchangers with baffles to enhance heat transfer, regions of flow normal to the rod axis or cross flow are encountered. A simple cross flow correlation is that of Zukauskas [52], who correlated laminar and turbulent flow pressure drop by equations of the form:

$$\Delta p = f \frac{NG_{max}^2}{2\rho} Z \tag{9.122}$$

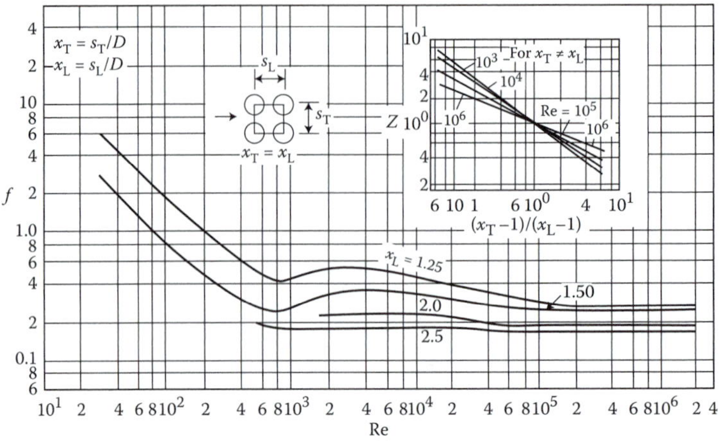

FIGURE 9.31 Friction factor (f) and the correction factor (Z) for use in Equation 9.122 for in-line tube arrangement. (From Zukauskas, A., *Adv. Heat Trans.*, 8:93, 1972.)

where f = friction factor, G_{max} = maximum mass flux, N = number of tube rows in the direction of flow, Z = a correction factor depending on the array arrangement.

Figures 9.31 and 9.32 provide the values of f and Z for various flow conditions as a function of Re = $G_{max}\ D/\mu$, where D = rod diameter.

A large number of more comprehensive but also more complex correlations exist. The various types of in-line and staggered arrays that have been investigated, together

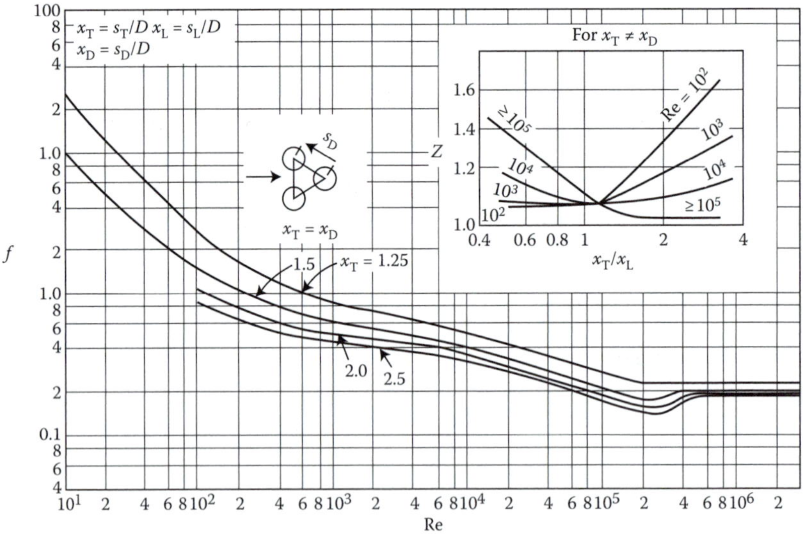

FIGURE 9.32 Friction factor (f) and the correction factor (Z) for use in Equation 9.122 for staggered tube arrangement. (From Zukauskas, A., *Adv. Heat Trans.*, 8:93, 1972.)

with the identification of their salient geometric characteristics, are illustrated in Figure 9.33. From the defining expression for the friction factor:

$$f \equiv \frac{2\Delta p}{\rho}\left(\frac{1}{V_{ref}^2}\right)\left(\frac{D_{ref}}{L}\right) \tag{9.123}$$

The correlations must adopt definitions for V_{ref}, D_{ref}, and D_{ref}/L; and in the literature on cross flow pressure drop, a wide range of definition sets have been

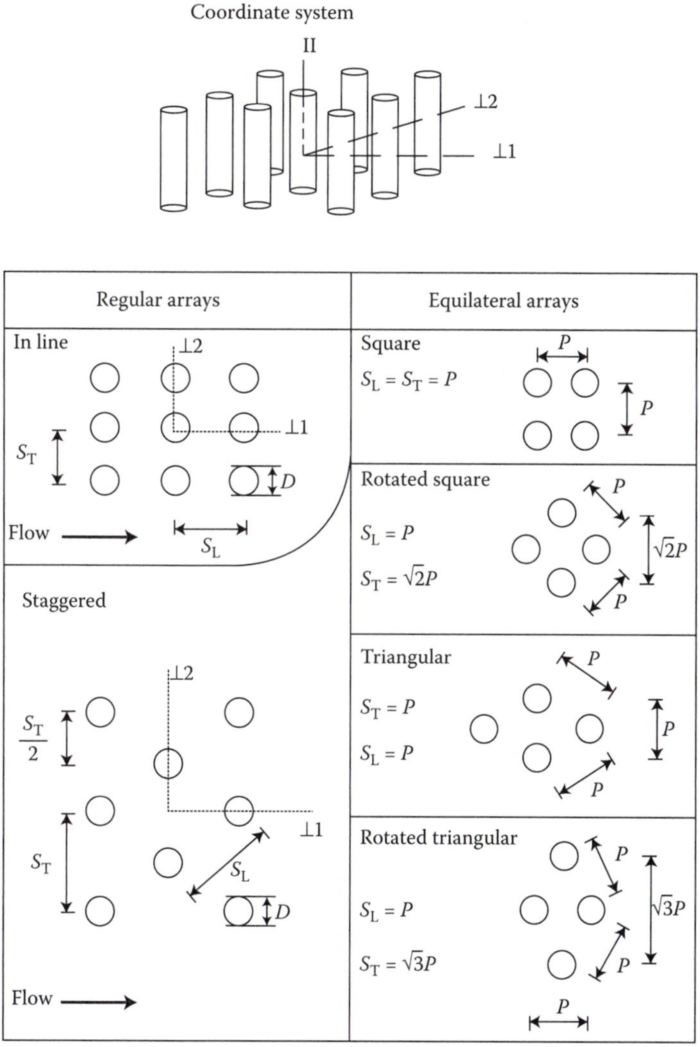

FIGURE 9.33 Regular rod array geometry and coordinate definitions. (Adopted from Ebeling-Koning, D., Robinson, J. T., and Todreas, N. E., *Nucl. Eng. Des.*, 91(1):29–40, 1986.)

adopted. For example, the velocity can be taken as any of the following velocities: V_{min}, V_{avg}, or V_{max}.

The reference diameter D_{ref} can be taken as: D (rod diameter), D_e, D_c (gap spacing), or D_V, where

$$D_V \equiv \text{volumetric hydraulic diamater} = \frac{4(\text{free volume of rod bundle})}{\text{friction surface area of rods}} \tag{9.124}$$

and the ratio D/L can be taken as: N (the number of major restrictions encountered by the flow) or with D or L individually identified with the selected characteristic length.

Figure 9.34 compares a number of correlations for the specific case of an equilateral triangular array of dimension $S_T/D = 1.25$. The parameter set V_{max}, D_V, and N, of the Gunter–Shaw correlation [16], is utilized to put all correlations on a consistent basis. The difference among these correlations exhibited in Figure 9.34 is typically less than the scatter among data from different experimenters. The Zukauskas correlation [53], which is a power law fit, appears the most comprehensive and also includes a correction for entrance effects. The Gunter–Shaw correlation is reasonably simple and covers a variety of geometries. It is fully detailed in Chapter 5, Volume II, where it is utilized to evaluate the distributed resistance for an in-line array in cross flow.

9.6.3.2 Across Wire-Wrapped Rod Bundles

A correlation has been presented by Suh and Todreas [48] based on their experimental data for cross flow over rod arrays with wire separators (also called displacers) parallel to the rod axes. Whereas in the wire-wrapped arrays the wires are arranged in helical fashion, numerical analyses of these arrays treat the wires as parallel to the rod axes in each axial control volume. For typical cases of control volumes with axial

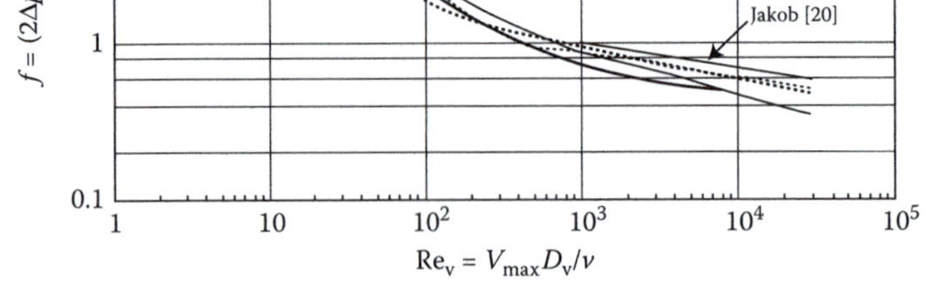

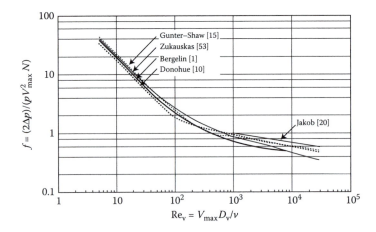

FIGURE 9.34 Friction factor correlations for equilateral-triangular-array bare tube (or rod) bundle with $S_T/D = 1.25$.

lengths small with respect to the wire lead length, this assumed parallel configuration of the wire and the rod is satisfactory. Figure 9.35 illustrates the geometry of the displacer and the rod in terms of the angle θ, which defines the position of the displacer with respect to the gap through which the cross flow passes. Because the displacer restricts the cross flow, the associated pressure drop is expected to be greater than that for a bare array and strongly dependent on the value of the angle θ.

The correlation for the friction factor is

$$f_{\text{displacer}} = \frac{f_{\text{bare}}}{E(\theta)} \tag{9.125}$$

The *Reynolds number* is

$$\text{Re}_{\text{v}} = \frac{V_{\text{max}} D_{\text{V}}}{\nu} \tag{9.126}$$

and the parameter $E(\theta)$ is

$$E_j(\theta) = \sum_{i=0}^{8} a_i^j \theta^i \tag{9.127}$$

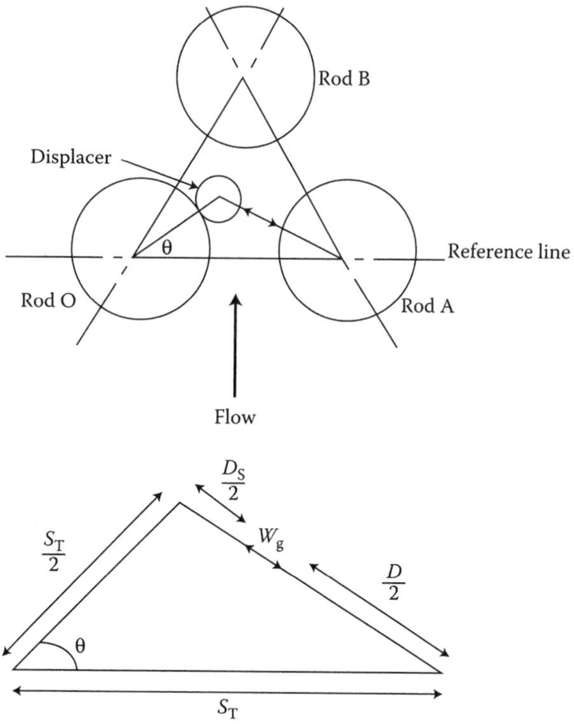

FIGURE 9.35 Geometry of displacer-rod bundle of equilateral triangular array.

TABLE 9.7
Regression Coefficients for Polynomial Equation 9.127 (for θ in Radians)

	a_i^j		
i	**Laminar**	**Transition**	**Turbulent**
0	−0.048629	−0.089567	−0.060247
1	−0.050047	−0.064236	−0.085772
2	2.893754	2.948909	2.534854
3	0.075660	0.086888	0.140045
4	−4.245446	−4.263047	−3.604788
5	−0.035919	−0.037395	−0.072517
6	2.213902	2.216716	1.869185
7	0.005437	0.004490	0.010781
8	−0.368285	−0.367479	−0.306346

where j = laminar, transition, turbulent for $Re_V < 500$; $500 \le Re_V \le 1000$; $Re_V > 1000$, respectively. The regression coefficients a_i^j are given in Table 9.7. It should be noted that this result has been evaluated for only one geometry, an equilateral triangular array of $S_T/D = S_L/D = 1.21$. The shape of $E(\theta)$, shown in Figure 9.36, illustrates that the displacer bundle friction factor increases as the displacer angular position approaches the gap.

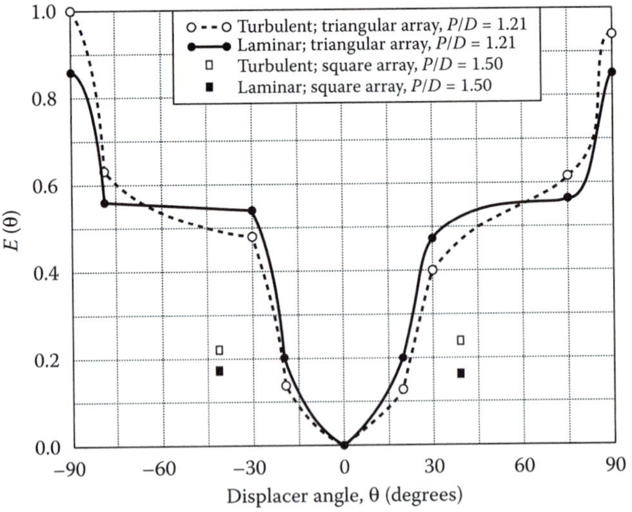

FIGURE 9.36 Parameter $E(\theta)$ as a function of displacer angle.

9.6.4 Form Losses for Abrupt Area Changes

Form losses in multichannel systems connected to headers are discussed in this section. This geometry is that of nuclear components such as fuel assemblies, steam generator tubing, and intermediate heat exchanger tubing fed from an inlet plenum and discharging to an outlet plenum.

In 1950, Kays [21] outlined an analytic procedure for calculation of pressure loss coefficients for incompressible flow through abrupt area changes in such a flow cross section. His results, shown in Figure 9.37, are applicable to entrance and exit losses between a large flow area and a restricted flow area such as the case at the inlet and

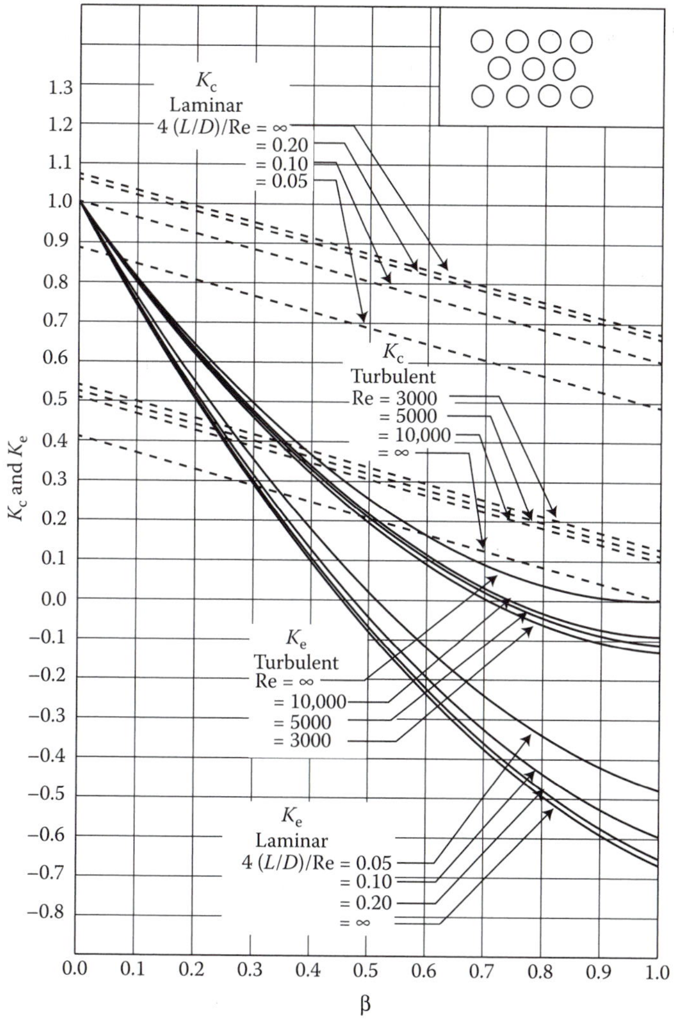

FIGURE 9.37 Loss coefficients for multitubular systems connected to entrance and exit headers. (From Kays, W. M. and London, A. L., *Compact Heat Exchangers*, 3rd Ed. McGraw-Hill, 1984.)

exit of the fuel assemblies,* but they are not applicable to flow between two channels with similar geometry. Hence, as shown in Figure 9.37, the loss coefficients do not approach zero if no area change is involved (i.e., an area ratio of 1), as should be the case at the intersection of identical geometry channels. The reason for Kays' results is that the velocity distributions in the flow areas are not identical on both sides of the abrupt change and hence indicate a pressure differential even when the area ratio is unity. For less drastic flow geometry changes, similar velocity distributions can be assumed on both sides of the geometry change. An outline of the method and illustrative results are given here.

An important caveat regarding Figure 9.37 is that the tubular array need not be restricted to the triangular geometry appearing in the top of Figure 9.37 taken from [23]. Rather, these results apply to the square array as well and, in fact, to any cross-sectional layout of the multiple channels. The confirmation of this observation will be presented in the derivation of these results which follows.

9.6.4.1 Method of Calculation

Consider a sudden expansion in the flow area (Figure 9.38). The behavior of this flow can be predicted by considering a momentum–force balance if the pressure on the downstream end of Section 2 in Figure 9.38 is known. The pressure on the downstream face may be taken as equal to the static pressure in the stream just prior to the expansion. Thus if we neglect wall friction, a force balance between Sections 2 and 3 in Figure 9.37 should yield the momentum change between Sections 2 and 3:

$$\int \rho \upsilon^2 \, dA_3 - \int \rho \upsilon^2 \, dA_2 = p_2 A_2 - p_3 A_3 \tag{9.128}$$

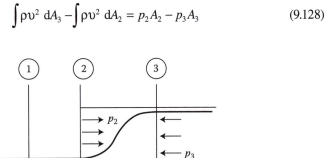

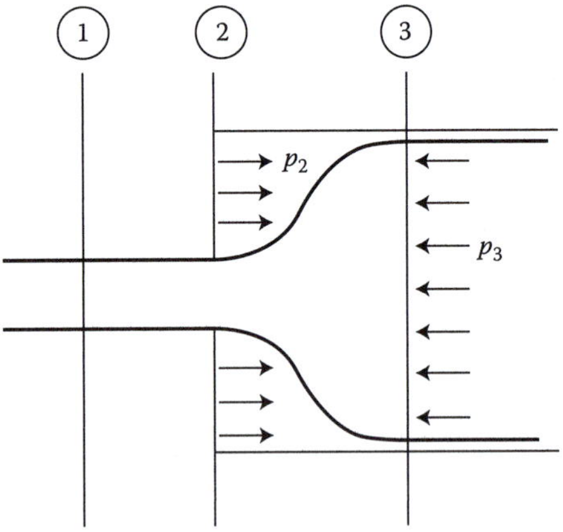

FIGURE 9.38 Flow past an expansion area.

* In a typical reactor, a significant pressure drop component is that associated with the flow entering the fuel assemblies from the lower plenum through the flow passages in the perforated support plate.

The pressure p_2 acts on all of A_2 because at the throat there is no change in the jet area. Note that

$$\int \rho \upsilon^2 \, dA_2 = \int \rho \upsilon^2 \, dA_1 \,(\text{upstream momentum}) \tag{9.129}$$

and

$$A_2 = A_3 \tag{9.130}$$

Furthermore, let K_d be defined as

$$K_d = \frac{1}{V_m^2 A} \int \upsilon^2 \, dA \tag{9.131}$$

where

$$V_m = \frac{1}{A} \int \upsilon \, dA \tag{9.132}$$

Assuming ρ is constant, Equations 9.129 through 9.131 can be used to recast Equation 9.128 in the form:

$$K_{d_3} \rho V_{m3}^2 A_3 - K_{d_1} \rho V_{m1}^2 A_1 = p_2 A_3 - p_3 A_3 \tag{9.133}$$

Now define the area ratio A_1/A_3 as

$$\beta = \frac{A_1}{A_3} \tag{9.134}$$

From the mass continuity equation

$$V_{m3} = \beta V_{m1} \tag{9.135}$$

Equation 9.133 can then be written in the form:

$$\frac{p_2 - p_3}{\rho} = -2(\beta K_{d_1} - \beta^2 K_{d_3}) \frac{V_{m1}^2}{2} \tag{9.136a}$$

This is the actual pressure change across the expansion which, as shown in Figure 9.21, is a pressure rise, that is, $p_3 > p_2$:

$$\frac{p_3 - p_2}{\rho} = 2(\beta K_{d_1} - \beta^2 K_{d_3}) \frac{V_{ml}^2}{2} \tag{9.136b}$$

From the conservation of the mechanical energy in the stream and uniform fluid velocity, the ideal pressure rise from flow deceleration can be obtained from

$$\frac{V_{m1}^2}{2} + \frac{p_2}{\rho} = \frac{V_{m3}^2}{2} + \frac{p_3}{\rho} \quad (9.137)$$

or

$$\left(\frac{p_3 - p_2}{\rho}\right)_{\substack{\text{recoverable or}\\ \text{ideal pressure}\\ \text{change}}} = (1-\beta^2)\frac{V_{m1}^2}{2} \quad (9.138)$$

The irrecoverable pressure loss is given by

$$(\Delta p)_{\text{loss}} = (p_3 - p_2)_{\text{ideal}} - (p_3 - p_2)_{\text{actual}} \quad (9.139)$$

Substituting the results of Equations 9.136b and 9.138 into Equation 9.139, we obtain

$$\frac{(\Delta p)_{\text{loss}}}{\rho} = [1 - 2\beta K_{d_1} + \beta^2(2K_{d_3} - 1)]\frac{V_{m1}^2}{2} \quad (9.140)$$

From the definition of the loss coefficient at an expansion:

$$K_e = \frac{(\Delta p)_{\text{loss}}}{\rho V_{m1}^2/2} \quad (9.141)$$

Thus

$$K_e = 1 - 2\beta K_{d_1} + \beta^2(2K_{d_3} - 1) \quad (9.142a)$$

Equation 9.142a derived by Kays [21] is general and applies to either single- or multiple-tube expansions. In particular, nonuniform cross-sectional velocity distributions as characterized by K_{d_1} and K_{d_3} can exist in either portion of the channel. If these velocity distributions are uniform, then $K_{d_1} = K_{d_3} = 1$ and Equation 9.142a reduces to the Borda–Carnot relation, which appears in Table 9.3 for the loss coefficient of a sudden expansion, that is

$$K_e = 1 - 2\beta + \beta^2 = (1-\beta)^2 \quad (9.143)$$

For multiple-tube systems, the following simplifications apply. If there are n small tubes, each of cross-sectional area A_1 discharging to a single large tube of area A_3 then:

$$\beta \equiv \frac{nA_1}{A_3} \quad (9.144)$$

From mass continuity Equation 9.144 also equals:

$$\frac{V_3}{V_1} = \frac{nA_1}{A_3} = \beta \tag{9.145a}$$

yielding

$$\frac{A_3}{A_1} = \frac{nV_1}{V_3} = \frac{n}{\beta} \tag{9.145b}$$

Hence, the ratio of Reynolds numbers in the large channel versus a small channel:

$$\frac{N_{\mathrm{Re}_3}}{N_{\mathrm{Re}_1}} = \frac{V_3}{V_1}\left(\frac{A_3}{A_1}\right)^{1/2} = \beta\left(\frac{n}{\beta}\right)^{1/2} = (n\beta)^{1/2} \tag{9.146}$$

utilizing Equations 9.145a and 9.145b.

Thus as the number of tubes increases, the Reynolds number in the large channel, N_{Re_3}, becomes much greater than in the small tubes. This justifies assuming the velocity distribution in the larger channel to be uniform ($K_{\mathrm{d}_3} = 1$) simplifying Equation 9.142a to the following form:

$$K_{\mathrm{e}} \cong 1 - 2K_{\mathrm{d}_1}\beta + \beta^2 \tag{9.142b}$$

The expansion loss coefficients in Figure 9.37 are evaluated from Equation 9.142b as illustrated in Example 9.10.

For abrupt contraction form loss, Kays [21] performed an analogous analysis but with key assumptions regarding conservation of mechanical energy and velocity distribution in the initial contraction and re-expansion regions. The result was

$$K_{\mathrm{c}} = \frac{1 - K_{\mathrm{cb}_1}\beta^2 C_{\mathrm{c}}^2 - 2C_{\mathrm{c}} + C_{\mathrm{c}}^2 2K_{\mathrm{d}_3}}{C_{\mathrm{c}}^2} - (1 - \beta^2) \tag{9.147a}$$

where

$$K_{\mathrm{cb}} = \frac{1}{AV_{\mathrm{avg}}^2}\int_0^A \upsilon^2\, \mathrm{d}A \tag{9.148}$$

Again assuming uniform velocity in the larger channel, $K_{\mathrm{cb}} \cong 1.0$, then Equation 9.147a reduces to Equation 9.147b (Kays also does directly evaluate K_{cb} for the circular-tube system only.):

$$K_{\mathrm{c}} \cong \frac{1 - 2C_{\mathrm{c}} + C_{\mathrm{c}}^2(2K_{\mathrm{d}_3} - 1)}{C_{\mathrm{c}}^2} \tag{9.147b}$$

Evaluating the constants C_c and K_{d_3} and employing them in Equation 9.147b results in the contraction loss coefficients presented in Figure 9.37.

9.6.4.2 Loss Coefficient Values

From Figure 9.37 observe the following coefficient value behavior:

a. Decreasing Reynolds number decreases the expansion coefficient and increases the contraction coefficient.
b. Laminar coefficients greatly exceed turbulent coefficients for contractions while the inverse holds for expansions
c. For expansions, negative coefficients result if K_{d_1} sufficiently exceeds K_{d_3} as Equation 9.142a dictates. This behavior is due to (1) the formulation of the K_e definition which includes both irrecoverable pressure loss due to irreversible free expansion and recoverable pressure rise due to deceleration upon expansion, and (2) the magnitude of the nonuniform velocity distribution in the smaller inlet tubes.

Example 9.10: Expansion Form Loss Coefficient Derivation

PROBLEM

Compute the expansion form loss coefficient for a circular tube multitubular system connected to an inlet flow header. For turbulent flow, use the semiempirical velocity distribution in circular tubes given by

$$\upsilon = V_m\left\{\sqrt{f}\left[2.15\log\left(\frac{y}{R}\right)+1.43\right]+1\right\} \tag{9.149}$$

where y = distance from the wall of a tube of radius R:

$$y = R - r$$

Use this velocity distribution for cases of Re = 10^4 and Re = ∞. Repeat for laminar flow.

SOLUTION

Substituting from Equation 9.149 into Equation 9.131 yields

$$K_{d_1} = 1.09068(f) + 0.05884\sqrt{f} + 1 \tag{9.150}$$

Equation 9.150 together with Equation 9.87 below with a different leading coefficient*

$$f = 0.196\ \text{Re}^{-0.2} \tag{9.87a}$$

* Kays selected the coefficient 0.196 versus the usual 0.184 smooth tube value as being more applicable to the heat exchanger surfaces being considered.

can be used to substitute for the respective parameters in Equation 9.142b to obtain the value of K_e in turbulent flow. Use Equation 9.142b versus Equation 9.142a since the system is multitubular, that is, take $K_{d_3} = 1$.

Examining limiting turbulent flow cases using Equation 9.142b for comparison with Figure 9.37 we observe

Taking $\beta = 1$,

- for $Re = \infty$ Equation 9.87a and Equation 9.150 yield $K_{d_1} = 1$ and hence $K_e = 1 - 2 + 1 = 0$, as plotted.
- for $Re = 10^4$: $K_{d_1} = 1.044$ using f from Equation 9.87a in Equation 9.150, and hence $K_e = -0.09$, as plotted.

Taking $\beta = 0$, $K_e = 1$ for all flow conditions as plotted.

For laminar flow, a parabolic velocity distribution can be assumed, and hence

$$K_{d_1} = 1.333, \quad \text{independent of Re} \tag{9.151}$$

Again for the multiple tube configuration $K_{d_3} = 1$ so that from Equation 9.142b:

$$K_e = 1 - 2.666\,\beta + \beta^2 \tag{9.152}$$

Taking $\beta = 1$,

$$K_e = 1 - 2.666 + 1 = -0.67 \quad \text{as plotted} \left(\text{for} \frac{f\,L/D}{\text{Re}} = \infty\right)$$

Taking $\beta = 0$, $K_e = 1$ as plotted.

Laminar and turbulent cases have been plotted in Figure 9.39. It is clear that these relations accurately predict no pressure loss if no area change is involved

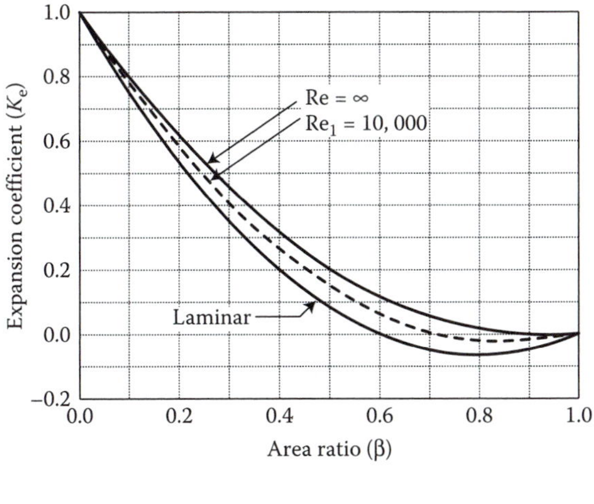

FIGURE 9.39 Variation of pressure loss coefficient due to sudden expansion (K_e) with the area ratio (β).

between channels of similar geometry. Note that because the pressure recovery due to acceleration has been accounted for, the negative values of K_e for $\beta = 0.6$–1.0 for the laminar case implies that there is additional pressure gain due to the velocity profile alone. This was also the case for Kays' original results, shown in Figure 9.37. Also note that the kinetic head $\upsilon^2/2$ is almost completely lost when the flow is discharged from a tube into a large area ($\beta = 0$).

PROBLEMS

9.1. *Emptying of a liquid tank* (Section 9.2)

Consider an emergency water tank, shown in Figure 9.40, that is designed to deliver water to a reactor following a loss of coolant event. The tank is prepressurized by nitrogen at 1.0 MPa. The water is discharged through a 0.2 m (inner diameter) pipe. What is the maximum flow rate delivered to the reactor if the flow is considered inviscid and the reactor pressure is

1. 0.8 MPa
2. 0.2 MPa

Answers:

1. $\dot{m} = 827.7$ kg/s
2. $\dot{m} = 1367.3$ kg/s

9.2. *Laminar flow velocity distribution and pressure drop in parallel plate geometry* (Section 9.4)

For flow between parallel flat plates:

1. Show that the momentum equation for fully developed, steady-state, constant density, and viscosity flow takes the form:

$$\frac{dp}{dx} = \mu \frac{d^2\upsilon_x}{dy^2}$$

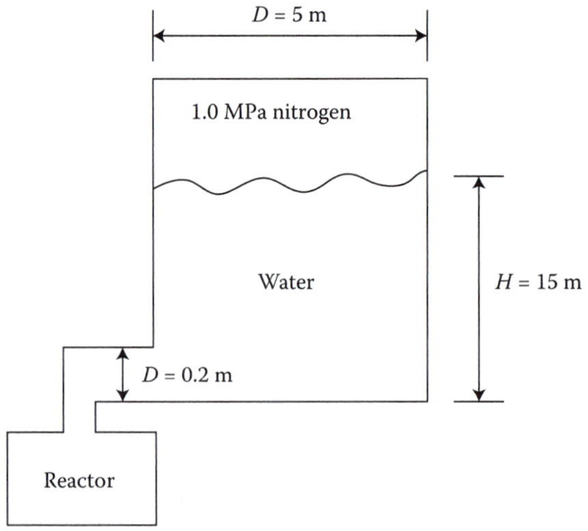

FIGURE 9.40 Emergency water tank.

2. For plate separation of $2y_0$, show that the velocity profile is given by

$$\upsilon_x(y) = \frac{3}{2}V_m\left[1-\left(\frac{y}{y_0}\right)^2\right]$$

3. Show that

$$-\frac{dp}{dx} = \frac{96}{Re}\frac{1}{4y_0}\frac{\rho V_m^2}{2}$$

9.3. *Velocity distribution in single-phase turbulent flow* (Section 9.5)
Consider a smooth circular flow channel of diameter 13.5 mm (for a hydraulic simulation of flow through a PWR assembly).
Operating conditions: Assume two adiabatic, fully developed flow conditions:
a. High flow (mass flow rate = 0.5 kg/s)
b. Low flow (mass flow rate = 1 g/s)
Properties (approximately those of pressurized water at 300°C and 15.5 MPa)
Density = 720 kg/m^3
Viscosity = 91 μPa s
1. For each flow condition, draw a quantitative sketch to show the velocity distribution based on Martinelli's formalism (Equations 9.80a through 9.80c).
2. Find the positions of interfaces between the laminar sublayer, the buffer zone, and the turbulent core.

Answers:
2a. Laminar layer: $0 \le y \le 3.2$ μm
Buffer zone: 3.2 μm $\le y <$ 19.2 μm
Turbulent core: 19.2 μm $\le y \le$ 6.75 mm
2b. Laminar flow

9.4. *Flow split in downflow* (Section 9.5)
High-pressure water at mass flow rate $\dot{m}$ enters a certain component having an inlet plenum and an outlet plenum, connected by two vertical tubes (Figure 9.41). These two tubes are identical except that one is heated and one is cooled. Is the mass flow rate in each tube the same?

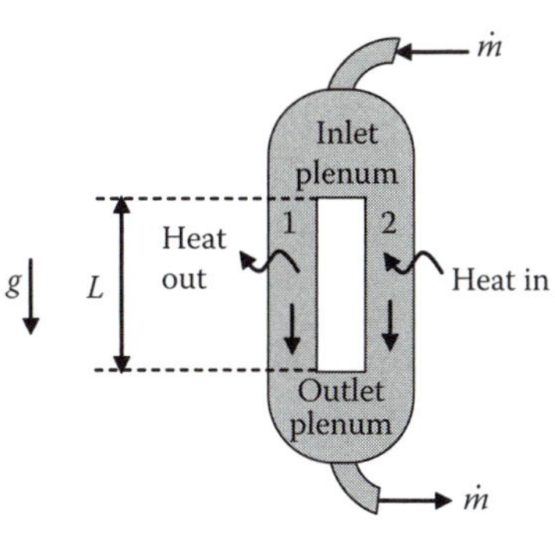

FIGURE 9.41 Parallel vertical channels.

If not, which tube has the higher mass flow rate? Support your answer with an analytic proof.

Assume the following:

- The density of water decreases as temperature increases.
- Downflow exists in both tubes.
- The form and acceleration terms in the momentum equation are negligible.
- The friction factor is the same in both tubes.
- Single-phase and steady-state conditions apply.

Answer:

The cooled channel has higher mass flow.

9.5. *Sizing of an orificing device* (Section 9.6)

In a hypothetical reactor, an orificing scheme is sought such that the core is divided into two zones. Each zone produces one-half of the total reactor power. However, zone 1 contains 100 assemblies, whereas zone 2 contains 80 assemblies.

It is desired to obtain equal average temperature rises in the two zones. Therefore, the flow in the assemblies of lower power (zone 1) is to be constricted by the use of orificing blocks, as shown in Figure 9.42.

Determine the appropriate diameter (D) of the flow channels in the orificing block. Assume negligible pressure losses in all parts of the fuel assemblies other than the fuel rod bundle and the orifice block. The flow in all the assemblies may be assumed to be fully turbulent. All coolant channels have smooth surfaces.

Data

Pressure drop across assembly	$\Delta p_A = 7.45 \times 10^5$ N/m^2
Total core flow rate	$\dot{m}_T = 17.5 \times 10^6$ kg/h
Coolant viscosity	$\mu = 2 \times 10^{-4}$ N s/m^2
Coolant density	$\rho = 0.8$ g/cm^3
Contraction pressure loss coefficient	$K_c = 0.5$
Expansion pressure loss coefficient	$K_e = 1.0$

Answer:

$D = 2.17$ cm

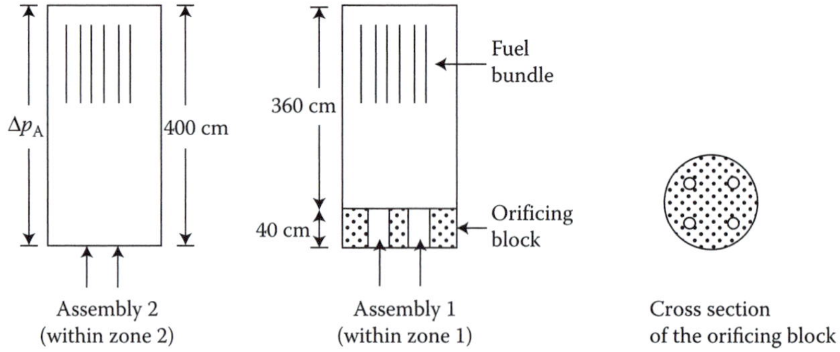

FIGURE 9.42 Orificing technique.

9.6. *Pressure drop features of a PWR core* (Section 9.6)
Consider a PWR core containing 50,952 fuel rods cooled with a total flow rate of 17.476 mg/s. Each rod has a total length of 3.876 m and a smooth outside diameter of 9.5 mm. The rods are arranged in a square array with pitch = 12.6 mm. The lower- and upper-end fittings are represented as a honeycomb grid spacer, with the thickness of individual grid elements being 1.5 mm. There are also five intermediate honeycomb grid spacers with thickness of 1 mm. Consider the upper and lower plenum regions to be entirely open.
1. What is the pressure drop for each of the following bundle regions (entrance, end fixtures, friction in the pin array between grids, gravity, grid spacers, and exit)?
2. What is the plenum-to-plenum pressure drop?

Properties
Water density = 720 kg/m^3
Water viscosity = 91 μPa s

Answers:
$\Delta p_{gravity} = 27.37$ kPa
$\Delta p_{friction} = 50.06$ kPa
$\Delta p_{entrance} + \Delta p_{exit} = 15.87$ kPa
$\Delta p_{spacers} = 26.07$ kPa
$\Delta p_{fittings} = 22.51$ kPa
$\Delta p_T = 141.88$ kPa

9.7. *Comparison of laminar and turbulent friction factors of water and sodium heat exchangers* (Sections 9.4 and 9.6)
Consider square arrays of vertical tubes utilized in two applications: a recirculation PWR steam generator and an intermediate heat exchanger for SFR service. In each case primary system liquid flows through the tubes, and secondary system liquid flows outside the tubes within the shell side.

TABLE 9.8
Operating and Geometric Conditions

Parameter	PWR Steam Generator	SFR Intermediate Heat Exchanger
System characteristics		
Primary fluid	Water	Sodium
Secondary fluid	Water	Sodium
P/*D*	1.5	1.5
D (cm)	1.0	1.0
Nominal shell-side properties		
Pressure (MPa)	5.5	0.202
Temperature (°C)	270	480
Density (kg/m^3)	767.9	837.1
Thermal conductivity (W/m °C)	0.581	88.93
Viscosity (kg/m s)	1.0×10^{-4}	2.92×10^{-4}
Heat capacity (J/kg K)	4990.0	1195.8

The operating and geometric conditions of both units are given in Table 9.8. Assume the wall heat flux is axially constant.

For the shell side of the tube array in each application answer the following questions:

1. Find the friction factor (f) for fully developed laminar flow at $Re_{De} = 10^3$.
2. Find the friction factor (f) for fully developed turbulent flow at $Re_{De} = 10^5$.
3. Can either of the above friction factors be found from the circular tube geometry using the equivalent diameter concept? Demonstrate and explain.
4. What length is needed to achieve fully developed laminar flow?
5. What length is needed to achieve fully developed turbulent flow?

Answers:

1. $f = 0.1198$
2. $f = 0.0194$
3. Only for turbulent case
4. Cannot be determined
5. $z = 0.466$ m to 0.746 m

REFERENCES

1. Bergelin, O. P., Brown, G. A., and Doberstein, S. C., Heat transfer and fluid friction during viscous flow across banks of tubes. IV. A study of the transition zone between viscous and turbulent flow. *Trans. ASME*, 74:953–960, 1952.
2. Bergles, A. E., Techniques to enhance heat transfer. In W. M. Rohsenow, J. P. Hartnett, and Y. I. Cho (eds.). Chapter 11 of *Handbook of Heat Transfer*, 3rd Ed. McGraw-Hill, 1998.
3. Bird, R. B., Stewart, W. E., and Lightfoot, E. N., *Transport Phenomena*. New York, NY: Wiley, 1960.
4. Cheng, S. K. and Todreas, N. E., Hydrodynamic models and correlations for bare and wire-wrapped hexagonal rod bundles: Bundle friction factors, subchannel friction factors and mixing parameters. *Nucl. Eng. Design*, 92(2):227–251, 1986.
5. Churchill, S. W., Empirical expressions for the shear stress in turbulent flow in commercial pipe. *AIChE, J.*, 19(2):375–376, 1973.
6. Colebrook, C. F., Turbulent flow in pipes with particular reference to the transition region between the smooth and rough pipe laws. *Proc. Inst. Civil Eng.*, 11:133, 1939.
7. Deissler, R. G. and Taylor, M. F., Analysis of axial turbulent flow and heat transfer through banks of rods or tubes. TID-7529. In: *Reactor Heat Transfer Conference*. Part 1, Book 2. USAEC, 1957.
8. De Stordeur, A. M., Drag coefficients for fuel elements spacers. *Nucleonics*, 19(6):74, 1961.
9. Diller, P., Todreas, N., and Hejzlar, P., Thermal hydraulic analysis for wire wrapped PWR cores. *Nucl. Eng. Des.*, 239:1461–1470, 2009.
10. Donohue, D. A., Heat transfer and pressure drop in heat exchangers. *Ind. Eng. Chem.*, 41:2499, 1949.
11. Ebeling-Koning, D., Robinson, J. T., and Todreas, N. E., Models for the fluid-solid interaction force for multidimensional single phase flow within tube bundles. *Nucl. Eng. Des.*, 91(1):29–40, 1986.
12. Filonenko, G. K., Hydraulic resistance in pipes. *Teploenergetica* 1(4):40–44, 1954.
13. Fleming, D. P. and Sparrow, E. M., Flow in the hydrodynamic entrance regions of ducts of arbitrary cross sections. *Trans. ASME J. Heat Transfer*, 91:345, 1969.

14. Fox, R. W. and McDonald, A. T., *Introduction to Fluid Mechanics*, 2nd Ed. New York, NY: Wiley, 1978.
15. Gunter, A. Y. and Shaw, W. A., A general correlation of friction factors for various types of surfaces in crossflow. *Trans. ASME*, 67:643, 1945.
16. Haaland, S. E., Simple and explicit formulas for the friction factor in turbulent pipe flow. *J. Fluids Eng.*, 105:89–90, 1983.
17. Heaton, H. S., Reynolds, W. C., and Kays, W. M., Heat transfer in annular passages: Simultaneous development of velocity and temperature fields in laminar flow. *Int. J. Heat Mass Transfer*, 7:763, 1964.
18. Idelchik, I. E., *Handbook of Hydraulic Resistance*, 2nd Ed. New York, NY: Hemisphere, 1986.
19. In, W. K., Oh, D. S., and Chun, T. H., Empirical and computational pressure drop correlations for PWR fuel spacer grids. *Nuc. Tech.*, 139(1):72–79, 2002.
20. Jakob, M., Heat transfer and flow resistance crossflow of gas over tube banks – discussions, *ASME Trans.*, 60:381–392, 1938.
21. Kays, W. M., Loss coefficients for abrupt changes in flow cross section with low Reynolds number flow in single and multiple-tube systems. *ASME Spring Meeting*, Washington, DC: April 12–14, 1950.
22. Kays, W. M. and Crawford, M. E., *Convective Heat and Mass Transfer*, 3rd Ed. New York, NY: McGraw-Hill, 1993.
23. Kays, W. M. and London, A. L., *Compact Heat Exchangers*, 3rd Ed. McGraw-Hill, New York, NY, 1984.
24. Langhaar, H. L. and No, P. H., Steady flow in the transition length of a straight tube. *J. Appl. Mech.,* Sect. 3, 9:55–58, 1942.
25. LeTourneau, B. W., Grimble, R. E., and Zerbe, J. E., Pressure drop for parallel flow through rod bundles. *Trans. ASME*, 79:483, 1957.
26. Lienhard IV, J. H. and Lienhard V, J. H., *A Heat Transfer Textbook*. Cambridge, MA: Phlogiston Press, 2008.
27. Malen, J., Todreas, N., Hejzlar, P., Ferroni, P., and Bergles, A., Thermal hydraulic design of a hydride-fueled inverted PWR core. *Nucl. Eng. Des.*, 239:1471–1480, 2009.
28. Manglik, R. M. and Bergles, A. E., Heat transfer and pressure drop correlations for twisted-tape inserts in isothermal tubes: Part I–Laminar flows. *J. Heat Transfer*, 115:881–889, 1993.
29. Manglik, R. M. and Bergles, A. E., Heat transfer and pressure drop correlations for twisted-tape inserts in isothermal tubes: Part II–Transition and turbulent flows. *J. Heat Transfer*, 115:890–896, 1993.
30. Manglik, R. M. and Bergles, A. E., Swirl flow heat transfer and pressure drop with twisted-tape inserts. *Adv. Heat Transfer*, 36:183 ff., 2002.
31. Martinelli, R. C., Heat transfer to molten metals. *Trans. ASME*, 69:947, 1947.
32. McEligot, D. M., Convective heat transfer in internal gas flows with temperature-dependent properties. In A. S. Majumdar and R. A. Mashelkar (eds.). *Advances in Transport Processes*, Vol. IV, pp.113–200. New York, NY: Wiley, 1986.
33. Moody, L. F., Friction factors for pipe flow. *Trans. ASME*, 66:671, 1944.
34. Nikiforova, A., Hejzlar, P., and Todreas, N. E., Lead-cooled flexible conversion ratio fast reactor. *Nucl. Eng. Design*, 239(12):2596–2611, 2009.
35. Pope, M. A., Hejzlar, P., and Driscoll, M. J., *Thermal Hydraulics of a 2400 MWth Supercritical CO_2-Direct Cycle GFR*. Cambridge: Massachusetts Institute of Technology, CANES Report MIT-ANP-TR-112, September 2006.
36. Petukhov, B. S., Heat transfer and friction in turbulent pipe flow with variable physical properties. In J. P. Hartnett and T. F. Irvine (eds.). *Advances in Heat Transfer*, Academic, New York, NY: 6:504–564, 1970.

37. Ravigururajan, T. S., A comparative study of thermal design correlations for turbulent flow in helical-enhanced tubes. *Heat Transfer Eng.*, 20(1):54–70, 1999.
38. Ravigururajan, T. S. and Bergles, A. E., Development and verification of general correlations for pressure drop and heat transfer in single-phase turbulent flow in enhanced tubes. *Exper. Thermal Fluid Sci.*, 13:55–70, 1996.
39. Rehme, K., Widerstandsbeiwerte von Gitterabstandshaltern für Reaktorbrennelemente. *AtomKernenergie*, 15(2):127–130, 1970.
40. Rehme, K., Laminarstromung in Stabbundden. *Chem. Ingenieur Technik*, 43:17, 1971.
41. Rehme, K., Pressure drop correlations for fuel element spacers. *Nucl. Technol.*, 17:15, 1973.
42. Rehme, K., Pressure drop performance of rod bundles in hexagonal arrangements. *Int. J. Heat Mass Transfer*, 15:2499, 1972.
43. Rehme, K., Simple method of predicting friction factors of turbulent flow in non-circular channels. *Int. J. Heat Mass Transfer*, 16:933, 1973.
44. Reichardt, H., Vollständige darstellung der turbulent geschwindigkeitsverteilung in glatten leitungen. *Z. Angew. Math. Mech.*, 31:208, 1951.
45. Reihman, T. C., *An Experimental Study of Pressure Drop in Wire Wrapped FFTF Fuel Assemblies*. BNWL-1207, September 1969.
46. Sparrow, E. M. and Loeffler, A. L. Jr., Longitudinal laminar flow between cylinders arranged in regular array. *AIChE J.*, 5:325, 1959.
47. Spengos, A. C., *Tests on Models of Nuclear Reactor Elements. IV. Model Study of Fuel Element Supports*, UMRI-2431-4-P, Univ. of Mich. Research Inst., Ann Arbor, 1959.
48. Suh, K. and Todreas, N. E., An experimental correlation of cross-flow pressure drop for triangular array wire-wrapped rod assemblies. *Nucl. Tech.*, 76:229, 1987.
49. Taylor, M. F., *Prediction of friction and heat-transfer coefficients with large variations in fluid properties*. NASA TM X-2145, December 1970.
50. Trupp, A. C. and Azad, R. S., Structure of turbulent flow in triangular rod bundles. *Nucl. Eng. Des.*, 32:47, 1975.
51. Webb, R. L. and Kim, N.-H., *Principles of Enhanced Heat Transfer*, 2nd Ed. New York, NY: Taylor and Francis, 2005.
52. Zukauskas, A., Heat transfer from tubes in crossflow. *Adv. Heat Transfer*, 8:93, 1972.
53. Zukauskas, A. and Ulinkskas, R., Banks of plain and finned tubes. *Heat Exchanger Design Handbook*, Vol.2. New York: Hemisphere, pp. 2.2.2.4–1–2.2.2.4–17, 1983.

10 Single-Phase Heat Transfer

10.1 FUNDAMENTALS OF HEAT TRANSFER ANALYSIS

10.1.1 Objectives of the Analysis

The objectives of heat transfer analysis for single-phase flows are generally (1) determination of the temperature field in the wall of the coolant channel so as to ensure that the operating temperatures are within the specified limits, and (2) determination of the parameters governing the heat transport rate at the channel walls including importantly the coolant temperature field. These parameters can then be used to choose materials and flow conditions that maximize heat transport in the process equipment.

Knowledge of the temperature field in the coolant leads to determination of the heat flux, $\vec{q}''(\mathrm{W/m^2})$, at the solid wall via Fourier's law for heat transfer. At any surface, this law states that

$$\vec{q}'' = -k\frac{\partial T}{\partial n}\vec{n} \tag{10.1}$$

where k is the thermal conductivity of the coolant (W/m K), and $\vec{n}$ is the unit vector perpendicular to the surface, so that $\partial T/\partial n$ is the temperature gradient in the direction of heat transfer (K/m). However, in engineering analyses, where only the second objective is desired, the heat flux is related to the bulk or mean temperature of the flow (T_b), via the so-called Newton law for heat transfer:

$$\vec{q}'' \equiv h(T_w - T_b)\vec{n} \tag{10.2}$$

where T_w is the wall temperature (K), and h is the heat transfer coefficient (W/m^2 K). Equation 10.2 is applied in the engineering analysis when it is possible to determine h for the flow conditions based on prior engineering experience. Often the heat transfer coefficient is a semiempirical function of the coolant properties and velocity as well as of the flow channel geometry.

10.1.2 Approximations to the Energy Equation

The general energy equation for single-phase flow is used in its temperature form:

$$\rho c_p \frac{DT}{Dt} = -\nabla \cdot \vec{q}'' + q''' + \beta T\frac{Dp}{Dt} + \Phi \tag{4.122a}$$

When the velocity $\vec{\upsilon}\,(\vec{r},t)$, pressure $p\,(\vec{r},t)$, and volumetric heat generation rate, q''', are known *a priori*, Equation 4.122a is used to specify the temperature field $T(\vec{r},t)$. The equations of state—$\rho(p,T)$, $c_p(p,T)$, and $\beta(T)$—as well as the constitutive relations for $q''(\rho, \upsilon, T)$ and $\Phi(\rho, \upsilon, T)$ are needed for the solution. Also needed are the initial values of $T\,(\vec{r},0)$ and $p\,(\vec{r},0)$ as well as the appropriate number of boundary conditions.

In general, as discussed in Section 9.1, the equation is first simplified to a form that is an acceptable approximation of the situation at hand. The common approximations are twofold:

1. The pressure term is negligible (in effect considering the phase incompressible).
2. The material properties are temperature independent and pressure independent.

The above two approximations are acceptable in forced convection flow analysis but should not be applied indiscriminately in natural-flow and mixed-flow analyses. When natural flow is important, the Boussinesq approximation is applied where the material properties are assumed to be temperature independent with the exception of the density in the gravity or buoyancy term in the momentum equation, which is assumed to vary linearly with temperature.

It is possible to ignore radiation heat transfer within the single-phase liquids and high-density gases so that the heat flux is due to conduction and convection alone. Using Equation 4.114, we get

$$\nabla \cdot \vec{q}'' = -\nabla \cdot k \nabla T$$

Thus for incompressible fluids and purely conductive heat flux, the energy Equation 4.122a is written as

$$\rho c_p \frac{DT}{Dt} = \nabla \cdot k \nabla T + q''' + \Phi \tag{10.3a}$$

For cylindrical geometry and a fluid with constant μ and k properties, Equation 10.3a takes the following form given in Table 4.12 when dh is taken as $c_p dT$ and radiation is neglected

$$\begin{aligned}
&\rho c_p \left(\frac{\partial T}{\partial t} + \upsilon_r \frac{\partial T}{\partial r} + \frac{\upsilon_\theta}{r}\frac{\partial T}{\partial \theta} + \upsilon_z \frac{\partial T}{\partial z} \right) \\
&= k \left(\frac{\partial^2 T}{\partial r^2} + \frac{1}{r}\frac{\partial T}{\partial r} + \frac{1}{r^2}\frac{\partial^2 T}{\partial \theta^2} + \frac{\partial^2 T}{\partial z^2} \right) + q''' \\
&\quad + \mu \left\{ 2\left[\left(\frac{\partial \upsilon_r}{\partial r} \right)^2 + \left(\frac{1}{r}\frac{\partial \upsilon_\theta}{\partial \theta} + \frac{\upsilon_r}{r} \right)^2 + \left(\frac{\partial \upsilon_z}{\partial z} \right)^2 \right] + \left(\frac{\partial \upsilon_z}{\partial \theta} + \frac{\partial \upsilon_\theta}{\partial z} \right)^2 \right. \\
&\quad \left. + \left(\frac{\partial \upsilon_r}{\partial z} + \frac{\partial \upsilon_z}{\partial r} \right)^2 + \left(\frac{1}{r}\frac{\partial \upsilon_r}{\partial \theta} + \frac{\partial \upsilon_\theta}{\partial r} - \frac{\upsilon_\theta}{r} \right)^2 \right\}
\end{aligned} \tag{10.3b}$$

If the dissipation energy, Φ, is also negligible, which is generally true unless the velocity gradients are very large, the terms between braces { } are neglected.

10.1.3 Dimensional Analysis

Consider the case of spatially and temporally constant pressure ($Dp/Dt = 0$) and no heat generation ($q''' = 0$) again. Also, let the heat flux be due to conduction alone, and let the material properties be temperature independent. Then Equation 10.3a takes the form

$$\rho c_p \frac{DT}{Dt} = k\nabla^2 T + \mu\phi' \tag{10.4}$$

where the viscosity has been factored out of the dissipation function Φ (i.e., $\Phi = \mu\phi'$); ϕ' is given in Table 4.12 as an explicit function of the velocity components.

Using the nondimensional parameters

$$\vec{\upsilon}^* = \frac{\vec{\upsilon}}{V}$$

$$x^* = \frac{x}{D_e}$$

$$t^* = \frac{tV}{D_e}$$

$$T^* = \frac{(T - T_o)}{(T_1 - T_o)}$$

where V, D_e, and $T_1 - T_o$ are convenient characteristic velocity, length, and temperature differences, respectively, in the system. Then Equation 10.4 can be cast in the form

$$\rho c_p \left[(T_1 - T_o)\frac{V}{D_e}\frac{\partial T^*}{\partial t^*} + (T_1 - T_o)\frac{V}{D_e}\vec{\upsilon}^* \cdot \nabla^* T^* \right] = \frac{k(T_1 - T_o)}{D_e^2}\nabla^{*2} T^* + \mu\left(\frac{V}{D_e}\right)^2 \phi^* \tag{10.5a}$$

or, by rearranging Equation 10.5a

$$\frac{\partial T^*}{\partial t^*} + \vec{\upsilon}^* \cdot \nabla^* T^* = \frac{1}{\text{Re Pr}}\nabla^{*2} T^* + \frac{\text{Br}}{\text{Re Pr}}\phi'^* \tag{10.5b}$$

where ∇^* and ∇^{*2} involve differentiation with respect to x^*:

$$\text{Re} \equiv \frac{\rho V D_e}{\mu}, \quad \text{the Reynolds number,} \tag{9.63}$$

$$\text{Pr} \equiv \frac{\mu c_p}{k}, \quad \text{the } \textit{Prandtl number} \tag{10.6}$$

$$\text{Br} \equiv \frac{\mu V^2}{k(T_1 - T_o)}, \quad \text{the } \textit{Brinkmann number} \tag{10.7}$$

Reynolds number is the ratio of inertial to viscous forces for the given flow conditions. Prandtl number is the ratio of molecular diffusivity of momentum to that of heat in a fluid. Brinkmann number is the ratio of heat production by viscous dissipation to heat transfer by conduction. In some analyses, Eckert number (Ec) is used instead of the ratio *Br/Pr*. Because

$$\text{Ec} = \frac{\text{Br}}{\text{Pr}} = \frac{V^2/c_p}{T_1 - T_o}, \quad \text{the } \textit{Eckert number} \tag{10.8}$$

it signifies the ratio of the dynamic temperature due to motion to the static temperature difference. The physical significance of several nondimensional groups of importance in single-phase heat transfer is given in Appendix G.

10.1.4 Thermal Conductivity

Thermal conductivity is the property relating the heat flux (rate of heat transfer per unit area) in a material to the temperature spatial gradient in the absence of radiation effects.

Thermal conductivities of engineering materials vary widely (Figure 10.1). The highest conductivities belong to the metals and the lowest to the gases. Even for the same material, the conductivity is a function of temperature and, in the case of gases, also of pressure. Figure 10.2 illustrates the pressure and temperature dependence of monoatomic substances. It may also be used to approximate the behavior of polyatomic substances.

For gases, an approximate relation for predicting the conductivity is given by the Eucken formula [3]:

$$k = \left(c_p + \frac{5R}{4M} \right) \mu \tag{10.9}$$

where R is the universal gas constant and M the molar mass.

For liquids and solids, the conductivity is more difficult to predict theoretically; however, for metals, thermal conductivity (k) is related to electrical conductivity (k_e) by

$$\frac{k}{k_e T} \equiv \text{L}, \quad \text{the } \textit{Lorentz number} \tag{10.10}$$

where T is the absolute temperature. (L is about 25×10^{-9} V^2/K^2 for pure metals at 0°C and increases only at a rate of about 10% per 100°C.) At very low temperatures (below −200°C), the metals become superconductors of electricity and heat.

10.1.5 Engineering Approach to Heat Transfer Analysis

For engineering analyses, the difference between the wall temperature and the bulk flow temperature is obtained by defining the heat transfer coefficient (h) through the nondimensional Nusselt number:

$$\mathrm{Nu} = f(\mathrm{Re}, \mathrm{Pr}, \mathrm{Gr}, \mu_w/\mu_b) \tag{10.11}$$

where

$$\mathrm{Nu} \equiv \frac{hD_{\mathrm{ref}}}{k}, \quad \text{the } \textit{Nusselt number} \tag{10.12a}$$

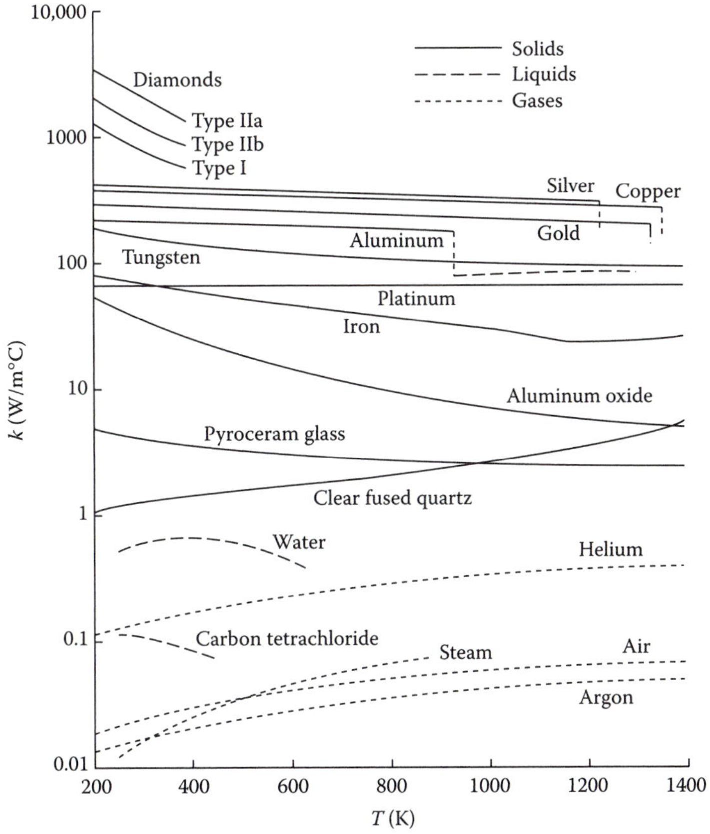

FIGURE 10.1 Variation of the thermal conductivity of various solids, liquids, and gases with temperature.* (From Çengel, Y. A., *Introduction to Thermodynamics and Heat Transfer*. New York, NY: McGraw-Hill, 1997.)

* For most gases at moderate pressures, the thermal conductivity is a function only of temperature. For pressure approaching the critical value, other sources for thermal conductivity values should be consulted.

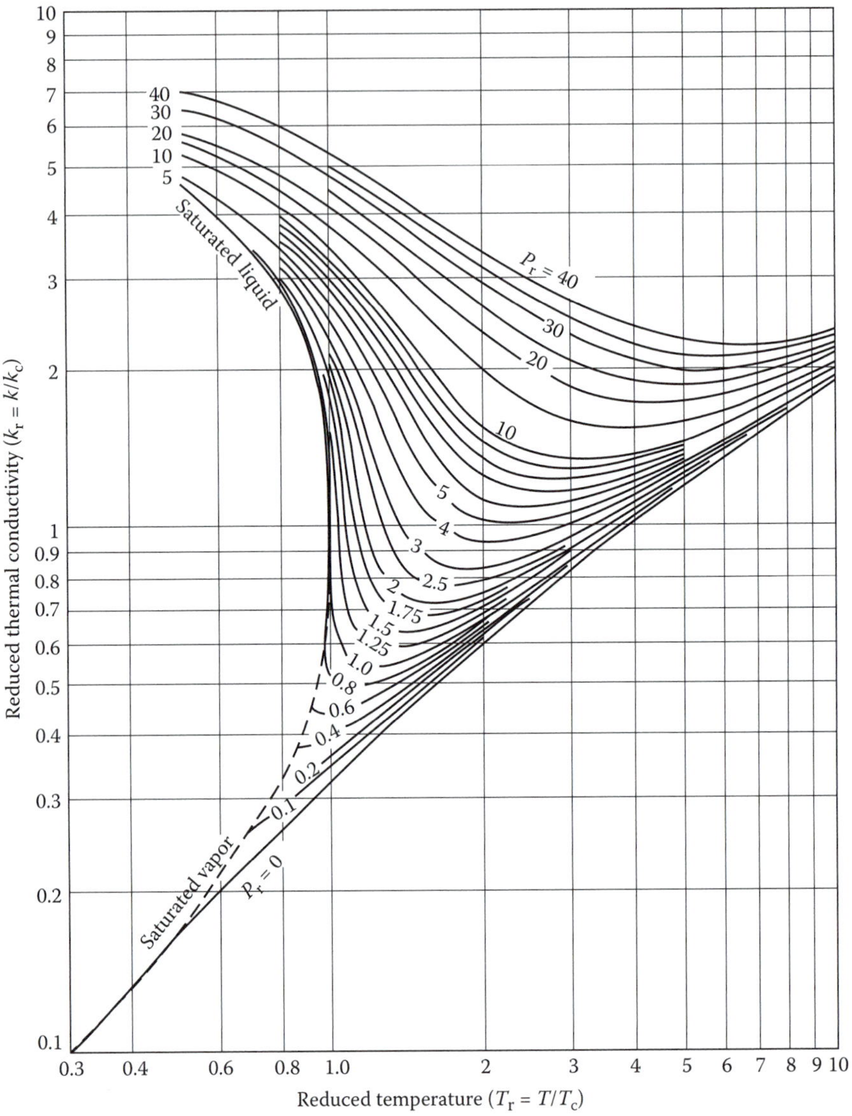

FIGURE 10.2 Thermal conductivity of monoatomic substances as a function of pressure and temperature where $p_r = p/p_c$. (From Bird, R. B., Stewart, W. E., and Lightfoot, E. N., *Transport Phenomena*. New York, NY: Wiley, 1960.)

$$\text{Gr} = \frac{\beta\, \Delta T g D_{\text{ref}}^3}{\nu^2}, \quad \text{the } \textit{Grashof number} \tag{10.12b}$$

and D_{ref} is an appropriate length or lateral dimension. (For external flows, the length dimension is usually used, whereas a lateral dimension is used for internal flows.)

TABLE 10.1
Representative Heat Transfer Parameters

Material	T (°C)	p (MPa)	k (W/m K) (Btu/h ft F)	Pr	Nu in a Tube Re = 10,000	Re = 100,000
Water	275	7.0	0.59 (0.35)	0.87	35	220
Gases						
Helium	500	4.0	0.31 (0.18)	0.67	32	195
CO_2	300	4.0	0.042 (0.024)	0.76	33	210
Sodium	500	0.3	52 (31)	0.004	5.5	8

The general form of Nu is obtained by a boundary layer analysis similar to that for the momentum equation (see Chapter 9). Experiments are often needed, particularly for turbulent flow, to define the numerical constants of Equation 10.11. The form of the Nu relation depends on the flow regime (laminar versus turbulent, external versus internal) and the coolant (metallic versus nonmetallic). At high values of Re, the heat transfer is aided by the presence of turbulent eddies, resulting in an increased heat transfer rate over the case of purely laminar flow. For metallic liquids, the molecular thermal conductivity is so high that the relative effect of turbulence is not as significant as in the case of nonmetallic flows.

Typical values of the heat transfer parameters associated with various coolants and processes are given in Tables 10.1 and 10.2, respectively. Boiling and condensation heat transfer processes and correlations are presented in Chapters 12 and 13.

TABLE 10.2
Typical Values of the Heat Transfer Coefficient for Various Processes

Process	Heat Transfer Coefficient (h) Btu/h ft^2 F	W/m^2 K
Natural convection		
Low-pressure gas	1–5	6–28
Liquids	10–100	60–600
Boiling water	100–2000	60–12,000
Forced convection in pipes		
Low-pressure gas	1–100	6–600
Liquids		
Water	50–2000	250–12,000
Sodium	500–5000	2500–25,000
Boiling water	500–10,000	2500–50,000
Condensation of steam	1000–20,000	5000–100,000

Example 10.1: Importance of Terms in the Energy Equation under Various Flow Conditions

PROBLEM

Consider the following two flow conditions in a pressurized water reactor steam generator on the primary side.

	Forced Flow	Natural Circulation
Flow per tube	1.184 kg/s	0.01184 kg/s
Characteristic temperature difference	15°C	25°C

For a tube of inner diameter 7/8 in. and an average temperature of 305°C, evaluate the various dimensionless parameters in Equation 10.6 and determine which terms are important under both flow conditions.

SOLUTION

The energy equation is

$$\frac{\partial T^*}{\partial t^*} + \vec{\upsilon}^* \cdot \nabla^* T^* = \frac{1}{\text{Re}\,\text{Pr}} \nabla^{*2} T^* + \frac{\text{Br}}{\text{Re}\,\text{Pr}} \phi^* \tag{10.6}$$

where

$$\text{Re} = \frac{\rho V D}{\mu}$$

$$\text{Pr} = \frac{\mu c_p}{k}$$

$$\text{Br} = \frac{\mu V^2}{k(T_\ell - T_o)}$$

Evaluating ρ, μ, c_p, and k for saturated water at 305°C, we find

$\rho = 701.9$ kg/m^3
$\mu = 8.9 \times 10^{-5}$ kg/m s
$k = 0.532$ W/m°C
$c_p = 5969$ J/kg°C

For the forced flow condition,

$$V = \frac{\dot{m}}{\rho A} = \frac{1.184 \text{ kg/s}}{(701.9 \text{ kg/m}^3)\dfrac{\pi}{4}\left[(0.875 \text{ in.})\dfrac{0.0254 \text{ m}}{\text{in.}}\right]^2} = 4.348 \text{ m/s}$$

so that

$$\text{Re} = \frac{(701.9 \text{ kg/m}^3)(4.348 \text{ m/s})(0.0222 \text{ m})}{8.90 \times 10^{-5} \text{kg/m s}} = 7.613 \times 10^5$$

$$\text{Pr} = \frac{(8.90 \times 10^{-5}\text{kg/m s})(5969\ \text{J/kg °C})}{0.532\text{W/m °C}} = 1.00$$

$$\text{Br} = \frac{(8.90 \times 10^{-5}\text{kg/m s})(4.348\ \text{m/s})^2}{(0.532\ \text{W/m °C})(15\text{°C})} = 2.11 \times 10^{-4}$$

Then

$$\frac{1}{\text{Re Pr}} = \frac{1}{\text{Pe}} = \frac{1}{(7.613 \times 10^5)(1.00)} = 1.314 \times 10^{-6}$$

$$\frac{\text{Br}}{\text{Re Pr}} = \frac{\text{Br}}{\text{Pr}} = (2.11 \times 10^{-4})(1.314 \times 10^{-6}) = 2.773 \times 10^{-9}$$

Thus in the forced-flow condition, these parameter values are sufficiently low that the conduction term and the dissipative term in the energy equation can often be ignored in the energy balance.

In the natural circulation condition, the velocity is two orders of magnitude lower than in the forced flow case, so that

$$\text{Re} = 7.613 \times 10^3$$

$$\text{Pr} = 1.00$$

$$\text{Br} = (2.11 \times 10^{-4})(1 \times 10^{-4})\left(\frac{15}{25}\right) = 1.266 \times 10^{-8}$$

Then

$$\frac{1}{\text{Pe}} = \frac{1}{\text{Re Pr}} = 1.314 \times 10^{-4}$$

$$\frac{\text{Br}}{\text{Pe}} = \frac{\text{Br}}{\text{Re Pr}} = (1.266 \times 10^{-8})(1.314 \times 10^{-4}) = 1.664 \times 10^{-12}$$

Again, neither term on the RHS is found to be large. However, the conduction term parameter has increased in value, whereas the dissipation term has decreased. The dissipation energy term can always be ignored, and the conduction term can often be ignored. It is not necessarily the case for all working fluids, however. In the case of sodium-cooled breeder reactors, for example, the high thermal conductivity of sodium results in a Pr number of the order 0.004. In natural circulation conditions, $1/\text{Pe} \approx 1$, and conduction may become important.

10.2 LAMINAR HEAT TRANSFER IN A PIPE

The development of the temperature profile in a boundary layer parallels the development of the velocity profile, as discussed in Chapter 9. For external flows, the thermal boundary layer is taken as the distance, δ_T, over which the temperature changes from the wall temperature to the stream temperature. For internal flows, the thermal boundary developing on the wall merges (at the center for symmetric

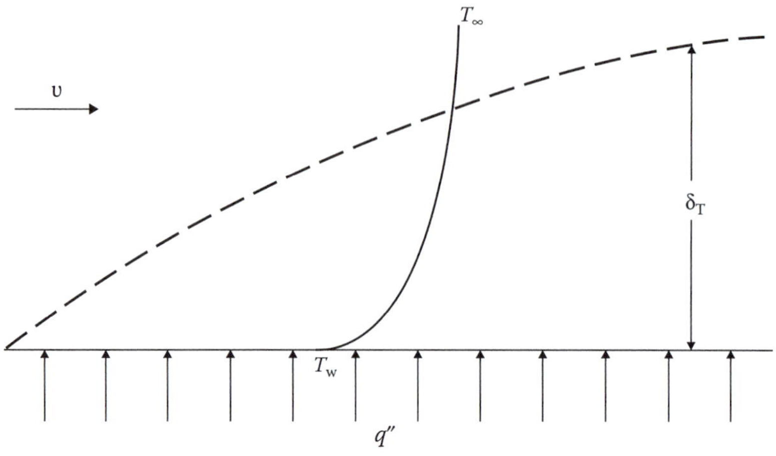

FIGURE 10.3 Thermal boundary layer for external flows.

channels) and provides thereafter the thermally developed region of the flow in the pipe. The temperature profiles for external and internal flows are illustrated in Figures 10.3 and 10.4, respectively.

10.2.1 Fully Developed Flow in a Circular Tube

Let us consider the fully developed flow region in a cylinder of radius R, with azimuthal symmetry. Two surface boundary conditions are of major importance—constant heat flux and constant wall temperature—because the solutions of these cases can be used to solve the cases of arbitrary wall temperature and heat flux distributions by superposition [56].

In the constant heat flux case, the wall and fluid temperatures increase linearly with length at the same rate. Because of the symmetry of the circular tube, the wall

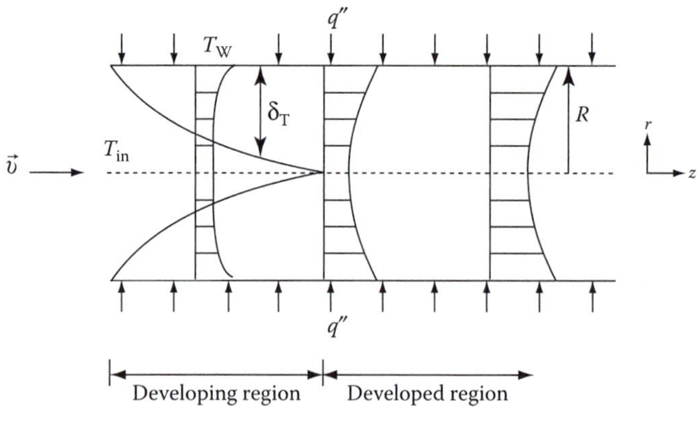

FIGURE 10.4 Thermal boundary layer for internal flow in pipe.

temperature is circumferentially constant at every cross section, though increasing axially, and the thermal properties of the wall material do not enter into any circumferential considerations. For a geometry that lacks circular symmetry, the analogous constant heat flux condition is more complex. Although the axial profile of the heat flux at any circumferential position can still be specified as constant, the conditions of circumferentially constant heat flux and temperature may be incompatible, as in the case of rod bundles.

The velocity characteristics of the fully developed flow conditions are

$$\upsilon_r = 0 \quad \text{and} \quad \frac{\partial \upsilon_z}{\partial z} = 0$$

Let T_m be the fluid mean temperature defined by

$$T_m(z) \equiv \frac{\int_o^R \rho \upsilon_z T 2\pi r \, dr}{\int_o^R \rho \upsilon_z 2\pi r \, dr} \tag{10.13}$$

The temperature lateral profile in a flow with fully developed heat transfer is independent of the axial distance, which implies the general condition:

$$\frac{\partial}{\partial z}\left(\frac{T_w - T}{T_w - T_m}\right) = 0 \tag{10.14a}$$

or

$$\left(\frac{\partial T_w}{\partial z} - \frac{\partial T}{\partial z}\right) - \left(\frac{T_w - T}{T_w - T_m}\right)\left(\frac{\partial T_w}{\partial z} - \frac{\partial T_m}{\partial z}\right) = 0 \tag{10.14b}$$

Since

$$\frac{T_w - T}{T_w - T_m} = f\left(\frac{r}{R}\right) \tag{10.14c}$$

Hence

$$\frac{\partial T_w}{\partial z} - \frac{\partial T}{\partial z} = f\left(\frac{r}{R}\right)\left(\frac{\partial T_w}{\partial z} - \frac{\partial T_m}{\partial z}\right) \tag{10.14d}$$

If the heat flux is constant, $\partial q_z''/\partial z = 0$ and the radial profile is constant, so that

$$\frac{\partial T_w}{\partial z} = \frac{\partial T_m}{\partial z} = \frac{\partial T}{\partial z}; \quad q_w''(z) = \text{constant} \tag{10.15}$$

whereas if the wall temperature is axially constant,

$$\frac{\partial T}{\partial z} = f\left(\frac{r}{R}\right)\frac{\partial T_{\mathrm{m}}}{\partial z}; \quad T_{\mathrm{w}}(z) = \text{constant} \tag{10.16}$$

In this case, $\partial q_z''/\partial z \neq 0$.

The steady-state temperature profile in the pipe is now derived for the constant heat flux case. As shown in Chapter 9, the velocity profile in this case is given by

$$\upsilon_{\mathrm{z}} = 2V_{\mathrm{m}}\left[1-\left(\frac{r}{R}\right)^2\right] \tag{9.53}$$

where

$$V_{\mathrm{m}} = \frac{\int_{\mathrm{o}}^{R} \rho\upsilon_{\mathrm{z}} 2\pi r \, \mathrm{d}r}{\int_{\mathrm{o}}^{\mathrm{R}} \rho 2\pi r \, \mathrm{d}r}$$

Neglecting the internal energy generation and the dissipation energy, at steady state the cylindrical (r, z) energy equation for fully developed conditions, Equation 10.3b reduces to

$$\rho c_{\mathrm{p}}\upsilon_{\mathrm{z}}\frac{\partial T}{\partial z} = \frac{\partial}{\partial z}k\frac{\partial T}{\partial z} + \frac{1}{r}\frac{\partial}{\partial r}\left(kr\frac{\partial T}{\partial r}\right) \tag{10.17}$$

In most applications, even a small velocity is sufficient to make the axial conduction heat transfer negligible, as

$$\frac{\partial}{\partial z}k\frac{\partial T}{\partial z} \ll \rho c_{\mathrm{p}}\upsilon_{\mathrm{z}}\frac{\partial T}{\partial z} \tag{10.18}$$

Substituting for υ_{z} from Equation 9.53 and applying the condition of Equation 10.18, Equation 10.17 can be written as

$$2\rho c_{\mathrm{p}}V_{\mathrm{m}}\left[1-\left(\frac{r}{R}\right)^2\right]\frac{\partial T}{\partial z} = \frac{1}{r}\frac{\partial}{\partial r}\left(kr\frac{\partial T}{\partial r}\right) \tag{10.19}$$

However, because $\partial T/\partial z$ is not a function of r, Equation 10.19 can be integrated to yield

$$2\rho c_{\mathrm{p}}V_{\mathrm{m}}\frac{\partial T}{\partial z}\left(\frac{r^2}{2} - \frac{r^4}{4R^2}\right) = kr\frac{\partial T}{\partial r} + C_1 \tag{10.20}$$

Applying the symmetry condition that at

$$r = 0 \quad \frac{\partial T}{\partial r} = 0 \tag{10.21}$$

we get $C_1 = 0$.

Integrating Equation 10.20 over r again, we get

$$2\rho c_p V_m \frac{\partial T}{\partial z}\left(\frac{r^2}{4} - \frac{r^4}{16R^2}\right) = kT + C_2 \tag{10.22}$$

Because at

$$r = R, \quad T = T_w, \tag{10.23}$$

$$C_2 = -kT_w + 2\rho c_p V_m \frac{\partial T}{\partial z}\left(\frac{3R^2}{16}\right) \tag{10.24}$$

and Equation 10.22 can be rearranged to give

$$T = T_w + \frac{2\rho c_p}{k} V_m \frac{\partial T}{\partial z}\left(\frac{r^2}{4} - \frac{r^4}{16R^2} - \frac{3R^2}{16}\right) \tag{10.25}$$

The heat flux at the wall, q''_w, is given by

$$q''_w = -k\left.\frac{\partial T}{\partial r}\right|_R = -2\rho c_p V_m \frac{\partial T}{\partial z}\left(\frac{R}{2} - \frac{R}{4}\right) \tag{10.26}$$

or

$$q''_w = -\left(2\rho c_p V_m \frac{\partial T}{\partial z}\right)\left(\frac{R}{4}\right) \tag{10.27}$$

Thus

$$\frac{\partial T}{\partial z} = -\frac{2q''_w}{\rho c_p V_m R} \tag{10.28}$$

Realizing that for the axially constant heat flux case $\partial T/\partial z = \partial T_m/\partial z$, Equation 10.28 could have been obtained from the energy balance for the cross section:

$$\rho V_m \pi R^2 \, c_p \left(\frac{\partial T_m}{\partial z}\right) = (-q''_w) 2\pi R \tag{10.29}$$

Note that the outward heat flux is given by $-k(\partial T/\partial r)$; hence for a heated channel $-q''_w$ is a positive number.

The mean temperature (T_m) can be evaluated using Equations 10.13 and 9.53 as

$$T_m - T_w = \frac{\int_o^R (T - T_w) 2V_m \left[1 - (r/R)^2\right] 2\pi r \, dr}{V_m \pi R^2}$$
$$= \frac{8\rho c_p V_m (\partial T/\partial z)}{R^2 k} \int_o^R \left[\frac{r^2}{4} - \frac{r^4}{16R^2} - \frac{3R^2}{16}\right]\left[1 - \left(\frac{r}{R}\right)^2\right] r \, dr$$
$$= -\frac{11}{48}\frac{\rho c_p V_m}{k}\left(\frac{\partial T}{\partial z}\right) R^2 \tag{10.30}$$

By combining Equations 10.29 and 10.30, we get

$$T_m - T_w = \frac{11}{24}\frac{R}{k} q''_w \tag{10.31}$$

From the definitions of h and Nu, we get

$$h = \frac{q''_w}{T_m - T_w} = \frac{24}{11}\frac{k}{R} \tag{10.32}$$

$$\text{Nu} = \frac{hD}{k} = \frac{h(2R)}{k} = \frac{48}{11} = 4.364 \tag{10.33}$$

The Nusselt number derived above is applicable to the constant heat flux case in a circular tube. The value of Nu depends on the boundary conditions as well as the geometry (Table 10.3). As Table 10.3 indicates for the case of a constant wall temperature heated tube, the Nusselt number is a constant of a different value, that is, Equation 3.66. In general, laminar flow conditions lead to Nusselt numbers that are constants, independent of flow velocity (or Re) and Pr.

10.2.2 Developed Flow in Other Geometries

In a manner similar to the tube flow with constant heat flux, the laminar flow heat transfer coefficient and the Nusselt numbers can be computed for other geometries. In all cases, the laminar Nusselt number is a constant independent of the flow velocity or Prandtl number. This case is also true for a constant axial wall temperature.

For geometries other than the round tube, a Nusselt number is defined using the concept of equivalent hydraulic diameter where

$$D_e \equiv \frac{4 \times \text{Flow area}}{\text{Wetted perimeter}} = \frac{4A_f}{P_w} \tag{10.34}$$

Table 10.4 provides the Nu values for various geometries.

TABLE 10.3
Nusselt Number for Laminar Fully Developed Velocity and Temperature Profiles in Tubes of Various Cross Sections

Cross-Sectional Shape	b/a	Nu^a q'' = Constant	Nu^a T_w = Constant
Circle	—	4.364	3.66
Square	1.0	3.61	2.98
Rectangle	1.43	3.73	3.08
Rectangle	2.0	4.12	3.39
Rectangle	3.0	4.79	3.96
Rectangle	4.0	5.33	4.44
Rectangle	8.0	6.49	5.60
Infinite parallel plates	∞	8.235	7.54
Infinite parallel plates (one insulated)	∞	5.385	4.86
Isosceles triangle	—	3.00	2.35

Source: From Kays, W. M. and Crawford, M. E., *Convective Heat and Mass Transfer*, 2nd Ed. New York, NY: McGraw-Hill, 1980.

[a] The constant heat rate solutions are based on constant *axial* heat rate but with constant *temperature* around the tube periphery. Nusselt numbers are averages with respect to tube periphery.

For flow parallel to a bundle of circular tubes, Sparrow et al. [55] solved the laminar problem for axially constant q''_w and circumferentially constant T_w (boundary condition A). Dwyer and Berry [16] solved the laminar case for axially and circumferentially constant q''_w (boundary condition B). Their results are given in Figure 10.5. For sufficiently large P/D values, Figure 10.5 demonstrates that the equivalent annulus approximation is good relative to the exact solution. The exact solution for the azimuthal variation of the temperature when P/D is less than 1.5 is described in Chapter 7, Volume II.

10.2.3 Developing Laminar Flow Region

In a pipe entry region, the heat transfer is more involved, as both the velocity and temperature profiles may vary axially. When the Prandtl number is higher than 5 [24], the velocity profile develops faster than the temperature profile. Hence, as in the

TABLE 10.4
Coefficients for Evaluation of Nusselt Numbers in Diabatic Flow

	Liquids $0.025 \leq \mu_b/\mu_w \leq 12.5$	Gases $0.27 \leq T_b/T_w \leq 2.7$
Heating ($T_w > T_b$)	$m = 0.11$	$m = 0.47$
Cooling ($T_w < T_b$)	$m = 0.25$	$m = 0$

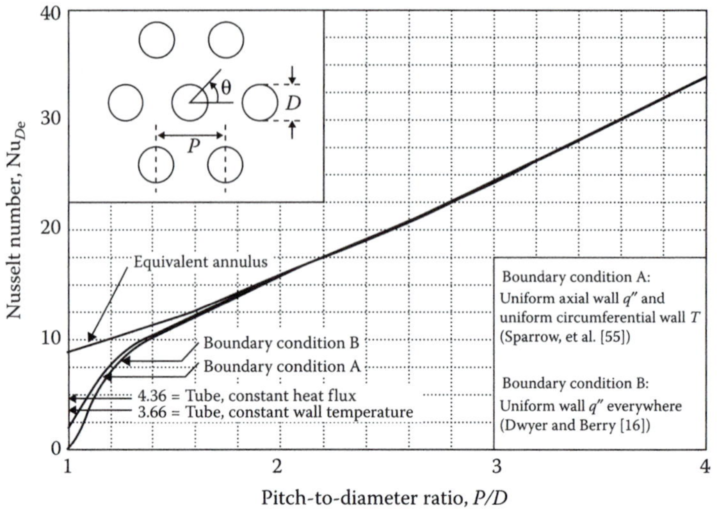

FIGURE 10.5 Nusselt numbers for fully developed laminar flow parallel to an array of circular tubes.

early work of Graetz, reported in 1885 [19], the parabolic velocity profile may be assumed to exist at the tube entrance with a uniform fluid temperature. When the Prandtl number is low, a tube with limited length may fully develop a thermal boundary layer while the velocity is still developing; hence, a slug (flat) velocity profile may be assumed.

Several solutions exist for simultaneous momentum and thermal laminar boundary layer development. Sparrow et al. [54] found the heat transfer coefficient for various Pr numbers and constant heat flux. Kays [25] solved the developing laminar flow problem, for Pr = 0.7, in a tube with uniform heat flux and uniform wall temperature (Figure 10.6). It is seen that in this region the Nu value is higher than the asymptotic value.

In general, the developing region (or thermal entry region) may be assumed to extend to a length ξ_T, such that

$$\frac{\xi_T}{D_e} = 0.05 \text{ Re Pr} \tag{10.35}$$

Bhatti and Savery [1] found that the thermal laminar developing length may be predicted by

$$\begin{aligned} \frac{\xi_T}{D_e} &= 0.1 \text{ Re Pr} && \text{for } 0.7 < \text{Pr} < 1 \\ &\simeq 0.004 \text{ Re} && \text{for Pr} = 0.01 \\ &\simeq 0.15 \text{ Re Pr} && \text{for Pr} > 5 \end{aligned} \tag{10.36}$$

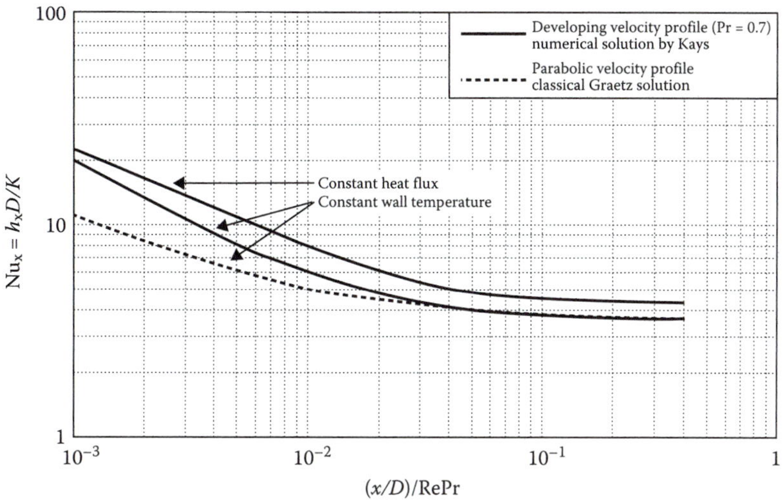

FIGURE 10.6 Local Nusselt number for simultaneous velocity and temperature development for laminar flow in a circular tube (Pr = 0.7). (From Kays, W. M. *Trans. ASME*, 77:1265, 1955.)

10.3 TURBULENT HEAT TRANSFER: MIXING LENGTH APPROACH

10.3.1 Equations for Turbulent Flow in Circular Coordinates

It was shown in Chapter 4 that the instantaneous values of velocity and temperature can be expanded into a time-averaged component and a fluctuating component. The steady-state incompressible fluid transport equations for mass, momentum, and energy can be given by inspection of Equations 4.135 through 4.137, respectively, as

$$\nabla \cdot [\rho \bar{\vec{\upsilon}}] = 0 \tag{9.68}$$

$$\nabla \cdot [\rho \bar{\vec{\upsilon}} \bar{\vec{\upsilon}}] = -\nabla p + \nabla \cdot [\bar{\bar{\tau}} - \rho \overline{\vec{\upsilon}' \vec{\upsilon}'}] + \rho \vec{g} \tag{9.69}$$

$$\nabla \cdot [\rho u^\circ \, \bar{\vec{\upsilon}}] = -\nabla \cdot \overline{q''} + [\overline{q''' - p\nabla \cdot \vec{\upsilon} + \Phi}] - \nabla \cdot \rho \overline{u^{\circ\prime} \vec{\upsilon}'} \tag{10.37a}$$

and by similar transformation of Equation 4.122a as

$$\nabla \cdot [\rho c_p \bar{\vec{\upsilon}} \, \bar{T}] = -\nabla \cdot \overline{q''} + \overline{q'''} + \bar{\Phi} - \nabla \cdot \rho c_p \overline{T' \vec{\upsilon}'} \tag{10.37b}$$

For a fluid with constant properties, the axial momentum equation can be expanded for the cylindrical coordinates as

$$\rho\left(\bar{\upsilon}_r \frac{\partial \bar{\upsilon}_z}{\partial r} + \frac{\bar{\upsilon}_\theta}{r}\frac{\partial \bar{\upsilon}_z}{\partial \theta} + \bar{\upsilon}_z \frac{\partial \bar{\upsilon}_z}{\partial z}\right) = -\frac{\partial p}{\partial z} + \frac{1}{r}\frac{\partial}{\partial r} r(\tau_{rz})_{\text{eff}} + \frac{1}{r}\frac{\partial}{\partial \theta}(\tau_{\theta z})_{\text{eff}} + \frac{\partial (\tau_{zz})_{\text{eff}}}{\partial z} + \rho \vec{g} \tag{10.38}$$

For fully developed flow, the r and θ direction momentum balances reduce to the condition of constant pressure in these directions. Also, $\bar{\upsilon}_r = \bar{\upsilon}_\theta = 0$.

The momentum fluxes $(\tau_{rz})_{eff}$, $(\tau_{\theta z})_{eff}$, and $(\tau_{zz})_{eff}$ incorporate contributions due to viscous effects and turbulent velocity fluctuations. In general, the effective stresses can be given by

$$(\tau_{rz})_{eff} = \mu\left(\frac{\partial \bar{\upsilon}_z}{\partial r} + \frac{\partial \bar{\upsilon}_r}{\partial z}\right) - \rho\overline{\upsilon_r'\upsilon_z'} \quad (10.39)$$

$$(\tau_{\theta z})_{eff} = \mu\left(\frac{\partial \bar{\upsilon}_\theta}{\partial z} + \frac{1}{r}\frac{\partial \bar{\upsilon}_z}{\partial \theta}\right) - \rho\overline{\upsilon_z'\upsilon_\theta'} \quad (10.40)$$

$$(\tau_{zz})_{eff} = \mu\left(2\frac{\partial \bar{\upsilon}_z}{\partial z}\right) - \rho\overline{\upsilon_z'\upsilon_z'} \quad (10.41)$$

where $\bar{\upsilon}_z, \bar{\upsilon}_r, \bar{\upsilon}_\theta$ is the time-averaged flow velocities and $\upsilon_z', \upsilon_r', \upsilon_\theta'$ is the fluctuating components of υ_z, υ_r, and υ_θ. The turbulent stress terms $\rho\overline{\upsilon_r'\upsilon_z'}$ and $\rho\overline{\upsilon_z'\upsilon_\theta'}$ are one-point correlations of mutually perpendicular velocity fluctuations and therefore have negative values. These correlation terms are called *Reynolds stresses*. The fluctuating components are typically small in magnitude relative to the time-averaged flow velocities.

For axial flow in axisymmetric geometries under fully developed conditions, the time-averaged transverse velocities $\bar{\upsilon}_r$ and $\bar{\upsilon}_\theta$ are zero. However, in nonaxisymmetric geometries these components exist even under fully developed conditions. They are caused by nonuniformities in wall turbulence and are called *secondary flows of the second kind*. They are in contrast to *secondary flows of the first kind*, which are produced by turning or skewing the primary flow, for example, flow in curved ducts. The secondary flow effects are small and much less studied than turbulent effects and are usually inferred from the pattern of shear stress distribution [57].

Neglecting the dissipation energy and internal heat generation, the energy equation (Equation 10.37b) can also be expressed as

$$\rho c_p\left(\bar{\upsilon}_r\frac{\partial \bar{T}}{\partial r} + \bar{\upsilon}_\theta\frac{1}{r}\frac{\partial \bar{T}}{\partial \theta} + \bar{\upsilon}_z\frac{\partial \bar{T}}{\partial z}\right) = -\left[\frac{1}{r}\frac{\partial r(q_r'')_{eff}}{\partial r} + \frac{1}{r}\frac{\partial (q_\theta'')_{eff}}{\partial \theta} + \frac{\partial (q_z'')_{eff}}{\partial z}\right] \quad (10.42)$$

The heat fluxes q_r'', q_θ'', and q_z'' can be given by

$$(q_r'')_{eff} = -\rho c_p \alpha\frac{\partial \bar{T}}{\partial r} + \rho c_p\overline{T'\upsilon_r'} \quad (10.43a)$$

$$(q_\theta'')_{eff} = -\rho c_p \alpha\frac{1}{r}\frac{\partial \bar{T}}{\partial \theta} + \rho c_p\overline{T'\upsilon_\theta'} \quad (10.43b)$$

$$(q_z'')_{\text{eff}} = -\rho c_p \alpha \frac{\partial \bar{T}}{\partial z} + \rho c_p \overline{T'\upsilon_z'} \tag{10.43c}$$

where $\bar{T}$ is the time-averaged temperature, and T' is its fluctuating component.

The foregoing equations can be developed further only when expressions are available relating the turbulent flux terms to the mean flow properties. Here, however, we treat the simpler case of fully developed flow in circular geometry where secondary flows do not exist and the velocity fluctuations can be related to the velocity gradients by the turbulent or eddy diffusivity for momentum (ε_M) and for heat (ε_H) such that

$$\overline{\upsilon_z'\upsilon_r'} = -\varepsilon_{M,r} \frac{\partial \bar{\upsilon}_z}{\partial r} \tag{10.44}$$

$$\overline{\upsilon_z'\upsilon_\theta'} = -\varepsilon_{M,\theta} \frac{1}{r} \frac{\partial \bar{\upsilon}_z}{\partial \theta} \tag{10.45}$$

$$\overline{T'\upsilon_r'} = -\varepsilon_{H,r} \frac{\partial \bar{T}}{\partial r} \tag{10.46}$$

$$\overline{T'\upsilon_\theta'} = -\varepsilon_{H,\theta} \frac{1}{r} \frac{\partial \bar{T}}{\partial \theta} \tag{10.47}$$

With the restrictions of fully developed flow,

$$\bar{\upsilon}_\theta = \bar{\upsilon}_r = 0 \tag{10.48}$$

$$\frac{\partial \bar{\upsilon}_z}{\partial z} = 0 \tag{10.49}$$

$$\tau_{zz} = 0 \tag{10.50}$$

For fully developed heat transfer, in general (see Equation 10.14a)

$$\frac{\partial}{\partial z}\left(\frac{T_w - \bar{T}(r)}{T_w - \bar{T}_m}\right) = 0 \tag{10.51}$$

To evaluate $\partial \bar{T}/\partial z$, we need to specify the axial boundary conditions at the wall.

Applying Equations 10.48 through 10.50 and neglecting gravity, Equations 10.38 and 10.42 become, respectively,

$$\frac{1}{r}\left(\frac{\partial(\tau_{rz} r)}{\partial r} + \frac{\partial \tau_{\theta z}}{\partial \theta}\right) = \frac{\partial p}{\partial z} \tag{10.52}$$

$$\frac{1}{r}\left[\frac{\partial}{\partial r}(q_r''r)+\frac{\partial q_\theta''}{\partial \theta}\right]+\frac{\partial q_z''}{\partial z}=-\rho c_p\bar{\upsilon}_z\frac{\partial \bar{T}}{\partial z} \quad (10.53)$$

where the subscript eff has been dropped for brevity.

For an axially constant wall heat flux, the fully developed flow features require that

$$\frac{\partial T_w}{\partial z}=\frac{\partial \bar{T}_m}{\partial z} \quad (10.54)$$

With this condition, Equation 10.51 reduces to a form similar to Equation 10.15 for laminar flow:

$$\frac{\partial \bar{T}}{\partial z}=\frac{\partial T_w}{\partial z}=\frac{\partial \bar{T}_m}{\partial z} \quad (10.55)$$

independent of r for $q_w''(z)=$ constant. The magnitude of $\partial\bar{T}/\partial z$ can be determined from an energy balance as

$$\frac{\partial \bar{T}}{\partial z}=\frac{\partial \bar{T}_m}{\partial z}=\frac{4q_w''}{V_m\rho c_p D_H}=\frac{q'}{\dot{m}c_p}=\text{constant} \quad (10.56)$$

For Equation 10.56 to be true, note that the axial heat conduction is neglected, that is

$$-k\frac{\partial^2 \bar{T}}{\partial z^2}=\frac{\partial q_z''}{\partial z}=0 \quad (10.57)$$

This condition is acceptable for practically all flows with reasonable velocities, so that $\rho c_p \upsilon_z > k(\partial\bar{T}/\partial z)$.

On the other hand, for a constant axial wall temperature boundary condition, $T_w(z)=$ constant. Hence

$$\frac{\partial T_w}{\partial z}=0 \quad (10.58)$$

and Equation 10.51 reduces to a form similar to Equation 10.16 for laminar flow:

$$\frac{\partial \bar{T}}{\partial z}=f\left(\frac{r}{R}\right)\frac{\partial \bar{T}_m}{\partial z} \quad (10.59)$$

Applying the conditions of Equations 10.44 through 10.47 to Equations 10.39 and 10.40, we get the shear stress in terms of the kinematic viscosity (ν) and eddy diffusivity of momentum ($\varepsilon_{M,r}$) and ($\varepsilon_{M,\theta}$):

$$(\tau_{rz})_{eff}=\rho(\nu+\varepsilon_{M,r})\frac{\partial \bar{\upsilon}_z}{\partial r} \quad (10.60)$$

$$(\tau_{\theta z})_{\text{eff}} = \rho(\nu + \varepsilon_{\text{M},\theta})\frac{\partial \bar{\upsilon}_z}{r\partial\theta} \tag{10.61}$$

Similarly, we get the heat flux in terms of the molecular thermal diffusivity (α) and the eddy diffusivity of heat ($\varepsilon_{\text{H,r}}$) and ($\varepsilon_{\text{H},\theta}$):

$$(q_r'')_{\text{eff}} = -\rho c_p[\alpha + \varepsilon_{\text{H,r}}]\frac{\partial \bar{T}}{\partial r} \tag{10.62}$$

$$(q_\theta'')_{\text{eff}} = -\rho c_p[\alpha + \varepsilon_{\text{H},\theta}]\frac{\partial \bar{T}}{r\partial\theta} \tag{10.63}$$

Of course, with circular geometry with azimuthal symmetry, the angular dependence is eliminated. Thus in round tubes, the applicable equations for axial momentum (Equation 10.52) and energy (Equation 10.53) are

$$\frac{1}{r}\frac{\partial}{\partial r}\left[\rho r(\nu + \varepsilon_{\text{M,r}})\frac{\partial \bar{\upsilon}_z}{\partial r}\right] = \frac{\partial p}{\partial z} \tag{10.64}$$

$$\frac{1}{r}\frac{\partial}{\partial r}\left[r(\alpha + \varepsilon_{\text{H,r}})\frac{\partial T}{\partial r}\right] = \bar{\upsilon}_z\frac{\partial \bar{T}}{\partial z} \tag{10.65}$$

10.3.2 Relation between ε_M, ε_H, and Mixing Lengths

The momentum and energy equations for fully developed flow (Equations 10.64 and 10.65) have the same form. The eddy diffusivity of momentum has the same dimensions as the eddy diffusivity of heat (both units are square meters per second or square feet per hour). There are no eddies of momentum and energy as such, but these diffusivities are related to the turbulent fluid eddy properties. Thus, there is reason to believe that these diffusivities are related. Indeed, in laminar flow both are zero, but in highly turbulent flow they may be so much higher than molecular diffusivities that they control the heat transfer.

For ordinary fluids, with $\text{Pr} \simeq 1$, it was often assumed that $\varepsilon_H = \varepsilon_M$, with a reasonable measure of success. However, it was noted by Jenkins [22] and Deissler [12] that in liquid metal flow an eddy may give up or gain heat during its travel owing to the high conductivity value. Hence $\varepsilon_H < \varepsilon_M$, but the exact value depends on the flow geometry and Reynolds number.

The mixing length theory has been extensively used to define ε_H and ε_M. The basic idea of the mixing length is similar to that of the mean free path of molecules in the kinetic theory of gases. Thus, a fluid eddy is assumed to travel a certain length (ℓ_M) perpendicular to the flow stream before losing its momentum. Thus

$$\upsilon_z' = \ell_M \frac{d\bar{\upsilon}_z}{dy} \tag{10.66}$$

where $y = R - r$.

Prandtl [44] first suggested that, for the parallel plate situation, υ'_z is of the same order as υ'_y, although by continuity consideration they should be of opposite sign. Thus he obtained

$$\overline{\upsilon'_z \upsilon'_y} = -\ell_M^2 \left| \frac{d\bar{\upsilon}_z}{dy} \right| \frac{d\bar{\upsilon}_z}{dy} \quad (10.67)$$

Hence by inspection of Equation 10.44, we get

$$\varepsilon_M = \ell_M^2 \left| \frac{d\bar{\upsilon}_z}{dy} \right| \quad (10.68)$$

Prandtl proposed that the mixing length is proportional to the distance from the wall, so that

$$\ell_M = Ky \quad (10.69)$$

where K is a universal constant, which was empirically found to be equal to 0.42 in the area adjacent to the wall.

Schlichting [48] reviewed the theories advanced for the mixing lengths since the original work of Prandtl. He suggested that for fully developed flow in ducts and pipes, the mixing length is well characterized by the Nikuradse formula:

$$\frac{\ell_M}{R} = 0.14 - 0.08\left(1 - \frac{y}{R}\right)^2 - 0.06\left(1 - \frac{y}{R}\right)^4 \quad (10.70)$$

From Dwyer's work [15], the following relation is recommended for liquid metals ($\Pr \ll 1$):

$$\frac{\varepsilon_H}{\varepsilon_M} \equiv \bar{\psi} = 1 - \frac{1.82}{\Pr\left(\varepsilon_M/\nu\right)_{max}^{1.4}} \quad (10.71)$$

where $(\varepsilon_M/\nu)_{max}$ is the maximum value in channel flow which for each geometry is a unique function of the Reynolds number. Various values of $(\varepsilon_M/\nu)_{max}$ are given for rod bundles in Figure 10.7. At low Reynolds numbers, $\bar{\psi}$ approaches zero because hot eddies lose heat in transit to the surrounding liquid metal. Equation 10.71 may produce negative values for $\bar{\psi}$ at very low values of Pr. Negative values are not permissible so in this case $\bar{\psi}$ should be set at zero. The contribution of turbulence to heat transfer would then be negligible, and a laminar heat transfer treatment would be adequate. At high Reynolds numbers, $\bar{\psi}$ approaches unity, as the eddies move so fast that they lose an insignificant amount of heat in transit.

The selection of ε_M and ε_H was comprehensively reviewed by Nijsing [40]. The mixing length theory was used extensively to define ε_M and ε_H and to solve for the temperature distribution. Advances in numerical computation, however, have shifted

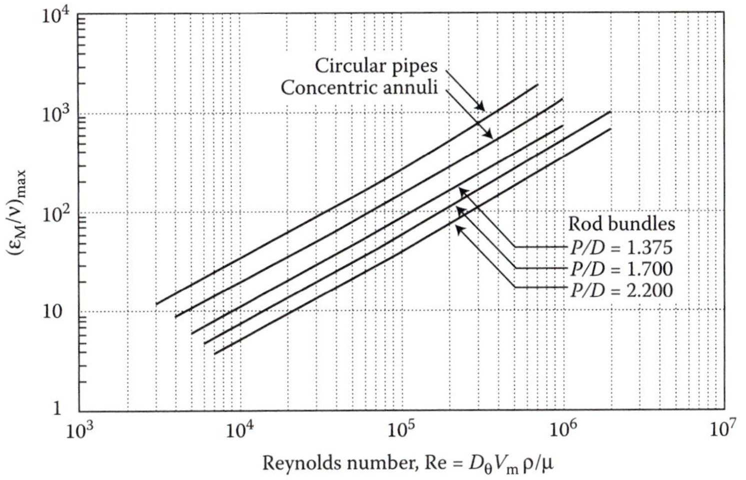

FIGURE 10.7 Values of $(\varepsilon_M/\nu)_{max}$ for fully developed turbulent flow of liquid metals through circular tubes, annuli, and rod bundles with equilateral triangular spacing. (From Dwyer, O. E., *AIChE J.*, 9:261, 1963.)

the treatment of turbulence to a more involved set of equations, for which more than one turbulent parameter (ε_M or ε_H) can exist in each transport equation. A brief description is given in Section 10.4.

10.3.3 Turbulent Temperature Profile

Analytic predictions of the temperature profile in turbulent flow date back to the simplest treatment of Reynolds (1874). Many refined treatments followed, notably those of Prandtl [43,44], von Karman [58], Martinelli [36], and Deissler [11,12]. These treatments described, using various approximations, the velocity profile and the eddy diffusivities ε_M and ε_H for fully developed flow.

Reynolds assumed that the entire flow field consists of a single zone that is highly turbulent so that molecular diffusivities can be neglected, that is, $\varepsilon_M \gg \nu$ and $\varepsilon_H \gg \alpha$. From Equations 10.60 and 10.62, we get in circular geometry where there is no azimuthal variation:

$$\frac{\left(\tau_{rz}\right)_{eff}}{\left(q''_r\right)_{eff}} = -\frac{\rho\varepsilon_{M,r}}{\rho c_p \varepsilon_{H,r}} \frac{d\bar{\upsilon}_z}{d\bar{T}} \tag{10.72}$$

Reynolds also assumed that $\varepsilon_{M,r}/\varepsilon_{H,r} = 1.0$ and that $\left(\tau_{rz}/q''_r\right)_{eff}$ is constant (i.e., equal to τ_w/q''_w) throughout the field. Therefore, integration of Equation 10.72 between the wall and the bulk or mean values gives

$$\int_{T_w}^{T_m} d\bar{T} = -\frac{q''_w}{\tau_w c_p} \int_{o}^{V_m} d\bar{\upsilon}_z$$

or

$$T_w - T_m = \frac{q''_w V_m}{\tau_w c_p} \tag{10.73}$$

By definition, the heat transfer coefficient is given by

$$h = \frac{q''_w}{T_w - T_m} \tag{10.74}$$

and the wall shear stress is related to the friction factor by

$$\tau_w = \frac{f}{4}\rho\frac{V_m^2}{2} \tag{9.60}$$

Equations 10.74 and 10.75, when substituted into Equation 10.73, lead to

$$\text{St} = \frac{\text{Nu}_D}{\text{Re}_D\ \text{Pr}} = \frac{h}{\rho c_p V_m} = \frac{f}{8} \tag{10.75}$$

where St is the dimensionless group called the *Stanton number.* This equation is known as the Reynolds analogy for momentum and heat transfer for fully developed turbulent flow and is valid only if $\text{Pr} \simeq 1.0$.

Prandtl assumed that the flow consists of two zones: a laminar sublayer where molecular effects dominate ($\varepsilon_M \ll \nu$ and $\varepsilon_H \ll \alpha$) and a turbulent layer where eddy diffusivities dominate ($\varepsilon_M \gg \nu$ and $\varepsilon_H \gg \alpha$). Again assuming $\varepsilon_M/\varepsilon_H = 1$ and a constant ratio of q''/τ in the turbulent layer, Prandtl obtained a modification of Equation 10.75 in which the ratio $f/8$ was multiplied by a term as noted in the following equation:

$$\text{St} = \frac{f}{8}\frac{1}{1+5\sqrt{f/8}(\text{Pr}-1)} \tag{10.76}$$

Further analogies of similar form were developed by subsequent investigators. Von Karman and Martinelli as discussed below are among the most prominent. A useful form of analogy which accommodates the Prandtl number effect (except at very low values) is that obtained for turbulent pipe flow from the Reynolds–Colburn analogy as demonstrated in the text of Lienhard and Lienhard [31], namely

$$\text{Nu}_D = \frac{(f/8)\text{Re}_D\ \text{Pr}}{1+12.8\sqrt{f/8}(\text{Pr}^{0.68}-1)} \tag{10.77}$$

This result was built upon by Petukhov [41] and then Gnielinski [17] to obtain their accurate correlations of Equations 10.95 and 10.96.

Von Karman [58] extended earlier treatments by assuming that three zones exist: a laminar sublayer, a buffer zone, and the turbulent core. He made assumptions similar to those of Prandtl but allowed molecular and eddy contributions in the buffer zone.

Martinelli [36] obtained solutions for the temperature profile in various geometries by using the assumption that the momentum and heat fluxes near the wall vary linearly with the distance from the wall (y). Thus

$$\frac{q''_{\mathrm{w}}}{\rho c_{\mathrm{p}}}\left(1-\frac{y}{R}\right)=(\alpha+\varepsilon_{\mathrm{H}})\frac{\mathrm{d}\bar{T}}{\mathrm{d}y} \tag{10.78}$$

$$\frac{\tau_{\mathrm{w}}}{\rho}\left(1-\frac{y}{R}\right)=(\nu+\varepsilon_{\mathrm{M}})\frac{\mathrm{d}\bar{\upsilon}_{\mathrm{z}}}{\mathrm{d}y} \tag{10.79}$$

If the velocity profile is empirically known, ε_{M} can be obtained from Equation 10.79. For a given relation of $\bar{\psi}=\varepsilon_{\mathrm{H}}/\varepsilon_{\mathrm{M}}$, Equation 10.78 can then be solved to obtain $\bar{T}(r)$.

Martinelli's results [36] for the temperature distribution in the three sublayers and for $\bar{\psi}=1$ are shown in Figure 10.8 for Re = 10,000 and 1,000,000 for various Prandtl numbers. It is interesting to note that for $\Pr \geq 1.0$, most of the temperature drop occurs in the laminar sublayer. For $\Pr \ll 1$ (e.g., in liquid metals) the temperature drop is more evenly distributed throughout the cross section. The higher Reynolds numbers appear to lead to more uniform temperature in the turbulent core sublayer. Kays and Leung [26] used various refinements for their approach and obtained numerical results for the Nusselt number dependence on Re in a uniformly heated round tube. Their results are shown in Figure 10.9. It is seen that for $\Pr \geq 0.5$ the change of Nu with Re is logarithmically linear. For $\Pr \leq 0.3$, the change of Nu with Re is more gradual, reflecting the influence of the high molecular thermal diffusivity.

Sleicher and Tribus [53] presented results for low Prandtl heat transfer solution with a constant wall temperature. Their results were used by Kays and Crawford [24] to plot the ratio of Nu for a constant heat flux, $(\mathrm{Nu})_{\mathrm{H}}$, to Nu for a constant wall temperature $(\mathrm{Nu})_{\mathrm{T}}$ (Figure 10.10). It is seen that the Nusselt number for constant q''_{w} is higher than that for constant T_{w} but that the difference is significant only for $Pr < 0.7$. Thus, for metallic fluids, the wall boundary condition significantly affects the turbulent Nu number, but for nonmetallic fluids the Nu number is practically independent of the boundary condition.

Example 10.2: Turbulent Heat Transfer in a Steam Generator Tube (Single-Phase Region Only)

PROBLEM

Consider primary system coolant flow in a typical steam generator tube immersed in a secondary system coolant. We are interested in determining the linear heat transfer rate from the primary side (bulk temperature T_{p}) to the single-phase region of the secondary side (bulk temperature T_{s}). The typical conditions of this

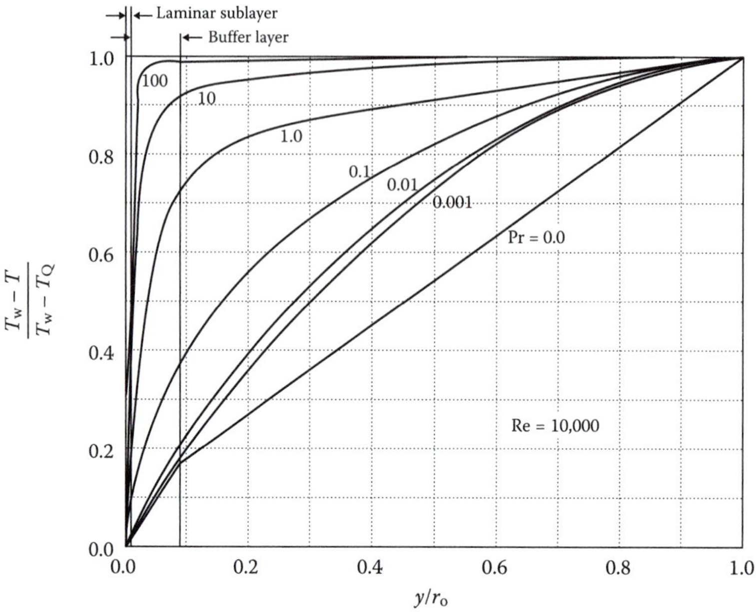

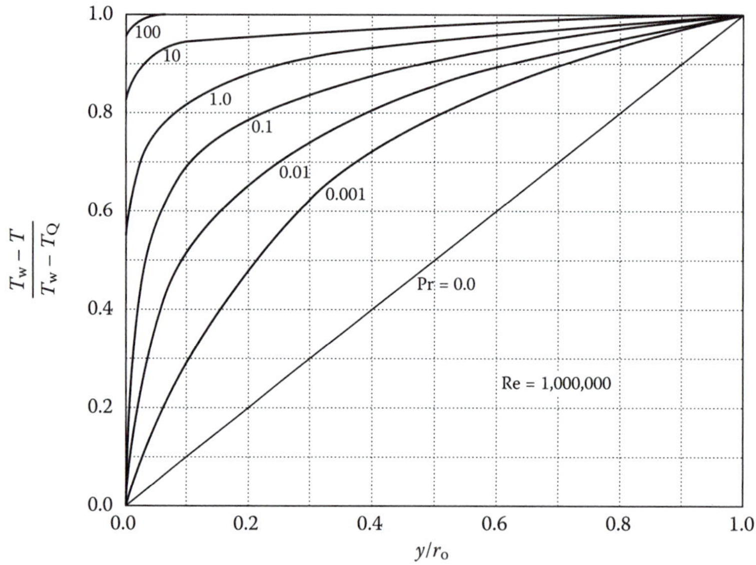

FIGURE 10.8 Martinelli's solution for the temperature distribution in a uniformly heated round tube. (From Martinelli, R. B., *Trans. ASME*, 69:947, 1947.)

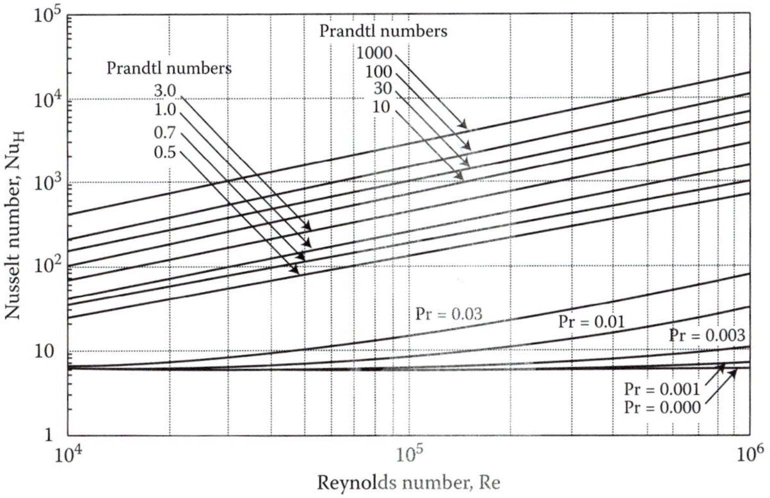

FIGURE 10.9 Nusselt number dependence on Reynolds number and Prandtl number for a uniformly heated round tube. (From Kays, W. M. and Leung, E. Y. *Int. J.H.M.T.*, 10:1533, 1963.)

region are described below. The tubes are made of Inconel ($k = 35$ W/m°C) with *I.D.* = 7/8 in. (2.22 cm) and *O.D.* = 1 in. (2.54 cm); the flow rate is 1.184 kg/s per tube. The average reactor coolant temperature is approximately 305°C, and the average secondary side temperature in the single-phase region is 280°C. Assume that the heat transfer coefficient on the outside of the tube is the same as that on the inside (not generally true as demonstrated later in Example 10.5).

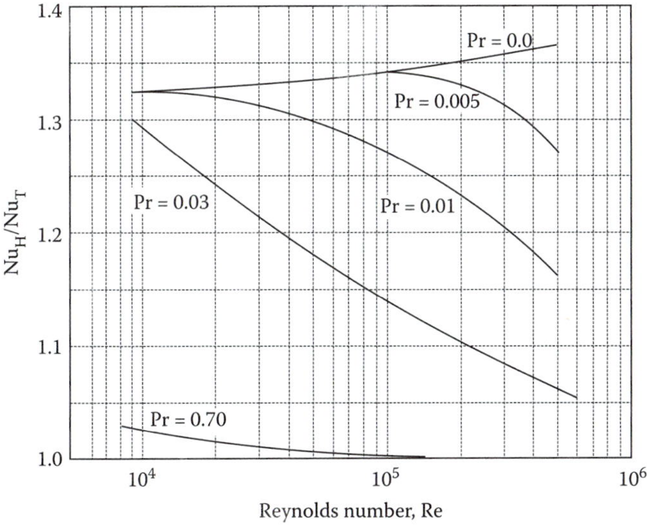

FIGURE 10.10 Ratio of Nusselt number for constant heat rate (Nu_H) to Nusselt number for constant surface temperature (Nu_T) for fully developed conditions in a circular tube. (From Sleicher, C. A. and Tribus, M., *Trans. ASME*, 79:789, 1957.)

SOLUTION

Using the concept of thermal resistance in series (see Sections 8.4.3 and 8.7.3), it can be seen that for a cylindrical geometry the heat transfer rate per unit length may be written as

$$q' = \frac{2\pi(T_p - T_s)}{1/(r_i h_i) + (\ell n(r_o/r_i)/k) + 1/(r_o h_o)}$$

From Example 10.1 we found that, under approximately the same conditions, $Re \approx 7.613 \times 10^5$, $Pr = 1$. We can therefore find an approximate value for the Nusselt number from Figure 10.9. (Note that it is not exactly correct, as the heat rejection along the length of a tube in a steam generator is not uniform.) However, Figure 10.9 indicates that, for $Re = 7.613 \times 10^5$ and $Pr = 1$, $Nu \approx 1200$. Hence,

$$h = Nu\frac{k}{D} = 1200\frac{0.532\ \text{W/m °C}}{(0.875\ \text{in.})(0.0254\ \text{m/in.})} = 2.87 \times 10^4\ \text{W/m}^2\ \text{°C}$$

Then the linear heat transfer rate is

$$q' = \frac{2\pi(25°\text{C})}{(1/((0.0111\ \text{m})(2.87 \times 10^4 \text{W/m}^2\ \text{°C}))) + ((\ell n\ (1/0.875))/35\ \text{W/m °C}) + \frac{1}{(0.0127\ \text{m})(2.87 \times 10^4 \text{W/m}^2 \text{°C})}}$$

$$q' = \frac{157.1}{3.14 \times 10^{-3} + 3.82 \times 10^{-3} + 2.74 \times 10^{-3}} = 1.62 \times 10^4 \text{W/m}$$

10.4 TURBULENT HEAT TRANSFER: DIFFERENTIAL APPROACH

10.4.1 Basic Models

The discussion in Section 10.3 focuses on representation of the fluctuation terms by algebraic relations involving the gradient of the mean velocity and temperature. It is the oldest and probably most popular approach to solving the time-averaged transport equations. The Prandtl mixing length hypothesis is the best-known method for the algebraic connection between the turbulent terms and the mean flow properties. However, this approach implies that the generation and dissipation of turbulence are in balance at each point. Therefore it does not allow for flow at a point to be influenced by turbulence generated at other points. In channel flow, turbulence is produced mainly near the walls and is transported to the bulk flow by diffusion. Turbulence generated by an obstruction is transported downstream, where the local velocity gradients are much smaller. The mixing length (and turbulent diffusivities) approach neglects the transport and diffusion of turbulence.

The differential approach to modeling turbulence was started during the 1940s and became intensive after 1960. Many reviews have appeared in this area, and the reader is referred to works by Launder and Spaulding [29], Rodi [47], and Bradshaw et al. [7] for a thorough review of the various possible modeling approaches. Here only brief descriptions of two popular models are presented: a one-equation model and a two-equation model. The mixing length approach of Section 10.3 is referred to as a zero-equation model, as it does not introduce any new transport equation.

It should be expected that in complex flow situations, for example, rapidly developing flow or recirculating flows, the mixing length approach is not suitable. However, it remains a useful tool for simple flows where ℓ_M can be well specified.

Prandtl was the first to suggest a one-equation model, which involves a transport equation for the turbulent kinetic energy per unit mass (k_t). The turbulent energy is defined by

$$k_t^2 \equiv \frac{1}{2}\left(\overline{\upsilon_x'^2} + \overline{\upsilon_y'^2} + \overline{\upsilon_z'^2}\right) \tag{10.80a}$$

or, using the tensor form

$$k_t^2 \equiv \frac{1}{2}\left(\overline{\upsilon_i' \upsilon_i'}\right) = \frac{1}{2}\left(\overline{\upsilon_i'^2}\right) \tag{10.80b}$$

Other one-equation models are based on the addition of a transport equation for the shear stresses or the turbulent viscosity.

Two-equation models for various parameters have also been introduced. The most popular model is the $k_t - \varepsilon_t$ model, where ε_t is the dissipation rate of the turbulent viscosity given by

$$\varepsilon_t = \nu\overline{\left(\frac{\partial \upsilon_i'}{\partial x_j}\right)^2} \tag{10.81}$$

where ν is the kinematic viscosity. In this model, the transport equations of both k_t and ε_t are solved to obtain a value for the turbulent viscosity (μ_t) defined by

$$\mu_t = \frac{C\rho k_t^2}{\varepsilon_t} \tag{10.82}$$

where C is a constant with a recommended value of 0.09 [20,23]. Note that subscripts t are not normally included in the literature on turbulence. Hence, the model above is referred to as simply the $k - \varepsilon$ model.

10.4.2 Transport Equations for the $k_T - \varepsilon_T$ Model

The time-averaged transport equations for mass, momentum, and energy may be written in tensor form as

$$\frac{\partial \rho}{\partial t} + \frac{\partial \rho \bar{\upsilon}_i}{\partial x_i} = 0 \tag{10.83}$$

$$\rho\left(\frac{\partial\bar{\upsilon}_i}{\partial t}+\bar{\upsilon}_j\frac{\partial\bar{\upsilon}_i}{\partial x_j}\right)=-\frac{\partial p}{\partial x_i}+\frac{\partial\tau_{ji}}{\partial x_j}-\frac{\overline{\partial\rho\upsilon_i'\upsilon_j'}}{\partial x_j}+\rho g_i \tag{10.84}$$

$$\begin{aligned}
\frac{\mathrm{D}k_\mathrm{t}}{\mathrm{D}t}=\frac{\partial}{\partial x_i}&\left[\overline{\upsilon_i'\left(\frac{\upsilon_j'\upsilon_j'}{2}+\frac{p}{\rho}\right)}\right] && \text{Diffusion term}\\
&-\overline{\upsilon_i'\upsilon_j'}\frac{\partial\bar{\upsilon}_i}{\partial x_j} && \text{Production by shear} = P_\mathrm{k}\\
&-\beta g_i\overline{\upsilon_i'T'} && \text{Production due to buoyancy} = G_\mathrm{k}\\
&-\nu\overline{\left(\frac{\partial\upsilon_i'}{\partial x_j}\right)^2} && \text{Viscous dissipation} = -\varepsilon_\mathrm{t}
\end{aligned} \tag{10.85}$$

where

$$\beta=-\frac{1}{\rho}\frac{\partial\rho}{\partial T}\bigg|_p \tag{4.117}$$

The term P_k represents the transfer of kinetic energy from the mean to the turbulent motion, and G_k represents an exchange between the kinetic and potential energies. In stable stratification, G_k is negative, whereas for unstable stratification, it is positive. The viscous dissipation term, ε_t, is always a sink term.

This energy equation is of little use given the number of new unknowns. To overcome this problem, it is assumed that the gradient of k_t controls the diffusion process:

$$\overline{\upsilon_i'\left(\frac{\upsilon_j'\upsilon_j'}{2}+\frac{p}{\rho}\right)}=\frac{\nu_\mathrm{t}}{\sigma_\mathrm{k}}\frac{\partial k_\mathrm{t}}{\partial x_i} \tag{10.86a}$$

where σ_k is the empirical diffusion constant with a recommended value of 1.0 for high Reynolds numbers and $\nu_\mathrm{t}=\mu_\mathrm{t}/\rho$.

Recalling that

$$-\overline{\upsilon_i'\upsilon_j'}=\nu_\mathrm{t}\left(\frac{\partial\bar{\upsilon}_i}{\partial x_j}+\frac{\partial\bar{\upsilon}_j}{\partial x_i}\right)-\frac{2}{3}\delta_{ij}k_\mathrm{t} \qquad \delta_{ij}\begin{array}{ll}=1 & i=j\\ =0 & i\neq j\end{array} \tag{10.86b}$$

$$-\overline{\upsilon_i'T'}=\frac{\nu_\mathrm{t}}{\sigma_\mathrm{t}}\frac{\partial\bar{T}}{\partial x_i} \tag{10.86c}$$

where σ_t = a turbulent Prandtl number ($= 1/\overline{\psi}$). Equation 10.85 can now be written as

$$\frac{Dk_t}{Dt} = \frac{\partial}{\partial x_i}\left(\frac{\nu_t}{\sigma_k}\frac{\partial k_t}{\partial x_i}\right) + \nu_t\left(\frac{\partial \overline{\upsilon_i}}{\partial x_j} + \frac{\partial \overline{\upsilon_j}}{\partial x_i}\right)\frac{\partial \overline{\upsilon_i}}{\partial x_j} - \frac{2}{3}k_t\delta_{ij}\frac{\partial \overline{\upsilon_i}}{\partial x_j} + \beta g_i \frac{\nu_t}{\sigma_t}\frac{\partial \overline{T}}{\partial x_i} - \varepsilon_t \qquad (10.87a)$$

An exact equation can be derived for ε_t from the Navier–Stokes equation, which again has been found to be of little use. A more useful semiempirical relation for ε_t may be given by

$$\rho\frac{D\varepsilon_t}{Dx_j} = C_1\frac{\varepsilon_t}{k_t}(P_k + G_k) - C_2\frac{\rho\varepsilon_t^2}{k_t} + \frac{\partial}{\partial x_j}\left[\left(\frac{\mu_t}{\sigma_t} + \mu\right)\frac{\partial \varepsilon_t}{\partial x_j}\right] \qquad (10.87b)$$

Here, the values of C_1 and C_2 are determined empirically. For grid turbulence, diffusion and $P_k + G_k$ are not important; thus, C_2 can be determined alone and is usually found to be between 1.8 and 2.0. Launder et al. [30] recommended that $C_1 = 1.92$. The value of the turbulent Prandtl number is $\sigma_t = 1.3$ [9]. Equations 10.82, 10.87a, and 10.87b comprise the full $k_t - \varepsilon_t$ model for turbulence.

10.4.3 One-Equation Model

When only the k_t equation is used, ε_t is calculated using the relation

$$\varepsilon_t = \frac{C^{3/4}k_t^{3/2}}{\ell_M} \qquad (10.88)$$

where $\ell_M = 0.42y$; y is the distance from the wall; and C is that of Equation 10.82 with a recommended value of 0.09.

10.4.4 Effect of Turbulence on the Energy Equation

In the energy equation, the transport of enthalpy due to turbulence can be described by the addition of a turbulent diffusivity (ε_H) such that

$$\varepsilon_H = \frac{\mu_t}{\rho\sigma_h} \qquad (10.89)$$

where σ_h is the turbulent Prandtl number for thermal energy transfer having a recommended value of 0.9 [9].

10.4.5 Summary

The preceding discussion presents simple approaches to modeling turbulence which is now the subject of intensive numerical activity. Work in the field of computational fluid dynamics (CFD) has intensified over the last few decades through extensive development of numerical methods and algorithms for problem solution. At the most ambitious level, all relevant length and timescales are resolved by direct numerical simulation (DNS). When the length scales to be resolved must be truncated due to practicality, only the large scales are numerically resolved, yielding a procedure called large eddy simulation (LES). Less numerically intense is the Reynolds-averaged Navier–Stokes (RANS) equation formulation utilizing a $k - \varepsilon$ or Reynolds stress model (RSM). The $k - \varepsilon$ model was briefly introduced in Section 10.4.2.

10.5 HEAT TRANSFER CORRELATIONS IN TURBULENT FLOW

A large amount of experimental and theoretical work has been done for the purpose of assessing the influence of various parameters on the heat transfer coefficient in the case of relatively simple geometries. These studies have shown the following:

1. The $h(x)$ value changes significantly in the entrance region (i.e., the region where the velocity and temperature profiles are still developing). When the entrance region is a small percentage of the whole channel (typical of LWR reactor channels), it can be neglected. If the entrance region is a large percentage of a channel (typical of SFR reactor channels), proper averaging is required. In any event, we can represent the heat transfer performance of a channel by the overall heat transfer coefficient of the channel (averaged for the whole length).
2. For the nonmetallic fluids, where Pr is ≥ 1, the laminar layer is very thin compared with the turbulent region (i.e., turbulent mechanism of heat transfer predominates) and the heat transfer coefficient is not very sensitive to boundary conditions. In the liquid metals case ($\mathrm{Pr} < 0.4$), where heat transfer by conduction is important, the heat transfer coefficient is sensitive to boundary conditions and channel shape. The prediction of heat transfer coefficients for these two ranges of Pr values is different, which makes it necessary to examine them separately.
3. Generally speaking, all heat transfer correlations are only accurate to 10–20%.

10.5.1 Nonmetallic Fluids—Smooth Heat Transfer Surfaces

10.5.1.1 Fully Developed Turbulent Flow

The application of greatest interest is the fully developed turbulent flow, mainly in long channels. In this section, we designate the asymptotic value (i.e., that of the fully developed flow) of the Nusselt number by Nu_∞.

Both experiment and theory show that for almost all nonmetallic fluids Nu_∞ is given by the equation

$$Nu_\infty = CRe^\alpha Pr^\beta \left(\frac{\mu_w}{\mu}\right)^\kappa \quad (10.90a)$$

where μ_w is the fluid viscosity at $T = T_w$; μ is the fluid viscosity at $T = T_b$; C, α, β, and κ are constants that depend on the fluid and the geometry of the channel.

When $T_w - T_b$ is not very large, $\mu_w \approx \mu$ and Equation 10.90a reduces to

$$Nu_\infty = CRe^\alpha Pr^\beta \quad (10.90b)$$

Let us now specialize these equations to geometries of interest.

10.5.1.1.1 Circular Tubes

For ordinary fluids, extensive use has been made of the following equations:

*i. Dittus–Boelter correlation [14] as introduced by McAdams [37]**: For cases when $\mu \approx \mu_w$ the Dittus–Boelter/McAdams correlation is the most universally used correlation. It is recommended by McAdams [38] for both heating and cooling† at moderate $T_w - T_b$ differences.

$$Nu_\infty = 0.023\ Re^{0.8}\ Pr^{0.4} \text{ when the fluid is heated} \quad (10.91)$$

for $0.7 < Pr < 100$, $Re > 10{,}000$, and $L/D > 60$. All fluid properties are evaluated at the mean bulk temperature when the correlation is used on a local basis. When used on an average basis, the arithmetic mean bulk temperature (i.e., the average of the bulk inlet and outlet temperature) is good for constant heat flux while the log mean bulk temperature applies for the constant wall temperature boundary condition.

ii. Colburn [10] correlations: For fluids with high viscosity, Colburn also unified the exponent of Pr for heating and cooling. The result is

$$St\ Pr^{2/3} = 0.023\ Re^{-0.2} \quad (10.92a)$$

where the *Stanton number* (St) = Nu/(RePr). Hence, Equation 10.92a is equivalent to

$$Nu_\infty = 0.023\ Re^{0.8}\ Pr^{1/3} \quad (10.92b)$$

The validity range is the same as for the Dittus–Boelter/McAdams equation and all properties in the St except for c_p are evaluated at the mean film temperature. The c_p value for St is evaluated at the bulk fluid temperature.

---

* This citing of the source of this classic equation versus the traditional paper [14] follows the suggestion of Winterton in his technical note tracing the history of the evolution of this equation. Winterton, R. H. S. *Int. J. Heat Mass Transf.*, 41(4–5):809–810, 1998.

† Other authors suggest Equation 10.91 with $Pr^{0.3}$ for cooling.

For the specific case of organic liquids, Silberberg and Huber [52] recommended

$$\text{Nu}_\infty = 0.015\ \text{Re}^{0.85}\ \text{Pr}^{0.3} \tag{10.93}$$

iii. Seider and Tate [51] correlations: The equation is

$$\text{Nu}_\infty = C\ \text{Re}^{0.8}\ \text{Pr}^{1/3}\left(\frac{\mu_w}{\mu}\right)^{-0.14} \tag{10.94}$$

where $C = 0.023$ for air and 0.027 for petroleum fractions.

Whereas μ_w is evaluated at the wall temperature, all other fluid properties are evaluated at the arithmetic mean bulk temperature for constant heat flux and the log mean bulk temperature for constant wall temperature. This equation is applicable for $0.7 < \text{Pr} < 120$ and $\text{Re} > 10{,}000$ and $L/D > 60$.

iv. Petukhov [41] correlations: Building from Equation 10.77, Petukhov proposed the following relation for fully developed turbulent flow in smooth pipes where all properties are evaluated at the bulk fluid temperature, T_b:

$$\text{Nu}_D = \frac{(f/8)\ \text{Re}_D\ \text{Pr}}{1.07 + 12.7\sqrt{f/8}(\text{Pr}^{2/3} - 1)} \tag{10.95}$$

valid for $10^4 < \text{Re}_D < 5 \times 10^6$, yielding

6% accuracy for $0.5 < \text{Pr} < 200$

10% accuracy for $200 < \text{Pr} < 2000$

v. Gnielinski [17] correlations: Gnielinski extended the results of the significant work done by Petukhov to the transition Reynolds number by making small adjustments of the terms of the Petukhov correlation to yield

$$\text{Nu}_D = \frac{(f/8)(\text{Re}_D - 1000)\ \text{Pr}}{1 + 12.7\sqrt{f/8}(\text{Pr}^{2/3} - 1)} \tag{10.96}$$

where

$$f = \frac{1}{(1.82\log_{10}\text{Re}_D - 1.64)^2} \tag{9.93}$$

and all properties are evaluated at T_b. Equation 10.97 is valid for $2300 \le \text{Re}_D \le 5 \times 10^6$ and as wide a range of Pr as the Petukhov Equation 10.95.

Gnielinski also proposed the following simplified equations, which are frequently used for convenience:

$$\text{Nu} = 0.0214(\text{Re}^{0.80} - 100)\text{Pr}^{0.4} \quad 0.5 < \text{Pr} < 1.5 \text{ (gases)} \tag{10.97a}$$

$$\mathrm{Nu} = 0.0120(\mathrm{Re}^{0.87} - 280)\,\mathrm{Pr}^{0.4} \quad 1.5 < \mathrm{Pr} < 500 \text{ (liquids)} \tag{10.97b}$$

The effect of pipe length flow development and fluid property variations were accommodated by application of two terms to Equation 10.97 yielding the following results for liquids and gases.

For liquids ($0.05 < \mathrm{Pr}/\mathrm{Pr}_{\mathrm{w}} < 20$):

$$\mathrm{Nu_D} = \left[\frac{(f/8)(\mathrm{Re_D} - 1000)\mathrm{Pr}}{1 + 12.7\sqrt{f/8}(\mathrm{Pr}^{2/3} - 1)}\right]\left[1 + \left(\frac{D}{L}\right)^{2/3}\right]\left(\frac{\mathrm{Pr}}{\mathrm{Pr_w}}\right)^{0.11} \tag{10.98a}$$

For gases ($0.5 < T_{\mathrm{b}} / T_{\mathrm{w}} < 1.5$):

$$\mathrm{Nu_D} = \left[\frac{(f/8)(\mathrm{Re_D} - 1000)\mathrm{Pr}}{1 + 12.7\sqrt{f/8}(\mathrm{Pr}^{2/3} - 1)}\right]\left[1 + \left(\frac{D}{L}\right)^{2/3}\right]\left(\frac{T_{\mathrm{b}}}{T_{\mathrm{w}}}\right)^{0.45} \tag{10.98b}$$

vi. Flows with significant variations in fluid properties: Following Lienhard and Lienhard [31], we select the proposal of Petukhov [41] that the effect of significant variations in fluid properties be correlated relative to the Nusselt number based on constant properties evaluated at the bulk temperature, T_{b}, using Equations 10.95 or 10.96 as

For liquids:

$$\frac{\mathrm{Nu}}{\mathrm{Nu}_{T_{\mathrm{b}}}} = \left(\frac{\mu_{\mathrm{b}}}{\mu_{\mathrm{w}}}\right)^m \tag{10.99a}$$

since viscosity change is the dominant effect.

For gases:

$$\frac{\mathrm{Nu}}{\mathrm{Nu}_{T_{\mathrm{b}}}} = \left(\frac{T_{\mathrm{b}}}{T_{\mathrm{w}}}\right)^m \tag{10.99b}$$

since three properties vary significantly with temperature, that is, viscosity, thermal conductivity, and density.

Table 10.4 presents correlations for the exponent m of Equations 10.99a and 10.99b.

vii. Gases for which compressibility effects can be neglected: For Mach number $\ll 1$, the preceding equations can be used. If the Mach number is on the order of 1 or higher, the above equations are used, with h defined as [2]

$$h \equiv \frac{q''_{\mathrm{w}}}{T_{\mathrm{w}} - T^{\mathrm{o}}} \tag{10.100}$$

where

$$T^{\circ} \equiv \left(1+\frac{\gamma-1}{2}M^2\right)T_{\mathrm{m}} \tag{10.101}$$

is called the stagnation temperature, M is the Mach number (see Appendix G), and T_{m} is the fluid mean temperature given by Equation 10.13.

10.5.1.1.2 Annuli and Noncircular Ducts

The same relations for the circular tube are used for annuli and noncircular ducts by employing the concept of the hydraulic diameter:

$$D_{\mathrm{e}} \equiv \frac{4A_{\mathrm{f}}}{P_{\mathrm{w}}} \tag{9.64}$$

for the Reynolds number.

Of course, use of the previous equations is justified only if the sections do not vary appreciably from circular. Such channels are square, rectangular not too far from square, and probably equilateral or nearly equilateral triangles. For geometries far from circular, this method may not be satisfactory [21], particularly in the regions of sharp corners.

10.5.1.1.3 Rod Bundles

For fully turbulent flow along rod bundles, Nu values may significantly deviate from the circular geometry because of the strong geometric nonuniformity of the subchannels. For this reason, Nu and h are expected to depend on the position of the rod within the bundle.

However, it is important to point out that, as stated earlier (Section 10.3.3), the value of Nu is insensitive to the boundary conditions for $\mathrm{Pr} > 0.7$. Furthermore, if the coolant area per rod is taken to define an equivalent annulus with zero shear at its outer boundary, it is found that the Nu predictions are accurate to within ±10% for $P/D \geq 1.12$. This point is illustrated for an interior pin in triangular geometry in Figure 10.11 for $\mathrm{Pr} > 0.7$ and in Figure 10.12 for $\mathrm{Pr} < 0.1$.

The usual way to represent the relevant correlation is to express the Nusselt number for fully developed conditions (Nu_{∞}) as a product of $(\mathrm{Nu}_{\infty})_{\mathrm{c.t.}}$ for a circular tube multiplied by a correction factor:

$$\mathrm{Nu}_{\infty} = \psi(\mathrm{Nu}_{\infty})_{\mathrm{c.t.}} \tag{10.102}$$

where $(\mathrm{Nu}_{\infty})_{\mathrm{c.t.}}$ is usually given by the Dittus–Boelter equation unless otherwise stated.

The problem is then formulated as the evaluation of Ψ.

i. Infinite array: Presser [45] suggested for this case,

$$\psi = 0.9090 + 0.0783\,P/D - 0.1283\ \mathrm{e}^{-2.4(P/D-1)} \tag{10.103a}$$

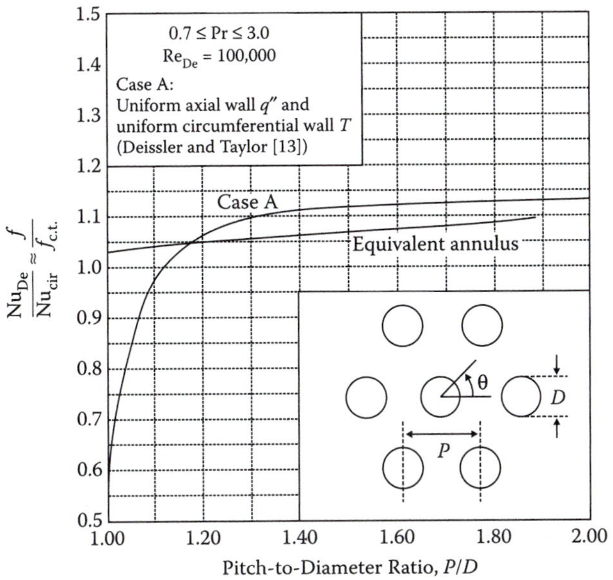

FIGURE 10.11 Fully developed turbulent flow parallel to a bank of circular tubes or rods. Reynolds number influence is small, and Nusselt number behavior is virtually the same as friction behavior.

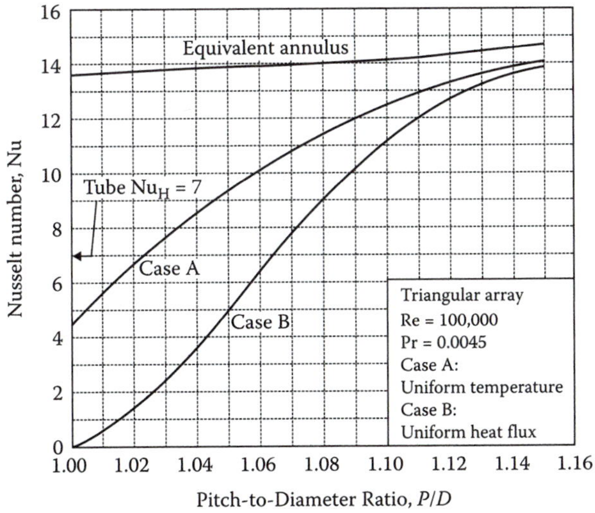

FIGURE 10.12 Variation of Nusselt number for fully developed turbulent flow with rod spacing for Prandtl number < 0.01. (From Nijsing, R., *Heat Exchange and Heat Exchangers with Liquid Metals.* AGRD-Ls-57012; AGRD Lecture Series No. 57 on Heat Exchangers by J. J. Ginoux, Lecture Series Director, 1972.)

for the triangular array and $1.05 \leq P/D \leq 2.2$, and

$$\psi = 0.9217 + 0.1478P/D - 0.1130\ \mathrm{e}^{-7(P/D-1)} \quad (10.103b)$$

for the square array and $1.05 \leq P/D \leq 1.9$.

In the particular case of water, Weisman [59] gave

$$\psi = 1.130\ P/D - 0.2609 \quad (10.104a)$$

for the triangular array and $1.1 < P/\mathrm{D} < 1.5$, and

$$\psi = 1.826\ P/D - 1.0430 \quad (10.104b)$$

for the square array and $1.1 \leq P/D \leq 1.3$, both for $(\mathrm{Nu}_\infty)_{\mathrm{c.t.}} = 0.023\ \mathrm{Re}^{0.8}\ \mathrm{Pr}^{0.333}$.

ii. Finite array: Markoczy [35] gave a general expression for Ψ valid for every rod within a finite lattice. The concept is as follows: Consider a fuel rod (R), as those in Figure 10.13, surrounded by J subchannels. Let the cross section of the flow area and the wetted perimeter of the jth subchannel be A_j and P_{wj}, respectively.

The general expression for Ψ is

$$\psi = 1 + 0.9120\ \mathrm{Re}^{-0.1}\mathrm{Pr}^{0.4}(1 - 2.0043\ \mathrm{e}^{-B}) \quad (10.105)$$

where

- The Reynolds number (Re) is based on the hydraulic diameter (D_e) and velocity characteristic of the subchannels surrounding the subject fuel rod (R). The hydraulic diameter is evaluated as

$$D_e = 4\frac{\sum_{j=1}^{J} A_j}{\sum_{j=1}^{J} P_{wj}} \quad (10.106a)$$

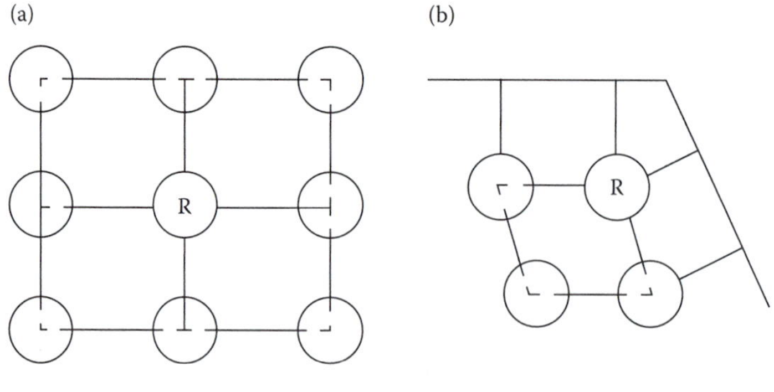

FIGURE 10.13 Interior rod (a) and edge (or corner) rod (b).

- The exponent B is given by

$$B = \frac{D_e}{D} \tag{10.106b}$$

where D is the diameter of the rod.

Note that for an interior rod the surrounding subchannels are those of an infinite array; and B, evaluated by Equations 10.106a and 10.106b, becomes (Figure 10.13)

$$B = \frac{2\sqrt{3}}{\pi}(P/D)^2 - 1 \tag{10.107a}$$

for triangular arrays, and

$$B = \frac{4}{\pi}(P/D)^2 - 1 \tag{10.107b}$$

for square arrays. The value of Ψ from Equation 10.105 successfully fits the experimental data for the following conditions with an average deviation between the correlation predictions and the experimental data of 7%:

$3 \times 10^3 \leq \text{Re} \leq 10^6$
$0.66 \leq \text{Pr} \leq 5.0$
$1 \leq P/D \leq 2.0$ for interior rods in triangular arrays
$1 \leq P/D \leq 1.8$ for interior rods in square arrays

10.5.1.2 Entrance Region Effect

In reality, h is not constant throughout the channel because (1) the entrance region where the temperature profile is still developing, and (2) there are changes of the properties due to the change in temperature. The effect of the changing temperature can be approximated by evaluating the physical constants at a representative temperature, which would be the mean value of the inlet and exit temperatures in the channel.

The case of the entrance region is more complicated because the entrance length is sensitive to Re, Pr, heat flux shape, and entrance conditions for the fluid. Within the entrance region, h decreases dramatically from a theoretical infinite value to the asymptotic value (Figure 10.14). In such a case, it is always possible to conservatively approximate the overall value of $\overline{\text{Nu}}$ by the fully developed value Nu_∞.

Generally, for circular geometry, at

$$\frac{L}{D_e} \gtrsim 60, \quad \text{Nu}_z \approx \text{Nu}_\infty \quad \text{for Pr} \ll 1.0 \tag{10.108a}$$

and at

$$\frac{L}{D_e} \gtrsim 40, \quad \text{Nu}_z = \text{Nu}_\infty \quad \text{for Pr} \geq 1.0 \tag{10.108b}$$

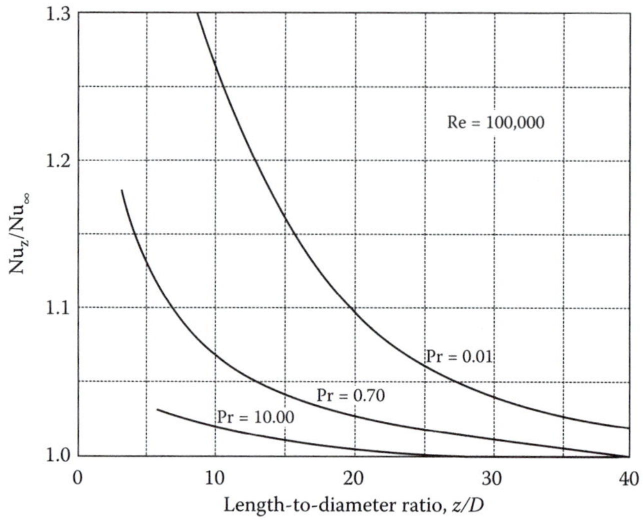

FIGURE 10.14 Local heat transfer coefficient variation with length to diameter of a tube at Re = 10^5. (From Kays, W. M. and Crawford, M. E., *Convective Heat and Mass Transfer*, 2nd Ed. New York, NY: McGraw-Hill, 1980.)

10.5.1.2.1 Overall Heat Transfer Coefficient

Several correlations have been proposed to include the entrance effects, which depend on the type of entrance condition. In general, the heat transfer rate is much higher in the entrance region.

1. For Re > 10,000 and 0.7 < Pr < 120 and square-edged entries, McAdams [38] recommended

$$\overline{\mathrm{Nu}} = \mathrm{Nu}_{\infty}[1 + (D_e/L)^{0.7}] \tag{10.109}$$

2. For tubes with bell-mouthed entry, McAdams [38] cited the relation of Latzko [28] for $L/D_e < 0.693\ \mathrm{Re}^{1/4}$:

$$\overline{\mathrm{Nu}} = 1.11\left[\frac{\mathrm{Re}^{1/5}}{(L/D_e)^{4/5}}\right]^{0.275} \mathrm{Nu}_{\infty} \tag{10.110a}$$

and for $L/D_e > 0.693\mathrm{Re}^{1/4}$:

$$\overline{\mathrm{Nu}} = \left(1 + \frac{A}{L/D_e}\right)\mathrm{Nu}_{\infty} \tag{10.110b}$$

where A is a function of Re defined as

$$A = 0.144\mathrm{Re}^{1/4} \tag{10.111}$$

Latzko's prediction was acceptable for $26{,}000 < \text{Re} < 56{,}000$, but with $A = 1.4$ compared with 1.83 to 2.22 predicted by Equation 10.111.

Correlations of the form of Equation 10.110b for gases are given for various types of flow entrance geometries in Table 10.5. For the particular case of superheated steam, McAdams et al. [39] predicted

$$\overline{\text{Nu}} = 0.0214\ \text{Re}^{0.8}\text{Pr}^{1/3}\left(1 + \frac{2.3}{L/D_e}\right) \tag{10.112}$$

10.5.1.2.2 Entrance Region Heat Transfer Coefficient

For short tubes, where the region of fully developed flow is a small percentage of the total length, the local value of the Nusselt number must be sought. Experimental data for gases have shown that [5]

1. For uniform velocity and temperature profile in the entrance,

$$\text{Nu(z)} = 1.5\left(\frac{z}{D_e}\right)^{-0.16} \text{Nu}_\infty \quad \text{for } 1 < \frac{z}{D_e} < 12 \tag{10.113a}$$

TABLE 10.5
Nusselt Number for Gas Flow in Tubes[a]

Type of Entrance	Illustration	*A*
Bellmouth	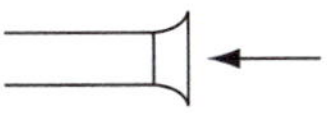	0.7
Bellmouth and one screen	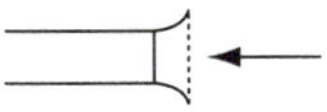	1.2
Long-calming section		1.4
Short-calming section	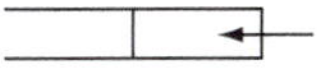	~3
45° Angle-bend entrance	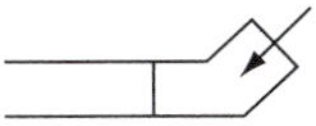	~5
90° Angle-bend entrance	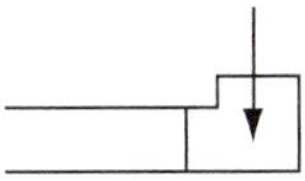	~7
Large orifice entrance (ratio of pipe I.D. to orifice diameter = 1.19)	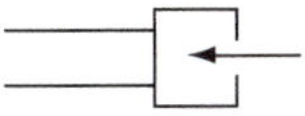	~16
Small orifice entrance (ratio of pipe I.D. to orifice diameter = 1.789)	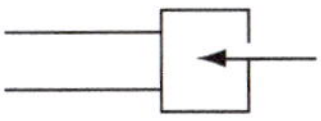	~7

Source: From Boelter, L. M. K., Young, G., and Iverson, H. W., *An Investigation of Aircraft Heaters*. XXVII, NACA-TN-1451, 1948.

[a] $\text{Nu} = \text{Nu}_\infty\left(1 + \frac{AD_e}{L}\right)$; for $\frac{L}{D_e} > 5$ and $\frac{D_e G}{\mu} > 17{,}000$.

$$\text{Nu}(z) = \text{Nu}_\infty \quad \text{for } \frac{z}{D_e} > 12 \tag{10.113b}$$

2. For abrupt entrance,

$$\text{Nu(z)} = \left(1 + \frac{1.2}{z/D_e}\right)\text{Nu}_\infty \quad \text{for } 1 < \frac{z}{D_e} < 40 \tag{10.114a}$$

$$\text{Nu}(z) = \text{Nu}_\infty \quad \text{for } \frac{z}{D_e} > 40 \tag{10.114b}$$

Gnielinski's accommodation of the entrance effect in his correlation has been presented in Equations 10.98a and 10.98b.

Example 10.3: Turbulent Heat Transfer in a Steam Generator Tube (Single-Phase Region Only): More Accurate Calculation

PROBLEM

In Example 10.2, the linear heat transfer rate of a tube in the single-phase region of the secondary side of the steam generator was based on the assumption that the shell side film coefficient was the same as the tube side coefficient. Using the information in Example 10.2 and the information below, determine the shell side heat transfer coefficient with one of the empirical relations presented in Section 10.5.1.

Shell side information
Lower shell shroud *I.D.* = 1.8 m
Secondary flow rate = 480 kg/s
Flow characteristics = assume saturated water at secondary side temperature of 280°C
Triangular tube array (not common practice for PWRs)
Total number of tubes = 3800

SOLUTION

We choose a correlation based on flow conditions that most closely approximate those of the steam generator. We must be sure, in particular, that the parameter space falls within the range of the correlation. To find the Re and Pr, first find the properties of water in the shell side:

Saturated water at 280°C:
$\rho = 750.7$ kg/m^3
$\mu = 9.75 \times 10^{-5}$ kg/m s
$k = 0.574$ W/m°C
$c_p = 5307$ J/kg°C

Now calculate the equivalent dimensions needed for the empirical formula. The area and equivalent diameter on the shell side are

$$A = \frac{\pi}{4}(1.8\text{ m})^2 - 3800\frac{\pi}{4}(0.0254\text{ m})^2 = 0.619\text{ m}^2$$

$$D_e = \frac{4A}{P_w} = \frac{4(0.619)}{\pi(1.8) + 3800\,\pi(0.0254)} = 0.008\text{ m}$$

The average P/D ratio can be found by setting the ratio of the tube to the total area for a single triangular cell as representative of the whole steam generator.

$$\frac{1/2\left(\pi/4D^2\right)}{1/2\left(\sqrt{3}/2P^2\right)} = \frac{3800(\pi/4)(0.0254\text{ m})^2}{\pi/4(1.8\text{ m})^2}$$

$$\frac{P}{D} = \sqrt{\frac{2\pi}{4\sqrt{3}}\frac{1}{3800}}\left(\frac{1.8}{0.0254}\right) = 1.1$$

The velocity on the shell side is

$$V = \frac{\dot{m}}{\rho A} = \frac{480\text{ kg/s}}{(750.7\text{ kg/m}^3)(0.619\text{ m}^2)} = 1.033\text{ m/s}$$

$$\text{Re} = \frac{\rho V D_e}{\mu} = \frac{(750.7\text{ kg/m}^3)(1.033\text{ m/s})(0.008\text{ m})}{9.75\times 10^{-5}\text{kg/m s}} = 6.363\times 10^4$$

$$\text{Pr} = \frac{9.75\times 10^{-5}(5307)}{0.574} = 0.901$$

Using the Weisman correlation (Equation 10.104) for water in triangular arrays, we find

$$(\text{Nu}_\infty)_{\text{c.t.}} = 0.023(6.363\times 10^4)^{0.8}(0.901)^{0.333} = 154.7$$

$$\psi = 1.130(1.1) - 0.2609 = 0.9821$$

Therefore

$$\text{Nu}_\infty = 154.7(0.9281) = 151.9$$

The shell side heat transfer coefficient is now:

$$h = \frac{\text{Nu}k}{D_e} = \frac{151.9(0.574)}{0.008}\text{W/m}^2\text{°C} = 1.09\times 10^4\text{W/m}^2\text{ °C}$$

and the linear heat transfer rate is

$$q' = \frac{2\pi(305-280)}{[1/(0.0111)(2.87\times 10^4)] + \{[\ell n(1/0.875)]/35\} + \{1/[(0.0127)(1.09\times 10^4)]\}}$$

$$q' = \frac{157.1}{3.14\times 10^{-3} + 3.82\times 10^{-3} + 7.22\times 10^{-3}} = 1.11\times 10^4\text{ W/m}$$

So the refinement in the calculation of the linear heat transfer rate, in this case, resulted in a decrease of about 33%.

10.5.2 Nonmetallic Fluids: Geometries for Enhanced Heat Transfer

The motivation for augmented heat transfer has been introduced in Section 9.5.6 where the associated increase in pressure drop for two geometric forms of augmentation—extended-ribbed surfaces and twisted tape inserts—has been presented in the form of enhanced friction factors. Here, the enhancement in heat transfer coefficient for these two geometries is presented.

10.5.2.1 Ribbed Surfaces

The adoption of the Brayton power conversion cycle for use with a primary liquid coolant creates the incentive for augmented heat transfer in gas/liquid nuclear heat exchangers on the low Nusselt number fluid side (e.g., carbon dioxide secondary side fluid). The Bergles and Ravigururajan correlation for friction factor is given as Equation 9.100a. These authors also developed a correlation for the Nusselt number in ribbed-surface tubes, Nu_a, and, consistent with their approach for the friction factor, they defined the Nu_a number with respect to the smooth tube: Thus, the computation of the heat transfer coefficient has to be performed according to the formula

$$h = \frac{\text{Nu}_a \times k}{D}$$

in which D is the inside diameter of the smooth tube (equal to the maximum inside diameter of the ribbed tube). The Ravigururajan–Bergles correlation [46] expresses the ratio of the augmented Nusselt number, Nu_a, to the smooth tube Nusselt number, Nu_{sm}, as

$$\frac{\text{Nu}_a}{\text{Nu}_{sm}} = \left\{1 + \left[2.64\ \text{Re}^{0.036}\left(\frac{e}{D}\right)^{0.212} \times \left(\frac{\delta}{D}\right)^{-0.21}\left(\frac{\alpha}{90}\right)^{0.29}\text{Pr}^{-0.024}\right]^7\right\}^{1/7} \tag{10.115a}$$

where Nu_{sm} is given by Petukhov and Popov [42] which, rewritten using the Fanning friction factor in place of the Darcy friction factor, becomes

$$\text{Nu}_{sm} = \frac{(f'_{sm}/2)\text{Re Pr}}{1 + 12.7\sqrt{\frac{f'_{sm}}{2}}\left(\text{Pr}^{2/3} - 1\right)} \tag{10.115b}$$

where f'_{sm} is the Fanning friction factor from Equation 9.100b. The Reynolds number shown in Equations 10.115a and 10.115b is that of the fluid assumed to flow in the smooth tube.

This correlation developed using air, water, hydrogen, and n-butyl alcohol data is applicable over the following fluid operational and geometric conditions:

$$0.66 \le \text{Pr} \le 37.6$$
$$3 \times 10^3 \le \text{Re} \le 5 \times 10^5$$

where

e rib height
D maximum inside tube diameter
δ rib separation
α helix angle (deg)
β contact angle profile (deg)

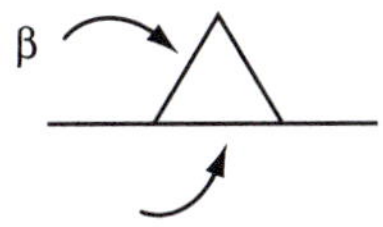

n number of shape corners facing the flow that characterize the rib profile (e.g., two for both triangular and rectangular profiles)

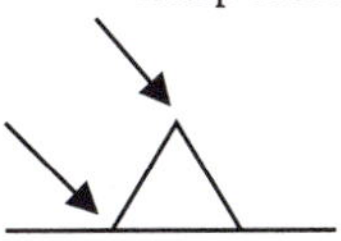

and

$$0.01 \le e/D \le 0.2$$
$$0.1 \le \delta/D \le 7.0$$
$$0.3 \le \alpha/90 \le 1.0$$

This Nusselt number correlation predicts 99% of the database upon which the correlation was based to within +50% and 69% within 20%.

Example 10.4: Heat Transfer Coefficient for an Extended Surface Intermediate Heat Exchanger Tube

PROBLEM

For the heat exchanger introduced in Example 9.6, calculate:

a. Heat transfer coefficient
b. Heat exchanger tube length

both in case of smooth tubes and in case of ribbed tubes.

For predicting the Nusselt number, use the correlations of Equations 10.115a and 10.115b developed by Ravigururajan and Bergles [46]. Compute the heat transfer rate through the lead-to-CO_2 heat exchanger with steel tubes using the log mean temperature difference: $\dot{Q} = UA\Delta T_{lm}$. Use the input parameters for the IHX given in Example 9.6 and Table 10.6.

TABLE 10.6
Useful Properties (Average) for Example 10.4

Pb density (kg/m^3)	10,160
Pb viscosity (Pa s)	1.7×10^{-3}
Pb thermal conductivity (W/m K)	15.6
Pb specific heat (J/kg K)	142.7
Pb velocity (m/s)	1.50
Steel thermal conductivity (W/m K)	26.0

SOLUTION

a. The overall heat transfer coefficient is given as

$$U = \frac{1}{(1 / h_{CO_2}) + (((D_i/2)\ln(D_o/D_i))/k_w) + (D_i/D_o h_{Pb})}$$

where

h_{CO_2} = heat transfer coefficient on the S-CO_2 side
h_{Pb} = heat transfer coefficient on the lead side
k_w = wall material thermal conductivity
D_i = inner radius of the IHX tube
D_o = outer radius of the IHX tube

The heat transfer coefficients, h_{CO_2} and h_{Pb}, are first calculated from the Nusselt number.

b. For smooth tubes,
Lead side:

$$V_{Pb} = \frac{\dot{m}}{\rho A_{total}} = \frac{\dot{m}}{\rho N\left(\sqrt{3}/2\ P^2 - \pi D_o^2/4\right)}$$

$$= \frac{21{,}700}{10{,}160 \times 10{,}000 \times \left[\sqrt{3}/2(1.23 \times 2.00 \times 10^{-2})^2 - (\pi(2.00 \times 10^{-2})^2/4)\right]}$$

$$= 1.02 \text{ m/s}$$

$$D_e = \frac{4A_{flow}}{P_w} = 4\frac{(\sqrt{3}/2)P^2 - (\pi D_o^2/4)}{\pi D_o}$$

$$= 4\frac{\sqrt{3}/2(1.23 \times 2.00 \times 10^{-2})^2 - ((\pi(2.00 \times 10^{-2})^2)/4)}{\pi \times 2.00 \times 10^{-2}} = 1.34 \times 10^{-2}\text{m}$$

$$Re_{Pb} = \left(\frac{\rho V D_e}{\mu}\right)_{Pb} = \frac{10{,}160 \times 1.02 \times 1.34 \times 10^{-2}}{1.70 \times 10^{-3}} = 8.17 \times 10^4$$

where D_e is the equivalent diameter on the shell side.

$$Pr_{Pb} = \left(\frac{c_p\mu}{k}\right)_{Pb} = \frac{142.7 \times 1.70 \times 10^{-3}}{15.6} = 1.56 \times 10^{-2}$$

The Nusselt number can be calculated using Equation 10.124 introduced in Section 10.5.3.4, which presents liquid metal heat transfer correlations for rod bundle geometry.

$$Nu_{Pb} = 4.0 + 0.33(1.23)^{3.8}\left(\frac{8.17 \times 10^4 \times 1.56 \times 10^{-2}}{100}\right)^{0.86} + 0.16(1.23)^{5.0} = 10.9$$

$$h_{Pb} = \left(\frac{Nu \times k}{D_e}\right)_{Pb} = \frac{10.9 \times 15.6}{1.34 \times 10^{-2}} = 1.27 \times 10^4 \text{ W/m}^2\text{K}$$

S-CO_2 side: For gases in circular smooth tubes, the correlation of Equation 10.115b for Nusselt number is used. The Re, Pr, f'_{sm}, and property values for the S-CO_2 side are given in Example 9.6.

$$Nu_{sm} = \frac{(f'_{sm}/2)\text{Re Pr}}{1 + 12.7\sqrt{f'_{sm}/2}\left(\text{Pr}^{2/3} - 1\right)} = \frac{(3.49 \times 10^{-3}/2)(3.607 \times 10^5)(0.75)}{1 + 12.7\sqrt{3.49 \times 10^{-3}/2}(0.75^{2/3} - 1)}$$
$$= 5.20 \times 10^2$$

$$h_{CO_2} = \left(\frac{Nu_{sm} \times k}{D_i}\right)_{CO_2} = \frac{5.20 \times 10^2 \times 5.77 \times 10^{-2}}{1.60 \times 10^{-2}} = 1.87 \times 10^3 \text{ W/m}^2\text{ K}$$

$$U = \frac{1}{(1/h_{CO_2}) + (((D_i/2)\ell n(D_o/D_i))/k_w) + (D_i/(D_o h_{Pb}))}$$
$$= \frac{1}{(1/1.87 \times 10^3) + ((1.60 \times 10^{-2}/2 \times \ell n(20.0/16.0))/26.0) + (1.60 \times 10^{-2}/2.00 \times 10^{-2} \times 1.27 \times 10^4)}$$
$$= 1.50 \times 10^3 \text{ W/m}^2\text{ K}$$

c. For enhanced tubes:
All of the conditions for the lead side are the same as for the smooth tube case. As for the S-CO_2 side, using Equation 10.115a, one obtains

$$Nu_a = Nu_{sm}\left\{1 + \left[2.64\text{Re}^{0.036}\left(\frac{e}{D}\right)^{0.212} \times \left(\frac{\delta}{D}\right)^{-0.21}\left(\frac{\alpha}{90}\right)^{0.29}\text{Pr}^{-0.024}\right]^7\right\}^{1/7}$$
$$= 5.20 \times 10^2\left\{1 + [2.64(3.607 \times 10^5)^{0.036}(0.022)^{0.212}]^7\right.$$
$$\left.[\times (1.25)^{-0.21}(0.3)^{0.29}(0.750)^{-0.024}]\right\}^{1/7}$$
$$= 6.72 \times 10^2$$

$$h_{CO_2} = \frac{Nu_a \times k}{D} = \frac{6.72 \times 10^2 \times 5.77 \times 10^{-2}}{1.60 \times 10^{-2}} = 2.42 \times 10^3 \text{W/m}^2 \text{ K}$$

$$U = \frac{1}{(1/h_{CO_2}) + (((D_i/2)\ln(D_o/D_i))/k_w) + (D_i/(D_o h_{Pb}))}$$

$$= \frac{1}{(1/2.42 \times 10^3) + ((1.60 \times 10^{-2}/2 \times \ln(20.0/16.0))/26.0) + (1.60 \times 10^{-2}/2.00 \times 10^{-2} \times 1.27 \times 10^4)}$$

$$= 1.84 \times 10^3 \text{ W/m}^2\text{K}$$

d. The length of the smooth and enhanced surface tubes can now be calculated by relating the total heat transfer rate, overall heat transfer coefficient, total heat transfer surface, and inlet and outlet fluid temperatures as

$$\dot{Q} = UA_T\Delta T_{lm}$$

The effective heat transfer surface can be calculated as

$$A_T = N(AL)_{tube} = N(2\pi rL)_{tube}$$

Log-mean temperature difference ΔT_{lm} is evaluated using the following equation:

$$\Delta T_{lm} = \frac{(T_{in,Pb} - T_{out,CO_2}) - (T_{out,Pb} - T_{in,CO_2})}{\ln(((T_{in,Pb} - T_{out,CO_2}))/((T_{out,Pb} - T_{in,CO_2})))} = 74.5°\text{C}$$

Smooth tubes:

$$A_T = \frac{\dot{Q}}{U\Delta T_{lm}} = \frac{300 \times 10^6}{1.50 \times 10^3 \times 74.5} = 2.68 \times 10^3 \text{ m}^2$$

$$L = \frac{A_T}{N(2\pi\, r)_{tube}} = \frac{2.86 \times 10^3}{10{,}000 \times 2 \times 3.14 \times 0.80 \times 10^{-2}} = 5.33 \text{ m}$$

Enhanced surface tubes:

$$A_T = \frac{\dot{Q}}{U\Delta T_{lm}} = \frac{300 \times 10^6}{1.84 \times 10^3 \times 74.5} = 2.19 \times 10^3 \text{ m}^2$$

$$L = \frac{A_T}{N(2\pi\, r)_{tube}} = \frac{2.19 \times 10^3}{10{,}000 \times 2 \times \pi \times 0.80 \times 10^{-2}} = 4.36 \text{ m}$$

Note that the heat transfer coefficient was calculated with respect to the inner radius of the tubes. Therefore, the inner radius is used to determine the tube length (or equivalently the heat transfer area).

10.5.2.2 Twisted Tape Inserts

The friction factors for twisted tape inserts for laminar and turbulent flow from Manglik and Bergles are given in Equations 9.101a and 9.102, respectively. The corresponding heat transfer correlations by these authors for laminar [33] and turbulent [34] flow are

Laminar flow:

$$\mathrm{Nu_m} = 4.612\left(\frac{\mu_b}{\mu_w}\right)^{0.14}\left\{\left[(1+0.0951\ \mathrm{Gz}^{0.894})^{2.5} + 6.413\times10^{-9}(Sw\times\mathrm{Pr}^{0.391})^{3.835}\right]^2 + 2.132\times10^{-14}(\mathrm{Re_a}\times\mathrm{Ra})^{2.23}\right\} \quad (10.116)$$

$$Sw = \left(\frac{\mathrm{Re}}{\sqrt{y}}\right)\frac{\pi}{[\pi-4(\delta/D)]}\left[1+\left(\frac{\pi}{2y}\right)^2\right]^{1/2} \quad (9.101\mathrm{d})$$

$$\mathrm{Re_a} = (V_a D/\nu), \quad V_a = (\dot{m}/\rho A_c), \quad \text{and} \quad A_c = [(\mu D^2/4) - \delta D] \quad (10.117\mathrm{a,b,c})$$

Turbulent flow:

$$\mathrm{Nu} = 0.023\ \mathrm{Re}^{0.8}\ \mathrm{Pr}^{0.4}\left[1+\frac{0.769}{y}\right]\left[\frac{\pi+2-(2\delta/D)}{\pi-(4\delta/D)}\right]^{0.2}\left[\frac{\pi}{\pi-(4\delta/D)}\right]^{0.8}\phi \quad (10.118\mathrm{a})$$

where

$$\phi = \left(\frac{\mu_b}{\mu_w}\right)^n \quad \text{or} \quad \left(\frac{T_b}{T_w}\right)^m \quad (10.118\mathrm{b})$$

$$n = \begin{cases} 0.18 & \text{liquid heating} \\ 0.30 & \text{liquid cooling} \end{cases} \quad \text{and} \quad m = \begin{cases} 0.45 & \text{gas heating} \\ 0.15 & \text{gas cooling} \end{cases}$$

Example 10.5: Heat Transfer Coefficient and Pressure Drop for a Twisted Tape Insert

PROBLEM

Under the constraints of fixed CO_2 flow rate per tube, fixed tube length and diameter and fixed operating conditions, the twisted tape insert defined in Example 9.7 proved to yield a larger friction pressure drop than that of the ribbed surface tube of Example 9.6, regardless of the twist ratio assumed for the tape. Now, replacing the ribbed surface tube design examined in Example 10.4 by tubes provided with twisted tapes:

- Determine the combination of tube length-to-twist ratio that minimizes the friction pressure drop across the heat exchanger.
- Compare the minimum pressure drop with that of the ribbed tube design of Example 10.4.

For the twisted tape design, total power transferred, outer tube diameter, tube thickness, number of tubes, flow rates, and inlet and outlet temperatures are the same as those of the ribbed surface tube design of Example 10.4. The tube inner diameter is equal to the maximum inner diameter of the ribbed tubes. Liquid Pb is assumed to flow on the shell side of the heat exchanger. Input data from this and previous examples are as follows:

Ribbed surface tubes:
Rib geometry: see Table 9.3 in Example 9.6;
L_{ribbed} (length of ribbed surface tubes) = 4.36 m

Twisted tape:
δ (twisted tape thickness) = 1 mm;
Common parameters (common to ribbed surface tube design and twisted tape design):
D_i (tube maximum inner diameter) = 16 mm;
D_o (tube outer diameter) = 20 mm;
Tube pitch-to-diameter ratio = 1.23;
N_{tubes} = 10,000;

$$\dot{Q}\ \left(\text{power to be transferred}\right) = 300\ \text{MW}_{th};$$

Total CO_2 total flow rate = 1594 kg/s;
h_{Pb} (heat transfer coefficient on the shell side) = 1.27×10^4 W/m² K

$$\begin{aligned} T_{in,CO_2} &= 397°C; \\ T_{out,CO_2} &= 591°C; \\ T_{in,Pb} &= 600°C; \\ T_{out,Pb} &= 503°C \end{aligned}$$

Average properties to be used for CO_2 are those given in Example 9.6, while those for Pb and stainless steel are given in Example 10.4. Neglect the wall-to-bulk viscosity difference in the heat transfer coefficient computation.

SOLUTION

Before examining the twisted tape design, let us calculate the pressure drop corresponding to the ribbed surface tubes of Example 10.4. In Example 9.6 the Darcy friction factor of the ribbed tubes and the CO_2 velocity in the tubes assumed to be smooth were found to be

$$\begin{aligned} f &= 1.69 \times 10^{-2} \\ V_{CO_2,smooth} &= 5.73 \text{ m/s} \end{aligned}$$

The pressure drop across the ribbed tube heat exchanger of Example 10.4 can be easily calculated as

$$\Delta p = f \frac{L_{ribbed}}{D_i} \frac{\rho V^2_{CO_2,smooth}}{2} = 1.69 \times 10^{-2} \frac{4.36}{1.60 \times 10^{-2}} \frac{138.5 \times 5.73^2}{2} = 10.48 \text{ kPa}$$

Now let us examine the twisted tape design. Both the tube length and twist ratio affect heat transfer performance and pressure drop of the heat exchanger. The combination of these parameters that minimizes the pressure drop can be found by relating them to the overall heat transfer coefficient and to the pressure drop, and by forcing the latter to have a minimum.

The heat balance across the heat exchanger can be expressed as

$$\dot{Q} = UA\Delta T_{lm} = U\left(N_{tubes}L\pi D_i\right)\Delta T_{lm} \tag{10.119}$$

where U is the overall heat transfer coefficient and ΔT_{lm} is the mean logarithmic temperature difference (calculated in Example 10.4 as 74.5°C). U can be expressed by means of the following relation:

$$U = \frac{1}{(1/h_{CO_2}) + (((D_i/2)\ln(D_o/D_i))/k_{steel}) + (D_i/D_o)(1/h_{Pb})} \tag{10.120}$$

After expressing the heat transfer coefficient on the tube side, h_{CO_2}, in terms of the corresponding Nusselt number (defined with respect to the empty tube), Equation 10.120 can be inserted into Equation 10.119 to find

$$\dot{Q} = \frac{1}{(D_i/\mathrm{Nu}_{CO_2}k) + (((D_i/2)\ln(D_o/D_i))/k_{steel}) + (D_i/D_o)(1/h_{Pb})}\left(N_{tubes}L\pi D_i\right)\Delta T_{lm} \tag{10.121}$$

The Nusselt number on the tube side, Nu_{CO_2}, can be expressed using the Bergles–Manglik correlation for the turbulent regime (Equation 10.118a), since in Example 9.6 the Reynolds number, for the tube assumed to be smooth, was calculated to be 3.607×10^5:

$$\mathrm{Nu}_{CO_2} = 0.023\,(\mathrm{Re}_{empty,CO_2})^{0.8}\,\mathrm{Pr}_{CO_2}^{0.4}\left[1+\frac{0.769}{y}\right]\left[\frac{\pi + 2 - (2\delta/D_i)}{\pi - (4\delta/D_i)}\right]^{0.2}\left[\frac{\pi}{\pi - (4\delta/D_i)}\right]^{0.8}\phi \tag{10.118a}$$

By introducing Equation 10.118a into Equation 10.121, we find

$$\dot{Q} = \frac{\left(N_{tubes}L\pi D_i\right)\Delta T_{lm}}{D_i/[0.023(\mathrm{Re}_{empty,CO_2})^{0.8}\mathrm{Pr}_{CO_2}^{0.4}(1+(0.769/y))((\pi + 2 - (2\delta/D_i))/(\pi - (4\delta/D_i)))^{0.2} (\pi/(\pi - (4\delta/D_i)))^{0.8}k] + [((D_i/2)\ln(D_o/D_i))/k_{steel}] + [(D_i/D_o)(1/h_{Pb})]} \tag{10.122}$$

where according to the problem statement, the wall-to-bulk viscosity correction coefficient, φ, was set equal to unity. In this equation, the tube length L and the twist ratio y are the only unknowns. Equation 10.122 can be rewritten using the numerical values of each parameter as

$$300\times10^6 = \frac{(10^4 L\pi 16\times10^{-3})74.5}{\begin{array}{r}16\times10^{-3}/\,[0.023(3.607\times10^5)^{0.8}(0.75)^{0.4}\left(1+(0.769/y)\right)((\pi+2-2\times0.0625)\\ /(\pi-4\times0.0625))^{0.2}(\pi/(\pi-4\times0.0625))^{0.8}5.77\times10^{-2}]\\ +[(16\times10^{-3}\ln(20/16))/(2\times26)]+[(16/20)(1/1.27\times10^4)]\end{array}}$$

After some manipulation, L in the equation above can be expressed as a function of y as

$$L = \frac{3.2546}{1+(0.769/y)} + 1.0547 \tag{10.123}$$

The friction pressure drop across the heat exchanger can be expressed following Equation 9.99 as

$$\begin{aligned}\Delta p &= f'\frac{L}{D_i}\frac{\rho V^2_{CO_2,\text{smooth}}}{2}\\ &= \left[\frac{4\times0.0791}{(\text{Re}_{CO_2\ \text{empty}})^{0.25}}\left(1+\frac{2.752}{y^{1.29}}\right)\left(\frac{\pi}{\pi-(4\delta/D_i)}\right)^{1.75}\left(\frac{\pi+2-(2\delta/D_i)}{\pi-(4\delta/D_i)}\right)^{1.25}\right]\frac{L}{D_i}\frac{\rho V^2_{CO_2\ \text{empty}}}{2}\\ &= \left[\frac{4\times0.0791}{(3.607\times10^5)^{0.25}}\left(1+\frac{2.752}{y^{1.29}}\right)\left(\frac{\pi}{\pi-4\times0.0625}\right)^{1.75}\left(\frac{\pi+2-2\times0.0625}{\pi-4\times0.0625}\right)^{1.25}\right]\\ &\quad\times\frac{L}{16\times10^{-3}}\frac{138.5\times5.73^2}{2} = 4223.48\left(1+\frac{2.752}{y^{1.29}}\right)L\end{aligned} \tag{10.124a}$$

where f' is given by Equation 9.102.

Now the expression for L given by Equation 10.123 can be introduced into Equation 10.124, the Δp relation above, to yield

$$\Delta p = 4223.48\left(1+\frac{2.752}{y^{1.29}}\right)\left(\frac{3.2546}{1+(0.769/y)} + 1.0547\right) \tag{10.124b}$$

Equation 10.124b is graphically represented in Figure 10.15, together with the tube length corresponding to each value of y.

The following observations can finally be drawn

- As expected, the larger the twist ratio, the smaller the pressure drop.
- The larger the twist ratio, the larger the tube length: this is due to the fact that, under the constraint of a constant heat load, the reduced heat transfer coefficient due to the lower swirl needs to be compensated for by means of a larger heat transfer area.

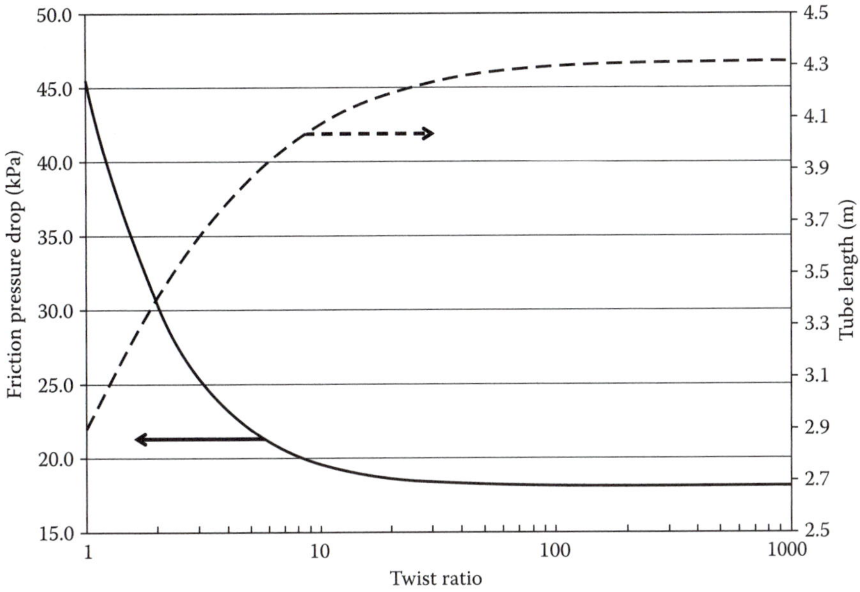

FIGURE 10.15 Relationship between tube length and twist ratio that minimizes friction pressure loss.

- The minimum attainable friction pressure drop for the twisted tape heat exchanger design is about 18 kPa, which is the friction pressure drop corresponding to tubes provided with straight tapes.
- The minimum attainable friction pressure drop for the twisted tape design is about 1.7 times larger than the friction pressure drop of the ribbed tube design.

10.5.3 Metallic Fluids—Smooth Heat Transfer Surfaces: Fully Developed Flow

The behavior of the *Nusselt number* (Nu) for liquid metals follows the relation

$$\mathrm{Nu}_{\infty} = A + B\,(\mathrm{Pe})^{\mathrm{C}} \tag{10.125}$$

where Pe is the *Peclet number*, that is, Pe = RePr. *A, B*, C are constants that depend on the geometry and the boundary conditions. The constant C is a number close to 0.8. The constant *A* reflects the fact that significant heat transfer by conduction in liquid metals occurs even as Re goes to zero.

10.5.3.1 Circular Tube

The following relations hold for the boundary conditions cited and fully developed flow conditions.

1. Constant heat flux along and around the tube, Lyon [32]:

$$\mathrm{Nu}_\infty = 7 + 0.025\ \mathrm{Pe}^{0.8} \tag{10.126a}$$

2. Uniform axial wall temperature and uniform radial heat flux, Seban and Shimazaki [50]:

$$\mathrm{Nu}_\infty = 5.0 + 0.025\ \mathrm{Pe}^{0.8} \tag{10.126b}$$

10.5.3.2 Parallel Plates

For fully developed flow, the following results from Seban [49] are applicable:

1. Constant heat flux through one wall only (the other is adiabatic):

$$\mathrm{Nu}_\infty = 5.8 + 0.02\ \mathrm{Pe}^{0.8} \tag{10.127}$$

2. For constant heat flux through both walls, a graphic correction factor for the heat transfer coefficient was supplied.

10.5.3.3 Concentric Annuli

For fully developed flow and the boundary condition of uniform heat flux in the inner wall when $D_2/D_1 > 1.4$:

$$\mathrm{Nu}_\infty = 5.25 + 0.0188\ \mathrm{Pe}^{0.8}\left(\frac{D_2}{D_1}\right)^{0.3} \tag{10.128}$$

If D_2/D_1 is close to unity, the use of Equation 10.127 was recommended by Seban [49]. More complex semiempirical formulations for the Nu_∞ of the form of Equation 10.125, where all constants are a function of annuli geometry, have been developed by Dwyer [15] for cases of either wall (inner or outer) subject to uniform heat flux and the other wall insulated.

10.5.3.4 Rod Bundles

As noted before, the nonuniformity of the subchannel shape creates substantial azimuthal variation of Nu. Also in finite rod bundles, the turbulent effects in a given subchannel affect adjacent subchannels differently depending on the location of the subchannels with respect to the duct boundaries. Therefore, the value of Nu is a function of position within the bundle. However, we provide here overall heat transfer correlations and defer the discussion of the variation of heat transfer with angle and location to Chapter 7, Volume II. Again, as shown in Figure 10.12, for $P/D > 1.1$, the equivalent annulus concept provides an acceptable answer.

The correlations considered here are as follows:
Kazimi and Carelli [27]:

$$\mathrm{Nu} = 4.0 + 0.33(P/D)^{3.8}(\mathrm{Pe}/100)^{0.86} + 0.16(P/D)^{5.0} \tag{10.129}$$

for $1.1 \le P/D \le 1.4$ and $10 \le \mathrm{Pe} \le 5000$.
Schad-modified [27]:

$$\mathrm{Nu} = [-16.15 + 24.96(P/D) - 8.55(P/D)^2]\,\mathrm{Pe}^{0.3} \tag{10.130a}$$

for $1.1 \le P/D \le 1.5$ and $150 \le \mathrm{Pe} \le 1000$.

$$\mathrm{Nu} = 4.496[-16.15 + 24.96(P/D) - 8.55(P/D)^2] \tag{10.130b}$$

for $\mathrm{Pe} \le 150$.
Graber and Rieger [18]:

$$\mathrm{Nu} = 0.25 + 6.2\left(P/D\right) + \left[-0.007 + 0.032\left(P/D\right)\right]\mathrm{Pe}^{0.8-0.024(P/D)} \tag{10.131}$$

for $1.25 \le P/D \le 1.95$ and $150 \le \mathrm{Pe} \le 3000$.
Borishanskii et al. [6]:

$$\begin{aligned}\mathrm{Nu} = {} & 24.15 \log[-8.12 + 12.76(P/D) - 3.65(P/D)^2] \\ & + 0.0174\,[1 - \exp(6 - 6P/D)][\mathrm{Pe} - 200]^{0.9}\end{aligned} \tag{10.132a}$$

for $1.1 \le P/D \le 1.5$ and $200 \le \mathrm{Pe} \le 2000$.

$$\mathrm{Nu} = 24.15 \log[-8.12 + 12.76(P/D) - 3.65(P/D)^2] \tag{10.132b}$$

for $1.1 < P/D < 1.5$ $\mathrm{Pe} \le 200$.

Experimental and predicted values of Nu in rod bundles with P/D values of 1.3 and 1.15 are shown in Figures 10.16 and 10.17, respectively. The comparison indicates that the correlations of Borishanskii and Schad-modified yield the best agreement over the entire range of P/D values. The Graber and Rieger correlation appears to significantly overpredict the heat transfer coefficient if extended beyond the published range of applicability $P/D \le 1.15$. The Kazimi and Carelli correlation underestimates Nu at high values of P/D.

Example 10.6: Typical Heat Transfer Coefficients of Water and Sodium

PROBLEM

Liquid flow velocity along rod bundles is limited principally to avoid rod vibration and possible cavitation at obstructions. Selecting a maximum velocity of 10 m/s,

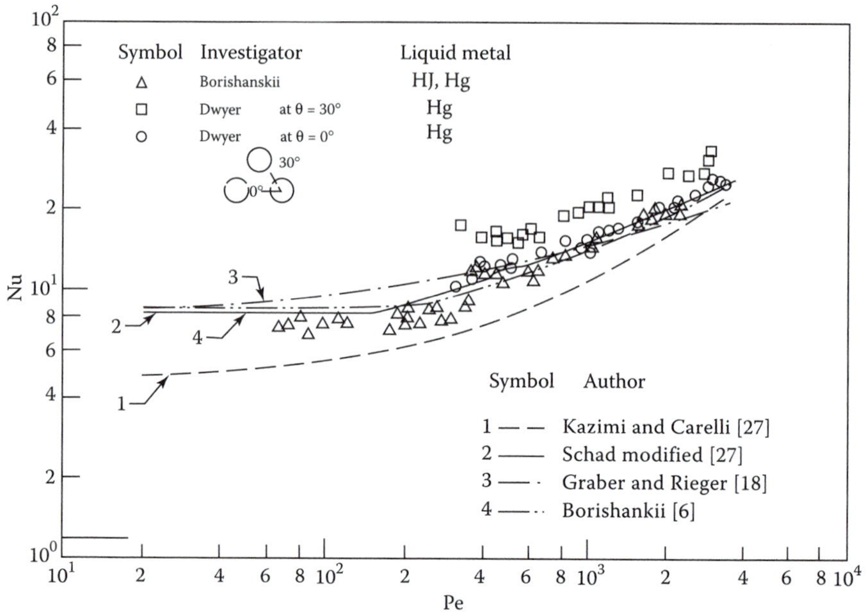

FIGURE 10.16 Comparison to predicted and experimental results of Nu for liquid metals in rod bundles for $P/D = 1.3$. (From Kazimi, M. S. and Carelli, M. D. *Heat Transfer Correlation for Analysis of CRBRP Assemblies.* Westinghouse Report, CRBRP-ARD-0034, 1976.)

compare the heat transfer coefficient of water (at an average coolant temperature for a PWR) to the heat transfer coefficient of sodium (at an average temperatures for an SFBR). Use geometric data for the PWR and SFBR in Table 1.3 and Appendix K with fluid property data in Table 10.7. How do the coefficients compare at a velocity of 1 m/s?

SOLUTION

1. For water at 10 m/s:

$$D_e = \frac{4A_f}{P_w} = \frac{4(P^2 - (\pi D^2/4))}{\pi D} = 11.78 \text{ mm}$$

$$\text{Nu}_\infty = \psi(\text{Nu}_\infty)_{\text{c.t.}}$$

$$(\text{Nu}_\infty)_{\text{c.t.Presser}} = 0.023\ \text{Re}^{0.8}\text{Pr}^{0.4} = 0.023\left(\frac{\rho V D_e}{\mu}\right)^{0.8}\left(\frac{c_p\mu}{k}\right)^{0.4}$$

$$= 0.023\left[\frac{(704)(10)(11.78\times10^{-3})}{8.69\times10^{-5}}\right]^{0.8}\cdot\left[\frac{(6270)8.69\times10^{-5}}{0.5}\right]^{0.4}$$

$$= 1447$$

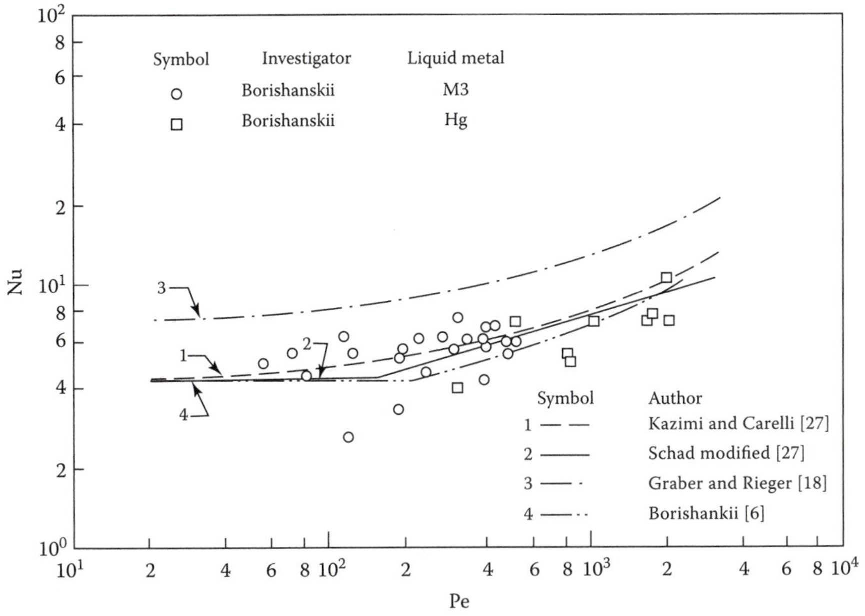

FIGURE 10.17 Comparison of predicted and experimental results of Nu for metals in rod bundles for $P/D = 1.15$. (Adopted from Kazimi, M. S. and Carelli, M. D., *Heat Transfer Correlation for Analysis of CRBRP Assemblies.* Westinghouse Report, CRBRP-ARD-0034, 1976.)

$$(Nu_{\infty})_{c.t.Weisman} = 0.023\ Re^{0.8}Pr^{0.333} = 1438$$

$$\psi_{Presser}(Eq.10.103b) = 0.9217 + 0.1478\left(\frac{0.0126}{0.0095}\right) - 0.1130\ e^{-7((0.0126/0.0095)-1)} = 1.11$$

$$\psi_{Weisman}(Eq.10.104b) = 1.826\left(\frac{0.0126}{0.0095}\right) - 1.0430 = 1.38$$

$$Nu_{\infty Presser} = 1447 \times 1.11 = 1606$$
$$Nu_{\infty Weisman} = 1438 \times 1.38 = 1984$$

TABLE 10.7
Property Data for Example 10.6

Parameter	Water (315°C)	Sodium (538°C)
k (W/m°C)	0.5	62.6
ρ (kg/m³)	704	817.7
μ (kg/m s)	8.69×10^{-5}	2.28×10^{-4}
c_p (J/kg°C)	6270	1254

However

$$\text{Nu} = \frac{hD_e}{k}$$

Therefore

$$\therefore h_{\text{Presser}} = \frac{k\ \text{Nu}}{D_e} = \frac{0.5(1606)}{11.78 \times 10^{-3}} = 68.2\text{k W/m}^2\,^\circ\text{C}$$

$$h_{\text{Weisman}} = \frac{0.5(1984)}{11.78 \times 10^{-3}} = 84.2\ \text{kW/m}^2\,^\circ\text{C}$$

2. For sodium at 10 m/s:

$$\text{Pe} = \text{RePr} = \left(\frac{\rho V D_e}{\mu}\right)\left(\frac{c_p \mu}{k}\right)$$

For a triangular array interior channel:

$$D_e = 4\frac{\left(\left(\sqrt{3}/4\right)P^2 - \pi\left(D^2/8\right)\right)}{\pi D/2}$$

From Tables 10.1 through 10.3, $P = 9.8$ mm and $D = 8.5$ mm:

$$\therefore D_e = 4\frac{[0.433(9.8)^2 - \pi(8.5)^2/8]}{\pi(8.5)/2} = 3.96\ \text{mm}$$

$$\text{Pe} = \left[\frac{(817.7)(10)(3.96 \times 10^{-3})}{2.28 \times 10^{-4}}\right]\left[\frac{(1254)(2.28 \times 10^{-4})}{62.6}\right] = 648.4$$

Using Equation 10.129:

$$\text{Nu} = 4 + 0.33\left(\frac{9.8}{8.5}\right)^{3.8}\left(\frac{648.4}{100}\right)^{0.86} + 0.16\left(\frac{9.8}{8.5}\right)^{5.0} = 7.15$$

$$h = \frac{k\ \text{Nu}}{D_e}$$

$$\therefore h = \frac{(62.6)(7.15)}{3.959 \times 10^{-3}} = 113.1\ \text{kW/m}^2\,^\circ\text{C}$$

3. For water at 1 m/s:
   Note that Re is still in the turbulent region.

$$h_{\text{Presser}} = (0.1)^{0.8}(68.2) = 10.8\ \text{kW/m}^2\,^\circ\text{C}$$

$$h_{\text{Weisman}} = (0.1)^{0.8}(84.2) = 13.3\ \text{kW/m}^2\,^\circ\text{C}$$

4 For sodium at 1 m/s:

$$\text{Pe} = 0.1 \times 648.4 = 64.84$$

$$\text{Nu} = 4 + 0.33\left(\frac{9.8}{8.5}\right)^{3.8}\left(\frac{64.84}{100}\right)^{0.86} + 0.16\left(\frac{9.8}{8.5}\right)^{5.0} = 4.72$$

$$h = \frac{k\text{Nu}}{D_e} = \frac{(62.6)(4.72)}{3.959 \times 10^{-3}} = 74.6 \text{ kW/m}^2{}^\circ\text{C}$$

In summary, at velocities of 1 and 10 m/s, the heat transfer coefficients (kW/m²°C) are

	h(kW/m² °C)		
	Water		Sodium
Velocity (m/s)	**Presser [45]**	**Weisman [59]**	**Kazimi and Carelli [27]**
10	68.2	84.2	113.1
1	10.8	13.3	74.6

PROBLEMS

10.1. *Derivation of Nusselt number for laminar flow in rectangular geometry* (Section 10.2)
Prove that the asymptotic Nusselt number for flow of a coolant with constant properties between two flat plates of infinite width, heated with uniform heat flux on both walls, is 8.235 for laminar flow (i.e., parabolic velocity distribution) and 12 for slug flow (i.e., uniform velocity distribution).

10.2. *Derivation of Nusselt number for laminar flow in circular and flat plate geometry* (Section 10.2)
Show that for a round tube, uniform wall temperature, and slug velocity conditions, the asymptotic Nusselt number for a round tube is 5.75, whereas for flow between two flat plates, it is 9.87.

10.3. *Derivation of Nusselt number for laminar flow in an equivalent annulus* (Section 10.2)
Derive the Nusselt number for slug flow in the equivalent annulus of an infinite rod array by solving the differential energy equation subject to the appropriate boundary conditions, that is

$$\frac{\partial^2 T}{\partial r^2} + \frac{1}{r}\frac{\partial T}{\partial r} = \frac{\upsilon \rho c_p}{k}\frac{\partial T}{\partial z}$$

Answer:

$$\text{Nu} = \frac{2(r_0^2 - r_i^2)^3}{r_i^2\left[r_0^4 \ell n(r_0/r_i) - (r_0^2 - r_i^2)(3r_0^2 - r_i^2)/4\right]}$$

10.4. *Estimating the effect of turbulence on heat transfer in SFBR fuel bundles* (Section 10.3)
Consider the fuel bundle of an SFBR whose geometry is described in Table 1.3. Using Dwyer's recommendations for the values of ε_m and ε_H

(Equation 10.71), estimate the ratio of $\varepsilon_H/\varepsilon_m$ for the sodium velocities in the bundles:

1. $\upsilon = 10$ m/s
2. $\upsilon = 1$ m/s.

Answers:

1. For $\upsilon = 10$ m/s:

$$\frac{\varepsilon_H}{\varepsilon_m} \text{ in core} = 0.510; \quad \left(\frac{\varepsilon_m}{\nu}\right)_{max} = 120$$

$$\frac{\varepsilon_H}{\varepsilon_m} \text{ in blanket} = 0.722; \quad \left(\frac{\varepsilon_m}{\nu}\right)_{max} = 180$$

2. For $\upsilon = 1$ m/s:

$$\frac{\varepsilon_H}{\varepsilon_m} \text{ in both core and blanket} = 0$$

10.5. *Reynolds analogy and equivalent diameter problem* (Section 10.3) Consider a uniformly heated tube (constant heat flux) of diameter 0.025 m with fluid flowing at an average velocity of 0.5 m/s. Find the fully developed heat transfer coefficient for two different fluids (Fluid A and Fluid B, whose properties are given in Table 10.8) by the following two procedures:

Procedure 1: Use only friction factor data. If you find this procedure not valid, state the reason.

Procedure 2: Select the relevant heat transfer correlation.

In summary, you are asked to provided four answers, that is,

	Fluid A	**Fluid B**
Procedure #1	$h = ?$	$h = ?$
Procedure #2	$h = ?$	$h = ?$

Answers:

Procedure #1:

Fluid A: $h = 5025$ W/m² K

Fluid B: h = cannot be computed

TABLE 10.8
Fluid Properties for Problem 10.5

Fluid Properties	**Fluid A**	**Fluid B**
k W/m°C	0.5	63
ρ kg/m³	700	818
μ kg/m s	8.7×10^{-5}	2.3×10^{-4}
c_p J/kg°C	6250	1250

Procedure #2:
Fluid A: $h = 4780$ W/m^2 K
Fluid B: $h = 22{,}000$ W/m^2 K

10.6. *Determining the temperature of the primary side of a steam generator* (Section 10.5)

Consider the flow of high-pressure water through the U-tubes of a PWR steam generator. There are 5700 tubes with outside diameter 19 mm, wall thickness 1.2 mm, and average length 16.0 m.

The steady-state operating conditions are:
Total primary flow through the tubes = 5100 kg/s
Total heat transfer from primary to secondary = 820 MW
Secondary pressure = 5.6 MPa (272°C saturation)

1. What is the primary temperature at the tube inlet?
2. What is the primary temperature at the tube outlet?

Use a Dittus–Boelter equation for the primary side heat transfer coefficient. Assume that the tube wall surface temperature on the secondary side is constant at 276°C.

Properties
For water at 300°C and 15 MPa
Density = 726 kg/m^3
Specific heat = 5.7 kJ/kg K
Viscosity = 92 μPa s
Thermal conductivity = 0.56 W/m K
For tube wall
Thermal conductivity = 26 W/m K

Hint: Consideration of the axial variation of the primary coolant bulk temperature is required.

Answers:
1. $T_{p,in} = 307.2$°C
2. $T_{p,out} = 279.0$°C

10.7. *Comparison of heat transfer characteristics of water and helium* (Section 10.5)

Consider a new design of a thermal reactor that requires square arrays of fuel rods. Heat is being generated uniformly along the fuel rods. Water and helium are being considered as single-phase coolants. Relevant coolant properties are given in Table 10.9. The design condition is that the maximum cladding surface temperature should remain below 316°C.

1. Find the minimum mass flow rate of water to meet the design requirement.
2. Would the required mass flow rate of helium be higher or lower than that of water?

TABLE 10.9
Coolant Properties for Problem 10.7

Coolant	ρ (kg/m^3)	c_p (kJ/kg K)	μ (kg/m s)	k (W/m K)
Water	735.3	5.317	0.955×10^{-4}	0.564
Helium	0.865	5.225	0.298×10^{-4}	0.230

Geometry of square array
P = pitch = 1.4 cm
D = fuel rod diameter = 1.09 cm
H = fuel height = 3.66 m
Operating conditions
Heat flux = 78.9 W/cm^2
Coolant inlet temperature = 260°C

Answers:
1. $\dot{m}_{water} = 0.451$ kg/s
2. $\dot{m}_{He} = 0.491$ kg/s

10.8. *Hydraulic and thermal analysis of the Emergency Core Spray System in a BWR* (Chapters 6 and 9, and Section 10.5)
The emergency spray system of a BWR delivers cold water to the core after a large-break loss of coolant accident has emptied the reactor vessel. The system comprises a large water pool, a pump, a spray nozzle, and connecting pipes (Figure 10.18). All pipes are smooth round tubes made of stainless steel with 10 cm internal diameter and 5 mm thickness. The pipe lengths are shown in Figure 10.18. Two sharp 90° elbows connect the vertical pipe to the horizontal pipe and the horizontal pipe to the spray nozzle. Each elbow has a form loss coefficient of 0.9. The spray nozzle has a total flow area of 26 cm^2 and a form loss coefficient of 15. The suction pipe in the pool has a sharp-edged entrance with a form loss coefficient of 0.5.
1. Calculate the pumping power required to deliver 50 kg/s of cold water to the core. (Assume steady-state and constant water properties. Do not neglect the acceleration terms in the momentum equation. Neglect entry region effects in calculating the friction factor. To calculate the irreversible term of the spray nozzle form loss, use the

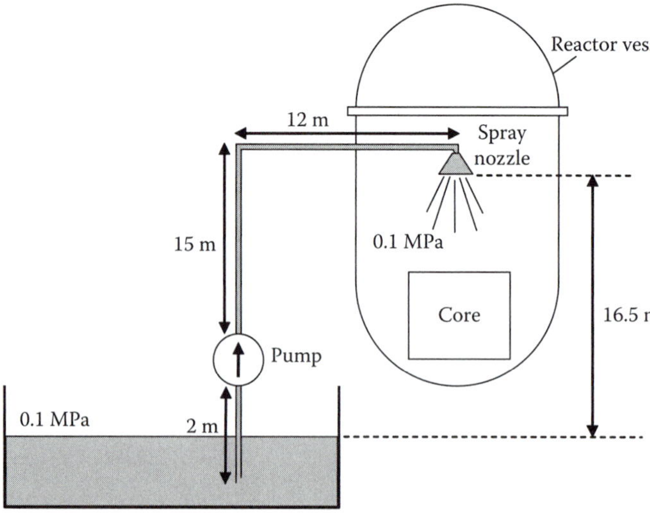

FIGURE 10.18 Emergency spray system.

value of the mass flux in the pipe. Neglect the vertical dimension of the pump. The isentropic efficiency of the pump is 80%).

2. The horizontal pipe leading to the spray nozzle is exposed to superheated steam at 200°C and 0.1 MPa. The length of the exposed section is 5 m. Estimate the heat transfer rate from the steam to the water inside the pipe. (Assume that the heat transfer coefficient on the outer surface of the pipe is 5000 W/m^{2}K. Neglect entry region effects in calculating the heat transfer coefficient within the pipe.)
3. In light of the results in (2), judge the accuracy of the constant property assumption made in calculating the pumping power in (1).

Relevant water, steam, and stainless steel properties are given in Table 10.10.

Answers:

1. $\dot{W}_p \approx 48$ kW
2. $\dot{Q} = 463$ kW

10.9. *Coolant selection for an advanced high-temperature reactor* (Chapters 8 and 9, and Section 10.5)

To improve the thermal–hydraulic performance of an advanced high-temperature reactor (AHTR), a vendor wishes to compare two alternative coolants, that is, a liquid metal (Na) and a liquid salt (LiF-BeF_2). In the AHTR core, the coolant flows inside 10 m long round channels arranged in a hexagonal lattice and surrounded by a solid fuel matrix. Consider the unit cell of this core (Figure 10.19).

TABLE 10.10
Property Values for Problem 10.8

Properties of Water at Room Temperature (25°C)

Property	Value
Density	997 kg/m^3
Viscosity	9×10^{-4} Pa s
Thermal conductivity	0.61 W/m K
Specific heat	4.2 kJ/kg K

Properties of Steam at 200°C and 0.1 MPa

Property	Value
Density	0.46 kg/m^3
Viscosity	2×10^{-5} Pa s
Thermal conductivity	0.03 W/m K
Specific heat	2.0 kJ/kg K

Properties of Stainless Steel

Property	Value
Density	8000 kg/m^3
Thermal conductivity	14 W/m K
Specific heat	0.47 kJ/kg K

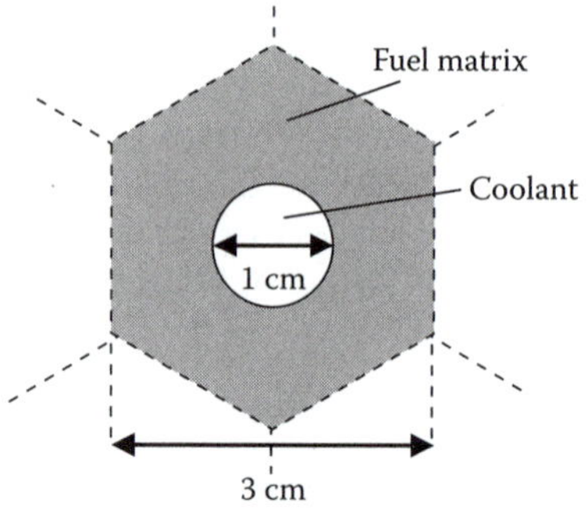

FIGURE 10.19 Cross-sectional view of the core unit cell.

1. The friction pressure drop in the coolant channel is to be limited to 200 kPa. Calculate the maximum allowable mass flow rate for the two candidate coolants. (Neglect surface roughness and entry effects.)
2. Calculate the pumping power for the mass flow rates computed in (1). (Assume $\eta_p = 100\%$.)
3. The coolant temperature at the channel inlet is 600°C. Assuming that the temperature in the fuel cannot exceed 1000°C, calculate the maximum allowable linear power for each coolant. (*Hint:* approximate the geometry of the fuel around the coolant channel as an equivalent annulus that conserves the fuel volume. Then solve the heat conduction equation for this annulus with a zero heat flux boundary condition. Assume axially and radially uniform heat generation rate within the fuel.)
4. In view of the above results, which coolant should the vendor select and why?

The properties for all materials in the system are given in Table 10.11.

Answers:

Liquid Na	Liquid LiF-BeF$_2$
(1) $\dot{m} = 0.357$ kg/s	$\dot{m} = 0.443$ kg/s
(2) $\dot{W}_p \approx 91$ W	$\dot{W}_p \approx 46$ W
(3) $q' = 9.4$ kW/m	$q' = 12.5$ kW/m

TABLE 10.11
Properties (All Properties Constant with Temperature) for Problem 10.9

Material	ρ (kg/m^3)	k (W/mK)	μ (Pa s)	c_p (J/kg K)
Liquid Na	780	60	1.7×10^{-4}	1300
Liquid LiF-BeF$_2$	1940	1	2.0×10^{-3}	2410
Fuel matrix	8530	6	/	500

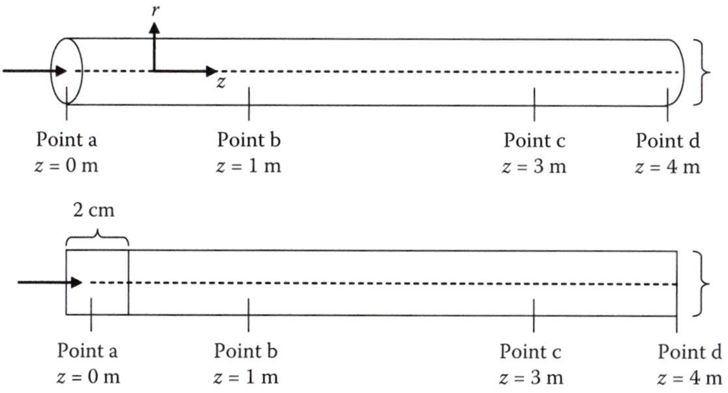

FIGURE 10.20 Round and square tubes (figure not to scale).

10.10. *Effect of geometry on single-phase heat transfer in straight tubes* (Chapter 9, Section 10.2 and Section 10.5)
Consider a smooth round tube (2 cm in diameter) and a smooth square tube (2 × 2 cm), each 4 m in length (shown in Figure 10.20). Each tube has a fluid flowing through it at the conditions given in Table 10.12.

TABLE 10.12
Fluid Properties for Problem 10.10

Parameter	Value
Density (ρ)	914 kg/m^3
Viscosity (μ)	9×10^{-5} Pa s
Specific heat (c_p)	5.8 kJ/(kg°C)
Thermal conductivity (k)	0.54 W/(m°C)
Mass flow rate ($\dot{m}$)	2.822×10^{-3} kg/s
Wall temperature	Constant throughout the length of the tube

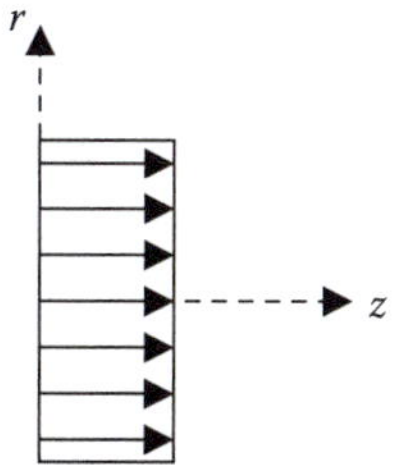

FIGURE 10.21 Velocity profile at Point a.

1. Determine if the flow is laminar or turbulent for both tubes. Justify your answer using calculations.
2. Given a uniform velocity profile at Point a (shown in Figure 10.21), sketch the velocity profiles at Points b and c for the round tube. Explain changes you see, if any, in the profiles at Points b and c.
3. Which tube has a higher heat transfer coefficient at Point c? Justify your answer using calculations. Give proper justification for the heat transfer correlation you choose.
4. Suppose that the flow rate triples. Which tube has a higher heat transfer coefficient at Point c?

Answers:

1. Laminar flow, Re = 1996 (round); Re = 1568 (square)
3. Round tube
4. Round tube

REFERENCES

1. Bhatti, M. S. and Savery, C. W., Heat Transfer in the entrance region of a straight channel: Laminar flow with uniform wall heat flux. *ASME* 76-HT-20, 1976. (Also in condensed form: *J. Heat Transfer*, 99:142, 1972.)
2. Bialokoz, I. G. and Saunders, O. A., Heat transfer in pipe flow at high speed. *Proc. Inst. Mech. Eng. (Lond.)*, 170:389, 1956.
3. Bird, R. B., Stewart, W. E., and Lightfoot, E. N., *Transport Phenomena*. New York, NY: Wiley, 1960.
4. Boelter, L. M. K., Young, G., and Iverson, H. W., *An Investigation of Aircraft Heaters*. XXVII, NACA-TN-1451, 1948.
5. Bonilla, C. F., *Heat Transfer*, Chapter 2. New York, NY: Interscience, 1964.
6. Borishanskii, V. M., Gotovskii, M. A., and Firsova, E. V., Heat transfer to liquid metals in longitudinally wetted bundles of rods. *Atom. Energy,* 27(6):1347–1350, 1969.
7. Bradshaw, P., Cebeci, T., and Whitelaw, J. H., *Engineering Calculation Methods for Turbulent Flow*. New York, NY: Academic Press, 1981.
8. Çengel, Y. A., *Introduction to Thermodynamics and Heat Transfer*. New York, NY: McGraw-Hill, 1997.
9. Chen, F. F., Domanus, H. M., Sha, W. T., and Shah, V. L., *Turbulence Modeling in the COMMIX Computer Code.* NUREG/CR-3504, ANL-83-65, April 1984.
10. Colburn, A. P., A method of correlating forced convection heat transfer data and a comparison with liquid friction. *Trans. AIChE*, 29:170, 1933.
11. Deissler, R. G., Investigation of turbulent flow and heat transfer in smooth tubes including the effects of variable physical properties. *Trans. ASME*, 73:101, 1951.
12. Deissler, R. G., Turbulent heat transfer and friction in the entrance regions of smooth passages. *Trans. ASME*, 77:1221, 1955.
13. Deissler, R. G. and Taylor, M. F., Analysis of axial turbulent flow and heat transfer through banks or rods or tubes. In: *Reactor Heat Transfer Conference*, 1956. TID-7529, Book 2, p. 416, 1957.
14. Dittus, F. W. and Boelter, L. M. K., Heat transfer in automobile radiators of the tubular type. University of California, Berkeley. *Publ. Eng.*, 2:443–461, 1930.
15. Dwyer, O. E., Eddy transport in liquid metal heat transfer. *AIChE J.*, 9:261, 1963.
16. Dwyer, O. E. and Berry, H. C., Laminar flow heat transfer for in-line flow through unbaffled rod bundles. *Nucl. Sci. Eng.*, 42:81, 1970.

17. Gnielinski, V., Neue gleichungen für den wärme- und den stoffübergang in turbulent durchströmten rohren und kanalen (New equations for heat and mass transfer in turbulent pipe and channel flow). *Int. Chem. Eng.*, 16:359–368, 1976.
18. Graber, H. and Rieger, M., Experimental study of heat transfer to liquid metals flowing in-line through tube bundles. *Prog. Heat Mass Transfer*, 7:151, 1973.
19. Graetz, L., Uber die warmeleitfahigkeit von flüssigkeiten. *Ann. Phys.*, 25:337, 1885.
20. Harlow, F. H. and Nakayama, P. I., *Transport of Turbulence Energy Decay.* Los Alamos Scientific Laboratory, Report LA-3854-UC-34, TID-4500, 1968.
21. Irvine, T. R., Non-circular convective heat transfer. In W. Ible (ed.). *Modern Developments in Heat Transfer*. New York, NY: Academic Press, 1963.
22. Jenkins, R., *Heat Transfer and Fluid Mechanics Institute*, pp. 147–158. Stanford University Press, Stanford, CA, 1951.
23. Jones, W. P. and Launder, B. E., The prediction of laminarization with a 2-equation model ot turbulence. *Int. J. Heat Mass Transfer*, 15:301, 1972.
24. Kays, W. M. and Crawford, M. E., *Convective Heat and Mass Transfer*, 2nd Ed. New York, NY: McGraw-Hill, 1980.
25. Kays, W. M., Numerical solutions for laminar flow heat transfer in circular tubes. *Trans. ASME*, 77:1265, 1955.
26. Kays, W. M. and Leung, E. Y., Heat transfer on annular passages. *Int. J.H.M.T.*, 10:1533, 1963.
27. Kazimi, M. S. and Carelli, M. D., *Heat Transfer Correlation for Analysis of CRBRP Assemblies.* Westinghouse Report, CRBRP-ARD-0034, 1976.
28. Latzko, H., Der Wärmeübergang an einer Turbulenten Flüssigkeits-Oder Gasstrom. *ZAMM*, 1:268, 1921.
29. Launder, B. E. and Spaulding, D. B., *Mathematical Models of Turbulence*. New York, NY: Academic Press, 1972.
30. Launder, B. E., Morse, A., Rodi, W., and Spaulding, D. B., The prediction of free shear flows—A comparison of the performance of six turbulent models. In: *Proceedings of NASA Conference on Free Shear Flows*, Langberg, 1972.
31. Lienhard IV, J. H. and Lienhard V, J. H., *A Heat Transfer Textbook*. Cambridge, MA: Phlogiston Press, 2008.
32. Lyon, R. N., Liquid metal heat transfer coefficients. *Chem. Eng. Prog.*, 47:75, 1951.
33. Manglik, R. M. and Bergles, A. E., Heat transfer and pressure drop correlations for twisted-tape inserts in isothermal tubes. I, Laminar flows. *J. Heat Transfer*, 115:881–889, 1993.
34. Manglik, R. M. and Bergles, A. E., Heat transfer and pressure drop correlations for twisted-tape inserts in isothermal tubes. II, Transition and turbulent flows. *J. Heat Transfer*, 115:890–896, 1993.
35. Markoczy, G., Convective heat transfer in rod clusters with turbulent axial coolant flow. 1. Mean value over the rod perimeter. *Wärme Stoffübertragung*, 5:204–212, 1972.
36. Martinelli, R. B., Heat transfer in molten metals. *Trans. ASME*, 69:947, 1947.
37. McAdams, W. H., *Heat Transmission*, 2nd Ed. New York, NY: McGraw-Hill, 1942.
38. McAdams, W. H., *Heat Transmission*, 3rd Ed. New York, NY: McGraw-Hill, 1954.
39. McAdams, W. H. Kennel, W. E., and Emmons, J. N., Heat transfer to superheated steam at high pressures addendum. *Trans ASME*, 72:421, 1950.
40. Nijsing, R., *Heat Exchange and Heat Exchangers with Liquid Metals.* AGRD-Ls-57012; AGRD Lecture Series No. 57 on Heat Exchangers by J. J. Ginoux, Lecture Series Director, 1972.
41. Petukhov, B. S., Heat transfer and friction in turbulent pipe flow with variable physical properties. In J. P. Hartnett and T. F. Irvine (ed.). *Advances in Heat Transfer*, Vol. 6, pp. 504–564. New York, NY, 1970.

42. Petukhov, B. S. and Popov, V. N., Theoretical calculation of heat exchange and frictional resistance in turbulent flow in tubes of an incompressible fluid with variable physical properties, *Teplofizika Vysokikh Temperature*, 1963.
43. Prandtl, L., Eine Beziehung zwischen Wärmeaustausch und Strömungswiderstand der Flüssigkeiten, *Z. Phys.*, 11:1072, 1910.
44. Prandtl, L., Über die Ausgebildete Turbulenz. *ZAMM*, 5:136, 1925.
45. Presser, K. H., *Wärmeübergang und Druckverlust an Reaktorbrennelementen in Form Langsdurchströmter Rundstabbündel*. Jul-486-RB, KFA, Julich, 1967.
46. Ravigururajan, T. S. and Bergles, A. E., Development and verification of general correlations for pressure drop and heat transfer in single-phase turbulent flow in enhanced tubes. *Exp. Therm. Fluid Sci.*, 13: 55–70, 1996.
47. Rodi, W., *Turbulence Models and Their Application to Hydraulics—A State of the Art Review*, International Association for Hydraulic Research, Delft, the Netherlands, 1980.
48. Schlichting, H., *Boundary Layer Theory*, 6th Ed. New York, NY: McGraw-Hill, 1968.
49. Seban, R. A., Heat transfer to a fluid flowing turbulently between parallel walls and asymmetric wall temperatures. *Trans. ASME*, 72:789, 1950.
50. Seban, R. A. and Shimazaki, T., Heat transfer to a fluid flowing turbulently in a smooth pipe with walls at constant temperature. ASME Paper 50-A-128, 1950.
51. Seider, E. N. and Tate, G. E., Heat transfer and pressure drop of liquids in tubes. *Ind. Eng. Chem.*, 28:1429, 1936.
52. Silberberg, M. and Huber, D. A., *Forced Convection Heat Characteristics of Polyphenyl Reactor Coolants*. AEC Report NAA-SR-2796, 1959.
53. Sleicher, C. A. and Tribus, M., Heat transfer in a pipe with turbulent flow and arbitrary wall-temperature distribution. *Trans. ASME*, 79:789, 1957.
54. Sparrow, E. M., Hallman, T. M., and Siegel, R., Turbulent heat transfer in the thermal entrance region of a pipe with uniform heat flux. *Appl. Sci. Res.*, A7:37, 1957.
55. Sparrow, E. M., Loeffler Jr., A. L., and Hubbard, H. A., Heat transfer to longitudinal laminar flow between cylinders. *J. Heat Transfer*, 83:415, 1961.
56. Tribus, M. and Klein, S. J., Forced convection from nonisothermal surfaces in heat transfer: A symposium. Engineering Research Institution, University of Michigan, Ann Arbor, MI, 1953.
57. Trup, A. C. and Azad, R. S., The structure of turbulent flow in triangular array rod bundles. *Nucl. Eng. Des.*, 32:47–84, 1975.
58. von Karman, T., The analogy between fluid friction and heat transfer. *Trans. ASME*, 61:701, 1939.
59. Weisman, J., Heat transfer to water flowing parallel to tube bundles. *Nucl. Sci. Eng.*, 6:79, 1959.
60. Winterton, R. H. S., Where did the Dittus and Boelter equation come from? *Int. J. Heat Mass Transf.*, 41(4–5):809–810, 1998.

11 Two-Phase Flow Dynamics

11.1 INTRODUCTION

In the present chapter the focus is on the hydraulics of two-phase flow, which includes phase configuration, pressure drop relations, and critical flow. The discussion in Chapter 5 indicates the wide range of possibilities for the selection of the sets of equations to be solved. The appropriate set of equations (more commonly called the *model*) for a particular two-phase flow system is influenced by the conditions of the system. The more detailed models can be laborious to solve because of the requirement for a large computing effort. The less detailed models introduce certain simplifying assumptions that are not always correct but can be solved more readily. Thus, it is appropriate to start the process of analyzing the two-phase flow systems by asking the following questions:

1. What is the number of flow dimensions that need to be represented? The difficulty of solving a multidimensional problem greatly exceeds that of a one-dimensional flow problem. In fact, there is relatively little information regarding the multidimensional effects on the flow hydraulics, and most of the existing information is related to one-dimensional channel flow.
2. What is the expected degree of mechanical equilibrium between the phases? Under conditions of high mass fluxes, the two phases can be expected to move at the same velocity. However, the relative velocity of one phase to the other is a complex function of the flow conditions and geometry under consideration.
3. What is the expected degree of thermal nonequilibrium in the flow? Thermal nonequilibrium is generally more important under transient conditions than under steady-state conditions. Even in steady-state conditions, however, the effects of nonequilibrium often must be included, as for subcooled boiling of liquids.

Once a model is selected, constitutive relations are needed to describe the rate of exchange of mass, momentum, and energy among the wall and the two phases. These relations depend on the flow configuration, or *flow regime*. They are often empirically determined correlations with broad error ranges compared to experimental data. The state equations and boundary and initial conditions are also needed to close the problem. The needed information for solving the two-phase flow equations is depicted in Figure 11.1.

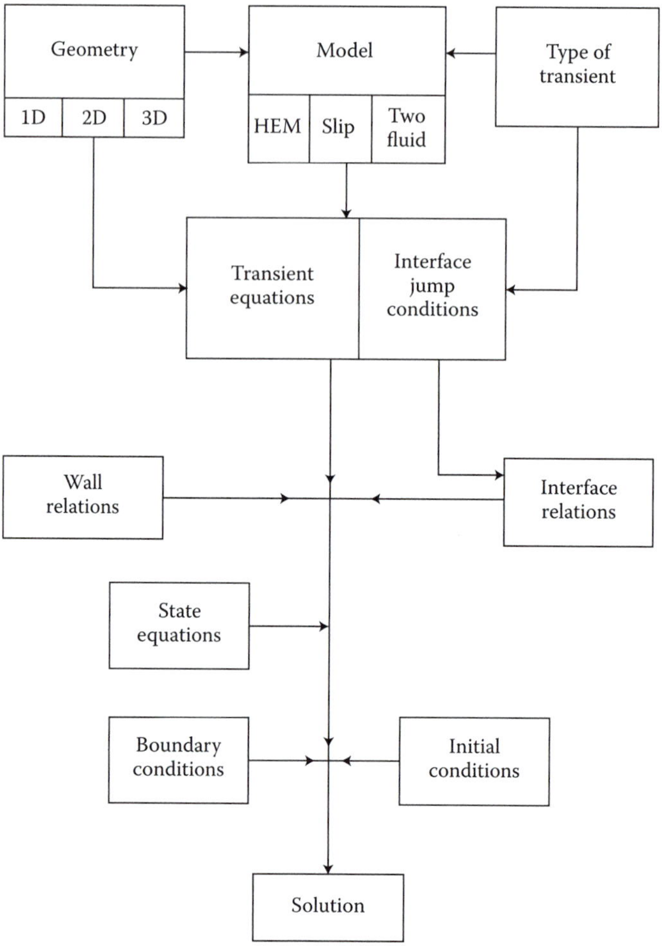

FIGURE 11.1 Information needs of two-phase flow analysis.

11.2 FLOW REGIMES

11.2.1 Regime Identification

The void distribution patterns of gas*–liquid flows have been characterized by many researchers. They did not always refer to the flow regimes with consistent terminology. However, the main flow regimes, the only ones addressed here, are sufficiently distinct. The void distribution depends on the pressure, channel geometry, gas and liquid flow rates, and the orientation of the flow with respect to gravity. Of major significance to the nuclear engineer are the cases of vertical upward co-current flow

* Indeed gas is the more general term, and many two-phase flow studies have been done with air–water as we discuss next. In reactor applications, the gas is typically a vapor, for example steam in particular, for water-cooled plants.

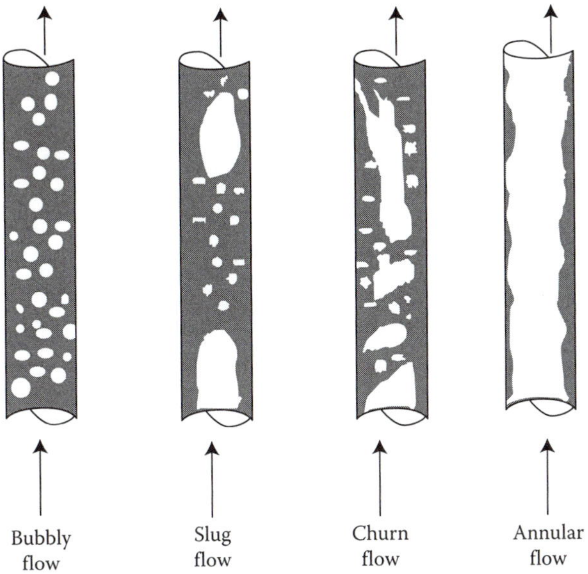

FIGURE 11.2 Flow patterns in vertical two-phase flow. With horizontal flow, the flow may stratify, creating additional patterns, as shown in Figure 11.3.

(as in a BWR fuel assembly) and horizontal co-current flow (as in the piping of an LWR during transients).

The flow regimes identified in vertical flows are the bubbly, slug, churn, and annular regimes (Figure 11.2). The bubbly regime is identified by the presence of dispersed gas bubbles in a continuous liquid phase. The bubbles can be of variable size and shape. Bubbles of 1 mm or less are spherical, but larger bubbles have variable shapes. Slug flow is distinguished by the presence of gas plugs (or large bubbles) separated by liquid slugs. The liquid film surrounding the gas plug usually moves downward. Several small bubbles may also be dispersed within the liquid. The churn flow is more chaotic but of the same basic character as the slug flow. Annular flow is recognizable from the presence of a continuous core of gas surrounded by an annulus of the liquid phase. If the gas flow in the core is sufficiently high, it may be carrying liquid droplets. In this case, an annular-dispersed flow regime is said to exist. Hewitt and Roberts [31] also suggested that the droplets can gather in clouds forming an wispy-annular regime. The liquid droplets are torn from the wavy liquid film, get entrained in the gas core, and can be de-entrained to join the film downstream of the point of their origin.

With horizontal flow, the flow may stratify, creating additional patterns, as shown in Figure 11.3.

11.2.2 Flow Regime Maps

With the flow regimes identified, the next step has been to create flow regime maps based on two independent flow variables where regions corresponding to these various flow regimes are identified. The flow and heat transfer correlations presented

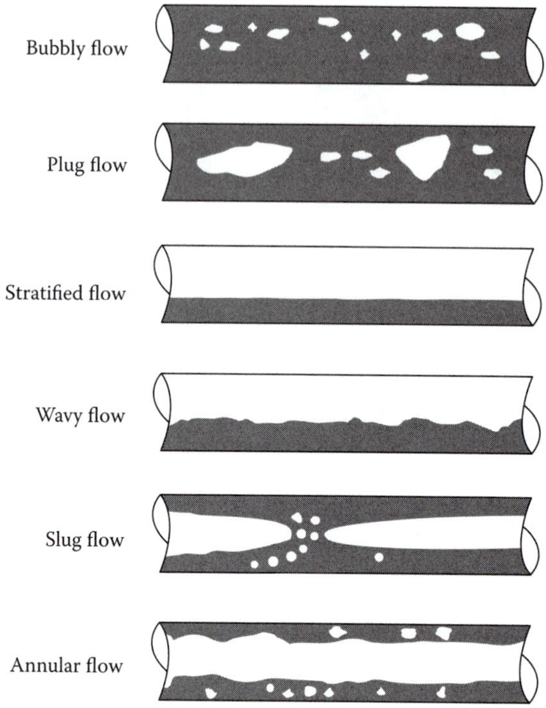

FIGURE 11.3 Flow patterns in horizontal two-phase flow.

later in this chapter and the next are generally applicable within a specific flow regime. Since the occurrence of a particular flow regime depends on multiple variables characterizing the fluid, geometry and pressure, such two-dimensional plots while suitable for particular applications, are not generally valid. In particular, the following cautions are advised

- Regime boundary lines should be considered broad transition bands since flow pattern transitions do not occur suddenly but over a range of the controlling variables, for example flow rates.
- Different maps are needed for different flow geometries (tubes, rod bundles), channel orientation (vertical, inclined, horizontal), and flow condition (adiabatic, diabatic).

Further, the use of the flow map approach in modern numerical analysis introduces several practical considerations:

- The sharp transition between regimes creates discontinuities in the constitutive relations which are coded into numerical tools.
- Most of the flow regimes have been defined in one-dimensional channels. The use of one-dimensional flow maps for three-dimensional flow analysis requires new approaches since there is more than a single velocity component

at each location. Two approaches can be considered: (1) The use of a static quantity such as the void fraction, to characterize the flow. One such flow map has been used in the RELAP5 [58] analysis code (Figure 11.4); and (2) The use of the total flow velocity as the characteristic velocity in the flow maps. Neither approach is without limitations. For example, the effect of the heat flux from the walls on the existence of a liquid or vapor region near the wall is not included in either approach.

11.2.2.1 Vertical Flow

The one-dimensional flow map of Hewitt and Roberts [31] (Figure 11.5) was developed on the basis of air–water data, obtained in a pipe of 31.2 mm diameter and at pressures varying from 0.14 to 0.54 MPa. It was found suitable for steam–water data in a pipe of 12.7 mm diameter at pressures of 3.45–6.90 MPa. It is based on the superficial liquid and vapor momentum fluxes:

$$\rho_\ell\{j_\ell\}^2 = \frac{G_m^2(1-x)^2}{\rho_\ell} \tag{11.1}$$

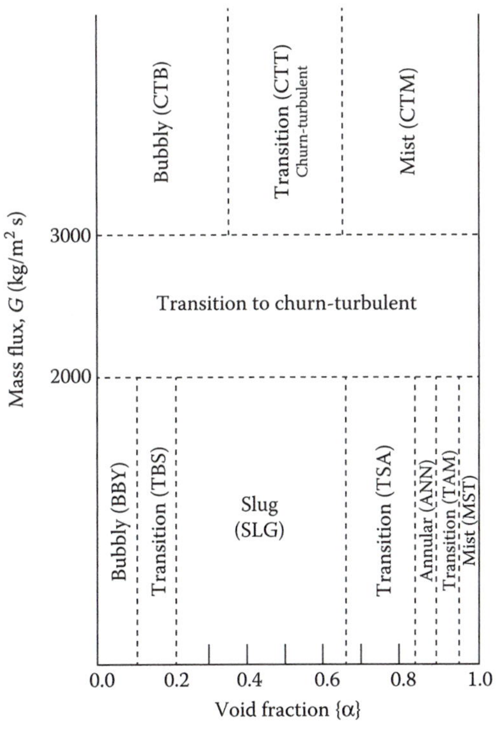

FIGURE 11.4 Vertical flow regime map of RELAP5. (From Ranson, V. H. et al., *RELAP5/MOD1 Code Manual*, Vols. 1 & 2, NUREG/CR-1826,EGG-2070, March 1982.)

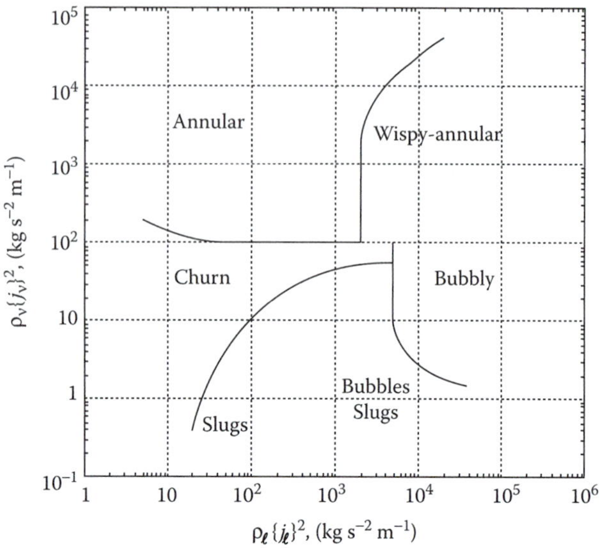

FIGURE 11.5 Hewitt and Roberts map for vertical upward flow. (From Hewitt, G. F. and Roberts, D. N., *Studies of Two-Phase Flow Patterns by Simultaneous X-Ray and Flash Photography*. AERE-M2159, 1969.)

and

$$\rho_v\{j_v\}^2 = \frac{G_m^2 x^2}{\rho_v} \tag{11.2}$$

Taitel and co-workers [67] compared several flow regime maps and found some discrepancies among them. They also developed their own map based on a theoretical analysis for the mechanisms contributing to the transition between regimes.

11.2.2.1.1 Bubbly to Slug

They postulated that the transition from bubbly to slug or churn flow is related to a maximum possible void fraction of 0.25 to prevent bubble coalescence to form slugs. The liquid and vapor velocities are related by

$$\upsilon_\ell = \upsilon_v - V_o \tag{11.3}$$

where V_o = rise velocity of large bubbles (5 mm < d < 20 mm), which was shown to be insensitive to bubble size by Harmathy [25] and is given by

$$V_o = 1.53\left(\frac{g(\rho_\ell - \rho_v)\sigma}{\rho_\ell^2}\right)^{1/4} \tag{11.4}$$

where σ = surface tension.

Using the definition of the volumetric fluxes and assuming uniform velocity distributions, we can obtain the following relation from Equations 11.3 and 11.4:

$$\frac{\{j_\ell\}}{\{1-\alpha\}} = \frac{\{j_v\}}{\alpha} - 1.53\left(\frac{g(\rho_\ell - \rho_v)\sigma}{\rho_\ell^2}\right)^{1/4} \tag{11.5}$$

For $\{\alpha\} = 0.25$, Equation 11.5 yields

$$\frac{\{j_\ell\}}{\{j_v\}} = 3 - 1.15\frac{[g(\rho_\ell - \rho_v)\sigma]^{1/4}}{\{j_v\}\rho_\ell^{1/2}} \tag{11.6}$$

Equation 11.6 is shown as line A in Figure 11.6. It should be noted that in an earlier paper by Taitel and Dukler [66] $\{\alpha\} = 0.3$ was chosen as the maximum permissible packing of bubbles to prevent coalescence, which yielded the transition limit of

$$\frac{\{j_\ell\}}{\{j_v\}} = 2.34 - 1.07\frac{[g(\rho_\ell - \rho_v)\sigma]^{1/4}}{\{j_v\}\rho_\ell^{1/2}} \tag{11.7}$$

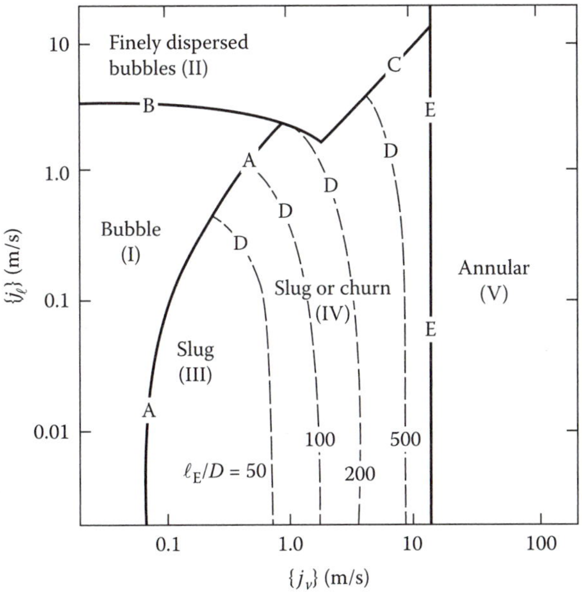

FIGURE 11.6 Flow regime map of Taitel et al. for air–water at 25°C and 0.1 MPa in 50 mm diameter tubes. (From Taitel, Y., Bornea, D., and Dukler, A. E., *AIChE J.*, 26:345, 1980.)

For small pipes, the rising bubbles can catch up to form a slug. Thus for very small channels the bubbly regime may not exist, in which case only slug flow would exist at low liquid and gas velocities and well-developed flow. This situation is shown in Figure 11.7 for air–water flow in a 25 mm diameter tube under the same pressure and temperature conditions as Figure 11.6.

The criterion for eliminating bubbly flow is that a deformable bubble rising at a velocity V_o approaches and coalesces with the slug or Taylor bubble rising at velocity V_b, that is

$$V_o \geq V_b$$

where

$$V_b = 0.35\sqrt{gD} \text{ (for } \rho_v \ll \rho_\ell) \tag{11.8}$$

Because V_o is given by Equation 11.4, this criterion yields the following relation for a small-diameter (no bubbly regime) system:

$$\left[\frac{\rho_\ell^2 g D^2}{(\rho_\ell - \rho_v)\sigma}\right]^{1/4} \leq 4.36 \tag{11.9}$$

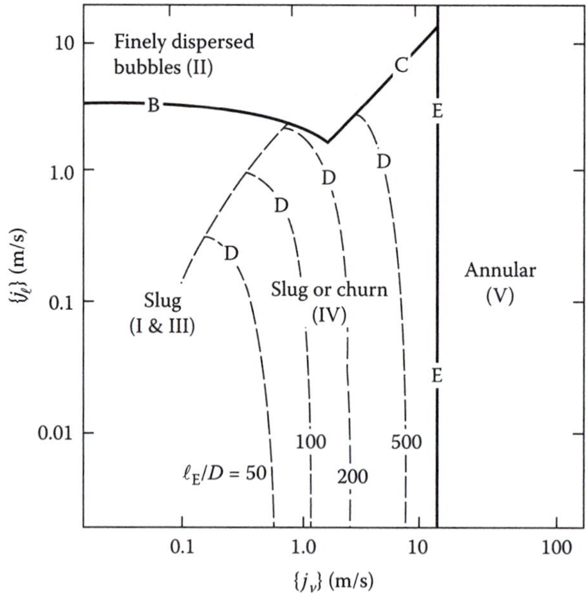

FIGURE 11.7 Flow regime map of Taitel et al. for air–water at 25°C and 0.1 MPa in 25 mm diameter tube. (From Taitel, Y., Bornea, D., and Dukler, A. E., *AIChE J.*, 26:345, 1980.)

Interestingly for the BWR channel, the criterion of Equation 11.9 yields a value less than 4.36, that is

$$\left[\frac{(737.3)^2 9.8(12.28\times10^{-3})^2}{(737.3-37.3)18.03\times10^{-3}}\right]^{1/4} = 2.82$$

which suggests the bubbly regime does not exist. Obviously, however, for the diabatic BWR channel, boiling on the fuel pin surfaces does release bubbles into the cooling channel but they soon form slugs for a finite length of the channel.

11.2.2.1.2 Finely Dispersed Bubbly to Bubbly/Slug Flow

For high liquid volumetric fluxes, bubble breakup would occur due to turbulent forces so that coalescence would also be prevented above a limit on the sum of $\{j_\ell\}$ and $\{j_v\}$ derived by Taitel et al. [67] as

$$\{j_\ell\}+\{j_v\} = 4\left\{\frac{D^{0.429}(\sigma/\rho_\ell)^{0.089}}{\nu_\ell^{0.072}}\left[\frac{g(\rho_\ell-\rho_v)}{\rho_\ell}\right]^{0.446}\right\} \qquad (11.10)$$

where ν_ℓ = liquid kinematic viscosity, and D = tube diameter. Equation 11.10 is shown as line B in Figures 11.6 and 11.7.

When small spherical bubbles are packed as tightly as possible together, the associated void fraction (for arrangement in a cubic lattice) is equal to 0.52. At high liquid velocities (when bubble breakup can occur), however, the relative velocity is zero, so from Table 5.2 or the one-dimensional relations of Figure 5.3, we get

$$\frac{\{\alpha\}}{1-\{\alpha\}} = \frac{\{\beta\}}{1-\{\beta\}}\frac{\upsilon_\ell}{\upsilon_v}$$

which, using Equation 5.59, reduces to

$$\{\alpha\} = \frac{\{j_v\}}{\{j_v\}+\{j_\ell\}} \quad \text{for } \upsilon_\ell \simeq \upsilon_v \qquad (11.11)$$

Thus the line,

$$0.52 = \frac{\{j_v\}}{\{j_v\}+\{j_\ell\}} \qquad (11.12)$$

defined as line C in Figures 11.6 and 11.7, is another line limiting the existence of the bubbly regime.

11.2.2.1.3 Slug to Churn Flow

The slug flow pattern develops from a bubbly flow pattern if enough small bubbles can coalesce to form Taylor bubbles. If the liquid slug between two Taylor bubbles is too small to be stable, the churn regime develops instead. Several mechanisms have been proposed for the onset of churn flow including flooding, which was suggested by Nicklin and Davidson [51]. Flooding is discussed later in this chapter in Section 11.2.3.

Taitel et al. [67] presented the alternative view that the churn regime is essentially a developing length region for slug flow. Thus they derived a maximum length (ℓ_ε) of the churn regime, given by

$$\frac{\ell_\varepsilon}{D} = 40.6\left(\frac{\{j\}}{\sqrt{gD}} + 0.22\right) \tag{11.13}$$

This equation shows the developing length to be dependent on $\{j\}/\sqrt{gD}$.

For conditions not allowing the existence of bubbly or annular flow, their criterion can be used to determine the length over which a churn regime can exist before changing to a slug flow. If a position along the tube length is shorter than the developing length, churn or slug flow may be observed. If the developing length is short relative to the position of interest along the tube, slug flow alone occurs. This situation is depicted in Figures 11.6 and 11.7 as lines D for given ℓ_ε/D.

In an earlier paper, Taitel and Dukler [66] had suggested that for $\ell_\varepsilon/D > 50$ churn flow would exist if

$$\{j_v\}/\{j\} \geq 0.85 \tag{11.14}$$

It should be noted that Equation 11.14 implies that churn flow exists at $\{j_v\}/\{j_\ell\} \geq 5.5$ whenever $\ell_\varepsilon/D > 50$. However, the more recent approach allows churn flow to exist even at lower values of $\{j_v\}/\{j_\ell\}$. For example, at $\{j_v\} = 1.0$ m/s, line D in Figure 11.7 for $\ell_\varepsilon/D = 100$ implies that churn flow can exist for $\{j_\ell\} \simeq 0.5$ m/s or up to $\{j_v\}/\{j_\ell\} \simeq 2.0$. Therefore, there is a substantial difference between the two approaches for defining the boundary between slug flow and churn flow.

11.2.2.1.4 Slug/Churn to Annular Flow

For transition from slug/churn flow to annular flow, Taitel et al. [67] argued that the gas velocity should be sufficient to prevent liquid droplets from falling and bridging between the liquid film. The minimum gas velocity to suspend a drop is determined from a balance between the gravity and drag forces:

$$\frac{1}{2}C_d\left(\frac{\pi d^2}{4}\right)\rho_v \upsilon_v^2 = \left(\frac{\pi d^3}{6}\right)g(\rho_\ell - \rho_v) \tag{11.15}$$

or

$$\upsilon_v = \frac{2}{\sqrt{3}}\left[\frac{g(\rho_\ell - \rho_v)d}{\rho_v C_d}\right]^{1/2} \tag{11.16}$$

The drop diameter is determined from the criterion shown by Hinze [34] for maximum stable droplet sizes:

$$d = \frac{K\sigma}{\rho_v \upsilon_v^2} \tag{11.17}$$

where K is the critical *Weber number*, which has a value between 20 and 30. Combining Equations 11.16 and 11.17, we get

$$\upsilon_v = \left(\frac{4K}{3C_d}\right)^{1/4} \frac{[\sigma g(\rho_\ell - \rho_v)]^{1/4}}{\rho_v^{1/2}} \tag{11.18}$$

If K is taken as 30 and C_d as 0.44 (note that υ_v is insensitive to the exact values because of the 1/4 power), and it is assumed that in annular flow the liquid film is thin so that $j_v \simeq \upsilon_v$, Equation 11.18 yields

$$\frac{\{j_v\}\rho_v^{1/2}}{[\sigma g(\rho_\ell - \rho_v)]^{1/4}} = 3.1 \tag{11.19}$$

Thus the transition to annular flow (Equation 11.19 depicted as line E in Figures 11.6 and 11.7) is independent of the tube diameter and the liquid flow rate.

The left-hand side group of Equation 11.19 is the vapor *Kutateladze number* (Ku), which represents the ratio of the gas kinetic head to the inertial forces acting on liquid capillary waves with a dimension of

$$\sqrt{\frac{\sigma}{g(\rho_\ell - \rho_v)}} \tag{11.20}$$

It is interesting that a similar criterion would be obtained for the initiation of annular flow if it were related to the onset of flow reversal (see Section 11.2.3).

11.2.2.2 Horizontal Flow

For horizontal flow, Mandhane et al. [43] proposed a map on the basis of the gas and liquid volumetric fluxes as shown in Figure 11.8. The range of data used to produce their results appears in Table 11.1. Taitel and Dukler [65] found similar map boundaries from a mechanistic approach.

Example 11.1: Flow Regime Calculations

PROBLEM

Consider a vertical tube of 17 mm I.D. and 3.8 m length. The tube is operated at steady state with the following conditions, which approximate the BWR fuel bundle condition.

Operating pressure, $p = 7.44$ MPa

Inlet temperature, $T_{in} = 275°C$
Saturation temperature, $T_{sat} \simeq 290°C$
Mass velocity, $G_m = 1700$ kg/m² s
Heat flux (constant), $q'' = 670$ kW/m²
Outlet quality, $x = 0.185$

1. Using the Hewitt and Roberts flow map, determine the flow regime at the exit.
2. Using the approach of Taitel et al. [67], determine if there is a transition from slug/churn flow to annular flow before the exit.

SOLUTION

Using $T_{sat} = 290°C$, the following fluid properties pertain

$$\rho_\ell = 732.33 \text{ kg/m}^3$$

$$\rho_v = 39.16 \text{ kg/m}^3$$

$$\sigma = 0.0167 \text{ N/m}$$

1. The coordinates for the Hewitt and Roberts flow map (Figure 11.5) are

$$\rho_\ell\{j_\ell\}^2 = \frac{[G_m(1-x)]^2}{\rho_\ell} = \frac{[1700(1-0.185)]^2}{732.33} = 2621.2 \text{kg/ms}^2$$

$$\rho_v\{j_v\}^2 = \frac{(G_m x)^2}{\rho_v} = \frac{[1700(0.185)]^2}{39.16} = 2525.8 \text{kg/ms}^2$$

Using these values with Figure 11.5, we find that the flow at the outlet is on the border between annular and wispy-annular.

2. The transition from slug/churn to annular flow in the approach of Taitel et al. [67] is predicted by Equation 11.19:

$$\frac{\{j_v\}\rho_v^{1/2}}{[\sigma g(\rho_\ell - \rho_v)]^{1/4}} = \text{Ku}_v = 3.1$$

At the outlet:

$$\frac{\sqrt{2525.8}}{[(0.0167)(9.8)(732.33 - 39.16)]^{1/4}} = 15.4 > 3.1$$

Thus at the exit the flow is already well into the annular regime.
The expression in the numerator of Equation 11.19 can be written as

$$\{j_v\}\rho_v^{1/2} = \frac{G_m x}{\rho_v^{1/2}}$$

TABLE 11.1
Parameter Range for the Flow Map of Mandhane et al.

Property	Parameter Range
Pipe inner diameter (mm)	12.7–165.1
Liquid density (kg/m^3)	705–1009
Gas density (kg/m^3)	0.80–50.5
Liquid viscosity (kg/m s)	3×10^{-4} to 9×10^{-2}
Gas viscosity (kg/m s)	10^{-5} to 2.2×10^{-5}
Surface tension (N/m)	24–103
Liquid superficial velocity (mm/s)	0.9–7310
Gas superficial velocity (m/s)	0.04–171

Source: From Mandhane, J. M., Gregory, G. A., and Aziz, K., *Int. J. Multiph. Flow,* 1:537, 1974.

Because the flow quality is linearly increasing from 0 at saturation condition to the exit value of $x = 0.185$, we see that Ku_v takes on values from 0 to 15.4. Thus, the transition from slug/churn flow to annular flow should take place at the location corresponding to $Ku_v = 3.1$.

11.2.3 Flooding and Flow Reversal

Flooding and flow reversal are basic phenomena encountered in several conditions in nuclear reactor thermal hydraulics, including flow regime transitions (described in

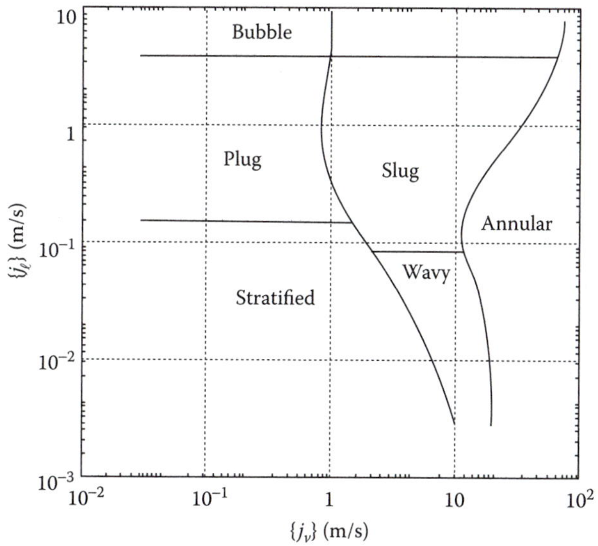

FIGURE 11.8 Flow map of Mandhane et al. [43] for horizontal flow. (From Mandhane, J. M., Gregory, G. A., and Aziz, K., *Int. J. Multiph. Flow,* 1:537, 1974.)

Section 11.2.2) and the rewetting of hot surfaces after a loss of coolant accident (LOCA) in an LWR. *Flooding* refers to the stalling of a liquid downflow by a sufficient rate of gas upflow, and *flow reversal* refers to the change in flow direction of liquid initially in co-current upflow with a gas as the gas flow rate is sufficiently decreased. Here, we review information on these phenomena as applied to simple tube geometry. Although several experiments have been carried out to predict the onset of flooding and flow reversal, no general adequate theories are available thus far. However, some empirical correlations have been developed.

Figure 11.9 shows a vertical tube with a liquid injection device composed of a porous wall. Initially there is no gas flow, and a liquid film flows down the tube at a constant flow rate. If gas is flowing upward at a low rate, a countercurrent flow takes place in the tube (tube 1). When the gas flow rate is increased, the film thickness remains constant and equal to the Nusselt-predicted thickness of a laminar film [32]. The nondimensional thickness is given by

$$\delta^* \equiv \delta\left[\frac{g\sin\theta}{v_\ell^2}\right]^{1/3} = 1.442\,\mathrm{Re}_\ell^{1/3} \tag{11.21a}$$

where

$$\mathrm{Re}_\ell \equiv \frac{4Q_\ell}{\pi D v_\ell} \tag{11.21b}$$

and where θ = inclination angle from the horizontal, v_ℓ = liquid kinematic viscosity, Q_ℓ = liquid volumetric flow rate, D = pipe diameter, and g = acceleration due to gravity. This relation is valid for liquid Reynolds number (Re_ℓ) up to 2000. For turbulent films, the thickness may be given as proposed by Bankoff and Lee [3] as

$$\delta^* = 0.304\,\mathrm{Re}_\ell^{7/12} \tag{11.22}$$

For a higher gas flow rate, the liquid film becomes unstable, large-amplitude waves appear, and the effective film thickness is increased. Droplets are torn from the crests of the waves and are entrained with the gas flow above the liquid injection level. Some of the droplets impinge on the wall and give rise to a liquid film (tube 2). Meanwhile, the pressure drop in the tube above the liquid injection level increases sharply. When the gas flow rate is increased further, the tube section below the liquid injection dries out progressively (tube 3), until the liquid downflow is completely prevented at a given gas flow rate. This point is called the *flooding condition* (tube 4). At higher gas flow rates, churn flow or annular flow appears in the upper tube.

Suppose we decrease the gas flow rate. For a given value the liquid film becomes unstable, and large-amplitude waves appear on the interface. The pressure drop increases, and the liquid film tends to fall below the liquid injection port. This point is called the *flow reversal phenomenon* (tube 5). When the gas flow rate decreases, the liquid film flows downward, and the upper tube dries out for a given gas flow rate (tube 6). It is noted that the change from upward to downward liquid flow and vice versa is preceded by a transition period of partial flow in both directions.

The flooding phenomenon has been described analytically by several authors. The onset of flooding has been explained by four mechanisms:

1. Occurrence of a standing wave on the liquid film
2. Interfacial instabilities between the two phases due to their relative velocities
3. No net flow in the liquid film
4. Inception of entrainment from the liquid film

The flow reversal phenomenon has been shown to be strongly connected to the hydrodynamics of hanging films. A review of the available experimental and analytic studies of countercurrent flow phenomena associated with flooding and flow reversal in vertical and horizontal flow has been given by Bankoff and Lee [3]. Because of the various definitions used for the onset of each phenomenon, the empirical relations should always be reviewed to ascertain their experimental base prior to application.

Experimental information on the pressure drop characteristics in vertical countercurrent flow transitions is given in Figure 11.10.

Two nondimensional numbers have been used extensively to correlate the transition conditions for two-phase systems: the Wallis number and the Kutateladze number. The former represents the ratio of inertial force to hydrostatic force on a bubble or drop of diameter *D*. The latter replaces the length scale *D* with the *Laplace constant* given as

$$\left[\frac{\sigma}{g(\rho_\ell - \rho_v)}\right]^{1/2} \tag{11.20}$$

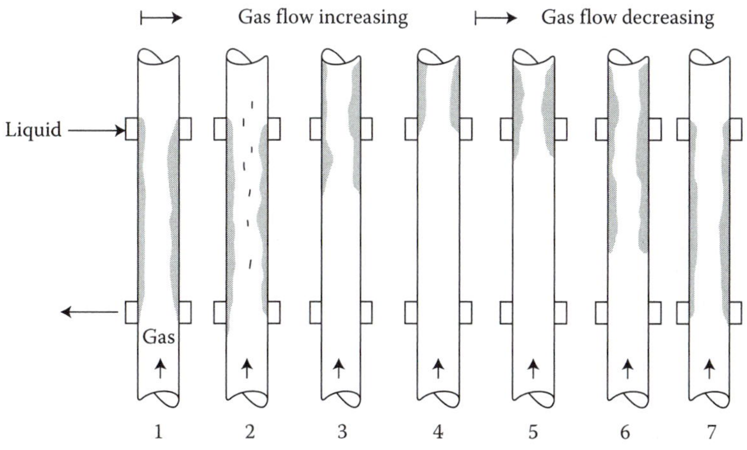

FIGURE 11.9 Flooding and flow reversal experiment (Tube 4—flooding; Tube 5—flow reversal).

The *Wallis number* is defined as

$$\{j_k^+\} \equiv \{j_k\}\left[\frac{\rho_k}{gD(\rho_\ell - \rho_v)}\right]^{0.5} \tag{11.23}$$

where k = ℓ for liquid and v for vapor.

The *Kutateladze number* (Ku) is defined as

$$\mathrm{Ku}_k \equiv \{j_k\}\left[\frac{\rho_k}{[g\sigma(\rho_\ell - \rho_v)]^{0.5}}\right]^{0.5} \tag{11.24}$$

Wallis [69] proposed correlations for flooding in vertical countercurrent flow transitions tubes that depend on the size of the channel:

$$\{j_v^+\}^{0.5} + m\{j_\ell^+\}^{0.5} = C \tag{11.25}$$

where m and C are constants that depend on the design of the channel ends. It is often found that $m = 1.0$ and $C = 0.9$ for round-edged tubes and 0.725 for sharp-edged tubes.

Porteous [54] gave the limit of flooding in a semitheoretical treatment as

$$\frac{\{j\}}{\sqrt{gD}} = 0.105\sqrt{\frac{\rho_\ell - \rho_v}{\rho_v}} \tag{11.26}$$

For flow reversal, Wallis [69] recommended

$$\{j_v^+\} = 0.5 \tag{11.27}$$

However, later Wallis and Kuo [70] modified this relation for flow reversal to reconcile it with the data, which showed less dependence on the tube diameter.

When analyzing their result for air–water flow in tubes of 6–309 mm diameter, Pushkin and Sorokin [56] proposed that flow reversal occurs for the *Kutateladze number* condition:

$$\mathrm{Ku}_v = 3.2 \tag{11.28}$$

They argued that this condition is also close to the flooding condition.

Whalley [71] recommended use of the Wallis flooding correlation (Equation 11.25) for tube diameters < 10 mm and the Pushkin and Sorokin proposal (Equation 11.28) for tube diameters > 10 mm.

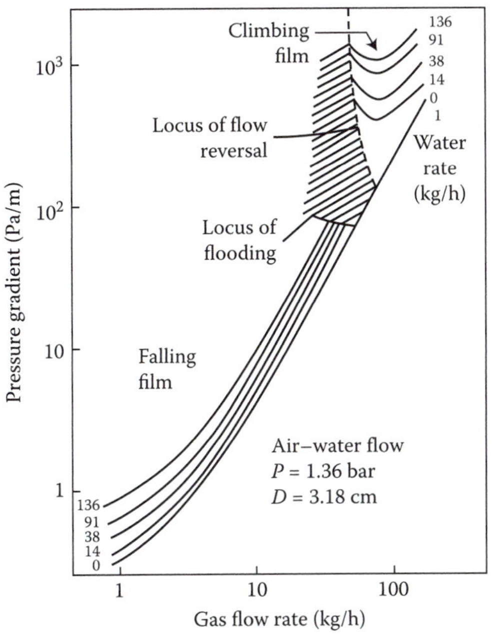

FIGURE 11.10 Pressure drop characteristics in vertical countercurrent flow transitions. (From Hewitt, G. F., Lacey, P. M. C., and Nicholls, B., *Transitions in Film Flow in a Vertical Flow*. AER-R4614, 1965.)

11.3 FLOW MODELS

As described in Chapter 5, many approaches can be applied to describe two-phase flow. The two degrees of freedom that can be specified by the various models involve equality or difference between the phasic velocities and the phasic temperatures. Specifically: (1) *unequal velocities* allow for not only potentially higher vapor velocity, but also the possibility of countercurrent flows, and (2) *thermal nonequilibrium*, which allows one or both of the two phases to have temperatures other than the saturation temperature despite the existence of the other phase. The equal velocity assumption is often referred to as the homogeneous flow assumption or the condition of mechanical equilibrium while the equal temperature assumption is referred to as the condition of thermal equilibrium.

Table 5.1 summarizes the possible models and the implied need for additional constitutive relations for each case. The main models can be summarized as follows:

1. The homogeneous equilibrium mixture model (HEM). Here, the two phases move with the same velocity and also exist at the same temperature (i.e., they are at the saturation temperature for the prevailing pressure). The mixture can then be treated as a single fluid. This model is particularly useful for high pressure and high-flow-rate conditions.
2. A thermal equilibrium mixture model with an algebraic relation between the velocities (or a slip ratio) of the two phases. A widely used such model

is the drift flux model. This model is different from the HEM model only in allowing the two phases to have different velocities that are related via a predetermined relation. This model is useful for low pressure and/or low-flow rate flows under steady-state or near-steady-state conditions.

3. The two-fluid (separate flow) model allows the phases to be in thermal non-equilibrium as well as to have unequal velocities. In this case, each phase has three independent conservation equations. The two-fluid model is needed for very fast transients when nonequilibrium conditions can be expected to be significant.

11.4 OVERVIEW OF VOID FRACTION AND PRESSURE LOSS CORRELATIONS

The use of the two-phase flow models above to establish the void fraction and pressure loss correlations is described next. In vertical flows the prediction of pressure gradients includes friction, acceleration, and gravitation terms which directly depend on fluid density. The choice of the correlations for the computation of these terms must be done very carefully. In fact, the processing of experimental adiabatic vertical flow data to obtain the friction pressure component requires estimating and eliminating the gravitational pressure component from each measured data point. Hence, when predicting pressure loss it is prudent to use frictional and gravitational correlations which have been consistently developed from the same data source. Many authors have published such related correlations. Where this is not the case, which is true for the most highly recommended friction gradient [22,23] and void fraction [55] correlations, it is suggested [30] that correlations of similar degrees of complexity be combined.

Table 11.2 summarizes the most widely used correlations for two-phase friction multiplier and void fraction. The major discriminating factors among these correlations are the range of fluids to which they are applicable, their pressure range, and, of course, the breadth and depth of the database upon which they were formulated and validated. Only the more accurate or broadly applicable correlations of Table 11.2 are presented in Sections 11.5 and 11.6 on void fraction and pressure loss.

11.5 VOID FRACTION CORRELATIONS

The void fraction is related to the slip ratio. Hence, void fraction prediction has been performed through modeling of the slip ratio under various conditions (HEM [10,11,59,73]) as well as by direct empirical correlation of the slip ratio [55]. The Zuber–Findlay and Chexal–Lellouche approaches are Drift Flux models (presented in Section 11.5.3). Coddington and Macian [16] presented a review of drift flux models which included these 2 and 11 other correlations. They found that the simpler correlations by Dix [17] and Bestion [6] gave comparable accuracy compared to other more complex correlations. Table 11.2 presents the range of validity of several void fraction correlations which are presented next.

TABLE 11.2
Principal Two-Phase Friction and Void Fraction Correlations

Correlation	Friction	Void Fraction	Fluid	Pressure Range (MPa)	Comment[a]
HEM, multiple models for various two-phase viscosity formulations	Yes	Yes	Any fluid		Good for high pressure, high velocity
Lockhart–Martinelli [42]	Yes	Yes	Air–water, -oil, -kerosene, -diesel fuel	0.12–0.36[b]	Accurate only for low pressure
Martinelli–Nelson [45]	Yes	Yes	Steam–water	0.1–0.18[b]	$(\mu_f/\mu_g) > 1000$ and $G < 100$ kg/m^2 s [29]
Thom [68]	Yes	Yes	Steam–water	0.1–20.7	
Baroczy [4]	Yes	Yes	Any fluid	1.0–13.8	Complicated graphical presentation
Chisholm [12] friction; [13] void	Yes	Yes	Any fluid	>3	$(\mu_f/\mu_g) > 1000$ and $G > 100$ kg/m^2 s [29]
Jones [39]	Yes	No	Steam–water		Empirical approximation to Martinelli–Nelson with synthesized flow rate effect
Friedel [22,23]	Yes	No	Any fluid	0.01–21.0 (single component) 0.10–17.1 (two components)	Applicable except for $(\mu_f/\mu_g) > 1000$
Chexal et al. [9]	Yes	No	Steam–water	0.172–9.65	
Premoli et al. [55]	No	Yes	Any fluid	0.5–8.8 steam–water	
Zuber and Findlay [73]	No	Yes	Any fluid	0.6–18.2	
Bestion [6]	Yes	Yes	Steam–water	>1 and $G > 10$ kg/m s	Closure laws for the 2-fluid CATHARE code
Chexal and Lellouche [10], [11], [59]	No	Yes	Steam–water Air–water Refrigerants	Steam–water 0.1–18 Air–water 0.1–0.70 Refrigerants 0.1–2.3	

[a] See Section 11.5.4 for comparative evaluation and recommended correlation use.
[b] Also included limited data between 3.4 and 20.6 MPa.

11.5.1 The Fundamental Void Fraction-Quality-Slip Relation

When time fluctuations in $\{\alpha\}$ can be ignored, for one-dimensional flow Equations 5.43a, 5.43b, and 5.47 can be used to derive the following relation between the slip ratio, quality, density ratio, and void fraction:

$$S = \frac{x}{1-x}\frac{\rho_\ell}{\rho_v}\frac{\{1-\alpha\}}{\{\alpha\}} \tag{5.49}$$

An explicit value of $\{\alpha\}$ can also be obtained as

$$\{\alpha\} = \frac{1}{1+((1-x)/x)(\rho_v/\rho_\ell)S} \tag{5.55}$$

From Equation 5.55, it is seen that the void fraction decreases with a higher slip ratio and higher density ratio ρ_v/ρ_ℓ. For steam–water flow, the void fraction at 6.9 MPa (1000 psi) is shown for various slip ratios in Figure 11.11a. It is seen that with $S = 1$ the void fraction approaches 50% even when the quality is only 5%, whereas for higher values of S the void fraction is less. For lower pressures the density ratio ρ_v/ρ_ℓ decreases, and the void fraction is higher for the same quality and slip ratio as Figure 11.11b illustrates.

The slip ratio itself is affected by the pressure (or density ratio) as well as by the profile of the void fraction within the cross-sectional area. As shown later, the macroscopic slip ratio can be obtained from a knowledge of the local phase velocity ratio (or microscopic slip) and the void distribution profile. The local slip can be obtained as a specified value or on the basis of a flow-regime-dependent approach.

11.5.2 Homogeneous Equilibrium Model

For the homogeneous model, the slip ratio is taken as 1.0. If thermal equilibrium is also assumed, the relation of void to quality is simply given by

$$\{\alpha\} = \frac{1}{1+((1-x)/x)(\rho_g/\rho_f)} \quad \text{(for HEM model)} \tag{11.29}$$

Note that $\{\beta\}$ for saturation conditions can be given by Equation 5.60 as

$$\{\beta\} = \frac{x/\rho_g}{x/\rho_g + (1-x)/\rho_f} = \frac{1}{1+(1-x)/x(\rho_g/\rho_f)} \tag{11.30}$$

Hence, it is clear that for the HEM model where $S = 1$:

$$\{\alpha\} = \{\beta\} \quad \text{for HEM} \tag{11.31}$$

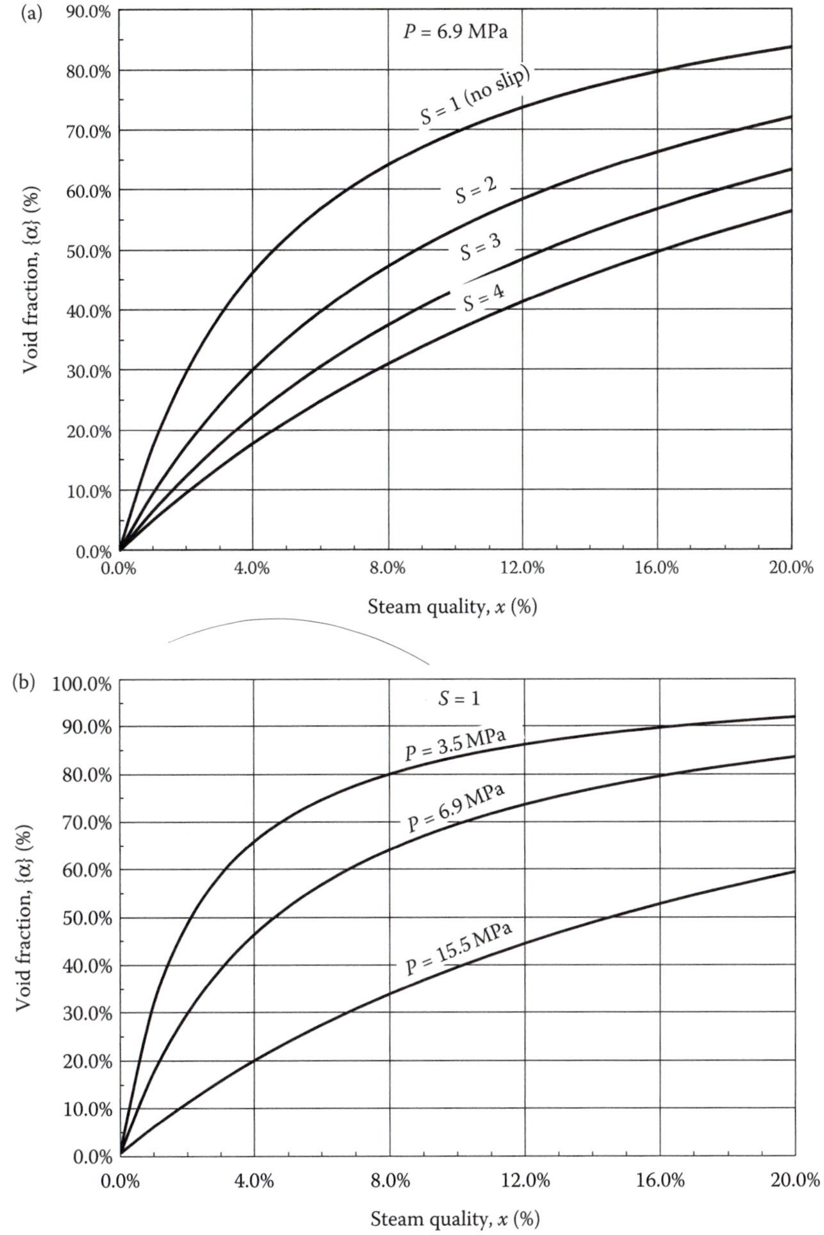

FIGURE 11.11 Variation of the void fraction with the quality for steam–water flow (using Equation 5.55): (a) for pressure of 6.9 MPa and (b) for slip ratio of unity.

Example 11.2: Void-Quality-Slip Calculations

PROBLEM

Consider two-phase water flow in a channel under saturated conditions for two cases:

$$T_{sat} = 100°C$$

$$T_{sat} = 270°C$$

If the flow quality x is equal to 0.1, for two cases of slip ratio ($S = 1$ [homogeneous flow] and $S = 2$) calculate:

1. Void fraction $\{\alpha\}$
2. Vapor flux $\{j_v\}$ in terms of the mass flux G
3. Volumetric fraction $\{\beta\}$
4. Average density $\{\rho\}$ (i.e., ρ_m)

SOLUTION

There are four cases to consider ($x = 0.1$ for all cases):

Case	T_{sat} (°C)	S
1	100	1
2	100	2
3	270	1
4	270	2

Fluid properties:
For $T_{sat} = 100°C$: $p_{sat} = 0.1$ MPa, $\rho_g = 0.5978$ kg/m^3, $\rho_f = 958.3$ kg/m^3
For $T_{sat} = 270°C$; $p_{sat} = 5.5$ MPa, $\rho_g = 28.06$ kg/m^3, $\rho_f = 767.9$ kg/m^3

1. To find $\{\alpha\}$, write Equation 5.55 assuming thermal equilibrium to obtain

$$\{\alpha\} = \frac{1}{1 + (1 - x)/x(\rho_g/\rho_f)S}$$

$$\text{Case 1:} \quad \{\alpha\} = \frac{1}{1 + ((1 - 0.1)/0.1)(0.5978/958.3)(1)} = 0.9944$$

$$\text{Case 2:} \quad \{\alpha\} = \frac{1}{1 + ((1 - 0.1)/0.1)(0.5978/958.3)(2)} = 0.9889$$

$$\text{Case 3:} \quad \{\alpha\} = \frac{1}{1 + ((1 - 0.1)/0.1)(28.06/767.9)(1)} = 0.7525$$

$$\text{Case 4:} \quad \{\alpha\} = \frac{1}{1 + ((1 - 0.1)/0.1)(28.06/767.9)(2)} = 0.6032$$

2. To find $\{j_v\}$, use

$$\{j_v\} = \{\alpha \upsilon_v\} = \frac{xG_m}{\rho_v} \tag{5.45; 5.43a}$$

Thus

$$\text{Cases 1 and 2:} \quad j_v = \frac{0.1}{0.5978} G_m = 0.1673\, G_m$$

$$\text{Cases 3 and 4:} \quad j_v = \frac{0.1}{28.06} G_m = 0.0036\, G_m$$

3. To find $\{\beta\}$, use

$$\{\beta\} = \frac{1}{1 + ((1-x)/x)(\rho_v/\rho_\ell)} \tag{5.60}$$

This equation is the same as Equation 5.55 for α with $S = 1$. Therefore we have, from solution 1:

Cases 1 and 2 $\{\beta\} = 0.9944$

Cases 3 and 4 $\{\beta\} = 0.7525$

4. To find $\{\rho\}$, use

$$\{\rho\} \equiv \rho_m = \{\alpha \rho_g\} + \{(1-\alpha)\rho_f\} = \{\alpha\}\rho_g + \{(1-\alpha)\}\rho_f \tag{5.50b}$$

Case 1: $\{\rho\} = (0.9944)0.5978 + (1 - 0.9944)958.3 = 5.961\ \text{kg/m}^3$

Case 2: $\{\rho\} = (0.9889)0.5978 + (1 - 0.9889)958.3 = 11.23\ \text{kg/m}^3$

Case 3: $\{\rho\} = (0.7525)28.06 + (1 - 0.7525)767.9 = 211.17\ \text{kg/m}^3$

Case 4: $\{\rho\} = (0.6032)28.06 + (1 - 0.6032)767.9 = 321.63\ \text{kg/m}^3$

11.5.3 Drift Flux Model

A more accurate approach to obtain the void fraction through evaluation of slip was suggested by Zuber and Findlay [73] by considering the average velocity of the vapor in the channel. To first define the drift flux, let us express the local vapor velocity as the sum of the two-phase volumetric velocity (j) and the local drift velocity of the vapor (υ_{vj}). Thus

$$\upsilon_v = j + \upsilon_{vj} \tag{11.32}$$

Hence,

$$j_v = \alpha \upsilon_v = \alpha j + \alpha(\upsilon_v - j) \tag{11.33}$$

When averaging Equation 11.33 over the flow area, we get

$$\{j_v\} = \{\alpha j\} + \{\alpha(\upsilon_v - j)\} \tag{11.34}$$

The second term on the right-hand side is defined as the drift flux. It physically represents the rate at which vapor passes through a unit area (normal to the channel axis) that is already traveling with the flow at a velocity j. The superficial velocity, $\{j_v\}$, is expressed as

$$\{j_v\} = C_0 \{\alpha\}\{j\} + \{\alpha\}V_{vj} \tag{11.35}$$

where the concentration parameter C_0 is defined by

$$C_0 \equiv \frac{\{\alpha j\}}{\{\alpha\}\{j\}} \tag{11.36}$$

and the effective drift velocity is

$$V_{vj} \equiv \{\alpha(\upsilon_v - j)\}/\{\alpha\} \tag{11.37}$$

The value of $\{\alpha\}$ can now be obtained from Equation 11.35 as

$$\{\alpha\} = \frac{\{j_v\}}{C_0\{j\} + V_{vj}} \tag{11.38}$$

The void fraction can be seen as due to two effects when Equation 11.38 is rewritten as

$$\frac{\{\alpha\}}{\{\beta\}} = \frac{1}{C_0 + (V_{vj}/\{j\})} \tag{11.39}$$

The C_0 term represents the global effect of planar nonuniform void and velocity profiles. If either the velocity or void profiles are uniform across the channel, then $C_0 = 1$, as can easily be shown from Equation 11.36. The $V_{vj}/\{j\}$ term represents the local relative velocity effect.

For homogeneous flow, neither local slip nor concentration profile effects are considered; therefore, $\upsilon_v = j$, $V_{vj} = 0$, and $C_0 = 1$, so from Equation 11.39 we get

$$\{\alpha\} = \{\beta\} \quad \text{for HEM} \tag{11.31}$$

For flow under slip conditions, several cases can be considered

1. At high total flow rate, the local effect term is negligible, as the relative velocity is negligible (i.e., $V_{vj} \approx 0$). This condition is implied by the old relations of Armand and Treshchev [1] and Bankoff [2], which were presented for steam–water systems and are of the form:

$$\{\alpha\} = K\{\beta\} \tag{11.40}$$

where the constant K is then defined by

$$K = \frac{1}{C_0} \tag{11.41}$$

For this case, Armand and Treshchev [1] suggested

$$K = 0.833 + 0.05\ell n(10p) \quad \text{for} \quad \beta < 0.9 \tag{11.42a}$$

where p is in MPa, while Bankoff [2] suggested that

$$K = 0.71 + 0.0001p(\text{psi}) \tag{11.42b}$$

2. For all flow regimes, Dix [17] suggested a general expression for C_0, given by

$$C_0 = \{\beta\}\left[1 + \left(\frac{1}{\{\beta\}} - 1\right)^b\right] \tag{11.43a}$$

where

$$b = \left(\frac{\rho_v}{\rho_\ell}\right)^{0.1} \tag{11.43b}$$

It is also useful to express Equation 11.38 in terms of the mass flow rate and quality:

$$\{\alpha\} = \frac{x\dot{m}/\rho_v A}{C_0((x/\rho_v) + (1-x)/\rho_\ell)(\dot{m}/A) + V_{vj}} \tag{11.44a}$$

where $\{j_v\}$ and $\{j\}$ have been expressed as given in Equations 5.43a and 5.44. Rearranging Equation 11.44a, we obtain obtain

$$\{\alpha\} = \frac{1}{C_0(1 + ((1-x)/x)(\rho_v/\rho_l)) + (V_{vj}\rho_v/xG)} \tag{11.44b}$$

Equating Equations 11.44b and 5.55, we can obtain an explicit relation for the slip ratio as

$$S = C_0 + \underbrace{\frac{(C_0 - 1)x\rho_\ell}{(1-x)\rho_v}}_{\text{Due to nonuniform void distribtuion}} + \underbrace{\frac{V_{vj}\rho_\ell}{(1-x)G}}_{\text{Due to local velocity differential between vapor and liquid}} \tag{11.45}$$

For uniform void or velocity distribution in the flow area, $C_0 = 1$. Zuber and Findlay [73] suggested that C_0 and V_{vj} are functions of the flow regime; $C_0 = 1.2$ for bubbly and slug flow, and $C_0 = 0$ for near zero void fraction and 1.0 for a high void fraction. The drift velocity (V_{vj}) for the bubbly and slug flow regimes can be given as

$$V_{vj} = (1 - \{\alpha\})^n V_\infty; \quad 0 < n < 3 \tag{11.46}$$

where V_∞ is the bubble rise terminal velocity in the liquid, as given in Table 11.3. It should be noted that in the bubbly regime the drift velocity is smaller than the terminal single bubble velocity because of the presence of the other bubbles. In the slug and churn flow, regimes V_{vj} and V_∞ are equal. Also note that when $n \neq 0$, calculation of $\{\alpha\}$ from Equation 11.38 requires iteration.

Ishii [37] extended the drift flux relations to annular flow. However, in annular flow, there is little difference between the vapor volumetric flow rate and the total volumetric flow rate, and V_{vj} is insignificant.

11.5.4 Chexal and Lellouche Correlation

This correlation, formulated in the drift flux approach, was originally developed at EPRI in 1992 [10], then subjected to a series of modifications leading to its most

TABLE 11.3
Values of n and V_∞ for Various Regimes

Regime	n	V_∞
Small bubbles ($d < 0.5$ cm)	3	$\frac{g(\rho_\ell - \rho_v)d^2}{18\mu_\ell}$
Large bubbles ($d < 2$ cm)	1.5	$1.53\left[\frac{\sigma g(\rho_\ell - \rho_v)}{\rho_\ell^2}\right]^{1/4}$
Churn flow	0	$1.53\left[\frac{\sigma g(\rho_\ell - \rho_v)}{\rho_\ell^2}\right]^{1/4}$
Slug flow (in tube of diameter D)	0	$0.35\sqrt{g\left(\frac{\rho_\ell - \rho_v}{\rho_\ell}\right)D}$

recent formulation, [11], available to the public in the form given in RELAP5 [59], which is presented below. It has the following desirable features:

- Applicable to all flow regimes (without the need for a flow regime map).
- Applicable to all channel inclinations (e.g., vertical, horizontal, inclined) and flow directions (e.g., co- and countercurrent, stagnant liquid situations).
- Applicable over a large range of channel sizes, pressures and mass fluxes.
- Continuous with x, G, p with continuous derivatives, so it can be used in codes without numerical problems. This is particularly important in applications where the fluid experiences more than one flow regime.
- Always predicts void fraction within the physically allowable range of 0–1.

The correlation also satisfies the following limits:

- Critical pressure limits: as $p \to p_{cr}$; $V_{vj} = 0$ and $C_0 = 1$
- High void fraction: as $\{\alpha\} \to 1$; $C_0 = 1$ and $V_{vj} = 0$

and was developed using experimental data covering the following ranges for steam–water mixtures:

$$0.01 \le \{\alpha\} \le 0.95$$

and for diabatic tests:

$$0.01 \le G \le 2100 \text{ kg/m}^2\text{s}$$

$$0.1 \le p \le 14.5 \text{ MPa}$$

while for adiabatic tests:

$$0.01 \le G \le 2550 \text{ kg/m}^2\text{s}$$

$$0.1 \le p \le 18 \text{ MPa}$$

The main disadvantage is that it is lengthy and requires iteration as some of its parameters depend on the void fraction itself. The application of the Chexal–Lellouche correlation is illustrated by a calculation which follows in Example 11.3.

The Chexal–Lellouche void model evaluates the void fraction from Equation 11.39.

Evaluation of C_0:

$$C_0 \equiv \frac{L}{K_0 + (1 - K_0)\alpha^r} \tag{11.47a}$$

where

$$L \equiv \frac{L_n}{L_d}$$

$$\begin{aligned} L_n &= 1 - \exp(-\alpha C_p) \quad \text{if } \alpha C_p < 170 \\ &= 1 \qquad\qquad\qquad\;\; \text{otherwise} \end{aligned}$$

$$L_d = 1 - \exp(-C_p) \quad \text{if } C_p < 170$$
$$= 1 \quad \text{otherwise}$$

$$C_p = \left| \frac{4p_{cr}^2}{p(p_{cr} - p)} \right|$$

$$p_{cr} = \text{critical pressure}$$

$$K_0 = B_1 + (1 - B_1)\left(\frac{\rho_g}{\rho_f}\right)^{0.25}$$

$$B_1 = \min\{0.8, A_1\}$$

$$A_1 = \frac{1}{1 + \exp\{\max[-85, \min(85, (-\text{Re}/60{,}000))]\}}$$

$$\text{Re} = \text{Re}_g \quad \text{if } \text{Re}_g > \text{Re}_f \text{ or } \text{Re}_g < 0$$
$$= \text{Re}_f \quad \text{otherwise}$$

$$\text{Re}_f = \frac{\rho_f j_f D_e}{\mu_f} \quad \text{(liquid superficial Reynolds number)}$$

$$\text{Re}_g = \frac{\rho_g j_g D_e}{\mu_g} \quad \text{(vapor/gas superficial Reynolds number)}$$

$$r = \frac{1 + 1.57(\rho_g/\rho_f)}{1 - B_1}$$

j_f and j_g: liquid and vapor/gas superficial velocities (positive if the phase moves upward, negative if downward; this sign convention determines the sign of Re_f and Re_g).

Evaluation of V_{vj}:

$$V_{vj} \equiv 1.41\left[\frac{(\rho_f - \rho_g)g\sigma}{\rho_f^2}\right]^{0.25} C_1 C_2 C_3 C_4 \tag{11.47b}$$

where

$$C_1 = (1 - \{\alpha\})^{B_1} \quad \text{if } \text{Re}_g \geq 0$$
$$= (1 - \{\alpha\})^{0.5} \quad \text{otherwise}$$

$$C_2 = 1 \quad \text{if } \frac{\rho_f}{\rho_g} \geq 18 \quad \text{and} \quad C_5 \geq 1$$

$$= 1 \quad \text{if } \frac{\rho_f}{\rho_g} \geq 18 \quad \text{and} \quad C_5 < 1 \text{ and } C_6 \geq 85$$

$$= \frac{1}{1-\exp(-C_6)} \quad \text{if } \frac{\rho_f}{\rho_g} \geq 18 \quad \text{and} \quad C_5 < 1 \text{ and } C_6 < 85$$

$$= 0.4757\left\{\ell n\left[\max\left(1.00001, \frac{\rho_f}{\rho_g}\right)\right]\right\}^{0.7} \quad \text{if } \frac{\rho_f}{\rho_g} < 18$$

$$C_5 = \left[150\left(\frac{\rho_g}{\rho_f}\right)\right]^{0.5}$$

$$C_6 = \frac{C_5}{1-C_5}$$

$$C_4 = 1 \quad \text{if } C_7 \geq 1$$

$$= \frac{1}{1-\exp(-C_8)} \quad \text{if } C_7 < 1$$

$$C_7 = \left(\frac{D_2}{D_e}\right)^{0.6}$$

$$D_2 = 0.09144 \text{ m (normalizing diameter)}$$

$$C_8 = \frac{C_7}{1-C_7}$$

$$C_3 = \max\left[0.50, 2\exp\left(\frac{|\text{Re}_f|}{300{,}000}\right)\right] \quad \text{if } j_f > 0 \quad \text{and} \quad j_g > 0$$

$$= 2\left(\frac{C_{10}}{2}\right)^{B_2} \quad \text{if } j_f < 0 \text{ and } j_g < 0, \quad \text{or} \quad \text{if } j_f < 0 \text{ and } j_g > 0$$

where

$$C_{10} = 2\exp\left[\frac{|\text{Re}_f|^{0.4}}{350{,}000}\right] - 1.7|\text{Re}_f|^{0.035}\exp\left[\frac{-|\text{Re}_f|}{60{,}000}\left(\frac{D_1}{D_e}\right)^2\right] + \left(\frac{D_1}{D_e}\right)^{0.1}|\text{Re}_f|^{0.001}$$

$$B_2 = \frac{1}{(1+0.05|\text{Re}_f/350{,}000|)^{0.4}}$$

$$D_1 = 0.0381 \text{ m (normalizing diameter)}$$

and D_e = equivalent diameter in meters.

Combining all terms, the Chexal–Lellouche void fraction correlation for flow in a vertical channel has the following form:

$$\{\alpha\} = \frac{\{\beta\}}{(L/(K_0 + (1-K_0)\alpha^{r})) + 1.41[((\rho_f - \rho_g)\sigma g)/\rho_f^2]^{0.25}((C_1C_2C_3C_4)/j)} \tag{11.48}$$

The desired void fraction value, α, can be determined using an iterative method from Equation 11.48, since some parameters on the RHS of Equation 11.48 depend on α.

A comparison of several void fractions follows next in Example 11.3.

11.5.5 Premoli Correlation

The correlation of Premoli et al. [55], usually known as the CISE correlation, is a correlation for α in terms of an empirical evaluation of the slip ratio. The slip ratio given as below is inserted into Equation 5.55 to obtain the void fraction:

$$\begin{aligned} S &= 1 + E_1\left(\frac{y}{1+yE_2} - yE_2\right)^{0.5} \quad \text{if } y \le \frac{1-E_2}{E_2^2} \\ &= 1 \quad \text{otherwise} \end{aligned} \tag{11.49}$$

where

$$y = \frac{\{\beta\}}{1-\{\beta\}} \quad \text{with } \beta = \frac{\rho_\ell x}{\rho_\ell x + \rho_v(1-x)} \tag{11.50a,b}$$

$$E_1 = 1.578\,\text{Re}^{-0.19}\left(\frac{\rho_\ell}{\rho_v}\right)^{0.22} \quad \text{and} \quad E_2 = 0.0273\,\text{We}\,\text{Re}^{-0.51}\left(\frac{\rho_\ell}{\rho_v}\right)^{-0.08} \tag{11.51a,b}$$

$$\text{Re} = \frac{GD_e}{\mu_\ell} \quad \text{and} \quad \text{We} = \frac{G^2D_e}{\sigma\rho_\ell} \tag{11.52a,b}$$

Here: x is the quality; ρ_v is the gas density (kg/m^3); ρ_ℓ is the liquid density (kg/m^3); D_e is the equivalent diameter (m); μ_ℓ is the liquid viscosity (N s/m^2); G is the total (liquid + gas) mass flux (kg/m^2 s); and σ is the surface tension (N/m).

The Premoli correlation is stated to be valid for upward flow of any fluid in adiabatic channels with the following operating conditions:

$$50 < G < 3500 \text{ kg/m}^2\text{s}$$

$$3 < D_e < 50 \text{ mm}$$

$$750 < \rho_f < 1000 \text{ kg/m}^3$$

$$3 < \rho_g < 42 \text{ kg/m}^3$$

$$0.018 < \sigma < 0.073 \text{ N/m}$$

$$300 < \mu_f < 1300 \text{ µPa s}$$

For steam–water mixtures, the viscosity range above would limit the correlation applicability to below about 0.1 MPa. However, based on the test results and subsequent experiments, the correlation authors* consider it reasonably applicable up to 8.8 MPa for steam–water mixtures.

The correlation is also stated to be valid in diabatic channels as long as the heat flux is below 1 MW/m^2 and the fluid entering the channel is saturated or slightly subcooled [55].

Note that the definition of the *Weber number*, We, is not the same as that used in the Friedel correlation for the two-phase friction multiplier (Equation 11.100e).

11.5.6 Bestion Correlation

The Bestion correlation was developed as part of the interfacial friction characterization performed for use in the CATHARE code. Hence, drift velocity correlations have been developed for PWR bundle, pipe, and annuli geometries [7]. The bundle correlation proposed was

$$C_0 = 1.0 \tag{11.53a}$$

as reported by Coddington and Macian [16]
and

$$V_{vj} = 0.188\sqrt{\frac{g d_h \Delta p}{\rho_v}} \tag{11.53b}$$

as originally proposed by Bestion in 1985 [6].

Subsequent formulations for V_{vj} have been proposed as Bestion [7] summarizes. The 1985 formulation above performs very well for PWR bundle data as Coddington and Macian demonstrate [16].

Example 11.3: Comparison of Void Fraction Correlations

PROBLEM

Consider vertical upflow of two-phase steam–water in a 20 mm diameter tube. Assume thermal equilibrium at 290°C (7.45 MPa). Find and plot the void fraction, α, versus flow quality, x, for $0.1 \le x \le 0.9$. Use three values of mass velocity:

$$G_1 = 3000 \text{ kg/m}^2 \text{ s}$$
$$G_2 = 1000 \text{ kg/m}^2 \text{ s}$$
$$G_3 = 100 \text{ kg/m}^2 \text{ s}$$

* A. Premoli, pers. comm., December 2009.

and four correlations for void fraction:

Homogeneous, $\{\alpha\} = \{\beta\}$
Drift flux (slug flow)
Chexal and Lellouche
Premoli et al.

How appropriate are these correlations for the flow conditions considered?

SOLUTION

Fluid properties:

$$v_g = 2.554 \times 10^{-2}\ \text{m}^3/\text{kg}$$
$$v_f = 1.366 \times 10^{-3}\ \text{m}^3/\text{kg}$$
$$\mu_f = 89.63\ \mu\text{Pa s}$$
$$\mu_g = 19.16\ \mu\text{Pa s}$$
$$\sigma = 0.0167\ \text{N/m}$$

Homogeneous Void Fraction Correlation

Use Equation 11.30:

$$\{\beta\} = \{\alpha\} = \frac{1}{1 + ((1-x)/x)(\rho_v / \rho_\ell)} = \frac{1}{1 + ((1-x)/x)(v_\ell / v_v)}$$

Independent of G, we can find the following values for $\{\alpha\}$:

x	{α}
0.1	0.68
0.3	0.89
0.5	0.95
0.7	0.98
0.9	0.99

Drift Flux

Use Equation 11.44b expressed for saturated conditions:

$$\{\alpha\} = \frac{1}{C_0(1 + ((1-x)/x)(\rho_g / \rho_f)) + (V_{vj}\rho_g / xG)}$$
$$= \frac{1}{C_0(1 + ((1-x)/x)(v_f / v_g)) + (V_{vj} / v_g xG)}$$

For $C_0 = 1.2$ and from Table 11.3 in saturated conditions:

$$V_{vj} = 0.35\sqrt{\frac{g(\rho_f - \rho_g)D}{\rho_f}} = 0.35\sqrt{\frac{g((1/v_f) - (1/v_g))D}{(1/v_f)}} = 0.15\text{m/s}$$

Using $g = 9.8$ m/s^2 and $D = 0.02$ m, we find the following values of $\{\alpha\}$:

	{α} (Drift Flux)				
	$x = 0.1$	$x = 0.3$	$x = 0.5$	$x = 0.7$	$x = 0.9$
G_1	0.56	0.74	0.79	0.81	0.83
G_2	0.54	0.73	0.78	0.81	0.82
G_3	0.40	0.63	0.71	0.75	0.76

Chexal and Lellouche

The EPRI correlation was structured in the mold of other drift flux void fraction correlations, taking the form:

$$\{\alpha\} = \frac{\{\beta\}}{C_0 + V_{vj}/\{j\}} \tag{11.39}$$

The calculation procedure for evaluating void fraction using this correlation has been presented in Section 11.5.4.

Evaluating G_1, G_2, and G_3 for the full range of qualities gives

	{α} (Chexal–Lellouche)				
	$x = 0.1$	$x = 0.3$	$x = 0.5$	$x = 0.7$	$x = 0.9$
G_1	0.61	0.83	0.91	0.95	0.99
G_2	0.60	0.82	0.91	0.95	0.98
G_3	0.35	0.64	0.77	0.85	0.91

Premoli et al.

Use Equation 11.49 for saturated conditions. Taking $D = 20$ mm, $G = 1000$ kg/m^{2}s and $x = 0.5$, we have

$$\text{Re} = \frac{GD_e}{\mu_l} = \frac{1000 \times 0.02}{0.00008963} = 223{,}139$$

and

$$\text{We} = \frac{G^2 D_e}{\sigma \rho_l} = \frac{1000^2 \times 0.02}{0.0167 \times 731.76} = 1637$$

This leads to

$$E_1 = 1.578\,\text{Re}^{-0.19}\left(\frac{\rho_f}{\rho_g}\right)^{0.22} = 1.578 \times (223{,}139)^{-0.19}\left(\frac{731.76}{39.18}\right)^{0.22} = 0.289$$

$$E_2 = 0.0273\text{WeRe}^{-0.51}\left(\frac{\rho_f}{\rho_g}\right)^{-0.08}$$

$$= 0.0273 \times 1637 \times (223{,}139)^{-0.51}\left(\frac{731.76}{39.18}\right)^{-0.08} = 0.0662$$

and finally

$$\{\beta\} = \frac{\rho_f x}{\rho_f x + \rho_g(1-x)} = \frac{731.76 \times 0.5}{731.76 \times 0.5 + 39.18 \times (1-0.5)} = 0.949$$

$$y = \frac{\{\beta\}}{1-\{\beta\}} = 18.68$$

Now, since

$$y < \frac{1-E_2}{E_2^2} = 213.2$$

the slip ratio is

$$S = 1 + E_1\left(\frac{y}{1+yE_2} - yE_2\right)^{0.5}$$

$$= 1 + 0.289\left(\frac{18.68}{1+18.68 \times 0.0662} - 18.68 \times 0.0662\right)^{0.5} = 1.77$$

Finally,

$$\{\alpha\} = \frac{1}{1+((1-x)/x)(S(\rho_g/\rho_f))}$$

$$= \frac{1}{1+((1-0.5)/0.5) \times 1.77 \times (39.18/731.76)} = 0.91$$

Evaluating G_1, G_2, and G_3 for the full range of qualities gives

	{α} (Premoli et al.)				
	$x = 0.1$	**$x = 0.3$**	**$x = 0.5$**	**$x = 0.7$**	**$x = 0.9$**
G_1	0.64	0.89	0.95	0.98	0.99
G_2	0.60	0.83	0.91	0.96	0.99
G_{3a}	0.56	0.78	0.87	0.92	0.97

SUMMARY

A comparison of all results using the models examined in this example is shown in Figures 11.12a and 11.12b.

Homogeneous model: This model is good for high-pressure, high-flow-rate conditions. Therefore, the $G_1 = 3000$ kg/m² s case is more appropriate for use with this model than G_2 or G_3, as this model is best used when $G > 2000$ kg/m² s.

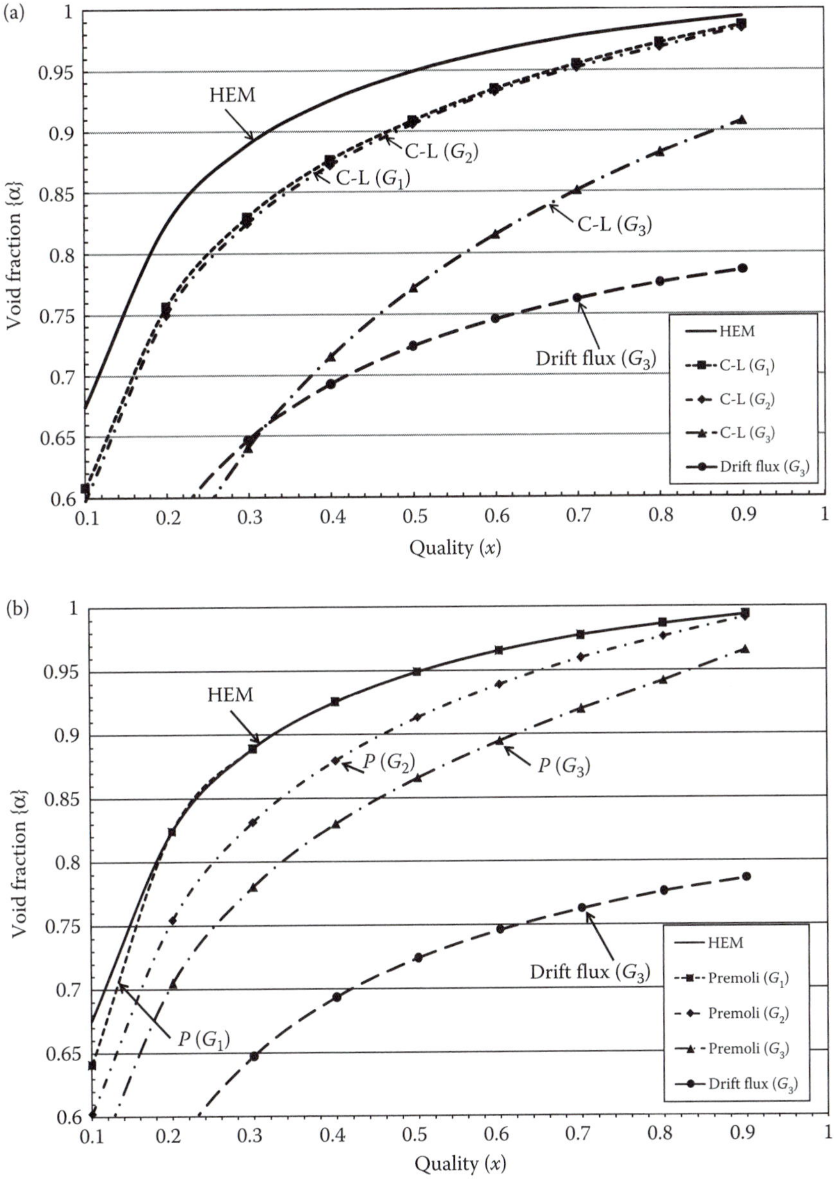

FIGURE 11.12 Variation of the void fraction with quality of Example 11.3. (a) Comparison of HEM, Chexal-Lellouche, and Drift Flux models, and (b) comparison of HEM, Premoli, and Drift Flux models.

Drift flux (slug flow) model: From Figure 11.5, based on the values of the two coordinates:

$$\rho_v\{j_v\}^2 = v_v(Gx)^2, \quad \rho_l\{j_l\}^2 = v_l[G(1-x)]^2$$

we find that for G_1 and G_2 most of the calculation region exhibits annular and wispy-annular flow. Hence, the slug flow assumptions used are not appropriate. However, for G_3 the values fall in the slug and churn flow regions, and therefore the slug flow assumptions are more appropriate. However, the predicted void fraction still falls well below the other correlations.

Chexal–Lellouche correlation: This correlation appropriately shows a strong dependence on both quality and mass flux. It approaches the HEM model at high mass flux (G_1 and G_2), and approaches the drift flux model for conditions in the slug flow regime (G_3); these are the trends that one would expect from an accurate correlation. Note that void fractions at G_1 and G_2 as predicted by this correlation are not appreciably different and therefore are difficult to distinguish in Figure 11.12a.

Premoli et al. correlation: While much simpler to apply, this correlation reproduces many of the features of Chexal–Lellouche. Premoli is a close match with the HEM model at G_1, and shows reduction in void fraction for a fixed quality as the mass flux is decreased (from G_1 to G_3). The G_2 case falls between these two extremes as Figure 11.12b illustrates.

11.6 PRESSURE–DROP RELATIONS

11.6.1 The Acceleration, Friction, and Gravity Components

Evaluation of the pressure drop in a two-phase flow channel can be accomplished using several models. Here we address the pressure drop prediction on the basis of mixture models, first using the HEM then allowing unequal phase velocities. This format follows the historical development of the insight into this area. It should not be surprising that a large degree of empiricism is involved, as two-phase flow is inherently chaotic.

The pressure gradient can be calculated using the momentum equation for a two-phase mixture flowing in the z direction of a one-dimensional channel:

$$\frac{\partial}{\partial t}(G_m A_z) + \frac{\partial}{\partial z}\left(\frac{G_m^2 A_z}{\rho_m^+}\right) = -\frac{\partial}{\partial z}(pA_z) + p\frac{\partial A_z}{\partial z} - \int_{P_z} \tau_w \, dP_z - \rho_m g \cos\theta A_z \quad (5.66)$$

where θ = the flow angle with the upward vertical and r is the channel radius at z, that is $r(z)$.

For steady state in a constant area channel ($dr/dz = 0$), it is possible to simplify Equation 5.66 to

$$-\frac{dp}{dz} = \frac{d}{dz}\left(\frac{G_m^2}{\rho_m^+}\right) + \frac{1}{A_z}\int_{P_z} \tau_w \, dP_z + \rho_m g \cos\theta \quad (11.54)$$

In this equation, the radial variation of p within the cross section is assumed negligible.

Equation 11.54 expresses the rate of change of the static pressure in the channel as the sum of three components due to acceleration, friction, and gravity:

$$-\frac{dp}{dz}=\left(\frac{dp}{dz}\right)_{acc}+\left(\frac{dp}{dz}\right)_{fric}+\left(\frac{dp}{dz}\right)_{grav} \tag{11.55}$$

where

$$\left(\frac{dp}{dz}\right)_{acc}=\frac{d}{dz}\left(\frac{G_m^2}{\rho_m^+}\right) \tag{11.56a}$$

$$\left(\frac{dp}{dz}\right)_{fric}=\frac{1}{A_z}\int_{P_z}\tau_w\, dP_z=\frac{\overline{\tau}_w P_z}{A_z} \tag{11.56b}$$

$$\left(\frac{dp}{dz}\right)_{grav}=\rho_m g\cos\theta \tag{11.56c}$$

where $\overline{\tau}_w$ is the circumferentially averaged wall shear stress. Note that dp/dz is negative for flow in the positive z direction, and $(dp/dz)_{fric}$ is always positive. The other two terms depend on the channel conditions. For heated channels ρ_m decreases as z increases, and $(dp/dz)_{acc}$ is positive. The $(dp/dz)_{grav}$ term is positive if $\cos\theta$ is positive.

To obtain the pressure drop, we integrate the pressure gradient equation, Equation 11.55, yielding

$$\Delta p=\Delta p_{acc}+\Delta p_{fric}+\Delta p_{grav} \tag{11.57}$$

where

$$\Delta p \equiv p_{in}-p_{out}=\int_{z_{in}}^{z_{out}}\left(-\frac{dp}{dz}\right)dz \tag{11.58}$$

$$\Delta p_{acc}=\left(\frac{G_m^2}{\rho_m^+}\right)_{out}-\left(\frac{G_m^2}{\rho_m^+}\right)_{in} \tag{11.59a}$$

$$\Delta p_{fric}=\int_{z_{in}}^{z_{out}}\frac{\overline{\tau}_w P_z}{A_z}\,dz \tag{11.59b}$$

$$\Delta p_{grav}=\int_{z_{in}}^{z_{out}}\rho_m g\cos\theta\, dz \tag{11.59c}$$

The dynamic density ρ_m^+ can be written in terms of the flow quality if we write the momentum flux of each phase in terms of the mass flux. From Equation 5.67, we have

$$\frac{G_m^2}{\rho_m^+} = \{\rho_v \alpha \upsilon_v^2\} + \{\rho_\ell (1-\alpha)\upsilon_\ell^2\} \tag{5.67}$$

Because the flow quality can be given by

$$xG_m = \{\rho_v \alpha \upsilon_v\} \quad \text{and} \quad (1-x)G_m = \{\rho_\ell (1-\alpha)\upsilon_\ell\} \qquad \text{(11.60a and b)}$$

then

$$\frac{G_m^2}{\rho_m^+} = \frac{x^2 G_m^2}{c_v \{\rho_v \alpha\}} + \frac{(1-x)^2 G_m^2}{c_\ell \{\rho_\ell (1-\alpha)\}} \tag{11.61}$$

where

$$c_v \equiv \frac{\{\rho_v \alpha \upsilon_v\}^2}{\{\rho_v \alpha \upsilon_v^2\}\{\rho_v \alpha\}} \tag{11.62a}$$

and

$$c_\ell \equiv \frac{\{\rho_\ell (1-\alpha)\upsilon_\ell\}^2}{\{\rho_\ell (1-\alpha)\upsilon_\ell^2\}\{\rho_\ell (1-\alpha)\}} \tag{11.62b}$$

For radially uniform velocity of each phase in the channel, $c_v = c_\ell = 1.0$. In that case, Equation 11.61 leads to

$$\frac{1}{\rho_m^+} = \frac{x^2}{\{\rho_v \alpha\}} + \frac{(1-x)^2}{\{\rho_\ell (1-\alpha)\}} \tag{11.63}$$

for uniform velocities. We ignore the radial distribution of the velocity in the development of the acceleration term in this section.

The friction pressure gradient for two-phase flow can be expressed in a general form similar to the single-phase flow:

$$\left(\frac{dp}{dz}\right)_{fric}^{TP} = \frac{\bar{\tau}_w P_w}{A_z} \equiv \frac{f_{TP}}{D_e}\left(\frac{G_m^2}{2\rho_m^+}\right) \tag{11.64}$$

where

$$D_e = \frac{4A_f}{P_w} \tag{9.64}$$

is the hydraulic or equivalent diameter.

The general approach for formulating the two-phase friction factor or the friction pressure gradient, $(dp/dz)^{TP}_{fric}$, is to relate them to friction factors and multipliers defined for a single-phase (either liquid or vapor) flowing at the same mass flux as the total two-phase mass flux and with a temperature corresponding to bulk conditions. If the single phase is liquid, the relevant parameters are $f_{\ell o}$ and $\phi^2_{\ell o}$, whereas if the single phase is vapor the parameters are f_{vo} and ϕ^2_{vo}. These parameters are related as

$$\left(\frac{dp}{dz}\right)^{TP}_{fric} = \phi^2_{\ell o}\left(\frac{dp}{dz}\right)^{\ell o}_{fric} = \phi^2_{vo}\left(\frac{dp}{dz}\right)^{vo}_{fric} \tag{11.65a,b}$$

so that

$$\phi^2_{\ell o} = \frac{\rho_\ell}{\rho^+_m}\frac{f_{TP}}{f_{\ell o}} \tag{11.66a}$$

$$\phi^2_{vo} = \frac{\rho_v}{\rho^+_m}\frac{f_{TP}}{f_{vo}} \tag{11.66b}$$

Typically, the "liquid-only" parameters are utilized in boiling channels, and the "vapor-only" multipliers are utilized for condensing channels. Therefore, in a two-phase boiling channl, the friction pressure gradient is given by

$$\left(\frac{dp}{dz}\right)^{TP}_{fric} = \phi^2_{\ell o}\frac{f_{\ell o}}{D_e}\left[\frac{G^2_m}{2\rho_\ell}\right] \tag{11.67}$$

11.6.2 Homogeneous Equilibrium Models

The Homogeneous equilibrium model (HEM) model implies that the velocity of the liquid equals that of the vapor, that the two velocities are uniform within the area, and that the two phases are in thermodynamic equilibrium. If

$$V_m \equiv \frac{G_m}{\rho_m} = \frac{[\rho_g\{\alpha\}\upsilon_g + \rho_f(1-\{\alpha\})\upsilon_f]}{[\rho_g\{\alpha\} + \rho_f(1-\{\alpha\})]} \tag{11.68}$$

we can see that for the HEM model (or any model with equal phase velocities):

$$\upsilon_g = \upsilon_f = V_m \tag{11.69}$$

Then, noting Equation 11.69 and substituting Equations 11.60a and 11.60b into Equation 11.63, and expressing the result for thermal equilibrium, we get

$$\begin{aligned}\frac{1}{\rho_m^+} &= \frac{x\{\rho_g\alpha\}V_m}{\{\rho_g\alpha\}G_m} + \frac{(1-x)\{\rho_f(1-\alpha)\}V_m}{\{\rho_f(1-\alpha)\}G_m}\\ &= \frac{xV_m + (1-x)V_m}{G_m} = \frac{V_m}{G_m} = \frac{1}{\rho_m}\end{aligned} \tag{11.70}$$

Equation 11.70 demonstrates that for the HEM model $\rho_m^+ = \rho_m$.

It is useful to note that we can use the equivalence of ρ_m^+ and ρ_m in the homogeneous equilibrium model to write

$$\frac{1}{\rho_m^+} = \frac{V_m}{G_m} = \frac{\{\alpha\}V_m + (1-\{\alpha\})V_m}{G_m} \tag{11.71}$$

Applying Equations 11.60a and 11.60b, Equation 11.71 can be written as

$$\frac{1}{\rho_m^+} = \frac{xG_m/\rho_g + (1-x)G_m/\rho_f}{G_m} \tag{11.72}$$

and finally

$$\frac{1}{\rho_m^+} = \frac{1}{\rho_m} = \frac{x}{\rho_g} + \frac{1-x}{\rho_f} \quad \text{(for HEM)} \tag{11.73}$$

From Equations 11.56a and 11.73, because G_m is constant for a constant area channel, the HEM acceleration pressure gradient is given by

$$\left(\frac{dp}{dz}\right)_{acc} = G_m^2 \frac{d}{dz}\left[\frac{1}{\rho_f} + \left(\frac{1}{\rho_g} - \frac{1}{\rho_f}\right)x\right] \tag{11.74a}$$

or in the expanded form:

$$\left(\frac{dp}{dz}\right)_{acc} = G_m^2\left[\frac{dv_f}{dz} + x\left(\frac{dv_g}{dz} - \frac{dv_f}{dz}\right) + (v_g - v_f)\frac{dx}{dz}\right] \tag{11.74b}$$

If v_g and v_f are assumed independent of z, that is, both the liquid and the gas are assumed incompressible, Equation 11.74b becomes

$$\left(\frac{dp}{dz}\right)_{acc} = G_m^2(v_g - v_f)\frac{dx}{dz} = G_m^2 v_{fg}\frac{dx}{dz} \tag{11.75}$$

When only the liquid compressibility is ignored, Equation 11.74b yields

$$\left(\frac{\mathrm{d}p}{\mathrm{d}z}\right)_{\mathrm{acc}} = G_{\mathrm{m}}^2\left(x\frac{\partial v_{\mathrm{g}}}{\partial p}\frac{\mathrm{d}p}{\mathrm{d}z} + v_{\mathrm{fg}}\frac{\mathrm{d}x}{\mathrm{d}z}\right) \quad (11.76)$$

For the HEM, Equation 11.64 for the friction pressure gradient reduces to

$$\left(\frac{\mathrm{d}p}{\mathrm{d}z}\right)_{\mathrm{fric}} = \frac{f_{\mathrm{TP}}}{D_{\mathrm{e}}}\left(\frac{G_{\mathrm{m}}^2}{2\rho_{\mathrm{m}}}\right) \quad (11.77)$$

Thus the total pressure drop when the gas compressibility is accounted for is obtained by substituting Equations 11.56c, 11.76, and 11.77 into Equation 11.55 and rearranging to get

$$-\left(\frac{\mathrm{d}p}{\mathrm{d}z}\right)_{\mathrm{HEM}} = \frac{(f_{\mathrm{TP}}/D_{\mathrm{e}})(G_{\mathrm{m}}^2/2\rho_{\mathrm{m}}) + G_{\mathrm{m}}^2 v_{\mathrm{fg}}(\mathrm{d}x/\mathrm{d}z) + \rho_{\mathrm{m}} g\cos\theta}{(1 + G_{\mathrm{m}}^2 x(\partial v_{\mathrm{g}}/\partial p))} \quad (11.78)$$

To evaluate the friction multiplier $\phi_{\ell o}^2$ given by Equation 11.66a, two approximations are possible for the two-phase friction factor and typical for elaborating the HEM:

1. f_{TP} is equal to the friction factor for liquid single-phase flow at the same mass flux G_{m} as the total two-phase mass flux:

$$f_{\mathrm{TP}} = f_{\ell o} \quad (11.79\mathrm{a})$$

2. f_{TP} has the same Re dependence as the single phase $f_{\ell o}$, so that

$$\frac{f_{\mathrm{TP}}}{f_{\ell o}} = \frac{C_1/\mathrm{Re}_{\mathrm{TP}}^n}{C_1/\mathrm{Re}_{\ell o}^n} = \left(\frac{\mu_{\mathrm{TP}}}{\mu_{\mathrm{f}}}\right)^n \quad (11.79\mathrm{b})$$

Turbulent flow conditions are usually assumed so that $C_1 = 0.316$ and $n = 0.25$; or $C_1 = 0.184$ and $n = 0.2$. Models for two-phase viscosity, μ_{TP}, are hypothetical forms which fit the limiting liquid or gas-only cases. In addition to those listed below, Cicchitti et al. [15] and Isbin [36] have also proposed models.

$$\text{McAdams et al. [47]:} \quad \frac{\mu_{\mathrm{TP}}}{\mu_{\mathrm{f}}} = \left[1 + x\left(\frac{\mu_{\mathrm{f}}}{\mu_{\mathrm{g}}} - 1\right)\right]^{-1} \quad (11.80\mathrm{a})$$

$$\text{Dukler et al. [18]:} \quad \frac{\mu_{\rm TP}}{\mu_{\rm f}} = \left[1+\beta\left(\frac{\mu_{\rm g}}{\mu_{\rm f}}-1\right)\right] \tag{11.80b}$$

A more complex form of relationship for viscosity also exists as follows:

$$\text{Beattie and Whalley [5]:} \quad \mu_{\rm TP} = \mu_{\rm f}(1-\{\beta\})(1+2.5\{\beta\}) + \mu_{\rm g}\{\beta\} \tag{11.80c}$$

where $\{\beta\}$ is the gas volume flow fraction (the void fraction in homogeneous flow) given by

$$\{\beta\} = \frac{\rho_{\rm f} x}{\rho_{\rm f} x + \rho_{\rm g}(1-x)} \tag{11.50b}$$

Note that all the above equations reduce to the proper single-phase viscosity in the extreme cases of $x = \{\beta\} = 1$ or $x = \{\beta\} = 0$.

The friction pressure drop multiplier, $\phi^2_{\ell o}$, has been defined by Equation 11.66a. Applying the approximation of Equation 11.79a, the two-phase multiplier can be expressed as

$$\phi^2_{\ell o} = \frac{\rho_\ell}{\rho_{\rm m}}\frac{f_{\rm TP}}{f_{\ell o}} = \frac{\rho_\ell}{\rho_{\rm m}} \tag{11.81}$$

For thermodynamic equilibrium Equation 11.81 reduces to

$$\phi^2_{\rm fo} = \left[1+x\left(\frac{\rho_{\rm f}}{\rho_{\rm g}}-1\right)\right] \tag{11.82}$$

Alternatively, if Equations 11.79b and 11.80a are used to define $f_{\rm TP}/f_{\ell o}$, we get

$$\phi^2_{\rm fo} = \frac{\rho_{\rm f}}{\rho_{\rm m}}\frac{f_{\rm TP}}{f_{\rm fo}} = \left[1+x\left(\frac{\rho_{\rm f}}{\rho_{\rm g}}-1\right)\right]\left[1+x\left(\frac{\mu_{\rm f}}{\mu_{\rm g}}-1\right)\right]^{-n} \tag{11.83}$$

Other similar relations can be obtained for the conditions 11.80b and 11.80c.

Table 11.4 gives representative values of $\phi^2_{\ell o}$ for steam/water mixtures at various pressures and qualities for the HEM models as well as the Friedel separated flow model to be discussed. It is seen that $\phi^2_{\ell o}$ increases with decreasing pressure and higher quality, and the inclusion of the viscosity effect is particularly important for high-quality conditions.

Finally, note that Armand and Treshchev [1] have a friction pressure gradient model that complements their void fraction model and performs well under low-quality ($x < 0.3$) conditions as in the BWR.

TABLE 11.4
Two-Phase Multiplier $\phi^2_{\ell o}$ of Various Models for Steam/Water Mixtures

p (MPa/psia)	G (kg/m² s)	$\phi^2_{\ell o}$ at Various Qualities (x) 0.0	0.1	0.2	0.5	0.8	1.0	Source
7.03/1020		1	2.7	4.3	8.3	11.8	14.0	Equation 11.83, $n = 0.25$
		1	2.1	4.1	10.4	16.6	20.7	Equation 11.82
	100	1	5.4	8.3	15.6	22.2	14.2	Friedel [22,23], with $D = 20$ mm
	3000	1	3.6	5.4	10.9	17.5	15.5	
5.09/738		1	3.9	6.4	12.9	18.5	21.9	Equation 11.83, $n = 0.25$
		1	3.0	6.0	14.9	23.8	29.8	Equation 11.82
	100	1	7.3	11.3	21.7	31.3	20.5	Friedel [22,23], with $D = 20$ mm
	3000	1	4.7	7.2	15.3	25.0	22.6	
2.01/291		1	8.2	14.4	29.7	42.9	51.0	Equation 11.83, $n = 0.25$
		1	8.5	17.0	42.5	67.0	85.0	Equation 11.82
	100	1	15.9	25.2	50.4	74.8	52.9	Friedel [22,23], with $D = 20$ mm
	3000	1	9.8	16.0	36.5	61.9	59.6	

11.6.3 Separate Flow Models

In the general two-fluid case, the velocities and the temperatures of the two phases may be different. However, in the separate flow model, thermodynamic equilibrium conditions (and hence equal temperatures) are assumed. The model is different from the homogeneous equilibrium case only in that the velocities are not equal. Thus, simple relations for the friction pressure gradient may be derived.

Ignoring the liquid compressibility and assuming radially uniform phasic densities and velocities, the acceleration pressure drop for a constant mass flux is defined by substituting from Equation 11.63 into Equation 11.56a to get

$$\begin{aligned}\left(\frac{\mathrm{d}p}{\mathrm{d}z}\right)_{\mathrm{acc}} &= G_{\mathrm{m}}^2 \frac{d}{\mathrm{d}z}\left[\frac{x^2 v_{\mathrm{g}}}{\{\alpha\}} + \frac{(1-x)^2 v_{\mathrm{f}}}{\{1-\alpha\}}\right] \\ &= G_{\mathrm{m}}^2 \left[\frac{2x v_{\mathrm{g}}}{\{\alpha\}} - \frac{2(1-x) v_{\mathrm{f}}}{\{1-\alpha\}}\right]\left(\frac{\mathrm{d}x}{\mathrm{d}z}\right) \\ &\quad + G_{\mathrm{m}}^2 \left[-\frac{x^2 v_{\mathrm{g}}}{\{\alpha\}^2} + \frac{(1-x)^2 v_{\mathrm{f}}}{\{1-\alpha\}^2}\right]\left(\frac{\mathrm{d}\alpha}{\mathrm{d}z}\right) \\ &\quad + G_{\mathrm{m}}^2 \frac{x^2}{\{\alpha\}}\frac{\partial v_{\mathrm{g}}}{\partial p}\left(\frac{\mathrm{d}p}{\mathrm{d}z}\right)\end{aligned} \tag{11.84}$$

Thus, the total static pressure drop is given by substituting Equations 11.56c, 11.67, and 11.84 into Equation 11.55 and rearranging to get

$$-\left(\frac{\mathrm{d}p}{\mathrm{d}z}\right)_{\mathrm{SEP}} = \left[1 + G_{\mathrm{m}}^2 \frac{x^2}{\{\alpha\}} \frac{\partial v_{\mathrm{g}}}{\partial p}\right]^{-1} \times \left\{ \phi_{\ell o}^2 \frac{f_{\ell o}}{D_{\mathrm{e}}} \left[\frac{G_{\mathrm{m}}^2}{2\rho_{\mathrm{f}}}\right] + G_{\mathrm{m}}^2 \left[\frac{2xv_{\mathrm{g}}}{\{\alpha\}} - \frac{2(1-x)v_{\mathrm{f}}}{\{1-\alpha\}}\right] \frac{\mathrm{d}x}{\mathrm{d}z} \right.$$
$$\left. + G_{\mathrm{m}}^2 \left[\frac{(1-x)^2}{\{1-\alpha\}^2} v_{\mathrm{f}} - \frac{x^2}{\{\alpha\}^2} v_{\mathrm{g}}\right] \frac{\mathrm{d}\alpha}{\mathrm{d}z} + \rho_{\mathrm{m}} g \cos\theta \right\} \tag{11.85}$$

Comparing Equations 11.78 and 11.85, it is seen that, because in the separate flow model the change in α is not uniquely related to the change in x, an additional term for the explicit dependence of the acceleration pressure drop on $\mathrm{d}\alpha/\mathrm{d}z$ appears in Equation 11.85.

Note that under high pressures the compressibility of the gas may also be ignored, that is

$$G_{\mathrm{m}}^2 \frac{x^2}{\{\alpha\}} \frac{\partial v_{\mathrm{g}}}{\partial p} \ll 1$$

This adjustment greatly simplifies the integration of Equations 11.78 and 11.85.

A large number of investigators (Table 11.2) have proposed correlations for two-phase friction multipliers and associated void fraction. These parameters then permit the two-phase pressure gradient to be evaluated by Equation 11.85, which can be subsequently integrated for a specified axial heat flux distribution to obtain the pressure drop. The most well-known of the correlations listed in Table 11.2 are presented next.

11.6.3.1 Lockhart–Martinelli Correlation

This approach [42] is built upon two basic assumptions:

1. The familiar single-phase pressure drop relations can be applied to each of the phases in the two-phase flow field.
2. The pressure gradients of the two phases are equal at any axial position.

In the previous discussion, the friction pressure drop was related to that of a single-phase flow at a mass flux equal to the total mixture mass flux. An alternative assumption is that the friction pressure gradient along the channel may also be predicted from the flow of each phase separately in the channel. Thus

$$\left(\frac{\mathrm{d}p}{\mathrm{d}z}\right)_{\mathrm{fric}}^{\mathrm{TP}} \equiv \phi_\ell^2 \left(\frac{\mathrm{d}p}{\mathrm{d}z}\right)_{\mathrm{fric}}^{\ell} \equiv \phi_{\mathrm{v}}^2 \left(\frac{\mathrm{d}p}{\mathrm{d}z}\right)_{\mathrm{fric}}^{\mathrm{v}} \tag{11.86}$$

where $(\mathrm{d}p/\mathrm{d}z)_{\mathrm{fric}}^{\ell}$ and $(\mathrm{d}p/\mathrm{d}z)_{\mathrm{fric}}^{\mathrm{v}}$ = pressure drops obtained when the liquid phase and the gas phase are assumed to flow alone in the channel at their actual flow rate, respectively.

By definition:

$$\left(\frac{\mathrm{d}p}{\mathrm{d}z}\right)_{\text{fric}}^{\ell} = \frac{f_\ell}{D_\mathrm{e}}\left[\frac{G_\mathrm{m}^2(1-x)^2}{2\rho_\ell}\right]; \quad \left(\frac{\mathrm{d}p}{\mathrm{d}z}\right)_{\text{fric}}^{\mathrm{v}} = \frac{f_\mathrm{v}}{D_\mathrm{e}}\left[\frac{G_\mathrm{m}^2 x^2}{2\rho_\mathrm{v}}\right] \tag{11.87a,b}$$

Hence,

$$\phi_\ell^2 = \frac{f_{\mathrm{TP}}}{f_\ell}\frac{\rho_\ell}{\rho_\mathrm{m}^+}\frac{1}{(1-x)^2} \tag{11.88a}$$

$$\phi_\mathrm{v}^2 = \frac{f_{\mathrm{TP}}}{f_\mathrm{v}}\frac{\rho_\mathrm{v}}{\rho_\mathrm{m}^+}\frac{1}{x^2} \tag{11.88b}$$

It is seen from Equations 11.66a and 11.88a that

$$\phi_{\ell o}^2 = \frac{f_{\mathrm{TP}}}{\rho_\mathrm{m}^+}\frac{\rho_\ell}{f_{\ell o}} = \phi_\ell^2\frac{f_\ell}{\rho_\ell}\frac{\rho_\ell}{f_{\ell o}}(1-x)^2 \tag{11.89}$$

but

$$\mathrm{Re}_{\ell o} = \mathrm{G_m}\frac{D_\mathrm{e}}{\mu_\ell} \quad \text{and} \quad \mathrm{Re}_\ell = \mathrm{G_m}(1-x)\frac{D_\mathrm{e}}{\mu_\ell}$$

The relation between the friction factor and the respective Re number can be given as

$$f_\ell \sim \left(\frac{\mu_\ell}{D_\mathrm{e}G_\ell}\right)^n, \quad f_\mathrm{v} \sim \left(\frac{\mu_\mathrm{v}}{D_\mathrm{e}G_\mathrm{v}}\right)^n, \quad f_{\ell o} \sim \left(\frac{\mu_\ell}{D_\mathrm{e}G_\mathrm{m}}\right)^n \tag{11.90}$$

where $n = 0.25$ or 0.2 depending on the correlation used to define the relation of Equation 11.79. Substituting from Equation 11.90 into Equation 11.89, we get

$$\phi_{\ell o}^2 = \phi_\ell^2\frac{[G_\mathrm{m}(1-x)D_\mathrm{e}/\mu_\ell]^{-n}}{(G_\mathrm{m}D_\mathrm{e}/\mu_\ell)^{-n}}(1-x)^2 = \phi_\ell^2(1-x)^{2-n} \tag{11.91}$$

Lockhart and Martinelli defined the parameter X such that

$$X^2 = \frac{(\mathrm{d}p/\mathrm{d}z)_{\text{fric}}^{\ell}}{(\mathrm{d}p/\mathrm{d}z)_{\text{fric}}^{\mathrm{v}}} \tag{11.92}$$

Note that from Equation 11.86, $X^2 = \phi_v^2 / \phi_\ell^2$. Substituting from Equations 11.87 and 11.90 into Equation 11.92, we obtain X^2 under thermal equilibrium conditions. When f is taken proportional to $\text{Re}^{-0.25}$:

$$X^2 = \left(\frac{\mu_f}{\mu_g}\right)^{0.25} \left(\frac{1-x}{x}\right)^{1.75} \left(\frac{\rho_g}{\rho_f}\right) \quad (11.93)$$

If f is taken to be proportional to $\text{Re}^{-0.2}$, we get

$$X^2 = \left(\frac{\mu_f}{\mu_g}\right)^{0.2} \left(\frac{1-x}{x}\right)^{1.8} \left(\frac{\rho_g}{\rho_f}\right) \quad (11.94)$$

Lockhart and Martinelli [42] suggested that ϕ_ℓ and ϕ_v as well as x can be correlated uniquely as a function of X. The graphic relation is shown in Figure 11.13 (where f and g are used in place of l and v). Their results were obtained from data on horizontal flow of adiabatic two-component systems at low pressure. An accurate

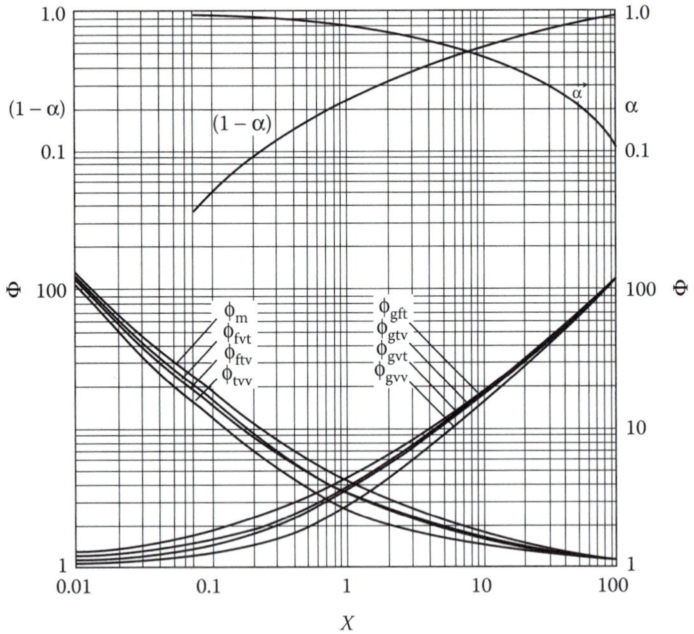

FIGURE 11.13 Martinelli model for pressure gradient ratios and void fractions. (From Lockhart, R. W. and Martinelli, R. C., *Chem Eng. Prog.*, 45(1):39–48, 1949.)

analytical representation of the curves shown in Figure 11.13 was suggested by Chisholm [12]:

$$\phi_\ell^2 = 1 + \frac{C}{X} + \frac{1}{X^2} \tag{11.95a}$$

$$\phi_v^2 = 1 + CX + X^2 \tag{11.95b}$$

$$1 - \alpha = \frac{X}{\sqrt{X^2 + CX + 1}} \tag{11.95c}$$

where C is given in Table 11.5, dependent on the phase flow being laminar (viscous) or turbulent. In practice, the scatter of the experimental results is as large if not larger than the variation in ϕ_ℓ^2 and ϕ_v^2 owing to the flow regime assumptions. Hence, only the value of $C = 20$ is usually used for all regimes (i.e., a turbulent-turbulent regime is used.)

Equations 11.95a and 11.95c imply that

$$\phi_\ell^2 = (1 - \{\alpha\})^{-2} \tag{11.96}$$

which was theoretically derived by Chisholm [12]. Chisholm's proposed correlation [13] listed in Table 11.2 is an extension of his earlier work with Sutherland [14].

11.6.3.2 Thom Correlation

Thom [68] dealt with a much greater steam–water database (that from the Cambridge University experimental program of 1951–59) than available to Martinelli and Nelson. He evaluated and correlated the acceleration pressure drop directly from the data. The friction pressure drop was deduced by subtracting the acceleration component from the measured total pressure drop in horizontal tubes. From the total and friction pressure drops, the mean value of the two-phase friction factor was derived. Finally, the gravity pressure drop was correlated in terms of the outlet quality for uniform wall heating and saturated inlet conditions of the test program.

Thus, Thom's correlated results are for pressure drop rather than pressure gradient components under conditions of uniform wall heating and saturated inlet

TABLE 11.5
Values of Constant C

Liquid–Gas	C
Turbulent-turbulent (tt)	20
Viscous-turbulent (vt)	12
Turbulent-viscous (tv)	10
Viscous-viscous (vv)	5

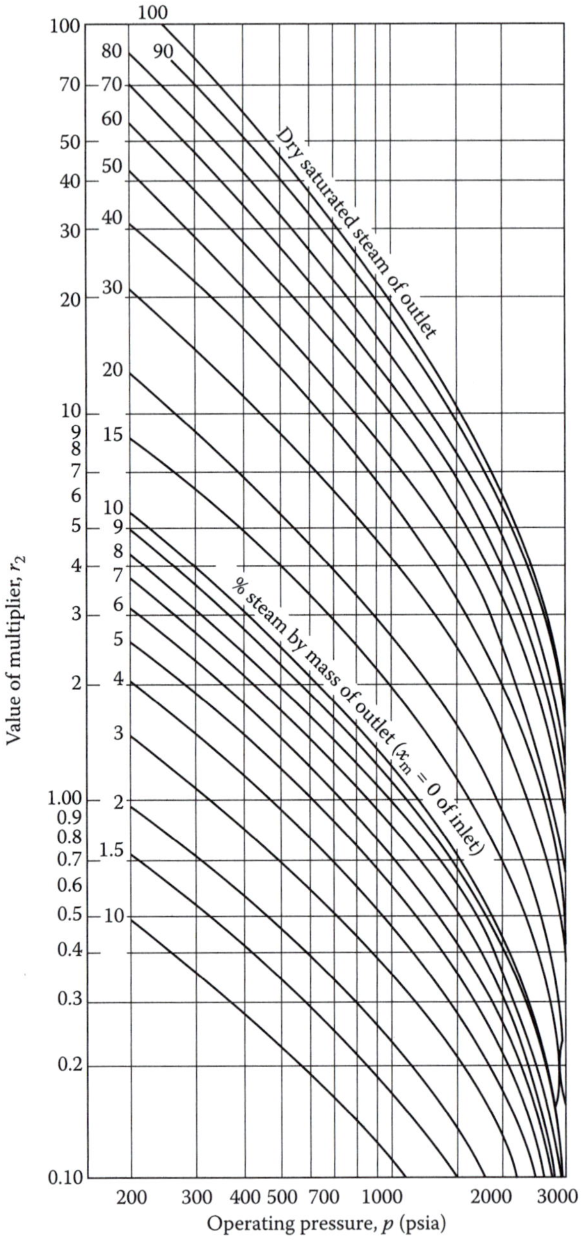

FIGURE 11.14 Thom's acceleration pressure drop multiplier (r_2). (From Thom, J. R. S., *Int.J. Heat Mass Transfer,* 7:709–724, 1964.)

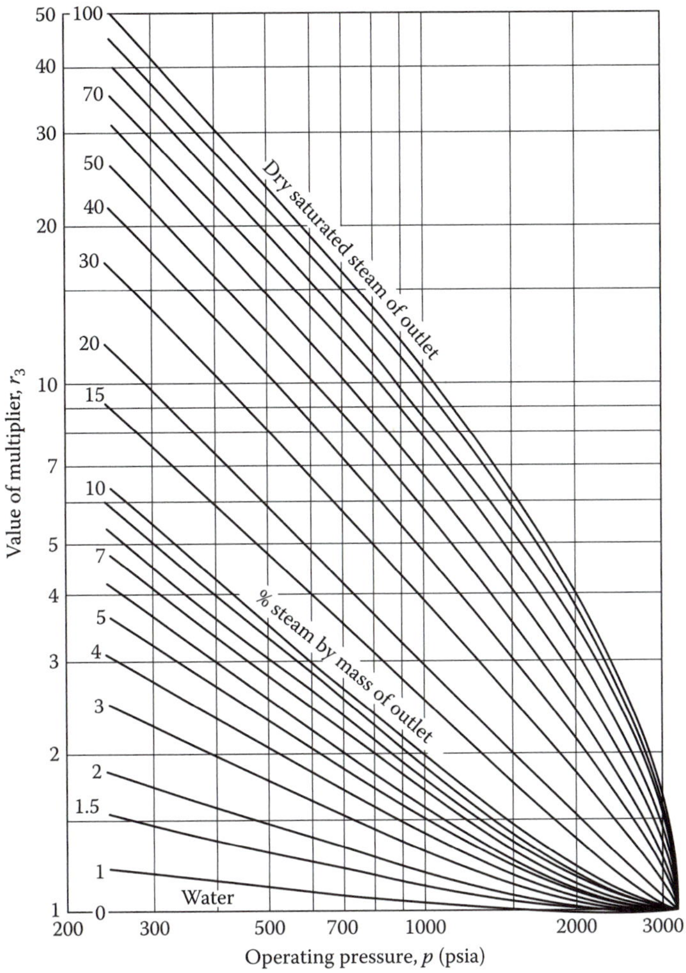

FIGURE 11.15 Thom's friction pressure drop multiplier (r_3). (From Thom, J. R. S., *Int. J. Heat Mass Transfer*, 7:709–724, 1964.)

conditions. The governing equation to which the correlated r factor results of Figures 11.14, 11.15, and 11.16, apply is Equation 11.57 expressed as follows:

$$\Delta p_{TP} = \Delta p_{acc} + \Delta p_{fric} + \Delta p_{grav}$$
$$= \frac{G_m^2}{\rho_\ell} r_2 + f_{\ell o} \frac{G_m^2}{2\rho_\ell} \frac{L}{D_e} r_3 + g\rho_\ell \cos\theta L r_4 \quad (11.97)$$

where θ is the flow angle with the upward vertical.

The definitions of r_2, r_3, (set in on-line) and r_4 are presented in Section 11.6.4.

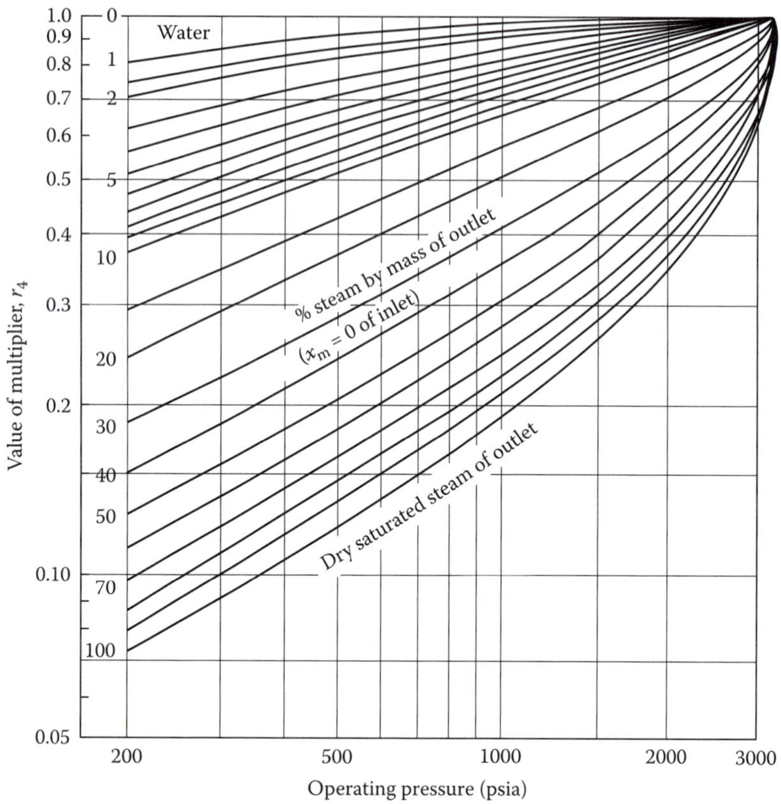

FIGURE 11.16 Thom's gravitational pressure drop multiplier (r_4). (From Thom, J. R. S., *Int. J. Heat Mass Transfer*, 7:709–724, 1964.)

11.6.3.3 Baroczy Correlation

Baroczy [4] included the influence of G_m on $\phi^2_{\ell o}$ for fluids other than steam–water. He generated two sets of curves. The first defines $\phi^2_{\ell o}$ at a reference $G_m = 1356$ kg/m^2 s (or 10^6 lb$_m$/ft^2·h) as a function of a fluid property index and at a given quality (Figure 11.17). The other set provides a multiplier correction factor (Ω) for various values of G as a function of the same property index (Figure 11.18). This multiplier correction factor is defined as

$$\phi^2_{\ell o}(G_m) \equiv \Omega \phi^2_{\ell o}(G_{ref}) \tag{11.98}$$

11.6.3.4 Friedel Correlation

The Friedel correlation [22,23] is a frequently recommended correlation for the frictional two-phase pressure gradient. It is written in terms of the total flow as liquid

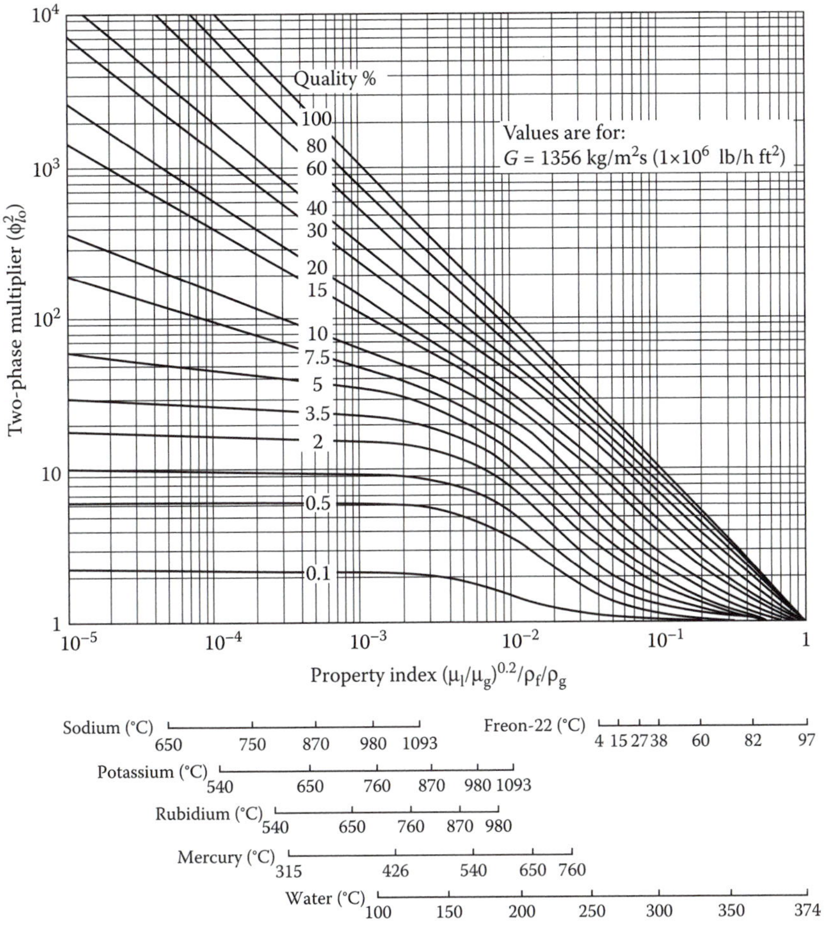

FIGURE 11.17 Baroczy two-phase friction multiplier ($\phi^2_{\ell o}$). (From Baroczy, C. J., *Chem. Eng. Prog. Symp. Ser.*, 62(44):232–249, 1966.)

only, that is, the two-phase multiplier, $\phi^2_{\ell o}$. For horizontal and vertical upflow, it has the following form*:

$$\phi^2_{\ell o} = E + \frac{3.24\,FH}{\mathrm{Fr}^{0.0454}\,\mathrm{We}^{0.035}} \tag{11.99}$$

$$E = (1-x)^2 + x^2\,\frac{\rho_\ell f_{vo}}{\rho_v f_{\ell o}} \tag{11.100a}$$

$$F = x^{0.78}(1-x)^{0.224} \tag{11.100b}$$

* The constant 3.24 of [22] in Equation 11.99 replaces the constant 3.21 of [23] (pers. comm., June 2009).

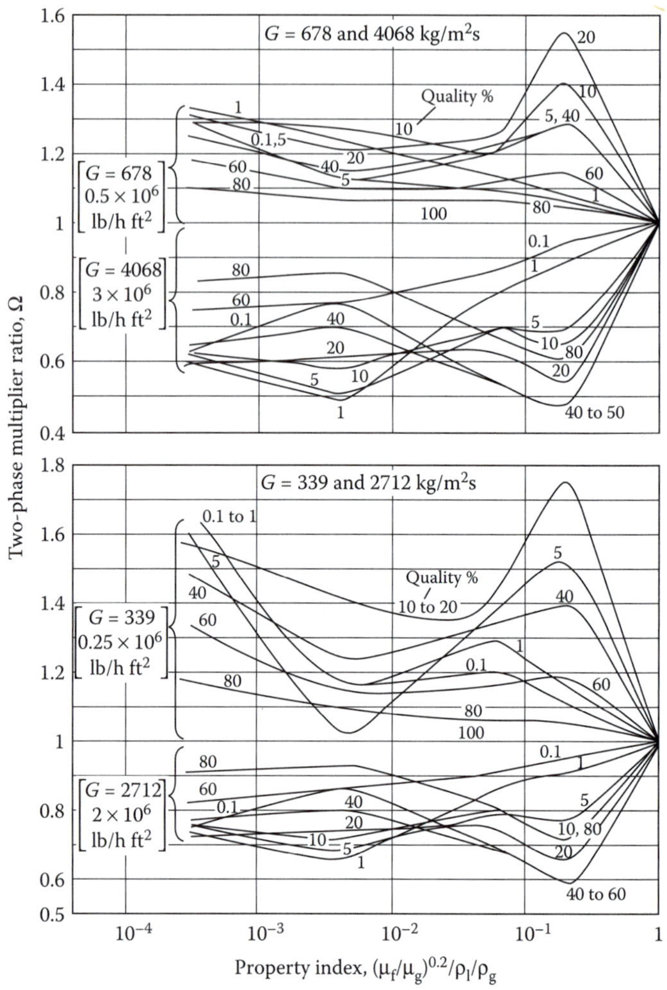

FIGURE 11.18 Baroczy mass flux correction factor. (From Baroczy, C. J., *Chem. Eng. Prog. Symp. Ser.*, 62(44):232–249, 1966.)

$$H = \left(\frac{\rho_\ell}{\rho_v}\right)^{0.91} \left(\frac{\mu_v}{\mu_\ell}\right)^{0.19} \left(1 - \frac{\mu_v}{\mu_\ell}\right)^{0.7} \tag{11.100c}$$

$$\text{Fr} = \frac{G_m^2}{g D_e \rho_m^2} \tag{11.100d}$$

$$\text{We} = \frac{G_m^2 D_e}{\sigma \rho_m} \tag{11.100e}$$

and

$$\rho_m = \left(\frac{x}{\rho_v} + \frac{1-x}{\rho_\ell} \right)^{-1} \quad \text{(for HEM)} \tag{11.73}$$

For the liquid-only and vapor-only friction factors, Friedel recommends different correlations depending on the channel geometry (circular, rectangular, annular). For circular geometries, he suggests

$$f_{jo} = \frac{64}{Re_j} \quad \text{if } Re_j \le 1055 \tag{11.101a}$$

$$= \left[0.86859\, \ell n \left(\frac{Re_j}{1.964\, \ell n\, Re_j - 3.8215} \right) \right]^{-2} \quad \text{if } Re_j > 1055 \tag{11.101b}$$

where the subscript "j" stands for "ℓ" or "v."

The parameters in the above equations are: x is the quality; ρ_v is the gas density (kg/m^3); ρ_ℓ is the liquid density (kg/m^3); f_{vo} and $f_{\ell o}$ are the Darcy friction factors for the total mass flux flowing with the gas and liquid properties, respectively; μ_v is the gas dynamic viscosity (Pa s); μ_ℓ is the liquid dynamic viscosity (Pa s); σ is the surface tension (N/m); and g is the acceleration due to gravity (= 9.81 m/s^2).

The correlation is applicable to vertical upflow and to horizontal flow. A very similar correlation is available for vertical downflow.

Example 11.4: Two-Phase Flow Pressure Gradient from the Friedel Correlation

PROBLEM

Consider vertical upflow of two-phase steam–water in a 20 mm diameter tube under thermal equilibrium at 290°C (7.45 MPa) as in Example 11.3. Find the two-phase pressure gradient from the Friedel correlation if the mass flux is 400 kg/m^2 s and the quality is 50%.

SOLUTION

The needed fluid properties are given in Example 11.3. Proceed to calculate the necessary parameters for insertion into Equation 11.99.

From Equation 11.73:

$$\rho_m = \left(\frac{0.5}{39.16} + \frac{0.5}{732.33} \right)^{-1} = 74.34\,\text{kg/m}^3$$

From Equation 11.100e:

$$We = \frac{(400)^2(0.02)}{(0.0167)(74.34)} = 2577.6$$

From Equation 11.100d:

$$\text{Fr} = \frac{(400)^2}{9.8(0.02)(74.34)^2} = 147.7$$

From Equation 11.100c:

$$H = \left(\frac{732.33}{39.16}\right)^{0.91}\left(\frac{19.16}{89.63}\right)^{0.19}\left(1 - \frac{19.16}{89.63}\right)^{0.7} = 9.06$$

From Equation 11.100b:

$$F = (0.5)^{0.78}(1 - 0.5)^{0.224} = 0.499$$

To evaluate the parameter E, the coefficients f_{vo} and f_{lo} are required which necessitates evaluating the Reynolds numbers as if the total flow were vapor and liquid only.

For the vapor, Equation 9.63 yields

$$\text{Re}_{\text{vo}} = \frac{G_{\text{m}}D_{\text{e}}}{\mu_{\text{v}}} = \frac{400(0.02)}{19.16 \times 10^{-6}} = 4.18 \times 10^5$$

Then for Equation 11.101b, the vapor-only friction factor is

$$f_{\text{vo}}(\text{Darcy}) = \left[0.86859\ell n\left(\frac{4.18 \times 10^5}{1.964\ell n(4.18 \times 10^5) - 3.8215}\right)\right]^{-2} = 0.0136$$

For the liquid, Equation 9.63 results in

$$\text{Re}_{\ell\text{o}} = \frac{G_{\text{m}}D_{\text{e}}}{\mu_{\ell}} = \frac{400(0.02)}{89.63 \times 10^{-6}} = 8.93 \times 10^4$$

Then from Equation 11.101b the liquid-only friction factor is

$$f_{\ell\text{o}} = \left[0.86859\ell n\left(\frac{8.93 \times 10^4}{1.964\ell n(8.93 \times 10^4) - 3.8215}\right)\right]^{-2} = 0.0184$$

yielding from Equation 11.100a

$$E = (1 - 0.5)^2 + (0.5)^2\left(\frac{732.33}{39.16}\right)\left(\frac{13.6 \times 10^{-3}}{18.4 \times 10^{-3}}\right) = 3.70$$

Hence, from Equation 11.99:

$$\phi_{\ell o}^2 = 3.70 + \frac{3.24(0.499)(9.06)}{(147.7)^{0.0454}(2577.4)^{0.035}} = 12.57$$

Finally, the desired two-phase friction gradient is given by

$$\left(\frac{dp}{dz}\right)_{\rm fric}^{\rm TP} = \left(\frac{dp}{dz}\right)_{\rm fric}^{\ell o} \phi_{\ell o}^2 \tag{11.65a}$$

where

$$\left(\frac{dp}{dz}\right)_{\rm fric}^{\ell o} = \frac{f_{\ell o} G_{\rm m}^2}{2D_{\rm e}\rho_\ell} = \frac{(0.0184)(400)^2}{2(0.02)(732.33)} = 100.5\,{\rm N/m^3 \ or \ Pa/m}$$

Hence,

$$\left(\frac{dp}{dz}\right)_{\rm fric}^{\rm TP} = 100.5(12.57) = 1263\,{\rm N/m^3 \ or \ Pa/m}$$

11.6.4 Two-Phase Pressure Drop

The two-phase pressure drop is evaluated by integration of the pressure gradient given by Equation 11.78 or 11.85 over the length L of the flow channel. Ignoring vapor compressibility, for the separated flow model of Equation 11.85 this integration* leads to

$$\Delta p = \frac{f_{\ell o}}{D_{\rm e}} \frac{G_{\rm m}^2}{2\rho_\ell} \int_0^L \phi_{\ell o}^2 {\rm d}z + G_{\rm m}^2 \left[\frac{(1-x)^2}{(1-\alpha)\rho_\ell} + \frac{x^2}{\alpha\rho_{\rm v}}\right]_{z=0}^{z=L} + \int_0^L g\cos\theta[\rho_{\rm v}\alpha + \rho_\ell(1-\alpha)]\,{\rm d}z \tag{11.102}$$

The procedure for evaluation of channel Δp using Equation 11.102 differs depending on whether the value of the inlet quality is zero or finite. Both cases are treated next for conditions of constant axial heat addition rate.

11.6.4.1 Pressure Drop for Zero Inlet Quality $x = 0$

For $\alpha = x = 0$ at $z = 0$ and constant heat addition rate over a length L, the vapor quality increases at a constant rate with distance z:

$$\frac{{\rm d}x}{{\rm d}z} = {\rm constant} = \frac{x_{\rm out}}{L}$$

* Integration of the acceleration component is most easily accomplished by treating only the first expression for the RHS of Equation 11.84.

Hence, Equation 11.102 can be written as

$$\Delta p = \frac{f_{\ell o}}{D_e}\frac{G_m^2}{2\rho_\ell}L\left[\frac{1}{x_{out}}\int_0^{x_{out}}\phi_{\ell o}^2\,dx\right] + \frac{G_m^2}{\rho_\ell}\left[\frac{(1-x_{out})^2}{(1-\alpha_{out})} + \frac{x_{out}^2\rho_\ell}{\alpha_{out}\rho_v} - 1\right]$$
$$+ \frac{L\rho_\ell g\cos\theta}{x_{out}}\int_0^{x_{out}}\left[1-\left(1-\frac{\rho_v}{\rho_\ell}\right)\alpha\right]dx \tag{11.103}$$

or

$$\Delta p = \frac{G_m^2}{\rho_\ell}r_2 + f_{\ell o}\frac{G_m^2}{2\rho_\ell}\frac{L}{D_e}r_3 + g\rho_\ell L\cos\theta r_4 \tag{11.97}$$

The parameter r_3 is a quality-averaged two-phase multiplier based on liquid flow. For this case of uniform axial heat addition rate, r_3 is equal to $\overline{\phi_{\ell o}^2}$.

The values of r_2, r_3, and r_4 are different for nonconstant heating situations and for different models for $\phi_{\ell o}^2$ and α. Values of r_2, r_3, and r_4 evaluated with the Thom correlation for uniform axial heat flux have been shown in Figures 11.14, 11.15, and 11.16, respectively. For the HEM, the r factor values can be analytically evaluated as demonstrated next for zero inlet quality and outlet quality x_{out}. Now, however, evaluate the total pressure drop using the HEM where the two-phase friction factor is taken equal to the single-phase friction factor. For this assumption, recall from Equation 11.82 that the two-phase liquid-only multiplier is

$$\phi_{fo}^2 = \left[1 + x\left(\frac{\rho_f}{\rho_g} - 1\right)\right] \tag{11.82}$$

Since

$$v_{fg} = v_g - v_f = \frac{1}{\rho_g} - \frac{1}{\rho_f} = \frac{\rho_f - \rho_g}{\rho_f\rho_g} \tag{11.104}$$

Equation 11.78, the relation for the HEM pressure gradient, becomes

$$-\left.\frac{dp}{dz}\right|_{HEM} = \frac{\dfrac{f_{TP}}{D_e}\dfrac{G_m^2}{2\rho_m} + G_m^2\left(\dfrac{\rho_f - \rho_g}{\rho_f\rho_g}\right)\dfrac{dx}{dz} + \rho_m g\cos\theta}{1 + G_m^2 x(\partial v_g/\partial p)} \tag{11.105}$$

In most cases of interest (within the riser following a high pre≪e boiler, or within coolant channels of a nuclear reactor), the operating pressure is sufficiently high that small, gradual changes in pressure do not significantly impact fluid properties, therefore we can say that when $p_{op} \gg \Delta p_{tot}$: $\partial v_g/\partial p \approx 0$ and ρ_f, ρ_g, μ_f need not be integrated over z. Hence, Equation 11.105 simplifies to

$$-\left.\frac{dp}{dz}\right|_{HEM} = \frac{f_{TP}}{D_e}\frac{G_m^2}{2\rho_m} + G_m^2\left(\frac{\rho_f - \rho_g}{\rho_f\rho_g}\right)\frac{dx}{dz} + \rho_m g\cos\theta \tag{11.106}$$

which upon integration over length L yields

$$\int_0^L\left(-\left.\frac{dp}{dz}\right|_{HEM}\right)dz = \int_0^L \frac{f_{TP}}{D_e}\frac{G_m^2}{2\rho_m}\,dz + \int_0^L G_m^2\left(\frac{\rho_f - \rho_g}{\rho_f\rho_g}\right)\frac{dx}{dz}\,dz + \int_0^L \rho_m g\cos\theta\,dz \tag{11.107}$$

Now treat each of these three pressure drop components.

1. Friction: Using the HEM model for the two-phase friction multiplier:

$$\int_0^L\left(-\left.\frac{dp}{dz}\right|_{fric}^{TP}\right)dz = \int_0^L -\left(\left.\frac{dp}{dz}\right|_{fric}^{\ell o}\right)\phi_{lo}^2\,dz$$

where

$$\left.\frac{dp}{dz}\right|_{fric}^{\ell o} = f\frac{L}{D_e}\frac{G_m^2}{2\rho_f},\quad \phi_{\ell o}^2 = \frac{f_{TP}}{f}\frac{\rho_f}{\rho_m}$$

and f is the Darcy friction factor calculated using, for example, the Moody diagram.
Integrating:

$$p_{in} - p_{out} = \int_0^L -\left(\left.\frac{dp}{dz}\right|_{fric}^{\ell o}\right)\times\phi_{\ell o}^2\,dz = -\left.\frac{dp}{dz}\right|_{fric}^{\ell o}\times\int_{x_i=0}^{x_{out}}\left(\frac{x\rho_f}{\rho_g}+1-x\right)\left(\frac{dz}{dx}\right)dx$$

For axially constant q'', dx/dz equals x_{out}/L. Thus,

$$\begin{aligned}
p_{in} - p_{out} &= \frac{fG_m^2}{D_e 2\rho_f}\times\int_{x_i=0}^{x_{out}}\left(\frac{x\rho_f}{\rho_g}+1-x\right)\frac{L}{x_{out}}dx \\
&= \frac{fG_m^2}{D_e 2\rho_f}\left[\frac{x_{out}^2}{2}\frac{\rho_f}{\rho_g} + x_{out} - \frac{x_{out}^2}{2}\right]\frac{L}{x_{out}} \\
&= f\frac{L}{D_e}\frac{G_m^2}{2\rho_f}\left[1+\frac{x_{out}}{2}\left(\frac{\rho_f-\rho_g}{\rho_g}\right)\right]
\end{aligned}$$

which, for

$$X_{\text{out}} \equiv x_{\text{out}}\left(\frac{\rho_{\text{f}}-\rho_{\text{g}}}{\rho_{\text{g}}}\right)$$

becomes

$$\Delta p_{\text{fric}} = f\frac{L}{D_{\text{e}}}\frac{G_{\text{m}}^2}{2\rho_{\text{f}}}\left[1+\frac{X_{\text{out}}}{2}\right] \tag{11.108}$$

2. Acceleration:

$$\begin{aligned}\Delta p_{\text{acc}} &= G_{\text{m}}^2\left(\frac{\rho_{\text{f}}-\rho_{\text{g}}}{\rho_{\text{f}}\rho_{\text{g}}}\right)\int_0^L \frac{\text{d}x}{\text{d}z}\text{d}z = G_{\text{m}}^2\left(\frac{\rho_{\text{f}}-\rho_{\text{g}}}{\rho_{\text{f}}\rho_{\text{g}}}\right)\int_0^{x_{\text{out}}}\text{d}x \\ &= x_{\text{out}}\left(\frac{\rho_{\text{f}}-\rho_{\text{g}}}{\rho_{\text{g}}}\right)\frac{G_{\text{m}}^2}{\rho_{\text{f}}} = \frac{G_{\text{m}}^2 X_{\text{out}}}{\rho_{\text{f}}}\end{aligned} \tag{11.109}$$

3. Gravity:

$$\Delta p_{\text{grav}} = \int_0^L \rho_{\text{m}} g\cos\theta\,\text{d}z$$

for a vertical riser.

$$\rho_{\text{m}} = \frac{1}{(x/\rho_{\text{g}})+((1-x)/\rho_{\text{f}})} = \frac{\rho_{\text{g}}\rho_{\text{f}}}{\rho_{\text{g}}+x(\rho_{\text{f}}-\rho_{\text{g}})} = \frac{\rho_{\text{f}}}{1+x((\rho_{\text{f}}-\rho_{\text{g}})/\rho_{\text{g}})}$$

$$\Delta p_{\text{grav}} = \rho_{\text{f}} g\int_0^{x_{\text{out}}}\frac{1}{1+x((\rho_{\text{f}}-\rho_{\text{g}})/\rho_{\text{g}})}\left(\frac{\text{d}z}{\text{d}x}\right)\text{d}x$$

Representing $(\rho_{\text{f}}-\rho_{\text{g}})/\rho_{\text{g}}$ as C_1 obtain

$$\begin{aligned}\Delta p_{\text{grav}} &= \frac{\rho_{\text{f}} gL}{x_{\text{out}}}\int_0^{x_{\text{out}}}\frac{1}{1+xC_1}\text{d}x = \frac{\rho_{\text{f}} gL}{x_{\text{out}}}\frac{1}{C_1}\ln(1+C_1 x)\Big|_0^{x_{\text{out}}} \\ &= \frac{\rho_{\text{f}} gL}{x_{\text{out}}}\frac{1}{C_1}\ln(1+C_1 x_{\text{out}}) \\ &= \frac{\rho_{\text{f}} gL}{x_{\text{out}}}\frac{\rho_{\text{g}}}{\rho_{\text{f}}-\rho_{\text{g}}}\ln\left(1+\frac{\rho_{\text{f}}-\rho_{\text{g}}}{\rho_{\text{g}}}x_{\text{out}}\right) \\ &= \frac{\rho_{\text{f}} gL}{X_{\text{out}}}\ln(1+X_{\text{out}})\end{aligned} \tag{11.110}$$

11.6.4.2 Pressure Drop for Non zero Inlet Quality

For this situation of positive x_{in} and uniform axial heat flux, the friction and the gravity losses must be expanded in the following manner if the tabulated results for $x_{in} = 0$ are to be used. For illustration, consider the case of an inlet quality equal to 0.1 and the exit quality equal to 0.7.

The desired pressure drop is expressed as the difference in pressure loss for two flow channels—one with $x_{in} = 0$ and $x_{out} = 0.7$ and the second with $x_{in} = 0$ and $x_{out} = 0.1$. Hence, following Equation 11.97 obtain

$$\Delta p = \frac{G_m^2}{\rho_\ell} r_2 \Big|_{0.1}^{0.7} + \frac{f_{\ell o} G_m^2}{2 D_e \rho_\ell}\left(L r_3\Big|_{0.0}^{0.7} - L_{0.1} r_3\Big|_{0.0}^{0.1}\right) + \rho_\ell g \cos\theta \left(L r_4\Big|_{0.0}^{0.7} - L_{0.1} r_4\Big|_{0.0}^{0.1}\right) \quad (11.111)$$

where

$$r_2 = X_{out};\quad r_3 = 1 + \frac{X_{out}}{2};\quad \text{and}\quad r_4 = \frac{\ell n(1 + X_{out})}{X_{out}}$$

with X_{out} evaluated using $x_{out} = 0.7$ or $x_{out} = 0.1$.

Also, in the above equation, while L is the total length of the flow channel, $L_{0.1}$ is the flow channel length required for the quality to attain the value of 0.1. This length is obtained by the following heat balance:

$$q''_w P_w L_{0.1} = \dot{m} h_{fg} \Delta x$$

where $\Delta x = 0.1$.

Hence,

$$L_{0.1} = \frac{\dot{m} h_{fg} 0.1}{q''_w P_w} \quad (11.112)$$

11.6.5 Relative Accuracy of Various Friction Pressure Loss Models

It is instructive to first examine the character of the pressure gradient data to appreciate the relative performance of homogeneous versus separated flow models. From the extensive data set of Gaspari et al. [24], let us extract a variable mass flux data set taken at a typical BWR pressure condition of 7.14 MPa. The friction pressure gradient (Pa/m) versus quality traces under these flow conditions are shown in Figure 11.19. The data are compared against the HEM correlation of Equation 11.83 and the Friedel correlation [22,23]. It is useful to comment on two characteristics of these plots: (1) the behavior of the data and the correlations at the all-vapor limit and (2) the relative accuracy of the correlations.

1. At the all-vapor limit, $x = 1$, $\phi_{vo}^2 = 1$ for the HEM case of Equation 11.79a.

$$\phi_{fo}^2 = \left(\frac{\rho_f}{\rho_g}\right)\left(\frac{\mu_g}{\mu_f}\right)^n \quad (11.113)$$

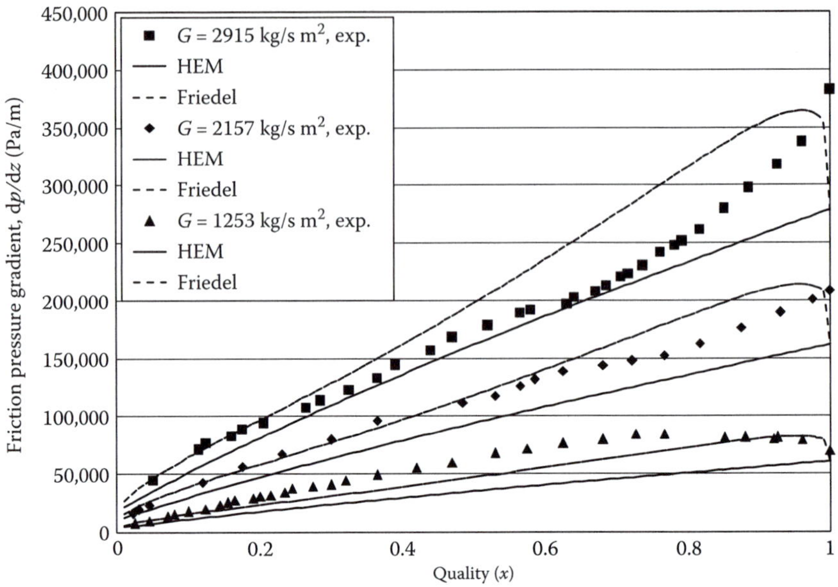

FIGURE 11.19 HEM and Friedel friction pressure gradient correlations compared to Gaspari et al. CISE data [24]. (HEM evaluated using Equation 11.83 with $n = 0.2$ and $f_{fo} = 0.184\ \text{Re}^{-0.2}$).

Hence, from Equation 11.65b:

$$\left.\frac{\mathrm{d}p}{\mathrm{d}z}\right|_{\text{fric}}^{\text{TP}} = \left.\frac{\mathrm{d}p}{\mathrm{d}z}\right|_{\text{fric}}^{\text{vo}}$$

that is, this HEM correlation should match the data identically at the all-vapor limit. Note that this is not the case for the HEM model of Equation 11.79b. However, as the mass flux increases sufficiently, it is hypothesized that the boundary layer thickness becomes comparable to the tube surface asperities such that the pressure gradient of the data becomes larger than that predicted by the HEM correlation and also the Friedel correlation.

2. Since homogeneous models are approximately linear, it is immediately obvious that these models generally well predict the high mass flux over most of the quality range but would fail to predict the low mass flux trace. This is reasonable since the high mass flux trace is mostly in mist, hence homogeneous flow, while the low mass flux trace is mostly in annular, hence separated, flow.

Consequently, various investigators have found the separated flow models to be better than the homogeneous model for $500 < G_m < 1500$ kg/m² s, whereas the homogeneous model appears to be better for $G_m \geq 2000$ kg/m² s. This finding is likely due to the fact that at a given quality, higher mass flow rates may lead to a well-mixed (dispersed) flow pattern characteristic of the homogeneous model.

Assessment of friction pressure loss models has been undertaken by a number of investigators. Central to the integrity of the assessment is the gathering and systematic qualification of the largest database possible. This has been performed on a continuing proprietary basis by G. Hewitt for the U.K. Engineering Science Data Unit. The results of the most recent evaluation [30] suggest the use of the Premoli and the Friedel correlations for general evaluation of void fraction and friction pressure gradient, respectively. Useful comparisons of correlations with qualified data are presented for specific ranges of pressure quality and mass flux in [30].

11.6.6 Pressure Losses across Singularities

Analysis of reactor flow systems involves evaluation of pressure losses across valves, throttling/metering devices (nozzles, venturis, and orifices), bends, and sudden contractions or expansions. To allow for such form losses, the homogeneous model would be expressed as

$$\left(\frac{dp}{dz}\right)_{\text{form}}^{\text{TP}} = \phi_{\ell o}^2 K\left[\frac{G_m^2}{2\rho_\ell}\right] \quad \text{(Adopted from Equation 11.64)}$$

where form loss coefficient K values for various geometries are listed in Table 9.4.

While the homogeneous model is effective for many configurations, nevertheless many specialized results exist for this range of configurations. These are best accessed by reference to handbook compilations of which that by Hewitt is comprehensive and clear [29].

Example 11.5: Pressure Drop Calculations in a Condensing Unit

PROBLEM

To predict the pressure drop in condensing equipment, it is possible to relate the friction pressure drop to an all-gas (single-phase) pressure drop by defining a new two-phase multiplier ϕ_{vo}^2 from Equations 11.77 and 11.66b as

$$\left(\frac{dp}{dz}\right)_{\text{friction}} = \phi_{vo}^2 \frac{f_{vo}}{D_e}\frac{G_m^2}{2\rho_v} \tag{11.114}$$

1. Using the HEM model, determine the multiplier in terms of the vapor density. Assume that the two-phase mixture viscosity is equal to the vapor viscosity.
2. Evaluate the pressure drop across a horizontal tube of length L and diameter D using the HEM approach. Assume axially uniform heat flux and the following conditions:
   $D = 20$ mm
   $L = 2$ m

$$f_{TP} = f_{go} = 0.005$$

$$p_{in} = 1.0 \text{ MPa } (\simeq 150 \text{ psi})$$

$$\dot{m} = 0.1 \text{kg/s}$$

Inlet equilibrium quality = 0.05
Exit equilibrium quality = 0.00

SOLUTION

1. From the problem statement and recognizing that $\rho_v = \rho_g$, we can write

$$\phi_{go}^2 = \frac{f_{TP}}{f_{go}} \frac{\rho_g}{\rho_m}$$

Using the HEM model, $\rho_m = \rho_m^+$, so Equation 11.73 gives

$$\frac{1}{\rho_m} = \frac{x}{\rho_g} + \frac{1-x}{\rho_f} = x v_g + (1-x) v_f$$

Thus

$$\phi_{go}^2 = \frac{f_{TP}}{f_{go}} \rho_g [x v_g + (1-x) v_f]$$

or

$$\phi_{go}^2 = \frac{f_{TP}}{f_{go}} \left[\frac{v_f}{v_g} + x \left(1 - \frac{v_f}{v_g} \right) \right]$$

2. The pressure gradient is found from:

$$-\frac{dp}{dz} = \left(\frac{dp}{dz}\right)_{acc} + \left(\frac{dp}{dz}\right)_{fric} + \left(\frac{dp}{dz}\right)_{gravity} \quad (11.55)$$

For a horizontal tube $(dp/dz)_{gravity} = 0$.

The pressure drop equation can be written in a manner analogous to Equation 11.59a and the integral of Equation 11.67 but in terms of saturated vapor conditions, as

$$\Delta p = p_{in} - p_{out} = G_m^2 \left[\left(\frac{1}{\rho_m}\right)_{out} - \left(\frac{1}{\rho_m}\right)_{in} \right] + \overline{\phi_{go}^2} \frac{f_{go} L}{D} \frac{G_m^2}{2\rho_g} \quad (11.115)$$

Now

$$\{\alpha\}_{in} = \frac{1}{1 + ((1-x)/x)(\rho_g/\rho_f)}$$

$$= \frac{1}{1 + ((1-0.05)/0.05)((1.1274 \times 10^{-3})/0.1943)} = 0.901$$

$$\rho_{in} = \alpha_{in}\rho_g + (1-\alpha_{in})\rho_f = \frac{\alpha_{in}}{v_g} + \frac{1-\alpha_{in}}{v_f} = \frac{0.901}{0.1943} + \frac{1-0.901}{1.1274 \times 10^{-3}}$$

$$\rho_{in} = 92.45 \text{ kg/m}^3$$

$$\rho_{out} = \rho_f = \frac{1}{v_f} = \frac{1}{1.1274 \times 10^{-3}} = 887\,\text{kg/m}^3$$

To find $\overline{\phi_{go}^2}$ we integrate the expression for ϕ_{go}^2, noting that $f_{TP} = f_{go}$, $x_{in} = 0.05$ to $x_{out} = 0$. Hence

$$\overline{\phi_{go}^2} = \frac{1}{x_{in} - x_{out}} \int_{x_{in}}^{x_{out}} \phi_{go}^2 \, dx$$

$$= \frac{1}{x_{in}} \left[\frac{v_f}{v_g} x_{in} + \frac{x_{in}^2}{2} \left(1 - \frac{v_f}{v_g} \right) \right]$$

$$= \frac{v_f}{v_g} + \frac{x_{in}}{2} \left(1 - \frac{v_f}{v_g} \right)$$

$$= \frac{1.1274 \times 10^{-3}}{0.1943} + \frac{0.05}{2} \left(1 - \frac{1.1274 \times 10^{-3}}{0.1943} \right) = 0.031$$

$$G_m = \frac{\dot{m}}{A} = \frac{\dot{m}}{(\pi/4)D^2} = \frac{0.1}{(\pi/4)(0.02)^2} = 318.3 \text{ kg/m}^2\,\text{s}$$

Substituting in Equation 11.115:

$$\Delta p = (318.3)^2 \left(\frac{1}{887} - \frac{1}{92.45} \right)$$
$$+0.031 \frac{(0.005)(2)}{0.02} 0.1943 \frac{(318.3)^2}{2} = -829.1\,\text{Pa}$$

Thus for condensation the pressure at the exit is larger than at the inlet for low pressure due to the large influence of the acceleration component at low pressure.

11.7 CRITICAL FLOW

11.7.1 Background

The critical flow rate is the maximum flow rate that can be attained by a compressible fluid as it passes from a high-pressure region to a low-pressure region. Although the flow rate of an incompressible fluid from the high-pressure region can be increased by reducing the receiving end pressure, a compressible fluid flow rate reaches a maximum for a certain (critical) receiving end pressure. This situation occurs for both single-phase flow of gases as well as two-phase gas–liquid flows. The observed velocity and pressure of a compressible fluid through a pipe from an upstream pressure (p_o) to a downstream (or back) pressure (p_b) increase as p_b is decreased until a limiting or critical value of p_b defined as p_{cr} is reached. The discharge velocity for all values of p_b below the critical pressure p_{cr} is constant.

The critical flow phenomenon has been studied extensively in single-phase and two-phase systems. It plays an important role in the design of two-phase bypass systems in steam turbine plants and of venting valves in the chemical and power industries. The critical flow conditions during a loss of coolant event from a nuclear power plant provided the motivation behind much of the experimental and theoretical studies of two-phase flows in recent years. Levy [41] presents a comprehensive review of the derivation and accuracy of the models presented in this chapter.

11.7.2 Single-Phase Critical Flow

It is useful to illustrate the analysis of critical flow by first considering single-phase flow in a one-dimensional horizontal tube which discharges fluid from a reservoir to the atmosphere. Take the fluid in the reservoir at pressure and temperature p_o and T_o and the atmospheric or back pressure relative to the tube discharge as p_b, as shown in Figure 11.20.

For a single-phase liquid under steady-state, incompressible, inviscid and adiabatic conditions, we can apply the Bernoulli equation of 9.15a to the flow geometry of Figure 11.20. Since the tube is horizontal and the velocity can be taken negligible at the inlet, we obtain

$$\frac{p_o}{\rho} = \frac{p_b}{\rho} + \frac{\upsilon_b^2}{2} \tag{11.116}$$

Hence

$$\upsilon_b = \sqrt{\frac{2}{\rho}(p_o - p_b)} \tag{11.117a}$$

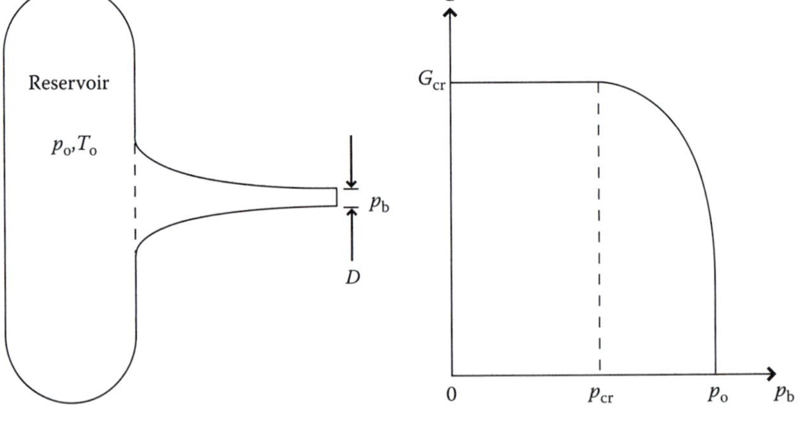

FIGURE 11.20 Critical flow behavior for a compressible fluid.

and

$$G_b = \rho \upsilon_b = \sqrt{2\rho(p_o - p_b)} \quad (11.117b)$$

Hence the mass flux increases as p_b decreases, reaching a maximum at $p_b = 0$, that is

$$G_{bmax} = \sqrt{2\rho p_o} \quad (11.118)$$

For a single-phase gas in horizontal flow under the same steady-state, inviscid, adiabatic, and hence isentropic conditions, from the conservation of energy principle, we obtain

$$0 = h_o - \left(h_b + \frac{\upsilon_b^2}{2} \right) \quad (11.119)$$

since the reservoir velocity is negligible.

Hence since for an ideal gas:

$$\mathrm{d}h = c_p \, \mathrm{d}T \quad (4.120b)$$

the exit velocity, υ_b, equals:

$$\upsilon_b = \sqrt{2(h_o - h_b)} = \sqrt{2c_p(T_o - T_b)} \quad (11.120a)$$

and

$$G_b = \rho_b \upsilon_b = \sqrt{2\rho_b^2 c_p T_o \left(1 - \frac{T_b}{T_o} \right)} \quad (11.120b)$$

For isentropic expansion of an ideal gas:

$$\frac{\rho_b}{\rho_o} = \left(\frac{p_b}{p_o} \right)^{1/\gamma} \quad \text{and} \quad \frac{T_b}{T_o} = \left(\frac{p_b}{p_o} \right)^{(\gamma-1)/\gamma} \quad (11.121a, b)$$

where $\gamma \equiv c_p/c_v$.

Hence, applying Equations 11.121a and 11.121b, Equation 11.120b can be written as

$$G_b = \rho_o \sqrt{2c_p T_o \left[\left(\frac{p_b}{p_o} \right)^{2/\gamma} - \left(\frac{p_b}{p_o} \right)^{(\gamma+1)/\gamma} \right]} \quad (11.122)$$

Again the mass flux increases as the p_b decreases. However, the mass flux of Equation 11.122 exhibits a maximum or critical value for the value of p_b equal to

$$(p_b)_{cr} = p_o\left(\frac{2}{\gamma+1}\right)^{\gamma/(\gamma-1)} \tag{11.123}$$

which can be obtained from differentiating G of Equation 11.122 with respect to the back pressure:

$$\frac{\partial G}{\partial p_b} = 0 \tag{11.124}$$

The corresponding velocity, $\upsilon_{b,cr}$, is obtained from Equation 11.122 by substituting for T_o using Equation 11.121b and replacing all $(p_b/p_o)_{cr}$ terms with the result of Equation 11.123. The resulting value of $\upsilon_{b,cr}$ is

$$\upsilon_{b,cr} = \sqrt{\gamma R_{sp} T_b} \tag{11.125}$$

Note $R_{sp} \equiv R/M$ is the specific gas constant, J/g K (specific value for each gas) where M = molar mass (g/mol) and $R \equiv$ universal gas constant (8.31447 J/mol K).

This velocity magnitude corresponds to the speed of sound, c, for a perfect gas, that is, in general:

$$c = \sqrt{\left(\frac{\delta p}{\delta \rho}\right)_s} = \sqrt{\gamma\left(\frac{\delta p}{\delta \rho}\right)_T} \tag{11.126}$$

and for a perfect gas using its equation of state:

$$p = \rho R_{sp} T \tag{11.127}$$

then

$$c = \sqrt{\gamma R_{sp} T} \tag{11.128}$$

Hence physically if p_b is reduced below p_{cr}, the change does not alter the mass flux in the manner dictated by Equation 11.122. This is because the information travels at the speed of sound, and no faster, in the direction opposite to the flow, and therefore the back pressure reduction cannot propagate up the channel from the exit position.

11.7.3 Two-Phase Critical Flow

The description of two-phase critical flow is affected by whether or not equilibrium between the two phases is maintained. Nonequilibrium conditions, in terms of both the temperature and the velocity differences between the two fluids, play a significant

TABLE 11.6
Principal Two-Phase Critical Flow Models

	Homogeneous (Slip Ratio = 1)	Nonhomogeneous (Slip Ratio > 1)
Thermal equilibrium (equal phasic temperatures)	HEM (Moody [50])	$S = (\rho_f/\rho_g)^{1/3}$ (Moody [49]) $S = (\rho_f/\rho_g)^{1/2}$ (Fauske [20])
Thermal nonequilibrium (unequal phasic temperatures)	Starkman [64] Henry [28]	Richter [60]

role in determining the flow rate at the exit. Hence, both thermal (phasic temperatures) and mechanical (phasic velocities) conditions must be considered. It is instructive to categorize the principal critical flow models with regard to their assumptions on thermal and mechanical equilibrium as illustrated in Table 11.6.

These most commonly used models are presented next, the first being the thermal equilibrium models.

11.7.3.1 Thermal Equilibrium Models

Following treatment of the single-phase gas in Equation 11.119, the enthalpy of the two-phase mixture undergoing a steady-state, adiabatic expansion under thermal equilibrium conditions can be expressed as

$$0 = h_o - \left[xh_g + (1-x)h_f + x\frac{\upsilon_g^2}{2} + (1-x)\frac{\upsilon_f^2}{2} \right] \tag{11.129}$$

where x = the flow quality under the assumption of thermodynamic equilibrium.

From the Second Law of Thermodynamics if the expansion is also inviscid, the entropy can also be written as

$$0 = s_o - [xs_g + (1-x)s_f] \tag{11.130}$$

so that

$$x = \frac{s_o - s_f}{s_g - s_f} \tag{11.131}$$

To obtain the mass flux (G), replace the phasic velocities υ_g and υ_f in Equation 11.129 with the relations involving G, x, ρ, and α of Equation 11.60a and b yielding

$$G = \sqrt{\frac{2[h_o - xh_g - (1-x)h_f]}{\left[(x^3/\rho_g^2\alpha^2) + ((1-x)^3/(\rho_f^2(1-\alpha)^2))\right]}} \tag{11.132a}$$

Now employing the fundamental void fraction, quality, slip ratio relation of Equation 5.55 for α, Equation 11.132a becomes

$$G = \sqrt{\rho''' 2[h_o - xh_g - (1-x)h_f]} \tag{11.132b}$$

where

$$\rho''' = \left\{ \left[\frac{x}{\rho_g} + \frac{(1-x)S}{\rho_f} \right]^2 \left[x + \frac{1-x}{S^2} \right] \right\}^{-1} \tag{11.132c}$$

and

$$S = \frac{\upsilon_g}{\upsilon_f} \tag{11.132d}$$

The mass flux G given in Equation 11.132b is a function of S, x and properties at the back pressure, p_b. From the reservoir state, h_o and s_o are known. Hence x is determined once p_b is fixed. Therefore with an assumed value of slip ratio, the mass flux is obtained as p_b is varied and reaches a maximum value. For the HEM model, the slip ratio would be unity. However, for separated flow, two prominent models have been proposed for the slip ratio which gives the maximum mass flux, that is G_{cr}. These are the models of Moody [49] and Fauske [20] for which the following S values were proposed:

$$\text{Moody model: } S = (\rho_f/\rho_g)^{1/3} \tag{11.133a}$$

$$\text{Fauske model: } S = (\rho_f/\rho_g)^{1/2} \tag{11.133b}$$

where ρ_f and ρ_g are densities at the critical back pressure.

The Moody model is based on maximizing the specific kinetic energy of the mixture with respect to the slip ratio:

$$\frac{\partial}{\partial S}\left[\frac{x\upsilon_g^2}{2} + \frac{(1-x)\upsilon_f^2}{2} \right] = 0 \tag{11.134a}$$

The Fauske model is based on maximizing the flow momentum with respect to the slip ratio:

$$\frac{\partial}{\partial S}[x\upsilon_g + (1-x)\upsilon_f] = 0 \tag{11.134b}$$

The critical mass velocity predicted by these models is available both in terms of reservoir stagnation conditions (p_o, h_o) and nozzle throat conditions (p_c, x_c). The former, the most convenient for design analysis, is illustrated in Figure 11.21 for the Moody model. An algebraic fit to the Moody model in terms of stagnation pressure

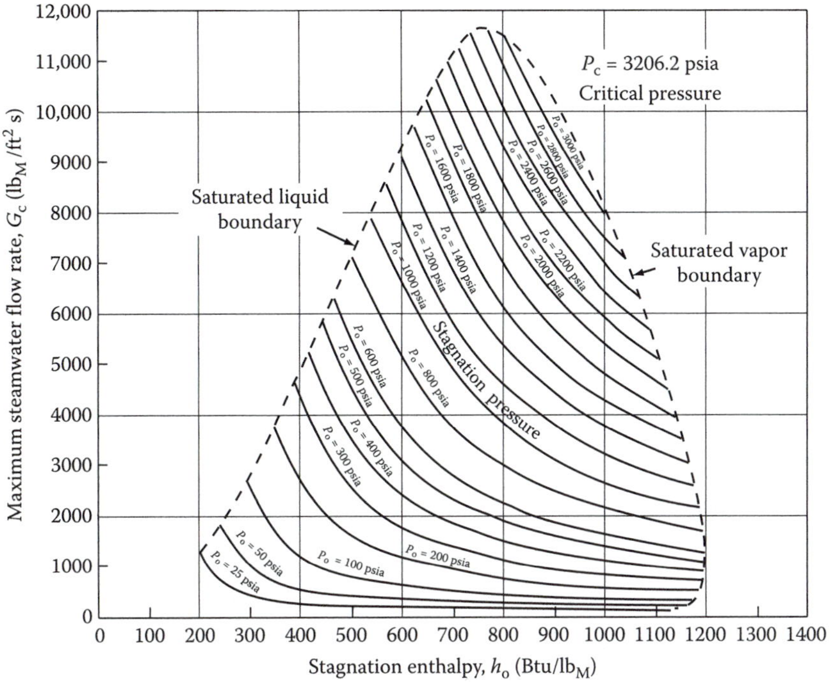

FIGURE 11.21 Maximum steam/water flow rate and local stagnation properties (Moody model). (From Lahey, Jr R. T. and Moody, F. J., *The Thermal-Hydraulics of a Boiling Water Nuclear Reactor*, 2nd Ed. American Nuclear Society, 1993.)

and specific enthalpy has been developed by McFadden et al. [48]. An analogous plot for the HEM model available in Moody [50] is also presented by Whalley [71].

Moody [49] included wall friction effects in the blowdown analysis for long pipes with significant values of $\bar{f}(L/D_H)$ where $\bar{f}$ is an average liquid friction factor over the flashing length, L. He modeled the pipe as an adiabatic length connected to a reservoir by an entrance section idealized as an isentropic nozzle. Moody presented the critical mass flux, G_{cr}, as a function of reservoir stagnation pressure, p_o (and alternately the pipe inlet static pressure p_1), and stagnation enthalpy, h_o, over the range of $\bar{f}(L/D)$ values from 0 to 100. Figure 11.22 presents these results for the case of $\bar{f}(L/D_H) = 3$. The results for $\bar{f}(L/D) = 0$ correspond to the isentropic results of Figure 11.21. Friction reduces the critical mass flux rate relative to the results of Figure 11.21, as Figure 11.22 illustrates. For example, at $h_o = 600$ Btu/lb and $p_0 = 1000$ psi, Figure 11.21 gives a critical mass flux $G_{cr} = 6700$ lb/ft^2 s whereas Figure 11.22 gives a $G_{cr} = 4000$ lb/ft^2 s.

A comparison of the predicted flow rate from various models by Edwards [19] suggests that the homogeneous flow prediction is good for pipe lengths greater than 300 mm and at pressures higher than 2.0 MPa. Moody's model overpredicts the data by a factor of 2, whereas Fauske's model falls in between. When the length of the tube is such that $L/D > 40$, the HEM model appears to do better than the other

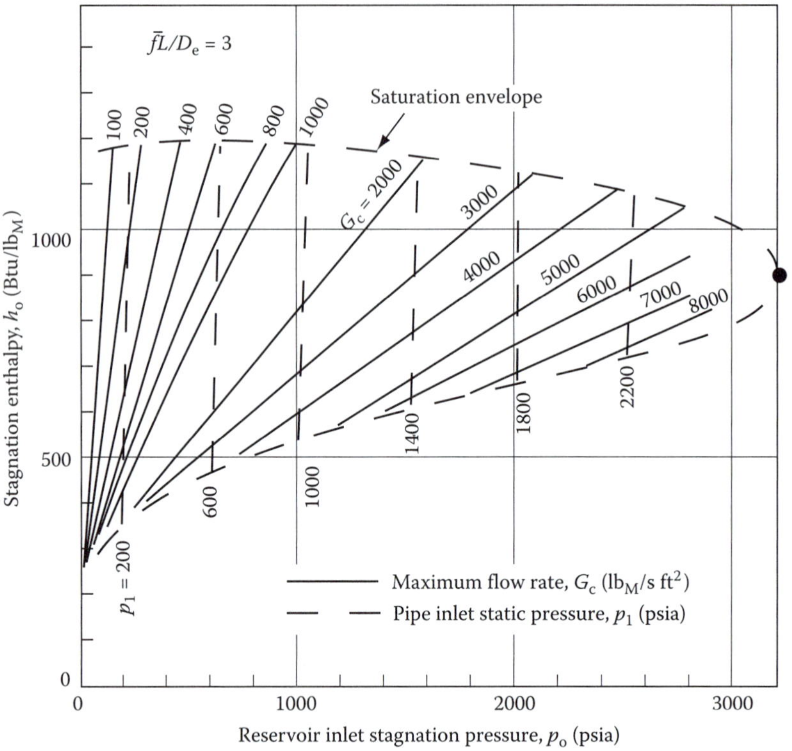

FIGURE 11.22 Pipe maximum steam/water discharge rate. (From Moody, F. J., *J. Heat Transfer*, 88:285, 1966.)

models. Generally, the predictability of critical two-phase flow remains uncertain. The results of one model appear superior for one set of experiments but not others.

11.7.3.2 Thermal Nonequilibrium Models

To attain thermal equilibrium, the fluid must spend adequate time in transit through the discharge flow path for the phase change mechanisms to take effect. This means that the fluid undergoing depressurization flashes sufficiently to increase quality until that of the equilibrium state is attained. For short pipe lengths or upstream fluid states far below saturation, the discharge fluid can remain in a metastable state producing a discharge flowrate much higher than for thermal equilibrium conditions.

When flashing occurs, a certain length of flow in a valve (or pipe) is needed before thermal equilibrium is achieved. In the absence of subcooling and noncondensable gases, the length to achieve equilibrium appears to be on the order of 0.1 m, as illustrated in Table 11.7 [20]. For flow lengths of less than 0.1 m, the discharge rate increases strongly with decreasing length as the degree of nonequilibrium increases, and more of the fluid remains in a liquid state. Figure 11.23

TABLE 11.7
Relaxation Length Observed in Various Critical Flow Experiments with Flashing Liquids

Source	*D* (mm)	*L/D*	*L* (mm)
Fauske (water)	6.35	~16	~100
Sozzi & Sutherland (water)	–12.7	~10	~127
Flinta (water)	–35	~3	~100
Uchida & Nariai (water)	4	~25	~100
Fletcher (Freon 11)	3.2	~33	~105
Van Den Akker et al. (Freon 12)	4	~22	90
Marviken data (water)	500	>0.33	<166

Source: From Fauske, H. K., *Plant Operations Prog.*, 4:132, 1985.

provides a demonstration of the critical pressure dependence on the length to diameter ratio (*L*/*D*).

Since metastable conditions are very difficult to predict, available homogeneous models either bound the nonthermal equilibrium discharge (frozen homogeneous flow) or model the actual but smaller than equilibrium quality empirically (Henry [26], Henry et al. [28]). A prevalent nonhomogeneous nonthermal equilibrium model is that of Richter [60]. These models will be discussed next.

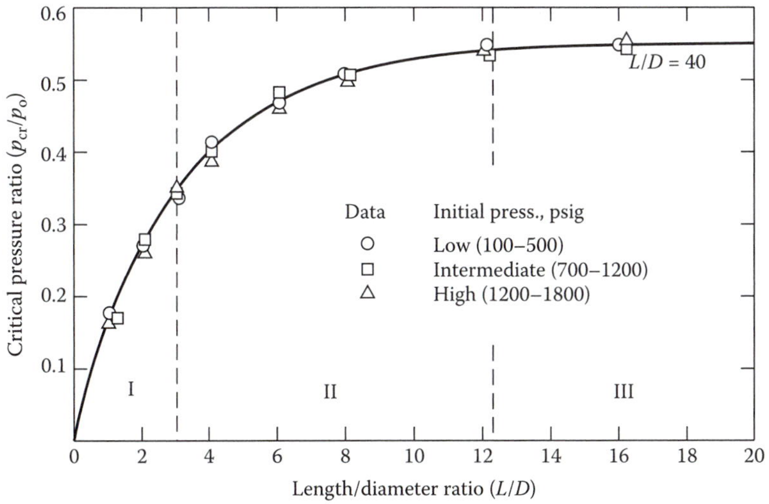

FIGURE 11.23 Critical or choked pressure ratio as a function of *L*/*D*. I, II, and III are the three regions mentioned in the text of the nonequilibrium models. (From Fauske, H. K., *Chem. Eng. Sym. Ser.*, 61:210, 1965.)

11.7.3.2.1 Frozen Homogeneous Flow

In this model, the initial liquid phase is constrained from flashing. The critical mass flux is given following Starkman et al. [64] and Henry and Fauske [27] as

$$G_c = \rho_c \left(\frac{2p_o x_e}{\rho_v} \frac{\gamma}{1+\gamma} \right)^{1/2} \tag{11.135a}$$

where

$$\frac{1}{\rho_c} = (1 - x_e)\frac{1}{\rho_v} + \frac{x_e}{\rho_v}\left(\frac{2}{1+\gamma}\right)^{\frac{1}{1-\gamma}} \tag{11.135b}$$

and x_e is the equilibrium quality, and γ is c_p/c_v of the vapor.

This result flows from the steady-flow energy equation, considering the enthalpy change of the vapor phase and assuming that the critical pressure ratio is the same as in single-phase flow.

11.7.3.2.2 Henry–Fauske Homogeneous Nonequilibrium Model

This model, developed by Henry [26] and Henry et al. [28], describes two-phase critical flow of initially saturated or subcooled liquid. For a constant area duct with sharp-edged inlet, the model is applied for $L/D \geq 12$. For $L/D < 12$, it is assumed that this entrance geometry creates a highly superheated liquid state which persists in a frozen condition. Beyond $L/D = 12$, the mixture takes on a thoroughly dispersed configuration and its quality relaxes exponentially to a long tube value x_{LT} which equals a constant N times the equilibrium quality, x_e, that is

$$x_{LT} = N x_e \tag{11.136}$$

where N is empirically established. An analogous model is proposed for smooth inlet geometries for $L/D \geq 0$.

Hence, the basic feature in this model is the concept that the actual flow quality may lag the equilibrium quality in the flow passage due to thermal nonequilibrium so that

$$\frac{dx}{dp} = N \frac{dx_e}{dp} \tag{11.137a}$$

where

$$N = C_\ell x_e \tag{11.137b}$$

(i.e., N itself is also a product of a constant times x_e for $x < 0.05$ and unity for $x \geq 0.05$).

The empirical constant, C_ℓ, which varies with flow geometry, is chosen to match the data. Since this model is empirically fitted to test data, its predictions are improved compared to the frozen homogeneous model and are higher than the HEM.

11.7.3.2.3 Richter Model

This most comprehensive nonthermal equilibrium model [60] also includes unequal liquid and vapor velocities. It is a two-fluid, five-equation model in which energy conservation is imposed through a mixture equation.

11.7.3.3 Practical Guidelines for Calculations

While the foregoing models are for thermal nonequilibrium conditions, Fauske [21] has provided the following convenient set of equations for calculating flashing flows involving both subcooled and saturated liquids including nonequilibrium and equilibrium flow regimes.

11.7.3.3.1 Equilibrium

For a sufficiently large value of L ($L \geq 0.1$ m), nucleation can be initiated in the pipe and the flow can attain the equilibrium condition. In that case, the critical mass flux is given as

$$G_{\text{cr}} = \frac{h_{\text{fg}}}{v_{\text{fg}}}\sqrt{\frac{1}{Tc_{\text{f}}}} \tag{11.138}$$

where h_{fg} = vaporization enthalpy (J/kg); v_{fg} = change in specific volume (m^3/kg); T = absolute temperature (K); and c_{f} = specific heat of the liquid (J/kg K).

When the properties are evaluated at p_{o}, the value of G_{cr} predicted by Equation 11.138 is called the equilibrium rate model (ERM). The ERM model is compared to experimental data and other models in Figure 11.24.

The effect of subcooling on the discharge rate is obtained by accounting for the increased single-phase pressure drop $[p_{\text{o}} - p_{\text{sat}}(T_{\text{o}})]$ resulting from the subcooling, where the subscript o refers to stagnation conditions. For example, for flow geometries where equilibrium rate conditions prevail for saturated inlet conditions ($L \geq 0.1$ m), the critical flow rate can be expressed as

$$G_{\text{cr}} \simeq \sqrt{2[p_{\text{o}} - p_{\text{sat}}(T_{\text{o}})]\rho_\ell + G_{\text{ERM}}^2} \tag{11.139}$$

If the subcooling is zero, that is $p_{\text{sat}}(T_{\text{o}}) = p_{\text{o}}$, the critical flow rate is approximated by Equation 11.138. Good agreement is illustrated between Equation 11.139 and various data including the large-scale Marviken data (Figures 11.25 and 11.26).

11.7.3.3.2 Nonequilibrium

In the absence of significant frictional losses, Fauske [21] proposed use of Equation 11.138 with the insertion of a nonequilibrium parameter, N, yielding

$$G_{\text{cr}} = \frac{h_{\text{fg}}}{v_{\text{fg}}}\sqrt{\frac{1}{NTc_{\text{f}}}} \tag{11.140}$$

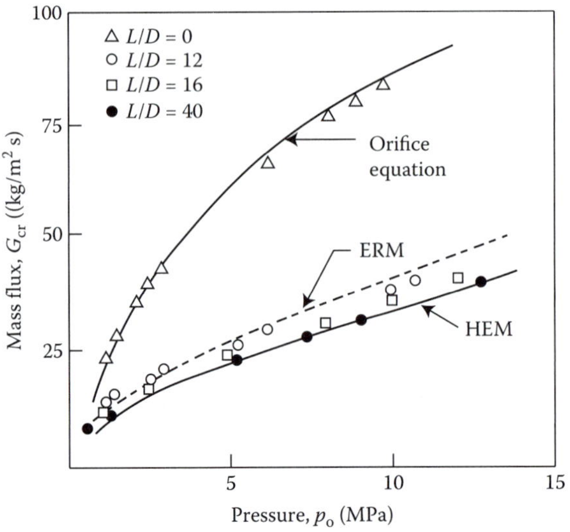

FIGURE 11.24 Typical flashing discharge data of initially saturated water and comparison with analytic models. (From Fauske, H. K., *Plant Operations Prog.*, 4:132, 1985.)

where N is given as

$$N = \frac{h_{fg}^2}{2\Delta p \rho_f K^2 v_{fg}^2 T c_f} + 10L \tag{11.141}$$

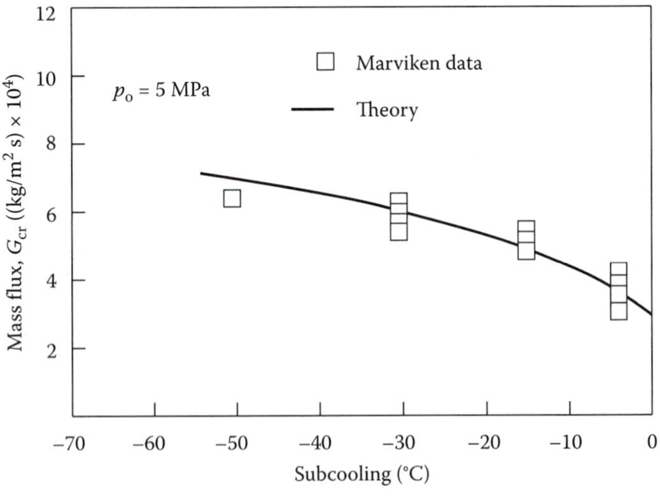

FIGURE 11.25 Comparison of typical Marviken data (D ranging from 200 to 509 mm and L ranging from 290 to 1809 mm) and calculated values based on Equation 11.139. (From Fauske, H. K., *Plant Operations Prog.*, 4:132, 1985.)

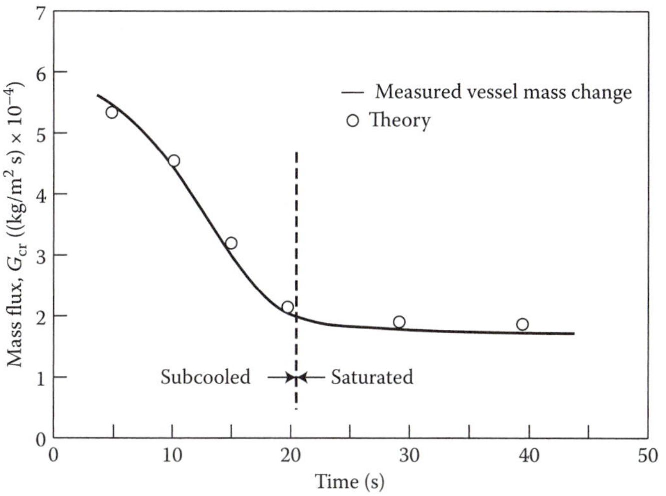

FIGURE 11.26 Comparison of Marviken test 4 ($D = 509$ mm and $L/D = 3.1$) and calculated values based on Equation 11.139. (From Fauske, H. K., *Plant Operations Prog.*, 4:132, 1985.)

where $\Delta p = p_o - p_b$ (Pa); K = discharge coefficient (0.61 for sharp edge); and L = length of tube in meters, ranging from 0 to 0.1 m.

These relations are alternately represented in terms of L/D as follows [20]:

- For $L/D = 0$ since the residence time is zero, flashing is effectively zero, Equation 11.140 reduces to the orifice equation, Equation 11.142, for discharge of incompressible liquid flow, that is

$$G_{cr} = 0.61\sqrt{2\rho_f(p_o - p_b)} \qquad (11.142)$$

- For $0 < L/D < 3$ (region I in Figure 11.23), it is possible to use Equation 11.142 with p_{cr} in place of p_b, that is

$$G_{cr} = 0.61\sqrt{2\rho_f(p_o - p_{cr})} \qquad (11.143)$$

where p_{cr} is obtained from Figure 11.23.

- For $3 < L/D < 12$ (region II in Figure 11.23), the flow is less than that predicted by Equation 11.143 using p_{cr} of Figure 11.23.
- For $40 > L/D > 12$ (region III in Figure 11.23), the flow can be predicted from Equation 11.143 and Figure 11.23.

Example 11.6: Critical Flow in a Tube of $L/D \simeq 3$

PROBLEM

Find the critical mass flux (G_{cr}) for water discharge from a pressurized vessel (5 MPa) through a tube with a diameter (D) = 509 mm, and length (L) = 1580 mm. Consider two cases for the water temperature: (1) saturated, and (2) with a 20°C subcooling. Use Figure 11.21 and the prescriptions of Equations 11.138 and 11.139. Compare the values obtained to those in Figure 11.24 and comment on the results.

SOLUTION

First, find *L/D*:

$$\frac{L}{D} = \frac{1580}{509} = 3.1$$

Therefore, the nonequilibrium case is appropriate here.

Saturated Condition

From Figure 11.23: $p_{cr} \simeq 0.37\ p_o = 0.37 \times 5$ MPa $= 1.85$ MPa (268.3 psia)

At $p = 5$ MPa, the saturated liquid enthalpy is

$$h_o(5\text{ MPa}) = 1154.23\text{ kJ/kg} = 496.26\text{ Btu/lb}$$

From Figure 11.21, we find for $h_o = 496$ Btu/lb and $p_{cr} = 268$ psi:

$$G_{cr} \simeq 2100\text{ lb/ft}^2\text{ s} \simeq 10253\text{ kg/m}^2\text{ s}$$

Now, we try the alternate set of equations. For saturated liquid, use Equation 11.138:

$$G_{cr} = \frac{h_{fg}}{v_{fg}}\sqrt{\frac{1}{Tc_f}}$$

$$h_{fg} = 1640.1\text{ kJ/kg}$$

$$v_{fg} = 0.03815\text{ m}^3\text{/kg}$$

$$T_{sat} = 263.99°\text{C} = 537.14\text{ K}$$

$$c_f \cong 5.0\text{ kJ/kg K}$$

Hence

$$G_{cr} = \frac{1640.1 \times 10^3}{0.03815}\sqrt{\frac{1}{(537.14)(5.0 \times 10^3)}} = 26{,}233\text{ kg/m}^2\text{ s}$$

Subcooled Condition

For 20°C subcooling, use Equation 11.139:

$$G_{cr} \simeq \sqrt{2[p_o - p(T_o)]\rho_\ell + G_{ERM}^2}$$

$$G_{ERM}^2 = (26{,}233\text{kg/m}^2\text{ s})^2$$

$$p_o = 5 \text{ MPa}$$

From $T_o = 243.99°C$, we get $p(T_o) = 3.585$ MPa and

$$\rho_\ell \simeq 809.4 \text{ kg/m}^3 \text{ (at } T_o \text{ and } p_o)$$

Hence

$$\begin{aligned} G_{cr} &\simeq \sqrt{2[5 \times 10^6 - 3.585 \times 10^6]809.4 + (26{,}233)^2} \\ &= 54{,}578 \text{ kg/m}^2 \text{ s} \end{aligned}$$

For $L/D = 3.1$, the value obtained using Figures 11.23 and 11.21 is lower than the value obtained using Equation 11.138. The flow rate for the subcooled water is higher than that for the saturated water owing to the larger fraction of single-phase liquid in the flow. The calculated values agree with those in Figure 11.25.

11.8 TWO-PHASE FLOW INSTABILITIES IN NUCLEAR SYSTEMS

Instabilities in two-phase systems have been classified in many ways. A general approach in nuclear systems has been to consider whether these instabilities are

- Thermal-hydraulic, caused by characteristics of the pressure drop variation with flow rate that are dependent on boiling or condensation in a diabatic flow at a constant power level.
- Thermal-hydraulics with neutronic feedback, where changes in the flow induce changes in the neutron energy spectrum, hence in the fission power level or distribution. This, in turn, causes changes in the flow characteristics.

Each of these classifications is discussed next.

11.8.1 Thermal-Hydraulic Instabilities

Two broad categories of thermal-hydraulic instabilities can be found in the literature: static and dynamic.

In static instabilities, which are more likely to occur at low pressure, changes in flow can occur without any need for external stimulus. They occur when the internal pressure drop change across a flow channel due to a flow increment becomes different from the external pressure drop increment required to induce the same increment in flow. The Ledinegg instability is the most dominant type of static instability. When the pressure drop across a heated channel is externally imposed, a small perturbation in flow can lead to a decrease in flow due to the interplay of the various components of the two-phase flow pressure drop. Another example of static instability is chugging or geysering which occurs if vapor formation in the channel can lead to expulsion of liquid.

In dynamic instabilities, due to expansion of the coolant in a heated channel or to coolant contraction in a cooled channel, changes in volumetric flowrate at the inlet cause much different volumetric flow changes at the exit. This leads to a time-dependent wave reflecting the effects of pressure drop adjustments to respond to the flow changes resulting in oscillatory changes in velocity or pressure at the inlet and the exit of a channel. Two types of instabilities exist in heated channels of nuclear systems: Acoustic Oscillations and Density Wave Oscillations. Acoustic oscillations have high frequencies, typically above 10 Hz, and can reach 1000 Hz depending on the acoustic velocity in the flowing mixture and the channel length. Density waves travel at much lower velocity than acoustic waves and produce oscillations with a frequency around 1 Hz for a 3.5 m long channel. Density wave oscillations in a small part of the core, such as in the hottest assembly, do not significantly alter the neutron spectrum, and the stability of that hot channel can be investigated without neutronic feedback.

11.8.1.1 Ledinegg Instabilities

The channel pressure drop-flow rate behavior, when extended over the entire flow rate range, is not a simple linearly increasing function for two-phase flow. In fact, for a wide range of pressure and heating rates, the $\Delta p - \dot{m}$ characteristic curve has a shape that can lead to multiple solutions and instabilities. Familiarity with these characteristics provides the physical basis essential to understanding and analyzing complex system behavior.

Here, we develop the channel pressure drop-flow rate relation in the friction-dominated regime (rather than a gravity-dominated regime, which is discussed in Chapters 3 and 4 of Volume II). Because heated channels are considered, the channel can be occupied mostly by the gas phase at sufficiently low-flow rates.

For a constant total power to the channel, the friction-dominated regime characteristic is shown in Figure 11.27. In this regime, pressure drop is given as

$$\Delta p_{\text{fric}} = \overline{\phi_{\ell o}^2} \frac{L}{D_e} f_{\ell o} \frac{\dot{m}^2}{2\rho_\ell A_f^2} = \overline{\phi_{go}^2} \frac{L}{D_e} f_{go} \frac{\dot{m}^2}{2\rho_g A_f^2} \quad \text{(from Equation 11.67)}$$

Point A is arbitrarily chosen at a high enough flow rate so that the channel is in liquid single-phase flow throughout its length. The region A–B is characterized principally by a pressure drop that decreases as $\dot{m}^{2-n}$ where n is the exponent of the flow rate in the governing expression for the friction factor, that is

$$f = \frac{C}{\text{Re}^n} = \frac{C}{(GD_e/\mu)^n} \quad (9.88)$$

Additional factors influencing this curve in the region A–B, but not shown in Figure 11.27, are through property (ρ and μ) variations with temperature and perhaps a change in f due to transition from turbulent to laminar flow conditions.

Prior but close to point B, boiling is initiated in the channel. At point B, vapor generation is sufficient to reverse the trend of decreasing Δp with decreasing $\dot{m}$. It occurs when the two-phase conditions lead to an increase in the friction pressure drop due to boiling that overcomes the decrease in pressure drop due to lower mass

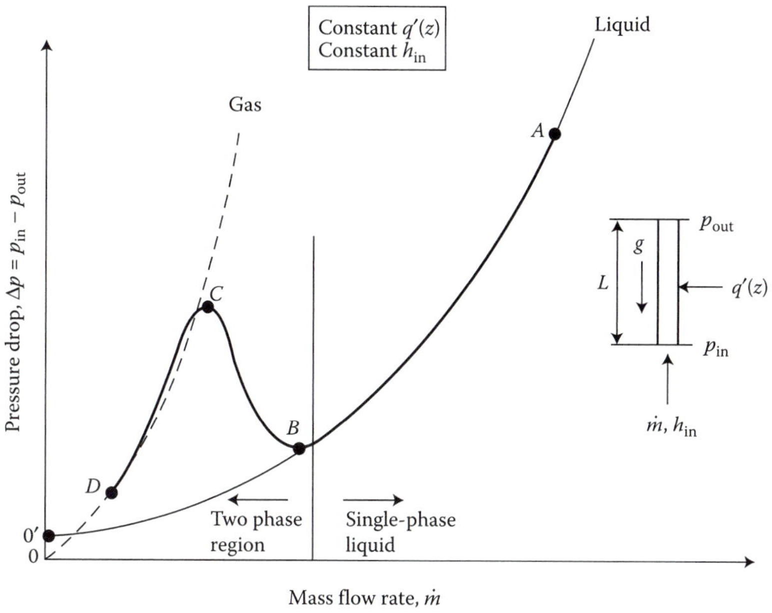

FIGURE 11.27 Pressure drop-flow rate characteristic for a heated channel. Curve 0′-liquid is for a liquid flowing adiabatically in the channel. Curve 0-gas is for a gas flowing adiabatically in the region.

flow. Therefore, this increase in total Δp occurs whenever boiling can be initiated at a relatively large flow rate and hence for relatively high-power input. At point C, where the flow rate is sufficiently low, the channel has considerable vapor flow so that the increase in f with decreasing $\dot{m}$ no longer governs the curve. Under these conditions, the region C–D is also characterized principally by a pressure drop, decreasing as $\dot{m}^{2-n}$. Here, however, the flow condition is mainly a single-phase vapor. Beyond point D, the gravity-dominated region is encountered.

In summary:

$$\text{In region } A \to B, \quad \overline{\phi_{\ell o}^2} = 1 \quad \text{and} \quad \Delta p_{\text{fric}} \propto \dot{m}^{2-n}$$

$$\text{In region } B \to C, \quad \overline{\phi_{\ell o}^2} \gg 1 \quad \text{and} \quad \Delta p_{\text{fric}} \propto \left(1 + \frac{\text{const}}{\dot{m}}\right)\dot{m}^{2-n}$$

$$\text{In region } C \to D, \quad \overline{\phi_{go}^2} = 1 \quad \text{and} \quad \Delta p_{\text{fric}} \propto \dot{m}^{2-n}$$

If the heat flux is low such that the flow rate is also sufficiently low when boiling occurs, the decrease in the gravity head may be larger than the increase in the friction pressure drop, leading to a decrease in the total pressure drop (Figure 11.28). This behavior was demonstrated experimentally by Rameau et al. [57].

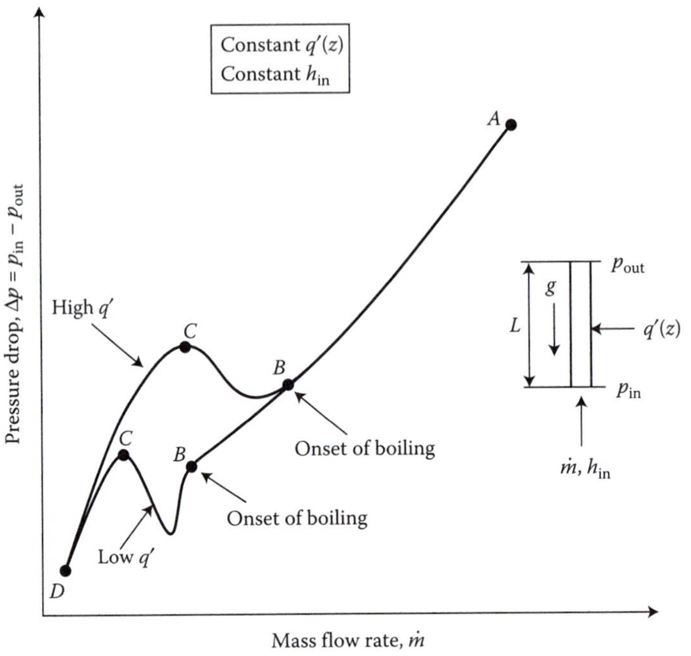

FIGURE 11.28 Pressure drop characteristics for the cases of high power and low power.

Let us now assume that an external pressure drop (Δp_{ex}) boundary condition is imposed on this heated channel; that is p_{in} and p_{out} are held constant. Three levels of Δp can be assumed (Figure 11.29):

$$\Delta p_{ex} > \Delta p_C \tag{11.144a}$$

$$\Delta p_B < \Delta p_{ex} < \Delta p_C \tag{11.144b}$$

$$\Delta p_{ex} < \Delta p_B \tag{11.144c}$$

For the intermediate case only, as indicated by intersection points 1, 2, and 3, multiple channel flow rates are possible. In this case, however, not all intersections are stable conditions. The criterion for stability can be developed by the following perturbation analysis. The fluid in a heated channel accelerates owing to the difference between the imposed external (or boundary) pressure drop (Δp_{ex}) and the intrinsic (mostly friction) pressure drop (Δp_f), as given by

$$I\frac{d\dot{m}}{dt} = \Delta p_{ex} - \Delta p_f \tag{11.145}$$

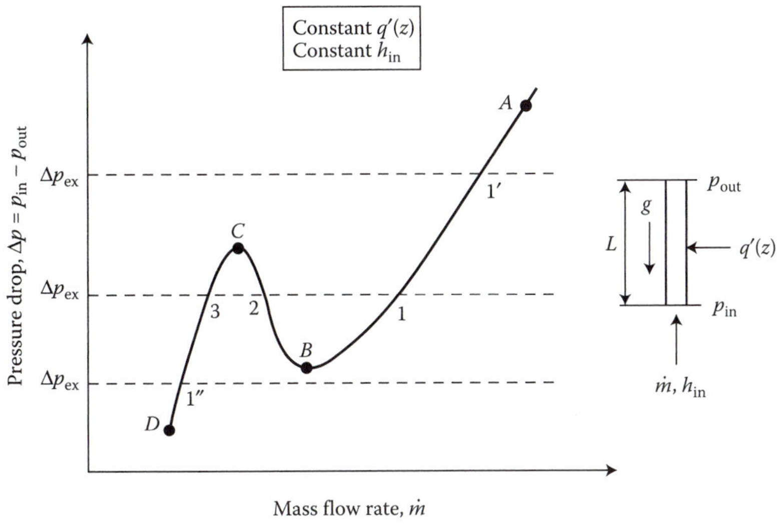

FIGURE 11.29 Stable and unstable operating conditions for prescribed external pressure drop conditions.

where $I = (L/Az)$, the geometric inertia of the fluid in the channel. The small perturbation equation for a small velocity change $\Delta\dot{m}$ is

$$I\frac{\partial\Delta\dot{m}}{\partial t} = \frac{\partial(\Delta p_{\text{ex}})}{\partial\dot{m}}\Delta\dot{m} - \frac{\partial(\Delta p_{\text{f}})}{\partial\dot{m}}\Delta\dot{m} \tag{11.146}$$

Expressing $\Delta\dot{m}$ as

$$\Delta\dot{m} = \varepsilon e^{\omega t} \tag{11.147}$$

Equation 11.146 leads to the result that

$$\omega = \frac{[(\partial(\Delta p_{\text{ex}}))/\partial\dot{m}] - [(\partial(\Delta p_{\text{f}}))/\partial\dot{m}]}{I} \tag{11.148}$$

The criterion of stability is that small perturbations should not grow with time. Hence, ω should be zero or negative. Equation 11.148 thus yields the following criterion for Ledinegg instability avoidance:

$$\frac{\partial(\Delta p_{\text{ex}})}{\partial\dot{m}} < \frac{\partial(\Delta p_{\text{f}})}{\partial\dot{m}} \tag{11.149}$$

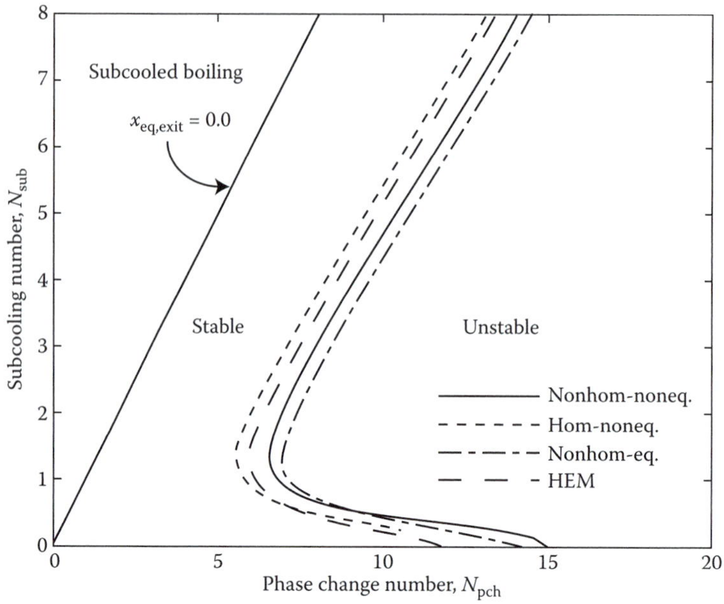

FIGURE 11.30 Comparison of stability boundaries using four different two-phase flow models at 5.0 MPa. (From Zhao, J., Saha, P., and Kazimi, M. S., *Nucl. Tech.*, 161:124–139, 2008.)

The external Δp_{ex} is represented in Figure 11.29 by a horizontal or zero-sloped line. Therefore, from Equation 11.149 a stable operating condition exists only for the positive-sloped region of the friction loss curve, that is segments *A–B* and *C–D*. We conclude that point 2 is not a stable operating point, whereas points 1′, 1, 3, and 1″ are stable operating points.

The behavior at point 2 can be explained physically by considering slight changes in $\dot{m}$ about that point. Suppose $\dot{m}$ is slightly decreased (increased). The friction pressure drop increases (decreases) more than the external Δp, further decelerating (accelerating) the mass flow rate. This deceleration (acceleration) causes the transition to point 3 (1) where the friction pressure drop is once again balanced by the imposed external pressure drop. Unlike point 2, however, points 1 and 3 are stable since they satisfy the stability criterion of Equation 11.149. This perturbation analysis of the well-known Ledinegg instability was presented by Maulbetsch and Griffith [46]. This instability is pronounced when the system pressure is low and therefore the liquid-to-vapor density ratio is high.

11.8.1.2 Density Wave Oscillations

Nondimensional groupings from the coupling of the single- and two-phase regions of a boiling channel were shown by Ishii and Zuber [38] to be important in determining the stability behavior of a heated channel. Saha [62] later added the nonequilibrium effects at low qualities, and showed experimentally that the boundary of stability can be given by two nondimensional numbers: the Subcooling number N_{sub} and the Phase Change Number, N_{pch}. They are defined as

$$N_{sub} = \frac{h_f - h_{in}}{h_{fg}} \times \frac{\rho_f - \rho_g}{\rho_g} \quad (11.150)$$

$$N_{pch} = \frac{(\rho_f - \rho_g)/\rho_f \rho_g}{h_{fg}} \times \frac{q'' P_h}{A_c} \times \frac{L}{\upsilon_{in}} \quad (11.151)$$

Note that the numbers are given as properties of the fluid and the boiling channel. The boiling channel has heated perimeter P_h, heated length L, and flow area A_c.

Figure 11.30 provides the stability boundary for a steam–water channel flow at 5 MPa. Additional analysis performed showed that the pressure level had little effect on the predicted stability boundary. On the other hand, the stability is greatly enhanced by a longer single-phase region (which for a given heat input implies a larger N_{sub} number).

When $N_{sub} > (N_{sub})_{cr}$, that is, at high subcooling, it is possible to use the stability criterion derived by Ishii and Zuber [38] to examine the single channel stability:

$$N_{pch} \le N_{sub} \frac{2\,(K_i + f_m/2D_e^* + K_e)}{1 + 0.5\,(f_m/2D_e^* + K_e)} \quad (11.152)$$

where K_i, K_e are inlet and exit form loss coefficients
f_m is the two-phase mixture friction factor
D_e is the hydraulic diameter
L is the length of the heated section
D_e^* is D_e/L,
$(N_{sub})_{cr}$ is given by

$$(N_{sub})_{cr} = 0.0022\ \mathrm{Pe}\left(\frac{A_c}{P_h L}\right)(N_{pch})_0 \quad \text{if Pe} < 70{,}000 \quad (11.153a)$$

$$= 154\left(\frac{A_c}{P_h L}\right)(N_{pch})_0 \quad \text{if Pe} > 70{,}000 \quad (11.153b)$$

$(N_{pch})_0$ is the value of N_{pch} for zero subcooling at the inlet and for most of the time can be approximated by $(N_{pch})_0 = x_e\,((\rho_f - \rho_g)/\rho_g)$ at the prevailing pressure.

In general, two approaches for the dynamic analysis have been followed: a frequency domain analysis of the stability of small perturbations in inlet conditions and an investigation of oscillations in the time-dependent behavior of a computer solution of the transport equations. The frequency domain equations are often linearized, and thus are considered approximate but adequate for instability onset predictions. However, the approach has the advantage of providing the ratio of the magnitude of sequential oscillations in a system, thus providing a measure for the level of stability margin when the ratio is less than one. On the other hand, a time domain analysis introduces the potential for numerical solution instability as well as the physical one. Thus, both approaches have been used without a clear favorite emerging.

Zhao et al. [72] examined the effects of non homogenous and non equilibrium modeling assumptions on the predicted boundary, and found that the HEM model tends to give a reasonable estimation for the boundary for thermal-hydraulic stability. Zhao et al. extended the approach to cover heating at supercritical conditions. Hu and Kazimi [35] developed a model to account for the flashing effects at low pressure.

11.8.2 Thermal-Hydraulic Instabilities with Neutronic Feedback

Density wave oscillations have been emphasized in analysis of nuclear reactor safety because they may affect the level of neutron moderation and hence could affect the power level in the core which might have an adverse impact on transients. Therefore, in BWRs, where the range of change in coolant density is larger than in PWRs, concerns about coupled oscillations have been part of the safety analysis from the early days of nuclear power. The in-phase flow oscillations of all channels in a core cause core-wide power oscillations which can lead to actuation of reactor scram, as happened in the LaSalle plant in the United States [52]. Concern about instabilities was raised earlier in Europe when several reactors underwent oscillations before actuation of reactor scrams as reported by March-Leuba [44]. For example, in 1978, the TVO-1 plant in Sweden reported oscillations while the total power remained little changed. In 1984, the Caorso BWR plant in Italy oscillated in an out-of-phase mode, where half the core increased in power while the other half decreased. Even if the reactor scram system is actuated only by total power indication, such out-of-phase (or regional) oscillations may not be terminated prior to damaging some of the fuel in the core. Thus, it is possible in principle to consider three types of coupled oscillations:

1. Small region (or hot channel) oscillations, in which no coupling to neutronics is needed for an analysis.
2. In-phase or whole core oscillations, in which the whole core behaves as a unit, and consideration of the loop external to the core is needed to capture changes in pressure drop across the core due to the oscillations.
3. Out-of-phase oscillations, where the total power and flow of the core can be assumed approximately constant, and a fixed pressure drop across the core would prevail.

Besides the concern about oscillations while the plant operates at or near full power, two other conditions have been studied

1. Oscillations of natural convection flow during BWR start up and partial load operation. Because the power to flow ratio is normally high at the maximum natural convection flow conditions, and in the absence of a pump to regulate the flow, stability of that mode of operation has been of concern. In fact, GE BWRs had to adjust the allowable operating region in BWR4 and BWR6 plants to avoid operation at low-flow conditions beyond stable limits.

2. Oscillations during startup of a BWR. Establishing the pressure at its maximum condition prior to raising the power level is one way to operate the plants and avoid this concern. On the other hand, raising the temperature in parallel with pressure avoids concerns about brittle fracture. However, if the power level is raised too quickly, boiling at low pressure (and even flashing due to exceeding the saturation temperature at the end of the chimney above the core) can lead to instabilities.

Good reviews of the various instabilities can be found by Bouré et al. [8], Saha and Zuber [63], Podowski [53], and Rizwan-Uddin [61].

PROBLEMS

11.1. *Methods of describing two-phase flow* (Section 11.2)

Consider vertical flow through a subchannel formed by fuel rods (12.5 mm outer diameter) arranged in a square array (pitch 16.3 mm). Neglect effects of spacers and unheated walls and assume the following:

1. Saturated water at 7.2 MPa (288°C); liquid density = 740 kg/m^3; vapor density = 38 kg/m^3.
2. For vapor fraction determination in the slug flow regime, use the drift flux representation (substituting subchannel hydraulic diameter for tube diameter).
3. Consider a simplified flow regime representation, including bubbly flow, slug flow, and annular flow. The slug-to-bubbly transition occurs at a vapor fraction of 0.15. The slug-to-annular transition occurs at a vapor fraction of 0.75.

Calculate and draw on a graph with axes the superficial vapor velocity, $\{j_v\}$ (ordinate, from 0 to 40 m/s); and the superficial liquid velocity, $\{j_\ell\}$ (abscissa, from 0 to 3 m/s).

Answers:

1. Slug-to-bubbly $\{j_v\} = 0.0235 + 0.22\{j_\ell\}$; Slug-to-annular $\{j_v\} = 0.966 + 9.0\{j_\ell\}$
2. $\{j_v\} = 34.2–19.5\{j_\ell\}$
3. $\{j_\ell\} = 1.51$ m/s
   $\{j_v\} = 4.78$ m/s

11.2. *Flow regime transitions* (Section 11.2)

Line B of the flow regime map of Taitel et al. [67] is given as Equation 11.10. Derive this equation assuming that, because of turbulence, a Taylor bubble cannot exist with a diameter larger than:

$$d_{\max} = k\left(\frac{\sigma}{\rho_\ell}\right)^{3/5}(\varepsilon)^{-2/5}$$

where $k = 1.14$

$$\varepsilon = \left|\frac{dp}{dz}\right|\frac{j}{\rho_m}$$

Use the friction factor $(f) = 0.046\ (jD/\nu)^{-0.2}$ and the fact that for small bubbles the critical diameter that can be supported by the surface tension is given by

$$d_{cr} = \left[\frac{0.4\sigma}{(\rho_\ell - \rho_g)g}\right]^{0.5}$$

11.3. *Regime map for vertical flow* (Section 11.2)
Construct a flow regime map based on coordinates $\{j_v\}$ and $\{j_\ell\}$ using the transition criteria of Taitel et al. [67] for the secondary side of the vertical steam generator.
Assume that the length of the steam generator is 3.7 m and the characteristic hydraulic diameter is the volumetric hydraulic diameter:

$$D_v = \frac{4(\text{net free volume})}{\text{friction surface}} = 0.134\,\text{m}$$

Operation conditions
Saturated water at 282°C
Liquid density = 747 kg/m^3
Vapor density = 34 kg/m^3
Surface tension = 17.6×10^{-3} N/m
Answers:

Bubbly-to-slug:	$\{j_\ell\} = 3.0\ \{j_v\} = 0.14$
Bubbly-to-dispersed bubbly:	$\{j_\ell\} + \{j_v\} = 5.554$
Disperse bubble-to-slug:	$\{j_\ell\} = 0.923\ \{j_v\}$
Slug-to-churn:	$\{j_\ell\} + \{j_v\} = 0.527$
Churn-to-annular:	$\{j_v\} = 1.78$

11.4. *Comparison of correlations for flooding* (Section 11.2)
1. It was experimentally verified that for tubes with a diameter larger than about 6.35 cm flooding is independent of diameter for air–water mixtures at low-pressure conditions. Test this assertion against the Wallis [69] correlation and the Pushkin and Sorokin [56] correlation (Equation 11.28) by finding the diameter at which they are equal at atmospheric pressure.
2. Repeat question 1 for a saturated steam–water mixture at 6.9 MPa.

Note: For the $j_\ell = 0$ condition, the terms flooding and flow reversal are effectively synonymous.

Answers:
1. $D = 3.17$ cm
2. $D = 2.08$ cm

The Taylor bubble diameter should be less than the tube diameter if the tube diameter is not to influence the flooding. At extremely low pressure situations, the limiting value of bubble diameter can reach 6.35 cm. Thus, the assertion that for $D > 6.35$ cm. Flooding is independent of diameter is applicable.

11.5. *Impact of slip model on the predicted void fraction* (Section 11.4)
In a water channel, the flow conditions are such that
Mass flow rate: $\dot{m} = 0.29$ kg/s
Flow area: $A = 1.5 \times 10^{-4}$ m^2
Flow quality: $x = 0.15$
Operating pressure: $p = 7.2$ MPa
Calculate the void fraction using (1) the HEM model, (2) Bankoff's slip correlation, and (3) the drift flux model using Dix's correlation for C_o and V_{vj} calculated assuming churn flow conditions.

Answers:
1. $\{\alpha\} = 0.775$
2. $\{\alpha\} = 0.631$
3. $\{\alpha\} = 0.703$

11.6. *Level swell in a vessel due to two-phase conditions* (Section 11.4)
Compute the level swell in a cylindrical vessel with volumetric heat generation under thermodynamic equilibrium and steady-state conditions. The vessel is filled with vertical fuel rods such that $D_e = 0.0122$ m. The collapsed water level is 2.13 m.
Operating conditions
No inlet flow to the vessel
$p = 5.516$ MPa
Q''' (volumetric heat source) $= 4.14 \times 10^6$ W/m^3
Water properties
$\sigma = 0.07$ N/m
$h_{fg} = 1.6 \times 10^6$ J/kg
$\rho_g = 28.14$ kg/m^3
$\rho_g = 767.5$ kg/m^3
Assumptions
Flow regime selected values from Table 11.3 *(n, V_∞)*
Bubbly large bubbles
Select appropriate transition for flow regimes so that there is a continuous shape for the α versus z curve. For these conditions, the following result for α versus z is known

$$\{\alpha\}V_\infty(1-\{\alpha\})^{n-1} = \frac{Q'''}{h_{fg}\rho_g}z$$

Answer:
$H_{swell} = 3.02$ m

11.7. *Void fraction evaluation using the EPRI correlation* (Section 11.4)
Using the EPRI correlation, evaluate the void fraction for a saturated water–steam mixture at a total mass flux, G, of 40 kg/m^2 s at 50% quality in a tube of diameter 20 mm at 7.45 MPa pressure. Use the fluid properties of Table 11.8.

Answer:
$\{\alpha\} = 0.62$

11.8. *Void, quality and pressure drop problem* (Sections 11.4 and 11.5)
Consider an adiabatic water channel 3 m long and 1 cm in diameter operating in homogeneous flow at 7.4 MPa pressure with a void (steam) distribution as shown in Figure 11.31. The total flow rate is 0.3 kg/s. Take the liquid viscosity at the operating conditions as 8.7×10^{-5} kg/m s.

TABLE 11.8
Fluid Properties for Problem 11.7

$\rho_f = 731.76\ \text{kg/m}^3$
$\rho_g = 39.18\ \text{kg/m}^3$
$\mu_f = 89.63\ \mu\text{Pa s}$
$\mu_g = 19.16\ \mu\text{Pa s}$
$\sigma = 0.0166\ \text{N/m}$

A. Find the values of $\{\alpha\}$, $\{\beta\}$ and x for the channel, that is, volume averaged values.
B. Which values of Part A would change if the local flow velocities of the two phases remain equal but the flow rate was reduced by a factor of 2 but the flow regime was unchanged?
C. Compute the pressure loss within the 3 m length.

Answers:
A. $\{\alpha\} = \{\beta\} = 0.041$ x = 0.0023
B. None
C. $\Delta p_{\text{total}} = 63.3$ kPa

11.9. *Calculation of a pipe's diameter for a specific pressure drop* (Section 11.5)
For the vertical plate riser shown in Figure 11.32, calculate the steam-generation rate and the riser diameter necessary for operation at the flow conditions given below:
Geometry: downcomer height = riser height = 4.6 m
Flow conditions:
T (steam) = 284.8°C
T (feed) = 226.7°C
Steam pressure = 6.7 MPa
Thermal power = 856 MW_{th} (from heater primary side to fluid in riser)
Assumptions
Homogenous flow model is applicable.

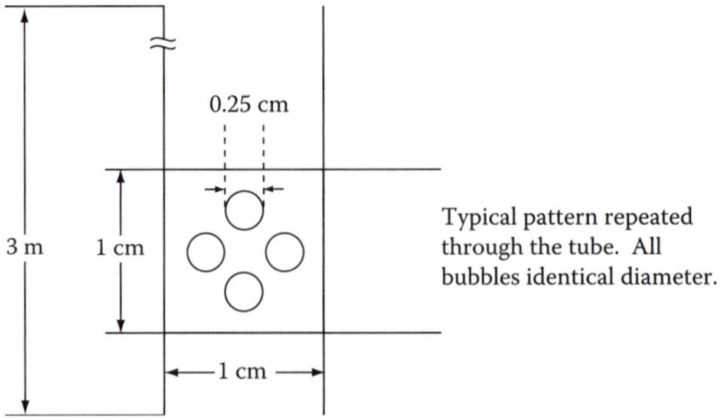

FIGURE 11.31 Void distribution.

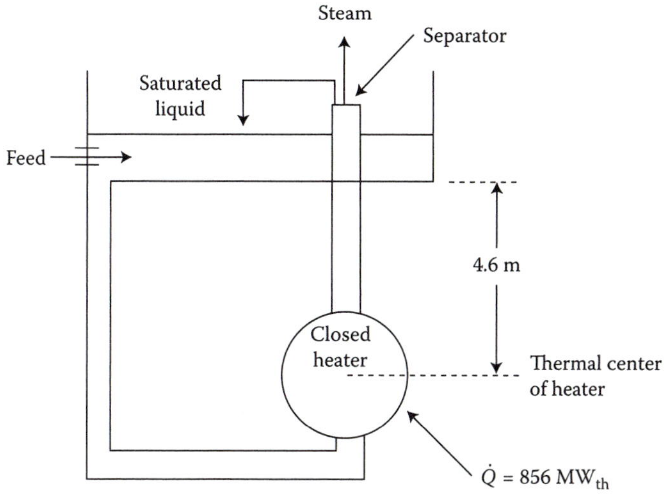

FIGURE 11.32 Vertical user schematic for Problem 11.9. Closed heater primary side not shown.

Friction losses in the downcomer, upper plenum, lower plenum, and heater are negligible.
Riser and downcomer are adiabatic.
Quality at heater exit (x_{out}) is 0.10.

Answers:
$D = 0.583$ m
Steam flow = 475 kg/s

11.10. *Flow dynamics of nanofluids* (Section 11.5)

Nanofluids are suspensions of solid nanoparticles in a base fluid (e.g., water), and are being investigated at MIT for their potential as coolants in nuclear systems. The flow behavior of nanofluids can be analyzed as a two-phase liquid/solid mixture. Since the particles are so small, they can be assumed to move homogeneously with the base fluid, that is the slip ratio is equal to one.

Consider a water-based nanofluid with alumina nanoparticles. It flows at steady state through a tube of 2.5 cm diameter. The nanofluid volumetric flowrate is 400 cm^3/s and the nanoparticle volumetric fraction is 0.05.

1. Find the mass flow rate of the nanoparticles and the total mass flow rate of the nanofluid.
2. Find the pressure gradient within the tube if the flow direction is vertically downward. To calculate the friction pressure gradient, assume fully developed flow, zero surface roughness and $f_{TP} = f_{\ell o}$
3. If the nanoparticles were made of a material denser than alumina, how would the components of the pressure gradient (friction, gravity, etc.) change? Assume that the nanofluid volumetric flow rate and the nanoparticle volumetric fraction are held at 400 cm^3/s and 0.05, respectively. The slip ratio is still equal to one. Provide a qualitative answer.

Properties

Water density 1 g/cm^3, viscosity 10^{-3} Pa s

Alumina density 4 g/cm^3, specific heat 780 J/kg K

Answers:

1. $\dot{m}_{np} = 80\,g/s \qquad \dot{m}_{tot} = 460\,g/s$
2. $(dp/dz)_{tot} = -\ 10.9$ kPa/m

11.11. *HEM pressure loss problem* (Section 11.5)

Consider a 3 m long water channel of circular cross-sectional area 1.5×10^{-4} m^2 operating at the following conditions:

$\dot{m} = 0.29$ kg/s

$p = 7.2$ MPa

Compute the pressure loss under homogeneous equilibrium assumptions for the following additional conditions:

a. Adiabatic channel with inlet flow quality of 0.15.
b. Uniform axial heat flux of sufficient magnitude to heat the entering saturated coolant ($x_{in} = 0$) to an exit quality of 0.15.

Answers:

a. $\Delta p_{total} = 36.6$ kPa
b. $\Delta p_{total} = 44.0$ kPa

11.12. *Analysis of a liquid metal-cooled reactor vessel* (Section 11.6)

Fast reactors have attracted renewed attention within the nuclear community because of their ability to consume the actinides from the LWR-spent fuel. Consider the vessel of a liquid–lead-cooled fast reactor, which is made of stainless steel with the dimensions shown in Figure 11.33. The top of the vessel is filled with a cover gas (nitrogen) whose operating temperature and pressure are 400°C and 0.5 MPa, respectively. The pressure outside the vessel is 0.1 MPa. Relevant material and fluid properties are given in Table 11.9.

1. A violent earthquake causes a 10 cm^2 crack in the vessel. Calculate the mass flow rate through the crack, immediately after the earthquake, for the following two cases:
   1. The crack occurs at the very bottom of the vessel.
   2. The crack occurs at the top of the vessel, that is in the cover gas region.

   (The flow through the crack can be treated as steady-state, adiabatic, and inviscid in both cases. The pressure outside the vessel

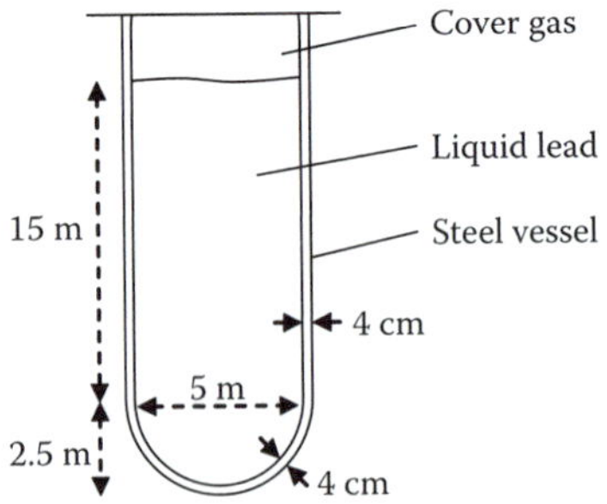

FIGURE 11.33 Schematic and dimensions of the vessel.

TABLE 11.9
Properties for Problem 11.12

Stainless steel	Maximum allowable stress intensity factor, $(S_m) = 138$ MPa at 400 °C, $\rho = 7500$ kg/m^3
Lead	$\rho = 10{,}500$ kg/m^3, $\mu = 1.6 \times 10^{-3}$ Pa s, boiling point 1750°C
Nitrogen	$c_p = 1039$ J/kg K, $R_{sp} = 297$ J/kg K, $\gamma = 1.4$

can be assumed constant at 0.1 MPa. Liquid lead can be treated as an incompressible liquid and nitrogen as a perfect gas.)

2. Which of the two cases considered in Part 1; would you judge more dangerous from a safety viewpoint?

Answers

1. $\dot{m} = 215$ kg/s
2. $\dot{m} = 0.766$ kg/s

REFERENCES

1. Armand, A. A. and Treshchev, G. G., Investigation of the resistance during the movement of steam-water mixtures in a heated boiler pipe at high pressures. 1947 *AERE Lib/Trans.*, 816, 1959.
2. Bankoff, S. G., A variable density single fluid model for two-phase flow with particular reference to steam-water flow. *J. Heat Transfer*, 82:265, 1960.
3. Bankoff, S. G. and Lee, S. C., *A Critical Review of the Flooding Literature.* NUREG/CR-3060 R2, 1983.
4. Baroczy, C. J., A systematic correlation for two-phase pressure drop. *Chem. Eng. Prog. Symp. Ser.,* 62(44):232–249, 1966.
5. Beattie, D. R. H. and Whalley, P. B., A simple two-phase frictional pressure drop calculations method. *Int. J. Multiph. Flow*, 8(1):83–87, 1982.
6. Bestion, D., *Interfacial Friction Determination for the 1-D-6 Equation 2 Fluid Model used in the Cathare Code.* European 2 phase Flow Group Meeting, Marchwood, 1985.
7. Bestion, D., The physical closure laws in the CATHARE code. *Nucl. Eng. Des.*, 124:229–245, 1990.
8. Bouré, J. A., Bergles, A. F., and Tang, L. S., Review of two-phase flow instability. *Nucl. Eng. Des.*, 25:165–192, 1973.
9. Chexal, B. et al., *Two-Phase Pressure Drop Technology for Design and Analysis.* Palo Alto, CA: Electric Power Research Institute, EPRI TR-113189, 1999.
10. Chexal, B., Lellouche, G., Horowitz, J., and Healzer, J., A void fraction correlation for generalized applications. *Prog. Nucl. Energy*, 27(4):255–295, 1992.
11. Chexal, B., Merilo, M., Maulbetsch, J., Horowitz, J., Harrison, J., Westacott, J., Peterson, C., Kastner, W., and Schmidt, H., *Void Fraction Technology for Design and Analysis.* Palo Alto, CA: Electric Power Research Institute, EPRI TR-106326, 1997.
12. Chisholm, D., A theoretical basis for the Lockhart-Martinelli correlation for two-phase flow. *Int. J. Heat Mass Transfer*, 10:1767–1778, 1967.
13. Chisholm, D., Pressure gradients due to friction during the flow of evaporating two-phase mixtures in smooth tubes and channels. *Int. J. Heat Mass Transfer*, 16:347–358, 1973.

14. Chisholm, D. and Sutherland, L. A., Prediction of pressure gradients in pipeline systems during two-phase flow. Paper 4, Presented at Symposium on Fluid Mechanics and Measurements in Two-Phase Flow Systems, University of Leeds, 1969.
15. Cicchitti, A. et al., Two-phase cooling experiments—pressure drop, heat transfer and burnout measurements. *Energ. Nucl.*, 7:407–425, 1960.
16. Coddington, P. and Macian, R., A study of the performance of void fraction correlations used in the context of drift-flux two-phase flow models. *Nucl. Eng. Des.*, 215:199–216, 2002.
17. Dix, G. E., *Vapor Void Fractions for Forced Convection with Subcooled Boiling at Low Flow Rates*. NEDO-10491. General Electric Company, 1971.
18. Dukler, A. E. et al., Pressure drop and hold-up in two-phase flow: Part A—A comparison of existing correlations. Part B—An approach through similarity analysis. Paper Presented at the AIChE Meeting, Chicago, 1962. Also *AIChE J.*, 10:38, 1964.
19. Edwards, A. R., *Conduction Controlled Flashing of a Fluid, and the Prediction of Critical Flow Rates in One Dimensional Systems*. UKAEA Report AHSB (5) R147, Risley, England, 1968.
20. Fauske, H. K., The discharge of saturated water through tubes. *Chem. Eng. Sym. Ser.*, 61:210, 1965.
21. Fauske, H. K., Flashing flows—some practical guidelines for emergency releases. *Plant Operations Prog.*, 4:132, 1985.
22. Friedel, L., Improved friction pressure drop correlations for horizontal and vertical two-phase flow. *European Two-phase Flow Group Meeting, Ispra, Italy*, Paper E2, 1979.
23. Friedel, L., Improved friction pressure drop correlations for horizontal and vertical two-phase pipe flow. *3R Int., J. Piping Eng. Pract.*, 18(7):485–491, July 1979.
24. Gaspari, G. P., Lombardi, C., and Peterlongo, G., Pressure drops in steam-water mixtures, round tubes—vertical upflow. CISE-R83, 1964.
25. Harmathy, T. Z., Velocity of large drops and bubbles in media of infinite or restricted extent. *AIChE J.*, 6:281, 1960.
26. Henry, R. E., The two-phase critical discharge of initially saturated or subcooled liquid. *Nucl. Sci. Eng.*, 41:336–342, 1970.
27. Henry, R. E. and Fauske, H. K., The two-phase critical heat flow of one-component mixtures in nozzles, orifices and short tubes. *J. Heat Transfer*, 93:179–187, 1971.
28. Henry, R. E., Fauske, H. K., and McComas, S. T., Two-phase flow at low qualities, I: Expermental; II. Analysis. *Nucl. Sci. Eng.*, 41:79–98, 1970.
29. Hewitt, G. F., Multiphase fluid flow and pressure drop. In *Heat Exchanger Design Handbook*, Vol. 2, 2.3.2–14, Begell House Inc., New York, NY, 2008.
30. Hewitt, G. F., Pressure gradient in upward adiabatic flows of gas-liquid mixtures in vertical pipes. Engineering Sciences Data Unit, Data Item 04006, Feb. 2004.
31. Hewitt, G. F. and Roberts, D. N., *Studies of Two-Phase Flow Patterns by Simultaneous X-Ray and Flash Photography*. AERE-M2159, 1969.
32. Hewitt, G. F. and Wallis, G. B., *Flooding and Associated Phenomena in Falling Film Flow in a Tube*. AERE-R, 4022, 1963.
33. Hewitt, G. F., Lacey, P. M. C., and Nicholls, B., *Transitions in Film Flow in a Vertical Flow*. AERE-R4614, 1965.
34. Hinze, J. V., Fundamentals of the hydrodynamic mechanisms of splitting in dispersion processes. *AIChE J.*, 1:289, 1955.
35. Hu, R. and Kazimi, M. S., Thermal hydraulic stability of natural circulation BWR under startup—Flashing Effects. Paper 9139, ICAPP '09, Tokyo, May 2009.
36. Isbin, H. S., Two phase steam-water pressure drops, *Chem Eng. Prog. Symp. Ser.*, 55:75–84, 1959.
37. Ishii, M., *One-Dimensional Drift Flux Model and Constitutive Equations for Relative Motion between Phases in Various Two-Phase Flow Regimes*. ANL-77-47, 1977.

38. Ishii, M. and Zuber, N., Thermally induced flow instability in two-phase mixtures. Paper B5.11, 4th Int. Heat Transfer Conf., Paris, 1973.
39. Jones, A. B., *Hydrodynamic Stability of a Boiling Channel.* KAPL-2170, Knolls Atomic Power Laboratory, 1961.
40. Lahey, R. T. Jr. and Moody, F. J., *The Thermal-Hydraulics of a Boiling Water Nuclear Reactor*, 2nd Ed. American Nuclear Society, La Grange Park, IL, 1993.
41. Levy, S., *Two-Phase Flow in Complex Systems.* Wiley & Sons, New York, NY, 1999.
42. Lockhart, R. W. and Martinelli, R. C., Proposed correlation of data for isothermal two-phase, two-component flow in pipes. *Chem Eng. Prog.*, 45(1):39–48, 1949.
43. Mandhane, J. M., Gregory, G. A., and Aziz, K., A flow pattern map for gas-liquid flow in horizontal pipes. *Int. J. Multiph. Flow*, 1:537, 1974.
44. March-Leuba, J. and M. Rey, J. M., Coupled thermo hydraulic-neutronic instabilities in boiling water nuclear reactors: A review of the state of the art. *Nucl. Eng. Des.*, 145:97–111, 1993.
45. Martinelli, R. C. and Nelson, D. B., Prediction of pressure drop during forced circulation boiling of water. *Trans. ASME*, 70:695–702, 1948.
46. Maulbetsch, J. S. and Griffith, P., *A Study of System Induced Instabilities in Forced Convection Flows with Subcooled Boiling*. MIT, Eng. Prof. Lab. Report 5382–35, 1965.
47. McAdams, W. H., et al., Vaporization inside horizontal tubes. II. Benzene-oil mixtures. *Trans. ASME*, 64:193–200, 1942.
48. McFadden, J. H. et al., *RETRAN-02, A Program for Transient Thermal Hydraulic Analysis of Complex Fluid Flow Systems*, EPRI NP-1850-CCMA, Vol. 1, Rev. 2, 1984.
49. Moody, F. J., Maximum two-phase vessel blowdown from pipes. *J. Heat Transfer*, 88:285, 1966.
50. Moody, F. J., Maximum discharge rate of liquid vapor mixtures from vessels. *Proc. ASME Symp. on Non-equilibrium Two-Phase Flows*, Vol. 27–36, American Society of Mechanical Engineers, New York, NY, 1975.
51. Nicklin, D. J. and Davidson, J. F., The onset of instability on two phase slug flow. In *Proceedings Symposium on Two Phase Flow.* Paper 4, Institute Mechanical Engineering, London, 1962.
52. NRC Bulletin 88–07 Supplement 1, Power Oscillations in Boiling Water Reactors. 1988.
53. Podowski, M. Z., Instabilities in boiling systems, 3rd Int. Top. Mtg. on Nuclear Plant Thermal Hydraulics and Operations, pp. 71–98, Seoul, Korea, 1988.
54. Porteous, A., Prediction of the upper limit of the slug flow regime. *Br. Chem. Eng.*, 14(9):117–119, 1969.
55. Premoli, A., Di Francesco, D., and Prina, A., A dimensionless correlation for determining the density of two-phase mixtures. *La Termotecnica*, 25(1):17–26, 1971. (First appeared as Premoli, A., Di Francesco, D., and Prina, A., An empirical correlation for evaluating two-phase mixture density under adiabatic conditions. *European Two-Phase Flow Group Meeting*, Milan, 1970.) Also personal communication with A. Premoli, Dec. 2009.
56. Pushkin, O. L. and Sorokin, Y. L., Breakdown of liquid film motion in vertical tubes. *Heat Transfer Sov. Res.* 1:56, 1969.
57. Rameau, B., Seiller, J. M., and Lee, K. W., Low heat flux flows sodium boiling experimental results and analysis. Presented at the ASME Winter Annual Meeting, New Orleans, 1984.
58. Ranson, V. H. et al., *RELAP5/MOD1 Code Manual*, Vols. 1 & 2, NUREG/CR-1826, EGG-2070, March 1982.
59. RELAP5–3D©, *Code Manual. Vol. IV: Models and Correlations.* Idaho National Laboratory, April 2005.
60. Richter, H. J., Separated two-phase flow model: Application to critical two-phase flow. *Int. J. Multiph. Flow*, 9(5):511–530, 1983.

61. Rizwan-uddin, Physics of density wave oscillations, Int. Conf. New Trends in Nuclear System Thermal Hydraulics, Pisa, May 30–June 2, 1994.
62. Saha, P., Thermally Induced Two-Phase Flow Instabilities, Including Effects of Thermal Non-equilibrium Between the Phases. Ph.D. thesis, Georgia Inst. of Tech., 1974.
63. Saha, P. and Zuber, N., An analytical study of thermally induced two-phase flow instabilities, including the effect of thermal non-equilibrium. *Int. J. Heat Mass Transfer*, 2:415–426, 1978.
64. Starkman, E. S. et al., Expansion of a very low quality two-phase fluid through a convergent-divergent nozzle. *J. Basic Eng.*, 86:247–256, 1964.
65. Taitel, Y. and Dukler, A. E., A model for predicting flow regimes transition in horizontal and near horizontal gas-liquid flow. *AIChE J.*, 22:47, 1976.
66. Taitel, Y. and Dukler, A. E., Flow regime transitions for vertical upward gas-liquid flow: A preliminary approach. Presented at the AIChE 70th Annual Meeting, New York, NY, 1977.
67. Taitel, Y., Bornea, D., and Dukler, A. E., Modelling flow pattern transitions for steady upward gas-liquid flow in vertical tubes. *AIChE J.*, 26:345, 1980.
68. Thom, J. R. S., Prediction of pressure drop during forced circulation boiling of water. *Int. J. Heat Mass Transfer*, 7:709–724, 1964.
69. Wallis, G. B., *One Dimensional, Two-Phase Flow.* McGraw-Hill, New York, NY, 1969.
70. Wallis, G. B. and Kuo, J. T., The behavior of gas-liquid interface in vertical tubes. *Int. J. Multiph. Flow*, 2:521, 1976.
71. Whalley, P. B., *Boiling, Condensation, and Gas-Liquid Flow*. Clarendon Press, Oxford, UK, 1987.
72. Zhao, J., Saha, P., and Kazimi, M. S., Core-wide (in-phase) stability of supercritical water reactors: II, Comparison with boiling water reactors. *Nucl. Tech.*, 161:124–139, 2008.
73. Zuber, N. and Findlay, J. A., Average volumetric concentration in two-phase flow systems. *J. Heat Transfer*, 87:453–468, 1965.

12 Pool Boiling

12.1 INTRODUCTION

Boiling heat transfer is the operating mode of heat transfer in BWR cores and may also occur in certain conditions in PWR cores. Furthermore, it is present in the steam generators and steam equipment of practically all nuclear plants. Therefore, it is of interest to the analysis of normal operating thermal conditions in reactor plants. Additionally, the provision of sufficient margin between the anticipated transient heat fluxes and the critical boiling heat fluxes is a major factor in the designs of LWR cores, as are the two-phase coolant thermal conditions under postulated loss of coolant events.

In this chapter and Chapter 13, the characteristics of boiling heat transfer are presented under nonflow or pool boiling conditions and flow boiling conditions, respectively. This chapter presents the fundamentals of nucleation, which transforms one phase into another, and liquid or vapor phase contact and stability on hot surfaces that are prerequisites to understanding both pool and flow boiling mechanisms. Finally, presentation of the condensation process has been placed at the end of this pool boiling discussion since laminar condensation heat transfer correlations have an analytic form analogous to those for film boiling.

The boiling process for both pool and flow boiling involves many phenomena, each of which has been described by many proposed models and correlations of varying accuracy. In this chapter, models and correlations that best describe the relevant physical processes have been selected for presentation. For results applicable to special conditions of fluid pressure and subcooling as well as surface geometry, orientation and conditions, specialized handbooks such as Hewitt [20] or Kandlikar et al. [25] and review articles cited in these sources should be consulted.

12.2 NUCLEATION

In order to describe the boiling process, we first determine the bubble radius that can exist in equilibrium with a superheated liquid in a liquid pool. Then we apply this result to nucleation in pool boiling by considering vapor trapping and retention in the heated surface microcavities and finally vapor growth out of these microcavities into the overhead liquid pool.

12.2.1 Equilibrium Bubble Radius

Let us consider a spherical vapor bubble of radius r in equilibrium within a superheated liquid at temperature T_ℓ and pressure p_ℓ. Since the liquid is superheated

$$T_\ell > T_{\text{sat}}(p_\ell) \quad \text{and} \quad p_\ell < p_{\text{sat}}(T_\ell)$$

as illustrated in Figure 12.1.

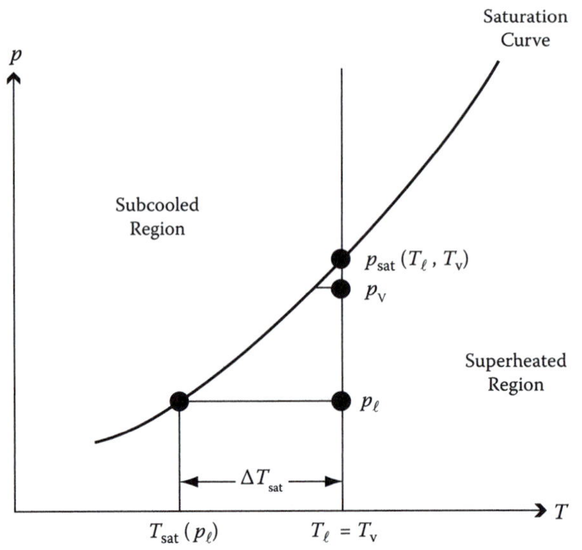

FIGURE 12.1 Vapor and temperature relationships for the equilibrium bubble in a superheated liquid pool.

The vapor and liquid pressures are related by the condition for mechanical equilibrium. Hence, by a force balance on the bubble hemisphere

$$\left(p_v - p_\ell\right)\pi r^2 = \sigma 2\pi r \tag{12.1a}$$

from which we obtain the Young–Laplace equation

$$p_v - p_\ell = \frac{2\sigma}{r} \tag{12.1b}$$

Thus, observe that mechanical equilibrium requires $p_v > p_\ell$. Also, since the vapor and liquid temperatures are equal

$$T_\ell = T_v \tag{12.2}$$

and T_ℓ is superheated, the vapor is also superheated. The vapor and liquid states at the bubble interface are also shown in Figure 12.1 along with two relevant saturation states

$$T_{sat} \text{ at } p_\ell \text{ that is } T_{sat}(p_\ell)$$
$$p_{sat} \text{ at } T_\ell \quad \text{and} \quad T_v \text{ that is } p_{sat}(T_\ell \text{ or } T_v)$$

To establish the equilibrium radius, r_e, in the desired form as a function of the liquid superheat, ΔT_{sat}, which is defined as

$$\Delta T_{sat} \equiv T_\ell - T_{sat}(p_\ell) \tag{12.3}$$

use the Clausius–Clapeyron equation which expresses the slope along the saturation curve as

$$\left.\frac{dp}{dT}\right|_{sat} = \frac{h_{fg}}{Tv_{fg}} \tag{12.4}$$

Assuming a constant slope over the ΔT interval $T_\ell - T_{sat}(p_\ell)$ yields

$$\frac{p_{sat}(T_\ell) - p_\ell}{T_\ell - T_{sat}(p_\ell)} = \frac{h_{fg}}{Tv_{fg}} \tag{12.5}$$

where T can be taken as any convenient value between $T_{sat}(p_\ell)$ and T_ℓ.

Now if we now assume

$$p_{sat}(T_\ell) \approx p_v$$

$$v_{fg} \approx v_g$$

and take

$$T = T_{sat}(p_\ell)$$

Then Equation 12.5 becomes

$$\frac{p_v - p_\ell}{\Delta T_{sat}} \simeq \frac{h_{fg}}{T_{sat}(p_\ell)v_g} \tag{12.6}$$

Finally, substituting Equations 12.6 into 12.1b leads to the desired result as

$$r_e \approx \frac{2\sigma T_{sat}(p_\ell)v_g}{h_{fg}\Delta T_{sat}} \tag{12.7}$$

This result can also be established on a more rigorous basis by an analysis which invokes the additional condition that the vapor and liquid chemical potentials are equal (Carey [6]):

$$\mu_v = \mu_\ell \tag{12.8a}$$

where

$$\mu = h - T_o s \tag{12.8b}$$

Example 12.1: The Equilibrium Bubble Radius

PROBLEM

For saturated water at 100°C, find the equilibrium radius size if the liquid superheat is about 5 K.

SOLUTION

Relevant physical properties of saturated water at 100°C from Table E.1 are

$$T_{sat} = 373.15 \text{ K}$$
$$\sigma = 58.78 \times 10^{-3} \text{ N/m or J/m}^2$$
$$v_g = 1.673 \text{ m}^3\text{/kg}$$
$$h_{fg} = 2.257 \times 10^6 \text{ J/kg}$$

$$r_e = \frac{2\sigma T_{sat} v_g}{h_{fg}\,\Delta T_{sat}} \tag{12.7}$$

$$= \frac{2(58.78 \times 10^{-3})373.15(1.673)}{2.257 \times 10^6(5)}$$

$$r_e = 6.5 \times 10^{-6} \text{ m}$$

12.2.2 Homogeneous and Heterogeneous Nucleation

The change of phase from liquid to vapor can occur via homogeneous or heterogeneous nucleation. In the former, no foreign bodies aid in the change of phase, but a number of liquid molecules that attain sufficiently high energy come together to form a nucleus of vapor. Therefore, homogeneous nucleation takes place only when the liquid temperature is substantially above the saturation temperature.

The naturally occurring embryo bubbles of vapor are of the order of molecular dimensions, so that $T_\ell - T_{sat}$ must be very high for homogeneous nucleation to occur. For water at atmospheric pressure the calculated superheat to initiate homogeneous nucleation is of the order of 220°C, a value that is much higher than what is measured in practice. Further discussion on homogeneous nucleation can be found in Blander and Katz [3].

For heterogeneous nucleation, which is of greater practical interest, microcavities (size ~10^{-3} mm) at the solid–liquid interfaces or on suspended bodies in the liquid serve as the nucleation sites. A solid surface contains a large number of microcavities with a wide distribution in size and shape. As well, some may contain air or vapor. The boiling process at the surface can begin if the coolant temperature near the surface is high enough that the preexisting vapor at the cavity site attains sufficient pressure to initiate the growth of a vapor bubble at that site. A major reduction in the superheat requirement is realized when dissolved gases are present in the surface microcavities, which reduces the required vapor pressure for bubble mechanical equilibrium. Equation 12.1b can then be generalized into the form

$$p_v + p_{gas} - p_\ell = \frac{2\sigma}{r} \tag{12.9}$$

The results for the equilibrium bubble radius can be applied to the heterogeneous nucleation process in both pool and flow (Chapter 13) boiling to characterize the size of microcavities active in the boiling process.

12.2.3 Vapor Trapping and Retention

The specific microcavities that serve as nucleation sites must be able to trap and retain vapor when the dry surface is flooded with liquid. The criteria governing these processes depend on the geometry of the liquid–vapor interface relative to the surface that is established by the wetting characteristic of the liquid–surface pair. This wettability or hydrophilicity is defined in terms of the contact angle, θ, which is the angle between the liquid–vapor interface and the solid surface measured through the liquid as illustrated in Figure 12.2a. Increased spreading of the liquid occurs for a decreasing value of the contact angle θ shown in Figure 12.2b.

Typical contact angles for water on brass are around 80–90 degrees; while on stainless steel, they are dependent on the presence of an oxide film, which reduces the contact angle hence enhancing the wettability. If the contact angle is <90°, the liquid wets the surface (e.g., ethanol on copper or lubricants on steel). If the contact angle is >90°, the liquid does not wet the surface (e.g., water on Teflon).

The criterion for trapping vapor in an idealized conical cavity is that the nose of the liquid front advancing in a cavity strikes the opposite cavity wall before the contact line reaches the bottom of the cavity as shown in Figure 12.3.

Hence, the criterion for trapping vapor in a conical cavity of half angle γ is

$$\theta > 2\gamma \tag{12.10}$$

The criterion for retaining the trapped vapor in the cavity when the system is cooled to room temperature between operating periods is that the curvature of the liquid–vapor interface is positive as Figure 12.4a illustrates, and hence the liquid pressure exceeds the vapor pressure. For the liquid–vapor interface in a conical cavity, the limiting interface orientation is an infinite radius of the liquid or the horizontal orientation relative to a vertically oriented cavity, as Figure 12.4b illustrates.

For a conical cavity of half-angle γ, this criterion requires that the contact angle be

$$\theta \geq \gamma + 90° \tag{12.11}$$

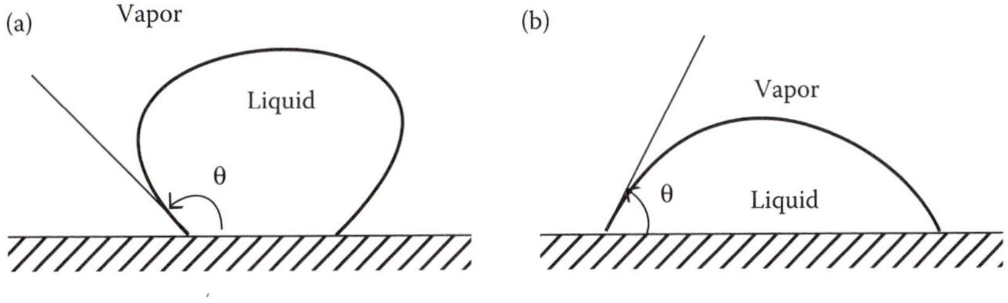

FIGURE 12.2 The contact angle θ. (a) poorly wetted fluid–solid pair; (b) well-wetted fluid–surface pair.

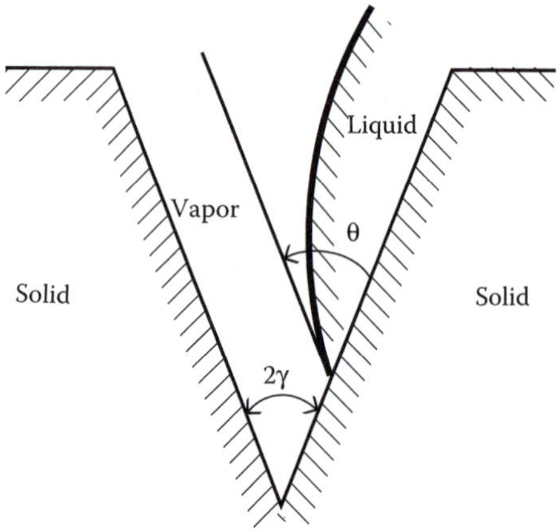

FIGURE 12.3 Criterion for trapping of vapor in a conical cavity of half-angle γ is $\theta > 2\gamma$. (Adopted from Carey, V., *Liquid-Vapor Phase-Change Phenomena*, 2nd Ed. New York: Taylor and Francis, 2008.)

which is characteristic of a poorly wetted surface. For this case the mechanical equilibrium equation is

$$p_\ell - p_v = \frac{2\sigma}{r} \tag{12.12}$$

so that $p_\ell > p_v$. For a system cooled to a temperature T at which both the liquid and vapor exist, the liquid is subcooled at pressure p_ℓ. The vapor is effectively saturated at p_v, which is below $T_{sat}(p_\ell)$ as illustrated in Figure 12.5.

The liquid–vapor interface recedes to the value of r consistent with mechanical equilibrium and the values of p_ℓ and p_v, that is

$$r = \frac{2\sigma}{p_\ell - p_v} \tag{12.13}$$

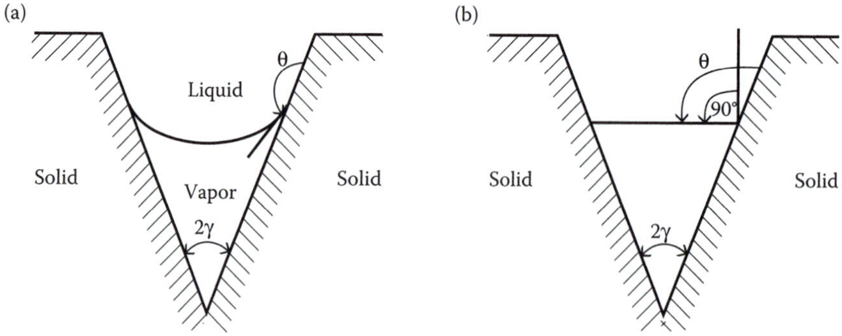

FIGURE 12.4 Retention of vapor in a conical cavity; (a) typical condition; (b) smallest contact angle for vapor retention, $\theta > \gamma + 90°$.

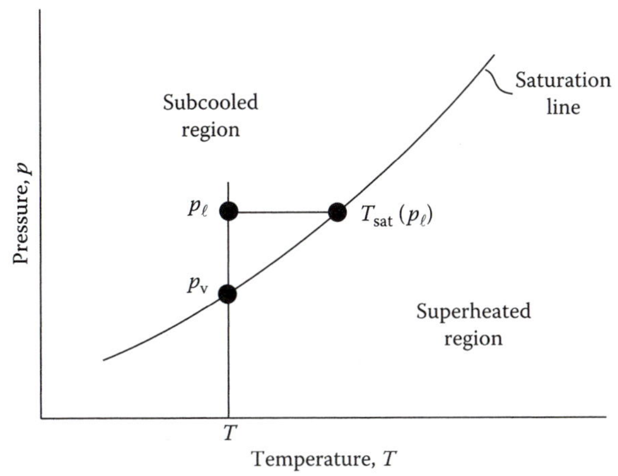

FIGURE 12.5 Liquid and vapor conditions for retaining vapor in a conical cavity.

In this state, the vapor in the cavity remains uncondensed at the state T, $p_v = p_{sat}(T)$ for the coolant conditions T, p_ℓ.

Should the contact angle be too small to satisfy the criterion of Equation 12.11 for a conical cavity, the presence of nonwetting inclusions or surface films would cause the criterion to be met. Conversely, the existence of reentrant cavities allows the required existence of a positive curvature of the liquid–vapor interface for values of θ much below the conical cavity criterion of Equation 12.11, as illustrated in Figure 12.6. Thus, if a liquid–surface pair has a contact angle θ that does not meet the criterion of Equation 12.11 for conical cavities, nonwetting inclusions or surface films deep in conical cavities and reentrant cavities can retain vapor by providing contact angles sufficient to maintain positive interface curvature.

12.2.4 Vapor Growth from Microcavities

In pool boiling, we envision the vapor embryo as growing with a changing radius of curvature, r, out of cavities of mouth radius R. The variation of R/r with embryo growth depends on the contact angle that the fluid makes with the surface.

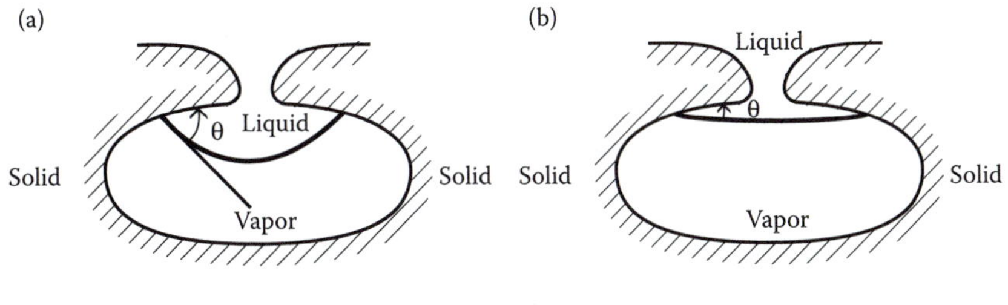

FIGURE 12.6 Retention of vapor in a reentrant cavity; (a) typical condition; (b) smallest contact angle θ for vapor retention.

For these cavities, two contact angle regimes hold. The first is

$$\theta \leq 90^\circ \quad (\text{well wetted}) \tag{12.14}$$

while the second is

$$\theta \geq 90^\circ \quad (\text{poorly wetted}) \tag{12.15}$$

The growth of the vapor embryo within, and then outside, the cavity is illustrated for these regimes in Figure 12.7a and b, respectively. In these figures, we illustrate the mouth region for a general cavity shape which has retained vapor. For the well-wetted case, this is either a conical cavity with nonwetting inclusions or films deep in the cavity or a reentrant cavity, while for the poorly wetted case ($\theta \geq 90^\circ$) this is simply a conical cavity without internal nonwetting inclusions or films (except for the region $90^\circ < \theta \leq 90^\circ + \gamma$, which also requires nonwetting inclusions or films deep in the cavity or a reentrant cavity).

For the well-wetted case, the interface radius of curvature, r, decreases, and hence R/r increases, as it moves up to the cavity mouth (positions 1 → 2). As the interface

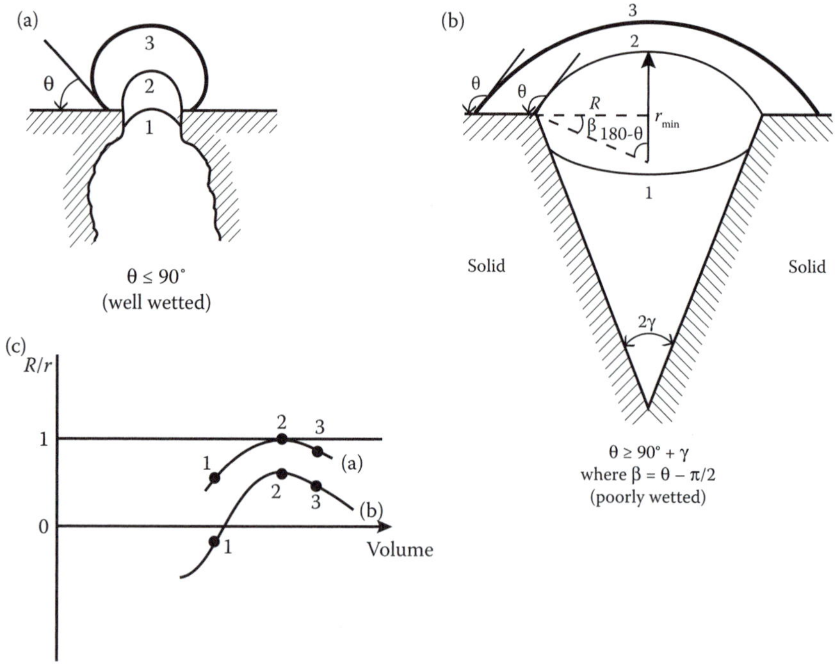

FIGURE 12.7 Bubble growth within and out of an idealized surface cavity for (a) a well wetted cavity, (b) a poorly wetted cavity and (c) the variation of bubble radius for both cavity types. (Adopted from Carey, V., *Liquid-Vapor Phase-Change Phenomena*, 2nd Ed. New York: Taylor and Francis, 2008.)

rotates at the mouth lip, r decreases to a minimum value equal to R and then increases again once outside the cavity (positions $2 \rightarrow 3$).

For the poorly wetted case, the interface starts concave into the liquid (hence r is taken negative) and r becomes more negative (hence R/r becomes less negative) as the embryo grows and the interface orientation moves to the cavity mouth (positions $1 \rightarrow 2$). At the mouth lip the interface reverses to be convex into the liquid, the ratio R/r becomes positive and reaches a positive maximum. For the interface just at the lip, r is a minimum and then increases once outside the cavity (positions $2 \rightarrow 3$). Figure 12.7c illustrates these R/r variations as the embryo increases in volume.

The minimum values of r for the two contact angle ranges are

$$\theta \leq 90^\circ \quad r_{\min} = R \tag{12.16a}$$

$$\theta \geq 90^\circ \quad r_{\min} = \frac{R}{\cos\beta} = \frac{R}{\sin\theta} \tag{12.16b}$$

This minimum value of r is important since it establishes the superheat necessary for the embryo to grow beyond the mouth of the cavity. Rewriting Equation 12.7 in terms of superheat, $T_\ell - T_{\text{sat}}(p_\ell)$ and expressing r_c as the value of $r_{\min}$ appropriate to the applicable fluid–surface pair contact angle, the minimum liquid superheat for bubble growth is established as

$$(\Delta T_{\text{sat}})_{\text{nucleation}} = T_\ell - T_{\text{sat}}(p_\ell) > \frac{2\sigma T_{\text{sat}}(p_\ell) v_{\text{g}}}{h_{\text{fg}}\, r_{\min}} \tag{12.17}$$

where $r_{\min}$ is given by Equation 12.16a or 12.6b.

The complete characterization of nucleation behavior of a surface for different combinations of fluid properties and surface conditions and nonidealized cavities is complicated but available through study of the extensive literature base already developed on this subject, which is elaborated in Carey [6].

12.2.5 Bubble Dynamics—Growth and Detachment

Subsequent to nucleation, the bubble dynamics processes include growth and detachment for which bubble departure diameter and frequency are well-studied parameters. Bubble growth is controlled in its early stage by the inertia of the surrounding liquid and in the later stage by the diffusion rate of heat to the vapor–liquid interface. Bubble departure dynamics depend on the balance between buoyancy and hydrodynamic drag forces promoting detachment and surface tension and liquid inertia forces resisting detachment. Correlations for each of these processes exist (see Carey [6] or Hewitt [19,20]), but they usually do not enter specifically into the relations for heat transfer rates and limiting conditions, which are of direct engineering interest and are presented next.

12.2.6 Nucleation Summary

This section has described the fundamentals of the nucleation process that underlies nucleate boiling heat transfer in both pool and flow boiling. The key factors applicable to boiling water in typical nuclear reactor cores and heat exchangers are

- Microcavities of size 0.5 to 1×10^{-3} mm at a solid–liquid interface, whether it is at a wall or on a suspended body in the fluid, serve as the nucleation sites for heterogeneous nucleation. These surfaces contain microcavities of a variety of shapes and sizes sufficient for nucleation to occur for wall superheat of order 2–5°C at 1.5 MPa or 0.06 to 0.1°C at 15.5 MPa.
- The sequence of processes that describes nucleation is the trapping, retention, and subsequent growth of vapor from the microcavities into bubbles which detach from the heated surface and move into the liquid pool or the flow stream.
- The wettability of the fluid–solid pair defined by the contact angle, θ, is an important parameter affecting these processes. For example, water on brass is a well-wetted pair, $\theta < 90°$, while water on Teflon is a poorly wetted pair, $\theta > 90°$. Oxides tend to be more easily wetted than metals.
- The prediction of the wall superheat required for nucleation in pool boiling makes direct use of the conditions for a spherical vapor bubble in equilibrium within a superheated liquid. For flow boiling this theory is adapted to consider the coolant temperature profile in the laminar sublayer adjacent to the heated solid surface (see Chapter 12).

12.3 THE POOL BOILING CURVE

The first determination of the heat transfer regimes of pool boiling was that of Nukiyama in 1934. The boiling curve, Figure 12.8, which remains useful for outlining the general features of pool boiling, is generally represented on a log–log plot of heat flux versus wall superheat ($T_w - T_{sat}$).

As can be seen starting with point B, the formation of bubbles leads to more effective heat transfer, and so the nucleate boiling heat flux can be one to two orders of magnitude higher than that of the single-phase natural convection heat flux but can be sustained only until a limiting condition (point C) is reached. However, higher heat fluxes (point C′ to F and beyond) lead to the formation of a continuous vapor film at the surface. Film boiling can also be established at lower heat fluxes if the surface temperature is sufficiently high, as along the curve D-E-C′. However, at low wall superheat (between C and D), the formation of the film is unstable, and this region is often called *transition boiling* (from nucleate to film boiling). Point C defines the critical heat flux or the departure from nucleate boiling condition. Point D is called the minimum stable *film boiling* (or Leidenfrost) temperature.

These heat transfer regimes and limiting conditions will be described next.

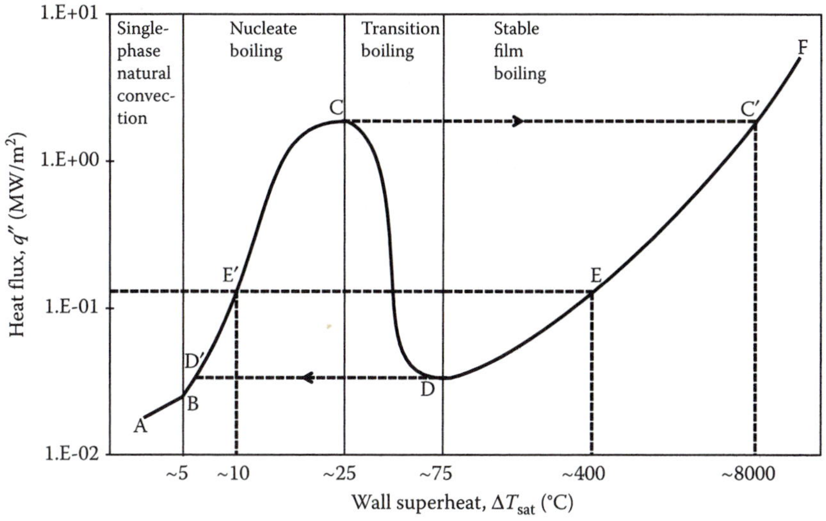

FIGURE 12.8 Typical pool boiling curve for saturated water at atmospheric pressure. (Adopted from Hewitt, G. F., In W. M. Rohsenow, J. P. Hartnett, and Y. Cho (eds.). *Handbook of Heat Transfer*, 3rd Ed. McGraw-Hill, 1998.)

12.4 HEAT TRANSFER REGIMES

12.4.1 Nucleate Boiling Heat Transfer (between Points B–C of the Boiling Curve of Figure 12.8)

From the earlier discussion on nucleation in Section 12.2, it is seen that boiling occurs at nucleation sites, and the number of sites that are activated depends on the surface profile and how well the liquid wets the surface and displaces air from the cavities.

Given these unknowns regarding active nucleation sites and the heat transfer mechanism at these sites, it is not surprising that complete theoretical correlations for nucleate boiling heat transfer have not been developed. The earliest and still useful result was presented by Rohsenow in 1952 [34]. He proposed that the form of the single-phase convective heat transfer be correlated for boiling as

$$\mathrm{Nu} = \frac{1}{C_{\mathrm{sf}}} \mathrm{Re}^{1-n}\, \mathrm{Pr}^{-m} \tag{12.18a}$$

with the following insights applied

- C_{sf} is a surface–fluid constant dependent on and tabulated for various surface–fluid combinations with typical values between 0.0025 and 0.015.
- The velocity in the Re is taken as the liquid velocity toward the surface sufficient to supply the vapor being produced, that is

$$\upsilon = \frac{q''}{h_{\mathrm{fg}}\rho_\ell} \tag{12.19}$$

- The length scale in both the Nu and Re is proportional to that of the spacing of vapor jets that arise from Rayleigh–Taylor instability of the vapor layer on a flat plate underlying a liquid pool. This jet spacing length scale is

$$L(\text{length scale}) = \left[\frac{\sigma}{(\rho_\ell - \rho_v)g}\right]^{1/2} \tag{12.20}$$

- The constants are taken as

$$n = 0.33 \quad \text{and} \quad m = 0.7$$

This value of n, as can be observed subsequently in Equation 12.18b, has been so chosen since it yields

$$q'' \propto \Delta T_{sat}^3 \tag{12.21}$$

which is the typical result for nucleate boiling.

Applying these four assumptions to Equation 12.18a yields

$$\frac{q''L}{\Delta T_{sat} k_\ell} = \frac{1}{C_{sf}}\left(\frac{q''}{h_{fg}}\frac{L}{\mu_\ell}\right)^{2/3}\left(\frac{\mu_\ell c_{p\ell}}{k_\ell}\right)^{-0.7} \tag{12.18b}$$

which reduces to*

$$\frac{c_{p\ell}\Delta T_{sat}}{h_{fg}} = C_{sf}\left\{\frac{q''}{\mu_\ell h_{fg}}\left[\frac{\sigma}{(\rho_\ell - \rho_v)g}\right]^{1/2}\right\}^{1/3}\left(\frac{\mu_\ell c_{p\ell}}{k_\ell}\right) \tag{12.18c}$$

Since the appropriate value of C_{sf} for a surface is often not known, more generalized methods have been proposed. Most useful among them are the correlations for saturated pool nucleate boiling of Stephan and Abdelsalam [43] and Cooper [11]. The more complicated result is from Stephan and Abdelsalam, which is applicable to all liquids and heater geometries:

$$\frac{q''d_d}{\Delta T_{sat} k_\ell} = 0.23\left(\frac{q''d_d}{k_f T_{sat}}\right)^{0.674}\left(\frac{\rho_v}{\rho_\ell}\right)^{0.297}\left(\frac{h_{fg}d_d^2}{\alpha_\ell^2}\right)^{0.371}\left(\frac{\rho_\ell - \rho_v}{\rho_\ell}\right)^{-1.73}\left(\frac{\alpha_\ell^2\rho_\ell}{\sigma d_d}\right)^{0.35} \tag{12.22a}$$

valid for the range $10^{-4} \le p/p_c \le 0.97$ where d_d is the bubble diameter at departure and α_ℓ is the liquid thermal diffusivity $(k_\ell/\rho_\ell c_{p\ell})$. For d_d, the authors use the relation

$$d_d = 0.146\theta\left[\frac{2\sigma}{g(\rho_f - \rho_g)}\right]^{1/2} \tag{12.22b}$$

where d_d is in meters and θ is the contact angle expressed in degrees.

---

* The Prandtl number exponent was altered by Rohsenow from the original published value of 1.7 to a value of unity for water only.

The simpler correlation of Cooper which employs reduced pressure, $p_R = p/p_{crit}$, surface roughness, ε, and molar mass, *M*, has the following dimensional form:

$$h = A\left[p_R^{(0.12-0.21\log_{10}\varepsilon)}\right]\left[-\log_{10} p_R\right]^{-0.55} M^{-0.5}\, q''^{2/3} \tag{12.23}$$

where h is in W/m² K
ε is in microns
q'' is in W/m²

The Cooper correlation applies to flat plates and horizontal cylinders for values of the constant A equal to 55 and 95, respectively, and for the following conditions:

- Water, refrigerants, and organic fluids
- $0.001 \le p_R \le 0.9$
- $2 \le M \le 200$
- $\varepsilon = 1$ micron if the surface condition is not known specifically

More mechanism-based correlations have also been proposed originating with the early work of Forster, first with Zuber [16], then with Greif [15]. The result of the Forster–Zuber effort has been adopted as the nucleate boiling component of the flow boiling correlation of Chen [8] that is presented in Chapter 13.

It is currently postulated that boiling heat transfer is the sum of heat transfer by three distinct mechanisms: bubble formation and departure, liquid microlayer evaporation, and microconvection of liquid moving with the bubbles and being replaced by cooler liquid reaching the surface. The formulation of a comprehensive treatment of these processes has been pursued through the numerical simulation of the boiling process by Dhir [14].

12.4.2 Transition Boiling (between Points C–D of the Boiling Curve of Figure 12.8)

This region of the boiling curve is obtainable only by controlling the surface temperature. If the surface heat flux is controlled, the wall temperature, on heat addition, follows the path C to C′ and upward toward F, and on heat reduction, traverses from F through C′ and E to D and then D′ to B, thus exhibiting hysteresis. In transition boiling, the liquid periodically contacts the heated surface leading to significant vapor formation that drives the liquid from the surface. Thus, the process is periodic and exhibits a mixed boiling mode with features of both nucleate and film boiling. It is usually reasonable to characterize this process by a linear interpolation between points C and D of a boiling curve on log–log coordinates. Auracher [1] presents a detailed review of transition boiling and its mechanisms.

12.4.3 Film Boiling (Between Points D–F of the Boiling Curve of Figure 12.8)

In this region, a continuous vapor film blankets the heater surface. Heat is transferred across this film and causes evaporation at the liquid–vapor interface. Interest exists

TABLE 12.1
Selected Film Boiling Heat Transfer Correlations

	Horizontal		Vertical
Flow Regime	**Cylinder**	**Plate**	**Plate or Cylinder**
Laminar	Bromley [4]	Berenson [2]	Hsu and Westwater [21]
	Sakurai and Shiotsu, 1992 [36]	Klimenko, 1981 [29]	
Turbulent	—	Klimenko, 1981 [29]	Hsu and Westwater, 1960 [21]

for film boiling heat transfer on horizontal surfaces where the film rises vertically under the action of buoyant forces. In all cases, laminar and turbulent regimes exist.

A vertical tube behaves as a vertical plate since its diameter is large compared to the vapor film thickness. However, for horizontal surfaces, the vapor flow pattern differs between the horizontal cylinder and the plate. For the tube, the vapor flows around the cylinder and departs from its top as bubbles while for the plate, vapor release is governed by the Taylor instability of the vapor–liquid interface.

Hence, for film boiling, we treat the three geometries listed in Table 12.1 for both laminar and turbulent conditions. The vapor–liquid configurations for the horizontal cylinder and the vertical flat plate are illustrated in Figure 12.9.

1. Horizontal flat plate—For laminar film boiling on a horizontal flat plate the Berenson [2] correlation given by Equation 12.24a is applicable:

$$h_c = 0.425\left\{\left[\frac{g(\rho_\ell - \rho_v)\rho_{vfm}k_{vfm}^3 h'_{fg}}{\mu_v \Delta T_{sat}}\right]\left[\frac{g(\rho_\ell - \rho_v)}{\sigma}\right]^{1/2}\right\}^{1/4} \quad (12.24a)$$

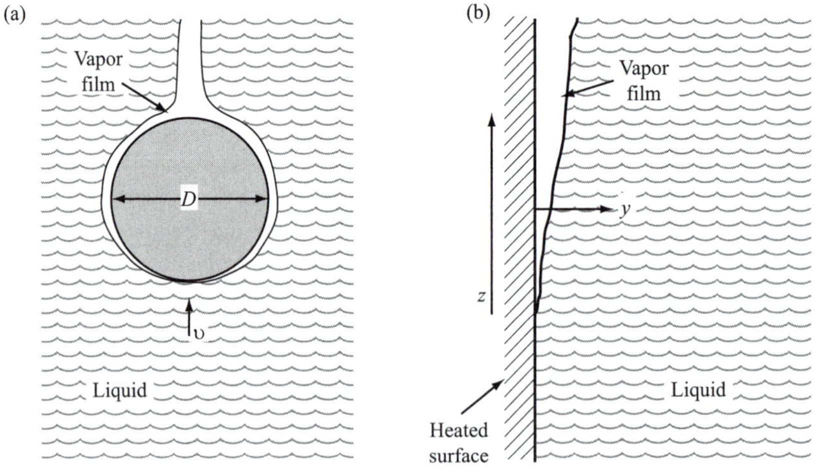

FIGURE 12.9 Film boiling (a) on a horizontal cylinder and (b) on a vertical flat plate.

where

$$h'_{fg} = h_{fg} + 0.5c_{pv}\Delta T_{sat} \tag{12.24b}$$

ρ_{vfm} and k_{vfm} are vapor film properties taken at the average of the hot wall and saturated liquid temperatures, the subscript c on the heat transfer coefficient signifies this is the convection heat transfer component, ρ_v, μ_v, and σ are properties at the saturated vapor temperature.

For laminar and turbulent film boiling, Klimenko extended the Berenson model to create correlations of the following form where the transition to turbulent vapor flow was proposed at $R > 10^8$:

$$\mathrm{Nu}_\lambda = C\sqrt{3}\,R^n\,\mathrm{Pr}_v^{1/3}\,f(\mathrm{Ja}) \tag{12.25a}$$

where

$$\mathrm{Nu}_\lambda = \frac{q''\lambda_D}{k_v(T_w - T_{sat})} \tag{12.25b}$$

$$\lambda_D = 2\pi\left(\frac{3\sigma}{\rho_\ell - \rho_v}\right)^{1/2} \quad \text{(the most dangerous Rayleigh–Taylor instability wavelength)} \tag{12.25c}$$

$$R = \frac{g\lambda_D^3}{(3)^{3/2}(\mu/\rho)_v^{3/2}}\left(\frac{\rho_\ell - \rho_v}{\rho_v}\right) \tag{12.25d}$$

$$\mathrm{Ja} = \text{Jacob number} = \frac{c_{pv}(T_w - T_{sat})}{h_{\ell v}} \tag{12.25e}$$

Table 12.2 lists values of parameters C, n, and f (Ja) for Equation 12.25.

2. Horizontal cylinder—For film boiling outside a horizontal cylinder of diameter D, the classical expression by Bromley [4] is

$$h_c = 0.62\left[\frac{g(\rho_\ell - \rho_v)\rho_{vfm}k_{vfm}^3 h'_{fg}}{D\mu_v\Delta T_{sat}}\right]^{1/4} \tag{12.26}$$

where the constant 0.62 differs slightly from the constant 0.73 applicable in the condensation case as given by Bromley et al. [5]. The value of h'_{fg} is given by Equation

TABLE 12.2
Parameters for Equation 12.25

	C	n	f
Laminar	0.19	1/3	1 for $\text{Ja} \geq 0.714$
			$0.89\ \text{Ja}^{-1/3}$ for $\text{Ja} < 0.714$
Turbulent	0.0086	1/2	1 for $\text{Ja} \geq 0.5$
			$0.71\ \text{Ja}^{-1/2}$ for $\text{Ja} < 0.5$

12.24b where the constant 0.5 is now replaced with 0.68. The correlation of Sakurai and Shiotsu [36] reproduced in Hewitt [19] covers a wide range of fluids, cylinder diameters, and subcoolings much better than that above of Bromley at the expense, however, of a much more complex formulation.

The physical properties in Equation 12.26 are evaluated at the same conditions as defined for Equation 12.24a.

3. Vertical plate—For film boiling on a vertical plate of height L, heat is conducted through the vapor film and causes evaporation at the liquid–vapor interface. For laminar flow of vapor in the film, the average value of the heat transfer coefficient over the plate of height L given by Hsu and Westwater [21] is

$$\bar{h}_{\text{cL}} = 0.943\left[\frac{g(\rho_\ell - \rho_\text{v})\rho_{\text{vfm}}k_{\text{vfm}}^3\, h'_{\text{fg}}}{L\mu_\text{v}\Delta T}\right]^{1/4} \tag{12.27a}$$

where

$$h'_{\text{fg}} = h_{\text{fg}} + 0.34\, c_{\text{pv}}\Delta T \tag{12.27b}$$

ΔT is the temperature difference from the solid surface to the bulk liquid. Physical properties are evaluated at the same conditions as defined for Equation 12.24a.

Note that this result for laminar film boiling on a vertical surface is analogous to the Nusselt film condensation result, Equation 12.41a, where the condensation of vapor to liquid is treated.

For a sufficiently long plate, the film eventually becomes turbulent. This occurs at a critical height L_o defined in terms of a critical film thickness, y^*, and Reynolds number Re*. This laminar to turbulent transition condition occurs at Re* = 100. These parameters, L_o and y^*, are defined as

$$L_\text{o} = \frac{\mu_\text{v}\,\text{Re}^*\, h'_{\text{fg}}\, y^*}{2k_\text{v}\Delta T} \tag{12.28a}$$

$$y^* = \left[\frac{2\mu_\text{v}^2\,\text{Re}^*}{g\rho_\text{v}(\rho_\ell - \rho_\text{v})}\right]^{1/2} \tag{12.28b}$$

and again

$$h'_{fg} = h_{fg} + 0.34\, c_{pv} \Delta T \tag{12.27b}$$

For this turbulent flow, the lower portion, L_o, is in laminar flow and the upper portion, $L - L_o$, is in turbulent flow. Hsu and Westwater [21] proposed that the average heat transfer coefficient over the entire plate height is given as

$$\frac{\bar{h}_{cT} L}{k_v} = \frac{2h'_{fg}\mu_v \,\text{Re}^*}{3k_v \Delta T} + \frac{B + \frac{1}{3}}{A}\left\{\left[\frac{2}{3}\left(\frac{A}{B + \frac{1}{3}}\right)(L - L_o) + \left(\frac{1}{y^*}\right)^2\right]^{3/2} - \left(\frac{1}{y^*}\right)^3\right\} \tag{12.29a}$$

where

$$A = \left[\frac{g(\rho_\ell - \rho_v)}{\rho_v}\right]\left(\frac{\bar{\rho}_v}{\mu_v \,\text{Re}^*}\right)^2 \tag{12.29b}$$

$$B = \frac{\mu_v + (f\rho_v \mu_v \,\text{Re}^* / 2\bar{\rho}_v) + (k_v \Delta T / h_{fg})}{(k_v \Delta T / h_{fg})} \tag{12.29c}$$

L_o, y^*, and h'_{fg} have been given above in Equations. 12.28a, 12.28b and 12.27b, $\text{Re}^* = 100$ and $\bar{\rho}_v$ is the average density of vapor in the laminar film, f is the Fanning friction factor.

In all cases at sufficiently high surface temperatures, for example, a surface temperature greater than the interface by about 540°C, enhancement of the heat transfer rate by radiation should be included. Typically, the convection component h_c is thus increased by a radiation component and the total h is expressed as

$$h = h_c + Jh_r \tag{12.30a}$$

where the constant J is nominally taken as 0.75 and the radiative heat transfer coefficient may be calculated by the following equation for parallel plate geometry:

$$h_r = \frac{\sigma}{(1/\varepsilon_s + 1/\varepsilon_\ell - 1)}\left(\frac{T_w^4 - T_{sat}^4}{T_w - T_{sat}}\right) \tag{12.30b}$$

where σ = Stefan–Boltzman constant, 56.7×10^{-9} W/m^{2}K^4
ε_s = emissivity of the solid surface
ε_ℓ = emissivity of the liquid surface (typically ≃1).

Example 12.2: Comparison of Heat Transfer Coefficients and Resulting Wall Temperature on a Horizontal Surface for Nucleate Boiling and Film Boiling

PROBLEM

Consider a flat surface in pool boiling with saturated water at 100°C. Compare the resultant wall temperatures in nucleate and film boiling for a surface heat flux of 1×10^5 W/m^2.

SOLUTION

The relevant physical properties of saturated water at 100°C from Table E.1 are

$$
\begin{aligned}
T_{sat} &= 100°C = 373.15 \text{ K}\\
h_{fg} &= 2.257 \times 10^6 \text{ J/kg}\\
\rho_f &= 958.3 \text{ kg/m}^3\\
\rho_g &= 0.598 \text{ kg/m}^3\\
\mu_f &= 283 \times 10^{-6} \text{ N s/m}^2\\
\mu_g &= 12.06 \times 10^{-6} \text{ N s/m}^2\\
k_f &= 0.681 \text{ W/m K}\\
k_g &= 24.9 \times 10^{-3} \text{ W/m K}\\
\sigma &= 58.78 \times 10^{-3} \text{ N/m}\\
c_{pf} &= 4.218 \times 10^3 \text{ J/kg K}\\
c_{pg} &= 2.034 \times 10^3 \text{ J/kg K}\\
p_{crit} &= 217.7 \text{ atm}
\end{aligned}
$$

The pressure ratio

$$p_R = \frac{1}{217.7} = 4.6 \times 10^{-3}$$

For nucleate boiling on a flat plate, use the correlation of Cooper since the problem conditions are within Cooper's stated range of applicability, that is

$$p_R > 0.001 \quad \text{and} \quad 2 < M < 200.$$

Take $\varepsilon = 1$ micron.
Hence,

$$
\begin{aligned}
h(\text{W/m}^2\text{K}) &= 55\left[p_R^{(0.12-0.21\log_{10}\varepsilon)}\right]\left[-\log_{10} p_R\right]^{-0.55} M^{-0.5}(q'')^{0.67}\\
&= 55\left[p_R^{[0.12-(0.21)0.4343\,\ell n\,\varepsilon]}\right][-0.4343\,\ell n\, p_R]^{-0.55} M^{-0.5}(q'')^{0.67}\\
&= 55\left[0.0046^{[0.12-(0.21)0.4343\,\ell n 1]}\right]\left[-0.4343\,\ell n 0.0046\right]^{-0.55}(18)^{-0.5}(q'')^{0.67}\\
&= 55[0.5242][0.6269][0.2357](q'')^{0.67}\\
&= 4.26(q'')^{0.67} \qquad (12.23)
\end{aligned}
$$

Now $h = q''/\Delta T_{sat}$

Hence,

$$\begin{aligned}\Delta T_{sat} &= 0.235(q'')^{0.33}\\ &= 0.235(1\times 10^5)^{0.33}\\ &= 10.5\ \text{K}\end{aligned}$$

For film boiling on a flat plate, use the correlation of Berenson:

$$h_c = 0.425\left\{\left[\frac{g(\rho_f-\rho_g)\rho_{vfm}k_{vfm}^3h'_{fg}}{\mu_v\Delta T_{sat}}\right]\left[\frac{g(\rho_f-\rho_g)}{\sigma}\right]^{1/2}\right\}^{1/4} \quad (12.24a)$$

where

$$\begin{aligned}h'_{fg} &= h_{fg} + 0.5c_{pg}\Delta T_{sat}\\ &= 2257\times 10^3 + 0.5(2.034\times 10^3)\Delta T_{sat}\end{aligned} \quad (12.24b)$$

and ρ_{vfm} and k_{vfm} are taken at the average of the hot wall temperature, assume to be 200°C and the saturated liquid at 100°C at 0.101 MPa.

Hence,

$$h_c\left(\frac{W}{m^2K}\right) = 0.425\left\{\left[\frac{9.81(958.3-0.598)0.523(28.9\times10^{-3})^3(2257\times10^3+1.017\times10^3\Delta T_{sat})}{12.06\times10^{-6}\Delta T_{sat}}\right]\right.$$

$$\left.\times\left[\frac{9.81(958.3-0.598)}{58.78\times10^{-3}}\right]^{1/2}\right\}^{1/4}$$

$$=0.425\left\{\left[\frac{9.83\times10^3(2257\times10^3+1.017\times10^3\Delta T_{sat})}{\Delta T_{sat}}\right]\left[4\times10^2\right]\right\}^{1/4}$$

Since

$$h = \frac{q''}{\Delta T_{sat}}$$

Then

$$q''\left(\frac{W}{m^2}\right) = 0.425\Delta T_{sat}\left[\frac{3.93\times10^6(2257\times10^3+1.017\times10^3\Delta T_{sat})}{\Delta T_{sat}}\right]^{1/4}$$

or $= 18.9\Delta T_{sat}^{3/4}(2257\times10^3+1.017\times10^3\Delta T_{sat})^{1/4}$

$= 733.3\Delta T_{sat}^{3/4}$ neglecting the h'_{fg} correction

So

$$\begin{aligned}\Delta T_{sat} &= \left(\frac{q''}{733.3}\right)^{4/3} = \left(\frac{1\times10^5}{733.3}\right)^{4/3}\\ &= (136.4)^{4/3}\\ &= 702.1\text{K}\end{aligned}$$

Without neglecting the h'_{fg} correction since the value of ΔT_{sat} is high and solving for ΔT_{sat}, we obtain

$$\Delta T_{sat} = 604.5 \text{ K}$$

Next, take ρ_{vfm} and k_{vfm} at the average of the hot wall temperature, 704.5°C and the saturated liquid at 100°C for the second iteration.

Hence,

$$h_c\left(\frac{\text{W}}{\text{m}^2\text{K}}\right) = 0.425\left\{\left[\frac{9.81(958.3-0.598)0.325(55.0\times10^{-3})^3(2257\times10^3+1.017\times10^3\Delta T_{sat})}{12.06\times10^{-6}\Delta T_{sat}}\right]\right.$$

$$\left.\times\left[\frac{9.81(958.3-0.598)}{58.78\times10^{-3}}\right]^{1/2}\right\}^{1/4}$$

$$= 0.425\left\{\left[\frac{4.21\times10^4(2257\times10^3+1.017\times10^3\Delta T_{sat})}{\Delta T_{sat}}\right]\left[4\times10^2\right]\right\}^{1/4}$$

Since

$$h = \frac{q''}{\Delta T_{sat}}$$

Then

$$q''\left(\frac{\text{W}}{\text{m}^2}\right) = 0.425\Delta T_{sat}\left[\frac{1.68\times10^7(2257\times10^3+1.017\times10^3\Delta T_{sat})}{\Delta T_{sat}}\right]^{1/4}$$

or

$$= 27.2\Delta T_{sat}^{3/4}(2257\times10^3+1.017\times10^3\Delta T_{sat})^{1/4}$$

$$= 1054.3\Delta T_{sat}^{3/4} \text{ neglecting the } h'_{fg} \text{ correction}$$

So

$$\Delta T_{sat} = \left(\frac{q''}{1054.3}\right)^{4/3} = \left(\frac{1\times10^5}{1054.3}\right)^{4/3}$$

$$= (94.8)^{4/3}$$

$$= 432.3 \text{ K}$$

Without neglecting the h'_{fg} correction and solving for ΔT_{sat}, we obtain

$$\Delta T_{sat} = 389.7 \text{ K}$$

Continuing the process above through five more iterations converges to

$$\Delta T_{sat} = 472.8 \text{ K}$$

12.5 LIMITING CONDITIONS ON THE BOILING CURVE

12.5.1 Critical Heat Flux (Point C of the Boiling Curve of Figure 12.8)

The critical heat flux (q''_{cr}) has been linked to the onset of fluidization of the pool (suspension of liquid by the vapor streams). This form of hydrodynamic instability has been the basis of numerous correlations since it was first suggested by Kutateladze in 1952 [30]. The general form of these critical heat flux correlations is

$$q''_{cr} = C_1 h_{fg} \rho_v \left[\frac{\sigma(\rho_\ell - \rho_v)g}{\rho_v^2} \right]^{1/4} \tag{12.31}$$

where

$$C_1 = 0.149$$

This equation form is plotted in Figure 12.10, where the lead constant has been optimized by various investigators (0.13 for Zuber [50], 0.16 for Kutateladze [30], and 0.18 for Rohsenow [35]) and differs slightly from the 0.149 value of Equation 12.31. This correlation with $C_1 = 0.149$ works well if the horizontal surface is reasonably large and no liquid enters from the sides of the plate.

Note that this critical flux correlation can be transformed to the Kutateladze number criterion for flooding by expressing the rate of energy supply to the horizontal heated surface A_h as

$$q''_{cr} A_h = h_{fg} \rho_v j_v A_h \tag{12.32}$$

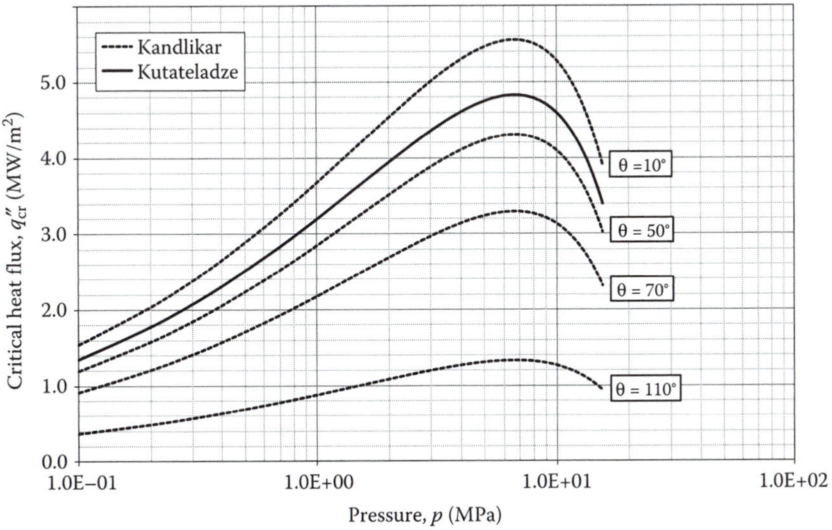

FIGURE 12.10 Effect of pressure on pool boiling CHF using the Kutateladze (Equation 12.31) and Kandlikar (Equation 12.34) correlations.

Substituting this result into Equation 12.31 yields the Kutateladze number criterion for flooding:

$$C_1 = j_v \left[\frac{\rho_v}{\left[g\sigma(\rho_\ell - \rho_v) \right]^{1/2}} \right]^{1/2} \tag{11.19, 11.24}$$

However, as Equation 11.19 illustrates, the applicable constant in the Kutateladze number is 3.1, an order of magnitude larger than the value of $C_1 = 0.149$.

For boiling on the outside of horizontal cylinders of large radii, the correlation of Equation 12.31 is used with the constant C_1 value of 0.116. For small radii, the constant is larger as determined by Sun and Lienhard [44] and equal to

$$C_1 = 0.116 + 0.3\exp\left[-3.44(R')^{1/2} \right] \quad \text{for } R' < 1 \tag{12.33a}$$

where

$$R' \equiv R \left[\frac{\sigma}{g(\rho_\ell - \rho_v)} \right]^{-1/2} \tag{12.33b}$$

and where R is the cylinder radius in meters.

As evident from Figure 12.10, the water critical heat flux seems to have a peak value at a pressure near 5.8 MPa. At low pressure, the effect of an increase in pressure is mainly to increase the vapor density, thus reducing the velocity and/or the flow area of the vapor for a given heat flux. Therefore, more vaporization is possible before the vapor flux reaches such a magnitude as to prevent the liquid from reaching the surface. On the other hand, because the heat of vaporization of water decreases with pressure, the heat flux associated with this critical vapor flux for fluidization starts to decrease at high pressures, when the relative changes in the vapor density with pressure become smaller.

Recently, Kandlikar [24] developed a critical heat flux correlation for pool boiling that includes the effects of contact angle (dynamic receding value in degrees), θ, surface orientation angle from horizontal in degrees, ϕ, and subcooling. The model is applicable to water, refrigerants, and cryogenic liquids and for orientations from the horizontal to 90 degrees (vertical).

For saturated pool boiling of pure liquids, the correlation is

$$q''_c = h_{fg}\rho_g \left(\frac{1+\cos\theta}{16} \right) \left[\frac{2}{\pi} + \frac{\pi}{4}(1+\cos\theta)\cos\phi \right]^{1/2} \left[\frac{\sigma(\rho_f - \rho_g)g}{\rho_g^2} \right]^{1/4} \tag{12.34}$$

and for subcooled conditions

$$q''_{c,sub} = q''_{c,sat} \left(1 + \frac{\Delta T_{sub}}{\Delta T_{sat}} \right) \tag{12.35}$$

12.5.2 Minimum Stable Film Boiling Temperature (Point D of the Boiling Curve of Figure 12.8)

For minimum stable film boiling, the hydrodynamic stability analysis of Berenson [2] leads to

$$j_v = C_2\left[\frac{\sigma g(\rho_\ell - \rho_v)}{(\rho_\ell + \rho_v)^2}\right]^{1/4} \tag{12.36}$$

where $C_2 = 0.09$. Berenson went on to derive a formula for the minimum wall temperature for a stable film (T^M) that depends only on the properties of the fluid, as given in Table 12.3. Henry [18] later suggested that the interface temperature on sudden contact between the liquid and the surface equals Berenson's minimum stable film boiling condition. Henry's correlation is recommended for highly subcooled water and for liquid metals. The correlation of Spiegler et al. [42], on the other hand, is based on the assumption that T^M should be related to the thermodynamic critical temperature. The Kalinin et al. [23] correlation requires the contact temperature between the solid surface and the liquid to reach a certain value relative to the thermodynamic critical temperature. Note that the Henry and Kalinin results introduce the effect of the surface properties that comports with the overwhelming existing evidence.

TABLE 12.3
Summary of Correlations for Prediction of Minimum Wall Temperature to Sustain Film Boiling T^M

Author	Correlation
Berenson [2]	$T_B^M - T_{sat} = 0.127\frac{\rho_{vfm} h_{fg}}{k_{vfm}}\left[\frac{g(\rho_\ell - \rho_v)}{\rho_\ell + \rho_v}\right]^{2/3}\left[\frac{\sigma}{g(\rho_\ell - \rho_v)}\right]^{1/2} \times\left[\frac{\mu_{fm}}{g(\rho_\ell - \rho_v)}\right]^{1/3}$ where subscript fm is for film (12.37a)[a]
Spiegler et al. [42]	$T_S^M = 0.84\,T_c$
Kalinin et al. [23]	$\frac{T_K^M - T_{sat}}{T_c - T_\ell} = 0.165 + 2.48\left[\frac{(\rho k c_p)_\ell}{(\rho k c_p)_w}\right]^{0.25}$ (12.38)
Henry [18]	$\frac{T_H^M - T_B^M}{T_B^M - T_\ell} = 0.42\left[\sqrt{\frac{(\rho k c_p)_\ell}{(\rho k c_p)_w}}\frac{h_{fg}}{c_{pw}(T_B^M - T_{sat})}\right]^{0.6}$ (12.39)

[a] Berenson's companion correlation for q'' is given by Equation 12.37b in Example 12.3.

Example 12.3: Minimum Film Boiling Temperature

PROBLEM

Compare the minimum film boiling temperature predicted by the Berenson and Henry correlations for saturated water at 100°C on a horizontal heated carbon steel surface of carbon content ≈0.5%.

SOLUTION

From Table 12.2, the relevant equation for Berenson is

$$T_B^M - T_{sat} = 0.127\frac{\rho_{vfm}h_{fg}}{k_{vfm}}\left[\frac{g(\rho_f - \rho_g)}{\rho_f + \rho_g}\right]^{2/3}\left[\frac{\sigma}{g(\rho_f - \rho_g)}\right]^{1/2}\left[\frac{\mu_{fm}}{g(\rho_f - \rho_g)}\right]^{1/3} \tag{12.37a}$$

and ρ_{vfm}, k_{vfm}, and μ_{fm} are taken at the average of the hot wall temperature, assumed to be 200°C and the saturated liquid at 100°C.

Thus,

$$\begin{aligned}T_B^M - T_{sat} &= 0.127\left[\frac{0.523(2.257\times10^6)}{28.9\times10^{-3}}\right]\left[\frac{9.8(958.3-0.598)}{958.3+0.598}\right]^{2/3}\\&\times\left[\frac{58.78\times10^{-3}}{9.8(958.3-0.598)}\right]^{1/2}\left[\frac{14.2\times10^{-6}}{9.8(958.3-0.598)}\right]^{1/3}\\&= 0.127(4.09\times10^7)(4.58)(2.50\times10^{-3})(1.15\times10^{-3})\\&= 68.4\text{ K}\end{aligned}$$

or

$$T_B^M = 441.5\text{ K }(168.4°\text{C})$$

Next iterate, taking ρ_{vfm}, k_{vfm}, and μ_{fm} at the average of the hot wall temperature, 168.4°C and the saturated liquid at 100°C.

Hence,

$$\begin{aligned}T_B^M - T_{sat} &= 0.127\left[\frac{0.544(2.257\times10^6)}{27.6\times10^{-3}}\right]\left[\frac{9.8(958.3-0.598)}{958.3+0.598}\right]^{2/3}\\&\times\left[\frac{58.78\times10^{-3}}{9.8(958.3-0.598)}\right]^{1/2}\left[\frac{13.6\times10^{-6}}{9.8(958.3-0.598)}\right]^{1/3}\\&= 0.127(4.45\times10^7)(4.58)(2.50\times10^{-3})(1.13\times10^{-3})\\&= 73.1\text{K}\end{aligned}$$

or

$$T_B^M = 446.2\text{ K }(173.1°\text{C})$$

Berenson also correlated the minimum stable heat flux value as

$$q''^M = 0.09\rho_{vfm}h_{fg}\left[\frac{g(\rho_\ell - \rho_v)}{\rho_\ell + \rho_v}\right]^{1/2}\left[\frac{\sigma}{g(\rho_\ell - \rho_v)}\right]^{1/4} \tag{12.37b}$$

Thus, for the conditions of this example

$$q'' = 0.09(0.544)(2.257 \times 10^6)\left[\frac{9.8(958.3 - 0.598)}{958.3 + 0.598}\right]^{1/2}\left[\frac{58.78 \times 10^{-3}}{9.8(958.3 - 0.598)}\right]^{1/4}$$

$$= 0.09(1.23 \times 10^6)(3.16)(0.05)$$

$$= 17.5 \times 10^3 \text{ W/m}^2$$

The Berenson results obtained are reflected in Figure 12.11.
Use Equation 12.39 from Henry:

$$\frac{T_H^M - T_B^M}{T_B^M - T_\ell} = 0.42\left[\sqrt{\frac{(\rho k c_p)_\ell}{(\rho k c_p)_w}}\frac{h_{fg}}{c_{pw}(T_B^M - T_{sat})}\right]^{0.6} \quad (12.39)$$

Obtain the needed wall surface properties for carbon steel with carbon content ≈0.5% from Table F.1 as

$$c_{pw} = 465 \text{ J/kg K}$$

$$\rho_w = 7833 \text{ kg/m}^3$$

$$\alpha = 1.47 \times 10^{-5} \text{ m}^2\text{/s}$$

Initially evaluate k_w from Table F.1 at the wall temperature of 173.1°C from the Berenson correlation as

$$k_w = 49.4 \text{ W/mK}$$

From above,

$$T_M^B = 446.2\text{ K} \quad \text{and} \quad T_B^M - T_f = 73.1\text{K}$$

Hence,

$$\frac{T_H^M - 446.2}{73.1} = 0.42\left\{\left[\frac{(958.3)(0.681)(4.218 \times 10^3)}{(7833)(49.4)(465)}\right]^{1/2}\frac{2.257 \times 10^6}{465(73.1)}\right\}^{0.6}$$

$$= 0.42\left[(0.0153)^{1/2}66.4\right]^{0.6}$$

$$= 1.486$$

$$T_H^M = 554.8\text{K } (281.7°\text{C})$$

From Table F.1, at this temperature the wall thermal conductivity is 45.6 W/m°C. Iterating increases the temperature T_H^M by only 2.7 K.

This difference in predicted minimum wall film boiling temperature reflects typical uncertainty in correlations for this parameter, that is

$$T_B^M = 446.2\text{K} \quad \text{vs} \quad T_H^M = 557.5\text{K}$$

12.6 SURFACE EFFECTS IN POOL BOILING

As discussed in Section 12.2, boiling on a surface depends on the activation of microcavities and the combined fluid/surface wetting characteristics. The microcavities are created by the fabrication process. In general, these characteristics affect the surface heat flux for given wall superheat.

Recently, investigation of pool boiling of dilute dispersions (<0.1% by volume) of nanoparticles by You et al. [49] and by Kim et al. [26] in water has revealed increases in critical heat flux and the minimum film boiling temperature. Note that the thermophysical properties of these dilute nanofluids are identical to those of pure water so that the change in boiling behavior must be attributed to changes in the boiling surface. The premature disruption of film boiling, and hence the higher minimum film boiling temperature, appears to be due to the increase in surface roughness and wettability. These in turn are due to nanoparticle deposition as reported by Kim et al. [26]. The linkage between improved wettability and CHF increase has been demonstrated by Kim et al. [27] for the prevalent CHF theories, particularly for the bubble interaction theory. This theory postulates the occurrence of DNB when the bubble density and departure frequency become high enough to lead to bubble radial coalescence and hence prevention of liquid access to the surface. However, insufficient data exist now to formulate a revised CHF correlation.

Example 12.4: Construction of the Pool Boiling Curve

PROBLEM

Draw the pool boiling curve for saturated water at 100°C. The saturation state physical properties are listed in Example 12.2.

Assume that boiling occurs on a flat horizontal surface with a conical cavity radius of 5 μm and a surface roughness ε of 1 μm.

SOLUTION

To produce a graph of the form of Figure 12.8, we will compute points B, C, D, and E and the equations of the lines representing the boiling regions connecting these points.

a. Point A—onset of nucleate boiling (Equation 12.7)

$$\Delta T_{sat} = \frac{2\sigma T_{sat} v_g}{h_{fg} r}$$

$$\Delta T_{sat} = \frac{2(58.78 \times 10^{-3})(373.15)(1.673)}{(2.257 \times 10^{6})(5 \times 10^{-6})} = 6.5\text{K}$$

b. Nucleate boiling region—the line between points A and B
Using the Cooper correlation (Equation 12.23), this line has been found in Example 12.2 as

$$\Delta T_{sat} = 0.235(q'')^{0.33}$$

or

$$q'' = 77.05\Delta T_{sat}^3 \quad \text{for} \quad q'' \text{ in W/m}^2 \text{ and } \Delta T_{sat} \text{ in K}$$

c. The critical condition—point C
Applying Equation 12.31 to this saturated water case, we obtain

$$q_c'' = 0.149 h_{fg}\rho_g \left[\frac{\sigma(\rho_f - \rho_g)g}{\rho_g^2}\right]^{1/4}$$

$$= 0.149(2257\times10^3)(0.598)\left[\frac{58.78\times10^{-3}(958.3-0.598)9.8}{(0.598)^2}\right]^{1/4} \quad (12.31)$$

$$= 0.149(1.35\times10^6)[6.27]$$

$$q_c'' = 1.26\times10^6 \text{ W/m}^2\text{K}$$

So from part b, this heat flux occurs at

$$\Delta T_{sat} = 0.235(q'')^{0.33}$$

$$= 0.235(1.26\times10^6)^{0.33}$$

$$\Delta T_{sat} = 24.2 \text{ K}$$

d. Point D—the minimum stable film boiling condition
Using the Berenson result of Table 12.1, this value has been calculated in Example 12.3 as

$$\Delta T_{sat} = 73.1 \text{ K}$$

Also in Example 12.3, the corresponding minimum heat flux was evaluated as

$$q''^{M} = 17.5\times10^3 \text{ W/m}^2$$

e. Film boiling region—the line from point D through points C′ and E
Use the Berenson result of Equations 12.24a and 12.24b to obtain the slope of this line.
Find h_c first:

$$h_c = 0.425\left\{\left[\frac{g(\rho_\ell - \rho_v)\rho_{vfm}k_{vfm}^3 h'_{fg}}{\mu_v\Delta T_{sat}}\right]\left[\frac{g(\rho_\ell - \rho_v)}{\sigma}\right]^{1/2}\right\}^{1/4} \quad (12.24a)$$

where

$$h'_{fg} = h_{fg} + 0.5c_{pv}\Delta T_{sat} \quad (12.24b)$$

Evaluating first h'_{fg}

$$h'_{fg} = 2.257 \times 10^6 + 0.5(2.034 \times 10^3)\Delta T_{sat}$$
$$= 2.257 \times 10^6 + 1.017 \times 10^3 \, \Delta T_{sat}$$

let us initially take $h'_{fg} = 2.257 \times 10^6$ for simplicity. This is reasonable, however, for ΔT_{sat} at or somewhat above ΔT_{sat}^{M}

$$h_c = 0.425\left\{\left[\frac{9.8(958.3-0.598)(0.544)(27.6\times10^{-3})^3(2.257\times10^6)}{12.06\times10^{-6}\Delta T_{sat}}\right]\right.$$
$$\left.\times\left[\frac{9.8(958.3-0.598)}{58.78\times10^{-3}}\right]^{1/2}\right\}^{1/4}$$

$$h_c = 0.425\left\{\left[\frac{2.01\times10^{10}}{\Delta T_{sat}}\right]\left[15.98\times10^4\right]^{1/2}\right\}^{1/4}$$
$$= 0.425\left(\frac{376.5}{\Delta T_{sat}^{1/4}}\right)(4.47)$$
$$h_c = \frac{715.3}{\Delta T_{sat}^{1/4}}$$

Since $q'' = h_c \Delta T_{sat}$.
Then the film boiling region is given as

$$q'' = 715.3\,\Delta T_{sat}^{3/4}$$

f. Summary—these results are displayed in Figure 12.11, which illustrates the pool boiling curve for saturated water at 100°C.

12.7 CONDENSATION HEAT TRANSFER

Condensation is a very effective means for heat transfer to a cold surface, for which the rate of heat transfer is influenced by vapor pressure, cold surface geometry, subcooling, and the presence of noncondensable gases. In the discussion presented here, a brief outline of these factors is provided, as pertinent to the operation and safety performance of a nuclear power plant. An extensive discussion of condensation can be found in specialized books such as Carey [6] or Kandlikar et al. [25].

Steam condensation plays a significant role in various components of nuclear power plants under normal conditions, for example, the steam condenser unit, the PWR pressurizer and the preheating equipment, thus influencing the plant overall thermodynamic efficiency. Due to low operating pressure of the condenser, the unit is relatively large, and inefficient condensation would make the condenser even larger. Condensation of steam in BWRs is an important phenomenon for accommodating plant transients, such as the discharge of steam into suppression pools when attempting to reduce the turbine load. During accidents, such as loss of cooling events, steam is also directed into the suppression pool of BWRs, and plays a role in

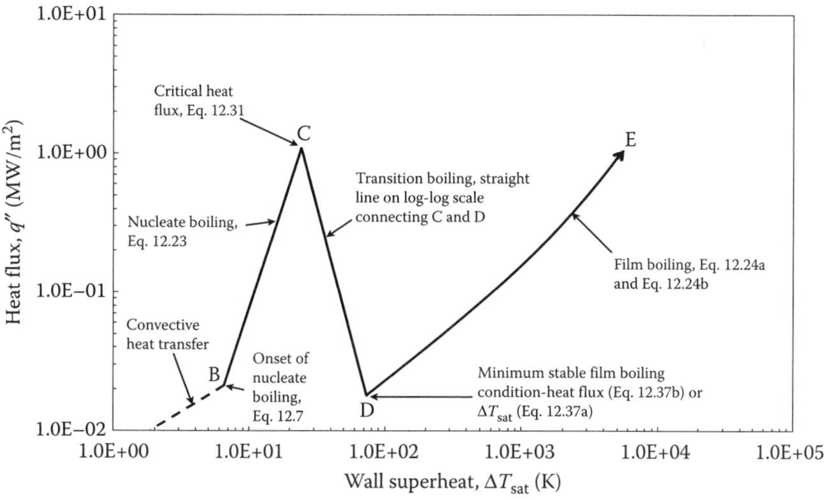

FIGURE 12.11 The pool boiling curve for saturated water at 100°C on a horizontal flat plate.

the containment response to loss of coolant events or more severe accidents. In PWRs, steam is ejected into the large containment atmosphere and condenses on the colder surfaces in the containment.

There are two main types of condensation: the homogenous one, taking place within the steam body itself, and the heterogeneous one, where the steam condenses on cold surfaces. Homogeneous condensation requires significant subcooling of the steam to form an embryo of a droplet and to enable the rejection of the phase change energy within the steam body away from the droplet forming location. Heterogeneous condensation requires less of a temperature difference since an embryo may be formed as adsorbed liquid at the surface and the heat is usually rejected within the cold body. Heterogeneous condensation can occur as droplet condensation when the drop does not wet the wall, or as filmwise condensation when wetting occurs. The rate of condensation heat transfer is higher when it occurs as droplet condensation. Some of the main correlations to estimate the rate of heat transfer by condensation and the main effects of geometry and other influencing factors follow.

12.7.1 Filmwise Condensation

12.7.1.1 Condensation on a Vertical Wall

On a vertical wall, the steam condenses and creates a liquid film, which flows downward by gravity. Based on a boundary layer treatment by Nusselt in 1916 for a constant wall temperature that accounts for momentum balance over the flowing film, it is possible to derive the heat transfer for a laminar film (Re < 1800) ignoring the temperature gradient and velocity gradient at the interface:

$$h(z) = \left[\frac{g(\rho_f - \rho_g)\rho_f k_f^3 h_{fg}}{4z\mu_f (T_{sat} - T_w)} \right]^{1/4} \tag{12.40}$$

where the subscript f refers to saturated liquid and g to saturated vapor, T_w is the wall temperature, and z is the downward distance on the vertical plate.

The average heat transfer coefficient over a length L becomes

$$\bar{h}_L = 1.13\left[\frac{g(\rho_f - \rho_g)\rho_f k_f^3 h_{fg}}{L\mu_f (T_{sat} - T_w)}\right]^{1/4} \quad \text{for } \mathrm{Re}_f \leq 1800 \tag{12.41a}$$

where

$$\mathrm{Re}_f = 4\frac{\dot{m}}{\mu_f} \quad \text{and } \dot{m} \text{ is the film flow rate per unit width.} \tag{12.41b}$$

As noted earlier, the result for flow condensation, Equation 12.41a, is analogous in form to that for laminar film boiling, Equation 12.26.

Solutions for film condensation heat transfer under uniform heat flux at the wall give approximately the same result for the mean heat transfer coefficient by Fujii [17].To account for the temperature gradient in the film and steam, it is possible to use a modified enthalpy to include all the enthalpy content of the change of phase:

$$h_{fg}^* = h_{fg} + 0.68 c_{pf}(T_{sat} - T_w) + c_{pg}(T_{steam} - T_{sat}) \tag{12.42}$$

which would replace h_{fg} in Equation 12.41a.

For turbulent film conditions, $\mathrm{Re}_f > 1800$, we have

$$\bar{h}_L = 0.0077 k_f \,\mathrm{Re}_f^{0.4}\left[\frac{g\rho_f^2(\rho_f - \rho_g)}{\mu_f^2}\right]^{1/3} \tag{12.43}$$

12.7.1.2 Condensation on or in a Tube

On a horizontal tube, laminar film condensation heat transfer as derived by Nusselt may be given by

$$\bar{h}_D = 0.728\left[\frac{g(\rho_f - \rho_g)\rho_f k_f^3 h_{fg}^*}{D\mu_f (T_{sat} - T_w)}\right]^{1/4} \quad \text{for } \mathrm{Re}_f < 2300 \tag{12.44}$$

where Re_f is given by Equation 12.41b, with $\dot{m}$ being the condensate rate per unit length of the tube. This follows the laminar film condensation on a vertical plate since the length of film flow around a tube is small enough that the Re number remains small, and the film stays laminar.

Comparing the laminar vertical plate (Equation 12.41a) and the horizontal tube (Equation 12.44) results, for the same fluid conditions:

$$\frac{\bar{h}_D}{\bar{h}_L} = \frac{0.728}{1.13}\left(\frac{L}{D}\right)^{1/4}$$

Hence, for typical geometries of interest where $L \gg D$, that is, $L \approx 1$ m, $D \approx 0.03$ m, the ratio of heat transfer coefficients is

$$\frac{\bar{h}_D}{\bar{h}_L} = \frac{0.728}{1.13}\left(\frac{1}{0.03}\right)^{1/4} = 1.55$$

When condensation occurs on a tube bundle containing N horizontal tubes in the same vertical plane, the Nusselt treatment for the laminar film can be used for the film dripping from one tube to the one below it, to derive an average coefficient given by Jakob [22]:

$$\bar{h}_{DN} = 0.728\left[\frac{g(\rho_f - \rho_g)\rho_f k_f^3 h_{fg}^*}{D\mu_f (T_{sat} - T_w)N}\right]^{1/4} \tag{12.45a}$$

that is

$$\bar{h}_{DN}/h_{D1} = (1/N)^{1/4} \tag{12.45b}$$

A better relation according to Collier and Thome [10] is given by

$$\bar{h}_{DN}/h_{D1} = (1/N)^{1/6} \tag{12.45c}$$

For laminar condensation inside a horizontal tube, Chato [7] recommends

$$\bar{h}_D = 0.555\left[\frac{g(\rho_f - \rho_g)\rho_f k_f^3 h_{fg}^*}{D\mu_f (T_{sat} - T_w)}\right]^{1/4} \tag{12.46}$$

which Chato found to apply for $Re_{vapor} \leq 35{,}000$ where

$$Re_{vapor} = \frac{G_{vapor} D}{\mu_v} \quad \text{(at the tube entrance.)}$$

For high-velocity conditions, Shah [37] recommends

$$\frac{h}{h_{\ell o}} = (1-x)^{0.8} + \frac{3.8(1-x)^{0.04} x^{0.76}}{(p/p_c)^{0.38}} \tag{12.47}$$

where x is the quality, p and p_c are the local and critical pressures, and the liquid-alone coefficient $h_{\ell o}$ can be obtained from the Dittus-Boelter/McAdams correlation:

$$h_{\ell o} = 0.023\left(\frac{k_l}{D}\right)\left(\frac{GD}{\mu_l}\right)^{0.8} \mathrm{Pr}_l^{0.4} \tag{10.91}$$

Shah [37] found agreement with data over a wide range of conditions:

$$11 \le G \le 211\,\text{kg/m}^2\text{s}$$
$$0 \le x < 1.0$$
$$0 \le \text{Pr}_\ell < 13.0$$

For $x = 1$, the correlation gives an unphysical result. Collier and Thome [10] recommend setting $x = 0.999$ when pure vapor ($x = 1$) enters the tube. They also find the Shah coefficient agreeable at $G > 200$ kg/m^{2}s.

Example 12.5: Condensation on a Horizontal Tube

PROBLEM

Saturated steam at 100°C is condensed on the outside of a horizontal tube of outside diameter 0.030 m and length 1 m. Assume a thin tube is made of highly conductive material. Hence, the wall temperature is constant throughout. Water at temperature 30°C flows in the tube at a flow rate such that the water to inside wall heat transfer coefficient is 1500 W/m^{2}K.

Find the condensation rate on the tube.

SOLUTION

The relevant physical properties of saturated steam at 100°C are listed in Example 12.2.

Let us use Nusselt's result of Equation 12.44 and confirm the flow is laminar on determination of the condensation rate per meter of tube, $\dot{m}$. Also the use of h^*_{fg} requires that a wall temperature be first assumed and iterated if necessary. Hence, let us first assume that the wall temperature is 93°C.

On one hand from Equation 12.42

$$\begin{aligned} h^*_{fg} &= h_{fg} + 0.68\,c_{pf}(T_{sat} - T_{wall}) \quad \text{since } T_{steam} = T_{sat} \\ &= 2.257 \times 10^6 + (4.218 \times 10^3)(100 - 93) \\ h^*_{fg} &= 2.286 \times 10^6 \text{ J/kg} \end{aligned}$$

$$\bar{h}_L = 0.728\left[\frac{g(\rho_f - \rho_g)\rho_f k_f^3 h^*_{fg}}{D\mu_f(T_{sat} - T_w)}\right] \tag{12.44}$$

$$\begin{aligned} \bar{h}_L &= 0.728\left[\frac{9.81(958.3 - 0.598)(958.3)(0.681)^3 2.286 \times 10^6}{(0.03)283 \times 10^{-6}(100 - T_w)}\right]^{1/4} \\ &= 0.728\left(\frac{76.56 \times 10^{16}}{100 - T_w}\right)^{1/4} \\ \bar{h}_L &= \frac{2.958 \times 10^4}{(100 - T_w)^{1/4}}\,\text{W/m}^2\text{K} \end{aligned}$$

Now

$$q' = \pi D\bar{h}_L(T_{sat} - T_w)$$
$$= \pi(0.03)\frac{2.958 \times 10^4}{(100 - T_w)^{1/4}}(100 - T_w)$$
$$q' = 0.279 \times 10^4(100 - T_w)^{3/4}\ \text{W/m} \tag{12.48}$$

On the other hand

$$q' = \pi D h_i(T_w - T_{water})$$
$$q' = \pi(0.03)(1500)(T_w - 30)$$
$$q' = 141.4(T_w - 30)\ \text{W/m} \tag{12.49}$$

Solving Equations 12.48 and 12.49 by trial and error for q' and T_w, we obtain

$$T_w = 95.1°\text{C}$$
$$q' = 9.18 \times 103\ \text{W/m}$$

This result could be iterated to bring the initial assumption for $T_w = 93°$C and this result of 95.1°C into closer correspondence, if desired. However, remaining with $T_w = 95.1°$C, the condensation rate is

$$\dot{m}_c = \frac{q'}{h_{fg}^*} = \frac{9.18 \times 10^3}{2.286 \times 10^6}$$
$$= 4.02 \times 10^{-3}\ \text{kg/ms}$$

and the Re_f can be confirmed to be in the laminar region since

$$Re_f = \frac{4\dot{m}}{\mu_f}$$
$$= \frac{4(4.02 \times 10^{-3})}{283 \times 10^{-6}}$$
$$Re_f = 56.8 < 2300 \tag{12.41b}$$

12.7.2 Dropwise Condensation

In dropwise condensation, droplets leave the surface at a given frequency, which creates sizable "fresh" areas for contact with steam. Thus, the droplet condensation results in higher heat transfer rate. Figure 12.12 gives the data of Takeyama and Shimizu [46] for steam condensation on the surface of a vertical copper surface. The data show that an order of magnitude difference is possible between the two modes of condensation at low subcooling of the surface.

12.7.3 The Effect of Noncondensable Gases

The presence of noncondensable gases greatly retards the condensation of steam on cold surfaces. This situation is encountered within reactor containments following a

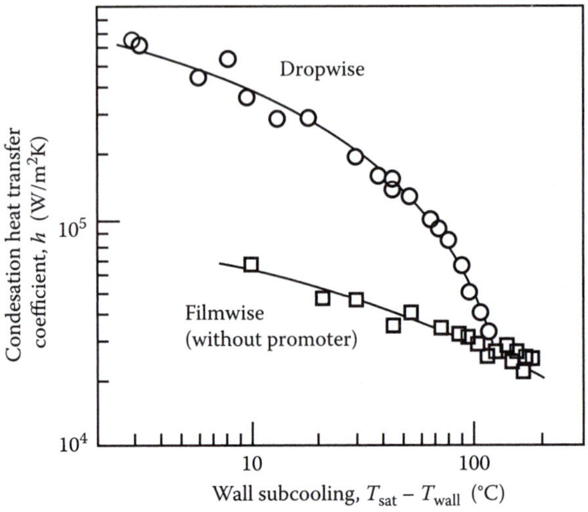

FIGURE 12.12 Data for condensation on a short vertical copper surface. (From Takeyama, T. and Shimizu, S., *Proceedings of the 5th International Heat Transfer Conference* 3:274, 1974.)

loss of coolant accident, or due to gradual leakage of air into the subatmospheric pressure condenser. The noncondensable gas arrives with the steam at the cold surface but is left by the condensing stream to form a boundary layer that retards the motion of the newly arriving steam toward the cold surface. The liquid film thermal resistance becomes smaller than the noncondensable gas layer resistance. Othmer [33] was first to point out that the presence of 0.5% mole fraction of air in a steam–air mixture reduces condensation over a 7.62 cm diameter copper tube by 50%. Uchida et al. [47] and Tagami [45] developed empirical relations for the effect of air, nitrogen, or argon on steam condensation. Their results are widely used for post-LOCA containment analysis. Their experiments involved condensation on the surface of a vertical cylinder of length 0.3 m and diameter 0.2 m. The pressure was allowed to vary between 1 and 3 atmospheres. Uchida et al.'s experiments were for small steam–air velocities, while Tagami's were for natural convection conditions. The results were correlated as

$$\bar{h}_{\text{Uchida}} = 380\left(\frac{1-y}{y}\right)^{0.7} \text{ W/m}^2\text{K} \tag{12.50}$$

$$\bar{h}_{\text{Tagami}} = 11.4 + 284\left(\frac{1-y}{y}\right) \text{W/m}^2\text{K} \tag{12.51}$$

where y is the air mass friction.

Kim and Corradini [28] showed that, for vertical plate condensation, waviness of the liquid film has a pronounced effect on the gas-phase resistance. Thus, it is difficult to produce an analytical approach to this effect in practical situations.

Dehbi et al. [12] conducted condensation experiments for a vertical cylinder under turbulent natural convection conditions of steam–air mixtures. The pressure varied between 1.5 and 4.5 atmospheres. Their results indicate that the pressure has a large effect while the wall length and wall subcooling are of secondary importance. They correlated the results of their experimental data as

$$\bar{h}_{\mathrm{L}} = \frac{L^{0.05}\left[(3.7+28.7\,p)-(2438+458.3\,p)\log(y)\right]}{\left(T_{\infty}-T_{\mathrm{w}}\right)^{0.25}} \tag{12.52}$$

where $\bar{h}_{\mathrm{L}}$ is the average condensation coefficient (W/m^2°C)
L is the length of vertical wall ($0.3 \le L < 3.5$ m)
p is pressure (atm) in the gas phase
T_{∞}, T_{w} are the bulk and average wall temperatures (°C), respectively
y is the air mass fraction in the gas phase

During forced flow condensation inside tubes, the heat transfer coefficient is higher than that for natural convection flow. The higher velocities of the gaseous core of the flow reduce the thickness of the condensate film. In addition, the bulk motion of the core sweeps any noncondensable gases. Hence, noncondensables have a lower effect on heat transfer than in nonflow conditions. Nevertheless, Vierow [48] found experimentally that 14% air inlet mass fraction can reduce the average heat transfer coefficient to one-seventh its pure steam value. The experiments were conducted for natural convection flow of steam/air using a vertical 22 mm diameter tube. Siddique et al. [38] measured local conditions for condensation in a vertical 46 mm diameter tube using air/steam and helium/steam mixtures under forced convection conditions. They showed that the local coefficient is quickly reduced as the fraction of noncondensable gases increases, so that the reduction factor from inlet to outlet in the 2.5 m tube was of the order of 5–10. Helium was more inhibiting for the same inlet mass fraction, but air was more inhibiting for the same molar inlet fraction.

Modeling the effects of noncondensables on heat transfer has followed two paths. Starting with Colburn and Hougen [9], the analogy of heat and mass transfer has been applied to model the heat transfer behavior in a vertical tube. Among the recent contributions to this approach are Siddique et al. [39] and Oh and Revankar [32], who provided an approach free of empiricism.

For a vertical plate or flow external to a pipe, and nonflow conditions, a boundary layer modeling approach is more common. Sparrow and Lin [40] considered laminar film condensation of saturated steam/air mixtures on a flat plate. They considered the liquid condensate layer and the steam/gas layer assuming constant properties. Minkowycz and Sparrow [31] extended this approach to include the effects of interfacial resistance, steam superheating, variable properties, and diffusion. The effect of air at a mass fraction of 0.1% was to reduce the heat transfer by 10%. This effect grew to 90% reduction at a mass fraction of 5%. Sparrowet al. [41] later included the effect of forced convection. Solutions with an increasingly more flexible description of the vapor and gas flows relying on numerical solutions of the coupled equations have followed, for example, Dehbi et al. [13].

PROBLEMS

12.1. *Comparison of liquid superheat required for nucleation in water and sodium* (Section 12.2)

1. Calculate the relation between superheat and equilibrium bubble radius for sodium at 1 atmosphere using Equation 12.7. Repeat for water at 1 atmosphere. Does sodium require higher or lower superheat than water?
2. Consider bubbles of 10 μm radius. Evaluate the vapor superheat and the bubble pressure difference for the bubbles of question 1.

Answers:

1. For sodium: $T_\ell - T_{\text{sat}} = 2.517 \times 10^{-4} \dfrac{1}{r_{\text{e}}}$

   For water: $T_\ell - T_{\text{sat}} = 3.25 \times 10^{-5} \dfrac{1}{r_{\text{e}}}$

2. $(T_\ell - T_{\text{sat}})_{\text{sodium}} = 25.2\,°\text{C}; \quad (T_\ell - T_{\text{sat}})_{\text{water}} = 3.25\,°\text{C}$

   $(p_{\text{v}} - p_\ell)_{\text{sodium}} = 22.6\,\text{kN/m}^2, \; (p_{\text{v}} - p_\ell)_{\text{water}} = 11.8\,\text{kN/m}^2$

12.2. *Nucleate boiling initiation and termination on a heat exchanger tube* (Chapters 8 and 10, and Section 12.2)

A heat exchanger tube is immersed in a water cooling tank at 290 K, as illustrated in Figure 12.13. Hot water (single phase, 550 K) enters the tube inlet and is cooled as it flows at 2 kg/s through the 316 grade stainless-steel tube (19 mm outside diameter and 15.8 mm inside diameter). Neglect entrance effects.

a. Compute the length from the inlet along the horizontal inlet length of the tube where nucleate boiling on the tube outside diameter is initiated.
b. Compute the length where nucleate boiling on the tube outside diameter is terminated.

The heat transfer coefficient between the outer tube wall and the water cooling tank is 500 W/m^{2}K for single-phase conditions and 5000 W/m^{2}K for nucleate boiling conditions. The wall superheat for incipient nucleation is 15°C for this configuration. Estimate and justify any additional information you need to execute the solution.

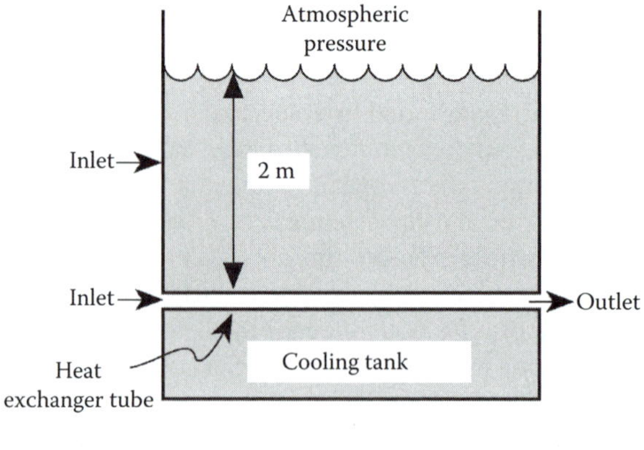

FIGURE 12.13 Cooling tank—heat exchange diagram.

Fluid properties of inlet water (assume they stay constant)
k, thermal conductivity = 0.5 W/m°C
ρ, density = 704 kg/m^3
μ, viscosity = 8.69×10^{-5} kg/ms
c_p, heat capacity = 6270 J/kg°C

Answers:

1. $x = 0$ m
2. $L = 35.16$ m

12.3. *Evaluation of pool boiling conditions at high pressure* (Section 12.4)

1. A manufacturer has free access on the weekends to a supply of 8000 amps of 440 V electric power. If his boiler operates at 3.35 MPa, how many 2.5 cm diameter, 2 m-long electric emersion heaters would be required to utilize the entire available electric power? He desires to operate at 80% of critical heat flux.
2. At what heat flux would incipient boiling occur? Assume that the water at the saturation temperature corresponds to 3.35 MPa. The natural convection heat flux is given by

$$q''_{NC} = 2.63(\Delta T)^{1.25}\ \text{kW/m}^2$$

$p = 3.35$ MPa
$T_{sat} = 240$°C
$h_{fg} = 1766$ kJ/kg
$\rho_\ell = 813$ kg/m^3
$\rho_v = 16.8$ kg/m^3
$\sigma = 0.0286$ N/m
$k = 0.0628$ W/m°C

Assume that the maximum cavity radius is very large. You may use the Rohsenow correlation for nucleate pool boiling taking $C_{s,f} = 0.0130$ in Equation 12.18c.

Answers:

1. $N = 6$ (based on $C_\ell = 0.18$ in Equation 12.31)
2. $q''_i = 4.19$ kW/m^2

12.4. *Shell and tube horizontal evaporator* (Chapters 9 and 10, and Section 12.4)

A shell-and-tube horizontal evaporator is to be designed with 30 tubes of 1 cm diameter. Inside the tubes water at 100 psia (690 kN/m^2) enters at one end at 130°C and leaves at the other end at 120°C. The water velocity υ in the tubes is 3 m/s. In the shell side, atmospheric pressure steam is generated at 100°C. Calculate

1. The length of the tubes
2. Rate of evaporation, kg/s
3. Rate of flow of the water on the tube side, kg/s
4. Pressure drop in the tubes on the water side

(Assume fully developed flow and neglect entrance and exit losses).

For the boiling side, use the Stephan and Abdelsalam correlation (Equation 12.22a).

Make your calculations for heat flux at mid-point where the liquid is at 125°C. Neglect thermal resistance of the thin tube wall.

Assume properties of liquid inside the tubes at 690 kPa in the range 120–130°C are the same as those at 1 atm and 100°C given in Table 12.4.

TABLE 12.4
Water Properties for Problem 12.4

	Liquid		Vapor
ρ (kg/m^3)	960		0.60
c_p (kJ/kg°C)	4.2		1.88
μ(kg/m s)	3×10^{-4}		1.3×10^{-5}
k(W/m°C)	0.68		0.025
σ (N/m)		0.06	
h_{fg} (kJ/kg)		2280	
Pr	1.9		0.97

Properties for water at $p = 1$ atm $T_{sat} = 100°C$

Answers:
1. $L = 0.138$ m
2. Rate of evaporation = 0.125 kg/s
3. Mass flow rate = 6.79 kg/s
4. $\Delta p = 554$ kPa

12.5. *Comparison of stable film boiling conditions in water and sodium* (Section 12.5)
Compare the value of the wall superheat required to sustain film boiling on a horizontal steel wall as predicted by Berenson's correlation to that predicted by Henry's correlation (Table 12.3).
1. Consider the cases of saturated water at (1) atmospheric pressure and (2) $p = 7.0$ MPa.
2. Consider the case of saturated sodium at atmospheric pressure.

Answers:
1. H_2O at 0.1 MPa:

$T_B^M - T_{sat} = 68.9°C$ $T_H^M - T_{sat} = 187°C$

H_2O at 7.0 MPa:

$T_B^M - T_{sat} = 1042°C$ $T_H^M - T_{sat} = 1443°C$
(too high in physical sense) (too high in physical sense)

2. Na at 0.1 MPa:

$T_B^M - T_{sat} = 30.6°C$ $T_H^M - T_{sat} = 411°C$
(too low in physical sense)

12.6. *Analysis of decay heat removal during a severe accident* (Chapters 3 and 4, and Section 12.5)
Extreme events in which the reactor core melts partially or completely are designated by nuclear engineers as "severe accidents." Consider a severe accident during which the core has completely melted, thus falling to the bottom of the pressure vessel. The situation is illustrated in Figure 12.14. The molten mixture of fuel (UO_2), fission products, clad (Zr), control

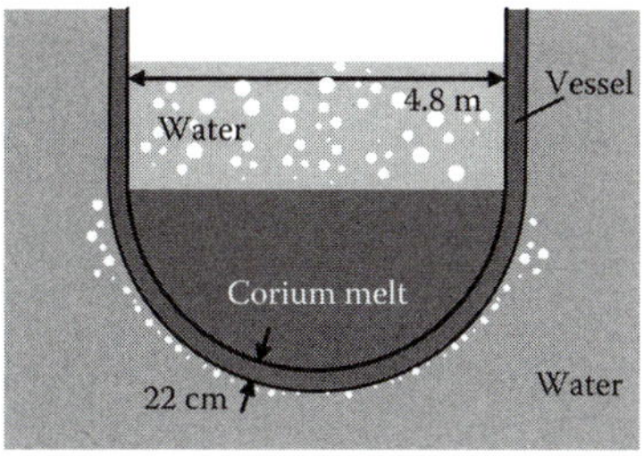

FIGURE 12.14 The lower head of the reactor vessel during a severe accident with complete melting of the core.

rod material (B_4C), and core-supporting structures (steel) is known in severe accident analysis as "corium." In the situation considered here, the corium melt fills the bottom of the vessel up to the junction of the hemispherical lower head with the cylindrical beltline region. There is water above the corium and water outside the vessel. The fuel decay heat is removed by boiling above the corium, and by conduction through the vessel wall. The whole system is at atmospheric pressure.

1. At normal operating conditions, the thermal power of this reactor is 3400 MW_{th}. Three hours after reactor shutdown, the corium melt is at 2000°C, the temperature on the outer surface of the vessel is 112°C, and the temperature of the water above the corium is 100°C. At this time, is the corium heating up, cooling down, or staying at steady temperature? (*Hint:* Assume that the temperature distribution within the corium melt is uniform.)
2. At a certain time during the accident, the heat flux at the upper surface of the corium melt is 200 kW/m^2. Calculate the average void fraction in the (stagnant) water above the corium melt. (Use the drift-flux model with $C_o = 1$ and the expression of V_{vj} for churn flow. Assume water at saturated conditions.)
3. With regard to question 2, would a HEM approach be acceptable? Why?

In solving question 1, use the following film boiling heat transfer correlation with convection (by Berenson) and radiation components (see Equation 12.30b). Table 12.5 lists materials properties that should be used in the solution of this problem.

Answers:

1. Heating up.
2. $\alpha \approx 0.381$
3. HEM is a poor approach since the liquid is nearly stagnant while the vapor is flowing upward.

12.7. *Void fraction and pressure drop in an isolation condenser* (Chapters 6, 9, 10, and 11, and Section 12.7)

A modern BWR uses an isolation condenser to remove the decay heat from the core following a feedwater pump trip. The isolation condenser receives 50 kg/s of saturated dry steam at 280°C and condenses it completely. The isolation condenser consists of 200 horizontal round tubes of 3 cm inner diameter and 12 m length. The condensing steam flows

TABLE 12.5
Materials Properties for Problem 12.4

Parameter	Value
T_{sat} (°C/K)	100 (373)
ρ_f (kg/m^3)	960
ρ_g (kg/m^3)	0.6
h_f (kJ/kg)	419
h_g (kJ/kg)	2675
c_{pf} [kJ/(kg°C)]	4.2
c_{pg} [kJ/(kg°C)]	2.1
μ_f (Pa s)	2.8×10^{-4}
μ_g (Pa s)	1.2×10^{-5}
k_f [W/(m°C)]	0.68
k_g [W/(m°C)]	0.02
σ (N/m)	0.06

Note: Corium—density: 8000 kg/m^3; specific heat: 530 J/kg°C; Emissivity: 0.5.
Vessel steel—density: 7500 kg/m^3; thermal conductivity: 30 W/m°C; saturated water at atmospheric pressure.

inside the tubes (see Figure 12.15). The tubes sit in a pool of water at atmospheric pressure.

1. Calculate the isolation condenser heat removal rate. (The properties of saturated water at 280°C are presented in Table 12.6.)
2. Using the simplified Chato correlation (Equation 12.46), estimate the temperature on the inner surface of the tubes. (Assume an axially uniform heat flux in the tubes.)
3. Sketch the axial profile of the void fraction in the tubes. (Assume linear variation of the quality in the tubes. Use HEM to calculate the void fraction.)
4. Calculate the acceleration, friction, gravity, and total pressure drops within the tubes. (Use the HEM approach with $f_{TP} = f_{\ell o}$ to calculate the friction pressure drop.)

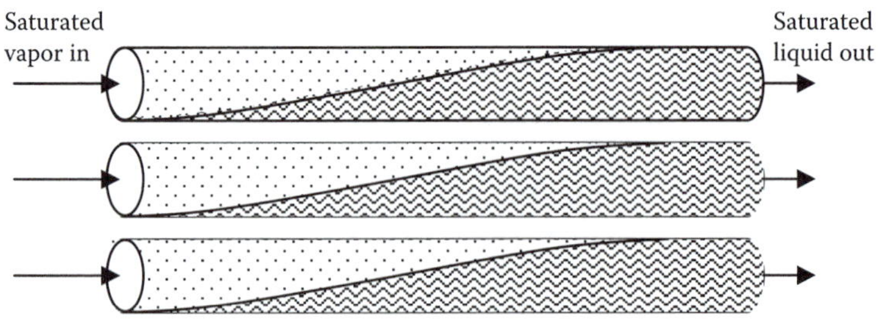

FIGURE 12.15 Isolation condenser tubes.

TABLE 12.6
Properties of Saturated Water at 280°C for Use in Problem 12.7

Parameter	Value
v_f (m^3/kg)	0.0013
v_g (m^3/kg)	0.03
h_f (kJ/kg)	1237
h_g (kJ/kg)	2780
μ_f (Pa s)	9.8×10^{-5}
μ_g (Pa s)	1.9×10^{-5}
k_f [W/(m°C)]	0.574
k_g [W/(m°C)]	0.061

5. How would the acceleration, friction, and gravity pressure drops within the tubes change, if the tubes were vertical and the steam flow were downward? (A qualitative answer is acceptable.)

Answers:

1. $\dot{Q} = 77.15\,\text{MW}_{\text{th}}$
2. $T_w \approx 216$ °C
4. $\Delta p_{\text{acc}} \approx -3594$ Pa
   $\Delta p_{\text{fric}} \approx 7060$ Pa
   $\Delta p_{\text{tot}} \equiv p_{\text{in}} - p_{\text{out}} \approx 3466$ Pa

REFERENCES

1. Auracher, H., Transition boiling in natural convection systems. In Dhir, V. and Bergles, A. E., (eds.). *Pool and External Flow Boiling*, pp. 219–236. New York, NY: ASME, 1992.
2. Berenson, P. J., Film boiling heat transfer from a horizontal surface. *J. Heat Transfer,* 83:351, 1961.
3. Blander, M. and Katz, J. L., Bubble nucleation in liquids. *AIChE J.*, 21:833, 1975.
4. Bromley, L. A., Heat transfer in stable film boiling. *Chem Eng. Prog.,* 46:221–227, 1950.
5. Bromley, L. A., Le Roy, N. R., and Robbers, J. A., Heat transfer in forced convection film boiling. *Ind. Eng. Chem.*, 49:1921–1928, 1953.
6. Carey, V., *Liquid-Vapor Phase-Change Phenomena*, 2nd Ed. Taylor and Francis, New York, NY, 2008.
7. Chato, J., Laminar condensation inside horizontal and inclined tubes. *J. ASHRAE,* 4:52–60, 1962.
8. Chen, J. C., A correlation for boiling heat transfer to saturated fluids in convective flow. *ASME paper 63-HT-34*, 1963, subsequently published as Chen, J. C. A correlation for boiling heat transfer in convection flow. *Ind. Eng. Chem. Process Des. Dev.,* 5:322, 1966.
9. Colburn, A. P. and Hougen, O. A., Design of cooler condensers for mixture of vapors with non-condensable gases. *Ind. Eng. Chem.,* 26:1178–1182, 1934.
10. Collier, J. G. and Thome, J. R., *Convective Boiling and Condensation*, 3rd Ed. Oxford: Clarendon Press, 1994.
11. Cooper, M. G., Saturation nucleate pool boiling—A simple correlation, Institute of Mechanical Engineers, London. *IchemE Symp. Ser.,* 86:786–793, 1984.

12. Dehbi, A., Golay, M., and Kazimi, M., Condensation experiments in Steam-Air and Steam-Air-Helium Mixtures under turbulent natural convection. *27th National Heat Transfer Conference (AIChE)*, Minneapolis, Minnesota, July 1991.
13. Dehbi, A., Golay, M., and Kazimi, M., A theoretical modeling of the effects of noncondensable gases on steam condensation under turbulent natural convection. *27th National Heat Transfer Conference (AIChE)*, Minneapolis, Minnesota, July 1991.
14. Dhir, V. J., Mechanistic prediction of nucleate boiling heat transfer—Achievable or a hopeless task. *J. Heat Transfer,* 128:1–12, 2006.
15. Forster, H. K. and Greif, R., Heat transfer to a boiling liquid—Mechanism and correlation. *ASME J. Heat Transfer,* 81:43–53, 1959.
16. Forster, H. K. and Zuber, N., Dynamics of vapor bubbles and boiling heat transfer. *AIChE J.*, 1:531–535, 1955.
17. Fujii, T. and Uehara, H., Laminar filmwise condensation on a vertical surface. *Int. J. Heat Mass Transfer,* 15:217–233, 1972.
18. Henry, R. E., A correlation for the minimum wall superheat in film boiling. *Trans. Am. Nucl. Soc.,* 15:420, 1972.
19. Hewitt, G. F., Boiling, Chapter 15 . In W. M. Rohsenow, J. P. Hartnett, and Y. Cho (eds.). *Handbook of Heat Transfer*, 3rd Ed. McGraw-Hill, 1998.
20. Hewitt, G. F., Exec. (ed.) *Heat Exchanger Design Handbook: Part 2, Fundamentals of Heat and Mass Transfer*. Begell House Inc., New York, NY, 2008.
21. Hsu, Y. Y. and Westwater, J. W., Approximate theory for film boiling on vertical surfaces. *AIChE Chem. Eng. Prog. Symp. Ser.,* 56(30), 15, AIChE, New York, NY, 1960.
22. Jakob, M., Heat transfer in evaporation and condensation I and II. *Mech. Eng.,* 58:643–660 (I), 729–739 (II), 1936.
23. Kalinin, E. et al., *Investigation of the Crisis of Film Boiling in Channels*. USSR: Moscow Aviation Institute, 1968.
24. Kandlikar, S. G., A theoretical model to predict pool boiling CHF incorporating effects of contact angle and orientation. *J. Heat Transfer,* 123:1071ff, 2001.
25. Kandlikar, S. G., Shoji, M., and Dhir, V. K., *Handbook of Phase Change: Boiling and Condensation*. Taylor and Francis, Philadelphia, PA, 1999.
26. Kim, H., DeWitt, G., McKrell, T., Buongiorno, J., and Hu, L., On the quenching of steel and zircaloy spheres in water-based nanofluids with alumina, silica and diamond nanoparticles. *Int. J. of Multiphase Flow*, 35:427–438, 2009.
27. Kim, S. J., Bang, I. C., Buongiorno, J., and Hu, L. W., Surface wettability change during pool boiling of nanofluids and its effect on critical heat flux. *Int. J. Heat Mass Transfer,* 50:4105–4116, 2007.
28. Kim, M. and Corradini, M., Modeling of condensation heat transfer in a reactor containment. *Nucl. Eng. Des.,* 118:193–212, 1990.
29. Klimenko, V. V., Film boiling on a horizontal plate—New correlation. *Int. J. Heat Mass Transfer,* 24, 69–79, 1981.
30. Kutateladze, S. S., *Heat Transfer in Condensation and Boiling*. AEC-TR-3770, U.S. Atomic Energy Commission, Oak Ridge, TN, 1959. (Originally published in Russian under same title by State Scientific and Technical Publ. of Literature and Machinery, Moscow, 1952.)
31. Minkowycz, W. J. and Sparrow, E. M., Condensation heat transfer in the presence of noncondensables, interfacial resistance, superheating, variable properties and diffusion. *Int. J. Heat Mass Transfer,* 9:1125–1144, 1966.
32. Oh, S. and Revankar, S., Experimental and theoretical investigation of film condensation with non-condensable gas. *Int. J. Heat Mass Transfer,* 49:2523, 2006.
33. Othmer, D., The condensation of steam. *Ind. Eng. Chem.,* 6(21): 577–583, 1929.
34. Rohsenow, W. M., A method of correlating heat transfer data for surface boiling of liquids. *Trans. ASME,* 74:969–976, 1952.

35. Rohsenow, W. M., Pool boiling. In G. Hetsroni (ed.). *Handbook of Multifluid Systems.* New York, NY: Hemisphere, 1982.
36. Sakurai, A. and Shiotsu, M., Pool film boiling heat transfer and minimum film boiling temperature. In V. K. Dhir and A. E. Bergles (eds.). *Pool and External Flow Boiling.* pp. 277–301, New York, NY: ASME, 1992.
37. Shah, M., A general correlation for heat transfer during film condensation inside pipes. *Int. J. Heat Mass Transfer,* 22:547–556, 1979.
38. Siddique, M., Golay, M. W., and Kazimi, M. S., Local heat transfer coefficients in a vertical tube in the presence of air, *7th Proc. of Nuclear Thermal Hydraulics*, ANS Winter Mtg., San Francisco, CA, November 1991, pp. 221–228; also appears in *Proceedings, of Symposium on Basic Aspects of Two Phase Flow and Heat Transfer*, ASME, 1992.
39. Siddique, M., Golay, M., and Kazimi, J., Theoretical modeling of forced convection condensation of steam in a vertical tube in the presence of a noncondensable gas. *Nucl. Tech.,* 106:202, 1994.
40. Sparrow, E. M. and Lin, S. H., Condensation heat transfer in the presence of a noncondensable gas. *J. Heat Transfer,* 86:430–436, 1964.
41. Sparrow, E. M., Minkowycz, W. J., and Saddy, M., Forced convection condensation in the presence of noncondensables and interfacial resistance. *Int. J. Heat Mass Transfer,* 10:1829–1845, 1967.
42. Spiegler, P., Hopenfeld, J., Silberberg, M., Bumpus, C. F. Jr., and Norman, A., Onset of stable film boiling and the foam limit. *Int. J. Heat Mass Transfer,* 6:987–994, 1963.
43. Stephan, K. and Abdelsalam, M., Heat transfer correlation for natural convection boiling. *Int. J. Heat Mass Transfer,* 23:73–87, 1980.
44. Sun, K. H. and Lienhard, J. H., The peak pool boiling heat flux on horizontal cylinders. *Int. J. Heat Mass Transfer,* 13:1425–1439, 1970.
45. Tagami, T., *Interim Report on Safety Assessments and Facilities Establishment Project for June 1965, No. 1*, Japanese Atomic Energy Research Agency, 1965.
46. Takeyama, T. and Shimizu, S., On the transition of dropwise condensation. *Proceedings of the 5th International Heat Transfer Conference* 3:274, 1974.
47. Uchida, H., Oyama, A., and Togo, Y., Evaluation of post-incident cooling systems of light water rectors, *Proceedings of the 3rd International Convention of Peaceful Uses of Atomic Energy,* 13:93–103, 1964.
48. Vierow, K., *Behaviour of Steam-Air Systems Condensing in Cocurrent Vertical Downflow.* MS Thesis, Dept. of Nuclear Eng., U. of California, Berkeley, 1990.
49. You, S. M., Kim, J. H., and Kim, K. H., Effect of nanoparticles on critical heat flux of water in pool boiling heat transfer. *App. Phys. Lett.,* 83(16), 3374ff, 20 Oct. 2003.
50. Zuber, N., *Hydrodynamic Aspects of Boiling Heat Transfer.* AECU-4439, AEC Technical Information Service Extension, Oak Ridge, TN, 1959. (Originally published as Ph.D. thesis under same title by University of California, Los Angeles, CA, 1959.)

13 Flow Boiling

13.1 INTRODUCTION

The regimes of heat transfer in a flowing system depend on a number of variables: mass flow rate, fluids employed, wall materials, geometry of the system, heat flux magnitude, and distribution. In nuclear applications, the two basic boundary conditions are the surface heat flux condition on the fuel elements surrounding the flow channel and the primary coolant side surface temperature condition on a steam generator tube or the surface temperature of other heat exchange equipment. For the fuel element, the axial heat flux distribution while ideally cosine shaped is altered significantly in LWRs, particularly BWRs, due to neutronic-thermal hydraulic interactions. For steam generators the once-through versus the U-tube design achieves a limited degree of secondary coolant superheat.

In principle for each of these cases, the same flow patterns and heat transfer regimes are encountered along the flow channel. However, their axial location and extent, and hence the wall temperature distributions, vary with the heat flux distribution.

The chapter first qualitatively describes the evolution of heat transfer regions and void fraction magnitude encountered in a heated boiling flow channel. Then the correlations for evaluating the heat flux magnitude or the wall temperature, including the channel behavior for top and bottom reflood following channel voiding upon a LOCA, are described. Finally, the critical heat flux condition is presented.

As advised in Chapter 12, the complex phenomena involved in the boiling process dictate that for results applicable to special conditions of fluid pressure and subcooling as well as surface geometry, orientation and condition, specialized handbooks such as Hewitt [36] or Kandlikar [42] and review articles cited in these sources should be consulted.

13.2 HEAT TRANSFER REGIONS AND VOID FRACTION/QUALITY DEVELOPMENT

Consider the boiling behavior in a tubular flow channel in vertical forced convection upflow subjected to an axially uniform heat addition. We first identify the heat transfer regions, then the void fraction development. In Section 13.3, specific correlations for each of the heat transfer regions are presented.

13.2.1 Heat Transfer Regions

The heat transfer regions and flow patterns, as well as the wall and fluid axial temperature variations for the heated flow channel, are illustrated in Figure 13.1. The

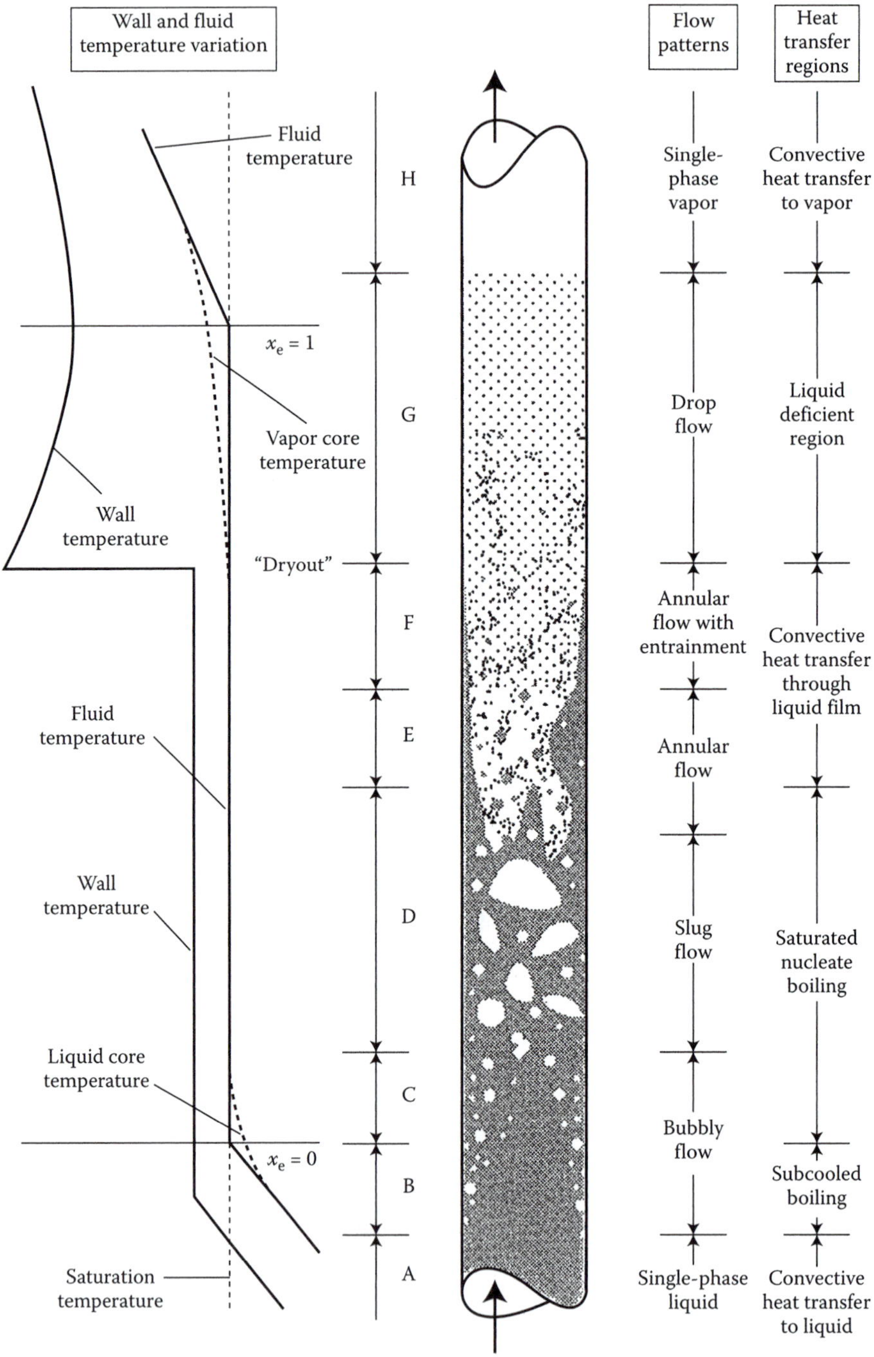

FIGURE 13.1 Regions of heat transfer in convective boiling in a flow channel with uniform wall heat flux. (Adopted from Collier, J. G. and Thome, J. R, *Convective Boiling and Condensation,* 3rd Ed. Oxford University Press, Oxford, UK, 1994.)

flow patterns have been described in Chapter 11. The regions labeled A through H in Figure 13.1 are further described in Figure 13.2, where their dependence on heat flux and quality is elaborated. For a modest magnitude of uniform wall heat flux, the axial transit up the flow channel would be represented by a horizontal line imposed in Figure 13.2. This transit would pass through each labeled region from A to H.

The liquid enters the channel in a subcooled state ($T_{in} < T_{sat}$), and rises in temperature due to heat addition. Then, at a certain height, the liquid near the wall becomes superheated and can nucleate a vapor bubble while the bulk liquid temperature may still be subcooled. (The equilibrium quality is still negative.) When *subcooled boiling* starts, the boiling process and the turbulence caused by the boiling enhance heat transfer and the wall temperature ceases to rise as fast as in the single-phase entry region. The bulk liquid temperature continues to rise until it reaches T_{sat}, thus starting a region of *saturated nucleate boiling*. Here, the bubbles become numerous and, as they detach, may start to agglomerate into larger bubbles, thus changing the flow pattern into slug or churn flow. In this regime, the wall temperature does not rise because the heat transfer coefficient is still high and the bulk fluid temperature is the saturation temperature, T_{sat}. In fact, due to the ever-increasing turbulence near the wall because of nucleation, the wall temperature decreases slightly. As boiling continues, the bubbles may merge totally into a vapor core in the tube while the liquid

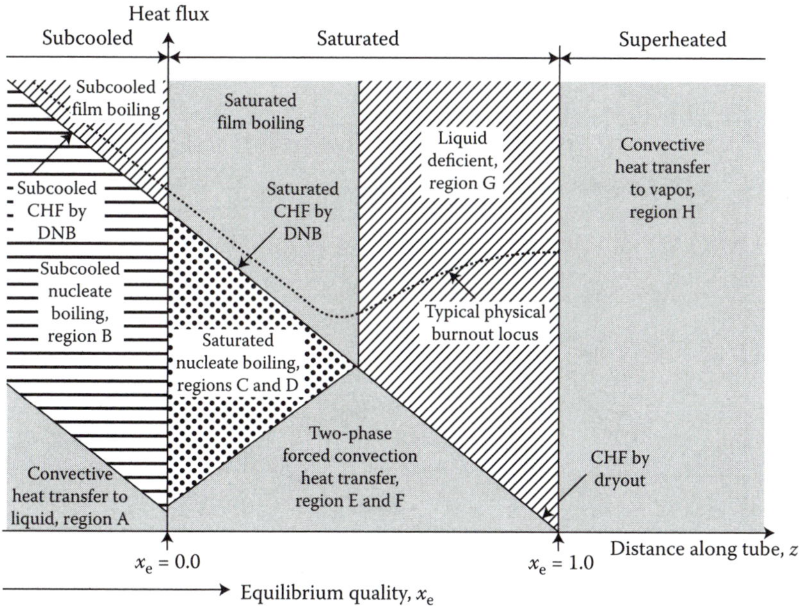

FIGURE 13.2 Dependence of two-phase forced convective heat transfer regimes on quality and heat flux. A–H = regions associated with axial regions of Figure 13.1 for conditions leading to CHF by Dryout. (From Collier, J, G. and Thome, J. R, *Convective Boiling and Condensation,* 3rd Ed. Oxford University Press, Oxford, UK, 1994.)

remains partially in a film on the walls and partially as entrained droplets in the vapor core. This annular flow pattern leads to a convection heat transfer mechanism through the liquid film. The wall temperature in regions E and F of Figure 13.1 may become insufficient to maintain the necessary wall superheat to form bubbles. Evaporation may continue only at the liquid–vapor interface (referred to as the nucleation suppression phenomenon). At some point, the liquid film on the wall becomes depleted owing to vapor entrainment and evaporation, and a *Dryout* condition occurs. Above that point, the flow is mostly vapor, with dispersed liquid droplets. This liquid-deficient heat transfer causes the wall temperature to rise abruptly, as vapor heat transfer is less efficient than liquid heat transfer. Thus, Dryout is a boiling crisis mechanism. The impingement of droplets on the wall may reduce the wall temperature right after the Dryout position. In the post-Dryout region, even before the thermodynamic equilibrium quality approaches unity, superheated vapor exists along with liquid droplets. With convective heat transfer mainly to vapor, the wall temperature again increases.

If the imposed heat flux is high, it is possible that the vapor-generation rate in the nucleate boiling regime becomes so high as to establish a gas film that separates the liquid from the wall, well before the occurrence of Dryout. This situation leads to DNB at the wall, similar to the CHF condition in pool boiling. The maximum heat flux that can be tolerated without establishing a vapor film is the CHF for these conditions.

The important dependence of the heat transfer coefficient on the vapor quality for both high- and low-heat flux conditions, shown in Figure 13.2, is specifically emphasized in Figure 13.3. It is seen that the higher the heat flux, the lower is the equilibrium quality at which the boiling inception and the CHF occur.

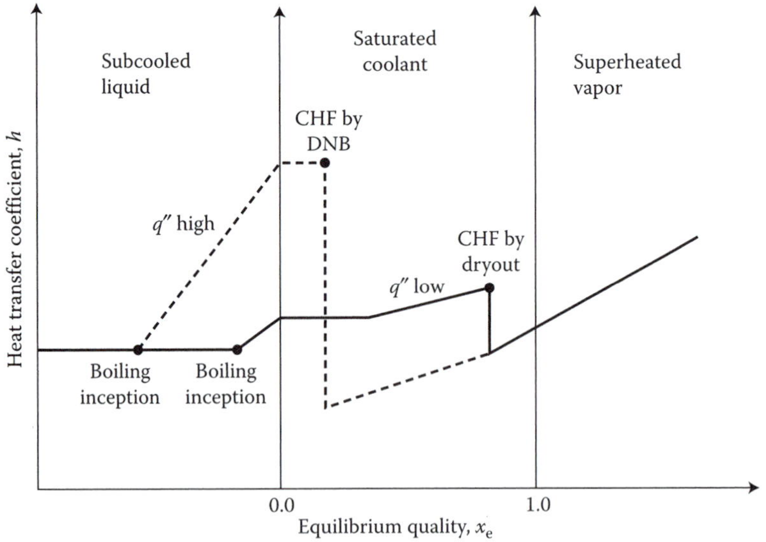

FIGURE 13.3 Possible variation of the heat transfer coefficient with quality. Pressure, flow rate, and heat flux are fixed.

The boundaries and void fraction behavior in the boiling channel will be discussed using the subregions defined in Figure 13.4. The parameters of interest are the locations of the transition points (Z_{ONB}, Z_D, Z_{OSB}, and Z_E)* as well as predictions of the quality and void fraction. The short summary that follows gives different models for the determination of these parameters.

The subcooled boiling region begins with the onset of nucleate boiling (at $Z = Z_{ONB}$) while the mean (or bulk) temperature is below the saturated temperature. However, for nucleation to occur, the fluid temperature near the wall must be somewhat higher than T_{sat}. Therefore, vapor bubbles begin to nucleate at the wall. Because most of the liquid is still subcooled, the bubbles do not detach but grow and collapse while attached to the wall, giving a small nonzero void fraction (Region I in Figure 13.4) that may be neglected. As the bulk of the coolant heats up, the bubbles can grow larger, and the possibility that they will detach from the wall surface into the flow stream increases. In Region II ($Z > Z_D$), the bubbles detach regularly and condense slowly as they move through the fluid, and the vapor voidage penetrates to the

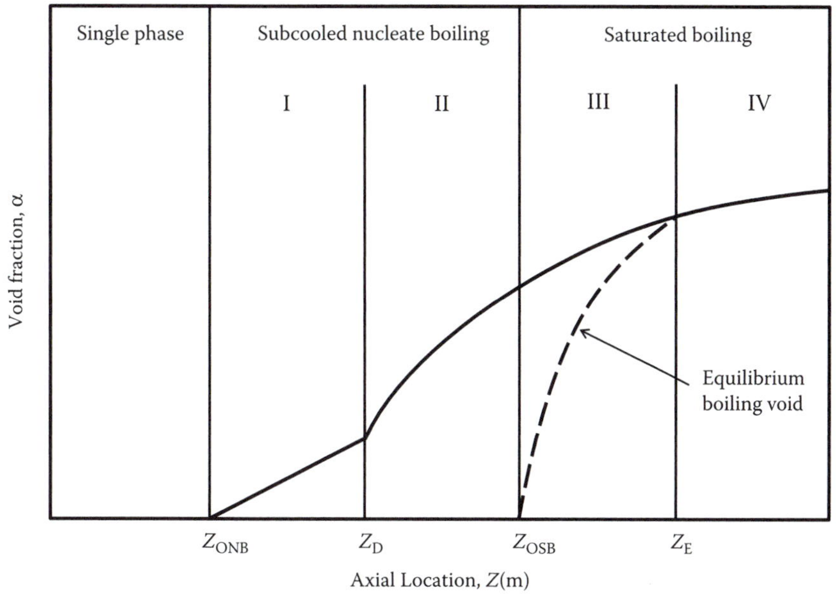

FIGURE 13.4 Development of area-averaged void fraction in a heated channel. Region I: void fraction is small and may be neglected. Region II: bubbles are significant; they are ejected from the wall into the bulk and partially collapse there. Region III: Bubbles do not condense, since thermal equilibrium exists in the channel. Region IV: Void fraction loses the subcooling history.

* Caution: axial locations designated by uppercase Z are measured from the channel inlet. For convenience with the cosine flux shape assumption, locations from the channel axial centerline are designated by lowercase z. Hence distinction exists between locations identified by Z and z, specifically $Z = L/2 + z$ where z can be negative or positive while Z is always positive.

fluid bulk. As Figure 13.4 indicates, the void fraction increases significantly. The next stage, Region III, is initiated when the bulk liquid becomes saturated at $Z = Z_{OSB}$, which is the onset of saturated boiling position. The void fraction continues to increase, approaching the thermal equilibrium condition (at $Z = Z_E$). This point marks the beginning of the region IV, where the thermodynamic nonequilibrium history is completely lost.

We next present approaches to predict locations Z_{ONB}, Z_D, Z_{OSB}, and Z_E.

13.2.1.1 Onset of Nucleate Boiling, Z_{ONB}

A criterion for boiling inception (i.e., $Z = Z_{ONB}$) in forced flow was developed by Bergles and Rohsenow [4] based on the suggestion of Hsu and Graham [37]. Their analyses were later confirmed in a more general derivation by Davis and Anderson [18]. The basic premise is that the liquid temperature due to the heat flux near the wall must be equal to the temperature associated with the required superheat for bubble stability (Figure 13.5).

The first possible equality occurs when the two temperatures tangentially make contact, which assumes that the wall has cavities of various sizes and the bubble grows at the cavity of radius r^*. The liquid temperature and the heat flux near the wall are related by

$$q'' = -k_\ell \frac{\partial T}{\partial r} \quad \text{or} \quad \frac{\partial T}{\partial r} = -\frac{q''}{k_\ell} \tag{13.1}$$

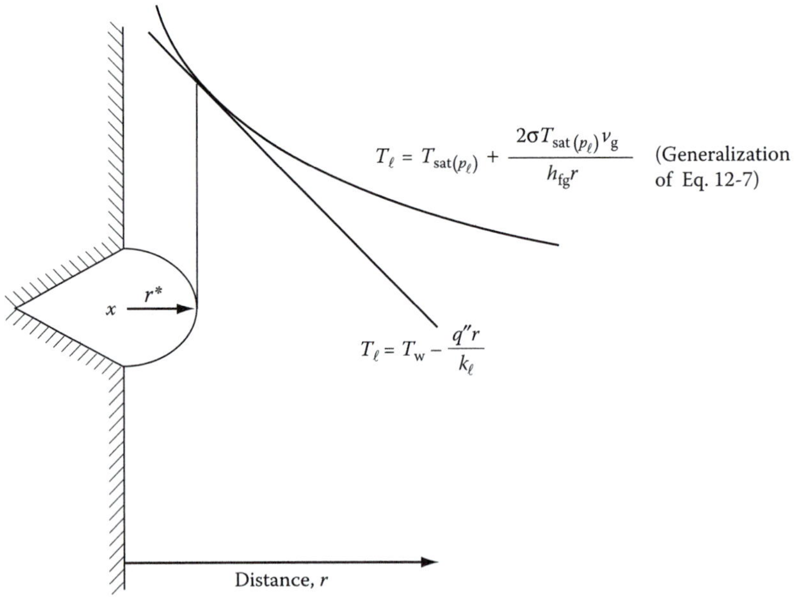

FIGURE 13.5 Critical cavity size for nucleation at a wall.

Equating this gradient to the gradient of the required liquid superheat (given in Equation 12.7) yields the cavity radius given by

$$r^*_{\text{tang}} = \sqrt{\frac{2\sigma T_{\text{sat}} v_{\text{g}} k_\ell}{h_{\text{fg}} q''}} \tag{13.2}$$

To support bubble nucleation, the liquid superheated boundary layer was assumed to extend to a thickness twice the critical cavity radius. This assumption, after some manipulation, leads to the following equation which expresses the required heat flux for onset of nucleation in terms of wall superheat assuming the required size surface cavity is present.

$$(q'')_{\text{i}} = \frac{k_\ell h_{\text{fg}}}{8\sigma T_{\text{sat}} v_{\text{g}}} (T_{\text{w}} - T_{\text{sat}})_{\text{i}}^2 \tag{13.3}$$

This is the form of the correlation proposed by Davis and Anderson [18] for systems at high-pressure or low-surface tension. It agrees well over a wide range of pressure and wall heat flux with the empirical dimensional correlation of Bergles and Rohsenow [4] proposed for water at a pressure between 0.1 and 13.6 MPa.

$$(q'')_{\text{i}} = 15.6\, p^{1.156} (T_{\text{W}} - T_{\text{sat}})_{\text{i}}^{2.3/p^{0.0234}} \tag{13.4}$$

where p is in psia; T_{W} and T_{sat} are in F; and q'' is in Btu/h ft^2. This correlation assumes the availability of large wall microcavities and ignores the surface finish effects. The incipient heat flux from Equation 13.4 is relatively low, as this relation was based on visual observation of the first nucleation occurrence rather than on observation of wall temperature response, which requires a significant number of nucleation bubbles.

If the wall heat flux is available, Equation 13.3 can be expressed as the required wall superheat for onset of nucleation, again assuming the required size of surface cavity is present yielding

$$\left(T_{\text{w}} - T_{\text{sat}}\right)_{\text{i}} = \left(\frac{8\sigma T_{\text{sat}} v_{\text{g}} q''}{k_\ell h_{\text{fg}}}\right)^{1/2} \tag{13.5}$$

Alternately, if the wall heat flux is not available, the required wall superheat can be expressed in terms of the difference between the incipient boiling wall temperature and the corresponding bulk fluid temperature by observing that at the point of incipience the heat flux is also given by the convection law, so that

$$(q'')_{\text{i}} = h_{\text{c}} (T_{\text{W}} - T_{\text{bulk}})_{\text{i}} \tag{13.6}$$

Hence substituting Equation 13.6 into Equation 13.3 yields the incipient wall superheat in the following form:

$$\frac{(T_W - T_{sat})_i^2}{(T_W - T_{bulk})_i} = \frac{1}{\Gamma} \tag{13.7a}$$

where

$$\Gamma \equiv \frac{k_\ell h_{fg}}{8\sigma T_{sat} v_g h_c} \tag{13.7b}$$

At low-heat fluxes, the point of tangency may be at a radius larger than the available size cavities in the wall. In this case, as in natural convection, Equation 13.3 would underpredict the required superheat and $(q'')_i$. Bjorge et al. [7] recommended that a maximum cavity for most surfaces in contact with water is $r_{max} = 10^{-6}$ m. When the liquid has a poor surface wetting ability, the apparent cavity size is larger than the actual size.

Rohsenow [50] showed that when the single-phase point of tangency occurs at the maximum available cavity radius, the convective heat transfer coefficient (h_c) is given by

$$h_c = \frac{k_\ell / r_{max}}{1 + \sqrt{1 + 4\Gamma(T_{sat} - T_{bulk})}} \tag{13.8}$$

If h_c is less than this magnitude, Equation 13.3 would not be reliable. When the liquid temperature T_{bulk} is at saturation, the value of h_c is $k_\ell/2r_{max}$, consistent with the assumption of a superheat boundary layer thickness equal to twice the cavity radius.

13.2.1.2 Bubble Departure, Z_D

The point at which bubbles can depart from the wall before they suffer condensation, also called the location of net vapor generation (Z_D), has been proposed to be either hydrodynamically controlled or thermally controlled. Among the early proposals for thermally controlled departure are those by Griffith et al. [25], Bowring [8], and Dix [20]. Griffith et al. [25] proposed

$$T_{sat} - T_{bulk,D} = \frac{q''}{5h_{\ell o}} \tag{13.9}$$

where $h_{\ell o}$ = the heat transfer coefficient of single-phase liquid flowing at the same total mass flow rate. In Equation 13.9, the unknown is $T_{bulk,D}$ where the subscript D refers to "departure" in the same manner as the subscript D in the location Z_D.

Bowring [8] proposed

$$T_{\text{sat}} - T_{\text{bulk,D}} = \frac{\eta q''}{G/\rho_{\text{f}}} \tag{13.10}$$

where $\eta = 0.94 + 0.00046\,p$ ($156 < p < 2000$ psia); T is in °F, G is in lb_m/h ft^2, ρ_f is in lb_m/ft^3, and q'' is in Btu/h ft^2.

Dix [20] proposed

$$T_{\text{sat}} - T_{\text{bulk,D}} = 0.00135\frac{q''}{h_{\ell o}}(\text{Re}_\ell)^{1/2} \tag{13.11}$$

The above three criteria are based on the assumption that at Z_D the wall heat flux is balanced by heat removal due to liquid subcooling.

Levy [45] introduced a hydrodynamically based model, assuming that the bubble detachment is primarily the result of drag (or shear) force overcoming the surface tension force. Staub [57] added the effect of buoyancy to the Levy model.

Later, Saha and Zuber [51] postulated that the hydrodynamic and the heat transfer mechanisms both may apply. Thus in the low-mass flow region, heat diffusion controls the condensation process and the departure process is heat-transfer-limited, signified by the *Nusselt number*:

$$\text{Nu} = \frac{q'' D_\text{e}}{k_\ell (T_{\text{sat}} - T_{\text{bulk,D}})} \tag{13.12}$$

whereas for high-flow rates both the heat transfer and the hydrodynamics are controlling, signified by the *Stanton number*:

$$\text{St} = \frac{q''}{G c_{p\ell}(T_{\text{sat}} - T_{\text{bulk,D}})} \tag{13.13}$$

The data from various sources were plotted against the *Peclet number* (Figure 13.6), where

$$\text{Pe} = \frac{\text{Nu}}{\text{St}} = \frac{G D_\text{e}\, c_{p\ell}}{k_\ell} \tag{13.14}$$

in rectangular, annular, and circular tubes as well as for some freon data. They developed the following criteria which are illustrated on Figure 13.6:

For $Pe < 7 \times 10^4$:

$$(\text{Nu})_\text{D} = 455 \quad \text{or} \quad T_{\text{sat}} - T_{\text{bulk,D}} = 0.0022\left(\frac{q'' D_\text{e}}{k_\ell}\right) \tag{13.15a}$$

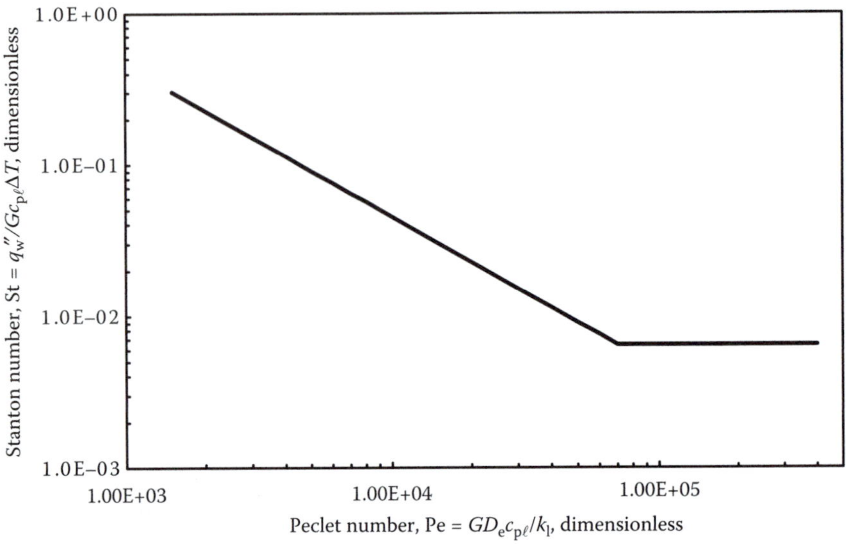

FIGURE 13.6 Bubble departure conditions as a function of the Peclet number. (Adopted from Saha, P. and Zuber, N., Point of net vapor generation and vapor void fraction in subcooled boiling. In *Proceedings 5th International Heat Transfer Conference*, pp. 175–179. Tokyo, 1974.)

For $\text{Pe} > 7 \times 10^4$:

$$(\text{St})_D = 0.0065 \quad \text{or} \quad T_{\text{sat}} - T_{\text{bulk,D}} = 154\left(\frac{q''}{G c_{p\ell}}\right) \tag{13.15b}$$

Since

$$\text{Nu} \equiv \text{St} \cdot \text{Re} \cdot \text{Pr} \equiv \text{St} \cdot \text{Pe} \tag{10.75}$$

Equation 13.15a becomes

$$(\text{St})_D = \frac{455}{\text{Pe}} \tag{13.15c}$$

Figure 13.6 illustrates the criteria expressed in terms of the Stanton vs. the Peclet numbers for Equations 13.15a and 13.15b.

The data used by Saha and Zuber [51] covered the following range of parameters for water: $p = 0.1$–13.8 MPa; $G = 95$–2760 kg/m^2 s; and $q'' = 0.28$–1.89 MW/m^2.

13.2.1.3 Onset of Saturated Boiling, Z_{OSB}

The point Z_{OSB} is the location at which the saturated liquid condition is achieved. The location of saturated liquid based on mixing cup enthalpy is simply obtained from a heat balance. Specifically, Z_{OSB} is determined from Equation 13.16 below:

$$h_f = h_{in} + \frac{1}{\dot{m}}\int_0^{Z_{OSB}} q'(Z)\,dZ \tag{13.16}$$

The representation of Equation 13.16 in Chapter 14 expresses the channel length as from $-L/2$ to $+ L/2$ for convenience in integrating the common expressions of $q'(z)$.

13.2.1.4 Location of Thermal Equilibrium, Z_E

This location is established as the position where the flow equilibrium quality, x_e, essentially attains the magnitude of the actual flow quality, x, analogous to the void behavior shown in Figure 13.4. Attempts to mechanistically determine the actual flow quality in the subcooled region have been reviewed by Lahey and Moody [44]. No completely satisfactory approach has been found, although the qualitative aspects of this region are well understood. The currently accepted approach for modeling this region is the profile-fit approach. The actual quality profile adopted here is that suggested by Levy [45] to be of the form

$$x(Z) = x_e(Z) - x_e(Z_D)\exp\left[\frac{x_e(Z)}{x_e(Z_D)} - 1\right] \tag{13.17}$$

which implies that the actual quality, x, asymptotically approaches the equilibrium quality, x_e with increasing Z.

Hence, while $x(Z) - x_e(Z)$ always has a finite value, the difference becomes infinitesimally small as Z increases since $x_e(Z_D)$ is negative, $x_e(Z)$ becomes larger and larger than $x_e(Z_D)$ as Z increases, and hence the term $\exp[x_e(Z)/x_e(Z_D) - 1] \Rightarrow 0$ as Z increases. The position Z_D can be obtained from Equation 13.17 for a given channel axial heat flux distribution and inlet subcooling upon selection of a finite (and typically small) value of $x(Z) - x_e(Z)$.

13.2.1.5 Void Fraction Profile, $\alpha(z)$

The void fraction may then be predicted from the void drift model (Equation 11.39) or any other void fraction model with $x(Z)$ available from Equation 13.17. Figure 13.4 illustrates this void fraction profile along a boiling channel.

$$\{\alpha(Z)\} = \frac{1}{C_O\left\{1 + \frac{1 - x(Z)}{x(Z)}\frac{\rho_v}{\rho_\ell}\right\} + \frac{V_{vj}\rho_v}{x(Z)G}} \tag{11.44b}$$

where C_o can be obtained from the suggestion of Dix [20]:

$$C_O = \{\beta\}\left[1+\left(\frac{1}{\{\beta\}}-1\right)^b\right] \tag{11.43a}$$

$$b=\left(\frac{\rho_v}{\rho_\ell}\right)^{0.1} \tag{11.43b}$$

and β = volumetric flow fraction of vapor.

Lahey and Moody [44] suggested that a general expression for V_{vj} may be given by

$$V_{vj} = 2.9\left[\left(\frac{\rho_\ell-\rho_v}{\rho_\ell^2}\right)\sigma g\right]^{0.25} \tag{13.18}$$

which is about twice the value of the bubble terminal velocity (V_∞) in a churn flow regime (see Table 11.3).

13.3 HEAT TRANSFER COEFFICIENT CORRELATIONS

The subcooled boiling (both single-phase convective and nucleate heat transfer components), saturated boiling (both nucleate and nucleation-suppressed convective heat transfer components), and the post-CHF heat transfer (the high-quality, liquid deficient region and the low-quality, flow film boiling region) processes will be sequentially described. The fundamentals of the nucleation phenomenon have already been presented in Chapter 12 to describe heat transfer in pool boiling. Here, the supplemental effect of liquid velocity on heat transfer in flow boiling will be described.

13.3.1 Subcooled Boiling Heat Transfer

This regime spans the range from boiling inception until the bulk coolant reaches the saturation temperature. Heat transfer occurs increasingly by nucleate boiling, h_{NB}, as the contribution of convection, h_c, in this situation single-phase convection, decreases. Hence

$$q'' = h_c(T_w - T_{bulk}) + h_{NB}(T_w - T_{sat}) \tag{13.19}$$

A procedure for calculating h_{NB} based on the Chen method for saturated boiling has been suggested by Collier and Thome [16]. For subcooled boiling, the F and S parameters from the Chen method (see Section 13.3.2) are adjusted by setting F to unity and calculating S with the quality x being set to zero. This method was found acceptable against experimental data of water, n-butyl alcohol and ammonia.

Figure 13.7 illustrates these two components of heat transfer and the transition between them with increasing wall superheat. In single-phase flow, the heat flux is proportional to $T_w - T_{bulk}$ while in nucleate boiling it is proportional to $(T_w - T_{sat})^3$ as Equation 12.21 illustrates. The locus of the points of onset of nucleation is approximately proportional to $(T_w - T_{sat})^2$. The total heat flux transfer, which is the sum of these components, is illustrated in Figure 13.7 by the dashed line. Figure 13.7 also illustrates the effect of different liquid velocity on the single-phase heat transfer component and hence the resultant subcooled boiling behavior.

The heat transfer in the early region of saturated boiling depends on the degree of bubble formation at the solid wall. So long as nucleation is present, it dominates the heat transfer rate. Hence, the correlations developed for estimating the heat flux associated with incipient boiling are still applicable in this region. Thus, the expressions for heat fluxes associated with subcooled nucleate boiling and low-quality saturated nucleate boiling are equal. However, when the flow quality is high, the liquid film becomes thin owing to evaporation and entrainment of droplets. The heat removal from the liquid film to the vapor core becomes so efficient that the nucleation within the film may be suppressed. Evaporation occurs mainly at the liquid film–vapor interface. Most correlations used for the annular heat transfer coefficient are found empirically.

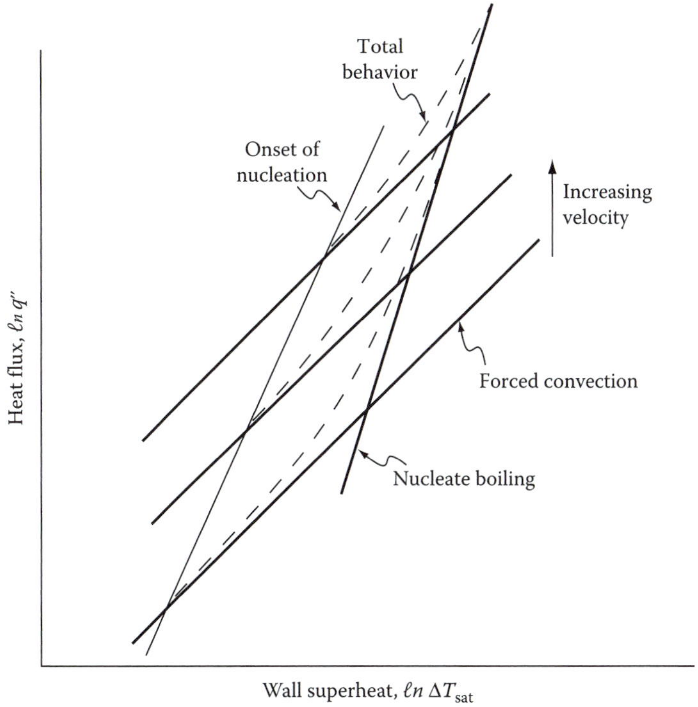

FIGURE 13.7 Superposition of forced convection and nucleate boiling heat transfer in subcooled boiling.

The heat transfer rate in the saturated boiling region is expressed analogously to Equation 13.19 as

$$q'' = (h_c + h_{NB})(T_w - T_{sat}) \tag{13.20}$$

because the bulk fluid temperature is at saturation conditions. The two-phase heat transfer coefficient ($h_{2\phi}$) is commonly expressed as the sum of the term due to nucleate boiling (h_{NB}) and the term due to convection heat transfer (h_c):

$$h_{2\phi} = h_{NB} + h_c \tag{13.21}$$

13.3.1.1 Multiple Author Correlations

A number of authors have utilized this two-term approach by formulating the two-phase heat transfer coefficient as a multiple of $h_{\ell o}$, the single-phase liquid heat transfer coefficient for the same total mass flux, that is

$$\frac{h_{2\phi}}{h_{\ell o}} = a_1 \frac{q''}{G h_{fg}} + a_2 X_{tt}^{-b} \tag{13.22}$$

where a_1, a_2, and b are empirical constants whose values in various correlations are summarized in Table 13.1 and

$$\frac{1}{X_{tt}} = \left(\frac{x}{1-x}\right)^{0.9} \left(\frac{\rho_f}{\rho_g}\right)^{0.5} \left(\frac{\mu_g}{\mu_f}\right)^{0.1} \tag{11.94}$$

Two simpler frequently used correlations for the nucleate boiling region (subcooled as well as saturated) for water at a pressure between 3.45 and 6.89 MPa are those of Jens and Lottes [41] and Thom et al. [58]. They are, respectively,

$$q'' = \frac{\exp(4p/6.2)}{(25)^4} \left(T_W - T_{sat}\right)^4 \tag{13.23a}$$

TABLE 13.1
Values of Constants for Saturated Flow Boiling Heat Transfer Coefficient in Equation 13.22

Author	a_1	a_2	b
Dengler and Addoms* [19]	0	3.5	0.5
Bennett et al.* [2]	0	2.9	0.66
Schrock and Grossman [53]	7400	1.11	0.66
Collier and Pulling [17]	6700	2.34	0.66

*Correlations appropriate only in the annular flow regime, hence $a_1 = 0$.

and

$$q'' = \frac{\exp(2p/8.7)}{(22.7)^2}\left(T_W - T_{sat}\right)^2 \tag{13.23b}$$

where q'' is in MW/m^2, p is in MPa, and T is in °C.

It should be noted that the lower slope of Thom et al.'s correlation is due to the limitation of the data they used to the early part of the boiling curve usually encountered in conventional boilers.

13.3.1.2 Chen Correlation

A popular composite correlation to cover the entire range of saturated boiling is that of Chen [13]. His correlation, which is widely used, is expressed in the form of Equation 13.21.

The convective part, h_c, is a modified Dittus–Boelter correlation given by

$$h_c = 0.023\left(\frac{G(1-x)D_e}{\mu_f}\right)^{0.8}(\mathrm{Pr_f})^{0.4}\frac{k_f}{D_e}F \tag{13.24}$$

The factor F accounts for the enhanced flow and turbulence due to the presence of vapor. F was graphically determined, as shown in Figure 13.8a. It can be approximated by

$$F = 1 \quad \text{for } \frac{1}{X_{tt}} < 0.1 \tag{13.25a}$$

$$F = 2.35\left(0.213 + \frac{1}{X_{tt}}\right)^{0.736} \quad \text{for } \frac{1}{X_{tt}} > 0.1 \tag{13.25b}$$

The nucleation part is based on the Forster–Zuber [22] equation with a suppression factor S:

$$h_{NB} = S(0.00122)\left[\frac{\left(k^{0.79}c_p^{0.45}\rho^{0.49}\right)_f}{\sigma^{0.5}\mu_f^{0.29}h_{fg}^{0.24}\rho_g^{0.24}}\right]\Delta T_{sat}^{0.24}\Delta p^{0.75} \tag{13.26}$$

where, $\Delta T_{sat} = T_W - T_{sat}$; $\Delta p = p(T_W) - p(T_{sat})$; and S is a function of the total Reynolds number as shown in Figure 13.8b. It can be approximated by

$$S = \frac{1}{1 + 2.53\times10^{-6}\,\mathrm{Re}^{1.17}} \tag{13.27}$$

where $\mathrm{Re} = \mathrm{Re}_\ell F^{1.25}$ and $\mathrm{Re}_\ell \equiv \dfrac{G(1-x)D}{\mu_\ell}$.

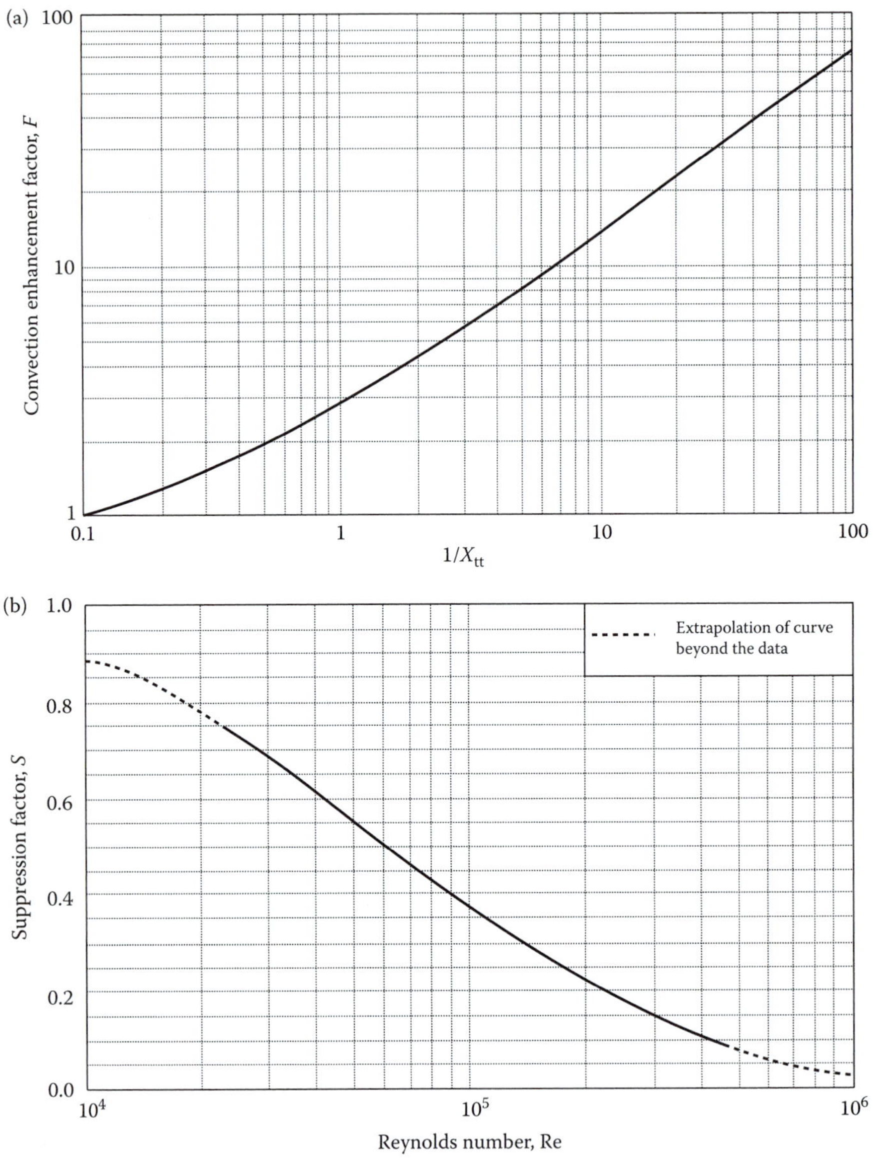

FIGURE 13.8 (a) Convection enhancement factor (F) used in Chen's correlation. (b) Suppression factor (S) used in Chen's correlation. (From Chen, J. C., *A Correlation for Boiling Heat Transfer to Saturated Fluids in Convective Flow.* ASME paper 63-HT-34, 1963 (first appearance). Chen, J. C., A correlation for boiling heat transfer in convection flow. *ISEC Process Des. Dev.*, 5:322, 1966 (subsequent archival publication).)

The range of water conditions of the original data was

Pressure = 0.17–3.5 MPa
Liquid inlet velocity = 0.06–4.5 m/s
Heat flux up to 2.4 MW/m^2
Quality = 0–0.7
Other tested fluids include methanol, cyclohexane, pentane, and benzene.

Chen's correlation has the advantage of being applicable over the boiling region prior to the DNB or Dryout location. It also has a lower-deviation error (11%) than the earlier correlations of Dengler and Addoms (38%) [19], Bennett et al. (32.6%) [2], and Shrock and Grossman (31.7%) [53] when tested for the range indicated above [13]. Subsequent data have extended the pressure range of the application to 6.9 MPa. A comparison of the Chen correlation to the Dengler and Addoms correlation as well as the Jens and Lottes correlation is shown in Figure 13.9.

Example 13.1 Heat Flux Calculation for Saturated Boiling

PROBLEM

Water is boiling at 7.0 MPa in a tube of an SFR steam generator. Using the Chen correlation, determine the heat flux at a position in the tube where the quality $x = 0.2$ and the wall temperature is 290°C.

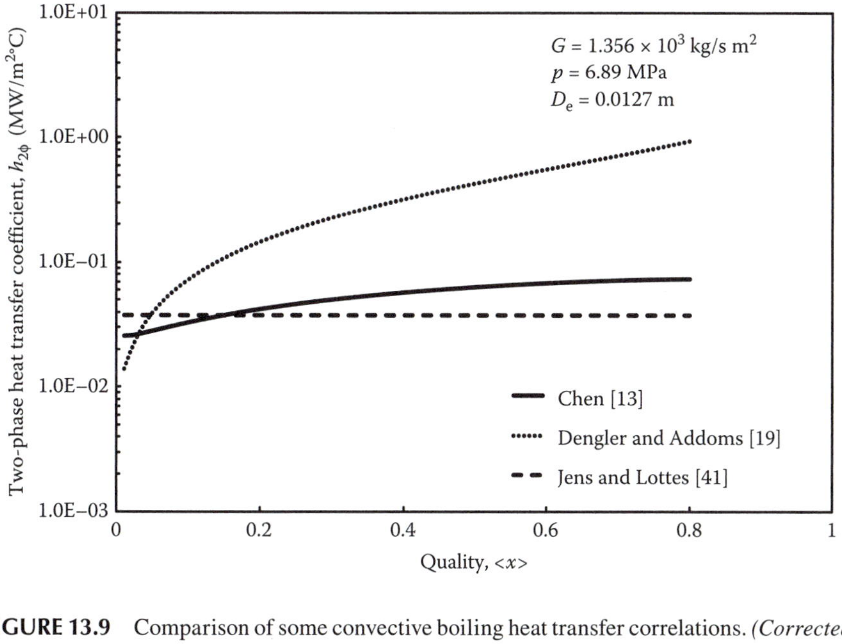

FIGURE 13.9 Comparison of some convective boiling heat transfer correlations. *(Corrected from* Lahey, R. T. and Moody, F. J., *The Thermal Hydraulic of Boiling Water Reactor*, p. 152. American Nuclear Society, 1977.)

Tube Flow Conditions

Diameter of tube (D) = 25 mm
Mass flow rate ($\dot{m}$) = 800 kg/h

SOLUTION

Required Water Properties

$\mu_f = 96 \times 10^{-6}$ N s/m^2
$\mu_g = 18.95 \times 10^{-6}$ N s/m^2
$c_{pf} = 5.4 \times 10^3$ J/kg K
$\rho_f = 740$ kg/m^3
$\rho_g = 36.5$ kg/m^3
$\sigma = 18.03 \times 10^{-3}$ N/m
$h_{fg} = 1513.6 \times 10^3$ J/kg
T_{sat} (7.0 MPa) = 284.64°C
$T_w = 290$°C
$\Delta T_{sat} = 5.36$ K
$k_f = 0.567$ W/m K

Pressure and Mass Flux

$$\Delta p_{sat} = p_{sat}(290°\text{C}) - p_{sat}(284.64°\text{C}) = 5.66 \times 10^5 \text{ Pa}$$

$$G = \frac{\dot{m}}{\frac{\pi}{4}D^2} = \frac{800\frac{1}{3600}}{\frac{\pi}{4}(0.025)^2} = 452.7 \text{ kg/m}^2\text{ s}$$

Chen Correlation

$$h_{2\phi} = h_{NB} + h_c \tag{13.21}$$

$$\begin{aligned} h_c &= 0.023\left[\frac{G(1-x)D}{\mu_f}\right]^{0.8}\left[\frac{\mu c_p}{k}\right]_f^{0.4}\left(\frac{k_f}{D}\right)F \\ &= 0.023\left[\frac{(452.7)(1-0.2)(0.025)}{96\times 10^{-6}}\right]^{0.8} \\ &\quad \times\left[\frac{(96\times 10^{-6})(5.4\times 10^3)}{0.567}\right]^{0.4}\left(\frac{0.567}{0.025}\right)F \end{aligned} \tag{13.24}$$

$$\begin{aligned} X_{tt} &= \left(\frac{1-x}{x}\right)^{0.9}\left(\frac{\rho_g}{\rho_f}\right)^{0.5}\left(\frac{\mu_f}{\mu_g}\right)^{0.1} \\ &= \left(\frac{1-0.2}{0.2}\right)^{0.9}\left(\frac{36.54}{740.0}\right)^{0.5}\left(\frac{96\times 10^{-6}}{18.95\times 10^{-6}}\right)^{0.1} = 0.9 \end{aligned}$$

$$F = 2.35\left(\frac{1}{X_{tt}} + 0.213\right)^{0.736} \quad \text{for } \frac{1}{X_{tt}} > 0.1 \tag{13.25b}$$

$$\therefore\ F = 2.87$$

$$\therefore\ h_c = 13{,}783 \text{ W/m}^2\text{ K}$$

$$h_{NB} = 0.00122\frac{\left(k^{0.79}c_p^{0.45}\rho^{0.49}\right)_f}{\sigma^{0.5}\mu_f^{0.29}h_{fg}^{0.24}\rho_g^{0.24}}\Delta T_{sat}^{0.24}\Delta p_{sat}^{0.75}S$$
$$= \left[0.00122\frac{(0.567)^{0.79}(5.4\times10^3)^{0.45}(740.0)^{0.49}}{(18.03\times10^{-3})^{0.5}(96\times10^{-6})^{0.29}(1513.6\times10^3)^{0.24}(36.5)^{0.24}}\right]$$
$$\times(5.36)^{0.24}(5.66\times10^5)^{0.75}S \tag{13.26}$$

To find *S*, use Figure 13.8b:

$$\text{Re}_\ell = \frac{G(1-x)D}{\mu_f} = \frac{(452.7)(1-0.2)(0.025)}{96\times10^{-6}} = 94{,}313$$

$$\text{Re} = \text{Re}_\ell F^{1.25} = 9.43\times10^4\,(2.87)^{1.25} = 3.52\times10^5$$

Therefore

$$S \simeq 0.12 \quad \text{and} \quad h_{NB} = 5309 \text{ W/m}^2\text{ K}$$

$$\text{So } h_{2\phi} = 5309 + 13{,}783 = 19{,}092 \text{ W/m}^2\text{ K}$$

and

$$q'' = h_{2\phi}\,\Delta T_{sat} = (19{,}092)(5.36) = 102.3 \text{ kW/m}^2$$

13.3.2 Bjorge, Hall, and Rohsenow Correlation

Bjorge et al. [7] also proposed that a nonlinear superposition of nucleate boiling and convection heat transfer may be used to produce a heat flux relation in the low-quality two-phase region as

$$q''^2 = q_c''^2 + (q''_{NB} - q''_{Bi})^2 \tag{13.28}$$

where q''_{Bi} is the value obtained from the fully developed boiling correlation at the incipient boiling point. It is subtracted to make $q'' = q''_c$ at the incipience of boiling. Figure 13.10 illustrates this superposition approach. As shown and as Equation 13.28 suggests, this correlation is proposed for subcooled as well as saturated boiling heat transfer. The fully developed nucleate boiling curve has a slope of about 3, that is,

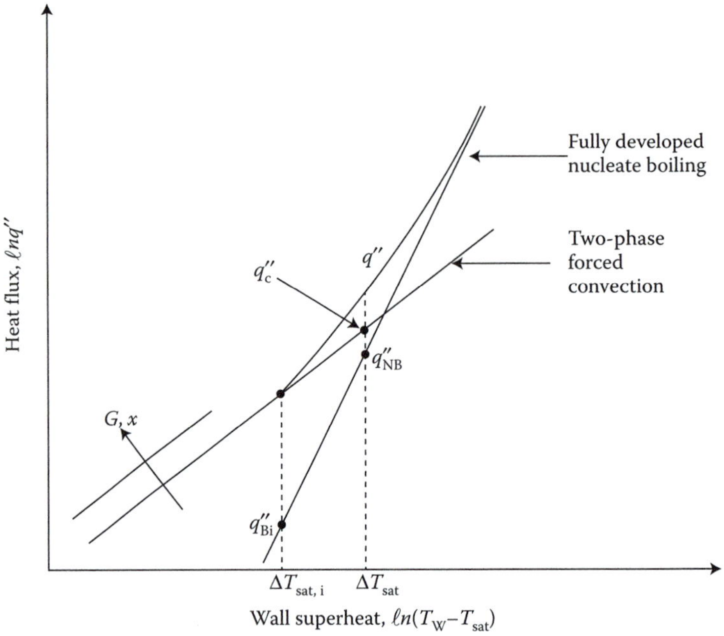

FIGURE 13.10 Superposition approach of Bjorge et al. (From Bjorge, R., Hall, G., and Rohsenow, W., *Int. J. Heat Mass Transfer*, 25(6):753–757, 1982.)

$q''_{NB} \propto \Delta T^3_{sat}$, so that Equation 13.28 can be written for subcooled and low-quality conditions as

$$q''^2 = q''^2_c + q''^2_{NB}\left\{1 - \left[\frac{(T_w - T_{sat})_i}{(T_w - T_{sat})}\right]^3\right\}^2 \tag{13.29}$$

where

$$(T_w - T_{sat})_i = \left(\frac{8\sigma T_{sat} v_g q''}{k_\ell h_{fg}}\right)^{1/2} \tag{13.3}$$

and

$$q''_{NB} = B_M \frac{(g)^{1/2} h_{fg}{}^{1/8} k_\ell{}^{1/2} \rho_\ell{}^{17/8} C_\ell^{19/8} \rho_v^{1/8}}{\sigma^{9/8}(\rho_\ell - \rho_v)^{5/8} T_{sat}^{1/8}} (T_w - T_{sat})^3 \tag{13.30}$$

where $B_M = 1.89 \times 10^{-14}$ in SI units (and 2.13×10^{-5} in British units).

For highly voided flow ($\alpha > 80\%$), Bjorge et al. [7] recommended a different correlation, given by

$$q'' = q_c'' + q_{NB}'' \left\{ 1 - \left[\frac{(T_w - T_{sat})_i}{(T_w - T_{sat})} \right]^3 \right\} \tag{13.31a}$$

where

$$q_c'' = \frac{F(X_{tt})}{F_2} \frac{k_\ell}{D} \mathrm{Re}_\ell^{0.9} \, Pr_\ell (T_w - T_{sat}) \tag{13.31b}$$

$$F(X_{tt}) = 0.15 \left[\frac{1}{X_{tt}} + 2.0 \left(\frac{1}{X_{tt}} \right)^{0.32} \right] \tag{13.31c}$$

and

$$F_2 = \mathrm{Pr}_\ell + 5\,ln(1 + 5\mathrm{Pr}_\ell) + 2.5\,ln(0.0031\,\mathrm{Re}_\ell^{0.812}) \quad \text{for } \mathrm{Re}_\ell > 1125 \tag{13.32a}$$

$$F_2 = 5\mathrm{Pr}_\ell + 5\,ln[1 + Pr_\ell(0.0964\,\mathrm{Re}_\ell^{0.585} - 1)] \quad 50 < \mathrm{Re}_\ell < 1125 \tag{13.32b}$$

$$F_2 = 0.707\,\mathrm{Pr}_\ell\,\mathrm{Re}_\ell^{0.5} \quad \mathrm{Re}_\ell < 50 \tag{13.32c}$$

Example 13.2 B–H–R (Bjorge et al.) Superposition Method

PROBLEM

Consider a coolant flowing in forced convection boiling inside a round tube. The following relations have been calculated from the liquid and vapor properties and flow conditions:

Forced convection:

$$q_c'' = h\,\Delta T_{sat}$$
$$h = 5.68 \text{ kW/m}^2\text{ K}$$

Fully developed nucleate boiling:

$$q_{NB}'' = \gamma_B\,\Delta T_{sat}^3$$
$$\gamma_B = 0.18 \text{ kW/m}^2\text{ K}^3$$

Incipient boiling condition:

$$q_i'' = \gamma_i\,\Delta T_{sat,i}^2$$
$$\gamma_i = 1.71 \text{ kW/m}^2\text{ K}^2$$

1. Determine the heat flux q'' when $\Delta T_{sat} = T_w - T_{sat} = 6.67$ K.
2. The above equations apply at the particular value of G_1. Keeping the heat flux q'' the same as above, how much must G be increased in order to stop nucleate boiling? What is the ratio of G_2/G_1? Assume the flow is turbulent.

SOLUTION

1. Assuming that the given wall superheat is sufficient to cause nucleate boiling at low-vapor quality, the heat flux is given by

$$q''^2 = (q''_c)^2 + (q''_{NB})^2 \left\{1 - \left[\frac{(\Delta T_{sat,i})}{\Delta T_{sat}}\right]^3\right\}^2 \tag{13.29}$$

Now we have to find $\Delta T_{sat,i}$ to calculate the heat flux. This quantity is the temperature difference between the wall and the saturated temperature for incipient boiling. For incipient boiling, the total heat flux equals that by forced convection and the incipient boiling condition is satisfied. Therefore, by considering the energy balance:

$$q''_i = q''_c$$
$$\gamma_i(\Delta T_{sat,i})^2 = h\Delta T_{sat,i}$$
$$\Delta T_{sat,i} = \frac{h}{\gamma_i} = \frac{5.68 \text{ kW/m}^2\text{ K}}{1.71 \text{ kW/m}^2\text{ K}^2} = 3.3 \text{ K}$$

The heat flux is therefore given by

$$q'' = \left\{\left[5.68(6.67)\right]^2 + \left[0.18(6.67)^3\right]^2\left[1 - \left(\frac{3.3}{6.67}\right)^3\right]^2\right\}^{1/2} \tag{13.29}$$

$$q'' = 60.37 \text{ kW/m}^2$$

2. The above heat transfer coefficient (h_c) is for a particular flow rate. We want to look for the flow rate that has a high enough forced convection heat transfer coefficient such that the wall temperature would be too low to allow incipient boiling for the given heat flux. Let the wall temperature in this case be T_{w2}.

To calculate the temperature $\Delta T_2 = T_{w2} - T_{sat}$

$$q''_i = \gamma_i \Delta T_{2,i}^2$$
$$\Delta T_{2,i} = \sqrt{\frac{q''_i}{\gamma_i}}$$
$$= \sqrt{\frac{60.37}{1.71}}$$
$$= 5.94 \text{ K}$$

The heat transfer coefficient of forced convection corresponding to ΔT_2 is

$$h_c = \frac{q''}{\Delta T_2} = \frac{60.37}{5.94} = 10.16\ \text{kW/m}^2\,\text{K}$$

From Equation 13.24, the forced convection heat transfer coefficient is expressed as

$$h_c = 0.023\left[\frac{(1-x)D}{\mu_\ell}\right]^{0.8} G^{0.8}\,\text{Pr}_\ell^{0.4}\,\frac{k_\ell}{D}F \propto G^{0.8}$$

So the new flow rate is obtained from

$$\frac{h_{c2}}{h_{c1}} = \left(\frac{G_2}{G_1}\right)^{0.8}$$

Hence

$$\frac{G_2}{G_1} = \left(\frac{h_{c2}}{h_{c1}}\right)^{1/0.8} = \left(\frac{10.16}{5.68}\right)^{1/0.8} = 2.07$$

13.3.3 Post-CHF Heat Transfer

Knowledge of the heat-transfer rate in the regime beyond the CHF is important in many reactor applications. In LWRs, an overpower transient or an LOCA may result in exposure of at least part of the fuel elements to CHF and post-CHF conditions. Once-through steam generators routinely operate with part of the length of their tubes in this heat-transfer region.

Mechanisms and correlations of CHF are discussed in Section 13.4. Here, heat transfer in the post-CHF regime is discussed.

Post-CHF heat transfer can be affected by two major factors:

1. The independent boundary condition. Is it the wall heat flux or the wall temperature? In the first case transition boiling is not obtained, whereas in the second case transition boiling may be realized.
2. The type of CHF (DNB versus Dryout). These two cases are illustrated in Figure 13.11 for vertical upflow.

In the post-DNB region (Figure 13.11a), film boiling creates a vapor film that separates the wall from the bulk liquid (inverted annular). The flow quality can

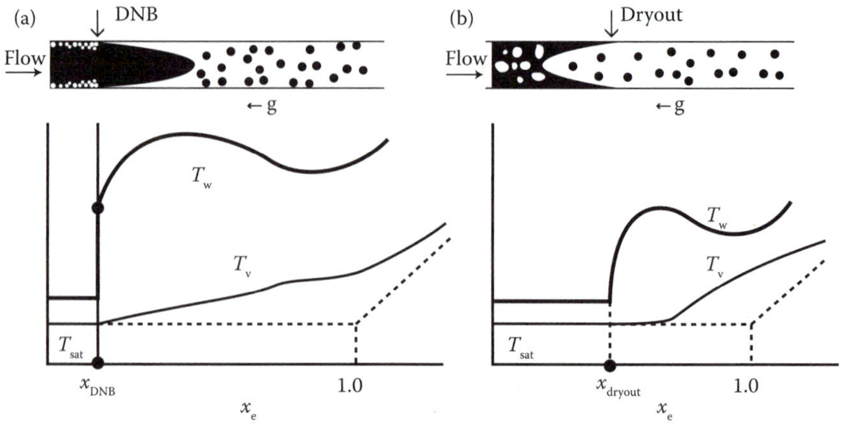

FIGURE 13.11 Post-CHF temperature distribution. (a) Post-DNB inverted annular, flow film boiling (IAFB), and (b) post-Dryout dispersed annular or liquid deficient flow film boiling (DFFB).

initially be relatively low. T_W is high even at low values of the thermodynamic quality because the post-CHF heat transfer coefficient is initially low due to the low vapor velocity. Further downstream in the dispersed droplet flow regime it is higher because of larger vapor velocity and also the drops receive heat from the vapor core since the vapor temperature well exceeds T_{sat}.

In the post-Dryout, dispersed annular flow or liquid deficient regime case (Figure 13.11b), the liquid exists only as droplets carried by the vapor, in which case the vapor receives the heat from the wall and only some of the droplets impinge on the wall. Most of the droplets receive heat from the vapor core when the vapor temperature exceeds T_{sat}.

Consequently, we present post-CHF heat transfer correlations for flow film boiling, both inverted annular flow applicable to post-DNB and dispersed annular flow applicable to post-Dryout as well as transition boiling (possible for either post-DNB or post-Dryout under controlled wall temperature conditions).

13.3.3.1 Both Film Boiling Regions (Inverted and Dispersed Annular Flow)

Recently Groeneveld et al. [31,32] have updated their earlier look-up table for fully developed film-boiling heat transfer which covers both regimes pictured in Figure 13.11.

The film boiling look-up table tabulates the fully developed film boiling heat transfer coefficient (kWm^{-2}K^{-1}) in terms of four prescribed parameters and a multiplier for tube diameter other than 8 mm covering the following ranges:

$0.1 \le p \le 20$ MPa
$0 < G \le 7000$ kg/m^2 s
$-0.2 \le x_e \le 2.0$
$50 \le T_w - T_{sat} \le 1200$ K

The data sets used to develop the look-up table spanned the tube diameter range

$$2.5 \le D \le 24.7 \text{ mm}$$

The table was based on synthesis of a number of correlations for various parameter regions. The table has three levels of shading applied to highlighted regions of uncertainty. Hence the predicted heat transfer coefficient is calculated as

$$h_{\text{pred}} = h_{\text{table}}\left[p, G, x_{\text{e}}, T_{\text{w}} - T_{\text{sat}}\right]\left(\frac{8}{D}\right)^{0.2}$$

with D in mm.

13.3.3.2 Inverted Annular Flow Film Boiling (Only)

In this case, illustrated in Figure 13.11a, as in pool boiling a thin film of vapor exists adjacent to the heated surface, but now an upward liquid velocity is also imposed.

The condition for the velocity to be important is obtained as the dominance of the dynamic pressure compared to the hydrostatic pressure or

$$\begin{aligned} \frac{\rho_\ell \upsilon^2}{2} &> \rho_\ell g(z \text{ or } D) \\ \text{or } \upsilon &> \sqrt{2g(z \text{ or } D)} \end{aligned} \tag{13.33}$$

where z applies to a vertical plate geometry and D to a horizontal cylinder of diameter D. The film boiling heat transfer coefficient on a vertical plate in pool boiling (no upward liquid velocity) has been given by Equation 12.27a (laminar) and Equation 12.29a (turbulent film) by Hsu and Westwater [38].

For the case of an upward liquid velocity υ across a tube of diameter D conforming to the criteria of Equation 13.33, the film boiling heat transfer coefficient is given by Bromley et al. [11] as

$$h_{\text{c}} = 2.7\sqrt{\frac{\rho_{\text{vfm}} k_{\text{vfm}} \upsilon h'_{\text{fg}}}{D\,\Delta T_{\text{sat}}}} \tag{13.34a}$$

where

$$h'_{\text{fg}} = h_{\text{fg}}\left(1 + 0.4\frac{\Delta T_{\text{sat}} c_{\text{pv}}}{h_{\text{fg}}}\right)^2 \tag{13.34b}$$

$$\Delta T_{\text{sat}} = T_{\text{w}} - T_{\text{sat}} \tag{13.34c}$$

Note that Equation 13.34a holds for $\upsilon > \sqrt{2gD}$ while Equation 12.26 is for the same geometry but for natural circulation velocities of magnitude $\upsilon < \sqrt{gD}$.

Finally, since the wall temperature is elevated in film boiling, radiative heat transfer can be important.

Hence

$$h_{\text{total}} = h_c + ah_r \tag{12.29}$$

where h_r has been given in Equation 12.30b and a is typically 3/4 but 7/8 when used with Equation 13.34a.

Whalley [64], in Example 9 of Chapter 25, gives detailed information on the calculation of heat transfer coefficients for film boiling on a tube with wall temperature of 600°C for cases with and without liquid flow and with and without the radiation correction. The heat transfer coefficient ranges from 9.3×10^4 W/m^{2}K for no υ and no h_r to 24.8×10^4 W/m^{2}K for both included with the velocity influence dominant.

13.3.3.3 Dispersed Annular or Liquid Deficient Flow Film Boiling (Only)

The transition into this region illustrated in Figure 13.11b occurs at Dryout. For this case of high G value, the wall temperature rapidly rises to a peak value and then falls. This wall temperature evolution from Dryout (point A) to the end of liquid drop evaporation (point B) is shown in Figure 13.12 from the results of Schmidt [52] illustrated in Collier and Thome [16] taken at $p = 16.9$ MPa, $G = 700$ kg/m^{2}s and heat flux from 290 to 700 kW/m^2. Increasing heat flux displaces point A to lower quality

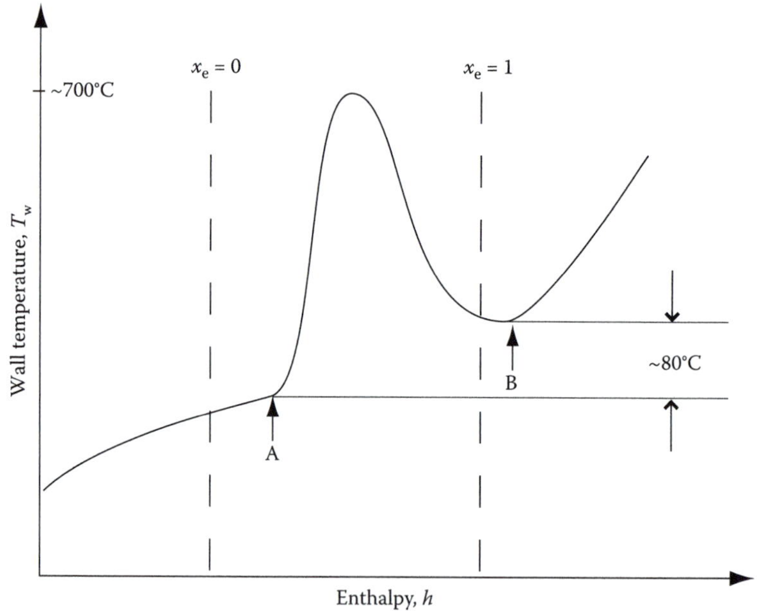

FIGURE 13.12 Wall temperature evolution in the liquid deficient region for heat flux of 700 kW/m^2 at 167 bar and 700 kg/m^2 s. (From Schmidt, K. R., Wärmetechnische Untersuchungen an hoch Kesselheizflächen. *Mitteilungen der Vereinigung der Grosskessel-bezitzer*, 391–401, December 1959.)

and increases the peak wall temperature as well as that at point B. For lower G values, a continuous increase in vapor temperature can easily result in a continuously increasing wall temperature.

Three types of correlations have been identified (Collier and Thome [16]) to calculate the heat transfer rate typically by calculating the wall temperature.

1. Empirical correlations which make no assumptions about the mechanism but attempt to relate the heat transfer coefficient (assuming the coolant is at the saturation temperature) and the independent variables.
2. Mechanistic correlations which recognize the departure from a thermodynamic equilibrium condition and attempt to calculate the true vapor quality and vapor temperature. A conventional single-phase heat transfer correlation is then used to calculate the heated wall temperature.
3. Semi-theoretical correlations which attempt to model individual hydrodynamic and heat transfer processes in the heated channel and relate them to the wall temperature.

13.3.3.3.1 *Empirical Correlations*

Groeneveld [27] compiled a bank of selected data from a variety of experimental post-Dryout studies in tubular, annular, and rod bundle geometries for steam–water flows and developed the following correlation:

$$\mathrm{Nu_g} = a\left\{\mathrm{Re_g}\left[x+\frac{\rho_\mathrm{g}}{\rho_\mathrm{f}}(1-x)\right]\right\}^{b}\mathrm{Pr_g^{\mathit{c}}}Y \quad (13.35)$$

where $\mathrm{Re_g} = GD/\mu_\mathrm{g}$

$$Y = \left[1-0.1\left(\frac{\rho_\mathrm{f}-\rho_\mathrm{g}}{\rho_\mathrm{g}}\right)^{0.4}(1-x)^{0.4}\right]^{d} \quad (13.36)$$

The coefficients a, b, c, and d are given in Table 13.2.

Slaughterbeck et al. [55,56] improved the correlation, particularly at low pressure. They recommended that the parameter Y be changed to

$$Y = (q'')^{\mathrm{e}}\left(\frac{k_\mathrm{g}}{k_\mathrm{cr}}\right)^{f} \quad (13.37)$$

where k_cr = conductivity at the thermodynamic critical point. Their recommended constants are also included in Table 13.2.

13.3.3.3.2 *Mechanistic Correlations*

This approach recognizes that heat transfer from the wall results in both heating of the vapor and some evaporation of the droplets from heat transfer from vapor. If all

TABLE 13.2
Constants in Empirical Post-CHF Correlations (Equations 13.35 through 13.37)

Author	a	b	c	d	e	f	No. of Points	% rms Error[a]
Groeneveld								
Tubes	1.09×10^{-3}	0.989	1.41	−1.15			438	11.5
Annuli	5.20×10^{-2}	0.688	1.26	−1.06			266	6.9
Slaughterbeck								
Tubes	1.16×10^{-4}	0.838	1.81		0.278	−0.508		12.0

Parameter	Range of Data: Tubes (Vertical and Horizontal)	Range of Data: Annuli (Vertical)
D_e (mm)	2.5–25.0	1.5–6.3
p (MPa)	6.9–21.8	3.5–10.1
G (kg/m^{2}s)	700–5300	800–4100
q''(kW/m^2)	120–2100	450–2250
x	0.1–0.9	0.1–0.9
Y	0.706–0.976	0.610–0.963

[a] rms = root mean square.

the heat from the wall went into first evaporating the droplets with the vapor remaining at the saturation temperature, then the situation would be that of complete thermodynamic equilibrium. On the other hand, if all the heat from the wall went into superheating the vapor, then there would be a complete departure from equilibrium. These bounding situations are illustrated in Figure 13.13. In general then, analyses are formulated in terms that partition the wall heat flux into components which evaporate the droplets and superheat the vapor. Hence, the ratio ε is created as

$$\varepsilon = \frac{q_f''(z)}{q''(z)} \tag{13.38a}$$

and

$$1 - \varepsilon = \frac{q_g''(z)}{q''(z)} \tag{13.38b}$$

where

$$q''(z) = q_f''(z) + q_g''(z) \tag{13.38c}$$

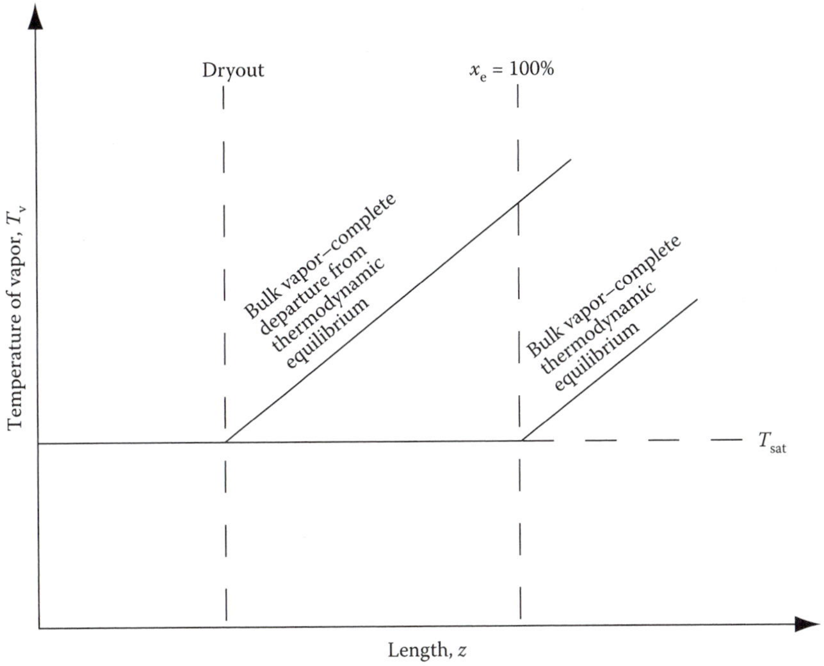

FIGURE 13.13 Bounding situations of thermodynamic equilibrium in post-Dryout heat transfer region.

Several correlations of this form have been proposed, notably that of Groeneveld and Delorme [28] and Chen et al. [14]. That of Chen et al. is the most comprehensive although involved.

13.3.3.3.3 Semitheoretical Models

These models extend the simplified tracking of the wall heat flux of the mechanistic models into a more comprehensive description of the multiple heat transfer mechanisms in the post-Dryout region. Six heat transfer mechanisms have been modeled

- From the wall to droplets
  1. Which impact and hence wet the wall
  2. Which enter the thermal boundary layer but do not wet the wall
  3. By radiation
- From the wall to bulk vapor
  4. By convective heat transfer
  5. By radiation
- From bulk vapor to droplets by convection

Collier and Thome [16] reviewed the evolution of models which have treated various of these mechanisms starting particularly with the work of Bennett et al. [3] and Iloeje et al. [39].

13.3.3.4 Transition Boiling

The existence of a transition boiling region in the forced convection condition as well as in pool boiling, under controlled wall temperature conditions, has been demonstrated experimentally.

Various attempts have been made to produce the correlations for the transition boiling region. Groeneveld and Fung [29] tabulated those available for forced convective boiling of water. In general, the correlations are valid only for the range of conditions of the data on which they were based. Figure 13.14 provides an example of the level of uncertainty in this region.

One of the earliest experimental studies of forced convection transition boiling was that by McDonough et al. [47], who measured heat transfer coefficients for water over the pressure range 5.5–13.8 MPa inside a 3.8 mm I.D. tube heated by NaK. They proposed the following correlation:

$$\frac{q''_{cr} - q''(z)}{T_w(z) - T_{cr}} = 4.15 \exp \frac{3.97}{p} \tag{13.39}$$

where q''_{cr} = CHF (kW/m²); $q''(z)$ = transition region heat flux (kW/m²); T_{cr} = wall temperature at CHF (°C); $T_w(z)$ = wall temperature in the transition region (°C); and p = system pressure (MPa).

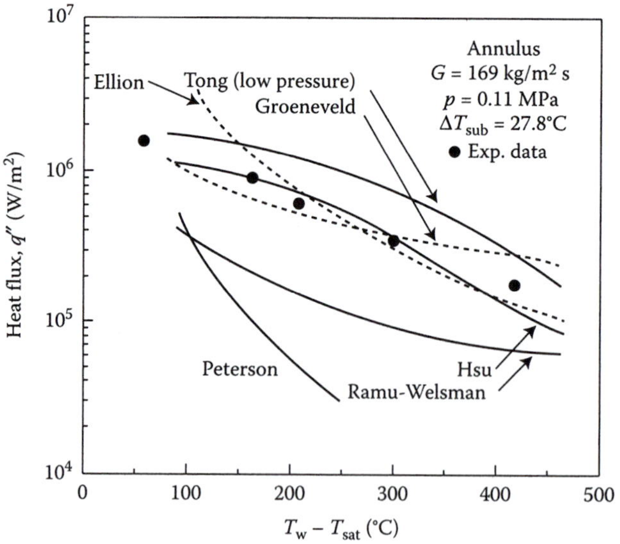

FIGURE 13.14 Comparison of various transition boiling correlations with Ellison's data. (Adopted from Groeneveld, D. C. et al., *Nucl. Eng. Des.*, 237:1909–1922, 2007.)

Tong [61] suggested the following equation for combined transition and stable film boiling at 6.9 MPa.

$$h_{tb} = 39.75\exp(-0.0144\Delta T) + 2.3\times10^{-5}\frac{k_g}{D_e}\exp\left(-\frac{105}{\Delta T}\right)\mathrm{Re}_f^{0.8}\,\mathrm{Pr}_f^{0.4} \quad (13.40)$$

where h_{tb} = the heat transfer coefficient for the transition region (kW/m^2 K); $\Delta T = T_w - T_{sat}$ (°C); D_e = equivalent diameter (m); k_g = vapor conductivity (kW/m K); Pr_f = liquid Pr; and $\mathrm{Re} = D_e G_m/\mu_f$.

Ramu and Weisman [48] later attempted to produce a single correlation for post-CHF and reflood situations. They proposed the transition boiling heat transfer coefficient:

$$h_{tb} = 0.5\,S\,h_{cr}\{\exp[-0.0140(\Delta T - \Delta T_{cr})] + \exp[-0.125(\Delta T - \Delta T_{cr})]\} \quad (13.41)$$

where h_{cr} (kW/m^2 °C) is the heat transfer coefficient, ΔT_{cr} (°C) is the wall superheat ($T_w - T_{sat}$) corresponding to the pool boiling CHF condition, and S = the Chen nucleation suppression factor.

Cheng et al. [15] suggested a simple correlation of the form

$$\frac{q''_{tb}}{q''_{cr}} = \left(\frac{T_w - T_{sat}}{\Delta T_{cr}}\right)^{-n} \quad (13.42)$$

They found that $n = 1.25$ fitted their low-pressure data acceptably well. A similar approach was adopted by Bjornard and Griffith [6], who proposed

$$q''_{tb} = \delta q''_{cr} + (1-\delta)q''_{min} \quad (13.43a)$$

$$\text{with } \delta = \left(\frac{T^M - T_w}{T^M - T_{cr}}\right)^2 \quad (13.43b)$$

where q''_{min} and T^M = heat flux and wall temperature, respectively, corresponding to the minimum heat flux for stable film boiling on the boiling curve; and q''_{cr} and T_{cr} = heat flux and wall temperature, respectively, at CHF. Some of the correlations for wall temperature for stable film boiling are given in Table 12.1.

13.3.4 Reflooding of a Core Which Has Been Uncovered

In an LOCA in an LWR, the core-primary system interaction progresses through key phases: (a) primary coolant blowdown due to depressurization leading to rapid but very limited core cooldown, (b) core heatup due to decay heat and stored energy release from the fuel uncompensated by coolant in the core, and finally (c) reintroduction of water into the core from water spray (BWRs) and/or bottom reflooding (PWRs). This coolant injection reverses core heatup which culminates in achievement of a peak cladding temperature with subsequent rapid cooldown. In large break LOCAs, the core uncovers completely* while in small break LOCAs the degree of core uncovery, if any, depends on the reactor design and the size of the assumed pipe break.

Core cladding temperature recovery involves the phenomena of quenching the temperature of the very hot cladding below the Leidenfrost point and the subsequent upward movement of the quenching front to cool the entire length of the fuel rod cladding.

BWRs employ both top spray and bottom reflood while PWRs employ only bottom reflood. This is because (1) the subassembly duct of the BWR being unheated provides a surface which acts as a heat sink which the spray can wet and run down toward the subassembly inlet, and (2) the fuel array size is limited so that cooling of the innermost fuel rods can occur by radiative heat transfer to the duct wall liquid film.

The channel flow configuration and quench front propagation for bottom reflooding can take two forms depending on the magnitude of the water-reflooding rate. At high-water flow rate, an inverted annular flow regime is created as in Figure 13.11a whereas at low-water flow rate the configuration resembles cocurrent upward annular flow, except that some sputtering occurs at the quench front.

Prediction of heat transfer rates and quench front progression is a difficult task. Among the difficulties are variation in liquid temperature within the film, heat transfer coefficient variation with axial position, particularly precooling due to water drops ahead of the quench front, axial conduction in the cladding, and the numerical value of the sputtering temperature which strongly depends on the surface physicochemical characteristics—the wall temperature between the quenched and hot cladding surfaces. Hence, no specific correlations for these parameters are presented, and the interested reader is referred to the literature for correlations relevant to specific configurations and conditions of interest. A useful overview is presented in Yadigaroglu et al. [65].

13.4 CRITICAL CONDITION OR BOILING CRISIS

The critical condition is used here to denote the situation in which the heat transfer coefficient of the two-phase flow substantially and abruptly deteriorates. Many terms have been used to denote the CHF conditions, including "boiling crisis" (mostly in non-English speaking countries) and "burnout" (preferred in Britain). The terminology critical condition is used here since it seems to best encompass the two mechanisms of rapid heat transfer coefficient deterioration—the DNB, observed at low-quality or even subcooled conditions, is best characterized by assessing the limiting heat flux and film Dryout, observed at moderately high quality, is best characterized by assessing the limiting channel power. The two phenomena are illustrated in Figure 13.15.

* The IRIS integral PWR is designed with a small, high-pressure spherical containment, which allows maintenance of core coverage with water throughout the accident progression.

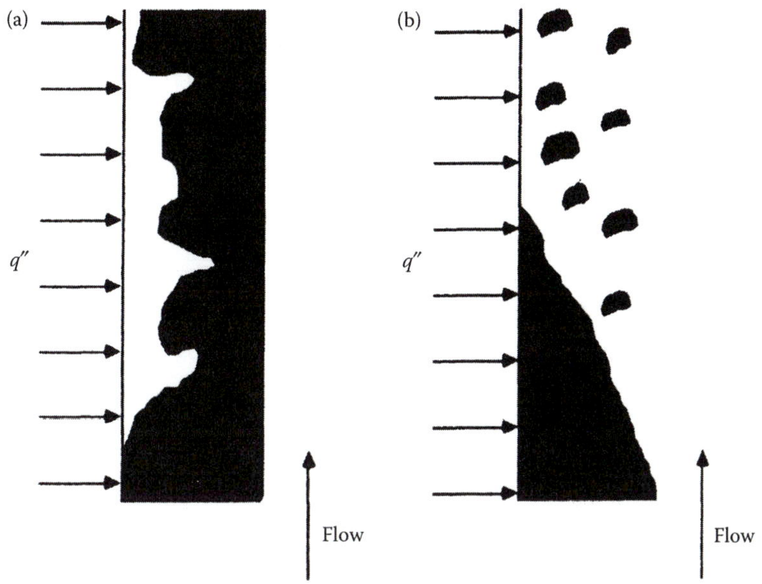

FIGURE 13.15 CHF mechanisms. (a) DNB. (b) Dryout. (Shaded flow channel area denotes liquid.)

In a system in which the heat flux or channel power is independently controlled such as a fuel assembly, the consequence of the critical condition is the rapid rise in the wall temperature. For systems in which the wall temperature is independently controlled, the occurrence of the critical condition implies a rapid decrease in the heat flux or channel power.

In assessing the operation of fuel relative to the critical condition, it is important to focus on the hot channel conditions. The hot channel flow rate is less than that of its neighbors while the degree of hot channel boiling is greater. This is a consequence of the condition of equal pressure drop which holds for an array of channels in parallel flow.

13.4.1 Critical Condition Mechanisms and Limiting Values

The existence of these separate mechanisms of the critical condition was demonstrated by Groeneveld's experiment [26] with the two test sections pictured in Figure 13.16—a uniformly heated channel and a channel with a short exit heat flux spike.

The CHF for the uniformly heated channel varies with quality as Figure 13.16 illustrates. The heat flux spike channel exhibits two behaviors:

1. At low quality, it reaches the critical condition when the heat flux at the spike, q''_{sp}, equals the CHF value for the uniformly heated channel, q''_{u}. Hence, this low-quality DNB behavior is local heat flux controlled.

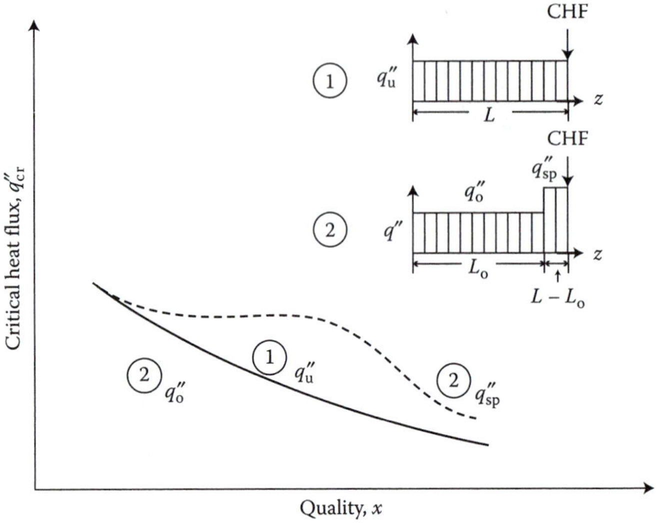

FIGURE 13.16 Effect of heat flux spike on q''_{cr}.

2. At high quality, the critical condition is reached in the spiked heat flux channel when its total power, $q''_o \pi D L_o + q''_{sp} \pi D(L - L_o)$, is effectively equal to the total power at the critical condition of the uniformly heated channel, $q''_u \pi DL$. Hence, this high-quality Dryout behavior is total power controlled. As a consequence, for high quality, the critical channel power should not be sensitive to the axial heat flux distribution for a fixed length. Keeys et al. [43] have demonstrated this conclusion as shown in Figure 13.17.

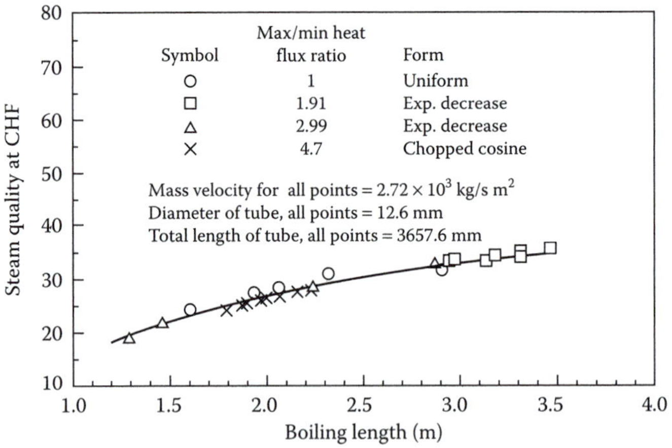

FIGURE 13.17 Effect of flux shape on critical quality. (From Keeys, R. K. F., Ralph, J. C., and Roberts, D. N. *Post Burnout Heat Transfer in High Pressure Steam-Water Mixtures in a Tube with Cosine Heat Flux Distribution.* AERE-R6411. U.K. Atomic Energy Research Establishment, Harwell, UK, 1971.)

The limiting magnitude of the heat flux and power for these two critical conditions is as follows:

a. For the DNB condition, the critical value of heat flux must be sufficiently high to at least cause the wall temperature to reach saturation ($T_w = T_{sat}$) while the bulk flow is still subcooled.
b. For the Dryout condition, the critical channel power $(\dot{q}_{cr})_{max}$ is at most that sufficient to achieve an equilibrium quality, $x_e = 1.0$.

These bounds, however, are far too broad compared to experiment. Hence, modeling and development of critical condition correlations have been required as presented next.

13.4.2 The Critical Condition Mechanisms

13.4.2.1 Models for DNB

Three types of general models have been proposed for DNB at modest velocities and subcooling:

- Flooding-like models: Here, the flow boiling crisis is essentially a hydrodynamic phenomenon. The liquid flows near the wall between jets of detaching vapor bubbles. As the heat flux increases the number of nucleation sites where the bubbles are produced also increases and so does the number of jets. If the vapor flux from the wall grows enough to blow off the liquid (separation of the liquid boundary layer), then the heated surface experiences a sudden temperature excursion.
- Bubble layer models: Here, bubbles detach from the heated wall where they are produced and form a bubble layer which is in contact with the subcooled liquid core and separated from the heated surface by a thin superheated liquid layer. Several different postulates have been suggested to relate this bubble layer to DNB. For example, DNB occurs when the void fraction in the bubble layer reaches a critical value (approximately 0.8) to prevent the subcooled core liquid from flowing through the bubble layer, reaching the heated surface and cooling it. Alternately, as the liquid from the subcooled core comes through the bubble layer to cool the wall, it quenches some vapor to make space for the bubbles produced at the heated wall. As a result, in the direction of the flow, the average enthalpy of the bubble layer increases and when it reaches a critical value the liquid flashes to void leading to DNB. Since it regards the boiling crisis as the result of the axial evolution of the bubble layer enthalpy, this model implies that DNB is not entirely a local phenomenon.
- Vapor blanket models: Here, the concept of bubble layer is replaced with that of a vapor blanket, made of elongated slugs of vapor that flow close to the wall over a thin underlying liquid layer. It is postulated that DNB occurs when the heat flux is large enough to dry the liquid layer when a vapor slug flows over it.

13.4.2.2 Model for Dryout

It is widely accepted that in high-steam quality channels (where the annular flow regime is dominant) the boiling crisis occurs as a result of the total evaporation (i.e., Dryout) of the liquid film in contact with the heated wall. Multiple models have been developed for a phenomenological description of this liquid film to enable prediction of the Dryout boiling crisis. Dryout can be considered as the result of competing hydrodynamic and thermal effects. The film is fed by deposition of liquid droplets from the vapor core. It is depleted by entrainment of droplets from the film (i.e., some liquid is sheared off the film by the high-velocity vapor flowing over it causing surface waves). Vapor bubbles can also be produced by nucleation if the surface heat flux is sufficiently high. Finally, some liquid evaporates at the film–vapor interface. Accurate modeling of these phenomena to allow calculation of the liquid film thickness along the channel has proven to be difficult.

13.4.2.3 Variation of the Critical Condition with Key Parameters

A heated tubular channel can be described by the following parameters:

System pressure, p
Inlet mass flow rate, $\dot{m}$ (or G for given A_F)
Inlet temperature, T_{in} (and hence inlet subcooling, $\Delta h_{sub,in}$)
Internal tube diameter, D_{in}
Tube length, L
Axial heat flux magnitude and distribution

The exit channel conditions, h_{out} or x_{out}, are variables dependent on the inlet conditions and applied heat flux via the heat balance equation

$$h_{out} = h_f - \Delta h_{sub,in} + \frac{4L}{GD} q'' \tag{13.44a}$$

$$x_{out} = x_{in} + \frac{1}{h_{fg}} \frac{4L}{GD} q'' \tag{13.44b}$$

Hence, for a uniform heat flux distribution, the CHF can be described in terms of the parameters

$$p, G, \Delta h_{sub,in}, D, L$$

or alternately since the CHF occurs at the exit applying the heat balance of Equation 13.44 to substitute

$$h_{out}$$

$$\text{or} \quad x_{out}$$

$$\text{for } \Delta h_{sub,in}$$

obtain

$$p, G, D, h_{\text{out}}, L$$

$$\text{or } p, G, D, x_{\text{out}}, L$$

Hence, let us now present the CHF behavior qualitatively for variations in p, $\Delta h_{\text{sub,in}}$ (or local quality x) and G with various other parameters controlled as described by Hewitt [36].

13.4.2.3.1 System Pressure (Figure 13.18a)

For fixed mass flux, tube diameter, and length, the CHF generally exhibits a peak value with increasing pressure. The detailed behavior varies depending upon whether the inlet condition or the exit conditions are held fixed. These trends for three such conditions are illustrated in Figure 13.18a. For fixed $\Delta h_{\text{sub,in}} = 0$ and $x_{\text{exit}} = 0$, the pressure peaks about 4 MPa with a slight second maximum for the former at about 18 MPa. For $T_{\text{f,in}} = 174°\text{C}$, a slight double humped behavior at 4 MPa and 12 MPa is observed.

13.4.2.3.2 Inlet Subcooling (Figure 13.18b)

For fixed pressure, mass flux, tube diameter, and length, the CHF increases nearly linearly with inlet subcooling. Figure 13.18b also further illustrates that at fixed inlet subcooling, increasing mass flux will yield a corresponding increase in CHF.

13.4.2.3.3 Mass Flux (Figure 13.18c)

For fixed pressure, tube diameter and length, the critical heat flux decreases nearly linearly with increasing exit quality for fixed mass flux. However, as the mass flux is increased, an important reversal in critical heat flux magnitude at fixed exit quality occurs. Specifically as Figure 13.18c illustrates, while critical heat flux increases with increasing mass flux in the DNB region due to a more efficient heat removal from the wall, the inverse behavior occurs at higher qualities presumably due to enhanced liquid entrainment from higher vapor velocity. These trends of CHF with mass flux at fixed exit quality are an important distinction between the behavior of PWR and BWR applications.

13.4.3 Correlations for the Critical Condition

Critical condition experiments have progressed from round tube geometry to annular geometry to small rod bundles and finally to full-scale bundles. In parallel, correlations for each of these geometries have been developed often with auxiliary recommended means to extend the applicability of the correlation to rod bundles. Literally, hundreds of correlations have been developed. Some are capable of predicting both DNB and Dryout, others DNB only, and others Dryout only. In this section, the most useful and popular of the tube and bundle geometry correlations will be presented sequentially for these three types of predictive capability. Annular correlations such as that by Barnett [1] have also been proposed and were of practical value since the corner subchannels are annulus shaped. However, since annular correlations are primarily of historic value they are not presented here.

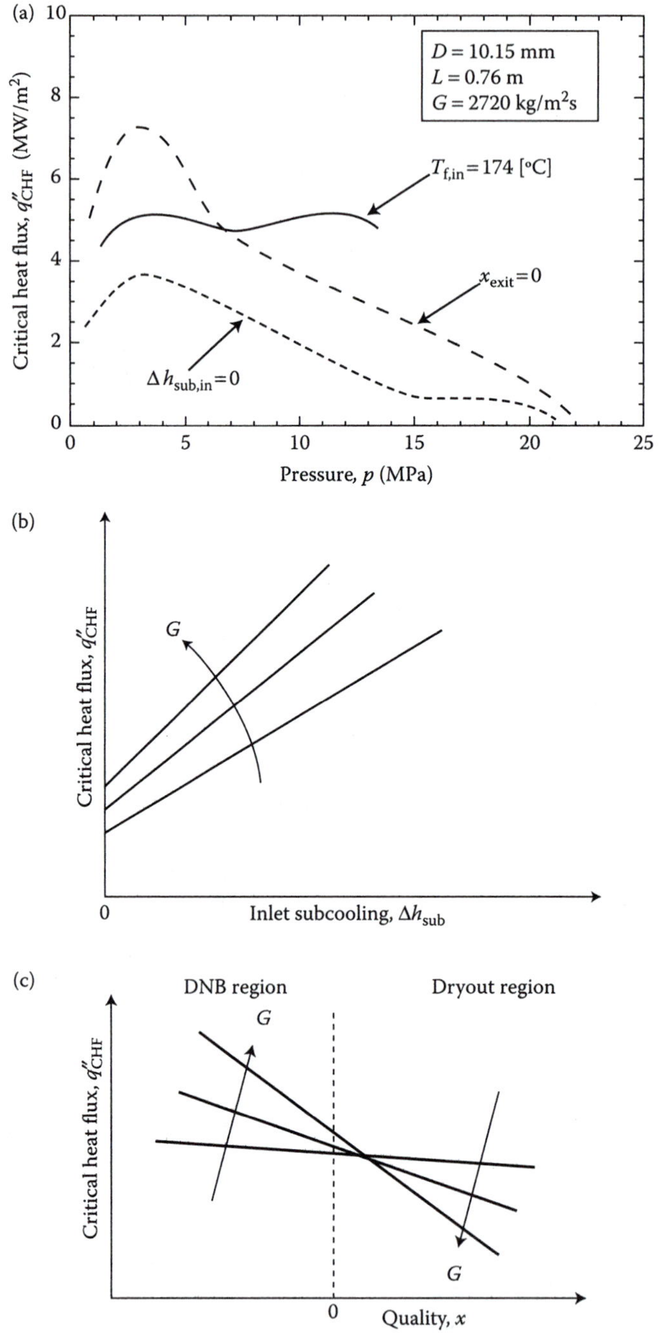

FIGURE 13.18 (a) Influence of pressure on the CHF (for $D = 10.15$ mm, $L = 0.76$ m, and $G = 2720$ kg/m^2 s). (b) Trend of CHF with inlet subcooling (constant p, D, L). (c) Trend of CHF with mass flux (constant p, D, L).

13.4.3.1 Correlations for Tube Geometry

13.4.3.1.1 Departure from Nucleate Boiling

The tube geometry correlations of general interest are all capable of predicting Dryout or DNB and Dryout. The Tong-68 correlation (given in [12]) is in the DNB-only category and is of the form

$$\frac{q''_{cr}}{h_{fg}} = C\frac{G^{0.4}\mu_f^{0.6}}{D^{0.6}} \tag{13.45a}$$

where

$$C = 1.76 - 7.433x_{exit} + 12.222x_{exit}^2 \tag{13.45b}$$

and q''_{cr} is the CHF, G is the mass flux, D is the inner tube diameter, h_{fg} is the latent heat, and μ_f is the dynamic viscosity of saturated liquid (SI units). The Tong correlation has also been presented in the form

$$\text{Bo} = \frac{C}{\text{Re}^{0.6}} \tag{13.46a}$$

where Bo and Re are *Boiling number* and *Reynolds number*, respectively. Tong showed that his correlation matched the then existing water tube data having uniform heat flux and pressure between 6.9 and 13.8 MPa to within ±25%. Celata et al. [12] modified the parameter C, together with a slight modification of the Reynolds number power, to give a more accurate prediction in the range of pressures below 5.0 MPa. This modification of the Tong correlation was based on data over the range ($2.2 < G < 40$ Mg m^{-2} s^{-1}, $0.1 < p < 5.0$ MPa, $2.5 < D < 8.0$ mm, $12 < L/D < 40$, $15 < T_{sub,ex} < 190$ K, $4.0 < q''_{cr} < 60.6$ MW m^{-2}). This new expression of the Tong correlation is

$$\text{Bo} = \frac{C}{\text{Re}^{0.5}} \tag{13.46b}$$

where

$C = (0.216 + 4.74 \times 10^{-2}p)\Psi$ [p in MPa]
$\Psi = 0.825 + 0.986x_{exit}$ if $x_{exit} > -0.1$;
$\Psi = 1$ if $x_{exit} < -0.1$;
$\Psi = 1/(2 + 30x_{exit})$ if $x_{exit} > 0$ (exit saturation conditions)

13.4.3.1.2 Dryout: The CISE-4 Correlation [23]

The CISE-4 correlation is a modification of CISE-3 in which the critical flow quality (x_c) approaches 1.0 as the mass flux decreases to 0. Unlike others, this correlation is based on the quality-boiling length concept and is restricted to BWR applications. The correlation was optimized in the flow range of $1000 < G < 4000$ kg/m^2 s. It has been suggested that the application of the CISE correlation in rod bundles is possible by including the ratio of heated-to-wetted perimeters in the correlation. The correlation is expressed by the following equations:

$$x_{cr} = \frac{D_e}{D_h}\left(a\frac{L_B}{L_B + b}\right) \tag{13.47a}$$

where

$$a = \frac{1}{1 + 1.481 \times 10^{-4}(1 - p/p_c)^{-3}G} \quad \text{if } G \le G^* \tag{13.47b}$$

and

$$a = \frac{1 - p/p_c}{(G/1000)^{1/3}} \quad \text{if } G \ge G^* \tag{13.47c}$$

where $G^* = 3375(1 - p/p_c)^3$; p_c = critical pressure (MPa); L_B = boiling length (distance from position of zero equilibrium quality to channel exit); and

$$b = 0.199(p_c/p - 1)^{0.4}\, GD^{1.4} \tag{13.47d}$$

where G is in kg/m²s and D is in meters (m).

The database for this correlation is

D = 0.0102–0.0198 m
L = 0.76–3.66 m
p = 4.96–6.89 MPa
G = 1085–4069 kg/m² s

13.4.3.1.3 DNB and Dryout

Three correlations will be presented here. Even though they are expressed in terms of heat flux, their developers state that they accurately predict the critical condition in the high-quality range, hence the occurrence of Dryout. Typically while heat flux approaches can be successfully used to predict the critical condition in the high-quality range, they are not as accurate as critical power (CP) approaches in predicting the axial location of the critical condition. Accurate knowledge of this axial location is of importance in predicting the maximum clad temperature before quenching following an LOCA.

13.4.3.1.3.1 Biasi Correlation [5] The Biasi correlation is a function of pressure, mass flux, flow quality, and tube diameter. The root-mean-square (rms) error of the correlation is 7.26% for more than 4500 data points, and 85.5% of all points are within ±10%. The correlation is capable of predicting both DNB and Dryout CHF conditions. The correlation is given by the following equations.

Use Equation 13.48b, below, for $G < 300$ kg/m² s; for higher G, use the larger of the two values. Hence

$$q''_{cr} = (2.764 \times 10^7)(100D)^{-n}G^{-1/6}[1.468F(p_{bar})G^{-1/6} - x]\,\text{W/m}^2 \quad (13.48a)$$

$$q''_{cr} = (15.048 \times 10^7)(100D)^{-n}\,G^{-0.6}H(p_{bar})[1 - x]\,\text{W/m}^2 \quad (13.48b)$$

where

$$F(p_{bar}) = 0.7249 + 0.099p_{bar}\exp(-0.032p_{bar}) \quad (13.48c)$$

$$H(p_{bar}) = -1.159 + 0.149p_{bar}\exp(-0.019p_{bar}) + 9p_{bar}(10 + p_{bar}^2)^{-1} \quad (13.48d)$$

Note: $p_{bar} = 10p$, when p is in MPa; and

$$n = 0.4 \text{ for } D \geq 0.01\text{ m}; \quad 0.6 \text{ for } D < 0.01\text{ m} \quad (13.48e)$$

The database for this correlation is

$D = 0.0030–0.0375$ m
$L = 0.2–6.0$ m
$p = 0.27–14$ MPa
$G = 100–6000$ kg/m² s
$x = 1/(1 + \rho_f/\rho_g)$ to 1

13.4.3.1.3.2 Bowring Correlation [9] The Bowring correlation contains four optimized pressure parameters. The rms error of the correlation is 7% for its 3800 data points. The correlation probably has the widest range of applicability in terms of pressure and mass flux. The correlation is described by the following equations in SI units:

$$q''_{cr} = \frac{A - Bh_{fg}x}{C}\,\text{W/m}^2 \quad (13.49a)$$

where

$$A = \frac{2.317(h_{fg}DG/4)F_1}{1 + 0.0143F_2D^{1/2}G} \quad (13.49b)$$

$$B = \frac{DG}{4} \quad (13.49c)$$

$$C = \frac{0.077F_3DG}{1 + 0.347F_4\left(\frac{G}{1356}\right)^n} \quad (13.49d)$$

$$p_R = 0.145p \text{ (where } p \text{ is in MPa)} \tag{13.49e}$$

$$n = 2.0 - 0.5p_R \tag{13.49f}$$

For $p_R < 1$ MPa:

$$\left.\begin{aligned} F_1 &= \{p_R^{18.942}\exp[20.89(1-p_R)]+0.917\}/1.917 \\ F_2 &= F_1/(\{p_R^{1.316}\exp[2.444(1-p_R)]+0.309\}/1.309) \\ F_3 &= \{p_R^{17.023}\exp[16.658(1-p_R)]+0.667\}/1.667 \\ F_4 &= F_3 p_R^{1.649} \end{aligned}\right\} \tag{13.49g}$$

For $p_R > 1$ MPa:

$$\left.\begin{aligned} F_1 &= p_R^{-0.368}\exp[0.648(1-p_R)] \\ F_2 &= F_1/\{p_R^{-0.448}\exp[0.245(1-p_R)]\} \\ F_3 &= p_R^{0.219} \\ F_4 &= F_3 p_R^{1.649} \end{aligned}\right\} \tag{13.49h}$$

The database for this correlation is

$D = 0.002–0.045$ m
$L = 0.15–3.7$ m
$p = 0.2–19.0$ MPa
$G = 136–18,600$ kg/m^2 s

13.4.3.1.3.3 The Groeneveld Correlation [33] A most useful and accurate current correlation is the CHF look-up table of Groeneveld. The Groeneveld CHF look-up table is basically a normalized data bank for a vertical 8 mm water-cooled tube. It provides the CHF, q''_{cr}, upon specification of the pressure, mass flux, and quality within the following ranges:

$0.1 \le p \le 21$ MPa
$0 < G \le 8000$ kg/m^{2}s
$3 < D \le 25$ mm
$-0.50 < x \le 0.90$

Various regions of the look-up table are shaded to differing degrees to specify the degree of uncertainty in the provided CHF value. The look-up table value for q''_{cr} is based on an 8 mm diameter tube and a uniform axial heat flux distribution. Table values are adjusted by multiplicative correction factors to account for a number of specific subchannel/tube and bundle conditions originally published in [30] as summarized in Table 13.3 [32].*

* The formulation of the heat flux distribution factor in Groeneveld et al. [32] contains two typographical errors: the correction factor is incorrectly shown as the reciprocal of that presented in Table 13.3 and Eq. 13.50c, and the parameter defining the two formulation ranges is specified in the nomenclature table, as flow quality (vapor weight fraction) instead of steam equilibrium quality as it should be.

For a subchannel or tube, therefore, table values are adjusted to the diameter and flux distribution of interest by means of correction factors that multiply the values of the look-up tables as

$$q''_{cr} = (q''_{cr})_{LUT} K_1 K_5 \tag{13.50a}$$

TABLE 13.3
Correction Factors for Groeneveld's CHF Look-Up Table

Factor	Form
K_1, Subchannel or tube-diameter	For $3 < D_e < 25$ mm: $K_1 = (8/D_e)^{1/2}$
Cross-section geometry Factor	For $D_e > 25$ mm: $K_1 = 0.57$
K_2, Bundle-geometry factor	$K_2 = \min\left[1, \left(\frac{1}{2} + \frac{2\delta}{D}\right)\exp\left(\frac{-(x_e)^{1/3}}{2}\right)\right]$ where δ = minimum rod spacing[a] = $P - D$
K_3, Mid-plane spacer factor for a 37-element bundle (CANDU)	See [32]
K_4, Heated-length factor	For $L/D_e > 5$: $K_4 = \exp\left[\left(\frac{D_e}{L}\right)\exp(2\alpha_{HEM})\right]$ $\alpha_{HEM} = \frac{x_e\rho_f}{[x_e\rho_f + (1 - x_e)\rho_g]}$
K_5, Axial flux distribution	For $x_e \le 0$: $K_5 = 1.0$ For $x_e > 0$: $K_5 = q''/q''_{BLA}$
K_6, Radial or circumferential Flux distribution factor	For $x_e > 0$: $K_6 = q''(z)_{max}/q''(z)_{avg}$[a] For $x_e \le 0$, $K_6 = 1.0$
K_7, Horizontal flow-orientation factor	See [32]
K_8, Vertical low-flow factor	$G < -400$ kg/m^2 s or $x_e << 0$: $K_8 = 1$ $-400 < G < 0$ kg/m^2 s: see [32] Minus sign refers to downward flow $G = 0$, $x_e = 0$ refer to pool boiling See [32] for further detail

Source: Groeneveld, D. C. et al., *International Topical Meeting on Nuclear Reactor Thermal Hydraulics (NURETH-10)*, Seoul, Korea, October 5–9, 2003b.

[a] Definition confirmed by D. Groeneveld (March 1, 2010, pers. comm.).

where $(q''_{cr})_{LUT}$ is the CHF from the look-up tables. The diameter correction factor, K_1, from Table 13.3 is

$$K_1 = \left(\frac{8}{D_e}\right)^{0.5} \tag{13.50b}$$

where D_e is the equivalent diameter under investigation, expressed in mm and the heat flux distribution correction factor, K_5, is

$$K_5 = \begin{cases} \dfrac{q''}{\overline{q''_{BLA}}} & \text{if } x_e > 0 \\ 1 & \text{if } x_e \le 0 \end{cases} \tag{13.50c}$$

where $\overline{q''_{BLA}}$ is the average heat flux from the onset of saturated boiling (OSB) to the location of interest, q'' is the heat flux at the location of interest, and x_e is the steam equilibrium quality.

For a bundle geometry, the factor K_2 would replace the tube geometry factor K_1.

The database upon which the table was developed contains more than 30,000 data points and provides CHF values at 24 pressures, 20 mass fluxes, and 23 qualities, covering the full range of conditions of practical interest. In addition, quoting Groeneveld [33], the 2006 CHF look-up table addresses several concerns with respect to previous CHF look-up tables raised in the literature. The major improvements of the 2006 CHF look-up table are

- An enhanced quality of the database (improved screening procedures, removal of clearly identified outliers, and duplicate data).
- An increased number of data in the database (an addition of 33 recent data sets).
- A significantly improved prediction of CHF in the subcooled region and the limiting quality region.
- An increased number of pressure and mass flux intervals (thus increasing the CHF entries by 20% compared to the 1995 CHF look-up table).
- An improved smoothness of the look-up table (the smoothness was quantified by a smoothness index).

A discussion of the impact of these changes on the prediction accuracy and table smoothness is presented in [33].

Example 13.3: Comparison of Tube Geometry Critical Heat Flux Correlations

PROBLEM

A vertical test tube in a high-pressure water boiling channel has the following characteristics:

$p = 6.89$ MPa $= 68.9$ bar
$D = 10.0$ mm

$L = 3.66$ m
$T_{in} = 204°C$

Using the Biasi and CISE-4 correlations, find the critical channel power for uniform heating at $G = 2000$ kg/m^2 s.

SOLUTION

Biasi correlation

Because $G > 300$ kg/m^2 s, check both Equations 13.48a and 13.48b. Use the larger q''_{cr}. In this case, Equation 13.48b is the larger.

$$q''_{cr} = 15.048 \times 10^7 (100D)^{-n} G^{-0.6} H(p_{bar})(1 - x) \tag{13.48b}$$

where

$$\begin{aligned} H(p_{bar}) &= -1.159 + 0.149 p_{bar} \exp(-0.019 p_{bar}) \\ &\quad + 9 p_{bar}/(10 + p_{bar}^2) \\ &= -1.159 + 0.149(68.9)\exp[-0.019(68.9)] \\ &\quad + 9(68.9)/[10 + (68.9)^2] \\ &= 1.744 \end{aligned} \tag{13.48d}$$

and $D = 0.01$ m, $n = 0.4$ from Equation 13.48e.

An expression is needed for $x = x_{cr}$, which is obtained from a heat balance:

$$q_{cr} = \dot{m} x h_{fg} + \dot{m} \Delta h_{sub}$$

Hence

$$x_{cr} = x(Z) = \frac{q_{cr}}{\dot{m} h_{fg}} - \frac{\Delta h_{sub}}{h_{fg}} = \frac{4 q''_{cr} Z}{G D h_{fg}} - \frac{\Delta h_{sub}}{h_{fg}}$$

For uniform heating, we take $Z = L$. Also

$$\Delta h_{sub} = 0.389 \times 10^6 \text{ J/kg}$$

$$h_{fg} = 1.51 \times 10^6 \text{ J/kg}$$

Thus, from Equation 13.48b

$$\begin{aligned} q''_{cr} &= 15.048 \times 10^7 \left[100(0.010)\right]^{-0.4} (2000)^{-0.6} (1.744) \\ &\quad \times \left[1 - \left(\frac{4 q''_{cr}(3.66)}{2000(0.01)(1.51 \times 10^6)} - \frac{0.386 \times 10^6}{1.51 \times 10^6} \right)\right] \end{aligned}$$

$$q''_{cr} = 3.423 \times 10^6 - 1.301q''_{cr} \rightarrow q''_{cr} = 1.488 \times 10^6 \text{ W/m}^2$$

$$q_{cr} = q''_{cr}\pi DL = (1.488 \times 10^6)\pi(0.01)(3.66) = 171.1 \text{ kW}$$

CISE-4 correlation

We need expressions for x_{cr} and L_B to use in Equation 13.47a. They come from energy balance considerations:

$$q''_{cr}\pi D(L - L_B) = G\frac{\pi D^2}{4}\Delta h_{sub}$$

$$\therefore L_B = L - \frac{GD\Delta h_{sub}}{4q''_{cr}}$$

$$q''_{cr}\pi DL_B = G\frac{\pi D^2}{4}x_{cr}h_{fg}$$

$$\text{Therefore } x_{cr} = \frac{4q''_{cr}L_B}{GDh_{fg}}$$

Thus Equation 13.47a becomes

$$\frac{4q''_{cr}}{GDh_{fg}}\left(L - \frac{GD\Delta h_{sub}}{4q''_{cr}}\right) = \frac{D_e}{D_h}\left[a\frac{(L - (GD\Delta h_{sub}/4q''_{cr}))}{L - (GD\Delta h_{sub}/4q''_{cr}) + b}\right]$$

or

$$\frac{4q''_{cr}}{GDh_{fg}} = \frac{D_e}{D_h}\left(\frac{a}{L - (GD\Delta h_{sub}/4q''_{cr}) + b}\right)$$

where $D_h = D_e = D = 0.01$ m; $p = 68.9$ bar $= 6.89$ MPa; $p_{cr} = 22.04$ MPa; and $G^* = 3375(1 - 6.89/22.04)^3 = 1096.17$ kg/m² s.

$$\therefore G > G^*$$

$$\therefore a = \frac{1 - p/p_c}{(G/1000)^{1/3}} = \frac{1 - 6.89/22.04}{(2000/1000)^{1/3}} = 0.54558$$

$$b = 0.199(p_c/p - 1)^{0.4}GD^{1.4}$$

$$= 0.199(22.04/6.89 - 1)^{0.4}(2000)(0.01)^{1.4} = 0.8839$$

Thus

$$\frac{4q''_{cr}}{(2000)(0.01)(1.51 \times 10^6)} = \frac{0.54558}{3.66 - \dfrac{(2000)(0.01)(0.389 \times 10^6)}{4q''_{cr}} + 0.8839}$$

$$1.30365 \times 10^{-7} q''_{cr} = \frac{0.54558}{4.544 - 1.976 \times 10^6 / q''_{cr}}$$

$$5.92366 \times 10^{-7} q''_{cr} = 0.54558 + 0.2576$$

$$q''_{cr} = 1356 \text{ kW/m}^2$$

$$q_{cr} = q''_{cr} \pi D L = (1{,}355{,}890.0)\pi(0.01)(3.66)$$

$$= 158.4 \text{ kW}$$

In this case, the CISE-4 correlation predicts a more conservative value for q_{cr} than the Biasi correlation.

13.4.3.2 Correlations for Rod Bundle Geometry

13.4.3.2.1 DNB: The W-3 Correlation

The most widely used correlation for evaluation of DNB conditions for PWRs is the W-3 correlation developed by Tong [59, 60]. The correlation may be applied to circular, rectangular, and rod-bundle flow geometries. The correlation has been developed for axially uniform heat flux, with a correcting factor for nonuniform flux distribution. This methodology for nonuniform heat flux conditions was developed by Tong [62] and independently by Silvestri [54]. Also cold wall and local spacer effects can be taken into account by specific factors which are presented after the nonuniform heat flux factor, F, is described. All W-3 correlation equations below are taken from the VIPRE-01 User's Manual [63].

13.4.3.2.1.1 Uniform Axial Heat Flux Formulation For a channel with axially uniform heat flux, the CHF, $q''_{cr,u}$, is given in SI units as

$$\begin{aligned} q''_{cr,u} = {} & \{(2.022 - 0.06238p) + (0.1722 - 0.01427p)\exp[(18.177 \\ & - 0.5987p)x_e]\}[(0.1484 - 1.596x_e + 0.1729x_e|x_e|)2.326G \\ & + 3271][1.157 - 0.869x_e][0.2664 + 0.8357 \\ & \times \exp(-124.1D_e)][0.8258 + 0.0003413(h_f - h_{in})] \end{aligned} \quad (13.51a)$$

where $q''_{cr,u}$ is in kW/m^2; p is in MPa; G is in kg/m^{2}s; h is in kJ/kg; and D_e is in m, and the correlation is valid in the ranges:

p (pressure) = 5.5–16 MPa
G (mass flux) = 1356–6800 kg/m^{2}s
D_e (equivalent diameter) = 0.015–0.018 m
x_e (equilibrium quality) = (minus)0.15–0.15
L (heated length) = 0.254–3.70 m
and the ratio of heated perimeter to wetted perimeter = 0.88–1.0

13.4.3.2.1.2 Nonuniform Axial Heat Flux Factor, F For predicting the DNB condition in a nonuniform heat flux channel, the following two steps are to be followed

1. The uniform CHF ($q''_{cr,u}$) is computed with the W-3 or other selected correlation, using the local reactor conditions.
2. The nonuniform DNB heat flux ($q''_{cr,n}$) distribution is then obtained for the prescribed flux shape by dividing $q''_{cr,u}$ by the F-factor, that is

$$q''_{cr,n} = q''_{cr,u}/F \tag{13.51b}$$

The factor F is presented in the form proposed by Lin et al. [46] except that for simplicity we take the lower limit of the integral below as zero versus the position of the onset of nucleation. This change introduces a negligible error for typical PWR conditions.

$$F = \frac{C\int_0^Z q''(Z')\exp[-C\cdot(Z-Z')]\,dZ'}{q''(Z)[1-\exp(-C\cdot Z)]} \tag{13.51c}$$

where Z = location of interest measured from core inlet; and C is an experimental coefficient describing the heat and mass transfer effectiveness at the bubble-layer/subcooled-liquid–core interface*:

$$C = 185.6\frac{[1-x_{eZ}]^{4.31}}{G^{0.478}}(\text{m}^{-1}) \tag{13.51d}$$

where G is in kg/m^{2}s and x_{eZ} is the equilibrium quality at the Z location of interest.

In British units, the CHF (Btu/h ft^2) for uniformly heated channels is given by

$$\begin{aligned} q''_{cr,u}/10^6 = {} & \{(2.022-0.0004302p)+(0.1722-0.0000984p) \\ & \times \exp[(18.177-0.004129p)x_e]\}[(0.1484-1.596x_e \\ & +0.1729x_e|x_e|)G/10^6+1.037](1.157-0.869x_e)[0.2664 \\ & +0.8357\exp(-3.151D_e)][0.8258+0.000794(h_f-h_{in})] \end{aligned} \tag{13.52a}$$

where q''_{cr} = DNB flux (Btu/h ft^2); p = pressure (psia); x_e = local equilibrium quality; D_e = equivalent diameter (in.); h_{in} = inlet enthalpy (Btu/lb); G = mass flux (lb$_m$/h ft^2); h_f = saturated liquid enthalpy (Btu/lb).

* This formulation based on the latest version of Tong's model supersedes his earlier version. Changes in DNB prediction caused by replacing the original model with that of Equation 13.51d are much smaller than the uncertainty in the W-3 DNB correlation itself.

The correlation is valid in the ranges:

p = 800–2300 psia
$G/10^6$ = 1.0–5.0 lb_m/h ft^2
D_e = 0.2– 0.7 in.
x_e = (minus)0.15–0.15
L = 10–144 in.

and the ratio of heated perimeter to wetted perimeter = 0.88–1.00.

The expression for the parameter C in Equation 13.52b in British units is

$$C = \frac{0.15[1 - x_{eZ}]^{4.31}}{(G/10^6)^{0.478}} \text{ in.}^{-1} \quad (13.52b)$$

where G is in lb/h ft^2.

The shape factor accounts for the difference in the amount of energy accumulated in the bubble layer up to the location of interest, for a uniform and nonuniform heated channel. To gain some physical insight on the shape factor, consider a channel with fixed mass flux, pressure, and equivalent diameter. Now consider the three heat flux profiles shown in Figure 13.19. Also, assume that in each case the inlet temperature is adjusted to obtain the same equilibrium quality at location z^*. Case 1 is the reference situation with uniform heating. In Case 2, a larger amount of energy than in Case 1 is supplied to the bubble layer before z^*; thus, it is clear that a small further heat supply at z^* (small relative to Case 1) may cause DNB. Hence, the shape factor is greater than one:

$$F = \frac{q''_{cr,u}}{q''_{cr,n}} > 1$$

Analogous reasoning would suggest that $F < 1$ in Case 3.

The shape factor conveys the so-called "memory effect" of the heating axial profile on the CHF. The quantity $1/C$, which shows up in the exponential of the shape

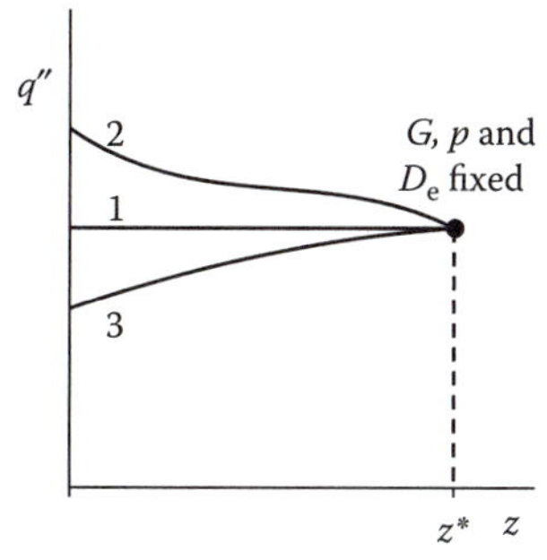

FIGURE 13.19 Nonuniform heat flux profiles.

factor, can be interpreted as a characteristic "memory length": it provides an estimate of the length of the upstream region that affects the CHF at a certain location. Note that C decreases as the quality increases therefore amplifying the memory effect. In the limit of very high qualities ($x_e > 0.2$), DNB is no longer the dominant CHF mechanism, and the occurrence of CHF can be considered as a global phenomenon, with little dependence on the local value of the heat flux.

13.4.3.2.1.3 Cold Wall Factor, $F_{cold\ wall}$ The wall correction factor, to account for the effect on CHF of the cold wall in a subchannel in which one side is a control rod thimble, is defined as

$$F_{\text{cold wall}} = \frac{q''_{\text{cr}}\big|_{D_h \neq D_e}}{q''_{\text{cr}}\big|_{\text{uniform},\, D_h}} \tag{13.53a}$$

Note carefully that Equation 13.53a specifies that if the cold wall factor is to be applied, the $q''_{\text{cr}}\big|_{\text{uniform},\, D_h}$ is based on hydraulic diameter D_h using the heated perimeter, P_h versus D_e using the wetted perimeter, P_w as in Equations 13.51a and 13.52a.

The correction factor is given by

$$F_{\text{cold wall}} = 1.0 - R_u[13.76 - 1.372e^{1.78x_e} - 4.732(G/10^6)^{-0.0535} - 0.0619(p/10^3)^{0.14} - 8.509(D_e)^{0.107}] \tag{13.53b}$$

where

$$R_u = 1 - \left(\frac{D_h}{D_e}\right) \tag{13.53c}$$

G is expressed in $\text{lb}_m/\text{h ft}^2$
p is expressed in psia
D_e is expressed in inches
D_h is hydraulic diameter based on heated perimeter

13.4.3.2.1.4 Grid Spacer Factors, F_{grid} Grids do affect CHF, probably by the enhancement of turbulence in the downstream flow field. Three grid spacer factors of the form

$$F_{\text{grid}} = \frac{q''_{\text{cr}}\big|_{\text{with grids}}}{q''_{\text{cr}}\big|_{\text{without grids}}} \tag{13.54a}$$

have been developed, one for each of the Westinghouse standard grid types. These are the simple or S grid, the mixing vane or R grid, and another mixing vane grid called the L-type. The R-grid factor and L-grid factor are identical except for a leading coefficient, F_g.

The S-grid factor is given by

$$F_{grid\,(S)} = 1.0 + 0.03(G/10^6)(TDC/0.019)^{0.35} \quad (13.54b)$$

where

G = mass flux, $lb_m/h\ ft^2$

TDC = turbulent cross flow mixing parameter (see Chapter 6, Volume II)

The L-grid or R-grid factor is given by

$$F_{grid\,(RL)} = F_g\{[1.445 - 0.0371L][p/225.896]^{0.5}[e^{(x_e+0.2)^2} - 0.73] + K_S \frac{G}{10^6}[TDC/0.019]^{0.35}\} \quad (13.54c)$$

where

L = heated length (ft)

G = mass velocity ($lb_m/h\ ft^2$)

p = pressure (psia)

K_S = grid spacing factor (an empirical constant dependent on grid design and spacing usually proprietary for a given design); Feng et al. [21] used a value of 0.066

TDC = turbulent crossflow mixing parameter

x_e = equilibrium quality

F_g = grid-type modifier; 1.0 for R-grids, 0.986 for L-grids

13.4.3.2.1.5 Forms of the W-3 Correlation—W-3S and W-3L To summarize, two forms of the Westinghouse CHF correlation have been described. These are

W-3S—uniform CHF equation with

- Standard nonuniform axial flux factor
- Unheated wall correction factor [if $\left(1 - \frac{D_h}{D_e}\right) \neq 0$]
- Simple S-grid spacer factor

W-3L—uniform CHF equation with

- Standard nonuniform axial flux factor
- Unheated wall correction factor [if $\left(1 - \frac{D_h}{D_e}\right) \neq 0$]
- L- and R-mixing vane grid factors

The VIPRE-01 User's Manual [63] more fully describes these correlations and their databases along with another form, the W-3C, based on earlier Westinghouse publications.

Example 13.4: Effect of Shape Factor in DNB Calculation

PROBLEM

Consider the interior hot subchannel of the PWR of Appendix K, assuming a cosine-shaped axial heat flux, and $q_0' = 44.62$ kW/m. Neglect cross-flow among subchannels and effects of the neutron flux extrapolation lengths.

Compute the axial variation of x_e, q'', $q''_{cr,u}$, $q''_{cr,n}$.

SOLUTION

The desired parameters including C are shown in Figures 13.20a through 13.20d. In this case, the axial power profile is not flat: given the physical interpretation of the shape factor, we expect $F < 1$ in the first half of the channel (increasing heat flux) and $F > 1$ in the second half of the channel (decreasing heat flux). Figure 13.20b confirms this expectation.

13.4.3.2.2 Dryout

In the high-vapor quality regions of interest to BWRs, two approaches have been taken to establish the required design margin. The first approach was to develop a limit line. That is a conservative lower envelope to the appropriate CHF data, such that virtually no data points fall below this line. The first set of limit lines used by the

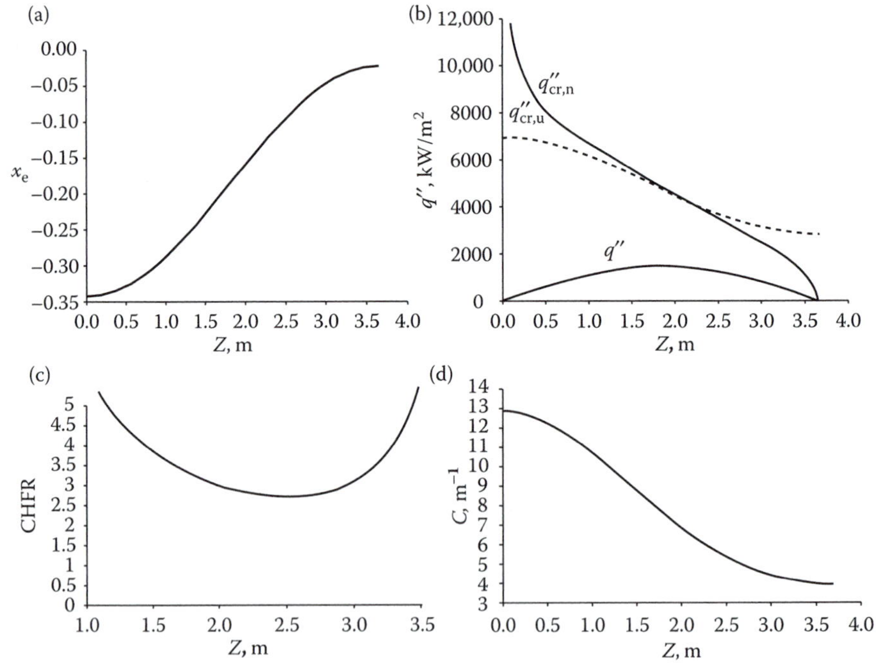

FIGURE 13.20 (a) Equilibrium quality as calculated in Example 13.4. (b) Actual and critical heat flux (CHF) as calculated in Example 13.4. (c) CHF ratio as calculated in Example 13.4. (d) Coefficient C as calculated in Example 13.4.

General Electric Company was based largely on single-rod annular boiling transition data having uniform axial heat flux. These design lines were known as the Janssen–Levy limit lines [40]. Some adjustments to the Janssen–Levy limit lines were found to be needed based on a subsequent experimental program in four- and nine-rod uniform axial heat flux bundles, and thus the Hench–Levy limit lines were developed [34].

These design curves were also constructed such that they fell below virtually all the data at each mass flux. Similar to the W-3 correlation for PWRs, these correlations have always been applied in BWR design with a margin of safety, in terms of an MCHFR. For the Hench–Levy limit lines the MCHFR was 1.9 for limiting transients. Obviously, although the limit line concept can be used for design purposes, it does not capture the axial heat flux effect; thus, the axial CHF location is normally predicted incorrectly. These correlations are also based on the assumption of local control of CHF conditions, which is invalid for high-quality flow.

The second approach, the total power concept, was developed in order to eliminate the undesirable features inherent in the local CHF hypothesis. A new correlation, known as the General Electric critical quality-boiling length (GEXL) correlation, was thus created [24].

13.4.3.2.2.1 The GEXL Correlation (General Electric Quality vs Boiling Length Correlation) The GEXL correlation [24] is based on a large amount of boiling crisis data taken in General Electric's ATLAS Heat Transfer Facility, which includes full-scale 49- and 64-rod data. The generic form of the GEXL correlation is

$$x_{cr} = x_{cr}(L_B, D_H, G, L, p, R) \tag{13.55}$$

where x_{cr} = bundle average critical quality; L_B = boiling length (distance from position of zero equilibrium quality to channel exit); D_H = heated diameter (i.e., four times the ratio of total flow area to heated rod perimeter); G = mass flux; L = total heated length; p = system pressure; and R = a parameter that characterizes the local peaking pattern with respect to the most limiting rod.

Unlike the limit line approach, the GEXL correlation is a "best fit" to the experimental data and is said to be able to predict a wide variety of data with a standard deviation of about 3.5% [24]. Additionally, the GEXL correlation is a relation between parameters that depend on the total heat input from the channel inlet to a position within the channel. The correlation line is plotted in terms of critical bundle radially averaged quality versus boiling length (Figure 13.21). The CP of a bundle is the value that leads to a point of tangency between the correlation line and a bundle operating condition. The critical quality-boiling length correlation leads to identification of a new margin of safety in BWR designs, that is the *critical power ratio* (CPR) of a channel:

$$\text{CPR} = \frac{\text{Critical power}}{\text{Operating power}} \tag{13.56}$$

The GEXL correlation has been kept as proprietary information.

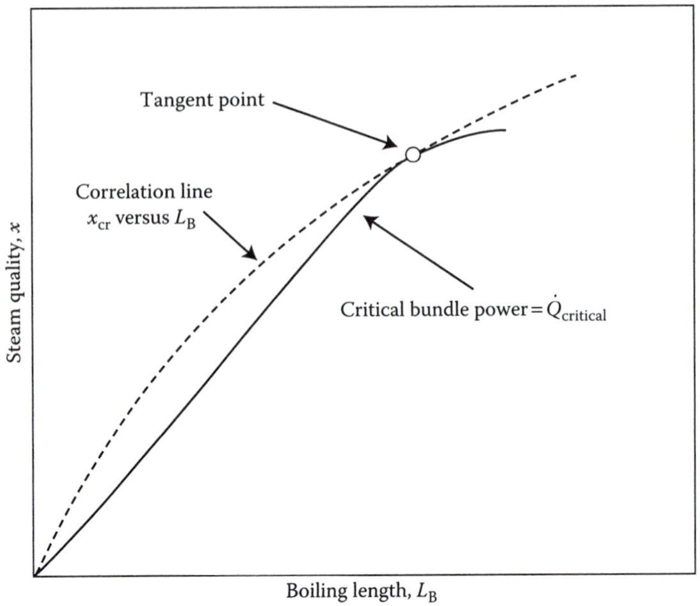

FIGURE 13.21 Correlation of critical conditions under Dryout.

13.4.3.2.2.2 The Hench–Gillis Correlation [35] The critical quality-boiling length Hench–Gillis correlation developed for EPRI and also called the EPRI-2 correlation is available. It was derived starting from an equation of the form of the CISE-2 correlation, and is based on publicly available Dryout data from BWR-type geometries.

$$x_c = \frac{AZ}{B+Z}(2-J)+F_p \tag{13.57a}$$

where $A = 0.50\ G^{-0.43}$

$B = 165 + 115\ G^{2.3}$

with G = mass flux, $Mlb_m/h\text{-}ft^2$

$$Z = \frac{\text{Boiling heat transfer area}}{\text{Bundle flow area}} = \frac{\pi D n L_B}{A_f}$$

with D = rod diameter

n = number of active rods in the bundle

L_B = boiling length

Pressure correction factor

$$F_p = 0.006 - 0.0157\left(\frac{p-800}{1000}\right) - 0.0714\left(\frac{p-800}{1000}\right)^2 \tag{13.57b}$$

where p is in psia.

The J-factor accounts for local peaking in the bundle, and is defined as follows: For corner rods:

$$J = J_1 - \frac{0.19}{G}(J_1 - 1)^2 \tag{13.57c}$$

For side rods:

$$J = J_1 - \frac{0.19}{G}(J_1 - 1)^2 - \left(\frac{0.07}{G+0.25} - 0.05\right) \tag{13.57d}$$

For central rods:

$$J = J_1 - \frac{0.19}{G}(J_1 - 1)^2 - \left(\frac{0.14}{G+0.25} - 0.10\right) \tag{13.57e}$$

J_1 is a weighted factor depending on the relative power factors, f_n, of the rods surrounding a given rod. Rods are weighted differently if they are in the same row (column) as a rod, or diagonally adjacent, as illustrated in Figure 13.22. The J_1 factors are calculated as follows:

For corner rods replace with:

$$J_1 = \frac{1}{16}\left(12.5 f_p + 1.5\sum_{i=1}^{2} f_i + 0.5 f_j\right) - \frac{nP(2S + D - P)}{64A}\left(4 f_p + \sum_{i=1}^{2} f_i\right) \tag{13.58a}$$

For side rods:

$$J_1 = \frac{1}{16}\left(11.0 f_p + 1.5\sum_{i=1}^{2} f_i + f_k + 0.5\sum_{j=1}^{2} f_j\right) - \frac{nP(2S + D - P)}{64A}\left(2 f_p + \sum_{i=1}^{2} f_i\right) \tag{13.58b}$$

For central rods:

$$J_1 = \frac{1}{4}\left(2.5 f_p + 0.25\sum_{i=1}^{4} f_i + 0.125\sum_{i=1}^{4} f_j\right) \tag{13.58c}$$

where

f_p = radial power factor for a given rod

f_i = radial power factor of rod adjacent to rod p in the same row or column (not including the rod adjacent to rod p in the same column, if p is a side rod)

f_j = radial power factor of rod adjacent to rod p on a diagonal line in the matrix

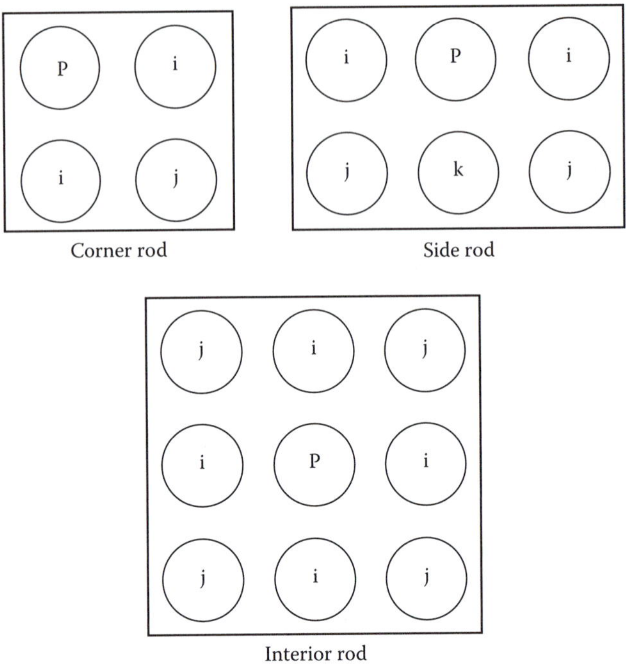

FIGURE 13.22 Adjacent rod weighting patterns for EPRI-2 correlation J-factors.

f_k = radial power factor of the rod adjacent to rod p and opposite the fuel channel wall when p is a side rod. (The weighted summations of rod power factors for corner and central rods do not include a k-rod.)

n = number of rods and water tubes in the fuel bundle

P = rod pitch

S = rod-to-fuel channel wall gap

A = total flow area in the bundle

D = fuel pin diameter

The Hench–Gillis correlation was derived from experimental data which did not include bundles with water rods. Therefore, no guidance was given for the calculation of the J-factor of a fuel rod facing water rods. A suggested solution consists of calculating the J-factor of that rod both as if it was a central or a side rod, and to use the resulting highest, most conservative, J-factor.

Example 13.5: The CPR Using the Hench–Gillis (EPRI-2) Correlation

PROBLEM

Considering the parameters of a BWR described in Appendix K, calculate the equilibrium quality and critical quality (using the Hench–Gillis correlation) as function of elevation for the rod-centered subchannel surrounding the "hot" fuel rod. Calculate also the CPR for that rod.

The radial power factors inside the fuel assembly are given in Figure 13.23, where the hot rod is the one printed in bold. For the sake of simplicity, assume a cosine heat flux distribution where $L_e = L$ and neglect cross-flow between subchannels. Use the following parameters for the rod-centered, lateral, hot subchannel.

$G = 1569.5$ kg m^{-2} s^{-1} (taken for simplicity as an average value between an edge and an interior coolant-centered subchannel)

$A_f = 1.42 \times 10^{-4}$ m^2

SOLUTION

The peak linear power for the hot rod is given in Appendix K as: $q_0' = 47.24$ kW/m.

According to the assumption of a cosine heat flux distribution, the total power generated by the fuel rod is $\dot{q} = \int_{-L/2}^{L/2} q_0' \cos\frac{\pi z}{L} dz = \frac{2Lq_0'}{\pi} = \frac{2(3.588)(47.24)}{\pi} =$ 107.9 kW.

The mass flowrate in the subchannel is

$$\dot{m} = GA_f = 1569.5 \cdot 1.42 \times 10^{-4} = 0.223 \text{ kg/s}$$

To obtain z_{OSB}, the location where bulk boiling starts to occur, Equation 14.31a can be applied.

$$z_{OSB} = \frac{L}{\pi}\sin^{-1}\left[-1 + \frac{2\dot{m}}{\dot{q}}(h_f - h_{in})\right] = \frac{3.588}{\pi}\sin^{-1}\left[-1 + \frac{2(0.223)}{107.9}(1274.4 - 1227.5)\right]$$

$$= 1.142 \sin^{-1}(-0.809) = -1.076 \text{ m}$$

0.92	1.04	1.08	1.10	1.08	1.10	1.08	1.04	0.92
1.04	0.77	0.98	0.79	0.71	0.80	0.99	0.77	1.04
1.08	0.98	0.76	0.92	0.99	0.98	0.79	0.99	1.08
1.10	0.79	0.91	1.07			0.98	0.80	**1.11**
1.08	0.71	0.98				0.99	0.71	1.08
1.10	0.80	0.98			1.07	0.92	0.79	1.10
1.08	0.99	0.79	0.98	0.98	0.91	0.76	0.98	1.07
1.04	0.78	0.99	0.80	0.71	0.79	0.98	0.77	1.03
0.92	1.04	1.08	1.08	1.08	1.10	1.08	1.03	0.92

FIGURE 13.23 Radial power factors to be used in Example 13.5. (From Greenspan, E., U. C. Berkeley, pers. comm., July, 2008.)

The axial distribution of the equilibrium quality may be obtained by applying Equation 14.33b as follows:

$$\begin{aligned} x_e(z) &= x_{e_{in}} + \frac{\dot{q}}{2\dot{m}h_{fg}}\left(\sin\frac{\pi z}{L} + 1\right) \\ &= \frac{1227.5 - 1274.4}{1496.4} + \frac{107.9}{2(0.223)(1496.4)}\left(\sin\frac{\pi z}{3.588} + 1\right) \\ &= -0.0314 + 0.162 + 0.162(\sin 0.876\,z) \end{aligned}$$

The mass flux for a subchannel is

$$G = 1569.5\ \text{kg/m}^2\ \text{s} \quad \text{or} \quad 1.157\ \text{Mlb}_\text{m}/\text{h ft}^2$$

The J-factor for the subject rod is determined from Equations 13.58b and 13.57d:

$$J_1 = \frac{1}{16}\left(11.0 f_P + 1.5\sum_{i=1}^{2} f_i + f_k + 0.5\sum_{j=1}^{2} f_j\right) - \frac{nP(2S + D - P)}{64A}\left(2f_P + \sum_{i=1}^{2} f_i\right) \tag{13.58b}$$

$$\begin{aligned} &= \frac{1}{16}\left[11.0(1.11) + 1.5(1.08 + 1.08) + 0.80 + 0.5(0.99 + 0.71)\right] \\ &\quad - \frac{76\cdot(0.01437)\left[2(0.00397) + 0.0112 - 0.01437\right]}{64(9.718\times10^{-3})}\left[2(1.11) + (1.08 + 1.08)\right] \\ &= 1.069 - 0.037 \\ &= 1.032 \end{aligned}$$

$$\begin{aligned} J &= J_1 - \frac{0.19}{G}(J_1 - 1)^2 - \left(\frac{0.07}{G + 0.25} - 0.05\right) \\ &= 1.032 - \frac{0.19}{1.157}(1.032 - 1)^2 - \left(\frac{0.07}{1.157 + 0.25} - 0.05\right) \\ &= 1.032 \end{aligned} \tag{13.57d}$$

The nominal pressure is

$$p = 7.14\ \text{MPa} \rightarrow 1035.6\ \text{psia}$$

So, from Equation 13.57b

$$\begin{aligned} F_p &= 0.006 - 0.0157\left(\frac{1035.6 - 800}{1000}\right) - 0.0714\left(\frac{1035.6 - 800}{1000}\right)^2 \\ &= -1.66\times10^{-3} \end{aligned}$$

The Hench–Gillis correlation gives the value for the critical quality:

$$x_c = \frac{AZ}{B+Z}(2-J) + F_p \tag{13.57a}$$

where
$A = 0.50\ G^{-0.43} = 0.5\ (1.157^{-0.43}) = 0.470$
$B = 165 + 115\ (1.157^{2.3}) = 326$

$$Z = \frac{\text{Boiling heat transfer area}}{\text{Bundle flow area}} = \frac{\pi D n L_B}{A_f} = \frac{\pi(0.0112)74 L_B}{9.718 \times 10^{-3}} = 267.9\ L_B$$

$$\therefore\ x_c = \frac{(0.470)267.9\ L_B}{326 + 267.9\ L_B}(2-1.032) - 1.66\times10^{-3} = \frac{121.9\ L_B}{326+267.9 L_B} - 1.66\times10^{-3}$$

The results for the critical quality and equilibrium quality are plotted in Figure 13.24, where the boiling length L_B is defined as

$$L_B = Z - Z_{OSB} \tag{13.59}$$

The same process can be iterated increasing the peak linear power until the equilibrium quality intersects the critical quality. This occurs at

$$q'_{o,cr} = 51.68\ \text{kW/m}$$

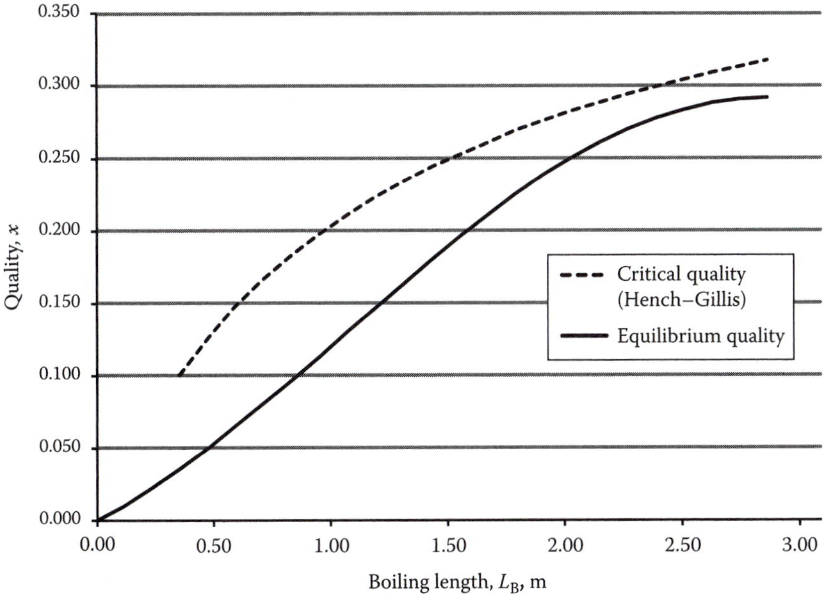

FIGURE 13.24 Critical quality and equilibrium quality from Example 13.5.

It is hence possible to calculate the CPR, as

$$CPR = \frac{\dot{q}_{cr}}{\dot{q}_{op}} = \frac{51.7}{47.2} = 1.10$$

13.4.3.2.3 Both DNB and Dryout

While the Biasi and Groeneveld correlations predict both low- and high-quality critical conditions for tubes, the only extensively used bundle geometry correlation with this capability is the EPRI-1 correlation.

13.4.3.2.3.1 The EPRI-1 Correlation [49] This correlation developed by Reddy and Fighetti at Columbia University for EPRI gives the CHF as a function of inlet subcooling, local quality, mass flux, and heat flux as

$$q''_{cr} = \frac{A - x_{in}}{C + \left(\dfrac{x_e - x_{in}}{q''_L}\right)} \tag{13.60}$$

where

$A = P_1 P_r^{P_2} G^{(P_5 + P_7 P_r)}$

$C = P_3 P_r^{P_4} G^{(P_6 + P_8 P_r)}$

q''_L = local heat flux (MBtu/h ft^2)

x_e = local equilibrium quality

x_{in} = inlet quality; $[(h_{inlet} - h_f)/h_{fg}]$

G = local mass velocity, Mlb$_m$/h ft^2

P_r = critical pressure ratio; system pressure/critical pressure and the Optimized Constants are

$P_1 = 0.5328$ $P_5 = -0.3040$

$P_2 = 0.1212$ $P_6 = 0.4843$

$P_3 = 1.6151$ $P_7 = -0.3285$

$P_4 = 1.4066$ $P_8 = -2.0749$

The correlation is based on experimental data spanning the range of parameters of Table 13.4.

The general form of the correlation (Equation 13.60) was obtained with data from normal matrix subchannels. Several correction factors were proposed for various conditions of practical interest.

a. BWR cold wall effect—for subchannels adjacent to the fuel channel interior wall.

$$q''_{cr} = \frac{AF_A - x_{in}}{CF_C + \left(\dfrac{x_e - x_{in}}{q''_L}\right)} \tag{13.61a}$$

where

$F_A = G^{0.1}$

$F_C = 1.183\, G^{0.1}$

TABLE 13.4
Parameter Range of EPRI Correlation

Pressure	200–2450 psia
Mass velocity	0.2–4.5 $Mlb_m/h\text{-}ft^2$
Quality	– 25.0 to +75.0%
Bundle geometries	3 × 3, 4 × 4, 5 × 5 with and without guide tubes
Heated length	30, 48, 66, 72, 84, 96, 144, 150, 168 in.
Rod diameters	PWR, BWR typical
Uniform axial power profiles	
Uniform and peaked radial power profiles	

The cold wall effect correction was developed from 661 corner channel CHF data points from 22 BWR-type test sections. It arises from the accumulation of liquid on the cold surface which retains channel liquid that otherwise would be available to the liquid film on the hot surface of the fuel rods. Hence the CHF is lowered as Equation 13.61a suggests, since the effect of F_c on the numerator overwhelms that of F_A on the denominator. The data covered the following parameter ranges:

Pressure	600–1500 psia
Mass flux	0.15–1.40 $Mlb_m/h\ ft^2$
Local quality	0.0–0.70

b. Grid spacer factor—a grid spacer factor, F_g, was developed to extend the applicability of the correlation to fuel assemblies with various types of grid spacers. It typically (but not in all cases of grid loss coefficient values) increases q''_{CHF}. The grid spacer factor is added to the correlation in the following manner:

$$q''_{cr} = \frac{A - x_{in}}{C F_g + \left(\dfrac{x_e - x_{in}}{q''_L}\right)} \tag{13.61b}$$

where
$F_g = 1.3–0.3\ C_g$
C_g = grid loss coefficient

The grid spacer correction factor was developed using Combustion Engineering data which included seven different types of grids, with loss coefficient values ranging from 0.815 to 2.0. The data consisted of seven groups, divided as follows:

Grid loss coefficient	0.815	0.885	1.08	1.25	1.38	1.48	2.00
Number of data points	1169	20	222	31	36	116	48

c. Nonuniform axial heat flux distribution factor—it is generally assumed that local CHF can depend, at least to some degree, on the heat flux distribution upstream of the point under consideration. The approach proposed by Bowring [10] to account for nonuniform axial heat flux profiles was adopted for use with the EPRI-1 correlation. The form of the correlation with the Bowring factor is

$$q''_{cr} = \frac{A - x_{in}}{C C_{nu} + \left(\frac{x_e - x_{in}}{q''_L}\right)} \tag{13.62}$$

where

$$C_{nu} = 1 + \frac{(Y - 1)}{(1 + G)} \tag{13.63a}$$

and Y is a measure of the nonuniformity of the axial heat flux profile defined as

$$Y = \frac{\int_0^Z \bar{q}''(Z)\,dZ}{\bar{q}''(Z)Z} \tag{13.63b}$$

where Z is taken from the start of the heated length and $\bar{q}''(Z)$ is the radially averaged heat flux at axial location Z. Bowring called the factor Y the Heat Balance Axial Heat Flux Parameter.

For clusters with a uniform axial heat flux, $Y = 1.0$.

This nonuniform axial heat flux form of the correlation was tested with 933 CHF data points from 23 test sections, having eight different nonuniform axial flux profiles. The data included symmetrical, top-peaked, bottom-peaked, spiked, and other irregular profiles. This modified correlation (as shown in Equation 13.62) predicted an average CHF ratio of 0.999 for the 933 data points, with a standard deviation of 6.92%. Nevertheless the base correlation (as given in Equation 13.60), without the C_{nu} correction, predicted an average CHF ratio of 1.00 for this same data, with a standard deviation of 8.56%. This demonstrates that the correlation can predict CHF for nonuniform axial flux profiles quite well, with or without the nonuniform correction factor.

Example 13.6: Comparison of Correlations for the Critical Condition in a PWR Hot Channel (Groeneveld Look-Up Table, W-3 and EPRI-1)

PROBLEM

Using the data referred to a PWR internal hot channel shown in Appendix K, calculate the heat flux at the distance from the core midplane: $z = 1$ m. Calculate then the CHF and the DNBR at that point using three different correlations: Groeneveld look-up table, W-3 and EPRI-1. For the sake of simplicity, consider a cosine axial heat flux distribution where $L_e = L$ and neglect cross-flow among subchannels.

SOLUTION

Adopting a cosine axial heat flux distribution, the heat flux for $z = 1$ m is

$$q''(z = 1\text{m}) = q_0''\cos\left(\frac{\pi z}{L}\right) = \frac{q_0'}{\pi D}\cos\left(\frac{\pi z}{L}\right) = \frac{44.62}{\pi(0.0095)}\cos\left(\frac{\pi(1)}{3.658}\right)$$
$$= 976.8 \text{ kW/m}^2$$

The equilibrium quality is given as

$$x_e(z = 1m) = x_{e_{in}} + \frac{\dot{q}}{2\dot{m}h_{fg}}\left(\sin\frac{\pi z}{L} + 1\right)$$
$$= \frac{h_{in} - h_f}{h_{fg}} + \frac{q_0' L/\pi}{AGh_{fg}}\left(\sin\frac{\pi z}{L} + 1\right)$$
$$= \frac{1300.6 - 1630.4}{2595.8 - 1630.4} + \frac{2(44.62)(3.658)}{\pi(2)(8.79\times 10^{-5})(3807)(2595.8 - 1630.4)}$$
$$\left(\sin\frac{\pi(1)}{3.658} + 1\right)$$
$$= -0.342 + 0.161(1.757)$$
$$= -0.059$$

Groeneveld look-up table

Interpolating the values from the Groeneveld look-up table, it is possible to obtain the CHF for $G = 3807$ kg m^{-2}s^{-1}, $p = 15.51$ MPa, and $x = -0.052$.

$$(q_c'')_{LUT} = 3567 \text{ kW/m}^2$$

The equivalent diameter D_e is

$$D_e = \frac{4A_f}{\pi D} = \frac{4(8.79 \times 10^{-5})}{\pi(0.0095)} = 0.0118 \text{ m}$$

The subchannel cross-section geometry factor K_1 is

$$K_1 = \sqrt{\frac{0.008}{D_e}} = \sqrt{\frac{0.008}{0.0118}} = 0.823$$

The bundle-geometry factor K_2 is

$$K_2 = \min\left[1, \left(\frac{1}{2} + \frac{2\delta}{D}\right)\exp\left(\frac{-(x_e)^{1/3}}{2}\right)\right]$$
$$= \min\left[1, \left(\frac{1}{2} + \frac{2(0.0031)}{0.0095}\right)\exp\left(-\frac{(-0.059)^{1/3}}{2}\right)\right] = 1$$

where

$$\delta = P - D = 0.0126 - 0.0095 = 0.0031$$

The heated length factor K_4 is

$$K_4 = \exp\left[\left(\frac{D_e}{L}\right)\exp\left(\frac{2x_e\rho_f}{\left[x_e\rho_f + (1 - x_e)\rho_g\right]}\right)\right]$$

$$= \exp\left\{\frac{0.0118}{3.658}\exp\left[\frac{2(-0.059)594.1}{(-0.059)594.1 + \left[1 - (-0.059)\right]102.1}\right]\right\}$$

$$= 1.0012$$

Since the equilibrium quality is negative, the axial flux distribution correction factor K_5 is equal to 1.

Hence the CHF is

$$q''_c = K_1K_2K_4K_5\left(q''_c\right)_{LUT} = (0.823)(1)(1.0012)(1)(3567)$$

$$= 2939\ \text{kW/m}^2$$

The DNBR for $z = 1$ m is hence

$$\text{DNBR} = \frac{q''_c}{q''} = \frac{2939}{976.8} = 3.01$$

W-3 correlation

The CHF according to the W-3 correlation for uniform axial heat flux is

$$q''_{cr,u} = \{(2.022 - 0.06238p) + (0.1722 - 0.01427p)\exp[(18.177$$
$$-0.5987p)x_e]\}[(0.1484 - 1.596x_e + 0.1729x_e|x_e|)2.326G$$
$$+3271][1.157 - 0.869x_e][0.2664$$
$$+0.8357\exp(-124.1D_e)][0.8258 + 0.0003413(h_f - h_{in})]$$

$$= \{(2.022 - 0.06238\cdot 15.51) + (0.1722 - 0.01427\cdot 15.51)\exp[(18.177$$
$$- 0.5987\cdot 15.51)(-0.059)]\}[(0.1484 - 1.596(-0.059)$$
$$+ 0.1729(-0.059)|-0.059|)2.326\cdot 3807 + 3271][1.157$$
$$- 0.869(-0.059)][0.2664$$
$$+ 0.8357\exp(-124.1\cdot 0.0118)][0.8258 + 0.0003413(1630.4 - 1300.6)]$$

$$= 2893\ \text{kW/m}^2$$

From Equation 13.51d

$$C = 186.5\frac{[1 - x_e]^{4.31}}{G^{0.478}} = 186.5\frac{[1 - (-0.059)]^{4.31}}{3807^{0.478}}$$
$$= 4.62$$

Inserting this value in Equation 13.51c expressed in terms of z versus Z obtain

$$F = \frac{C\int_{-L/2}^{z} q''(z')\exp[-C(z - z')]\,dz'}{q''(z)\left\{1 - \exp\left[-C\left(z + \frac{L}{2}\right)\right]\right\}}$$
$$= \frac{4.64q_0''\int_{-3.588/2}^{1}\cos\left(\frac{\pi z'}{3.658}\right)\exp[-4.62(1 - z')]\,dz'}{q_0''\cos\left(\frac{\pi}{3.658}\right)\left\{1 - \exp\left[-4.62\left(1 + \frac{3.658}{2}\right)\right]\right\}}$$
$$= 7.07\int_{-1.829}^{1}\cos(0.859z')\exp[-4.62(1 - z')]\,dz'$$

Integrating this function numerically:

$$F = 1.17$$

The CHF is

$$q''_{cr} = \frac{q''_{cr,u}}{F} = \frac{2893}{1.17} = 2473\ \text{kW/m}^2$$

The DNBR for $z = 1$ m is hence

$$\text{DNBR} = \frac{q''_{cr}}{q''} = \frac{2473}{976.8} = 2.53$$

EPRI-1 correlation

The EPRI-1 correlation requires the following units to be converted into the British system:

$$G = 3807\ \text{kg/m}^2\ \text{s} \quad \text{or} \quad 2.807\ \text{Mlb}_m/\text{h ft}^2$$

$q''(z = 1\ \text{m}) = 976.8\ \text{kW/m}^2$ or 0.3096 M Btu/h ft^2

$$x_{in} = \frac{h_{in} - h_f}{h_{fg}} = \frac{1300.6 - 1630.4}{2595.8 - 1630.4} = -0.342$$

$$P_r = \frac{15.51}{22.04} = 0.704$$

$$A = P_1 P_r^{P_2} G^{(P_5 + P_7 P_r)} = 0.5328\left(0.704^{0.1212}\right)\left(2.807^{(-0.3040-0.3285\cdot 0.704)}\right) = 0.294$$

$$C = P_3 P_r^{P_4} G^{(P_6 + P_8 P_r)} = 1.6151\left(0.704^{1.4066}\right)\left(2.807^{(0.4843-2.0749\cdot 0.704)}\right) = 0.360$$

The nonuniform axial heat flux distribution factor can be evaluated as follows, where z is taken from the core axial centerline:

$$Y(z = 1\,\text{m}) = \frac{\int_{-L/2}^{z} \bar{q}''(z^*)\,dz^*}{\bar{q}''(z)(z + L/2)} = \frac{\bar{q}_0'' \int_{-L/2}^{z} \cos\frac{\pi z^*}{L}\,dz^*}{(z + L/2)\bar{q}_0''\cos\frac{\pi z}{L}} = \frac{\bar{q}_0''\frac{L}{\pi}\left[\sin\frac{\pi z^*}{L}\right]_{-L/2}^{z}}{(z + L/2)\bar{q}_0''\cos\frac{\pi z}{L}}$$

$$= \frac{\frac{L}{\pi}\left[\sin\frac{\pi z}{L} + 1\right]}{(z + L/2)\cos\frac{\pi z}{L}} = \frac{\frac{3.658}{\pi}\left[\sin\frac{\pi}{3.658} + 1\right]}{(1 + 3.658/2)\cos\frac{\pi}{3.658}} = 1.11$$

$$C_{nu} = 1 + \frac{Y - 1}{1 + G} = 1 + \frac{1.11 - 1}{1 + 2.807} = 1.0289$$

$$q''_{cr} = \frac{A - x_{in}}{C \cdot C_{nu} + \left(\frac{x_e - x_{in}}{q''_L}\right)} = \frac{0.294 - (-0.342)}{0.360(1.0289) + \left(\frac{-0.059 - (-0.342)}{0.3096}\right)}$$

$$= 0.495\ \text{MBtu/h ft}^2 \quad \text{or} \quad 1562\,\text{kW/m}^2$$

The DNBR for $z = 1$ m is hence

$$\text{DNBR} = \frac{q''_{cr}}{q''} = \frac{1562}{976.8} = 1.60$$

In summary, the results for this PWR example are

Correlation	DNBR at 1 m above core midplane
Groeneveld	3.01
W-3	2.53
EPRI-1	1.60

However, these numerical DNBR values do not reflect the margin in power to the critical condition (DNB). The determination of this margin for the value of DNBR from any correlation is discussed in Section 13.4.4.

Example 13.7: Comparison of Correlations for the Critical Condition in a BWR Hot Channel (Groeneveld Look-Up Table, Hench–Gillis, and EPRI-1)

PROBLEM

Using the data referred to a BWR hot channel shown in Appendix K, calculate the equilibrium quality and heat flux at the distance from the core midplane: $z = 1$ m.

Calculate the critical quality at that point using the Hench–Gillis correlation and the CHF using the Groeneveld look-up table and the EPRI-1 correlation. For the sake of simplicity, consider a cosine axial heat flux distribution where $L_e = L$ and neglect cross-flow among different channels.

Use the following parameters, which refer to a lateral rod-centered subchannel:

$$G = 1569.5 \text{ kg m}^{-2} \text{ s}$$

$A_f = 1.42 \times 10^{-4} \text{ m}^2$

SOLUTION

The equilibrium quality is given by

$$x_e(z = 1\,\text{m}) = x_{e_{in}} + \frac{\dot{q}}{2\dot{m}h_{fg}}\left(\sin\frac{\pi z}{L} + 1\right)$$

$$= \frac{h_{in} - h_f}{h_{fg}} + \frac{2q_0'L/\pi}{2AGh_{fg}}\left(\sin\frac{\pi z}{L} + 1\right)$$

$$= \frac{1227.5 - 1274.4}{1496.4} + \frac{2(47.24)(3.588)}{\pi(2)(1.42\cdot 10^{-4})(1569.5)(1496.4)}\left(\sin\frac{\pi(1)}{3.588} + 1\right)$$

$$= -0.0314 + 0.162(1.768) = 0.255$$

Adopting a cosine axial heat flux distribution, the heat flux for $z = 1$ m is

$$q''(z = 1\,\text{m}) = q_0''\cos\left(\frac{\pi \cdot z}{L}\right) = \frac{q_0'}{\pi D}\cos\left(\frac{\pi \cdot z}{L}\right) = \frac{47.24}{\pi \cdot 0.0112}\cos\left(\frac{\pi \cdot 1}{3.588}\right)$$

$$= 860.0 \text{ kW/m}^2$$

Hench–Gillis correlation

The same values used in this Example have been used in Example 13.5. The Hench–Gillis correlation leads to the following equation:

$$x_c = \frac{121.9\,L_B}{326 + 267.9L_B} - 1.66 \times 10^{-3}$$

The boiling length is

$$L_B = z - z_{OSB}$$

Where $z_{OSB} = -1.076$ m, as calculated in Example 13.5. Hence

$$x_c(z = 1) = \frac{121.9\,(1 + 1.076)}{326 + 267.9(1 + 1.076)} - 1.66 \times 10^{-3} = 0.285$$

This value is greater than the equilibrium quality, which is 0.255, so according to the Hench–Gillis correlation there is no Dryout condition for $z = 1$ m.

The CPR at $z = 1$ m can be calculated by increasing the power until the equilibrium quality reaches the critical quality. This occurs at a power of 123.0 kW (based on a numerical calculation) compared to the Appendix K hot rod power of 107.9 kW, so the CPR = 1.14 at z = 1 m.

Groeneveld look-up table

Interpolating the values from the Groeneveld look-up table, it is possible to obtain the CHF which refers to the desired *G*, *p*, and *x*:

$$(q''_c)_{LUT} = 3094\ \text{kW/m}^2$$

The geometric correction factor K_1 is

$$K_1 = \sqrt{\frac{0.008}{D_H}} = \sqrt{\frac{0.008}{0.0162}} = 0.70$$

The bundle-geometry factor K_2 is

$$K_2 = \min\left[1, \left(\frac{1}{2} + \frac{2\delta}{D}\right)\exp\left(\frac{-(x_e)^{1/3}}{2}\right)\right]$$

$$= \min\left[1, \left(\frac{1}{2} + \frac{2 \cdot 0.00317}{0.0112}\right)\exp\left(-\frac{(0.255)^{1/3}}{2}\right)\right] = 0.776$$

where

$$\delta = P - D = 0.01437 - 0.0112 = 0.00317\ \text{m}$$

The heated length factor K_4 is

$$K_4 = \exp\left[\left(\frac{D_e}{L}\right)\exp\left(\frac{2x_e\rho_f}{\left[x_e\rho_f + (1 - x_e)\rho_g\right]}\right)\right]$$

$$= \exp\left\{\frac{0.0162}{3.588}\exp\left[\frac{2(0.255)737.3}{(0.255)737.3 + \left[1 - (0.255)\right]37.3}\right]\right\} = 1.026$$

To determine the axial flux distribution factor K_5, we must calculate the average heat flux from z_{OSB} to $z = 1$.

$$\bar{q}''_{BLA} = \frac{\int_{-1.076}^{1} q''(z^*)dz^*}{(1 + 1.076)} = \frac{q''_0\int_{-1.076}^{1} \cos\frac{\pi z^*}{L}dz^*}{(1 + 1.076)} = \frac{q''_0\frac{L}{\pi}\left[\sin\frac{\pi z^*}{L}\right]_{-1.076}^{1}}{(1 + 1.076)}$$

$$= \frac{\dfrac{q_0'}{\pi D}\dfrac{L}{\pi}\left[\sin\dfrac{\pi}{L} - \sin\dfrac{\pi(-1.076)}{L}\right]}{(1 + 1.076)}$$

$$= \frac{\dfrac{47.24}{\pi(0.0112)}\dfrac{3.588}{\pi}\left[\sin\dfrac{\pi}{3.588} - \sin\dfrac{\pi(-1.076)}{3.588}\right]}{(1 + 1.076)} = 1165\ \text{kW/m}^2$$

Hence

$$K_5 = \frac{q''}{\bar{q}''_{\text{BLA}}} = \frac{860.0}{1165} = 0.74$$

So the CHF is

$$q_c'' = K_1K_2K_4K_5\left(q_c''\right)_{\text{LUT}} = (0.70)(0.776)(1.026)(0.74)(3094)$$
$$= 1276\ \text{kW/m}^2$$

This value is greater than the local heat flux, which is 860.0 kW m^{-2}, so according to the Groeneveld look-up table there is no Dryout condition for $z = 1$ m.

The CPR at $z = 1$ m can be calculated by increasing the power until the heat flux reaches the CHF. This occurs at a power of 141.4 kW (based on a numerical calculation) compared to the Appendix K hot rod power of 107.9 kW, so the CPR = 1.31 at z = 1 m.

EPRI-1 correlation

The EPRI-1 correlation requires the following units to be converted into the British system:

$$G = 1569.5\ \text{kg m}^{-2}\ \text{s} \quad \text{or} \quad 1.157\ \text{Mlb}_m/\text{h ft}^2$$

$$q''(z = 1\ \text{m}) = 860.0\ \text{kW/m}^2 \quad \text{or} \quad 0.273\ \text{MBtu/h ft}^2$$

$$x_{\text{in}} = \frac{h_{\text{in}} - h_f}{h_{fg}} = \frac{1227.5 - 1274.4}{1496.4} = -0.0314$$

$$P_r = \frac{7.14}{22.04} = 0.324$$

$$A = P_1P_r^{P_2}G^{(P_5 + P_7P_r)} = 0.5328 \times 0.324^{0.1212} \times 1.157^{(-0.3040-0.3285\cdot 0.324)} = 0.438$$

$$C = P_3P_r^{P_4}G^{(P_6 + P_8P_r)} = 1.6151 \times 0.324^{1.4066} \times 1.157^{(0.4843-2.0749\cdot 0.324)} = 0.322$$

The nonuniform axial heat flux distribution factor can be evaluated as follows:

$$Y(z = 1\text{m}) = \frac{\int_{-L/2}^{z} \bar{q}''(z^*)\,dz^*}{\bar{q}''(z)(z + L/2)} = \frac{\bar{q}_0''\int_{-L/2}^{z} \cos\dfrac{\pi z^*}{L}\,dz^*}{(z + L/2)\bar{q}_0''\cos\dfrac{\pi z}{L}} = \frac{\bar{q}_0''\dfrac{L}{\pi}\left[\sin\dfrac{\pi z^*}{L}\right]_{-L/2}^{z}}{(z + L/2)\,\bar{q}_0''\cos\dfrac{\pi z}{L}}$$

$$= \frac{\frac{L}{\pi}\left[\sin\frac{\pi z}{L} + 1\right]}{(z + L/2)\cos\frac{\pi z}{L}} = \frac{\frac{3.588}{\pi}\left[\sin\frac{\pi}{3.588} + 1\right]}{\left(1 + 3.588/2\right)\cos\frac{\pi}{3.588}} = 1.13$$

$$C_{nu} = 1 + \frac{Y - 1}{1 + G} = 1 + \frac{1.13 - 1}{1 + 1.157} = 1.06$$

Since this hot channel is adjacent to the fuel channel wall, the correction factors in Equation 13.61 should be applied

$$F_A = G^{0.1} = 1.157^{0.1} = 1.015$$

$$F_C = 1.183G^{0.1} = 1.183 \times 1.157^{0.1} = 1.200$$

$$q''_{cr} = \frac{AF_A - x_{in}}{C \times C_{nu} F_C + \left(\frac{x_e - x_{in}}{q''_L}\right)} = \frac{0.438(1.015) + 0.0314}{0.322(1.06)(1.200) + \left(\frac{0.255 + 0.0314}{0.273}\right)}$$

$$= 0.326 \text{ MBtu/h ft}^2 \quad \text{or} \quad 1028 \text{ kW/m}^2$$

This value is greater than the local heat flux, which is 860.0 kW/m², so according to the EPRI-1 correlation there is no Dryout condition for $z = 1$ m.

The CPR at $z = 1$ m can be calculated by increasing the power until the heat flux reaches the CHF. This occurs at a power of 130.6 kW (based on a numerical calculation) compared to the Appendix K hot rod power of 107.9 kW, so the CPR = 1.21 at $z = 1$ m.

In summary, the results for this BWR example are

Correlation	CPR at 1 m above core midplane
Hench–Gillis	1.14
Groeneveld	1.31
EPRI-1	1.21

13.4.4 Design Margin in Critical Condition Correlation

With CHF and CP correlations now presented in Section 13.4.3, let us examine how they are applied in design and the comparative design margins among these correlations.

13.4.4.1 Characterization of the Critical Condition

Table 2.3 introduced the characterization of the critical condition for PWRs (subcooled to very low-quality operation conditions) and BWRs (moderate quality operating conditions) as respectively:

For PWRs:

$$\text{CHFR} \equiv \text{DNBR} = \left.\frac{q''_{cr}}{q''_{op}}\right|_{\text{for any z}} \tag{13.64a}$$

where the critical condition is caused by DNB.

For BWRs:

$$\text{CPR} = \left.\frac{\dot{q}_{\text{cr}}}{\dot{q}_{\text{op}}}\right|_{\text{for full channel length}} \tag{13.64b}$$

where the critical condition is caused by film Dryout.

For a channel with a cosine axial heat flux shape, Figure 13.25a illustrates the operating condition parameters $q''_{\text{op}}(z)$ for increasing operating power as well as the critical condition parameters q''_{cr} for the PWR. The analogous illustration of $\dot{q}_{\text{op}}$ and $\dot{q}_{\text{c}}$ for the BWR in terms of x_{cr} and x_{e} is presented in Figure 13.25b.

13.4.4.2 Margin to the Critical Condition

The thermal margin that exists at the operating condition (100% power) for PWRs is the minimum value of the DNBR, that is, the MDNBR. This minimum occurs in the upper half of the channel for a cosine axial flux shape as illustrated in Figure 13.26a. The relationship between the DNBR and the operating power is illustrated in Figure 13.26b. The slope of the curve of Figure 13.26b is directly dependent on the specific correlation plotted. Such slopes vary significantly among popular available correlations, which introduce implications to be elaborated upon below in Section 13.4.4.3.

The thermal margin that exists at the operating condition for BWR is the CPR directly. The relationship between the CPR and the operating power is illustrated in Figure 13.27. As for DNBR correlations, the slope of the curve of Figure 13.27 is directly dependent on the specific correlation, now a correlation of the Dryout phenomenon, which is plotted.

Both the DNBR and the CPR are greater than unity at the 100% power operating condition as Figures 13.26b and 13.27 illustrate. For the BWR, the CPR gives the operating margin directly in terms of power. For the PWR, however, the DNBR is not a margin in channel power but rather a ratio of critical to local heat flux.

13.4.4.3 Comparison of Various Correlations

For the hot channel conditions of the typical BWR and PWR (see Appendix K), Figures 13.28, 13.29a, and 13.29b illustrate various DNBR and Dryout correlations.

	PWR	BWR
p (MPa)	15.51	7.14
T_{IN} (°C)	293.1	278.3
G[a] (kg/m^2 s)	3770.5	1569.5
L (m)	3.658	3.588
q''_0 (kW/m^2)	1495	1343

[a] The hot channel G value has been taken as an average value between an edge and an interior coolant-centered subchannel.

For BWR conditions, the dispersion between multiple correlations as shown in Figure 13.28 represents differences among the correlations in terms of their physical representation of the Dryout phenomenon. Conversely, however, for PWR conditions,

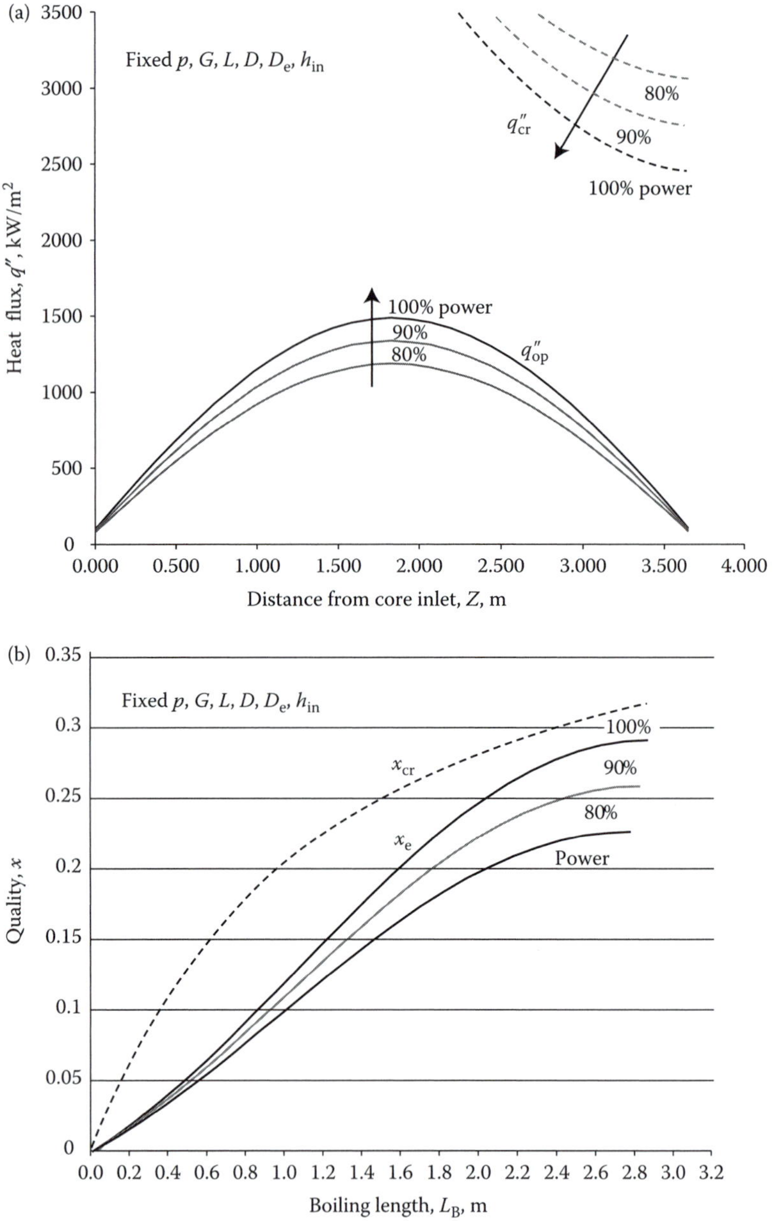

FIGURE 13.25 (a) Critical condition correlation and operating conditions under DNB. (For PWR interior hot channel of Appendix K using the Groeneveld look-up table in Groeneveld, D. C. et al., *Nucl. Eng. Des.*, 237:1909–1922, 2007.) (b) Critical condition correlation and operating conditions under Dryout. (For BWR side hot channel of Appendix K used in Example 13.5, calculated using the Hench–Gillis correlation in Hench, J. E. and Gillis, J. C. *Correlation of Critical Heat Flux Data for Application to Boiling Water Reactor Conditions*, EPRI NP-1898, June 1981.)

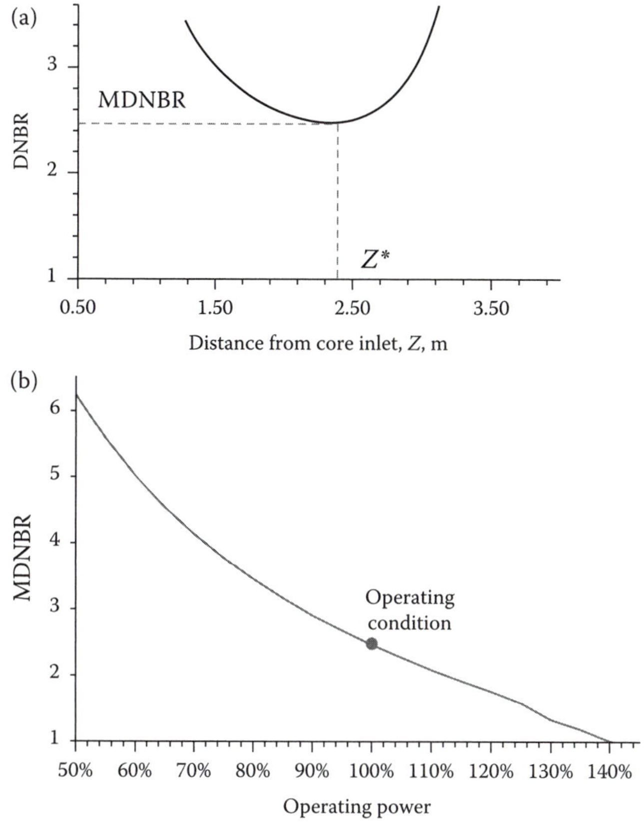

FIGURE 13.26 (a) The definition of MDNBR. (Referred to PWR interior hot channel of Appendix K using the Groeneveld look-up table in Groeneveld, D. C. et al., *Nucl. Eng. Des.*, 237:1909–1922, 2007.) (b) The relationship between DNBR and operating power. (Referred to PWR interior hot channel of Appendix K using the Groeneveld look-up table in Groeneveld, D. C. et al., *Nucl. Eng. Des.*, 237:1909–1922, 2007.)

the dispersion between multiple correlations as shown in Figure 13.29a represents differences both in their physical representation of the DNB phenomenon as well as their selected correlation form. All DNBR correlations predict reasonably the same channel power at the occurrence of the critical condition, that is, at MDNBR = 1.0, and the modern EPRI-sponsored CPR correlations, that is, Hench–Gillis (EPRI-2) and Reddy and Fighetti (EPRI-1) do so at CPR = 1.0. Hence, while all modern correlations of each class effectively predict the critical condition equally well, the margin to DNB in terms of power at operation conditions less than critical cannot be ascertained by the DNBR value. This margin is expressed as the CPR. Figure 13.29b which is derived from Figure 13.29a demonstrates that DNB correlations exhibit drastically different margins to DNB, that is, CPRs at the same value of DNBR. Interestingly, only the EPRI-I correlation gives a close match of margins represented by CPR (Figure 13.28) and by MDNBR (Figure 13.29b).

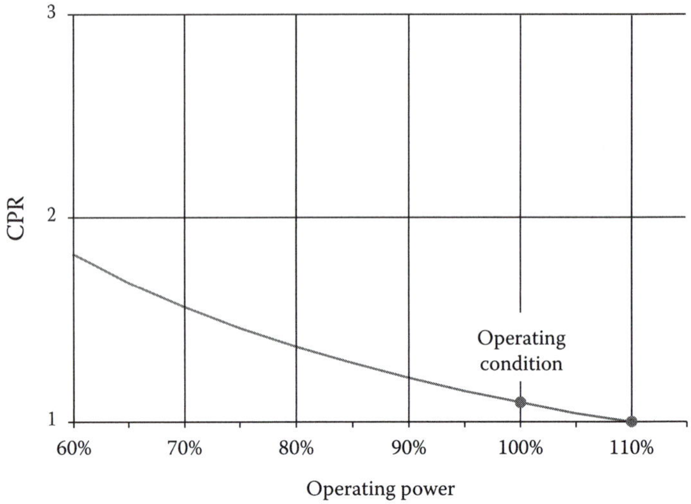

FIGURE 13.27 The relationship between CPR and operating power (for BWR side hot channel of Appendix K used in Example 13.5, calculated using the Hench–Gillis correlation [38]).

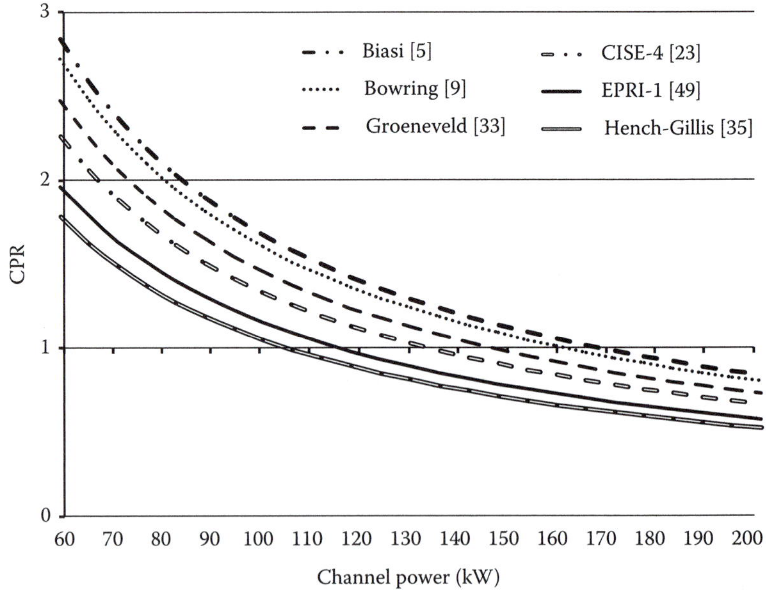

FIGURE 13.28 Dryout correlation comparison for a typical BWR hot channel condition (Appendix K), in a 12.3 mm diameter tube.

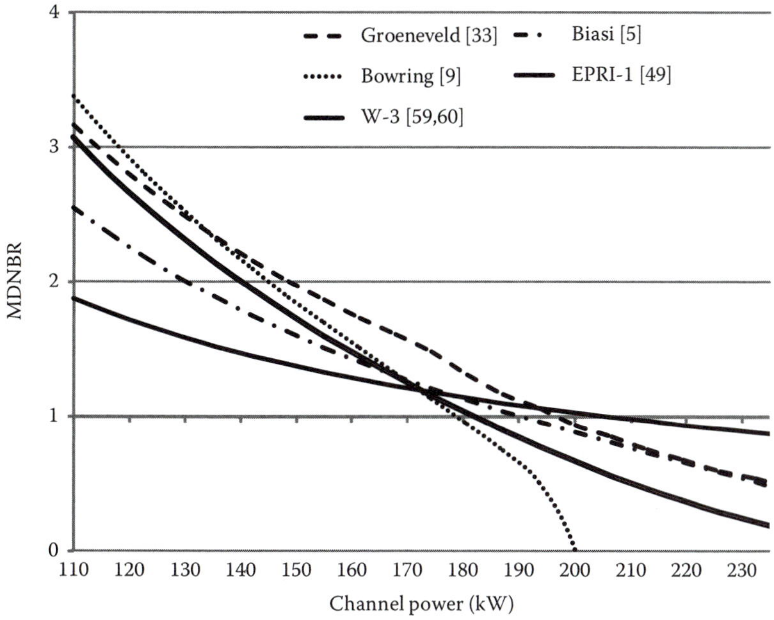

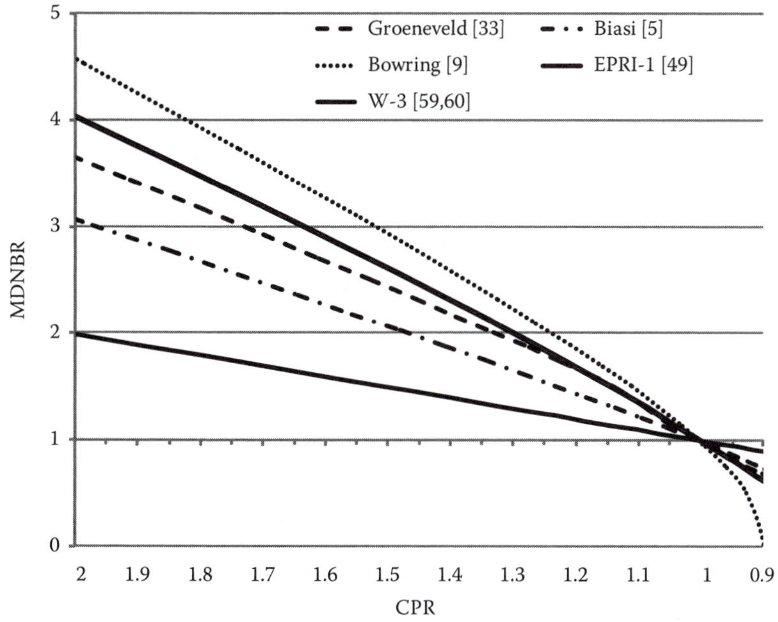

FIGURE 13.29 (a) DNBR correlation comparison for typical Appendix K PWR hot channel conditions in a 11.8 mm diameter tube. (b) DNBR correlation comparison for typical Appendix K PWR hot channel conditions in a 11.8 mm diameter tube.

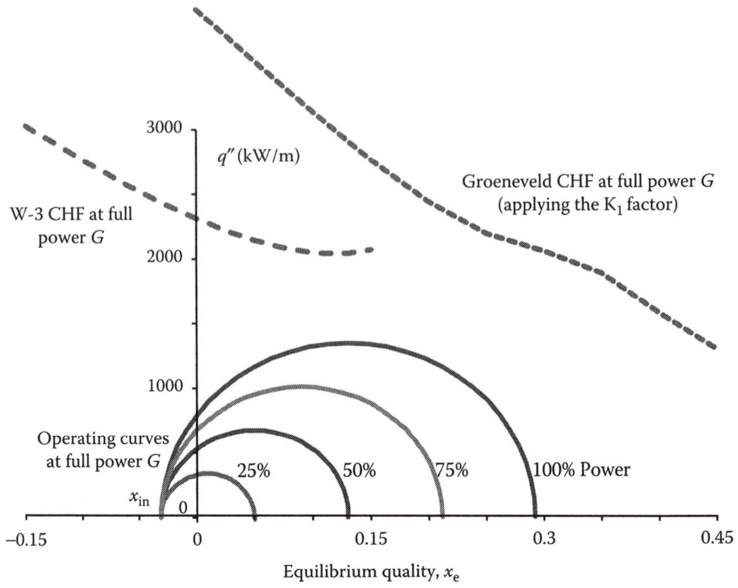

FIGURE 13.30 Trajectory of the startup of the BWR hot channel of Appendix K on a heat flux quality map (for $p = 7.14$ MPa full power $G = 1569.5$ kg m^{-2}s^{-1}).

13.4.4.4 Design Considerations

It is often asked how the BWR hot channel avoids experiencing DNB as the reactor power is raised to operating conditions upon startup. No doubt this channel experiences subcooled and then saturated boiling upon startup, but apparently the magnitude of the local channel heat flux during ascension to full power is less than the DNB value. Let us demonstrate that this is the case.

Figure 13.30 presents the Groeneveld [33] CHF limits and the W-3 CHF limits at the BWR operating conditions. As Figure 13.30 illustrates, the intersection of the operating curve and the critical condition limit occurs in the positive quality region at larger than 100% power. Note that in the negative or about zero quality region, the two curves are reasonably separated. This is a consequence of the small degree of inlet subcooling and the high value of CHF around the zero quality region for the 7.2 MPa operating BWR condition.

PROBLEMS

13.1. *Nucleation in pool and flow boiling* (Section 13.2)

A platinum heat surface has conical cavities of uniform size, R, of 10 microns.

1. If the surface is used to heat water at 1 atm in pool boiling, what is the value of the wall superheat and surface heat flux required to initiate nucleation?
2. If the same surface is now used to heat water at 1 atm in forced circulation, what is the value of the wall superheat required to initiate nucleate boiling? What is the surface heat flux required to initiate nucleation?

Answers:
1. $\Delta T_{sat} = 3.3$ K
   Rohsenow method: $q'' = 4682$ W/m^2 using $C_{sf} = 0.013$
   Stephan and Abdelsalam method: $q'' = 2849$ W/m^2
   Cooper method: $q'' = 1029$ W/m^2
2. $\Delta T_{sat} = 6.5$ K
   $q'' = 221$ kW/m^2

13.2. *Factors affecting incipient superheat in a flowing system* (Section 13.2)
1. Saturated liquid water at atmospheric pressure flows inside a 20 mm diameter tube. The mass velocity is adjusted to produce a single-phase heat transfer coefficient equal to 10 kW/m^2 K. What is the incipient boiling heat flux? What is the corresponding wall superheat?
2. Provide answers to the same questions for saturated liquid water at 290°C, flowing through a tube of the same diameter, and with a mass velocity adjusted to produce the same single-phase heat transfer coefficient.
3. Provide answers to the same questions if the flow rate in the 290°C case is doubled.

Answers:
1. $q_i'' = 1.92 \times 10^4$ W/m^2
   $T_W - T_{sat} = 1.92$°C
2. $q_i'' = 229.8$ W/m^2
   $T_W - T_{sat} = 0.023$°C
3. $q_i'' = 697$ W/m^2
   $T_W - T_{sat} = 0.04$°C

13.3. *Heat transfer problems for a BWR channel* (Section 13.3)
Consider a heated tube operating at BWR pressure conditions with a cosine heat flux distribution.
Relevant conditions are as follows:

Channel geometry
$D = 11.20$ mm
$L = 3.588$ m

Operating conditions
$p = 7.14$ MPa
$T_{in} = 278.3$°C
$G = 1625$ kg/m^{2}s
$q'_{max} = 47.24$ kW/m
1. Find the axial position where the equilibrium quality, x_e, is zero.
2. What is the axial extent of the channel where the actual quality is zero? That is, this requires finding the axial location of boiling incipience. (It is sufficient to provide a final equation with all parameters expressed numerically to determine this answer without solving for the final result.)
3. Find the axial location of maximum wall temperature assuming the heat transfer coefficient given by the Thom et al., correlation for nuclear boiling heat transfer (Equation 13.23b).
4. Find the axial location of maximum wall temperature assuming the heat transfer coefficient is not constant but varies as is calculated by relevant correlations. Here, you are not asked for the exact location,

but whether the location is upstream or downstream from the value from Part 3.

Answer:
1. $Z_{x_e=0} = 0.61$ m (from the bottom of channel)
2. $Z_{ONB} = 0.14$ m (see Example 14.5a for final equation)
3. 1.79 m
4. Upstream

13.4. *Thermal parameters in a heated channel in two-phase flow* (Sections 13.3 and 13.4)
Consider a 3 m long water channel of circular cross-sectional area 1.5×10^{-4} m^2 operating at the following conditions:

$$\dot{m} = 0.29\,\text{kg/s}$$
$$p = 7.2\,\text{MPa}$$
$$h_{in} = \text{saturated liquid}$$
$$q'' = \text{axially uniform}$$
$$x_{exit} = 0.15$$

Compute the following:
1. Fluid temperature
2. Wall temperature using Jens and Lottes correlation
3. Determine CPR using the Groeneveld Look-up Table

Answers:
1. 287.7°C
2. 294.3°C
3. CPR = 2.75

13.5. *CHF calculation with the Bowring correlation* (Section 13.4)
Using the data of Example 13.3, calculate the minimum critical heat flux ratio using the Bowring correlation.

Answer:
MCHFR = 2.13

13.6. *Boiling crisis on the vessel outer surface during a severe accident* (Section 13.4)
Consider water boiling on the outer surface of the vessel where water flows in a hemispherical gap between the surface of the vessel and the vessel insulation (Figure 13.31). The gap thickness is 20 cm. The system is at atmospheric pressure.
1. The water inlet temperature is 80°C and the flow rate in the gap is 300 kg/s. The heat flux on the outer surface of the vessel is a uniform 350 kW/m^2 in the hemispherical region ($0 \le \theta \le 90°$) and zero in the beltline region ($\theta > 90°$). If a boiling crisis occurred in this system, what type of boiling crisis would it be (DNB or Dryout)?
2. At what angle θ within the channel would you expect the boiling crisis to occur first and why?

Answers:
1. DNB
2. $\theta = 90°$

13.7. *Calculation of MCPR for a BWR hot channel* (Section 13.4)

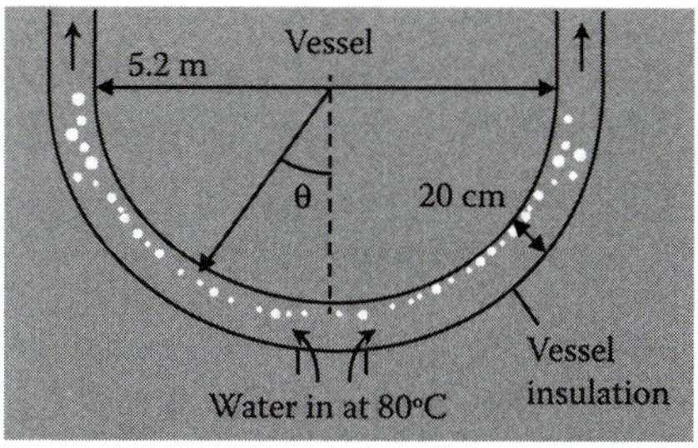

FIGURE 13.31 Two-phase flow in the gap between the outer surface of the vessel and the vessel insulation.

Consider a BWR channel operating at 100% power at the conditions noted below. Using the Hench–Gillis correlation (Equation 13.62) determine the MCPR at 100% power.

Operating conditions

$$q'(z) = q'_{\text{ref}} \exp\left[-\alpha\left(\frac{z}{L} + \frac{1}{2}\right)\right]\cos\left(\frac{\pi z}{L}\right)$$

where $z = 0$ defines the channel midplane

$q'_{\text{ref}} = 104.75$ kW/m $p = 7.14$ MPa

$\alpha = 1.96$ $T_{\text{in}} = 278.3°\text{C}$

$G = 1569.5$ kg/m^{2}s

Channel conditions
$L = 3.588$ m
$P = 14.73$ mm
$D = 11.20$ mm
$A_{\text{fch}} = 1.42 \times 10^{-4}$ m^2
$A_{\text{f}} = 9.718 \times 10^{-3}$ m^2

Hench–Gillis correlation
for simplicity assume
$J = 1.032$
$F_{\text{p}} = -1.66 \times 10^{-3}$
from Example 13.5

Answer:
MCPR = 1.173

REFERENCES

1. Barnett, P. G., *A Correlation of Burnout Data for Uniformly Heated Annuli and Its Use for Predicting Burnout in Uniformly Heat Rod Bundles*. AEEW-R-463, U.K. Atomic Energy Authority, Winfrith, UK, 1966.
2. Bennett, J. A. et al., Heat transfer to two-phase gas liquid systems. I. Steam/water mixtures in the liquid dispersed region in an annulus. *Trans. Inst. Chem. Eng.*, 39:113, 1961; and AERE-R-3519, 1959.

3. Bennett, A. W., Hewitt, G. F., Kearsey, H. A., and Keeys, R. K. F., Heat transfer to steam-water mixtures flowing in uniformly heated tubes in which the critical heat flux has been exceeded. Paper 27 presented at *Thermodynamics and Fluid Mechanics Convention, IMechE, Bristol*, Bristol, UK, 27–29 March, 1968. See also AERE-R-5373, 1967.
4. Bergles, A. E. and Rohsenow, W. M., The determination of forced convection surface boiling heat transfer. *J. Heat Transfer,* 86:363, 1964.
5. Biasi, L. et al., Studies on burnout. Part 3. *Energy Nucl.*, 14:530, 1967.
6. Bjornard, T. A. and Griffith, P., PWR blowdown heat transfer. In O. C. Jones and G. Bankoff (eds.). *Symposium on Thermal and Hydraulic Aspects of Nuclear Reactor Safety,* Vol. 1. New York, NY: ASME, 1977.
7. Bjorge, R., Hall, G., and Rohsenow, W., Correlation for forced convection boiling heat transfer data. *Int. J. Heat Mass Transfer*, 25(6):753–757, 1982.
8. Bowring, R. W., *Physical Model Based on Bubble Detachment and Calculation of Steam Voidage in the Subcooled Region of a Heated Channel.* Report HPR10. Halden, Norway: Halder Reactor Project, 1962.
9. Bowring, R. W., *A Simple but Accurate Round Tube, Uniform Heat Flux Dryout Correlation over the Pressure Range 0.7 to 17 MPa.* AEEW-R-789. U.K. Atomic Energy Authority, Winfrith, UK, 1972.
10. Bowring, R. W., A new mixed flow cluster Dryout correlation for pressures in the range 0.6–15.5 MN/m^2 (90–2250 psia)—for use in a transient blowdown code, *Proceedings of the IME Meeting on Reactor Safety*, Paper C217, Manchester, 1977.
11. Bromley, L. A., Le Roy, N. R., and Robbers, J. A., Heat transfer in forced convection film boiling. *Ind. Eng. Chem.*, 49:1921–1928, 1953.
12. Celata, G. P., Cumo, M., and Mariani, A., Assessment of correlations and models for the prediction of CHF in water subcooled flow boiling. *Int. J. Heat Mass Transfer*, 37(2):237–255, 1994.
13. Chen, J. C., *A Correlation for Boiling Heat Transfer to Saturated Fluids in Convective Flow.* ASME paper 63-HT-34, 1963 (first appearance). Chen, J. C. A correlation for boiling heat transfer in convection flow. *ISEC Process Des. Dev.,* 5:322, 1966 (subsequent archival publication).
14. Chen, J. C., Sundaram, R. K., and Ozkaynak, F. T., *A phenomenological correlation for post CHF Heat Transfer*. Lehigh University, NUREG-0237, Washington, DC, 1977.
15. Cheng, S. C., Ng, W. W., and Heng, K. T., Measurements of boiling curves of subcooled water under forced convection conditions. *Int. J. Heat Mass Transfer*, 21:1385, 1978.
16. Collier, J. G. and Thome, J. R., *Convective Boiling and Condensation,* 3rd Ed. Oxford University Press, Oxford, UK, 1994.
17. Collier, J. G. and Pulling, D. J., *Heat Transfer to Two-Phase Gas–Liquid Systems*.
18. Davis, E. J. and Anderson, G. H., The incipience of nucleate boiling in forced convection flow. *AIChE J.,* 12:774, 1966.
19. Dengler, C. E. and Addoms, J. N., Heat transfer mechanism for vaporization of water in a vertical tube. *Chem. Eng. Prog. Symp. Ser.*, 52:95, 1956.
20. Dix, G. E., Vapor void fraction for forced convection with subcooled boiling at low flow rates. Report NADO-10491. General Electric, San Jose, CA, 1971.
21. Feng, D, Hejzlar, P., and Kazimi, M. S., Thermal-hydraulic design of high-power-density annular fuel in PWRs. *Nucl. Tech.*,160:16–44, 2007.
22. Forster, K. and Zuber, N., Dynamics of vapor bubbles and boiling heat transfer. *AIChE J.*, 1:531, 1955.
23. Gaspari, G. P. et al., A rod-centered subchannel analysis with turbulent (enthalpy) mixing for critical heat flux prediction in rod clusters cooled by boiling water. In *Proceedings of the 5th International Heat Transfer Conference*, Tokyo, Japan, September 3–7, 1974, CONF-740925, 1975.

24. *General Electric BWR Thermal Analysis Basis (GETAB) Data: Correlation and Design Applications*. NADO-10958, San Jose, CA, 1973.
25. Griffith, P. J., Clark, A., and Rohsenow, W. M., Void volumes in subcooled boiling systems. ASME paper 58-HT-19, New York, NY, 1958.
26. Groeneveld, D. C., Effect of a heat flux spike on the downstream Dryout behavior. *J. Heat Transfer*, 96C:121, 1974. Also see Groeneveld, D. C. *The Effect of Short Flux Spikes on the Dryout Power.* AECL-4927, Chalk River, Ontario, 1975.
27. Groeneveld, D. C., *Post-Dryout Heat Transfer at Reactor Operating Conditions*. Paper presented at the National Topical Meeting on Water Reactor Safety, American Nuclear Society, AECL-4513, Salt Lake City, UT, 1973.
28. Groeneveld, D. C. and Delorme, G. G. J., Prediction of the thermal non-equilibrium in the post-Dryout regime. *Nucl. Eng. Des.*, 36:17–26, 1976.
29. Groeneveld, D. C. and Fung, K. K., Heat transfer experiments in the unstable post-CHF region. Presented at the *Water Reactor Safety Information Exchange Meeting*, Washington, DC. 1977.
30. Groeneveld, D. C., Cheng, S. C., and Doan, T., AECL-UO critical heat flux lookup table. *Heat Transfer Eng.*, 7(1–2):46–62, 1986.
31. Groeneveld, D. C., Leung, L. K. H., Vasic, A. Z., Guo, Y. J., and Cheng, S. C., A look-up table for fully developed film-boiling heat transfer. *Nucl. Eng. Des.*, 225:83–97, 2003a.
32. Groeneveld, D. C., Leung, L. K. H., Guo, Y., Vasic, A., El Nakla, M., Peng, S. W., Yang, J., and Cheng, S. C., Look-up tables for predicting CHF and film boiling heat transfer: Past, present, and future. *10th International Topical Meeting on Nuclear Reactor Thermal Hydraulics (NURETH-10)*, Seoul, Korea, October 5–9, 2003b.
33. Groeneveld, D. C. et al., The 2006 CHF look-up table. *Nucl. Eng. Des.*, 237:1909–1922, 2007.
34. Healzer, J. M., Hench, J. E., Janssen, E., and Levy, S., *Design Basis for Critical Heat Flux Condition in Boiling Water Reactors.* APED-5286, General Electric, San Jose, CA, 1966.
35. Hench, J. E. and Gillis, J. C., *Correlation of Critical Heat Flux Data for Application to Boiling Water Reactor Conditions*, EPRI NP-1898, Campbell, CA, June 1981.
36. Hewitt, G. F., Exec. Ed., *Heat Exchanger Design Handbook: Part 2* (Section 2.7.3 by Collier, J. G., and Hewitt, G. F.) *Fundamentals of Heat and Mass Transfer.* Begell House Inc., New York, NY, 2008.
37. Hsu, Y. Y. and Graham, R. W., Analytical and experimental study of thermal boundary layer and ebullition cycle. NASA Technical Note TNO-594, 1961.
38. Hsu, Y. Y. and Westwater, J. W., Approx. theory for film boiling on vertical surfaces, *AIChE Chem. Eng. Prog. Symp. Ser.*, 56(30):15, AIChE, New York, NY, 1960.
39. Iloeje, O. C., Plummer, D. N., Rohsenow, W. M., and Griffith, P., A study of wall rewet and heat transfer in dispersed vertical flow. MIT, Department of Mechanical Engineering Report 72718-92, Cambridge, MA, 1974.
40. Janssen, E. and Levy, S., *Burnout Limit Curves for Boiling Water Reactors.* APED-3892, General Electric, San Jose, CA, 1962.
41. Jens, W. H. and Lottes, P. A., *Analysis of Heat Transfer, Burnout, Pressure Drop and Density Data for High Pressure Water.* Report ANL-4627, Argonne, IL, 1951.
42. Kandlikar, S. G., Shoji, M., and Dhir, V., *Handbook of Phase Change: Boiling and Condensation.* Taylor and Francis, Philadelphia, PA, 1999.
43. Keeys, R. K. F., Ralph, J. C., and Roberts, D. N., *Post Burnout Heat Transfer in High Pressure Steam-Water Mixtures in a Tube with Cosine Heat Flux Distribution.* AERE-R6411. U.K. Atomic Energy Research Establishment, Harwell, UK, 1971.
44. Lahey, R. T. and Moody, F. J., *The Thermal Hydraulic of Boiling Water Reactor.* American Nuclear Society, 1977, p. 152.
45. Levy, S., Forced convection subcooled boiling prediction of vapor volumetric fraction. *Int. J. Heat Mass Transfer*, 10:951, 1967.

46. Lin, W.-S., Pei, B.-S., and Lee, C.-H., Bundle critical power predictions under normal and abnormal conditions in pressurized water reactors, *Nucl. Tech.*, 98:354ff, 1992.
47. McDonough, J. B., Milich, W., and King, E. C., An experimental study of partial film boiling region with water at elevated pressures in a round vertical tube. *Chem. Eng. Prog. Sym. Ser.,* 57:197, 1961.
48. Ramu, K. and Weisman, J., A method for the correlation of transition boiling heat transfer data. Paper B.4.4. In *Proceedings of the 5th International Heat Transfer Meeting*, Tokyo, 1974.
49. Reddy, D. G. and Fighetti, C. F., Parametric study of CHF data, volume 2: A generalized subchannel CHF correlation for PWR and BWR fuel assemblies. Electric Power Research Institute (EPRI) Report NP-2609, prepared by Heat Transfer Research Facility, Department of Chemical Engineering, Columbia University, NY, January 1983.
50. Rohsenow, W. M. Boiling. In W. M. Rohsenow, J. P. Hartnett and E. N. Ganic (eds.). *Handbook of Heat Transfer Fundamentals*, 2nd Ed. New York, NY: McGraw-Hill, 1985.
51. Saha, P., and Zuber, N., Point of net vapor generation and vapor void fraction in subcooled boiling. In *Proceedings of the 5th International Heat Transfer Conference*, Tokyo, 1974, pp. 175–179.
52. Schmidt, K. R., Wärmetechnische Untersuchungen an hoch belasteten Kesselheizflächen. *Mitteilungen der Vereinigung der Grosskessel-bezitzer*, 63:391–401, December 1959.
53. Schrock, V. E. and Grossman, L. M., Forced convection boiling in tubes. *Nucl. Sci. Eng.*, 12:474, 1962.
54. Silvestri, M., On the burnout equation and on location of burnout points. *Energia Nucleare*, 13(9):469–479, 1966.
55. Slaughterbeck, D. C., Ybarrondo, L. J., and Obenchain, C. F., Flow film boiling heat transfer correlations—Parametric study with data comparisons. Paper presented at the *National Heat Transfer Conference*, Atlanta, 1973.
56. Slaughterback, D. C., Vesely, W. E., Ybarrondo, L. J., Condie, K. G., and Mattson, R. J., Statistical regression analyses of experimental data for flow film boiling heat transfer. Paper presented at *ASME-AIChE Heat Transfer Conference*, Atlanta, 1973.
57. Staub, F. The void fraction in subcooled boiling—Prediction of the initial point of net vapor generation. *J. Heat Transfer*, 90:151, 1968.
58. Thom, J. R. S., Walker, W. M., Fallon, T. A., and Reising, G. F. S., Boiling in subcooled water during flow up heated tubes or annuli, 2nd Ed. *Proc. Inst. Mech. Engrs.*, 180:226, 1965.
59. Tong, L. S., *Boiling Crisis and Critical Heat Flux.* USAEC Critical Review Series, Report TID-25887, 1972.
60. Tong, L. S., Heat transfer in water cooled nuclear reactors. *Nucl. Eng. Des.*, 6:301, 1967.
61. Tong, L. S., Heat transfer mechanisms in nucleate and film boiling. *Nucl. Eng. Des.*, 21:1, 1972.
62. Tong, L. S. et al., Influence of axially non-uniform heat flux on DNB. *AIChE Chem. Eng. Prog. Symp. Ser.*, 62(64):35–40, 1966.
63. *VIPRE-01: A Thermal-Hydraulic Code for Reactor Cores, Vol. 2: User's Manual (Rev. 2), Appendix D: CHF and CPR Correlations. NP-2511-CCM-*$_A$, Research Project 1584–1, Prepared by Batelle, Pacific Northwest Laboratories for Electric Power Research Institute, July, 1985.
64. Whalley, P. B., *Boiling, Condensation, and Gas-Liquid Flow.* Oxford: Clarendon Press, 1987.
65. Yadigaroglu, G., Nelson, R. A., Teschendorff, V., Murao, Y., Kelley, J., and Bestion, D., Modeling of reflooding. *Nucl. Eng. Des.*, 145:1–35, 1993.

14 Single Heated Channel
Steady-State Analysis

14.1 INTRODUCTION

Solutions of the mass, momentum, and energy equations of the coolant in a single channel are presented in this chapter. The channel is generally taken as a representative coolant subchannel within an assembly, which is assumed to receive coolant only through its bottom inlet. The fuel and clad heat transport equations are also solved under steady-state conditions for the case in which they are separable from the coolant equations.

We start with a discussion on one-dimensional transient transport equations of the coolant with radial heat input from the clad surfaces. It is assumed that the flow area is axially uniform, although form pressure losses due to local area changes (e.g., spacers) can still be accounted for. Solutions for the coolant equations are presented for steady state in this chapter and for transient conditions in Chapter 2, Volume II.

14.2 FORMULATION OF ONE-DIMENSIONAL FLOW EQUATIONS

Consider the coolant as a mixture of liquid and vapor flowing upward in a heated channel. The positions Z_{ONB}, Z_D, Z_{OSB}, and Z_E have been defined in Figure 13.4. Figure 14.1 illustrates these locations and the associated relations between thermal equilibrium and actual qualities, x_e and x, respectively.* Based on the two-phase mixture equations to be developed next, we will present examples that consider equilibrium (both thermal and mechanical conditions) and nonequilibrium flow conditions.

Following the basic concepts discussed in Chapter 5, the radially averaged coolant flow equations can be derived by considering the flow area at any axial position as the control area. For simplicity, we consider the mixture equations of a two-phase system rather than deal with each fluid separately.

14.2.1 Nonuniform Velocities

The one-dimensional mass, momentum, and energy transport equations have been derived in Chapter 5 and can be written as follows:

* As noted in Chapter 13, axial locations designated by uppercase Z are measured from the channel inlet. For convenience with the cosine flux shape assumption, locations from the channel axial centerline are designed by lowercase z. Specifically, $Z = L/2 + z$ where z can be negative or positive while Z is always positive.

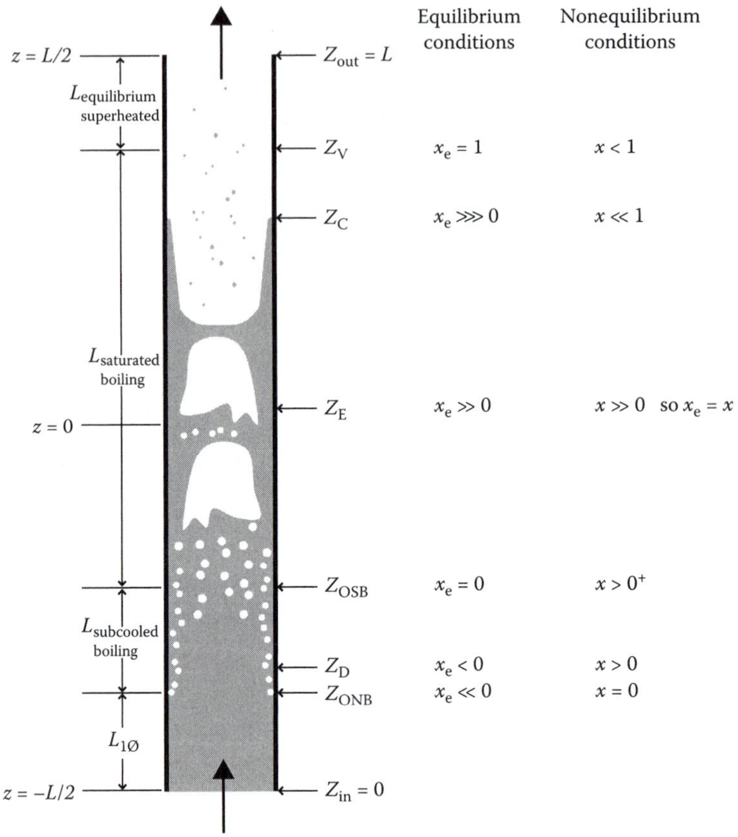

FIGURE 14.1 Heated flow channel illustrating thermal equilibrium and actual quality relationships.

Mass:

$$\frac{\partial}{\partial t}(\rho_m A_z) + \frac{\partial}{\partial z}(G_m A_z) = 0 \tag{5.63}$$

Momentum:

$$\frac{\partial}{\partial t}(G_m A_z) + \frac{\partial}{\partial z}\left(\frac{G_m^2}{\rho_m^+} A_z\right) = -\frac{\partial(pA_z)}{\partial z} + p\frac{\partial A_z}{\partial z} - \int_{P_z} \tau_w \, dP_z - \rho_m g \cos\theta A_z \tag{5.66}$$

where

$$\rho_m = \{\rho_v \alpha\} + \{\rho_\ell (1-\alpha)\} \tag{5.50b}$$

$$G_m = \{\rho_v \alpha \upsilon_{vz}\} + \{\rho_\ell (1-\alpha)\upsilon_{\ell z}\} \tag{5.40c}$$

$$\frac{1}{\rho_m^+} \equiv \frac{1}{G_m^2}\{\rho_v \alpha \upsilon_{vz}^2 + \rho_\ell (1-\alpha)\upsilon_{\ell z}^2\} \tag{5.67}$$

θ = angle of the z direction with the upward vertical

$$\frac{1}{A_z}\int_{P_z} \tau_w \mathrm{d}P_z = \frac{\bar{\tau}_w P_z}{A_z} = \left(\frac{\mathrm{d}p}{\mathrm{d}z}\right)_{\text{fric}} \tag{11.56b}$$

In the mass and momentum relations above, the liquid and vapor velocities can have a nonuniform radial distribution across the flow channel. The friction pressure gradient for two-phase flow can be related to the wall shear stress and the momentum flux by terms analogous to the single-phase flow case:

$$\left(\frac{\mathrm{d}p}{\mathrm{d}z}\right)_{\text{fric}}^{\text{TP}} = \frac{\bar{\tau}_w P_w}{A_z} = \frac{f_{\text{TP}}}{D_e}\left(\frac{G_m^2}{2\rho_m^+}\right) \tag{11.64}$$

Energy:

$$\frac{\partial}{\partial t}[(\rho_m h_m - p)A_z] + \frac{\partial}{\partial z}(G_m h_m^+ A_z) = q_m''' A_z - q_w'' P_w + \frac{G_m}{\rho_m}\left[F_{wz}''' + \frac{\partial p}{\partial z}\right]A_z \tag{5.160}$$

where

$$h_m = \frac{\rho_v \alpha h_v + \rho_\ell (1-\alpha) h_\ell}{\rho_m} \tag{5.72}$$

$$h_m^+ = \frac{\rho_v \alpha h_v \upsilon_{vz} + \rho_\ell (1-\alpha) h_\ell \upsilon_{\ell z}}{G_m} \tag{5.73}$$

$$F_{wz}''' = \frac{1}{A_z}\int_{P_z} \tau_w \mathrm{d}P_z \tag{5.143}$$

From Equation 11.54:

$$F_{wz}''' = \left(\frac{\partial p}{\partial z}\right)_{\text{fric}}$$

For a vertical constant area channel, under the assumptions of $p_v \simeq p_\ell \simeq p$ and negligible q_m''', the mass, momentum, and energy equations take the following form for a two-phase mixture:

$$\frac{\partial \rho_m}{\partial t} + \frac{\partial}{\partial z}(G_m) = 0 \tag{14.1}$$

$$\frac{\partial G_{\rm m}}{\partial t}+\frac{\partial}{\partial z}\left(\frac{G_{\rm m}^2}{\rho_{\rm m}^+}\right)=-\frac{\partial p}{\partial z}-\frac{f_{\rm TP}G_{\rm m}|G_{\rm m}|}{2D_{\rm e}\rho_{\rm m}}-\rho_{\rm m}g\cos\theta \tag{14.2}$$

$$\frac{\partial}{\partial t}(\rho_{\rm m}h_{\rm m}-p)+\frac{\partial}{\partial z}(h_{\rm m}^+G_{\rm m})=\frac{q''P_{\rm h}}{A_{\rm z}}+\frac{G_{\rm m}}{\rho_{\rm m}}\left(\frac{\partial p}{\partial z}+\frac{f_{\rm TP}G_{\rm m}|G_{\rm m}|}{2D_{\rm e}\rho_{\rm m}}\right) \tag{14.3a}$$

The absolute value notation is used with $G_{\rm m}$ in the momentum and energy equations to account for the friction force change in direction depending on the flow direction.

The rearrangement of Equation 14.3a leads to

$$\frac{\partial}{\partial t}(\rho_{\rm m}h_{\rm m})+\frac{\partial}{\partial z}(h_{\rm m}^+G_{\rm m})=\frac{q''P_{\rm h}}{A_{\rm z}}+\frac{\partial p}{\partial t}+\frac{G_{\rm m}}{\rho_{\rm m}}\left(\frac{\partial p}{\partial z}+\frac{f_{\rm TP}G_{\rm m}|G_{\rm m}|}{2D_{\rm e}\rho_{\rm m}}\right) \tag{14.3b}$$

In the above equations, all parameters are functions of time and axial position.

Although the specific enthalpy energy equations are used here, there is no fundamental difficulty in using the internal energy instead. However, when the pressure, p, can be assumed to be a constant (in both time and space), the energy equation (Equation 14.3b) can be mathematically manipulated somewhat more easily. For numerical solutions, the particular form of the energy equation may have more significance.

14.2.2 Uniform and Equal Phase Velocities

The momentum and energy equations can be simplified by combining each with the continuity equation. The mass equation can be written as

$$\frac{\partial}{\partial t}\rho_{\rm m}+\frac{\partial}{\partial z}\rho_{\rm m}\upsilon_{\rm m}=0 \tag{14.4}$$

where

$$\upsilon_{\rm m}\equiv\frac{G_{\rm m}}{\rho_{\rm m}} \tag{11.68}$$

If the two-phase velocities are radially uniform and equal (for the homogenous two-phase flow model), then

$$\upsilon_{\rm v}=\upsilon_{\ell}=\upsilon_{\rm m} \tag{11.69}$$

For uniform velocity across the channel, Equation 5.67 yields

$$\rho_{\rm m}^+=\frac{G_{\rm m}^2}{G_{\rm v}\upsilon_{\rm v}+G_{\ell}\upsilon_{\ell}} \tag{14.5}$$

If the uniform phase velocities are also equal

$$\rho_m^+ = \frac{G_m^2}{(G_v + G_\ell)\upsilon_m} = \rho_m \tag{5.74}$$

The left-hand side of the momentum equation (Equation 14.2) can be simplified if it is assumed that the vapor and liquid velocities are equal to the following form applying Equation 14.4:

$$\begin{aligned}\frac{\partial}{\partial t}(\rho_m \upsilon_m) + \frac{\partial}{\partial z}(\rho_m \upsilon_m \upsilon_m) &= \rho_m \frac{\partial \upsilon_m}{\partial t} + \upsilon_m \frac{\partial \rho_m}{\partial t} + \upsilon_m \frac{\partial(\rho_m \upsilon_m)}{\partial z} + \rho_m \upsilon_m \frac{\partial \upsilon_m}{\partial z} \\ &= \rho_m \frac{\partial \upsilon_m}{\partial t} + \rho_m \upsilon_m \frac{\partial \upsilon_m}{\partial z}\end{aligned} \tag{14.6}$$

Now substituting Equation 14.6 into the momentum equation (Equation 14.2), we obtain

$$\rho_m \frac{\partial \upsilon_m}{\partial t} + G_m \frac{\partial \upsilon_m}{\partial z} = -\frac{\partial p}{\partial z} - \frac{f_{TP} G_m |G_m|}{2\rho_m D_e} - \rho_m g \cos\theta \tag{14.7}$$

Similarly, the left-hand side of the energy equation (Equation 14.3a) for the case of radially uniform and equal velocities (where $h_m^+ = h_m$ from Equation 5.75) can be written as

$$\begin{aligned}\frac{\partial}{\partial t}(\rho_m h_m - p) + \frac{\partial}{\partial z}(\rho_m \upsilon_m h_m) &= \rho_m \frac{\partial h_m}{\partial t} - \frac{\partial p}{\partial t} + h_m \frac{\partial \rho_m}{\partial t} \\ &\quad + h_m \frac{\partial \rho_m \upsilon_m}{\partial z} + \rho_m \upsilon_m \frac{\partial h_m}{\partial z}\end{aligned} \tag{14.8}$$

Again, applying the continuity condition of Equation 14.4 now to Equation 14.8 and combining the result with Equation 14.3b gives

$$\rho_m \frac{\partial h_m}{\partial t} + G_m \frac{\partial h_m}{\partial z} = \frac{q'' P_h}{A_z} + \frac{\partial p}{\partial t} + \frac{G_m}{\rho_m}\left(\frac{\partial p}{\partial z} + \frac{f_{TP} G_m |G_m|}{2 D_e \rho_m}\right) \tag{14.9}$$

14.3 DELINEATION OF BEHAVIOR MODES

Before illustrating solutions of the equations presented in the last section, it is advantageous to identify the hydrodynamic characteristics of interest. These characteristics tend to influence the applicability of the simplifying assumptions used to reduce the complexity of the equations.

To begin with, it should be mentioned that the underlying assumption of the discussion to follow is that the axial variation in the laterally variable local conditions may be represented by the axial variation of the bulk conditions. It is true only when the flow is fully developed (i.e., the lateral profile is independent of axial location). For single-phase turbulent flow, the development of the flow is achieved within a length equal to 25–40 times the channel hydraulic diameter. For two-phase flow in a heated channel, the flow does not reach a "developed" state owing to changing vapor quality and distribution along the axial length.

Another important flow feature is the degree to which the pressure field is influenced by density variation in the channel and the connecting system. For "forced convection" conditions, the flow is meant to be unaffected, or only mildly affected, by the density change along the length of the channel. Hence, buoyancy effects can be neglected. For "natural convection" the pressure gradient is governed by density changes with enthalpy, and, therefore, the buoyancy head should be described accurately. When neither the external pressure head nor the buoyancy head governs the pressure gradient independently, the convection is termed "mixed." Table 14.1 illustrates the relation between the convection state and the appropriate boundary conditions in a calculation.

Lastly, as discussed in Chapter 1, Volume II, the conservation equations of coolant mass and momentum may be solved for a single channel under boundary conditions of (1) specified inlet pressure and exit pressure, (2) specified inlet flow and exit pressure, or (3) specified inlet pressure and outlet velocity. For cases 2 and 3 boundary conditions, the unspecified boundary pressure is uniquely obtained. However, for case 1 boundary conditions, more than one inlet flow rate may satisfy the equations. Physically, this is possible because density changes in a heated channel can create several flow rates at which the integrated pressure drops are identical, particularly if boiling occurs within the channel. This subject is discussed in Section 14.5.2.

14.4 THE LWR CASES ANALYZED IN SUBSEQUENT SECTIONS

In the following sections, pressure, temperature, enthalpy, quality, and void fraction conditions in single heated channels will be determined under equilibrium and

TABLE 14.1
Consideration of Flow Conditions in a Single Channel

	Flow Conditions[b]	
Inlet Boundary Condition[a]	**Forced Convection (Buoyancy Can Be Neglected)**	**Natural Convection (Buoyancy Is Dominant)**
Pressure	√	√
Flow rate or velocity	√	This boundary condition cannot be applied

[a] Exit pressure is prescribed for both cases.

[b] Conditions in which buoyancy effects are neither negligible nor dominant lead to "mixed convection" in the channel.

nonequilibrium assumptions for a sinusoidal axial heat flux distribution. Both thermal and mechanical equilibrium and nonequilibrium conditions are included. Recall that mechanical equilibrium refers to the case where the velocities of the two phases are equal, and hence the slip ratio S is unity. This is also referred to as homogeneous flow. Thermal equilibrium refers to the case that the phasic temperatures are equal. Hence, the fully equilibrium flow condition is also called the homogeneous equilibrium model (HEM), for example, homogeneous for the mechanical condition, equilibrium for the thermal condition.

Examples will illustrate typical PWR and BWR conditions. Both core average and hot interior channels are treated. Recall as mentioned in Section 13.4, the conditions in the hot subchannel of the core are most limiting since the flow rate is the lowest and the heat addition the greatest. The reactor parameters for the LWRs utilized in these examples are tabulated in Appendix K.

Table 14.2 summarizes the heated channel cases analyzed in the remainder of this chapter.

14.5 STEADY-STATE SINGLE-PHASE FLOW IN A HEATED CHANNEL

14.5.1 Solution of the Energy Equation for a Single-Phase Coolant and Fuel Rod (PWR Case)

The radial temperature distribution across a fuel rod surrounded by flowing coolant was derived in Chapter 8. Here, the axial distribution of the coolant temperature is derived and linked with the radial temperature distribution within the fuel rod to yield the axial distribution of temperatures throughout the fuel rod.

At steady state, the energy equation (Equation 14.9) in a channel with constant axial flow area (hence $G =$ constant) reduces for single phase to

$$G\frac{d}{\mathrm{d}z}h = \frac{q''P_\mathrm{h}}{A_\mathrm{z}} + \frac{G}{\rho}\left(\frac{\mathrm{d}p}{\mathrm{d}z} + f\frac{G|G|}{2D_\mathrm{e}\rho}\right) \tag{14.10a}$$

TABLE 14.2
Heated Channel Cases Analyzed

	Energy Equation Solution	Momentum Equation Solution
Single phase (PWR case and GFR case)	Sections 14.5.1 and 14.5.2 Example 14.1	Section 14.5.3 Example 14.2
Two phase (BWR case)		
Thermal and mechanical equilibrium	Section 14.6.1 Example 14.3	Section 14.6.2 Example 14.4
Thermal and mechanical nonequilibrium	Section 14.7.1	Section 14.7.2
Prescribed wall heat flux	Example 14.5	Example 14.7
Prescribed coolant temperature (PWR once-through steam generator case)	Example 14.6	

The partial derivative terms are changed to full derivative terms since z is the only spatial variable in Equation 14.10a.

Neglecting the energy terms due to the pressure gradient and friction dissipation, Equation 14.10a yields

$$GA_z \frac{d}{dz} h = q'' P_h \tag{14.10b}$$

or

$$\dot{m} \frac{d}{dz} h = q'(z) \tag{14.10c}$$

For a given mass flow rate ($\dot{m}$), the coolant enthalpy rise depends on the axial variation of the heat generation rate. In nuclear reactors, the local heat generation depends on the distribution of both the neutron flux and the fissile material. The neutron flux is affected by the moderator density, absorbing materials (e.g., control rods), and the local concentration of the fissile and fertile nuclear materials. Thus, a coupled neutronic-thermal hydraulic analysis is necessary for a complete design analysis.

For the purpose of illustrating the general solution methods and some essential features of the axial distribution, we apply the following simplifying assumptions:

1. The variation of $q'(z)$ is sinusoidal:

$$q'(z) = q'_o \cos \frac{\pi z}{L_e} \tag{14.11}$$

where q'_o = axial peak linear heat generation rate, and L_e = core length plus extrapolation length at the core bottom and top boundaries. The axial power profile is representative of what may be the neutron flux shape in a homogenous reactor core when the effects of neutron absorbers within the core or

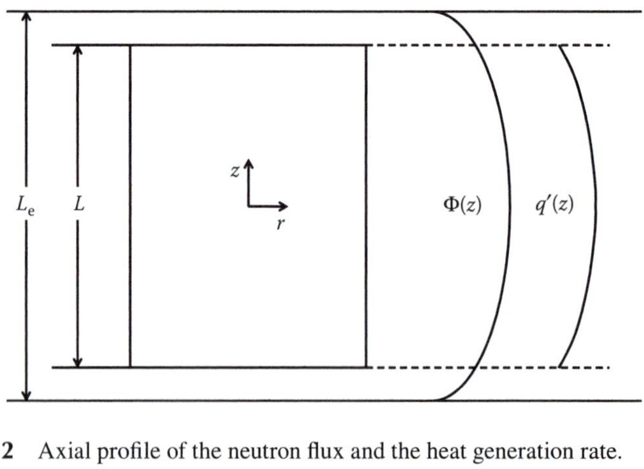

FIGURE 14.2 Axial profile of the neutron flux and the heat generation rate.

axially varying coolant/moderator density due to change of phase are neglected. In real reactors, the axial neutron flux shape cannot be given by a simple analytic expression but is generally less peaked than the sinusoidal distribution.

Note that whereas the extrapolated neutron flux extends from $z = -L_e/2$ to $z = +L_e/2$, the heat generation rate is confined to the actual heated length between $z = -L/2$ to $z = +L/2$.

2. The change in the physical properties of the coolant, fuel, gap bond, and cladding can be ignored. Thus, the coolant heat capacity and fuel, gap, and cladding thermal conductivities are constant, independent of z.
3. The coolant remains in the liquid phase. Subcooled boiling, which is substantial in a PWR and increases the heat transfer coefficient above the single-phase value, is neglected here. Further, the heat transfer coefficient and the channel mass flow rate are taken to be constant.

14.5.1.1 Coolant Temperature

Equation 14.10c can be integrated over the axial length:

$$\dot{m}\int_{h_{in}}^{h_m(z)} dh = q_o' \int_{-L/2}^{z} \cos\left(\frac{\pi z}{L_e}\right) dz \tag{14.12}$$

Equation 14.12 for single-phase flow with constant c_p may be written as

$$\dot{m}c_p \int_{T_{in}}^{T_m(z)} dT = q_o' \int_{-L/2}^{z} \cos\left(\frac{\pi z}{L_e}\right) dz \tag{14.13}$$

The subscript "m" denotes the mixed mean or bulk coolant condition at axial location z. The result of integrating Equation 14.13 may be expanded into the form

$$T_m(z) - T_{in} = \frac{q_o'}{\dot{m}c_p}\frac{L_e}{\pi}\left(\sin\frac{\pi z}{L_e} + \sin\frac{\pi L}{2L_e}\right) \tag{14.14}$$

This equation specifies the bulk temperature of the coolant as a function of height. The exit temperature of the coolant at $z = L/2$ is given by

$$T_{out} = T_{in} + \left(\frac{q_o'}{\dot{m}c_p}\right)\left(\frac{2L_e}{\pi}\right)\sin\frac{\pi L}{2L_e} \tag{14.15}$$

When the neutronic extrapolation height (L_e) can be approximated by the physical core height (L), the equation simplifies to

$$T_{out} = T_{in} + \frac{2q_o'L}{\pi \dot{m}c_p} \tag{14.16}$$

14.5.1.2 Cladding Temperature

The axial variation of the outside cladding temperature can be determined by considering the heat flux at the cladding outer surface:

$$h\left[T_{co}(z) - T_m(z)\right] = q''_{co}(z) = \frac{q'(z)}{P_h} \tag{14.17}$$

where $P_h = 2\pi R_{co}$, and h = cladding-coolant heat transfer coefficient.

When combining Equations 14.11 and 14.17:

$$T_{co}(z) = T_m(z) + \frac{q'_o}{2\pi R_{co} h}\cos\left(\frac{\pi z}{L_e}\right) \tag{14.18}$$

Eliminating $T_m(z)$ with Equation 14.14 gives

$$T_{co}(z) = T_{in} + q'_o\left[\frac{L_e}{\pi \dot{m} c_p}\left(\sin\frac{\pi z}{L_e} + \sin\frac{\pi L}{2L_e}\right) + \frac{1}{2\pi R_{co} h}\cos\frac{\pi z}{L_e}\right] \tag{14.19}$$

The maximum cladding surface temperature can be evaluated by the condition

$$\frac{dT_{co}}{dz} = 0 \tag{14.20}$$

and

$$\frac{d^2 T_{co}}{dz^2} < 0 \tag{14.21}$$

The condition of Equation 14.20 leads to

$$\tan\left(\frac{\pi z_c}{L_e}\right) = \frac{2\pi R_{co} L_e h}{\pi \dot{m} c_p} \tag{14.22a}$$

or equivalently

$$z_c = \frac{L_e}{\pi}\tan^{-1}\left(\frac{2\pi R_{co} L_e h}{\pi \dot{m} c_p}\right) \tag{14.22b}$$

Because all the quantities in the arc-tangent are positive, z_c is a positive value. Hence, the maximum cladding temperature occurs at z_c such that $0 < z_c < L/2$.

The second derivative is given by

$$\frac{d^2T_{co}}{dz^2} = -q_o'\left[\frac{\pi}{L_e \dot{m} c_p}\sin\left(\frac{\pi z}{L_e}\right) + \frac{\pi^2}{L_e^2 2\pi R_{co} h}\cos\left(\frac{\pi z}{L_e}\right)\right] \quad (14.23)$$

and yields a negative value when z is positive.

14.5.1.3 Fuel Centerline Temperature

It is possible to extrapolate this approach to determine the maximum temperature in the fuel rod itself. First, by combining Equations 8.152 and 14.14, we find that the fuel centerline temperature, T_{cl} (corresponding to T_{max} in Equation 8.152), is given by

$$T_{cl}(z) = T_{in} + q_o'\left\{\frac{L_e}{\pi \dot{m} c_p}\left(\sin\frac{\pi z}{L_e} + \sin\frac{\pi L}{2L_e}\right) + \left[\frac{1}{2\pi R_{oo} h} + \frac{\delta_o}{2\pi R_{oo} k_o} + \frac{1}{2\pi k_c}\ell n\left(\frac{R_{co}}{R_{ci}}\right) + \frac{1}{2\pi R_g h_g} + \frac{1}{4\pi k_f}\right]\cos\frac{\pi z}{L_e}\right\} \quad (14.24)$$

where k_f, k_c, and k_o = the thermal conductivities of the fuel, cladding, and oxide film, respectively; h_g = gap conductance; R_{co} and R_{ci} = outer and inner clad radii, respectively; $R_g = (R_f + R_{ci})/2$, R_{oo} = outside oxide film radius, and δ_o = oxide film thickness. By differentiating the last equation, we find the position of maximum fuel temperature as

$$z_f = \frac{L_e}{\pi}\tan^{-1}\left\{\frac{L_e/\pi \dot{m} c_p}{\left[(1/4\pi k_f) + (1/2\pi k_c)\ell n\left(R_{co}/R_{ci}\right) + (1/2\pi R_g h_g) + (1/2\pi R_{oo} h) + (\delta_o/2\pi R_{oo} k_o)\right]}\right\} \quad (14.25)$$

The maximum fuel centerline temperature is found by substituting the position z_f into Equation 14.24. Note that again z_f is expected to be a positive quantity. An illustration of the axial variation of T_m, T_{co}, T_{ci}, and T_{cl} is given in Figure 14.3.

Example 14.1: Determination of the Maximum Fuel Temperature for a Single-Phase Coolant

PROBLEM

For the interior hot subchannel of a PWR with parameters as described in Appendix K, determine the coolant temperature as a function of height and the maximum fuel and cladding temperatures. Neglect extrapolation lengths and cross-flow among subchannels. Assume that the heat transfer coefficients and thermal conductivities remain constant at the values listed in Appendix K.

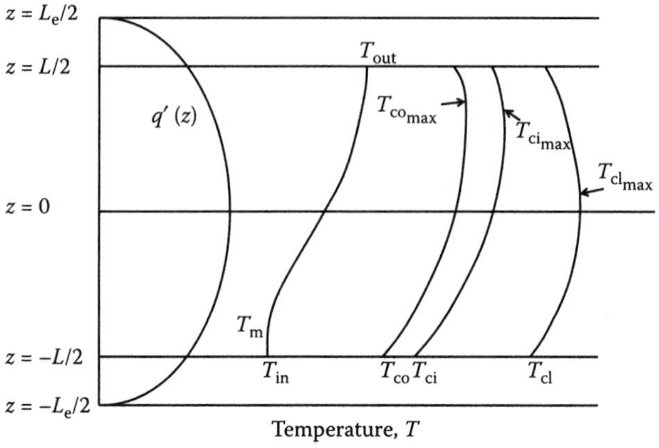

FIGURE 14.3 Axial variation of the bulk coolant temperature (T_m), the clad temperatures (T_{co} and T_{ci}), and the fuel centerline temperature (T_{cl}).

SOLUTION

Given the sinusoidal heat addition over the heated length (L) = 3.658 m, Equation 14.14 can be used to obtain the coolant temperature, $T_m(z)$:

$$T_m(z) = T_{in} + \frac{q'_o}{\dot{m}c_p}\frac{L_e}{\pi}\left(\sin\frac{\pi z}{L_e} + \sin\frac{\pi L}{2L_e}\right) \quad (14.14)$$

$$= 293.1 + \left(\frac{44.62}{(0.335)(5.742)}\right)\left(\frac{3.658}{\pi}\right)\left(\sin\frac{\pi z}{3.658} + 1\right)$$

$$= 320.1 + 27.01\sin\frac{\pi z}{3.658}$$

Equation 14.22b may be utilized to determine the position at which the maximum cladding temperature occurs

$$z_c = \frac{L_e}{\pi}\tan^{-1}\left(\frac{2\pi R_{co}L_e h}{\pi\dot{m}c_p}\right) \quad (14.22b)$$

$$= \frac{3.658}{\pi}\tan^{-1}\left(\frac{2\pi(0.00475)(3.658)(34.0)}{\pi(0.335)(5.742)}\right)$$

$$= 0.641\,\text{m}$$

Substituting this value into Equation 14.19 gives the cladding surface maximum temperature:

$$T_{co}(z) = T_{in} + q'_o\left[\frac{L_e}{\pi\dot{m}c_p}\left(\sin\frac{\pi z}{L_e} + \sin\frac{\pi L}{2L_e}\right) + \frac{1}{2\pi R_{co}h}\cos\frac{\pi z}{L_e}\right] \quad (14.19)$$

$$= 293.1 + 44.62\left\{\frac{3.658\left[\sin\left(\pi\, 0.641/3.658\right) + 1\right]}{\pi(0.335)(5.742)} + \frac{\cos\left(\pi(0.641)/3.658\right)}{\pi(0.0095)(34)}\right\}$$

$$= 293.1 + 44.62(0.922 + 0.840) = 371.7\,°\text{C}$$

To obtain the maximum fuel centerline temperature, Equation 14.25 is applied to determine the position, z_f, where this maximum temperature occurs. First we determine the fuel pellet radius, the effective gap radius and the clad internal radius:

$$\begin{aligned}
R_f &= \frac{8.192}{2} = 4.096\ \text{mm} \\
R_g &= 4.096 + \frac{0.0826}{2} = 4.14\ \text{mm} \\
R_{ci} &= 4.096 + 0.0826 = 4.18\ \text{mm} \\
R_{co} &= 4.75\ \text{mm} \quad \text{for } \delta_c = 0.57\ \text{mm} \\
R_{oo} &= 4.80\ \text{mm} \quad \text{for } \delta_o = 0.05\ \text{mm}
\end{aligned}$$

$$\begin{aligned}
z_f &= \frac{L_e}{\pi}\tan^{-1}\left\{\frac{(L_e/\pi\dot{m}c_p)}{\left[(1/4\pi\bar{k}_f) + (1/2\pi k_c)\ell n\left(R_{co}/R_{ci}\right) + (1/2\pi R_g h_g) + (1/2\pi R_{co} h) + (\delta_o/2\pi R_{oo} k_o)\right]}\right\} \\
&= \frac{3.658}{\pi}\tan^{-1}\left(\frac{(3.658/\pi(0.335)(5.742))}{1/4\pi(0.002163) + \left(\ell n(4.75/4.18)/2\pi(0.01385)\right) + 10^3/2\pi(4.14)(5.7)} \right. \\
&\qquad \left. + 10^3/2\pi(4.75)(34) + 5\times10^{-2}/2\pi(4.80)2\times10^{-3}\right) \\
&= \frac{3.658}{\pi}\tan^{-1}\left\{\frac{0.605}{36.79 + 1.47 + 6.74 + 0.985 + 0.829}\right\} = \frac{3.658}{\pi}\tan^{-1}\left\{\frac{0.605}{46.81}\right\} \\
&= 0.015\ \text{m}
\end{aligned} \tag{14.25}$$

Substituting z_f into Equation 14.24 gives the maximum fuel centerline temperature as

$$\begin{aligned}
T_{cl} &= 293.1 + 44.62\left[0.605\left(\sin\frac{\pi(0.015)}{3.658} + 1\right) + 46.81\left(\cos\frac{\pi(0.015)}{3.658}\right)\right] \\
&= 2409\,°\text{C}
\end{aligned}$$

14.5.2 Solution of the Energy Equation for a Single-Phase Coolant with Roughened Cladding Surface (Gas Fast Reactor)

This reactor case uses the same conservation equations as those for the single-phase PWR coolant case in Section A. However, for the metallic clad gas fast reactor (GFR) the peak cladding temperature has been reduced by roughening the clad surface over an appropriate length of the fuel rod. This appropriate length is established for a given degree of roughness by a trade-off between the achievable cladding temperature reduction and the increased channel pressure drop. The degree of roughness enhances the turbulent heat transfer, thus increasing the heat transfer coefficient.

Hence Equation 14.18 is modified, this time by an enhanced h value over the roughened portion of the fuel rod cladding surface.

14.5.3 Solution of the Momentum Equation to Obtain Single-Phase Pressure Drop

Consider an upward vertical channel, as shown in Figure 14.1. Under steady-state conditions, Equation 14.2 may be written for single-phase flow as

$$\frac{d}{dz}\left(\frac{G^2}{\rho_\ell}\right) = -\frac{dp}{dz} - f\frac{G|G|}{2D_e\rho_\ell} - \rho_\ell g \tag{14.26}$$

Note that the partial derivative terms are changed to the full derivative terms, since z is the only spatial variable in Equation 14.26.

The exact solution of the momentum equation requires identifying the temperature dependence of the properties of the fluid, such as ρ_ℓ and μ_ℓ (which are present in Equation 14.26 through Reynolds number dependence of f). Determining the pressure drop is then accomplished by integrating Equation 14.26, which can be written after rearrangement as

$$p_{in} - p_{out} = \left(\frac{G^2}{\rho_\ell}\right)_{out} - \left(\frac{G^2}{\rho_\ell}\right)_{in} + \int_{z_{in}}^{z_{out}} \frac{fG|G|}{2D_e\rho_\ell}dz + \int_{z_{in}}^{z_{out}} \rho_\ell g\, dz \tag{14.27}$$

Because the acceleration term is an exact differential, it depends on the endpoint conditions only, whereas the friction and gravity terms are path dependent.

Under conditions of single-phase liquid flow, it is possible to assume that the physical property change along the heated channel is negligible, thereby decoupling the momentum equation from the energy equation. Hence, ρ_ℓ is taken as constant. Since the flow area has already been taken axially constant, the mass flux, G, is constant. The acceleration pressure drop is negligible. Hence, Equation 14.27 can be approximated by

$$p_{in} - p_{out} = \frac{fG|G|}{2D_e\rho_\ell}(z_{out} - z_{in}) + \rho_\ell g(z_{out} - z_{in}) \tag{14.28}$$

In practice, "average" properties, evaluated at the center of the channel, are used. The accuracy of such an evaluation is good for single-phase liquid flow. However, for gas flow or two-phase flow, the radial and axial variation in fluid properties cannot always be ignored, and a proper average would then have to be defined by performing the integration of Equation 14.27 both radially and axially.

If the flow area varies axially, G is not a constant for constant mass flow rate and Equation 14.26 must include the extra term $p(dA_z/dz)$ in accordance with

Equation 5.66 and the discussion of Chapter 9 about pressure changes due to flow area changes.

Furthermore, if p_{in} and p_{out} are taken to be the pressures in the plena at the channel extremities, the entrance and exit pressure losses (see Chapter 9, Section 9.6.4) should be added to the right-hand side of Equation 14.28.

Example 14.2: Calculation of Friction Pressure Drop in a Bare Rod PWR Assembly (for Fully Equilbrium Conditions Thus Neglecting Subcooled Boiling)

PROBLEM

For a 3411 MW_{th} PWR with parameters as presented in Appendix K, determine the pressure drop across an interior subchannel. You may neglect spacer, inlet, and exit losses; that is, consider a bare rod PWR fuel assembly. Also ignore subcooled boiling, if it exists.

SOLUTION

Equation 14.28 is utilized. Fluid properties are evaluated at the center of the core, where the average temperature is given by

$$T_{avg} = \frac{293.1 + 326.8}{2} = 310\,°C$$

From Appendix K:

$$\rho_\ell = 704.88\ \text{kg/m}^3$$
$$c_p = 5.742\ \text{kJ/kg K}$$
$$\mu_\ell = 84.60\ \mu\text{Pa s}$$

To determine f, the Reynolds number must first be determined, which is a function of the subchannel equivalent diameter (D_e):

$$D_e = \frac{4(\text{flow area})}{\text{Wetted perimeter}} = \frac{4(87.9\ \text{mm}^2)}{\pi(9.5\ \text{mm})} = 11.8\ \text{mm} = 0.0118\ \text{m}$$

Thus,

$$\text{Re} = \frac{GD_e}{\mu} = \frac{(3807)(0.0118)}{84.6 \times 10^{-6}} = 5.13 \times 10^5$$

Using the McAdams correlation (Equation 9.87) to obtain the friction factor (f):

$$f = 0.184(\text{Re})^{-0.20} = 0.01327$$

Substituting f, D_e, G, ρ_ℓ, and z into Equation 14.28, the pressure drop is obtained as

$$\begin{aligned} p_{in} - p_{out} &= \frac{fG|G|}{2D_e\rho_\ell}(z_{out} - z_{in}) + \rho_\ell g(z_{out} - z_{in}) \\ &= \frac{(0.01327)(3807)^2}{2(0.0118)(704.88)}(3.876) + (704.88)(9.81)(3.876) \\ &= 44{,}812 + 26{,}802 = 71{,}615\,\text{Pa} = 0.07\,\text{MPa} \end{aligned} \tag{14.28}$$

14.6 STEADY-STATE TWO-PHASE FLOW IN A HEATED CHANNEL UNDER FULLY EQUILIBRIUM (THERMAL AND MECHANICAL) CONDITIONS

14.6.1 Solution of the Energy Equation for Two-Phase Flow (BWR Case with Single-Phase Entry Region)

The liquid coolant may undergo boiling as it flows in a heated channel under steady-state conditions, as it does in BWR assemblies. The coolant state undergoes a transition to a two-phase state at a given axial length. Thus, the initial channel length (Figure 14.4) can be called the nonboiling length and is analyzed by the equations described in Section 14.5. The *boiling length* requires the development of new equations in terms of enthalpy because phase change occurs at isothermal conditions. Subcooled boiling does occur in BWRs, but because the channel is mostly in

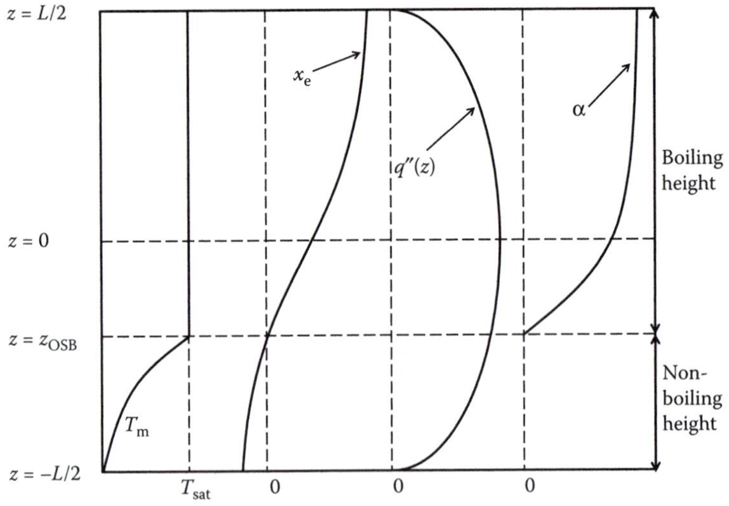

FIGURE 14.4 Coolant temperature, equilibrium quality, and void fraction for sinusoidal heat addition in a single channel under equilibrium conditions. (*Note:* z_{OSB} is a negative value.)

saturated boiling, the subcooled boiling impact on the pressure drop and the temperature field under steady-state conditions tends to be small and is neglected here. Hence, only equilibrium conditions are considered in this section.

Thus, we assume the coolant temperature remains constant after the saturation temperature is reached. This location is the position z_{OSB}, and its placement depends on the axial variation of the linear heat generation rate. For the typical BWR flux shape with bottom entry control rods, this saturated condition is reached at about one-third of the fuel assembly length from the inlet. Thus, the axial coolant temperature variation is expressed only up to the axial position at which the coolant mixed mean temperature, T_m, achieves saturation. Above that location, T_m remains constant but the quality keeps increasing. Further, for a BWR, the heat transfer coefficient changes markedly between the single- and two-phase channel regions. Hence, the outside clad temperature within these two axial coolant regions must be represented by separate equations of the form of Equation 14.18, each with its appropriate heat transfer coefficient, h.

Figure 14.4 illustrates this behavior along with the associated axial profiles of the equilibrium quality and void fraction. For illustration purposes, the axial heat flux distribution has been taken as sinusoidal. The more BWR appropriate bottom-peaked axial heat flux shape is prescribed for computation in a problem exercise at the end of this chapter.

In order to determine z_{OSB} and these quality and void profiles, start with Equation 14.12 expressed for a two-phase mixture and integrating from the inlet to any position along the length:

$$h^+(z) - h_{in} = \frac{P_h}{G_m A_z} \int_{-L/2}^{z} q''(z)\,dz \tag{14.29}$$

where $h^+(z) = h(z)$ since the flow condition is homogeneous.

The onset of saturated boiling, OSB, is determined to occur when the bulk coolant enthalpy reaches the saturated liquid enthalpy value. Equation 14.29 can then be used to determine Z_{OSB}, the demarcation between the boiling length and the nonboiling length. For simplicity, we have assumed thermodynamic equilibrium, $h(z_{OSB}) \equiv h_f$ (the saturated liquid enthalpy) at the local pressure.

For the case of sinusoidal heat generation (Equation 14.11), and L_e taken equal to L, Equation 14.29 yields

$$h_f - h_{in} = \frac{P_h q''_o L}{G_m A_z \pi}\left(\sin\frac{\pi z_{OSB}}{L} + 1\right) \tag{14.30a}$$

However, because

$$\frac{2}{\pi} q''_o P_h L = \frac{2}{\pi} q'_o L = \dot{q} \quad \text{and} \quad G_m A_z = \dot{m}$$

Equation 14.30a takes the form

$$h_{\text{f}} - h_{\text{in}} = \frac{\dot{q}}{2\dot{m}}\left(\sin\frac{\pi z_{\text{OSB}}}{L} + 1\right) \tag{14.30b}$$

The boiling boundary (z_{OSB}) is determined from Equation 14.30b as

$$z_{\text{OSB}} = \frac{L}{\pi}\sin^{-1}\left[-1 + \frac{2\dot{m}}{\dot{q}}(h_{\text{f}} - h_{\text{in}})\right] \tag{14.31a}$$

Applying Equation 14.29 from the inlet to the outlet also yields

$$h_{\text{out}} - h_{\text{in}} = \frac{P_{\text{h}} q_{\text{o}}''}{G_{\text{m}} A_{\text{z}}} \int_{-L/2}^{L/2} \cos\frac{\pi z}{L}\,\text{d}z = \frac{2P_{\text{h}} q_{\text{o}}'' L}{G_{\text{m}} A_{\text{z}} \pi} = \frac{\dot{q}}{\dot{m}} \tag{14.32}$$

which, when combined with Equation 14.31a, yields

$$z_{\text{OSB}} = \frac{L}{\pi}\sin^{-1}\left[-1 + 2\left(\frac{h_{\text{f}} - h_{\text{in}}}{h_{\text{out}} - h_{\text{in}}}\right)\right] \tag{14.31b}$$

When the axial variation of pressure is small with respect to the inlet pressure, h_{f} and h_{g} can be assumed to be axially constant. Therefore, the quality at any axial position can be predicted from Equation 5.53. Inserting $h^{+}(z)$ from Equation 14.29 into Equation 5.53, obtain the result for $x_{\text{e}}(z)$ expressed in Equation 14.33a as follows:

$$x_{\text{e}}(z) = x_{\text{e}_{\text{in}}} + \frac{P_{\text{h}}}{G_{\text{m}} A_{\text{z}} h_{\text{fg}}} \int_{-L/2}^{z} q''(z)\,\text{d}z \tag{14.33a}$$

$$x_{\text{e}}(z) = x_{\text{e}_{\text{in}}} + \frac{\dot{q}}{2\dot{m} h_{\text{fg}}}\left(\sin\frac{\pi z}{L} + 1\right) \tag{14.33b}$$

Note that $x_{\text{e}_{\text{in}}}$ is negative under normal conditions because the liquid enters the core in a subcooled state. Once the quality axial distribution is known, it is possible to predict the axial void fraction distribution from Equation 5.55, written for fully equilibrium conditions as

$$\alpha(z) = \frac{1}{1 + [(1 - x_{\text{e}})/x_{\text{e}}](\rho_{\text{g}}/\rho_{\text{f}})} \tag{14.34}$$

where $x_{\text{e}}(z)$ is given by Equation 14.33b.

Example 14.3: Calculation of Axial Distribution of Multiple Parameters in a BWR

PROBLEM

Consider a BWR as described in Appendix K. For core-averaged conditions determine the location where bulk boiling starts to occur, and the axial profiles of the enthalpy, $h_m(z)$, equilibrium quality $x_e(z)$, and void fraction $\alpha(z)$ in the core. Assume fully (thermal and mechanical) equilibrium conditions. Neglect effects of the neutron flux extrapolation lengths.

SOLUTION

The exit enthalpy (h_{out}) may be evaulated as follows:

$$h_{out} = h_f + x_{e_{out}} h_{fg} = 1274.4 + (0.146)(2770.8 - 1274.4) = 1492.9 \text{ kJ/kg}$$

To obtain z_{OSB}, the location where bulk boiling starts to occur, Equation 14.31b may be applied, yielding

$$z_{OSB} = \frac{L}{\pi}\sin^{-1}\left[-1 + 2\left(\frac{h_f - h_{in}}{h_{out} - h_{in}}\right)\right] \tag{14.31}$$

$$= \frac{3.588}{\pi}\sin^{-1}\left[-1 + \frac{2(1274.4 - 1227.5)}{(1492.9 - 1227.5)}\right] = -0.803 \text{ m}$$

Hence,

$$Z_{OSB} = \frac{L}{2} + z_{OSB} = \frac{3.588}{2} - 0.803 = 0.991 \text{ m}$$

Now from a heat balance across the core

$$\frac{\dot{q}}{\dot{m}} = h_{out} - h_{in} = 1492.9 - 1227.5 \tag{14.32}$$

$$= 265.4 \text{ kJ/kg}$$

Note that if we were considering conditions in the interior hot subchannel versus the core average conditions, then the following values for $\dot{q}$ and $\dot{m}$ would apply

$$\dot{q} = q'_o \frac{2}{\pi} L$$

$$= 47.24\left(\frac{2}{\pi}\right)3.588 = 107.9 \text{ kW}$$

$$\dot{m} = 0.175 \text{ kg/s} \quad (\text{Table } K\text{–1})$$

hence

$$\left(\frac{\dot{q}}{\dot{m}}\right)_{\text{hot interior channel}} = 616.6(\text{kJ/kg})$$

One may obtain $h(z)$ by integrating Equation 14.29:

$$\begin{aligned} h(z) &= h_{\text{in}} + \frac{P_h L q_o''}{\dot{m}\pi} \int_{-L/2}^{z} \cos\frac{\pi z}{L} \\ &= h_{\text{in}} + \frac{\dot{q}}{2\dot{m}}\left(\sin\frac{\pi z}{L} + 1\right) \\ &= 1227.5 + \frac{265.4}{2}\left(\sin\frac{\pi z}{L} + 1\right) \end{aligned}$$

or

$$h(z) = 1360.2 + 132.7\sin\left(\frac{\pi z}{3.588}\right)$$

The axial distribution of quality may then be obtained by applying Equation 14.33b as follows:

$$\begin{aligned} x_e(z) &= x_{e_{\text{in}}} + \frac{\dot{q}}{2\dot{m}h_{fg}}\left(\sin\frac{\pi z}{L} + 1\right) \\ &= \frac{1227.5 - 1274.4}{1496.4} + \frac{265.4}{2(1496.4)}\left(\sin\frac{\pi z}{3.588} + 1\right) \end{aligned} \tag{14.33b}$$

or

$$x_e(z) = 0.0573 + 0.0887\sin\frac{\pi z}{3.588}$$

The void fraction axial distribution can be obtained by inserting Equation 14.33b into Equation 14.33c yielding

$$\begin{aligned} \alpha(z) &= \frac{1}{1 + [(1 - x_e(z))/x_e(z)](\rho_g/\rho_f)} \\ &= \frac{1}{1 + [(0.9427 - 0.0887\sin(z/3.588))/(0.0573 + 0.0887\sin(z/3.588))](37.3/737.3)} \end{aligned} \tag{14.34}$$

The values of $h(z)$, $x_e(z)$, and $\alpha(z)$ are plotted in Figures 14.5 through 14.7, respectively.

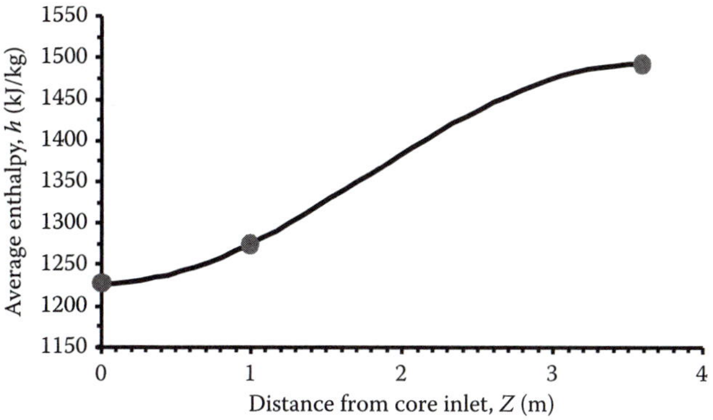

FIGURE 14.5 Axial distribution of average enthalpy in the core (Example 14.3).

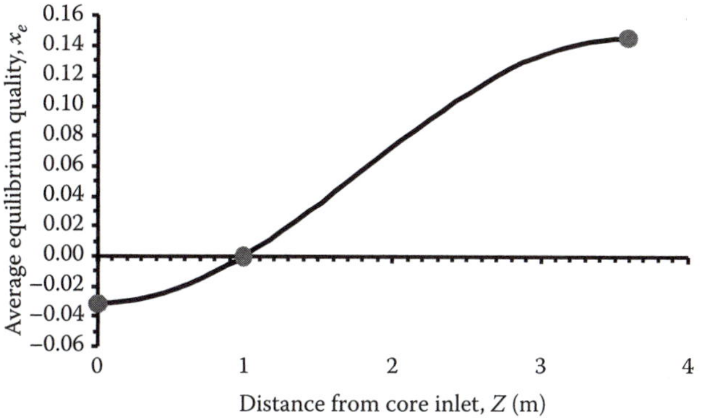

FIGURE 14.6 Axial distribution of average equilibrium quality in the core (Example 14.3).

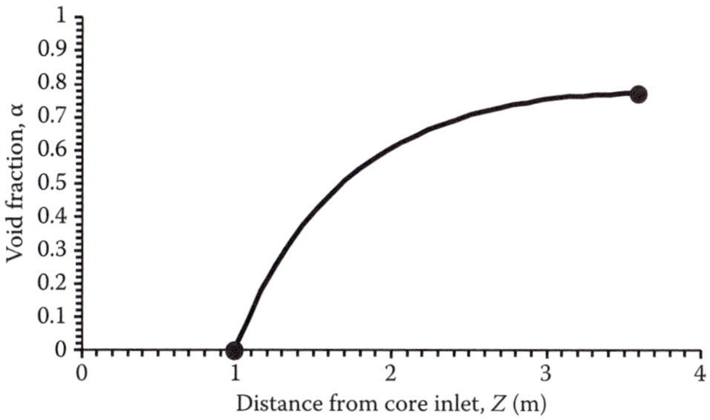

FIGURE 14.7 Axial distribution of average void fraction in the core (Example 14.3).

14.6.2 Solution of the Momentum Equation for Fully Equilibrium Two-Phase Flow Conditions to Obtain Channel Pressure Drop (BWR Case with Single-Phase Entry Region)

The steady-state pressure drop across a vertical channel for the thermal and mechanical equilibrium assumption will be obtained through the integration of Equation 14.2 over the liquid and the two-phase mixture lengths, and then applying the equilibrium assumptions. First, the integration of Equation 14.2 yields

$$p_{\text{in}} - p_{\text{out}} = \left(\frac{G_{\text{m}}^2}{\rho_{\text{m}}^+}\right)_{\text{out}} - \left(\frac{G_{\text{m}}^2}{\rho_{\text{m}}^+}\right)_{\text{in}} + \int_{z_{\text{in}}}^{z_{\text{OSB}}} \rho_\ell g \,\text{d}z + \int_{z_{\text{OSB}}}^{z_{\text{out}}} \rho_{\text{m}} g \,\text{d}z \tag{14.35}$$

$$+ \frac{f_{\ell o} G_{\text{m}} |G_{\text{m}}| (z_{\text{OSB}} - z_{\text{in}})}{2D_{\text{e}}\rho_\ell} + \frac{\overline{\phi_{\ell o}^2} f_{\ell o} G_{\text{m}} |G_{\text{m}}| (z_{\text{out}} - z_{\text{OSB}})}{2D_{\text{e}}\rho_\ell}$$

$$+ \sum_i \left(\phi_{\ell o}^2 K \frac{G_{\text{m}} |G_{\text{m}}|}{2\rho_\ell}\right)_i$$

where the friction pressure drop multiplier depends on the axial variation of the flow quality, system pressure, and flow rate. Note that ρ_ℓ is used in Equation 14.35 for both the single- and two-phase regions—because the value of the liquid density which is initially subcooled and then saturated is not significantly different. To evaluate the boiling height and the two-phase friction multiplier, the energy equation needs to be solved as illustrated earlier. Since the channel area is constant and we are referring to the steady-state condition, the mass flux is constant, but in the entry single-phase region, it is liquid while in the boiling region it is two phase. For simplicity, the mass flux is designated as G_{m}, mixture mass flux, throughout the channel.

Equation 14.35 is commonly written as

$$\Delta p = \Delta p_{\text{acc}} + \Delta p_{\text{grav}} + \Delta p_{\text{fric}} + \Delta p_{\text{form}} \tag{14.36}$$

where

$$\Delta p = p_{\text{in}} - p_{\text{out}} \tag{14.37}$$

$$\Delta p_{\text{acc}} = \left(\frac{G_{\text{m}}^2}{\rho_{\text{m}}^+}\right)_{\text{out}} - \left(\frac{G_{\text{m}}^2}{\rho_{\text{m}}^+}\right)_{\text{in}} \tag{14.38a}$$

$$\Delta p_{\text{gravity}} = \int_{z_{\text{in}}}^{z_{\text{OSB}}} \rho_\ell g \, dz + \int_{z_{\text{OSB}}}^{z_{\text{out}}} \rho_{\text{m}} g \, dz \tag{14.38b}$$

$$\Delta p_{\text{fric}} = \left[(z_{\text{OSB}} - z_{\text{in}}) + \overline{\phi_{\ell o}^2} (z_{\text{out}} - z_{\text{OSB}}) \right] \frac{f_{\ell o} G_{\text{m}} |G_{\text{m}}|}{2 D_{\text{e}} \rho_\ell} \tag{14.38c}$$

$$\Delta p_{\text{form}} = \sum_i \left(\phi_{\ell o}^2 K \frac{G_{\text{m}} |G_{\text{m}}|}{2 \rho_\ell} \right)_i \tag{14.38d}$$

Let us now evaluate these components for the sinusoidal heat input case further assuming the critical condition is not reached in the channel. Note that the two-phase friction multiplier values plotted in Chapter 11 were all for axially uniform heat addition in the channels.

14.6.2.1 Δp_{acc}

From Equations 14.38a and 11.63, we get

$$\Delta p_{\text{acc}} = \left\{ \left[\frac{(1 - x_{\text{e}})^2}{(1 - \alpha)\rho_{\text{f}}} + \frac{x_{\text{e}}^2}{\alpha \rho_{\text{g}}} \right]_{\text{out}} - \frac{1}{\rho_\ell} \right\} G_{\text{m}}^2 \tag{14.39}$$

For high flow rate conditions when boiling does not occur $(\rho_{\text{m}}^+)_{\text{out}} \simeq \rho_\ell$ and $\Delta p_{\text{acc}} \simeq 0$. For low flow rate conditions, the term ρ_{m}^+ decreases, approaching ρ_{g} when the total flow is evaporated. It implies that as G_{m} decreases, the difference between outlet- and inlet-specific volume terms in Equation 14.39, for example,

$$\left(\frac{1}{\rho_{\text{m}}^+} \right)_{\text{out}} - \left(\frac{1}{\rho_\ell} \right)$$

increases. The net effect as illustrated in Example 14.4 is that Δp_{acc} exhibits a maximum as G_{m} increases.

While Δp_{acc} can be calculated directly from 14.39 in terms of $x_{\text{e,out}}$ and α_{out}, the Δp_{grav} and Δp_{fric} components involve the integration of $\alpha(z)$ and $x_{\text{e}}(z)$, respectively. Each of these parameters and their integrations as well as the evaluation of Δp_{acc} can be formulated in terms of the quantities

$$x_{\text{avg}} = \frac{x_{\text{e}_{\text{in}}} + x_{\text{e}_{\text{out}}}}{2} \tag{14.40a}$$

$$\Delta x_{\text{rise}} = x_{\text{e}_{\text{out}}} - x_{\text{e}_{\text{in}}} \tag{14.40b}$$

which we express next for this sinusoidal heat input case.

First express $x_e(z)$ from Equation 14.33b for this heat input distribution as

$$x_e(z) = x_{e_{in}} + \frac{\dot{q}}{2\dot{m}h_{fg}} + \frac{\dot{q}}{2\dot{m}h_{fg}}\left(\sin\frac{\pi z}{L}\right)$$
$$= x_{e_{in}} + \left(\frac{x_{e_{out}} - x_{e_{in}}}{2}\right) + \left(\frac{x_{e_{out}} - x_{e_{in}}}{2}\right)\sin\left(\frac{\pi z}{L}\right)$$
$$= x_{ave} + \frac{\Delta x_{rise}}{2}\sin\left(\frac{\pi z}{L}\right) \quad (14.41)$$

The void fraction is next given for this heat flux distribution by substituting Equation 14.41 into Equation 5.55 yielding

$$\alpha(z) = \frac{x_{ave} + (\Delta x_{rise}/2)\sin(\pi z/L)}{(1-\rho_g/\rho_f)[x_{ave} + \Delta x_{rise}/2\sin(\pi z/L)] + \rho_g/\rho_f} \quad (14.42)$$

or

$$\alpha(z) = \frac{x_{ave} + (\Delta x_{rise}/2)\sin(\pi z/l}{x' + x''\sin(\pi z/L)}$$

where

$$x' \equiv x_{ave} + \frac{\rho_g}{\rho_f}(1 - x_{ave}) \quad (14.43a)$$

$$x'' \equiv \left(1 - \frac{\rho_g}{\rho_f}\right)\frac{\Delta x_{rise}}{2} \quad (14.43b)$$

Note that at low pressure $\rho_g/\rho_f \ll 1$; therefore, $x' \approx x_{ave}$ and $x'' \approx (\Delta x_{rise}/2)$.

14.6.2.2 Δp_{grav}

From Equation 14.38b, the gravity pressure drop is given by

$$\Delta p_{grav} = \rho_\ell g(z_{OSB} - z_{in}) + \int_{z_{OSB}}^{z_{out}} [\rho_f - \alpha(\rho_f - \rho_g)]\, g\, dz$$
$$= \rho_\ell g(z_{out} - z_{in}) - (\rho_f - \rho_g)g\int_{z_{OSB}}^{z_{out}} \frac{x_{ave} + (\Delta x_{rise}/2)\sin(\pi z/L)}{x' + x''\sin(\pi z/L)}\, dz \quad (14.44)$$

If $|x''| > |x'|$, which is generally the case at very low pressure:

$$\Delta p_{\text{grav}} = \rho_{\text{f}} g L - (\rho_\ell - \rho_{\text{g}}) g \left\{ \frac{\Delta x_{\text{rise}}}{2x''}(z_{\text{out}} - z_{\text{OSB}}) + \left(x_{\text{ave}} - \frac{\Delta x_{\text{rise}} x'}{2x''} \right) \frac{L}{\pi} \frac{1}{(x''^2 - x'^2)^{1/2}} \right\}$$
$$\times \left[\ln \frac{x' \tan(\pi z/2L) + x'' - (x''^2 - x'^2)^{1/2}}{x' \tan(\pi z/2L) + x'' + (x''^2 - x'^2)^{1/2}} \right]_{z_{\text{OSB}}}^{z_{\text{out}}} \tag{14.45}$$

If $|x'| > |x''|$, which is typically the case at BWR pressure (7.14 MPa):

$$\Delta p_{\text{grav}} = \rho_\ell g L - (\rho_{\text{f}} - \rho_{\text{g}}) g \left\{ \frac{\Delta x_{\text{rise}}}{2} \frac{(z_{\text{out}} - z_{\text{OSB}})}{x''} + \left(x_{\text{ave}} - \frac{\Delta x_{\text{rise}} x'}{2x''} \right) \right.$$
$$\left. \times \frac{L}{\pi} \frac{2}{(x'^2 - x''^2)^{1/2}} \left[\tan^{-1} \frac{x' \tan(\pi z/2L) + x''}{(x'^2 - x''^2)^{1/2}} \right]_{z_{\text{OSB}}}^{z_{\text{out}}} \right\} \tag{14.46}$$

14.6.2.3 Δp_{fric}

Using the HEM friction pressure drop multiplier where the effect of viscosity is neglected

$$\phi_{\ell o}^2 = \left(\frac{\rho_{\text{f}}}{\rho_{\text{g}}} - 1 \right) x_{\text{e}} + 1.0 \tag{11.82}$$

The friction pressure drop is obtained from Equation 14.38c obtaining $\overline{\phi_{\ell o}^2}$ from integration of Equation 11.82 as

$$\Delta p_{\text{fric}} = \frac{f_{\ell o} G_{\text{m}} |G_{\text{m}}|}{2 D_{\text{e}} \rho_\ell} \left\{ (z_{\text{OSB}} - z_{\text{in}}) + \int_{z_{\text{OSB}}}^{z_{\text{out}}} \left[\left(\frac{\rho_{\text{f}}}{\rho_{g}} - 1 \right) x_{\text{e}} + 1.0 \right] dz \right\}$$
$$= \frac{f_{\ell o} G_{\text{m}} |G_{\text{m}}|}{2 D_{\text{e}} \rho_\ell} \left\{ (z_{\text{out}} - z_{\text{in}}) + \left(\frac{\rho_{\text{f}}}{\rho_{\text{g}}} - 1 \right) \int_{z_{\text{OSB}}}^{z_{\text{out}}} \left[x_{\text{ave}} + \frac{\Delta x_{\text{rise}}}{2} \sin\left(\frac{\pi z}{L} \right) \right] dz \right\} \tag{14.47}$$

Performing the integration of Equation 14.47

$$\Delta p_{\text{fric}} = \frac{f_{\ell o} G_{\text{m}} |G_{\text{m}}|}{2 D_{\text{e}} \rho_\ell} \left\{ L + \left(\frac{\rho_{\text{f}}}{\rho_{\text{g}}} - 1 \right) \cdot \left[x_{\text{ave}} (z_{\text{out}} - z_{\text{OSB}}) \right. \right.$$
$$\left. \left. + \frac{\Delta x_{\text{rise}}}{2} \frac{L}{\pi} \left(\cos \frac{\pi z_{\text{OSB}}}{L} - \cos \frac{\pi z_{\text{out}}}{L} \right) \right] \right\} \tag{14.48}$$

14.6.2.4 Δp_{form}

The form losses due to abrupt geometry changes, for example, spacers, can be evaluated using the homogeneous multiplication factor, so that Equation 14.38d becomes

$$\Delta p_{\text{form}} = \sum_i \frac{G_m^2 K_i}{2\rho_\ell}\left[1+\left(\frac{\rho_\ell}{\rho_g}-1\right)x_{e,i}\right] \quad (14.49)$$

where K_i = the single-phase pressure loss coefficients at location "i."

Example 14.4: Determination of Pressure Drop in a Bare Rod BWR Bundle for Fully Equilibrium Conditions

PROBLEM

For the BWR conditions of Example 14.3 (sinusoidal axial power distribution and exit quality of 14.6%), calculate the pressure drops Δp_{acc}, Δp_{fric}, Δp_{grav}, and Δp_{total} for an interior channel neglecting fuel rod spacers for the flow rate range from $\dot{m}/4$ to $4\dot{m}$ where $\dot{m}$ is the nominal flow rate for an interior subchannel. Assume fully equilibrium conditions.

SOLUTION

For this array, the pressure loss is equal in each subchannel and hence characteristic of the rod bundle loss. Hence, treating an interior subchannel, calculate the pressure drop components sequentially as follows:

1. Δp_{acc}: Assuming homogenous equilibrium conditions, the void fraction at the top of the core is given by

$$\alpha = \frac{1}{1+[(1-x_e)/x_e](\rho_g/\rho_f)}$$
$$= \frac{1}{1+[(1-0.146)/0.146](37.3/737.3)} = 0.772$$

The inlet and outlet qualities have been calculated in Example 14.3 as

$$x_{e_{in}} = \frac{h_{in}-h_f}{h_{fg}} = \frac{1227.5-1274.4}{1496.4} = -0.031$$
$$x_{e_{out}} = 0.146$$
$$\Delta p_{acc} = \left\{\left[\frac{(1-x_e)^2}{(1-\alpha)\rho_f}+\frac{x_e^2}{\alpha\rho_g}\right]_{out} - \left(\frac{1}{\rho_\ell}\right)_{in}\right\}G_m^2 \quad (14.39)$$
$$= \left\{\left[\frac{(1-0.146)^2}{(1-0.772)(737.3)}+\frac{(0.146)^2}{0.772(37.3)}\right] - \frac{1}{754.6}\right\}(1625)^2$$
$$= 9.9\ \text{kPa}$$

2. Δp_{fric}: To determine the single-phase friction factor $f_{\ell o}$ the Reynolds number, which is a function of the subchannel equivalent diameter (D_e), must first be calculated

$$D_e = \frac{4(\text{flow area})}{\text{wetted perimeter}} = \frac{4(1.08\times10^{-4})}{\pi(0.0112)} = 0.01228\ \text{m}$$

Thus

$$\text{Re} = \frac{G_m D_e}{\mu} = \frac{(1625)(0.01228)}{9.28\times10^{-5}} = 2.15\times10^{5}$$

where μ is calculated at $T_{avg} = 282.2°\text{C}$ and $p = 7.14$ MPa.

$$f_{\ell o} = \frac{0.184}{\text{Re}^{0.2}} = \frac{0.184}{(2.15\times10^{5})^{0.2}} = 0.0158$$

For the friction component

$$\Delta p_{\text{fric}} = \frac{f_{\ell o} G_m |G_m|}{2 D_e \rho_\ell}\left\{L + \left(\frac{\rho_f}{\rho_g} - 1\right)\cdot\left[x_{ave}\left(z_{out} - z_{OSB}\right) + \frac{\Delta x_{rise}}{2}\frac{L}{\pi}\left(\cos\frac{\pi z_{OSB}}{L} - \cos\frac{\pi z_{out}}{L}\right)\right]\right\} \tag{14.48}$$

where

$$z_{out} = \frac{3.588}{2} = 1.794\ \text{m};\quad Z_{out} = 3.588\ \text{m}$$

$$L = 3.588\ \text{m}$$

$$z_{OSB} = -0.803\ \text{m (from Example 14.3)};$$

$$Z_{OSB} = L/2 + z_{OSB} = 1.794 - 0.803 = 0.991\ \text{m}$$

$$G_m = 1625\ \text{kg/m}^2\text{s}$$

$$\rho_g = 37.3\ \text{kg/m}^3;\quad \rho_f = 737.3\ \text{kg/m}^3$$

$$\rho_\ell \approx \rho_\ell\left(T = 286.1°\text{C}, p = 7.14\,\text{MPa}\right) = 739\ \text{kg/m}^3$$

$$x_{ave} = \frac{x_{in} + x_{out}}{2} = \frac{-0.0313 + 0.146}{2} = 0.057$$

$$\frac{\Delta x_{rise}}{2} = \frac{x_{out} - x_{in}}{2} = \frac{0.146 - (-0.0313)}{2} = 0.089$$

$$f_{\ell o} = 0.0158;\quad D_e = 0.01228\ \text{m}$$

Thus,

$$\Delta p_{\text{fric}} = 0.0158\frac{(1625)^2}{2(0.01228)(739)}\left\{3.588 + \left(\frac{737.3}{37.3} - 1\right)\left[0.057(1.794 + 0.803)\right.\right.$$

$$\left.\left. + 0.089\left(\frac{3.588}{\pi}\right)\left(\cos\frac{-0.803\,\pi}{3.588} - \cos\frac{1.794\,\pi}{3.588}\right)\right]\right\}$$

$$= 18.0\text{ kPa}$$

3. Δp_{grav}: Let us now calculate the gravity term. First determine x' and x''.

$$x' \equiv x_{\text{ave}} + \frac{\rho_g}{\rho_f}(1 - x_{\text{ave}}) = 0.057 + \frac{37.3}{737.3}(1 - 0.057) = 0.105 \quad (14.43a)$$

$$x'' \equiv \left(1 - \frac{\rho_g}{\rho_f}\right)\frac{\Delta x_{\text{rise}}}{2} = \left(1 - \frac{37.3}{737.3}\right)(0.089) = 0.0845 \quad (14.43b)$$

$$\therefore x' > x''$$

Hence, for the gravity component

$$\Delta p_{\text{grav}} = \rho_\ell g L - (\rho_f - \rho_g) g\left\{\frac{\Delta x_{\text{rise}}(z_{\text{out}} - z_{\text{OSB}})}{2x''} + \left(x_{\text{ave}} - \frac{\Delta x_{\text{rise}} x'}{2x''}\right) \times \frac{L}{\pi}\frac{2}{(x'^2 - x''^2)^{1/2}}\right.$$

$$\left.\left[\tan^{-1}\frac{x'\tan(\pi z_{\text{out}}/2L) + x''}{(x'^2 - x''^2)^{1/2}} - \tan^{-1}\frac{x'\tan(\pi z_{\text{OSB}}/2L) + x''}{(x'^2 - x''^2)^{1/2}}\right]\right\}$$

$$\therefore \Delta p_{\text{grav}} = (739)(9.81)(3.588) - (737.3 - 37.3)(9.81) \times \left\{\frac{0.089(1.794 + 0.803)}{0.0845}\right.$$

$$+ \left[0.057 - \frac{0.089(0.105)}{0.0845}\right] \times \left[\frac{3.588}{\pi}\frac{2}{(0.105^2 - 0.0845^2)^{1/2}}\right]$$

$$\times \left\{\tan^{-1}\left[\frac{0.105\tan[\pi(1.794)/2(3.588)] + 0.0845}{(0.105^2 - 0.0845^2)^{1/2}}\right]\right.$$

$$\left.\left. - \tan^{-1}\left[\frac{0.105\tan[-\pi(0.803)/2(3.588)] + 0.0845}{(0.105^2 - 0.0845^2)^{1/2}}\right]\right\}\right\}$$

$$\therefore \Delta p_{\text{grav}} = 15.5\text{ kPa} \quad (14.46)$$

4. Δp_{tot}: The total pressure drop is given by the sum of the three components:

$$\Delta p_{\text{tot}} = \Delta p_{\text{acc}} + \Delta p_{\text{fric}} + \Delta p_{\text{grav}} = 9.9 + 18.0 + 15.5 = 43.4\text{ kPa}$$

The pressure drop components for various flow rates have been calculated using the same procedure as above and are plotted in Figure 14.8. The corresponding values of x_{out}, x', and x'' are shown in Figure 14.9.

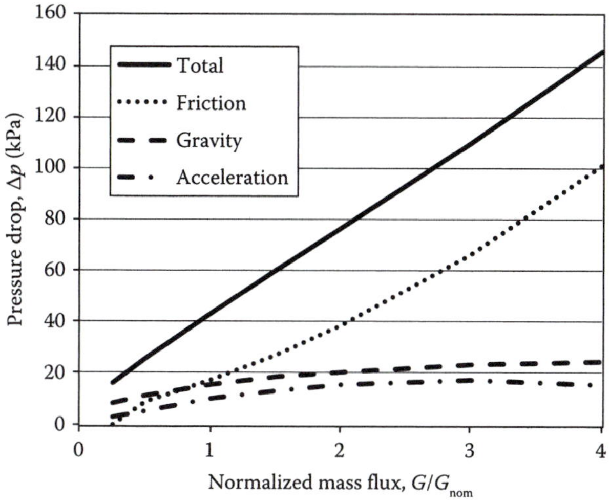

FIGURE 14.8 Pressure drop components in the BWR interior channel of Example 14.4. The mass flux at the nominal condition is $G_{nom} = 1625$ kg/m^2 s.

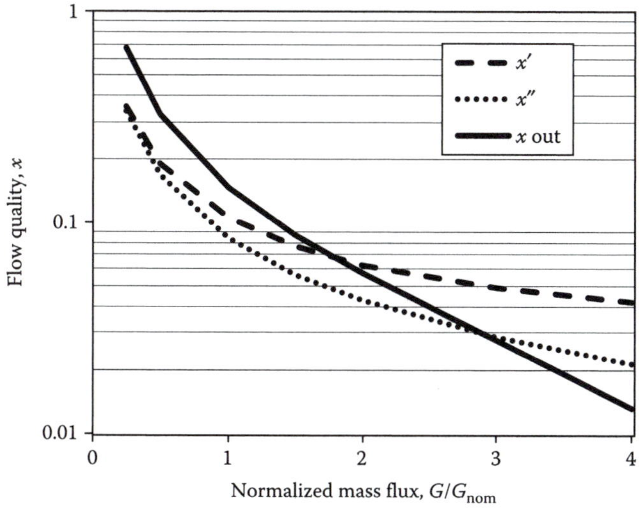

FIGURE 14.9 Effect of the mass flux on the flow quality for Example 14.4.

14.7 STEADY-STATE TWO-PHASE FLOW IN A HEATED CHANNEL UNDER NONEQUILIBRIUM CONDITIONS

The calculation of thermal parameters and the pressure drop in a heated channel allowing for nonequilibrium conditions (both unequal velocity and unequal phase temperatures) is discussed in this section. Essentially four flow regions may exist over the entire length of the subchannel. The existence of these flow regions depends on the heat flux and the inlet conditions. The flow enters as single-phase liquid and may undergo subcooled boiling, bulk boiling, and possibly Dryout before exiting the channel. In a properly designed LWR fuel assembly, the critical heat flux or condition is not encountered, whereas it is encountered in the tube of a once-through steam generator.

14.7.1 Solution of the Energy Equation for Nonequilibrium Conditions (BWR and PWR Cases)

Two axial thermal boundary conditions are of interest: prescribed wall heat flux (BWR case) and prescribed coolant temperature (PWR once-through steam generator case).

14.7.1.1 Prescribed Wall Heat Flux

This condition is characteristic of a subchannel in a BWR reactor core fuel assembly.

We sequentially determine the axial locations of Figure 14.1 that define the transitions between heat transfer regions as follows:

a. Point of onset or initiation of nucleate boiling, Z_{ONB}
   The axial position where boiling initiates, Z_{ONB}, can be determined using single-phase and nucleate boiling heat transfer correlations as follows. Until subcooled boiling occurs, the temperature of the wall is governed by the single-phase heat transfer coefficient. Once subcooled boiling starts, the wall temperature is governed by a subcooled nucleate boiling correlation. At the location where these two temperatures intersect, boiling will be initiated as shown in the graph of wall temperature versus wall height, z^* (Figure 14.10). This location determined by the selected single phase and nucleate boiling heat transfer correlations will generally be different from the locations determined from the Bergles and Rohsenow [1] as well as the Davis and Anderson [4] ONB correlations (Equations 13.4 and 13.3, respectively).
b. Point of transition from onset of nucleation to effective subcooled flow boiling, z_D.
   This point is the point of bubble detachment, also called the location of net vapor generation. Let

$$(\Delta T_{sub})_{z_D} = T_{sat} - T_m(z_D) \tag{14.50}$$

then following the correlation of Saha and Zuber [13] as introduced in Chapter 13:

$$(\Delta T_{sub})_{z_D} = 0.0022\left(\frac{q''D_e}{k_\ell}\right) \quad \text{for Pe} \leq 70{,}000 \tag{13.15a}$$

---

* Since a cosine flux shape is assumed, it is convenient to define the procedure for finding the axial locations of Figure 14.1 in terms of z's and then convert them to Z positions by the relation $Z = L/2 + z$ of footnote of Section 14.2.

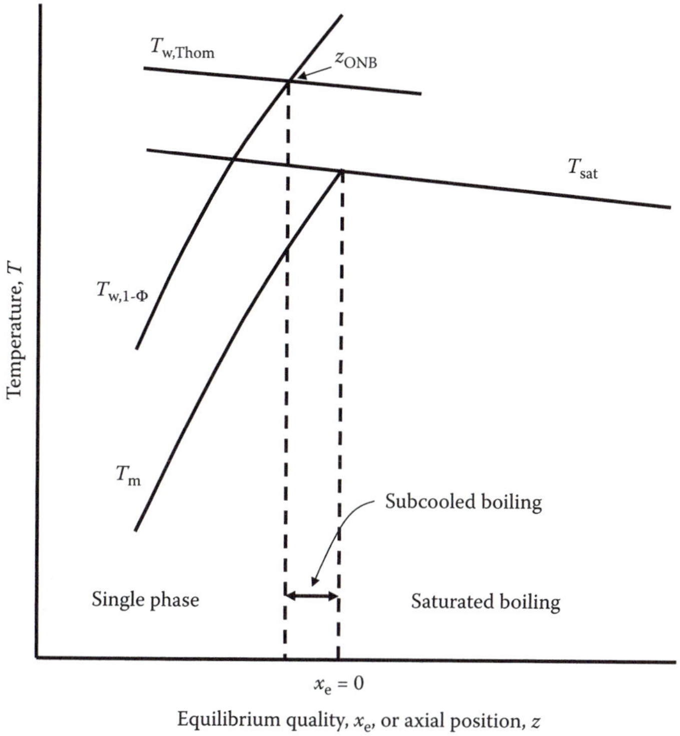

FIGURE 14.10 Prediction of the location z_{ONB} for a sinusoidal axial heat flux distribution. (For subcooled boiling heat transfer we adapt the Thom correlation from Thom, J. R. S., Walker, W. M., Fallon, T. A., and Reising, G. F. S., *Proc. Instn. Mech. Engrs.*, 180:226,1965.)

$$(\Delta T_{sub})_{z_D} = 154\left(\frac{q''}{G_m c_{p\ell}}\right) \quad \text{for Pe} > 70{,}000 \tag{13.15b}$$

where

$\text{Pe} \equiv$ Peclet number $= G_m D_e c_{p\ell}/k_\ell$

$c_{p\ell}$ and k_ℓ = heat capacity and thermal conductivity, respectively, of water (bulk)

T_m = bulk temperature of water

From an energy balance written up to the location z_D:

$$\int_{-L/2}^{z_D} q'' \cos\frac{\pi z}{L}\,dz = \frac{c_{p\ell} G_m D_e}{4}\left[T_m(z_D) - T_{in}\right] \tag{14.51}$$

where $T_m(z_D)$ is obtained depending on the value of Pe from Equation 13.15a or 13.15b.

c. Point of transition from subcooled to bulk boiling, z_{OSB}
At this point, the fluid is saturated. The energy balance of Equation 14.31a written for a sinusoidal heat generation case, which yields z_{OSB}, is applicable for equilibrium as well as nonequilibrium conditions.

d. Point of achievement of thermal equilibrium, z_E
Following the discussion of Chapter 13, z_E is obtained from Equation 13.17 from Levy [11] on selection of a finite (and typically small) value of $x(z) - x_e(z_E)$.

$$x(z) = x_e(z_E) - x_e(z_D)\exp\left[\frac{x_e(z_E)}{x_e(z_D)} - 1\right] \tag{13.17}$$

e. Point of CHF or CPR, z_C
Obtained from a suitable critical condition correlation from Section 13.4.

f. Point of single-phase vapor, z_V
Subsequent to the CHF or CPR location, mist flow exists with film boiling until the flow attains essentially the single-phase vapor condition. For thermal nonequilibrium, the vapor in the two-phase region is heated by the wall, but the liquid phase is heated principally by contact with the vapor versus direct impact on the walls. Hence, the vapor can become highly superheated with liquid droplets still present.

In summary then, six regions of tube length can be separately considered

- Single phase $0 \le z \le z_{ONB}$
- Wall bubbles $z_{ONB} \le z \le z_D$
- Subcooled boiling $z_D \le z \le z_{OSB}$
- Bulk boiling (actual quality greater than equilibrium quality) $z_{OSB} \le z \le z_E$
- Bulk boiling (from achievement of equilibrium quality to the critical condition location) $z_E \le z \le z_C$
- Mist flow $z_C \le z \le z_V$

For simplicity with little loss of accuracy in the computation of channel wall temperature and pressure loss, we can condense these six regions to four:

- Single phase $0 \le z \le z_{ONB}$
- Subcooled boiling $z_{ONB} \le z \le z_{OSB}$
- Bulk boiling $z_{OSB} \le z \le z_C$
- Mist flow $z_C \le z \le z_V$

Example 14.5: Thermal Conditions in a BWR Channel

PROBLEM

Consider the hot interior subchannel of the BWR of Appendix K. Assume sinusoidal axial heat generation. Neglect cross-flow among subchannels and effects of the neutron flux extrapolation lengths. Calculate the values of the following parameters:

a. Z_{ONB}, axial location in the channel where boiling initiates considering nonequilibrium channel conditions
b. Z_D, point of bubble detachment
c. Z_{OSB}, point of transition from subcooled to bulk boiling
d. Z_E, point of achievement of thermal equilibrium
e. Z_C, critical condition location

where by convention all Z locations above are measured from the channel inlet.

SOLUTION

a. Z_{ONB}: To determine T_w single phase, we must determine if flow is laminar or turbulent:

$$D_e = \frac{4A_f}{\pi D} = \frac{4 \cdot 1.08 \times 10^{-4}}{\pi \cdot 0.0112} = 0.0123 \text{ m}$$

$$\text{Re} = \frac{\rho \upsilon D_e}{\mu} = \frac{GD_e}{\mu} = \frac{1625 \cdot 0.0123}{9.28 \times 10^{-5}} = 2.15 \cdot 10^5$$

where μ is calculated at $T_{avg} = 282.2°\text{C}$ and $p = 7.14$ MPa.

Since the flow is turbulent, we can use the Dittus–Boelter/McAdams correlation (Equation 10.91) to determine T_w, single phase. Since

$$T_{w,D\text{-}B} = T_m + \frac{q''(z)}{h_{D\text{-}B}}$$

And using Equation 14.14 to find T_m as a function of height (taking $L = L_e$):

$$T_m(z) = T_{in} + \frac{q_o'}{\dot{m}c_p}\frac{L}{\pi}\left(\sin\frac{\pi z}{L} + 1\right)$$

Combining to find T_w as a function of $q''(z)$ and z:

$$T_{w,D\text{-}B} = T_{in} + \frac{q_o'}{\dot{m}c_p}\frac{L}{\pi}\left(\sin\frac{\pi z}{L} + 1\right) + \frac{q_0'' \cos(\pi z/L)}{h_{D\text{-}B}}$$

where

$$\text{Nu} = 0.023\ \text{Re}^{0.8}\ \text{Pr}^{0.4} = 0.023 \cdot (2.15 \times 10^5)^{0.8}(1)^{0.4} = 424.3 \qquad (10.91)$$

$$h_{D\text{-}B} = \frac{k_\ell \text{Nu}}{D_e} = \frac{0.579(424.3)}{0.0123} = 2.00 \times 10^4\ \text{W/m}^2\text{K}$$

For the hot interior subchannel

$$q'_0 = 47.24 \text{ kW/m}$$

$$\dot{m} = 0.175 \text{ kg/s}$$

$$q''_0 = \frac{q'_0}{\pi D} = \frac{47.24}{\pi(0.0112)} = 1343 \text{ kW/m}^2$$

$$\dot{q} = \frac{2q'_0 L}{\pi} = \frac{2(47.24)(3.588)}{\pi} = 107.9 \text{ kW}$$

$$\frac{\dot{q}}{\dot{m}} = 616.6 \text{ kJ/kg}$$

The Thom correlation for subcooled boiling heat transfer is

$$q'' = \frac{\exp(2p/8.7)}{(22.7)^2}\left(T_{w,\text{Thom}} - T_{sat}\right)^2 \quad (13.23b)$$

where q'' is in MW/m², p is in MPa, and T is in °C.

Thus, $T_{w,\text{Thom}}$ is

$$T_{w,\text{Thom}} = \sqrt{\frac{q''_{\text{MW/m}^2}(22.7)^2}{\exp(2p/8.7)}} + T_{sat} = 22.7\sqrt{\frac{q''_0\cos(\pi z/L)}{10^6\exp(2p/8.7)}} + T_{sat}$$

Now, setting $T_{w,\text{Thom}} = T_{w,\text{D-B}}$:

$$22.7\sqrt{\frac{q''_0\cos(\pi z/L)}{10^6\exp(2p/8.7)}} + T_{sat} = T_{in} + \frac{q'_0}{\dot{m}c_p}\frac{L}{\pi}\left(\sin\frac{\pi z}{L} + 1\right) + \frac{q''_0\cos(\pi z/L)}{h_{\text{D-B}}}$$

Inserting all the numerical values

$$22.7\sqrt{\frac{1343\cos(\pi z/3.588)}{10^3\exp(2\cdot 7.14/8.7)}} + 287 = 278.3 + \frac{47.24}{0.175(5.42)}\frac{3.588}{\pi}$$
$$\times\left(\sin\frac{\pi z}{3.588} + 1\right) + \frac{1343\times 10^3\cos(\pi z/3.588)}{2.00\times 10^4}$$

$$11.58\sqrt{\cos\left(\frac{\pi z}{3.588}\right)} + 287 = 278.3 + 56.88\left(\sin\frac{\pi z}{3.588} + 1\right) + 67.15\cos\left(\frac{\pi z}{3.588}\right)$$

The desired numerical solution of this equation is $z_{ONB} = -1.577$ m. The location of the onset of nucleate boiling, calculated from the bottom of the core, is

$$Z_{ONB} = z_{ONB} + L/2 = -1.577 + 3.588/2 = 0.217 \text{ m}$$

b. Z_D: The Peclet number is

$$\text{Pe} = \text{Re Pr} = 2.15 \times 10^5 \cdot (1) = 2.15 \times 10^5$$

Since $\text{Pe} > 7 \times 10^4$, then

$$\left(\Delta T_{sub}\right)_{Z_D} = 154\left(\frac{q''}{G_m c_{p\ell}}\right) \quad (13.15b)$$

Now for a sinusoidal axial heat flux,

$$\int_{-L/2}^{z_D} q_o'' \cos\frac{\pi z}{L} dz = \frac{c_{p\ell} G_m D_e}{4}\left[T_m(z_D) - T_{in}\right] \quad (14.51)$$

Integrating and applying Equation 13.15b,

$$q_o''\frac{L}{\pi}\left(\sin\frac{\pi z_D}{L} + 1\right) = \frac{c_{p\ell} G_m D_e}{4}\left[T_{sat} - 154\left(\frac{q_o'' \cos(\pi z_D/L)}{G_m c_{p\ell}}\right) - T_{in}\right]$$

$$1343 \times 10^3 \frac{3.588}{\pi}\left(\sin\frac{\pi z_D}{3.588} + 1\right) = \frac{5.42 \times 10^3 (1625)(0.0123)}{4}$$
$$\times\left[287 - 154\left(\frac{1343 \times 10^3 \cos(\pi z_D/3.588)}{(1625) 5.42 \times 10^3}\right) - 278.3\right]$$

$$1.534 \times 10^6\left(\sin\frac{\pi z_D}{3.588} + 1\right) = 2.71 \times 10^4\left[8.7 - 23.48 \cos\frac{\pi z_D}{3.588}\right]$$

The desired numerical solution for this equation is $z_D = -1.474$ m. Hence

$$Z_D = z_D + \frac{L}{2} = -1.474 + \frac{3.588}{2} = 0.320 \text{ m}$$

c. Z_{OSB}: The Z_{OSB} location is determined by Equation 14.33b, imposing the equilibrium quality equal to zero.

$$x_e(z_{OSB}) = x_{e_{in}} + \frac{\dot{q}}{2\dot{m} h_{fg}}\left(\sin\frac{\pi z_{OSB}}{L} + 1\right) = 0 \quad (14.33b)$$

$$\frac{1227.5 - 1274.4}{1496.4} + \frac{616.6}{2(1496.4)}\left(\sin\frac{\pi z_{OSB}}{3.588} + 1\right) = 0$$

$$-0.0313 + 0.206\left(\sin\frac{\pi z_{OSB}}{3.588} + 1\right) = 0$$

$$z_{OSB} = \frac{3.588}{\pi}\sin^{-1}\left(\frac{0.0313}{0.206} - 1\right) = -1.156\,\text{m}$$

Hence

$$Z_{OSB} = z_{OSB} + L/2 = -1.156 + 3.588/2 = 0.638\ \text{m}$$

d. Z_E: Rearranging Equation 13.17, the following equation is obtained

$$x_e(z_D)\exp\left[\frac{x_e(z)}{x_e(z_D)} - 1\right] = x_e(z) - x(z) \quad (13.17)$$

Now $x_e(z)$ is available from Equation 14.33 as

$$x_e(z) = x_{e_{in}} + \frac{\dot{q}}{2\dot{m}h_{fg}}\left(\sin\frac{\pi z}{L} + 1\right) \quad (14.33)$$

where

$$\frac{\dot{q}}{\dot{m}}\frac{1}{2h_{fg}} = \frac{616.6 \cdot 10^3}{2 \cdot 1496.4 \cdot 10^3} = 0.206$$

and

$$x_{e_{in}} = \frac{h_{in} - h_f}{h_{fg}} = \frac{1227.5 - 1274.4}{1496.4} = -0.0313$$

so $x_e(z_E)$ and $x_e(z_D)$ can be expressed using Equation 14.33.
Taking $z_D = -1.474$ m from part (b) and assuming $x(z) - x_e(z)$ is small, say 0.001, Equation 13.17 can be expressed in terms of $x_e(z_E)$ as follows:

$$\left[x_{e_{in}} + \frac{\dot{q}}{2\dot{m}h_{fg}}\left(\sin\frac{\pi z_D}{L} + 1\right)\right]\exp\left[\frac{x_{e_{in}} + \dot{q}/2\dot{m}h_{fg}\left(\sin\pi z_E/L + 1\right)}{x_{e_{in}} + \dot{q}/2\dot{m}h_{fg}\left(\sin\pi z_D/L + 1\right)} - 1\right] = -0.001 \quad (14.52)$$

$$\left[-0.0313 + 0.206\left(\sin\frac{\pi(-1.474)}{3.588} + 1\right)\right]$$
$$\exp\left[\frac{-0.0313 + 0.206\left(\sin\pi z_E/3.588 + 1\right)}{-0.0313 + 0.206\left(\sin\pi(-1.474)/3.588 + 1\right)} - 1\right] = -0.001$$

$$-0.0233 \cdot \exp\left[\frac{-0.0313 + 0.206\left(\sin\dfrac{\pi z_E}{3.588} + 1\right)}{-0.0233} - 1\right] = -0.001$$

$$z_E = \frac{3.588}{\pi}\sin^{-1}\left[\frac{-0.0233\left(\ell n\dfrac{0.001}{0.0233} + 1\right) + 0.0313}{0.206} - 1\right] = -0.742\ \text{m}$$

Hence

$$Z_E = z_E + L/2 = -0.742 + 3.588/2 = 1.052\ \text{m}$$

e. Z_C: In Example 13.5, the Hench–Gillis [10] correlation has been applied to the hot channel of the BWR of Appendix K, showing that the equilibrium quality is always lower than the critical quality, and so no Dryout occurs.

14.7.1.2 Prescribed Coolant Temperature

The inlet and outlet of the hot fluid and the inlet of the cold fluid are prescribed. These conditions are characteristic of a unit cell (primary side tube and the surrounding annulus of secondary coolant) of a steam generator.

We will consider a tube of secondary coolant heated by primary coolant flowing concurrently in a surrounding annulus. For this once-through PWR steam generator, the secondary coolant is heated to superheat conditions. Hence, the critical condition occurs within the tube length. However, in contrast to the prescribed wall heat flux situation, here the wall temperature excursion is bounded by the primary and secondary coolant temperatures existing at the location of the critical condition.

Example 14.6: Wall Temperature Calculation in a Once-through Steam Generator

PROBLEM

A once-through steam generator of a new design has the high-pressure primary coolant in the shell side and the lower-pressure secondary coolant in the tubes. A single tube, 8.5 m long, with its associated annular primary system flow region and several system parameters are presented in Figure 14.11. The inlet temperatures are prescribed for both coolants while the outlet temperature is prescribed for secondary coolant only. Table 14.3 summarizes the saturation properties for primary and secondary fluids.

Develop a numerical algorithm to calculate the tube wall temperature and the secondary void fraction axial profile through the entire length of the steam generator. Assume that

- Tube walls have zero thickness, that is, zero thermal resistance.
- The heat transfer coefficient on the shell side is axially constant. (Use saturated water properties to calculate it, even if the primary coolant is subcooled.)

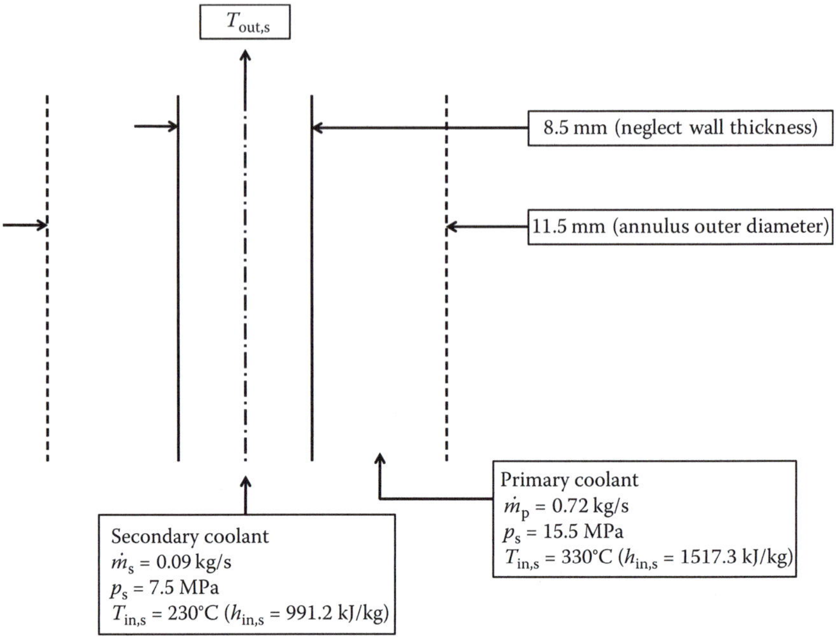

FIGURE 14.11 A primary coolant tube and its surrounding annulus of secondary coolant.

SOLUTION

The logic for solving the problem is the following. The resulting algorithm can be written with any numerical code, for example, MATLAB®.

The tube-annulus system has to be partitioned in a number of axial zones, for example, 1000, each having length $\Delta z = 8.5/1000 = 8.5 \times 10^{-3}$ m. The general idea is to solve the heat balance of a zone and use the results of this balance to calculate the conditions at the inlet of the downstream zone, and repeat this procedure all the way up the steam generator. However, for a given axial zone, the heat balance can be written correctly only once the heat transfer modes (single-phase liquid-forced convection, subcooled boiling, etc.) characterizing that zone are known. Each zone has two sides: the primary side, which operates in single-phase liquid-forced convection, and the secondary side, for which the heat

TABLE 14.3
Relevant Properties of Saturated Water and Steam

p (MPa)	T_{sat} (°C)		h (kJ/kg)	k (W/mK)	c_p (kJ/kgK)	$\mu \times 10^6$ (Pa s)	ρ (kg/m³)	Pr (--)	$\sigma \times 10^3$ (N/m)
7.5	290.5	Water	1292.9	0.564	5.50	89.45	730.88	0.8723	16.54
		Steam	2765.2	0.065	5.61	19.17	39.45	1.6545	
15.5	344.8	Water	1629.9	0.458	8.95	68.33	594.38	1.3353	4.67
		Steam	2596.1	0.121	14.00	23.11	101.92	2.6739	

transfer mode has to be determined. Therefore, the thermal hydraulic parameters in each zone must first be calculated

$$G_p = \frac{\dot{m}_p}{\frac{\pi}{4}\left(D_{\text{annulus}}^2 - D_{\text{tube}}^2\right)} = 15{,}279 \text{ kg/s m}^2$$

$$G_s = \frac{\dot{m}_s}{\frac{\pi}{4} D_{\text{tube}}^2} = 1586 \text{ kg/s m}^2$$

$$D_{eq,p} = \frac{\pi\left(D_{\text{annulus}}^2 - D_{\text{tube}}^2\right)}{\pi\left(D_{\text{annulus}} + D_{\text{tube}}\right)} = 0.0030 \text{ m}$$

$$D_{eq,s} = D_{\text{tube}} = 0.0085 \text{ m}$$

$$\text{Re}_{f,p} = \frac{G_p D_{eq,p}}{\mu_{f,p}} = 6.708 \times 10^5$$

$$\text{Re}_{f,s} = \frac{G_s D_{eq,s}}{\mu_{f,s}} = 1.507 \times 10^5$$

$$(x_e)_{s,1} = \frac{h_{in,s} - h_{f,s}}{h_{fg,s}} = -0.2049 \text{ (secondary fluid equilibrium quality at tube inlet or node 1)}$$

To calculate the energy transferred to the secondary coolant, the mechanisms of heat transfer in the secondary coolant along the length of the steam generator must be known. These mechanisms can be identified, per the problem statement, as

- Single-phase liquid-forced convection
- Subcooled nucleate boiling
- Saturated nucleate boiling
- Transition boiling
- Film boiling
- Single-phase vapor-forced convection

Because these modes of heat transfer have significantly varied heat transfer coefficients, the location where each mechanism takes effect must be determined, and, therefore, the location of the following transition points must be calculated

- Onset of subcooled nucleate boiling
- Onset of saturated nucleate boiling
- Onset of boiling crisis (CHF)
- Onset of superheated steam

First, the heat transfer mode on the shell side is single-phase liquid-forced convection. The corresponding heat transfer coefficient* which, per problem statement, is constant throughout the whole tube height can be calculated using a popular variant of the Dittus–Boelter/McAdams equation (Equation 10.91) with the Prandtl number taken as $Pr^{0.3}$ for cooling:

$$H_p = Nu_p \frac{k_{f,p}}{D_{e,p}} = \left(0.023\ Re_{f,p}^{0.8}\, Pr_{f,p}^{0.3}\right)\frac{k_{f,p}}{D_{e,p}} = 1.76 \times 10^5\ \text{W/m}^2\text{K}$$

In the first axial zone on the tube side, the bulk of the secondary fluid is subcooled but, depending on the wall temperature (unknown), the heat transfer regime can be single-phase forced convection or subcooled boiling. Since this is not known *a priori*, single-phase forced convection can be assumed, and this assumption then verified. The heat transfer coefficient on the tube side can be calculated using the Dittus–Boelter equation for heating:

$$H_{s,1} = Nu_s \frac{k_{f,s}}{D_{e,s}} = \left(0.023\ Re_{f,s}^{0.8}\, Pr_{f,s}^{0.4}\right)\frac{k_{f,s}}{D_{e,s}} = 2.01 \times 10^4\ \text{W/m}^2\text{K}$$

The wall temperature at the generic *i*th axial zone can be obtained from the heat balance:

$$q''_{\text{tube inside}} = q''_{\text{tube outside}} \Rightarrow H_{s,i}(T_{w,i} - T_{s,i}) = H_p(T_{p,i} - T_{w,i})$$

where, at the first axial location, $T_{s,i} = T_{s,in}$ and $T_{p,i} = T_{p,in}$ are known. Thus,

$$T_{w,1} = \frac{H_p T_{p,in} + H_{s,1} T_{s,in}}{H_p + H_{s,1}} = 319.7\,°\text{C}$$

Now we need to verify whether this temperature is consistent with the assumption made, that is, single-phase liquid-forced convection on the tube side of the first axial zone. Let us use the Davis–Anderson correlation to calculate the wall temperature required for nucleate boiling incipience (ONB):

$$T_{w,inc} = T_{sat,s} + \sqrt{\frac{8 q''_1 \sigma_s T_{sat,s}}{k_{f,s} h_{fg,s} \rho_{g,s}}} = 292.5°\text{C} \quad (13.3)$$

Since $T_{w,1} > T_{w,inc}$, subcooled boiling exists already in the first axial zone. The wall temperature needs to be reevaluated using a heat transfer correlation for subcooled boiling. For example, the Chen correlation [3] can be used for the iterative† computation of the secondary side heat transfer coefficient:

$$H_{s,i} = (H_{s,FC})_i + (H_{s,NB})_i$$

* In this problem, since both the heat transfer coefficient and enthalpy appear, they are differentiated by use of the symbols H and h, respectively.

† The computation of $H_{s,i}$ has to be done iteratively since T_{wi} is unknown, that is, by varying T_{wi} until: $H_{s,i}(T_{w,i} - T_{sat,s}) = H_p(T_{p,i} - T_{w,i})$.

where the forced convection contribution is given by

$$(H_{s,c})_i = F_i \frac{k_{f,s}}{D_e} 0.023 \left\{ \frac{G_s [1-(x)_s] D_e}{\mu_{f,s}} \right\}_i^{0.8} Pr_{f,s}^{0.4} \quad (13.24)$$

while the nucleate boiling contribution is given by

$$(H_{s,NB})_i = S(0.00122) \left(\frac{k_f^{0.79} C_{pf}^{0.45} \rho_f^{0.49}}{\sigma^{0.5} \mu_f^{0.29} h_{fg}^{0.24} \rho_g^{0.24}} \right)_s \left(T_{w,i} - T_{sat,s} \right)^{0.24} \left[p_{sat}(T_{w,i}) - p_{sat}(T_{sat,s}) \right]^{0.75}$$

$$(13.26)$$

The enhanced turbulence factor, F_i, is given by

$$F_i = 1 \quad \text{for} \left(\frac{1}{X_{tt}} \right) \leq 0.1 \quad (13.25a)$$

$$= 2.35 \left[0.213 + \left(\frac{1}{X_{tt}} \right)_i \right]^{0.736} \quad \text{for } \frac{1}{X_n} > 0.1 \quad (13.25b)$$

with

$$\left(\frac{1}{X_{tt}} \right)_i = \left[\frac{(x_e)_{s,i}}{1-(x_e)_{s,i}} \right]^{0.9} \left(\frac{\rho_f}{\rho_g} \right)_s^{0.5} \left(\frac{\mu_g}{\mu_f} \right)_s^{0.1} \quad (11.94)$$

while the boiling suppression factor, S, is given by

$$S = \frac{1}{1 + 2.53 \times 10^{-6} \left\{ F^{1.25} \frac{G_s [1-(x_e)_s] D_{tube}}{\mu_{f,s}} \right\}_i^{1.17}} \quad (13.27)$$

For the first axial zone $T_{p,i} = T_{p,in}$ and it follows that $T_{w,i} = 317.0°C$.

Knowing the wall temperature allows calculation of the power that is transferred from the primary to the secondary fluid along the i-th axial zone:

$$\dot{Q}_i = q_i''(\pi D_{tube} \Delta z) = H_p (T_{p,i} - T_{w,i})(\pi D_{tube} \Delta z)$$

which, for the first axial zone, is

$$\dot{Q}_1 = 60{,}830 \ \Delta z = 60{,}830 \times (8.5 \times 10^{-3}) = 517.6 \text{ W}$$

The enthalpies of the fluids at the inlet of axial zone $i + 1$ are

$$h_{p,i+1} = h_{p,i} - \frac{\dot{Q}_i}{\dot{m}_p}$$

$$h_{s,i+1} = h_{s,i} + \frac{\dot{Q}_i}{\dot{m}_s}$$

These relations are to be applied to each axial zone starting from the second one using, as input, known data referred to the upstream axial zone. Thus, for $i = 2$:

$$h_{s,2} = h_{s,in} + \frac{\dot{Q}_1}{\dot{m}_s} = 991.2 + \frac{517.6 \times 10^{-3}}{0.09} = 996.95 \text{ kJ/kg}$$

and

$$(x_e)_{s,2} = \frac{h_{s,2} - h_{f,s}}{h_{fg,s}} = \frac{996.95 - 1292.9}{1472.3} = -0.2010$$

Particularly, when $h_{s,i} = h_{f,s}$, then "i" is the onset of saturated boiling (OSB) location.

The calculation of the void fraction requires knowing the steam flow quality x which, in subcooled boiling, is effectively zero upstream of the onset of significant void or bubble departure location, z_D. Whether z_D has been reached must therefore be checked in each axial zone, until OSV is detected, using the Saha–Zuber correlation, Equation 13.15. Since

$$\begin{aligned} Pe_s &= Pr_s Re_s = \left(\frac{\mu_f c_{pf}}{k_f}\right)_s Re_s = \left[\frac{(89.45 \times 10^{-6})5500}{0.564}\right] 1.507 \times 10^5 \\ &= 1.315 \times 10^5 > 7 \times 10^4 \end{aligned}$$

the Saha–Zuber correlation establishes that Z_D occurs at the 24th axial location where

$$T_{w,i} \geq T_{sat,s} - 154\left(\frac{q_i''}{G_s c_{pf,s}}\right)$$

When applied to the 24th axial zone, this inequality is verified, since

$$308.9 > 290.5 - 154\left[\frac{2.1321 \times 10^6 (328.2 - 316.1)}{1586 \times 5500}\right] = 244.9°\text{C}$$

which means that bubbles depart at the 24th node.

Beyond z_D, the steam flow quality is calculated using the Levy model:

$$(x)_{s,i} = (x_e)_{s,i} - (x_e)_{s,i=D} \exp\left[\frac{(x_e)_{s,i}}{(x_e)_{s,i=D}} - 1\right] \tag{13.17}$$

Thus, for the 24th and 25th axial zones:

$$(x)_{s,24} = 0$$

$$(x)_{s,25} = -0.1142 + 0.1178 \exp\left[\frac{-0.1142}{-0.1178} - 1\right] = 5.6 \times 10^{-5}$$

The void fraction can then be calculated as

$$\alpha_{s,i} = \frac{1}{1 + \frac{1-(x)_{s,i}}{(x)_{s,i}} \left(\frac{\rho_g}{\rho_f}\right)_s S_i}$$

where the slip ratio, S, can be obtained, for example, using the Premoli correlation [12]:

$$\begin{aligned} S_i &= 1 + E_1 \left(\frac{y_i}{1 + y_i E_2} - y_i E_2\right)^{0.5} && \text{if } y_i \le \frac{1-E_2}{E_2^2} \\ &= 1 && \text{otherwise} \end{aligned} \tag{11.49}$$

where

$$E_1 = 1.578\ \mathrm{Re}_{f,s}^{-0.19} \left(\frac{\rho_f}{\rho_g}\right)_s^{0.22} = 0.3113$$

$$E_2 = 0.0273 \frac{G_s^2 D_{tube}}{(\sigma \rho_f)_s} \mathrm{Re}_{f,s}^{-0.51} \left(\frac{\rho_f}{\rho_g}\right)_s^{-0.08} = 0.0874$$

and for $i = 2$:

$$y_2 = \frac{\beta_2}{1-\beta_2} = 0.0016 < \frac{1-E_2}{E_2^2} = 119$$

$$\beta_2 = \frac{\rho_{f,s}(x)_{s,2}}{\rho_{f,s}(x)_{s,2} + \rho_{g,s}\left[1-(x)_{s,2}\right]} = 0.0016$$

$$S_2 = 1 + E_1 \left(\frac{y_2}{1 + y_2 E_2} - y_2 E_2\right)^{0.5} = 1.0119$$

Thus,

$$\alpha_{s,2} = \frac{1}{1 + \frac{1 - 8.6 \times 10^{-5}}{8.6 \times 10^{-5}} \left(\frac{1.368 \times 10^{-3}}{2.533 \times 10^{-2}}\right) 1.0119} = 0.0016$$

By repeating these calculations at each axial zone, we find that the secondary fluid axial profile reaches $h_{f,s}$ at about 0.485 m from the inlet, that is, at axial zone 58, which is therefore the OSB axial location. From this point up to the CHF point, the heat transfer regime is forced convection-saturated boiling, for which the Chen correlation, Equation 13.24, can be used for the iterative* computation of the secondary side heat transfer coefficient:

$$H_{s,i} = (H_{s,FC})_i + (H_{s,NB})_i$$

* The computation of $H_{s,i}$ has to be done iteratively since T_{wi} is unknown, that is, by varying T_{wi} until: $H_{s,i}\,(T_{w,i} - T_{sat,s}) = H_p\,(T_{p,i} - T_{w,i})$.

where the forced convection contribution is given by

$$(H_{s,FC})_i = F_i \frac{k_{f,s}}{D_{tube}} 0.023 \left\{ \frac{G_s [1-(x)_s] D_{tube}}{\mu_{f,s}} \right\}_i^{0.8} Pr_{f,s}^{0.4} \tag{13.24}$$

while the nucleate boiling contribution is given by

$$(H_{s,NB})_i = 0.00122\, S_i \left(\frac{k_f^{0.79} c_{pf}^{0.45} \rho_f^{0.49}}{\sigma^{0.5} \mu_f^{0.29} h_{fg}^{0.24} \rho_g^{0.24}} \right)_s (T_{w,i} - T_{sat,s})^{0.24} \left[p_{sat}(T_{w,i}) - p_{sat}(T_{sat,s}) \right]^{0.75} \tag{13.26}$$

The enhanced turbulence factor, F_i, is given by

$$F_i = 1 \qquad \text{if } \left(\frac{1}{X_{tt}} \right) \le 0.1 \tag{13-25a}$$

$$= 2.35 \left[0.213 + \left(\frac{1}{X_{tt}} \right)_i \right]^{0.736} \qquad \text{otherwise} \tag{13-25b}$$

with

$$\left(\frac{1}{X_{tt}} \right)_i = \left[\frac{(x_e)_{s,i}}{1-(x_e)_{s,i}} \right]^{0.9} \left(\frac{\rho_f}{\rho_g} \right)_s^{0.5} \left(\frac{\mu_g}{\mu_f} \right)_s^{0.1} \tag{11.94}$$

while the boiling suppression factor, S_i, is given by

$$S_i = \frac{1}{1 + 2.53 \times 10^{-6} \left\{ F^{1.25} \frac{G_s [1-(x_e)_s] D_{tube}}{\mu_{f,s}} \right\}_i^{1.17}} \tag{13.27}$$

The knowledge of $H_{s,i}$ and $T_{w,i}$ allows calculating $h_{s,i}$, from which $(x)_{s,i}$ and $\alpha_{s,i}$ can easily be determined.

To calculate where CHF occurs, two correlations are used: Biasi [2] and CISE-4 [7].

At each node, q''_{cr} is calculated using the Biasi correlation:

$$q''_{cr} = (2.764 \times 10^7)(100D)^{-n} G^{-1/6} [1.468 F(p_{bar}) G^{-1/6} - x] \text{ W/m}^2 \tag{13.48a}$$

$$q''_{cr} = (15.048 \times 10^7)(100D)^{-n} G^{-0.6} H(p_{bar})[1 - x] \text{ W/m}^2 \tag{13.48b}$$

where

$$F(p_{bar}) = 0.7249 + 0.099p_{bar}\exp(-0.032p_{bar}) \quad (13.48c)$$

$$H(p_{bar}) = -1.159 + 0.149p_{bar}\exp(-0.019p_{bar}) + 9p_{bar}(10 + p_{bar}^2)^{-1} \quad (13.48d)$$

Note: $p_{bar} = 10p$, when p is in MPa; and

$$n = \begin{cases} 0.4 \text{ for } D \geq 0.01 \\ 0.6 \text{ for } D < 0.01 \end{cases} \quad (13.48e)$$

When the condition

$$q'' > q''_{cr}$$

is satisfied, CHF occurs. For the Biasi correlation, CHF occurs at 2.49 m.

At each node, x_{cr} is also calculated

$$x_{cr} = \frac{D_e}{D_h}\left(a\frac{L_B}{L_B + b}\right) \quad (13.47a)$$

where

$$a = \frac{1}{1 + 1.481 \times 10^{-4}(1 - p/p_c)^{-3}G} \quad \text{if } G \leq G^*$$

and

$$a = \frac{1 - p/p_c}{(G/1000)^{1/3}} \quad \text{if } G \geq G^*$$

where $G^* = 3375(1 - p/p_c)^3$; p_c = critical pressure (MPa); L_B = boiling length (distance from position of zero equilibrium quality to channel exit); and

$$b = 0.199(p_c/p - 1)^{0.4}\, GD^{1.4}$$

When the condition

$$x > x_{cr}$$

is satisfied, CHF occurs. For the CISE-4 corrrelation, CHF occurs at 2.25 m. This elevation is taken as the point where CHF occurs because it is less than the elevation calculated using the Biasi correlation, and thus is more conservative. Since the heat transfer coefficient decreases significantly at this location, the wall temperature profile exhibits a discontinuity as shown in Figure 14.12.

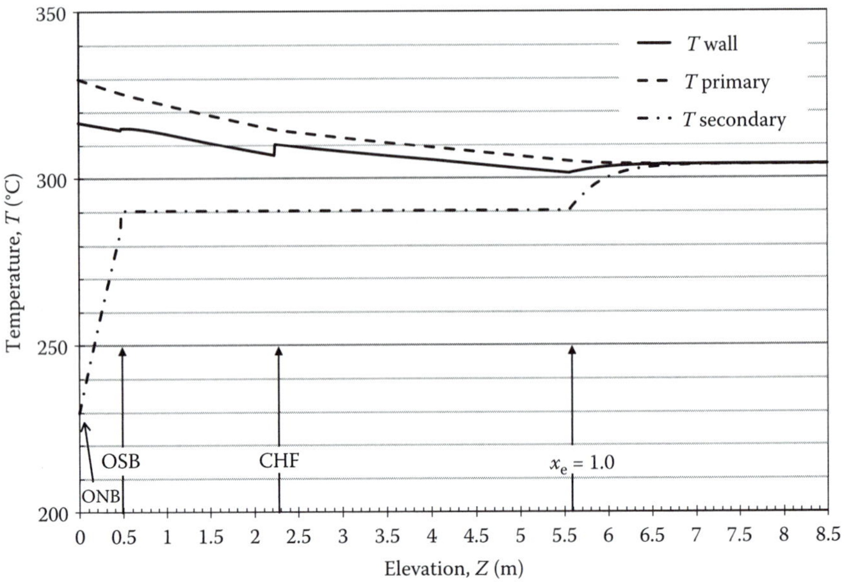

FIGURE 14.12 Axial temperature profiles for Example 14.6.

To calculate post-CHF heat transfer, the Groeneveld film boiling look-up tables [8,9] are used to determine the appropriate heat transfer coefficient inside the tube. The heat tranfser coefficient is interpolated at each node and the heat flux is calculated iteratively to satisfy the heat balance equation. The resulting axial heat flux profile is shown in Figure 14.14.

Eventually, the enthalpy of the secondary steam exceeds $h_{s,g}$ and superheated steam is produced. This occurs at an elevation of 4.83 m; however, the Groeneveld film boiling heat transfer lookup table is still used to find the heat transfer coefficient at each axial node above this point since the look-up table is valid to an equilibrium quality of 2.0. Using this method, the secondary steam is calculated to exit the steam generator at about 14°C superheat. However, since the associated equilibrium quality is only 1.05, it is most likely that saturated liquid droplets are entrained in the superheated steam flow.

The resulting data are shown in Figures 14.12 through 14.14, which show the axial temperature, secondary coolant void fraction, and axial heat flux profiles, respectively, obtained through the methodology described. Note that Figure 14.12 indicates that at the prescribed steam generator flow conditions, the last 1–2 m do not achieve the effective heat transfer.

14.7.2 Solution of the Momentum Equation for Channel Nonequilibrium Conditions to Obtain Pressure Drop (BWR Case)

For the computation of pressure drop, we can adopt the results of Section 14.6.2 but with the following changes in parameters to reflect the nonequilibrium assumption which is made in this section.

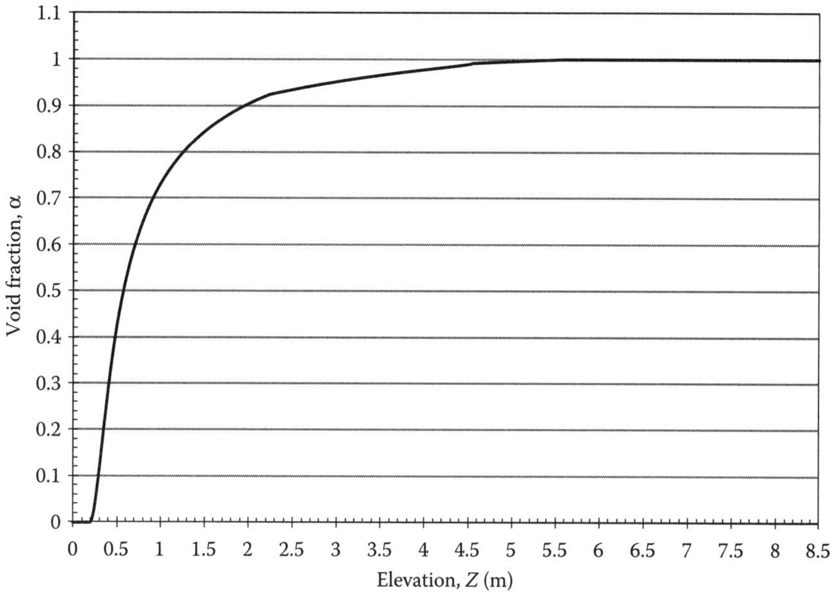

FIGURE 14.13 Secondary coolant void fraction profile for Example 14.6.

- For all pressure drop components, replace
  ρ_f with ρ_ℓ
  ρ_g with ρ_v
  Z_{OSB} with Z_{ONB}
  and for α use a separated flow correlation such as Premoli [12] (Equation 11.49).
- For the friction component additionally replace $\phi_{\ell o}$ with a separated flow correlation such as Friedel [5,6] (Equation 11.99).

Note that the correlations of Premoli and Friedel cover the whole two-phase region so that it is not necessary to explicitly introduce the locations of two-phase phenomena beyond Z_{ONB} into these equations.

Example 14.7: Determination of Pressure Drop in a Bare Rod BWR Bundle for Nonequilibrium Conditions

PROBLEM

For the average BWR conditions of Example 14.3, calculate the pressure drops Δp_{acc}, Δp_{grav}, Δp_{fric}, and Δp_{total} for a bare rod bundle. Consider nonequilibrium conditions, same power for all the rods and neglect cross-flow among subchannels.

SOLUTION

According to our assumptions, the pressure loss is equal in every subchannel. We will treat an interior subchannel, whose mass flux density G is 1625 kg/m²s (from Appendix K) and whose equivalent diameter D_e is 0.01228 m (from Example 14.4).

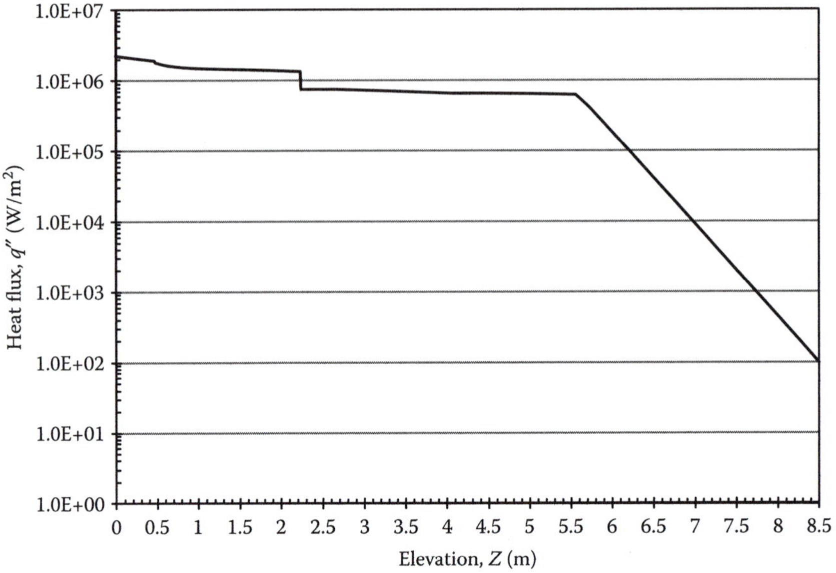

FIGURE 14.14 Axial heat flux profile for Example 14.6.

Applying the procedure used in part (a) of Example 14.5 to core-averaged conditions in an interior subchannel, Z_{ONB} and Z_D are determined as

$$Z_{ONB} = 0.466 \text{ m}$$

$$Z_D = 0.619 \text{ m}$$

Nondetached bubbles exist on the wall between positions Z_{ONB} and Z_D. We neglect their effect on pressure drop over this short length and use the position Z_D in the following calculations as the start of the two-phase condition.

The actual quality profile can be determined applying Equation 13.17 to the equilibrium quality profile calculated in Example 14.3.

$$x(z) = x_e(z) - x_e(z_D)\exp\left[\frac{x_e(z)}{x_e(z_D)} - 1\right] \tag{13.17}$$

$$x_e(z) = 0.0573 + 0.0887 \sin\frac{\pi z}{3.588}$$

Calculating z_D from the core midplane,

$$z_D = Z_D - L/2 = 0.619 - 3.588/2 = -1.175 \text{ m}$$

Hence,

$$x_e(z_D) = 0.0573 + 0.0887\sin\frac{\pi z_D}{3.588} = 0.0573 + 0.0887\sin\frac{\pi(-1.175)}{3.588} = -0.0187$$

The actual quality profile is

$$x(z) = 0.0573 + 0.0887\,\sin\frac{\pi z}{3.588} + 0.0187\exp\left[\frac{0.0573+0.0887\sin\dfrac{\pi z}{3.588}}{-0.0187} - 1\right]$$

Using this equation, the actual outlet quality is very close to the outlet equilibrium quality:

$$x_{out} = x\left(z = \frac{L}{2}\right) = 0.146$$

The pressure drop calculations are based on Equation 11.102:

$$\Delta p = \Delta p_{acc} + \Delta p_{fric} + \Delta p_{grav} = G_m^2\left[\frac{(1-x)^2}{(1-\alpha)\rho_\ell} + \frac{x^2}{\alpha\rho_v}\right]_{\alpha,x_z=0}^{\alpha,x_z=L}$$

$$+\frac{f_{\ell o}}{D_e}\frac{G_m^2}{2\rho_\ell}\int_0^L \phi_{\ell o}^2\,dz + \int_0^L g\cos\theta\,[\rho_v\alpha + \rho_\ell(1-\alpha)]\,dz \qquad (11.102)$$

1. Δp_{acc}: The acceleration component in Equation 11.102 is

$$\Delta p_{acc} = \left\{\left[\frac{(1-x)^2}{(1-\alpha)\rho_\ell} + \frac{x^2}{\alpha\rho_v}\right]_{out} - \left[\frac{1}{\rho_\ell}\right]_{in}\right\}G_m^2$$

At the inlet:

$$\rho_{\ell,in} = 754.6\ \text{kg/m}^3$$

At the outlet:

$$\rho_{\ell,out} = \rho_\ell\,(T = 286.1°\text{C},\ p = 7.14\ \text{MPa}) = 739\ \text{kg/m}^3$$

$$\mu_{\ell,out} = \mu_\ell\,(T = 286.1°\text{C},\ p = 7.14\ \text{MPa}) = 9.12 \times 10^{-5}\ \text{kg/m s}$$

$$\rho_{v,out} \approx \rho_g = 37.3\ \text{kg/m}^3$$

The Premoli correlation can be used to calculate α_{out}:

$$\beta = \frac{\rho_\ell x}{\rho_\ell x + \rho_v(1-x)} = \frac{739 \cdot 0.146}{739 \cdot 0.146 + 37.3(1-0.146)} = 0.772$$

$$y = \frac{\beta}{1-\beta} = 3.387$$

$$\text{Re} = \frac{GD_e}{\mu_\ell} = \frac{1625(0.01228)}{9.12 \times 10^{-5}} = 2.19 \times 10^5 \quad (11.52a)$$

$$\text{We} = \frac{G^2 D_e}{\sigma \rho_\ell} = \frac{1625^2(0.01228)}{17.77 \times 10^{-3} \cdot 739} = 2470 \quad (11.52b)$$

$$E_1 = 1.578\,\text{Re}^{-0.19}\left(\frac{\rho_\ell}{\rho_v}\right)^{0.22} = 1.578\left(2.19 \times 10^5\right)^{-0.19}\left(\frac{739}{37.3}\right)^{0.22} = 0.294$$

(11.51a)

$$E_2 = 0.0273\,\text{We}\,\text{Re}^{-0.51}\left(\frac{\rho_\ell}{\rho_v}\right)^{-0.08}$$

$$= 0.0273 \cdot 2470 \cdot \left(2.19 \times 10^5\right)^{-0.51}\left(\frac{739}{37.3}\right)^{-0.08} = 0.100 \quad (11.51b)$$

The slip factor is given, according to the following condition, by the Premoli correlation:

$$(y = 3.387) \le \left(\frac{1-E_2}{E_2^2} = 90\right)$$

$$S = 1 + E_1\left(\frac{y}{1+yE_2} - yE_2\right)^{0.5} \quad (11.49)$$

$$= 1 + 0.294\left(\frac{3.387}{1+3.387 \cdot 0.100} - 3.387 \cdot 0.100\right)^{0.5} = 1.435$$

The void fraction can be determined using the void fraction–quality–slip relation:

$$\alpha_{out} = \frac{1}{1 + \frac{(1-x_{out})}{x_{out}}\frac{\rho_v}{\rho_\ell}S} = \frac{1}{1 + \frac{(1-0.146)}{0.146}\frac{37.3}{739}1.435} = 0.702 \quad (5.55)$$

The acceleration pressure drop component is hence

$$\Delta p_{acc} = \left\{ \left[\frac{(1-x)^2}{(1-\alpha)\rho_\ell} + \frac{x^2}{\alpha\rho_v} \right]_{out} - \left[\frac{1}{\rho_\ell} \right]_{in} \right\} G_m^2$$

$$= \left\{ \left[\frac{(1-0.146)^2}{(1-0.702)739} + \frac{0.146^2}{0.702 \cdot 37.3} \right] - \left[\frac{1}{754.6} \right] \right\} 1625^2$$

$$= 7.4 \text{ kPa}$$

2. Δp_{fric}: The friction term integral of Equation 11.102 may be divided in two components, one for the single-phase lower portion of the channel (from z_{in} to effectively z_D) and one for the boiling length (from effectively z_D to z_{out}):

$$\Delta p_{fric} = \left(\frac{z_D - z_{in}}{\rho_{\ell,sub}} + \frac{\int_{z_D}^{z_{out}} \phi_{\ell o}^2 \, dx}{\rho_{\ell,boil}} \right) \frac{f_{\ell o} G_m |G_m|}{2D_e}$$

The two-phase frictional multiplier $\phi_{\ell o}^2$ can be determined using the Friedel correlation:

$$\phi_{\ell o}^2 = E + \frac{3.24FH}{\text{Fr}^{0.0454}\text{We}^{0.035}} \quad (11.99)$$

$$E = (1-x)^2 + x^2 \frac{\rho_\ell f_{vo}}{\rho_v f_{\ell o}} \quad (11.100a)$$

$$F = x^{0.78}(1-x)^{0.224} \quad (11.100b)$$

$$H = \left(\frac{\rho_\ell}{\rho_v} \right)^{0.91} \left(\frac{\mu_v}{\mu_\ell} \right)^{0.19} \left(1 - \frac{\mu_v}{\mu_\ell} \right)^{0.7} \quad (11.100c)$$

$$\text{Fr} = \frac{G^2}{g D_e \rho_m^2} \quad (11.100d)$$

$$\text{We} = \frac{G^2 D_e}{\sigma \rho_m} \quad (11.100e)$$

And

$$\rho_m = \left(\frac{x}{\rho_v} + \frac{1-x}{\rho_\ell} \right)^{-1} \quad (11.73)$$

The two friction factors are calculated as follows:

$$\mathrm{Re}_\ell = \frac{GD_e}{\mu_\ell} = \frac{1625(0.01228)}{9.12\times10^{-5}} = 2.19\times10^5$$

$$\mathrm{Re}_v = \frac{GD_e}{\mu_v} = \frac{1625(0.01228)}{1.90\times10^{-5}} = 1.05\times10^6$$

$$f_{jo} = \left[0.86859\ \ell n\left(\frac{\mathrm{Re}_j}{1.964\ \ell n\ \mathrm{Re}_j - 3.8215}\right)\right]^{-2} \quad \text{if } \mathrm{Re}_j > 1055 \qquad (11.100h)$$

where "j" is "v" or "ℓ."

$$f_{\ell o} = \left[0.86859\ \ell n\left(\frac{2.19\times10^5}{1.964\ \ell n\ 2.19\times10^5 - 3.8215}\right)\right]^{-2} = 0.0154$$

$$f_{vo} = \left[0.86859\ \ell n\left(\frac{1.05\times10^6}{1.964\ \ell n\ 1.05\times10^6 - 3.8215}\right)\right]^{-2} = 0.0116$$

The Friedel correlation can be integrated numerically assuming, for the sake of simplicity, constant densities, viscosities and surface tensions, and adopting the previously calculated actual quality distribution. The result is

$$\int_{z_D}^{z_{out}} \phi_{\ell o}^2\, dx = 9.32\ \mathrm{m}$$

Since the boiling length is about 3 m [$z_{out} - z_D = 3.588/2 - (-1.175) = 2.97$ m], this result demonstrates that the friction pressure drop in this region is about three times that of the liquid-only case.

For simplicity, we use two constant averaged values for the liquid density in the subcooled region and in the boiling length. The former is calculated at the core-averaged temperature and pressure, while the latter is assumed as equal to the outlet density, neglecting temperature and pressure differences throughout the boiling length:

$$\rho_{\ell,sub} \approx \rho_\ell\,(T = 282.2°\mathrm{C},\ p = 7.14\ \mathrm{MPa}) = 747.2\ \mathrm{kg/m^3}$$

$$\rho_{\ell,boil} \approx \rho_{\ell,out} = 739\ \mathrm{kg/m^3}$$

Hence, the friction pressure drop is

$$\Delta p_{fric} = \left(\frac{z_D - z_{in}}{\rho_{\ell,sub}} + \frac{\int_{z_D}^{z_{out}} \phi_{\ell o}^2\, dx}{\rho_{\ell,boil}}\right)\frac{f_{\ell o} G_m |G_m|}{2D_e}$$

$$= \left[\frac{(-1.175 + 1.794)}{747.2} + \frac{9.32}{739}\right]\frac{0.0154(1625^2)}{2(0.01228)}$$

$$= 22.3 \text{ kPa}$$

3. Δp_{grav}: The gravity term of Equation 11.102 may be divided in two components, one for the single-phase lower portion of the channel (from z_{in} to effectively z_D) and one for the boiling length (from effectively z_D to z_{out}):

$$\Delta p_{gravity} = \rho_{\ell,sub} g(z_D - z_{in}) + \int_{z_D}^{z_{out}} [\rho_{\ell,boil} - \alpha(\rho_{\ell,boil} - \rho_v)] g dz$$

The numerical integration of this equation, using the previously determined profiles of x and α and the density values used for the calculation of the friction component, gives

$$\Delta p_{gravity} = \rho_{\ell,sub} g(z_D - z_{in}) + \int_{z_D}^{z_{out}} [\rho_{\ell,boil} - \alpha(\rho_{\ell,boil} - \rho_v)] g \, dz$$

$$= (747.2)(9.81)(-1.175 + 1.794) + \int_{-1.175}^{1.794} \left[739 - \alpha(z)(739 - 37.3)\right] 9.81 \, dz$$

$$= 16.7 \text{ kPa}$$

4. Δp_{tot}: From the sum of the three components

$$\Delta p_{tot} = \Delta p_{acc} + \Delta p_{fric} + \Delta p_{grav} = 7.4 + 22.3 + 16.7 = 46.4 \text{ kPa}$$

The results of this example based on nonequilibrium compared to those of Example 14.4 based on equilibrium for the same interior subchannel conditions are presented in Table 14.4.

The nonequilibrium assumption yields a slip ratio, S, greater than unity that decreases the void fraction for given quality and density conditions. Hence,

TABLE 14.4
Comparison of Results of Examples 14.4 and 14.7

	Example 14.4 (Equilibrium)	Example 14.7 (Nonequilibrium)
Δp_{acc} (kPa)	9.9	7.4
Δp_{fric} (kPa)	18.0	22.3
Δp_{grav} (kPa)	15.5	16.7
Δp_{total} (kPa)	43.4	46.4

at the subchannel outlet, the slip ratio and void fraction for nonequilibrium conditions are 1.435 and 0.702, respectively, (Example 14.7) compared to 1.0 and 0.772 (Example 14.4) for the imposed exit quality and pressure of 14.6% and 7.14 MPa. (Pressure loss in the channel is neglected.) As a consequence, the component Δp_{acc} is decreased while Δp_{grav} is increased for nonequilibrium compared to equilibrium conditions, as Table 14.4 demonstrates. The friction component is increased for nonequilibrium vs. equilibrium conditions primarily due to the increased slip for the nonequilibrium case.

PROBLEMS

14.1. *Heated channel power limits* (Section 14.5)
How much power can be extracted from a PWR with the geometry and operating conditions of Examples 14.1 and 14.2 and a cosine radial power distribution if:
1. The coolant exit temperature is to remain subcooled.
2. The maximum clad temperature is to remain below saturation conditions.
3. The fuel maximum temperature is to remain below the melting temperature of 2400°C (ignore sintering effects).

Answers:
1. $\dot{Q} = 3228\ \text{MW}_{\text{th}}$
2. $\dot{Q} = 2220\ \text{MW}_{\text{th}}$
   $\dot{Q} = 3415\ \text{MW}_{\text{th}}$

14.2. *Specification of power profile for a given clad temperature* (Section 14.6)
Consider a nuclear fuel rod whose cladding outer radius is a. Heat is transferred from the fuel rod to coolant with constant heat transfer coefficient h. The coolant mass flow rate along the rod is $\dot{m}$. Coolant-specific heat c_p is independent of temperature.
It is desired that the temperature of the outer surface of the fuel rod t (at radius a) be constant, independent of distance Z from the coolant inlet to the end of the fuel rod.
Derive a formula showing how the linear power of the fuel rod q' should vary with Z if the temperature at the outer surface of the fuel rod is to be constant.

Answer:

$$q'(Z) = q'_o \exp[-2\pi haZ/\dot{m}c]$$

14.3. *Pressure drop-flow rate characteristic for a fuel channel* (Section 14.6)
A designer is interested in determining the effect of a 50% channel blockage on the downstream-clad temperatures in a BWR core. The engineering department proposes to assess this effect experimentally by inserting a prototypical BWR channel containing the requisite blockage in the test loop sketched in Figure 14.15 and running the centrifugal pump to deliver prototypic BWR pressure and single-channel flow conditions.

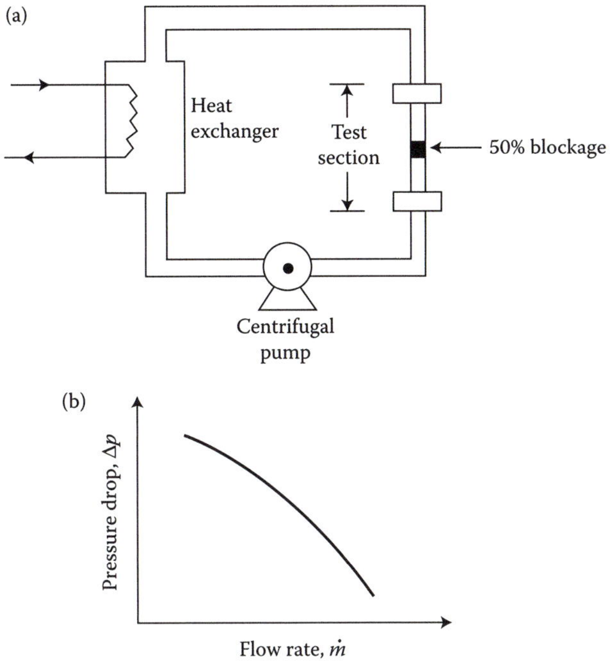

FIGURE 14.15 Test loop (a) and pump characteristic curve (b).

Is the test plan acceptable to you? If not, what changes would you propose and why?

Answer:
It is not possible to specify both flow and pressure conditions to be identical to BWR conditions. A suitable modification to the test loop must be accomplished.

14.4. *Pressure drop in a two-phase flow channel* (Section 14.6)
For the BWR conditions given below, using the HEM model determine the acceleration, frictional, gravitational, and total pressure drop (ignore the spacer pressure drop).
Compare the frictional pressure drop obtained using the HEM model to that obtained from the Thom approach (Eq 11.97).
Geometry: Consider a vertical tube of 17 mm I.D. and 3.8 m length.
Operating Conditions: Steady-state conditions at
Operating pressure = 7550 kPa
Inlet temperature = 275°C
Mass flux = 1700 kg/m^2 s
Heat flux (axially constant) = 670 kW/m^2

Answers:
The pressure drops (HEM model) are as follows:
$\Delta p_{\text{acc}} = 12.5$ kPa
$\Delta p_{\text{fric}} = 16.0$ kPa
$\Delta p_{\text{grav}} = 14.1$ kPa
$\Delta p_{\text{total}} = 42.6$ kPa

Friction pressure drop (Thom): $\Delta p_{fric} = 18.0$ kPa

14.5. *Critical heat flux for PWR channel* (Section 14.6)
Using the conditions of Examples 14.1 and 14.2, determine the minimum critical heat flux ratio (MCHFR) using the W-3 correlation.

Answer:
MCHFR = 2.41

14.6. *Nonuniform linear heat rate of a BWR* (Section 14.6)
For the hot subchannel of a BWR-5, calculate the axial location where bulk boiling starts to occur. Assuming the exit quality is 29.2%, determine the axial profiles of the hot channel enthalpy, $h_m(z)$, and thermal equilibrium quality, $x_e(z)$. Assume thermodynamic equilibrium and the following axial heat generation profile:

$$q' = q'_{ref} \exp\left[-\alpha\left(\frac{z}{L} + \frac{1}{2}\right)\right]\cos\left(\frac{\pi z}{L}\right) \quad -\frac{L}{2} < z < \frac{L}{2}$$

Parameters

$$q'_{max} = 47.24 \text{ kW/m}$$
$$\alpha = 1.96$$
$$T_{in} = 278.3°\text{C}$$
$$p = 7.14 \text{ MPa}$$
$$L = 3.588 \text{ m}$$
$$x_{out} = 0.292$$

Answer:
$z_{OSB} = -1.29$ m

14.7. *Thermal hydraulic analysis of a pressure tube reactor* (Chapters 3, 8, 9, 10, 11, and Section 14.6)
Consider the light water-cooled and moderated pressure tube reactor shown in Figure 14.16. The fuel and coolant in the pressure tube are

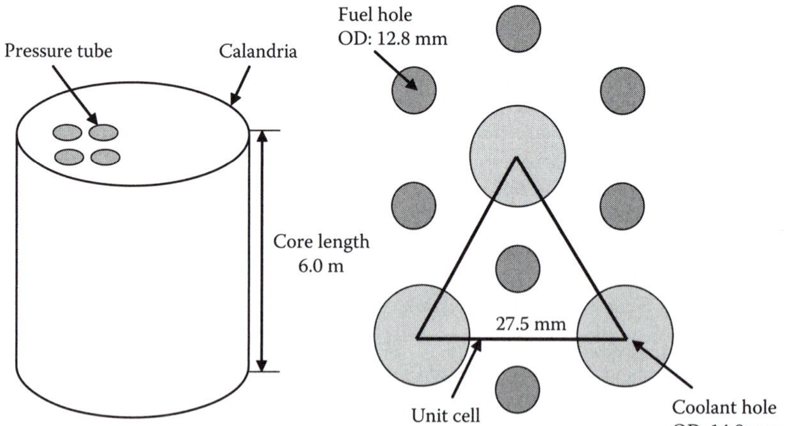

FIGURE 14.16 Calandria with pressure tubes and unit cell in the pressure tube.

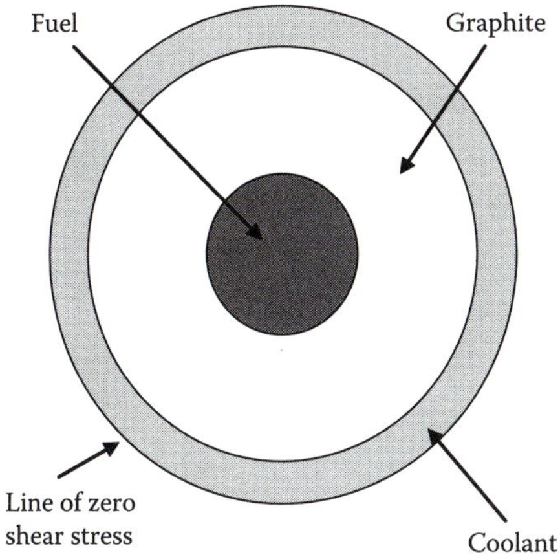

FIGURE 14.17 Equivalent Annuli Model (not to scale).

within a graphite matrix. Each pressure tube consists of a graphite matrix that has 24 fuel holes and 12 coolant holes. Part of the graphite matrix and a unit cell are also shown in Figure 14.16. An equivalent annuli model for thermal analysis is shown in Figure 14.17. Consider the fuel (although composed of fuel particles in each fuel hole) as operating at a uniform volumetric heat generation rate in the r, θ plane. Operating conditions and property data are in Tables 14.5 and 14.6.

Assumptions

- HEM (Homogeneous Equilibrium Model) for two phase flow analysis is valid.
- Cosine axial heat flux (neglect extrapolation).

  Saturated coolant data at 6.89 MPa

  Density: $\rho_f = 742.0$ kg/m3, $\rho_g = 35.94$ kg/m^3

  Enthalpy: $h_f = 1261.6$ kJ/kg, $h_{fg} = 1511.9$ kJ/kg

  Saturated temperature at $p = 6.89$ MPa, $T_{sat} = 284.86$°C

TABLE 14.5
Operating Data for Problem 14.7

	Units	Data		Units	Data
Reactor system			Fuel Hole		
Core thermal power	MW_{th}	2000	Fuel hole diameter	mm	12.8
Number of pressure tubes		740	Mass of UC	kg/hole	2.3
Core radius	m	8.5			
Core length	m	6	Coolant Hole		
			Coolant hole diameter	mm	14.8
Primary system			Coolant flow rate	kg/s	1.4
Pressure	MPa	6.89			
Inlet coolant temperature	°C	245	Unit cell pitch	mm	27.5

TABLE 14.6
Property Data for Problem 14.7

Parameters	Fuel	Graphite	Coolant
Thermal conductivity, k (W/mK)	7	23	0.59
Dynamic viscosity, μ (Pa s)			101×10^{-6}
Specific heat, c_p (J/kgK)			5.0×10^3
Coolant inlet enthalpy, h_{in} (kJ/kg)			1062.3
Single phase density, ρ (kg/m^3)			776.3

Questions

1. What is the radial peaking factor assuming an axial cosine and radial Bessel function flux shape (neglect extrapolation length)?
   For the following questions, assume that the total power of the fuel hole is 260 (kW/fuel hole) in the hot channel.
2. What is the coolant exit temperature in the hot channel?
3. What is the coolant exit enthalpy in the hot channel?
4. What is the exit void fraction in the hot channel?
5. What is the nonboiling length in the hot channel?
6. What is the fuel centerline temperature at the position where bulk boiling starts in the hot channel?
7. What is the pressure drop in the hot channel? To simplify your calculation, assume a uniform heat flux value that provides total power equivalent to the cosine shape heat flux distribution.

Answers:

1. $P_R = 2.32$
2. $T_{m,exit} = 319.3°C$
3. $h_{exit} = 1433.7$ kJ/kg
4. $\alpha_{exit} = 0.726$
5. $L_{NB} = 3.14$ m
6. $T_{CL}(z_{OSB}) = 1213.1°C$
7. $\Delta P_{tot} = 527.5$ kPa

14.8. *Maximum clad temperature for a SFBR* (Chapters 3, 10, and Section 14.6) Derive the relationship between the physical and extrapolated axial lengths for a SFBR core such that the maximum clad temperature occurs at the core outlet during steady-state operating conditions. This relationship describes the truncation of the assumed sinusoidal thermal flux variation along the core axis. Ignore the reactor blankets and assume the following remain constant along the axial length of the core:

i. Heated perimeter of channels
ii. Mass flux of coolant
iii. Coolant specific heat
iv. Film heat transfer coefficient

Answer:

$$L \leq \frac{2L_e}{\pi} \tan^{-1}\left(\frac{2L_e h_{film} P_h}{\pi \dot{m} c_p} \right)$$

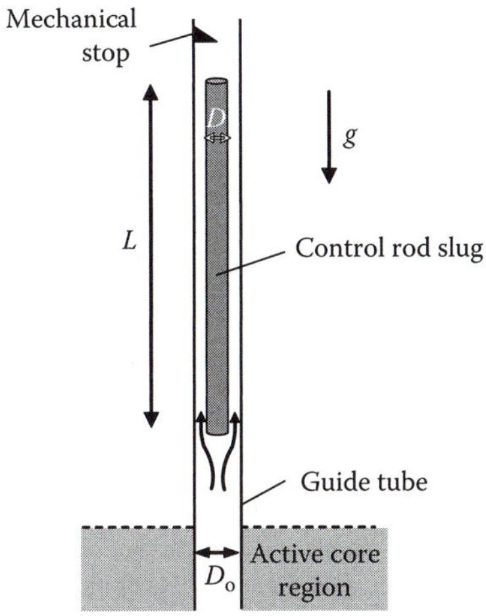

FIGURE 14.18 The flow-levitated control rod system. (There is a mechanical stop to ensure that the control rod is not ejected by the flow.)

14.9. *Flow-levitated control rod in a PWR* (Chapters 8, 9, 10 and Section 14.5)
An engineer has recently proposed a novel control rod design based on flow levitation, to be used in PWRs. The control rod consists of a slug of absorbing material (boron carbide, B_4C) that can slide within a guide tube (see Figure 14.18). During normal operating conditions, the control rod is held out of the core by the coolant flow. If the coolant flow suddenly decreases, the control rod falls back into the core. The idea is to create a scram system that would automatically (passively) shut down the reactor in case of loss of flow and loss of coolant accidents. The control rod slug diameter is $D = 2$ cm and its length is $L = 3.7$ m. The guide tube diameter is $D_o = 2.6$ cm. The properties of B_4C and coolant are reported in Table 14.7.

TABLE 14.7
Property Data for Problem 14.9

Properties of Liquid Water[a]	
Density	730 kg/m^3
Viscosity	9×10^{-5} Pa s
Specific heat	5500 J/kgK
Thermal conductivity	0.56 W/mK
Properties of Boron Carbide	
Density	2500 kg/m^3
Specific heat	950 J/kgK
Thermal conductivity	35 W/mK

[a] Assumed constant over the temperature and pressure range of interest.

i. Calculate the minimum flow rate through the guide tube required for control rod levitation. Assume turbulent flow in the annulus between the guide tube and the control rod slug. Use a friction factor equal to 0.017, and form loss coefficients equal to 0.25 and 1.0 for the flow contraction and expansion at the bottom and top of the control rod slug, respectively. Ignore the presence of the stop. Assume steady state.

ii. Now consider the operation of this control rod at reduced flow conditions, when it is fully inserted in the core. In this situation, the control rod is exposed to a high neutron flux, and so significant heat is generated due to (n, α) reactions on the B_4C. Because the absorption cross section of B_4C is so high, the neutrons only penetrate a few microns beneath the control rod surface. In fact, for all practical purposes, we can assume that the volumetric heat generation rate within the control rod is zero, and describe the situation simply by means of a surface heat flux. Calculate the maximum temperature in the control rod, assuming that the heat flux at its surface has a cosine axial shape with an average of 80 kW/m^2. At the conditions of interest, here the coolant flow rate, inlet temperature, and pressure are 0.3 kg/s, 284°C, and 15.5 MPa, respectively.

Answers:

i. $\dot{m} = 0.54$ kg/s

ii. $T_{o,max} = 298.3$°C

14.10. *Thermal behavior of a plate fuel element following a loss of coolant* (Chapters 8, 10, and Section 14.6)

A reactor fuel assembly of the MIT research reactor is made up of plate elements as shown in Figure 14.19 (only 5 of 13 elements are shown). Suppose the flow channel between plates 2 and 3 is blocked at the inlet (Figure 14.20). What is the axial location of the maximum fuel temperature in plate 3? Solve this in the following steps: (Steps A and B can be solved independently of each other).

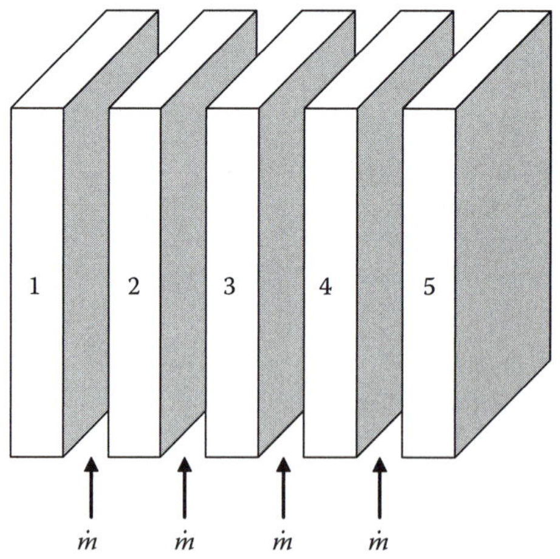

FIGURE 14.19 MIT research reactor fuel plate elements. (Only 5 of 13 elements are shown.)

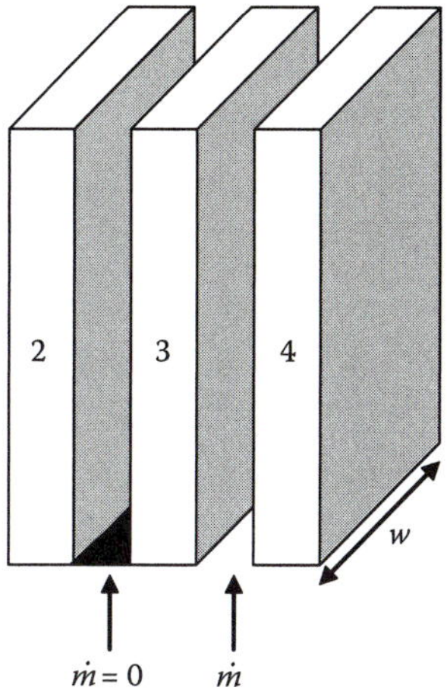

FIGURE 14.20 Fuel plate elements with flow blockage between 2 and 3.

A. Find $T_w(z)$ where T_w is the element 3 surface temperature on the cooled side (RHS).
B. Find $T_{\text{Fuel}_{\text{LHS}}}(z) - T_w(z)$ where $T_{\text{Fuel}_{\text{LHS}}}(z)$ is the element 3 surface temperature on the insulated side (LHS).
C. Find the axial location of the maximum $T_{\text{Fuel}_{\text{LHS}}}(z)$.

In solving this problem, you can make the following assumptions:

- All heat transfer through the fuel element is radial, that is, there is no axial heat transfer within the fuel element.

- All of the energy generated in plate 3 flows radially to the right to the coolant channel between elements 3 and 4, that is, the left side of element 3 has an insulated boundary (see Figure 14.21).
- For simplicity, we neglect the clad and take the elements as only composed of fuel a metallic fuel.
- Assume the flow is fully developed.

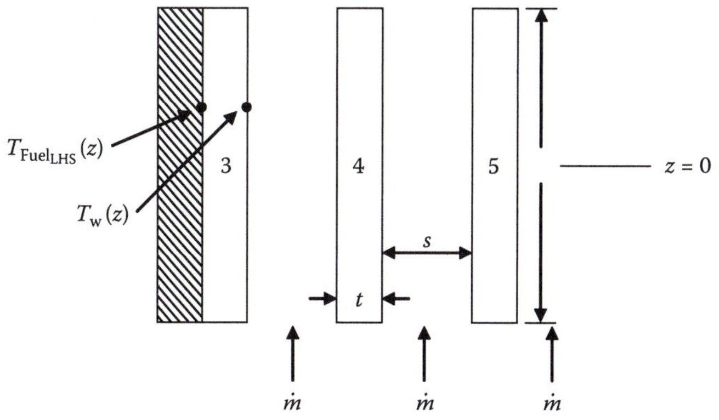

FIGURE 14.21 Geometry of fuel plate elements.

Operating Conditions:
$p = 55$ psi (0.379 MPa)
$T_{\text{inlet}} = 123.8$ F (51°C)
$\dot{m} = 0.32 \text{kg/s}$

$$q''' = 8.54 \times 10^5 \cos\left(\frac{\pi z}{L}\right) \text{kW/m}^3$$

Geometry:
$L = 23$ in. (58.42 cm)
$s = 0.098$ in. (0.249 cm)
$t = 0.030$ in. (0.0762 cm)
$w = 2.082$ in. (5.288 cm)
Properties:
Water: $c_p = 4.181$ kJ/kg K

$\rho = 987.2$ kg/m^3
$k = 0.644$ W/mK
$\mu = 544 \times 10^{-6}$ kg/m s
Pr = 3.597
Fuel:
$k = 41.2$ W/mK

Answers:

A. $$T_w(z) = \frac{3q_o'''wtL}{2\pi \dot{m} c_p}\left[\sin\left(\frac{\pi z}{L}\right) + 1\right]$$

B. $$T_{\text{fuel}_{\text{LHS}}}(z) - T_w(z) = \frac{q'''(z)t^2}{2k_{\text{fuel}}}$$

C. $z = 14.8$ cm

REFERENCES

1. Bergles, A. E. and Rohsenow, W. M., The determination of forced convection surface boiling heat transfer. *J. Heat Transfer,* 86:363, 1964.
2. Biasi, L. et al., Studies on burnout. Part 3: *Energy Nucl.*, 14:530, 1967.
3. Chen, J. C., *A Correlation for Boiling Heat Transfer to Saturated Fluids in Convective Flow*. ASME paper 63-HT-34, 1963 (first appearance). Chen, J. C. A correlation for boiling heat transfer in convection flow. *ISEC Process Des. Dev.,* 5:322, 1966 (subsequent archival publication).
4. Davis, E. J. and Anderson, G. H., The incipience of nucleate boiling in forced convection flow. *AIChE J.,* 12:774, 1966.
5. Friedel, L., Improved friction pressure drop correlations for horizontal and vertical two-phase flow. *European Two-phase Flow Group Meeting, Ispra, Italy*, Paper E2, 1979.
6. Friedel, L., Improved friction pressure drop correlations for horizontal and vertical two-phase pipe flow. *3R Int. J. Piping Eng. Pract.*, 18(7):485–491, July 1979.
7. Gaspari, G. P. et al., A rod-centered subchannel analysis with turbulent (enthalpy) mixing for critical heat flux prediction in rod clusters cooled by boiling water. In *Proceedings 5th International Heat Transfer Conference*, Tokyo, Japan, 3–7 September 1974, CONF-740925, 1975.
8. Groeneveld, D. C., Leung, L. K. H., Vasic, A. Z., Guo, Y. J., and Cheng, S. C., A look-up table for fully developed film-boiling heat transfer. *Nucl. Eng. Des.*, 225:83–97, 2003a.

9. Groeneveld, D. C., Leung, L. K. H., Guo, Y., Vasic, A., El Nakla, M., Peng, S. W., Yang, J., and Cheng, S. C., Look-up tables for predicting CHF and film boiling heat transfer: Past, present, and future. *10th International Topical Meeting on Nuclear Reactor Thermal Hydraulics (NURETH-10),* Seoul, Korea, October 5–9, 2003b.
10. Hench, J. E. and Gillis, J. C., *Correlation of Critical Heat Flux Data for Application to Boiling Water Reactor Conditions*, EPRI NP-1898, June 1981.
11. Levy, S., Forced convection subcooled boiling prediction of vapor volumetric fraction. *Int. J. Heat Mass Transfer,* 10:247, 1967.
12. Premoli, A., Di Francesco, D., and Prina, A., A dimensionless correlation for determining the density of two-phase mixtures. *La Termotecnica*, 25(1):17–26, 1971. (First appeared as Premoli, A., Di Francesco, D., and Prina, A., An empirical correlation for evaluating two-phase mixture density under adiabatic conditions. *European Two-Phase Flow Group Meeting*, Milan, 1970.)
13. Saha, P. and Zuber, N., Point of net vapor generation and vapor void fraction in subcooled boiling. Paper B4.7. In *Proceedings of the 5th International Heat Transfer Conference,* Tokyo, 1974.
14. Thom, J. R. S., Walker, W. M., Fallon, T. A., and Reising, G. F. S., Boiling in subcooled water during flow up heated tubes or annuli, *Proc. Instn. Mech. Engrs.*, 180:226, 1965.

Appendix A: Selected Nomenclature

Coordinates

$\vec{i}, \vec{j}, \vec{k}$	Unit vectors along coordinate axes
$\vec{n}$	Unit normal vector directed outward from control surface
$\vec{r}$	Unit position vector
x, y, z	Cartesian
r, θ, z	Cylindrical
r, θ, ϕ	Spherical
u, v, w	Velocity in Cartesian coordinates
$\upsilon_x, \upsilon_y, \upsilon_z$	
$\upsilon_r, \upsilon_\theta, \upsilon_z$	Velocity in cylindrical coordinates

Extensive and Specific Properties

		Dimensions in Two Unit Systems[a]	
Symbol	**Description**	**ML θ T**	**FE Plus ML θ T**
C, c	General property		Varies
E, e	Total energy (internal + kinetic + potential)		E, EM^{-1}
H, h	Enthalpy		E, EM^{-1}
H°, h°	Stagnation enthalpy		E, EM^{-1}
M, m	Total mass	*M*	
S, s	Entropy		E, EM^{-1}
U, u	Internal energy		E, EM^{-1}
U°, u°	Stagnation internal energy		E, EM^{-1}
V, v	Volume	L^3, L^3M^{-1}	

[a] M: mass; θ: temperature; F: force; E: energy; L: length; T: time; –: dimensionless

Dimensionless Numbers

Symbol	Description	Location Where Symbol First Appears[b]
Br	Brinkmann number	E10.7
Ec	Eckert number	E10.8
Fr	Froude number	E9.36

continued

Dimensionless Numbers (continued)

Symbol	Description	Location Where Symbol First Appears[b]
Gr	Grashof number	E10.12b
Ku	Kutateladze number	E11.19
L	Lorentz number	E10.10
M	Mach number	E10.100
M_{ij}	Mixing Stanton number	Vol. II
Nu	Nusselt number	E10.12
Pe	Peclet number	E10.125
Pr	Prandtl number	E10.6
Re	Reynolds number	E9.63
St	Stanton number	E10.75
$\{y_k^+\}$	Wallis number	E11.23
We	Weber number	E11.17

[a] Dimension: M: mass; θ: temperature; F: force; E: energy; L: length; T: time; –: dimensionless

[b] Appearance in Text: A: Appendix; E: Equation; Ex: Example; F: Figure; T: Table; for example, E10-6: Equation 10.6

General Notation

Symbol	Description	Dimension in Two Unit Systems[a] ML θ T	FE Plus ML θ T	Location Where Symbol First Appears[b]
	General English Notation			
A, A_f	General area or flow area	L^2		F2.1
A	Atomic mass number	–		E3.1
	Availability function		E	E4.27
	Constant	–		E8.16
A_s	Projected frontal area of spacer	L^2		E9.111
A_v	Avogadro's number	M^{-1}		E3.7b
	Unobstructed flow area in channels	L^2		E9.111
A_{fb}	Sum of the area of the total fluid–solid interface and the area of the fluid	L^2		Vol. II
A_{fs}	Area of the total fluid–solid interface within the volume	L^2		Vol. II
a	Atomic fraction	–		E3.16a
	Half thickness of plate fuel	L		F8.10
B	Buildup factor	–		E3.52
B_d	Discharge burnup		EM^{-1}	E2.24
B_1	Single batch loaded core burnup		EM^{-1}	E2.25
B_{uc}	Operating cycle burnup		EM^{-1}	E2.25

General Notation (continued)

Symbol	Description	Dimension in Two Unit Systems[a]: ML θ T	FE Plus ML θ T	Location Where Symbol First Appears[b]
C	Tracer concentration	–		Vol. II
	Constant for friction factor			E9.105a
C_D	Nozzle coefficient	–		E9.20b
C_d	Drag coefficient	–		E11.15
C_o	Concentration factor in drift flux model for nonuniform void distribution	–		E11.36
C_s	Spacer drag coefficient	–		E9.110
C_v	Modified drag coefficient	–		E9.111
C_1	Temperature-cladding half thickness parameter	θ/L		E8.53a
C_2	Temperature parameter	θ		E8.53b
c	Isentropic speed of sound in the fluid	LT^{-1}		E11.126
	Constant	–		E8.27
c_p	Specific heat, constant pressure		$EM^{-1}\theta^{-1}$	E4.115
c_R	Hydraulic resistance coefficient	–		Vol. II
c_v	Specific heat, constant volume		$EM^{-1}\theta^{-1}$	E8.1
D	Diameter, rod diameter	L		T1.3
	Tube diameter	L		F9.7
	Nozzle diameter	L		F9.6
D_e	Equivalent diameter	L		E2.44
D_H	Heated diameter	L		E13.55
D_s	Wire spacer diameter	L		F1.15
D_V	Volumetric hydraulic diameter	L		E9.124
D_{ft}	Wall-to-wall distance of hexagonal bundle	L		F1.15
d	Diameter of liquid drop	L		E11.17
	Day	T		E2.24
E	Neutron kinetic energy		E	E3.1
	Constant	–		E8.22a
E_c	Energy level below which energy loss by neutrons is negligible		E	E3.47
$\bar{E}$	Effective energy		E	E3.61
F	Function	–		T3.6
	Modification factor of Chen's correlation	–		E13.25a
	Looseness of bundle packing	–		Vol. II
	Hot spot factor	–		Vol. II
	Distance across hexagon flats	L		T1.3
	Ratio of decay power to total reactor power	–		E3.74a

continued

General Notation (continued)

Symbol	Description	Dimension in Two Unit Systems[a]: ML θ T	FE Plus ML θ T	Location Where Symbol First Appears[b]
	Constant	–		E8.22a
FLR	Forced loss rate	–		E2.35
$\vec{F}$; $\vec{f}$	Force, force per unit mass		F, FM^{-1}	E4.18
F_Q	Heat flux hot channel factor	–		Vol. II
F_{ix}	Total drag force in the control volume i in the x direction		F	Vol. II
F_{iz}	Subchannel circumferentially averaged force for vertical flow over the solid surface in the control volume i		F	Vol. II
$F_{V_{DC}}$	Void factor for annular pellet cooled on inner and outer surfaces (dual cooling)	–		E8.87
$F_{V_{IC}}$	Void factor for annular pellet cooled on inner surface only	–		E8.101
$F_{V_{OC}}$	Void factor for annular pellet cooled on outside surface only	–		E8.101
$F_{\Delta h}$	Coolant enthalpy rise hot channel factor	–		Vol. II
$F_{\Delta T}$	Coolant temperature rise hot channel factor	–		Vol. II
f	Moody friction factor	–		E9.39e
	Function	–		E8.16
	Fission gas release fraction	–		E8.25
	Fraction of coolant flowrate	–		F6.32
$f(E)$	Fraction of neutron energy loss in a collision	–		E3.57
f_{IHM}	Mass fraction of initial heavy atoms	–		E2.11
f_{TP}	Two-phase friction factor	–		E11.79a
$f_{c.t.}$	Friction factor in a circular tube	–		E9.107
f_j	Subfactor of x_j	–		Vol. II
$f_{j,y}$	Subfactor relative to parameter x_j affecting the property y	–		Vol. II
f_{tr}	Friction factor for the transverse flow	–		Vol. II
f'	Fanning friction factor in the fuel	–		E9.62
G	Mass flux	$ML^{-2}T^{-1}$		E5.38a
g	Distance from rod surface to array flow boundary	L		F1.12
	Function	–		E8.16
g, s, l	States of substance	–		E3.76
$\vec{g}$ or g	Acceleration due to gravity	LT^{-2}		E4.23
gad	Gadolinium	–		E8.22a
H	Axial lead of wire wrap	L		E9.120a
	Grid strap height	L		E9.115b

General Notation (continued)

Symbol	Description	Dimension in Two Unit Systems[a]: ML θ T	FE Plus ML θ T	Location Where Symbol First Appears[b]
	Head (pump)	L		Vol. II
	Heated length	L		T1.3
	Heat of formation			E3.76
HTR	High temperature regenerator	–		F6.32
h	Wall heat transfer coefficient	L	$EL^{-2}\,\theta^{-1}T^{-1}$	E2.8
	Dimensionless irradiation defect annealing parameter	–		E8.22a
	Dimensionless pump head	–		Vol. II
h_g	Gap conductance		$EL^{-2}\,\theta^{-1}T^{-1}$	E8.136
h_m^+	Flow mixing-cup enthalpy		EM^{-1}	E5.52
I	Geometric inertia of the fluid	L^{-1}		E11.145
	Energy flux		$EL^{-2}T^{-1}$	E3.48
$\dot{I}$	Irreversibility or lost work		ET^{-1}	E4.47
$I(\vec{r})$	Indicator function	–		Vol. II
I_p	Prompt neutron lifeline	T		E3.65
J	Total number of neighboring subchannels	–		Vol. II
J_o	Bessel function	–		E3.35
$\vec{J}$	Generalized surface source or sink for mass, momentum, and energy			E4.50
$\vec{j}$ or j	Volumetric flux (superficial velocity)		LT^{-1}	E5.42
j_{AX}	Flux of species A diffusing through a binary mixture of A and B due to the concentration gradient of A	$ML^{-2}T^{-1}$		Vol. II
K	Total form loss coefficient	–		E9.39d
K_G	Form loss coefficient in transverse direction	–		Vol. II
k	Thermal conductivity	–	$EL^{-1}\,\theta^{-1}T^{-1}$	E4.114
	Neutron multiplication factor	–		E3.65
L	Flashing length	L		F11.22
	Heated or active fuel length	L		E2.4
	Length	L		E9.39e
	Length scale	L		E12.20
	Liters	L^3		T1.8
L_A	Availability factor			E2.32a
L_B	Boiling length	L		E13.59
L_C	Capacity factor	–		T2.4
L_{CB}	Plant unit capability factor	–		E2.29

continued

General Notation (continued)

Symbol	Description	Dimension in Two Unit Systems[a]: ML θ T	FE Plus ML θ T	Location Where Symbol First Appears[b]
L_{NB}	Nonboiling length	L		Vol. II
LTR	Low temperature regenerator	–		F6.32
ℓ	Axial dimension	L		E9.12
	Transverse length	L		Vol. II
ℓ_M, ℓ_H	Mixing length	L		E10.66
M	Atomic density	ML^{-3}		E3.7b
	Mass	M		T2.4
	Molecular mass (per mole)	M		E2.14
MCPR	Minimum critical power ratio	–		T2.3
MDNBR	Minimum departure from nucleate boiling ratio	–		T2.3
$\dot{m}$	Mass flow rate	MT^{-1}		E2.9
N	Atomic density	L^{-3}		E2.14
	Number of subchannels	–		T1.4
N_p	Total number of rods	–		T1.4
N_H, N_w, N_p	Transport coefficient of lumped subchannel	–		Vol. II
N_{ps}	Number of rods along a side	–		T1.5
N_{rings}	Number of rings in a rod bundle	–		T1.5
N_{rows}	Number of rows of rods	–		T1.4
N_γ	Photon density	L^{-3}		E3.48
n	Index number	–		T1.5
	Integer number	–		E2.6
	Number of batches	–		T2.4
$\vec{n}$	Unit vector	L		E2.3
P	Decay power		ET^{-1}	E3.68
	Perimeter	L		E5.64
	Pitch	L		F1.12
	Porosity	–		E8.17
	Power rating		ET^{-1}	E2.30
P_I	Pitch (between flow channels—inverted array)	L		F2.17
P_P	Pitch (between pins—pin array)	L		F2.17
P_R	Power peaking factor	–		Vol. II
P_w	Wetted perimeter	L		E9.64
ΔP	Clearance on a per-pin basis	L		Vol. II
p	Pressure		FL^{-2}	E2.9
p_i	Surface contact pressure		FL^{-2}	E8.141
$\hat{p}$	$\hat{p} = p + \rho g z$		FL^{-2}	Vol. II

General Notation (continued)

Symbol	Description	Dimension in Two Unit Systems[a]: ML θ T	FE Plus ML θ T	Location Where Symbol First Appears[b]
Δp^{+}	$\dfrac{\Delta p}{\rho * gL} - 1$	–		Vol. II
Q	Heat	ML^2T^{-2}	E	E4.19
	Temperature of dimensionless annealing parameter	θ		E8.22a
	Volumetric flow rate	L^3T^{-1}		E2.9
Q^{M}	Specific power		$ET^{-1}M^{-1}$	E2.2
Q'''	Power density		$ET^{-1}L^{3}$	E2.10a
$\dot{Q}$	Core power		ET^{-1}	E2.6
	Heat flow		ET^{-1}	E7.2a
	Heat-generation rate		ET^{-1}	E5.124
q	Rate of heat transfer in a flow path		ET^{-1}	F2.1
$\dot{q}$	Rate of energy generated in a pin		ET^{-1}	E2.5
q'	Linear heat-generation rate		$ET^{-1}L^{-1}$	E2.4
q'_{o}	Peak linear heat-generation rate		$ET^{-1}L^{-1}$	E14.11
q'_{rb}	Linear heat generation of equivalent, dispersed heat source		$ET^{-1}L^{-1}$	Vol. II
$\vec{q}'', q''$	Heat flux, surface heat flux		$ET^{-1}L^{-2}$	E2.3
q''_{cr}	Critical heat flux		$ET^{-1}L^{-2}$	F2.4
q''_{x}	Heat flux at a surface		$ET^{-1}L^{-2}$	E2.3
q'''	Volumetric heat-generation rate		$ET^{-1}L^{-3}$	E2.3
q'''_{rb}	Equivalent dispersed heat source		$ET^{-1}L^{-3}$	Vol. II
R	Proportionality constant for hydraulic resistance	$L^{-4}M^{n}\theta^{n}$		Vol. II
	Radius	L		E2.7
	Universal gas constant (per mole)		$E\theta^{-1}$	T8.9
R_{SP}	Specific gas constant		$E\theta^{-1}M^{-1}$	E11.125
RR	Reaction rate in a unit volume	$T^{-1}L^{-3}$		E3.6
	Recirculation ratio	–		E1.1
$\vec{R}$	Distributed resistance		F	Vol. II
r	Enrichment	–		T2.4
	Radius	L		E2.50
r_{e}	Equilibrium vapor bubble radius for nucleation	L		E12.7
r_{p}	Core-average reload enrichment	–		E2.27
	Pressure ratio	–		E6.97

continued

General Notation (continued)

Symbol	Description	Dimension in Two Unit Systems[a] ML θ T	FE Plus ML θ T	Location Where Symbol First Appears[b]
S	Number of photons	–		E3.54
	Slip ratio	–		E5.48
	Suppression factor of Chen's correlation	–		E13.27
	Surface area	L^2		E2.3
S_T	Pitch	L		F9.33
S_{ij}	Open flow area in the transverse direction	L^2		Vol. II
$\dot{S}_{gen}$	Rate of entropy generation		$E\theta^{-1}$	E4.25b
s_{ij}	Gap within the transverse direction	L		Vol. II
T	Magnitude of as-fabricated clearance or tolerance in an assembly	L		Vol. II
	Temperature	θ		F2.1
T_c	Operating cycle length	T		T2.4
T_o	Reservoir temperature	θ		E4.27
T_{res}	Fuel in-core residence time	T		E2.24
T_s	Surroundings temperature	θ		E4.25a
t	Temperature (K)/1000	θ		E8.27
	Time	T		E3.67
	Tonne (metric ton)	M		T2.4
t_g	Equivalent gap thickness	L		T2.7
	Spacer thickness	L		F1.12
t_s	Time after shutdown	T		E3.65
t^*	Peripheral spacer thickness	L		ATJ.2
t'	Time after occurrence of fission	T		E3.67
U	Internal energy (extensive)		E	E3.7
	Overall heat transfer coefficient		$EL^{-2}\theta^{-1}T^{-1}$	F2.1
u	Internal energy (intensive)		EM^{-1}	E3.7
V	Mean velocity	LT^{-1}		E2.9
	Volume	L^3		E2.3
V_m	Mean velocity	LT^{-1}		E9.51
V_o	Rise velocity of large bubbles	LT^{-1}		E11.3
V_∞	Bubble rise velocity	LT^{-1}		E11.46
V_{vj}, υ_{vj}	Local drift velocity of vapor	LT^{-1}		E11.46; E11.32
υ	Velocity in the lateral direction	LT^{-1}		Vol. II
	Velocity of a point	LT^{-1}		E2.9
$\vec{\upsilon}$	Velocity vector	LT^{-1}		T4.2
$\upsilon_r, \vec{\upsilon}_r$	Relative velocity of the fluid with respect to the control surface	LT^{-1}		E4.15
$\vec{\upsilon}_s$	Velocity of the control volume surface	LT^{-1}		E4.11

General Notation (continued)

Symbol	Description	Dimension in Two Unit Systems[a]: ML θ T	FE Plus ML θ T	Location Where Symbol First Appears[b]
W	Work	ML^2T^{-2}	FL	E4.19
W_u	Useful work	ML^2T^{-2}	FL	E4.29
W_{ij}	Transverse mass flow rate per unit length of channel	$MT^{-1}L^{-1}$		Vol. II
W'^D_{ij},	Transverse mass flow rate per unit length due to turbulent fluctuations	$MT^{-1}L^{-1}$		Vol. II
W'^M_{ij},	Hypothetical transverse mass flow rate per unit length for momentum transfer	$MT^{-1}L^{-1}$		Vol. II
W'^H_{ij}	Hypothetical transverse mass flow rate per unit length for energy transfer	$MT^{-1}L^{-1}$		Vol. II
W^{*M}_{ij}	Hypothetical transverse mass flow rate for molecular and turbulent momentum transfer	$MT^{-1}L^{-1}$		Vol. II
W^{*H}_{ij}	Hypothetical transverse mass flow rate for molecular and turbulent energy transfer	$MT^{-1}L^{-1}$		Vol. II
X	Dimensionless radius	–		Vol. II
	Parameter for evaluation of effective fuel conductivity	–		E8.23b
X^2	Lockart-Martinelli parameter	–		E11.92
x	Flow quality	–		E1.1
x_{cr}	Critical quality at CHF	–		E13.47a
x_{st}	Static quality	–		E5.22
$\Delta x'$	Transverse length	L		Vol. II
Y_i	Preference for downflow in a channel			Vol. II
Z	Atomic number	–		E3.1
	Axial location above core inlet	L		F13.4
Z_D	Bubble departure location	L		F13.4
Z_E	Location of thermal equilibrium	L		F13.4
Z_{ONB}	Onset of nucleate boiling location	L		F13.4
Z_{OSB}	Onset of saturated boiling location	L		F13.4
z	Axial position	L		E2.4
z_c	Position of maximum temperature in cladding	L		E14.22b
z^L_{ij}	Laminar mixing length	L		Vol. II
z^T_{ij}	Turbulent mixing length in COBRA	L		Vol. II

[a] Dimension: M: mass; θ: temperature; F: force; E: energy; L: length; T: time; –: dimensionless

[b] Appearance in Text: A: Appendix; E: Equation; Ex: Example; F: Figure; T: Table; For example, E10.6: Equation 10.6

continued

General Notation (continued)

Symbol	Description	Dimension in Two Unit Systems[a] ML θ T	FE Plus ML θ T	Location Where Symbol First Appears[b]
	General Greek Symblos			
α	Confidence level	–		Vol. II
	Dimensionless angular speed (pump)	–		Vol. II
	Dimensionless parameter	–		P2.3
	General function			E3.43
	Linear thermal expansion coefficient	θ^{-1}		E8.23b
	Local void fraction	–		E5.11a
	Radius ratio	–		E8.76
	Thermal diffusivity	L^2T^{-1}		E10.43
α_k	Phase density function	–		E5.1
β	Constant	–		E8.5
	Delayed neutron fraction	–		E3.65
	Dimensionless torque (pump)	–		Vol. II
	Direction angle	–		Vol. II
	Fuel to fuel plus gap and cladding area ratio	–		E2.46
	Mixing parameter	–		Vol. II
	Parameter to express non-uniform spatial heat generation rate	–		E8.76
	Pressure loss parameter	–		T6.9
	Ratio of flow area	–		T9.1
	Thermal volume expansion coefficient	θ^{-1}		E4.117
	Volumetric fraction of vapor or gas	–		E5.51
Γ	Volumetric vaporization rate	$MT^{-1}L^{-3}$		T5.1
γ	Gamma ray	–		E3.48
	% of energy deposited in the fuel rods	–		E2.11
	Porosity	–		Vol. II
	Specific heat ratio (c_p/c_v)	–		T6.3
δ	Thickness of heat-exchanger tube	L		Vol. II
	Thickness of liquid film in annular flow	L		E11.21a
δR	Extrapolated radius	L		T3.6
δL	Extrapolated length	L		T3.6
δ^*	Dimensionless film thickness	–		E11.21a
δ_c	Cladding thickness	L		F8.10
δ_g	Gap between fuel and cladding	L		E2.11
δ_T	Thickness of temperature boundary layer	L		F10.3
ε	$\ell\, dw/dy$ in COBRA code	L^2T^{-1}		Vol. II
	Ratio of frontal grid area to the bundle flow area away from the grid	–		E9.114b
	Surface emissivity	–		E8.137a

General Notation (continued)

Symbol	Description	Dimension in Two Unit Systems[a] ML θ T	FE Plus ML θ T	Location Where Symbol First Appears[b]
ε_H	Eddy diffusivity of energy	L^2T^{-1}		E10.46
ε_M	Momentum diffusivity	L^2T^{-1}		E9.72
ζ	Thermodynamic efficiency (or effectiveness)	–		E6.31
η	Porosity correction factor	–		E8.22b
	Pump efficiency	–		E2.9
η_s	Isentropic efficiency	–		E6.37
η_{th}	Thermal efficiency	–		E6.38
φ	Reynolds number partition factor	–		E9.119
Λ	Mean neutron generation time	T		E3.65
λ	Decay constant	–		E3.65
	Length along channel until fluid reaches saturation	L		Vol. II
λ_c	Taylor instability wavelength	L		E12.25c
λ_{tr}	Transport mean free path of neutrons	L		T3.6
μ	Attenuation coefficient	L^{-1}		E3.51
	Dynamic viscosity	$ML^{-1}T^{-1}$	FTL^{-2}	E4.84
μ_a	Absorption coefficient	L^{-1}		E3.49
$\mu_o(E)$	Average cosine of the scattering angle for neutrons of energy *E*			E3.41
μ'	Bulk viscosity	$ML^{-1}T^{-1}$	FTL^{-2}	E4.84
$\hat{\mu}$	Estimated mean of distribution	–		Vol. II
ν	Dimensionless volumetric flow rate (pump)	–		Vol. II
	Kinematic viscosity (μ/ρ)	L^2T^{-1}		E9.31a
	Time it takes fluid packet to lose its subcooling	T		Vol. II
	Volume fraction	–		E2.22
π	Torque		FL	Vol. II
θ	Conductivity integral		$E\theta^{-1}L^{-1}$	E8.11
	Film temperature drop	θ		Vol. II
	Position angle	–		E3.24a
	Two-phase multiplier for mixing	–		Vol. II
θ^*	Influence coefficient	–		Vol. II
ρ	Density	ML^{-3}		E2.11
	Step reactivity change	–		E3.65
ρ_m^+	Two-phase momentum density	ML^{-3}		E5.66
Σ	Macroscopic cross section	L^{-1}		E3.11

continued

General Notation (continued)

Symbol	Description	Dimension in Two Unit Systems[a] ML θ T	FE Plus ML θ T	Location Where Symbol First Appears[b]
σ	Microscopic cross section	L^2		E3.7a
	Normal stress component		FL^{-2}	F4.8
	Stefan–Boltzman constant		$ET^{-1}L^{-2}\theta^{-4}$	E8.137a
	Surface tension		FL^{-1}	E11.4
$\hat{\sigma}$	Estimated standard deviation of distribution	–		Vol. II
τ	Shear stress component		FL^{-2}	F4.8
	Time after startup	T		E3.68
	Time constant	T		Vol. II
τ_s	Operational time	T		E3.69a
υ	Velocity	$L\theta^{-1}$		E4.2
Φ	Dissipation function		$EL^{-2}T^{-1}$	E4.107
	Neutron flux	$L^{-2}T^{-1}$		E3.11
ϕ	Azimuthal angle	–		T4.7
	Generalized volumetric source or sink for mass, momentum or energy	Property c units $\theta^{-1}M^{-1}$		E4.50
	Relative humidity	–		E7.19
ϕ_g^2	Two-phase frictional multiplier based on pressure gradient of gas or vapor flow	–		E11.86
ϕ_ℓ^2	Two-phase frictional multiplier based on pressure gradient of liquid flow	–		E11.86
$\phi_{\ell o}^2$	Two-phase friction multiplier based on all-liquid-flow pressure gradient	–		E11.65a
ϕ_{vo}^2	Two-phase frictional multiplier based on all-vapor flow pressure gradient	–		E11.65b
ξ	Development length	L		E10.35
	Effectiveness of regenerator	–		E6.113, E6.116
	Logarithmic energy decrement	–		E3.45
	Mole percent of PuO_2 in the oxide	–		E8.26
χ	Energy per fission deposited in the fuel		E	E3.9
	2.00 – O/M	–		T8.4
ψ	Correction factor	–		E10.102
	Force field per unit mass of fluid		FLM^{-1}	E4.21
	Ratio of eddy diffusivity of heat to momentum	–		E10.71
$\bar{\omega}$	Vorticity	T^{-1}		E9.4b
	Subscripts			
1	Energy group number			E3.12
1, 2, 3	Interior, edge, corner subchannels			T1.4

General Notation (continued)

Symbol	Description	Dimension in Two Unit Systems[a] ML θ T	FE Plus ML θ T	Location Where Symbol First Appears[b]
	Fuel structure density zones			E8.103
0.98	98% theoretical density			Ex8.4
0.95	95% theoretical density			E8.22a
0.88	88% theoretical density			Ex8.4
25	Uranium 235 isotope			E3.18c
28	Uranium 238 isotope			E3.18c
A	Area			Vol. II
	Annulus			Vol. II
	Side A (of fuel plate)			F8.12
AF	Atmospheric flow			E6.68
ACT	Actual			E2.30
a	Absorption			T3.3
	Air			E7.1
acc	Acceleration			T9.1
ann	Annular			E8.89
avg	Average			F2.1
B,b	Boiling			?
B	Buoyancy			Vol. II
	Side B (of fuel plate)			F8.12
b	Bulk			E2.8
C	Cold, cold leg or side			F2.1
	Coolant channel			E2.43a
	Critical condition			E13.17
CD	Condensor			E6.87
CP	Compressor			E6.102
c	Cladding			E8.38
	Conduction			E4.113
	Containment			E7.1
	Contraction			T9.1
	Convection			E13.19
	Coolant			T2.7
	Core			Vol. II
	Critical point (thermodynamic)			E9.31b
ci	Cladding inside			F8.10
cii	Inner surface of annular pellet inner cladding			T8.6
cio	Outer surface of annular pellet inner cladding			T8.6
cl	Center Line			F8.22
$c\ell$	Cladding (outside diameter)			F2.20

continued

General Notation (continued)

Symbol	Description	Dimension in Two Unit Systems[a] ML θ T	FE Plus ML θ T	Location Where Symbol First Appears[b]
co	Cladding outside			E2.7
cr	Critical flow			E11.123
c.m.	Control mass			E4.17c
c.v.	Control volume			E4.30
c.t.	Circular tube			E9.107
DC	Dual (surface) cooled			E8.89
d	Downflow			Vol. II
	Discharge			E2.24
EA	Equivalent annulus			E2.50
EFPP	Effective full power period			T2.6
EO	Outage extension			T2.6
EQUIL	Equilibrium void distribution			Vol. II
e	Electrical			T2.4
	Equilibrium			E5.53
	Expansion			T9.1
	Extrapolated			E3.35
eff	Effective			E2.30
$e\ell$	Elastic			E3.39
ex	External			Vol. II
exit, ex, e	Indicating the position of flowing exit			E1.1
FCR	Fuel to coolant area ratio			E2.45
FE	Iron			E3.63
FGC	Fuel plus gap plus cladding			E2.46
FO	Forced outage			T2.6
f	Fission			E3.1
	Fluid			Vol. II
	Saturated liquid			E5.53
f, F	Fuel			E2.22; E2.43c
fb	Fuel bearing region			E2.42a
ff	Fissile atoms			E2.14
fg	Difference between fluid and gas			T6.3
fi	Flow without spacers			E9.105a
fo	Fuel outside surface			E2.7
fric	Friction			T9.1
ft	Distance across hexagonal flats			E3.33
g	Gap between fuel and cladding			F2.20
	Saturated vapor			E5.53
grav	Gravity			E11.5
H	Hot leg or side			F2.1
H, h	Heated, hot			F2.1
HEM	Homogeneous equilibrium model			E11.78

General Notation (continued)

Symbol	Description	Dimension in Two Unit Systems[a] ML θ T	FE Plus ML θ T	Location Where Symbol First Appears[b]
HM	Heavy metal			T2.4
HX	Heat exchanger			E6.104
h, (heater)	Denoting the selected subchannel control volume			Vol. II
	Index of streams into or out of the control volume			E4.30a
	Inner surface			Vol. II
	Pressurizer heater			F7.10
I	Idle outage			T2.6
I, IC	Inverted channel			F2.18
i	Index number			E3.1
iℓ	Inelastic			E3.58
j	Adjacent subchannel control volume of the subchannel control volume *i*			Vol. II
	Index of isotopes			E3.8b
	Index of properties			Vol. II
K	Number of energy groups			E3.11
k	Index of phase			E5.1
	Index number			E3.11
L	Laminar			E9.105a
LHS	Left-hand side			E8.44b
ℓ	Liquid phase in a two-phase flow			T5.1
ℓo	Liquid only			E11.65a
M	Temperature of a quantity of interest			Vol. II
MO	Maintenance outage			T2.6
m	Log mean (temperature difference)			F2.1
	Mixture			E5.38a
	Temperature difference of a quantity of interest			Vol. II
max	Maximum			E8.34
NB	Nucleate boiling			E13.19
NOM	Nominal			Vol. II
n	Index number			E2.6
	Nuclear core			E7.2d
nf	Nonfissile atoms			E2.14
O	Oxygen			E2.17
OC	Outside (surface) cooled			E8.89
o	Indicating initial value			E8.5
	Outer surface			Vol. II

continued

General Notation (continued)

		Dimension in Two Unit Systems[a]		
Symbol	Description	ML θ T	FE Plus ML θ T	Location Where Symbol First Appears[b]
	Reservoir conditions			E4.27
P	Pump			E6.48
PO	Planned outage			T2.6
p	Pins (total number)			T1.4
	Pins (along a side)			T1.5
	Pore			E8.17
	Primary			E6.49
R	Rated			Vol. II
	Reactor			E6.53
RHS	Right-hand side			E8.43
RO	Refueling outage			T2.6
r	Enrichment			T2.4
	Radial coordinate			E8.51
	Radiation			E4.113
rb	Equivalent or disperse			Vol. II
	Rod bundle			Vol. II
ref	Reference			T1.5
res	Restructured (fuel)			E8.107
SC	Standard channel			E2.43
SEP	Separated flow model			E11.85
SG	Steam generator			E6.54
SP	Single phase			Vol. II
SQ	Square			E2.51b
s	Scattering			E3.39
	Secondary			E6.49
	Shutdown			E3.70
	Sintered			E8.128
	Slug flow			Vol. II
	Solid			E8.17
	Spacer			E9.110
	Surface, interface			E4.11
	Wire spacer			F1.15
s	Isentropic			E6.10b
sat	Saturated			E12.3
st	Static			E5.22
	Structures			E7.2a
T,t	Turbine			T6.5
T	Total			E2.43a
TD	Theoretical density			E2.53b
TP	Two phase			E11.77

General Notation (continued)

Symbol	Description	Dimension in Two Unit Systems[a] ML θ T	FE Plus ML θ T	Location Where Symbol First Appears[b]
th	Thermal			E2.6
tr	Transverse flow			Vol. II
t′	Time after occurrence of fission			E3.67
U	Uranium			E2.17
UO	Unplanned outage			T2.6
u	Upflow			Vol. II
	Useful			E4.28
V	Volume			E2.1
v	Cavity (void)			E8.63
	Vapor or gas phase in a two-phase flow			T5.1
υ_j	Local vapor drift			E11.32
v_o	Vapor only			E11.65b
w	Wall			E2.3
x, y, z	Cartesian spatial coordinates			E8.4
ϕ	Neutron flux			Vol. II
ρ	Fuel density			Vol. II
∞	Free stream			F10.3
	Superscripts			
25	Uranium 235 isotope			E3.16a
D	Hot-spot factor resulting from direct contributors			Vol. II
i	Intrinsic			Vol. II
j	Index of isotopes			E3.6
o	Denoting nominal			Vol. II
	Stagnation			T4.2
S	Hot-spot factor resulting from statistical contributors			Vol. II
t	Turbulent effect			E4.132
$\rightarrow$	Vector			E2.3
—	Spatial average			E1.1
′ ″ ‴	Per unit length, surface area, volume, respectively			E2.7
′	Denoting perturbation			E4.125
TP	Two phase			E11.64
*	Denoting the velocity or enthalpy transported by the diversion crossflow			Vol. II
	Reference			Vol. II
=	Tensor			E4.8
–, ~	Averaging (time)			E4.124; E5.6

continued

General Notation (continued)

Symbol	Description	Dimension in Two Unit Systems[a] ML θ T	FE Plus ML θ T	Location Where Symbol First Appears[b]
	Special Symbols			
Δ	Change in, denoting increment			E2.9
∇	Gradient			E4.2
δ	Change in, denoting increment			T3.6
< >	Volumetric averaging			E2.7
{ }	Area averaging			E3.22
≡	Defined as			
≃, ≈, ~	Approximately equal to			E3.38

Appendix B
Physical and Mathematical Constants

Avogadro's number (A_v)	0.602252×10^{24} molecules/g mol
	2.731769×10^{26} molecules/lbm mol
Barn	10^{-24}cm^2, 1.0765×10^{-27} ft^2
Boltzmann's constant ($k = \overline{R}/A_v$)	1.38054×10^{-16} erg/°K
	8.61747×10^{-5} eV/°K
Curie	3.70×10^{10} dis/s
Electron charge	4.80298×10^{-10} esu, 1.60210×10^{-19} Coulomb
Faraday's constant	9.648×10^{4} coulombs/mol
g_c Conversion factor	1.0 g cm^2/erg s^2, 32.17 lbm ft/lb$_f$ s^2, 4.17×10^{8} lbm ft/lb$_f$ h^2, 0.9648×10^{18} amu cm^2/MeV s^2
Gravitational acceleration (standard)	32.1739 ft/s^2, 980.665 cm/s^2
Joule's equivalent	778.16 ft-lb$_f$/Btu
Mass–energy conversion	1 amu = 931.478 MeV = 1.41492×10^{-13} Btu = 4.1471×10^{-17} kw h
	1 g = 5.60984×10^{26} MeV = 2.49760×10^{7} kw h = 1.04067 Mwd
	1 lbm = 2.54458×10^{32} MeV = 3.86524×10^{16} Btu
Mathematical constants	$e \equiv 2.71828$
	$\pi \equiv 3.14159$
	$ln\ 10 \equiv 2.30259$
Molecular volume	22413.6 cm^3/g mol, 359.0371 ft^3/lbm mol, at 1 atm and 0°C
Neutron energy	0.0252977 eV at 2200 m/s, 1/40 ev at 2187.017 m/s
Planck's constant	6.6256×10^{-27} erg s, 4.13576×10^{-15} eV s
Rest masses	
Electron	5.48597×10^{-4} amu, 9.10909×10^{-28} g, 2.00819×10^{-30} lbm
Neutron	1.0086654 amu, $1.6748228 \times 10^{-24}$ g, 3.692314×10^{-27} lbm
Proton	1.0072766 amu, 1.672499×10^{-24} g, 3.687192×10^{-27} lbm
Stephan–Boltzmann constant	5.67×10^{-12} W/cm^{2}K^4

Universal gas constant ($\bar{R}$)	1545.08 ft-lb_f/lbm mol °R
	1.98545 cal/g mol °K
	1.98545 Btu/lbm mol °R
	8.31434×10^7 erg/g mol °K
Velocity of light	2.997925×10^{10} cm/s, 9.83619×10^8 ft/s

Source: Adopted from El-Wakil, M. M. *Nuclear Heat Transport.* Scranton, PA: International Textbook Company. 1971.

REFERENCE

El-Wakil, M. M. *Nuclear Heat Transport.* Scranton, PA: International Textbook Company. 1971.

Appendix C
Unit Systems

Table	Unit	References
C.1	Length	El-Wakil (1971)
C.2	Area	El-Wakil (1971)
C.3	Volume	El-Wakil (1971)
C.4	Mass	El-Wakil (1971)
C.5	Force	Bird et al. (1960)
C.6	Density	El-Wakil (1971)
C.7	Time	El-Wakil (1971)
C.8	Flow	El-Wakil (1971)
C.9	Pressure, momentum flux	Bird et al. (1960)
C.10	Work, energy, torque	Bird et al. (1960)
C.11	Power	El-Wakil (1971)
C.12	Power density	El-Wakil (1971)
C.13	Heat flux	El-Wakil (1971)
C.14	Viscosity, density × diffusivity, concentration × diffusivity	Bird et al. (1960)
C.15	Thermal conductivity	Bird et al. (1960)
C.16	Heat-transfer coefficient	Bird et al. (1960)
C.17	Momentum, thermal or molecular diffusivity	Bird et al. (1960)
C.18	Surface tension	—

Source: Adopted from Bird, R. B., Stewart, W. E., and Lightfoot, E. N. *Transport Phenomena.* New York, NY: Wiley, 1960; El-Wakil, M. M. *Nuclear Heat Transport*. Scranton, PA: International Textbook Company, 1971.

Relevant SI Units for Conversion Tables

Quantity	Name	Symbol
SI Base Units		
Length	meter	m
Time	second	s
Mass	kilogram	kg
Temperature	kelvin	°K
Amount of matter	mole	mol
Electric current	ampere	A
SI Derived Units		
Force	newton	$N = kg\ m/s^2$
Energy, work, heat	joule	$J = N\ m = kg\ m^2/s^2$

continued

Relevant SI Units for Conversion Tables (continued)

Quantity	Name	Symbol
Power	watt	W = J/s
Frequency	hertz	$Hz = s^{-1}$
Electric charge	coulomb	C = A s
Electric potential	volt	V = J/C
	Allowed Units (To Be Used with SI Units)	
Time	minute	min
	hour	h
	day	d
Plane angle	degree	°
	minute	′
	second	″
Volume	liter	l
Mass	tonne	t = 1000 kg
	atomic mass unit	$U \approx 1.66053 \times 10^{-27}$ kg
Fluid pressure	bar	$bar = 10^5$ Pa
Temperature	degree Celsius	°C
Energy	electron volt	$eV \approx 1.60219 \times 10^{-19}$ J

C.1 UTILIZATION OF UNIT CONVERSION TABLES

The column and row units correspond to each other, as illustrated below. Given a quantity in units of a row, multiply by the table value to obtain the quantity in units of the corresponding column.

Example

How many meters is 10 cm?

Answer: We desire the quantity in units of meters (the column entry) and have been given the quantity in units of centimeters (the row entry), that is, 10 cm. Hence meters = 0.01 cm = 0.01 (10) = 0.1

	Columns			
Rows	**a**	**b**	**c**	**d**
	Centimeters	**Meters**		
a = centimeters		0.01		
b = meters				
c				
d				

TABLE C.1
Length

	Centimeters (cm)	Meters[a] (m)	Inches (in.)	Feet[b] (ft)	Miles	Microns (μ)	Angstroms (Å)
cm	1	0.01	0.3937	0.03281	6.214×10^{-6}	10^{4}	10^{8}
m	100	1	39.37	3.281	6.214×10^{-4}	10^{6}	10^{10}
in.	2.540	0.0254	1	0.08333	1.578×10^{-5}	2.54×10^{4}	2.54×10^{8}
ft	30.48	0.3048	12	1	1.894×10^{-4}	0.3048×10^{6}	0.3048×10^{10}
Miles	1.6093×10^{5}	1.6093×10^{3}	6.336×10^{4}	5.280×10^{3}	1	1.6093×10^{9}	1.6093×10^{13}
Microns	10^{-4}	10^{-6}	3.937×10^{-5}	3.281×10^{-6}	6.2139×10^{-10}	1	10^{4}
Ångstroms	10^{-8}	10^{-10}	3.937×10^{-9}	3.281×10^{-10}	6.2139×10^{-14}	10^{-4}	1

[a] SI units.
[b] English units.

TABLE C.2
Area

	cm^2	m^{2a}	in.^{2}	ft^{2b}	$Mile^2$	Acre	Barn
cm^2	1	10^{-4}	0.155	1.0764×10^{-3}	3.861×10^{-11}	2.4711×10^{-8}	10^{24}
m^2	10^4	1	1.550×10^3	10.764	3.861×10^{-7}	2.4711×10^{-4}	10^{28}
in.^{2}	6.4516	6.4516×10^{-4}	1	6.944×10^{-3}	2.491×10^{-10}	1.5944×10^{-7}	6.4516×10^{24}
ft^2	929	0.0929	144	1	3.587×10^{-8}	2.2957×10^{-5}	9.29×10^{26}
$Mile^2$	2.59×10^{10}	2.59×10^6	4.0144×10^{11}	2.7878×10^7	1	640	2.59×10^{34}
Acre	4.0469×10^7	4.0469×10^3	6.2726×10^6	4.356×10^4	1.5625×10^{-3}	1	4.0469×10^{31}
Barn	10^{-24}	10^{-28}	1.55×10^{-25}	1.0764×10^{-27}	3.861×10^{-35}	2.4711×10^{-32}	1

[a] SI units.
[b] English units.

TABLE C.3
Volume

	cm³	Liters	m³[a]	in.³	ft³[b]	Cubic Yards	U.S. (liq.) Gallons	Imperial Gallons
cm^3	1	10^{-3}	10^{-6}	0.06102	3.532×10^{-5}	1.308×10^{-6}	2.642×10^{-4}	2.20×10^{-4}
Liters	10^3	1	10^{-3}	61.02	0.03532	1.308×10^{-3}	0.2642	0.220
m^3	10^6	10^3	1	6.102×10^4	35.31	1.308	264.2	220.0
in.^{3}	16.39	0.01639	1.639×10^{-5}	1	5.787×10^{-4}	2.143×10^{-5}	4.329×10^{-3}	3.605×10
ft^3	2.832×10^4	28.32	0.02832	1728	1	0.03704	7.481	6.229
Cubic yards	7.646×10^5	764.6	0.7646	4.666×10^4	27.0	1	202.0	168.2
U.S. gallons	3.785×10^3	3.785	3.785×10^{-3}	231.0	0.1337	4.951×10^{-3}	1	0.8327
Imperial gallons	4.546×10^3	4.546	4.546×10^{-3}	277.4	0.1605	5.946×10^{-3}	1.201	1

[a] SI units.
[b] English units.

TABLE C.4
Mass

	Grams (g)	Kilograms (kg)[a]	Pounds (lbm)[b]	Tons (short)	Tons (long)	Tons (metric)	Atomic Mass Units (amu)
g	1	0.001	2.2046×10^{-3}	11.102×10^{-6}	9.842×10^{-7}	10^{-6}	6.0225×10^{23}
kg	1×10^{3}	1	2.2046	11.102×10^{-3}	9.842×10^{-4}	10^{-3}	6.0225×10^{26}
lbm	453.6	0.4536	1	5.0×10^{-4}	4.464×10^{-4}	4.536×10^{-4}	2.7318×10^{26}
Tons (short)	9.072×10^{5}	907.2	2.0×10^{3}	1	0.8929	0.9072	5.4636×10^{29}
Tons (long)	1.016×10^{6}	1016	2.240×10^{3}	1.12	1	1.016	6.1192×10^{29}
Tons (metric)	10^{6}	1000	2.2047×10^{3}	1.1023	0.9843	1	6.0225×10^{29}
amu	1.6604×10^{-24}	1.6604×10^{-27}	3.6606×10^{-27}	1.8303×10^{-30}	1.6343×10^{-30}	1.6604×10^{-30}	1

[a] SI units.
[b] English units.

TABLE C.5
Force

	g cm s^{-2} (Dynes)	kg m s^{2}[a] (Newtons)	lbm ft s^{-2} (Poundals)	lbf[b]
Dynes	1	10^{-5}	7.2330×10^{-5}	2.2481×10^{-6}
Newtons	10^{5}	1	7.2330	2.2481×10^{-1}
Poundals	1.3826×10^{4}	1.3826×10^{-1}	1	3.1081×10^{-2}
lbf	4.4482×10^{5}	4.4482	32.1740	1

[a] SI units.
[b] English units.

TABLE C.6
Density

	g/cm^3	kg/m^3[a]	lbm/in.^{3}	lbm/ft^3[b]	lbm/U.S. gals	lbm/Imp gals
g/cm^3	1	10^{3}	0.03613	62.43	8.345	10.02
kg/m^3	10^{-3}	1	3.613×10^{-5}	0.06243	8.345×10^{-3}	0.01002
lbm/in.^{3}	27.68	2.768×10^{4}	1	1.728×10^{3}	231	277.4
lbm/ft^3	0.01602	16.02	5.787×10^{-4}	1	0.1337	0.1605
lbm/U.S. gals	0.1198	119.8	4.329×10^{-3}	7.481	1	1.201
lbm/Imp gals	0.09978	99.78	4.605×10^{-3}	6.229	0.8327	1

[a] SI units.
[b] English units.

TABLE C.7
Time

	Microseconds (μs)	Seconds (s)	Minutes (min)	Hours (h)	Days (d)	Years (yr)
μs	1	10^{-6}	1.667×10^{-8}	2.778×10^{-10}	1.157×10^{-11}	3.169×10^{-14}
s	10^{6}	1	1.667×10^{-2}	2.778×10^{-4}	1.157×10^{-5}	3.169×10^{-8}
min	6×10^{7}	60	1	1.667×10^{-2}	6.944×10^{-4}	1.901×10^{-6}
h	3.6×10^{9}	3.6×10^{3}	60	1	0.04167	1.141×10^{-4}
d	8.64×10^{10}	8.64×10^{4}	1440	24	1	2.737×10^{-3}
yr	3.1557×10^{13}	3.1557×10^{7}	5.259×10^{5}	8.766×10^{3}	365.24	1

TABLE C.8
Flow

	cm^3/s	ft^3/min	U.S. gal/min	Imperial gal/min
cm^3/s	1	0.002119	0.01585	0.01320
ft^3/min	472.0	1	7.481	6.229
U.S. gal/min	63.09	0.1337	1	0.8327
Imperial gal/min	75.77	0.1605	1.201	1

TABLE C.9
Pressure, Momentum Flux

	$g\ cm^{-1}\ s^{-2}$ (Dyne cm^{-2})	Pascal $kg\ m^{-1}\ s^{-2}$[a] (Newtons m^{-2})	$lbm\ ft^{-1}\ s^{-2}$ (Poundals ft^{-2})	$lbf\ ft^{-3}$	lbf $in.^{-3}$[b] (psia)	Atmospheres (atm)	mm Hg	in. Hg
Dyne cm^{-2}	1	10^{-1}	6.7197×10^{-2}	2.0886×10^{-3}	1.4504×10^{-5}	9.8692×10^{-7}	7.5006×10^{-4}	2.9530×10^{-5}
Newtons m^{-2}	10	1	6.7197×10^{-1}	2.0886×10^{-2}	1.4504×10^{-4}	9.8692×10^{-6}	7.5006×10^{-3}	2.9530×10^{-4}
Poundals ft^{-2}	1.4882×10^{1}	1.4882	1	3.1081×10^{-2}	2.1584×10^{-4}	1.4687×10^{-5}	1.1162×10^{-2}	4.3945×10^{-4}
lbf ft^{-2}	4.7880×10^{2}	4.7880×10^{1}	32.1740	1	6.9444×10^{-3}	4.7254×10^{-4}	3.5913×10^{-1}	1.4139×10^{-2}
psia	6.8947×10^{4}	6.8947×10^{3}	4.6330×10^{3}	144	1	6.8046×10^{-2}	5.1715×10^{1}	2.0360
atm	1.0133×10^{6}	1.0133×10^{5}	6.8087×10^{4}	2.1162×10^{3}	14.696	1	760	29.921
mm Hg	1.3332×10^{3}	1.3332×10^{2}	8.9588×10^{1}	2.7845	1.9337×10^{-2}	1.3158×10^{-3}	1	3.9370×10^{-2}
in. Hg	3.3864×10^{4}	3.3864×10^{3}	2.2756×10^{3}	7.0727×10^{1}	4.9116×10^{-1}	3.3421×10^{-2}	25.400	1

[a] SI units.
[b] English units.

TABLE C.10
Work, Energy, Torque

	$g\ cm^2$ (Ergs)	$kg\ m^2\ s^{-2}$[a] (Absolute Joules)	$lb_m\ ft^2\ s^{-2}$ (ft-poundals)	ft lb_f	cal	Btu[b]	hp-h	kw-h
$g\ cm^2\ s^{-2}$	1	10^{-7}	2.3730×10^{-6}	7.3756×10^{-8}	2.3901×10^{-8}	9.4783×10^{-11}	3.7251×10^{-14}	2.7778×10^{-14}
$kg\ m^2\ s^{-2}$	10^7	1	2.3730×10^{1}	7.3756×10^{-1}	2.3901×10^{-1}	9.4783×10^{-4}	3.7251×10^{-7}	2.7778×10^{-7}
$lbm\ ft^2\ s^{-2}$	4.2140×10^{5}	4.2140×10^{-2}	1	3.1081×10^{-2}	1.0072×10^{-2}	3.9942×10^{-5}	1.5698×10^{-8}	1.1706×10^{-8}
ft lb_f	1.3558×10^{7}	1.3558	32.1740	1	3.2405×10^{-1}	1.2851×10^{-3}	5.0505×10^{-7}	3.7662×10^{-7}
Thermochemical calories[c]	4.1840×10^{7}	4.1840	9.9287×10^{1}	3.0860	1	3.9657×10^{-3}	1.5586×10^{-6}	1.1622×10^{-6}
British thermal units	1.0550×10^{10}	1.0550×10^{3}	2.5036×10^{4}	778.16	2.5216×10^{2}	1	3.9301×10^{-4}	2.9307×10^{-4}
Horsepower hours	2.6845×10^{13}	2.6845×10^{6}	6.3705×10^{7}	1.9800×10^{6}	6.4162×10^{5}	2.5445×10^{3}	1	7.4570×10^{-3}
Absolute kilowatt-hours	3.6000×10^{13}	3.6000×10^{6}	8.5429×10^{7}	2.6552×10^{6}	8.6042×10^{5}	3.4122×10^{3}	1.3410	1

[a] SI units.
[b] English units.
[c] This unit, abbreviated cal. is used in chemical thermodynamic tables. To convert quantities expressed in International Steam Table calories (abbreviated I.T. cal) to this unit, multiply by 1.000654.

TABLE C.11
Power

	Ergs/s	Joule/s watt[a]	kw	Btu/h[b]	hp	eV/s
Ergs/s	1	10^{-7}	10^{-10}	3.412×10^{-7}	1.341×10^{-10}	6.2421×10^{11}
Joule/s	10^{7}	1	10^{-3}	3.412	1.341×10^{-3}	6.2421×10^{18}
Kw	10^{10}	10^{3}	1	3412	1.341	6.2421×10^{21}
Btu/h	2.931×10^{6}	0.2931	2.931×10^{-4}	1	3.93×10^{-4}	1.8294×10^{18}
Hp	7.457×10^{9}	745.7	0.7457	2.545×10^{3}	1	4.6548×10^{21}
eV/s	1.6021×10^{-12}	1.6021×10^{-19}	1.6021×10^{-22}	5.4664×10^{-19}	2.1483×10^{-22}	1

[a] SI units.
[b] English units.

TABLE C.12
Power Density

	Watt/cm^3, kw/lit[a]	cal/s cm^3	Btu/h in.^{3}	Btu/h ft^3[b]	MeV/s cm^3
Watt/cm^3, kw/lit	1	0.2388	55.91	9.662×10^4	6.2420×10^{12}
cal/s cm^3	4.187	1	234.1	4.045×10^5	2.613×10^{13}
Btu/h in^3	0.01788	4.272×10^{-3}	1	1728	1.1164×10^{11}
Btu/h ft^3	1.035×10^{-5}	2.472×10^{-6}	5.787×10^{-4}	1	6.4610×10^7
MeV/s cm^3	1.602×10^{-13}	3.826×10^{-14}	8.9568×10^{-12}	1.5477×10^{-8}	1

[a] SI units.
[b] English units.

TABLE C.13
Heat Flux

	Watt/cm^2[a]	cal/s cm^2	Btu/h ft^2[b]	MeV/s cm^2
Watt/cm^2	1	0.2388	3170.2	6.2420×10^{12}
cal/s cm^2	4.187	1	1.3272×10^4	2.6134×10^{13}
Btu/h ft^2	3.155×10^{-4}	7.535×10^{-5}	1	1.9691×10^9
MeV/s cm^2	1.602×10^{-13}	3.826×10^{-14}	5.0785×10^{-10}	1

[a] SI units.
[b] English units.

TABLE C.14
Viscosity Density × Diffüsivity, Concentration × Diffusivity

	$g\ cm^{-1}\ s^{-1}$ (poises)	$kg\ m^{-1}\ s^{-1}$[a]	$lbm\ ft^{-1}\ s^{-1}$[b]	$lbf\ s\ ft^{-2}$	Centipoises	$lbm\ ft^{-1}\ h^{-1}$
$g\ cm^{-1}\ s^{-1}$	1	10^{-1}	6.7197×10^{-2}	2.0886×10^{-3}	10^{-2}	2.4191×10^{2}
$kg\ m^{-1}\ s^{-1}$	10	1	6.7197×10^{-1}	2.0886×10^{-2}	10^{3}	2.4191×10^{3}
$lbm\ ft^{-1}\ s^{-1}$	1.4882×10^{1}	1.4882	1	3.1081×10^{-2}	1.4882×10^{3}	3.60×10^{3}
$lbf\ s\ ft^{-2}$	4.7880×10^{2}	4.7880×10^{1}	32.1740	1	4.7880×10^{4}	1.1583×10^{5}
Centipoises	10^{-2}	10^{-3}	6.7197×10^{-4}	2.0886×10^{-5}	1	2.4191
$lbm\ ft^{-1}\ h^{-1}$	4.1338×10^{-3}	4.1338×10^{-4}	2.7778×10^{-4}	8.6336×10^{-6}	4.1338×10^{-1}	1

[a] SI units.
[b] English units.

TABLE C.15
Thermal Conductivity

	$g\ cm\ s^{-3}\ K^{-1}$ (Ergs $s^{-1}\ cm^{-1}\ K^{-1}$)	$kg\ m\ s^{-3}\ K^{-1}$[a] (Watts $m^{-1}\ K^{-1}$)	$lbm\ ft\ s^{-3}\ °F^{-1}$	$lbf\ s^{-1}\ °F^{-1}$	$cal\ s^{-1}\ cm^{-1}\ K^{-1}$	$Btu\ h^{-1}\ ft^{-1}\ °F^{-1}$[b]
$g\ cm\ s^{-3}\ K^{-1}$	1	10^{-5}	4.0183×10^{-5}	1.2489×10^{-6}	2.3901×10^{-8}	5.7780×10^{-6}
$kg\ m\ s^{-3}\ K^{-1}$	10^{5}	1	4.0183	1.2489×10^{-1}	2.3901×10^{-3}	5.7780×10^{-1}
$lbm\ ft\ s^{-3}\ °F^{-1}$	2.4886×10^{4}	2.4886×10^{-1}	1	3.1081×10^{-2}	5.9479×10^{-4}	1.4379×10^{-1}
$lbf\ s^{-1}\ °F^{-1}$	8.0068×10^{5}	8.0068	3.2174×10^{1}	1	1.9137×10^{-2}	4.6263
$cal\ s^{-1}\ cm^{-1}\ K^{-1}$	4.1840×10^{7}	4.1840×10^{2}	1.6813×10^{3}	5.2256×10^{1}	1	2.4175×10^{2}
$Btu\ h^{-1}\ ft^{-1}\ °F^{-1}$	1.7307×10^{5}	1.7307	6.9546	2.1616×10^{-1}	4.1365×10^{-3}	1

[a] SI units.
[b] English units.

TABLE C.16
Heat Transfer Coefficient

	$g\ s^{-3}\ K^{-1}$	$kg\ s^{-3}\ K^{-1}$[a] (Watts $m^{-2}\ K^{-1}$)	$lbm\ s^{-3}\ °F^{-1}$	$lbf\ ft^{-1}\ s^{-1}\ °F^{-1}$	$cal\ cm^{-2}\ s^{-1}\ K^{-1}$	Watts $cm^{-2}\ K^{-1}$	$Btu\ ft^{-2}\ h^{-1}\ °F^{-1}$[b]
$g\ s^{-3}\ K^{-1}$	1	10^{-3}	1.2248×10^{-3}	3.8068×10^{-5}	2.3901×10^{-8}	10^{-7}	1.7611×10^{-4}
$kg\ s^{-3}\ K^{-1}$	10^{3}	1	1.2248	3.8068×10^{-2}	2.3901×10^{-5}	10^{-4}	1.7611×10^{-1}
$lbm\ s^{-3}\ °F^{-1}$	8.1647×10^{2}	8.1647×10^{-1}	1	3.1081×10^{-2}	1.9514×10^{-5}	8.1647×10^{-5}	1.4379×10^{-1}
$lbf\ ft^{-3}\ s^{-3}\ °F^{-1}$	2.6269×10^{4}	2.6269×10^{1}	32.1740	1	6.2784×10^{-4}	2.6269×10^{-3}	4.6263
$cal\ cm^{-2}\ s^{-1}\ K^{-1}$	4.1840×10^{7}	4.1840×10^{4}	5.1245×10^{4}	1.5928×10^{3}	1	4.1840	7.3686×10^{3}
Watts $cm^{-2}\ K^{-1}$	10^{7}	10^{4}	1.2248×10^{4}	3.8068×10^{2}	2.3901×10^{-1}	1	1.7611×10^{3}
$Btu\ ft^{-2}\ h^{-1}\ °F^{-1}$	5.6782×10^{3}	5.6782	6.9546	2.1616×10^{-1}	1.3571×10^{-4}	5.6782×10^{-4}	1

[a] SI units.
[b] English units.

TABLE C.17
Momentum, Thermal or Molecular Diffusivity

	$cm^2\ s^{-1}$	$m^2\ s^{-1}$[a]	$ft^2\ h^{-1}$[b]	Centistokes
$cm^2\ s^{-1}$	1	10^{-4}	3.8750	10^2
$m^2\ s^{-1}$	10^4	1	3.8750×10^4	10^6
$ft^2\ h^{-1}$	2.5807×10^{-1}	2.5807×10^{-5}	1	2.5807×10^1
Centistokes	10^{-2}	10^{-6}	3.8750×10^{-2}	1

[a] SI units.
[b] English units.

TABLE C.18
Surface tension

	N/m[a]	Dyne/cm	lbf/ft[b]
N/m	1	10^3	6.852×10^{-2}
Dyne/cm	0.001	1	6.852×10^{-5}
lbf/ft	14.594	1.4594×10^4	1

[a] SI units.
[b] English units.

REFERENCES

Bird, R. B., Stewart, W. E., and Lightfoot, E. N. *Transport Phenomena.* New York, NY: Wiley, 1960.

El-Wakil, M. M. *Nuclear Heat Transport.* Scranton, PA: International Textbook Company, 1971.

Appendix D
Mathematical Tables

D.1 BESSEL FUNCTION*

Some useful derivatives and integrals of Bessel functions are given in Tables D.l and D.2.

TABLE D.1
Derivatives of Bessel Functions

$$\frac{\mathrm{d}J_0(x)}{\mathrm{d}x} = -J_1(x)$$

$$\frac{\mathrm{d}Y_0(x)}{\mathrm{d}x} = -Y_1(x)$$

$$\frac{\mathrm{d}I_0(x)}{\mathrm{d}x} = I_1(x)$$

$$\frac{\mathrm{d}K_0(x)}{\mathrm{d}x} = -K_1(x)$$

$$\frac{\mathrm{d}J_\upsilon(x)}{\mathrm{d}x} = J_{\upsilon-1}(x) - \frac{\upsilon}{x}J_\upsilon(x) = -J_{\upsilon+1}(x) + \frac{\upsilon}{x}J_\upsilon(x) = \frac{1}{2}[J_{\upsilon-1}(x) - J_{\upsilon+1}(x)]$$

$$\frac{\mathrm{d}Y_\upsilon(x)}{\mathrm{d}x} = Y_{\upsilon-1}(x) - \frac{\upsilon}{x}Y_\upsilon(x) = -Y_{\upsilon+1}(x) + \frac{\upsilon}{x}Y_\upsilon(x) = \frac{1}{2}[Y_{\upsilon-1}(x) - Y_{\upsilon+1}(x)]$$

$$\frac{\mathrm{d}I_\upsilon(x)}{\mathrm{d}x} = I_{\upsilon-1}(x) - \frac{\upsilon}{x}I_\upsilon(x) = I_{\upsilon+1}(x) + \frac{\upsilon}{x}I_\upsilon(x) = \frac{1}{2}[I_{\upsilon-1}(x) - I_{\upsilon+1}(x)]$$

$$\frac{\mathrm{d}K_\upsilon(x)}{\mathrm{d}x} = -K_{\upsilon-1}(x) - \frac{\upsilon}{x}K_\upsilon(x) = -K_{\upsilon+1}(x) + \frac{\upsilon}{x}K_\upsilon(x) = -\frac{1}{2}[K_{\upsilon-1}(x) - K_{\upsilon+1}(x)]$$

$$\frac{\mathrm{d}x^\upsilon J_\upsilon(x)}{\mathrm{d}x} = x^\upsilon J_{\upsilon-1}(x)$$

$$\frac{\mathrm{d}x^{-\upsilon} J_\upsilon(x)}{\mathrm{d}x} = -x^{-\upsilon} J_{\upsilon+1}(x)$$

$$\frac{\mathrm{d}x^\upsilon Y_\upsilon(x)}{\mathrm{d}x} = x^\upsilon Y_{\upsilon-1}(x)$$

$$\frac{\mathrm{d}x^{-\upsilon} Y_\upsilon(x)}{\mathrm{d}x} = -x^{-\upsilon} Y_{\upsilon+1}(x)$$

$$\frac{\mathrm{d}x^\upsilon I_\upsilon(x)}{\mathrm{d}x} = x^\upsilon I_{\upsilon-1}(x)$$

$$\frac{\mathrm{d}x^{-\upsilon} I_\upsilon(x)}{\mathrm{d}x} = x^{-\upsilon} I_{\upsilon+1}(x)$$

$$\frac{\mathrm{d}x^\upsilon K_\upsilon(x)}{\mathrm{d}x} = -x^\upsilon K_{\upsilon-1}(x)$$

$$\frac{\mathrm{d}x^{-\upsilon} K_\upsilon(x)}{\mathrm{d}x} = -x^{-\upsilon} K_{\upsilon+1}(x)$$

* Adopted from El-Wakil, M. M. *Nuclear Heat Transport.* Scranton, PA: International Textbook Company, 1971.

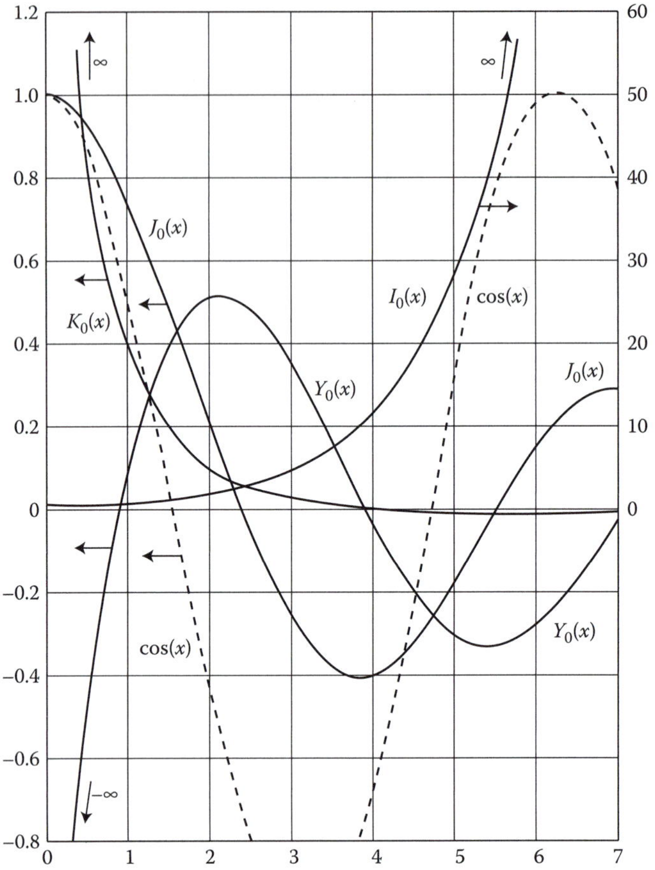

FIGURE D.1 The four Bessel functions of zero order.

TABLE D.2
Integrals of Bessel Functions

$\int J_1(x)\mathrm{d}x = -J_0(x) + C$	$\int Y_1(x)\mathrm{d}x = -Y_0(x) + C$
$\int I_1(x)\mathrm{d}x = I_0(x) + C$	$\int K_1(x)\mathrm{d}x = -K_0(x) + C$
$\int x^{\upsilon} J_{\upsilon-1}(x)\mathrm{d}x = x^{\upsilon} J_{\upsilon}(x) + C$	$\int x^{-\upsilon} J_{\upsilon+1}(x)\mathrm{d}x = -x^{-\upsilon} J_{\upsilon}(x) + C$

The Bessel and modified Bessel functions of zero and first order are tabulated in Table D.3 for positive values of x up to $x = 4.0$. Some roots are given below:

Roots of $J_0(x)$: $x = 2.4048, 5.5201, 8.6537, 11.7915, \ldots$

Roots of $J_1(x)$: $x = 3.8317, 7.0156, 10.1735, 13.3237, \ldots$

Roots of $Y_0(x)$: $x = 0.8936, 3.9577, 7.0861, 10.2223, \ldots$

Roots of $Y_1(x)$: $x = 2.1971, 5.4297, 8.5960, 11.7492, \ldots$

TABLE D.3
Some Bessel Functions

X	$J_0(x)$	$J_1(x)$	$Y_0(x)$	$Y_1(x)$	$I_0(x)$	$I_1(x)$	$K_0(x)$	$K_1(x)$
0	1.0000	0.0000	$-\infty$	$-\infty$	1.000	0.0000	∞	∞
0.05	0.9994	0.0250	−1.979	−12.79	1.001	0.0250	3.114	19.91
0.10	0.9975	0.0499	−1.534	−6.459	1.003	0.0501	2.427	9.854
0.15	0.9944	0.0748	−1.271	−4.364	1.006	0.0752	2.030	6.477
0.20	0.9900	0.0995	−1.081	−3.324	1.010	0.1005	1.753	4.776
0.25	0.9844	0.1240	−0.9316	−2.704	1.016	0.1260	1.542	3.747
0.30	0.9776	0.1483	−0.8073	−2.293	1.023	0.1517	1.372	3.056
0.35	0.9696	0.1723	−0.7003	−2.000	1.031	0.1777	1.233	2.559
0.40	0.9604	0.1960	−0.6060	−1.781	1.040	0.2040	1.115	2.184
0.45	0.9500	0.2194	−0.5214	−1.610	1.051	0.2307	1.013	1.892
0.50	0.9385	0.2423	−0.4445	−1.471	1.063	0.2579	0.9244	1.656
0.55	0.9258	0.2647	−0.3739	−1.357	1.077	0.2855	0.8466	1.464
0.60	0.9120	0.2867	−0.3085	−1.260	1.092	0.3137	0.7775	1.303
0.65	0.8971	0.3081	−0.2476	−1.177	1.108	0.3425	0.7159	1.167
0.70	0.8812	0.3290	−0.1907	−1.103	1.126	0.3719	0.6605	1.050
0.75	0.8642	0.3492	−0.1372	−1.038	1.146	0.4020	0.6106	0.9496
0.80	0.8463	0.3688	−0.0868	−0.9781	1.167	0.4329	0.5653	0.8618
0.85	0.8274	0.3878	−0.0393	−0.9236	1.189	0.4646	0.5242	0.7847
0.90	0.8075	0.4059	−0.0056	−0.8731	1.213	0.4971	0.4867	0.7165
0.95	0.7868	0.4234	0.0481	−0.8258	1.239	0.5306	0.4524	0.6560
1.0	0.7652	0.4401	0.0883	−0.7812	1.266	0.5652	0.4210	0.6019
1.1	0.6957	0.4850	0.1622	−0.6981	1.326	0.6375	0.3656	0.5098
1.2	0.6711	0.4983	0.2281	−0.6211	1.394	0.7147	0.3185	0.4346
1.3	0.5937	0.5325	0.2865	−0.5485	1.469	0.7973	0.2782	0.3725
1.4	0.5669	0.5419	0.3379	−0.4791	1.553	0.8861	0.2437	0.3208
1.5	0.4838	0.5644	0.3824	−0.4123	1.647	0.9817	0.2138	0.2774
1.6	0.4554	0.5699	0.4204	−0.3476	1.750	1.085	0.1880	0.2406
1.7	0.3690	0.5802	0.4520	−0.2847	1.864	1.196	0.1655	0.2094
1.8	0.3400	0.5815	0.4774	−0.2237	1.990	1.317	0.1459	0.1826
1.9	0.2528	0.5794	0.4968	−0.1644	2.128	1.448	0.1288	0.1597
2.0	0.2239	0.5767	0.5104	−0.1070	2.280	1.591	0.1139	0.1399
2.1	0.1383	0.5626	0.5183	−0.0517	2.446	1.745	0.1008	0.1227
2.2	0.1104	0.5560	0.5208	−0.0015	2.629	1.914	0.0893	0.1079

continued

TABLE D.3 (continued)
Some Bessel Functions

X	$J_0(x)$	$J_1(x)$	$Y_0(x)$	$Y_1(x)$	$I_0(x)$	$I_1(x)$	$K_0(x)$	$K_1(x)$
2.3	0.0288	0.5399	0.5181	0.0523	2.830	2.098	0.0791	0.0950
2.4	0.0025	0.5202	0.5104	0.1005	3.049	2.298	0.0702	0.0837
2.5	−0.0729	0.4843	0.4981	0.1459	3.290	2.517	0.0623	0.0739
2.6	−0.0968	0.4708	0.4813	0.1884	3.553	2.755	0.0554	0.0653
2.7	−0.1641	0.4260	0.4605	0.2276	3.842	3.016	0.0493	0.0577
2.8	−0.1850	0.4097	0.4359	0.2635	4.157	3.301	0.0438	0.0511
2.9	−0.2426	0.3575	0.4079	0.2959	4.503	3.613	0.0390	0.0453
3.0	−0.2601	0.3391	0.3769	0.3247	4.881	3.953	0.0347	0.0402
3.2	−0.3202	0.2613	0.3071	0.3707	5.747	4.734	0.0276	0.0316
3.4	−0.3643	0.1792	0.2296	0.4010	6.785	5.670	0.0220	0.0250
3.6	−0.3918	0.0955	0.1477	0.4154	8.028	6.793	0.6175	0.0198
3.8	−0.4026	0.0128	0.0645	0.4141	9.517	8.140	0.0140	0.0157
4.0	−0.3971	−0.0660	−0.0169	0.3979	11.302	9.759	0.0112	0.0125

D.2 DIFFERENTIAL OPERATORS*

TABLE D.4
Summary of Differential Operations Involving the ∇-Operator in Rectangular Coordinates[a] (*x, y, z*)

$$(\nabla \cdot \vec{\upsilon}) = \frac{\partial \upsilon_x}{\partial x} + \frac{\partial \upsilon_y}{\partial y} + \frac{\partial \upsilon_z}{\partial z} \tag{A}$$

$$(\nabla^2 s) = \frac{\partial^2 s}{\partial x} + \frac{\partial^2 s}{\partial y^2} + \frac{\partial^2 s}{\partial z^2} \tag{B}$$

$$(\bar{\bar{\tau}} : \nabla \vec{\upsilon}) = \tau_{xx}\left(\frac{\partial \upsilon_x}{\partial x}\right) + \tau_{yy}\left(\frac{\partial \upsilon_y}{\partial y}\right) + \tau_{zz}\left(\frac{\partial \upsilon_z}{\partial z}\right) + \tau_{xy}\left(\frac{\partial \upsilon x}{\partial y} + \frac{\partial \upsilon_y}{\partial x}\right) + \tau_{yx}\left(\frac{\partial \upsilon_y}{\partial z} + \frac{\partial \upsilon_z}{\partial y}\right) + \tau_{zx}\left(\frac{\partial \upsilon_z}{\partial x} + \frac{\partial \upsilon x}{\partial z}\right) \tag{C}$$

$$[\nabla s]_x = \frac{\partial s}{\partial x} \tag{D}$$

$$[\nabla s]_y = \frac{\partial s}{\partial y} \tag{E}$$

$$[\nabla s]_z = \frac{\partial s}{\partial z} \tag{F}$$

continued

* Adopted from Bird, R. B., Stewart, W. E., and Lightfoot, E. N. *Transport Phenomena*. New York, NY: Wiley, 1960.

TABLE D.4 (continued)
Summary of Differential Operations Involving the ∇-Operator in Rectangular Coordinates[a] (x, y, z)

$$\begin{cases} [\nabla \times \vec{\upsilon}]_x = \dfrac{\partial \upsilon_z}{\partial y} - \dfrac{\partial \upsilon_y}{\partial z} & (G) \\ [\nabla \times \vec{\upsilon}]_y = \dfrac{\partial \upsilon_x}{\partial z} - \dfrac{\partial \upsilon_z}{\partial x} & (H) \\ [\nabla \times \vec{\upsilon}]_z = \dfrac{\partial \upsilon_y}{\partial x} - \dfrac{\partial \upsilon_x}{\partial y} & (I) \end{cases}$$

$$\begin{cases} [\nabla \cdot \bar{\bar{\tau}}]_x = \dfrac{\partial \tau_{xx}}{\partial x} + \dfrac{\partial \tau_{xy}}{\partial y} + \dfrac{\partial \tau_{xz}}{\partial z} & (J) \\ [\nabla \cdot \bar{\bar{\tau}}]_y = \dfrac{\partial \tau_{xy}}{\partial x} + \dfrac{\partial \tau_{yy}}{\partial y} + \dfrac{\partial \tau_{yz}}{\partial z} & (K) \\ [\nabla \cdot \bar{\bar{\tau}}]_z = \dfrac{\partial \tau_{xz}}{\partial x} + \dfrac{\partial \tau_{yz}}{\partial y} + \dfrac{\partial \tau_{zz}}{\partial z} & (L) \end{cases}$$

$$\begin{cases} [\nabla^2 \vec{\upsilon}]_x = \dfrac{\partial^2 \upsilon_x}{\partial x^2} + \dfrac{\partial^2 \upsilon_x}{\partial y^2} + \dfrac{\partial^2 \upsilon_x}{\partial z^2} & (M) \\ [\nabla^2 \vec{\upsilon}]_y = \dfrac{\partial^2 \upsilon_y}{\partial x^2} + \dfrac{\partial^2 \upsilon_y}{\partial y^2} + \dfrac{\partial^2 \upsilon_y}{\partial z^2} & (N) \\ [\nabla^2 \vec{\upsilon}]_z = \dfrac{\partial^2 \upsilon_z}{\partial x^2} + \dfrac{\partial^2 \upsilon_z}{\partial y^2} + \dfrac{\partial^2 \upsilon_z}{\partial z^2} & (O) \end{cases}$$

$$\begin{cases} [\vec{\upsilon} \cdot \nabla \vec{\upsilon}]_x = \upsilon_x \dfrac{\partial \upsilon_x}{\partial x} + \upsilon_y \dfrac{\partial \upsilon_x}{\partial y} + \upsilon_z \dfrac{\partial \upsilon_x}{\partial z} & (P) \\ [\vec{\upsilon} \cdot \nabla \vec{\upsilon}]_y = \upsilon_x \dfrac{\partial \upsilon_y}{\partial x} + \upsilon_y \dfrac{\partial \upsilon_y}{\partial y} + \upsilon_z \dfrac{\partial \upsilon_y}{\partial z} & (Q) \\ [\vec{\upsilon} \cdot \nabla \vec{\upsilon}]_z = \upsilon_x \dfrac{\partial \upsilon_z}{\partial x} + \upsilon_y \dfrac{\partial \upsilon_z}{\partial y} + \upsilon_z \dfrac{\partial \upsilon_z}{\partial z} & (R) \end{cases}$$

[a] Operations involving the tensor τ are given for symmetrical τ only.

TABLE D.5
Summary of Differential Operations Involving the ∇-Operator in Cylindrical Coordinates[a] (r, θ, z)

$$(\nabla \cdot \vec{\upsilon}) = \frac{1}{r}\frac{\partial}{\partial r}(r\upsilon_r) + \frac{1}{r}\frac{\partial \upsilon_\theta}{\partial \theta} + \frac{\partial \upsilon_z}{\partial z} \qquad (A)$$

$$(\nabla^2 s) = \frac{1}{r}\frac{\partial}{\partial r}\left(r\frac{\partial s}{\partial r}\right) + \frac{1}{r^2}\frac{\partial^2 s}{\partial \theta^2} + \frac{\partial^2 s}{\partial z^2} \qquad (B)$$

continued

TABLE D.5 (continued)

Summary of Differential Operations Involving the ∇-Operator in Cylindrical Coordinates[a] (r, θ, z)

$$(\bar{\bar{\tau}}:\nabla\vec{\upsilon}) = \tau_{rr}\left(\frac{\partial \upsilon_r}{\partial r}\right) + \tau_{\theta\theta}\left(\frac{1}{r}\frac{\partial \upsilon_\theta}{\partial \theta} + \frac{\upsilon_r}{r}\right) + \tau_{zz}\left(\frac{\partial \upsilon_z}{\partial z}\right) + \tau_{r\theta}\left[r\frac{\partial}{\partial r}\left(\frac{\upsilon_\theta}{r}\right) + \frac{1}{r}\frac{\partial \upsilon_r}{\partial \theta}\right] + \tau_{\theta z}\left(\frac{1}{r}\frac{\partial \upsilon_z}{\partial \theta} + \frac{\partial \upsilon_\theta}{\partial z}\right) + \tau_{rz}\left(\frac{\partial \upsilon_z}{\partial r} + \frac{\partial \upsilon_r}{\partial z}\right) \quad (C)$$

$$[\nabla s]_r = \frac{\partial s}{\partial r} \quad (D)$$

$$[\nabla s]_\theta = \frac{1}{r}\frac{\partial s}{\partial \theta} \quad (E)$$

$$[\nabla s]_z = \frac{\partial s}{\partial z} \quad (F)$$

$$[\nabla\times\vec{\upsilon}]_r = \frac{1}{r}\frac{\partial \upsilon_z}{\partial \theta} - \frac{\partial \upsilon_\theta}{\partial z} \quad (G)$$

$$[\nabla\times\vec{\upsilon}]_\theta = \frac{\partial \upsilon_r}{\partial z} - \frac{\partial \upsilon_z}{\partial r} \quad (H)$$

$$[\nabla\times\vec{\upsilon}]_z = \frac{1}{r}\frac{\partial}{\partial r}(r\upsilon_\theta) - \frac{1}{r}\frac{\partial \upsilon_r}{\partial \theta} \quad (I)$$

$$[\nabla\cdot\bar{\bar{\tau}}]_r = \frac{1}{r}\frac{\partial}{\partial r}(r\tau_{rr}) + \frac{1}{r}\frac{\partial}{\partial \theta}\tau_{r\theta} - \frac{1}{r}\tau_{\theta\theta} + \frac{\partial \tau_{rz}}{\partial z} \quad (J)$$

$$[\nabla\cdot\bar{\bar{\tau}}]_\theta = \frac{1}{r}\frac{\partial \tau_{\theta\theta}}{\partial \theta} + \frac{\partial \tau_{r\theta}}{\partial r} + \frac{2}{r}\tau_{r\theta} + \frac{\partial \tau_{\theta z}}{\partial z} \quad (K)$$

$$[\nabla\cdot\bar{\bar{\tau}}]_z = \frac{1}{r}\frac{\partial}{\partial r}(r\tau_{rz}) + \frac{1}{r}\frac{\partial \tau_{\theta z}}{\partial \theta} + \frac{\partial \tau_{zz}}{\partial z} \quad (L)$$

$$[\nabla^2\vec{\upsilon}]_r = \frac{\partial}{\partial r}\left(\frac{1}{r}\frac{\partial}{\partial r}(r\upsilon_r)\right) + \frac{1}{r^2}\frac{\partial^2 \upsilon_r}{\partial \theta^2} - \frac{2}{r^2}\frac{\partial \upsilon_\theta}{\partial \theta} + \frac{\partial^2 \upsilon_r}{\partial z^2} \quad (M)$$

$$[\nabla^2\vec{\upsilon}]_\theta = \frac{\partial}{\partial r}\left(\frac{1}{r}\frac{\partial}{\partial r}(r\upsilon_\theta)\right) + \frac{1}{r^2}\frac{\partial^2 \upsilon_\theta}{\partial \theta^2} - \frac{2}{r^2}\frac{\partial \upsilon_r}{\partial \theta} + \frac{\partial^2 \upsilon_\theta}{\partial z^2} \quad (N)$$

$$[\nabla^2\vec{\upsilon}]_z = \frac{1}{r}\frac{\partial}{\partial r}\left(r\frac{\partial \upsilon_z}{\partial r}\right) + \frac{1}{r^2}\frac{\partial^2 \upsilon_z}{\partial \theta^2} + \frac{\partial^2 \upsilon_z}{\partial z^2} \quad (O)$$

$$[\vec{\upsilon}\cdot\nabla\vec{\upsilon}]_r = \upsilon_r\frac{\partial \upsilon_r}{\partial r} + \frac{\upsilon_\theta}{r}\frac{\partial \upsilon_r}{\partial \theta} - \frac{\upsilon_\theta^2}{r} + \upsilon_z\frac{\partial \upsilon_r}{\partial z} \quad (P)$$

$$[\vec{\upsilon}\cdot\nabla\vec{\upsilon}]_\theta = \upsilon_r\frac{\partial \upsilon_\theta}{\partial r} + \frac{\upsilon_\theta}{r}\frac{\partial \upsilon_\theta}{\partial \theta} + \frac{\upsilon_r\upsilon_\theta}{r} + \upsilon_z\frac{\partial \upsilon_\theta}{\partial z} \quad (Q)$$

$$[\vec{\upsilon}\cdot\nabla\vec{\upsilon}]_z = \upsilon_r\frac{\partial \upsilon_z}{\partial r} + \frac{\upsilon_\theta}{r}\frac{\partial \upsilon_z}{\partial \theta} + \upsilon_z\frac{\partial \upsilon_z}{\partial z} \quad (R)$$

[a] Operations involving the tensor τ are given for symmetrical τ only.

TABLE D.6

Summary of Differential Operations Involving the ∇-Operator in Spherical Coordinates[a] (r, θ, ϕ)

$$(\nabla\cdot\vec{\upsilon}) = \frac{1}{r^2}\frac{\partial}{\partial r}(r^2\upsilon_r) + \frac{1}{r\sin\theta}\frac{\partial}{\partial\theta}(\upsilon_\theta \sin\theta) + \frac{1}{r\sin\theta}\frac{\partial\upsilon_\phi}{\partial\phi} \tag{A}$$

$$(\nabla^2 s) = \frac{1}{r^2}\frac{\partial}{\partial r}\left(r^2\frac{\partial s}{\partial r}\right) + \frac{1}{r^2\sin\theta}\frac{\partial}{\partial\theta}\left(\sin\theta\frac{\partial s}{\partial\theta}\right) + \frac{1}{r^2\sin^2\theta}\frac{\partial^2 s}{\partial\phi^2} \tag{B}$$

$$\begin{aligned}(\bar{\bar{\tau}}:\nabla\vec{\upsilon}) = {} & \tau_{rr}\left(\frac{\partial\upsilon_r}{\partial r}\right) + \tau_{\theta\theta}\left(\frac{1}{r}\frac{\partial\upsilon_\theta}{\partial\theta} + \frac{\upsilon_r}{r}\right)\\ & + \tau_{\phi\phi}\left(\frac{1}{r\sin\theta}\frac{\partial\upsilon_\phi}{\partial\phi} + \frac{\upsilon_r}{r} + \frac{\upsilon_\theta\cot\theta}{r}\right)\\ & + \tau_{r\theta}\left(\frac{\partial\upsilon_\theta}{\partial r} + \frac{1}{r}\frac{\partial\upsilon_r}{\partial\theta} - \frac{\upsilon_\theta}{r}\right) + \tau_{r\phi}\left(\frac{\partial\upsilon_\phi}{\partial r} + \frac{1}{r\sin\theta}\frac{\partial\upsilon_r}{\partial\phi} - \frac{\upsilon_\phi}{r}\right)\\ & + \tau_{\theta\phi}\left(\frac{1}{r}\frac{\partial\upsilon_\phi}{\partial\theta} + \frac{1}{r\sin\theta}\frac{\partial\upsilon_\theta}{\partial\phi} - \frac{\cot\theta}{r}\upsilon_\phi\right)\end{aligned} \tag{C}$$

$$[\nabla s]_r = \frac{\partial s}{\partial r} \tag{D}$$

$$[\nabla s]_\theta = \frac{1}{r}\frac{\partial s}{\partial\theta} \tag{E}$$

$$[\nabla s]_\phi = \frac{1}{r\sin\theta}\frac{\partial s}{\partial\phi} \tag{F}$$

$$[\nabla\times\vec{\upsilon}]_r = \frac{1}{r\sin\theta}\frac{\partial}{\partial\theta}(\upsilon_\phi\sin\theta) - \frac{1}{r\sin\theta}\frac{\partial\upsilon_\theta}{\partial\phi} \tag{G}$$

$$[\nabla\times\vec{\upsilon}]_\theta = \frac{1}{r\sin\theta}\frac{\partial\upsilon_r}{\partial\phi} - \frac{1}{r}\frac{\partial}{\partial r}(r\upsilon_\phi) \tag{H}$$

$$[\nabla\times\vec{\upsilon}]_\phi = \frac{1}{r}\frac{\partial}{\partial r}(r\upsilon_\theta) - \frac{1}{r}\frac{\partial\upsilon_r}{\partial\theta} \tag{I}$$

$$[\nabla\cdot\bar{\bar{\tau}}]_r = \frac{1}{r^2}\frac{\partial}{\partial r}(r^2\tau_{rr}) + \frac{1}{r\sin\theta}\frac{\partial}{\partial\theta}(\tau_{r\theta}\sin\theta) + \frac{1}{r\sin\theta}\frac{\partial\tau_{r\phi}}{\partial\phi} - \frac{\tau_{\theta\theta}+\tau_{\phi\phi}}{r} \tag{J}$$

$$[\nabla\cdot\bar{\bar{\tau}}]_\theta = \frac{1}{r^2}\frac{\partial}{\partial r}(r^2\tau_{r\theta}) + \frac{1}{r\sin\theta}\frac{\partial}{\partial\theta}(\tau_{\theta\theta}\sin\theta) + \frac{1}{r\sin\theta}\frac{\partial\tau_{\theta\phi}}{\partial\phi} + \frac{\tau_{r\theta}}{r} - \frac{\cot\theta}{r}\tau_{\phi\phi} \tag{K}$$

$$[\nabla\cdot\bar{\bar{\tau}}]_\phi = \frac{1}{r^2}\frac{\partial}{\partial r}(r^2\tau_{r\phi}) + \frac{1}{r}\frac{\partial\tau_{\theta\phi}}{\partial\theta} + \frac{1}{r\sin\theta}\frac{\partial\tau_{\phi\phi}}{\partial\phi} + \frac{\tau_{r\phi}}{r} + \frac{2\cot\theta}{r}\tau_{\theta\phi} \tag{L}$$

$$[\nabla^2\vec{\upsilon}]_r = \nabla^2\upsilon_r - \frac{2\upsilon_r}{r^2} - \frac{2}{r^2}\frac{\partial\upsilon_\theta}{\partial\theta} - \frac{2\upsilon_\theta\cot\theta}{r^2} - \frac{2}{r^2\sin\theta}\frac{\partial\upsilon}{\partial\phi} \tag{M}$$

$$[\nabla^2\vec{\upsilon}]_\theta = \nabla^2\upsilon_\theta - \frac{2}{r^2}\frac{\partial\upsilon_r}{\partial\theta} - \frac{r_\theta}{r^2\sin^2\theta} - \frac{2\cos\theta}{r^2\sin^2\theta}\frac{\partial\upsilon_\phi}{\partial\phi} \tag{N}$$

$$[\nabla^2\vec{\upsilon}]_\phi = \nabla^2\upsilon_\phi - \frac{\upsilon_\phi}{r^2\sin^2\theta} + \frac{2}{r^2\sin\theta}\frac{\partial\upsilon_r}{\partial\phi} + \frac{2\cos\theta}{r^2\sin^2\theta}\frac{\partial\upsilon_\theta}{\partial\phi} \tag{O}$$

continued

TABLE D.6 (continued)
Summary of Differential Operations Involving the ∇-Operator in Spherical Coordinates[a] (*r*, θ, ϕ)

$$\begin{cases} [\vec{\upsilon}\cdot\nabla\vec{\upsilon}]_r = \upsilon_r \dfrac{\partial \upsilon_r}{\partial r} + \dfrac{\upsilon_\theta}{r}\dfrac{\partial \upsilon_r}{\partial \theta} + \dfrac{\upsilon_\phi}{r \sin\theta}\dfrac{\partial \upsilon_r}{\partial \phi} - \dfrac{\upsilon_\theta^2 + \upsilon_\phi^2}{r} & (P) \\ [\vec{\upsilon}\cdot\nabla\vec{\upsilon}]_\theta = \upsilon_r \dfrac{\partial \upsilon_\theta}{\partial r} + \dfrac{\upsilon_\theta}{r}\dfrac{\partial \upsilon_\theta}{\partial \theta} + \dfrac{\upsilon_\phi}{r \sin\theta}\dfrac{\partial \upsilon_\theta}{\partial \phi} + \dfrac{\upsilon_r\upsilon_\theta}{r} - \dfrac{\upsilon_\phi^2 \cot\theta}{r} & (Q) \\ [\vec{\upsilon}\cdot\nabla\vec{\upsilon}]_\phi = \upsilon_r \dfrac{\partial \upsilon_\phi}{\partial r} + \dfrac{\upsilon_\theta}{r}\dfrac{\partial \upsilon_\phi}{\partial \theta} + \dfrac{\upsilon_\phi}{r \sin\theta}\dfrac{\partial \upsilon_\phi}{\partial \phi} + \dfrac{\upsilon_\phi\upsilon_r}{r} + \dfrac{\upsilon_\theta\upsilon_\phi \cot\theta}{r} & (R) \end{cases}$$

[a] Operations involving the tensor τ are given for symmetrical τ only.

REFERENCES

Bird, R. B., Stewart, W. E., and Lightfoot, E. N. *Transport Phenomena.* New York, NY: Wiley, 1960.

El-Wakil, M. M. *Nuclear Heat Transport.* Scranton, PA: International Textbook Company, 1971.

Appendix E
Thermodynamic Properties

TABLE E.1
Saturation State Properties of Steam and Water

Temp. (°C)	Pressure (MPa)	Specific Volume (m^3/kg)		Specific Enthalpy (kJ/kg)		Water					Steam			
		Water	Steam	Water	Steam	c_{pf} (kJ/kg K)	σ (N/m)	μ_f (Pa s)	k_f (W/m K)	$(Pr)_f$ (--)	c_{pg} (kJ/kg K)	μ_g (Pa s)	k_g (W/m K)	$(Pr)_g$ (--)
0.01	0.00061165	1.000E-03	205.990	0.0006	2500.9	4.2199	0.07565	1.791E-03	0.56104	13.4726	1.884	9.216E-06	1.707E-02	1.0174
10	0.0012282	1.000E-03	106.300	42.0	2519.2	4.1955	0.07422	1.306E-03	0.58000	9.4471	1.895	9.461E-06	1.762E-02	1.0173
20	0.0023393	1.002E-03	57.757	83.9	2537.4	4.1844	0.07274	1.002E-03	0.59842	7.0036	1.906	9.727E-06	1.823E-02	1.0171
30	0.004247	1.004E-03	32.878	125.7	2555.5	4.1801	0.07119	7.974E-04	0.61546	5.4155	1.918	1.001E-05	1.889E-02	1.0165
40	0.0073849	1.008E-03	19.515	167.5	2573.5	4.1796	0.06960	6.530E-04	0.63058	4.3280	1.931	1.031E-05	1.960E-02	1.0158
50	0.012352	1.012E-03	12.027	209.3	2591.3	4.1815	0.06794	5.468E-04	0.64355	3.5531	1.947	1.062E-05	2.037E-02	1.0148
60	0.019946	1.017E-03	7.667	251.2	2608.8	4.1851	0.06624	4.664E-04	0.65435	2.9829	1.965	1.094E-05	2.119E-02	1.0141
70	0.031201	1.023E-03	5.040	293.1	2626.1	4.1902	0.06448	4.039E-04	0.66309	2.5521	1.986	1.126E-05	2.207E-02	1.0134
80	0.047414	1.029E-03	3.405	335.0	2643.0	4.1969	0.06267	3.543E-04	0.66999	2.2196	2.012	1.159E-05	2.301E-02	1.0136
90	0.070182	1.036E-03	2.359	377.0	2659.5	4.2053	0.06082	3.144E-04	0.67525	1.9580	2.043	1.193E-05	2.402E-02	1.0146
100	0.10142	1.044E-03	1.672	419.2	2675.6	4.2157	0.05891	2.817E-04	0.67909	1.7490	2.080	1.227E-05	2.510E-02	1.0169
110	0.14338	1.052E-03	1.209	461.4	2691.1	4.2283	0.05696	2.547E-04	0.68169	1.5798	2.124	1.261E-05	2.625E-02	1.0209
120	0.19867	1.060E-03	0.891	503.8	2705.9	4.2435	0.05497	2.321E-04	0.68319	1.4413	2.177	1.296E-05	2.747E-02	1.0269
130	0.27028	1.070E-03	0.668	546.4	2720.1	4.2615	0.05293	2.129E-04	0.68370	1.3270	2.239	1.330E-05	2.877E-02	1.0353
140	0.36154	1.080E-03	0.508	589.2	2733.4	4.2826	0.05086	1.965E-04	0.68330	1.2318	2.311	1.365E-05	3.014E-02	1.0463
150	0.47616	1.091E-03	0.392	632.2	2745.9	4.3071	0.04874	1.825E-04	0.68204	1.1522	2.394	1.399E-05	3.160E-02	1.0602
160	0.61823	1.102E-03	0.307	675.5	2757.4	4.3354	0.04659	1.702E-04	0.67996	1.0854	2.488	1.434E-05	3.313E-02	1.0768
170	0.79219	1.114E-03	0.243	719.1	2767.9	4.3678	0.04441	1.596E-04	0.67705	1.0293	2.594	1.468E-05	3.475E-02	1.0961
180	1.0028	1.127E-03	0.194	763.1	2777.2	4.4050	0.04219	1.501E-04	0.67332	0.9822	2.713	1.503E-05	3.645E-02	1.1183
190	1.2552	1.142E-03	0.156	807.4	2785.3	4.4474	0.03995	1.418E-04	0.66875	0.9429	2.844	1.537E-05	3.824E-02	1.1433
200	1.5549	1.157E-03	0.127	852.3	2792.0	4.4958	0.03768	1.343E-04	0.66331	0.9104	2.990	1.572E-05	4.011E-02	1.1712

210	1.9077	1.173E-03	0.104	897.6	2797.3	4.5512	0.03538	1.276E-04	0.65697	0.8840	3.150	1.606E-05	4.209E-02	1.2022
220	2.3196	1.190E-03	0.086	943.6	2800.9	4.6146	0.03307	1.215E-04	0.64965	0.8632	3.329	1.641E-05	4.417E-02	1.2368
230	2.7971	1.209E-03	0.072	990.2	2802.9	4.6876	0.03074	1.160E-04	0.64131	0.8476	3.529	1.677E-05	4.638E-02	1.2756
240	3.3469	1.230E-03	0.060	1037.6	2803.0	4.7719	0.02839	1.109E-04	0.63185	0.8372	3.754	1.713E-05	4.873E-02	1.3192
250	3.9762	1.252E-03	0.050	1085.8	2800.9	4.8701	0.02604	1.061E-04	0.62119	0.8319	4.011	1.750E-05	5.126E-02	1.3687
260	4.6923	1.276E-03	0.042	1135.0	2796.6	4.9856	0.02369	1.017E-04	0.60924	0.8321	4.308	1.788E-05	5.403E-02	1.4252
270	5.503	1.303E-03	0.036	1185.3	2789.7	5.1230	0.02134	9.750E-05	0.59591	0.8382	4.656	1.828E-05	5.711E-02	1.4901
280	6.4166	1.333E-03	0.030	1236.9	2779.9	5.2889	0.01899	9.351E-05	0.58115	0.8510	5.073	1.870E-05	6.061E-02	1.5651
290	7.4418	1.366E-03	0.026	1290.0	2766.7	5.4931	0.01666	8.966E-05	0.56496	0.8717	5.582	1.915E-05	6.471E-02	1.6523
300	8.5879	1.404E-03	0.022	1345.0	2749.6	5.7504	0.01436	8.590E-05	0.54743	0.9023	6.220	1.965E-05	6.965E-02	1.7548
310	9.8651	1.448E-03	0.018	1402.2	2727.9	6.0848	0.01209	8.217E-05	0.52875	0.9456	7.045	2.021E-05	7.584E-02	1.8772
320	11.284	1.499E-03	0.015	1462.2	2700.6	6.5373	0.00986	7.841E-05	0.50920	1.0066	8.159	2.085E-05	8.391E-02	2.0270
330	12.858	1.561E-03	0.013	1525.9	2666.0	7.1863	0.00770	7.454E-05	0.48907	1.0952	9.753	2.161E-05	9.494E-02	2.2194
340	14.601	1.638E-03	0.011	1594.5	2621.8	8.2080	0.00563	7.043E-05	0.46851	1.2339	12.236	2.255E-05	1.109E-01	2.4882
350	16.529	1.740E-03	0.009	1670.9	2563.6	10.1160	0.00367	6.588E-05	0.44737	1.4896	16.692	2.382E-05	1.360E-01	2.9246
360	18.666	1.895E-03	0.007	1761.7	2481.5	15.0040	0.00188	6.033E-05	0.42572	2.1262	27.356	2.572E-05	1.815E-01	3.8770
370	21.044	2.215E-03	0.005	1890.7	2334.5	45.1550	0.00039	5.207E-05	0.42504	5.5317	96.598	2.968E-05	3.238E-01	8.8526
373.94	22.062	2.995E-03	0.003	2064.9	2105.1	∞	0	4.854E-05	∞		∞	4.58E-05	∞	

Source: From Lemmon, E.W., McLinden, M.O., and Friend, D.G. Thermophysical properties of fluid systems, in *NIST Chemistry WebBook*, *NIST Standard Reference Database Number 69*, P.J. Linstrom and W.G. Mallard (Eds), National Institute of Standards and Technology, Gaithersburg MD, 20899, http://webbook.nist.gov (retrieved August 13, 2010).

TABLE E.2
Thermodynamic Properties of Dry Saturated Steam: In Pressure Increments

		Specific Volume (m^3/kg)		Enthalpy (kJ/kg)			Entropy (kJ/kg K)		
Pressure (MPa)	Temp. (°C)	Saturated Liquid	Saturated Vapor	Saturated Liquid	Evap.	Saturated Vapor	Saturated Liquid	Evap.	Saturated Vapor
0.00061248	2.86E-02	1.000E-03	205.73	7.8916E-02	2500.8	2500.9	2.8665E-04	9.1547	9.1550
0.10	100.00	1.044E-03	1.6718	419.2	2256.4	2675.6	1.3072	6.0469	7.3541
0.25	127.41	1.067E-03	7.187E-01	535.3	2181.2	2716.5	1.6072	5.4452	7.0524
0.50	151.83	1.093E-03	3.748E-01	640.1	2108.0	2748.1	1.8604	4.9603	6.8207
0.75	167.75	1.111E-03	2.555E-01	709.2	2056.4	2765.6	2.0195	4.6641	6.6836
1.00	179.88	1.127E-03	1.944E-01	762.5	2014.6	2777.1	2.1381	4.4469	6.5850
1.25	189.81	1.141E-03	1.570E-01	806.6	1978.5	2785.1	2.2337	4.2737	6.5074
1.50	198.29	1.154E-03	1.317E-01	844.6	1946.4	2791.0	2.3143	4.1287	6.4430
1.75	205.73	1.166E-03	1.134E-01	878.2	1917.0	2795.2	2.3845	4.0032	6.3877
2.00	212.38	1.177E-03	9.959E-02	908.5	1889.8	2798.3	2.4468	3.8922	6.3390
2.25	218.41	1.187E-03	8.872E-02	936.2	1864.3	2800.5	2.5029	3.7925	6.2954
2.50	223.95	1.197E-03	7.995E-02	961.9	1840.0	2801.9	2.5543	3.7015	6.2558
2.75	229.08	1.207E-03	7.273E-02	985.9	1817.0	2802.8	2.6016	3.6178	6.2194
3.00	233.85	1.217E-03	6.666E-02	1008.3	1794.9	2803.2	2.6455	3.5401	6.1856
3.25	238.33	1.226E-03	6.151E-02	1029.6	1773.5	2803.1	2.6867	3.4673	6.1540
3.50	242.56	1.235E-03	5.706E-02	1049.8	1752.8	2802.6	2.7254	3.3989	6.1243
3.75	246.56	1.244E-03	5.318E-02	1069.1	1732.8	2801.9	2.7620	3.3343	6.0963
4.00	250.35	1.253E-03	4.978E-02	1087.5	1713.3	2800.8	2.7968	3.2728	6.0696
4.25	253.98	1.261E-03	4.676E-02	1105.2	1694.3	2799.5	2.8299	3.2142	6.0441
4.50	257.44	1.270E-03	4.406E-02	1122.2	1675.7	2797.9	2.8615	3.1582	6.0197
4.75	260.75	1.278E-03	4.164E-02	1138.7	1657.5	2796.2	2.8918	3.1045	5.9963
5.00	263.94	1.286E-03	3.945E-02	1154.6	1639.6	2794.2	2.9210	3.0527	5.9737

5.25	267.01	1.295E-03	3.746E-02	1170.1	1621.9	2792.0	2.9491	3.0027	5.9518
5.50	269.97	1.303E-03	3.564E-02	1185.1	1604.6	2789.7	2.9762	2.9545	5.9307
5.75	272.82	1.311E-03	3.398E-02	1199.7	1587.5	2787.2	3.0024	2.9077	5.9101
6.00	275.58	1.319E-03	3.245E-02	1213.9	1570.7	2784.6	3.0278	2.8623	5.8901
6.25	278.26	1.327E-03	3.104E-02	1227.8	1554.0	2781.8	3.0524	2.8182	5.8706
6.50	280.86	1.336E-03	2.973E-02	1241.4	1537.5	2778.9	3.0764	2.7752	5.8516
6.75	283.38	1.344E-03	2.851E-02	1254.7	1521.1	2775.8	3.0997	2.7333	5.8330
7.00	285.83	1.352E-03	2.738E-02	1267.7	1504.9	2772.6	3.1224	2.6924	5.8148
7.25	288.21	1.360E-03	2.632E-02	1280.4	1488.9	2769.3	3.1446	2.6523	5.7969
7.50	290.54	1.368E-03	2.533E-02	1292.9	1473.0	2765.9	3.1662	2.6131	5.7793
7.75	292.80	1.376E-03	2.440E-02	1305.2	1457.1	2762.3	3.1874	2.5746	5.7620
8.00	295.01	1.385E-03	2.353E-02	1317.3	1441.4	2758.7	3.2081	2.5369	5.7450
8.25	297.16	1.393E-03	2.270E-02	1329.2	1425.7	2754.9	3.2284	2.4998	5.7282
8.50	299.27	1.401E-03	2.192E-02	1340.9	1410.1	2751.0	3.2483	2.4634	5.7117
8.75	301.33	1.410E-03	2.119E-02	1352.5	1394.5	2747.0	3.2678	2.4275	5.6953
9.00	303.34	1.418E-03	2.049E-02	1363.9	1379.0	2742.9	3.2870	2.3921	5.6791
9.25	305.32	1.427E-03	1.983E-02	1375.1	1363.6	2738.7	3.3058	2.3573	5.6631
9.50	307.25	1.435E-03	1.920E-02	1386.2	1348.2	2734.4	3.3244	2.3229	5.6473
9.75	309.14	1.444E-03	1.860E-02	1397.2	1332.8	2730.0	3.3427	2.2889	5.6316
10.00	311.00	1.453E-03	1.803E-02	1408.1	1317.4	2725.5	3.3606	2.2554	5.6160
10.25	312.82	1.461E-03	1.749E-02	1418.8	1302.1	2720.9	3.3784	2.2221	5.6005
10.50	314.60	1.470E-03	1.697E-02	1429.4	1286.7	2716.1	3.3959	2.1892	5.5851
10.75	316.36	1.479E-03	1.647E-02	1440.0	1271.3	2711.3	3.4132	2.1565	5.5697
11.00	318.08	1.489E-03	1.599E-02	1450.4	1255.9	2706.3	3.4303	2.1242	5.5545
11.25	319.77	1.498E-03	1.553E-02	1460.8	1240.5	2701.3	3.4471	2.0922	5.5393
11.50	321.43	1.507E-03	1.509E-02	1471.1	1225.0	2696.1	3.4638	2.0603	5.5241
11.75	323.07	1.517E-03	1.467E-02	1481.3	1209.5	2690.8	3.4803	2.0287	5.5090
12.00	324.68	1.526E-03	1.426E-02	1491.5	1193.9	2685.4	3.4967	1.9972	5.4939
12.25	326.26	1.536E-03	1.387E-02	1501.5	1178.4	2679.9	3.5129	1.9659	5.4788

continued

TABLE E.2 (continued)
Thermodynamic Properties of Dry Saturated Steam: In Pressure Increments

		Specific Volume (m^3/kg)		Enthalpy (kJ/kg)			Entropy (kJ/kg K)		
Pressure (MPa)	Temp. (°C)	Saturated Liquid	Saturated Vapor	Saturated Liquid	Evap.	Saturated Vapor	Saturated Liquid	Evap.	Saturated Vapor
12.50	327.81	1.546E-03	1.350E-02	1511.6	1162.7	2674.3	3.5290	1.9348	5.4638
12.75	329.35	1.556E-03	1.313E-02	1521.6	1147.0	2668.6	3.5449	1.9038	5.4487
13.00	330.85	1.567E-03	1.278E-02	1531.5	1131.2	2662.7	3.5608	1.8728	5.4336
13.25	332.34	1.577E-03	1.244E-02	1541.4	1115.3	2656.7	3.5765	1.8419	5.4184
13.50	333.80	1.588E-03	1.211E-02	1551.3	1099.2	2650.5	3.5921	1.8111	5.4032
13.75	335.25	1.599E-03	1.179E-02	1561.1	1083.2	2644.3	3.6077	1.7803	5.3880
14.00	336.67	1.610E-03	1.149E-02	1571.0	1066.9	2637.9	3.6232	1.7495	5.3727
14.25	338.07	1.621E-03	1.119E-02	1580.8	1050.5	2631.3	3.6386	1.7187	5.3573
14.50	339.45	1.633E-03	1.090E-02	1590.6	1034.0	2624.6	3.6539	1.6879	5.3418
14.75	340.81	1.645E-03	1.061E-02	1600.4	1017.3	2617.7	3.6693	1.6570	5.3263
15.00	342.16	1.657E-03	1.034E-02	1610.2	1000.5	2610.7	3.6846	1.6260	5.3106
15.25	343.48	1.670E-03	1.007E-02	1620.0	983.5	2603.5	3.6998	1.5949	5.2947
15.50	344.79	1.682E-03	9.811E-03	1629.9	966.2	2596.1	3.7151	1.5637	5.2788
15.75	346.08	1.696E-03	9.557E-03	1639.8	948.8	2588.6	3.7304	1.5322	5.2626
16.00	347.35	1.709E-03	9.309E-03	1649.7	931.1	2580.8	3.7457	1.5006	5.2463
16.25	348.61	1.724E-03	9.067E-03	1659.7	913.1	2572.8	3.7611	1.4687	5.2298
16.50	349.85	1.738E-03	8.830E-03	1669.7	894.9	2564.6	3.7765	1.4365	5.2130
16.75	351.08	1.754E-03	8.598E-03	1679.8	876.4	2556.2	3.7921	1.4039	5.1960
17.00	352.29	1.769E-03	8.371E-03	1690.0	857.5	2547.5	3.8077	1.3710	5.1787
17.25	353.49	1.786E-03	8.148E-03	1700.3	838.2	2538.5	3.8235	1.3376	5.1611
17.50	354.67	1.803E-03	7.929E-03	1710.8	818.5	2529.3	3.8394	1.3037	5.1431
17.75	355.84	1.821E-03	7.714E-03	1721.3	798.4	2519.7	3.8555	1.2693	5.1248

18.00	356.99	1.840E-03	7.502E-03	1732.1	777.7	2509.8	3.8718	1.2343	5.1061
18.25	358.13	1.860E-03	7.292E-03	1743.0	756.5	2499.5	3.8884	1.1984	5.0868
18.50	359.26	1.881E-03	7.086E-03	1754.1	734.7	2488.8	3.9053	1.1617	5.0670
18.75	360.37	1.903E-03	6.881E-03	1765.5	712.2	2477.7	3.9225	1.1241	5.0466
19.00	361.47	1.927E-03	6.677E-03	1777.2	688.8	2466.0	3.9401	1.0855	5.0256
19.25	362.56	1.952E-03	6.475E-03	1789.1	664.6	2453.7	3.9581	1.0455	5.0036
19.50	363.64	1.979E-03	6.273E-03	1801.4	639.4	2440.8	3.9767	1.0041	4.9808
19.75	364.70	2.008E-03	6.070E-03	1814.1	612.9	2427.0	3.9958	0.9610	4.9568
20.00	365.75	2.040E-03	5.865E-03	1827.2	585.1	2412.3	4.0156	0.9158	4.9314
20.25	366.79	2.075E-03	5.658E-03	1840.9	555.6	2396.5	4.0362	0.8682	4.9044
20.50	367.81	2.113E-03	5.446E-03	1855.3	523.9	2379.2	4.0579	0.8174	4.8753
20.75	368.83	2.156E-03	5.226E-03	1870.7	489.5	2360.2	4.0811	0.7624	4.8435
21.00	369.83	2.206E-03	4.996E-03	1887.6	451.0	2338.6	4.1064	0.7015	4.8079
21.25	370.82	2.267E-03	4.749E-03	1906.6	406.9	2313.5	4.1352	0.6319	4.7671
21.50	371.79	2.347E-03	4.473E-03	1929.5	353.6	2283.1	4.1698	0.5483	4.7181
21.75	372.75	2.466E-03	4.145E-03	1959.8	283.2	2243.0	4.2158	0.4384	4.6542

Source: From Lemmon, E.W., McLinden, M.O., and Friend, D.G. Thermophysical properties of fluid systems, in *NIST Chemistry WebBook, NIST Standard Reference Database Number 69*, P.J. Linstrom and W.G. Mallard (Eds), National Institute of Standards and Technology, Gaithersburg MD, 20899, http://webbook.nist.gov (retrieved August 13, 2010).

TABLE E.3
Thermodynamic Properties of Dry Saturated Steam: In Temperature Increments

		Specific Volume (m^3/kg)		Enthalpy (kJ/kg)			Entropy (kJ/kg K)		
Temp. (°C)	Pressure (MPa)	Saturated Liquid	Saturated Vapor	Saturated Liquid	Evap.	Saturated Vapor	Saturated Liquid	Evap.	Saturated Vapor
5	0.00087258	1.000E-03	147.0	21.0	2489.08	2510.1	0.0763	8.9485	9.0248
10	0.0012282	1.000E-03	106.3	42.0	2477.2	2519.2	0.1511	8.7487	8.8998
15	0.0017058	1.001E-03	77.88	63.0	2465.3	2528.3	0.2245	8.5558	8.7803
20	0.0023393	1.002E-03	57.76	83.9	2453.5	2537.4	0.2965	8.3695	8.6660
25	0.0031699	1.003E-03	43.34	104.8	2441.7	2546.5	0.3672	8.1894	8.5566
30	0.004247	1.004E-03	32.88	125.7	2429.8	2555.5	0.4368	8.0153	8.4520
35	0.005629	1.006E-03	25.21	146.6	2417.9	2564.5	0.5051	7.8466	8.3517
40	0.0073849	1.008E-03	19.52	167.5	2406.0	2573.5	0.5724	7.6831	8.2555
45	0.009595	1.010E-03	15.25	188.4	2394.0	2582.4	0.6386	7.5247	8.1633
50	0.012352	1.012E-03	12.03	209.3	2382.0	2591.3	0.7038	7.3710	8.0748
55	0.015762	1.015E-03	9.564	230.3	2369.8	2600.1	0.7680	7.2218	7.9898
60	0.019946	1.017E-03	7.667	251.2	2357.6	2608.8	0.8313	7.0768	7.9081
65	0.025042	1.020E-03	6.194	272.1	2345.4	2617.5	0.8937	6.9360	7.8296
70	0.031201	1.023E-03	5.040	293.1	2333.0	2626.1	0.9551	6.7989	7.7540
75	0.038595	1.026E-03	4.129	314.0	2320.6	2634.6	1.0158	6.6654	7.6812
80	0.047414	1.029E-03	3.405	335.0	2308.0	2643.0	1.0756	6.5355	7.6111
85	0.057867	1.032E-03	2.826	356.0	2295.3	2651.3	1.1346	6.4088	7.5434
90	0.070182	1.036E-03	2.359	377.0	2282.5	2659.5	1.1929	6.2852	7.4781
95	0.084608	1.040E-03	1.981	398.1	2269.5	2667.6	1.2504	6.1647	7.4151

100	0.10142	1.044E-03	1.672	419.2	2256.4	2675.6	1.3072	6.0469	7.3541
105	0.1209	1.047E-03	1.418	440.3	2243.1	2683.4	1.3633	5.9319	7.2952
110	0.14338	1.052E-03	1.209	461.4	2229.7	2691.1	1.4188	5.8193	7.2381
115	0.16918	1.056E-03	1.036	482.6	2216.0	2698.6	1.4737	5.7091	7.1828
120	0.19867	1.060E-03	8.912E-01	503.8	2202.1	2705.9	1.5279	5.6012	7.1291
125	0.23224	1.065E-03	7.700E-01	525.1	2188.0	2713.1	1.5816	5.4954	7.0770
130	0.27028	1.070E-03	6.680E-01	546.4	2173.7	2720.1	1.6346	5.3918	7.0264
135	0.31323	1.075E-03	5.817E-01	567.7	2159.2	2726.9	1.6872	5.2900	6.9772
140	0.36154	1.080E-03	5.085E-01	589.2	2144.2	2733.4	1.7392	5.1901	6.9293
145	0.41568	1.085E-03	4.460E-01	610.6	2129.2	2739.8	1.7907	5.0919	6.8826
150	0.47616	1.091E-03	3.925E-01	632.2	2113.7	2745.9	1.8418	4.9953	6.8371
155	0.5435	1.096E-03	3.465E-01	653.8	2098.0	2751.8	1.8924	4.9002	6.7926
160	0.61823	1.102E-03	3.068E-01	675.5	2081.9	2757.4	1.9426	4.8065	6.7491
165	0.70093	1.108E-03	2.724E-01	697.2	2065.6	2762.8	1.9923	4.7143	6.7066
170	0.79219	1.114E-03	2.426E-01	719.1	2048.8	2767.9	2.0417	4.6233	6.6650
175	0.8926	1.121E-03	2.166E-01	741.0	2031.7	2772.7	2.0906	4.5335	6.6241
180	1.0028	1.127E-03	1.938E-01	763.1	2014.2	2777.2	2.1392	4.4448	6.5840
185	1.1235	1.134E-03	1.739E-01	785.2	1996.2	2781.4	2.1875	4.3572	6.5447
190	1.2552	1.142E-03	1.564E-01	807.4	1977.9	2785.3	2.2355	4.2704	6.5059
195	1.3988	1.149E-03	1.409E-01	829.8	1959.0	2788.8	2.2832	4.1846	6.4678
200	1.5549	1.157E-03	1.272E-01	852.3	1939.7	2792.0	2.3305	4.0997	6.4302
205	1.7243	1.165E-03	1.151E-01	874.9	1919.9	2794.8	2.3777	4.0153	6.3930
210	1.9077	1.173E-03	1.043E-01	897.6	1899.7	2797.3	2.4245	3.9318	6.3563
215	2.1058	1.181E-03	9.468E-02	920.5	1878.8	2799.3	2.4712	3.8488	6.3200
220	2.3196	1.190E-03	8.609E-02	943.6	1857.3	2800.9	2.5177	3.7663	6.2840
225	2.5497	1.199E-03	7.840E-02	966.8	1835.3	2802.1	2.5640	3.6843	6.2483

continued

TABLE E.3 (continued)

Thermodynamic Properties of Dry Saturated Steam: In Temperature Increments

Temp. (°C)	Pressure (MPa)	Specific Volume (m^3/kg) Saturated Liquid	Saturated Vapor	Enthalpy (kJ/kg) Saturated Liquid	Evap.	Saturated Vapor	Entropy (kJ/kg K) Saturated Liquid	Evap.	Saturated Vapor
230	2.7971	1.209E-03	7.150E-02	990.2	1812.7	2802.9	2.6101	3.6027	6.2128
235	3.0625	1.219E-03	6.530E-02	1013.8	1789.4	2803.2	2.6561	3.5214	6.1775
240	3.3469	1.230E-03	5.971E-02	1037.6	1765.4	2803.0	2.7020	3.4403	6.1423
245	3.6512	1.240E-03	5.465E-02	1061.5	1740.7	2802.2	2.7478	3.3594	6.1072
250	3.9762	1.252E-03	5.008E-02	1085.8	1715.1	2800.9	2.7935	3.2786	6.0721
255	4.3229	1.264E-03	4.594E-02	1110.2	1688.9	2799.1	2.8392	3.1977	6.0369
260	4.6923	1.276E-03	4.217E-02	1135.0	1661.6	2796.6	2.8849	3.1167	6.0016
265	5.0853	1.289E-03	3.875E-02	1160.0	1633.5	2793.5	2.9307	3.0354	5.9661
270	5.503	1.303E-03	3.562E-02	1185.3	1604.4	2789.7	2.9765	2.9539	5.9304
275	5.9464	1.318E-03	3.277E-02	1210.9	1574.3	2785.2	3.0224	2.8720	5.8944
280	6.4166	1.333E-03	3.015E-02	1236.9	1543.0	2779.9	3.0685	2.7894	5.8579
285	6.9147	1.349E-03	2.776E-02	1263.2	1510.5	2773.7	3.1147	2.7062	5.8209
290	7.4418	1.366E-03	2.556E-02	1290.0	1476.7	2766.7	3.1612	2.6222	5.7834
295	7.9991	1.385E-03	2.353E-02	1317.3	1441.4	2758.7	3.2080	2.5371	5.7451
300	8.5879	1.404E-03	2.166E-02	1345.0	1404.6	2749.6	3.2552	2.4507	5.7059
305	9.2094	1.425E-03	1.993E-02	1373.3	1366.1	2739.4	3.3028	2.3629	5.6657
310	9.8651	1.448E-03	1.834E-02	1402.2	1325.7	2727.9	3.3510	2.2734	5.6244
315	10.556	1.472E-03	1.685E-02	1431.8	1283.3	2715.1	3.3998	2.1818	5.5816
320	11.284	1.499E-03	1.547E-02	1462.2	1238.4	2700.6	3.4494	2.0878	5.5372

325	12.051	1.528E-03	1.418E-02	1493.5	1190.8	2684.3	3.5000	1.9908	5.4908
330	12.858	1.561E-03	1.298E-02	1525.9	1140.1	2666.0	3.5518	1.8904	5.4422
335	13.707	1.597E-03	1.185E-02	1559.5	1085.9	2645.4	3.6050	1.7856	5.3906
340	14.601	1.638E-03	1.078E-02	1594.5	1027.3	2621.8	3.6601	1.6755	5.3356
345	15.541	1.685E-03	9.769E-03	1631.5	963.4	2594.9	3.7176	1.5586	5.2762
350	16.529	1.740E-03	8.802E-03	1670.9	892.7	2563.6	3.7784	1.4326	5.2110
355	17.57	1.808E-03	7.868E-03	1713.7	812.9	2526.6	3.8439	1.2941	5.1380
360	18.666	1.895E-03	6.949E-03	1761.7	719.8	2481.5	3.9167	1.1369	5.0536
365	19.821	2.017E-03	6.012E-03	1817.8	605.1	2422.9	4.0014	0.9483	4.9497
370	21.044	2.215E-03	4.954E-03	1890.7	443.8	2334.5	4.1112	0.6900	4.8012

Source: From Lemmon, E.W., McLinden, M.O., and Friend, D.G. Thermophysical properties of fluid systems, in *NIST Chemistry WebBook*, *NIST Standard Reference Database Number 69*, P.J. Linstrom and W.G. Mallard (Eds), National Institute of Standards and Technology, Gaithersburg MD, 20899, http://webbook.nist.gov (retrieved August 13, 2010).

TABLE E.4
Thermodynamic Properties of Superheated Steam

Pressure (MPa)	Temperature (°C) 200	400	600	800	1000
0.1					
v	2.1439	3.0621	3.9752	4.8871	5.7986
ρ	0.46645	0.32658	0.25156	0.20462	0.17246
h	2875.4	3278.5	3705.6	4160.2	4642.6
s	7.8294	8.5391	9.0937	9.5621	9.9739
2.0					
v		0.15121	0.19961	0.24674	0.29342
ρ		6.6131	5.0097	4.0528	3.4081
h		3248.3	3690.7	4151.5	4637
s		7.1292	7.7043	8.179	8.5936
4.0					
v		0.073431	0.098859	0.12292	0.14652
ρ		13.618	10.115	8.1352	6.8248
h		3214.5	3674.9	4142.3	4631.2
s		6.7714	7.3705	7.8523	8.2697
6.0					
v		0.047419	0.065265	0.081648	0.09756
ρ		21.088	15.322	12.248	10.25
h		3178.2	3658.7	4133.1	4625.4
s		6.5432	7.1693	7.6582	8.0786
8.0					
v		0.034344	0.048463	0.061011	0.073079
ρ		29.117	20.634	16.39	13.684
h		3139.4	3642.4	4123.8	4619.6
s		6.3658	7.0221	7.5184	7.9419
10.0					
v		0.026436	0.038378	0.048629	0.05839
ρ		37.827	26.057	20.564	17.126
h		3097.4	3625.8	4114.5	4613.8
s		6.2141	6.9045	7.4085	7.8349
12.0					
v		0.021106	0.031651	0.040375	0.048599
ρ		47.38	31.594	24.768	20.577
h		3052	3608.9	4105.1	4608
s		6.0764	6.8054	7.3173	7.7467

TABLE E.4 (continued)
Thermodynamic Properties of Superheated Steam

Pressure (MPa)	Temperature (°C) 200	400	600	800	1000
14.0					
v		0.01724	0.026845	0.034479	0.041605
ρ		58.003	37.252	29.003	24.035
h		3002.3	3591.8	4095.8	4602.1
s		5.9459	6.7191	7.2391	7.6716
16.0					
v		0.014281	0.023238	0.030058	0.036361
ρ		70.021	43.034	33.269	27.502
h		2947.6	3574.4	4086.3	4596.3
s		5.8179	6.6421	7.1703	7.606
18.0					
v		0.011916	0.020431	0.026619	0.032282
ρ		83.924	48.945	37.566	30.977
h		2886.4	3556.8	4076.9	4590.5
s		5.6883	6.572	7.1089	7.5476
20.0					
v		0.00995	0.018185	0.023869	0.02902
ρ		100.5	54.991	41.895	34.459
h		2816.9	3539	4067.5	4584.7
s		5.5525	6.5075	7.0531	7.495
22.0					
v		0.008256	0.016347	0.02162	0.026352
ρ		121.13	61.175	46.253	37.948
h		2735.8	3521	4058	4578.9
s		5.4051	6.4473	7.002	7.447

Note: Units: v in m³/kg, ρ in kg/m³, h in kJ/kg, s in kJ/kg K.

Source: From Lemmon, E.W., McLinden, M.O., and Friend, D.G. Thermophysical properties of fluid systems, in *NIST Chemistry WebBook, NIST Standard Reference Database Number 69*, P.J. Linstrom and W.G. Mallard (Eds), National Institute of Standards and Technology, Gaithersburg MD, 20899, http://webbook.nist.gov (retrieved August 13, 2010).

TABLE E.5
Thermodynamic Properties of Helium

Pressure (MPa)	Temperature (°C) 0	200	400	600	800	1000
0.1						
v	5.6026	9.7025	13.803	17.903	22.003	26.103
ρ	0.17849	0.10307	0.072451	0.055858	0.045449	0.03831
h	1424	2462.6	3501.2	4539.8	5578.4	6617
s	27.512	30.365	32.196	33.547	34.618	35.506
2.0						
v	0.28658	0.49409	0.70166	0.90928	1.1169	1.3246
ρ	3.489427	2.0239228	1.4251917	1.0997712	0.8953353	0.7549449
h	1430.2	2468.5	3506.8	4545.2	5583.6	6622
s	21.319	24.172	26.002	27.352	28.423	29.311
4.0						
v	0.14474	0.24838	0.35209	0.45585	0.55963	0.66343
ρ	6.9089402	4.0260891	2.8401829	2.1937041	1.7868949	1.507318
h	1436.7	2474.7	3512.7	4550.8	5589	6627.3
s	19.882	22.733	24.563	25.914	26.984	27.871
6.0						
v	0.097459	0.16648	0.23558	0.30471	0.37387	0.44305
ρ	10.260725	6.0067275	4.2448425	3.2818089	2.6747265	2.2570816
h	1443.2	2480.8	3518.5	4556.4	5594.4	6632.6
s	19.043	21.893	23.722	25.072	26.143	27.03
8.0						
v	0.07382	0.12554	0.17732	0.22914	0.28099	0.33286
ρ	13.546464	7.9655887	5.6395218	4.3641442	3.5588455	3.0042661
h	1449.6	2487	3524.4	4562.1	5599.9	6637.9
s	18.447	21.297	23.126	24.476	25.546	26.433
10.0						
v	0.059637	0.10097	0.14237	0.1838	0.22527	0.26675
ρ	16.768114	9.9039319	7.0239517	5.4406964	4.4391175	3.7488285
h	1456.1	2493.2	3530.4	4567.7	5605.4	6643.2
s	17.986	20.835	22.664	24.013	25.083	25.97
12.0						
v	0.050183	0.084595	0.11907	0.15358	0.18812	0.22267
ρ	19.927067	11.82103	8.3984211	6.5112645	5.3157559	4.4909507
h	1462.6	2499.4	3536.3	4573.4	5610.9	6648.5
s	17.61	20.458	22.286	23.635	24.705	25.592

TABLE E.5 (continued)
Thermodynamic Properties of Helium

Pressure (MPa)	Temperature (°C) 0	200	400	600	800	1000
14.0						
v	0.043429	0.072899	0.10243	0.13199	0.16158	0.19119
ρ	23.026089	13.717609	9.7627648	7.5763315	6.1888848	5.2303991
h	1469	2505.7	3542.2	4579.1	5616.4	6653.8
s	17.292	20.14	21.967	23.316	24.386	25.272
16.0						
v	0.038364	0.064128	0.089946	0.1158	0.14168	0.16758
ρ	26.066104	15.593812	11.117782	8.6355786	7.0581592	5.9672992
h	1475.4	2511.9	3548.1	4584.8	5621.9	6659.2
s	17.017	19.864	21.691	23.039	24.109	24.995
18.0						
v	0.034425	0.057306	0.08024	0.10321	0.12621	0.14922
ρ	29.048656	17.45018	12.462612	9.6889836	7.9233024	6.7015145
h	1481.8	2518.1	3554.1	4590.5	5627.4	6664.6
s	16.775	19.621	21.448	22.796	23.865	24.751
20.0						
v	0.031273	0.051849	0.072476	0.093139	0.11383	0.13453
ρ	31.976465	19.286775	13.797671	10.736641	8.7850303	7.4332863
h	1488.2	2524.3	3560.1	4596.3	5632.9	6669.9
s	16.558	19.404	21.23	22.578	23.647	24.533

Note: Units: *v* in m^3/kg, ρ in kg/m^3, *h* in kJ/kg, *s* in kJ/kg K.

Source: From Lemmon, E.W., McLinden, M.O., and Friend, D.G. Thermophysical properties of fluid systems, in *NIST Chemistry WebBook, NIST Standard Reference Database Number 69*, P.J. Linstrom and W.G. Mallard (Eds), National Institute of Standards and Technology, Gaithersburg MD, 20899, http://webbook.nist.gov (retrieved August 13, 2010).

TABLE E.6
Thermodynamic Properties of CO_2

Pressure (MPa)	Temperature (°C)				
	0	200	400	600	800
0.1					
v	0.50573	0.88113	1.2547	1.6278	2.0008
ρ	1.9773	1.1349	0.79702	0.61432	0.49979
h	484.87	668.22	880.07	1111.5	1356.7
s	2.6633	3.1627	3.5349	3.8355	4.0882
2.0					
v	0.021926	0.043869	0.063497	0.082727	0.10181
ρ	45.608	22.795	15.749	12.088	9.822
h	460	661.43	876.9	1109.9	1356
s	2.0341	2.5882	2.967	3.2697	3.5233
4.0					
v		0.021538	0.031718	0.041496	0.051131
ρ		46.429	31.528	24.099	19.558
h		654.21	873.64	1108.3	1355.3
s		2.4455	2.8315	3.1363	3.3909
6.0					
v		0.014107	0.021135	0.027758	0.03424
ρ		70.889	47.315	36.026	29.205
h		646.95	870.46	1106.8	1354.6
s		2.3569	2.7504	3.0574	3.3128
8.0					
v		0.010401	0.01585	0.020893	0.025797
ρ		96.142	63.09	47.864	38.764
h		639.66	867.36	1105.3	1354
s		2.2905	2.6916	3.0007	3.257
10.0					
v		0.008188	0.012685	0.016777	0.020733
ρ		122.13	78.833	59.607	48.232
h		632.39	864.34	1103.8	1353.4
s		2.2362	2.645	2.9562	3.2134
12.0					
v		0.006722	0.010579	0.014035	0.017358
ρ		148.77	94.523	71.25	57.609
h		625.17	861.4	1102.4	1352.8
s		2.1896	2.6063	2.9195	3.1776

TABLE E.6 (continued)
Thermodynamic Properties of CO_2

Pressure	Temperature (°C)				
(MPa)	0	200	400	600	800
14.0					
v		0.005684	0.009079	0.012079	0.014949
ρ		175.93	110.14	82.789	66.895
h		618.05	858.55	1101.1	1352.3
s		2.1484	2.573	2.8882	3.1471
16.0					
v		0.004915	0.007958	0.010613	0.013143
ρ		203.46	125.66	94.22	76.088
h		611.08	855.79	1099.8	1351.8
s		2.1114	2.5436	2.8608	3.1206
18.0					
v		0.004326	0.007089	0.009475	0.011739
ρ		231.16	141.07	105.54	85.188
h		604.31	853.11	1098.6	1351.4
s		2.0776	2.5174	2.8365	3.097
20.0					
v		0.003864	0.006396	0.008566	0.010616
ρ		258.82	156.35	116.74	94.195
h		597.79	850.54	1097.4	1350.9
s		2.0465	2.4935	2.8145	3.0758

Note: Units: v in m^3/kg, ρ in kg/m^3, h in kJ/kg, s in kJ/kg K.

Source: From Lemmon, E.W., McLinden, M.O., and Friend, D.G. Thermophysical properties of fluid systems, in *NIST Chemistry WebBook*, *NIST Standard Reference Database Number 69*, P.J. Linstrom and W.G. Mallard (Eds), National Institute of Standards and Technology, Gaithersburg MD, 20899, http://webbook.nist.gov (retrieved August 13, 2010).

TABLE E.7
Thermodynamic Properties of Sodium

	Saturation Properties			Temperature (K)—Superheated Vapor					
Temperature (K) Pressure (MPa)	Sat. Liquid	Sat. Vapor	Pressure (MPa)	600	800	1000	1200	1400	1600
400.0									
2.596E-10			1.0000E-06						
v	1.0878E-03	8.0461E+11	*v*	2.1651E+08	2.8929E+08	3.6164E+08	4.3395E+08	5.0628E+08	5.7860E+08
ρ	919.27	1.243E-12	ρ	4.619E-09	3.457E-09	2.765E-09	2.304E-09	1.975E-09	1.728E-09
h	246.66	4757.05	*h*	4915.4	5103.3	5284.3	5465.1	5645.9	5826.7
s	2.9243	14.1575	*s*	11.399	11.714	11.909	12.040	12.138	12.224
800.0									
9.906E-04			1.00E-05						
v	1.2072E-03	2.9114E+05	*v*		2.8917E+07	3.6161E+07	4.3395E+07	5.0628E+07	5.7860E+04
ρ	828.35	3.435E-06	ρ		3.458E-08	2.765E-08	2.304E-08	1.975E-08	1.728E-05
h	769.22	4966.28	*h*		5101.9	5284.1	5465.1	5645.9	5826.9
s	3.8338	9.0953	*s*		10.879	11.067	11.207	11.305	11.3919
1200.0									
1.487E-01			1.00E-04						
v	1.3670E-03	2.5359E+03	*v*		2.8797E+06	3.6147E+06	4.3391E+06	5.0628E+06	5.7860E+06
ρ	731.52	3.943E-04	ρ		3.473E-07	2.767E-07	2.305E-07	1.975E-07	1.728E-07
h	1272.63	5110.51	*h*		5088.0	5282.7	5464.8	5645.8	5826.7
s	4.3458	7.5415	*s*		10.031	10.248	10.374	10.473	10.559
1600.0									
1.787E+00			1.00E-03						
v	1.5981E-03	2.5072E+02	*v*			3.5996E+05	4.3352E+05	5.0612E+05	5.7854E+05
ρ	625.73	3.989E-03	ρ			2.778E-06	2.307E-06	1.976E-06	1.729E-06

h	1812.04	5217.17	*h*			5268.9	5461.7	5644.8	5826.3
s	4.7281	6.8630	*s*			9.397	9.539	9.639	9.726
2000.0									
7.850E+00			1.0000E-02						
v	1.9847E-03	6.0516E+01	*v*			3.4670E+04	4.2970E+04	5.0459E+04	5.7777E+04
ρ	503.85	1.652E-02	ρ			2.884E-05	2.327E-05	1.982E-05	1.731E-05
h	3858.32	5272.64	*h*			5146.80	5423.40	5634.80	5821.80
s	5.0635	6.4802	*s*			8.456	8.685	8.801	8.891
2400.0									
2.089E+01			1.0000E-01						
v	2.9894E-03	1.7318E+01	*v*				3.9955E+03	4.9068E+03	5.7039E+03
ρ	334.52	5.774E-02	ρ				2.503E-04	2.038E-04	1.753E-04
h	4204.30	5077.61	*h*				5202.10	5.54E+03	5779.40
s	5.4467	6.1662	*s*				7.689	7.913	8.036
2503.7									
2.556E+01			1.0000E+00						
v	4.5662E-03	4.5662E-03	*v*						5.1235E-01
ρ	219.00	2.190E+02	ρ						1.952E+00
h	4294.00	4294.00	*h*						5480.8
s	5.7412	5.9126	*s*						7.187

Note: Units: *v* in m^3/kg, ρ in kg/m^3, *h* in kJ/kg, *s* in kJ/kg K.

Source: From J. K. Fink, *Tables of Thermodynamic Properties of Sodium,* ANL-CEN-RSD-82-4, Argonne National Laboratory, Argonne, IL, 1982.

TABLE E.8
Thermodynamic Transport Properties of Liquid Sodium at Atmospheric Pressure

Temperature (°C)	Density (kg/m^3)	Enthalpy (kJ/kg)	Specific Heat (kJ/kg K)	Thermal Conductivity (W/m K)	Viscosity (Pa s)	Surface Tension (N/m)
100	888.62	209.69	1.3822	89.276	6.808E-04	0.2005
125	880.76	244.12	1.3727	87.364	6.042E-04	0.1979
150	872.91	278.31	1.3632	85.503	5.430E-04	0.1952
175	865.05	312.25	1.3537	83.692	4.933E-04	0.1926
200	857.19	345.95	1.3442	81.930	4.521E-04	0.1900
225	849.33	379.42	1.3347	80.216	4.176E-04	0.1873
250	841.48	412.67	1.3264	78.549	3.882E-04	0.1847
275	833.62	445.70	1.3181	76.928	3.631E-04	0.1821
300	825.76	478.53	1.3099	75.352	3.413E-04	0.1795
325	817.90	511.18	1.3016	73.819	3.222E-04	0.1769
350	810.05	543.64	1.2954	72.329	3.054E-04	0.1742
375	802.19	575.94	1.2894	70.880	2.905E-04	0.1716
400	794.33	608.08	1.2834	69.471	2.772E-04	0.1690
425	786.48	640.07	1.2774	68.102	2.652E-04	0.1664
450	778.62	671.94	1.2731	66.770	2.545E-04	0.1638
475	770.76	703.69	1.2688	65.476	2.447E-04	0.1613
500	762.90	735.33	1.2646	64.217	2.358E-04	0.1587
525	755.05	766.89	1.2603	62.993	2.276E-04	0.1561
550	747.19	798.36	1.2581	61.802	2.202E-04	0.1535
575	739.33	829.77	1.2561	60.644	2.133E-04	0.1510
600	731.47	861.13	1.2541	59.518	2.069E-04	0.1484
625	723.62	892.45	1.2521	58.421	2.010E-04	0.1458
650	715.76	923.75	1.2520	57.354	1.955E-04	0.1433
675	707.90	955.03	1.2520	56.315	1.904E-04	0.1407
700	700.04	986.32	1.2520	55.302	1.856E-04	0.1382
725	692.19	1017.63	1.2520	54.316	1.812E-04	0.1356
750	684.33	1048.96	1.2541	53.354	1.770E-04	0.1331
775	676.47	1080.33	1.2563	52.416	1.730E-04	0.1306
800	668.61	1111.77	1.2586	51.500	1.693E-04	0.1281
825	660.76	1143.27	1.2608	50.606	1.658E-04	0.1255
850	652.90	1174.86	1.2652	49.732	1.625E-04	0.1230

Source: From J. K. Fink and L. Leibowitz, by *High Temperature and Materials Science* Vol. 35, 65–103, 1996.

REFERENCES

Fink, J. K. *Tables of Thermodynamic Properties of Sodium,* ANL-CEN-RSD-82-4, Argonne National Laboratory, Argonne, IL, 1982.

Fink, J. K. and Leibowitz, L. A consistent assessment of the thermophysical properties of sodium, *High Temperature and Materials Science* Vol. 35, 65–103, 1996.

Lemmon, E.W., McLinden, M.O., and Friend, D.G. Thermophysical properties of fluid systems, in *NIST Chemistry WebBook, NIST Standard Reference Database Number 69*, P.J. Linstrom and W.G. Mallard (Eds), National Institute of Standards and Technology, Gaithersburg MD, 20899, http://webbook.nist.gov (retrieved August 13, 2010).

Appendix F
Thermophysical Properties of Some Substances

Figures F.1 through F.4 presenting the thermophysical properties of some fluids have been adopted from Poppendiek, H. F. and Sabin, C. M. *Some Heat Transfer Performance Criteria for High Temperature Fluid Systems* (American Society of Mechanical Engineers, 75-WA/HT-103, 1975) except for sodium, which is from *Liquid Metals Handbook* (NAV EXOS P-733 Rev., Lyon, R. N., ed., Office of Naval Research, Washington, DC, June 1952.)

Caution: The property values are presented in a comparative manner, which may not yield the accuracy needed for detailed assessments. In such cases, recent tabulated property listings should be consulted.

Note: For dynamic and kinematic fluid viscosities, see Figures 9.10 and 9.11, respectively. For thermal conductivity of engineering materials, see Figure 10.1.

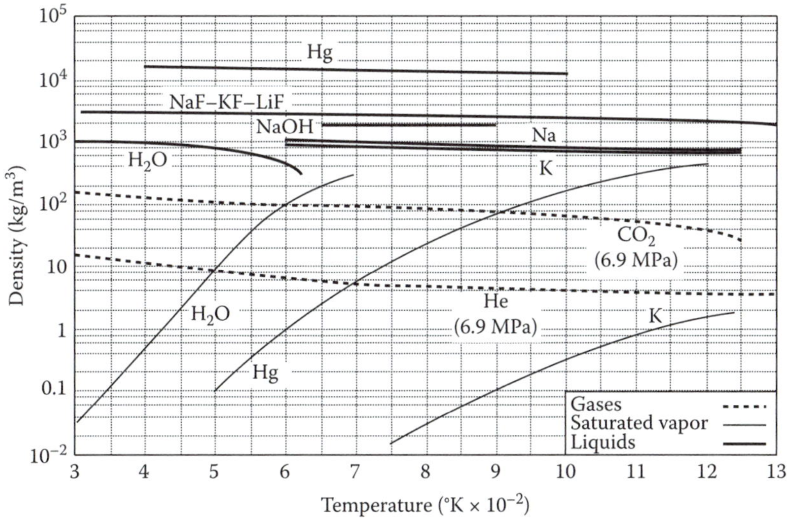

FIGURE F.1 Density versus temperature.

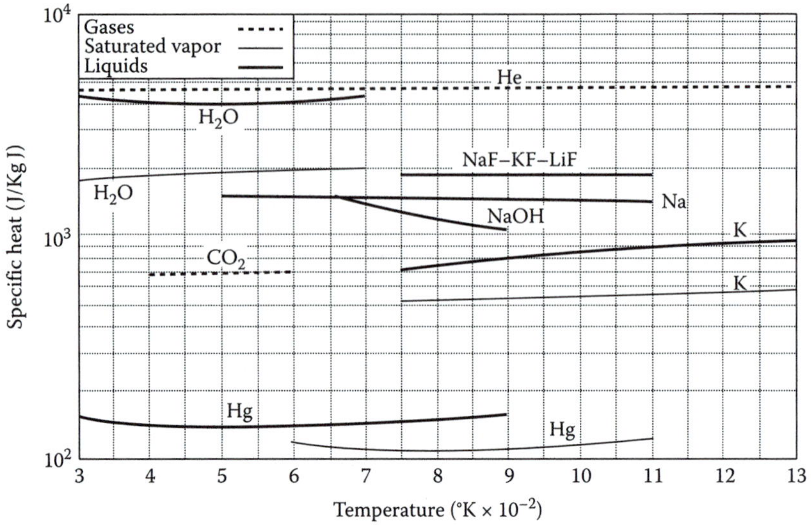

FIGURE F.2 Specific heat versus temperature.

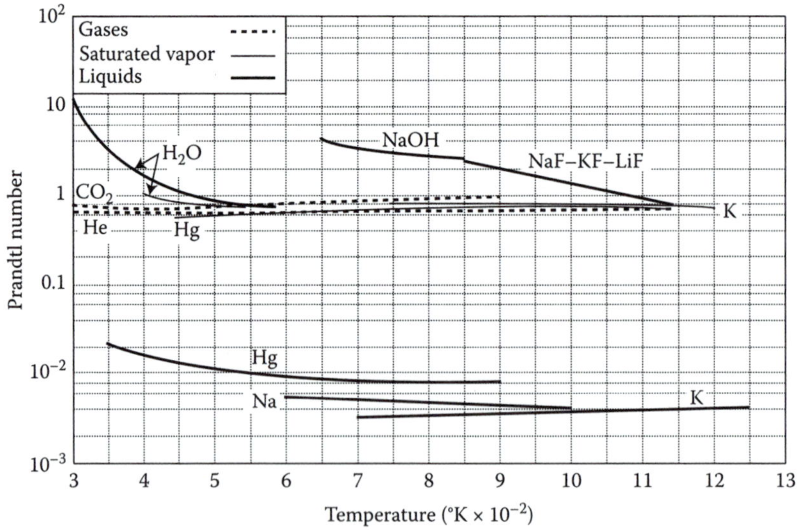

FIGURE F.3 Prandtl number versus temperature.

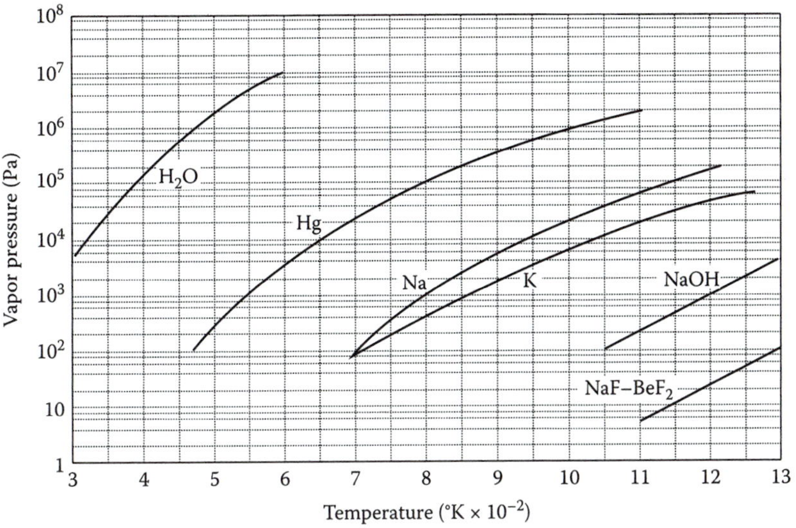

FIGURE F.4 Vapor pressure versus temperature.

TABLE F.1
Thermal Properties of Solids: Metals

	Properties at 20°C				k (W/m °C)				
Metal	ρ (kg/m^3)	c_p (J/kg °C)	k (W/m °C)	α (m^2/s)	100°C	200°C	300°C	400°C	600°C
Aluminum, pure	2707	896	204	8.42×10^{-5}	206	215	229	249	
Duralumin, 94–96 Al, 3–5 Cu	2787	883	164	6.68×10^{-5}	182	194			
Lead	11,370	130	34.6	2.34×10^{-5}	33.4	31.5	29.8		
Iron, pure	7897	452	72.7	2.03×10^{-5}	67.5	62.3	55.4	48.5	39.8
Iron, wrought, C < 0.5%	7849	460	58.9	1.63×10^{-5}	57.1	51.9	48.5	45.0	36.4
Iron, cast, C ≈ 4%	7272	419	51.9	1.70×10^{-5}					
Carbon steel, C ≈ 0.5%	7833	465	53.7	1.47×10^{-5}	51.9	48.5	45.0	41.5	34.6
Carbon steel, C = 1.5%	7753	486	36.4	0.97×10^{-5}	36.3	36.3	34.6	32.9	31.2
Nickel steel, 10%	7945	460	26.0	0.72×10^{-5}					
Nickel steel, 30%	8073	460	12.1	0.33×10^{-5}					
Nickel steel, 50%	8266	460	13.8	0.36×10^{-5}					
Nickel steel, 70%	8506	460	26.0	0.67×10^{-5}					
Nickel steel, 90%	8762	460	46.7	1.16×10^{-5}					
Chrome steel, 1%	7865	460	60.6	1.67×10^{-5}	55.4	51.9	46.7	41.5	36.4
Chrome steel, 5%	7833	460	39.8	1.11×10^{-5}	38.1	36.4	36.4	32.9	29.4
Chrome steel, 10%	7785	460	31.2	0.87×10^{-5}	31.2	31.2	29.4	29.4	31.2
Cr–Ni steel, 18% Cr, 8% Ni	7817	460	16.3	0.44×10^{-5}	17.3	17.3	19.0	19.0	22.5
Ni–Cr steel, 20% Ni, 15% Cr	7865	460	14.0	0.39×10^{-5}	15.1	15.1	16.3	17.3	19.0
Manganese steel, 2%	7865	460	38.1	1.05×10^{-5}	36.4	36.4	36.4	34.6	32.9
Tungsten steel, 2%	7961	444	62.3	1.76×10^{-5}	58.9	53.7	48.5	45.0	36.4
Silicon steel, 2%	7673	460	31.2	0.89×10^{-5}					

Copper, pure	8954	383	386	11.2×10^{-5}	379	374	369	364	353
Bronze, 75 Cu, 25 Sn	8660	343	26.0	0.86×10^{-5}					
Brass, 70 Cu, 30 Zn	8522	385	111	3.41×10^{-5}	128	144	147	147	
German silver, 62 Cu 15 Ni, 22 Zn	8618	394	24.9	0.73×10^{-5}	31.2	39.8	45.0	48.5	
Constantan, 60 Cu, 40 Ni	8922	410	22.7	0.61×10^{-5}	22.2	26.0			
Magnesium, pure	1746	1013	171	9.71×10^{-5}	168	163	158		
Molybdenum	10,220	251	123	4.79×10^{-5}	118	114	111	109	106
Nickel, 99.9% pure	8906	446	90.0	2.27×10^{-5}	83.1	72.7	64.0	58.9	
Silver, 99.9% pure	10,520	234	407	16.6×10^{-5}	415	374	362	360	
Tungsten	19,350	134	163	6.27×10^{-5}	151	142	133	126	113
Zinc, pure	7144	384	112	4.11×10^{-5}	109	106	100	93.5	
Tin, pure	7304	227	64.0	3.88×10^{-5}	58.9	57.1			

Source: Adopted from Eckert, E. R. G. and Drake, Jr., R.M. *Heat and Mass Transfer*. New York, NY: McGraw-Hill, 1959.

REFERENCES

Collier J. G. *Convective Boiling and Condensation*, 2nd Ed., New York, NY: McGraw-Hill, 1981.

Eckert, E. R. G. and Drake, Jr., R.M. *Heat and Mass Transfer*. New York, NY: McGraw-Hill, 1959.

Poppendiek, H. F. and Sabin, C. M. *Some Heat Transfer Performance Criteria for High Temperature Fluid Systems*. American Society of Mechanical Engineers, 75-WA/HT-103, 1975.

Appendix G

Dimensionless Groups of Fluid Mechanics and Heat Transfer

Name	Notation	Formula	Interpretation in Terms of Ratio
Biot number	Bo	$\frac{hL}{k_s}$	Surface conductance ÷ internal conduction of solid
Cauchy number	Ca	$\frac{V^2}{B_s/\rho}=\frac{V^2}{a^2}$	Inertia force ÷ compressive force = (Mach number)^2
Eckert number	Ec	$\frac{V^2}{c_p \Delta T}$	Temperature rise due to energy conversion ÷ temperature difference
Euler number	Eu	$\frac{\Delta p}{\rho V^2}$	Pressure force ÷ inertia force
Fourier number	Fo	$\frac{kt}{\rho c_p L^2}=\frac{\alpha t}{L^2}$	Rate of conduction of heat ÷ rate of storage of energy
Froude number	Fr	$\frac{V^2}{gL}$	Inertia force ÷ gravity force
Graetz number	Gz	$\frac{D}{L}\cdot\frac{V\rho c_p D}{k}$	Re Pr ÷ *(L/D)*; heat transfer by convection in entrance region ÷ heat transfer by conduction
Grashof number	Gr	$\frac{g\beta \Delta T L^3}{\nu^2}$	Buoyancy force ÷ viscous force
Knudsen number	Kn	$\frac{\lambda}{L}$	Mean free path of molecules ÷ characteristic length of an object
Lewis number	Le	$\frac{\alpha}{D_c}$	Thermal diffusivity ÷ molecular diffusivity
Mach number	M	$\frac{V}{a}$	Macroscopic velocity ÷ speed of sound
Nusselt number	Nu	$\frac{hL}{k}$	Temperature gradient at wall ÷ overall temperature difference
Péclet number	Pé	$\frac{V\rho c_p D}{k}$	(Re Pr); heat transfer by convection ÷ heat transfer by conduction

continued

(continued)

Name	Notation	Formula	Interpretation in Terms of Ratio
Prandtl number	Pr	$\frac{\mu c_p}{k} = \frac{v}{\alpha}$	Diffusion of momentum ÷ diffusion of heat
Reynolds number	Re	$\frac{\rho VL}{\mu} = \frac{VL}{v}$	Inertia force ÷ viscous force
Schmidt number	Sc	$\frac{\mu}{\rho D_c} = \frac{v}{D_c}$	Diffusion of momentum ÷ diffusion of mass
Sherwood number	Sh	$\frac{h_D L}{D_c}$	Mass diffusivity ÷ molecular diffusivity
Stanton number	St	$\frac{h}{V\rho c_p} = \frac{h}{c_p G}$	Heat transfer at wall ÷ energy transported by stream
Stokes number	Sk	$\frac{\Delta p L}{\mu V}$	Pressure force ÷ viscous force
Strouhal number	SI	$\frac{L}{tV}$	Frequency of vibration ÷ characteristic frequency
Weber number	We	$\frac{\rho V^2 L}{\sigma}$	Inertia force ÷ surface tension force

Appendix H
Multiplying Prefixes

tera	T	10^{12}
giga	G	10^{9}
mega	M	10^{6}
kilo	k	10^{3}
hecto	h	10^{2}
deca (deka)	da	10^{1}
deci	d	10^{-1}
centi	c	10^{-2}
milli	m	10^{-3}
micro	μ	10^{-6}
nano	n	10^{-9}
pico	p	10^{-12}
femto	f	10^{-15}
atto	a	10^{-18}

Appendix I
List of Elements

Atomic Number	Symbol	Name	Atomic Weight[a]	Atomic Number	Symbol	Name	Atomic Weight[a]
1	H	Hydrogen	1.00794 (7)	36	Kr	Krypton	83.80
2	He	Helium	4.002602 (2)	37	Rb	Rubidium	85.4678 (3)
3	Li	Lithium	6.941 (2)	38	Sr	Strontium	87.62
4	Be	Beryllium	9.01218	39	Y	Yttrium	88.9059
5	B	Boron	10.811 (5)	40	Zr	Zirconium	91.224 (2)
6	C	Carbon	12.011	41	Nb	Niobium	92.9064
7	N	Nitrogen	14.0067	42	Mo	Molybdenum	95.94
8	O	Oxygen	15.9994 (3)	43	Tc	Technetium	[98]
9	F	Fluorine	18.998403	44	Ru	Ruthenium	101.07 (2)
10	Ne	Neon	20.179	45	Rh	Rhodium	102.9055
11	Na	Sodium	22.98977	46	Pd	Palladium	106.42
12	Mg	Magnesium	24.305	47	Ag	Silver	107.8682 (3)
13	Al	Aluminum	26.98154	48	Cd	Cadmium	112.41
14	Si	Silicon	28.0855 (3)	49	In	Indium	114.82
15	P	Phosphorus	30.97376	50	Sn	Tin	118.710 (7)
16	S	Sulfur	32.066 (6)	51	Sb	Antimony	121.75 (3)
17	Cl	Chlorine	35.453	52	Te	Tellurium	127.60 (3)
18	Ar	Argon	39.948	53	I	Iodine	126.9045
19	K	Potassium	39.0983	54	Xe	Xenon	131.29 (3)
20	Ca	Calcium	40.078 (4)	55	Cs	Cesium	132.9054
21	Sc	Scandium	44.95591	56	Ba	Barium	137.33
22	Ti	Titanium	47.88 (3)	57	La	Lanthanum	138.9055 (3)
23	V	Vanadium	50.9415	58	Ce	Cerium	140.12
24	Cr	Chromium	51.9961 (6)	59	Pr	Praseodymium	140.9077
25	Mn	Manganese	54.9380	60	Nd	Neodymium	144.24 (3)
26	Fe	Iron	55.847 (3)	61	Pm	Promethium	[145]
27	Co	Cobalt	58.9332	62	Sm	Samarium	150.36 (3)
28	Ni	Nickel	58.69	63	Eu	Europium	151.96
29	Cu	Copper	63.546 (3)	64	Gd	Gadolinium	157.25 (3)
30	Zn	Zinc	65.39 (2)	65	Tb	Terbium	158.9254
31	Ga	Gallium	69.723 (4)	66	Dy	Dysprosium	162.50 (3)
32	Ge	Germanium	72.59 (3)	67	Ho	Holmium	164.9304
33	As	Arsenic	74.9216	68	Er	Erbium	167.26 (3)
34	Se	Selenium	78.96 (3)	69	Tm	Thulium	168.9342
35	Br	Bromine	79.904	70	Yb	Ytterbium	173.04 (3)

continued

(continued)

Atomic Number	Symbol	Name	Atomic Weight[a]	Atomic Number	Symbol	Name	Atomic Weight[a]
71	Lu	Lutetium	174.967	89	Ac	Actinium	227.0278
72	Hf	Hafnium	178.49 (3)	90	Th	Thorium	232.0381
73	Ta	Tantalum	180.9479	91	Pa	Protactinium	231.0359
74	W	Tungsten	183.85 (3)	92	U	Uranium	238.0289
75	Re	Rhenium	186.207	93	Np	Neptunium	237.0482
76	Os	Osmium	190.2	94	Pu	Plutonium	[244]
77	Ir	Iridium	192.22 (3)	95	Am	Americium	[243]
78	Pl	Platinum	195.08 (3)	96	Cm	Curium	[247]
79	Au	Gold	196.9665	97	Bk	Berkelium	[247]
80	Hg	Mercury	200.59 (3)	98	CI	Californium	[251]
81	Tl	Thallium	204.383	99	Es	Einsteinium	[252]
82	Pb	Lead	207.2	100	Fm	Fermium	[257]
83	Bi	Bismuth	208.9804	101	Md	Mendelevium	[258]
84	Po	Polonium	[209]	102	No	Nobelium	[259]
85	Al	Astatine	[210]	103	Lr	Lawrencium	[260]
86	Rn	Radon	[222]	104	Rf	Rutherfordium[b]	[261]
87	Fr	Francium	[223]	105	Ha	Hahnium[b]	[262]
88	Ra	Radium	226.0254	106	Unnamed	Unnamed	[263]

Source: From Walker, F. W., Miller, D. G., and Feiner, F. *Chart of the Nuclides*, 13th Ed. Revised 1983, General Electric Co. Schenectady, NY.

[a] Values in parentheses are the uncertainty in the last digit of the stated atomic weights. Values without a quoted error in parentheses are considered to be reliable to ±1 in the last digit except for Ra, Ac, Pa, and Np, which were not given by Walker et al. (1983). Brackets indicate the most stable or best known isotope.

[b] The names of these elements have not been accepted because of conflicting claims of discoverer.

REFERENCE

Walker, F. W., Miller, D. G., and Feiner, F. *Chart of the Nuclides*, 13th Ed. Revised 1983, General Electric Co. Schenectady, NY.

Appendix J
Square and Hexagonal Rod Array Dimensions

J.1 LWR FUEL BUNDLES: SQUARE ARRAYS

Tables J.1 and J.2 present formulas for determining axially averaged unit subchannel and overall bundle dimensions, respectively.

The presentation of these formulas as axially averaged values is arbitrary and reflects the fact that grid-type spacers occupy a small fraction of the axial length of a fuel bundle. Therefore this fraction (δ) has been defined as

$$\delta = \frac{\text{Total axial length of grid spacers}}{\text{Axial length of the fuel bundle}}$$

Formulas applicable at the axial grid locations or between grids are easily obtained from Tables J.1 and J.2 by taking $\delta = 1$ or $\delta = 0$, respectively.

Determination of precise dimensions for an LWR assembly would require knowledge of the specific grid configuration used by the manufacturer. Typically, the grid strap thickness at the periphery is slightly enhanced. Here it is taken as thickness t^*. It should also be carefully noted that these formulas are based on the assumptions of a rectangular grid, and no support tabs or fingers. The dimension (g) is the spacing from rod surface to the flow boundary of the assembly. In the grid plane, the segment along g open for flow is $g - t/2$.

J.2 LMFBR FUEL BUNDLES: HEXAGONAL ARRAYS

Tables J.3 and J.4 summarize the formulas for determining axially averaged unit subchannel and overall bundle dimensions, respectively. Fuel pin spacing is illustrated as being performed by a wire wrap. In practice, both grids and wires are used as spacers in LMFBR fuel and blanket bundles. The axially averaged dimensions in these tables are based on averaging the wires over one lead length.

TABLE J.1
Square Arrays: Axially Averaged Unit Subchannel Dimensions for Ductless Assembly

	$i \equiv 1$ Interior	$i \equiv 2$ Edge	$i \equiv 3$ Corner
	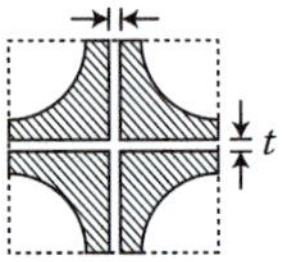	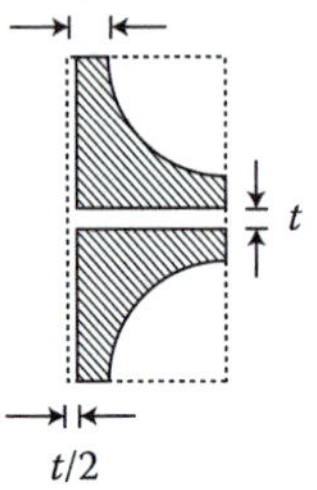	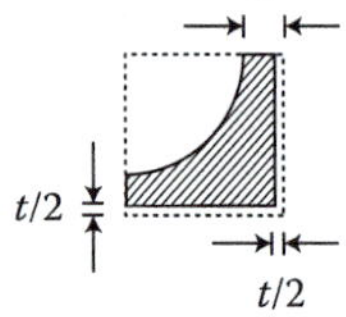
Area			
A_{fi} (area for flow without spacer)	$P^2 - \frac{\pi D^2}{4}$	$P\left(\frac{D}{2}+g\right)-\frac{\pi D^2}{8}$	$\left(\frac{D}{2}+g\right)^2-\frac{\pi D^2}{16}$
A_{fi_s} (average area for flow with spacer)	$A_{fi} - (2Pt - t^2)\delta$	$A_{f2} - \left[\left(\frac{D}{2}+g-t^*\right)t+\frac{Pt^*}{2}\right]\delta$	$A_{f3} - \left[2\left(\frac{D}{2}+g\right)t^* - t^{*2}\right]$
P_{wi} (wetted perimeter without spacer and duct)	πD	$\frac{\pi}{2}D$	$\frac{\pi}{4}D$
P_{wi_s} (average wetted perimeter including spacer but without duct)	$\pi D + 4(P - t)\delta$	$\frac{\pi}{2}D+\left[\left(\frac{D}{2}+g-t^*\right)2+(P-t)\right]\delta$	$\frac{\pi}{4}D+\left(\frac{D}{2}+g-t^*\right)2\delta$

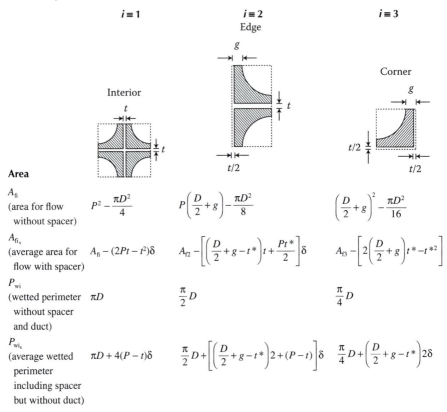

TABLE J.2
Square Arrays: Axially Averaged Overall Dimensions for Ductless Assembly

Assuming grid spacer around N_p rods of thickness t and that $g = \frac{P-D}{2}$

1. Total area inside square (A_T):

$$A_T = D_\ell^2$$

where D_ℓ = length of one side of the square.

2. Total average cross-sectional area for flow (A_{fT}):

$$A_{fT} = D_\ell^2 - N_p \frac{\pi}{4} D^2 - N_p \frac{t}{2} P4\delta + N_p \left(\frac{t}{2}\right)^2 4\delta$$

$$A_{fT} = D_\ell^2 - \left[N_p \frac{\pi}{4}(D^2) + 2\delta\left(\sqrt{N_p}\right)\left(D_\ell t\right) - N_p t^2 \delta \right]$$

Because

$$D_\ell = \sqrt{N_p} P \text{ and } \quad t^* = t$$

where

D = rod diameter

N_p = number of rods

t = interior spacer thickness

t^* = peripheral spacer thickness

3. Total average wetted perimeter (P_{wT})

$$P_{wT} = N_p \pi D + 4\sqrt{N_p} D_\ell \delta - 4N_p t\delta$$

4. Equivalent hydraulic diameter for overall square (D_{eT})

$$D_{eT} = \frac{4A_{fT}}{P_{wT}} = \frac{4D_\ell^2 - \left[N_p \pi D^2 + 8\sqrt{N_p} D_\ell t\delta - 4N_p t^2 \delta \right]}{N_p \pi D + 4\sqrt{N_p} D_\ell \delta - 4N_p t\delta}$$

TABLE J.3
Hexagonal Array with Wire Wrap Spacer: Axially Averaged Unit Subchannel Dimensions

	$i \equiv 1$	$i \equiv 2$	$i \equiv 3$
	Interior 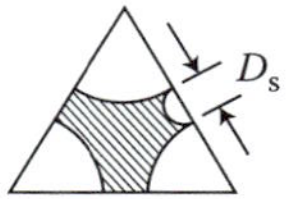	Edge 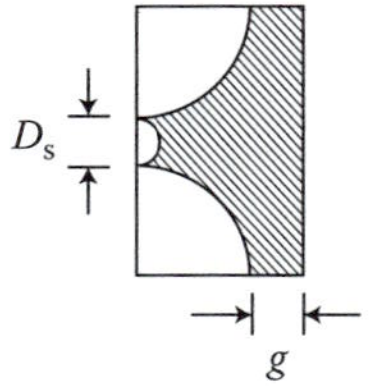	Corner 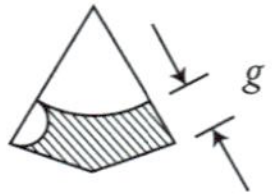
Area			
A_{fi} (area for flow without wire wrap spacers)	$\frac{1}{2}P\left(\frac{\sqrt{3}}{2}P\right)-\frac{\pi D^2}{8}=\frac{\sqrt{3}}{4}P^2-\frac{\pi D^2}{8}$	$P\left(\frac{D}{2}+g\right)-\frac{\pi D^2}{8}$	$\left[\frac{1}{\sqrt{3}}\left(\frac{D}{2}+g\right)^2-\frac{\pi D^2}{24}\right]$
A_{fi_s} (area for flow including wire wrap spacers)	$A_{f1}-\left(\frac{3}{6}\right)\frac{\pi}{4}D_s^2=A_{f_1}\frac{\pi D_s^2}{8}$ (three spacers traverse cell per unit lead; each traverse is 60° of 360°)	$A_{f2}-\left(\frac{2}{4}\right)\frac{\pi D_s^2}{4}$ (two spacers traverse cell per unit lead; each traverse is 90° of 360°)	$A_{f3}-\left(\frac{1}{6}\right)\frac{\pi}{4}D_s^2$ (one spacers traverses cell per unit lead; each traverse is 60° of 360°)
P_{wi_s} (wetted perimeter including wire wrap spacers)	$\frac{\pi D}{2}+\frac{\pi D_s}{2}$ (three spacers traverse cell per unit lead; each traverse is 60° of 360°)	$\frac{\pi D}{2}+P+\frac{\pi D_s}{2}$ (two spacers traverse cell per unit lead; each traverse is 90° of 360°)	$\frac{\pi}{6}(D+D_s)+\frac{2}{\sqrt{3}}\left(\frac{D}{2}+g\right)$ (one spacers traverses cell per unit lead; each traverse is 60° of 360°)

Note: $g \equiv$ spacer diameter plus wall-pin clearance, if any.

TABLE J.4

Hexagonal Array: Axially Averaged Overall Dimensions Assuming Wire Wrap Spacers Around Each Rod

1. Total area inside hexagon (A_{hT})

$$A_{hT} = D_{ft} D_{\ell} + 2\left(\frac{1}{2}\right) D_{ft} D_{\ell} \sin 30^{\circ} = \frac{\sqrt{3}}{2} D_{ft}^2$$

Because

$$D_{\ell} = \frac{D_{ft}}{2} \frac{1}{\cos 30^{\circ}} = \frac{\sqrt{3}}{3} D_{ft}$$

where

D_{ℓ} = length of one side of the hexagon

D_{ft} = distance across flats of the hexagon; for a bundle considering clearances or tolerances between rods and duct

Now

$$D_{\ell} = (N_{ps} - 1)(D + D_s) + \frac{2\sqrt{3}}{3}\left(\frac{D}{2} + g\right)$$

$$D_{ft} = 2\left[\left(\sqrt{\frac{3}{2}}\right) N_{rings}(D + D_s) + \frac{D}{2} + g\right]$$

D = rod diameter

D_s = wire wrap diameter

g = rod to duct spacing

N_p = number of rods

N_{ps} = number of rods along a side

N_{rings} = number of rings

2. Total cross-sectional area for flow (A_{fT})

$$A_{ft} = A_{hT} - N_p \frac{\pi}{4}(D^2 + D_s^2)$$

3. Total wetted perimeter (P_{wT})

$$P_{wT} = 6D_{\ell} + N_p \pi D + N_p \pi D_s = 2\sqrt{3} D_{ft} + N_p \pi (D + D_s)$$

4. Equivalent diameter for overall hexagonal array (D_{eT})

$$D_{eT} = \frac{4A_{fT}}{P_{wT}} = \frac{2\sqrt{3} D_{ft}^2 - N_p \pi \left(D^2 + D_s^2\right)}{2\sqrt{3} D_{ft} + N_p \pi \left(D + D_s\right)}$$

Appendix K

Parameters for Typical BWR-5 and PWR Reactors

TABLE K.1
Key Characteristics of the Nine Mile Point 2 General Electric BWR-5 with GE11 Fuel and of the Seabrook Station PWR

Parameters	Units	BWR	Sources	PWR	Sources
Reactor General Parameters		**Nine Mile Point 2, GE BWR-5**		**Seabrook Station Reactor**	
Thermal power, $\dot{Q}_{th}$	MWth	3323	A	3411	G
Net electric power, $\dot{Q}_e$	MWe	1062	B	1148	B
Efficiency, η	%	32.0	(1)	33.7	(1)
Nominal pressure, p	MPa	7.14	A	15.51	G
Steam dome pressure, p_{dome}	MPa	7.03	A	—	—
Total core pressure drop, Δp_{core}	MPa	0.171	A	0.197	G
Final feedwater temperature	°C	215.6	A	—	—
Core inlet temperature, T_{in}	°C	278.3	A	293.1	G
Core exit temperature, T_{exit}	°C	286.1	(2)	326.8	(12)
Core average exit quality, x	%	14.6	C	—	—
Total steam flow rate, $\dot{m}_{steam}$	kg/s	1798	A	—	—
Core coolant flow rate, $\dot{m}_{core}$	kg/s	13671	A	17476[a]	G
Number of assemblies, N_a	—	764	B	193	G
Active core equivalent diameter	m	4.75	B	3.37	B
Coolant mass in primary circuit	t	260	B	354	B
Fuel enrichment (initial core), r	%	0.7/1.8/2.2	B	1.6/2.4/3.1	B
Fuel enrichment (reloads), r	%	3.5	B	3.1/3.4/4.2[c]	B
Number of loops	—	2	B	4	B
Cycle length	months	16[d]	B	12[d]	B
Average discharge burnup	MWd/tU	32300	B	33000	B
Fuel inventory	tHM	141	B	89	(4)
	t(UO_2)	160	(4)	101	G
Average core power density	kW_{th}/L	52.3	(3)	104.5	(3)
Average core specific power	kW_{th}/kg_{HM}	23.6	(15)	38.3	(15)

continued

TABLE K.1 (continued)
Key Characteristics of the Nine Mile Point 2 General Electric BWR-5 with GE11 Fuel and of the Seabrook Station PWR

Parameters	Units	BWR	Sources	PWR	Sources
Reactor General Parameters		**Nine Mile Point 2, GE BWR-5**		**Seabrook Station Reactor**	
Configuration	—	9×9	D	17×17	G
Fuel rods per assembly, N_{rods}	—	74	D	264	G
Number of part length fuel rods	—	8	D	—	—
Number of full length fuel rods	—	66	D	—	—
Number of water rods, N_{wr}	—	2	E	—	—
Channel width, l_{ch}	mm	134.1 (inside)	E	214.0	G
		138.6 (outside)	H		
Assembly pitch, l	mm	152.4	B	215.0	G
Core average flow rate per assembly, $\dot{m}_a$	kg/s	15.4	(5)	89.8	(13)
Assembly flow area, A_{fa}	m^2	9.718×10^{-3}[b]	(6)	2.444×10^{-2}	(14)
Core average assembly mass flux, G_a	$kg\ m^{-2}\ s^{-1}$	1584[b]	(7)	3675.4	G
Fuel Rods		**GE11, 9 × 9 fuel**		**Seabrook Station Reactor**	
Pellet percent of theoretical density	—	97	D	95	G
Rod-to-rod pitch, P	mm	14.37	D	12.6	G
Fuel rod outside diameter, D	mm	11.20	D	9.5	G
Cladding thickness, t_{clad}	mm	0.71	D	0.572	G
Fuel-cladding gap (cold), t_{gap}	mm	0.09	(8)	0.0826	G
Fuel pellet diameter, D_f	mm	9.60	D	8.192	G
Fuel pellet length, L_f	mm	10	D	9.8	G
Diameter of water rods, D_{wr}	mm	24.9	E	—	—
Total fuel rod height	m	4.09	D	3.876	G
Heated fuel height, L	m	3.588	D	3.658	G
Part length rod length	m	2.286	D	—	—
% of energy deposited in the fuel rods	%	96.5	I	97.4	G
Peak LHGR, q'_0	kW/m	47.24	D	44.62	G
Core average LHGR, $\langle q' \rangle$	kW/m	17.6	F	17.86	G
Core average subchannel flow rate, $\dot{m}_{ct}$	kg/s	0.175 (interior)	(9)	0.335 (interior)	(9)
		0.134 (edge)		0.159 (edge)	
		0.0922 (corner)		0.0759 (corner)	
Subchannel flow area, A_{fch}	m^2	1.08×10^{-4} (int)	(10)	8.79×10^{-5} (int)	(10)
		8.83×10^{-5} (edg)		4.27×10^{-5} (edg)	
		6.70×10^{-5} (cor)		2.07×10^{-5} (cor)	

continued

TABLE K.1 (continued)
Key Characteristics of the Nine Mile Point 2 General Electric BWR-5 with GE11 Fuel and of the Seabrook Station PWR

Parameters	Units	BWR	Sources	PWR	Sources
Core average subchannel mass flux, G_{ch}	kg m^{-2} s^{-1}	1625 (int) 1514 (edg) 1378 (cor)	(11)	3807 (int) 3734 (edg) 3661 (cor)	(11)

[a] Effective flow rate for heat transfer (total minus: flow through thimble tubes, leakage from barrel-baffle into core, head cooling flow, leakage to the vessel outlet nozzle).
[b] Assuming all the fuel rods being full-length rods.
[c] Many PWRs now reload about 4.5%
[d] Typically USA BWRs and PWRs now operate on 24- and 18-month cycle lengths.

References

A Nine Mile Point 2 Nuclear Plant, Updated Safety Analysis Report, Table 4.4-1, USAR Revision 8, Oswego, New York, November 1995.
B *2009 World Nuclear Industry Handbook*, Wilmington Media, Kent, United Kingdom, 2009.
C Typical BWR value.
D Anonymous, Fuel design data, *Nuclear Engineering International*, 52, 638, Sciences Module, p.32, Sept 2007.
E General Electric Company, Retransmittal of Response to Request for Additional Information (RAI) for ESBWR Pre-application Review, San Jose, CA, 2003.
F Watford, G. A., GE 10 × 10 Advanced BWR fuel design, ANS topical meeting on Advances in Nuclear Fuel Management II, Myrtle Beach, SC, 1997.
G Ferroni, P., Hejzlar, P., and Todreas N., Compilation of Thermal–hydraulic and geometric data of the Seabrook Nuclear Power Plant, Unpublished, 2006.
H BWR/6, General Description of a Boiling Water Reactor, General Electric Co., San Jose, CA, revised 1980.
I Author's best estimate.

Calculations

(1) $$\eta = \frac{\text{Net electric power}}{\text{Thermal power}}$$

(2) $T_{exit} = T_{sat}\,(p = 7.03\ MPa)$. Saturation temperature of water at the steam dome pressure.

(3) $$\text{Average core power density} = \frac{\text{Thermal power}}{L\dfrac{\pi}{4}\left(\text{active core equivalent diameter}\right)^2}$$

(4) $$t\left(UO_2\right) = \frac{t\left(HM\right)}{f_{HM}}$$

The heavy metal fraction is given by Equation 2.21:

$$f_{HM} = \frac{rM_{ff} + (1-r)M_{nf}}{rM_{ff} + (1-r)M_{nf} + M_{O_2}} \tag{2.21}$$

The values in Table K.1 are obtained using the average reload enrichments for BWRs and PWRs.

continued

TABLE K.1 (continued)
Key Characteristics of the Nine Mile Point 2 General Electric BWR-5 with GE11 Fuel and of the Seabrook Station PWR

(5) $\dot{m}_a = \frac{(1-0.10-0.04)\times \dot{m}_{core}}{N_a} = \frac{0.86\times \dot{m}_{core}}{N_a} = \frac{0.86\times 13671}{764} = 15.39\,\text{kg/s},$

assuming 10% core bypass flow and 4% flow through the water rods.

(6) $A_{fa} = (l_{ch})^2 - N_{wr}\frac{\pi(D_{wr})^2}{4} - N_{rods}\frac{\pi(D)^2}{4}$

$= 0.1341^2 - 2\frac{\pi 0.0249^2}{4} - 74\frac{\pi 0.0112^2}{4} = 9.718\times 10^{-3}\ \text{m}^2$

(7) $G_a = \frac{\dot{m}_a}{A_{fa}} = \frac{15.39}{9.718\times 10^{-3}} = 1584\frac{\text{kg}}{\text{m}^2\ \text{s}}$

(8) $t_{gap} = \frac{D - D_f - 2t_{clad}}{2} = \frac{11.20 - 9.60 - 2\times 0.71}{2} = 0.09\ \text{mm}$

(9) Calculated using equation 4.114 (*Nuclear Systems Vol. II,* p. 154), using $n = 0.2$ and considering for BWR four different types of channels (50 interior channels, 32 edge, 4 corner and 1 near water rods) and for PWR five types (156 interior, 64 edge, 4 corner, 4 near the instrumentation tube, 96 near the 24 guide tubes). From source *G*, the instrumentation tube diameter is 12.29 mm and the guide tube diameter is 11.58 mm.

(10) $A_{fch} = P^2 - \frac{\pi D^2}{4}$

(11) $G_{ch} = \frac{\dot{m}_{ch}}{A_{fch}}$

(12) $T_{exit} = T_{in} + \Delta T_{core} = 293.1 + 33.7 = 326.8°\text{C}$, where $\Delta T_{core} = 33.7°\text{C}$ from source *G*.

(13) $\dot{m}_a = G_a A_{fa} = 3675.4 \times 2.444 \times 10^{-2} = 89.83\ \text{kg/s}$

Note: $\dot{m}_a N_a = 89.83 \times 193 = 17365\,\text{kg/s} \cong 99\%\ \dot{m}_{core}$. The two core flows are different because of the water flow between the assemblies.

(14) Using the data about guide tubes and the instrumentation tube cited in (9):

$$A_{fa} = l_{ch}^2 - \frac{\pi}{4}\left(N_{gt}D_{gt}^2 + N_{it}D_{it}^2 + N_{rods}D^2\right)$$
$$= 0.214^2 - \frac{\pi}{4}\left(24 \times 0.01158^2 + 1 \times 0.01229^2 + 264 \times 0.0095^2\right)$$
$$= 2.444 \times 10^{-2}\ \text{m}^2$$

(15) $Q^M = \frac{\dot{Q}_{th}}{f_{HM}M_{fc}}$

where $\dot{Q}_{th}$ = reactor thermal power
f_{HM} = heavy metal fraction of core fuel material
M_{fc} = mass of fuel material in the core

TABLE K.2
Properties of the Light Water Reactors in Table K.1

	Symbol	Unit	BWR	PWR
Coolant Properties				
Density	ρ at T_{avg}	kg/m^3	—	704.88
	ρ_{in}	kg/m^3	754.65	740.29
	ρ_f	kg/m^3	737.30	594.12
	ρ_g	kg/m^3	37.320	102.06
Viscosity	μ at T_{avg}	$kg\ m^{-1}\ s^{-1}$	—	8.460×10^{-5}
	μ_f	$kg\ m^{-1}\ s^{-1}$	9.075×10^{-5}	6.829×10^{-5}
	μ_g	$kg\ m^{-1}\ s^{-1}$	1.902×10^{-5}	2.312×10^{-5}
Specific heat	c_p at T_{avg}	$kJ\ kg^{-1}\ K^{-1}$	—	5.742
	c_{pf}	$kJ\ kg^{-1}\ K^{-1}$	5.428	8.976
	c_{pg}	$kJ\ kg^{-1}\ K^{-1}$	5.423	14.05
Specific enthalpy	h_{in}	kJ/kg	1227.5	1300.6
	h at T_{avg}	kJ/kg	—	1393.3
	h_f	kJ/kg	1274.4	1630.4
	h_g	kJ/kg	2770.8	2595.8
Other Properties				
Clad-coolant heat transfer coefficient	h	$kW\ m^{-2}\ °C^{-1}$	—	34
Pellet-clad gap conductance	h_{gap}	$kW\ m^{-2}\ °C^{-1}$	—	5.7
Cladding (Zr) thermal conductivity	k_c	$W\ m^{-1}\ K^{-1}$	13.85	13.85
Fuel (UO_2) thermal conductivity	k_f	$W\ m^{-1}\ K^{-1}$	2.163	2.163
Helium gap bond thermal conductivity	k_{He}	$15.8 \times 10^{-6} T^{0.79}\ W\ cm^{-1}\ K^{-1}$ (Equation 8.140) with T in Kelvin		

Index

Note: n = Footnote

A

C

D

E

G

H

R

T

W